$s^2$ defined : p. 27
$\mu$, $\sigma$ defined : p. 39
Reason for N-1 : p. 44
statistic p. 70
average of mean ~ pp 80-83
STANDARD ERROR = $\frac{dev}{\sqrt{N}}$ see p. 77

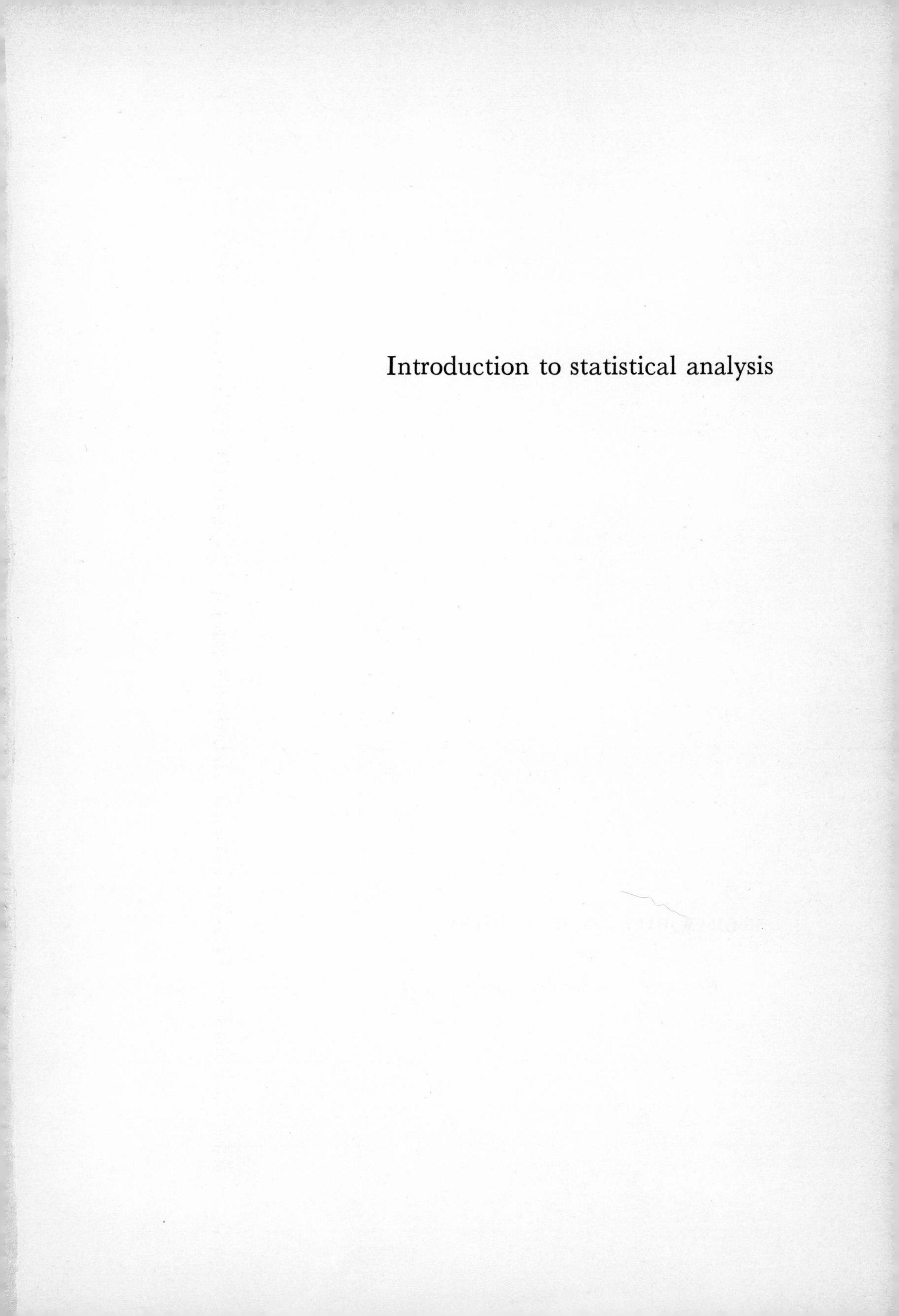

# Introduction to statistical analysis

**McGRAW-HILL BOOK COMPANY**

*New York San Francisco St. Louis Toronto London Sydney*

*Third Edition*

# INTRODUCTION TO STATISTICAL ANALYSIS

Wilfrid J. Dixon

*Professor of Biomathematics*
*University of California, Los Angeles*

Frank J. Massey, Jr.

*Professor of Biostatistics*
*University of California, Los Angeles*

**TO:**

Eva, Janet, Kitty

Mildred, Frank, Laura

**INTRODUCTION TO STATISTICAL ANALYSIS**



Printed in the United States of America.

Library of Congress catalog card number: 68-17180

**07-017070-3** 890 KPKP 7543

# Preface

This textbook is written for a basic course in statistics to be taken by students from all fields in which statistics finds application. We have attempted to present the fundamental concepts of the subject in a manner which will show the student how general is the application of the statistical method. It is intended that interested students continue this type of training in courses giving special applications in their own fields after one, two, or three quarters of this course.

We have found that the contents of this revised text can easily be covered in a one-year course having either three lectures per week or two lectures with one laboratory per week. For shorter courses the following topics are suggested: Chapters 1 to 8 plus selections from any of Chapters 10, 11, 13, 17, and 20. Except for parts of Chapter 20 the only mathematical ability assumed of the student is a knowledge of algebraic addition, subtraction, and multiplication. We feel that the topic of probability is more meaningful for students with a minimum of mathematical background if it is presented late in the year course. With students who have the equivalent of two years of high school algebra some teachers may wish to present Chapter 20 quite early, and it has been prepared with this in mind. It may be desirable to introduce part or all of Chapter 20 immediately following Chapter 4. We have avoided conventional gambling games, dice problems, etc., in Chapter 20 and have stressed the statistical applications of the theory.

Many changes have been made to improve sequence or to change emphasis, and several sections presenting new or different methods have been added.

The principal changes from the earlier editions are: (1) A reorganization of the material which is now in Chapters 6 to 8; (2) various topics on efficiency and alternate tests and estimates have been collected together in Chapter 9; (3) analysis-of-variance topics have been added; (4) Chapter 16 contains an expanded discussion of problems of normality and transformations. The analysis-of-variance material is divided between Chapters 10 and 15 to make it simpler to defer some of these topics until regression (Chapter 11) and analysis-of-covariance (Chapter 12) methods have been introduced.

At the end of Chapter 2 we have included data on 12 variables for 200 people chosen from the Los Angeles Heart Study, by courtesy of Dr. John M. Chapman, UCLA. These data are used in exercises in following chapters. The availability of real data is important to students, and the teacher may wish to assign other exercises based on these data.

Following Chapter 20 we have included a section of "General Comments" which provides notes for extended reading either into various areas of application or into a study in greater depth of special topics.

The tables have again been expanded. A number of changes in format have been made. These include a new table (A-9*e*) for fourfold contingency tables, a short table of common logarithms (A-14*b*), additional required sample sizes in Table A-12, a new table of the studentized range (A-8*a*), and an extended table of the rank correlation coefficient (A-30*c*).

The concepts of distribution, sample, and population are introduced early. The elementary descriptive procedures of statistics are introduced as they are needed in the development of the ideas of sampling, tests of hypotheses, and design of experiments. The analysis of variance is introduced sufficiently early for its inclusion in a one-semester or two-term course. Nonparametric statistics have been included because of their wide applicability and because of their validity under general conditions. The sampling distributions of the various statistics are introduced by means of experimental sampling. Experimental verifications of tabled distributions have been carried out by comparing percentiles of observed sampling distributions with the mathematical results. The sampling experiments indicated at the end of the chapters are organized so that computations on the samples drawn are used in several following class exercises.

The authors wish to express their appreciation to Professor E. S. Pearson for permission to reprint from *Biometrika* parts of Tables A-7, A-8, A-9, A-13, A-18, and A-30; to S. K. Banerjee for permission to reprint from *Sankhyā* Table A-25; to the RAND Corporation for permission to print the random-number tables; to A. Hald and John Wiley & Sons for permission to copy certain percentiles forming part of Table A-7*c*. We are indebted to Sir Ronald A. Fisher, Frank Yates, and to Messrs. Oliver & Boyd, Ltd., Edinburgh, for permission to reprint, in part, Table III from their book

*Statistical Tables for Biological, Agricultural and Medical Research.* For other tables in the Appendix we are indebted to C. Colcord, L. S. Deming, C. Eisenhart, M. W. Hastay, L. A. Knowler, R. F. Link, F. Mosteller, E. G. Olds, F. Swed, W. A. Wallis, J. E. Walsh, and E. K. Yost.

We wish to take this opportunity to express our appreciation to the many friends and colleagues who have made helpful criticisms and suggestions on the earlier editions. In the preparation of the third edition, thanks are due to M. Tarter and K. A. Brownlee for comments on Chapters 10 and 15, and to William Shonick and Hung Teh Chen for preparing answers to many of the exercises, as well as for valuable suggestions. Appreciated assistance with manuscript preparation was provided by Mary Ash and Jean Angle.

*Wilfrid J. Dixon*
*Frank J. Massey, Jr.*

# Contents

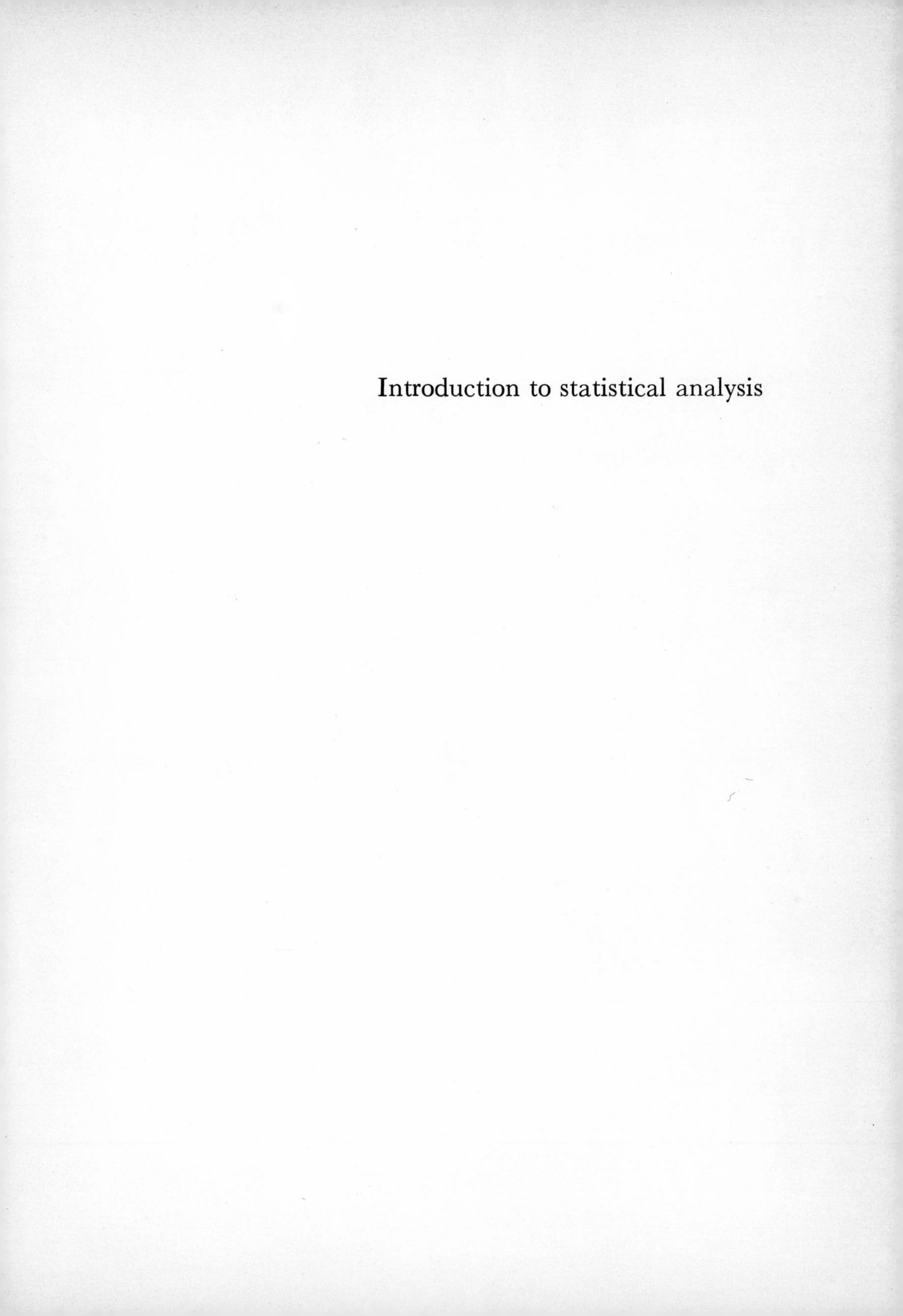

Introduction to statistical analysis

**CHAPTER ONE**

# Introduction

The term *statistics* pertains to a listing of facts, to systematic methods of arranging and describing the data, and finally to a science of inferring generalities from specific observations. The emphasis in this book will be on the problem of inference, but in order to build up the necessary background the first few chapters will be of a descriptive nature. The laws of physical, biological, and social science have their proof in statistical facts. The study of statistics here will not stop with the mere description of an existing situation but will continue on to the study of procedures of scientific inference and proof.

## TYPES OF PROOF 1-1

You, no doubt, have heard the statement "You can prove anything with statistics." We shall determine just what sort of things can be proved by statistics, and furthermore, just what we mean by "proof." Natural and physical *laws* are *hypotheses* which have been subjected to various tests and have become accepted, or, as some say, proved. The proof of a hypothesis is the *testing* of the hypothesis. If the tests show the hypothesis satisfactory it is accepted; if the tests show the hypothesis unsatisfactory it is rejected. When have we tested a hypothesis sufficiently to reject it? The standard procedure will be to collect information in the form of numerical observations and to base our decision on these observations. For example, if someone tosses a coin 100 times and obtains heads every time, he may feel that the hypothesis that he has an unbiased coin is no longer acceptable, and he will reject that hypothesis. It is well known that it is possible for this result to occur with a true coin, but if we were to demand that we be completely positive before making a decision, we could never decide that the coin was biased, even if it had two heads, unless we were allowed to examine it. We would not be able to accept the hypothesis of gravity as a law until all the apples of all time had fallen. The procedures of *statistical inference* will make it possible (under certain assumptions) for us to state just what the probability is that we will accept false hypotheses or reject true hypotheses. We will never know for sure in any particular case, of course, whether the hypothesis is true or not. This *statistical proof* is the basic form of proof used in the investigations of all sciences. We must make a distinction here between the methods of statistical proof and the methods of mathematical proof. *Mathematical proof* is available only within the framework of mathematics itself and cannot be applied outside that field. A hypothesis in mathematics may be declared false by the presentation of a single example which violates it. However, a single example which does not agree with the hypothesis cannot usually by itself cause us to reject a hypothesis outside of mathematics.

The methods of proof used to develop the statistical procedures presented in this book will in a few cases be *mathematical proofs* in the adaptation of various formulas. For the most part, however, the development will be that of statistical proof, or *experimentation*. These statistical procedures can all be developed by mathematical means, although it will not be possible in most cases to present these mathematical developments because of the limited mathematical background assumed of the reader of this text. The development by experimentation will serve a double purpose, for here we are interested not only in the results of the experiments, but also in a study of the process of proof by experiment.

## 1-2 GENERALITY OF APPLICATIONS OF STATISTICS

There does not exist a theory of statistics applicable only to economics, only to medicine, or only to education. There is a *general* theory of statistics which is applicable to any field of study in which observations are made. Statistical procedures now form an important part of all fields of science, and procedures which have been developed for use in one field have almost invariably found important application in a number of other fields. There are, however, statistical procedures which are more frequently used in one field than in another. We shall concentrate on those procedures which are most widely used.

We need not look far to find problems using statistical ideas. The concept of *average* is used in referring to a man of average height, to a ballplayer's batting average, or to the average number of cigarettes smoked per day. We use the single figure of an average to represent the whole group of men, or the ballplayer's performance in all his games, or the many different speeds on the trip between towns. We use the single quantity, the average, to describe one characteristic of the group. For example, in saying that the average height of a group of men is 69 inches we do not mean that all men in the group are 69 inches tall; the average describes the whole *group* of men, so that if we were to pick a man at random from this group and were asked what height we would *expect* him to be, we would give as our estimate 69 inches. Only if all individuals in the group were exactly the same height could we *guarantee* that the chosen one would have the same height as the average.

Another concept with which we are all familiar is that of *dispersion*, or variability. A teacher may say that one class is more uniform in ability than another class. An engineer may say that one batch of electric light bulbs is more variable in quality than another batch. A textile worker may say that one type of yarn is more variable in breaking strength than it should be. A manufacturer who wishes to use mass production must reduce the variability in the dimensions of parts if any shaft is to fit into any sleeve. The manufacturer of propellant for a missile must produce propellant of sufficiently uniform burning time that he can adjust the total propellant burning time. If the propellant burns too rapidly the missile explodes before it leaves the ground; if it burns too slowly the missile cannot be launched.

Another commonly used statistical description of data is that of *correlation*, or association. The teacher may say of his class that the faster readers are also better at arithmetic and the slower readers poorer at arithmetic. As an example of a negative relationship, a teacher may say of a particular class that the older children read more slowly and the younger children more rapidly.

These concepts of proof, average, dispersion, and correlation will be developed more fully, and their combined use in various types of statistical investigations will be illustrated in the ensuing chapters.

A major area of statistical theory is concerned with the design of experiments and the efficient collection of information to aid in this design. Many experimental materials are so expensive that it is essential that the desired information be obtained with a minimum number of observations. Statistics is also concerned with problems which arise from the necessity of designing experiments to investigate several factors at the same time, either because of the great length of time required for the experiment or because of the difficulty in reproducing the experimental treatments or conditions. In such cases statistical methods must be used to separate the effects of the separate treatments.

Whenever anything is measured numerically, even though the attempt to make an assessment results in numbers no more refined than simple counting, there arises the desire for judging the significance of the data and for making the maximum use of the information gathered. These are the principal problems with which statistical methods are concerned.

CHAPTER TWO

# Distributions

It is necessary in the investigation of any phenomenon, whether it is the study of forces of attraction by the physicist, the study of the effects of anxiety by the psychologist, the study of radiation effects on animals by the biologist, or any other research, to observe and record some characteristic of the objects under consideration.

## OBSERVATIONS 2-1

We must have observations of some form, even if they are of a very rudimentary sort. It is perhaps easiest to think of *measurements* of such simple properties as length, weight, volume. For such properties it is generally easy to establish a measuring stick to compare objects or individuals. For example, we may mark off a pole in units of equal length and use it to show that an object 60 inches long is twice as long as another 30 inches long.

Let us consider, for example, such a characteristic as intelligence. It is possible to construct an examination which will indicate whether one individual has *more* than another of something that might be called intelligence, but this scale does not enable us to say that a particular individual is *twice* as intelligent as some other individual. The physical sciences also deal with quantities for which only a *scale* is established, for example, the common property temperature. We may say that today is hotter than yesterday, but we cannot decide that today is twice as hot as yesterday.

*Scores* are often applied to indicate quality, such as the scores applied to butter or to meat. Here there is not even the comparability between individuals that is available when a scale is established. Scores are compiled from an arbitrary assignment of a certain number of points for various characteristics such as flavor, appearance, or consistency. The term "score" is correctly used with reference to a football score. Here various points have been assigned for a touchdown, a conversion after touchdown, a field goal, and a safety. The team obtaining the highest score wins the game. There is not always general agreement, however, that the team with the highest score played the best football, nor is it possible to say that a team making 14 points played twice as well as the team making 7 points.

A *rank* is used merely to indicate where a particular individual stands with respect to the others which have been observed at the same time. For example, we may decide that this painting is the best, this is next best, etc.

Many statistical procedures can be applied alike to numbers which are true measurements, readings from a scale, scores, or ranks. In some cases the procedure must be different. We shall avoid the use of the term "measurement" when referring to scores or ranks. Instead, *observation* will represent any sort of numerical recording of information.

It will be necessary to distinguish between *continuous* and *discrete* observations. A continuous observation is one for which successive refinements of our measuring stick or scale will give more precise observations. Discrete observations always result in one of a particular set of values. In a five-man committee the number of aye votes recorded must always be exactly 0, 1, 2, 3, 4, or 5. Other examples of discrete observations would be the number of bolts in a box, the number of people entering a department store in a day, the number of $\beta$ rays counted in 1 second, the number of petals on a flower, etc. Examples of observations which are usually continuous are length, weight, amount of current, or velocity.

It is important to recognize that the interpretation of some statistical numbers is different when they are computed on measurements or scales than when they are computed on scores or ranks. In the following section the observations used as an illustration are the heights of boys. Remember that the procedures developed are equally applicable to measurements in general. Height is chosen for illustration, since it is a familiar measurement. Height is a continuous variable which is recorded in discrete units.

## 2-2 HISTOGRAM

A *histogram* is a picture of a number of observations.

EXAMPLE 2-1. Suppose we have measured the heights of five boys in a certain school and have recorded them to the nearest inch as 68, 72, 66, 67,

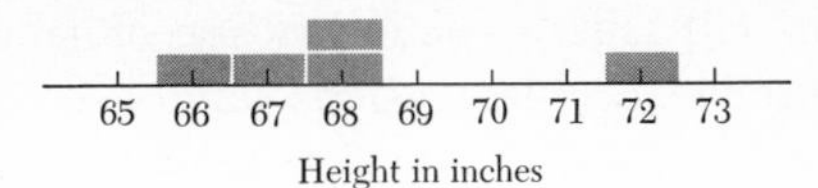


**Figure 2-1** Histogram of boys' heights. $N = 5$; 1-inch interval.

68. We then represent them by *equal units of area* above a scale marked with those heights (Fig. 2-1).

EXAMPLE 2-2. If we measure another 75 boys we get a picture from which we can form at sight an idea of how the heights are distributed at various values along a line marked off in inches (Fig. 2-2).

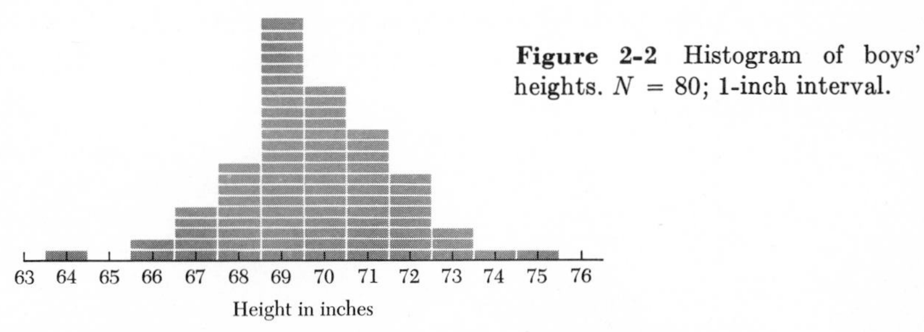


**Figure 2-2** Histogram of boys' heights. $N = 80$; 1-inch interval.

It is not difficult to see from this histogram that the average height is around 69 or 70 inches and that a small proportion of the boys are less than 66 inches or more than 73 inches tall.

This picture, or histogram, is particularly useful when a large number of measurements have been made. The histogram of the first five measurements was not so informative.

The histogram is usually presented without the lines showing the individual observations, and the *frequencies* are not indicated for each case but are instead shown as a scale along the left. Note that each observation is represented by an equal amount of area, and since in this example the intervals are of equal width, observations are also represented by equal units on the vertical axis.

As more and more measurements are made, the histogram approaches more and more closely the distribution of the heights of all the boys in this school. Although a histogram of only five measurements does not

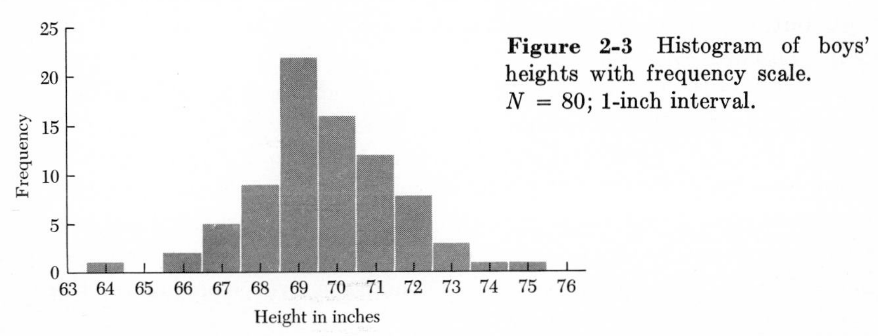


**Figure 2-3** Histogram of boys' heights with frequency scale. $N = 80$; 1-inch interval.

suggest the form the distribution will take, some indication of its final form is present before a very large number of observations have been added to the distribution.

Although the heights were recorded as a discrete variable, we have drawn the histogram to reflect the fact that height is a continuous measurement. The above description of how to draw a histogram will thus apply to discrete measurements as they occur and will apply to continuous measurements if they are grouped into a number of categories (for example, to the nearest inch). Even if the measurements had been made to a much greater precision, they could be represented by the above histogram by combining all measurements between 68.5 and 69.5 inches into the group at 69 inches, all measurements between 69.5 and 70.5 into the group at 70, and so on. The

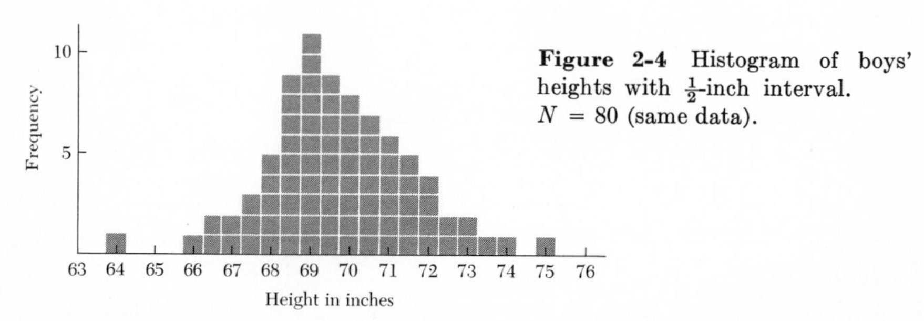


**Figure 2-4** Histogram of boys' heights with $\frac{1}{2}$-inch interval. $N = 80$ (same data).

measurements can, of course, be grouped into any length interval. If we had assumed that the measurements were made to the nearest half inch, or if we had grouped, say, all measurements between 68.25 and 68.75 at 68.5, those between 68.75 and 69.25 at 69, and so on, we would have obtained a histogram similar in appearance to that above, but the data would have been grouped into about twice as many groups. We have made each rectangle half as wide to indicate that we are now reading to the nearest half inch, and we have increased the height of each rectangle to hold the area the same for each observation. Also, notice that the total area will be 80 units of area, to represent 80 observations.

It is important to record observations to the proper degree of precision for the study being attempted. To provide a picture of the data in general it is important to leave enough groups of data to discern the shape of the distribution, but it is also important not to have so many that too much detail obscures the picture.

Either of the histograms above would usually be considered satisfactory. For most studies 8 to 20 groups would be satisfactory.

If more and more observations were taken, and the histograms were constructed by grouping the measurements into intervals of less and less width, we can imagine that the histograms would approach a smooth curve which would represent more truly the distribution of continuous measurements. Figure 2-5 illustrates such a smooth curve. Here again, the total

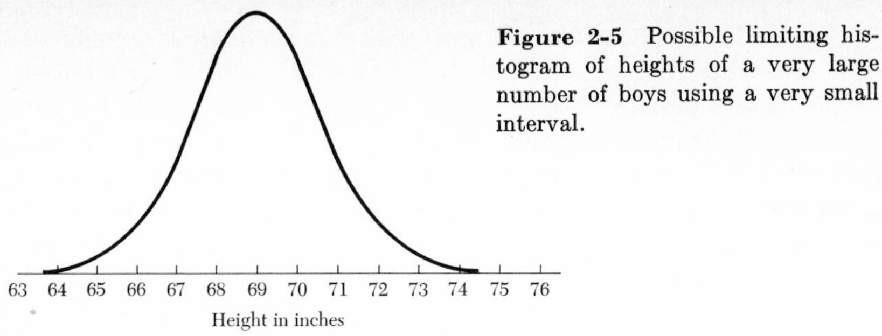


**Figure 2-5** Possible limiting histogram of heights of a very large number of boys using a very small interval.

area is equal to the number of observations. Often, in order to have comparable pictures for different numbers of observations, the *proportion* is indicated in place of the frequencies. This leaves the shapes of the distributions the same and changes only the scale at the left. If Figs. 2-1 to 2-4 were changed in this manner, the scale in Fig. 2-3 would be marked $\frac{5}{80}$ instead of 5 and $\frac{10}{80}$ instead of 10, etc. The same change would be necessary for Fig. 2-4. Below is a list of the 80 measurements of height (to the nearest tenth of an inch) from which the histograms in Figs. 2-1 to 2-4 were drawn, with the measurements arranged in order of size:

| | | | | | | | |
|---|---|---|---|---|---|---|---|
| 64.2 | 67.9 | 68.6 | 69.1 | 69.5+ | 70.2 | 71.0 | 72.0 |
| 65.8 | 67.9 | 68.7 | 69.1 | 69.6 | 70.3 | 71.0 | 72.1 |
| 66.3 | 68.2 | 68.7 | 69.1 | 69.7 | 70.4 | 71.1 | 72.2 |
| 66.6 | 68.2 | 68.8 | 69.2 | 69.8 | 70.4 | 71.2 | 72.4 |
| 66.9 | 68.3 | 68.9 | 69.3 | 69.9 | 70.5− | 71.4 | 72.5− |
| 67.0 | 68.4 | 68.9 | 69.3 | 69.9 | 70.5+ | 71.4 | 72.8 |
| 67.4 | 68.5− | 68.9 | 69.4 | 70.0 | 70.6 | 71.4 | 73.0 |
| 67.5− | 68.5+ | 69.0 | 69.5− | 70.0 | 70.6 | 71.6 | 73.4 |
| 67.6 | 68.6 | 69.0 | 69.5− | 70.1 | 70.9 | 71.7 | 74.2 |
| 67.8 | 68.6 | 69.0 | 69.5+ | 70.1 | 70.9 | 71.8 | 74.9 |

Although it is difficult to make measurements to greater precision than tenths of an inch, the observer was requested when a boy's measurement was near a half inch to indicate whether it was above or below and indicate that decision with a plus sign or a minus sign. Thus we can decide in which group to place these measurements. For example, 68.5− will go in the lower group and 68.5+ in the upper group.

## PERCENTILES 2-3

The term *percentile* will be used in reference to a distribution of observations. For example, the 10th percentile, $P_{10}$, is defined as the value below which 10 per cent of the distribution of values will fall. In Example 2-2 we would obtain $P_{10} = 67.5$ inches, since 10 per cent of the boys were shorter than 67.5 inches.

**Cumulative-distribution polygon.** A *cumulative-distribution polygon* is constructed of line segments connecting points plotted for the *cumulative* percentage of observations at each division point of the horizontal scale. The 80 measurements of height are tabulated in Table 2-1. For each height interval the observed frequency and the percentage are listed. Cumulations of these frequencies and percentages are given in the right-hand columns. The cumulative percentages are plotted in Fig. 2-6, with the points connected to form the polygon.

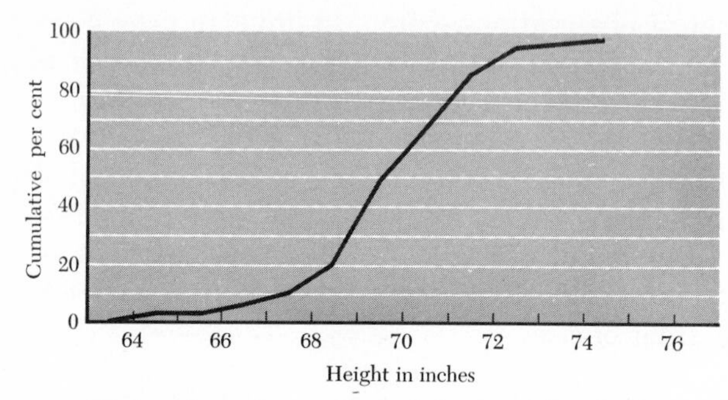


**Figure 2-6** Cumulative percentage polygon of heights of boys. $N = 80$.

Note that the value of $P_{10} = 67.5$ inches or the value for any other percentile can be obtained by reference to the cumulative-distribution polygon. Note in Fig. 2-6, that if we locate 10 per cent on the left-hand scale, draw a line horizontally to the polygon, and then draw a line down to the lower scale from the point of intersection with the polygon, we can read

**Table 2-1** Heights of 80 boys, measurements to nearest inch

| *Height, in. (mid-point)* | *Frequency* | *Per cent* | CUMULATION THROUGH INTERVAL *Frequency* | CUMULATION THROUGH INTERVAL *Per cent* |
|---|---|---|---|---|
| 76 | | | | |
| 75 | 1 | 1.25 | 80 | 100.00 |
| 74 | 1 | 1.25 | 79 | 98.75 |
| 73 | 3 | 3.75 | 78 | 97.50 |
| 72 | 8 | 10.00 | 75 | 93.75 |
| 71 | 12 | 15.00 | 67 | 83.75 |
| 70 | 16 | 20.00 | 55 | 68.75 |
| 69 | 22 | 27.50 | 39 | 48.75 |
| 68 | 9 | 11.25 | 17 | 21.25 |
| 67 | 5 | 6.25 | 8 | 10.00 |
| 66 | 2 | 2.50 | 3 | 3.75 |
| 65 | | | 1 | 1.25 |
| 64 | 1 | 1.25 | 1 | 1.25 |
| 63 | | | 0 | 0 |
| *Total* | 80 | 100.00 | | |

the value, 67.5 inches. In this manner we find, for example, $P_{50} = 69.6$. Rarely is it necessary to report percentiles to more accuracy than can be obtained easily from a graph such as Fig. 2-6.

If we wish to know the *percentile rank* for any particular height, that is, the percentage of individuals who are shorter than this particular height, we need only refer to the polygon in Fig. 2-6. For example, suppose we wish to find the percentile rank of 67.5 inches. The answer is, of course, 10, as we computed above and observed from this same polygon. The percentile rank of 70 inches is obtained by drawing a line vertically from the point marked 70 on the lower scale, extending this line up to the polygon, and then drawing a line horizontally to the left-hand scale. Here we read 59 per cent.

**Frequency table.** The most widely used method for the numerical reporting of data is the *frequency table.* The first two columns of Table 2-1 represent a frequency table. The third column has been added to indicate the percentage at each height. The fourth and fifth columns indicate a cumulation of the frequencies and the computation for a cumulative-distribution polygon.

## GLOSSARY

In the development of any phase of science the student will find that certain words are used with specific meanings. The following words introduced thus far have special import in statistics:

average
continuous
correlation
cumulative distribution
discrete
dispersion
experiment
frequency table
histogram
hypothesis
induction
inference
law
measurement
percentile
percentile rank
proof
rank
scale
score

## DISCUSSION QUESTIONS

**1.** What information can be obtained from a frequency distribution?

**2.** Distinguish between percentile and percentile rank.

**3.** What is the interpretation of the area under a histogram? What feature of a cumulative-distribution polygon has a similar interpretation?

**4.** What is an advantage in having a large number of intervals? What is a disadvantage?

**5.** Give two examples each of scores, ranks, and measurements other than those mentioned in the text.

**6.** Define each of the terms in the Glossary. For each, give the statistical meaning and compare this definition with that in a dictionary.

## PROBLEMS

*Section* 2-1

**1.** Classify the following observations as continuous or discrete; if they are continuous, classify them as measurements, scales, scores, or ranks:

(*a*) A ball-bearing has a diameter of 3.25 millimeters.
(*b*) A cable has a breaking strength of 9,750 pounds.
(*c*) There are 23 students attending class.
(*d*) A mark on a history examination is 75.
(*e*) A toss of 10 coins results in 6 heads.
(*f*) There are seven petals on a flower.
(*g*) A relative-humidity reading is 55 per cent.
(*h*) Jerry is thirteenth in his class.
(*i*) A man swims 100 yards in 73 seconds.
(*j*) Receipts for the day are $416.75.
(*k*) The guinea pig is 10 months old.
(*l*) A man's yearly income is $3,600.
(*m*) A man's blood pressure (systolic) is 120.
(*n*) The half-life of a sample of an element is 6.8 minutes.
(*o*) A river is 3 feet above flood stage.

*Sections* 2-2 *and* 2-3

**2.** Read from Fig. 2-6 the percentage of boys who have heights between 67.5 and 71.5 inches.

**3.** The following table shows adult male age distributions of a community in Maine and of a sample from that community. Draw a histogram and a cumulative-distribution polygon for each distribution. Read $P_{10}$, $P_{50}$, $P_{75}$, and $P_{95}$ from the polygon.

| | AGE GROUP | | | | | | | | |
|---|---|---|---|---|---|---|---|---|---|
| | 21–24 | 25–34 | 35–44 | 45–54 | 55–64 | 65–74 | 75–84 | 85–94 | |
| *Community* | 128 | 400 | 406 | 410 | 307 | 314 | 110 | 21 | 2,096 |
| *Sample* | 20 | 60 | 83 | 138 | 68 | 68 | 20 | 2 | 459 |

**4.** Construct the cumulative-distribution polygons for the data in Fig. 2-4. Read from the polygons values of $P_{10}$, $P_{25}$, $P_{40}$, $P_{50}$, $P_{60}$, $P_{75}$, $P_{90}$.

**5.** A group of 50 boys and a group of 50 girls were given a test in arranging different-shaped blocks. The completion times are recorded in the frequency tables below. Draw a histogram and a cumulative-distribution polygon for each set. Read the values of $P_{25}$, $P_{50}$, and $P_{75}$ from each polygon.

| | TIME, SECONDS | | | | | | | | | | | |
|---|---|---|---|---|---|---|---|---|---|---|---|---|
| | 51 | 50 | 49 | 48 | 47 | 46 | 45 | 44 | 43 | 42 | 41 | 40 |
| *Girls* | | 1 | 1 | 1 | 3 | 9 | 9 | 12 | 6 | 3 | 4 | 1 |
| *Boys* | 1 | 2 | 4 | 3 | 7 | 12 | 10 | 6 | 2 | 3 | | |

Problems 6 to 19 refer to data in Tables 2-2*a* and 2-2*b*. Each table contains observations made on 100 men. These men were survivors of a group who had an initial examination in 1952 and were reexamined in 1962. The variables listed are as follows:

$A$ = age in 1952
$B$ = a socioeconomic-status scale with the high end coded as 1 and the low end as 5
$C$ = systolic blood pressure in millimeters of mercury measured in 1952 by method $A$
$D$ = diastolic blood pressure in millimeters of mercury measured in 1952 by method $A$
$E$ = systolic blood pressure in millimeters of mercury measured in 1962 by a second method $B$
$F$ = diastolic blood pressure in millimeters of mercury measured in 1962 by method $B$
$G$ = blood cholesterol measured in 1952
$H$ = blood cholesterol measured in 1962
$I$ = height in inches measured in 1952
$J$ = weight in pounds measured in 1952
$K$ = pulse rate in beats per minute measured in 1962
$L$ = a dichotomous variable coded as 1 if a coronary incident occurred between 1952 and 1962

All measurements were to the nearest integer. Note that a number of men in the original sample died between 1952 and 1962, some from coronary incidents and some from other causes, and are not included in the table. Thus no association should be made between the last variable and the physical measurements. Also, comparisons such as that of cholesterol in 1952 and 1962 must be conditional upon survival.

**6.** Using intervals 20–24, 25–29, 30–34, . . . , collect the ages $A$ in Table 2-2*a* into a frequency table and draw the histogram and cumulative-distribution polygon. Read the values of $P_{10}$, $P_{50}$, $P_{75}$ from the polygon.

**7.** Using intervals 20–24, 25–29, 30–34, . . . , collect the ages $A$ in Table 2-2*b* into a frequency table and draw the histogram and cumulative-distribution polygon. Read the values of $P_{10}$, $P_{50}$, $P_{75}$ from the polygon.

**8.** Using intervals 90–99, 100–109, 110–119, . . . , collect the 1952 systolic-blood-pressure recordings $C$ in Table 2-2*a* into a frequency table and draw the histogram and cumulative-distribution polygon. Read the values of $P_{10}$, $P_{50}$, $P_{75}$ from the polygon.

**9.** Using intervals 90–99, 100–109, 110–119, . . . , collect the 1952 systolic-blood-pressure recordings $C$ in Table 2-2*b* into a frequency table and draw the histogram and cumulative-distribution polygon. Read the values of $P_{10}$, $P_{50}$, $P_{75}$ from the polygon.

**10.** Using intervals 50–54, 55–59, 60–64, . . . , collect the 1952 diastolic-blood-pressure readings $D$ in Table 2-2*a* into a frequency table and draw the histogram and cumulative-distribution polygon. Read the values of $P_{10}$, $P_{50}$, $P_{75}$ from the polygon.

**11.** Using intervals 50–54, 55–59, 60–64, . . . , collect the 1952 diastolic-blood-pressure readings $D$ in Table 2-2*b* into a frequency table and draw the histogram and cumulative-distribution polygon. Read the values of $P_{10}$, $P_{50}$, $P_{75}$ from the polygon.

**12.** Using intervals 125–149, 150–174, 175–199, . . . , collect the 1952 cholesterol readings $G$ in Table 2-2*a* into a frequency table and draw the histogram and cumulative-distribution polygon. Read the values of $P_{10}$, $P_{50}$, $P_{75}$ from the polygon.

**13.** Using intervals 125–149, 150–174, 175–190, . . . , collect the 1952 cholesterol readings $G$ in Table 2-2*b* into a frequency table and draw the histogram and cumulative-distribution polygon. Read the values of $P_{10}$, $P_{50}$, $P_{75}$ from the polygon.

**14.** Using 1-inch intervals, collect the height determinations $I$ in Table 2-2*a* into a frequency table and draw the histogram and cumulative-distribution polygon. Read the values of $P_{10}$, $P_{50}$, $P_{75}$ from the polygon.

**Table 2-2a** Data from L. A. Heart Study (Group A)

| *A* | *B* | *C* | *D* | *E* | *F* | *G* | *H* | *I* | *J* | *K* | *L* |
|---|---|---|---|---|---|---|---|---|---|---|---|
| 44 | 4 | 124 | 80 | 140 | 70 | 254 | 165 | 70 | 190 | 84 | 0 |
| 35 | 5 | 110 | 70 | 114 | 74 | 240 | 209 | 73 | 216 | 84 | 0 |
| 41 | 3 | 114 | 80 | 164 | 90 | 279 | 270 | 68 | 178 | 72 | 0 |
| 31 | 3 | 100 | 80 | 118 | 84 | 284 | 274 | 68 | 149 | 78 | 0 |
| 61 | 3 | 190 | 110 | 150 | 58 | 315 | 208 | 68 | 182 | 72 | 1 |
| 61 | 1 | 130 | 88 | 140 | 76 | 250 | 173 | 70 | 185 | 60 | 0 |
| 44 | 3 | 130 | 94 | 160 | 88 | 298 | 209 | 68 | 161 | 84 | 0 |
| 58 | 3 | 110 | 74 | 148 | 90 | 384 | 238 | 67 | 175 | 76 | 0 |
| 52 | 1 | 120 | 80 | 110 | 68 | 310 | 281 | 66 | 144 | 70 | 0 |
| 52 | 2 | 120 | 80 | 108 | 58 | 337 | 269 | 67 | 130 | 72 | 0 |
| 52 | 4 | 130 | 80 | 146 | 76 | 367 | 300 | 69 | 162 | 80 | 0 |
| 40 | 2 | 120 | 90 | 152 | 96 | 273 | 194 | 68 | 175 | 90 | 0 |
| 49 | 3 | 130 | 75 | 148 | 68 | 273 | 234 | 66 | 155 | 68 | 0 |
| 34 | 3 | 120 | 80 | 114 | 66 | 314 | 274 | 74 | 156 | 80 | 0 |
| 37 | 3 | 115 | 70 | 120 | 82 | 243 | 223 | 65 | 151 | 64 | 0 |
| 63 | 3 | 140 | 90 | 176 | 78 | 341 | 269 | 74 | 168 | 84 | 0 |
| 28 | 4 | 138 | 80 | 154 | 78 | 245 | 221 | 70 | 185 | 96 | 0 |
| 40 | 3 | 115 | 82 | 126 | 80 | 302 | 272 | 69 | 225 | 84 | 0 |
| 51 | 4 | 148 | 110 | 142 | 84 | 302 | 219 | 69 | 247 | 80 | 1 |
| 33 | 5 | 120 | 70 | 136 | 64 | 386 | 285 | 66 | 146 | 80 | 0 |
| 37 | 5 | 110 | 70 | 136 | 84 | 312 | 251 | 71 | 170 | 80 | 1 |
| 33 | 3 | 132 | 90 | 136 | 78 | 302 | 222 | 69 | 161 | 72 | 0 |
| 41 | 3 | 112 | 80 | 114 | 68 | 394 | 306 | 69 | 167 | 92 | 0 |
| 38 | 3 | 114 | 70 | 126 | 78 | 358 | 252 | 69 | 198 | 80 | 0 |
| 52 | 3 | 100 | 78 | 116 | 76 | 336 | 295 | 70 | 162 | 60 | 0 |
| 31 | 5 | 114 | 80 | 104 | 68 | 251 | 239 | 71 | 150 | 66 | 0 |
| 44 | 3 | 110 | 80 | 104 | 74 | 322 | 227 | 68 | 196 | 100 | 1 |
| 31 | 3 | 108 | 70 | 122 | 74 | 281 | 249 | 67 | 130 | 80 | 0 |
| 40 | 4 | 110 | 74 | 128 | 58 | 336 | 240 | 68 | 166 | 84 | 1 |
| 36 | 3 | 110 | 80 | 106 | 76 | 314 | 271 | 73 | 178 | 60 | 0 |
| 42 | 3 | 136 | 82 | 152 | 88 | 383 | 229 | 69 | 187 | 68 | 0 |
| 28 | 3 | 124 | 82 | 118 | 72 | 360 | 407 | 67 | 148 | 76 | 0 |
| 40 | 5 | 120 | 85 | 130 | 90 | 369 | 261 | 71 | 180 | 60 | 0 |
| 40 | 3 | 150 | 100 | 124 | 66 | 333 | 206 | 70 | 172 | 70 | 0 |
| 35 | 2 | 100 | 70 | 114 | 68 | 253 | 196 | 68 | 141 | 84 | 0 |
| 32 | 3 | 120 | 80 | 130 | 82 | 268 | 199 | 68 | 176 | 72 | 0 |
| 31 | 3 | 110 | 80 | 120 | 64 | 257 | 191 | 71 | 154 | 84 | 0 |
| 52 | 2 | 130 | 90 | 174 | 108 | 474 | 310 | 69 | 145 | 72 | 0 |
| 45 | 3 | 110 | 80 | 142 | 84 | 391 | 357 | 69 | 159 | 84 | 1 |
| 39 | 5 | 106 | 80 | 100 | 68 | 248 | 208 | 67 | 181 | 78 | 0 |
| 40 | 1 | 130 | 90 | 122 | 72 | 520 | 325 | 68 | 169 | 100 | 1 |
| 48 | 3 | 110 | 70 | 120 | 66 | 285 | 254 | 66 | 160 | 68 | 1 |
| 29 | 5 | 110 | 70 | 94 | 60 | 352 | 295 | 66 | 149 | 72 | 0 |
| 56 | 3 | 141 | 100 | 124 | 72 | 428 | 276 | 65 | 171 | 72 | 1 |
| 53 | 3 | 90 | 55 | 147 | 78 | 334 | 204 | 68 | 166 | 70 | 0 |
| 47 | 3 | 90 | 60 | 110 | 62 | 278 | 208 | 69 | 121 | 84 | 0 |
| 30 | 3 | 114 | 76 | 90 | 54 | 264 | 239 | 73 | 178 | 68 | 0 |
| 64 | 3 | 140 | 90 | 142 | 90 | 243 | 209 | 71 | 171 | 72 | 1 |
| 31 | 2 | 130 | 88 | 130 | 78 | 348 | 299 | 72 | 181 | 60 | 0 |
| 35 | 2 | 120 | 88 | 128 | 88 | 290 | 223 | 70 | 162 | 80 | 0 |

**Table 2-2a** *(Continued)*

| *A* | *B* | *C* | *D* | *E* | *F* | *G* | *H* | *I* | *J* | *K* | *L* |
|---|---|---|---|---|---|---|---|---|---|---|---|
| 65 | 1 | 130 | 90 | 120 | 60 | 370 | 367 | 65 | 153 | 88 | 1 |
| 43 | 2 | 122 | 82 | 162 | 74 | 363 | 294 | 69 | 164 | 68 | 0 |
| 53 | 1 | 120 | 80 | 112 | 60 | 343 | 264 | 71 | 159 | 72 | 0 |
| 58 | 2 | 138 | 82 | 112 | 54 | 305 | 224 | 67 | 152 | 62 | 1 |
| 67 | 1 | 168 | 105 | 158 | 76 | 365 | 295 | 68 | 190 | 80 | 1 |
| 53 | 2 | 120 | 80 | 132 | 74 | 307 | 283 | 70 | 200 | 60 | 0 |
| 42 | 5 | 134 | 90 | 136 | 86 | 243 | 206 | 67 | 147 | 72 | 0 |
| 43 | 5 | 115 | 75 | 132 | 80 | 266 | 220 | 68 | 125 | 72 | 0 |
| 52 | 4 | 110 | 75 | 126 | 70 | 341 | 270 | 69 | 163 | 72 | 0 |
| 68 | 5 | 110 | 80 | 152 | 76 | 268 | 191 | 62 | 138 | 75 | 0 |
| 64 | 5 | 105 | 68 | 164 | 68 | 261 | 220 | 66 | 108 | 72 | 0 |
| 46 | 3 | 138 | 90 | 130 | 70 | 378 | 273 | 67 | 142 | 75 | 0 |
| 41 | 3 | 120 | 90 | 130 | 79 | 279 | 236 | 70 | 212 | 80 | 0 |
| 58 | 3 | 130 | 90 | 154 | 68 | 416 | 230 | 68 | 188 | 96 | 0 |
| 50 | 1 | 160 | 110 | 194 | 108 | 261 | 240 | 66 | 145 | 80 | 0 |
| 45 | 3 | 100 | 70 | 136 | 84 | 332 | 260 | 67 | 144 | 84 | 0 |
| 59 | 3 | 156 | 90 | 156 | 82 | 337 | 264 | 67 | 158 | 80 | 0 |
| 56 | 2 | 120 | 92 | 110 | 64 | 365 | 265 | 65 | 154 | 72 | 0 |
| 59 | 3 | 126 | 96 | 120 | 80 | 292 | 196 | 67 | 148 | 64 | 0 |
| 47 | 1 | 110 | 80 | 166 | 90 | 304 | 293 | 67 | 155 | 88 | 0 |
| 43 | 1 | 95 | 70 | 98 | 62 | 341 | 194 | 69 | 154 | 66 | 0 |
| 37 | 3 | 120 | 74 | 124 | 70 | 317 | 220 | 74 | 184 | 70 | 0 |
| 27 | 3 | 100 | 60 | 118 | 70 | 296 | 241 | 67 | 140 | 72 | 0 |
| 44 | 5 | 110 | 80 | 128 | 70 | 390 | 249 | 66 | 167 | 60 | 0 |
| 41 | 3 | 120 | 80 | 120 | 60 | 274 | 245 | 69 | 138 | 76 | 0 |
| 33 | 4 | 120 | 80 | 130 | 82 | 355 | 278 | 68 | 169 | 68 | 0 |
| 29 | 3 | 115 | 80 | 130 | 76 | 225 | 174 | 70 | 186 | 88 | 0 |
| 24 | 3 | 120 | 80 | 114 | 68 | 218 | 199 | 69 | 131 | 76 | 0 |
| 36 | 3 | 108 | 66 | 106 | 56 | 298 | 218 | 67 | 160 | 72 | 0 |
| 23 | 2 | 110 | 78 | 122 | 74 | 178 | 229 | 66 | 142 | 88 | 0 |
| 47 | 5 | 120 | 80 | 116 | 68 | 341 | 272 | 70 | 218 | 64 | 1 |
| 26 | 2 | 110 | 75 | 112 | 64 | 274 | 227 | 70 | 147 | 64 | 0 |
| 45 | 3 | 130 | 90 | 166 | 80 | 285 | 253 | 65 | 161 | 96 | 0 |
| 41 | 4 | 164 | 110 | 158 | 88 | 259 | 258 | 66 | 245 | 78 | 0 |
| 55 | 3 | 125 | 88 | 182 | 108 | 266 | 212 | 67 | 167 | 94 | 0 |
| 34 | 4 | 110 | 80 | 112 | 70 | 214 | 220 | 67 | 139 | 66 | 1 |
| 51 | 2 | 110 | 75 | 162 | 98 | 267 | 312 | 66 | 150 | 80 | 0 |
| 58 | 2 | 120 | 80 | 130 | 74 | 256 | 171 | 67 | 175 | 64 | 0 |
| 51 | 3 | 118 | 88 | 176 | 90 | 273 | 209 | 64 | 123 | 72 | 0 |
| 35 | 3 | 110 | 75 | 114 | 80 | 348 | 313 | 72 | 174 | 80 | 0 |
| 34 | 1 | 118 | 78 | 124 | 78 | 322 | 250 | 69 | 192 | 72 | 0 |
| 26 | 3 | 120 | 70 | 114 | 58 | 267 | 226 | 70 | 140 | 80 | 0 |
| 25 | 5 | 110 | 75 | 118 | 68 | 270 | 232 | 74 | 195 | 64 | 0 |
| 44 | 3 | 100 | 70 | 109 | 58 | 280 | 209 | 65 | 144 | 78 | 0 |
| 57 | 2 | 130 | 85 | 148 | 66 | 320 | 245 | 69 | 193 | 80 | 0 |
| 67 | 5 | 110 | 80 | 120 | 68 | 320 | 258 | 64 | 134 | 72 | 0 |
| 59 | 4 | 160 | 90 | 162 | 76 | 330 | 321 | 63 | 144 | 52 | 0 |
| 62 | 4 | 130 | 88 | 132 | 76 | 274 | 241 | 69 | 179 | 62 | 0 |
| 40 | 3 | 140 | 90 | 122 | 80 | 269 | 215 | 63 | 111 | 64 | 0 |
| 52 | 3 | 120 | 90 | 122 | 62 | 269 | 245 | 68 | 164 | 64 | 0 |

**Table 2-2b** Data from L. A. Heart Study (Group B)

| *A* | *B* | *C* | *D* | *E* | *F* | *G* | *H* | *I* | *J* | *K* | *L* |
|---|---|---|---|---|---|---|---|---|---|---|---|
| 28 | 2 | 130 | 80 | 118 | 58 | 135 | 153 | 67 | 168 | 96 | 0 |
| 34 | 1 | 115 | 80 | 128 | 79 | 403 | 289 | 69 | 175 | 76 | 0 |
| 43 | 2 | 122 | 78 | 106 | 74 | 294 | 270 | 68 | 173 | 80 | 0 |
| 38 | 3 | 125 | 80 | 138 | 80 | 312 | 231 | 71 | 158 | 84 | 0 |
| 45 | 1 | 110 | 80 | 120 | 78 | 311 | 310 | 69 | 154 | 72 | 0 |
| 26 | 3 | 120 | 84 | 136 | 90 | 222 | 269 | 72 | 214 | 72 | 0 |
| 35 | 3 | 112 | 90 | 124 | 86 | 302 | 244 | 67 | 176 | 60 | 0 |
| 51 | 2 | 120 | 90 | 138 | 70 | 269 | 201 | 70 | 262 | 76 | 0 |
| 55 | 2 | 120 | 82 | 132 | 76 | 311 | 229 | 71 | 181 | 75 | 0 |
| 45 | 4 | 130 | 90 | 154 | 82 | 286 | 249 | 73 | 143 | 80 | 0 |
| 69 | 4 | 160 | 90 | 150 | 70 | 370 | 191 | 67 | 185 | 72 | 1 |
| 58 | 5 | 140 | 90 | 182 | 98 | 403 | 229 | 66 | 140 | 76 | 0 |
| 64 | 5 | 190 | 100 | 148 | 70 | 244 | 210 | 66 | 187 | 84 | 0 |
| 70 | 5 | 190 | 112 | 158 | 72 | 353 | 201 | 66 | 163 | 82 | 0 |
| 27 | 2 | 112 | 78 | 116 | 62 | 252 | 293 | 68 | 164 | 70 | 0 |
| 53 | 4 | 155 | 104 | 136 | 98 | 453 | 290 | 66 | 170 | 88 | 0 |
| 28 | 2 | 115 | 65 | 114 | 54 | 260 | 276 | 66 | 150 | 70 | 0 |
| 29 | 1 | 110 | 80 | 104 | 66 | 269 | 248 | 68 | 141 | 82 | 0 |
| 23 | 2 | 120 | 82 | 128 | 72 | 235 | 240 | 65 | 135 | 84 | 0 |
| 40 | 2 | 120 | 78 | 126 | 79 | 264 | 203 | 71 | 135 | 84 | 0 |
| 53 | 3 | 160 | 90 | 160 | 62 | 420 | 389 | 67 | 141 | 60 | 0 |
| 25 | 5 | 110 | 75 | 116 | 76 | 235 | 253 | 69 | 148 | 62 | 0 |
| 63 | 3 | 130 | 90 | 156 | 80 | 420 | 304 | 69 | 160 | 68 | 1 |
| 48 | 4 | 110 | 78 | 122 | 62 | 277 | 206 | 71 | 180 | 68 | 1 |
| 36 | 5 | 120 | 80 | 108 | 66 | 319 | 268 | 70 | 157 | 80 | 0 |
| 28 | 3 | 120 | 86 | 124 | 74 | 386 | 300 | 70 | 189 | 72 | 1 |
| 57 | 3 | 110 | 68 | 110 | 40 | 353 | 262 | 71 | 166 | 66 | 0 |
| 39 | 4 | 110 | 80 | 114 | 70 | 344 | 270 | 66 | 175 | 60 | 0 |
| 52 | 3 | 130 | 90 | 188 | 92 | 210 | 208 | 65 | 172 | 60 | 1 |
| 51 | 3 | 140 | 90 | 158 | 82 | 286 | 273 | 67 | 134 | 88 | 0 |
| 37 | 5 | 120 | 90 | 110 | 78 | 260 | 189 | 67 | 188 | 84 | 0 |
| 28 | 4 | 110 | 75 | 126 | 76 | 252 | 280 | 67 | 149 | 72 | 0 |
| 44 | 1 | 120 | 90 | 136 | 76 | 336 | 323 | 72 | 175 | 72 | 0 |
| 35 | 4 | 100 | 70 | 128 | 74 | 216 | 165 | 66 | 126 | 72 | 0 |
| 41 | 4 | 100 | 65 | 98 | 58 | 208 | 164 | 69 | 165 | 68 | 0 |
| 29 | 4 | 120 | 80 | 114 | 80 | 352 | 321 | 68 | 160 | 84 | 0 |
| 46 | 2 | 125 | 90 | 118 | 82 | 346 | 371 | 63 | 155 | 68 | 0 |
| 55 | 1 | 148 | 90 | 214 | 98 | 259 | 265 | 71 | 140 | 72 | 0 |
| 32 | 3 | 100 | 70 | 112 | 72 | 290 | 278 | 70 | 181 | 60 | 0 |
| 40 | 3 | 125 | 90 | 157 | 85 | 239 | 226 | 67 | 178 | 80 | 0 |
| 61 | 3 | 154 | 80 | 196 | 66 | 333 | 325 | 66 | 141 | 64 | 0 |
| 29 | 5 | 100 | 60 | 108 | 66 | 173 | 201 | 69 | 143 | 64 | 0 |
| 52 | 3 | 110 | 80 | 126 | 74 | 253 | 218 | 70 | 139 | 70 | 0 |
| 25 | 3 | 120 | 80 | 118 | 86 | 156 | 218 | 67 | 136 | 60 | 0 |
| 27 | 3 | 110 | 70 | 110 | 60 | 156 | 208 | 67 | 150 | 72 | 0 |
| 27 | 3 | 130 | 90 | 110 | 80 | 208 | 270 | 69 | 185 | 70 | 0 |
| 53 | 3 | 130 | 80 | 138 | 66 | 218 | 258 | 73 | 185 | 78 | 0 |
| 42 | 3 | 120 | 80 | 110 | 68 | 172 | 307 | 68 | 161 | 72 | 0 |
| 64 | 3 | 115 | 75 | 144 | 72 | 357 | 289 | 67 | 180 | 80 | 0 |
| 27 | 2 | 130 | 90 | 136 | 86 | 178 | 188 | 74 | 198 | 64 | 0 |

**Table 2-2b** *(Continued)*

| *A* | *B* | *C* | *D* | *E* | *F* | *G* | *H* | *I* | *J* | *K* | *L* |
|---|---|---|---|---|---|---|---|---|---|---|---|
| 55 | 1 | 110 | 70 | 146 | 58 | 283 | 200 | 70 | 128 | 68 | 1 |
| 33 | 5 | 120 | 80 | 160 | 92 | 275 | 310 | 67 | 177 | 84 | 0 |
| 58 | 1 | 110 | 80 | 138 | 72 | 187 | 202 | 71 | 224 | 72 | 0 |
| 51 | 3 | 110 | 80 | 118 | 70 | 282 | 294 | 71 | 160 | 72 | 0 |
| 37 | 3 | 134 | 80 | 142 | 78 | 282 | 290 | 71 | 181 | 88 | 0 |
| 47 | 3 | 120 | 70 | 98 | 64 | 254 | 272 | 65 | 136 | 63 | 0 |
| 49 | 3 | 120 | 90 | 156 | 86 | 273 | 243 | 71 | 245 | 80 | 0 |
| 46 | 1 | 150 | 104 | 174 | 94 | 328 | 368 | 71 | 187 | 76 | 0 |
| 40 | 2 | 110 | 70 | 146 | 77 | 244 | 199 | 70 | 161 | 78 | 1 |
| 26 | 3 | 110 | 80 | 112 | 78 | 277 | 280 | 74 | 190 | 72 | 0 |
| 28 | 1 | 110 | 70 | 132 | 68 | 195 | 230 | 73 | 180 | 90 | 0 |
| 23 | 2 | 108 | 68 | 118 | 64 | 206 | 261 | 71 | 165 | 80 | 0 |
| 52 | 3 | 125 | 90 | 164 | 78 | 327 | 299 | 65 | 147 | 108 | 0 |
| 42 | 3 | 110 | 75 | 114 | 58 | 246 | 254 | 67 | 146 | 80 | 0 |
| 27 | 2 | 115 | 88 | 118 | 78 | 203 | 198 | 70 | 182 | 104 | 0 |
| 29 | 1 | 120 | 80 | 112 | 78 | 185 | 226 | 72 | 187 | 84 | 0 |
| 43 | 2 | 122 | 78 | 106 | 52 | 224 | 196 | 66 | 128 | 60 | 0 |
| 34 | 3 | 110 | 70 | 108 | 74 | 246 | 255 | 68 | 140 | 80 | 0 |
| 40 | 2 | 120 | 80 | 98 | 68 | 227 | 200 | 67 | 163 | 89 | 0 |
| 28 | 3 | 115 | 70 | 124 | 66 | 229 | 270 | 70 | 144 | 88 | 0 |
| 30 | 3 | 115 | 70 | 110 | 52 | 214 | 216 | 71 | 150 | 68 | 0 |
| 34 | 1 | 108 | 75 | 118 | 78 | 206 | 204 | 71 | 137 | 100 | 0 |
| 26 | 2 | 120 | 80 | 106 | 64 | 173 | 183 | 67 | 141 | 72 | 0 |
| 34 | 3 | 135 | 90 | 160 | 98 | 248 | 308 | 70 | 141 | 72 | 0 |
| 35 | 3 | 100 | 70 | 110 | 72 | 222 | 248 | 73 | 190 | 76 | 0 |
| 34 | 3 | 110 | 60 | 114 | 66 | 230 | 246 | 73 | 167 | 84 | 0 |
| 45 | 3 | 100 | 70 | 112 | 60 | 219 | 233 | 69 | 159 | 64 | 0 |
| 47 | 3 | 120 | 80 | 124 | 76 | 239 | 269 | 67 | 157 | 72 | 0 |
| 54 | 3 | 134 | 90 | 168 | 100 | 258 | 276 | 66 | 170 | 64 | 0 |
| 30 | 2 | 110 | 80 | 120 | 80 | 190 | 206 | 69 | 132 | 72 | 0 |
| 29 | 1 | 110 | 80 | 172 | 96 | 252 | 288 | 68 | 155 | 80 | 0 |
| 48 | 3 | 110 | 80 | 128 | 76 | 253 | 261 | 72 | 178 | 80 | 0 |
| 37 | 3 | 120 | 88 | 132 | 82 | 172 | 231 | 71 | 168 | 80 | 0 |
| 43 | 3 | 138 | 94 | 162 | 104 | 320 | 412 | 65 | 159 | 76 | 1 |
| 31 | 3 | 124 | 90 | 124 | 64 | 166 | 188 | 67 | 160 | 84 | 0 |
| 48 | 2 | 140 | 85 | 132 | 78 | 266 | 296 | 71 | 165 | 72 | 0 |
| 34 | 3 | 110 | 80 | 122 | 78 | 176 | 231 | 72 | 194 | 70 | 0 |
| 42 | 3 | 130 | 80 | 126 | 58 | 271 | 288 | 70 | 191 | 84 | 1 |
| 49 | 1 | 122 | 94 | 166 | 106 | 295 | 276 | 73 | 198 | 80 | 0 |
| 50 | 2 | 130 | 85 | 118 | 74 | 271 | 230 | 68 | 212 | 102 | 1 |
| 42 | 4 | 120 | 80 | 154 | 94 | 259 | 333 | 64 | 147 | 108 | 0 |
| 50 | 3 | 140 | 95 | 144 | 68 | 178 | 224 | 68 | 173 | 72 | 1 |
| 60 | 5 | 160 | 100 | 138 | 82 | 317 | 291 | 72 | 206 | 80 | 0 |
| 27 | 5 | 124 | 80 | 116 | 78 | 192 | 219 | 70 | 190 | 72 | 0 |
| 29 | 2 | 120 | 85 | 114 | 76 | 187 | 266 | 68 | 181 | 72 | 0 |
| 29 | 2 | 110 | 70 | 108 | 70 | 238 | 253 | 72 | 143 | 60 | 0 |
| 49 | 3 | 112 | 78 | 116 | 68 | 283 | 253 | 64 | 149 | 82 | 0 |
| 49 | 4 | 100 | 70 | 104 | 59 | 264 | 243 | 70 | 166 | 68 | 0 |
| 50 | 1 | 128 | 92 | 148 | 84 | 264 | 325 | 70 | 176 | 96 | 0 |
| 31 | 2 | 105 | 68 | 98 | 54 | 193 | 180 | 67 | 141 | 64 | 0 |

**15.** Using 1-inch intervals, collect the height determinations $I$ in Table 2-2*b* into a frequency table and draw the histogram and cumulative-distribution polygon. Read the values of $P_{10}$, $P_{50}$, $P_{75}$ from the polygon.

**16.** Using intervals 101–115, 116–130, 131–145, . . . , collect the weight determinations $J$ in Table 2-2*a* into a frequency table and draw the histogram and cumulative-distribution polygon. Read the values of $P_{10}$, $P_{50}$, $P_{75}$ from the polygon.

**17.** Using intervals 101–115, 116–130, 131–145, . . . , collect the weight determinations $J$ in Table 2-2*b* into a frequency table and draw the histogram and cumulative-distribution polygon. Read the values of $P_{10}$, $P_{50}$, $P_{75}$ from the polygon.

**18.** Using intervals 50–54, 55–59, . . . , collect the pulse-rate determinations $K$ in Table 2-2*a* into a frequency table and draw the histogram and cumulative-distribution polygon. Read the values of $P_{10}$, $P_{50}$, $P_{75}$ from the polygon.

**19.** Using intervals 50–54, 55–59, . . . , collect the pulse-rate determinations $K$ in Table 2-2*b* into a frequency table and draw the histogram and cumulative-distribution polygon. Read the values of $P_{10}$, $P_{50}$, $P_{75}$ from the polygon.

**20.** The table below gives the distribution of the number of hours before onset of illness after eating at a banquet. Draw the histogram and the cumulative-distribution polygon. Read the values of $P_{10}$, $P_{50}$, $P_{75}$ from the polygon.

| *Hours before onset* | *Frequency* |
|---|---|
| 25 | 1 |
| 24 | 0 |
| 23 | 0 |
| 22 | 1 |
| 21 | 0 |
| 20 | 2 |
| 19 | 0 |
| 18 | 1 |
| 17 | 1 |
| 16 | 1 |
| 15 | 3 |
| 14 | 6 |
| 13 | 8 |
| 12 | 11 |
| 11 | 7 |
| 10 | 3 |
| 9 | 4 |
| 8 | 2 |
| 7 | 1 |
| 6 | 0 |
| 5 | 0 |
| 4 | 0 |
| 3 | 2 |
| 2 | 1 |
| 1 | 0 |

21. The table below gives the distribution of number of visits for well-baby care for a group of 246 families. Draw a histogram and cumulative-distribution polygon. Read the values of $P_{10}$, $P_{50}$, $P_{75}$ from the polygon.

| *No. of visits* | 9 | 8 | 7 | 6 | 5 | 4 | 3 | 2 | 1 | 0 |
|---|---|---|---|---|---|---|---|---|---|---|
| *Frequency* | 1 | 2 | 4 | 7 | 28 | 56 | 53 | 29 | 29 | 37 |

22. The table below gives the distribution of the number of calls required to obtain complete interviews from 646 participants in a study. Draw the histogram and cumulative-distribution polygon. Read the values of $P_{10}$, $P_{50}$, $P_{75}$ from the polygon.

| *No. of calls* | 10 | 9 | 8 | 7 | 6 | 5 | 4 | 3 | 2 | 1 |
|---|---|---|---|---|---|---|---|---|---|---|
| *Frequency* | 3 | 0 | 2 | 4 | 10 | 43 | 48 | 72 | 163 | 301 |

23. The table below gives the distribution of the number of people in families for a group of 434 families which had penicillin prophylaxis and 407 families which did not. The 434 families had been chosen at random from the 841 families available. Draw a histogram and cumulative-distribution polygon for each distribution and read the values of $P_{10}$, $P_{50}$, $P_{75}$ from each polygon.

| | NO. IN FAMILY | | | | | | | | | | | | | | | | | |
|---|---|---|---|---|---|---|---|---|---|---|---|---|---|---|---|---|---|---|
| | 22 | 16 | 15 | 14 | 13 | 12 | 11 | 10 | 9 | 8 | 7 | 6 | 5 | 4 | 3 | 2 | 1 | |
| *Prophylaxis* | 1 | 1 | 3 | 5 | 5 | 12 | 16 | 17 | 28 | 56 | 52 | 72 | 86 | 43 | 29 | 7 | 1 | 434 |
| *Comparison* | 0 | 1 | 2 | 1 | 6 | 7 | 11 | 22 | 44 | 36 | 57 | 62 | 62 | 51 | 28 | 16 | 1 | 407 |

24. The table below gives determination of serum cholesterol in milligrams per 100 milliliters for a group of males. Draw the histogram and cumulative-distribution polygon. Read the values of $P_{10}$, $P_{50}$, $P_{75}$ from the polygon.

| *Interval* | *Frequency* |
|---|---|
| 450–479 | 25 |
| 420–449 | 29 |
| 390–419 | 49 |
| 360–389 | 111 |
| 330–359 | 201 |
| 300–329 | 253 |
| 270–299 | 277 |
| 240–269 | 242 |
| 210–239 | 176 |
| 180–209 | 82 |
| 150–179 | 49 |
| 120–149 | 8 |

25. The table gives reported incomes of single-person households where the individual was 60 years of age or over. Draw a histogram and cumulative-distribution polygon. Read values of $P_{10}$, $P_{50}$, $P_{75}$ from the polygon.

| *Income, nearest $1,000* | *Frequency, thousands* |
|---|---|
| 1,000 | 2,000 |
| 2,000 | 585 |
| 3,000 | 235 |
| 4,000 | 165 |
| 5,000 | 115 |
| 6,000 | 30 |
| 7,000 | 25 |
| 8,000 | 15 |
| 9,000 | 10 |
| 10,000 | 8 |
| 11,000 | 6 |
| 12,000 | 4 |
| 13,000 | 2 |
| *Total* | 3,200 |

26. The table below gives distributions by sex of sweat chloride concentrations for carriers of the cystic fibrosis gene and for noncarrier controls. Draw a histogram and cumulative-distribution polygon for each. Read values of $P_{10}$, $P_{50}$, $P_{75}$ from each polygon.

| | CF CARRIERS | | CONTROLS | |
|---|---|---|---|---|
| *Sweat chloride, mEq/liter* | *Male* | *Female* | *Male* | *Female* |
| 91–100 | 1 | 1 | 0 | 2 |
| 81–90 | 0 | 1 | 0 | 0 |
| 71–80 | 4 | 1 | 4 | 2 |
| 61–70 | 5 | 1 | 4 | 3 |
| 51–60 | 11 | 17 | 10 | 10 |
| 41–50 | 8 | 14 | 16 | 27 |
| 31–40 | 16 | 14 | 23 | 23 |
| 21–30 | 10 | 15 | 28 | 24 |
| 11–20 | 7 | 5 | 16 | 8 |
| 1–10 | 1 | 0 | 0 | 2 |

27. In the following table are reasons given by 220 patients for referral to a hospital emergency service. Draw a histogram-type chart, putting the reason on the horizontal axis and the percentage of patients on the vertical axis (the figure drawn in such cases where the observation is not a measurement is called a *bar graph*). Why,

although it is possible to draw a cumulative-distribution polygon, would such a curve be of little value?

| *Reason* | *Frequency* |
|---|---|
| Patients' choice | 73 |
| No private physician | 61 |
| Physician unavailable | 45 |
| Proper hospital unit closed | 23 |
| Administrative | 13 |
| Convenience | 4 |
| Cost | 1 |

CHAPTER THREE

# Introduction to measures of central value and dispersion

In order to investigate the character of a distribution it is useful to have various words and measurements which will serve to describe the distribution. In Chap. 2 we drew pictures (histograms, polygons, etc.) of distributions, but it is not always possible or convenient to do so (sometimes it is not even desirable). In this chapter we shall define certain measurements which are most commonly used to describe a distribution. In Chaps. 4 to 8 we shall see how these particular measurements are used, and later, in Chap. 9, we shall examine certain alternative measurements and explain how a choice is made among them.

## 3-1 MEASURES OF CENTRAL VALUE: ARITHMETIC MEAN

*Central value* refers to the location of the *center* of the distribution. One of the intuitive notions of the center, or average, of a distribution is the *arithmetic mean,* which is defined as the sum of all the observations divided by the number of observations. If each observation is considered as having a unit mass and being distributed along an axis (as in a histogram), then the arithmetic mean is located at the centroid, or center of gravity, of the distribution. The equal amount of area allotted to each observation would have equal mass, so that if the histogram were a sheet of metal, it would balance on a pivot under the arithmetic mean. We shall use a bar above the measurement symbol to denote the value of the arithmetic mean. Thus $\bar{X}$ will represent the mean of a set of observations designated as $X$'s.

If there is a small number of observations, we shall add them directly and divide the total by the number of observations. Suppose we let $X_1$ be the value of the first observation, $X_2$ be the value of the second observation, $X_3$ be the value of the third, and so on. If there are four observations $X_1$, $X_2$, $X_3$, $X_4$, then we shall have

$$\bar{X} = \frac{X_1 + X_2 + X_3 + X_4}{4}$$

If there are 100 observations, then

$$\bar{X} = \frac{X_1 + X_2 + X_3 + \cdots + X_{100}}{100}$$

There will be many circumstances in which we need to add several numbers. Whenever we use the symbol $\Sigma$ (Greek letter capital sigma), we shall understand that it is read "the sum of" and that it tells us to add certain expressions. For example, if there are four observations $X_1$, $X_2$, $X_3$, $X_4$, then

$$\sum_{i=1}^{4} X_i = X_1 + X_2 + X_3 + X_4 \qquad \text{and} \qquad \bar{X} = \frac{\sum_{i=1}^{4} X_i}{4}$$

For the sum of the numbers $X_1, \ldots, X_{20}$ we write $\sum_{i=1}^{20} X_i$; this is read as "the sum of the 20 observations $X_1$ up to $X_{20}$." The subscript $i$ attached to the $X$ refers to the $i$th observation, and $X_i$ is the value of the $i$th observation. The symbol $\sum_{i=1}^{20}$ tells us to substitute $i = 1$, then $i = 2$, $i = 3, \ldots, i = 20$ into the expression immediately following the $\Sigma$ and then to add the 20 numbers obtained. Frequently, if there is no chance of ambiguity, the symbol is abbreviated to $\Sigma X_i$ or $\Sigma X$, and it is understood in either case that we are to add the 20 observations $X_1, \ldots, X_{20}$.

Following are several examples of situations in which it is convenient to use the symbol $\Sigma$. These examples are typical of the uses we shall make of this notation. Verbal statements for the summation precede each example.

1. The sum of squares of the four observations $X_1$, $X_2$, $X_3$, $X_4$:

$$\sum_{i=1}^{4} X_i^2 = X_1^2 + X_2^2 + X_3^2 + X_4^2$$

2. The sum of the products of pairs of values $a_i$, $X_i$:

$$\sum_{i=1}^{6} a_iX_i = a_1X_1 + a_2X_2 + a_3X_3 + a_4X_4 + a_5X_5 + a_6X_6$$

3. The sum of products of a constant $a$ times each of six numbers is equal to the constant times the sum of the six numbers:

$$\sum_{i=1}^{6} aX_i = aX_1 + aX_2 + aX_3 + aX_4 + aX_5 + aX_6$$

$$= a(X_1 + X_2 + X_3 + X_4 + X_5 + X_6) = a \sum_{i=1}^{6} X_i$$

Similarly, if $a$ is a constant, then $\sum_{i=1}^{N} aX_i = a \sum_{i=1}^{N} X_i$.

4. The sum of six numbers, each of which is the number 2:

$$\sum_{i=1}^{6} 2 = 2 + 2 + 2 + 2 + 2 + 2 = 6 \times 2 = 12$$

5. The sum of $N$ numbers, each number having the value $a$:

$$\sum_{i=1}^{N} a = a + a + \cdots + a = Na$$

6. The arithmetic mean is equal to the sum of the $N$ observations divided by the number of observations:

$$\bar{X} = \frac{X_1 + X_2 + \cdots + X_N}{N} = \frac{\sum_{i=1}^{N} X_i}{N}$$

7. The sum of products of pairs of observations. We shall use a particular example where $f_i$ refers to the frequency of observations of value $X_i$:

$$\sum_{i=1}^{4} f_iX_i = f_1X_1 + f_2X_2 + f_3X_3 + f_4X_4$$

8. The sum of products, with each product $\frac{1}{3}$ times the square of the observation:

$$\sum_{i=1}^{12} \tfrac{1}{3}X_i^2 = \tfrac{1}{3}X_1^2 + \tfrac{1}{3}X_2^2 + \cdots + \tfrac{1}{3}X_{12}^2$$

9. The sum of four numbers, each of which is the deviation of the observation from 3. This is equal to the sum of the four observations minus 4 times 3:

$$\sum_{i=1}^{4} (X_i - 3) = (X_1 - 3) + (X_2 - 3) + (X_3 - 3) + (X_4 - 3)$$
$$= X_1 + X_2 + X_3 + X_4 - 4 \times 3 = \left(\sum_{i=1}^{4} X_i\right) - 12$$

10. The sum of deviations of $N$ observations from the number $a$ is equal to the sum of the $N$ observations minus $N$ times $a$:

$$\sum_{i=1}^{N} (X_i - a) = \left(\sum_{i=1}^{N} X_i\right) - Na$$

**Computation procedure.** If the data are arranged in a frequency table, the approximate computation of the value of $\bar{X}$ can be shortened as follows. Let $X_1$ be the value of the mid-point of the first interval, and let $f_1$ be the number of observations that fall in this interval. Now, instead of adding $X_1$ to itself $f_1$ times, we can obtain the sum by multiplying $f_1$ by $X_1$. In the same way we find that the sum of all observations in the second interval is $f_2X_2$, in the third interval $f_3X_3$, and so on. If there are $k$ intervals, the sum of all the observations can be written as

$$f_1X_1 + f_2X_2 + \cdots + f_kX_k = \sum_{i=1}^{k} f_iX_i$$

The total number of observations is

$$f_1 + f_2 + f_3 + \cdots + f_k = \sum_{i=1}^{k} f_i = N$$

Thus we have

$$\bar{X} = \frac{\sum_{i=1}^{k} f_iX_i}{\Sigma f_i}$$

Referring to the observations on boys' heights in Chap. 2, there was one boy whose height was in the interval 63.5 to 64.5. The mid-point of this interval is 64. Here, then $X_1 = 64$ and $f_1 = 1$; $X_2 = 65$, $f_2 = 0$; $X_3 = 66$,

$f_3 = 2$; $X_4 = 67, f_4 = 5$, etc. Furthermore,

$$\Sigma f_i X_i = (64 \times 1) + (65 \times 0) + (66 \times 2) + (67 \times 5) + \cdots$$
$$= 64 + 0 + 132 + 335 + \cdots$$

and

$$\Sigma f_i = 1 + 0 + 2 + 5 + \cdots$$

In Table 3-1 is an example of the computation for finding $\bar{X}$ from the frequency table of the heights given in Chap. 2.

**Table 3-1**

| *Mid-point of interval* $X_i$ | *Frequency* $f_i$ | $f_i X_i$ |
|---|---|---|
| 75 | 1 | 75 |
| 74 | 1 | 74 |
| 73 | 3 | 219 |
| 72 | 8 | 576 |
| 71 | 12 | 852 |
| 70 | 16 | 1,120 |
| 69 | 22 | 1,518 |
| 68 | 9 | 612 |
| 67 | 5 | 335 |
| 66 | 2 | 132 |
| 65 | 0 | 0 |
| 64 | 1 | 64 |
| *Total* | 80 | 5,577 |

$$N = \Sigma f_i = 80$$
$$\Sigma f_i X_i = 5{,}577$$
$$\bar{X} = \frac{\Sigma f_i X_i}{\Sigma f_i} = \frac{5{,}577}{80} = 69.7 \text{ inches}$$

**Median.** Another measure of the "center" of a distribution is the *median,* the point on the scale of observations on each side of which there are equal areas under the histogram. This point is also the 50th percentile, $P_{50}$. Thus the median is the middle observation if there is an odd number of cases and, by convention, is the mean of the two central observations if there is an even number of cases. Alternatively, the value of the median could be defined as the quantity read from a cumulative-distribution polygon as described on page 10. The median is discussed further in Chap. 9.

## 3-2 MEASURES OF DISPERSION: VARIANCE

An important concept in statistics is that any average does not in itself give a clear picture of a distribution. The graphs of Fig. 3-1 show a group of distributions all having the same arithmetic mean yet obviously differing in general appearance.

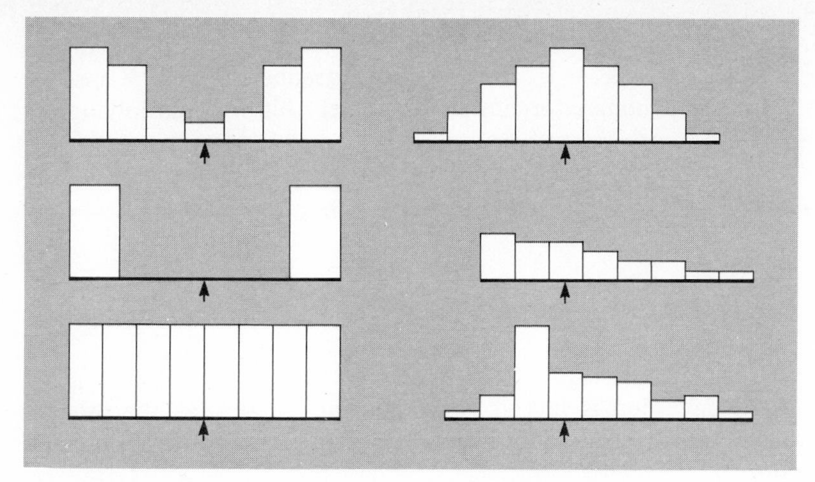

**Figure 3-1** Different distributions with equal mean values.

Another type of measure which helps to clarify the shape of the distribution is one that indicates how the observations are spread out from the average. Such a measure could be called a measure of *dispersion*, spread, or variability. It is usually desirable to have a measure which will be large if the observations are distant from the mean and small if they are close to the mean.

At first glance the sum of the deviations of the observations from the mean,

$$\sum_{i=1}^{N} (X_i - \bar{X})$$

this is sum of deviations

may seem to be a good measure for this purpose, but on further examination we see that its value is always equal to zero.

For example, the arithmetic mean of the numbers 2, 3, 5, 8 is 4.5, so the deviations from the mean are $-2.5$, $-1.5$, $+.5$, $+3.5$. The total of these numbers is zero.

This objection may be overcome by squaring the deviations before they are added. The *variance* is defined as the sum of squares of the deviations of the observations from $\bar{X}$ divided by one less than the total number of observations. In symbols, $X_1 - \bar{X}$ is the deviation of the first observation from the mean, $X_2 - \bar{X}$ is the deviation of the second observation from the mean, and so on. Thus the variance, which we shall represent symbolically by $s^2$, is

$$s^2 = \frac{(X_1 - \bar{X})^2 + (X_2 - \bar{X})^2 + \cdots + (X_N - \bar{X})^2}{N - 1} = \frac{\sum_{i=1}^{N} (X_i - \bar{X})^2}{N - 1}$$

We shall find subsequently that the division by $N - 1$ in place of $N$ is necessary in order to use $s^2$ for the various functions it will perform in a number of statistical methods. Many books introduce $s^2$ with $N$ instead of $N - 1$ in the denominator and later adjust this by multiplying by $N/(N - 1)$.

The *standard deviation* is defined as the positive square root of the variance. Thus the standard deviation is given by

$$s = \sqrt{\frac{\sum_{i=1}^{N} (X_i - \bar{X})^2}{N - 1}}$$

The standard deviation is in the same units of measure as the original observations and can be visually interpreted as a horizontal-scale unit on a histogram.

The variance is almost the mean of the squares of the deviations (if there were an $N$ in the denominator, it would be the arithmetic mean of these quantities). Occasionally you will see the variance referred to as the *mean-square deviation* and the standard deviation referred to as the *root-mean-square deviation.*

An algebraically equivalent formula for $s^2$ that is often simpler to use for computations is

$$s^2 = \frac{\Sigma X_i^2 - (\Sigma X_i)^2/N}{N - 1}$$

where we use the more convenient notation $\Sigma$ for $\sum_{i=1}^{N}$.

The equivalence of the two formulas for $s^2$ given above may be shown as follows:

$$\begin{aligned}
\Sigma(X_i - \bar{X})^2 &= \Sigma(X_i^2 - 2X_i\bar{X} + \bar{X}^2) \\
&= (X_1^2 - 2X_1\bar{X} + \bar{X}^2) + (X_2^2 - 2X_2\bar{X} + \bar{X}^2) \\
&\qquad + \cdots + (X_N^2 - 2X_N\bar{X} + \bar{X}^2) \\
&= (X_1^2 + X_2^2 + X_3^2 + \cdots + X_N^2) \\
&\qquad - 2\bar{X}(X_1 + X_2 + \cdots + X_N) \\
&\qquad + (\bar{X}^2 + \bar{X}^2 + \cdots + \bar{X}^2) \\
&= \Sigma X_i^2 - 2\frac{\Sigma X_i}{N}\Sigma X_i + N\left(\frac{\Sigma X_i}{N}\right)^2 \\
&= \Sigma X_i^2 - \frac{(\Sigma X_i)^2}{N}
\end{aligned}$$

If the observations are arranged in a frequency table, some of the work can be shortened in a manner similar to that employed in the computa-

**Table 3-2a**

| $X_i$ | $f_i$ | $f_iX_i$ | $f_iX_i^2$ |
|---|---|---|---|
| 75 | 1 | 75 | 5,625 |
| 74 | 1 | 74 | 5,476 |
| 73 | 3 | 219 | 15,987 |
| 72 | 8 | 576 | 41,472 |
| 71 | 12 | 852 | 60,492 |
| 70 | 16 | 1,120 | 78,400 |
| 69 | 22 | 1,518 | 104,742 |
| 68 | 9 | 612 | 41,616 |
| 67 | 5 | 335 | 22,445 |
| 66 | 2 | 132 | 8,712 |
| 65 | 0 | 0 | 0 |
| 64 | 1 | 64 | 4,096 |
| *Total* | 80 | 5,577 | 389,063 |

$$N = \Sigma f_i = 80$$

$$\bar{X} = \frac{\Sigma f_iX_i}{N} = \frac{5{,}577}{80} = 69.7$$

$$s^2 = \frac{\Sigma f_iX_i^2 - (\Sigma f_iX_i)^2/N}{N-1}$$

$$= \frac{389{,}063 - (5{,}577)^2/80}{79}$$

$$= \frac{276.4}{79} = 3.50$$

tion of the mean. If $f_i$ observations have the value $X_i$, then instead of adding the square of the $X_i$, $f_i$ times, we multiply the square of the $X_i$ by $f_i$ to give $f_iX_i^2$. To facilitate the work an additional column $f_iX_i^2$ may be added to the frequency table. Then the numbers in this column can be added to give $\Sigma f_iX_i^2$. If we already have a column of $f_iX_i$, we can get the $f_iX_i^2$ by multiplying $f_iX_i$ by $X_i$.

The square root of $s^2 = 3.50$ can be obtained from Table A-33 (in the Appendix). Direct reference to Table A-33 gives $s = 1.87$ correct to three figures. Square roots may also be computed by the method given at the foot of Table A-32.

The computations for $\Sigma X_i^2$ may be checked by replacing each $X_i$ by $X_i + 1$ and computing $\Sigma(X_i + 1)^2$. The check consists in noting that

$$\Sigma(X_i + 1)^2 = \Sigma X_i^2 + 2\Sigma X_i + N$$

Some other measures of dispersion are the *range, semiquartile range, mean deviation,* and distance between $P_{07}$ and $P_{93}$. The range is the distance between the smallest and largest observations and will be denoted by $w$. The semiquartile range is the distance between $P_{25}$ and $P_{75}$. The mean deviation is the mean of the deviations of the observations from $\bar{X}$ where the deviation is taken as the positive distance between the observation and $\bar{X}$. While each of these measures describes dispersion, they are not substitutes for each other. For example, although no formula relates the range and the standard deviation, it is impossible for the range to be much less than two standard deviations and if you compute a standard deviation to be less than, say, one-eighth of the range, you should recheck your arithmetic.

## 3-3 EFFECT OF UNIFORM CHANGE IN THE OBSERVATIONS

A few types of manipulation of the observations are frequently required, and we shall study them in some detail. In particular, adding or subtracting a constant to each observation and multiplying or dividing each observation by a constant have very specific effects upon the arithmetic mean, variance, and standard deviation.

Suppose we add 5 units to each observation. Then, instead of $X_1$, $X_2, \ldots, X_N$, we have the new numbers $X_1 + 5, X_2 + 5, \ldots, X_N + 5$. The mean of these new numbers will be

$$\text{New mean} = \frac{(X_1 + 5) + (X_2 + 5) + \cdots + (X_N + 5)}{N}$$

$$= \frac{1}{N}\left(\sum_{i=1}^{N} X_i + 5N\right) = \bar{X} + \frac{5N}{N} = \bar{X} + 5$$

The variance of the new numbers will be

$$\text{New variance} = \frac{\sum_{i=1}^{N} [X_i + 5 - (\bar{X} + 5)]^2}{N - 1}$$

$$= \frac{\sum_{i=1}^{N} (X_i + 5 - \bar{X} - 5)^2}{N - 1} = \frac{\sum_{i=1}^{N} (X_i - \bar{X})^2}{N - 1} = s^2$$

We see that the mean is increased by 5 units but that the variance and standard deviation are unchanged. If the increase (or decrease) in each observation is $a$ units instead of 5, the new arithmetic mean will be $a$ units larger (or smaller) than the old mean, and the variance and standard deviation will not be changed.

*Thus adding (or subtracting) a constant from a group of observations will add (or subtract) the same constant to the arithmetic mean but will not change the variance or standard deviation.*

Suppose we multiply each observation by a constant, say, $c$. Then the new observations are $cX_1, cX_2, cX_3, \ldots, cX_N$. The arithmetic mean of these quantities is $c$ times the arithmetic mean of the original observations:

$$\text{New mean} = \frac{\Sigma cX_i}{N} = \frac{c\Sigma X_i}{N} = c\bar{X}$$

The variance of the new numbers* is $c^2s^2$, and the new standard deviation is $cs$. If the observations are divided by $c$ instead of being multiplied, then the new mean is $\bar{X}/c$, the new variance is $s^2/c^2$, and the new standard deviation is $s/c$.

*Thus multiplying (or dividing) each observation by a constant will multiply (or divide) the arithmetic mean by the same constant, will multiply (or divide) the variance by the square of the constant, and will multiply (or divide) the standard deviation by the constant.*

A special case is the subtraction of $\bar{X}$ from each observation. Subtracting $\bar{X}$ from each observation also subtracts $\bar{X}$ from the mean. The new mean is $\bar{X} - \bar{X} = 0$. These new observations

$$X_1 - \bar{X}, X_2 - \bar{X}, \ldots, X_N - \bar{X}$$

have the same variance as before, since the subtraction of a constant from each observation does not affect the variance. Now suppose that each of these new observations is divided by the standard deviation $s$. Then we shall have a new set of numbers,

$$\frac{X_1 - \bar{X}}{s}, \frac{X_2 - \bar{X}}{s}, \frac{X_3 - \bar{X}}{s}, \ldots, \frac{X_N - \bar{X}}{s}$$

These new numbers still have mean 0 (dividing by $s$ affects the mean by dividing it by $s$, but $0/s$ is still 0). The variance will be divided by $s^2$. Thus the variance of the new set of numbers is $s^2/s^2 = 1$, and the new standard deviation is $\sqrt{1} = 1$. Observations transformed in this way are called *standardized observations*. This transformation is often carried out for scores or grades, since they are often arbitrarily constructed. In dealing with such observations it is sometimes desirable to have (for comparison of different grades or for quick reference) scores which can be easily compared. Standardization as described above may serve this purpose. A further step to obtain, say, a mean of 500 and a standard deviation of 100 is often taken. Such scores are referred to as *standard scores*.

To accomplish this we change each score by the formula

$$Z_i = 500 + 100\,\frac{X_i - \bar{X}}{s}$$

At the end of the last section we saw that $(X_i - \bar{X})/s$ has mean 0, and we

$$\text{* New variance} = \frac{\Sigma(cX_i - c\bar{X})^2}{N-1} = \frac{\Sigma c^2(X_i - \bar{X})^2}{N-1} = \frac{c^2\Sigma(X_i - \bar{X})^2}{N-1} = c^2s^2$$

$$\text{New standard deviation} = \sqrt{c^2s^2} = cs$$

note that the $Z_i$ will have a mean 500 units higher than zero. The $(X_i - \bar{X})/s$ was seen to have a standard deviation of 1, and the standard deviation of $100(X_i - \bar{X})/s$ is $100 \times 1 = 100$.

We have already shown that the addition of a constant to each score does not affect the standard deviation of a distribution. Therefore $Z_i$ will have a mean of 500 and a standard deviation of 100.

***Coding.** If the mid-points of the intervals are equally spaced, they may be replaced by the numbers 0, 1, 2, 3, . . . , or the 0 may replace an $X_i$ near the center of the distribution and the numbers run . . . , −3, −2, −1, 0, 1, 2, 3, . . . This is called *coding* of the observations. Where the $X_i$'s are large or where the number of observations is large, coding may save considerable time in the computation of $\bar{X}$ and $s^2$.

In coding we subtract the score coded as zero from each observation and divide the results by the length of the class interval $i$. As we have seen, the mean and standard deviation of the original observations are given by

$$\bar{X} = i\bar{x} + X_0 \qquad \text{and} \qquad s_X = is_x$$

where $\bar{x}$ = mean of coded scores
$X_0$ = score coded as zero
$s_X$ = standard deviation of original observations
$s_x$ = standard deviation of coded observations

* This section and later sections are marked with an asterisk to indicate that there will be no loss of continuity in omitting them.

**Table 3-2b**

| *Coded score* $x_i$ | $f_i$ | $f_ix_i$ | $f_ix_i^2$ |
|---|---|---|---|
| 5 | 1 | 5 | 25 |
| 4 | 1 | 4 | 16 |
| 3 | 3 | 9 | 27 |
| 2 | 8 | 16 | 32 |
| 1 | 12 | 12 | 12 |
| 0 | 16 | 0 | 0 |
| −1 | 22 | −22 | 22 |
| −2 | 9 | −18 | 36 |
| −3 | 5 | −15 | 45 |
| −4 | 2 | − 8 | 32 |
| −5 | 0 | 0 | 0 |
| −6 | 1 | − 6 | 36 |
| *Total* | 80 | −23 | 283 |

$$N = \Sigma f_i = 80$$

$$\bar{x} = \frac{\Sigma f_i x_i}{N} = \frac{-23}{80} = -.3$$

$$\bar{X} = i\bar{x} + X_0 = 1(-.3) + 70 = 69.7$$

$$s_x^2 = \frac{\Sigma f_i x_i^2 - (\Sigma f_i x_i)^2/N}{N-1}$$

$$= \frac{283 - (-23)^2/80}{79}$$

$$= \frac{276.4}{79} = 3.50$$

$$s_X = is_x = 1 \times 1.87 = 1.87$$

The computation given in Table 3-2*a* appears as in Table 3-2*b* if it is coded with $i = 1$ and $X_0 = 70$.

## GLOSSARY

| | | |
|---|---|---|
| coded score | median | standard score |
| dispersion | range | variance |
| mean | standard deviation | |

## DISCUSSION QUESTIONS

**1.** How can the sum of deviations from the mean be used as a check on arithmetic?

**2.** Is it possible for $\bar{X}$ and $s^2$ to have the same numerical value?

**3.** Distinguish between a measurement on an individual and a numerical quantity describing (measuring) a distribution. Give several examples of each.

**4.** Do the values of $\bar{X}$ and $s^2$ completely describe a distribution? Can you think of other measures analogous to $s^2$?

**5.** Suppose a group of scores are changed to $Z$ scores. What happens to the percentile rank of an individual? What happens to the deviation of an individual score from the mean?

**6.** Suppose each measurement in a distribution is multiplied by 2. What happens to the mean of the distribution? What happens to the variance of the distribution? What happens to the standard deviation of the distribution? What happens to each of the three if 4 is added to each measurement?

**7.** Suppose a set of observations has mean $\bar{X}$ and variance $s^2$. What happens to the mean and variance if each score is divided by $s$ and then $\bar{X}$ is subtracted from each quotient? Is the result the same if you first subtract $\bar{X}$ from each score and then divide the difference by $s$?

**8.** Give an example of a situation in which the median might be a more appropriate descriptive measure than the mean.

**9.** Is there any difference between $\Sigma X_i^2$ and $(\Sigma X_i)^2$?

## CLASS EXERCISE

Select from Table A-2 three groups of five numbers each. Start at some random place in the table, so that each student will have a different group of numbers. Select also three groups of five numbers each from Table A-23. Perform the necessary computations to obtain $\bar{X}$, $s^2$, and $s$ for each of the groups.

The results of this exercise will be used several times in later chapters to provide experimental verification of various theoretical statements. Therefore care should be taken in drawing the observations and the computations should be checked.

A possible worksheet for recording the data and results is shown as Table 3-3. Columns $A$, $B$, $C$ are for the samples from Table A-2, and the next column will be used for the combination of these three samples. Columns $D$, $E$, $F$ are for the samples from Table A-23, and the next column will be used for the combination of these three samples. The lines below the $s$ values will be used for computations in later chapters.

**Table 3-3** Data sheet for sampling experiment

| | From table A-2 | | | | From table A-23 | | | |
|---|---|---|---|---|---|---|---|---|
| | *A* | *B* | *C* | | *D* | *E* | *F* | |
| 1 | | | | | | | | |
| 2 | | | | | | | | |
| 3 | | | | | | | | |
| 4 | | | | | | | | |
| 5 | | | | | | | | |
| $\Sigma X_i$ | | | | | | | | |
| $\bar{X}$ | | | | | | | | |
| $\Sigma X_i^2$ | | | | | | | | |
| $\Sigma(X_i + 1)^2$ | | | | | | | | |
| $\Sigma X_i^2 + 2\Sigma X_i + N$ | | | | | | | | |
| $\Sigma X_i^2 - \frac{(\Sigma X_i)^2}{N}$ | | | | | | | | |
| $s^2$ | | | | | | | | |
| $s$ | | | | | | | | |
| $t = \sqrt{N}\frac{\bar{X} - 0}{s}$ | | | | | | | | |
| $\bar{X} + \frac{t_{.05}s}{\sqrt{N}}$ | | | | | | | | |
| $\bar{X} + \frac{t_{.95}s}{\sqrt{N}}$ | | | | | | | | |
| Median | | | | | | | | |
| Range | | | | | | | | |
| Midrange | | | | | | | | |
| $\frac{s^2}{\chi_{.05}^2/df}$ | | | | | | | | |
| $\frac{s^2}{\chi_{.95}^2/df}$ | | | | | | | | |

**Figure 3-2** Work sheet for sampling experiments.

Section ______________________ Date ______________________

Experiments for investigation of sampling distribution of various statistics; $\bar{X}$, $s^2$, $\bar{X}_1 - \bar{X}_2$, $t$, etc.

| Observed frequency (~~IIII~~ II) | Observation (midpoint of interval) | Code | $f$ | $xf$ | $x^2f$ | Check $(x+1)^2f$ | Cum. per cent to upper boundary |
|---|---|---|---|---|---|---|---|
| | | 15 | | | | | |
| | | 14 | | | | | |
| | | 13 | | | | | |
| | | 12 | | | | | |
| | | 11 | | | | | |
| | | 10 | | | | | |
| | | 9 | | | | | |
| | | 8 | | | | | |
| | | 7 | | | | | |
| | | 6 | | | | | |
| | | 5 | | | | | |
| | | 4 | | | | | |
| | | 3 | | | | | |
| | | 2 | | | | | |
| | | 1 | | | | | |
| | | 0 | | | | | |
| | | −1 | | | | | |
| | | −2 | | | | | |
| | | −3 | | | | | |
| | | −4 | | | | | |
| | | −5 | | | | | |
| | | −6 | | | | | |
| | | −7 | | | | | |
| | | −8 | | | | | |
| | | −9 | | | | | |
| | | −10 | | | | | |
| | | −11 | | | | | |
| | | −12 | | | | | |
| | | −13 | | | | | |
| | | −14 | | | | | |
| | | −15 | | | | | |
| Totals | | | | | | | |

$\Sigma f_i(X_i + 1)^2 = \Sigma f_i X_i^2 + 2\Sigma f_i X_i + N$

______ = ______ + ______ + ______

Observed mean $= \dfrac{\Sigma X_i}{N} =$

Observed variance $= \dfrac{\Sigma X_i^2 - \dfrac{(\Sigma X)^2}{N}}{N-1}$

$=$ ______________

Theoretical mean $=$

Theoretical variance $=$

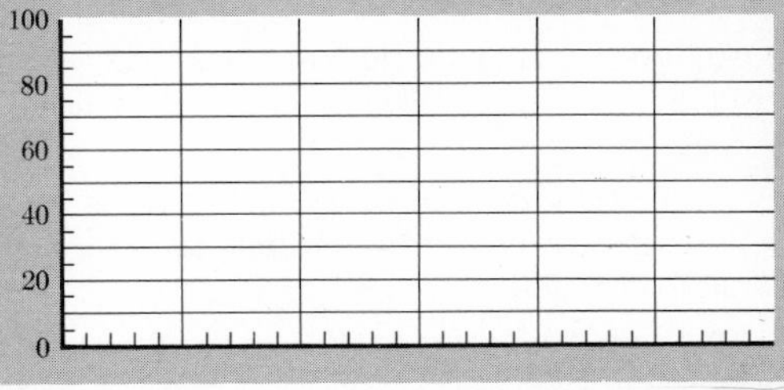

100

80

60

40

20

0

## PROBLEMS

**1.** Compute the values of $\bar{X}$, $s^2$, and $s$ for the data in the table, where the values for $X$ represent the mid-points of intervals.

| $X$ | 15 | 25 | 35 | 45 | 55 | 65 | 75 | 85 | 95 | 105 |
|---|---|---|---|---|---|---|---|---|---|---|
| $f$ | 1 | 5 | 12 | 18 | 21 | 19 | 10 | 7 | 6 | 1 |

In Probs. 2 to 24 compute the values of $\bar{X}$, $s^2$, $s$, the median, and the range for the data given in the indicated problems of Chap. 2.

**2.** Prob. 3
**3.** Prob. 5
**4.** Prob. 6
**5.** Prob. 7
**6.** Prob. 8
**7.** Prob. 9
**8.** Prob. 10
**9.** Prob. 11
**10.** Prob. 12
**11.** Prob. 13
**12.** Prob. 14
**13.** Prob. 15
**14.** Prob. 16
**15.** Prob. 17
**16.** Prob. 18
**17.** Prob. 19
**18.** Prob. 20
**19.** Prob. 21
**20.** Prob. 22
**21.** Prob. 23
**22.** Prob. 24
**23.** Prob. 25
**24.** Prob. 26

**25.** Compute the values of $\bar{X}$, $s^2$, and $s$ for the following five observations:

96 103 98 100 96

**26.** Sketch histograms of two distributions having the same mean and total area such that one distribution has a large variance and the other has a small variance.

**27.** Sketch a histogram of a distribution where the mean is equal to the median. Is there any general class of distributions for which the mean and median are equal?

**28.** The weights of a number of packages of frozen peas are given as follows:

| | | | | | | | |
|---|---|---|---|---|---|---|---|
| 16.1 | 15.9 | 15.8 | 16.3 | 16.2 | 16.0 | 16.1 | 16.0 |
| 16.0 | 16.1 | 16.0 | 15.9 | 16.1 | 16.0 | 16.0 | 15.9 |

(*a*) Make a frequency table. (*b*) Draw a histogram. (*c*) Find the mean. (*d*) Find the median. (*e*) Find the range. (*f*) Find the variance. (*g*) Find the standard deviation.

**29.** The following observations are the yields in pounds of hops. Find the mean, variance, and standard deviation.

| | | | | | | | | | | |
|---|---|---|---|---|---|---|---|---|---|---|
| 3.4 | 4.4 | 4.8 | 4.5 | 5.1 | 4.6 | 5.5 | 4.7 | 3.5 | 3.6 | 4.2 |
| 4.8 | 3.4 | 4.3 | 5.0 | 3.6 | 3.5 | 5.3 | 4.7 | 5.4 | 2.2 | 4.0 |
| 4.6 | 3.0 | 5.3 | 2.6 | 4.3 | 5.0 | 5.8 | 3.1 | 2.7 | 4.8 | 4.0 |
| 3.6 | 5.0 | 3.0 | 3.2 | 3.4 | 5.6 | 5.3 | 5.8 | 4.2 | 4.6 | 3.7 |
| 6.0 | 6.2 | 5.0 | 6.8 | 6.0 | 6.5 | 4.8 | 6.6 | 7.0 | 7.4 | 5.6 |

**30.** What would be the effect on $\bar{X}$ of subtracting 40 from each observation in Prob. 5, Chap. 2? What would be the effect on $s^2$? On $s$? On the median? On the range?

**31.** Subtract 64 from each observation in Table 3-2*a*. Compute the new mean, variance, standard deviation, median, and range and compare the values computed with those for Table 3-2*a*.

**32.** Find the $Z$ score for $X = 80$ in Prob. 1 above.

CHAPTER FOUR

# Population and sample

Any set of individuals (or objects) having some common observable characteristic constitutes a *population*, or *universe*. Any subset of a population is a *sample* from that population. The examples of *inference* presented here are primarily for *random* samples because the sampling technique is appealing and widely used and because of the simplicity of certain formulas. For a random sample the individuals are chosen in turn from the population, with each remaining individual having an equal chance of being chosen on the next draw. If the individuals selected are returned to the population before the next sample member is selected, the process is called sampling *with replacement;* if the next sample member is selected from a population that does not include previously selected members, the sampling is called sampling *without replacement.*

The term "population" may refer either to the individuals measured or to the measurements themselves. There is then a distribution of the measurements of a sample, which we actually observe and study, and a distribution of the measurements of the population, which may exist but is usually not in observed or recorded form. One of the most important problems in statistics is to decide what information about the distribution of the population can be inferred from a study of the sample.

A third type of distribution consists of the distribution of a measurement made on each of all possible samples of a fixed size which could be taken from a universe. For example, if we take all possible samples of 10 students from a given school and compute the mean height of each sample, we will have a large number of means. These means form a distribution, which we call the *sampling distribution* of the mean of samples of size 10.

EXAMPLE 4-1. A study is to be made of the heights of the men in a certain city. A sample of 200 men is chosen, and their heights are recorded. Here the universe consists of all men in the city. The sample consists of the 200 men chosen. The measurable characteristic is height. We can form a distribution of the heights observed in the sample, but presumably the distribution of the universe is not available. If we compute the mean of the heights of the 200 men in the sample, we can think of this mean as a single observation from the distribution of means of all possible samples of 200 men.

EXAMPLE 4-2. An investigation is being undertaken to test the effects of a particular type of drug on tetanus. A group of rats infected with tetanus is treated with the drug, and then the proportion of rats recovering within a specified time interval is observed. The universe consists of all rats which will be, or could be, infected with tetanus and then treated. The sample consists of the group of rats actually used in the investigation, and the characteristic is either recovery or failure to recover within the specified time interval. The universe here cannot be observed, since we could not perform this experiment on every rat.

EXAMPLE 4-3. In a survey of the manual abilities of children (aged 12) in a certain locality, a manual-skills test was given to 200 children. The universe would be all 12-year-old children in the locality sampled from, and the sample would be the 200 children taking the test.

EXAMPLE 4-4. A field of wheat has an area of 4,000 acres. A sample of 30 plots of size 1 square rod each is chosen, and the yields of these plots are measured. The population consists of the 640,000 individual square-rod plots, the sample is the 30 chosen, and the measurement is the yield of a plot.

EXAMPLE 4-5. A survey is made to determine the opinion of the people in a certain city on a proposed pension law. A sample of 500 people is questioned, and the number of favorable opinions is recorded. Here, as in example 2, the measurement is of a discrete type; we may record a 1 if the response is favorable and a 0 if it is unfavorable.

EXAMPLE 4-6. It is desired to estimate the mean score of students in a particular school on a standardized test. A sample of 10 students is chosen to take the test. The measurement is the student's score on the test.

EXAMPLE 4-7. Twelve seeds of a certain type of hybrid corn are planted and the yields observed. Here the population does not actually exist; it consists of all seeds of this type of corn. Probably very little seed actually exists, and whether or not much will exist in the future may depend on the outcome of this experiment.

## 4-1 POPULATION, OR UNIVERSE

We shall use the words "population" and "universe" interchangeably. A universe can be finite (children in a city, cows in a geographical region, electric light bulbs manufactured during a day), or it can be infinite (all points on a line, all heights between 5 feet and 6 feet, the time required for rats to run a maze).

In many cases a universe will be finite but so large that we treat it as though it were infinite (a boxcar of grain, the apples in an orchard). Many populations are so scattered as to be inaccessible as a whole. For example, in studying some characteristic of all students graduated from a university it might be impractical to contact every graduate but quite reasonable to contact a sample; only if the people were collected in some way might it be reasonable to measure every individual in the population. Similarly, if it were desired to study some characteristic of railway workers, the population would be so disperse as to make any complete survey very difficult.

Most of our procedures will be based on the assumption of an infinite population. Some modifications of formulas are necessary in case the population is not large in relation to the size of the sample. Generally, however, such corrections for finite population size can be ignored if the population has, say, several hundred individuals and the size of the sample is not more than 5 per cent of the number of individuals in the population.

A different type of universe is obtained if we consider the problem of measuring the length of a field. If we measure it a large number of times, we shall get a distribution of measurements (not all alike owing to errors in measurement). The set of all possible measurements of the field which have been, or will be, or could be taken can be thought of as forming a universe. Obviously, we can never observe the entire universe. Physicists in measuring the attraction of gravity or chemists in performing very accurate weighing have this type of problem.

The various descriptive terms and measurements previously mentioned are applied to the distribution of the universe, as well as to other distributions. We shall be especially interested in the *mean* and *variance*. We shall denote the arithmetic mean of the universe by the Greek letter $\mu$ (mu) and the variance of the universe by $\sigma^2$ (sigma squared). The *standard deviation* of the universe is then denoted by $\sigma$.

Any measurable characteristic of the universe is called a *parameter*. $\mu$ and $\sigma$ are examples of parameters.

## SAMPLE 4-2

The size of a sample, usually represented by the letter $N$, is the number of individuals in the sample. A sample may be any size from $N = 1$ to the number of items in the universe. The measurements on the individuals in the sample will form a distribution which will have a mean, denoted by $\bar{X}$, and a variance, denoted by $s^2$. Presumably, $\bar{X}$ and $s^2$, which we can actually measure, should give us some information about $\mu$ and $\sigma^2$, whose values are usually not known. $\bar{X}$ and $s^2$ are different from sample to sample, while $\mu$

and $\sigma^2$ are constant; that is, they have particular values for a particular universe.

A value computed entirely from the sample is called a *statistic*. $\bar{X}$, $s^2$, the median, and the range of the sample are examples of statistics.

The method of choosing a sample is an important factor in determining what use can be made of the sample. Generally, if we know the probability of each individual in the population being selected for the sample, then the reliability of the sample can be assessed. In particular, for random samples this probability is constant. If a random sample was desired but some individuals in the universe are more likely to be chosen than others, the sample is said to be *biased*. It has been found that subjective methods of picking individuals for a sample often (perhaps usually) lead to biased samples, apparently owing primarily to preferences of the person making the selections. To prevent such bias, and to avoid the criticism of bias even if the sample is not biased, some nonsubjective method of choosing a sample, such as random sampling, should be employed.

**Random sampling.** In a random sample every individual in the population has an equal and independent chance of being chosen for the sample. Technically, for sampling with replacement every individual chosen should be measured and returned to the population before another selection is made. This means, of course, that an individual can be chosen twice in the same sample. For example, to choose a random sample of five cards out of a full deck you should choose one card, record its face value, and return it to the deck, shuffle the deck thoroughly, draw another card, record its value, and replace it in the deck, and so on. If the population is large in relation to the sample size, very little error will result from the procedure of not returning each individual to the population. In practice the individual is rarely returned, and in fact it is frequently impossible to do so, as in a case where the individual is changed or destroyed by the examination. In many cases care is actually taken to prevent the same individual from appearing twice in a sample. Note in Sec. 4-5, however, that where there is a very small population we return each individual before drawing another.

## 4-3 RANDOM NUMBERS

Suppose that we have a finite population from which we wish to draw a random sample of $N$ individuals. One method of doing this would be to assign a number to each member of the population, put a set of numbered tags corresponding to the individuals into a box, and draw $N$ tags from the box. The numbers on these $N$ tags will correspond to the individuals to be selected. This is a satisfactory method except for the labor involved. Very thorough mixing is required. Experience has shown that "card shuffling" often provides inadequate mixing.

We can shorten this process by use of a table of *random numbers*, Table A-1 in the Appendix. Such a table consists of numbers chosen in a fashion similar to drawing numbered tags out of a box. The table is so made that all numbers, 0, 1, . . . , 9 appear with approximately the same frequency; by combining numbers in pairs we have the numbers from 00 to 99; by using the numbers three at a time we have numbers from 000 to 999, and so on. Such a table could be constructed as follows. Number 10 tags 0, 1, 2, 3, 4, 5, 6, 7, 8, 9 and place them in a box. Draw one and record its value, return it and draw another, and so on. We shall use the numbers from Table A-1 to select random samples.

The table should be entered in a random manner. One way is to close your eyes and place a finger on one page of the table. Either the digits under your finger can be used, or these digits may be used to locate other digits in the table. For example, (29, 32) selected in this way can be taken to indicate the use of the 29th number in the 32d column of the table. This procedure may be repeated until the required number of random digits is obtained. Since the digits are thoroughly mixed in the table, the entire sample may be taken in a group once a location has been determined randomly. The table is large enough to make it very unlikely that there will be much duplication in different samples.

Suppose we wish to select at random 20 items from a universe of 400 items. First we assign numbers to the 400 items in the universe. Three-place numbers are needed (400 of them), so we agree to use three columns in the table. If the number in the table is 400 or less we take the corresponding item from the universe. If the number is more than 400 we skip it. If a number comes up twice, we skip it the second time (thus the sampling is without replacement). We continue reading numbers until the desired number of items, 20 in this case, have been chosen for the sample.

For large populations this procedure has the drawback that the universe has to be numbered, and it must be relatively easy and inexpensive to find any particular individual. Modern card files and machines such as punched-card sorters may make this type of sampling practical for extremely large populations.

Sometimes, when the population cannot be easily numbered, some sort of geographical numbering can be made. For example, to sample from a field of some agricultural product we might divide the field into a large number of small rectangles by lines running north and south and lines running east and west. The rectangles are then numbered, and some of them are chosen in the manner described above.

Another example might be the choice of a sample for determining the consumer-price index. The counties in the United States could be numbered and a sample chosen from them (actually types of sampling other than random are ordinarily used in surveys of this sort).

Suppose we wish to investigate the effect of two drugs on mice. A random-number table could be used to select from a group of 100 mice a random sample of 50 mice to receive the first drug. The remainder of the mice could be given the second drug. Sometimes other methods are preferred (we shall discuss some of them later); for example, we might wish to guard against all the first 50 being males and all the second 50 being females.

## 4-4 DESIGN OF EXPERIMENTS

The method of choosing a sample is called the *design of the experiment.* We have focused on the design where the observations are chosen at random, but this sampling design is by no means always the best. For some fields of application other designs will give more precise information. It is perhaps more appropriate to study these designs in detail in specialized courses after you have obtained a good general knowledge of statistical application. Hence we shall not emphasize any one design (except random sampling) to any great extent.

The so-called "experiment" is usually a situation in which the population does not actually exist and the purpose of the investigation is to establish something about such a population if it were to exist. Many agricultural, biological, medical, psychological, educational, and chemical experiments are of this type. Examples 4-2 and 4-7 are of this sort, as are problems of measuring the effects of fertilizers or drugs, the effects of teaching methods, the results of a particular method of chemical analysis, and so on. In situations of this type specialized designs, where outside conditions are controlled or where groups of individuals are chosen in particular ways, may give more information than random sampling. In fact, it is frequently difficult in experimental work to ascertain that we have actually sampled in a particular manner, and considerable theory is concerned with this problem.

A second situation occurs when there is actually an existing population and we wish to determine by a "survey" some characteristic of that population. Examples 4-1, 4-3, and 4-4 illustrate this type of situation. Frequently a statistical problem will involve both these situations, as in Example 4-2. In surveys a sampling design involving "stratification" of the population is often used and will, if used correctly, give more precise information about the population than random sampling. For stratification the population is subdivided into several parts, or strata, and the number of observations in the sample is apportioned among these strata. Frequently the proportion of the sample to be taken from each of the strata is fixed to be the same as the proportion of the population in that strata. Stratification methods have been widely used in sampling human populations, as in public-opinion polls and market surveys.

Very misleading results can occur if samples are not taken correctly. Probably the most publicized example is the case of the news magazine which attempted to predict the presidential election of 1936 by taking a large sample entirely from a few strata, completely ignoring the other strata. Of course, in public-opinion surveys (as in many other practical situations) there is further difficulty in actually measuring the individual. An inexperienced interviewer may easily obtain false responses from an obliging interviewee or no response from certain people.

Most of the discussion in this book is restricted to random sampling. We shall base inferences and conclusions on the assumption that the sampling has been carried out randomly.

## SAMPLING DISTRIBUTIONS 4-5

Any statistic (measurement on a sample) has a *sampling distribution.* Here we shall examine the sampling distribution of $\bar{X}$ and $s^2$. Suppose we take from a universe *all possible* samples of, say, size 10. On each sample we perform the necessary computations to find $\bar{X}$ and $s^2$. We should then have a universe of means (consisting of all these values of $\bar{X}$) and a universe of $s^2$. These have distributions which are called, respectively, the *sampling distribution of the mean* and the *sampling distribution of the variance* (sometimes abbreviated to *distribution of mean* and *distribution of variance*). By examination of the sampling distribution of the mean we can tell how frequently the mean $\bar{X}$ will fall in any interval we wish to discuss. For example, we can tell how frequently it will fall within two units of the universe mean $\mu$, between $X = 7$ and $X = 8$, etc. Let us actually record the sampling distributions of the mean and variance in a simple case and see what they look like. Remember that in practical work it would be essentially impossible to record such distributions, and we are doing it here to show how these statistics behave in an actual situation. The distributions for our analyses are available from theoretical considerations.

EXAMPLE 4-8. Suppose our universe consists of six rabbits. The weights of the rabbits in ounces are

$$X_1 = 11 \qquad X_2 = 16 \qquad X_3 = 12 \qquad X_4 = 15 \qquad X_5 = 16 \qquad X_6 = 14$$

Now suppose we take samples of size $N = 2$. There are 36 different samples possible, and they are listed in Table 4-1. Beside each one is recorded the $\bar{X}$ and the $s^2$ for that particular sample. Notice that it is possible for an individual to be chosen more than once in the same sample. In practice the universe will be so large that the chances of duplication will be very small.

Here our sampling distribution of the mean consists of the 36 values of $\bar{X}$. The sampling distribution of the variance consists of the 36 values of $s^2$. Table 4-1 shows all possible samples of size 2, with mean and variance for each. The first observation is recorded along the top, and the second observation is recorded on the left. In each box of the main part of the table the first number is the mean and the second is the variance of the two observations in the sample. To make comparisons easier, Fig. 4-1 depicts these two distributions with the distribution of the universe. Since we have the entire universe of six items, we can compute the mean:

$$\mu = \frac{11 + 16 + 12 + 15 + 16 + 14}{6} = \frac{84}{6} = 14$$

The population variance, defined with $N$ in the denominator, is

$$\sigma^2 = \frac{11^2 + 16^2 + 12^2 + 15^2 + 16^2 + 14^2 - (84)^2/6}{6} = \frac{22}{6} = \frac{11}{3}$$

We also compute the mean and variance for the distribution of $\bar{X}$ and for the distribution of $s^2$ (using 36, not 35, in the denominator for the variance). The computation is shown in Table 4-2. Note that the mean of the sampling distribution of $\bar{X}$ has the same value as the mean $\mu$ of the universe. Note, also that the mean of the $s^2$ distribution has the same value as the variance $\sigma^2$ of the universe. This second equality occurs only if we use $N - 1$ in the computation of the *sample* variance. Thus we can say that "on the average" the sample variance is equal to the variance of the universe. If we had used $N$ instead of $N - 1$ in the sample variance, the mean of the sampling distribution would not equal universe variance $\sigma^2$. If the entire universe is available, the use of $N$ (not $N - 1$) provides the correct value of $\sigma^2$.

A statistic whose sampling distribution has its mean equal to a population parameter is called an *unbiased estimate* of that parameter; thus for random samples with replacement $\bar{X}$ is an unbiased estimate of $\mu$ and $s^2$ is an unbiased estimate of $\sigma^2$.

**Table 4-1**

| SECOND OBSERVATION \ FIRST OBSERVATION | 11 | 16 | 12 | 15 | 16 | 14 |
|---|---|---|---|---|---|---|
| 11 | 11 / 0 | 13.5 / 12.5 | 11.5 / .5 | 13 / 8 | 13.5 / 12.5 | 12.5 / 4.5 |
| 16 | 13.5 / 12.5 | 16 / 0 | 14 / 8 | 15.5 / .5 | 16 / 0 | 15 / 2 |
| 12 | 11.5 / .5 | 14 / 8 | 12 / 0 | 13.5 / 4.5 | 14 / 8 | 13 / 2 |
| 15 | 13 / 8 | 15.5 / .5 | 13.5 / 4.5 | 15 / 0 | 15.5 / .5 | 14.5 / .5 |
| 16 | 13.5 / 12.5 | 16 / 0 | 14 / 8 | 15.5 / .5 | 16 / 0 | 15 / 2 |
| 14 | 12.5 / 4.5 | 15 / 2 | 13 / 2 | 14.5 / .5 | 15 / 2 | 14 / 0 |

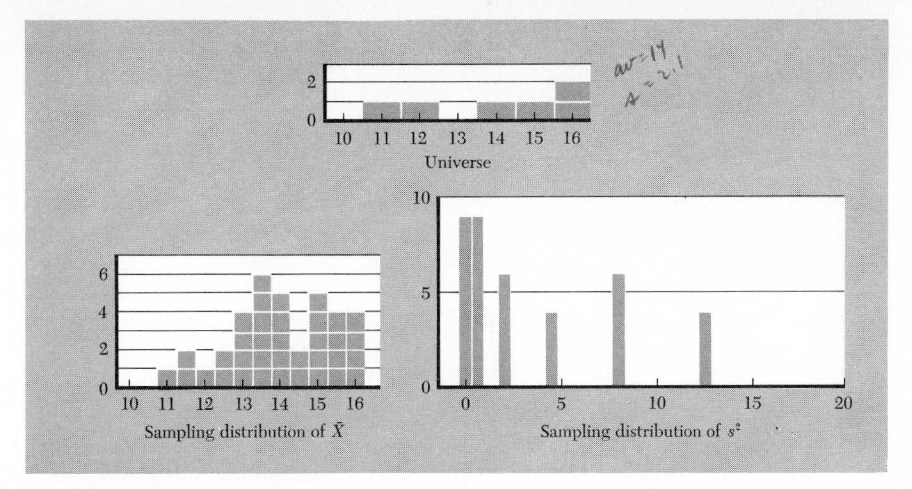


**Figure 4-1** Distribution of universe and sampling distributions.

Denote the variance of the universe by $\sigma^2$ and the variance of the sampling distribution of the mean by $\sigma_{\bar{X}}{}^2$. In the example above $\sigma^2/\sigma_{\bar{X}}{}^2 = 2$, which is $N$, the size of the sample. This is always true for random samples with replacement and can be written as $\sigma_{\bar{X}}{}^2 = \sigma^2/N$. Correspondingly, $\sigma_{\bar{X}} = \sigma/\sqrt{N}$. It is common practice to refer to the standard deviation of a sampling distribution of a statistic as the *standard error* of that statistic. Thus $\sigma_{\bar{X}}$ is the standard error of $\bar{X}$ and for random samples with replacement is numerically equal to $\sigma/\sqrt{N}$.

The variance of the distribution of $s^2$ does not have such a nice relationship to $\sigma^2$.

**Table 4-2**

| $\bar{X}$ | $f$ | $f\bar{X}$ |
|---|---|---|
| 16 | 4 | 64 |
| 15.5 | 4 | 62 |
| 15 | 5 | 75 |
| 14.5 | 2 | 29 |
| 14 | 5 | 70 |
| 13.5 | 6 | 81 |
| 13 | 4 | 52 |
| 12.5 | 2 | 25 |
| 12 | 1 | 12 |
| 11.5 | 2 | 23 |
| 11 | 1 | 11 |
| *Total* | 36 | 504 |

Mean $\bar{X} = \frac{504}{36} = 14$
Var. $\bar{X} = \frac{66}{36} = \frac{11}{6}$

| $s^2$ | $f$ | $f(s^2)$ |
|---|---|---|
| 12.5 | 4 | 50 |
| 8 | 6 | 48 |
| 4.5 | 4 | 18 |
| 2 | 6 | 12 |
| .5 | 8 | 4 |
| 0 | 8 | 0 |
| *Total* | 36 | 132 |

Mean $s^2 = \frac{132}{36} = \frac{11}{3}$

The sampling distribution of $\bar{X}$ shows the chance that a sample mean $\bar{X}$ may have certain values. In Example 4-8 the chance that $\bar{X} = 14$ is $\frac{5}{36}$ (5 cases in 36). As another example, we compute the chance that $\bar{X} = 13$, 13.5, 14, or 14.5 to be $\frac{17}{36}$. Compare this result with the chance that a single observation $X$ is equal to 13, 13.5, 14, or 14.5. We compute

$$\frac{0 + 0 + 1 + 0}{6} = \frac{1}{6} = \frac{6}{36}$$

The four values chosen for $X$ are near $\mu = 14$, and we can see that the mean of a sample of two observations has a greater chance than a single observation of being close to the population mean. This result is also indicated by the smaller variance $\sigma^2/2$ of the sampling distribution of $\bar{X}$ for $N = 2$.

It is a lengthy task to write down the sampling distribution for even such a small universe and such a very small sample size. For a universe of 1,000 and a sample of size 20 it would be impossible to enumerate the many billion individual items in the sampling distribution. We can approximate the sampling distribution in cases of this type by drawing a large number (perhaps several hundred) of random samples and recording the distribution observed. However, this is a tedious method of obtaining information or "proving" conjectures. Sometimes we can fall back on the authority of statements of "proven" mathematical theorems. These theorems will have certain assumptions, and in applying them we must keep the assumptions, as well as conclusions, in mind. Usually these assumptions will involve the distribution of the population and the size of the sample, as well as the method of drawing the sample.

EXAMPLE 4-9. With these limitations in our methods in mind, let us draw a large number of samples from some universe and attempt to approximate the sampling distribution of the mean $\bar{X}$ and the variance $s^2$. We take two sample sizes $N = 10$ and $N = 40$ to illustrate the effect of sample size.

We shall see now the use of another random-number table. This table, Table A-2, has items chosen at random from a universe which has a *normal* distribution. This normal distribution is quite important in statistical method and will be discussed in detail in the next chapter. Random-number Table A-1 was taken from a uniform distribution, with all numbers 0, 1, 2, 3, 4, 5, 6, 7, 8, 9 being equally frequent. In Table A-2 numbers with values near zero occur fairly frequently, and numbers farther and farther away from zero occur less and less frequently. The numbers in the universe (several pages of Table A-2) have mean 0 and variance $\sigma^2 = 1$ ($\sigma = 1$). Suppose we take 200 samples of $N = 10$ and 200 of $N = 40$ and record the values of $\bar{X}$ and $s^2$ for each sample. We can then obtain some idea of the general shape of the sampling distributions of $\bar{X}$ and $s^2$.

The method of choosing samples should be fairly random. First we pick a two-digit number from Table A-1; the number is (3,6). We start on page 3, column 6, of Table A-2 and record the numbers in 200 sets of 10. It is not necessary to mix them up, since this was done before putting them into the table. We put these sets of 10 together, four at a time for 50 samples of 40 each. Then we take 150 more samples of 40 each.

Table 4-3 shows the observed means and variances corresponding to the indicated mid-points of class intervals. Figure 4-2 contains the histograms for the four frequency tables.

Some of the things we should look for in these sampling distributions are as follows:

1. The general shape of the curves.

2. The mean of the sampling distributions. From the statements in Example 4-8 we expect the sampling distribution of the mean $\bar{X}$ to have an average approximately equal to $\mu$ (zero in this example) and the sampling distribution of the variance to have an average of approximately $\sigma^2$ (1 in this

**Table 4-3** Observed values of $\bar{X}$ and $s^2$ from a normal population with $\mu = 0$ and $\sigma^2 = 1$

| VALUES OF $\bar{X}$ | | | | VALUES OF $s^2$ | | | |
|---|---|---|---|---|---|---|---|
| $N = 10$ | | $N = 40$ | | $N = 10$ | | $N = 40$ | |
| *Mid-point* | *Freq.* | *Mid-point* | *Freq.* | *Mid-point* | *Freq.* | *Mid-point* | *Freq.* |
| .75 | 2 | .425 | 3 | .10 | 3 | .45 | 3 |
| .65 | 3 | .375 | 0 | .30 | 18 | .55 | 16 |
| .55 | 7 | .325 | 9 | .50 | 33 | .65 | 16 |
| .45 | 13 | .275 | 6 | .70 | 28 | .75 | 21 |
| .35 | 13 | .225 | 8 | .90 | 29 | .85 | 27 |
| .25 | 19 | .175 | 11 | 1.10 | 26 | .95 | 37 |
| .15 | 22 | .125 | 21 | 1.30 | 20 | 1.05 | 28 |
| .05 | 24 | .075 | 18 | 1.50 | 18 | 1.15 | 22 |
| −.05 | 26 | .025 | 30 | 1.70 | 10 | 1.25 | 12 |
| −.15 | 19 | −.025 | 36 | 1.90 | 6 | 1.35 | 12 |
| −.25 | 17 | −.075 | 16 | 2.10 | 3 | 1.45 | 6 |
| −.35 | 12 | −.125 | 14 | 2.30 | 1 | | |
| −.45 | 11 | −.175 | 10 | 2.50 | 2 | | |
| −.55 | 5 | −.225 | 9 | 2.70 | 1 | | |
| −.65 | 4 | −.275 | 3 | 2.90 | 1 | | |
| −.75 | 3 | −.325 | 4 | 3.10 | 1 | | |
| | | −.375 | 0 | | | | |
| | | −.425 | 2 | | | | |
| *Total* | 200 | | 200 | | 200 | | 200 |
| Mean $\bar{X}$ | .009 | | .019 | Mean $s^2$ | 1.001 | | .945 |
| Var $\bar{X}$ | .1035 | | .0258 | Var $s^2$ | .2983 | | .0579 |
| Std. dev. $\bar{X}$ | .322 | | .161 | Std. dev. $s^2$ | .546 | | .241 |

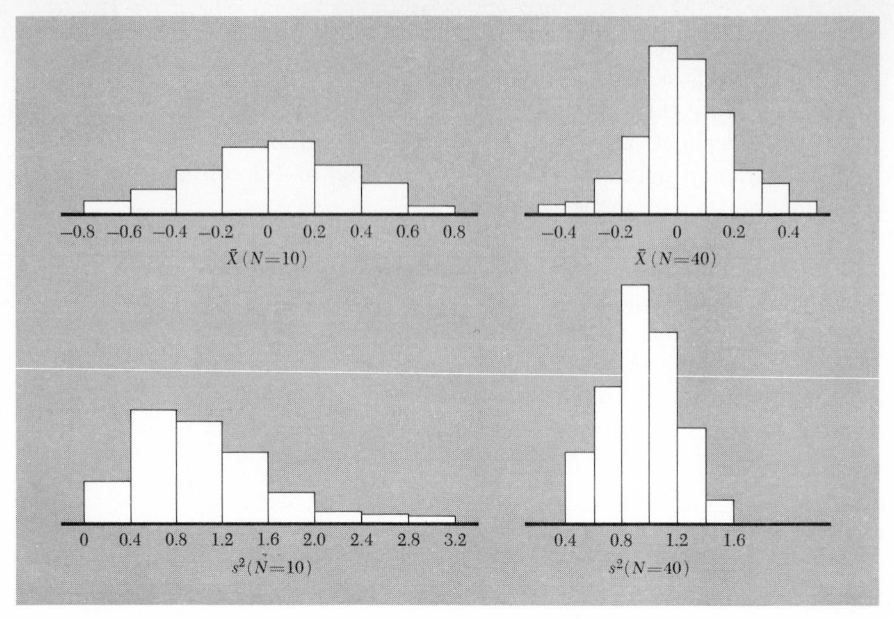


**Figure 4-2** Results of a sampling experiment.

example). We do not expect exact values because we do not have the entire sampling distribution—only 200 observations from it.

3. The variance of the sampling distributions of $\bar{X}$. Again from the previous examples, we expect the sampling distributions of the mean to have variances approximately equal to $\sigma^2/N$ (which is $\frac{1}{10} = .1$ and $\frac{1}{40} = .025$ for $N = 10$ and $N = 40$, respectively).

We also record for future reference the percentiles $P_{05}$, $P_{10}$, $P_{20}$, $P_{30}$, $P_{50}$, $P_{70}$, $P_{80}$, $P_{90}$, $P_{95}$ of the observed distribution for $s^2$ for $N = 10$.

The values of the percentiles for the complete sampling distribution (obtained by a mathematical process) are also recorded so that the comparison can be seen. The theoretical distribution is called the $\chi^2/df$ (chi square over degrees of freedom) distribution.

**Table 4-4**

| | $P_{05}$ | $P_{10}$ | $P_{20}$ | $P_{30}$ | $P_{50}$ | $P_{70}$ | $P_{80}$ | $P_{90}$ | $P_{95}$ |
|---|---|---|---|---|---|---|---|---|---|
| *Observed* | .28 | .39 | .52 | .64 | .92 | 1.23 | 1.44 | 1.70 | 1.97 |
| *Theoretical* | .369 | .463 | .598 | .710 | .927 | 1.18 | 1.36 | 1.63 | 1.88 |

As we proceed we shall find many occasions to approximate sampling distributions experimentally, and we shall also study several sampling distributions in enough detail to make use of mathematically obtained tables of percentiles.

In Sec. 4-5 the example of the 36 possible samples of size $N = 2$ taken from a population of size 6 assumed that a single individual could be selected twice in the same sample. In most practical sampling problems the population is so large in relation to the sample size that the chance of getting the same individual twice is so small that consideration of this possibility can be neglected. However, to see the necessary modification in the formulas if an individual cannot be chosen more than once in the same sample, we consider the 30 possible samples of size $N = 2$ if the first individual chosen for the sample is not returned to the population before the second is chosen. If the samples were recorded as in Table 4-1, with the first observation recorded along the top and the second on the left, the table would appear as before, except there would be no entries in the squares along the diagonal running from the upper left corner to the lower right corner of the table. Therefore the sampling distributions of $\bar{X}$ and $s^2$ recorded in Table 4-2 would be altered by removal of the six means $\bar{X} = 11, 16, 12, 15, 16, 14$ and six zero values of $s^2$; Table 4-5 shows the resulting sampling distributions of $\bar{X}$ and $s^2$. Note that, as in the case of sampling with replacement, the mean of the sampling distribution of $\bar{X}$ is $\mu$. However, the variance of the sampling distribution of $\bar{X}$ is not the same as before. The new formula is

$$\text{var } \bar{X} = \frac{\sigma^2}{N}\left(1 - \frac{N-1}{N_p - 1}\right) = \frac{\sigma^2}{N}\left(\frac{N_p - N}{N_p - 1}\right)$$

where $N_p$ = size of population
$N$ = size of sample

**Table 4-5**

| $\bar{X}$ | $f$ | $f\bar{X}$ |
|---|---|---|
| 16 | 2 | 32 |
| 15.5 | 4 | 62 |
| 15 | 4 | 60 |
| 14.5 | 2 | 29 |
| 14 | 4 | 56 |
| 13.5 | 6 | 81 |
| 13 | 4 | 52 |
| 12.5 | 2 | 25 |
| 12 | 0 | 0 |
| 11.5 | 2 | 23 |
| 11 | 0 | 0 |
| *Total* | 30 | 420 |

Mean $\bar{X} = \frac{420}{30} = 14$

$$\text{Var } \bar{X} = \frac{44}{30} = \frac{8.8}{6} = 1.467$$

| $s^2$ | $f$ | $fs^2$ |
|---|---|---|
| 12.5 | 4 | 50 |
| 8 | 6 | 48 |
| 4.5 | 4 | 18 |
| 2.0 | 6 | 12 |
| .5 | 8 | 4 |
| 0 | 2 | 0 |
| *Total* | 30 | 132 |

Mean $s^2 = \frac{132}{30} = 4.4$

When $N$ is small compared with $N_p$, the term $(N - 1)/(N_p - 1)$ can be neglected and the approximation $\sigma^2/N$ is adequate. For the illustrative population

$$\text{var } \bar{X} = \frac{11/3}{2}\left(\frac{6 - 2}{6 - 1}\right) = 1.467$$

The mean $s^2$ is also changed and is now given by

$$\frac{N_p}{N_p - 1}\sigma^2$$

For a large population this is very close to $\sigma^2$, as we obtained in the case of sampling with replacement. If you are sampling without replacement from a finite population, you should bear in mind that the variance of $\bar{X}$ and the mean $s^2$ are different from those for sampling with replacement or in sampling from an infinite population.

For the illustrative population the mean variance is

$$\text{Mean } s^2 = \frac{6}{6 - 1}\left(\frac{11}{3}\right) = 4.4$$

## 4-7 SAMPLING EXPERIMENTS

One of the main goals of this book is to impart an understanding of statistics through sampling experiments. In these experiments a sampling distribution is approximated by drawing a large number of samples from some universe. In the hope of encouraging more actual sampling experiments we shall examine here (and at other places in the text) some methods of performing such experiments. Such experiments can be and have been used as a means of studying sampling distribution in cases where mathematical results are not available. In fact, it is not unusual to perform such investigations on electronic computers, which can produce many samples per second.

The use of tables of random numbers (Table A-1) or tables of observations drawn from a specified normal population (Table A-2) is convenient and fast. However, it is often easier to visualize the mechanism of forming a sampling distribution by drawing samples from a visible population rather than from a row of numbers on a page. Drawing from an actual population may also serve to eliminate any mistrust of the arrangement of the numbers in the table.

To illustrate a situation in which individuals in a population either have or do not have a certain characteristic we use a box of beads and a paddle (see Fig. 4-3). The paddle is a plastic scoop with five rows of eight holes each. By dipping the paddle into the box of beads we can obtain five samples of $N = 8$, eight samples of $N = 5$, or one sample of $N = 40$ individuals. The beads are small plastic spheres in two (or more) colors. Thus, for example, if

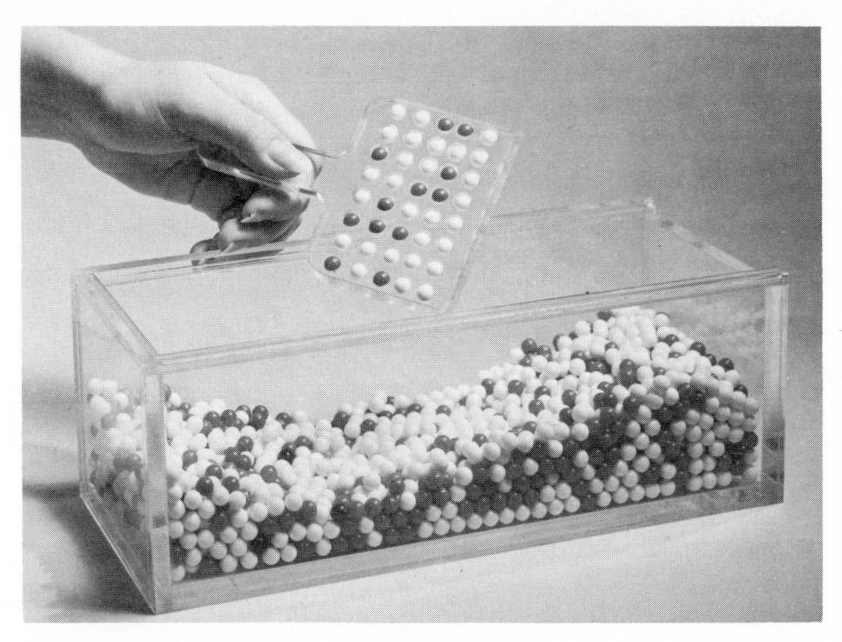

**Figure 4-3** Bead box used in sampling experiments.

there are 1,000 white beads and 500 red beads in the box, we have a population in which two-thirds of the individuals have some characteristic and one-third does not. Such a box has an advantage over tossing coins, not only in time and effort, but in the possibility of constructing (by altering the proportions of red and white beads in the box) any desired type of population.

Another device that can be employed is a box of tags (representing individuals), each tag having on it a number (representing the measurement on the individual). The tags may be small round paper tags with metal rims, such as those used in marking prices on merchandise. If there is a large number of tags in the box (the universe), samples of 10 or so can be drawn at once and all 10 values recorded before the tags are returned to the box. If the number of tags in the box is small it is better to return each observed tag before drawing another. Frequently the distribution of numbers on the tags is made up to be an approximately normal distribution (as described in Chap. 5). Table A-22 gives six possible sets of numbers for tags. Sets 1, 2, 3, 4, 5 represent approximately normal distributions with means and standard deviations as marked. Set 6 is a uniform distribution in which each measurement has the same frequency. For example, to construct a universe from set 1, take 200 tags, mark 1 of them 5, 3 of them 4, 10 of them 3, and so on.

For the experiments in this chapter only one such set is used. However, in later chapters we shall study experiments using several sets at once. To keep them separate, the sets are recorded in different colors as noted in the table.

**Models and assumptions.** It should be emphasized that tables of random numbers or prepared boxes of tags or beads are *ideal* populations, where no assumptions need be made. In using Table A-2, for example, we may be sure that the population is normally distributed and that observations are drawn at random from this population. Such ideal populations are *models*, used only to demonstrate certain mathematical theorems. In practical problems, where assumptions must be made concerning a population, we should be reasonably certain that such assumptions are satisfied before using statistical procedures based on these assumptions, or else we should use statistical procedures which are little affected by the truth of the assumptions that have been made.

## GLOSSARY

biased sample
design of experiment
parameter
population
random numbers
random sample
sample
sample size
sampling distribution
statistic
survey
unbiased estimate
universe

## DISCUSSION QUESTIONS

**1.** What three types of distributions were presented in this chapter? How do these three types differ?

**2.** Explain why the statement "Every individual in the population has an equal chance of being in the sample" is not the same as saying "Every measurement in the population has an equal chance of being in the sample."

**3.** Distinguish between a parameter and a statistic. Are these constants? What are two uses of a statistic?

**4.** Why do we not measure parameters directly?

**5.** How many random samples of size 3 could be taken from a population of 100 individuals? From a population of 1,000 individuals? How many samples of size $N = 100$ could be taken from a population of 1 million individuals?

**6.** Discuss methods of obtaining random samples (except for replacement) from the following populations. Give for each a possible measurement which might be made on individuals in the sample.

(*a*) Students in a school
(*b*) Oranges on a tree
(*c*) People in a city
(*d*) Fir trees in a forest
(*e*) Houses in a county
(*f*) A day's production of electric light bulbs from a factory
(*g*) The wheat in a freight car

(*h*) Children in a certain hospital during 1956
(*i*) All graduates of some college
(*j*) The fish in a river

**7.** What happens to the variance of the sampling distribution of the mean $\sigma_{\bar{X}}^2$ if the sample size is doubled? What happens to the standard deviation of the sampling distribution of the mean if the sample size is doubled?

**8.** If you observe a *single* sample mean $\bar{X} = 12$, how can there be a distribution of means?

**9.** What information do you get from the sampling distribution of any statistic?

**10.** What might be meant by the statement "In a sample of size $N = 40$ you can count on $\bar{X}$ being only half as far from the population mean as in a sample of size $N = 10$"? Does this mean that in every sample of size 40 the sample mean is only half as far from $\mu$ as the mean of any sample of size 10?

**11.** Is a sample mean ever equal, usually equal, or always equal to the population mean?

**12.** Suppose that a single measurement is made of the length of a field. Can this be thought of as a sample? If so, from what population? Is it possible to estimate the population variance from these data?

**13.** We sometimes try to determine whether two school classes are significantly different on the basis of the results of some tests. In what populations are we interested?

**14.** If a sample is obtained by selecting every tenth item, what possible bias could result? Give examples. Why is this not a random sample?

**15.** Suppose it is desired to estimate the mean family size in a certain town. Would the recording of the family size of a random selection of high school students be a reasonable way to obtain data?

## CLASS EXERCISES

**1.** Collect the observed values of $\bar{X}$ and $s^2$ for the samples of size $N = 5$ drawn from Table A-2 (Class Exercise, Chap. 3) into distributions. Draw histograms and compute the mean $\bar{X}$, the mean $s^2$, and the variance of $\bar{X}$. Compare these values with the theoretical values.

**2.** Proceed as in Exercise 1 for the samples drawn from Table A-23.

**3.** Collect the three samples of size 5 from Table A-2 into one sample of size 15 and compute the $\bar{X}$ and $s^2$ for this sample. Then collect the results of all students into a distribution of values of $\bar{X}$ for $N = 15$ and into a distribution of values of $s^2$ for $N = 15$. Compute the mean $\bar{X}$, the mean $s^2$, and the variance of $\bar{X}$ and compare these numbers with the theoretical values.

**4.** Compute for each of the samples drawn from Table A-2 the values of the median and of the range $w$. Collect the results into distributions and compute the mean median, the mean range, and the variance of the medians. Compare the estimated sampling distribution of the mean with that of the median. Compare the estimated sampling distribution of $s^2$ with that of $w$.

**5.** Consider the observations 1, 2, 3, 4, 5 as a population. Draw a histogram and find the mean and variance of the population. From this population draw all possible samples, with replacement, of size 2. Find the mean and variance of the sampling distribution of the means and the mean of the sampling distribution of the variance. Compare these values with the mean and variance of the universe.

**6.** Consider the population of Exercise 5 and proceed as described there, except sample without replacement.

**7.** Suppose a population consists of the four measurements $X_1 = 2$, $X_2 = 4$, $X_3 = 6$, $X_4 = 8$. How many random samples of size 3 could be chosen from this population? List these samples and compute the mean and median for each s mple. Form the sampling distribution of the mean and the sampling distribution of the median for samples of size 3 from this population. Which of these distributions has the smaller variance? How do the means of these two distributions compare with the population mean? Does the variance of each of these distributions equal $\sigma^2/N$? If you had a single sample of three observations, which measurement, the mean or the median, would be more likely to be close to the population mean? Do you think this is a general rule, or do you think that the answer to this last question would depend upon the form of the distribution of the population?

## PROBLEMS

*Section* 4-3

**1.** Use Table A-1 to select 10 individuals from the 80 measurements given on page 9. Describe exactly how you entered the table and how you decided which individuals to take for the sample. Compute the mean $\bar{X}$ and the variance $s^2$ for the sample you selected and compare them with $\mu$ and $\sigma^2$.

**2.** Use Table A-1 to select 10 individuals from the distribution of $s^2$ for samples of size 10 recorded in Table 4-3. Describe exactly how you entered the table and how you decided which individuals to take for the sample. Find the mean of the ten $s^2$ values selected and compare it with $\sigma^2$.

*Section* 4-5

**3.** Suppose 500 samples of size 20 each are to be taken from Table A-2 and $\bar{X}$ and $s^2$ recorded for each. Approximately what values would you expect to find for the mean $\bar{X}$, the variance of the $\bar{X}$'s, and the mean of the 500 values of $s^2$?

**4.** A population of tags is formed with the distribution of numbers in Table 2-1. Suppose 1,000 random samples of five tags each are taken from the population and $\bar{X}$ and $s^2$ are recorded for the numbers of each sample. Approximately what values would you expect to find for the mean $\bar{X}$, the variance of the $\bar{X}$'s, and the mean of the $s^2$'s?

**5.** Suppose in a population of 1,000 tags 500 are marked 1 and 500 are marked 0. Verify $\mu = .5$, $\sigma^2 = .25$. Suppose 100 random samples of 10 tags are taken from the population and $\bar{X}$ is recorded for each sample. Approximately what values would you expect to find for the mean of the $\bar{X}$'s and the variance of the $\bar{X}$'s?

**6.** In Chap. 2, Probs. 5 to 26 give results on observed distributions. Assume each set of data to be a random sample, and for each set for which you have done the arithmetic use $s^2$ in place of $\sigma^2$ to find an estimated standard error of the mean. (Note values of means and standard deviations on page 592.)

**7.** The mean $\bar{X} = .009$ for the 200 samples of size 10 in Table 4-3 could be interpreted as the mean of a single random sample of size 2,000. What is its standard error? Correspondingly, in the same table, what is the standard error of the observed value .019?

**8.** A population consists of 1,000 tags marked 1 and 2,000 tags marked 0. Using the 1's and 0's as measurements, find $\mu$ and $\sigma$. A random sample of 25 tags is selected with replacement. Find the standard error of the sample mean. How large a sample would be needed to reduce the standard error of the mean to .001?

**9.** A large population of people includes 20 per cent with blue eyes. If blue-eyed people are given the score 1 and the rest are scored 0, find the standard error of the mean score of a random sample of 100 people.

**10.** A population of men has a height distribution with $\sigma = 3$ inches. How large a sample is needed for the standard error of the mean height to equal .5 inches?

**11.** A population consists of 1,000 men, and the distribution of heights of these men has $\sigma = 3$ inches. Find the standard error of a mean height for a random sample of 50 men selected without replacement. Find the standard error of the mean height of a random sample of 50 with replacement. Compare this result with the first result.

**12.** How much of an increase in sample size is needed to decrease a standard error of a mean by 50 per cent? By 75 per cent? By 87.5 per cent? By 93.75 per cent?

CHAPTER FIVE

# The normal distribution

One of the most important frequency distributions in statistics is the *normal distribution.* Many of the procedures in later chapters will be based on a knowledge of this distribution. Its physical appearance is that of a symmetrical bell-shaped curve, extending infinitely far in both positive and negative directions. Its use is the same as that of any other distribution curve. The relative frequency with which a variable will take on values between two points is the area under the curve between the two points on the horizontal axis. Not all symmetrical bell-shaped distributions are normal distributions; the term "normal distribution" refers to the fact that the area under the curve is distributed in a specified manner which will be discussed later.

The word "normal" should not be read as "usual" or as the opposite of "abnormal." It is true that a number of statistics have, for large sample sizes, sampling distributions which are normal, but there is no natural reason for us to expect that every population which is sampled has a normal distribution.

## 5-1 EQUATION OF THE NORMAL CURVE

The equation of the normal curve is

$$Y = \frac{1}{\sigma\sqrt{2\pi}} e^{-\frac{1}{2}\left(\frac{X-\mu}{\sigma}\right)^2}$$

where $\pi$ = a constant (approximately 3.1416)

$e$ = a constant (approximately 2.7183)

$\sigma$ = a parameter (equal to the standard deviation of the distribution; different normal curves may have different standard deviations and thus different values for $\sigma$, but for a given distribution $\sigma$ is constant)

$\mu$ = a parameter (equal to the mean of the distribution; different normal curves may have different means, hence different values for $\mu$, but for a given distribution $\mu$ is constant)

$X$ = abscissa, measurement or score marked on horizontal axis

$Y$ = ordinate, height of curve corresponding to an assigned value of $X$

The total area between this curve and the $X$ axis is 1 square unit. Thus the area under the curve between the points, say, $X = a$ and $X = b$ is equal to the expected proportion of cases that lie between the two points. It is very important, therefore, to be able to find the area under sections of the curve.

It is sometimes desirable to construct a normal-distribution curve which has the same area as a given histogram. In this case the equation is

$$Y = \frac{Ni}{\sigma\sqrt{2\pi}} e^{-\frac{1}{2}\left(\frac{X-\mu}{\sigma}\right)^2}$$

where $N$ = number of cases observed

$i$ = length of class interval used to draw the histogram

The area under such a curve is $Ni$ square units. If the histogram is drawn with the proportion of observations in each interval as the height of that interval, then the $N$ is replaced by 1 and the area under the curve is $i$ square units.

The curves in Fig. 5-1 indicate the shape of the normal curve. The three curves each have the same mean $\mu$ and the same standard deviation $\sigma$ but different total areas $Ni$. The area under each of the three curves in Fig. 5-1 has the same proportional distribution to the total area under the curve. For

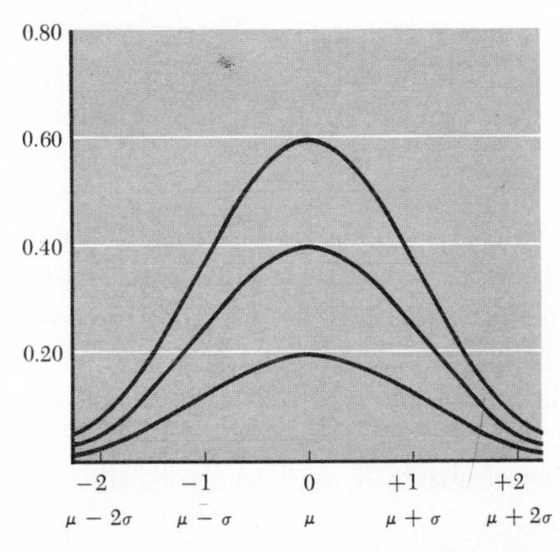


**Figure 5-1** "Normal" frequency distributions with equal means and standard deviations but unequal population sizes.

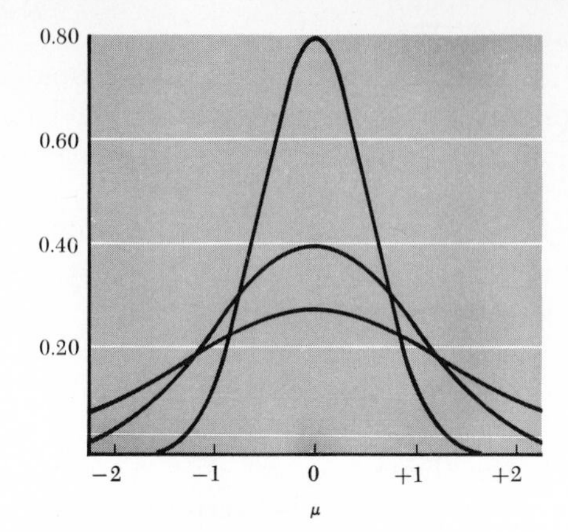


**Figure 5-2** "Normal" frequency distributions with equal means and population sizes but unequal standard deviations. Note that the scale marked on the abscissa corresponds to standard deviation units for the middle curve only.

example, the area between $\mu$ and $\mu + \sigma$ is in each case 34.13 per cent of the area under the curve.

We shall usually draw the normal curves with total area 1. This is standard practice in describing a distribution, and the area under the curve between any two values is also the proportion of the total area.

Figure 5-2 shows a comparison of normal curves with the same area. The three normal curves in Fig. 5-2 each have the same mean $\mu$ but different standard deviations. A curve with area concentrated closely about the mean has a smaller value of $\sigma$, while a curve with area less concentrated about the mean has a larger value of $\sigma$.

Normal curves all extend infinitely far both to the right and to the left. They are all symmetrical about $X = \mu$, that is, about the mean. They are all concave downward for $X$ within $1\sigma$ of the mean and concave upward for $X$ farther than $1\sigma$ from the mean. The proportion of area under a normal curve between any two values is completely determined by $\mu$ and $\sigma$.

## 5-2 AN EXAMPLE OF THE NORMAL DISTRIBUTION

One of the most useful applications of the normal distribution is in the sampling distribution of the mean. To demonstrate in an actual case we shall construct a universe and approximate the sampling distribution of the mean experimentally.

EXAMPLE 5-1. Suppose our universe contains an equal frequency of quantities ranging from 60 to 69. To correspond to the scores in our experiment we take 300 tags, 30 labeled 60, 30 labeled 61, and so on, to 69. We place the tags in a box, mix them thoroughly, draw a sample of two tags ($N = 2$), and compute the mean of the scores on the two tags. We repeat this process 200 times (200 samples of size $N = 2$), record the results in a frequency table (Table 5-1), and draw a histogram (Fig. 5-3). The same procedure is carried out for samples of size 4, 8, and 16. Note that each of

**Table 5-1** Frequency distribution of observed means of observations drawn from a uniform population with $\mu = 64.5$ and $\sigma^2 = 8.25$

| UNIVERSE $N = 1$ | | $N = 2$ | | $N = 4$ | | $N = 8$ | | $N = 16$ | |
|---|---|---|---|---|---|---|---|---|---|
| $X$ | $f$ | $\bar{X}$ | $f$ | $\bar{X}$ | $f$ | $\bar{X}$ | $f$ | $\bar{X}$ | $f$ |
| 69 | 30 | 68.75 | 9 | 68.50 | 3 | 67.250 | 2 | 66.1875 | 3 |
| 68 | 30 | 67.75 | 12 | 67.75 | 5 | 66.625 | 4 | 65.8750 | 5 |
| 67 | 30 | 66.75 | 23 | 67 | 4 | 66 | 26 | 65.5625 | 21 |
| 66 | 30 | 65.75 | 37 | 66.25 | 21 | 65.375 | 36 | 65.2500 | 27 |
| 65 | 30 | 64.75 | 33 | 65.50 | 38 | 64.750 | 40 | 64.9375 | 24 |
| 64 | 30 | 63.75 | 38 | 64.75 | 35 | 64.125 | 43 | 64.6250 | 25 |
| 63 | 30 | 62.75 | 22 | 64 | 53 | 63.500 | 30 | 64.3125 | 35 |
| 62 | 30 | 61.75 | 13 | 63.25 | 25 | 62.875 | 15 | 64 | 26 |
| 61 | 30 | 60.75 | 13 | 62.50 | 8 | 62.250 | 3 | 63.6875 | 18 |
| 60 | 30 | | | 61.75 | 8 | 61.625 | 1 | 63.3750 | 15 |
| | | | | | | | | 63.0625 | 1 |
| *Total* | 300 | | 200 | | 200 | | 200 | | 200 |
| Mean $\bar{X}$ | | | 64.66 | | 64.63 | | 64.57 | | 64.58 |
| Var $\bar{X}$ | | | 4.173 | | 1.868 | | 1.105 | | .509 |
| $s_{\bar{X}}$ | | | 2.043 | | 1.367 | | 1.051 | | .714 |
| $Ns_{\bar{X}}^2$ | | | 8.35 | | 7.47 | | 8.84 | | 8.14 |

the four values of $Ns_{\bar{X}}^2$ is approximately equal to $\sigma^2$. If we had taken all possible samples, all four would equal $\sigma^2$.

The top histogram in Fig. 5-3 is the frequency distribution of the universe. The four lower histograms are for means with sample sizes 2, 4, 8, 16, respectively. The approximation of the bell-shaped curve may be noted. The universe does not have a normal distribution, but the sampling distribution of the means tends to be more normal than the universe. The following theorem states the situation more explicitly.

THEOREM. *If the distribution of the universe has a finite variance, the sampling distribution of the means of random samples will be approximately normal if the sample size is sufficiently large.*

For most practical examples the universe will have a finite variance, and thus this condition is not very restrictive. The condition "sufficiently large" is more troublesome, since it depends upon the distribution of the universe, which is usually unknown. If, however, the distribution of the universe is exactly normal, then the sampling distribution of the mean of any size sample, even as small as $N = 1$ or $N = 2$, will be exactly normal. If the population is reasonably symmetrical and not too disperse, a moderate sample size will usually make the distribution approximately normal.

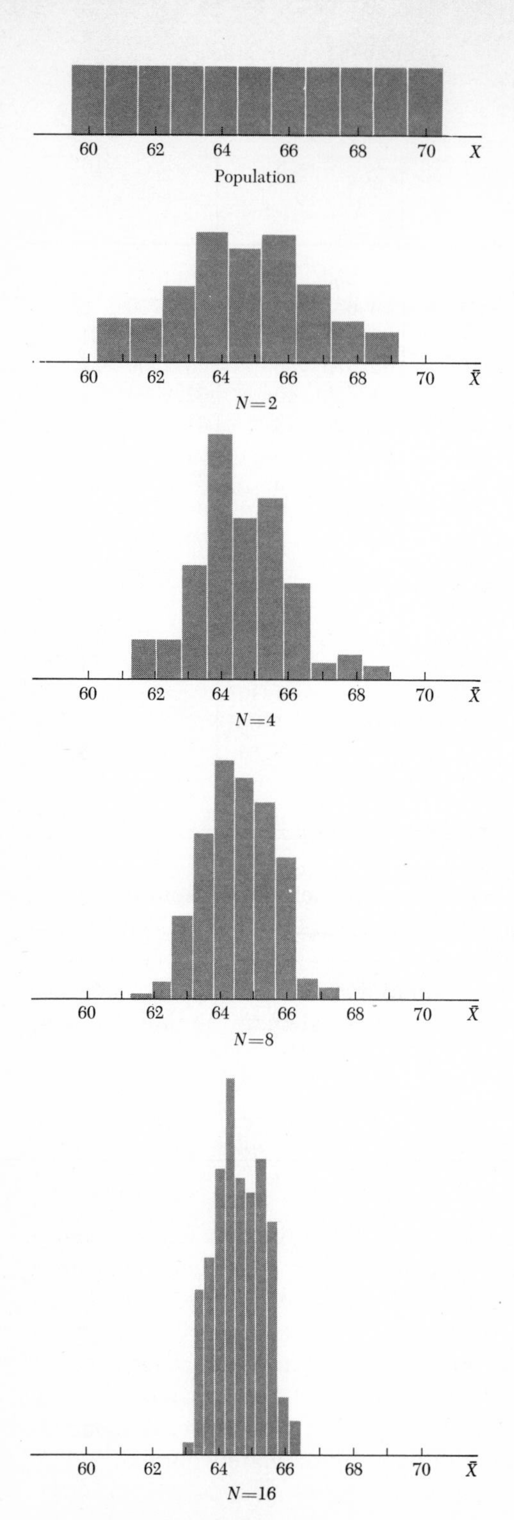


**Figure 5-3** Results of sampling experiments.

Our concept of "approximately normal" is so far largely intuitive. We say that a histogram is approximately normal when the areas of the rectangles are approximately equal to the corresponding areas under a normal curve. Section 5-5 discusses a simple method of examining a set of data for marked deviations from normality.

For the above examples it can be seen that the mean of the sample means falls fairly close to the mean of the universe, $\mu$, in every case. This agrees reasonably with the assertion in Chap. 4 that for random samples the sample mean is an unbiased estimate of the population mean; if we had several thousand samples instead of 200 we would anticipate that the mean of the means would be even closer to $\mu$. If we use a statistic which is an unbiased estimate of a parameter, we shall estimate the parameter correctly "on the average." This does not mean that we are likely to be close to the correct value of the parameter. The chance of being close depends not only upon the center of the sampling distribution, but also upon the dispersion of the distribution. The standard deviation is frequently used to measure dispersion, and in the case of an unbiased statistic a small standard deviation indicates a large chance of being close.

## THE CUMULATIVE NORMAL DISTRIBUTION 5-3

The cumulative-distribution function gives the relative frequency with which observations are expected to be smaller than (to the left of) any specified value. This relative frequency is the area under the distribution curve from minus infinity to the specified value.

In the case of the normal distribution the areas under the curve from the extreme left (minus infinity) to various points are given in Table A-4. This table is entered with the value of $z = (X - \mu)/\sigma$ and gives the cumulated area. Note that $\mu$ and $\sigma$ are constant and $X$ is the variable; however,

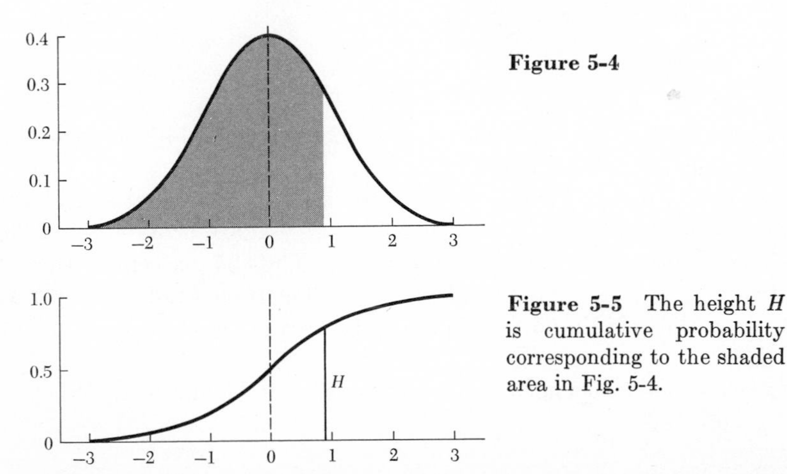


**Figure 5-4**

**Figure 5-5** The height $H$ is cumulative probability corresponding to the shaded area in Fig. 5-4.

the table must be entered by way of the deviation of $X$ from the mean measured in $\sigma$ units. The advantage of this method is that the same table can be used for any value of $\mu$ and $\sigma$.

The height $H$ indicated in Fig. 5-5 is numerically equal to the area shaded under the normal curve in Fig. 5-4 and thus gives the frequency with which observations less than or equal to any given value of $X$ will occur.

## 5-4 AREAS UNDER THE NORMAL CURVE

The area under the normal curve between the points $X = a$ and $X = b$, where $a$ is less than $b$ ($a < b$), gives the probability with which cases fall between $a$ and $b$. To find this area first find the area from minus infinity to $X = b$, then find the area from minus infinity to $X = a$, and subtract the second area from the first.

EXAMPLE 5-2. In a normal distribution $\mu = 30$ and $\sigma = 5$. What is the probability that a case falls between 20 and 35?

*Answer.* $X = 20$ gives $\dfrac{X - \mu}{\sigma} = \dfrac{20 - 30}{5} = -2$

$X = 35$ gives $\dfrac{X - \mu}{\sigma} = \dfrac{35 - 30}{5} = 1$

Table A-4 gives the area to the left of $-2$ to be .0228 and the area to the left of 1 to be .8413. Thus the area (proportion of cases) between $X = 20$ and $X = 35$ is $.8413 - .0228 = .8185$; that is, 81.85 per cent of cases are between 20 and 35 (see Fig. 5-6).

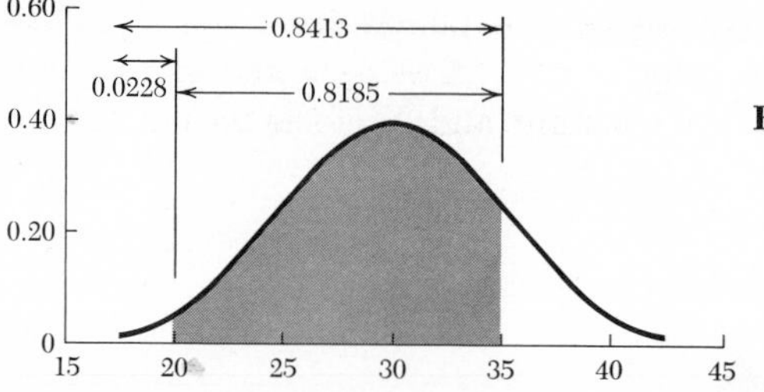


**Figure 5-6** Normal curve areas.

EXAMPLE 5-3. Between what values does the central 95 per cent of all the area under the normal curve lie?

*Answer.* Table A-4 shows that 2.5 per cent (half of 5 per cent) of the area is below $(X - \mu)/\sigma = -1.96$ (or $X - \mu = -1.96\sigma$, or $X = \mu - 1.96\sigma$) and 2.5 per cent lies above $(X - \mu)/\sigma = 1.96$. Thus 95 per cent of the area lies between $X = \mu - 1.96\sigma$ and $X = \mu + 1.96\sigma$. If, for example, $\mu = 7$ and $\sigma = 2$, then the middle 95 per cent of the area is between $X = 3.08$ and $X = 10.92$.

EXAMPLE 5-4. In a normal distribution $\mu = 163$ and $\sigma = 12$. Where are $P_{10}$, $P_{50}$, and $P_{95}$?

*Answer.* $P_{10}$ is a point having 10 per cent of the area below it. Table A-4 shows this point to be $(X - 163)/12 = -1.28$ [or $X = 163 + 12(-1.28)$], and thus $P_{10} = 148$. Similarly, $P_{50} = 163$ and $P_{95} = 183$.

EXAMPLE 5-5. Samples of 25 items were taken from a normal distribution with $\mu = 50$ and $\sigma = 10$. What proportion of the sample means would lie between 48 and 52?

*Answer.* From Chap. 4 we know that the mean of the sampling distribution of the $\bar{X}$'s is the same as the mean of the universe; thus $\mu_{\bar{X}} = 50$. We also know that $\sigma_{\bar{X}}^2 = \sigma^2/N = \frac{100}{25} = 4$, so that $\sigma_{\bar{X}} = 2$. As indicated in Sec. 5-2, the sampling distribution of the mean of samples from a normal universe is a normal distribution. Thus we can make use of Table A-4 with $\mu = 50$ and $\sigma = 2$. The area below 48 is obtained by entering Table A-4 with the value $(X - \mu)/\sigma = (48 - 50)/2 = -1$. The area below 52 is obtained by using $(X - \mu)/\sigma = (52 - 50)/2 = 1$. These areas are .1587 and .8413, and the area between 48 and 52 is $.8413 - .1587 = .6826$. Hence 68 per cent of sample means may be expected to lie between 48 and 52. If we choose a single sample, we have a 68 per cent chance of finding the $\bar{X}$ between 48 and 52.

EXAMPLE 5-6. Between what two points does the central 50 per cent of the cases in a normal distribution fall? The central 99 per cent?

*Answer.* $(X - \mu)/\sigma = -.674$ has 25 per cent of the area below it, and $(X - \mu)/\sigma = .674$ has 75 per cent of the area below it. Thus 50 per cent of the area lies between $\mu - .674\sigma$ and $\mu + .674\sigma$. The middle 99 per cent falls between $\mu - 2.576\sigma$ and $\mu + 2.576\sigma$. These central areas can also be obtained directly from Table A-3.

## NORMAL-PROBABILITY PAPER 5-5

By changing nonuniformly the vertical scale on the graph of the cumulative-normal-distribution curve, it is possible to have the cumulative-normal-distribution curve take on the shape of a straight line. You can visualize this by thinking of the curve as plotted on an elastic sheet which is then stretched out to make the curve a straight line. Figure 5-7 shows the

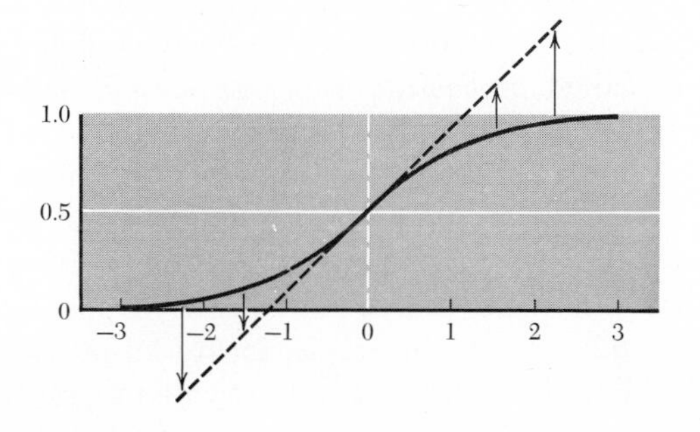


**Figure 5-7** Transformation to straighten a cumulative normal curve.

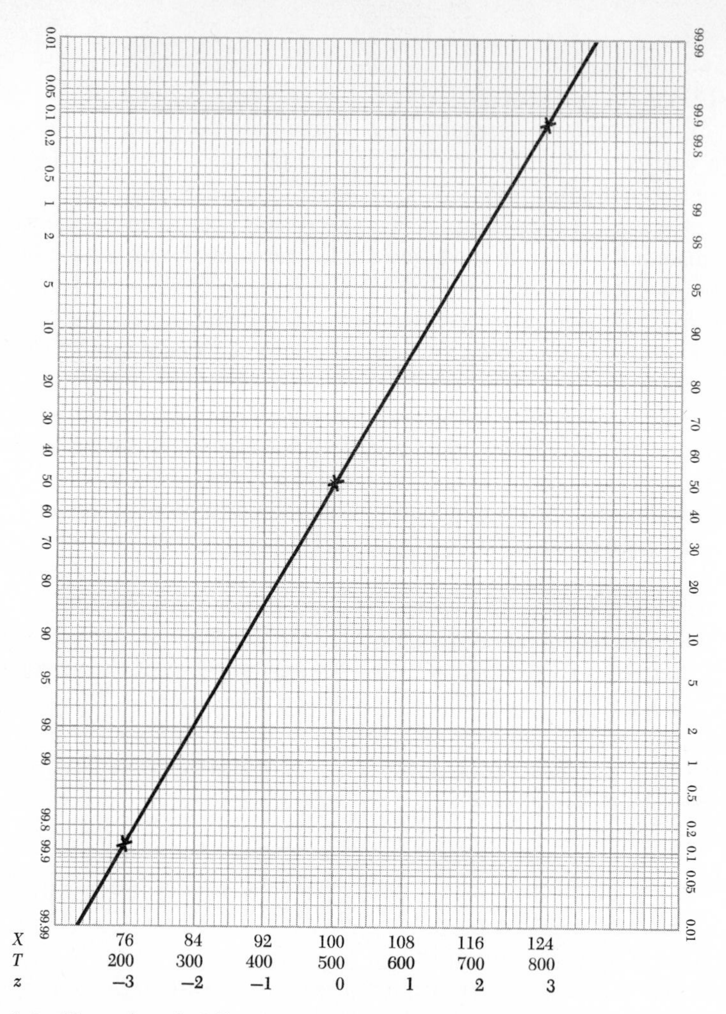


**Figure 5-8** Normal probability paper with cumulative normal curve drawn.

cumulative-normal-distribution curve and indicates the stretching necessary: to the left of the mean the curve should be pulled down, and to the right of the mean it should be pulled up until it coincides with the dotted line.

There is special graph paper called *probability paper* or *normal-probability paper* which is scaled so that any cumulative-normal-distribution curve is a straight line. Figure 5-8 shows a sheet of this type of graph paper. The horizontal axis is marked for a mean of 100 and a standard deviation of 8 units. The line drawn represents the cumulative normal distribution with $\mu = 100$ and $\sigma = 8$. Note that percentiles can be read directly from the right-hand scale of this graph. To find $P_{20}$, for example, locate 20 on the vertical scale and draw a horizontal line through this point extended to meet

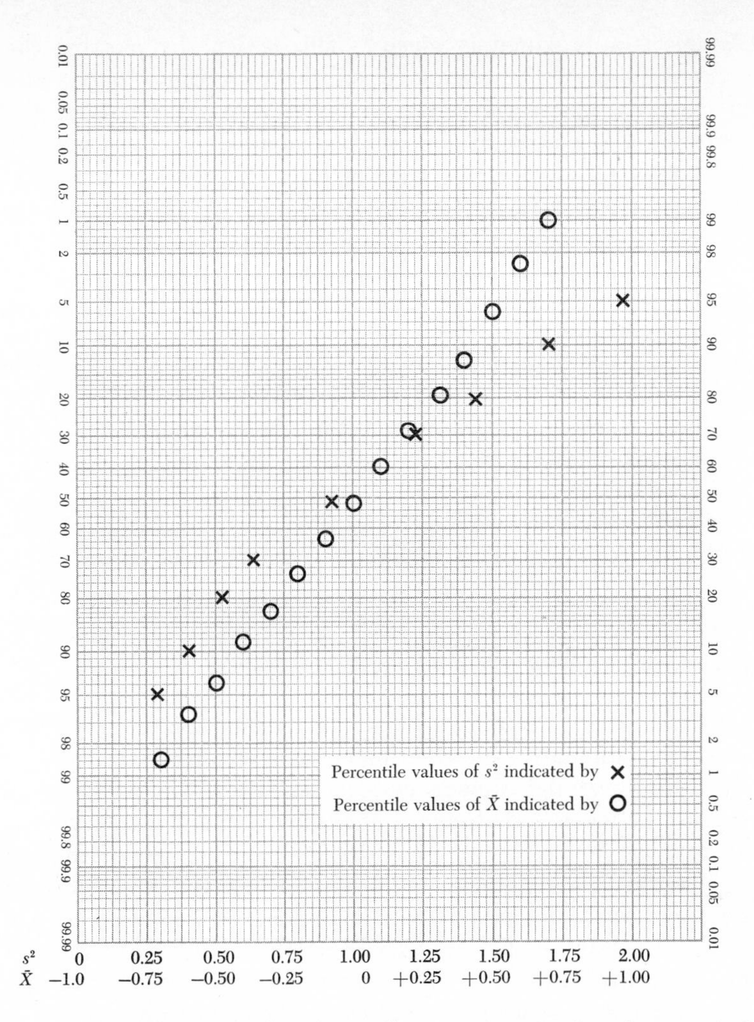


**Figure 5-9** Results of sampling experiments plotted on normal probability paper.

the cumulative-distribution line. Then read the horizontal scale directly below this intersection.

The values recorded in Table A-4 can also be read from this paper by using $z$ scores on the horizontal scale. At least two percentile values must be known to draw the line, but once the line is drawn any percentile can be read from it. Thus if we want the value of $P_{70}$ we find 70 on the vertical scale, move horizontally to the line, and read the $z$ score directly below the intersection. Verify from Fig. 5-8 that this procedure gives the same value, $z = .525$, as Table A-4. To find the percentile rank of $z = -1.5$, we find the height of the line above $z = -1.5$ to be 7.

Probability paper may also be used to indicate whether the distribution of a given sample is approximately normal. This is done by plotting the cumulative distribution of the sample upon the probability paper and then noting how closely this curve approximates a straight line. If the curve is approximately a straight line the distribution is approximately normal. If it deviates considerably from a straight line the distribution is not normal. The fact that the sample distribution gives a curve which is not a straight line is an indication that the population from which it came is not normal.

Figure 5-9 shows plots on normal-probability paper of the cumulative curves for 200 observations of $\bar{X}$ and $s^2$ for samples of size 10. These results are recorded in Table 4-3. Note from the curves that the distribution of $\bar{X}$ is approximately a straight line, while that of $s^2$ is markedly curved. This gives empirical evidence, or proof, of our statement that the sampling distribution of the mean is approximately normal and indicates that the distribution of $s^2$ is not normal.

Normal scores can be standardized to any mean and standard deviation by marking the horizontal scale as in Fig. 5-8, where $T$ provides scores corresponding to $\mu = 500$ and $\sigma = 100$.

## 5-6 FITTING A NORMAL CURVE TO A HISTOGRAM

There are various methods of fitting a normal distribution to some given data. We can draw a cumulative-distribution polygon on normal-probability paper and then draw a straight line as close as possible to the polygon. As noted in Sec. 5-5, the mean and standard deviation of the fitted curve can then be read from the graph. This procedure has the advantage of an easy visual check of normality. Alternatively, we could find the equation of a normal distribution with the same mean and standard deviation as the given data and then draw the normal curve over the histogram to visually check the fit. Note that a normal curve can be fitted to any histogram. Whether or not the curve accurately pictures the frequency distribution is another matter. The "fit" may be good or bad.

If $\bar{X}$ is the mean and $s$ the standard deviation of the sample, the equation of the curve is

$$Y = \frac{iN}{s}\left[\frac{1}{\sqrt{2\pi}}e^{-\frac{1}{2}\left(\frac{X-\bar{X}}{s}\right)^2}\right]$$

The value of the factor in brackets is given in Table A-3. This table is entered with values of $(X - \bar{X})/s$, and the ordinate column gives the value of $\frac{1}{\sqrt{2\pi}}e^{-\frac{1}{2}\left(\frac{X-\bar{X}}{s}\right)^2}$. This value must be multiplied by $iN/s$ to give the height of the curve for a histogram of total area $iN$.

The example in Table 5-2 shows the method of finding points on the curve. We have drawn a normal curve with the same mean and standard

**Table 5-2**

| $X$ | $\dfrac{X-\bar{X}}{s}$ | $\dfrac{1}{\sqrt{2\pi}} e^{-\frac{1}{2}\left(\frac{X-\bar{X}}{s}\right)^2}$ From *Table A-3* | $Y = \dfrac{iN}{s} \times$ *col. 3* $= 62.5 \times$ *col. 3* |
|---|---|---|---|
| .50 | 3 | .004 | .3 |
| .42 | 2.5 | .018 | 1.1 |
| .34 | 2 | .054 | 3.4 |
| .26 | 1.5 | .130 | 8.1 |
| .18 | 1 | .242 | 15.1 |
| .10 | .5 | .352 | 22.0 |
| .02 | 0 | .399 | 24.9 |
| −.06 | − .5 | .352 | 22.0 |
| −.14 | −1 | .242 | 15.1 |
| −.22 | −1.5 | .130 | 8.1 |
| −.30 | −2 | .054 | 3.4 |
| −.38 | −2.5 | .018 | 1.1 |
| −.46 | −3 | .004 | .3 |

deviation as the observed distribution of means for samples of size 40 as recorded in Table 4-3. For this distribution we have $\bar{X} = .02$, $s = .16$, $i = .05$, $N = 200$, $iN/s = 62.5$.

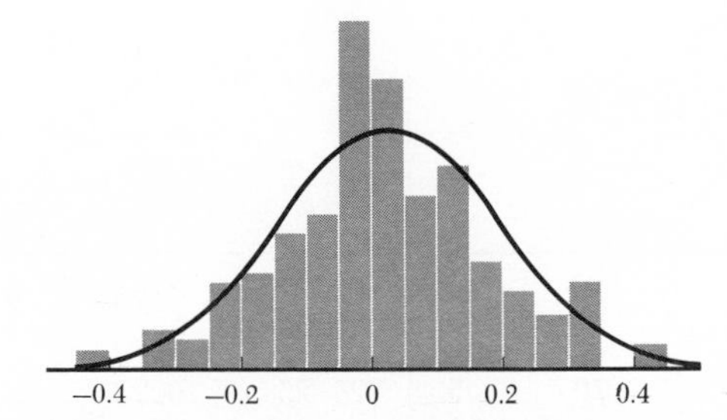


**Figure 5-10** A histogram with fitted normal curve.

Enough values of $X$ should be chosen to permit sketching a smooth curve to the accuracy desired. In this example the mean was taken for one value of $X$, and then points were taken on either side of the mean at intervals of $.5s$. The computation is simplified by this procedure, but we must compute the $X$ values in the first column in Table 5-2 to use for plotting the points on the curve. For example, when $(X - \bar{X})/s = 2$, we have $X = .34$.

## SAMPLING FROM A DICHOTOMOUS (YES-OR-NO) POPULATION 5-7

Frequently a population can be divided into two groups on the basis of some characteristic. For example, a population of animals may be divided

**Table 5-3**

| $X$ | $f$ | $fX$ | $fX^2$ |
|---|---|---|---|
| 1 | 1,000 | 1,000 | 1,000 |
| 0 | 1,000 | 0 | 0 |
| *Total* | 2,000 | 1,000 | 1,000 |

$$\mu = \frac{1{,}000}{2{,}000} = .5$$

$$\sigma^2 = \frac{1{,}000 - (1{,}000)^2/2{,}000}{2{,}000} = .25$$

$$\sigma = \sqrt{.25} = .5$$

into males and females, a population of voters may be divided into those voting for a Republican candidate and those voting for other candidates (or those voting yes on a proposition and those voting no), a population of diseased people may be divided into those who recover and those who do not, a population of missiles may be divided into those which are defective and those which are not, and so on.

We can record observations of this type in a manner similar to that used for continuous measurements if we assign the number one ($X = 1$) to each member of the population belonging to one group (say, group $A$) and the number zero ($X = 0$) to the members belonging to the other group (say, group $B$). If we take a sample from the population, the sum of these assigned numbers $\sum_{i=1}^{N} X_i$, which is the sum of 1's and 0's, will be the number in the sample belonging to group $A$ and the mean $\bar{X}$ of the 1's and 0's will be

**Table 5-4** Observed proportions of red beads in samples of sizes 10, 20, and 40 from a population containing 50 per cent red beads ($\mu = .5$ and $\sigma^2 = .25$)

| $N = 10$ | | | $N = 20$ | | | $N = 40$ | | |
|---|---|---|---|---|---|---|---|---|
| *No. of red* | *Prop.* ($\bar{X}$) | $f$ | *No. of red* | *Prop.* ($\bar{X}$) | $f$ | *No. of red* | *Prop.* ($\bar{X}$) | $f$ |
| 10 | 1 | 0 | 16 | .80 | 1 | 28–29 | .7125 | 2 |
| 9 | .9 | 0 | 15 | .75 | 1 | 26–27 | .6625 | 5 |
| 8 | .8 | 7 | 14 | .70 | 5 | 24–25 | .6125 | 9 |
| 7 | .7 | 11 | 13 | .65 | 10 | 22–23 | .5625 | 16 |
| 6 | .6 | 22 | 12 | .60 | 12 | 20–21 | .5125 | 24 |
| 5 | .5 | 20 | 11 | .55 | 16 | 18–19 | .4625 | 24 |
| 4 | .4 | 18 | 10 | .50 | 15 | 16–17 | .4125 | 10 |
| 3 | .3 | 14 | 9 | .45 | 13 | 14–15 | .3625 | 8 |
| 2 | .2 | 6 | 8 | .40 | 10 | 12–13 | .3125 | 2 |
| 1 | .1 | 2 | 7 | .35 | 7 | | | |
| 0 | 0 | 0 | 6 | .30 | 5 | | | |
| | | | 5 | .25 | 3 | | | |
| | | | 4 | .20 | 2 | | | |
| Mean $\bar{X}$ | | .493 | | | .500 | | | .503 |
| Var $\bar{X}$ | | .0291 | | | .0161 | | | .0074 |
| $N$ Var $\bar{X}$ | | .291 | | | .322 | | | .296 |

the proportion of the sample belonging to group $A$. As noted earlier, the sampling distribution of $\bar{X}$ from any population with a finite variance will be approximately normal if the sample size is sufficiently large. The variance of the sampling distribution is $\sigma^2/N$. These results apply to our present problem.

EXAMPLE 5-7. We construct a population of 1,000 red beads and 1,000 white beads and assign the number 1 to the red beads and the number 0 to the white beads. The population distribution is recorded in Table 5-3 and shows the computation of the mean and variance.

To approximate the sampling distribution of $\bar{X}$ we take 100 samples of size $N = 10$ each and record the number $\Sigma X_i$ and the proportion $\bar{X}$ of red beads for each. The experiment is repeated with 100 samples of size $N = 20$ and 100 samples of size $N = 40$. The results are in Table 5-4. Note that the mean of the $\bar{X}$'s is in each case close to $\mu = .5$ and the variance is about equal to $\sigma^2/N = .25/N$.

The three distributions in Table 5-4 are pictured in Fig. 5-11 as histograms with normal curves sketched over them.

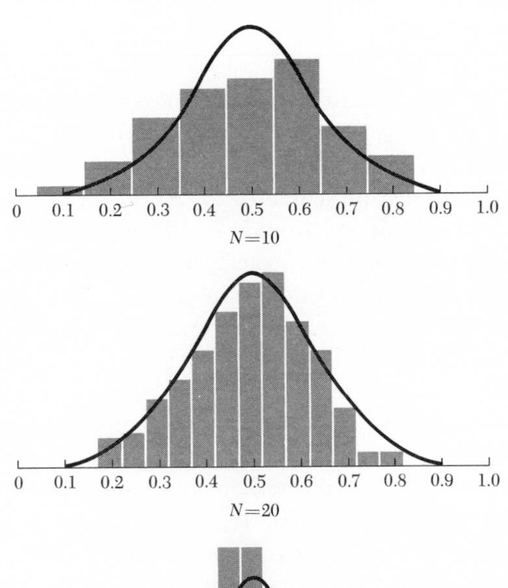


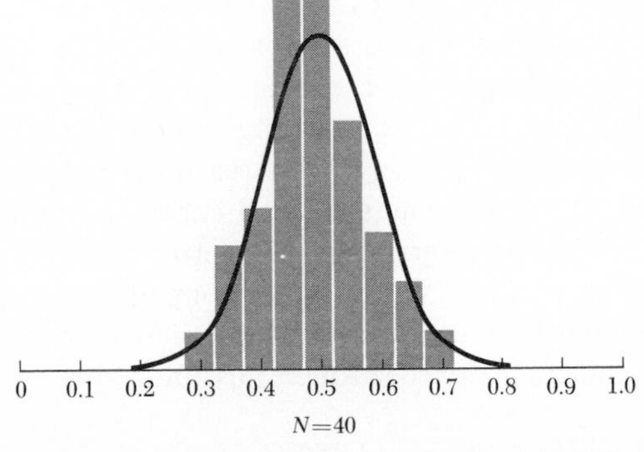


**Figure 5-11** Results of sampling experiments.

**Table 5-5**

| $p$ | $\sigma^2 = p(1-p)$ | $\sigma = \sqrt{p(1-p)}$ | $1/\sigma$ |
|---|---|---|---|
| .1 or .9 | .09 | .30 | 3.33 |
| .2 or .8 | .16 | .40 | 2.50 |
| .3 or .7 | .21 | .46 | 2.18 |
| .4 or .6 | .24 | .49 | 2.04 |
| .5 | .25 | .50 | 2 |

| $N$ | $.5/\sqrt{N}$ |
|---|---|
| 10 | .1581 |
| 15 | .1291 |
| 20 | .1118 |
| 25 | .1000 |
| 40 | .0791 |
| 50 | .0707 |
| 100 | .0500 |
| 200 | .03536 |
| 300 | .02887 |
| 400 | .02500 |
| 500 | .02236 |
| 1,000 | .01581 |
| 5,000 | .00707 |
| 10,000 | .00500 |
| 1,000,000 | .00050 |

Formulas giving $\mu$ and $\sigma^2$ for a population containing a proportion $p$ of 1's and a proportion $1 - p$ of 0's are $\mu = p$ and $\sigma^2 = p(1 - p)$. This result was given for $p = \frac{1}{2}$ in Table 5-3. Table 5-5 records values of $p$ and $\sigma^2 = p(1 - p)$, $\sigma = \sqrt{p(1 - p)}$ taken from Table A-28. Note that there is little variation in $\sigma$ for proportions between .3 and .7. Therefore, if $p$ is between .3 and .7, even though we do not know exactly where, we may use $\sigma = .5$, which will be at least as large as $\sigma$ and not be in error by more than 10 per cent. Table 5-5 also gives values of $.5/\sqrt{N}$ for several values of $N$.

Let us apply these formulas for $\mu$ and $\sigma^2$, together with the assumption that the sample proportion $\bar{X}$ is approximately normal, in an example.

EXAMPLE 5-8. Suppose a sample of $N = 100$ people is taken from a population containing 50 per cent men and 50 per cent women. What is the chance that the proportion of men in the sample will be between .48 and .52, that is, that the number of men will be 48, 49, 50, 51, or 52?

*Answer.* $\mu = .5, \sigma = \sqrt{.5 \times .5} = .5$, and $\sigma_{\bar{X}} = .50/\sqrt{100} = .05$. Since we are using a continuous distribution to approximate the distribution of observed proportions, we shall obtain better accuracy if we consider the number of men to be between 47.5 and 52.5, a range of 5 units corresponding to the five quantities 48, 49, 50, 51, 52. The reason for this correction can easily be seen from Fig. 5-11, where normal curves have been drawn over the discrete distributions. By referring to the pictures for $N = 10$ and $N = 20$ in Fig. 5-11 (where the data have not been grouped into larger intervals) we can see that the chance of any particular value of $\bar{X}$ is represented by a rectangle which extends halfway to the next possible value for $\bar{X}$.

Since it is the area of these rectangles that we wish to approximate, we must find the area under the normal curve which corresponds to the rectangles. In the figure for $N = 10$ the rectangle for $\bar{X} = .5$ extends from .45 to .55, and in the figure for $N = 20$ the rectangle for $\bar{X} = .5$ extends from .475 to .525. Returning to our example, we note that the five rectangles corresponding to 48, 49, 50, 51, 52 for $N = 100$ will extend from 47.5 to 52.5 or for the proportion $\bar{X}$ from .475 to .525. The normal deviates $z$ for .475 and for .525 are

$$\frac{\bar{X} - \mu}{\sigma_{\bar{X}}} = \frac{.475 - .5}{.05} = -.5 \qquad \text{and} \qquad \frac{\bar{X} - \mu}{\sigma_{\bar{X}}} = \frac{.525 - .5}{.05} = .5$$

Table A-3 gives the area between $z = -.5$ and $z = .5$ to be .3829, which is the chance that $\bar{X}$ will fall between .475 and .525 or that the number of men will be 48, 49, 50, 51, 52.

As an additional example consider a sample of $N = 25$ from the same population. Let us find the chance that the number of men will be 12 or 13, that is, either $\bar{X} = .48$ or $\bar{X} = .52$. We consider that the number of men will fall between 11.5 and 13.5, or the proportion $\bar{X}$ will be between .46 and .54. The values of $z$ corresponding to $\bar{X} = .46$ and .54 are $-.4$ and .4, which from Table A-3 gives a .3108 chance that $\bar{X}$ will be .48 or .52. For larger values of $N$ the chance that $\bar{X}$ takes on some value from .48 to .52 increases. For example, for $N = 400$ the chance is .6047.

## NORMAL POPULATIONS 5-8

Many practical problems have statistical answers based on the assumption that the distribution of the population is normal. As noted before, the truth of this assumption may be checked by plotting the sample cumulative-percentage points on normal-probability paper. Frequently in practice the variable of interest is not distributed normally. It may be possible, however, to modify the measuring device in such a way that the distribution is at least approximately normal.

Often a research worker has sufficient data and enough experience with his material to be able to specify the type of measurement transformation which will give a normally distributed variable. Sometimes there are general classes of problems for which a standard change of variable is employed. For example, in dosage-response or sensitivity experiments it often happens that the logarithm of the dosage concentration is approximately normally distributed. Data can be plotted on normal-probability paper with a logarithmic scale on one axis. Samples of graph paper are shown on pages 430 to 437. Various types of frequently used transformations are discussed in Sec. 16-2.

## GLOSSARY

cumulative normal distribution
curve fitting
normal curve
normal-probability paper

## DISCUSSION QUESTIONS

**1.** Suppose we multiplied by 6 every number in Table A-2 and then added 20 to each new number. Describe the resulting distribution. Is it still a normal distribution?

**2.** What is a normal distribution? How many normal curves are there?

**3.** How many parameters determine a normal distribution?

**4.** Is there always 95 per cent of the area under a normal curve between $X = -1.96$ and $X = 1.96$?

**5.** Is a normal curve always bell-shaped? Is a bell-shaped distribution curve always a normal curve?

**6.** What use of the standard deviation has been emphasized in this chapter?

**7.** How do you interpret the area under a section of a normal curve?

**8.** What is a method of checking to see whether or not a sample is approximately normally distributed?

**9.** Can a sample be exactly normally distributed?

**10.** Why might we wish to fit a normal curve to a histogram? Is there any other way to accomplish the same purpose?

**11.** Explain the difference between the standard deviation of the sampling distribution of single observations and the standard deviation of the sampling distribution of means. Why is the standard deviation of means especially useful?

**12.** What property does normal-probability paper have?

**13.** What distributions can you expect to be approximately normal? Describe experiments to prove your answer. Is the word "prove" used correctly here?

**14.** Define each term in the Glossary.

## CLASS EXERCISES

**1.** Using a box of 500 white beads and 1,000 red beads, draw samples of size $N = 20$. Form an estimated sampling distribution of the number of red beads in a sample, as was done in Sec. 5-7. Verify that the mean of the sampling distribution is two-thirds of 20. Plot the observed cumulative distribution on normal-probability paper and verify that the distribution is approximately normal.

**2.** Plot on normal-probability paper the estimated sampling distribution of $\bar{X}$ of the samples of size $N = 5$ from Table A-2 drawn for the Class Exercise in Chap. 3, and verify that the distribution is approximately normal (actually, since the population is normal, the sampling distribution of $\bar{X}$ is exactly normal).

**3.** Perform the operations described in Exercise 2 with the samples from Table A-23.

**4.** Perform the operations described in Exercise 2 with the samples of size 15 from Table A-2.

**5.** Draw on normal-probability paper the observed distributions of $s^2$ for samples of size 5 and of size 15 from Table A-2. Verify that the distribution for samples of size 15 is closer to a normal distribution than the distribution for samples of size 5.

*Section* 5-3

**1.** Read values from Table A-4 for the following areas under a normal curve:

(*a*) The area to the left of $\mu + \sigma$, or $z = 1$

(*b*) The area to the left of $\mu + 2\sigma$, or $z = 2$

(*c*) The area to the left of $\mu - 2\sigma$, or $z = -2$

Read also values of $z$ such that the percentile rank or cumulative area have the following values:

(*d*) A percentile rank equal to 2.5 per cent

(*e*) A percentile rank equal to 50 per cent

(*f*) A percentile rank equal to 95 per cent

(*g*) A percentile rank equal to 99.5 per cent

**2.** In a normal distribution with $\mu = 0$ and $\sigma = 1$ what are the values of $P_{10}$, $P_{25}$, $P_{75}$, $P_{97.5}$?

*Section* 5-4

**3.** In a normal distribution with $\mu = 30$ and $\sigma = 5$ find (*a*) the area below 24, (*b*) the area between 24 and 36, (*c*) the area between 30 and 40, (*d*) the area above 37, (*e*) the point that has 95 per cent of the area below it, (*f*) the two points containing the middle 90 per cent of the area.

**4.** In a normal distribution with $\mu = 47.6$ and $\sigma = 16.2$ find (*a*) the chance that a single observation will be larger than 50 (*b*) two points such that a single observation has a 97 per cent chance of falling between them, (*c*) $P_{10}$, $P_{30}$, and $P_{99}$, (*d*) standard ($z$) scores for 63.8 and 39.5.

**5.** In a normal population with $\mu = 100$ and $\sigma = 12$ (*a*) what is the chance that a sample of 36 items has a mean less than 102? (*b*) What are two points such that the chance of a mean of 36 items, lying between the two points, is .95? (*c*) Answer (*b*) for .99 instead of .95.

**6.** In a normally distributed population of heights with $\mu = 68$ inches and $\sigma = 3$ inches what proportion of the population has heights between 65 inches and 71 inches? Below (shorter than) 62 inches? Above (taller than) 74 inches? What heights include the middle 99 per cent of the population? The upper 25 per cent? The lower 30 per cent?

**7.** A test is standardized with mean 80 and standard deviation 5. If the population which was standardized was normal, what proportion of the population would have grades over 90? Less than 75?

**8.** The mean life of stockings used by an army was 40 days, with a standard deviation of 8 days. Assume the life of the stockings follows a normal distribution. If 1 million pairs are issued, how many would need replacement before 35 days? After 46 days?

*Section* 5-5

**9.** For a normal distribution with $\mu = 43$ and $\sigma = 15$ use normal-probability paper to draw the cumulative-frequency curve and read from this curve the values of $P_{10}$, $P_{50}$, $P_{75}$, $P_{90}$, $P_{99}$.

**10.** Plot on normal-probability paper the observed cumulative distributions of $\bar{X}$ and $s^2$ from Class Exercise 1, Chap. 4. Check to see whether or not they agree in mean, variance, and type of distribution as stated in the text for the sampling distributions of $\bar{X}$ and $s^2$.

**11.** For the data in Prob. 29, Chap. 3, draw the cumulative-frequency distribution on normal-probability paper. Does it appear that these data come from a normal population?

**12.** Plot on normal-probability paper the normal cumulative distribution with $\mu = 0$ and $\sigma = 1$. Read from the curve the area between 0 and 2. The area between $-1.5$ and 2.5. Check these values against those in Table A-4.

**13.** Problems 5 to 26, Chap. 2, present distributions of samples. Plot a number of these on normal-probability paper. Read $P_{50}$ from those plotted and compare them with the corresponding values of the means (note values of $\bar{X}$ and $s$ on page 592). For plotted distributions which appear to be approximately normal read estimated standard deviations from the graph and compare each estimated value with the corresponding value of $s$.

*Section* 5-6

**14.** Fit a normal curve to the histogram of the following test-score data. Draw both the histogram and the fitted curve on the same set of coordinate axes.

| *Test score* | 95 | 90 | 85 | 80 | 75 | 70 | 65 | 60 |
|---|---|---|---|---|---|---|---|---|
| *Frequency* | 2 | 3 | 5 | 6 | 8 | 12 | 8 | 6 |

*Section* 5-7

**15.** In Class Exercise 1 above, $p = \frac{2}{3}$. What are the mean and variance of the population if the red beads are labeled 1 and the white beads 0?

**16.** In a box of 1,000 nuts 500 have a left-hand thread and 500 have a right-hand thread. What is the chance that a sample of 25 will contain fewer than 10 nuts with a left-hand thread? Fewer than 8? Between 10 and 15 inclusive?

**17.** A random sample of 3,600 voters is questioned. With labels of 1 for yes and 0 for no, what is the chance that $\bar{X}$ will differ from the population proportion by less than .05? By less than .01? (Assume the approximate value $\sigma = .5$.)

**18.** Table 5-4 has three approximate sampling distributions of $\bar{X}$ for dichotomous data. Plot each on normal-probability paper and verify the approximate normality. Read the means and standard deviations from the graphs and compare them with the means and standard deviations computed from the data in the table.

CHAPTER SIX

# Statistical inference: estimation and tests

*Statistical inference* is any procedure by which one generalizes to a population from the data obtained in a sample. A television poll of a sample of households estimates the total number of people in the country watching a program, the national health survey estimates from a sample of several hundred people per week the proportion of people in the country who are ill, a geneticist tests a theory that traits are inherited from both parents by observation of a sample of children, a physician tests the value of a new drug in relation to a standard drug on a sample group of patients, and so on.

## ESTIMATION AND TESTS OF HYPOTHESES 6-1

The process by which a statistic computed from one or more samples is used to approximate the value of a parameter is called *estimation*. Sometimes the statistic and parameter are computed in the same way. This is the case with $\bar{X}$ and $\mu$, and we use $\bar{X}$ as an estimate of $\mu$. As noted in Chaps. 4 and 5, the sampling distribution of $\bar{X}$ gives explicit information which can be used to describe the reliability of $\bar{X}$ as an estimate of $\mu$. If a random sample is taken $\bar{X}$ will be an unbiased estimate of $\mu$ and there is a high probability that $\bar{X}$ will be within a few standard deviation units of $\mu$. In the same way, $s^2$ is an unbiased estimate of $\sigma^2$ and $s$ is an estimate of $\sigma$, although it is biased.

A hypothesis typically arises in the form of speculation concerning observed phenomena of nature or man. Thus examples of hypotheses are that men are taller than women, that aspirin cures a headache, that smog kills people, and that tall parents have tall children. If the speculation is translated into a statement concerning the distribution of a defined population or the distributions of several defined populations, that statement is a *statistical hypothesis.*

A procedure which details how a sample is to be inspected so that we may conclude that it either agrees reasonably with the hypothesis or does not agree with the hypothesis will be called a *test* of the hypothesis; it is a decision rule which tells us to *accept* the hypothesis for certain types of samples and to *reject* it for other types. Decision rules are seldom infallible, and hypotheses which are actually true may be rejected, and alternatively, hypotheses which are actually false may be accepted. The statistical hypothesis under test is often referred to as the *null hypothesis.*

A statistical test is sometimes referred to as *statistical proof.* Outside fields such as mathematics or philosophical logic, statistical proof is usually the only sort available.

Examples of hypotheses which can be subjected to statistical tests are as follows:

1. This group of observations is a sample from a population with mean equal to a specified value:

(*a*) These electric light bulbs are a standard quality (average length of life $\mu$ equals some specified value $\mu_0$).

(*b*) The average number of bacteria killed by test drops of a germicide is equal to some number.

(*c*) The average intelligence of this class is equal to the average intelligence of all students.

2. This group of observations is a sample from a population with variance $\sigma^2$:

(*a*) This class is just as variable in intelligence as the usual class.

(*b*) This machine produces axles of more uniform diameter than the allowable variation $\sigma^2$.

(*c*) The growing period for this hybrid of corn is more variable than the growing period for other hybrids.

3. These two groups are from populations with the same mean ($\mu_1 = \mu_2$):

(*a*) Method $A$ is better than method $B$ for teaching algebra.

(*b*) Steel made by method $A$ has a mean hardness greater than steel made by method $B$.

(*c*) Penicillin is more effective than streptomycin in the treatment of disease $X$.

We shall examine a number of statistical hypotheses and corresponding methods of proof. The examples given are simple in nature, are well established historically, and have well-accepted (as well as otherwise satisfactory) decision rules. Hypotheses of other types, such as those involving more than two means and variances, those involving proportions, and those concerning association, will be introduced in Chaps. 8 and 10.

The preferred scientific approach is to state a hypothesis which is of interest and then to choose an optimum statistical technique to test this hypothesis. For the present, however, we shall examine particular techniques with examples of their applications.

## CONFIDENCE-INTERVAL ESTIMATE OF THE MEAN 6-2

Since a statistic measured on one sample can rarely be expected to be exactly equal to a parameter, it is important that an estimate be accompanied by a statement which describes the precision of the estimate. *Confidence intervals* provide a method of stating both how close the value of a statistic is likely to be to the value of a parameter and the chance of its being that close.

For random samples with replacement the sampling distribution of $\bar{X}$ has mean $\mu$ and standard deviation (or *standard error*) equal to $\sigma/\sqrt{N}$. Also, at least for large samples, the sampling distribution of $\bar{X}$ is normal in shape, and we may use Table A-4 to find the probability that a sample mean will be within any given distance of $\mu$. For example, in 95 per cent of the samples we will observe a mean between $\mu - 1.96\sigma/\sqrt{N}$ and $\mu + 1.96\sigma/\sqrt{N}$. This is exactly equivalent to saying that in 95 per cent of the samples the interval between $\bar{X} - 1.96\sigma/\sqrt{N}$ and $\bar{X} + 1.96\sigma/\sqrt{N}$ will include the value of $\mu$. This can be written with inequality symbols as

$$\bar{X} - \frac{1.96\sigma}{\sqrt{N}} < \mu < \bar{X} + \frac{1.96\sigma}{\sqrt{N}}$$

which is read as "the value of $\bar{X} - 1.96\sigma/\sqrt{N}$ is less than $\mu$, which in turn is less than $\bar{X} + 1.96\sigma/\sqrt{N}$."

EXAMPLE 6-1. Suppose that $\sigma$ is known to be equal to 2.80 and, with the objective of estimating $\mu$, we collect $N = 16$ random observations obtaining, say, $\bar{X} = 15.70$. On substituting these values into the above inequality we obtain

$$15.70 - 1.96\frac{2.80}{\sqrt{16}} < \mu < 15.70 + 1.96\frac{2.80}{\sqrt{16}}$$

or $\quad 15.70 - 1.37 < \mu < 15.70 + 1.37$

or $\quad 14.33 < \mu < 17.07$

This interval is called a 95 per cent *confidence interval* for estimating $\mu$, and the numbers 14.33 and 17.07 are called *confidence limits* for $\mu$. The percentage level, 95 per cent, is called the *confidence level.*

The interval quoted depends on the value of $\bar{X}$ obtained in the particular sample. If we obtained a sample mean equal to 15.20, the interval would extend from 13.83 to 16.57; if $\bar{X} = 17$, the interval extends from 15.63 to 18.37; and so on. Since we cannot guarantee that all such intervals include $\mu$, we cannot guarantee that this interval (14.33 to 17.07), or any particular interval, does include $\mu$. We can guarantee, however, that in the long run 95 per cent of the intervals obtained in this way will include $\mu$.

Figure 6-1 shows the sampling distribution of $\bar{X}$'s if, in fact, $\mu = 15$, $\sigma = 2.80$, and $N = 16$. Also, it presents the result of a sampling experiment in which 2 out of 20, or 10 per cent, of the sample 95 per cent confidence intervals failed to cover $\mu$. Remember that the 95 per cent confidence level refers to *all possible* samples, and may not apply exactly to any experiment with a small number of trials.

The choice of 95 per cent in the above discussion is arbitrary. We could have selected 99 per cent confidence in place of 95 per cent by replacing the factor 1.96 by 2.58, or we could have selected 99.8 per cent confidence by using the factor 3, and so on. A longer interval will have a higher probability of including $\mu$. Thus with fixed sample size higher confidence can be obtained, but the statement will lose precision. Conversely, a more precise estimate may be stated with less chance that it is correct.

It will be convenient to denote by $z_\alpha$ the $100\alpha$ percentile obtained from the normal distribution, Table A-4. In this notation $z_{.025} = -1.96$, $z_{.005} = -2.58$, $z_{.975} = 1.96$, $z_{.995} = 2.58$, and so on. The $100(1 - \alpha)$ per cent confidence interval then can be written as

$$\bar{X} + \frac{z_{\frac{1}{2}\alpha}\sigma}{\sqrt{N}} < \mu < \bar{X} + \frac{z_{1-\frac{1}{2}\alpha}\sigma}{\sqrt{N}}$$

and, for example, the 99 per cent confidence interval could appear as

$$\bar{X} - \frac{2.58\sigma}{\sqrt{N}} < \mu < \bar{X} + \frac{2.58\sigma}{\sqrt{N}}$$

Since the normal distribution is symmetrical, $z_{\frac{1}{2}\alpha}$ is always the negative of $z_{1-\frac{1}{2}\alpha}$, and we may write confidence limits for $\mu$ in the form $\bar{X} \pm z_{1-\frac{1}{2}\alpha}\sigma/\sqrt{N}$ ($15.70 \pm 1.37$ in the example). The minus sign gives the lower limit and the plus sign gives the upper limit. It is important to state clearly that such intervals are confidence intervals, since many research reports use a similar presentation for $\bar{X} \pm s$ (mean $\pm$ standard deviation) or $\bar{X} \pm s/\sqrt{N}$ (mean $\pm$ standard error).

The population standard deviation $\sigma$ is required in order to estimate $\mu$ as just shown. If the sample value of $s$ is substituted for $\sigma$, the factor 1.96 will no longer ensure a 95 per cent chance for the confidence interval to include $\mu$. This point will be discussed further in Sec. 6-3.

The word "confidence" is used here instead of "probability" because of the commonly accepted interpretation that "probability" should refer to a possible future happening, whereas once a sample is taken, the interval either does or does not include the parameter value; that is, it has 100 per

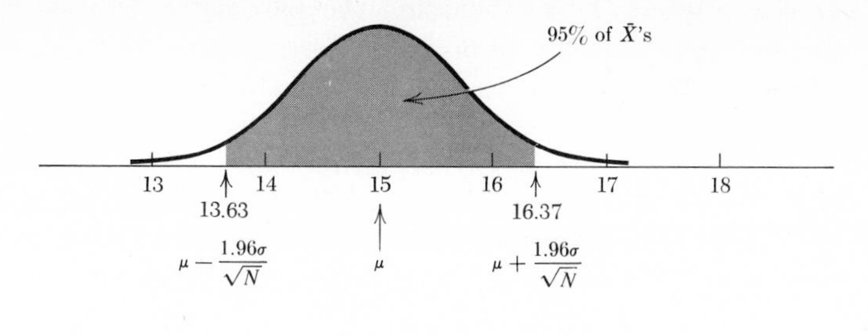


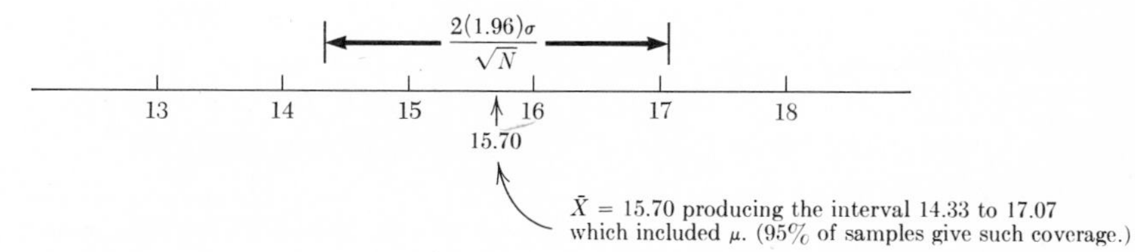


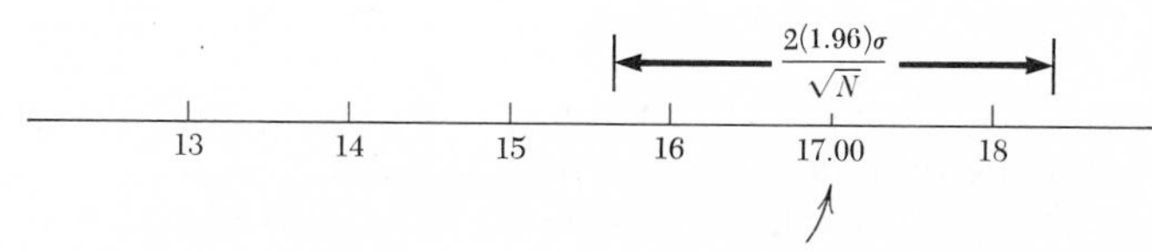


$\bar{X}$ = 17.00 producing the interval 15.63 to 18.37 which did not include $\mu$. (5% of samples similarly fail.)

Results of 20 samples. (Two of the 20 intervals did not include $\mu$. The 5% value applies to all possible samples—not necessarily to any particular set of 20.)

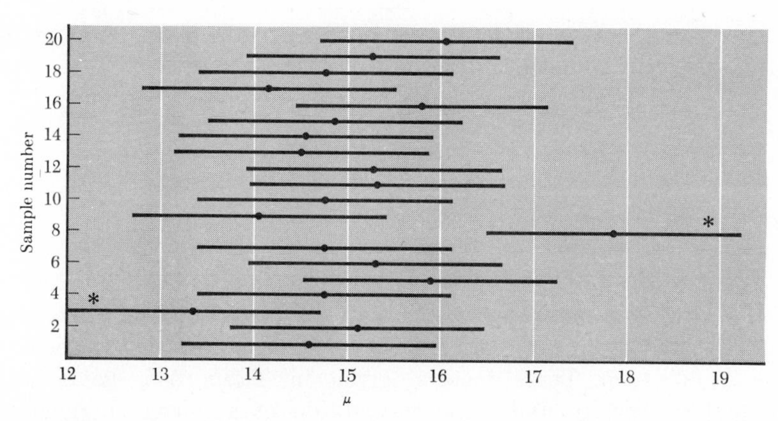


**Figure 6-1**

cent or 0 per cent probability of including the parameter. However, the frequency interpretation "95 per cent of interval estimates will be correct" is quite appropriate. Also, the probability statement refers to different intervals coming from different samples, and not to different values of $\mu$. Thus the statement "$\mu$ has a 95 per cent chance of being between $\bar{X} - 1.96\sigma/\sqrt{N}$ and $\bar{X} + 1.96\sigma/\sqrt{N}$" is considered incorrect, or at least misleading, since it may be understood to imply a varying $\mu$.

**Sample size required for estimation.** Since larger samples provide estimates of greater precision, we can attain any stated precision by taking samples sufficiently large.

We estimated $\mu$ as being between $\bar{X} + z_{\frac{1}{2}\alpha}\sigma/\sqrt{N}$ and $\bar{X} + z_{1-\frac{1}{2}\alpha}\sigma/\sqrt{N}$. If we specify that the maximum allowable length of a $100(1 - \alpha)$ per cent confidence interval be a certain length, say, $2d$, then $d = z_{1-\frac{1}{2}\alpha}\sigma/\sqrt{N}$, and the required sample size is found to satisfy this equation; that is,

$$N = \left(\frac{z_{1-\frac{1}{2}\alpha}\sigma}{d}\right)^2$$

With this sample size we can state that the confidence interval will be of length $2d$, or that the precision of the estimate is $\pm d$.

EXAMPLE 6-2. Suppose that a population has standard deviation $\sigma = 3$ inches and we wish to estimate $\mu$ to the nearest half inch, that is, with a 95 per cent confidence interval of length 1 inch. Using $d = .5$, $\sigma = 3$, $z_{.975} = 1.96$, we require that $.5 = 1.96 \times 3/\sqrt{N}$ or that we have

$$N = \left(\frac{1.96 \times 3}{.5}\right)^2 = 139$$

observations from the population. Accordingly, if we take $N = 139$ observations, the 95 per cent confidence interval will be $\bar{X} - .5 < \mu < \bar{X} + .5$.

Table A-12*e* gives necessary sample sizes corresponding to various values of $k = d/\sigma$ and of $\alpha$.

## 6-3 ESTIMATION OF $\mu$ WHEN $\sigma$ IS UNKNOWN

If $\sigma$ is unknown and we wish to replace $\sigma$ with $s$ in calculating confidence intervals for $\mu$, we must also replace the factors $z_{\frac{1}{2}\alpha}$ and $z_{1-\frac{1}{2}\alpha}$ with factors $t_{\frac{1}{2}\alpha}$ and $t_{1-\frac{1}{2}\alpha}$ obtained from the percentiles of the sampling distribution of $t = (\bar{X} - \mu)/(s/\sqrt{N})$. These factors, given in Table A-5, have been obtained mathematically under the assumption of a normal population distribution. Each line of Table A-5 refers to a different distribution identified by the number of degrees of freedom, which for the case under discussion is $df = N - 1$.

Percentiles read from this table will be denoted by $t_\alpha\ (df)$. The number in the parentheses references the number of degrees of freedom, and $100\alpha$ is the value of the percentile rank. Thus for a sample of size $N = 10$ we would read $t_{.95}(9) = 1.833$, $t_{.975}(9) = 2.262$, and so on. The distributions are symmetrical about zero, and the lower percentiles are the negatives of those recorded. For example, $t_{.05}(9) = -1.833$ and $t_{.025}(9) = -2.262$.

Values from this table are interpreted as follows. For a random sample of size $N$ from a normal population with mean $\mu$ the probability that $\bar{X}$ will be between $\mu + t_{\frac{1}{2}\alpha}(df)\ s/\sqrt{N}$ and $\mu + t_{1-\frac{1}{2}\alpha}(df)\ s/\sqrt{N}$ is $100(1 - \alpha)$ per cent. Thus a $100(1 - \alpha)$ per cent confidence interval for $\mu$ is given by

$$\bar{X} + t_{\frac{1}{2}\alpha}(df)\frac{s}{\sqrt{N}} < \mu < \bar{X} + t_{1-\frac{1}{2}\alpha}(df)\frac{s}{\sqrt{N}}$$

Example 6-3. If for a random sample of $N = 10$ men from a population assumed to be normally distributed we obtain a mean height $\bar{X} = 68.2$ inches and a standard deviation $s = 2.9$ inches, we can then state a 95 per cent confidence interval estimate of $\mu$ to be

$$68.2 - 2.262\frac{2.9}{\sqrt{10}} < \mu < 68.2 + 2.262\frac{2.9}{\sqrt{10}}$$

or

$$68.2 - 2.07 < \mu < 68.2 + 2.07 \qquad \text{or} \qquad 66.13 < \mu < 70.27$$

The $t$ distributions are symmetrical, but the sampling distribution of $t = (\bar{X} - \mu)/(s/\sqrt{N})$ has greater dispersion than the normal distribution of $z = (\bar{X} - \mu)/(\sigma/\sqrt{N})$. If $N$ is very large, $s$ can be expected to be very close to $\sigma$ and the $t$ distribution can be expected to be very close to normal. These characteristics can be verified from Table A-5; percentiles are farther from zero for small $df$ and are close to the percentiles of the normal distribution as given in Table A-4 for large $df$.

Figure 6-2 shows $t$-distribution curves for $df = 2$ and $df = 4$, corresponding to $N = 3$ and $N = 5$, and a normal distribution curve with $\mu = 0$ and $\sigma = 1$.

The area under the curves represents probability, and by inspection of the curves we see that to include any given percentage of the area an interval must be longer for smaller samples. Values from Table A-5 indicate that to include 95 per cent of the area the interval for the $N = 3$ curve must extend from $-4.30$ to 4.30, for the $N = 5$ curve from $-2.78$ to 2.78, and for the normal distribution from $-1.96$ to 1.96.

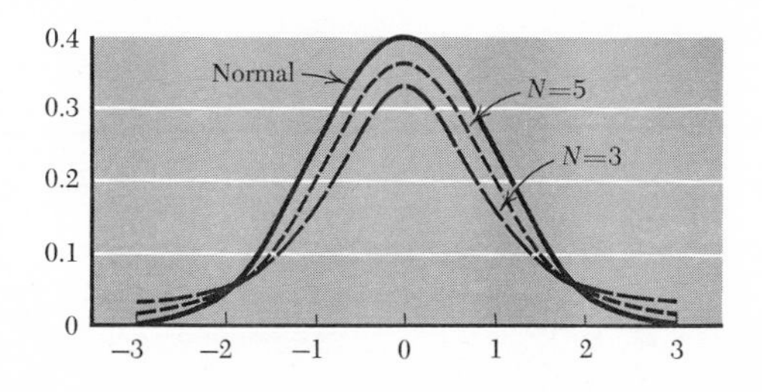


**Figure 6-2** Normal and $t$ distributions with $df = 2$ and $df = 4$.

## 6-4 TEST OF HYPOTHESIS ON THE MEAN

The sampling distribution of $\bar{X}$ is used as the basis for testing a hypothesis about $\mu$. The values of the statistic $\bar{X}$ which we agree should cause rejection of a hypothesis will be called a *critical region* for the test, and the probability that the statistic $\bar{X}$ will be in the critical region when the hypothesis is true will be called the *level of significance* of the test. Generally we shall use a small level of significance and accordingly provide protection against rejecting a true hypothesis. If the value of the statistic is observed to be in the critical region, we shall reject the hypothesis or say that "The result is significant at the specified level." This specialized use of the word "significant" applies to the rejection of a statistical hypothesis. A statistically significant departure from a hypothesis may or may not be of practical interest in a particular study. If the sample size is large, statistical significance may occur for a difference which is of marginal practical importance. For example, a difference of 1 pound in weight may be of great importance in infants but negligible in adults. Each of these would show as significant if the sample were sufficiently large.

Suppose the hypothesized mean $\mu$ of a population is stated to be $\mu_0$. The critical region will be defined as values of $\bar{X}$ sufficiently far from $\mu_0$ that such values would be unlikely to occur by chance if $\mu_0$ is actually correct. Such values are defined in terms of $z = (\bar{X} - \mu_0)/(\sigma/\sqrt{N})$ if $\sigma$ is known with the critical region defined to be values of $z$ less than $z_{\frac{1}{2}\alpha}$ or greater than $z_{1-\frac{1}{2}\alpha}$. We then have a test with a level of significance equal to $\alpha$. Alternatively, if $\sigma$ is not known and the population is normal, then the critical region is defined as values of $t = (\bar{X} - \mu_0)/(s/\sqrt{N})$ less than $t_{\frac{1}{2}\alpha}(df)$ or greater than $t_{1-\frac{1}{2}\alpha}(df)$.

The test of a statistical hypothesis should include the following information:

- Statement of the hypothesis and assumptions
- Statement of the level of significance chosen
- The test statistic and critical region
- Presentation of any computation
- A full statement of the conclusions

Example 6-4. Suppose we know that a population of men has heights with standard deviation $\sigma = 3$ inches and a hypothesis is stated that the population mean is 67 inches, that is, $\mu_0 = 67$, so that the hypothesis can be written $H\colon \mu = 67$ inches. We shall test the hypothesis at the 5 per cent level of significance using a random sample of $N = 25$ men. The above steps then are summarized as follows:

- $H\colon \mu = 67$, $\sigma = 3$ is known, $N = 25$
- $\alpha = .05$
- $z = \dfrac{\bar{X} - 67}{\sigma/\sqrt{N}} = \dfrac{\bar{X} - 67}{3/\sqrt{25}} = \dfrac{\bar{X} - 67}{.6}$

and we reject if $z \leq -1.96$ or if $z \geq 1.96$.

• Suppose the observations result in $\bar{X} = 68$, so that

$$z = \frac{68 - 67}{.6} = 1.67$$

• This value of $z$ is between $-1.96$ and $1.96$ and therefore does not lead to rejection of the hypothetical value $\mu = 67.0$ at the 5 per cent level of significance. We sometimes say "The result is not significant," or "There is not sufficient evidence to reject the hypothesis." We shall avoid a wording that implies that the hypothesis has been proved to be true. Additional discussion on this point is given in the next section.

In this example the critical region corresponds to a 5 per cent level of significance. Any other desired level could be used. Thus for $\alpha = .01$ we would set the critical limits at $\mu_0 \pm 2.58\sigma/\sqrt{N}$, for $\alpha = .10$ we would set the critical limits at $\mu_0 \pm 1.64\sigma/\sqrt{N}$, and so on.

Since results in professional journals are often given in the form of the mean and standard error, $\bar{X} \pm s/\sqrt{N}$, we may choose an appropriate coefficient $t_{1-\frac{1}{2}\alpha}$ to obtain intervals for any selected significance level. Also, many articles report a "$P$ value" or a "probability level." This usually indicates the smallest value of $\alpha$ that would lead to rejection of the hypothesis.

## SECOND TYPE OF ERROR 6-5

In testing a null hypothesis the level of significance is the probability of rejecting a true hypothesis. We may also make the error of accepting a hypothesis that is not true. Thus four possible situations exist:

1. The hypothesis is true and it is accepted.
2. The hypothesis is true and it is rejected.
3. The hypothesis is false and it is accepted.
4. The hypothesis is false and it is rejected.

It is clear that situations 1 and 4 result in correct decisions, while situations 2 and 3 apply to incorrect decisions. These concepts are presented in Table 6-1, where the rows represent decisions made on the basis of particular observations in the sample and the columns represent the truth or falsity of the hypothesis.

**Table 6-1**

| | | HYPOTHESIS | |
|---|---|---|---|
| | | *Actually true* | *Actually false* |
| DECISION | *To accept* | Correct | $\beta$ error or type II error |
| | *To reject* | $\alpha$ error or type I error | Correct |

We refer to the error made if a true hypothesis is rejected as an *$\alpha$ error*, or *type I error*. Accepting a false hypothesis we call a *$\beta$ error*, or *type II error*. The letters $\alpha$ and $\beta$ will also be used alone to denote, respectively, probabilities related to these errors; they represent numerical values between zero and unity, while "$\alpha$ error" and "$\beta$ error" are the names of the errors. The letter $\alpha$ also represents the level of significance of the decision rule or test.

Table 6-2 illustrates the roles of $\alpha$ and $\beta$ as conditional probabilities in the decision table. Note that $\alpha$ can be interpreted as a frequency only when the hypothesis is true, and similarly, $\beta$ is so interpreted only for a false hypothesis. Either error is undesirable, of course, and we wish both $\alpha$ and $\beta$ to be small. The three quantities $\alpha$, $\beta$, and $N$ are interrelated, so that generally both $\alpha$ and $\beta$ can be made small only by increasing $N$.

**Table 6-2**

| | | HYPOTHESIS | |
|---|---|---|---|
| | | *Actually true* | *Actually false* |
| DECISION | *To accept* | $1 - \alpha$ | $\beta$ |
| | *To reject* | $\alpha$ | $1 - \beta$ |
| | | 1 | 1 |

**Power of test for $H$: $\mu = \mu_0$.** In Sec. 6-4 a test of the hypothesis $H$: $\mu = \mu_0$ was presented in which the desired level of significance was obtained by adjusting the critical region. However, even for a fixed critical region, and consequently a fixed value of $\alpha$, the numerical value of $\beta$ varies according to possible true values of $\mu$. The value of $\beta$ is large if $\mu$ is close to $\mu_0$ and small if $\mu$ is very different from $\mu_0$, which reflects the fact that alternatives which are only slightly different from the hypothesis will be difficult to discover and large differences will be easier to recognize.

As an illustration, consider Example 6-4, where the population standard deviation $\sigma$ is 3 inches and the hypothesis $H$: $\mu = 67$ inches is to be tested with a 5 per cent level of significance and a sample of $N = 25$ observations. The standard error of $\bar{X}$ is equal to $\sigma/\sqrt{N} = .6$, and the critical region consists of values of $\bar{X}$ smaller than 65.82 and values of $\bar{X}$ larger than 68.18.

Figure 6-3 shows the sampling distribution of $\bar{X}$ with the assumption that $\mu$ is equal to 67. If the value 67 is correct, there is only a 5 per cent chance that a single $\bar{X}$ will fall in the critical region.

We now suppose that some specified alternative to the hypothesis $H$: $\mu = 67$ is actually true and use the techniques of Chap. 5 to compute a value of $\beta$. We accept the hypothesis $\mu = 67$ if $\bar{X}$ is between 65.82 and 68.18; thus, for example, if the mean is actually 68, we would compute $z_1 = (65.82 - 68)/.6 = -3.63$ and $z_2 = (68.18 - 68)/.6 = .30$. We refer to Table A-4 for these two values of $z$ and find an area or probability equal to

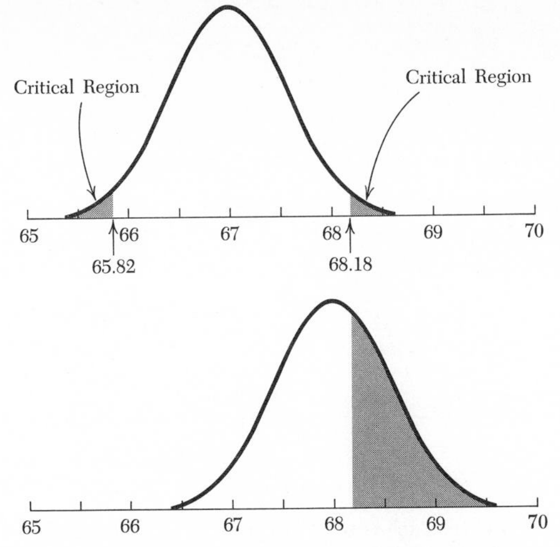


**Figure 6-3** Sampling distributions of $\bar{X}$ if $\mu = 67$ and if $\mu = 68$, respectively, for $N = 25$ and $\sigma = 3$.

$.62 - .00 = .62$ between $z_1 = -3.63$ and $z_2 = .30$. Thus for the possible alternative $\mu = 68$ we have $\beta = .62$. Figure 6-3 illustrates the areas discussed. This computation was repeated for several possible values of $\mu$, and the results are given in Table 6-3. Note the symmetry of the computations and results.

If the values of $\beta$ are plotted against the alternative values of $\mu$, the resulting curve is called the *operating characteristic* of the test. It is more conventional to plot $1 - \beta$ against the alternative $\mu$ to construct a *power curve*. The power curve for this example is presented in Fig. 6-4. The same curve can be used for any $\mu_0$, $\sigma$, and $N$ with the alternative scale provided in Fig. 6-4.

Once the power curve has been drawn, it can be used to read the value of $1 - \beta$ for a given alternative $\mu$, or conversely, an alternative $\mu$ can be read for given $1 - \beta$. For example, for $1 - \beta = .8$ or $\beta = .2$ we read on the

**Table 6-3** Values of $\beta$ for a test of $H: \mu = 67$, $\sigma = 3$, $N = 25$ with $\alpha = .05$

| $\mu$ | $z_1$ | $z_2$ | *Probability of accepting* ($\beta$) | *Power* ($1 - \beta$) |
|---|---|---|---|---|
| 68.5 | −4.47 | −.53 | .29 | .71 |
| 68 | −3.63 | .30 | .62 | .38 |
| 67.5 | −2.80 | 1.13 | .87 | .13 |
| 67 | −1.96 | 1.96 | .95 | .05 |
| 66.5 | −1.13 | 2.80 | .87 | .13 |
| 66 | −.30 | 3.63 | .62 | .38 |
| 65.5 | .53 | 4.47 | .29 | .71 |

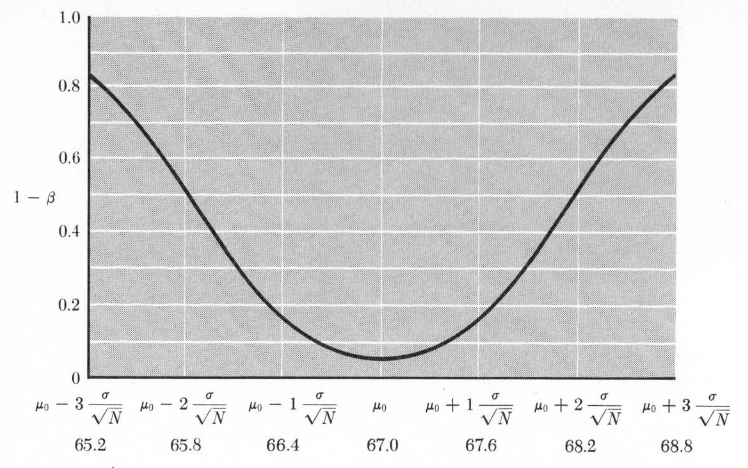


**Figure 6-4** Power curve of the test of the hypothesis $H:\mu = 67$ given $\sigma = 3.0$, $N = 25$ with $\alpha = .05$.

horizontal scale $\mu = \mu_0 + 2.8\sigma/\sqrt{N}$. For given $\sigma = 10$, $\mu_0 = 50$, $N = 16$ we find for the alternative $\mu = 55$, which is $(55 - 50)/(10/\sqrt{16}) = 2$ standard deviation units to the right of $\mu_0$ on the horizontal scale, that $\beta = .50$.

The power curve of this test appears close to a straight line if it is drawn on normal-probability paper. Several power curves, for different levels of significance, are so drawn in Table A-12*a*. Inspection of these curves for any fixed alternative shows that as $\alpha$ is increased the value of the power $1 - \beta$ increases, and consequently $\beta$ decreases. Thus a gain in power can be obtained by accepting a lower level of protection from the $\alpha$ error.

**Sample size and power.** If, with a fixed sample size, we decrease the level of significance $\alpha$, the acceptance region becomes larger, and correspondingly, $\beta$ becomes larger. However, by increasing the sample size it is possible to decrease both $\alpha$ and $\beta$. In Example 6-4, where we had $\sigma = 3$ inches and the hypothesis $H$: $\mu = 67$ inches was tested with $\alpha = .05$, the critical region used consisted of values of $\bar{X} < 67 - 1.96\sigma/\sqrt{N}$ and values of $\bar{X} > 67 + 1.96\sigma/\sqrt{N}$. We used $N = 25$, and the critical region consisted

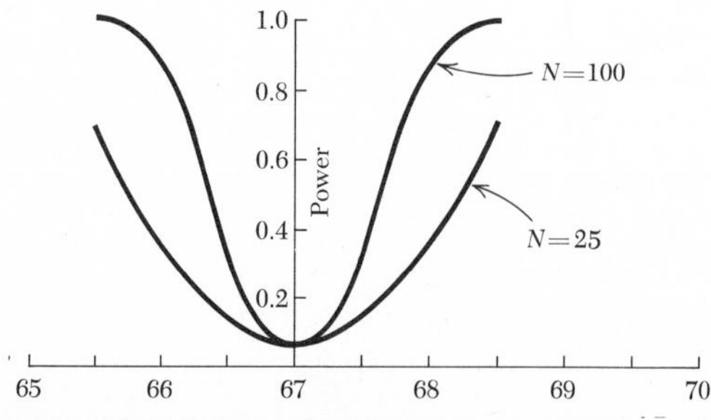


**Figure 6-5** Power of single sample test of $H:\mu = 67$ for $N = 25$ and $N = 100$.

**Table 6-4** A comparison of the probability of accepting $H: \mu = 67$ for sample sizes $N = 25$ and $N = 100$. In both cases $\sigma = 3$ and $\alpha = .05$.

| | | | $N = 100$ | | $N = 25$ |
|---|---|---|---|---|---|
| $\mu$ | $z_1$ | $z_2$ | *Probability of accepting* ($\beta$) | *Power* $(1 - \beta)$ | *Power* $(1 - \beta)$ |
| 68.5 | −6.97 | −3.04 | .00 | 1.00 | .71 |
| 68 | −5.30 | −1.37 | .09 | .91 | .38 |
| 67.5 | −3.63 | .30 | .62 | .38 | .13 |
| 67 | −1.96 | 1.96 | .95 | .05 | .05 |
| 66.5 | −.30 | 3.63 | .62 | .38 | .13 |
| 66 | 1.37 | 5.30 | .09 | .91 | .38 |
| 65.5 | 3.04 | 6.97 | .00 | 1.00 | .71 |

of values of $\bar{X}$ less than 65.82 and values of $\bar{X}$ greater than 68.18. If a sample of size $N = 100$ were used, the 5 per cent level critical region would be values of $\bar{X} < 66.41$ and $\bar{X} > 67.59$. We know that this shorter acceptance region, coupled with a smaller $\sigma/\sqrt{N}$ will result in a smaller $\beta$. If we repeat the calculations for $\mu = 68$, we find that:

$$\bar{X} = 66.41 \text{ translates to } z_1 = (66.41 - 68)/.3 = -5.30 \text{ and}$$
$$\bar{X} = 67.59 \text{ translates to } z_2 = (67.59 - 68)/.3 = -1.37$$

From Table A-4 we find $\beta = .09 - .00 = .09$. Calculations corresponding to those in Table 6-3 have been recorded in Table 6-4, with the values of $1 - \beta$ from Table 6-3 copied for comparison.

Clearly, as expected, the larger sample size has resulted in an improved test. However, recognize that the increase in power required an additional 75 observations. In practical problems a compromise must be made between the cost of observations and the desire to have small values of $\alpha$ and $\beta$. Figure 6-5 presents the two power curves for this example corresponding to $N = 25$ and $N = 100$.

## ONE-SIDED TESTS 6-6

In the previous sections we used critical regions that consisted of extreme values both larger and smaller than the value specified by the hypothesis. We have assumed that both positive and negative alternatives were to be protected against. However, situations exist where only large (or small) alternative values are of interest, and it is natural that a critical region be set up of only large (or small) values. For example, a new drug is to be used only if it differs from some standard in a positive way, that is, if it produces a better result.

The desired single alternative may be represented, in some situations, by a value larger than that specified by the hypothesis and in other cases by a value smaller. The seller of cereal, for instance, may prefer that packages barely meet the labeled amount, while the buyer would be willing to accept more—at the same price, of course.

We use the notation $H: \mu \leq \mu_0$, read as "$\mu$ less than or equal to $\mu_0$," to indicate that we shall reject the hypothesis $\mu = \mu_0$ only if alternative values larger than $\mu_0$ are supported by our test. Accordingly, we shall set up a critical region consisting of large values of $\bar{X}$. In the format used the procedure is as follows:

- $H: \mu \leq \mu_0$, with $\sigma$ known and a random sample of size $N$ (the alternative is some value greater than $\mu_0$).
- State $\alpha$.
- Reject if $\bar{X}$ is greater than $\mu_0 + z_{1-\alpha}\sigma/\sqrt{N}$, which is equivalent to rejection if $z = (\bar{X} - \mu_0)/(\sigma/\sqrt{N})$ is greater than $z_{1-\alpha}$.
- Draw the sample and compute $\bar{X}$ and $z$.
- State the conclusions.

## 6-7 COMMENTS ON INFERENCE

**Confidence intervals and critical regions.** In testing the hypothesis $H: \mu = \mu_0$ the critical region used,

$$\bar{X} < \mu_0 + \frac{z_{\frac{1}{2}\alpha}\sigma}{\sqrt{N}} \qquad \text{and} \qquad \bar{X} > \mu_0 + \frac{z_{1-\frac{1}{2}\alpha}\sigma}{\sqrt{N}}$$

has striking similarities to the confidence interval estimate of $\mu$ using $\bar{X}$,

$$\bar{X} + \frac{z_{\frac{1}{2}\alpha}\sigma}{\sqrt{N}} < \mu < \bar{X} + \frac{z_{1-\frac{1}{2}\alpha}\sigma}{\sqrt{N}}$$

and, in fact, values of $\mu$ which would be acceptable by that test would be values in the confidence interval and vice versa.

This direct connection between tests and confidence intervals will apply in some other analyses, but there are tests for which the corresponding confidence interval is not optimum and there are confidence intervals for which the corresponding test does not have maximum power. For a fixed level of significance in a test we should be concerned about minimizing $\beta$. For a fixed level of confidence we should be concerned about the length of the interval. The test of a hypothesis has two associated probability statements: the chance that the null hypothesis is rejected if true and the chance that the null hypothesis is not rejected when some alternative is true. The single probability for a confidence interval is the chance that the computed interval covers the population parameter.

**Sensitivity and specificity.** Two descriptive situations may help to clarify the meaning of the decision process, one in terms of a medical diagnosis and the other in terms of a stock-purchase problem. In the diagnosis problem we have a patient who does or does not have a disease, say, tuberculosis. We could label as a hypothesis the assertion that the patient does have tuberculosis. We also have a symptom (severe cough) or laboratory measurement (size of welt after skin test) which is under scrutiny as a possible diagnostic tool such that if the symptom is present we decide to diagnose the patient as having tuberculosis (clearly, things are not so simple in actual practice). Conventionally, symptoms implying disease are said to be positive and other symptoms are said to be negative. Table 6-5 is set up with this terminology. In this situation $\alpha$ is called a false-negative rate and $\beta$ a false-positive rate. The value of $1 - \alpha$ is also called the *sensitivity* of the diagnostic test; if $1 - \alpha$ is large, the test recognizes or is sensitive to the disease. In the same terminology, the value of $1 - \beta$ is called the *specificity* of the test; if $1 - \beta$ is large the test is generally positive only for the disease or it is specific to that disease.

**Table 6-5**

| | | HYPOTHESIS | |
|---|---|---|---|
| | | *Actually true (patient does have tuberculosis)* | *Actually false (patient does not have tuberculosis)* |
| DECISION | *Test is positive (patient is diagnosed as sick)* | $1 - \alpha$ = proportion of diseased patients diagnosed as sick | $\beta$ = proportion of well patients diagnosed as sick |
| | *Test is negative (patient is diagnosed as well)* | $\alpha$ = proportion of diseased patients diagnosed as well | $1 - \beta$ = proportion of well patients diagnosed as well |

We can also illustrate the concept by presenting an example concerning the purchase of stock. A particular stock is, by hypothesis, to rise in price, and if it is purchased now it could be sold for a profit. Of course, it may not rise after it is purchased, or alternatively, we may not purchase the stock and find that our hypothesis was true, and we have lost the potential income. The decision to buy or not to buy may be based on such factors as the previous years' sales, profits, acquisition of new equipment, or changes in company policy. We shall not attempt to assign values of $\alpha$ and $\beta$ but list only the consequences of the action in Table 6-6.

**Minimizing cost.** Sometimes, but not always, it is possible to put comparable *cost* figures on the errors and agree to minimize the total cost.

**Table 6-6**

| | | HYPOTHESIS | |
|---|---|---|---|
| | | *True* (*stock price will rise*) | *False* (*stock price will not rise*) |
| DECISION | *Buy stock* | Makes profit | No gain (or loss if stock drops) |
| | *Do not buy stock* | Loses potential gain (from this stock) | No loss (from this stock) |

Thus if the cost of making an $\alpha$ error is $C_1 = \$5$ and the cost of making a $\beta$ error is $C_2 = \$100$, we would set up a test in which $\beta$ was very small. If we know the probability that the hypothesis is true to be $p$, then the average cost of error or loss is

$$p\alpha C_1 + (1 - p)\beta C_2$$

We can then define an optimum test as one which minimizes the average cost of error. Of course, we must know $p$, $C_1$, and $C_2$ to use this criterion.

If $p$ is not known exactly, but a range of possible values of $p$ is known, a possible criterion of a good test is that it protect against the most unfavorable possibility for $p$; that is, we choose a test which minimizes the maximum average loss. A decision rule of this type is called a *minimax* decision. This has considerable appeal, but there are problems in the assignment of values to $C_1$ and $C_2$.

If $p$, $C_1$, and $C_2$ are unknown (a common case), it is usual to minimize $\beta$ for given $\alpha$ and given $N$, as we have done in this chapter.

**Best tests against a specified alternative.** If we are testing in a normal population the hypothesis $H: \mu = \mu_0$ with preset level of significance $\alpha$ and sample size $N$ and wish to be protected against an alternative value of $\mu$ which is greater than $\mu_0$, then the one-sided critical region consisting of values of $\bar{X} > \mu_0 + z_{1-\frac{1}{2}\alpha}\sigma/\sqrt{N}$ gives a maximum power $1 - \beta$ in comparison with other tests with the same assumptions, level of significance, and sample size $N$. Thus the one-sided test is best against an alternative $\mu > \mu_0$. However, for this region, the value of the power is small if $\mu < \mu_0$, and consequently the one-sided test is undesirable if such values of $\mu$ are of interest. The two-sided test gives a power curve (Fig. 6-4) which rises on both sides of the hypothetical value $\mu = \mu_0$. If we restrict our choice to tests that have flat power curves at $\mu_0$ but rise most steeply near that value, the two-tail test defined in Sec. 6-4 is best.

## GLOSSARY

$\alpha$
$\alpha$ error
accept hypothesis
$\beta$
$\beta$ error
confidence interval
confidence level
confidence limits
critical region
decision rule
degrees of freedom
estimation
hypothesis
inference
level of confidence
level of significance
null hypothesis
one-sided test
operating-characteristic curve
*P* value
power
proof
reject hypothesis
sensitivity
significance
specificity
statistical hypothesis
statistical inference
statistical proof
statistical test
*t* distribution
test of hypothesis
two-sided test
type I error
type II error

## DISCUSSION QUESTIONS

**1.** Define each of the terms in the glossary.

**2.** Describe an experiment to verify the proportion of possible confidence intervals which actually cover the mean of a normal population.

**3.** What are the two kinds of inference we are studying? What is a statistical inference?

**4.** Does our use of the term "statistical proof" agree with the definition of the word "proof" in the dictionary?

**5.** What are the two types of errors which we might make in testing hypotheses? Why, in a particular problem, can we make only one? Would it be an advantage if we knew which error we might make?

**6.** How can the chances of making both types of error be made very small?

**7.** To which type of error does the level of significance refer?

**8.** Is the chance of making a $\beta$ error equal to 1 minus the chance of making an $\alpha$ error? Explain.

**9.** Distinguish between assumption and hypothesis.

**10.** What assumptions are made about the population when an experiment is carried out and analyzed by use of a *t* test?

**11.** Why in testing a one-sided hypothesis is the rejection region entirely on one side of the hypothetical value?

**12.** Would there be any objection to using the rejection region

$$\mu_0 - \frac{.125\sigma}{\sqrt{N}} < \bar{X} < \mu_0 + \frac{.125\sigma}{\sqrt{N}}$$

to test the hypothesis $\mu = \mu_0$ at the 10 per cent level of significance?

## CLASS EXERCISES

**1.** Compute 90 per cent confidence limits for $\mu$ from the value of $\bar{X}$ observed in each of the three samples of size 5 from Table A-2 (computed for the Class Exercise in Chap. 3). Assume $\sigma = 1$. Count, for all students in the class, the number of these intervals which actually cover $\mu$. Note that the intervals change location from sample to sample, but that all the intervals are of the same length.

**2.** Suppose that you believe you are sampling from the population represented in Table A-2 but you actually are sampling from the population in Table A-23. You wish to test the hypothesis that the population mean is zero at the 5 per cent level of significance. Assuming that $\sigma^2 = 1$, determine a test and a critical region. Then, using the samples from Table A-23, approximate the chance, or relative frequency, of accepting your false hypothesis (that is, the $\beta$ error).

**3.** For each of the samples of size $N = 5$ drawn from Table A-2 compute the value of $t = \dfrac{(\bar{X} - 0)\sqrt{5}}{s}$ and verify that less than 95 per cent fall between $-1.96$ and $1.96$ ($\mu - 1.96\sigma$ and $\mu + 1.96\sigma$). Compute $P_{95}$ for the observed distribution of $t$ values and compare it with $t_{.95}(4)$ from Table A-5.

**4.** For each of the samples of size $N = 5$ drawn from Table A-23 find 90 per cent confidence limits for $\mu$ (that is, $\bar{X} + t_{.05}(4)\ s/\sqrt{5}$, $\bar{X} + t_{.95}(4)\ s/\sqrt{5}$). What proportion of these observed intervals covered the population mean?

## PROBLEMS

*Section* 6-2

**1.** A population of people has incomes with a standard deviation of $\sigma = \$500$. If a random sample of 25 people resulted in $\bar{X} = \$4{,}800$, estimate $\mu$ by a 95 per cent confidence interval. Repeat to get a 99 per cent confidence interval for $\mu$.

**2.** A random sample of 25 men had systolic blood pressure measured resulting in $\bar{X} = 130$ millimeters of mercury. If $\sigma = 10$ millimeters of mercury, estimate $\mu$ by a 95 per cent confidence interval.

**3.** If it is known that the variance of the length of life of electric light bulbs is 2,500, and if we obtain a mean life of 500 hours for a sample of 25 bulbs, determine a 95 per cent confidence interval estimate of the population mean.

**4.** A sample of 40 observations from a population with $\sigma = 3$ inches has $\bar{X} = 64.2$ inches. Give 90 per cent confidence limits for $\mu$. Give 99.9 per cent confidence limits for $\mu$.

**5.** What size of sample should be taken from a population with $\sigma = 3$ inches to give a 90 per cent confidence interval for $\mu$ of length 1 inch? A 99 per cent confidence interval of length .5 inch?

**6.** If a population has $\sigma = \$500$, how many observations would be needed to estimate $\mu$ within \$100 with 95 per cent confidence?

*Section* 6-3

**7.** A random sample of 25 people from a population showed incomes with a mean $\bar{X} = \$4{,}800$ and a standard deviation $s = \$500$. Estimate $\mu$ with a 95 per cent confidence interval. What assumption did you make about the population?

**8.** A sample of 10 ball bearings showed $\bar{X} = 1.004$ centimeters and $s = .003$. Give a 95 per cent confidence interval for $\mu$.

**9.** A fertilizer-mixing machine produced ten 100-pound bags with percentages of nitrate as follows: 9, 12, 11, 10, 11, 9, 11, 12, 9, 10. Give a 90 per cent confidence interval for $\mu$.

**10.** Using the data in Prob. 5, Chap. 2, give 95 per cent confidence limits for the mean time for boys and 95 per cent confidence limits for the mean time for girls. Can you be 95 per cent confident that both estimates are correct?

**11.** Assume that the data in Probs. 6 to 26, Chap. 2, are from random samples. Assuming that $X$ has a normal distribution for each set, estimate $\mu$ by a 95 per cent confidence interval.

*Section* 6-4

**12.** We wish to test the hypothesis that the mean IQ of a student body is 100. Using $\sigma = 15$, $\alpha = .01$, and a sample of 100 students, determine the critical region for a two-sided test.

**13.** Suppose in a sample of 100 people the mean height $\bar{X}$ was observed to be 67 inches. If the hypothetical height of the population is 66 inches and $\sigma = 3$ inches, would you reject $\mu = 66$ with $\alpha = .05$? With $\alpha = .01$? With $\alpha = .001$? Use a two-sided test.

**14.** Determine, for $\alpha = .10$, $N = 36$, $\sigma = 5$, values of $\bar{X}$ which would cause rejection of $\mu = 50$, using a two-sided test.

**15.** In which of the following cases would you reject the hypothetical value of $\mu$, using a two-sided test?

(*a*) $N = 25$, $\sigma = 2$, $\alpha = .05$, $\mu = 50$, $\bar{X} = 51$
(*b*) $N = 100$, $\sigma = 2$, $\alpha = .01$, $\mu = 67$, $\bar{X} = 66.5$
(*c*) $N = 2{,}500$, $\sigma = 2$, $\alpha = .01$, $\mu = 1.60$, $\bar{X} = 1.70$
(*d*) $N = 3{,}600$, $\sigma = .5$, $\alpha = .01$, $\mu = .50$, $\bar{X} = .51$
(*e*) $N = 4$, $\sigma = .5$, $\alpha = .05$, $\mu = .50$, $\bar{X} = .75$

**16.** A machine is set to turn out ball bearings having a radius of 1 centimeter. A sample of 10 ball bearings produced by this machine has a mean radius of 1.004 centimeters, with $s = .003$. Is there reason to suspect the machine is turning out ball bearings having a mean radius different from 1 centimeter?

**17.** A fertilizer-mixing machine is set to give 10 pounds of nitrate for every 100 pounds of fertilizer. Ten 100-pound bags are examined. The percentages of nitrate are 9, 12, 11, 10, 11, 9, 11, 12, 9, 10. Is there reason to believe that the mean is not equal to 10 per cent?

**18.** Two varieties of wheat are each planted in ten localities with differences in yield as follows: 2, 4, 2, 2, 3, 6, 2, 2, 4, 3. Test the hypothesis that the population mean difference is zero, using a test at $\alpha = .01$ level.

*Section* 6-5

**19.** We wish to test the hypothesis that the mean weight of a population of people is 140 pounds. Using $\sigma = 15$ pounds, $\alpha = .05$, and a sample of 36 people, find (*a*) the values of $\bar{X}$ which would lead to rejection of the hypothesis and (*b*) the value of $\beta$ if $\mu = 150$ pounds. Use a two-sided test.

**20.** (*a*) Find the two-sided critical regions for the hypothesis $H$: $\mu = 67$ given $\sigma = 3$ and using $N = 100$ and $N = 25$ and $\alpha = .01$. (*b*) For these examples compute values of $\beta$, draw power curves, and compare the results with Table 6-4 and Fig. 6-4. (*c*) Read from the curves values of $1\text{-}\beta$ for $\mu = 67.25$ and for $\mu = 67.75$.

**21.** The hypothesis $H$: $\mu = 100$ is to be tested with $\alpha = .05$. The population standard deviation is known to be $\sigma = 10$. (*a*) Would a sample of size $N = 100$ result in a value of $\beta$ less than .2 if, in fact, $\mu = 110$? (*b*) How large a sample would be required so that $\beta = .01$ if, in fact, $\mu = 110$?

*Section* 6-6

**22.** A cigarette manufacturer wished to advertise that the nicotine content of his cigarettes had an average less than 24 milligrams. A laboratory made five determinations of the nicotine content in milligrams as 26, 28, 22, 23, 29. (*a*) Would the manufacturer feel safe at the 5 per cent level in making the above claim for his product? Carry out the solution, stating assumptions and conclusions in the same general fashion as in the text. (*b*) How could the manufacturer make an $\alpha$ error (type I error)? What is the chance of his doing so? What would be the consequences of his making an error of this kind? (*c*) How could the manufacturer make a $\beta$ error (type II error)? What would be the consequences of such an error?

**23.** A manufacturer of flashlight batteries claims that his batteries will operate continuously for an average life of at least 20 hours with a certain type of bulb. A government testing laboratory experimented with five batteries and obtained the following burning times: 19, 18, 22, 20, 17. (*a*) Would the laboratory feel safe at the 5 per cent level in saying that the manufacturer's claim is correct? Carry out the solution, stating assumptions and conclusions in the same general fashion as in the text. (*b*) How could the laboratory make an $\alpha$ error (type I error)? What is the chance of its doing so? What would be the consequences of such an error? (*c*) How could the laboratory make a $\beta$ error (type II error)? What would be the consequences of such an error?

**24.** Repeat Prob. 20 with a one-sided test.

**25.** Repeat Prob. 21 for the hypothesis $H: \mu \leq 100$.

CHAPTER SEVEN

# Inference: single population

The general ideas concerning estimations and tests of hypotheses were illustrated in the previous chapter for the population mean. An expanded presentation of this problem will be given in this chapter, with the application extended to proportions and variances.

## THE MEAN 7-1

It will be useful to construct a general outline for the test of a hypothesis, indicating the origin of the hypothesis and the consequences of the conclusions. This outline will be provided for the test of the following hypothesis.

HYPOTHESIS. *Population mean is equal to a specified number when $\sigma$ is known. Two-sided test.*

1. State the experimental goal.
2. State the statistical hypothesis and alternatives: $H: \mu = \mu_0, \sigma$ known, alternatives $\mu > \mu_0$ or $\mu < \mu_0$.
3. Decide on the level of significance $\alpha$, the chance of rejecting the hypothesis if it is true.

4. Decide upon the statistic to be used for testing this hypothesis, in this case, $z = (\bar{X} - \mu_0)/(\sigma/\sqrt{N})$.

5. Find the sampling distribution of the statistic under the assumption that the hypothesis is true. Here assume $z = (\bar{X} - \mu_0)/(\sigma/\sqrt{N})$ has a normal distribution with mean 0 and variance 1.

6. Decide upon a critical region, those values of the statistic which would cause us to reject the hypothesis. The chance that the statistic will fall in the critical region if the hypothesis is true should be $\alpha$. For this we find in Table A-4 the values of $z_{\frac{1}{2}\alpha}$ and $z_{1-\frac{1}{2}\alpha}$ such that a proportion of $\frac{1}{2}\alpha$ of the cases will be larger than $z_{1-\frac{1}{2}\alpha}$ and a proportion $\frac{1}{2}\alpha$ of the cases will be smaller than $z_{\frac{1}{2}\alpha}$. Then the critical region consists of all values of $z$ below $z_{\frac{1}{2}\alpha}$ and above $z_{1-\frac{1}{2}\alpha}$. If $\alpha = .05$, then the critical region is below $z_{.025} = -1.96$ and above $z_{.975} = 1.96$. If the observed statistic $(\bar{X} - \mu_0)/(\sigma/\sqrt{N})$ is less than $-1.96$ or greater than 1.96, we reject the hypothesis. If $\alpha = .01$, the points are $z_{\frac{1}{2}\alpha} = -2.58$ and $z_{1-\frac{1}{2}\alpha} = 2.58$.

7. Compute the value of the statistic from the sample and see whether or not it falls in the critical region.

8. Make a statement of acceptance or rejection of the hypothesis. If the statistic has a value in the critical region we reject; if not, we accept.

9. State the consequences of the experimental findings in light of the acceptance or rejection of the statistical hypothesis.

Example 7-1

1. A standard intelligence examination has been given for several years with an average score of 80 and a standard deviation of 7. A group of 25 students are taught with special emphasis on reading skill. If the 25 students obtained a mean grade of 83 on the examination, is there reason to believe that the special emphasis changes the results on the test? We shall test the hypothesis that the population mean of students taught by the new method is 80.

2. $H: \mu = 80, \sigma = 7$, alternatives $\mu > 80$ or $\mu < 80$.

3. $\alpha = .05$ (this choice is arbitrary, of course, and is set by the experimenter before the data are examined).

4. $z = (\bar{X} - \mu_0)/(\sigma/\sqrt{N})$

5. Assume that $z$ has normal distribution with mean 0 and variance 1.

6. The critical region for $\alpha = .05$ is $z < -1.96$ and $z > 1.96$.

7. $z = \dfrac{83 - 80}{7/\sqrt{25}} = \dfrac{3}{7/\sqrt{25}} = \dfrac{3}{1.40} = 2.14.$

8. The value of $z$ is greater than 1.96, so we reject the hypothesis.

9. We have sufficient evidence to conclude that the scores of the students taught by the special method do not still have the same mean of 80 from a test at the 5 per cent level of significance. We reason that if the mean of the population of students taught by the new method is 80, then there is less than a 5 per cent chance that a mean of 25 scores would be as far away

as 83. Thus there is sufficient reason to doubt the statement that the mean is 80.

HYPOTHESIS. *Population mean is equal to a specified number when $\sigma$ is known. One-sided test, alternative $\mu > \mu_0$.*

Many problems arise in which we are chiefly interested in the possibility that the mean may be too large. For example, we may wish to test whether or not a particular type of fertilizer increases the yield of wheat. If it does not, we continue using the old type of fertilizer. Another example is a chemical additive to raise the melting point of lead. If the melting point is not raised, we would not add the chemical. We usually state the hypothesis in such cases as $\mu \leq \mu_0$ to take advantage of the strength of the rejecting statement. Thus if we reject we have a certain degree of assurance that the mean $\mu$ is larger than $\mu_0$. In computations we use $\mu_0$ as the hypothetical mean. The less-than-or-equal-to sign reminds us that values of $\bar{X}$ which are less than or equal to $\mu_0$ would not be cause for rejection. If the true mean $\mu$ does equal $\mu_0$, the chance of rejecting the hypothesis is controlled at the value $\alpha$. If the true mean $\mu$ is actually less than $\mu_0$, the chance of rejecting the hypothesis is less than $\alpha$.

Our procedure for testing a hypothesis of this sort is as follows:

1. State the experimental goal.
2. State the hypothesis: $H: \mu \leq \mu_0$, alternative $\mu > \mu_0$.
3. Choose $\alpha$.
4. Use $z = (\bar{X} - \mu_0)/(\sigma/\sqrt{N})$ as a statistic to test the hypothesis. Note that we shall not want to reject if $z$ happens to be negative.
5. Assume that the sampling distribution of $z = (\bar{X} - \mu_0)/(\sigma/\sqrt{N})$ is normal, with mean 0 and variance 1.
6. Here the critical region will be $z$ greater than $z_{1-\alpha}$ (that is, $z > z_{1-\alpha}$), where $z_{1-\alpha}$ is the value (from Table A-4) above which a proportion $\alpha$ of the sample values will fall if the hypothesis is true.
7. Compute the value of $z = (\bar{X} - \mu_0)/(\sigma/\sqrt{N})$, and reject the hypothesis if $z > z_{1-\alpha}$ and accept it if $z < z_{1-\alpha}$.
8. State the statistical conclusion.
9. State the experimental conclusion.

EXAMPLE 7-2

1. A certain manufacturing process produces 12.3 units per hour. This yield has a variance $\sigma^2 = 2$. A new process is suggested which is expensive to install but would be worthwhile if production could be increased to an average of at least 13 units per hour. In order to decide whether or not to make the change, 14 new machines are tested. They produce a mean of 13.3 units per hour. The hypothesis is that the mean yield from new machines is less than or equal to 13. If we have sufficient reason to doubt this, we shall buy the machines.

2. $H: \mu \leq 13, \sigma^2 = 2$, alternative $\mu > 13$.

3. $\alpha = .25$. This indicates that there is a 25 per cent chance of profiting from the new machines when they produce at a rate no greater than 13 units per hour. The manufacturer is willing to take a risk of this size.

4. $z = \dfrac{\bar{X} - \mu_0}{\sigma/\sqrt{N}} = \dfrac{\bar{X} - 13}{1.4/\sqrt{14}}$.

5. Assume $z$ has a normal distribution.

6. Reject the hypothesis if $z > z_{.75} = .67$.

7. $z = \dfrac{13.3 - 13}{1.4/\sqrt{14}} = \dfrac{.3}{1.4/3.74} = .80$.

8. $z$ is greater than .67, so we reject the hypothesis.

9. Rejection of the hypothesis $\mu \leq 13$ leads to a decision to purchase the new machines.

HYPOTHESIS. *Population mean is equal to a specified number when $\sigma$ is known. One-sided test, alternative $\mu < \mu_0$.*

The procedure for testing this hypothesis is the same as in the previous case, except that the critical region will be $z$ less than $z_\alpha$.

HYPOTHESIS. *Population mean is equal to a specified constant when $\sigma$ is not known. Two-sided test.*

1. State the experimental goal.
2. State the hypothesis: $H: \mu = \mu_0$, alternative $\mu < \mu_0$ or $\mu > \mu_0$.
3. Choose $\alpha$.
4. Use the statistic $t = (\bar{X} - \mu_0)/(s/\sqrt{N})$.
5. If the universe has a normal distribution and if the hypothesis is true, the statistic $t$ has a $t(N - 1)$ distribution.
6. The critical region is $t < t_{\frac{1}{2}\alpha}(N - 1)$ and $t > t_{1-\frac{1}{2}\alpha}(N - 1)$, where these values are read from Table A-5.
7. Compute the value of $t$ from the sample and reject the hypothesis if the observed $t$ is less than $t_{\frac{1}{2}\alpha}(N - 1)$ or if it is greater than $t_{1-\frac{1}{2}\alpha}(N - 1)$, and otherwise accept the hypothesis.
8. State the statistical conclusion.
9. State the experimental conclusion.

EXAMPLE 7-3

1. A certain type of rat shows a mean gain in weight of 65 grams during the first 3 months of life. Twelve rats were fed a particular diet from birth until age 3 months, and the following weight gains were observed: 55, 62, 54, 58, 65, 64, 60, 62, 59, 67, 62, 61. Is there reason at the 5 per cent level of significance to believe that the diet causes a change in the amount of weight gained? We test the hypothesis that the mean weight gain is 65. Rejection will allow the conclusion of an effect on mean weight gain.

2. $H: \mu = 65$, alternative $\mu < 65$ or $\mu > 65$.

3. $\alpha = .05$.

4. $t = \dfrac{\bar{X} - 65}{s/\sqrt{12}}$.

5. We assume this statistic has a $t$ distribution with 11 degrees of freedom.

6. We reject if $t < -2.20$ or if $t > 2.20$.

7. $t = \dfrac{\bar{X} - 65}{s/\sqrt{12}} = \dfrac{60.75 - 65}{3.8406/3.4641} = -3.83$.

8. $t$ is less than $-2.20$, so we reject the hypothesis.

9. We conclude that the diet under study has affected the weight gain of rats.

HYPOTHESIS. *Population mean is equal to a specified number when $\sigma$ is not known. One-sided test, alternative $\mu > \mu_0$.*

The modifications of the above procedure necessary to test this hypothesis follow:

2. $H: \mu \leq \mu_0$, alternative $\mu > \mu_0$.

6. The critical region is $t > t_{1-\alpha}(N - 1)$. We are interested in rejecting our hypothesis only when $t$ is larger than $t_{1-\alpha}(N - 1)$ in order to provide the chosen level of significance $\alpha$.

EXAMPLE 7-4

1. Injection of a certain type of hormone into hens is said to increase the mean weight of eggs .3 ounce. A sample of 30 eggs has an arithmetic mean .4 ounce above the preinjection mean and a value of $s$ equal to .20. Is this enough reason to accept the statement that the mean increase is at least .3 ounce? We test the hypothesis that the mean is not greater than .3 with the intention of concluding a greater mean effect if we reject the hypothesis.

2. $H: \mu \leq .3$, alternative $\mu > .3$.

3. Choose $\alpha = .05$.

4. $t = \dfrac{\bar{X} - .3}{s/\sqrt{N}}$.

5. We assume that this statistic has a $t$ distribution with

$$N - 1 = 30 - 1 = 29 \text{ degrees of freedom}$$

6. Reject if $t > t_{.95}(29) = 1.70$.

7. $t = \dfrac{.4 - .3}{.20/\sqrt{30}} = .5\sqrt{30} = 2.75$.

8. $t$ is larger than 1.70, so we reject the hypothesis that $\mu \leq .3$.

9. We can agree with the statement that the mean increase is more than .3 ounce.

HYPOTHESIS. *Population mean is equal to a specified constant when $\sigma$ is not known. One-sided test, alternative $\mu < \mu_0$.*

This is the same as the preceding case, except that the critical region is $t < t_\alpha(N - 1)$.

## 7-2 PROPORTIONS: ESTIMATION AND TESTS

**Estimation of a proportion.** The method used for estimating means may also be used to estimate the proportion of individuals in a population having a certain characteristic, for example, to estimate from a sample of registered voters the proportion of all registered voters in favor of a certain candidate.

In Sec. 5-7 we noted that a proportion of a sample having a stated characteristic could be treated as a special case of a mean $\bar{X}$ by assigning the score 1 to the individuals having the characteristic and the score 0 to those not having the characteristic. The mean of these scores is the statistic $\bar{X}$, with mean $p$, the proportion in the population, and standard deviation $\sigma/\sqrt{N} = \sqrt{p(1 - p)/N}$, and which for large sample sizes has an approximately normal sampling distribution. The fact that the standard deviation depends on the unknown $p$ is troublesome, and further attention will be given to this point later. For this chapter we shall be satisfied with noting that (1) whatever the value of $p$, the quantity $p(1 - p)$ is never larger than .25; so if we use $.5/\sqrt{N}$ instead of $\sigma/\sqrt{N}$, we shall be using a number equal to or larger than the true standard deviation, and (2) if $N$ is moderately large, there is little error in replacing $\sqrt{p(1 - p)/N}$ by $\sqrt{\bar{X}(1 - \bar{X})/N}$. These approximations seem more reasonable if we note how little $\sqrt{p(1 - p)}$ changes even for considerable changes in $p$. This point was discussed in Sec. 5-7.

Using either of the procedures, we can now form approximate confidence intervals for $p$ in the same manner as for the mean $\mu$ in Sec. 6-3. As an example using the observed $\bar{X}$ in the standard-deviation formula, suppose a sample of 500 people had been interviewed and 200 of them stated they were in favor of a certain candidate for president. Approximate 95 per cent confidence limits for the population proportion in favor of the candidate are given by

$$\frac{200}{500} - 1.96\sqrt{\frac{.4 \times .6}{500}} < p < \frac{200}{500} + 1.96\sqrt{\frac{.4 \times .6}{500}}$$

or

$$.357 < p < .443$$

The use of .5 as an approximate value for the standard deviation will give an interval $.356 < p < .444$, which is little different from the interval using the observed $\bar{X}$.

Further attention is given to the estimation of proportions in Chap. 13.

**Test of hypotheses on proportions**

HYPOTHESIS. *Population proportion is equal to a specified proportion.*

The procedure used in Sec. 6-4 for testing a hypothesis specifying $\mu$ can be used, with the modification that $\sigma = \sqrt{p(1-p)}$, to test a hypothesis specifying a value of $p$ in a dichotomous population. The test proceeds as follows:

1. State the experimental goal.
2. $H: p = p_0$, alternative $p < p_0$ or $p > p_0$.
3. Choose $\alpha$.
4. Use the statistic $z = \dfrac{\bar{X} - p_0}{\sqrt{p_0(1-p_0)/N}}$.
5. Assume that $z$ has a normal distribution with mean 0 and variance 1.
6. Use Table A-4 to find for a two-sided test $z_{\frac{1}{2}\alpha}$ and $z_{1-\frac{1}{2}\alpha}$ or for one-sided tests either $z_\alpha$ or $z_{1-\alpha}$.
7. Compute $\bar{X}$ and $z$.
8. State the statistical conclusion.
9. State the experimental conclusion.

EXAMPLE 7-5

1. Suppose, for example, that we wish to test, at the 5 per cent level of significance, the hypothesis that in a certain population of flies grown in a laboratory 50 per cent are male and 50 per cent are female. The sex for a sample of 100 flies will be determined. We test the hypothesis of equal percentages of males and females with rejection for significant deviation in either direction.
2. $H: p = .5$, alternative $p < .5$ or $p > .5$.
3. $\alpha = .05$.
4. Use the statistic $z = \dfrac{\bar{X} - .5}{\sqrt{.5 \times .5/100}}$.
5. Assume a normal sampling distribution for $z$.
6. Reject if $z \leq -1.96$ or if $z \geq 1.96$.
7. The random sample of $N = 100$ flies from the population was observed to consist of 40 males and 60 females. Thus $\bar{X} = .4$ and

$$z = \frac{.40 - .50}{\sqrt{.5 \times .5/100}} = \frac{-.10}{.05} = -2.$$

8. Since $-2$ is less than $-1.96$, we conclude that we have statistical evidence at the 5 per cent level that the proportion of male flies is not .50.
9. We conclude a significant imbalance in numbers of each sex.

## THE STANDARD DEVIATION AND VARIANCE 7-3

It is often of great importance to study the variability of some measurement. For example, the problem of teaching a class which is homogeneous in

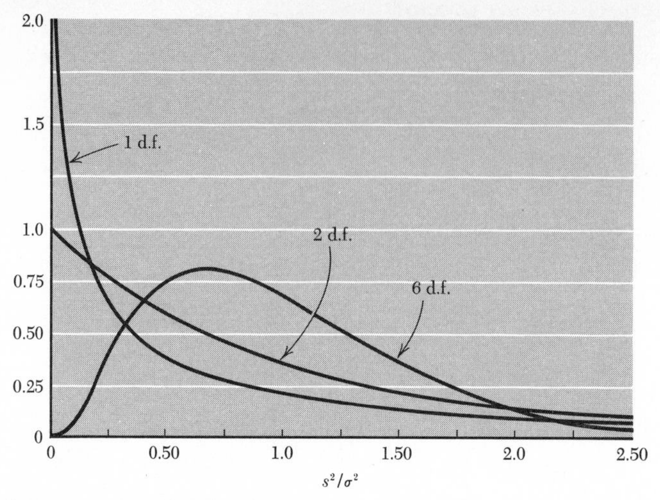


**Figure 7-1** $\chi^2/df$ distributions with 1, 2, and 6 degrees of freedom.

ability is much different from the problem of teaching a class in which abilities differ widely, even though both classes are of similar ability on the average. In the manufacture of links for a chain, not only should the links have a sufficient strength on the average, but it is essential that the variability about that average be small.

**Confidence-interval estimates of $\sigma^2$ and $\sigma$.** The estimation of $\sigma^2$ by confidence limits is based on the sampling distribution of $s^2$ (the sampling distribution of $s^2/\sigma^2$ was introduced in Chap. 4 where percentiles were experimentally obtained). These percentiles can be obtained more precisely from tables. The percentiles of the sampling distribution of $s^2/\sigma^2$ are in Table A-6*b* and are called the percentiles of the $\chi^2/df$ distribution, read "chi square over degrees of freedom." The graphs in Fig. 7-1 show the shapes of several $\chi^2/df$ distributions. The percentile values are different for different sample sizes, and Table A-6*b* gives, for samples of size 10, $P_{05} = .369$, $P_{95} = 1.88$. These values are in the columns headed $P_{05}$ and $P_{95}$ and on the line stating the number used in the denominator of $s^2$. This number, called *degrees of freedom*, is $N - 1$. Therefore 90 per cent of the samples of size $N = 10$ would be such that

$$.369 < \frac{s^2}{\sigma^2} < 1.88$$

or equivalently, such that

$$\frac{1}{1.88} < \frac{\sigma^2}{s^2} < \frac{1}{.369}$$

so that, multiplying by $s^2$, we obtain

$$\frac{s^2}{1.88} < \sigma^2 < \frac{s^2}{.369}$$

and have 90 per cent confidence limits for the estimation of $\sigma^2$. Suppose for a sample of 10 items the sample variance is computed to be 300. The limits for 90 per cent confidence are $\frac{300}{.369}$ and $\frac{300}{1.88}$. These limits are 813 and 160. In general, the 90 per cent confidence limits are

$$\frac{s^2}{P_{05}} > \sigma^2 > \frac{s^2}{P_{95}}$$

where the percentile values indicated are obtained from the $\chi^2/df$ distribution in Table A-6*b*. If confidence limits are desired for other than 90 per cent confidence, different percentile values must be used.

Confidence limits for $\sigma$ may be obtained directly from the confidence limits for $\sigma^2$ by taking the square root of the three terms in the inequality. For our example we have

$$\sqrt{\frac{s^2}{.369}} > \sigma > \sqrt{\frac{s^2}{1.88}}$$

$$\sqrt{813} > \sigma > \sqrt{160}$$

and obtaining the square roots from Table A-33, we have

$$28.5 > \sigma > 12.6$$

Table A-6*e* gives percentiles of the sampling distribution of $\sigma/s$, and these when multiplied by $s$ give confidence limits for $\sigma$ directly.

We can also specify the required sample size to estimate $\sigma^2$ correctly within a certain percentage. For example, if we use a 95 per cent confidence interval, we obtain from Table A-6*b* the 2.5 and the 97.5 percentiles of $s^2/\sigma^2$. If the limits are to be within, say, 10 per cent of $\sigma^2$, we need to locate approximately the value .90 in the 2.5 column and approximately 1.10 in the 97.5 column. By reference to the table we can see that about 740 or 750 degrees of freedom are required. Therefore a sample variance with about 750 degrees of freedom is required. The sample size is given by $N = df + 1$. Table A-6*c* gives required sample sizes for estimating $\sigma^2$ or $\sigma$ with specified accuracy and level of confidence.

**Tests of hypotheses concerning $\sigma$ or $\sigma^2$.** Suppose the hypothesis is made that the variance $\sigma^2$ of a normal population has some fixed value, say $\sigma_0^2$. Our procedure to test this statement will be to draw a random sample from the population and compute the value of $s^2$ for this sample. If the observed $s^2$ is markedly smaller or markedly larger than the supposed variance $\sigma_0^2$, we shall reject that value as a reasonable value for $\sigma^2$. The criterion of "how much larger or how much smaller" will be based on the sampling distribution of $s^2/\sigma_0^2$. If, *and only if*, the hypothesis of $\sigma^2 = \sigma_0^2$ is true, the sampling distribution of $s^2/\sigma_0^2$ is a $\chi^2/df$ distribution. By using this distribution, therefore, we can specify the level of significance $\alpha$ and control the chance of rejecting the hypothesis when it is actually true.

*One-sided alternatives.* Usually statistical problems are concerned with the rejection of hypotheses on variances only when the sample indicates that the parameter involved is actually larger than the hypothetical value. For example, it might be satisfactory to manufacture an article with variance 5 but unsatisfactory if the variance is much larger than 5. If the variance is much greater than 5, we would wish to retool our machinery, but if the variance is 5 *or less than* 5, we would wish to continue production without change. To test this hypothesis we would reject only if $s^2/5$ is significantly large. To indicate this situation we state as a hypothesis $\sigma^2 \leq 5$, where the less-than-or-equal-to sign reminds us to reject only if $s^2/5$ is large. We use $\sigma^2 = 5$ as a hypothetical value and use the sampling distribution of $s^2/5$, that is, $\chi^2/df(N - 1)$, to determine a critical region corresponding to a given level of significance $\alpha$. Then, if $\sigma^2 = 5$, we have a chance $\alpha$ of saying that $\sigma^2$ is larger than 5, and if $\sigma^2$ is actually less than 5, we have an even smaller chance than $\alpha$ of saying that $\sigma^2$ is larger. Situations such as this concern "one-sided" alternatives; that is, values on only one side of $\sigma_0^2$ are cause for rejection.

The outline below indicates steps in testing a one-sided hypothesis. The hypothesis $\sigma^2 \geq \sigma_0^2$ is tested in a similar fashion, except that it is rejected only if $s^2/\sigma_0^2$ is small. The hypothesis $\sigma^2 = \sigma_0^2$ can, of course, be tested by using a two-sided rejection region.

HYPOTHESIS. *Population variance is equal to a specified number. One-sided hypothesis, alternative $\sigma^2 > \sigma_0^2$.*

1. State the experimental hypothesis.
2. $H: \sigma^2 \leq \sigma_0^2$. Alternative $\sigma^2 > \sigma_0^2$.
3. Choose the level of significance $\alpha$.
4. The statistic used to test this hypothesis is $\chi^2/df = s^2/\sigma_0^2$.
5. If we assume that the observations are a random sample from a normally distributed population which does have $\sigma^2 = \sigma_0^2$, then percentiles of the sampling distribution of this statistic are as given in Table A-6*b*.
6. The critical region is $s^2/\sigma_0^2 > \chi^2_{1-\alpha}/df(N - 1)$, where $\chi^2_{1-\alpha}/df(N - 1)$ is the value which has a proportion of $1 - \alpha$ of the cases to the left. For example, if $\alpha = .05$ and $N = 10$, then $\chi_{.95}{}^2/df(9) = 1.88$.
7. Compute the statistic $s^2/\sigma_0^2$.
8. Reject the hypothesis if the value is in the critical region or accept the hypothesis if the value is not in the critical region.
9. State the experimental conclusion.

Example 7-6

1. A certain type of light bulb whose burning time is measured in hours has a variance of 10,000. A sample of 20 new-type light bulbs is observed to have $s^2 = 12{,}000$. Is this reason to believe, at the 5 per cent level of significance, that the variance of the new type is more than 10,000? We test the hypothesis of variance less than or equal to 10,000 to decide whether a significant increase in variance has occurred.

2. $H: \sigma^2 \leq 10{,}000$. This is stated in this fashion so that if we reject we have sufficient reason to believe that $\sigma^2 > 10{,}000$.

3. $\alpha = .05$.

4. Use $s^2/10{,}000$.

5. If the assumptions above hold, the sampling distribution of this statistic is $\chi^2/df(19)$.

6. Reject if $(s^2/10{,}000) > \chi_{.95}{}^2/df(19) = 1.59$.

7. Here $s^2/10{,}000 = 1.2$.

8. We do not have justification at the 5 per cent level of significance to reject the hypothesis.

9. There is not sufficient evidence to conclude that there has been an increase in the variance of light bulbs.

**Values of $\beta$ for tests involving a variance.** In the test of the hypothesis $H: \sigma^2 \leq \sigma_0{}^2$ against the alternative $\sigma^2 > \sigma_0{}^2$ we accept if

$$\frac{s^2}{\sigma_0{}^2} < \chi^2_{1-\alpha}/df$$

where $\chi^2_{1-\alpha}/df$ is read from Table A-6*b* for $df = N - 1$. If $\sigma_0{}^2$ is the correct value; that is, if the sample is actually taken from a population having variance $\sigma_0{}^2$, then the probability that the inequality will hold is $1 - \alpha$, and the test has level of significance equal to $\alpha$. If an alternative value of $\sigma^2$, say, $\sigma_1{}^2$, is correct, then $s^2/\sigma_1{}^2$ has the $\chi^2/df$ distribution, and the probability of accepting $\sigma_0{}^2$ can be found by noting that

$$\frac{s^2}{\sigma_0{}^2} < \chi^2_{1-\alpha}/df$$

is equivalent to

$$\frac{s^2}{\sigma_1{}^2} < \frac{\sigma_0{}^2}{\sigma_1{}^2}\,\chi^2_{1-\alpha}/df$$

Table A-6*b* gives the probability of this occurrence.

Example 7-7. Suppose we are testing the hypothesis $H: \sigma^2 = 9$ with alternative $\sigma^2 > 9$ for $N = 10$ and $\alpha = .05$. We shall accept $\sigma^2 = 9$ if $s^2/9$ is less than or equal to 1.88, with risk $\alpha = .05$ of rejecting a true hypothesis. Now, suppose that the sample is taken from a population with $\sigma^2 = 18$, that is, from a population with a variance twice as large. We compute the value of $\beta$ (the chance of accepting $\sigma^2 = 9$) for this alternative. Note that the statement that $s^2/9 < 1.88$ is equivalent to the statement that $s^2/18 < \frac{1}{2} \times 1.88 = .94$. Since $s^2/18$ has a $\chi^2/df(9)$ distribution under the alternative hypothesis, we can use the line for 9 degrees of freedom in Table A-6*b*. There is .51 chance that $s^2/18$ will be less than .94, so the chance of accepting the hypothesis $H: \sigma^2 = 9$ when actually $\sigma^2 = 18$ is $\beta = .51$.

EXAMPLE 7-8. Suppose we are testing the hypothesis $H: \sigma^2 \leq 10{,}000$ with $N = 20$ and $\alpha = .05$, as in Example 7-6. We shall accept this hypothesis if $s^2/10{,}000$ is less than 1.59. Now, suppose that actually $\sigma^2 = 30{,}000$, three times the hypothetical value. Note that $s^2/10{,}000 < 1.59$ is equivalent to $s^2/30{,}000 < 1.59 \times \frac{1}{3} = .53$. The chance that this inequality holds is read from Table A-6*b* for $df = 20 - 1 = 19$ and is approximately .05. Therefore for the alternative $\sigma^2 = 30{,}000$ our test gives $\beta = .05$. If we wish to find the alternative $\sigma^2$ which corresponds to a particular value of $\beta$, say, $\beta = .50$, we solve for the $\sigma^2$ which makes $1.59(10{,}000/\sigma^2) = .965$, the 50th percentile of the $\chi^2/df$ distribution with 19 degrees of freedom. Here

$$\sigma^2 = 1.59 \times 10{,}000/.965 = 16{,}500$$

## GLOSSARY

| | | |
|---|---|---|
| $\chi^2$ | estimation of mean | test of mean |
| $\chi^2/df$ distribution | estimation of proportion | test of proportion |
| $\chi^2/df$ statistic | estimation of standard deviation | test of standard deviation |

## DISCUSSION QUESTIONS

**1.** Define each of the terms in the Glossary.

**2.** Under what conditions is the sampling distribution of $s^2$ a $\chi^2/df$ distribution?

**3.** Explain the reasoning and assumptions used in forming a confidence interval for estimating $\sigma^2$ from $s^2$.

## CLASS EXERCISES

**1.** Compute 90 per cent confidence limits for $\sigma^2$ from the sample variances computed for the three samples of size 5 from Table A-2. Collect the results of all the students and count the number of these intervals that actually cover $\sigma^2$. Note that the intervals not only change position from sample to sample, but they also change in length.

**2.** From the observed values of $s^2$ for samples of size 5 from Table A-2 verify that if you were testing the hypothesis $\sigma^2 = 1$ at the 5 per cent level of significance you would have rejected this hypothesis 5 per cent of the time.

## PROBLEMS

*Section* 7-1

**1.** A population has mean weight equal to 150 pounds. A random sample of 50 people from the population had, after a 60-day diet, a mean weight equal to 145 pounds and a standard deviation equal to 10 pounds. Is the sample mean significantly below the original population mean? What other data on the 50 people would give better information?

**2.** It was thought that the mean protein intake of a population was at least 65 grams per day. One hundred randomly chosen individuals kept 21-day diet records, and their average daily gram intake was recorded. These 100 numbers had a mean equal to 61.9 with a standard deviation of 15.9. State a hypothesis and analyze the data.

**3.** Use the first 10 observations for systolic blood pressure, column $C$ in Table 2-2$b$, to test the hypothesis that the mean is 110 millimeters of mercury. Repeat with the first 20 observations.

*Section* 7-2

**4.** Complete the table showing the standard error of the mean of a random sample of size $N$ from a population having a proportion $p$ of 1's and a proportion $1 - p$ of 0's.

| $p$ | $N = 25$ | $N = 100$ | $N = 400$ | $N = 900$ | $N = 2{,}500$ | $N = 10{,}000$ |
|---|---|---|---|---|---|---|
| .5 | | | | | | |
| .4 | | | | | | |
| .3 | | | | | | |
| .2 | | | | | | |
| .1 | | | | | | |
| .05 | | | | | | |
| .01 | | | | | | |
| .001 | | | | | | |

**5.** How large a random sample is required for the standard error of the sample proportion to be less than .01, regardless of the size of $p$?

**6.** If a population proportion is known to be less than .10, how large a sample is required for the standard error of the sample proportion to be less than .01?

**7.** The first 200 babies born in January were 120 boys and 80 girls. Estimate by an 80 per cent confidence interval the population proportion of male births.

**8.** Of 1,000 people treated with a new drug, 200 showed an allergic reaction. Estimate with a 90 per cent confidence interval the proportion of the sample population that would show an allergic reaction.

**9.** What size sample should be taken from a population to estimate $p$ to the nearest 1 per cent with 98 per cent confidence?

**10.** What values of $\bar{X}$ for a sample $N = 400$ from a population containing a proportion $p$ of 1's and a proportion $1 - p$ of 0's would be in the critical regions for tests of the hypotheses $p = .50$, $p = \frac{1}{3}$, $p = .25$ if a two-sided test with $\alpha = .01$ were used? If a one-sided test with $\alpha = .05$ were used?

**11.** Would an observed proportion of .30 in a sample of $N = 900$ be in the critical region for a two-sided test of the hypothesis $p = \frac{1}{3}$ at the 5 per cent level of significance?

*Section* 7-3

**12.** Suppose it is desired to estimate the variance of the number of seeds of a certain plant when grown in a new environment. For 20 plants $\Sigma(X_i - \bar{X})^2 = 6{,}000$. What is the 90 per cent confidence-interval estimate of $\sigma^2$?

**13.** In Table 3-2 the mean of a sample of 80 observations is $\bar{X} = 69.7$. If $\sigma^2 = 3.50$, estimate $\mu$ with a 95 per cent confidence interval. Using $s^2 = 3.50$, estimate $\sigma^2$ with a 95 per cent confidence interval.

**14.** What sample size should be taken so that a 90 per cent confidence interval for the variance will not be in error by more than 20 per cent, that is, so that $s^2/\sigma^2$ is approximately between .80 and 1.20?

**15.** What values of $s^2$ from a sample of $N$ observations from a normal population would cause you to reject the hypothetical value $\sigma_0{}^2$ in each of the following cases:

(a) $N = 21$, $\sigma_0{}^2 = 100$, $\alpha = .01$, two-sided test
(b) $N = 50$, $\sigma_0{}^2 \leq 15$, $\alpha = .05$, one-sided test
(c) $N = 12$, $\sigma_0{}^2 = 20$, $\alpha = .05$, two-sided test
(d) $N = 61$, $\sigma_0{}^2 = 30$, $\alpha = .05$, two-sided test
(e) $N = 38$, $\sigma_0{}^2 \geq .016$, $\alpha = .10$, one-sided test

**16.** If a sample of 25 observations has $s^2 = 12.6$, would you accept or reject at the 5 per cent level of significance the hypothesis $\sigma^2 = 20$? The hypothesis $\sigma^2 \geq 20$?

**17.** For Prob. 15 compute the value of $\beta$ if the true value of $\sigma^2$ for the corresponding five cases is (a) 50, (b) 20, (c) 10, (d) 15, (e) .004.

**18.** A one-sided hypothesis $\sigma^2 \leq 16$ is to be tested at the 5 per cent level of significance. What size sample should be used to give $\beta = .50$ if the sample is actually drawn from a population with $\sigma^2 = 32$?

CHAPTER EIGHT

# Inference: two populations

In many research studies the chief interest is in comparing two groups rather than in comparing one group to a known value. Two groups may be used to compare teaching methods, to compare two drugs, or to compare two production methods. Even when a new drug or treatment is being compared to a standard, it is usually necessary to study both the new and old regimen with two samples collected under the same circumstances. Comparison of two groups is usually in terms of their mean values. However, we shall first introduce methods for comparing variances, since this technique will be of use in the later test of means, where assumptions are made about equality of variances.

## SAMPLING DISTRIBUTION OF $F$ 8-1

Suppose we have two random samples drawn from a normal population. We shall let $X_{1i}$ be the $i$th observation in the first sample and $X_{2i}$ be the $i$th observation in the second sample. Let $N_1$ be the number of observations in the first sample and $N_2$ the number in the second. Then the means and variances of the two samples are as follows:

$$\bar{X}_1 = \frac{\Sigma X_{1i}}{N_1} \quad \text{and} \quad s_1^2 = \frac{\Sigma(X_{1i} - \bar{X}_1)^2}{N_1 - 1}$$

$$\bar{X}_2 = \frac{\Sigma X_{2i}}{N_2} \quad \text{and} \quad s_2^2 = \frac{\Sigma(X_{2i} - \bar{X}_2)^2}{N_2 - 1}$$

For samples from the same normal population, the statistic $F$ given by the formula

$$F = \frac{s_1^2}{s_2^2}$$

has a sampling distribution called the *F distribution*. There are two sample variances involved and two sets of degrees of freedom, $N_1 - 1$ in the numerator of the $F$ and $N_2 - 1$ in the denominator. Each pair of degrees of freedom determines an $F$ distribution, and to indicate which is intended we shall write $F(N_1 - 1, N_2 - 1)$, where the first number in parentheses is the number of degrees of freedom in the numerator and the second is the number of degrees of freedom in the denominator.

Table A-7 gives critical values of $F(N_1 - 1, N_2 - 1)$. This table must be entered with three values: the appropriate percentage level, the number of degrees of freedom in the numerator, and the number of degrees of freedom in the denominator. Two separate tables, A-7*a* and A-7*b*, are given for cumulative proportions of 95 per cent and 99 per cent ($F_{.95}$, $F_{.99}$). These tables are also labeled 5% and 1%, and these percentages refer to the proportion of the area under the curves to the right of the values given in the tables. The number of degrees of freedom in the numerator of the $F$ ratio is at the top of the tables, and the number of degrees of freedom in the denominator is in the left-hand column. For example, if $N_1 - 1 = 10$ and $N_2 - 1 = 12$, we could read from the tables that 1 per cent of the area under the $F(10,12)$ curve is to the right of 4.30, 5 per cent is to the right of 2.75 and so on. The chance that $F = s_1^2/s_2^2$ will be larger than 2.75 is .05, and so on. Note that the distribution of $F(10,12)$ is different from the distribution of $F(12,10)$, and correspondingly different numbers appear in the table under 12 and across from 10 from those appearing under 10 and across from 12.

Table A-7*c* lists these percentiles, as well as a number of other percentiles. The percentiles for each distribution are given in a single group. For example, suppose we wish to find percentiles for the ratio of two variances, the variance in the numerator having 4 degrees of freedom and the variance in the denominator having 8 degrees of freedom. We enter the

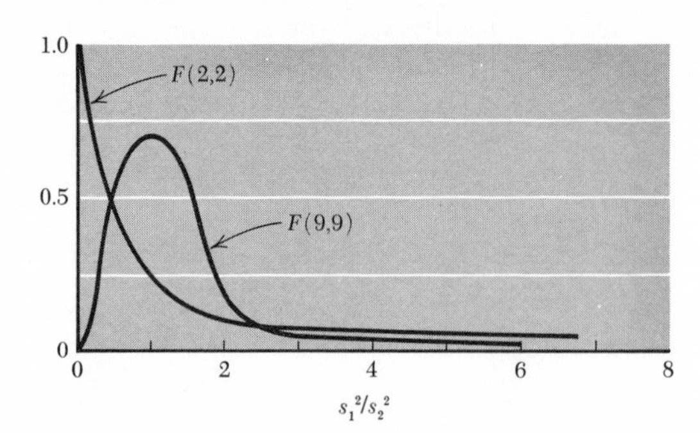


**Figure 8-1** Two $F$ distributions.

table in the column headed $\nu_1 = 4$, and, moving down the left-hand side of the table to $\nu_2 = 8$, we can read $F_{.95}(4,8) = 3.84$. Similarly, $F_{.50}(4,8) = .915$, $F_{.10}(4,8) = .253$, $F_{.50}(12,15) = .989$, $F_{.95}(2,\infty) = 3.00$, etc. Two $F$-distribution curves are sketched in Fig. 8-1.

It might be noted that $\infty$ degrees of freedom corresponds to a case where the value of $s^2$ is equal to $\sigma^2$, and the $F(N-1, \infty)$ distribution is the same as the distribution of $s^2/\sigma^2$, a $\chi^2/df$ distribution. Therefore the critical values of $F(N-1, \infty)$ are the same as those of $\chi^2/df(N-1)$ in Table A-6b.

## TESTS OF HYPOTHESES CONCERNING TWO VARIANCES 8-2

The $F$ distribution can be used to test the hypothesis that the variances $\sigma_1^2$ and $\sigma_2^2$ of two normally distributed populations are equal. It is not necessary to assume that the two populations have equal means. To test this hypothesis our procedure is to take a random sample from each population and compute the ratio of the sample variances, $F = s_1^2/s_2^2$, rejecting the hypothesis if the observed value is unusually large or unusually small. Percentiles of the $F$ distribution read from Table A-7 can be used to define a critical region, and for $\alpha = .05$ this region will consist of values of $F > F_{.975}(N_1 - 1, N_2 - 1)$ and values of $F < F_{.025}(N_1 - 1, N_2 - 1)$. Suppose that the first sample has $N_1 = 13$ cases and the second sample has $N_2 = 11$ cases. Then if $\sigma_1^2 = \sigma_2^2$, there is a 2.5 per cent chance that $F = s_1^2/s_2^2$ will be less than $F_{.025}(12,10) = .296$ and a 2.5 per cent chance that $F = s_1^2/s_2^2$ will be greater than $F_{.975}(12,10) = 3.62$. If we agreed to reject the hypothesis if either of these results occurs, we would test the hypothesis with a 5 per cent level of significance.

HYPOTHESIS. *Population variances are equal. Two-sided test. Alternative* $\sigma_1^2 < \sigma_2^2$ *or* $\sigma_1^2 > \sigma_2^2$.

1. State the experimental goal.
2. $H: \sigma_1^2 = \sigma_2^2$, alternative $\sigma_1^2 < \sigma_2^2$ or $\sigma_1^2 > \sigma_2^2$.
3. Choose $\alpha$.
4. Use the statistic $F = s_1^2/s_2^2$ to test the hypothesis.
5. If the observations are random samples from normal populations and if the hypothesis is true, then the sampling distribution of this statistic is $F(N_1 - 1, N_2 - 1)$.
6. Reject if $F > F_{1-\frac{1}{2}\alpha}(N_1 - 1, N_2 - 1)$ or if $F < F_{\frac{1}{2}\alpha}(N_1 - 1, N_2 - 1)$.
7. Compute $F$ from the sample.
8. Reject or accept the hypothesis.
9. State the experimental conclusion.

EXAMPLE 8-1

1. We wish to test whether the yields of two types of wheat have the same variances. We observe $s_1^2 = 17$ for the weights of 16 plots of one type

of wheat and $s_2^2 = 29$ for the weights of 12 plots of the second type. Is there reason to doubt at the 10 per cent level of significance that the population variances are equal? We hypothesize equal variances in yield to investigate the differences in variance.

2. $H: \sigma_1^2 = \sigma_2^2$, alternative $\sigma_1^2 < \sigma_2^2$ or $\sigma_1^2 > \sigma_2^2$.
3. $\alpha = .10$.
4. Use $F = s_1^2/s_2^2 = \frac{17}{29} = .586$.
5. If the assumptions are satisfied, the sampling distribution of this statistic is $F(15,11)$.
6. Reject if $F > 2.72$ or if $F < .398$. Note that these values are 95 and 5 per cent values, respectively, of $F(15,11)$.
7. $F$ is not in the critical region.
8. We accept the hypothesis.
9. We do not have sufficient evidence to reject the hypothesis of equal variance.

**One-sided alternatives.** We can also test hypotheses of the type $\sigma_1^2 \leq \sigma_2^2$. The procedure is similar to that above, except that we would reject only if the ratio $s_1^2/s_2^2$ is larger than $F_{1-\alpha}(N_1 - 1, N_2 - 1)$.

**Power of the $F$ test.** If we are testing the hypothesis that two populations have equal variances $H$: $\sigma_1^2 = \sigma_2^2$, we accept the hypothesis if $F = s_1^2/s_2^2$ is between two percentiles of the $F$ distribution:

$$F_{\frac{1}{2}\alpha}(N_1 - 1, N_2 - 1) < \frac{s_1^2}{s_2^2} < F_{1-\frac{1}{2}\alpha}(N_1 - 1, N_2 - 1)$$

If $\sigma_1^2$ is not equal to $\sigma_2^2$, then the quantity $\dfrac{s_1^2/\sigma_1^2}{s_2^2/\sigma_2^2}$ has the $F$ distribution, with $N_1 - 1$ and $N_2 - 1$ degrees of freedom. In order to find the value of $\beta$ we change the above inequality by multiplying each term by $\sigma_2^2/\sigma_1^2$, obtaining:

$$\frac{\sigma_2^2}{\sigma_1^2} F_{\frac{1}{2}\alpha}(N_1 - 1, N_2 - 1) < \frac{s_1^2/\sigma_1^2}{s_2^2/\sigma_2^2} < \frac{\sigma_2^2}{\sigma_1^2} F_{1-\frac{1}{2}\alpha}(N_1 - 1, N_2 - 1)$$

The chance of this occurrence can be read from Table A-7*c*.

EXAMPLE 8-2. Suppose we are testing the hypothesis that $\sigma_1^2 = \sigma_2^2$, using samples of size $N_1 = 16$ and $N_2 = 12$, with $\alpha = .10$, as in Example 8-1, and that actually $\sigma_1^2 = 2\sigma_2^2$; that is, the first variance is twice as large as the second. We accept $\sigma_1^2 = \sigma_2^2$ if $s_1^2/s_2^2$ is between .398 and 2.72. For the second inequality above we evaluate $\frac{1}{2} \times .398$ and $\frac{1}{2} \times 2.72$ for reference to Table A-7*c* with 15 and 11 degrees of freedom. We find the chance to be $.68 - .00 = .68$.

## 8-3 ESTIMATION OF VARIANCES USING SEVERAL SAMPLES

If two or more samples are from the same population or from different populations having equal variances $\sigma^2$, then the variances of the several samples, one from each population, can be pooled or averaged to give an estimate of $\sigma^2$. The formula for this estimate if there are $k$ samples is

$$s_p^2 = \frac{(n_1 - 1)s_1^2 + (n_2 - 1)s_2^2 + \cdots + (n_k - 1)s_k^2}{n_1 + n_2 + \cdots + n_k - k}$$

This $s_p^2$ is an unbiased estimate of $\sigma^2$. If the populations have normal distributions but not necessarily the same means, the sampling distribution of $s_p^2/\sigma^2$ is a $\chi^2/df$ distribution, with $n_1 + n_2 + \cdots + n_k - k$ degrees of freedom. In this case 90 per cent confidence limits for $\sigma^2$ are

$$\frac{s_p^2}{\chi_{.95}^2/df(\Sigma n_i - k)} < \sigma^2 < \frac{s_p^2}{\chi_{.05}^2/df(\Sigma n_i - k)}$$

Another method of estimating $\sigma^2$ is to group all the observations into one sample and compute the variance of this single set of numbers. This procedure is valid only if the separate populations have the same mean. Otherwise this estimate will tend to be somewhat larger than $\sigma^2$. As an example, consider a case where one population has mean 0 and a second population has mean 10, and each has variance $\sigma^2 = 1$. Individuals in a sample from the first population would cluster around 0, while individuals in a sample from the second population would cluster around 10. If the two samples were combined into one group, the mean would be about 5, and the variance of the single set would be approximately 25, much larger than $\sigma^2 = 1$. This is an extreme case, of course, but it indicates that grouping several samples into a single sample to estimate variance may be considerably less reliable than averaging the several sample variances.

Still another method of estimating $\sigma^2$ is discussed in Sec. 9-7. The variance of the means of several samples each of size $N$ is computed. This variance is an estimate of $\sigma_{\bar{X}}^2 = \sigma^2/N$. Thus we may estimate $\sigma^2$ by multiplying the variance of the means by $N$. Again this method is valid if the samples come from populations having the same mean, but the estimate tends to be too large if the population means are unequal.

Note that if $X_{ij}$, where $i = 1, 2, \ldots, k$ and $j = 1, 2, \ldots, n_i$, is the $j$th observation in the $i$th sample, if $\bar{X}_{i\cdot}$ is the mean of the $i$th sample, and if $\bar{X}$ is the mean of all observations, then

$$\sum_i \sum_j (X_{ij} - \bar{X})^2 = (n_1 + n_2 + \cdots + n_k - k)\, s_p^2 + \sum_i n_i(\bar{X}_{i\cdot} - \bar{X})^2$$

This identity, used in analysis-of-variance techniques in Chap. 10 and later, displays the variation in a group of numbers as the sum of two components, the within-sample or pooled variability and the variation of the sample means. If we restrict our observation to a single subgroup the variability is of the order of $s_p^2$ in size, while if we select observations from all groups the variability can be considerably larger if the subgroups have different means.

## 8-4 TESTS OF HYPOTHESES CONCERNING THE MEANS OF TWO POPULATIONS

Suppose we have two samples, one from each of two populations. We might wish to know whether the means of the two populations are equal. We denote the two population means by $\mu_1$ and $\mu_2$ and the two population variances by $\sigma_1^2$ and $\sigma_2^2$. To compare $\mu_1$ and $\mu_2$ we take a sample of $N_1$ observations from the first population and a sample of $N_2$ observations from the second population, and compute the respective means $\bar{X}_1$ and $\bar{X}_2$. Then, to test the hypothesis that $\mu_1 - \mu_2 = 0$ we use the statistic $\bar{X}_1 - \bar{X}_2$ and reject the hypothesis if this difference is significantly far from zero. The sampling distribution of $\bar{X}_1 - \bar{X}_2$ has mean $\mu_1 - \mu_2$ and variance $(\sigma_1^2/N_1) + (\sigma_2^2/N_2)$ and is, for large sample sizes, approximately normal. Thus, if $\mu_1 - \mu_2 = 0$, we may assume that the sampling distribution of the statistic

$$z = \frac{\bar{X}_1 - \bar{X}_2}{\sqrt{(\sigma_1^2/N_1) + (\sigma_2^2/N_2)}}$$

is approximately as given in Table A-4, and this table can be used to determine a critical region.

Consider samples of sizes $N_1 = N_2 = 1$ from Table A-2, so $\mu_1 = \mu_2 = 0$ and $\sigma_1 = \sigma_2 = 1$, with $\bar{X}_1 - \bar{X}_2$ computed for each pair of samples. The results for 200 such pairs of samples are recorded in Table 8-1. Note that the mean of the differences $\bar{X}_{1i} - \bar{X}_{2i}$ agrees approximately with the theoretical

**Table 8-1** Results of 200 observations of differences $\bar{X}_{1i} - \bar{X}_{2i}$ of means of samples of sizes $N_1 = N_2 = 1$ from normal populations with $\mu_1 = \mu_2 = 0$ and $\sigma_1 = \sigma_2 = 1$

| *Midpoint of* $\bar{X}_1 - \bar{X}_2$ | *Frequency* |
|---|---|
| 3.5 | 3 |
| 3 | 4 |
| 2.5 | 4 |
| 2 | 14 |
| 1.5 | 22 |
| 1 | 18 |
| .5 | 25 |
| 0 | 31 |
| − .5 | 24 |
| −1 | 16 |
| −1.5 | 20 |
| −2 | 10 |
| −2.5 | 7 |
| −3 | 2 |
| *Total* | 200 |

| | *Observed* | *Theoretical* |
|---|---|---|
| Mean $(\bar{X}_1 - \bar{X}_2)$ | .1125 | $\mu_1 - \mu_2 = 0$ |
| Var $(\bar{X}_1 - \bar{X}_2)$ | 1.978 | $(\sigma_1^2/1) + (\sigma_2^2/1) = 2$ |

value zero, and the observed variance is also close to the value

$$\frac{\sigma_1^2}{N_1} + \frac{\sigma_2^2}{N_2} = \frac{1}{1} + \frac{1}{1} = 2$$

given by the formula. To test agreement of the observed mean with the theoretical mean, we can compute $z = \dfrac{.1125 - 0}{\sqrt{2}/\sqrt{200}} = 1.125$ and see that the difference is not significant at the 5 per cent level.

HYPOTHESIS. *Means of two populations are equal assuming they have the same $\sigma$ when $\sigma$ is known. Two-sided test.*

1. State the experimental goal.
2. $H: \mu_1 = \mu_2$, given $\sigma_1 = \sigma_2 = \sigma$, with the value of $\sigma$ known; alternative $\mu_1 < \mu_2$ or $\mu_1 > \mu_2$.
3. Decide on $\alpha$.
4. Use the statistic $z = \dfrac{\bar{X}_1 - \bar{X}_2}{\sigma\sqrt{(1/N_1) + (1/N_2)}}$.
5. If the population is normally distributed the sampling distribution of $z$ is normal. If the sample sizes $N_1$ and $N_2$ are large the sampling distribution of $z$ is approximately normal even if the population is not normally distributed.
6. The critical region is $z < z_{\frac{1}{2}\alpha}$ and $z > z_{1-\frac{1}{2}\alpha}$.
7. Complete the computation.
8. State the statistical conclusion.
9. State the experimental conclusion.

EXAMPLE 8-3

1. Two astronomers recorded observations on a certain star. The 12 observations obtained by the first astronomer have a mean reading of 1.20 and the 8 observations obtained by the second astronomer have a mean of 1.15. Past experience has indicated that each astronomer obtains readings with a variance of about .40. Does the difference between the two results seem reasonable? We hypothesize equal means for the two observers and will conclude a difference only if the observed means differ significantly at the 1 per cent level.
2. $H: \mu_1 = \mu_2$, given $\sigma^2 = .40$, $\sigma = \sqrt{.40} = .6325$; alternative $\mu_1 < \mu_2$ or $\mu_1 > \mu_2$.
3. Choose $\alpha = .01$.
4. Use $z = \dfrac{\bar{X}_1 - \bar{X}_2}{\sigma\sqrt{(1/N_1) + (1/N_2)}}$.
5. We assume that $z$ has a normal distribution with mean 0 and variance 1.
6. Reject if $z < -2.58$ or $z > 2.58$.

7. $z = \dfrac{1.20 - 1.15}{.6325\sqrt{\frac{1}{12} + \frac{1}{8}}} = .17.$

8. $z$ is between $-2.58$ and $2.58$, so we accept the hypothesis.

9. The observed difference does not seem unreasonably large. We conclude that a difference in the observers has not been shown.

HYPOTHESIS. *Mean of one population is equal to the mean of another population when $\sigma$ is known. One-sided test. Alternative $\mu_1 > \mu_2$.*

Problems involving inequality of the means of two populations often arise. As in Sec. 7-1, the statement of the hypothesis is in the form $H$: $\mu_1 \leq \mu_2$, where we consider the equality $\mu_1 = \mu_2$ to be in question, and the less-than-or-equal-to sign is to remind us that we will reject only if $\bar{X}_1$ is significantly larger than $\bar{X}_2$.

1. State the experimental goal.
2. $H$: $\mu_1 \leq \mu_2$, given $\sigma_1 = \sigma_2 = \sigma$, with the value of $\sigma$ known; alternative $\mu_1 > \mu_2$.
3. Choose $\alpha$.
4. Use $z = \dfrac{\bar{X}_1 - \bar{X}_2}{\sigma\sqrt{(1/N_1) + (1/N_2)}}$.
5. We assume that this statistic has a normal distribution with mean 0 and unit variance.
6. We would not wish to reject the hypothesis if $\bar{X}_1$ turns out to be less than $\bar{X}_2$, so for a critical region take $z > z_{1-\alpha}$.
7. Compute the statistic.
8. Reject or accept the hypothesis.
9. State the experimental conclusion.

HYPOTHESIS. *Two populations have the same mean when $\sigma$ is not known. Two-sided test.*

1. State the experimental goal.
2. $H$: $\mu_1 = \mu_2$, given $\sigma_1 = \sigma_2 = \sigma$, with the value of $\sigma$ unknown; alternative $\mu_1 < \mu_2$ or $\mu_1 > \mu_2$.
3. Choose $\alpha$.
4. For a statistic to test this hypothesis use

$$t = \frac{\bar{X}_1 - \bar{X}_2}{s_p\sqrt{(1/N_1) + (1/N_2)}}$$

where $s_p^2$ is the pooled mean-square estimate of $\sigma^2$ given by

$$s_p^2 = \frac{\Sigma X_{1i}^2 - [(\Sigma X_{1i})^2/N_1] + \Sigma X_{2i}^2 - [(\Sigma X_{2i})^2/N_2]}{N_1 + N_2 - 2}$$

$$= \frac{(N_1 - 1)s_1^2 + (N_2 - 1)s_2^2}{N_1 + N_2 - 2}$$

where $\Sigma X_{1i}^2$ = sum of squares in first sample
$\Sigma X_{2i}^2$ = sum of squares in second sample
$\Sigma X_{1i}$ = sum of observations in first sample
$\Sigma X_{2i}$ = sum of observations in second sample
Here $s_p^2$ is the same pooled estimate discussed in Sec. 8-3.

5. If both populations have normal distributions with the same mean and the same variance, then this statistic has the $t(N_1 + N_2 - 2)$ distribution, with percentiles as recorded in Table A-5.

6. The critical region consists of values of $t < t_{\frac{1}{2}\alpha}(N_1 + N_2 - 2)$ or $t > t_{1-\frac{1}{2}\alpha}(N_1 + N_2 - 2)$.

7. Compute $t$.

8. Reject or accept the hypothesis.

9. State the experimental conclusion.

EXAMPLE 8-4

1. Two new types of rations are fed to pigs, and it is desired to test whether one or the other of the types is better. A sample of 12 pigs is fed ration $A$ and a sample of 12 pigs is fed ration $B$. The gains in weight are as follows:

| *Ration A* | 31 | 34 | 29 | 26 | 32 | 35 | 38 | 34 | 30 | 29 | 32 | 31 |
|---|---|---|---|---|---|---|---|---|---|---|---|---|
| *Ration B* | 26 | 24 | 28 | 29 | 30 | 29 | 32 | 26 | 31 | 29 | 32 | 28 |

We test for equal weight gains in the two groups and will conclude which ration is better.

2. $H$: $\mu_1 = \mu_2$, $\sigma_1 = \sigma_2 = \sigma$, with $\sigma$ unknown; alternative $\mu_1 < \mu_2$ or $\mu_1 > \mu_2$.

3. $\alpha = .05$.

4. Use $t = \dfrac{\bar{X}_1 - \bar{X}_2}{s_p\sqrt{(1/N_1) + (1/N_2)}}$.

5. If the two populations have normal distributions with the same mean and the same variance, then this statistic has a $t$ distribution with $12 + 12 - 2 = 22$ degrees of freedom.

6. Reject if $t < -2.07$ or $t > 2.07$.

7. For ration $A$, $\bar{X}_1 = 31.75$ and $\Sigma X_{1i}^2 - \dfrac{(\Sigma X_{1i})^2}{12} = 112.25$.

For ration $B$, $\bar{X}_2 = 28.6667$ and $\Sigma X_{2i}^2 - \dfrac{(\Sigma X_{2i})^2}{12} = 66.64$.

$$s_p^2 = \frac{112.25 + 66.64}{12 + 12 - 2} = 8.131$$

$$s_p = \sqrt{8.131} = 2.85$$

$$t = \frac{\bar{X}_1 - \bar{X}_2}{s_p\sqrt{\frac{1}{12} + \frac{1}{12}}} = \frac{3.083}{2.85/\sqrt{6}} = 2.65.$$

8. $t$ is larger than 2.07, so we reject the hypothesis.

9. Rejection of the statistical hypothesis permits the choice of ration $A$ as the better feed.

HYPOTHESIS. *Mean of one population is equal to the mean of a second population when $\sigma$ is unknown. One-sided test. Alternative $\mu_1 > \mu_2$.*

This problem is a combination of those in previous examples. We use the same test statistic as above, but we shall reject only if $t$ is larger than $t_{1-\alpha}(N_1 + N_2 - 2)$.

EXAMPLE 8-5

1. Two types of paint are to be tested. Paint I is somewhat cheaper than paint II. The test consists in giving scores to the paints after they have been exposed to certain weather conditions for a period of 6 months. Five samples of each type of paint are scored as follows:

| *Paint* I | 85 | 87 | 92 | 80 | 84 |
|---|---|---|---|---|---|
| *Paint* II | 89 | 89 | 90 | 84 | 88 |

We should like to adopt paint I, the cheaper one, unless we have definite reason to believe that paint II is better. If we state as a hypothesis that $\mu_2 \leq \mu_1$, our chance of adopting paint II when the paints are equally good will be controlled at $\alpha$.

2. $H$: $\mu_2 \leq \mu_1$, alternative $\mu_2 > \mu_1$.

3. $\alpha = .05$.

4. $$t = \frac{\bar{X}_1 - \bar{X}_2}{s_p \sqrt{(1/N_1) + (1/N_2)}}.$$

5. Assume that this statistic has a $t$ distribution with $N_1 + N_2 - 2$ degrees of freedom.

6. Reject if $t < t_{.05}(8)$, that is, if $t < -1.86$. We reject here if $t$ is negative ($t < -1.86$), since this would lead us to believe that $\mu_2 > \mu_1$.

7. $$\bar{X}_1 = 85.6 \text{ and } \Sigma X_{1i}^2 - \frac{(\Sigma X_{1i})^2}{5} = 77.2.$$

$$\bar{X}_2 = 88.0 \text{ and } \Sigma X_{2i}^2 - \frac{(\Sigma X_{2i})^2}{5} = 22.0.$$

$$s_p^2 = \frac{77.2 + 22.0}{8} = 12.4$$

$$s_p = 3.52$$

$$t = \frac{85.6 - 88.0}{3.52 \sqrt{\frac{1}{5} + \frac{1}{5}}} = -1.08$$

8. We accept the hypothesis.

9. There is not sufficient evidence to conclude a difference in paints I and II.

HYPOTHESIS. *Two populations have the same mean (variances not equal).*

Suppose that we wish to compare the means of two normally distributed populations having variances $\sigma_1^2$ and $\sigma_2^2$. As before, we denote the observations by $X_{1i}$ and $X_{2i}$ and the sample sizes by $N_1$ and $N_2$. The sampling distribution of the statistic

$$z = \frac{(\bar{X}_1 - \bar{X}_2) - (\mu_1 - \mu_2)}{\sqrt{(\sigma_1^2/N_1) + (\sigma_2^2/N_2)}}$$

is normal with mean 0 and variance 1. This is approximately true for sufficiently large values of $N_1$ and $N_2$ even if the populations are not normal. For this hypothesis we take $\mu_1 - \mu_2 = 0$.

If the values of $\sigma_1$ and $\sigma_2$ are known they are substituted into the formula, and observed values of $z$ can be compared with the normal distribution in Table A-4. If the values of $\sigma_1$ and $\sigma_2$ are not known but the experimenter feels that there is evidence that each population is normally distributed, then he may wish to use the statistic obtained by substituting the observed $s_1$ and $s_2$ for $\sigma_1$ and $\sigma_2$ to obtain the statistic

$$t = \frac{(\bar{X}_1 - \bar{X}_2) - (\mu_1 - \mu_2)}{\sqrt{(s_1^2/N_1) + (s_2^2/N_2)}}$$

which, if the assumptions of normality are correct, has approximately a $t$ distribution with

$$df = \frac{[(s_1^2/N_1) + (s_2^2/N_2)]^2}{\dfrac{(s_1^2/N_1)^2}{N_1} + \dfrac{(s_2^2/N_2)^2}{N_2}}$$

This value of $df$ will not be an integer, and interpolation in the $t$ table may be necessary for more accurate levels of significance; however, usually the closest value in the table is sufficient.

If there is evidence that $\sigma_1$ and $\sigma_2$ are not equal, a more convenient sampling design and analysis sometimes results from pairing the observations from the two populations, as described in the next section.

## PAIRING OBSERVATIONS 8-5

**Extraneous factors.** In sampling from two populations it sometimes happens that extraneous factors cause a significant difference in means, even though there may be no difference in the effects we are trying to measure.

Conversely, extraneous factors can mask or obscure a real difference. For example, in an experiment to test which of two types ($A$ or $B$) of fertilizers is the better, two plots of wheat are planted at each of 10 experiment stations. One of the two plots has fertilizer $A$ and the other fertilizer $B$. If the average yield of the 10 having type $A$ is compared with the average of the 10 having type $B$, part of the difference observed (if there is any) may be due to the different types of soil or different weather conditions at the different stations instead of the different fertilizers. Or the fertilizers may cause a difference, which is obscured by the other factors.

An experimental design that sometimes overcomes part of this difficulty is the observation of *pairs*. We try to make sure that the two members of any pair are alike in other respects. This is an ideal, of course, and we are limited by the availability of pairs which are similar and by our ability to choose similar pairs. In the example each pair of plots would have approximately the same types of soil, weather conditions, and so on.

Suppose we let $X_{1i}$ be the first member of the $i$th pair and $X_{2i}$ be the second member. We have, say, $N$ pairs of observations,

$$(X_{11},X_{21}),\ (X_{12},X_{22}),\ (X_{13},X_{23}),\ \ldots\ ,\ (X_{1N},X_{2N})$$

If we take differences $d_i = X_{1i} - X_{2i}$, we shall have a set of $N$ observations, each of which is a difference between two original observations. There may be certain extraneous factors affecting some of the individuals, but we hope (and assume) that they affect each member of any one pair in exactly the same manner. We also assume that the effect is essentially one of increasing (or decreasing) each of the means by some constant, so that the subtraction has removed the effect.

We wish to test the hypothesis $H\colon \mu_1 = \mu_2$. This hypothesis specifies that there is no difference in the mean treatment effects within pairs. There may well be differences in treatment effect from pair to pair, but we hope that these have been subtracted out by our use of $X_{1i} - X_{2i}$. To test the hypothesis we note that if the hypothesis is true the differences $X_{1i} - X_{2i}$ would come from a set of numbers with mean 0. Thus we can test the hypothesis that the differences come from a population with mean 0. This is done as in Sec. 7-1 (a $t$ test with $N - 1$ degrees of freedom.)

If there are no extraneous effects, then we actually lose information by pairing. With only $N - 1$ degrees of freedom in the estimate of $\sigma^2$, we accept larger differences in $\bar{X}_1 - \bar{X}_2$ than we might accept if we had $2N - 2$ degrees of freedom. Of course, if there are likely to be extraneous effects (as in the example above), we may prefer to pair and take the risk of some loss in power. This loss results in an increase in the probability of accepting the hypothesis when it is false. The increase is slight, however, if the number of pairs is moderately large, say, greater than 10. The level of significance is not affected. This point is discussed further in Chap. 14.

HYPOTHESIS. *Population mean difference is equal to a specified number.*

EXAMPLE 8-6

1. In a study of learning ability 10 boys and 10 girls were chosen at random from a freshman class. Scores were obtained as measurements of their ability to learn nonsense syllables. It was decided that pairing the subjects on the basis of IQ would effect a reduction in population variance and thus make it easier to discover any differences in ability. The observations and analysis are recorded in Table 8-2.

2. $H$: $\mu_1 = \mu_2$; that is, the mean learning score for boys equals mean learning score for girls when paired on the basis of IQ; alternative $\mu_1 < \mu_2$ or $\mu_1 > \mu_2$.

3. $\alpha = .01$.

4. $t = \dfrac{\bar{d} - 0}{s_d/\sqrt{N}} = \dfrac{-4.4}{\sqrt{128.7}/\sqrt{10}} = -1.2$ (note that $\bar{X}_1 - \bar{X}_2 = \bar{d}$ and that $s_d^2$ is the variance of the differences).

5. If the populations have normal distributions, then this statistic has a $t$ distribution with $10 - 1 = 9$ degrees of freedom; $t_{.995}(9) = 3.25$.

6. Reject if $t < -3.25$ or $t > 3.25$.

7. $t = -1.2$.

8. We accept the hypothesis.

9. This experiment does not demonstrate a difference for boys and girls in ability to learn nonsense syllables when an adjustment is made for the effect of IQ (care should be taken not to claim, without further information about $\beta$, that the experiment has "proved" that there is no difference in mean ability).

**Table 8-2** Data observed on pairs of individuals

| | PAIR NO. | | | | | | | | | |
|---|---|---|---|---|---|---|---|---|---|---|
| | 1 | 2 | 3 | 4 | 5 | 6 | 7 | 8 | 9 | 10 |
| *Boys* | 28 | 18 | 22 | 27 | 25 | 30 | 21 | 21 | 20 | 27 |
| *Girls* | 19 | 38 | 42 | 25 | 15 | 31 | 22 | 37 | 30 | 24 |
| *Difference* | 9 | −20 | −20 | 2 | 10 | −1 | −1 | −16 | −10 | 3 |

$\Sigma d_i = -44 \qquad \Sigma d_i^2 = 1{,}352 \qquad \bar{d} = -4.4 \qquad s_d^2 = \dfrac{1{,}352 - (44)^2/10}{9} = 128.7$

**Pairing of observations with unequal variances.** Suppose two populations have quite different unknown variances $\sigma_1^2$ and $\sigma_2^2$ and we wish to compare their means $\mu_1$ and $\mu_2$. We could take a random sample of size $N$ from each population and pair the individuals in the order obtained. Then the difference of observations in each pair could be obtained. These differences can be considered to be a sample of size $N$ from a population having mean $\mu_1 - \mu_2$ and standard deviation $\sigma = \sqrt{\sigma_1^2 + \sigma_2^2}$. Thus the methods of Sec. 7-1 can be used to estimate $\mu_1 - \mu_2$ or to test the hypothesis $H: \mu_1 - \mu_2 = 0$.

**Pairing of dependent observations.** Suppose two measurements are made on a single individual, say, before and after a treatment, and we are interested in the change observed. The observations cannot be considered as random samples from a population corresponding to the "before" situation and a population for the "after" situation. However, the differences can be analyzed as above. Other methods for analyzing dependent observations are given in Chaps. 11 and 12.

EXAMPLE 8-7

1. A certain stimulus is to be tested for its effect on blood pressure. Twelve men have their blood pressure measured before and after the stimulus. The results are shown in Table 8-3. Is there reason to believe that the stimulus would, on the average, raise the mean systolic blood pressure? The pairing was natural here, since two observations are made on the same individual. The sample consists of 12 individuals with two measurements on each. To test for evidence that the stimulus raises the blood pressure we hypothesize no change but set up a one-sided test.

2. $H$: $\mu_2 \leq \mu_1$ (or mean increase no more than zero), alternative $\mu_2 > \mu_1$.

3. $\alpha = .05$.

4. $t = \dfrac{\bar{X}_2 - \bar{X}_1}{s_d/\sqrt{12}}$.

**Table 8-3** Data observed before and after for 12 men

| MEN | *Before* | *After* | INCREASE $d_i$ | $d_i^2$ |
|---|---|---|---|---|
| 1 | 120 | 128 | +8 | 64 |
| 2 | 124 | 131 | +7 | 49 |
| 3 | 130 | 131 | +1 | 1 |
| 4 | 118 | 127 | +9 | 81 |
| 5 | 140 | 132 | −8 | 64 |
| 6 | 128 | 125 | −3 | 9 |
| 7 | 140 | 141 | +1 | 1 |
| 8 | 135 | 137 | +2 | 4 |
| 9 | 126 | 118 | −8 | 64 |
| 10 | 130 | 132 | +2 | 4 |
| 11 | 126 | 129 | +3 | 9 |
| 12 | 127 | 135 | +8 | 64 |
| *Total* | | | 22 | 414 |

5. If the populations have normal distributions and the treatment has no effect on blood pressure, then this statistic has a $t$ distribution with $12 - 1 = 11$ degrees of freedom.

6. Reject if $t > 1.80$.

7. We compute

$$s_d^2 = \frac{414 - [(22)^2/12]}{11} = 33.97$$

$$s_d = 5.83$$

$$t = \frac{1.833}{5.83/\sqrt{12}} = \frac{1.833}{1.68} = 1.09$$

8. We accept the hypothesis.

9. We do not have sufficient evidence to conclude that the stimulus increases blood pressure.

## CONFIDENCE LIMITS FOR THE DIFFERENCE BETWEEN TWO MEANS 8-6

If we have two populations with means $\mu_1$ and $\mu_2$, and $\sigma_1^2 = \sigma_2^2 = \sigma^2$, we can estimate the difference between $\mu_1$ and $\mu_2$ by the following confidence limits when $\sigma$ is known:

$$\bar{X}_1 - \bar{X}_2 + z_{\frac{1}{2}\alpha}\sigma\sqrt{\frac{1}{N_1} + \frac{1}{N_2}} < \mu_1 - \mu_2 < \bar{X}_1 - \bar{X}_2 + z_{1-\frac{1}{2}\alpha}\sigma\sqrt{\frac{1}{N_1} + \frac{1}{N_2}}$$

If $\sigma$ is not known, $z_{\frac{1}{2}\alpha}$ and $z_{1-\frac{1}{2}\alpha}$ must be replaced by $t_{\frac{1}{2}\alpha}$ and $t_{1-\frac{1}{2}\alpha}$ and $\sigma$ by $s_p$. In this case the number of degrees of freedom is $N_1 + N_2 - 2$.

If the observations of two samples are paired, confidence limits for a mean difference are as follows:

$$\bar{d} + t_{\frac{1}{2}\alpha}\frac{s_d}{\sqrt{N}} < \mu_1 - \mu_2 < \bar{d} + t_{1-\frac{1}{2}\alpha}\frac{s_d}{\sqrt{N}}$$

Here $s_d$ is the standard deviation of the differences (as in Sec. 8-5), and the $t$ values are read from Table A-5 for $\alpha$ and $N - 1$ degrees of freedom.

## GLOSSARY

difference in means
$F$ distribution
$F$ statistic
pairing
pooled variance

## DISCUSSION QUESTIONS

1. What assumptions lead to the sampling distribution of $s_1^2/s_2^2$ being an $F$ distribution?

2. How many $F$ distributions are represented in Table A-7c?

3. What hypothesis did we test with an $F$ distribution?

4. If $X_1$ and $X_2$ are random observations from a population, is the standard error of $X_1 - X_2$ larger or smaller than the standard deviation of $X_1$?

5. If it is not necessary to pair observations and you decide to pair, is the level of significance different? Discuss the effect of pairing on your chance of making a correct inference both in the case where you should pair and in the case where it is not necessary.

6. Describe the three situations in which pairing may be useful and give an example of each from a field of application.

## CLASS EXERCISES

1. Collect a number of pairs of observations from adjacent columns of Table A-2 and obtain the difference of the observations in each pair. If you had taken all possible pairs, what would the mean difference and standard deviation of differences be? Compare these values with the mean and standard deviation of your observed differences.

2. Subdivide the class into two groups on some arbitrary basis, such as sex, age, or school major, and record the heights of the students in each group. Test the hypothesis $H: \sigma_1^2 = \sigma_2^2$ and test the hypothesis $H: \mu_1 = \mu_2$. Discuss the reasonableness of the assumption behind the tests for this situation and the conclusions reached.

3. For the six samples drawn for the Class Exercise, Chap. 3, compute $s_p^2$; also compute the variance of all 30 observations. Collect the results of all the students and verify the statement that $s_p^2$ is an unbiased estimate of $\sigma^2$, but that the variance of the combined groups is generally much larger than $\sigma^2$.

4. Draw three samples of size $N = 5$ from Table A-24. Take one of these samples and one of the samples you have previously drawn from Table A-2 and compute the value of $\dfrac{\bar{X}_1 - \bar{X}_2}{s_p \sqrt{(1/N_1) + (1/N_2)}}$. Repeat this for the other pairs of samples. For how many of these experiments is the hypothesis that the two populations have the same mean rejected at the 5 per cent level of significance?

Table A-24 records observations from a normal universe with mean 0 and variance 2. Here the two populations do have equal means, and this experiment will indicate the effect of unequal variances upon an experiment.

## PROBLEMS

*Section* 8-1

1. Read from Table A-7 the values of $F_{.99}(3,4)$, $F_{.99}(4,3)$, $F_{.50}(3,4)$, $F_{.50}(4,3)$, $F_{.50}(4,4)$, $F_{.05}(11,10)$, $F_{.999}(10,10)$.

2. Compare $F_{.95}(3, \infty)$ with the 95th percentile of the $\chi^2/df$ distribution with 3 degrees of freedom.

3. How do values of $F_\alpha(N_1,N_2)$ and $F_{1-\alpha}(N_2,N_1)$ compare?

*Section* 8-2

4. For each of the following cases test the hypothesis $H: \sigma_1 = \sigma_2$:
(*a*) $N_1 = N_2 = 16$, $s_1^2 = 50$, $s_2^2 = 16$, $\alpha = .05$
(*b*) $N_1 = 41$, $N_2 = 13$, $s_1^2 = 15.6$, $s_2^2 = 6.3$, $\alpha = .01$
(*c*) $N_1 = 60$, $N_2 = 120$, $s_1^2 = 8$, $s_2^2 = 17$, $\alpha = .02$

5. A tire manufacturer wished to test two types of tires. Fifty tires of type $A$ had a mean life of 24,000 miles, with $s^2 = 6{,}250{,}000$. Forty tires of type $B$ had a mean life of 26,000 miles with $s^2 = 9{,}000{,}000$. Is there a significant difference between the two sample variances?

6. In order to time events accurately it is important to have a timepiece which gains or loses a constant amount, so that appropriate allowance may be made. The following data, given in seconds, show the daily gain or loss for two timepieces. Is one significantly more variable than the other?

| $A$ | −46 | −37 | −51 | −50 | −38 | −46 | −42 | −43 | −40 | −39 |
|---|---|---|---|---|---|---|---|---|---|---|
| $B$ | 96 | 88 | 96 | 102 | 92 | 98 | 103 | 94 | 91 | 97 |

7. A mixture is considered thoroughly mixed if samples selected randomly from it are very much the same. Eight samples each from mixtures produced by methods $A$ and $B$ have the variances $s_A{}^2 = 24.2$ and $s_B{}^2 = 37.3$. Are methods $A$ and $B$ equally effecive?

8. Compute for Prob. 4 the value of $\beta$ if the true value of $\sigma_1{}^2/\sigma_2{}^2$ for the corresponding three cases is (*a*) 2, (*b*) .5, (*c*) .2, instead of the hypothetical value $\sigma_1{}^2/\sigma_2{}^2 = 1$.

9. Compute for two samples of size 10 each the chance that the hypothesis of equal variances will be rejected if one variance is actually twice the other. Use $\alpha = .05$.

*Section* 8–3

10. Suppose two samples from the same population with 12 and 35 observations, respectively, have $s_1{}^2 = 16.0$ and $s_2{}^2 = 18.0$. Pool these values and estimate $\sigma^2$ with a 95 per cent confidence interval.

11. Suppose 10 samples of 9 observations each have variances 23.5, 30.6, 29.3 27.5, 27.5, 26.3, 29.8, 30.7, 22.3, 26.5. Form a pooled estimate of $\sigma^2$ and use it to give 90 per cent confidence limits for $\sigma^2$.

12. Three samples of 20 observations each have means $\bar{X}_1 = 10$, $\bar{X}_2 = 20$, $\bar{X}_3 = 30$ and variances $s_1{}^2 = 4$, $s_2{}^2 = 6$, $s_3{}^2 = 7$. Compute $s_p{}^2$ and the variance of all 60 observations.

*Section* 8-4

13. Twenty plots of ground were planted with corn. On 10 plots a new type of phosphorus fertilizer was applied. The yields of the 20 plots are recorded below. Is there a significant difference between the yields at the 5 per cent level?

| *Control* | 6.1 | 5.8 | 7.0 | 6.1 | 5.8 | 6.4 | 6.1 | 6.0 | 5.9 | 5.8 |
|---|---|---|---|---|---|---|---|---|---|---|
| *Phosphorus* | 5.9 | 5.7 | 6.1 | 5.8 | 5.9 | 5.6 | 5.6 | 5.9 | 5.7 | 5.6 |

14. On an examination a class of 18 students had a mean of 70 with $s = 6$. Another class of 21 had a mean of 77 with $s = 8$ on the same examination. Is there reason to believe that one class is significantly better than the other? Consider the students as samples from one population. What might the population be?

15. The following data give readings in foot-pounds of the impact strength on two kinds of insulating material. Determine whether there is any difference in uniformity or in mean strength between the two kinds of material.

| $A$ | 1.25 | 1.16 | 1.33 | 1.15 | 1.23 | 1.20 | 1.32 | 1.28 | 1.21 |
|---|---|---|---|---|---|---|---|---|---|
| $B$ | 1.01 | .89 | .97 | .95 | .94 | 1.02 | .99 | 1.06 | .98 |

**16.** A group of 50 boys and a group of 50 girls were given a test in arranging different-shaped blocks. The times are recorded in the frequency table below. Analyze the data. If this was an industrial experiment, give a 90 per cent confidence interval for the mean saving in time if girls perform the task rather than boys.

| | TIME, SECONDS | | | | | | | | | | | |
|---|---|---|---|---|---|---|---|---|---|---|---|---|
| | 40 | 41 | 42 | 43 | 44 | 45 | 46 | 47 | 48 | 49 | 50 | 51 |
| *Girls* | 1 | 4 | 3 | 6 | 12 | 9 | 9 | 3 | 1 | 1 | 1 | |
| *Boys* | | | 3 | 2 | 6 | 10 | 12 | 7 | 3 | 4 | 2 | 1 |

**17.** A tire manufacturer wished to test two types of tires. Fifty tires of type $A$ had a mean life of 24,000 miles with $s^2 = 6{,}250{,}000$. Forty tires of type $B$ had a mean life of 26,000 miles with $s^2 = 9{,}000{,}000$. Is there a significant difference between the two sample means?

**18.** How large samples of size $N_1 = N_2 = N$ should be taken from each of two populations which have standard deviations $\sigma_1 = \sigma_2 = 3$ inches for the standard error of the difference in means to be as small as .5 inch?

*Section* 8-5

**19.** The following data give paired yields of two varieties of wheat. Each pair was planted in a different locality. Test the hypothesis that the mean yields are equal. Find a 90 per cent confidence interval for the difference in the mean yields.

| *Variety* I | 45 | 32 | 58 | 57 | 60 | 38 | 47 | 51 | 42 | 38 |
|---|---|---|---|---|---|---|---|---|---|---|
| *Variety* II | 47 | 34 | 60 | 59 | 63 | 44 | 49 | 53 | 46 | 41 |

Explain why pairing is necessary in this problem.

**20.** An experiment was performed with seven hop plants. One half of each plant was pollinated and the other half was not pollinated. The yield of the seed of each hop plant is tabulated as follows:

| *Pollinated* | .78 | .76 | .43 | .92 | .86 | .59 | .68 |
|---|---|---|---|---|---|---|---|
| *Unpollinated* | .21 | .12 | .32 | .29 | .30 | .20 | .14 |

(*a*) Determine at the 5 per cent level whether the pollinated half of the plant gives a higher yield in seed than the nonpollinated half. State the assumptions and hypotheses to be tested and carry through the computations to make a decision. (*b*) How can the experimenter make an $\alpha$ error? What are the consequences of his doing so? (*c*) How can the experimenter make a $\beta$ error? What are the consequences of his doing so? (*d*) Give 90 per cent confidence limits for the difference in mean yields.

CHAPTER NINE

# Efficiency and various statistics

In the earlier chapters most of the discussion was confined to the use of the arithmetic mean and the variance in procedures for estimation and tests of hypotheses. In this chapter we shall discuss more fully the reasons for that choice and several alternative statistics which are useful under certain conditions. The choice of procedure will sometimes depend on which estimate is the easiest or quickest to compute in a particular situation, and, of course, the substitution of one method for another will also depend on the relative sampling accuracy of the substitute method. The less efficient statistics may be quite adequate for the following purposes:

1. For *preliminary* or *pilot studies* which are based on very few observations from which rough estimates of population parameters are made for use in the principal study to follow;

2. For *rapid computation*, perhaps performed by slide rule or by mental calculation for use in surveying a large number of data in search of portions of the data which may require more precise or extensive analysis;

3. For *simple computation*, where there is no time to perform and check extensive computations (the simpler computation will generally introduce fewer errors arising from the computation itself).

In connection with the third point the authors have conducted classroom sampling experiments involving both $s$ and the range for small samples where the efficiency of the range is high. The computed values were collected into distributions for comparison of the variances, that is, to investigate the sampling accuracy of the range as an estimate of $\sigma$. When these values were collected into distributions before the students had checked their computations, more often than not the variance of the estimate obtained from the range was smaller than the variance of the distribution of $s$ values.

## 9-1 *COMPARISON OF MEAN AND MEDIAN

When reference is made to an "average," it seems to be generally assumed that the arithmetic mean is intended. However, other measures are sometimes used, and therefore any statement of average should indicate whether or not it is the arithmetic mean. The *median*, $P_{50}$, is often quoted in place of the mean as an average. The median is the middle value in size in a group of observations; for the distribution 10, 12, 14, 18, 19, the median is 14. If there is an even number of observations we define the median to be the mean of the two middle values. For the distribution 10, 12, 14, 18, 19, 21 the median is $(14 + 18)/2 = 16$. This score is central in the sense that half the observations have values which are larger and half which are smaller. In many cases there will be little difference between the mean and the median, but sometimes they may differ widely. Consider the situation in which a majority of salaries are at a comparatively low level and there are a few very large salaries; here the mean salary may be twice as large as the median salary. It is obvious in such a case that a report giving the average salary may be misleading unless we know whether the mean or median is being used, and many studies report both.

If a measure of central value is being chosen only to describe a set of numbers, a choice between the mean and the median is mainly determined by the interpretation put on the concept of "central." Thus, for example, in buying a large number of bags of grain the mean weight of 100 bags is as useful to the buyer as the weights of the individual bags. However, in a discussion of amounts of money spent by families for milk it may be more useful (from the over-all health viewpoint, if not from the dairy's) to know a median, which specifically pinpoints the dividing line between the upper and lower 50 per cent of the families, rather than the mean, which indicates the total amount spent.

There is also the question of the relative effort or difficulty in finding the mean value or the median value. If we have an unarranged collection of numbers it will probably be as easy to compute the mean (and the variance also) as to arrange the numbers in a frequency table. However, if the numbers are arranged in order of size or in a frequency table, the median can be found by inspection in several seconds, while computation of the mean would perhaps still require several minutes. If the median is used, measurement of every item in the sample may sometimes be avoided. In observing heights of small groups of people, the middle person can often be noted by visual inspection, and the single measurement of his height would determine the median of the group.

The value of an increase in the ease of computation of the median over the computation of the mean should be considered only in the light of the whole problem. The effort required to set up an experiment, to collect the observations, and to interpret the results is usually great and is often quite expensive. The actual computation (at least for techniques described in this book) seldom requires more than a few hours at the outside, and though this may sound tedious, it comprises a small part of the total time spent. Actually, most of the computations discussed here can be carried out in much less time.

It is also important in choosing the mean or median to consider the possibility that a statistic (measured on a sample) will be used to estimate a parameter (measured on a population). Of course, it seems reasonable to use a sample median to estimate a population median and a sample $\bar{X}$ to estimate a population $\mu$. However, if we suppose that the population mean and population median are approximately equal, we will then wish to decide whether to use the sample mean or the sample median as an estimate of $\mu$. We can base this choice on the standard errors of these two statistics, as discussed in the next section. Alternatively, the lengths of the confidence intervals for estimating the mean and the variance could be used to compare the various measures of mean. The statistic which yields the shortest confidence interval could be termed the "best," and a rating of any other statistic could be obtained by comparing the length of its confidence interval with that of the best estimate.

## *STATISTICAL EFFICIENCY 9-2

In random sampling from a normal population (as well as from many other populations) the sample mean $\bar{X}$ is an *efficient estimate* of $\mu$ in the sense that it has a standard error smaller than that of any other unbiased estimate of $\mu$. For example, the sample median has a standard error larger than $\sigma/\sqrt{N}$. The *relative efficiency* of another estimate is defined by dividing its variance into the variance of the efficient estimate. This quotient, represented here by the symbol $E$, is generally less than 1 but could be equal to 1 if an estimate equivalent to the efficient statistic is used.

Generally an increased sample size will decrease the standard error or variance of a statistic, and an inefficient statistic computed with a larger sample size may give as precise an estimate (as small a standard error) as the efficient statistic computed with a smaller sample size. For a number of examples the relative efficiency is the ratio of these sample sizes.

Note that if we have an estimate of the population mean which is 50 per cent efficient, then the ratio of the variances is .50. Therefore the variance of the sampling distribution of the other estimate is twice as large as the variance of the sampling distribution of $\bar{X}$, or equal to $2\sigma^2/N$.

**Mean and median.** We noted in Chap. 4 that the sampling distribution of $\bar{X}$ has variance $\sigma^2/N$. If the population is normal the variance of the sampling distribution of medians is given by $\sigma^2/(NE)$, where values of $E$, called the *relative efficiency* of the median compared with the mean, are given in Table A-8*b*(4). Thus for samples of size 4 the variance of the median is $\sigma^2/(4 \times .838) = \sigma^2/3.35$, and for samples of size 10 it is $\sigma^2/7.23$. For large sample sizes the value $E = .637$ should be used.

Although the variances for median and mean are different for the same sample size, they may be made equal if different sample sizes are used. If we equate the two variances, using $N_1$ as the size of the sample from which the median is computed and $N_2$ as the equivalent sample size using the mean, we have for large values of $N_1$ and $N_2$

$$\frac{\sigma^2}{.637N_1} = \frac{\sigma^2}{N_2}$$

Since the same $\sigma^2$ appears on both sides of the equation, we can find the sample size $N_1$ which corresponds to any value of $N_2$ by solving the equation

$$.637N_1 = N_2$$

In particular, if we take $N_1 = 1{,}000$ we obtain $N_2 = 1{,}000 \times .637 = 637$. Therefore, if we measure precision by the variance or standard error of the sampling distribution we obtain the same precision from a sample mean based on 637 observations as we do from a sample median based on 1,000 observations.

Note that for normal populations the $\bar{X}$ distribution has a smaller variance and thus a smaller spread than the median distribution; consequently (this is the important point), a single observed mean has a greater chance of being close to $\mu$ than does a single observed median. This is no guarantee that for *every* sample the mean will be closer than the median to the population mean.

Approximate determination or verification of the relative efficiency can be obtained from sampling experiments such as those described in Chaps. 4 and 5. To illustrate, we take 200 samples, each of size $N = 10$, from Table A-2 and record the median of each sample in Table 9-1, using the same class interval as in the experiment recorded in Table 4-3. The values of $\bar{X}$ are copied from Table 4-3, and the slightly greater spread of the median distribution can be noted. The observed variance of the 200 medians is .1398, which can be compared with the value from the formula $\sigma^2/(NE) = 1/(10 \times .723) = .138$. The ratio of the observed variance of means to the observed variance of medians is $.1035/.1398 = .740$, which agrees fairly well with the value .723 recorded in Table A-8*b*(4).

**Table 9-1** Observed values of $\bar{X}$ and medians for 200 samples of size 10, each from a normal population with $\mu = 0$ and $\sigma^2 = 1$

| *Mid-point* | *Frequency of $\bar{X}$'s* | *Frequency of medians* |
|---|---|---|
| .95 | 0 | 1 |
| .85 | 0 | 1 |
| .75 | 2 | 3 |
| .65 | 3 | 4 |
| .55 | 7 | 7 |
| .45 | 13 | 8 |
| .35 | 13 | 21 |
| .25 | 19 | 12 |
| .15 | 22 | 21 |
| .05 | 24 | 23 |
| −.05 | 26 | 23 |
| −.15 | 19 | 14 |
| −.25 | 17 | 17 |
| −.35 | 12 | 13 |
| −.45 | 11 | 9 |
| −.55 | 5 | 9 |
| −.65 | 4 | 6 |
| −.75 | 3 | 5 |
| −.85 | 0 | 2 |
| −.95 | 0 | 1 |

| | *Theoretical* | *Observed* |
|---|---|---|
| Mean $\bar{X}$ | $\mu = 0$ | .009 |
| Var $\bar{X}$ | $\sigma^2/N = .1$ | .1035 |
| Mean med. | $\mu = 0$ | −.013 |
| Var med. | $\sigma^2/(NE) = .1383$ | .1398 |
| Efficiency | $.1/.1383 = .723$ | .740 |

## *DATA-HANDLING EQUIPMENT 9-3

Statistical efficiency in the sense of minimum variance is only one consideration in a research analysis. In the discussion of the comparative merits of the mean and median reference was made to the ease of computing the median. If the sample is large it actually may not be simpler unless one has equipment to arrange the observations in order. The type of equipment available will play a part in the choice of the statistical estimator to be used.

The treatment of census data involving millions of observations was adapted to automatic equipment about 70 years ago. The procedure used then, and modified in various ways since, is that of registering data on cards by punching holes in the cards in various positions. By means of the positions of the holes in the cards the machines can automatically sort and count the cards. These machines are available in most universities. The basic machines to be described are the punch, verifier, sorter, tabulator, and reproducer.

The *punch* automatically punches a hole in successive columns of the card as the information to be recorded is "typed" on the keyboard. The card has 80 columns in which data can be recorded and 12 positions in each column. A single hole is punched in each column to indicate a number, and a combination of two holes is punched for letters of the alphabet. Several columns can be used to record numbers of several digits. The cards feed automatically into the machine and are advanced and stacked automatically, so that the operator records the data on the cards as fast as the keys are depressed. An efficient operator can punch several hundred cards per hour.

The punches are checked for accuracy in a *verifier*, where another operator "repunches" the cards. The machine indicates discrepancies automatically.

The *sorter* can be set to scan a single column on each of a large stack of cards and separate the original stack into 12 different stacks of cards according to the position punched in that column. The sorter scans a particular column of each card in turn and electrically controls gates to the 12 *pockets*. The cards pass into the pocket corresponding to the hole punched. The sorter is usually equipped with a set of counters which tally the number of cards having each position punched without the necessity of having the cards go into the separate pockets. The sorter leafs through the stack of cards at the rate of 450 to 1,200 cards per minute.

The *tabulator* performs the functions of printing and addition. Any information punched on the cards can be printed. The printing operation is performed by a set of 88 to 120 type bars, which print a whole line at a time and will print about 100 lines per minute. The tabulator can add several different sequences of numbers at the same time, up to 80 digits. These totals are printed automatically at intervals which can be determined by the cards themselves. The tabulator can perform a large number of specialized operations, many of which can be controlled by the cards or the numbers recorded on the cards as they pass automatically through the machine.

The *reproducer* reproduces punches from one set of cards into another set. Certain columns can be selected for reproduction in any of the columns of the new set. The reproducer can automatically verify the reproducing it is performing.

Various other machines perform specialized operations. The operations to be discussed in this chapter rely mainly on the use of a sorter for very large numbers of observations. Data may be placed on cards for analysis by the methods to be discussed here, but very frequently the observations have been made and entered on punched cards for other purposes. Routine recordings of engineering measurements on manufactured items are frequently maintained. Information on production and sales is regularly recorded in this fashion. Employees' time cards and information cards often contain a large amount of information. Police and hospital records are often available on punched cards. Punched cards are also used by labor unions, universities, and state bureaus. Most departments of the Federal government record data in this manner. These files of cards are sometimes available for special researches.

These comments on computing machinery are intended only to indicate some of the effects the machines may have on statistical procedures. Many of these functions are now carried out by electronic computing machinery.

## *OTHER ESTIMATES OF CENTRAL VALUE 9-4

**Percentile estimates of $\mu$: large samples.** In symmetrical populations the median of the random sample is an unbiased estimate of $\mu$, the population mean, and we have noted that for a normal population the large sample efficiency of the median is .637. Thus by use of a single percentile, $P_{50}$, we can obtain the same precision from 200 items as from an $\bar{X}$ based on $200 \times .637 = 128$ items.

We can obtain higher efficiencies from the mean of several percentile values. If we wish to pick the two percentile values whose mean would give the highest efficiency possible, we could construct the sampling distributions of each suggested combination and pick that pair whose sampling distribution has the smallest dispersion. This investigation can be made mathematically, and the result is $(P_{28.6} + P_{71.4})/2$. There is only a slight change in efficiency from $(P_{25} + P_{75})/2$; its efficiency is .808. The latter is easier to remember and will be suggested for use here. The efficiency can, of course, be increased by obtaining more percentile values. The efficiencies in Table A-8*a*(1) are for the percentile estimates obtained from the mean of the indicated percentiles. For example, we read from this table that 93 per cent of the available information for estimating the mean is obtained by observing five percentile values.

**Other estimates of $\mu$: small samples.** Three alternative estimates of central value will be considered for small samples. Since the median of an odd number of observations is the observation with an equal number of observations on either side and the median of an even number of observations is the mean of the two middle values, the efficiency of the median is higher for even sample sizes than for odd sample sizes. For larger sample sizes the efficiency levels off and approaches .637. The variance and the efficiency of the median are listed in Table A-8*b*(4).

Several presentations in this and later sections assume the observations to be arranged in order of size and labeled $X_1, X_2, \ldots, X_N$, with $X_1$ the smallest of the $N$ observations, $X_2$ the second smallest, and so on. Observations arranged in this way are called *order* statistics.

The *midrange* $M_r = (X_1 + X_N)/2$ is the mean of the largest and smallest observations and estimates the mean of the population. The efficiencies relative to the sample mean are shown in Table A-8*b*(4). It is probably not advisable to use the midrange for samples of more than five observations, since for larger sample sizes its efficiency drops below that of the median.

It was noted above that in large samples the most efficient estimate to

**Table 9-2** Comparison of estimates with $N = 7$, $\mu = 2$, $\sigma = 1$.
Observations: .676, 1.113, 1.707, 1.847, 2.167, 2.713, 3.445.

| | *Formula* | *Computation* | *Result* | *Eff.* |
|---|---|---|---|---|
| Mean | $\frac{\Sigma X_i}{N}$ | 13.668/7 | 1.953 | 1. |
| Midrange | $\frac{1}{2}(X_1 + X_7)$ | $\frac{0.676 + 3.445}{2}$ | 2.060 | .654 |
| Median | $X_4$ | 1.847 | 1.847 | .679 |
| Mean best two | $\frac{1}{2}(X_2 + X_6)$ | $\frac{1.113 + 2.713}{2}$ | 1.913 | .849 |

be obtained from two percentiles is the mean of $P_{28.6}$ and $P_{71.4}$. Except for very small samples, the median and midrange are computed from observations which are not near these percentiles, so we may expect that two observations may be selected whose mean is more efficient. Table A-8*b*(4) indicates for each sample size the best choice of two observations for estimating the mean of a normal population. For example, in a sample of size $N = 10$, the statistic $(X_3 + X_8)/2$ is more efficient than $(X_1 + X_{10})/2$, or $(X_4 + X_7)/2$, or any other such estimate based on two observations.

Table 9-2 illustrates the computation of these statistics on seven observations ($N = 7$) from a normal population arranged in order of size, $X_1 = 0.676$, $X_2 = 1.113$, and so on. The data were taken from Table A-23, which are observations from a normal population with $\mu = 2$ and $\sigma = 1$.

## 9-5 *ESTIMATES OF DISPERSION

**Percentile estimates of dispersion: large samples.** For a large number of observations the percentile estimate $P_{93} - P_{07}$ is sometimes used as an estimate of dispersion. This estimate is the most efficient if only two values are used; the efficiency is .65. A third percentile is of little additional value in estimating dispersion. The four percentiles which are the most efficient are $P_{03}$, $P_{15}$, $P_{85}$, $P_{97}$. Here we compute $P_{97} + P_{85} - P_{15} - P_{03}$. With these four values the efficiency is increased to .80. Other results in addition to these are contained in Table A-8*a*(2).

The values for the coefficients in the percentile estimates for the standard deviation [Table A-8*a*(2)] are obtained from the normal distribution. For example, the difference between $P_{07}$ and $P_{93}$ for a normal distribution is from $\mu - 1.476\sigma$ to $\mu + 1.476\sigma$ as read from Table A-4; this difference is $2.952\sigma$. The coefficient $.3388 = 1/2.952$. We divide our estimate of $2.952\sigma$ by 2.952 (or multiply by .3388) to obtain an estimate of $\sigma$. The coefficient for the second estimate is $.1714 = 1/5.834$, where 5.834 is obtained as for the first estimate, $(2 \times 1.881) + (2 \times 1.036) = 5.834$.

The efficiencies of the estimates of the standard deviation are obtained by comparing these estimates with an estimate obtained from the sample variance. For the large sample sizes considered here we can state, as we did for estimates of the mean, that equal precision is available from an estimate which is 50 per cent efficient by examining a sample twice as large. For example, if we consider a sample of 300 items and compute the percentile estimate of $\sigma$ based on four percentiles, we may note that equivalent precision could be obtained from the computation of $s$ for a sample of $300 \times .80 = 240$ observations.

An example of the use of this procedure is as follows. Suppose the percentiles of a distribution of values are

| *Per cent* $\alpha$ | $P_\alpha$ | |
|---|---|---|
| 97 | 45.2 | |
| 85 | 39.2 | |
| | | 84.4 |
| 15 | 25 | |
| 3 | 18.6 | |
| | | 43.6 |
| | | 40.8 |

We add $P_{97}$ and $P_{85}$ and subtract the sum of $P_{15}$ and $P_{03}$. The estimate of $\sigma$ is $40.8 \times .1714 = 6.99$.

For estimation of the mean we obtain high efficiencies by finding percentiles which are near the center. For estimating the standard deviation the best percentiles are closer to the end of the distribution. However, we are often interested in estimating both the mean and standard deviation; sample percentile values can be used for estimating both and are given in Table A-8$a$(3). These values are not the best for each but are a compromise, giving fairly high efficiencies for both the mean and standard deviation with percentiles that are easy to remember. The values of $K$ are the multipliers for the estimate of $\sigma$.

As an example of the use of these values, suppose that 572 observations are arranged in order. If we wish to use six percentiles to estimate the mean and standard deviation, we note from Table A-8$a$(3) the percentages 05, 15, 40, 60, 85, and 95. We multiply each by $\frac{572}{100}$ and obtain 29, 86, 229, 343, 486, 543. We find the value of observation number 29, that is, $P_{05}$, observation number 86, and so on. We add these and divide by 6 to obtain an estimate of the mean $\mu$. We add the last three and subtract the first three and multiply by $K$ (in this case .1704) to obtain our estimate of $\sigma$. These estimates are of the same precision in sampling from a normal population as an $\bar{X}$ obtained from a sample of $572 \times .89 = 509$ items and an $s$ obtained from a sample of $572 \times .80 = 458$ items.

**Small sample estimates of dispersion.** The mean of the sampling distribution of $s^2 = \dfrac{\Sigma(X_i - \bar{X})^2}{N - 1}$ is equal to the population variance $\sigma^2$ for random samples from any population. Usually we shall use $s$ to estimate $\sigma$, and it should be noted that for normal populations a correction factor can be used to make $s$ an unbiased estimate of $\sigma$. For $N = 2, 3, 4$ we use $1.253s$, $1.128s$, and $1.085s$, respectively. For larger values of $N$ we can use the estimate $\left[1 + \dfrac{1}{4(N-1)}\right] s$. These correction factors are listed in Table A-8$b$(1). They are to be used only for samples from approximately normal populations; other factors apply for other population distributions.

**The range.** The range $w$ is the distance between the largest and smallest observations. In order-statistics notation, $w = X_N - X_1$; for example, the range of the observations 10, 12, 14, 18, 19, 21 is $w = 21 - 10 = 11$. Usually the range is used as a measure of dispersion only for a small number of observations. There is some objection to the use of the range, since it can be greatly affected by a single value which may be removed a considerable distance from all other values observed.

The range must be adjusted to be an unbiased estimate of $\sigma$, since the mean of its sampling distribution is not equal to $\sigma$. The correction factors $K_1$ for the range vary with $N$ and are given, for normal populations, in Table A-8$b$(1). For example, we can see from the table that for samples of size 10 the range must be multiplied by .325 to give an unbiased estimate of $\sigma$. In other words, the mean range is $1/.325 = 3.08$ standard deviations.

We have recorded in Table 9-3 the values of the range $w$ and the values of $s$ observed for a number of samples from a normal population having $\sigma^2 = 1$. The mean of the observed values of $w$ is 3.0425 standard deviations, which agrees fairly well with the correct value 3.08 noted above. When the values of $w$ are multiplied by .325, the observed mean is .984, which is close to the theoretical $\sigma = 1$. The values of $s$ after being multiplied by $1 + 1/(4 \times 9) = 1.028$ to remove the bias have a mean of 1.015, which is also close to the theoretical value.

Even in very large samples the range is often used as a very rough estimate of the standard deviation. For normal populations the entire sample for a wide range of sample sizes will be contained in an interval of about four to six standard deviations. For example, in the distribution given in the first column of Table 4-3 the entire distribution is contained in an interval of length 1.6 units. We may state a rough estimate of the standard deviation as about $1.6/4 = .4$ or $1.6/6 = .27$. In this case the standard deviation is actually .322.

After adjusting the estimates to remove bias we can compare the relative efficiencies of the various estimates. Strictly speaking, the most efficient estimate of $\sigma^2$ is $\Sigma(X - \mu)^2/N$. However, $\mu$ is generally unknown, so no

**Table 9-3** Observed values of $w$, $kw$, and $cs$ for 200 samples of size 10, each from a normal population with $\mu = 0$ and $\sigma = 1$.
($k$ is chosen so that the mean of the sampling distribution of $kw$ is $\sigma$ and for samples of size 10 is $k = .325$. $c$ is chosen so that the mean of the sampling distribution of $cs$ is $\sigma$, and for samples of size 10 is $c = 1.028$.)

| DISTRIBUTION OF $w$ | | DISTRIBUTION OF $kw$ AND $cs$ | | |
|---|---|---|---|---|
| *Mid-point* | *Freq.* | *Mid-point* | *Freq. of* $.325w$ | *Freq. of* $1.028s$ |
| 6.0 | 1 | 1.9 | 2 | 1 |
| 5.5 | 1 | 1.7 | 0 | 2 |
| 5.0 | 1 | 1.5 | 12 | 9 |
| 4.5 | 11 | 1.3 | 25 | 29 |
| 4.0 | 24 | 1.1 | 46 | 65 |
| 3.5 | 36 | .9 | 64 | 55 |
| 3.0 | 50 | .7 | 44 | 31 |
| 2.5 | 42 | .5 | 7 | 7 |
| 2.0 | 29 | .3 | 0 | 1 |
| 1.5 | 5 | .1 | 0 | 0 |
| *Total* | 200 | | 200 | 200 |

| | *Theoretical* | *Observed* |
|---|---|---|
| Mean $w$ | 3.08 | 3.0425 |
| Mean $.325w$ | 1.000 | .984 |
| Var $.325w$ | .0671 | .0685 |
| Mean $1.028s$ | 1.000 | 1.015 |
| Var $1.028s$ | .0570 | .0645 |

further consideration will be given to this estimate other than to state that the relative efficiency of $s^2$ with respect to this statistic is $1 - 1/N$. The other estimates of dispersion are compared for relative efficiency with the unbiased estimate based on $s$ as mentioned above. These efficiencies are given in Table A-8$b$(1). We can see that the efficiency of the mean deviation for samples of size 10 or less is .89 or greater. The efficiency of the range estimate is high for very small sample sizes, .955 for $N = 5$, but decreases to .850 for $N = 10$ and to .700 for $N = 20$ for a sampling from normal populations. The efficiency of the range estimate approaches zero as the sample size increases indefinitely.

Referring again to the sampling distribution reported in Table 9-3, we see that the ratio of the variances of the distribution of $1.028s$ and $.325w$ is $.0645/.0685 = .942$, which, although larger than the efficiency of the range

$E = .850$ noted above, still indicates that the estimate based on $s$ has a smaller variance than the estimate based on $w$.

**Mean deviation.** The *mean deviation* is computed by finding the mean of the absolute deviations from the mean or median. To estimate $\sigma$ we must multiply the mean deviation for large samples by $\sqrt{\pi/2} = 1.253$. The efficiency of the mean deviation for large samples is .88.

The mean deviation is easy to compute for small samples if the deviations are taken from the median. Table A-8*b*(2) indicates the computation of the sum of these deviations. The multiplier to convert this sum to an unbiased estimate of $\sigma$ and the efficiency of this estimate are also included. Instead of computing the mean of the differences indicated in the column headed "Computation," we have included that factor in the multiplier to simplify computation.

The efficiencies of the range and mean deviation are less than the efficiencies for estimates simpler to compute than the mean deviation. For sample sizes of 5 to 10 these estimates employ the difference of the sum of the two largest values and the sum of the two smallest values. For larger sample sizes three or four of the largest and smallest values are used. A *modified linear estimate*, given in Table A-8*b*(3), indicates the values to use in computing an estimate of $\sigma$ which will give the highest efficiency for an estimate of this type. The coefficient is such that this statistic will give an unbiased estimate of $\sigma$.

**Best linear estimate of $\sigma$.** The estimation of $\sigma$ by a linear function of the observations requires the use of different coefficients for the different observations if it is to be the best estimate in the sense of being the most efficient unbiased linear estimate. For samples of size 2 or 3 this estimate is the same as above. However, beginning with samples of size 4, a greater efficiency is obtained by computing the estimates as given in Table A-8*b*(6). The efficiency of each of these estimates is also indicated.

The computation of the estimates of dispersion given in this section is illustrated in Table 9-4 for the same data used in Table 9-2.

**Table 9-4** Further comparisons for $N = 7$ and $\sigma = 1$; observations: .676, 1.113, 1.707, 1.847, 2.167, 2.713, 3.445

| | *Computation* | *Est. of* $\sigma$ | *Eff.* |
|---|---|---|---|
| Range | $.370(3.445 - .676)$ | 1.025 | .911 |
| Mean dev. | $.2031(3.445 + 2.713 + 2.167 - 1.707 - 1.113 - .676)$ | .981 | .920 |
| Modified linear est. | $.2370(3.445 + 2.713 - 1.113 - .676)$ | 1.035 | .967 |
| Best linear est. | $.2778(3.445 - 0.676) + .1351(2.713 - 1.113) + .0625(2.167 - 1.707)$ | 1.014 | .989 |

For this particular sample the best linear estimate actually was the closest to $\sigma$ of the four statistics, but for other samples it may not be the closest. The higher efficiency for the best linear estimate indicates that with all possible samples the values of this statistic will vary less (have a smaller standard deviation) than the values of the other statistics.

## *SUBSTITUTE $t$ AND $F$ RATIOS 9-6

The inefficient estimates given in the previous sections can be combined to form ratios similar to the $t$ ratio $(\bar{X} - \mu)/(s/\sqrt{N})$.

**Mean over range.** If the arithmetic mean is retained in the numerator and $s$ in the denominator is replaced by the estimate $K_1 w$ obtained from the range $w$, we could form the ratio $(\bar{X} - \mu)/(K_1 w/\sqrt{N})$ and use this statistic to test hypotheses about $\mu$. It is simpler to compute $\tau_1 = (\bar{X} - \mu)/w$. Percentiles of the sampling distribution of $\tau_1$ are given in Table A-8*c*(1) for sample sizes up to 20. Since the range has high efficiency for small sample sizes, it can be expected that $\tau_1$ will have high efficiency as a substitute for $t$.

The range may also be used as a substitute for the $t$ ratio for differences between means. Percentiles for the statistic $\tau_d = \dfrac{\bar{X}_1 - \bar{X}_2}{\frac{1}{2}(w_1 + w_2)}$ are given in Table A-8*c*(2) for samples of equal size up to 20.

**Midrange over range.** The use of the midrange as an estimate of $\mu$ and the range as an estimate of $\sigma$ has the advantage that both are computed from the same two observations. A substitute $t$ ratio is $\tau_2 = (M_r - \mu)/w$. It should be noted that the efficiency of the midrange decreases rapidly for sample sizes greater than 5. For small sample sizes the efficiency of the midrange is larger than that of the median. A comparison of the power function for this test with the power function for the $t$ test for smaller sample sizes shows that an additional observation will compensate for the loss in performance of this statistic in samples up to size 8 at the 1 per cent level of significance and in samples up to size 6 at the 5 per cent level of significance. For larger sample sizes the loss in performance is greater. Percentiles of the distribution of $\tau_2$ are given in Table A-8*c*(3) for samples of size 10 or less.

**Substitute $F$ Ratio.** The ratio of two ranges can be used as substitute for the ratio of two variances. Although it is somewhat less efficient for normal populations, it may be desirable to use such a substitute if the available time is seriously limited. Percentiles for the distribution of the ratio of two ranges are given in Table A-8*d* for the respective sample sizes $N_1$ and $N_2$ less than or equal to 15. The hypothesis $\sigma_1 = \sigma_2$ is rejected if $w_1/w_2$ is significantly large or small. For example, if two samples of sizes $N_1 = N_2 = 5$ are used to test the hypothesis $\sigma_1 = \sigma_2$ at the 5 per cent level of significance, we may use the critical region $w_1/w_2 < .317$ and $w_1/w_2 > 3.16$. For the one-sided test $\sigma_1 \leq \sigma_2$ the 5 per cent critical region is $w_1/w_2 > 2.57$.

## 9-7 *ESTIMATES WITH INEFFICIENT STATISTICS

Statistics such as the median and range, although somewhat inefficient as estimators for samples from normal populations, often have advantages in time or effort and may be used to form confidence-interval estimates in a similar manner to those based on $\bar{X}$ and $s^2$. This section discusses four examples of inefficient statistics in confidence-interval estimates of parameters. For normal populations these methods will usually give longer confidence intervals than those based on $\bar{X}$ and $s^2$. In particular cases, however, the speed of computation or the use of simpler sampling designs may make them advantageous.

EXAMPLE 9-1. *Range as an estimate of* $\sigma$. For small samples from a normal population percentiles of the distribution of the range divided by the population standard deviation, $w/\sigma$, are recorded in Table A-8*b*(1). We can form a confidence-interval estimate of $\sigma$ by stating that the observed range divided by $\sigma$ is between certain percentile limits. For example, suppose that in a sample of five observations we find $w = 15.3$. The 2.5 and 97.5 per cent limits of $w/\sigma$ for $N = 5$ are .85 and 4.20. Thus we estimate $\sigma$ by the interval $.85 < 15.3/\sigma < 4.20$, which gives $3.64 < \sigma < 18$.

EXAMPLE 9-2. *Confidence interval for* $\mu$ *based on medians.* It is sometimes convenient to take a number of small samples, compute an inefficient statistic for each, and average the values for all the samples. Suppose we measure the medians of 200 samples of five observations each. If our original population is normal the variance of the population of medians for samples of size 5 is $\sigma^2/(EN) = \sigma^2/(.697 \times 5) = \sigma^2/3.48$, so we can estimate the unknown $\sigma^2$ by multiplying the observed variance of the medians by 3.48. Suppose, however, that the original population variance is known to be 100; we can then use the 200 medians to estimate the population mean $\mu$. The mean of the 200 medians can be assumed to have an approximately normal sampling distribution with mean equal to the population mean $\mu$ and a variance which is $\sigma^2/3.48$ divided by 200. The confidence limits for $\mu$ are

$$\bar{X} + z_{\frac{1}{2}\alpha}\sqrt{\frac{28.7}{200}} < \mu < \bar{X} + z_{1-\frac{1}{2}\alpha}\sqrt{\frac{28.7}{200}}$$

EXAMPLE 9-3. *Estimates of* $\mu$ *and* $\sigma^2$ *based on means.* If a collection of $\bar{X}$'s of samples of size $N$ are available, they may be used to estimate the mean and variance of the original population (in this example only the estimation of the variance is inefficient). Suppose we have 500 samples of four observations each from a population and the means $\bar{X}_1, \bar{X}_2, \ldots, \bar{X}_{500}$

have been recorded. The mean $\bar{X} = \sum_{1}^{500} \bar{X}_i/500$ is the mean of a sample of $500 \times 4 = 2{,}000$ observations, and we can estimate $\mu$ by the interval

$$\bar{X} + z_{\frac{1}{2}\alpha} \frac{\sigma}{\sqrt{2{,}000}} < \mu < \bar{X} + z_{1-\frac{1}{2}\alpha} \frac{\sigma}{\sqrt{2{,}000}}$$

If, for example, $\bar{X} = 70$ and $\sigma = 10$, then 95 per cent confidence limits are

$$70 - 1.96 \frac{10}{\sqrt{2{,}000}} < \mu < 70 + 1.96 \frac{10}{\sqrt{2{,}000}}$$

$$69.56 < \mu < 70.44$$

If $\sigma^2$ is not known, we compute the variance $s_{\bar{X}}{}^2$ of the 500 values $\bar{X}_i$, which is an estimate of $\sigma^2/N = \sigma^2/4$, the variance of $\bar{X}$'s for samples of size 4. We refer to Table A-6*b* to form a confidence-interval estimate for $\sigma^2$. Reference is made to the table for 499 degrees of freedom. Suppose we observe $s_{\bar{X}}{}^2 = 2.30$ and thus estimate $\sigma^2 = 4 \times 2.30 = 9.20$. A 90 per cent confidence-interval estimate of $\sigma^2$ is

$$.898 < \frac{s_{\bar{X}}{}^2}{\sigma^2/4} < 1.11$$

or

$$\frac{4 \times 2.30}{1.11} < \sigma^2 < \frac{4 \times 2.30}{.898}$$

EXAMPLE 9-4. *Confidence interval for $\sigma$ based on mean range.* The mean of the ranges of a number of small samples can be used to estimate $\sigma$ as follows. Suppose that we have 200 samples of size $N = 4$ from a normal population and have recorded the range $w_i$ of each. An unbiased estimate of $\sigma$ is $.486\bar{w} = .486 \sum_{1}^{200} w_i/200$. The factor .486 is obtained from Table A-8*b*(1) for the range of four observations. Note also from Table A-8*b*(1) that the variance of the range estimate $.486w$ is $.183\sigma^2$. Now, $.486\bar{w}$ is the mean of 200 observations from the population of range estimates and thus has a variance $.183\sigma^2/200$. Since $\bar{w}$ is the mean of a fairly large sample of ranges, we can assume that it is normally distributed; hence we can compute confidence limits for $\sigma$ from

$$.486\bar{w} + z_{\frac{1}{2}\alpha} \sqrt{\frac{.183\sigma^2}{200}} < \sigma < .486\bar{w} + z_{1-\frac{1}{2}\alpha} \sqrt{\frac{.183\sigma^2}{200}}$$

which yields

$$\frac{.486\bar{w}}{1 + z_{1-\frac{1}{2}\alpha} \sqrt{\frac{.183}{200}}} < \sigma < \frac{.486\bar{w}}{1 + z_{\frac{1}{2}\alpha} \sqrt{\frac{.183}{200}}}$$

If the observed $.486\bar{w} = 21.3$, the observed 95 per cent confidence limits for $\sigma$ are

$$\frac{21.3}{1 + 1.96\sqrt{\frac{.183}{200}}} < \sigma < \frac{21.3}{1 - 1.96\sqrt{\frac{.183}{200}}}$$

or

$$20.1 < \sigma < 22.6$$

## 9-8 *TOLERANCE LIMITS

We have found that confidence limits may be determined so that the interval between these limits will cover a population parameter with a certain confidence, that is, a certain proportion of the time. Sometimes it is desirable to obtain an interval which will cover a fixed *portion of the population distribution* with a specified confidence. These intervals are called *tolerance intervals,* and the end points of such intervals are called *tolerance limits.* For example, it is of considerable interest to a manufacturer to estimate what proportion of manufactured articles will have dimensions in a given range. Tolerance intervals may be of the form $\bar{X} \pm Ks$, where $K$ is determined so that the interval will cover a proportion $P$ of the population with confidence $\gamma$. Confidence limits for $\mu$ are also in the form $\bar{X} \pm ks$. However, we determined $k$ so that the confidence interval would cover the population mean $\mu$ a certain proportion of the time. It is obvious that the interval must be longer to cover a large portion of the distribution than to cover just the single value $\mu$. Table A-16 gives values of $K$ for $P = .75, .90, .95, .99, .999$ and $\gamma = .75, .90, .95, .99$ and for many different sample sizes $N$. For example, if $\bar{X} = 14$ and $s = 1.5$, $N = 18$, we can estimate that the interval $\bar{X} \pm Ks = 14 \pm (3.702 \times 1.5) = 14.0 \pm 5.553$, or the interval 8.447 to 19.553 will contain 99 per cent of the population with confidence 95 per cent. The values for $K$ in Table A-16 are computed with the assumption that observations are from normal populations.

## 9-9 *CONTROL CHARTS

In this section we shall describe a simple graphical device which is useful in keeping track of production quality, the *quality-control chart.* "Control" here implies that some static condition exists, not that we are actually improving or changing the quality. If the quality does improve or change significantly we say that we no longer have control, and we depend upon our device to detect the change. In the past most of the applications of this method have been in engineering; however, the techniques are being adopted in many other fields. We shall discuss examples of the use of several quality-control charts.

**Control chart for means and ranges.** Suppose that an article which should have size $\mu$ is being produced. We shall take a sample of $N$ of these articles every day and compute the mean of the observations, which we shall record on a graph as in Fig. 9-1. On the vertical axis we have a scale for observed values of $\bar{X}$. On the horizontal axis we have a scale for days, or the time at which the sample was taken. A solid horizontal line is drawn through the assumed mean, and two parallel dotted lines are drawn, one above and one below the solid line. We shall observe the mean of a sample every day and plot it on the graph. If the point is between the dotted lines we say that the manufacturing process is in control and take no further action. If the point is above the upper or below the lower dotted line we say that the procedure is out of control, and we attempt to find the factor which is causing the extreme observations. In the example in the figure the process was out of control on the fifth day.

We wish the chance of asserting lack of control (a change in mean) to be very small when the static condition still exists. We control this chance by the distance between the dotted lines. If we assumed a normal distribution of means, we could draw the lines at $\mu + 1.96\sigma/\sqrt{N}$ and $\mu - 1.96\sigma/\sqrt{N}$ for a 5 per cent chance of making such a mistake on any one trial. In practice it is customary to use $3\sigma/\sqrt{N}$ as control lines. For a normal distribution these control lines give a .27 per cent chance of deciding erroneously that there is a lack of control. The chance of this kind of error corresponds to the level of significance in testing hypotheses.

Usually another control chart to record dispersion is kept simultaneously with the control chart for means. The dispersion is usually measured by the sample standard deviation $s$ or by the sample range $R$. Because of the speed and ease with which it can be computed, $R$ is more widely used. Actually, for small samples $R$ is fairly efficient. A control chart for $R$ has the same general appearance as one for $\bar{X}$. The range is scaled on the vertical axis and days or time of observation are scaled on the horizontal. There are upper and lower control lines, as before. Each sample range is plotted as it is observed, and the process is inspected if the sample range is outside the control lines.

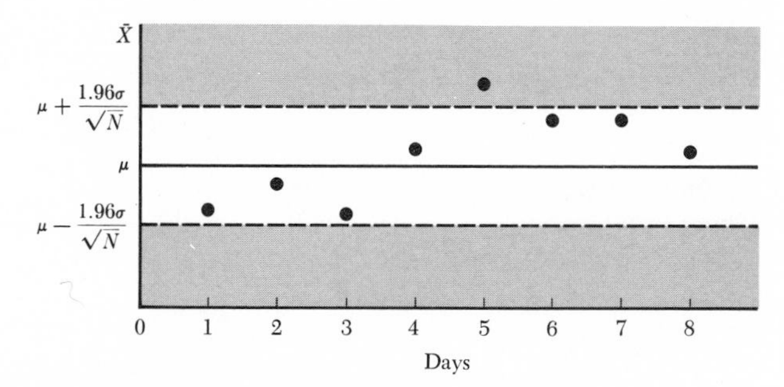


**Figure 9-1** 95% control chart for means.

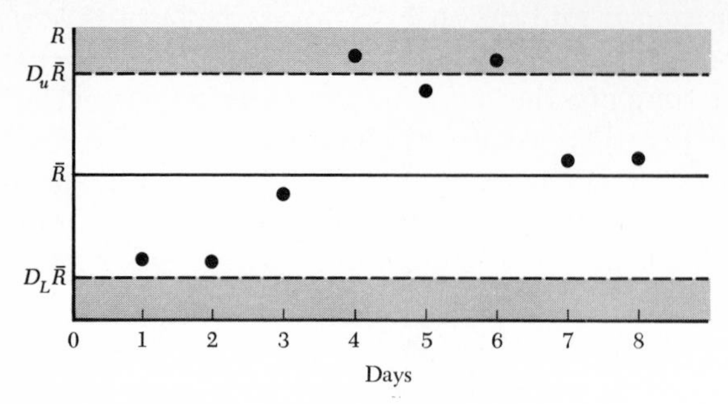


**Figure 9-2** 99% control chart for ranges.

The control lines for the range are found from Table 9-5. If $\bar{R}$ denotes the standard or hypothetical range,* then the control limits are $D_L\bar{R}$ and $D_U\bar{R}$, where $D_L$ and $D_U$ are read across from the sample size $N$. These multipliers are computed to give a 1 per cent chance that the sample range will exceed $D_U$ or be less than $D_L$. They are the .5 and the 99.5 percentiles of the sampling distribution of the statistic $R$, and $\bar{R}$ is the mean of this distribution. These values are based on the assumption that the measurements have a normal distribution.

Frequently $\sigma$ is not known, and the mean range $\bar{R}$ is used to measure $3\sigma/\sqrt{N}$ limits for the control chart for means. Control limits $\mu \pm A_2\bar{R}$ can be used instead of $\mu \pm 3\sigma/\sqrt{N}$, with values of $A_2$ as given in Table 9-5.

* We have used the bar above a symbol to represent the mean value as obtained from a sample. This use of $\bar{R}$ here is inconsistent but is standard notation in quality-control work. The sample range is denoted by $w$ elsewhere in this book, since $w$ is the usual notation in other than quality-control applications.

**Table 9-5** Multipliers for control charts

| $N$ | $D_L$ | $D_U$ | $A_2$ |
|---|---|---|---|
| 2 | .009 | 3.52 | 1.88 |
| 3 | .100 | 2.58 | 1.02 |
| 4 | .185 | 2.26 | .73 |
| 5 | .254 | 2.09 | .58 |
| 6 | .308 | 1.97 | .48 |
| 7 | .351 | 1.90 | .42 |
| 8 | .386 | 1.84 | .37 |
| 9 | .415 | 1.79 | .34 |
| 10 | .441 | 1.76 | .31 |
| 11 | .463 | 1.72 | .29 |
| 12 | .482 | 1.70 | .27 |
| 13 | .498 | 1.68 | .25 |
| 14 | .511 | 1.66 | .24 |
| 15 | .524 | 1.64 | .22 |

The factor $A_2$ includes the $3/\sqrt{N}$, and also a factor which, in a normal population, is such that $A_2\bar{R} = 3\sigma/\sqrt{N}$.

Frequently the mean of a large number of samples is treated as $\mu$, and the mean range of these samples is treated as $\bar{R}$. Then the control lines are drawn and the charts are used to indicate whether the production has been in control.

EXAMPLE 9-5. In 1965 a department in a store made weekly reports of its overtime pay. For the entire year the mean monthly overtime pay was 21.3 per cent. Every month the range of the four weekly reports was computed. The mean of these 12 ranges was 12.1 per cent.

Control limits on monthly ($N = 4$) mean overtime and range are

$\bar{X}$: $21.3 \pm (.73 \times 12.1)$ — 12.5 to 30.1

$R$: $.185 \times 12.1$ to $2.26 \times 12.1$ — 2.20 to 27.3

During 1966 the $\bar{X}$ and $R$ values in the table were recorded. $\bar{X}$ is out of control for January, March, and September. $R$ is out of control for May.

| | *Jan.* | *Feb.* | *Mar.* | *Apr.* | *May* | *June* | *July* | *Aug.* | *Sept.* | *Oct.* | *Nov.* | *Dec.* |
|---|---|---|---|---|---|---|---|---|---|---|---|---|
| $\bar{X}$ | 12.1 | 15.8 | 10.2 | 18.6 | 19.2 | 12.6 | 13.8 | 12.6 | 31.0 | 18.6 | 20.3 | 28.6 |
| $R$ | 13.6 | 12.3 | 6.5 | 26.0 | 29.0 | 17.1 | 12.3 | 10.3 | 13.6 | 12.3 | 23.0 | 19.1 |

Other uses of control charts in management are control of output and quality of clerical help, sales performance in various fields, production by skilled labor, and so on.

## GLOSSARY

best linear estimate of $\sigma$
control chart
efficiency
efficient estimate
mean deviation
mean over range
median
midrange
midrange over range
order statistics
percentile estimate
range
statistical efficiency
substitute $F$ ratio
substitute $t$ ratio
$\tau_1$
$\tau_2$
$\tau_d$
tolerance interval
tolerance limits

## DISCUSSION QUESTIONS

**1.** Define each of the terms in the Glossary.

**2.** Give examples of situations in which it might be practical to take a larger number of observations and use an inefficient estimating statistic. Situations of this type might occur if it is easy and inexpensive to obtain observations and results are needed in a short period of time, or if it is easy to order the individuals and difficult to measure any particular one.

**3.** Is it easier to find the median or the mean?

4. State why the critical values for $N = 2$ for $\tau_1$ are equal to the critical values for $N = 2$ for $\tau_2$.

5. Discuss the comparative frequencies with which the confidence intervals using $t$ and those using $\tau_1$ cover the population mean.

6. Give an example of a situation in which the observations may easily be ordered before the actual measurements are made.

7. Describe a procedure for selecting random samples from a population recorded on cards by using random numbers which are punched on cards.

8. Explain the difference between tolerance intervals and confidence intervals.

9. Why are the control limits for $R$ given in this chapter better than $\bar{R} \pm 3\sigma_{\bar{R}}$?

## CLASS EXERCISES

1. From the observed distribution of the median as recorded in Class Exercise 4, Chap. 4, estimate the efficiency of the median compared with the mean for samples of size 5 from a normal population. Compare this estimated efficiency with the theoretical efficiency .697.

2. Compute the midrange for the samples of size 5 drawn from Table A-2 (Class Exercise, Chap. 3). Collect the computed values into a distribution and compute the mean and variance. Compare the results with the theoretical values given in Table A-8*b*(4). Estimate the efficiency of the midrange for samples of size 5 from a normal population.

3. Compute the following statistics (if they have not already been computed) for the six samples of five observations each used in previous class exercises: the median, the midrange, the range, the mean deviation, $\tau_1$, $\tau_2$, and $\tau_d$ (for pairs of samples). Collect the computed values from all the students and estimate from distributions of these statistics (*a*) the efficiencies of the median and midrange as estimates of the population mean by comparison with sample mean and (*b*) the efficiencies of the range and mean deviation as estimates of the population $\sigma$ by comparison with the sample standard deviation $s$. (*c*) Compare the performance of $\tau_1$, $\tau_2$, $\tau_d$ as substitutes for $t$ by comparing the lengths of confidence intervals from these substitutes with the lengths of confidence intervals obtained by using $t$. If distributions of the sample means, sample standard deviations, and $t$ values both for single samples and for pairs of samples have not been constructed previously, they should be prepared for comparison with the distributions described above.

## PROBLEMS

*Section* 9-2

1. For what size sample will the sampling distribution of the mean have the same dispersion as the sampling distribution of the median of 20 observations?

2. If data collected for determining the median cost 10 cents per observation and data collected for computing the mean cost 15 cents per observation, would it be more economical to take a sample of 100 cases and measure $\bar{X}$ or to take a sample sufficiently large to provide a median as precise as $\bar{X}$?

3. An experimenter has the choice of finding $\bar{X}$ for 163 observations or the median for 200. Which will give him the more precise estimate of $\mu$?

*Section* 9-4

**4.** If $P_{25}$ and $P_{75}$ are observed in a sample of 1,000 observations from a normal population, how can an unbiased estimate of $\mu$ be constructed?

**5.** The selection of percentiles 17, 50, 83 from a large body of data yields the values 127, 136, 147. Estimate the mean and state the efficiency of the estimate.

*Section* 9-5

**6.** For the data given at the end of Sec. 2.2 find (*a*) the median, (*b*) the midrange, (*c*) the range.

**7.** Compute for the data of Prob. 29, Chap. 3, (*a*) the median, (*b*) the midrange, (*c*) the range.

**8.** The data give readings in foot-pounds of the impact strength on two kinds of insulating material. For each sample compute the median, the midrange, and the range. Assuming normality, compute an unbiased estimate of $\sigma$ based on the range for each sample.

| *A* | 1.25 | 1.16 | 1.33 | 1.15 | 1.23 | 1.20 | 1.32 | 1.28 | 1.21 |
|---|---|---|---|---|---|---|---|---|---|
| *B* | 1.01 | .89 | .97 | .95 | .94 | 1.02 | .99 | 1.06 | .98 |

**9.** A sample of 1,000 cases has $P_{93} = 16.1$ and $P_{07} = 10.3$. Estimate $\mu$ and $\sigma$. On the assumption that the efficiency of $.3388\ (P_{93} - P_{07})$ relative to $s$ is $E = .65$ and the variances for $.3388\ (P_{93} - P_{07})$ and $s$ are approximately $\sigma^2/(2NE)$ and $\sigma^2/(2N)$, respectively, for what size sample would $s$ be as precise as the percentile estimate?

**10.** Suppose percentiles were obtained from a distribution of 3,000 items as in the following tabulation. If you were asked to compute estimates of the mean with only four of these values, which would you use? Compute your estimate of the mean. Compute in a similar manner an estimate of the standard deviation based on two items. If you were asked to estimate both the mean and standard deviation from the same four values, which would you select? Compute estimates of the mean and standard deviation from these values.

How would you use all 19 percentiles to estimate $\mu$? How could you use all 19 percentiles to estimate $\sigma$? Find the multipliers to make the estimate of $\sigma$ unbiased.

| *P* | 95 | 90 | 85 | 80 | 75 | 70 | 65 | 60 | 55 | 50 | 45 | 40 | 35 | 30 | 25 | 20 | 15 | 10 | 05 |
|---|---|---|---|---|---|---|---|---|---|---|---|---|---|---|---|---|---|---|---|
| *X* | 150 | 139 | 130 | 126 | 120 | 116 | 112 | 107 | 104 | 100 | 97 | 91 | 88 | 84 | 80 | 75 | 69 | 62 | 50 |

**11.** If percentiles 07 and 93 of a distribution of data give the values $-27$ and 58, what is an estimate of the standard deviation? What is the efficiency of that estimate? Can the mean be estimated from these values?

*Section* 9-6

**12.** Two types of paint are to be tested. Paint I is somewhat cheaper than paint II. Samples of size 5 for each paint were given scores after they had been exposed to the weather for a period of 6 months. Analyze the following scores using $\tau_d$.

| *Paint* I | 85 | 87 | 92 | 80 | 84 |
|---|---|---|---|---|---|
| *Paint* II | 89 | 89 | 90 | 84 | 88 |

**13.** Using the ratio of ranges, test for differences in dispersion with the following observations of weight gain on 12 pigs fed ration $A$ and 12 pigs fed ration $B$:

| | | | | | | | | | | | | |
|---|---|---|---|---|---|---|---|---|---|---|---|---|
| *Ration A* | 31 | 34 | 29 | 26 | 32 | 35 | 38 | 34 | 30 | 29 | 32 | 31 |
| *Ration B* | 26 | 24 | 28 | 29 | 30 | 29 | 32 | 26 | 31 | 29 | 32 | 28 |

**14.** Twenty plots of ground were planted with corn. On 10 plots a new type of phosphorus fertilizer was applied. Analyze using $\tau_d$, estimating the difference in means by a 95 per cent confidence interval, and test the hypothesis that the new type of fertilizer is no better in mean value than the old.

| | | | | | | | | | | |
|---|---|---|---|---|---|---|---|---|---|---|
| *Control* | 6.1 | 5.8 | 7.0 | 6.1 | 5.8 | 6.4 | 6.1 | 6.0 | 5.9 | 5.8 |
| *Phosphorus* | 5.9 | 5.7 | 6.1 | 5.8 | 5.9 | 5.6 | 5.6 | 5.9 | 5.7 | 5.6 |

**15.** Analyze Prob. 17, Chap. 6, using $\tau_1$. Give a 95 per cent confidence interval for the mean and test the machine setting. Repeat with $\tau_2$. Compare these results with the solution using $t$.

**16.** The following data give paired yields of two varieties of wheat. Each of the 10 pairs was planted in a different locality. Analyze the results using $\tau_1$. Give a 95 per cent confidence interval for the mean difference and test the hypothesis of mean difference equal to zero. Repeat with $\tau_2$.

| | PAIR NO. | | | | | | | | | |
|---|---|---|---|---|---|---|---|---|---|---|
| | 1 | 2 | 3 | 4 | 5 | 6 | 7 | 8 | 9 | 10 |
| *Variety* I | 45 | 32 | 58 | 57 | 60 | 38 | 47 | 51 | 42 | 38 |
| *Variety* II | 47 | 34 | 60 | 59 | 63 | 44 | 49 | 53 | 46 | 41 |

Explain why pairing is necessary in this problem.

**17.** Given measurements on the percentage of $Na_2O$ in soda ash as 40.32, 40.37, 40.27, 40.35, 40.30, determine 90 per cent confidence limits for the population mean $\mu$ using (*a*) $\bar{X}$ and $s$, (*b*) $\bar{X}$ and $w$, (*c*) midrange and $w$.

*Section* 9-7

**18.** In Table 9-1 the mean median of 200 samples of size 10 each was observed to be $-.013$. Use this result to estimate $\mu$ with a 99 per cent confidence interval, assuming $\sigma^2 = 1$.

**19.** In Table 4-3 are recorded $\bar{X}$'s of samples of size 10 and 40. Using the $\bar{X}$'s of samples of size 10, estimate $\mu$ with a 95 per cent confidence interval and estimate $\sigma^2$ (not $\sigma_{\bar{X}}^2$) with a 90 per cent confidence interval. Repeat for the $\bar{X}$'s of samples of size 40.

*Section* 9-8

**20.** For the data in Prob. 17, Chap. 6, determine an interval such that you have 99 per cent confidence that it will cover 90 per cent of the population.

**21.** If a certain machined part is accurate within $\pm .2$ inch it is usable. A random sample of 10 such parts showed deviations from the exact size of $+.08$, $-.11$, $+.04$, $-.01$, $-.07$, $+.04$, $-.02$, $+.03$, $+.12$, $-.10$. About what proportion of the population sampled from can you be 95 per cent confident of finding between $-.20$ and $.20$?

*Section* 9-9

**22.** In the manufacturing of parts the following data were obtained for the daily number defective for a production of 100 parts per day. Construct a control chart with the 20 readings in the first four columns and see whether or not the production goes out of control after that point.

| | | | | | | | |
|---|---|---|---|---|---|---|---|
| 22 | 25 | 15 | 26 | 31 | 22 | 17 | 26 |
| 23 | 20 | 28 | 32 | 43 | 18 | 16 | 36 |
| 21 | 16 | 29 | 26 | 18 | 24 | 28 | 42 |
| 17 | 14 | 26 | 33 | 26 | 24 | 32 | 36 |
| 38 | 26 | 25 | 30 | 21 | 16 | 18 | 34 |

**23.** (*a*) Construct a control chart for $\bar{X}$ for the following data on the blowing time of fuses, with samples of size 5 taken every hour for 12 hours. (*b*) If these are the first data taken on this product, would you say that the process seemed to be under control and hence that the mean and range from these data could be used for future control? (*c*) If it is known that previous control existed with a mean and standard deviation about equal to these sample values, would these data justify some action on the part of the engineer in charge at any time? Each set of five has been arranged vertically in order of magnitude.

| | | | | | | | | | | | |
|---|---|---|---|---|---|---|---|---|---|---|---|
| 42 | 42 | 19 | 36 | 42 | 51 | 60 | 18 | 15 | 69 | 64 | 61 |
| 65 | 45 | 24 | 54 | 51 | 74 | 60 | 20 | 30 | 109 | 91 | 78 |
| 75 | 68 | 80 | 69 | 57 | 75 | 72 | 27 | 39 | 113 | 93 | 94 |
| 78 | 72 | 81 | 77 | 59 | 78 | 95 | 42 | 62 | 118 | 109 | 109 |
| 87 | 90 | 81 | 84 | 78 | 132 | 138 | 60 | 84 | 153 | 112 | 136 |

CHAPTER TEN

# Analysis of variance

In this chapter we shall explore procedures for testing for differences among the means of two or more populations. In Sec. 8-3 it was noted that if means of subgroups are greatly different, the variance of the combined groups is much larger than the variances of the separate groups. The analysis of differences in means will be based on this fact.

Circumstances frequently make it necessary to design an experiment in such a way that several variables or populations can be studied simultaneously. If we wish to investigate the differences among five means by performing a $t$ test on each pair of means, there would be ten $t$ values to compute. It is not good statistical procedure to do this. First, if we use a 5 per cent critical value and then test a number of means in this manner, the level of significance will be much larger than 5 per cent for testing the hypothesis of no difference in means as a group, since, even though all samples are from the same population, 5 per cent of the $t$ values will exceed the critical value on the average. It can be shown that the chance that some one or more of 10 independent $t$ values will exceed $t_{.95}$ is .40. If the above hypothesis that all five means are equal is rejected when some one of the ten $t$ values exceeds $t_{.95}$, this hypothesis will be rejected with a chance which may be many times the level of significance .05. Exact statements are difficult to make owing to the possible dependence of the various tests.

In addition to the disadvantage of a materially increased level of significance $\alpha$, there is the loss of precision in estimating the variance if we use only those measurements in the two groups being compared. This can be improved by using a pooled variance, but the $\alpha$ still remains indefinite.

The examples in the next section illustrate some types of experimental design which are analyzed with analysis-of-variance methods. As will be seen in the following sections, the analysis rests on a separation of the variance of all the observations into parts, each of which measures variability attributable to some specific source, such as internal variation of the several populations or variations from one population to another, and the expression "analysis of variance" refers to this breakdown of the sample variance. Note, however, that the basic question concerns a comparison of the means of the several populations, and the parts of the sample variance are analyzed for this purpose. We shall see that there are a number of ways of completing the analysis, depending on the specific conclusions desired; for example, we may desire an omnibus test of the hypothesis that all populations have equal means, or we may wish to compare individual populations to find if any one has a significantly greater mean.

## EXAMPLES AND DISCUSSION OF SEVERAL PROBLEMS 10-1

**Fixed-constants model: model I.** *Single classification.* The simplest application of analysis-of-variance procedures is to the problem of estimating or testing hypotheses about the means $\mu_1, \mu_2, \ldots, \mu_k$ of $k$ populations and is referred to as a *single-classification problem.* All individuals are classified into exactly one of the $k$ populations or categories. The case $k = 2$ was discussed in detail in Chap. 8, where the difference $\mu_1 - \mu_2$ was estimated by a confidence interval and where the hypothesis $\mu_1 = \mu_2$ was tested.

It should be emphasized that the $k$ categories exhaust the cases in which there is interest and are not merely a sample from a larger number of categories; that is, the conclusion is to be made only for these $k$ categories (such cases are discussed below). Examples of the classification of individuals of a population on a single variable are as follows:

1. We have five ($k = 5$) teaching methods and a group of school children. We imagine five hypothetical populations, each consisting of all the children taught by one of the methods, and are interested in comparing the mean amount of material learned by each group (as measured by an examination) using the respective five methods.

2. We have three cities ($k = 3$) and are interested in comparing the mean incomes of the inhabitants of each of the three cities. In this case the populations actually exist at present. In the previous example the populations may not ever exist except in the mind of the experimenter.

3. We have four varieties of wheat ($k = 4$) and imagine a population of yields of each variety. We are interested in comparing the mean yields of the four varieties.

**Table 10-1** Two-way classification

| | | FIRST-VARIABLE CATEGORIES $c$ | | | |
|---|---|---|---|---|---|
| | | 1 | 2 | 3 | |
| SECOND-VARIABLE CATEGORIES $r$ | 1 | $\mu_{11}$ | $\mu_{21}$ | $\mu_{31}$ | $\mu_{\cdot 1}$ |
| | 2 | $\mu_{12}$ | $\mu_{22}$ | $\mu_{32}$ | $\mu_{\cdot 2}$ |
| | 3 | $\mu_{13}$ | $\mu_{23}$ | $\mu_{33}$ | $\mu_{\cdot 3}$ |
| | 4 | $\mu_{14}$ | $\mu_{24}$ | $\mu_{34}$ | $\mu_{\cdot 4}$ |
| | | $\mu_{1\cdot}$ | $\mu_{2\cdot}$ | $\mu_{3\cdot}$ | $\mu_{\cdot\cdot}$ |

*Two classifications.* In this example individuals are categorized on the basis of two characteristics. Each individual will belong to exactly one category for each variable. One of the characteristics is arbitrarily called the first variable and has, say, $c$ categories, and the other variable is called the second and has, say, $r$ categories. In Table 10-1 the first variable is placed at the top and is also referred to as a column variable. Thus there are $c$ categories for the first variable and $c$ columns in the table. The second variable appears as rows in the table, and thus the table has $r$ rows corresponding to the $r$ categories of the second variable. Table 10-1 represents the case $c = 3$ and $r = 4$. We may consider each of the $c \times r = 3 \times 4 = 12$ cells as separate populations, each individual belonging to one and only one of the 12. We are interested in how the means of these 12 populations differ from one another. In the table the means of the separate populations are labeled $\mu_{ij}$, with the first subscript referring to the column and the second to the row. For example, $\mu_{23}$ is the mean of the population of individuals belonging to category 2 of the column variable of classification and to category 3 of the row variable of classification. On the right side of the table are recorded $\mu_{\cdot j}$, defined as the mean of $\mu_{ij}$ for that row, for example, $\mu_{\cdot 1} = (\mu_{11} + \mu_{21} + \mu_{31})/3$, and at the bottom of the table $\mu_{i\cdot}$, for example, $\mu_{1\cdot} = (\mu_{11} + \mu_{12} + \mu_{13} + \mu_{14})/4$, defined as the mean of the $\mu_{ij}$ for that column. The mean of all 12 values $\mu_{11}, \mu_{12}, \ldots, \mu_{34}$ is written as $\mu_{\cdot\cdot}$ in the lower right-hand corner of the table. If the 12 populations are of equal size, than $\mu_{\cdot 1}$ is the mean of the population consisting of all individuals in the first row (category 1 of variable 2), $\mu_{2\cdot}$ is the mean of the population in the second column, etc.

As in the first example, it is supposed that the $c$ and $r$ categories exhaust all categories of interest for the first and second variables, respectively, and neither is considered as a sample of categories for the particular variable; this is still a fixed-constants, or model I, situation.

Examples of the classification of individuals by two characteristics are as follows:

1. We have the children of three schools ($c = 3$) and four teaching methods ($r = 4$) to investigate. Note here that for the fixed-constants model we suppose that we are interested in only three schools and, specifi-

cally, do not plan to infer from our results anything about a larger group of schools. We may wish here to compare the schools, to compare the teaching methods, or possibly to see whether there is some interacting effect which a particular school has with a particular method.

2. We have three cities ($c = 3$) and have divided the employed people into groups by sex ($r = 2$). We may wish to compare the mean income for men with the mean income for women, to compare the mean incomes for the three cities, and also to see whether there is some interaction, such as a particular city having an unusually large or small mean income for either men or women, but not for both.

3. We have four varieties of wheat ($c = 4$) and three sources of fertilizer ($r = 3$) and are interested in discovering whether a combination of a particular variety of wheat with fertilizer from a particular source will result in a larger mean yield than the other combinations. Or, as before, we may also be interested in whether some one source of fertilizer is better with all four varieties of wheat.

*Several classifications.* The previous example can be extended to cases where individuals are classified on more than two variables. For example, we may consider all people in three cities classified as to the city in which they live, sex, and income over or under $6,000 per year. The original population is divided into 12 subpopulations: men in city 1 earning under $6,000, men in city 1 earning over $6,000, women in city 1 earning under $6,000, etc. We may wish to consider whether the mean percentage of income spent on medicine was the same for the three cities, for the two income groups, or for the two sexes, or possibly whether some interaction was present.

**Components of variance, model II.** *Single classification.* A situation which appears to be similar to the single-variable model I problem occurs if the $k$ populations examined are actually not the entire set of populations of interest, but rather a sample from the larger group. This is referred to as a *components-of-variance model,* and although the arithmetic of the analysis is the same as in the fixed-constants case, the interpretations differ. Here, in addition to the problem of comparing the means $\mu_1, \ldots, \mu_k$ of the chosen populations, we have the problem of inference beyond them to the class of populations sampled from.

Examples of this situation are as follows:

1. We select five schools ($k = 5$) at random from a group of 50 schools and choose students randomly from each. We are interested in estimating means or testing hypotheses about the means of the 50 schools.

2. We have a group of 200 cities and choose samples of people from 10 of the cities ($k = 10$). We are interested in estimating the mean or variance of the mean incomes of the 200 cities.

3. We have a fertilizer which may be used for 10 types of wheat. We

choose three types ($k = 3$) from the 10 and wish to estimate the mean yield if the fertilizer was used on all 10 types. Note that since there is a single fertilizer we still have only one variable of classification.

*Components with two variables; mixed model.* Components-of-variance situations also occur if there are two variables of classification. If the categories for both variables are samples from larger groups of categories, we refer to it as a components-of-variance situation. However, if all categories are used for one variable and the categories for the other are a sample of categories, we refer to the situation as a *mixed-model* problem. Examples of the two situations are as follows:

1. Components of variance: We have a group of 50 schools and 12 age groups of students. We choose five schools ($c = 5$) and two age groups ($r = 2$) and measure samples of students in the two age groups from each of the five schools. Our measurements might be learning skill, or money spent for comics, or growth rate, etc. We are interested in an inference to the entire group of 50 schools and to all 12 age groups. For example, we may wish to determine whether the mean learning skill for all 50 schools is above a given norm, basing our inference on the data from the five schools. (When the categories are defined according to the values of a scaled variable, e.g., age or income, a sampling design leading to regression analysis, introduced in Chaps. 11 and 15, would likely be more informative than the components-of-variance design and analysis.)

2. Components of variance: We have a group of 200 cities and 15 income categories of workers. We choose samples of workers from 20 cities ($c = 20$) and from three chosen income categories ($r = 3$) and measure the amount of education for each worker. We are interested in an inference comparing the mean years of education of workers in the 200 cities and an inference comparing the mean years of education of workers in the 15 income categories. We may also be interested in whether there is evidence of an interaction of city and income level; it may be that in some cities a particular income classification has a mean level of education which is unusually high or low.

3. Mixed model: We have exactly four types of wheat ($c = 4$) and 60 general locations in a state in which to grow wheat. We choose five locations at random ($r = 5$) and grow each of the four types of wheat at each of the five locations. Since all the types of wheat are present in our experiment, but only a sample of locations, we have a mixed-model experiment.

Probably in most scientific investigations an attempt is made to consider factors other than the one specifically under study. The methods illustrated above, which arrange the factors into general classifications, are often useful for this purpose, since they permit a study at each category level of each variable. This type of experimentation is contrary to the method of holding all factors constant except for the one under study. To illustrate this point consider the following examples.

Suppose we wish to compare two methods of teaching a certain subject.

We select test groups which have the same proportion of boys and girls, who are equal in intelligence, of the same age, and so on. An experiment conducted by one teacher or in one school, however, is subject to the criticism that this teacher may be efficient with a certain method or that the superiority of a certain method may be due to the particular school involved. It is clear that an experiment of this sort should involve several schools and several teachers in each school. Depending on the problem being studied, it may be better to include more schools, even with one teacher per school. Thus to investigate a particular phenomenon we do not really wish to hold constant all other variables, but to show that the phenomenon exists independent of the other variables.

Suppose we wish to investigate the effect of several treatments on a group of white rats. For example, suppose four groups of rats are given four different concentrations of an injection and their performance in running a maze is observed. If we can determine that there is a significant difference in the performance of the different groups but we discover in a study of the records that there is also a significant difference in the ages of the rats in the four groups, this experiment must be declared inconclusive since the observed differences may be due to the different ages rather than to the different treatment. By designing the experiment to balance the age groups we may minimize this possibility. If sufficient animals are available it may be desirable to also balance on training, sex, previous experiments, etc.

In comparing the yield of several hybrids of corn we must consider the important variables such as planting time, soil fertility, or amount of fertilizer. Either these variables must be controlled, or the experiment must be designed in such a way that the effects of planting time, fertility, and so on can be estimated and separated from the effect on yield of the difference among the hybrids themselves. It will probably be necessary to use several planting times and to plant the corn with, say, different amounts of fertilizer in order to estimate these effects.

If we wish to decide whether some additional treatment makes soil pipe more resistant to corrosion, we should like to be sure that the improvement, if any exists, is valid for several different types of soil before this treatment is generally adopted.

A study of various ways of treating delinquent children should include cases from various educational, economic, regional groups, and so on.

In cases where the important variables can be held constant, as in the laboratory, simpler statistical procedures are possible. When the circumstances are such that these variables cannot be controlled, a statistical procedure must be developed which will take into account these various changes. However, even in cases where the important variables can be controlled, it may be prohibitively expensive to do so. Furthermore, there is always the possibility that the values at which the variables have been fixed cause a different result to occur in our experiment from that which might occur for other values of these variables. Also, it is sometimes very difficult to create the same experimental situation every time we wish to repeat our experiment on one or two more items.

Time is frequently the chief factor in causing us to design an experiment which will study a number of treatments and materials simultaneously. It often takes a whole year to perform one experiment in biology or agriculture. In many sciences a whole generation or several generations are required, so that it is important to have a single experiment as complete as possible.

An important point to bear in mind in designing experiments is that in using a mean $\bar{X}$ of observations to estimate a population mean $\mu$ our precision depends on the size of the variance $\sigma^2/N$ of the sampling distribution of $\bar{X}$. If $\sigma^2/N$ is small, either because $N$ is large or because $\sigma^2$ is small, then our estimate will be more precise. The population $\sigma^2$ consists of two parts, the actual variability of the members of the population and the variability introduced by lack of complete precision in measuring the sampled members of the population. This second portion of $\sigma^2$ may be reduced by refining the measuring instruments or improving the experimental apparatus. From an economic point of view, careful consideration should be given to the appropriate balance between the number of observations and the precision of each. In some cases we may wish to reduce $\sigma^2$, and in some cases it is simpler and less expensive to increase $N$.

This chapter presents only an introduction to statistical methods applicable to experiments of the type described above. The analyses will cover the following topics:

Single variable of classification:

1. Fixed constants, model I:
   (*a*) Test of hypothesis of equal means, Sec. 10-2
   (*b*) Contrasts or comparisons, Sec. 10-3
2. Components-of-variance model, model II, Chap. 15

Two variables of classification:

1. Fixed constants, model I:
   (*a*) Test of hypothesis of equal means in terms of row, column, and interaction effects, Secs. 10-4 and 10-5
   (*b*) Contrasts or comparisons, Sec. 10-6
2. Components-of-variance model, model II, Chap. 15

Three variables of classification, Latin square, Chap. 15.

## 10-2 SINGLE VARIABLE OF CLASSIFICATION: MODEL I

In this design every individual belongs to one and only one of $k$ distinct populations, with means $\mu_1, \mu_2, \ldots, \mu_k$. To make inferences concerning the sizes of these means we shall develop a test of the hypothesis that *all* $k$ means are equal. In the next section tests for and estimates of more detailed comparisons among the means will be developed, such as estimates of $\mu_1 - \mu_2$, $\mu_1 - \mu_3$.

A random sample of individuals will be taken from each population—$n_1$ from the first population, $n_2$ from the second, and so on, to $n_k$ individuals from the $k$th population. We shall compute from the means $\bar{X}_{1\cdot}$, $\bar{X}_{2\cdot}$, . . . , $\bar{X}_{k\cdot}$ of the samples an estimate of the variance $\sigma^2$ of each of the $k$ populations to compare with the pooled variance $s_p^2$ obtained from the variances of the $k$ samples.

For convenience we shall refer to the different categories, which may represent different schools, classes, hybrids, fertilizers, stages of learning, dosages of penicillin, economic groups, number of children in family, age groups, color of eyes, metals of different atomic-weight classes, different classes of polymers, and so on. It might be noted that if the categories represent different treatments given to individuals, the populations differ only in treatment given. In this case we could select a number of individuals and randomly assign an equal number to each of the categories.

To simplify the presentation further, only three categories ($k = 3$) will be considered. The changes necessary to include more groups will be pointed out. Suppose we have observed five individuals at random from each category. $X_{ij}$ is the measurement on the $j$th individual chosen from the $i$th category. The data may be arranged as in Table 10-2 and the totals and averages computed as indicated. Thus $T_{1+}$ is the total of all the observations in the first column, $T_{2+}$ the total for the second column, and $T_{3+}$ the total for the third column. $T_{++}$ is the total of all the column totals, which, of course, is equal to the total of all the observations.

The hypothesis to be tested is that there is no difference in means among the categories (the three groups of five are all drawn from populations

**Table 10-2**

| | CATEGORY *A* | *B* | *C* | | |
|---|---|---|---|---|---|
| | $X_{11}$ | $X_{21}$ | $X_{31}$ | | |
| | $X_{12}$ | $X_{22}$ | $X_{32}$ | | |
| | $X_{13}$ | $X_{23}$ | $X_{33}$ | | |
| | $X_{14}$ | $X_{24}$ | $X_{34}$ | | |
| | $X_{15}$ | $X_{25}$ | $X_{35}$ | | |
| *Total* | $T_{1+}$ | $T_{2+}$ | $T_{3+}$ | $T_{++}$ | *Grand total* |
| *Mean* | $\bar{X}_{1\cdot}$ | $\bar{X}_{2\cdot}$ | $\bar{X}_{3\cdot}$ | $\bar{X}_{\cdot\cdot}$ | *Grand mean* |

$$T_{1+} = X_{11} + X_{12} + X_{13} + X_{14} + X_{15}, \text{ etc.}$$
$$T_{++} = T_{1+} + T_{2+} + T_{3+}$$
$$\bar{X}_{1\cdot} = \frac{T_{1+}}{5}, \ldots \qquad \bar{X}_{\cdot\cdot} = \frac{T_{++}}{15}$$

with the same mean). The choice of the level of significance should be made before the experiment is conducted. We shall estimate the population $\sigma^2$ in two ways and then compare these two estimates:

1. We can form a pooled estimate of $\sigma^2$, as was done in Chap. 8:

$$s_p^2 = \frac{\sum_{j=1}^{n_1} (X_{1j} - \bar{X}_{1\cdot})^2 + \sum_{j=1}^{n_2} (X_{2j} - \bar{X}_{2\cdot})^2 + \sum_{j=1}^{n_3} (X_{3j} - \bar{X}_{3\cdot})^2}{n_1 + n_2 + n_3 - 3} \tag{1}$$

where $n_1$, $n_2$, $n_3$ are the number of cases in category $A$, $B$, $C$, respectively, and, for the data above, $n_1 = n_2 = n_3 = 5$.

2. We can compute the variance of the means directly and multiply by the sample size $n$, as was done in Chap. 8. The variance of the means is an estimate of $\sigma^2/n$. Therefore the variance of the means multiplied by $n$ is an estimate of $\sigma^2$. This estimate is

$$s_M^2 = n \frac{\sum_{i=1}^{k} (\bar{X}_{i\cdot} - \bar{X}_{\cdot\cdot})^2}{k - 1} \tag{2}$$

where $k$ = number of categories or treatments ($k = 3$ in this case)
$n$ = number of measurements in each category ($n = 5$ in this case)

When the number of observations is not the same in all categories, this estimate is

$$s_M^2 = \frac{\sum_{i=1}^{k} n_i (\bar{X}_{i\cdot} - \bar{X}_{\cdot\cdot})^2}{k - 1} \tag{2'}$$

These two quantities, $s_p^2$ and $s_M^2$, may be tested for significant difference by using the $F$ ratio. If the groups are from populations having unequal means, estimate (2) or (2′) will usually be considerably larger than estimate (1). We wish to reject the hypothesis of no difference in means if the observed means are more disperse than we would expect when they are all obtained from the same population. However, we wish to reject this hypothesis only if these means are *significantly* more disperse. Therefore we always place this second estimate in the numerator and reject the hypothesis of equal means if the $F$ ratio obtained exceeds the critical value of the $F$ table for $k - 1$ and $\Sigma n_i - k$ degrees of freedom (for this example, 2 and 12 degrees of freedom). This use of the $F$ table has been justified mathematically for normal populations.

As was noted in Chap. 8, an alternative computing formula for $s_p^2$ is

$$s_p^2 = \frac{\sum_j X_{1j}^2 - \frac{(\Sigma X_{1j})^2}{n_1} + \sum_j X_{2j}^2 - \frac{(\Sigma X_{2j})^2}{n_2} + \sum_j X_{3j}^2 - \frac{(\Sigma X_{3j})^2}{n_3}}{n_1 + n_2 + n_3 - 3}$$

or, if we rearrange the terms and replace the total of the $i$th column $\sum_j X_{ij}$ by $T_{i+}$,

$$s_p^2 = \frac{\sum_i \sum_j X_{ij}^2 - \left(\frac{T_{1+}^2}{n_1} + \frac{T_{2+}^2}{n_2} + \frac{T_{3+}^2}{n_3}\right)}{\sum_i n_i - 3}$$

The numerator is referred to as the *within-groups sum of squares,* and the pooled variance $s_p^2$ is sometimes called the *within-groups mean square* or *within-groups variance.* The denominator is referred to as the *number of degrees of freedom in* $s_p^2$. The general formula for $s_p^2$ if there are $k$ categories can be written as

$$s_p^2 = \frac{\Sigma\Sigma X_{ij}^2 - \sum \frac{T_{i+}^2}{n_i}}{\Sigma n_i - k}$$

In a similar manner the estimates (2) or (2′) can be rearranged to give as a computing formula

$$s_M^2 = \frac{\frac{T_{1+}^2}{n_1} + \frac{T_{2+}^2}{n_2} + \frac{T_{3+}^2}{n_3} - \frac{T_{++}^2}{N}}{3 - 1}$$

where $T_{++} = \Sigma\Sigma X_{ij}$ = total of all observations
$N = \Sigma n_i$ = total number of observations

The numerator is called the *sum of squares for means* (or *between categories*), and $s_M^2$ is called the *mean square for* (or *between*) *categories.* The computing formula for $s_M^2$ if there are $k$ categories is

$$s_M^2 = \frac{\sum_{i=1}^{k} \frac{T_{i+}^2}{n_i} - \frac{T_{++}^2}{N}}{k - 1}$$

The number of degrees of freedom for $s_M^2$ is $k - 1$.

Note that if the numerators of $s_p^2$ and $s_M^2$ are added the result is

$$\sum \sum X_{ij}^2 - \frac{T_{++}^2}{N}$$

which, when divided by the degrees of freedom $N - 1$, is exactly the computing formula for the variance of the entire set of observations. The breakdown of the total sum of squares into two parts can be conveniently recorded in a table, as shown in Table 10-3. The last column of Table 10-3 will be discussed later in this section.

The computations are simple to perform. The following data are fictitious and are introduced only to illustrate the computation. The data

**Table 10-3** Analysis of variance for one variable

| | Sum of squares | Degrees of freedom | Mean square | Estimate of |
|---|---|---|---|---|
| Means | $\sum \frac{T_{i+}^2}{n_i} - \frac{T_{++}^2}{N}$ | $k - 1$ | $s_M^2$ | $\sigma^2 + n\sigma_m^2$ |
| Within | $\sum \sum X_{ij}^2 - \sum \frac{T_{i+}^2}{n_i}$ | $N - k$ | $s_p^2$ | $\sigma^2$ |
| *Total* | $\sum \sum X_{ij}^2 - \frac{T_{++}^2}{N}$ | $N - 1$ | | |

are presented in four categories to illustrate the changes necessary in the formulas for a different number of categories. The hypothesis of no difference in categories (equal means) will be tested at the 5 per cent level of significance.

**Table 10-4**

| | CATEGORY | | | | |
|---|---|---|---|---|---|
| | *A* | *B* | *C* | *D* | |
| | 7 | 6 | 8 | 7 | |
| | 2 | 4 | 4 | 4 | |
| | 4 | 6 | 5 | 2 | |
| | | | | 5 | |
| *Total* | 13 | 16 | 17 | 18 | 64 |

**Computation.** Category means:

$$\left[\frac{(13)^2}{3} + \frac{(16)^2}{3} + \frac{(17)^2}{3} + \frac{(18)^2}{4}\right] - \frac{(64)^2}{13} = 319 - 315.08 = 3.92$$

Within:

$$(7)^2 + (2)^2 + (4)^2 + (6)^2 + (4)^2 + (6)^2 + (8)^2 + (4)^2 + (5)^2 + (7)^2 + (4)^2 + (2)^2 + (5)^2 - \left[\frac{(13)^2}{3} + \frac{(16)^2}{3} + \frac{(17)^2}{3} + \frac{(18)^2}{4}\right] = 37$$

Total:

$$(7)^2 + (2)^2 + (4)^2 + (6)^2 + (4)^2 + (6)^2 + (8)^2 + (4)^2 + (5)^2 + (7)^2 + (4)^2 + (2)^2 + (5)^2 - \frac{(64)^2}{13} = 356 - 315.08 = 40.92$$

The analysis-of-variance table (Table 10-5) has been almost universally accepted as the particular form for reporting the computation results.

**Table 10-5** Analysis of variance

| | *Sum of squares* | *Degrees of freedom* | *Mean square* | *F ratio* |
|---|---|---|---|---|
| Category means | 3.92 | 3 | 1.31 | $F = \frac{1.31}{4.11} = .32$ |
| Within | 37.00 | 9 | 4.11 | $F_{.95}\ (3,9) = 3.86$ |
| *Total* | 40.92 | 12 | | |

The number of degrees of freedom among categories in this table is $k - 1$, in this case 3, or one less than the number of categories. The degrees of freedom for the pooled estimate of the variance (sum of squares within groups) is $\Sigma n_i - k$, in this case 9. The number of degrees of freedom for the total is $N - 1$, in this case 12. Note that the degrees of freedom for the two estimates of variance total to $N - 1$. The row of totals in the analysis-of-variance table is used only for computation of other numbers in the table. The numbers in the mean-square column are obtained by dividing the corresponding sum of squares by the degrees of freedom. The two mean squares obtained are the two estimates of variance $s_M^2$ and $s_p^2$. The mean square obtained from the means is always placed in the numerator of the $F$ ratio, since we wish to declare the means significantly different only if they are significantly more spread out than would be expected for samples from the same population. For the particular data above the hypothesis of equal means would be accepted, since the observed $F$ is less than $F_{.95}\ (3,9) = 3.86$.

**Hypothesis and assumptions.** The hypothesis we are testing with the above analysis of variance is that the samples are from populations with the same mean, that is, $\mu_1 = \mu_2 = \mu_3 = \mu_4$. The computations except for the test of significance gives valid results for estimating the variances if the samples are randomly chosen from populations having approximately equal variances. However, the test of significance using the $F$ distribution in the analysis of variance above is known to be valid if the observations are from normally distributed populations with equal variances. Investigation has shown that the results of the analysis are changed very little by moderate violations of the assumptions of normal distribution and equal variance. If a considerable amount of experimentation is involved these assumptions can be checked from the data gathered.

OUTLINE OF ANALYSIS-OF-VARIANCE TEST
FOR SINGLE VARIABLE OF CLASSIFICATION

1. State the experimental goal.
2. $H: \mu_1 = \mu_2 = \cdots = \mu_k$. The means of the $k$ categories are all equal.

3. Choose the level of significance $\alpha$.

4. The statistic used is $F$, the ratio of the mean square for means to the mean square for within groups.

5. On the assumption that the observations are randomly selected from normal populations with $\sigma_1^2 = \sigma_2^2 = \cdots = \sigma_k^2$ (homogeneous variance) and that the hypothesis is true, the distribution of $F$ is $F(k - 1, \Sigma n_i - k)$, as given in Table A-7.

6. The critical region is $F > F_{1-\alpha}(k - 1, \Sigma n_i - k)$.

7. Compute $F$ and see whether or not it falls in the critical region.

8. Accept or reject the hypothesis.

9. State the conclusions for the experiment.

Often it can be assumed from past experience or from an examination of the data that the assumption of homogeneous variance is satisfied. If the variances are not homogeneous it may be possible to transform the observations so that this assumption will be satisfied (see Sec. 16-2).

If the hypothesis $\mu_1 = \mu_2 = \cdots = \mu_k$ is true, both $s_p^2$ and $s_M^2$ are unbiased estimates of $\sigma^2$, the variance of the individual populations. If, in fact, the hypothesis is not true and some of the $\mu_i$'s are unequal, then if we assume for simplicity that there is an equal number $n$ of observations from each population, $s_M^2$ will have a sampling distribution with mean equal to $\sigma^2 + n\sigma_m^2$, where $\sigma_m^2 = \sum_{i=1}^{k} (\mu_i - \mu.)^2/(k - 1)$ and $\mu. = \Sigma\mu_i/k$. However, the within-groups variance $s_p^2$ is still an unbiased estimate of $\sigma^2$. These mean values are listed in the last column of Table 10-3. If the $n_i$'s are not equal, $n\sigma_m^2$ is replaced by $\Sigma n_i(\mu_i - \mu.)^2/(k - 1)$.

The value of $\sigma_m^2$ may be estimated by subtracting the within-groups variance from the mean square for means and dividing by $n$. Thus $(s_M^2 - s_p^2)/n$ is an unbiased estimate of $[(\sigma^2 + n\sigma_m^2) - \sigma^2]/n = \sigma_m^2$.

To illustrate the meaning of $\sigma_m^2$, consider 5 observations from each of six populations, each having $\sigma^2 = 1$. Suppose the means of the six populations are $\mu_1 = \mu_2 = \mu_3 = 0$ and $\mu_4 = \mu_5 = \mu_6 = 2$. Then $\mu. = 1$, and $\sigma_m^2 = \frac{6}{5} = 1.20$. Therefore the mean square for means $s_M^2$ estimates $\sigma^2 + n\sigma_m^2 = 1 + (5 \times 1.20) = 7$, and $s_p^2$ estimates $\sigma^2 = 1$. In the usual situation the $\mu_i$'s are not known, and we estimate their dispersion from the formula $(s_M^2 - s_p^2)/n$.

## 10-3 INDIVIDUAL COMPARISONS IN THE ONE-VARIABLE CASE

If the hypothesis $\mu_1 = \mu_2 = \cdots = \mu_k$ is rejected by the methods of the previous section we conclude that there are differences among the means. The test, however, indicates very little about the nature of the differences. The procedure introduced in this section provides a method for examining in a single experiment a wide variety of possible differences, including those suggested by the data itself. It is based on the range of the $k$ observed means, that is, the difference between the smallest and largest observed means.

In the preceding section the total sum of squares was separated into two parts, one involving the variance of the observed means and the other the within-groups sum of squares. The two portions were used in a test of the hypothesis that the several populations have equal means. If the observed means are close together we accept the hypothesis; if they are significantly dispersed we reject the hypothesis. In the test we used the variance of the observed means $\bar{X}_{1\cdot}, \bar{X}_{2\cdot}, \ldots, \bar{X}_{k\cdot}$ to measure their dispersion.

The range $w$ is also a measure of dispersion and can be used to measure the dispersion of the $k$ means. A test of the hypothesis of equal means can be made by comparing the range of the means to the within-groups sum of squares. This can be done with the test statistic $q = w/(s_p/\sqrt{n})$, where $s_p^2$ is the pooled or within-groups mean square and $n$ is the size of each sample. The denominator $s_p/\sqrt{n}$ is an estimate of $\sigma/\sqrt{n}$, the standard deviation of $\bar{X}$'s for samples of size $n$ from the same population.

Several percentiles of the sampling distribution of the statistic $q$ are given in Table A-18$a$, where $k$ is the number of means and $\nu$ is the number of degrees of freedom in $s_p^2$. The heading in Table A-18$a$ is $q = w/s$, which is the ratio for $k$ single observations ($n = 1$) from a normal population. The statistic is divided by the factor $\sqrt{n}$ to apply it to the means of $n$ observations each.

Suppose we have four observations ($n = 4$) from each of three populations ($k = 3$); then the pooled variance has $12 - 3 = 9$ degrees of freedom. We find from Table A-18$a$ that $q = w/(s_p/\sqrt{4})$ has a 95 per cent chance of being less than 3.95 and a 99 per cent chance of being less than 5.43. An $\alpha = .05$ critical region for testing the hypothesis that the three populations have equal means consists of values of $q$ greater than 3.95, and an $\alpha = .01$ critical region consists of values of $q$ greater than 5.43. If, for example, the observed means are $\bar{X}_{1\cdot} = 2.25$, $\bar{X}_{2\cdot} = 4.00$, $\bar{X}_{3\cdot} = 4.50$ and the pooled variance is $s_p^2 = 4.41$, then the observed $q = \dfrac{2.25}{2.10/2} = 2.14$, and this is less than 3.95. Thus we do not reject the hypothesis of equal means at the 5 per cent level of significance.

**Extension to other hypotheses.** The preceding analysis was presented as a test of the hypothesis that the $k$ population means $\mu_1, \mu_2, \ldots, \mu_k$ are equal. We may also use the $q$ table to estimate or to test simultaneously hypotheses about all possible differences of the means of the type $\mu_1 - \mu_2$, $\mu_1 - \mu_3$, $\mu_4 - \mu_2$, and so on, as well as all possible linear expressions of the form $a_1\mu_1 + a_2\mu_2 + \cdots + a_k\mu_k$, where $a_1 + a_2 + \cdots + a_k = 0$, and where for convenience we shall make the sum of the positive $a$'s equal to 1.

Such forms are called *contrasts* or *comparisons* among the means $\mu_1$, $\mu_2$, . . . , $\mu_k$. Tests for contrasts are based on the following theorem:

THEOREM. *For random samples from k normal populations with the same variance the chance that all comparisons simultaneously satisfy*

$$\frac{-q_{1-\alpha}s_p}{\sqrt{n}} < (a_1\bar{X}_{1\cdot} + a_2\bar{X}_{2\cdot} + \cdots + a_k\bar{X}_{k\cdot}) - (a_1\mu_1 + a_2\mu_2 + \cdots + a_k\mu_k) < \frac{q_{1-\alpha}s_p}{\sqrt{n}}$$

*is equal to* $1 - \alpha$, *where the value of* $q_{1-\alpha}$ *is read from Table* A-18*a*.

For example, consider the observations on three populations given above: $k = 3$, $n = 4$, $\bar{X}_{1\cdot} = 2.25$, $\bar{X}_{2\cdot} = 4.00$, $\bar{X}_{3\cdot} = 4.50$, and $s_p^2 = 4.41$. For 95 per cent confidence in our statements we use $q_{.95} = 3.95$ from Table A-18 for $k = 3$ and 9 degrees of freedom. Using the notation $q_{1-\alpha}(k,\nu)$, we have $q_{.95}(3,9) = 3.95$. We first compute

$$\frac{q_{1-\alpha}s_p}{\sqrt{n}} = \frac{3.95 \times 2.10}{\sqrt{4}} = 4.15$$

We may now say with 95 per cent confidence of being correct in all statements that

$$\bar{X}_{1\cdot} - \bar{X}_{2\cdot} - 4.15 < \mu_1 - \mu_2 < \bar{X}_{1\cdot} - \bar{X}_{2\cdot} + 4.15$$
$$-5.90 < \mu_1 - \mu_2 < 2.40$$

and

$$\bar{X}_{1\cdot} - \bar{X}_{3\cdot} - 4.15 < \mu_1 - \mu_3 < \bar{X}_{1\cdot} - \bar{X}_{3\cdot} + 4.15$$
$$-6.40 < \mu_1 - \mu_3 < 1.90$$

and

$$\bar{X}_{2\cdot} - \bar{X}_{3\cdot} - 4.15 < \mu_2 - \mu_3 < \bar{X}_{2\cdot} - \bar{X}_{3\cdot} + 4.15$$
$$-4.65 < \mu_2 - \mu_3 < 3.65$$

and

$$\frac{\bar{X}_{1\cdot} + \bar{X}_{2\cdot}}{2} - \bar{X}_{3\cdot} - 4.15 < \frac{\mu_1 + \mu_2}{2} - \mu_3 < \frac{\bar{X}_{1\cdot} + \bar{X}_{2\cdot}}{2} - \bar{X}_{3\cdot} + 4.15$$
$$-5.53 < \frac{\mu_1 + \mu_2}{2} - \mu_3 < 2.77$$

and

$$\frac{\bar{X}_{1\cdot} + \bar{X}_{3\cdot}}{2} - \bar{X}_{2\cdot} - 4.15 < \frac{\mu_1 + \mu_3}{2} - \mu_2 < \frac{\bar{X}_{1\cdot} + \bar{X}_{3\cdot}}{2} - \bar{X}_{2\cdot} + 4.15$$
$$-4.77 < \frac{\mu_1 + \mu_3}{2} - \mu_2 < 3.53$$

and

$$\frac{\bar{X}_{2\cdot} + \bar{X}_{3\cdot}}{2} - \bar{X}_{1\cdot} - 4.15 < \frac{\mu_2 + \mu_3}{2} - \mu_1 < \frac{\bar{X}_{2\cdot} + \bar{X}_{3\cdot}}{2} - \bar{X}_{1\cdot} + 4.15$$

$$-2.15 < \frac{\mu_2 + \mu_3}{2} - \mu_1 < 6.15$$

and so on, for as many contrasts as we can, or wish to, write down.

This is a most useful result, since it makes possible confidence statements for differences suggested by the data in contrast to a particular difference stated in a hypothesis before data are collected. Frequently exploratory collections of data can be used to suggest hypotheses that may be tested by a future experiment, which is then carried out to decide whether to reject or to accept the hypotheses. Using the $q$ statistic in this case makes it possible to look for unexpected results in the data and then to use the same data to establish significance.

Computations for a number of contrasts can be recorded conveniently in tabular form. The coefficients $a_1, a_2, \ldots, a_k$ corresponding to $\bar{X}_{1\cdot}$, $\bar{X}_{2\cdot}, \ldots, \bar{X}_{k\cdot}$ are listed for each contrast of interest in a table like Table 10-6. Although only six contrasts are listed, we note again that any others may be added to the table with no loss of confidence in the final conclusions, provided the $q$ statistic is used. The numerical results for the example discussed above are given in Table 10-7*a*. In this example there is no evidence, with the $q$ statistic at the 5 per cent level, to indicate rejection of a zero value for any of the six contrasts listed.

Provided that we are estimating only one particular contrast and no other, we may use the $t$ statistic. Suppose the single contrast chosen in

**Table 10-6** Examples of contrasts among three groups

| $\bar{X}_{1\cdot}$ | $\bar{X}_{2\cdot}$ | $\bar{X}_{3\cdot}$ | *Confidence limits* | *Population contrast* |
|---|---|---|---|---|
| $a_1$ | $a_2$ | $a_3$ | | |
| 1 | −1 | 0 | $(\bar{X}_{1\cdot} - \bar{X}_{2\cdot}) \pm \frac{qs_p}{\sqrt{n}}$ | $\mu_1 - \mu_2$ |
| 1 | 0 | −1 | $(\bar{X}_{1\cdot} - \bar{X}_{3\cdot}) \pm \frac{qs_p}{\sqrt{n}}$ | $\mu_1 - \mu_3$ |
| 0 | 1 | −1 | $(\bar{X}_{2\cdot} - \bar{X}_{3\cdot}) \pm \frac{qs_p}{\sqrt{n}}$ | $\mu_2 - \mu_3$ |
| $\frac{1}{2}$ | $\frac{1}{2}$ | −1 | $\left(\frac{\bar{X}_{1\cdot} + \bar{X}_{2\cdot}}{2} - \bar{X}_{3\cdot}\right) \pm \frac{qs_p}{\sqrt{n}}$ | $\frac{\mu_1 + \mu_2}{2} - \mu_3$ |
| $\frac{1}{2}$ | −1 | $\frac{1}{2}$ | $\left(\frac{\bar{X}_{1\cdot} + \bar{X}_{3\cdot}}{2} - \bar{X}_{2\cdot}\right) \pm \frac{qs_p}{\sqrt{n}}$ | $\frac{\mu_1 + \mu_3}{2} - \mu_2$ |
| −1 | $\frac{1}{2}$ | $\frac{1}{2}$ | $\left(\frac{\bar{X}_{2\cdot} + \bar{X}_{3\cdot}}{2} - \bar{X}_{1\cdot}\right) \pm \frac{qs_p}{\sqrt{n}}$ | $\frac{\mu_2 + \mu_3}{2} - \mu_1$ |

**Table 10-7a** Numerical example for contrasts among three groups

| 2.25 | 4.00 | 4.50 | *Confidence limits* | *Population contrast* |
|---|---|---|---|---|
| 1 | −1 | 0 | $-1.75 \pm 4.15$ | $\mu_1 - \mu_2$ |
| 1 | 0 | −1 | $-2.25 \pm 4.15$ | $\mu_1 - \mu_3$ |
| 0 | 1 | −1 | $-\ .50 \pm 4.15$ | $\mu_2 - \mu_3$ |
| $\frac{1}{2}$ | $\frac{1}{2}$ | −1 | $-1.38 \pm 4.15$ | $\frac{\mu_1 + \mu_2}{2} - \mu_3$ |
| $\frac{1}{2}$ | −1 | $\frac{1}{2}$ | $-\ .62 \pm 4.15$ | $\frac{\mu_1 + \mu_3}{2} - \mu_2$ |
| −1 | $\frac{1}{2}$ | $\frac{1}{2}$ | $2.00 \pm 4.15$ | $\frac{\mu_2 + \mu_3}{2} - \mu_1$ |

advance of the experiment is the third one listed in Table 10-7*a*. We could then state the 95 per cent confidence limits for the single comparison $\mu_2 - \mu_3$ as $-.50 \pm (2.26 \times 2.10)\sqrt{\frac{2}{4}}$ or $-.50 \pm 3.36$.

The $q$ statistic makes it possible to examine a wide variety of comparisons simultaneously with a stated probability that *all* estimates are correct. If a certain single comparison is of major interest, a shorter confidence interval may be obtained for this comparison by using percentiles of the $t$ distribution from Table A-5 to form the interval

$$t_{\frac{1}{2}\alpha} s_p \sqrt{\frac{\Sigma a_i^2}{n}} < (a_1\bar{X}_1 + a_2\bar{X}_2 + \cdots + a_k\bar{X}_k) - (a_1\mu_1 + a_2\mu_2 + \cdots + a_k\mu_k) < t_{1-\frac{1}{2}\alpha} s_p \sqrt{\frac{\Sigma a_i^2}{n}}$$

However, the level of confidence applies only to this single statement

**Table 10-7b** Numerical example for contrasts among three groups

| 2.25 | 4.00 | 4.50 | *Confidence limits using F* | *Confidence limits using q* | *Population contrast* |
|---|---|---|---|---|---|
| 1 | −1 | 0 | $-1.75 \pm 4.33$ | $-1.75 \pm 4.15$ | $\mu_1 - \mu_2$ |
| 1 | 0 | −1 | $-2.25 \pm 4.33$ | $-2.25 \pm 4.15$ | $\mu_1 - \mu_3$ |
| 0 | 1 | −1 | $-\ .50 \pm 4.33$ | $-\ .50 \pm 4.15$ | $\mu_2 - \mu_3$ |
| $\frac{1}{2}$ | $\frac{1}{2}$ | −1 | $-1.38 \pm 3.75$ | $-1.38 \pm 4.15$ | $\frac{\mu_1 + \mu_2}{2} - \mu_3$ |
| $\frac{1}{2}$ | −1 | $\frac{1}{2}$ | $-\ .62 \pm 3.75$ | $-\ .62 \pm 4.15$ | $\frac{\mu_1 + \mu_3}{2} - \mu_2$ |
| −1 | $\frac{1}{2}$ | $\frac{1}{2}$ | $2.00 \pm 3.75$ | $2.00 \pm 4.15$ | $\frac{\mu_2 + \mu_3}{2} - \mu_1$ |

and not to repeated single statements. Further discussion is devoted to tests involving independent (orthogonal) contrasts in Chap. 15.

Some writers advocate making multiple comparisons of the type discussed in this section with a much larger value of $\alpha$. This, of course, improves the chance of recognizing cases where some one of the hypotheses tested is not true, but it should be noted that it also increases the chance of saying a true hypothesis is false.

Multiple comparisons can also be based on the $F$ statistic by use of the following theorem:

THEOREM. *For random samples from $k$ normal populations with the same variance there is probability $1 - \alpha$ that all comparisons simultaneously satisfy*

$$-L < (a_1\bar{X}_1 + a_2\bar{X}_2 + \cdots + a_k\bar{X}_k) - (a_1\mu_1 + a_2\mu_2 + \cdots + a_k\mu_k) < L$$

*where* $L^2 = (k - 1)F_{1-\alpha}\,(k - 1,\nu)\, s_p{}^2 \left(\frac{a_1{}^2}{n_1} + \frac{a_2{}^2}{n_2} + \cdots + \frac{a_k{}^2}{n_k}\right)$

*and the value of $F_{1-\alpha}\,(k - 1,\nu)$ with $\nu = \Sigma n_i - k$ is read from Table* A-7.

The results of using this method for the contrasts of Table 10-7*a* are presented in Table 10-7*b*. Here $k = 3$, $n = 4$, $\nu = 12 - 3 = 9$, and with $\alpha = .05$ we compute

$$L^2 = 2 \times 4.26 \times 4.41(\tfrac{1}{4} + \tfrac{1}{4}) = 18.79$$

for the first three contrasts and

$$L^2 = 2 \times 4.26 \times 4.41(.25/4 + .25/4 + 1/4) = 14.09$$

for the last three contrasts.

The confidence limits using the $q$ statistic are repeated from Table 10-7*a* for comparison. It can be noted that the $F$ procedure resulted arithmetically in shorter confidence intervals for some comparisons at the expense of others and it is true in general that a comparison of two means will be improved with the use of $q$. Note that $L^2$ allows unequal sample sizes $n_1, n_2, \ldots, n_k$, whereas the $q$ statistic is presented here for samples of equal size.

## TWO VARIABLES OF CLASSIFICATION: SINGLE OBSERVATION 10-4

This section will present the analysis of an experiment designed to study populations of individuals classified by two characteristics. Experiments can be conducted in such a way that several variables may be studied in the same experiment. For each variable a number of categories or levels may be chosen for study. If an equal number of observations is made for all possible combinations of levels (one level from each variable), the experiment is called a *factorial experiment.* In an experiment with two (or more) variables it sometimes happens that all combinations of both variables cannot be studied. As another example we might have three experimental teaching methods to test and 10 groups of students from which to choose. Assignment of the methods to six of the groups of students could be made at random, with two replications of each method. This is called a completely randomized design. If three groups of students at each of five schools were assigned the three methods randomly in each school, the experiment would be called a *randomized-block design.* The schools in this example are referred to as blocks. Other examples of experiments with several variables are as follows.

In studying the yield of different grain hybrids we may wish to investigate at the same time the effect of varying fertility on the yields of these hybrids.

The sociologist in studying family size may wish to investigate the effects of city size and region of the country. Average size of family could be computed for five categories of city size in each of, say, six regions of the country and the results investigated for significant differences in size of family for different city sizes independent of region of the country and for significant differences in size of family for different regions of the country independent of city size.

An experiment in teaching methods might be designed with two variables of classification, different classes and different schools, and the average score obtained on an examination might be observed.

Often one of the variables of classification is important as a control; we may wish to confirm that this variable does not affect the measurements. For example, the amount of eye pigment in *Drosophila* might be measured for each of several hybrids for three or four different conditions of environment (food, temperature).

In this section we shall consider only the case of a factorial experiment with one observation made for each combination of levels. Our observations can be recorded as in the 12 cells of Table 10-8. The computation is illustrated for data for four treatments $A$, $B$, $C$, $D$ and three varieties $a$, $b$, $c$.

$T_{1+}$ is the total of the observations in the first column, $T_{+1}$ is the total of those in the first row, and so on, and $T_{++}$ is the total of all the observations.

**Table 10-8**

| SECOND VARIABLE | FIRST VARIABLE $A$ | $B$ | $C$ | $D$ | *Total* | *Mean* |
|---|---|---|---|---|---|---|
| $a$ | $X_{11}$ | $X_{21}$ | $X_{31}$ | $X_{41}$ | $T_{+1}$ | $\bar{X}_{\cdot 1}$ |
| $b$ | $X_{12}$ | $X_{22}$ | $X_{32}$ | $X_{42}$ | $T_{+2}$ | $\bar{X}_{\cdot 2}$ |
| $c$ | $X_{13}$ | $X_{23}$ | $X_{33}$ | $X_{43}$ | $T_{+3}$ | $\bar{X}_{\cdot 3}$ |
| *Total* | $T_{1+}$ | $T_{2+}$ | $T_{3+}$ | $T_{4+}$ | $T_{++}$ | |
| *Mean* | $\bar{X}_{1\cdot}$ | $\bar{X}_{2\cdot}$ | $\bar{X}_{3\cdot}$ | $\bar{X}_{4\cdot}$ | | $\bar{X}_{\cdot\cdot}$ |

The $\bar{X}_{i\cdot}$'s are the means of the columns, the $\bar{X}_{\cdot j}$'s the means of the rows, $\bar{X}_{\cdot\cdot}$ is the grand mean, $c$ is the number of columns, and $r$ is the number of rows.

The variation of the observations in this table is caused not only by the population variance and the basic experimental errors that are always present, but also by differences which may be due to a difference in treatments (first variable of classification) or a difference in varieties (second variable of classification). The development of the analysis-of-variance procedure for this two-way table of observations will be illustrated numerically.

**Table 10-9a**

| | $A$ | $B$ | $C$ | $D$ | $T_{+i}$ |
|---|---|---|---|---|---|
| $a$ | 7 | 6 | 8 | 7 | 28 |
| $b$ | 2 | 4 | 4 | 4 | 14 |
| $c$ | 4 | 6 | 5 | 3 | 18 |
| $T_{i+}$ | 13 | 16 | 17 | 14 | 60 |

In the previous case of a single variable of classification an estimate of the variance was obtained from the means of the columns. Here we shall also estimate the variance from the row means. The sum of squares we shall use in the estimate of the population variance obtained from column means is computed as before:

$$\frac{(13)^2}{3} + \frac{(16)^2}{3} + \frac{(17)^2}{3} + \frac{(14)^2}{3} - \frac{(60)^2}{12} = 303.33 - 300 = 3.33$$

Similarly, the sum of squares for estimating the population variance from the row means is

$$\frac{(28)^2}{4} + \frac{(14)^2}{4} + \frac{(18)^2}{4} - \frac{(60)^2}{12} = 326 - 300 = 26.00$$

The total sum of squares is

$$(7)^2 + (2)^2 + (4)^2 + (6)^2 + \cdots + (3)^2 - \frac{(60)^2}{12} = 336 - 300 = 36.00$$

Note that these formulas are all similar. The denominator in every term is the number of items which have been added into the total which appears in the numerator.

The analysis of variance is shown in Table 10-10. The residual sum of squares is obtained by subtracting the sums of squares for row means and column means from the total sum of squares. Since there are three row means, there are 2 degrees of freedom for rows. There are four columns and thus 3 degrees of freedom for column means. The degrees of freedom for residual is obtained by subtraction, $11 - 2 - 3 = 6$. We can form an $F$

**Table 10-10** Analysis of variance

| | *Sum of squares* | *Degrees of freedom* | *Mean square* |
|---|---|---|---|
| Row means | 26.00 | 2 | 13.00 |
| Column means | 3.33 | 3 | 1.11 |
| Residual | 6.67 | 6 | 1.11 |
| *Total* | 36.00 | 11 | |

ratio to test the difference among rows for significance by comparing the estimate of variance from row means with the residual variance, which is an estimate of variance independent of the differences in means of both rows and columns:

$$F = \frac{13.00}{1.11} = 11.7 \qquad F_{.95}(2,6) = 5.14$$

We then reject the hypothesis of no difference among means of different rows.

A test can also be made for the difference among means of the different columns independent of any difference in rows.

$$F = \frac{1.11}{1.11} = 1.00 \qquad F_{.95}(3,6) = 4.76$$

We accept the hypothesis of no difference in means of the columns.

The above example indicates the computing procedure for the analysis of variance for a two-way table. The following analysis of the same data, although more lengthy, will perhaps make the analysis easier to understand. First the table of data is bordered with row and column means and the deviation of these means from the grand mean, as in Table 10-9*b*. The estimate of variance from column means is

$$\frac{r\Sigma(\bar{X}_{i\cdot} - \bar{X}_{\cdot\cdot})^2}{c - 1} = \frac{3}{3}[(-.667)^2 + (.333)^2 + (.667)^2 + (-.333)^2] = 1.11$$

**Table 10-9b**

| | *A* | *B* | *C* | *D* | $T_{+j}$ | $\bar{X}_{\cdot j}$ | $\bar{X}_{\cdot j} - \bar{X}_{\cdot\cdot}$ |
|---|---|---|---|---|---|---|---|
| *a* | 7 | 6 | 8 | 7 | 28 | 7.0 | 2.0 |
| *b* | 2 | 4 | 4 | 4 | 14 | 3.5 | −1.5 |
| *c* | 4 | 6 | 5 | 3 | 18 | 4.5 | − .5 |
| $T_{i+}$ | 13 | 16 | 17 | 14 | 60 | | |
| $\bar{X}_{i\cdot}$ | 4.333 | 5.333 | 5.667 | 4.667 | | 5.00 | |
| $\bar{X}_{i\cdot} - \bar{X}_{\cdot\cdot}$ | −.667 | .333 | .667 | −.333 | | | |

as was also obtained from the computing formulas and given in the analysis-of-variance table. The estimate of variance from row means is

$$\frac{c\Sigma(\bar{X}_{\cdot j} - \bar{X}_{\cdot\cdot})^2}{r - 1} = \frac{4}{2}[(2)^2 + (-1.5)^2 + (-.5)^2] = 13$$

The residual sum of squares will be obtained by computing the sum of squares of all the observations after the differences in row and column means have been "removed." We can now modify the values of our observations so that all row means are the same by adding or subtracting the same amount from each value in a row. Thus we subtract 2 from each value in the first row, add 1.5 to each value in the second row, and add .5 to each value in the third row. The results are shown in Table 10-9*c*. Note that the column means and the grand mean are unchanged.

**Table 10-9c**

| | $A$ | $B$ | $C$ | $D$ | $T_{+j}$ | $\bar{X}'_{\cdot j}$ |
|---|---|---|---|---|---|---|
| $a$ | 5 | 4 | 6 | 5 | 20 | 5 |
| $b$ | 3.5 | 5.5 | 5.5 | 5.5 | 20 | 5 |
| $c$ | 4.5 | 6.5 | 5.5 | 3.5 | 20 | 5 |
| $T_{i+}$ | 13 | 16 | 17 | 14 | 60 | |
| $\bar{X}_{i\cdot}$ | 4.333 | 5.333 | 5.667 | 4.667 | | 5 |

Now we modify the values in Table 10-9*c* in a similar manner, so that the column means will be the same. To do so we add .667 to each observation in the first column, subtract .333 from each observation in the second column, subtract .667 in the third column, and add .333 in the fourth column. The results are shown in Table 10-9*d*. The remaining variation of the values in the table is not due to differences in row and column means, since all row and column means are now the same. Computing the variance of the values corrected for differences in means, we obtain

$$(5.667)^2 + (4.167)^2 + (5.167)^2 + (3.667)^2 + \cdots + (3.833)^2 - \frac{(60)^2}{12} = 6.67$$

This is recorded in the analysis-of-variance table, Table 10-10, as the

**Table 10-9d**

| | $A$ | $B$ | $C$ | $D$ | $T_{+j}$ | $\bar{X}'_{\cdot j}$ |
|---|---|---|---|---|---|---|
| $a$ | 5.667 | 3.667 | 5.333 | 5.333 | 20 | 5 |
| $b$ | 4.167 | 5.167 | 4.833 | 5.833 | 20 | 5 |
| $c$ | 5.167 | 6.167 | 4.833 | 3.833 | 20 | 5 |
| $T_{i+}$ | 15 | 15 | 15 | 15 | 60 | |
| $\bar{X}'_{i\cdot}$ | 5 | 5 | 5 | 5 | | 5 |

residual sum of squares. If the variables are additive (see below), the residual sum of squares measures experimental errors not explained by differences in row means or differences in column means. Note that the value just obtained and the two sums of squares for means add up to the total sum of squares: $6.67 + 3.33 + 26. = 36.00$.

The 12 residual values in Table 10-9*d* have been corrected for three row means, which themselves must average $\bar{X}..$, and four column means, which themselves must average $\bar{X}..$, resulting in a loss of 2 and 3 degrees of freedom in addition to a loss of 1 degree of freedom for the grand mean. The degrees of freedom for this sum of squares of residuals is $(c - 1)(r - 1) = 6$, and this sum of squares divided by the degrees of freedom is an unbiased estimate of $\sigma^2$ which is independent of any differences there might be in means for "treatments" or "varieties." In the above numerical example the $X_{ij}$ were "corrected" as follows:

$$X_{ij} - (\bar{X}_{i\cdot} - \bar{X}..) - (\bar{X}_{\cdot j} - \bar{X}..)$$

The relation which we noted in the numerical example above is true in general:

$$\sum_j \sum_i (X_{ij} - \bar{X}..)^2 = \sum_j \sum_i [X_{ij} - (\bar{X}_{i\cdot} - \bar{X}..) - (\bar{X}_{\cdot j} - \bar{X}..) - \bar{X}..]^2 + r \sum_i (\bar{X}_{i\cdot} - \bar{X}..)^2 + c \sum_j (\bar{X}_{\cdot j} - \bar{X}..)^2$$

where the values of $i$ run from 1 to $c$ and the values of $j$ from 1 to $r$. The proof of this equation can be obtained from the computing form of each of the four terms in the equation as follows. As discussed in Sec. 10-2, the sum on the left can be written as

$$S_T = \sum \sum X_{ij}^2 - \frac{T_{++}^2}{cr}$$

and the last two terms on the right can be written as

$$S_r = \frac{\Sigma T_{+j}^2}{c} - \frac{T_{++}^2}{cr} \quad \text{and} \quad S_c = \frac{\Sigma T_{i+}^2}{r} - \frac{T_{++}^2}{cr}$$

The new term on the right is first written as

$$S_R = \sum_i \sum_j (X_{ij} - \bar{X}_{i\cdot} - \bar{X}_{\cdot j} + \bar{X}..)^2$$

and then, after some manipulation, as

$$S_R = \sum_i \sum_j X_{ij}^2 - \frac{\Sigma T_{i+}^2}{r} - \frac{\Sigma T_{+j}^2}{c} + \frac{T_{++}^2}{cr}$$

It can be seen that the computing forms of the three sums of squares on the right add up to the computing form of the sum of squares on the left. Thus the residual sum of squares $S_R$ can be found by computing the total sum of squares $S_T$ and subtracting from it $S_c$ and $S_r$, the sums of squares for the column and row means.

**Hypothesis and assumptions.** Assumptions for this case consider the population means for the 12 cells in the above layout to be of the following sort. We suppose the mean for the first cell to be $\mu_{11} = \mu + r_1 + c_1$, where $\mu$ is the same for all cells, $r_1$ is the same for all cells in row 1, and $c_1$ is the same for all cells in column 1. The mean for the second cell in the first row is $\mu_{21} = \mu + r_1 + c_2$, and so on, to the fourth cell in the third row, where the mean is $\mu_{43} = \mu + r_3 + c_4$. As an example, consider the means for the 12 cells as illustrated in Table 10-11, for which $\mu = 50$ and the row "effects" for the means are $r_1 = 0$, $r_2 = 5$, $r_3 = -5$, the column "effects" for the means are $c_1 = 7$, $c_2 = -2$, $c_3 = 4$, $c_4 = -9$. These effects will always be taken so that their sum is zero. This is always possible, since $\mu$ can be changed if necessary. We select one observation from each population as indicated in Table 10-11, that is, one from a normal population with mean 57, one from a normal population with mean 48 and so on. In this example, neither population row means nor population column means are equal. If, for example, there were no column effects, the second number added to 50 in Table 10-11 would be zero in every cell. Since in practice we do not know the population

**Table 10-11**

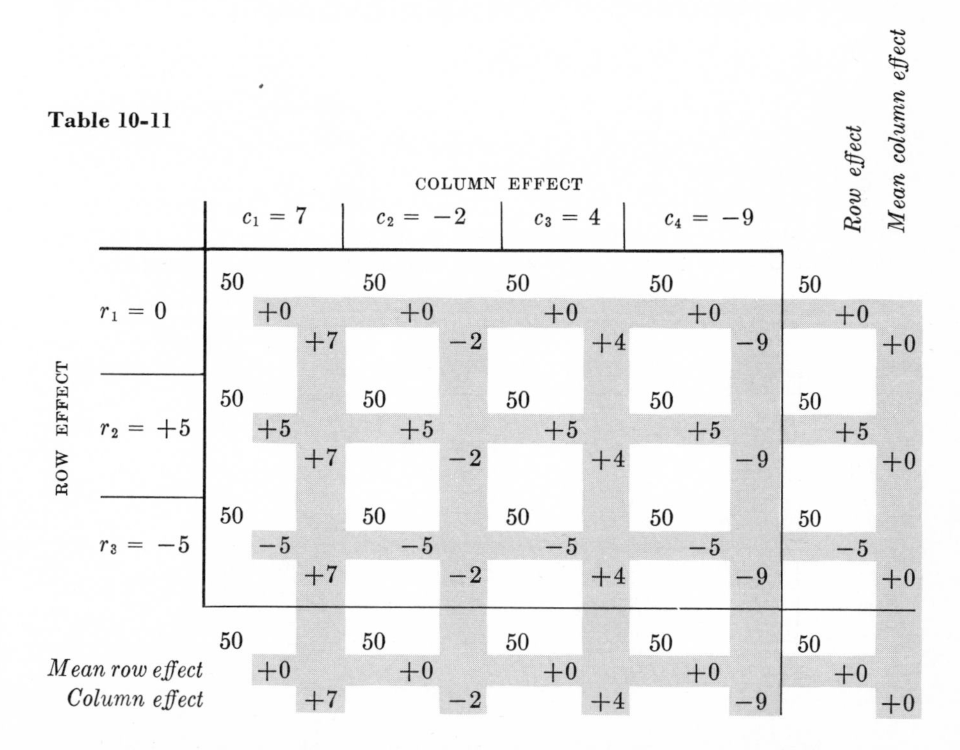

| ROW EFFECT | COLUMN EFFECT $c_1 = 7$ | $c_2 = -2$ | $c_3 = 4$ | $c_4 = -9$ | *Row effect* / *Mean column effect* |
|---|---|---|---|---|---|
| $r_1 = 0$ | 50 +0 +7 | 50 +0 −2 | 50 +0 +4 | 50 +0 −9 | 50 +0 +0 |
| $r_2 = +5$ | 50 +5 +7 | 50 +5 −2 | 50 +5 +4 | 50 +5 −9 | 50 +5 +0 |
| $r_3 = -5$ | 50 −5 +7 | 50 −5 −2 | 50 −5 +4 | 50 −5 −9 | 50 −5 +0 |
| *Mean row effect* / *Column effect* | 50 +0 +7 | 50 +0 −2 | 50 +0 +4 | 50 +0 −9 | 50 +0 +0 |

means for any of the cells, we shall use the analysis of variances to test for differences in the $c$'s and in the $r$'s.

It is assumed that the variance $\sigma^2$ of each cell population is the same.

The computation in Table 10-10, and immediately following, tests the following two hypotheses:

HYPOTHESIS 1. *There are no column effects;* $c_1 = c_2 = c_3 = c_4 = 0$.

HYPOTHESIS 2. *There are no row effects;* $r_1 = r_2 = r_3 = 0$.

When the experiment is designed as above, we may test the hypothesis stating that all $c_i$'s are zero, independent of whether or not there are row effects. Similarly, we may test the hypothesis stating that all $r_j$'s are zero, independent of whether or not there are column effects. We must assume, however, that these effects are *additive*, that is, that the mean of the cell in the $i$th column and $j$th row can be written as $\mu_{ij} = \mu + r_j + c_i$. This is an assumption, since any arbitrary set of 12 cell means written in a $3 \times 4$ table may not satisfy this condition. One interpretation of additivity is that the row and column effects do not *interact*, that is, do not have an effect in combination different from the sum of their separate effects.

Interaction occurs, for example, in studies of mixtures of chemicals where one chemical may react with the other and produce a much larger (or smaller) effect than the sum of the effects of each chemical administered separately. In pharmacological studies enhancement, or positive interaction, is referred to as *synergism*. Negative interaction is called *antagonism*. It might be noted that analyses of variance are frequently made without full assurance that the variables do not interact. Such analyses may lead to useful results, but caution should be exercised, as little is known about the effects of interaction on the results of the analysis.

Table 10-12 shows the analysis-of-variance table for two-way classification with single observation, including the population value estimated by the computed mean square. $\sigma^2$ is the variance of each of the $rc$ populations;

**Table 10-12** Analysis of variance for two-way classification with single observation

| | *Sum of squares* | *Degrees of freedom* | *Mean square* | *Estimate of* |
|---|---|---|---|---|
| Column means | $\sum \frac{T_{i+}^2}{r} - \frac{T_{++}^2}{rc} = S_c$ | $c - 1$ | $\frac{S_c}{(c-1)}$ | $\sigma^2 + r\sigma_c^2$ |
| Row means | $\sum \frac{T_{+j}^2}{c} - \frac{T_{++}^2}{rc} = S_r$ | $r - 1$ | $\frac{S_r}{(r-1)}$ | $\sigma^2 + c\sigma_r^2$ |
| Remainder | $S_T - S_c - S_r = S_R$ | $(c-1)(r-1)$ | $\frac{S_R}{(c-1)(r-1)}$ | $\sigma^2$ |
| *Total* | $\sum\sum X_{ij}^2 - \frac{T_{++}^2}{rc} = S_T$ | $rc - 1$ | | |

$\sigma_c^2$ and $\sigma_r^2$ are $\Sigma c_i^2/(c-1)$ and $\Sigma r_j^2/(r-1)$, respectively. If there is a significant column (or row) effect, $\sigma_c^2$ (or $\sigma_r^2$) may be estimated as shown at the end of Sec. 10-2.

### OUTLINE OF ANALYSIS-OF-VARIANCE TEST FOR TWO VARIABLES OF CLASSIFICATION, SINGLE OBSERVATION

1. State the experimental goal.

2. HYPOTHESIS 1. *The c column effects are zero.* The test of this hypothesis is made independent of the row effects.

   HYPOTHESIS 2. *The r row effects are zero.* The test of this hypothesis is made independent of the column effects.

3. Choose the level of significance.

4. The statistic used is $F$. For hypothesis 1 the ratio of the mean square for column means to the residual mean square; for hypothesis 2 the ratio of the mean square for row means to the residual mean square.

5. Assuming the observations are randomly selected from normal populations with homogeneous variance and that the row and column effects are additive, percentiles of the $F$ distribution are given in Table A-7.

6. The critical region for hypothesis 1 is

$$F > F_{1-\alpha}[c-1,(r-1)(c-1)]$$

The critical region for hypothesis 2 is

$$F > F_{1-\alpha}[r-1,(r-1)(c-1)]$$

7. Compute $F$ ratios.
8. Accept or reject the hypotheses.
9. State the conclusions for the experiment.

## TWO VARIABLES OF CLASSIFICATION: REPEATED MEASUREMENTS 10-5

The best estimate of experimental error (population $\sigma^2$) for use in investigating differences in means is obtained from measurements repeated under similar conditions. Suppose that in the study of yield of corn two variables of classification are treatments and varieties, and for every combination of treatment and variety the corn is planted in several plots. We say the experiment has been *replicated,* or repeated. This replication will enable us to analyze the data more fully. The analysis is a combination of the two procedures presented thus far. The data will appear as in Table 10-13 if there are two rows, three columns, and three replications.* First, this set

* Only the case of an equal number of replications in each cell will be considered here. The analysis must be modified for an unequal number of replications.

**Table 10-13**

| | $A$ | $B$ | $C$ |
|---|---|---|---|
| $a$ | $X_{111}$ | $X_{211}$ | $X_{311}$ |
| | $X_{112}$ | $X_{212}$ | $X_{312}$ |
| | $X_{113}$ | $X_{213}$ | $X_{313}$ |
| $b$ | $X_{121}$ | $X_{221}$ | $X_{321}$ |
| | $X_{122}$ | $X_{222}$ | $X_{322}$ |
| | $X_{123}$ | $X_{223}$ | $X_{323}$ |

**Table 10-14**

| | $A$ | $B$ | $C$ |
|---|---|---|---|
| $a$ | $T_{11+}$ | $T_{21+}$ | $T_{31+}$ |
| $b$ | $T_{12+}$ | $T_{22+}$ | $T_{32+}$ |

**Table 10-15**

| | $A$ | $B$ | $C$ | |
|---|---|---|---|---|
| $a$ | 4 | 2 | 5 | |
| | 7 | 3 | 6 | |
| | 5 | 2 | 4 | |
| $b$ | 9 | 8 | 10 | |
| | 8 | 7 | 8 | |
| | 8 | 5 | 7 | |
| | | | | 108 |

CELL TOTALS

| | $A$ | $B$ | $C$ | *Total* |
|---|---|---|---|---|
| $a$ | 16 | 7 | 15 | 38 |
| $b$ | 25 | 20 | 25 | 70 |
| *Total* | 41 | 27 | 40 | 108 |

of data can be considered to be samples of size 3 each from six populations and can be analyzed as in Sec. 10-2 of this chapter. That is, we can test the hypothesis that the means of the six populations are equal by comparing the variance of the observed means in the six cells with the within-groups variance. Also we may wish to look into the type of differences in the means, using the methods either of Sec. 10-3 or of Sec. 10-4. In this section we shall continue the approach of Sec. 10-4, and in the next section we shall analyze a similar problem with contrasts as in Sec. 10-3. We now record the six totals as in Table 10-14 and consider these data to be the single observations classified in a two-way table, as in Sec. 10-4, proceeding with the computations as performed there, except that each denominator in the computing formulas is the number of observations in each total, and not merely the number of

**Table 10-16** Analysis of variance

| | *Sum of squares* | *Degrees of freedom* | *Mean square* | *F ratio* |
|---|---|---|---|---|
| Between six means | 78.67 | 5 | 15.73 | 10.9 |
| Within | 17.33 | 12 | 1.44 | $F_{.95}(5,12) = 3.11$ |
| *Total* | 96.00 | 17 | | |

rows or columns. Table 10-15 shows an example of data collected as above. The analysis of variance, carried out as in Sec. 10-2, is recorded in Table 10-16.

We compute

Sum of squares for total:

$$(4)^2 + (7)^2 + (5)^2 + (9)^2 + (8)^2 + (8)^2 + (2)^2 + \cdots + (7)^2 - \frac{(108)^2}{18}$$
$$= 744 - 648 = 96$$

Sum of squares for subtotals from six populations as described above:

$$\frac{(16)^2}{3} + \frac{(25)^2}{3} + \frac{(7)^2}{3} + \frac{(20)^2}{3} + \frac{(15)^2}{3} + \frac{(25)^2}{3} - \frac{(108)^2}{18}$$
$$= 726.67 - 648 = 78.67$$

The significant $F$ ratio indicates that there is evidence that the means of the six populations are unequal. We now analyze the cell totals as in Sec. 10-4, recording the results in Table 10-17.

We compute

Sum of squares for rows:

$$\frac{(38)^2}{9} + \frac{(70)^2}{9} - \frac{(108)^2}{18} = 704.89 - 648 = 56.89$$

Note that the totals 38 and 70 are each totals of 9 of the original observations. The total 108 is the sum of all 18 original observations.

Sum of squares for columns:

$$\frac{(41)^2}{6} + \frac{(27)^2}{6} + \frac{(40)^2}{6} - \frac{(108)^2}{18} = 668.33 - 648 = 20.33$$

The sum of squares for interaction in Table 10-17 is the same as the residual sum of squares in the two-way analysis of variance of single observations and is obtained as in Sec. 10-4,

$$78.67 - 56.89 - 20.33 = 1.45$$

**Table 10-17** Analysis of variance

| | *Sum of squares* | *Degrees of freedom* | *Mean square* |
|---|---|---|---|
| Row means | 56.89 | 1 | 56.89 |
| Column means | 20.33 | 2 | 10.17 |
| Interaction | 1.45 | 2 | .72 |
| *Subtotal* | 78.67 | 5 | |
| Within groups | 17.33 | 12 | 1.44 |
| *Total* | 96.00 | 17 | |

The degrees of freedom are also determined as in Secs. 10-2 and 10-4. The subtotal sum of squares corresponds to the sum of squares for means in Table 10-16.

The term "interaction" is used in this analysis in place of "residual," since we have a further estimate of the variance, the within groups mean square, to use in a test for those differences among means which cannot be accounted for by constant shifts in row means and column means, as was illustrated in the previous section. These remaining effects are called interactions.

Note that we have a breakdown of the total sum of squares into four parts: (1) the within-groups sum of squares, which measures the variances of the six individual populations, (2) the sum of squares of row means, which measures the variability from one row to another, (3) the analogous sum of squares for column means, and (4) the sum of squares for interaction, which measures the lack of additivity of row and column effects.

From this point on the analysis definitely depends on whether we are interested in main effects, the differences in row (or column) means averaged over all columns (or rows), or whether we are interested in which combination of row and column (main effect plus interaction) produces a very large or very small result. As an example of a situation in which main effects are of interest, suppose that the two rows $a$ and $b$ refer to two proposed smog devices and the three columns $A$, $B$, and $C$ refer to three automobile models. Suppose also that the observations are the number of miles before necessary replacement of the device and that there are 18 cars in all in the experiment, three of each model with each device. If a single device is to be used for all car models, then our interest would be in the row main effect, that is, in the device producing the best effect averaged over cars. It if is considered possible to supply different devices to different car models, then our interest is not in main effects, but in which device is best for each model. If a single device and a single model are to be chosen for best results in the future, then our interest is in the best combination of device and model. In this example it should be noted that interest in the row main effect implies the assumption that there are the same number of cars of each model. Also, important economic questions such as the price of the device in relation to size or life of car model, cost of adjustment or replacement, or cost of smog damage are not included in the analysis.

In order to introduce these considerations into the analysis we temporarily leave our example to give the relationship between the numbers in the analysis-of-variance table and the various parameters to be estimated. The statements of the assumptions and tests of hypotheses underlying our problem are as follows:

1. We assume that the observations in the $i$th column and $j$th row are random samples from a normal population with mean $\mu + c_i + r_j + I_{ij}$, where $\mu$ is the same for all cells, $c_i$ is the same for all cells in the $i$th column,

$r_j$ is the same for all cells in the $j$th row, and where the $I_{ij}$'s may be different for each $i$ and $j$. We can suppose with no loss in generality that $\sum_i c_i = 0$, $\sum_j r_j = 0$, and $\sum_i I_{ij} = 0$ for each $j$, and $\sum_j I_{ij} = 0$ for each $i$, since if these sums are not zero we may make them zero by redefining $\mu$.

2. We assume that the variance $\sigma^2$ is the same for each of the normal populations.

Although we are introducing as assumptions normality and equal variance (both of which may be tested), the fact that the cell means are written in terms of a grand mean, row, column, and interaction effects implies no assumption, since this places no restriction on the means of the individual populations. In Sec. 10-4, where the $I_{ij}$'s are assumed to be zero, a restriction was imposed.

The various statistics in the analysis-of-variance table (Table 10-17) are unbiased estimates of certain functions of the population means as recorded in Table 10-18. We suppose there are $n$ observations from each cell of a two-way classification of $c$ columns and $r$ rows. $X_{ije}$ is the $e$th observation from the population in the $i$th column and $j$th row. As before, $\sigma^2$ is the variance of each of the $rc$ populations, $\sigma_c^2 = \Sigma c_i^2/(c-1)$ and $\sigma_r^2 = \Sigma r_j^2/(r-1)$ and $\sigma_I^2 = \Sigma\Sigma I_{ij}^2/(c-1)(r-1)$. We see from Table 10-18 that if we are interested in the inequality of the $c_i$'s we can compare the mean square for columns with the within-groups mean square and, if the ratio is much larger than unity, conclude that the $c_i$'s are not equal; that is, $\sigma_c^2$ is not equal to zero. Similarly, we divide the mean square for rows by the within-groups mean square, to test for inequality of the row effects, and divide the interaction mean square by the within-groups mean square, to test for inequality of the $I_{ij}$'s.

**Table 10-18** Analysis of variance with estimates

| | *Sum of squares* | *Degrees of freedom* | *Mean square* | *Estimate of* |
|---|---|---|---|---|
| Column means | $rn\Sigma(\bar{X}_{i..} - \bar{X}_{...})^2 = S_c$ | $c-1$ | $\dfrac{S_c}{(c-1)}$ | $\sigma^2 + rn\sigma_c^2$ |
| Row means | $cn\Sigma(\bar{X}_{.j.} - \bar{X}_{...})^2 = S_r$ | $r-1$ | $\dfrac{S_r}{(r-1)}$ | $\sigma^2 + cn\sigma_r^2$ |
| Interaction | $S_s - S_c - S_r = S_I$ | $(c-1)(r-1)$ | $\dfrac{S_I}{(c-1)(r-1)}$ | $\sigma^2 + n\sigma_I^2$ |
| *Subtotal* | $n\Sigma\Sigma(\bar{X}_{ij.} - \bar{X}_{...})^2 = S_s$ | $rc-1$ | | |
| Within | $S_T - S_s = S_w$ | $rc(n-1)$ | $\dfrac{S_w}{rc(n-1)}$ | $\sigma^2$ |
| *Total* | $\Sigma\Sigma\Sigma(X_{ije} - \bar{X}_{...})^2 = S_T$ | $rcn-1$ | | |

With these general results in mind we return to the example in Table 10-17. To look for interaction, we divide .72 by 1.44, getting $F = .72/1.44 = .5$ which gives no indication that $\sigma_I^2$ is greater than zero. For the row effects $F = 56.89/1.44 = 39.5$, which, since it is larger than $F_{.95}(1,12) = 4.75$, is significant at the $\alpha = .05$ level. For the columns $F = 10.17/1.44 = 7.06$, which is also significant at the $\alpha = .05$ level. Thus we have no evidence of lack of additivity but significant differences in both row and column effects.

Note in Table 10-18 that if $\sigma_I^2 = 0$, then both the lines labeled "Interaction" and "Within" give unbiased estimates of $\sigma^2$. This indicates that a combination of the two may improve the estimate of $\sigma^2$. A suggested procedure which has some theoretical justification (see the References at the end of the text) is to average, or pool, interaction and within sums of squares if the ratio of the interaction and within mean squares is less than twice the 50th percentile of the $F$ distribution. Using this rule, we would pool in our example, since the ratio is .50, which is less than

$$2F_{.50}(2,12) = 2 \times .735 = 1.470$$

To pool we add the sum of squares for "Interaction" to the sum of squares for "Within" to get a "Residual" sum of squares and then add the respective degrees of freedom to get the degrees of freedom for "Residual." For our example the residual mean square is $18.78/14 = 1.34$. This residual mean square is now used as a denominator in testing for row or column main effects. The results are shown in Table 10-19.

To test for row effects we compute

$$F = \frac{56.89}{1.34} = 42.5 \qquad F_{.95}(1,14) = 4.60$$

This value is significant, and we reject the hypothesis of equal row means. To test for column effects we compute

$$F = \frac{10.17}{1.34} = 7.59 \qquad F_{.95}(2,14) = 3.74$$

This is also significant at the 5 per cent level.

**Table 10-19** Analysis of variance

| | *Sum of squares* | *Degrees of freedom* | *Mean square* |
|---|---|---|---|
| Rows | 56.89 | 1 | 56.89 |
| Columns | 20.33 | 2 | 10.17 |
| Residual | 18.78 | 14 | 1.34 |
| *Total* | 96.00 | 17 | |

It should be emphasized that the level of significance in the above analysis will apply where it has been agreed before collecting data to perform a single test. If we proceed beyond the first test, the chance of rejecting at least one true hypothesis will increase, perhaps on the order of the size of $1 - (1 - \alpha)^h$, where $h$ tests are performed at the $\alpha$ level of significance. This formula applies to completely unrelated tests and is known to be an upper bound for this case. Instead of performing several tests on the data as above and in Table 10-17 and then proceeding to estimate the differences in the means of the several cell populations, the procedures outlined in Secs. 10-3 and 10-6 are recommended.

## TWO-BY-TWO FACTORIAL DESIGN: INDIVIDUAL COMPARISONS 10-6

As noted in Sec. 10-5, a factorial experiment has an equal number of observations (replications) for every combination of categories in the variables of classification. This type of experiment, as well as many other types, may be analyzed by setting up comparisons as in Sec. 10-3. We illustrate this for the case where there are two variables with two categories each. This case is called a two-by-two ($2 \times 2$) factorial experiment, where $2 \times 2$ denotes the number of categories in each variable. In this terminology a $3 \times 4 \times 5$ factorial experiment would involve three variables, the first having three categories, the second having four categories, and the third having five categories. The example in Sec. 10-4 is a $3 \times 4$ factorial experiment, and the example in Sec. 10-5 is a $2 \times 3$ factorial experiment replicated three times.

As an example, suppose that we have two populations, say, men and women, and two methods of teaching, say, $A$ and $B$. We use each teaching method on three men and three women (12 people in all) and record a final examination score for each. The data are recorded in Table 10-20 in the same manner as in Sec. 10-5, and the corresponding analysis of variance is in Table 10-21. Table 10-22 shows some of the possible contrasts among the observed means. As in Sec. 10-3, these contrasts are in the form

$$a_{11}\bar{X}_{11\cdot} + a_{12}\bar{X}_{12\cdot} + a_{21}\bar{X}_{21\cdot} + a_{22}\bar{X}_{22\cdot}$$

where $a_{11} + a_{12} + a_{21} + a_{22} = 0$ and the sum of the positive $a_{ij}$'s is unity. This table shows how the cell means can be used to estimate the average difference between men and women for the two teaching methods (main effects) and also the degree to which a particular method is good or poor for either sex. If a single teaching method is to be used for both men and women, the experimenter is interested in main effects. If a different method could be used for men from that for women, the experimenter would be interested in which method is best for men and which method is best for women. If the experimenter is interested in the best teaching method for *either* men or women, then he wishes to find that combination having the largest mean.

**Table 10-20** Data for 2 × 2 factorial

| | A | B |
|---|---|---|
| *Men* | 5 | 5 |
| | 8 | 5 |
| | 9 | 6 |
| *Women* | 11 | 7 |
| | 14 | 8 |
| | 15 | 6 |

| | A | B | Total |
|---|---|---|---|
| *Men* | 22 | 16 | 38 |
| *Women* | 40 | 21 | 61 |
| *Total* | 62 | 37 | 99 |

**Table 10-21** Analysis of variance

| | Sum of squares | Degrees of freedom | Mean square | F ratio |
|---|---|---|---|---|
| Rows | $\frac{5{,}165}{6} - 816.75 = 44.08$ | 1 | 44.08 | 17.63 |
| Columns | $\frac{5{,}213}{6} - 816.75 = 52.08$ | 1 | 52.08 | 20.83 |
| Interaction | $110.25 - 44.08 - 52.08 = 14.09$ | 1 | 14.09 | 5.64 |
| *Subtotal* | $\frac{2{,}781}{3} - 816.75 = 110.25$ | 3 | | |
| Within | $130.25 - 110.25 = 20.00$ | 8 | 2.50 | |
| *Total* | $947 - 816.75 = 130.25$ | 11 | | |

In Table 10-21 all three $F$ ratios are larger than $F_{.95}(1,8) = 5.32$. Therefore, if the experiment is designed to investigate any one of the three effects, row, column, or interaction, we may declare significance at the 5 per cent level. However, it is not clear that we can state significance at the 5 per cent level for all three simultaneously.

We may look at any or all contrasts in terms of the $q$ statistic. Some of these contrasts are listed in Table 10-22. For example, line 1 shows the contrast $\frac{1}{2}\bar{X}_{11\cdot} + \frac{1}{2}\bar{X}_{21\cdot} - \frac{1}{2}\bar{X}_{12\cdot} - \frac{1}{2}\bar{X}_{22\cdot}$, which measures the mean difference for men and women. We note that the contrasts in lines 2, 5, 6, 7, and 10 are significant, since the confidence limits do not cover zero. These contrasts are column (teaching) main effects, difference in methods for women, difference for sex for method $A$, difference between method $B$ for men and method $A$ for women, and difference for method $A$ for women and the average of the three other cells. Some interpretations are as follows:

From contrast 2, if a single teaching method is to be used for both men and women, method $A$ should be used.

From contrasts 4 and 5, if different methods may be used for men and for women, for method $A$ it can be said from contrast 5 that this experiment demonstrates that women perform better than men. From contrast 4 we can see that this is not demonstrated for method *B*.

From contrasts 6 and 8, if we can choose to train only men or only women, then only in the case of the women is it conclusive that method *A* will obtain superior results.

From contrasts 5, 6, and 7 together, the combination of women and method *A* results in significantly higher examination scores than any of the other three combinations.

From contrast 10, the combination of women and method *A* is better than the average of the other three combinations. In this example this result could be inferred from contrasts 5, 6, and 7 taken together.

**Table 10-22** Individual comparisons in a 2 × 2 factorial experiment

| *Contrast* | $\bar{X}_{11\cdot} = \frac{22}{3}$ | $\bar{X}_{21\cdot} = \frac{16}{3}$ | $\bar{X}_{12\cdot} = \frac{40}{3}$ | $\bar{X}_{22\cdot} = \frac{21}{3}$ | $\Sigma a_{ij}\bar{X}_{ij\cdot} \pm q_{.95}\frac{s}{\sqrt{n}}$ | *Estimate of* |
|---|---|---|---|---|---|---|
| 1 | $\frac{1}{2}$ | $\frac{1}{2}$ | $-\frac{1}{2}$ | $-\frac{1}{2}$ | $-3.83 \pm 4.13$ | $r_1 - r_2$, row main effect, difference for men and women |
| 2 | $\frac{1}{2}$ | $-\frac{1}{2}$ | $\frac{1}{2}$ | $-\frac{1}{2}$ | $4.17 \pm 4.13$ | $c_1 - c_2$, column main effect, difference for methods |
| 3 | $\frac{1}{2}$ | $-\frac{1}{2}$ | $-\frac{1}{2}$ | $\frac{1}{2}$ | $-2.17 \pm 4.13$ | $2I_{11}$, interaction effect |
| 4 | 1 | −1 | 0 | 0 | $2.00 \pm 4.13$ | Difference of methods for men |
| 5 | 0 | 0 | 1 | −1 | $6.33 \pm 4.13$ | Difference of methods for women |
| 6 | 1 | 0 | −1 | 0 | $-6.00 \pm 4.13$ | Difference for men and women using method *A* |
| 7 | 0 | 1 | −1 | 0 | $-8.00 \pm 4.13$ | Difference of method *B* for men and *A* for women |
| 8 | 0 | 1 | 0 | −1 | $-1.67 \pm 4.13$ | Difference for men and women using method *B* |
| 9 | 1 | 0 | 0 | −1 | $.33 \pm 4.13$ | Difference of method *A* for men and *B* for women |
| 10 | $-\frac{1}{3}$ | $-\frac{1}{3}$ | 1 | $-\frac{1}{3}$ | $6.78 \pm 4.13$ | Difference of method *A* for women and average of three other cells |

$q_{.95}(4,8) = 4.53 \qquad s = \sqrt{2.50} = 1.58 \qquad \frac{qs}{\sqrt{n}} = \frac{4.53 \times 1.58}{\sqrt{3}} = 4.13$

## 10-7 *WEIGHTED MEANS

The preceding examples have implicitly treated the individual cell means as having equal importance, but there are cases where it is desirable to give them unequal weights. For example, suppose an existing population is subdivided, or stratified, into a number $k$ of possibly unequally sized subgroups and a random sample is taken from each subgroup. If the strata are of sizes $M_1, M_2, \ldots, M_k$ and the sample means are $\bar{X}_1, \bar{X}_2, \ldots, \bar{X}_k$, then an appropriate (unbiased) estimate of the population mean would be $\bar{X}_s = \Sigma M_j\bar{X}_j/\Sigma M_i$. The individual $\bar{X}_j$'s are said to be *weighted* by $w_j = M_j/\Sigma M_i$ to give the *weighted mean* $\bar{X}_s$.

If we denote the standard deviations in the population strata by $\sigma_1, \sigma_2, \ldots, \sigma_k$ and the sample sizes by $n_1, n_2, \ldots, n_k$, then the variance of the sampling distribution of $\bar{X}_s$ is given by

$$\operatorname{var} \bar{X}_s = w_1^2 \frac{\sigma_1^2}{n_1} + w_2^2 \frac{\sigma_2^2}{n_2} + \cdots + w_k^2 \frac{\sigma_k^2}{n_k}$$

and the standard error of $\bar{X}_s$ is the square root of this expression. Thus an unbiased estimate of the over-all population mean can be obtained by sampling from strata of known sizes, and if values of $\sigma_i^2$ are known, its standard error can be computed. If the $n_i$'s are not small, $s_i^2$ could be used in place of the $\sigma_i^2$ in the above formula. If the $\sigma_i^2$'s are approximately equal to, say, $\sigma^2$, it is appropriate to use the pooled or within groups variance from the analysis of variance table as an estimate of $\sigma^2$.

We should clearly separate the use of stratification with the objective of more precisely estimating the over-all population mean from the case where the existing separate strata are of interest. In the former case we attempt, by use of prior knowledge or insight, to obtain strata which have $\sigma_i$ as small as possible. Typically, they are small if the subgroup means are quite unequal, and sometimes inequality of means is a simpler criterion for stratification.

It is of interest that for specified total sample size and with equal $\sigma_i^2$'s the value of $\sigma_{\bar{X}_s}^2$ is a minimum if the sample sizes $n_i$ are proportional to $M_i$, the strata sizes. Such an experimental design is called a *representative sample.* If the $\sigma_i^2$'s are not equal, then values of $n_i$ proportional to the product $M_i\sigma_i$ make the standard error of $\bar{X}_s$ a minimum. Alternatively, if the subgroups are of interest separately, then the individual $n_i$'s should not be small. If the $n_i$'s are proportional to $\sigma_i^2$, the standard errors of the strata means, $\bar{X}_i$, are equal.

As an example, consider 60,000 adults in a city distributed into three age groups with $M_1 = 10{,}000$ persons between 20 and 39 years of age, $M_2 = 20{,}000$ between 40 and 59, and $M_3 = 30{,}000$ older than 59. Denote the mean incomes of the people in these strata as $\mu_1$, $\mu_2$, and $\mu_3$, respectively, and suppose the strata standard deviations of income are $\sigma_1 = \sigma_2 = \sigma_3 = \$50$.

The over-all population mean is

$$\mu = \frac{10{,}000\mu_1 + 20{,}000\mu_2 + 30{,}000\mu_3}{60{,}000}$$

and the means $\bar{X}_1$, $\bar{X}_2$, and $\bar{X}_3$ of samples of sizes $n_1$, $n_2$, and $n_3$ then give as an estimate of $\mu$ the statistic

$$\bar{X}_s = \frac{10{,}000\bar{X}_1 + 20{,}000\bar{X}_2 + 30{,}000\bar{X}_3}{60{,}000}$$

which has variance given by

$$\begin{aligned}\operatorname{var} \bar{X}_s &= \left(\frac{10{,}000}{60{,}000}\right)^2 \frac{2{,}500}{n_1} + \left(\frac{20{,}000}{60{,}000}\right)^2 \frac{2{,}500}{n_2} + \left(\frac{30{,}000}{60{,}000}\right)^2 \frac{2{,}500}{n_3} \\ &= \frac{2{,}500}{36}\left(\frac{1}{n_1} + \frac{4}{n_2} + \frac{9}{n_3}\right)\end{aligned}$$

For the case $n_1 = n_2 = n_3 = 100$,

$$\operatorname{var} \bar{X}_s = \frac{2{,}500}{36}\left(\frac{1}{100} + \frac{4}{100} + \frac{9}{100}\right) = 9.72$$

and the estimate $\bar{X}_s$ has a standard error equal to \$3.12.

To carry the example further let us suppose that the 300 individuals in the sample form a representative sample with $n_1 = 50$, $n_2 = 100$, and $n_3 = 150$. The variance of the estimate is

$$\frac{2{,}500}{36}\left(\frac{1}{50} + \frac{4}{100} + \frac{9}{150}\right) = \frac{25}{3} = \$8.33$$

and the standard error is \$2.89. This is an improvement over the equal-sample-size case.

It is of interest to note that we cannot evaluate the advantage of the representative sample over a simple random sample of size $n = 300$ from the 60,000 people unless we have some knowledge of the real differences in the strata means. In the case $\mu_1 = \mu_2 = \mu_3$ the stratification helps least, and a random sample of size 300 gives an estimate with standard error $\sigma/\sqrt{n} = 50/\sqrt{300} = \$2.90$, the same value as for the representative sample. If $\mu_1 - \mu = \$100$ and $\mu_2 - \mu = \$100$ and $\mu_3 - \mu = -\$100$, then the population standard deviation is \$112 (compared with the standard deviation of \$50 in the strata), giving a standard error of $\$112/\sqrt{300} = \$6.50$ for the random sample estimate.

The use of weighted means also are of interest in cases where two or more populations being compared are themselves distributed differently with

**Table 10-23**

| Age | Frequency in population A | Frequency in population B | Sample size A | Sample size B | Sample mean income A | Sample mean income B |
|---|---|---|---|---|---|---|
| 20–39 | 10,000 | 30,000 | 100 | 100 | $400 | $500 |
| 40–59 | 20,000 | 20,000 | 100 | 100 | 600 | 600 |
| 60–79 | 30,000 | 10,000 | 100 | 100 | 800 | 700 |

respect to a stratifying variable. Thus in Table 10-23 the age distribution of city $A$ is much different from that of city $B$, and the mean income in the two cities would be estimated by use of different weights. For city $A$ the estimated mean income is

$$\bar{X}_A = (\tfrac{1}{6} \times 400) + (\tfrac{2}{6} \times 600) + (\tfrac{3}{6} \times 800) = \$667$$

and for city $B$ it is

$$\bar{X}_B = (\tfrac{3}{6} \times 500) + (\tfrac{2}{6} \times 600) + (\tfrac{1}{6} \times 700) = \$567$$

While these values \$667 and \$567 appropriately estimate the means of the populations, they also reflect the different age distributions, and it might be desired to obtain comparable values which reflected only the differences in the means specific to the ages. This can be done by choosing any convenient or reasonable single (standard) age distribution and comparing the weighted (standardized or adjusted) means obtained. For example, if we used the population in the first city $A$ as a standard we would have standardized means equal, respectively, to $\bar{X}'_A = \$667$ and $\bar{X}'_B = \$633$. Note that the analysis-of-variance method for the two variables of classification situation described in Secs. 10-4 and 10-5 compares the city means by use of equal weights on the age strata. The standardized population used above did not use equal weights for the age strata.

Once a standard distribution has been chosen, the hypothesis of equal population standardized means can be tested. First we note that the difference between two independent standardized means has, as in Sec. 8-4, a variance equal to the sum of the variances of the standardized means. Thus, if the within cell standard deviations are all equal to $\sigma$, the variance of $\bar{X}'_A - \bar{X}'_B$ is given by

$$\begin{aligned}\text{var}\,(\bar{X}'_A - \bar{X}'_B) &= \left(\frac{1}{6}\right)^2 \frac{\sigma^2}{100} + \left(\frac{2}{6}\right)^2 \frac{\sigma^2}{100} + \left(\frac{3}{6}\right)^2 \frac{\sigma^2}{100} \\ &\quad + \left(\frac{1}{6}\right)^2 \frac{\sigma^2}{100} + \left(\frac{2}{6}\right)^2 \frac{\sigma^2}{100} + \left(\frac{3}{6}\right)^2 \frac{\sigma^2}{100} \\ &= \frac{28}{3{,}600}\sigma^2\end{aligned}$$

If $\sigma$ = \$50 the standard error of the difference is \$4.42. In this case the observed difference, $\bar{X}'_A - \bar{X}'_B = 667 - 633 = \$34$, is $z = 34/4.42 = 7.7$ standard errors from zero. In testing the hypothesis of no population difference this result would be classified as highly significant. If $\sigma^2$ is unknown the pooled variance may be used.

## GLOSSARY

adjusted mean
analysis of variance
between means
interaction
mean square
replication
representative sample
residual
standardized mean
strata
variable of classification
weighted mean
within groups

## DISCUSSION QUESTIONS

**1.** State in your own words the assumptions about the population which are made in the use of the factorial experiment. Do you have any way to check these assumptions? Of what significance is it to an experimenter if the assumptions do not appear to be true?

**2.** Give examples of situations in fields of application where analysis of variance might be used.

**3.** Define each term in the Glossary.

**4.** What conclusions are made when you reject a hypothesis using the analysis-of-variance technique? When you accept a hypothesis?

**5.** If you reject a hypothesis using analysis of variance, what further analysis of the data can be made?

## CLASS EXERCISES

**1.** For Class Exercise 3, Chap. 8, you computed the value of $s_p^2$ and the total sum of squares for an analysis-of-variance single-classification experiment with $k = 6$, $n = 5$, $N = 30$. Now compute the mean square for categories $s_M^2$. Collect the values of $s_p^2$ for all the students into a distribution and verify that the mean $s_p^2$ is approximately equal to $\sigma^2$, which is 1 for the populations sampled (Tables A-2 and A-23). Collect the values of $s_M^2$ into a distribution and verify that the mean $s_M^2$ is $\sigma^2 + n\sigma_m^2$. Here $\mu_1 = \mu_2 = \mu_3 = 0$, $\mu_4 = \mu_5 = \mu_6 = 2$, $\mu_. = 1$, $\sigma_m^2 = \frac{6}{5}$, and so the mean $s_M^2$ should be 7.

Form the quotient of $s_M^2$ and $s_p^2$ for each group and collect the results for all the students into a distribution. Compare the 95th percentile of the observed distribution with $F_{.95}(5,24)$. Why is there a large discrepancy? Observe the frequency with which you reject the hypothesis of equal means for the six populations at the 5 per cent level of significance.

**2.** Group one sample of size $N = 5$ from Table A-2 with one sample of size $N = 5$ from Table A-23 (as recorded in the Class Exercise, Chap. 3). Then compute the values of $s_p^2$ and $s_M^2$ for each of the three groups ($k = 2, n = 5$), and collect the results of the class into distributions of $s_p^2$ and of $s_m^2$. Verify, as in Exercise 1, whether the mean $s_p^2$ is equal to the population variance and whether the mean $s_M^2$ is equal to $\sigma^2 + n\sigma_m^2$.

Form the quotient of $s_M^2$ and $s_p^2$ for each group and collect the results of all the students (three from each student) into a distribution. Compare the 95th percentile of the observed distribution with $F_{.95}(1,8)$. Why is there a large discrepancy? Observe the frequency with which you reject the hypothesis of equal means if actually $\mu_1 = 0$, $\mu_2 = 2$ (as is the case in this experiment).

**3.** Compute, on the three samples of size 5 each drawn from Table A-2, the values of $s_p^2$ and $s_M^2$. Here $k = 3$, $n = 5$, $N = 15$. Collect the results of all the students and verify that the mean $s_p^2$ and the mean $s_M^2$ are both equal to $\sigma^2$.

Form the quotient of $s_M^2$ and $s_p^2$ for each group and collect these results into a distribution. Compare the 95th percentile of this observed distribution with $F_{.95}(2,12)$. If there is a large number of cases in your experiment you can expect fairly close agreement, otherwise only approximate agreement.

## PROBLEMS

*Section* 10-2

**1.** The students in three classes in an elementary statistics course obtained total scores as in the tabulation. Is there a significant difference in the scores received by students meeting at different times of day? State completely the hypothesis you are testing and your conclusions.

| *8 o'clock* | | *10 o'clock* | | *2 o'clock* | |
|---|---|---|---|---|---|
| 121 | 122 | 97 | 131 | 134 | 162 |
| 117 | 141 | 145 | 143 | 89 | 128 |
| 145 | 126 | 119 | 107 | 108 | 133 |
| 108 | 145 | 139 | 86 | 88 | 93 |
| 142 | 114 | 143 | 94 | 146 | 118 |
| 154 | 136 | 133 | 164 | 153 | 126 |
| 115 | 151 | 149 | 139 | 130 | 127 |
| 81 | 105 | 107 | 151 | 144 | 150 |
| 122 | 103 | 154 | 141 | 125 | 138 |
| 127 | 108 | 102 | 131 | 111 | 119 |
| | | 108 | 65 | 87 | 142 |
| | | 131 | 141 | | |

**2.** Each of the sets of observations is a random sample drawn from a normal population. Test for equality of means by the analysis of variance. State the hypotheses and assumptions. Indicate how an $\alpha$ error could be made. Indicate how a $\beta$ error could be made.

| | | | | | | | | |
|---|---|---|---|---|---|---|---|---|
| *A* | 49 | 42 | 47 | 76 | 69 | 58 | | |
| *B* | 49 | 44 | 50 | 58 | 70 | | | |
| *C* | 44 | 57 | 34 | 48 | 50 | | | |
| *D* | 58 | 54 | 64 | 60 | 53 | 64 | 52 | 42 |

3. Random samples of 100 people each from three cities gave weekly income figures as summarized below, with $X_{ij}$ the income for the $j$th person from the $i$th city. Set up an analysis-of-variance table. State and test the standard hypothesis from the table.

| *City* | *Sample size* | $\sum_j X_{ij}$ | $\sum_j X_{ij}^2$ |
|---|---|---|---|
| I | 100 | \$10,000 | 1,056,300 |
| II | 100 | \$11,200 | 1,596,000 |
| III | 100 | \$9,800 | 1,100,000 |

4. A fertilizer was tested at two concentrations on 16 plots, with the resulting yields given below. Set up an analysis-of-variance table and test the hypothesis of equality of means.

| | | | | | | | | |
|---|---|---|---|---|---|---|---|---|
| *Concentration* 1 | 18 | 17 | 16 | 15 | 14 | 13 | 13 | 12 |
| *Concentration* 2 | 21 | 20 | 19 | 18 | 16 | 16 | 14 | 14 |

*Section* 10-3

5. Using the data in Prob. 3, set up a comparison between each pair of cities.

6. Verify for the example in Prob. 4 (where there are only two populations and consequently only one comparison which can be made) that the $q$ test leads to the same conclusion as the analysis-of-variance-table method.

7. Samples of 25 students each were taken from five schools and given an examination on current events. The mean scores for the five samples were $\bar{X}_{1\cdot} = 63$, $\bar{X}_{2\cdot} = 72$, $\bar{X}_{3\cdot} = 60$, $\bar{X}_{4\cdot} = 80$, $\bar{X}_{5\cdot} = 92$ with $s_p^2 = 26.01$. Give a set of interval estimates for all differences of pairs of means for the five schools such that we can be 95 per cent confident that all are correct.

8. Using the data in Table 2-2*a* (or Table 2-2*b*), stratify into subgroups by age as follows: men under 40 in group 1, men 40 to 49 in group 2, men 50 to 59 in group 3, and men 60 and over in group 4. (*a*) Perform an analysis of variance on systolic blood pressure (1952 readings). (*b*) Perform an analysis of variance on diastolic blood pressure (1952 readings). (*c*) Perform an analysis of variance on blood cholesterol (1952 readings).

9. The following experiment was designed to determine the relative merit of four different feeds in regard to the gain in weight of pigs. Analyze the data, comparing all pairs among the four feeds. Indicate how each of the two types of error might be made and give the consequence of each error.

Twenty pigs are divided at random into four lots of five pigs in each and each lot is given a different feed. The weight gain in pounds by each of the pigs for a fixed length of time is given in the table.

| | | | | | |
|---|---|---|---|---|---|
| *Feed A* | 133 | 144 | 135 | 149 | 143 |
| *Feed B* | 163 | 148 | 152 | 146 | 157 |
| *Feed C* | 210 | 233 | 220 | 226 | 229 |
| *Feed D* | 195 | 184 | 199 | 187 | 193 |

**10.** Set up an analysis-of-variance table for height for the two samples in Table 2-2*a* and Table 2-2*b*. State a hypothesis and draw a conclusion.

**11.** Set up an analysis-of-variance table for weight for the two samples in Table 2-2*a* and Table 2-2*b*. State a hypothesis and draw a conclusion.

**12.** Using the data in Table 2-2*a* (or Table 2-2*b*), stratify into subgroups by age and height as follows: men less than 50 years of age and 66 inches or less in height in subgroup 1, men less than 50 years of age and over 66 inches in height in subgroup 2, men 50 years or over in age and 66 inches or less in height in subgroup 3, and men 50 years or over in age and over 66 inches in height in subgroup 4. (*a*) Perform an analysis of variance on systolic blood pressure (1952 readings). (*b*) Perform an analysis of variance on diastolic blood pressure (1952 readings). (*c*) Perform an analysis of variance on cholesterol (1952 readings). (*d*) Perform an analysis of variance of the 1952 blood pressure gradient (systolic minus diastolic).

In each case above set up several contrasts which might be of interest and use the multiple comparison analysis based on the $F$ statistic to state conclusions. (Note that this problem anticipates the analysis in later sections.)

**13.** Repeat Prob. 12, except stratify on age as before and stratify also on weight with men weighing less than or equal to 170 pounds in one stratum and men weighing over 170 pounds in the other.

**14.** Using the height-weight data in Table 2-2*a*, stratify the 100 men into groups according to height less than or equal to 63 inches, 64 to 66 inches, 67 to 69 inches, and over 69 inches. (*a*) Set up an analysis-of-variance table for weight; do the arithmetic and draw a conclusion. (*b*) Set up an analysis-of-variance table for systolic blood pressure (1952); do the arithmetic and draw a conclusion.

**15.** Samples of people from four height classifications had the observed weights given below. Construct the analysis-of-variance table. Compare the mean weights of each pair of samples by the $F$-statistic-comparison method.

| *Height classification* | *Sample size* | *Observed weights* |
|---|---|---|
| 60 | 3 | 110 135 120 |
| 62 | 4 | 120 140 130 135 |
| 64 | 2 | 150 145 |
| 70 | 3 | 170 185 160 |

**16.** The values given are the responses to the three specified dosages of a drug. Using the $q$-statistic method, estimate the difference between mean values for the first and second dosage, between the mean values for the second and third dosage, and the difference of these differences.

| *Dose, grams* | $n_i$ | *Mean response* | *Standard deviation* |
|---|---|---|---|
| 1 | 20 | 12.3 | 4.1 |
| 2 | 20 | 14.1 | 4.8 |
| 4 | 20 | 16.0 | 3.9 |

*Section* 10-4

**17.** The drained weight in ounces of frozen apricots was measured for various types of sirup and various concentrations of sirup. The original weights of the apricots were the same. Differences in drained weight would be attributable to differences in concentration or type of sirup. Analyze the data.

| SIRUP CONCENTRATION | SIRUP COMPOSITION All sucrose | $\frac{2}{3}$ sucrose, $\frac{1}{3}$ corn sirup | $\frac{1}{2}$ sucrose, $\frac{1}{2}$ corn sirup | All corn sirup |
|---|---|---|---|---|
| 30 | 28.80 | 28.21 | 29.28 | 29.12 |
| 40 | 29.12 | 28.64 | 29.12 | 30.24 |
| 50 | 29.76 | 30.40 | 29.12 | 28.32 |
| 65 | 30.56 | 29.44 | 28.96 | 29.60 |

*Section* 10-5

**18.** Below are results of computations on the observations of the life of four brands of automobile tires used under five different road conditions. Four tires of each brand were used for each type of road. Complete the analysis of variance, stating hypotheses and conclusions. Explain the reason for including different road conditions in the experiment.

| | *Sum of squares* |
|---|---|
| Tires | 190.1 |
| Roads | 200.2 |
| *Subtotal* | 500.4 |
| *Total* | 804.6 |

**19.** An agency wished to determine whether five makes of automobiles would average the same number of miles per gallon. A random sample of three cars of each make was taken from each of three cities, and each car had a test run with 1 gallon of gasoline. The table records the number of miles traveled. (*a*) Why were three cities used instead of just one city? (*b*) What populations are sampled from? (*c*) How would you go about getting such a random sample of three cars from a city? (*d*) What assumptions are made about the populations, and what hypotheses can be tested? (*e*) Perform the analysis of variance and state fully your conclusions.

| CITIES | AUTOMOBILES *A* | *B* | *C* | *D* | *E* |
|---|---|---|---|---|---|
| *Los Angeles* | 20.3 | 19.5 | 22.1 | 17.6 | 23.6 |
| | 19.8 | 18.6 | 23.0 | 18.3 | 24.5 |
| | 21.4 | 18.9 | 22.4 | 18.2 | 25.1 |
| *San Francisco* | 21.6 | 20.1 | 20.1 | 19.5 | 17.6 |
| | 22.4 | 19.9 | 21.0 | 19.2 | 18.3 |
| | 21.3 | 20.5 | 19.8 | 20.3 | 18.1 |
| *Portland* | 19.8 | 19.6 | 22.3 | 19.4 | 22.1 |
| | 18.6 | 18.3 | 22.0 | 18.5 | 24.3 |
| | 21.0 | 19.8 | 21.6 | 19.1 | 23.8 |

**20.** Subdivide the data in Table 2-2*a* (or 2-2*b*) by a two-variable age-height classification, with ages 20 to 39, 40 to 49, 50 and over as one variable of classification and heights 60 to 67, 68 to 69, and 70 and over as the second. Using the first five individuals in each cell, perform an analysis of variance (*a*) for cholesterol and (*b*) for systolic blood pressure.

*Section* 10-7

**21.** A population of 10,000 households is subdivided geographically into three strata containing 1,000, 4,000, and 5,000 households, respectively. Random samples of size 200 each were taken from each stratum, and the distance to work was measured. The means were $\bar{X}_1 = 10$ miles, $\bar{X}_2 = 20$ miles, and $\bar{X}_3 = 4$ miles, and the pooled variance was $s_p^2 = 5.76$. Show how these numbers would fit into an analysis-of-variance table. Estimate the population mean and indicate the accuracy expected.

**22.** The students at a university consist of 8,000 freshmen, 6,000 sophomores, 6,000 juniors, and 5,000 seniors. A random sample of 100 students from each class is collected to determine the amount of outside employment done by the students. If the sample means are $\bar{X}_1 = 10$ hours, $\bar{X}_2 = 14$ hours, $\bar{X}_3 = 16$ hours, and $\bar{X}_4 = 10$ hours, estimate the over-all mean. If $\sigma_1 = \sigma_2 = \sigma_3 = \sigma_4 = 10$ hours, how accurate is the result? How much would be gained if a representative sample were chosen instead of equal sample sizes in each stratum?

CHAPTER ELEVEN

# Regression and correlation

A *regression problem* considers the frequency distributions of one variable when another is held fixed at each of several levels. A *correlation problem* considers the joint variation of two measurements, neither of which is restricted by the experimenter. Examples of regression problems can be found in the study of the yields of crops grown with different amounts of fertilizer, the length of life of certain animals exposed to different amounts of radiation, the hardness of plastics which are heat-treated for different periods of time. In these problems the variation in one measurement is studied for particular levels of the other variable selected by the experimenter. Examples of correlation problems are found in the study of the relationship between IQ and school grades, blood pressure and metabolism, or height of cornstalk and yield. In these examples both variables are observed as they naturally occur, neither variable being fixed at predetermined levels. For convenience in terminology the examples of the regression problem and the correlation problem treated in this chapter are stated in terms of the variables height and weight.

## REGRESSION 11-1

Suppose that we wish to make a study of the distribution of the weights of a population of men with relation to the heights of the men. We shall

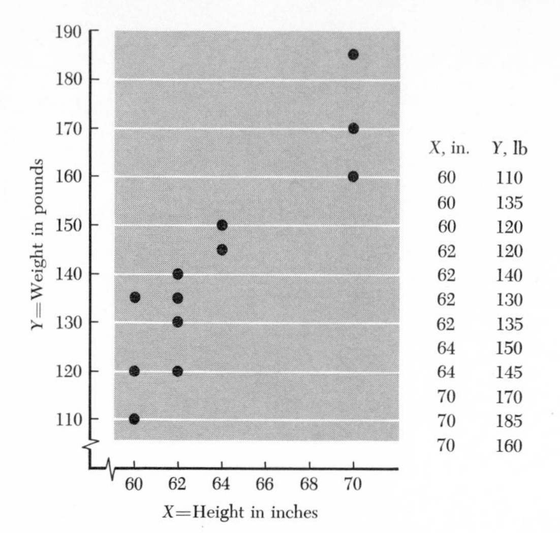


| X, in. | Y, lb |
|---|---|
| 60 | 110 |
| 60 | 135 |
| 60 | 120 |
| 62 | 120 |
| 62 | 140 |
| 62 | 130 |
| 62 | 135 |
| 64 | 150 |
| 64 | 145 |
| 70 | 170 |
| 70 | 185 |
| 70 | 160 |

**Figure 11-1** Weights plotted against heights for selected men. ($N = 12$)

subdivide the population into groups of men of approximately equal height and examine the relationships between the distributions of weights in the various groups.

For any chosen height there is a distribution of weights. This distribution has a mean, the mean weight of all men who have this height, and a variance, the variance of weights of all men who have this height. We shall define the *regression of weight on height* as the mean of the distribution of weights of all men who have the given height. In this usage the distribution of weights depends upon the particular height chosen, and we shall speak here of the weight as a dependent variable and the height as an independent variable.

It is convenient to introduce symbols, and we shall designate $Y$ as the dependent measurement and $X$ as the independent measurement. For any individual in the population there is, then, a pair of measurements, $X$ and $Y$. The mean of any distribution of weights $Y$ for given height $X$ will be denoted by $\mu_{y \cdot x}$ and the variance of this distribution by $\sigma^2_{y \cdot x}$. These are parameters. They are constant for any fixed height but may vary between distributions of weights for different heights. The mean of weights of all men of all heights will be denoted by $\mu_y$ and the variance of these weights by $\sigma_y{}^2$. We shall assume that $\sigma^2_{y \cdot x}$ is constant for all values of $X$.

To study the distributions of weights we shall select several heights and make several random observations of men having these heights. Suppose, for example, we decide to choose heights of $X = 60$, 62, 64, 70 inches, and we make the observations in the tabulation. These observations have been plotted on the graph in Fig. 11-1. Each point represents one man; his weight is read on the vertical scale and his height is read on the horizontal scale. The graph with the points plotted is called a *scatter diagram*.

In many important applications of regression theory the regression curve is a straight or approximately straight line for the range of $X$ values under consideration. In a case of this sort we say that there is *linear regression*. We shall assume for our examples that the regression curve is a straight line. A test of this hypothesis using the observed data is given in Sec. 11-5.

If there is linear regression we may write the formula for the mean $Y$ when $X$ is given as

$$\mu_{y\cdot x} = A + B(X - \bar{X}) \tag{1}$$

where $A$ and $B$ are parameters,* $X$ is any given value (height for our example), and $\bar{X}$ is the mean of the chosen $X$ values of the individuals in the sample. $B$ is the *slope* of the line, the amount that $\mu_{y\cdot x}$ changes when $X$ changes by one unit.

## ESTIMATION OF PARAMETERS 11-3

The regression equation (1) is in terms of the population parameters. We shall use the data in a sample to estimate these parameters. The unbiased estimates of $A$ and $B$, whose sampling distributions have minimum variance, are $\bar{Y}$ and $b$, where

$$b = \frac{\Sigma X_iY_i - \Sigma X_i \Sigma Y_i/N}{\Sigma X_i^2 - (\Sigma X_i)^2/N} \tag{2}$$

If we let $\bar{Y}_x$ denote our estimate of the mean $Y$ when $X$ is given, we have

$$\bar{Y}_x = \bar{Y} + b(X - \bar{X}) \tag{3}$$

An unbiased estimate of $\sigma^2_{y\cdot x}$ is $s^2_{y\cdot x}$, where

$$s^2_{y\cdot x} = \frac{1}{N-2} \sum \{Y_i - [\bar{Y} + b(X_i - \bar{X})]\}^2 \tag{4}$$

or the algebraically equivalent formula

$$s^2_{y\cdot x} = \frac{N-1}{N-2} (s_y^2 - b^2 s_x^2) \tag{5}$$

where $s_x^2$ and $s_y^2$ are the variances of the observed $X$ values and of the observed $Y$ values, respectively. It can be seen from formula (4) that $s^2_{y\cdot x}$ is a mean-square deviation of sample points from the estimated regression line. The value of $s_{y\cdot x}$ is often called the *standard error of estimate.*

* These parameters, called *regression coefficients,* are frequently denoted in the literature by $\alpha$ and $\beta$. We do not use $\alpha$ and $\beta$, since these symbols have been used for chances of error in testing hypotheses.

For the example above we have computed

$$\Sigma X_i = 766 \qquad \bar{X} = 63.83$$
$$\Sigma Y_i = 1{,}700 \qquad \bar{Y} = 141.67$$
$$\Sigma X_i Y_i = 109{,}380$$
$$\Sigma X_i^2 = 49{,}068$$
$$\Sigma Y_i^2 = 246{,}100$$

$$s_x^2 = \frac{49{,}068 - (766)^2/12}{11} = \frac{171.67}{11} = 15.61$$

$$s_y^2 = \frac{246{,}100 - (1{,}700)^2/12}{11} = \frac{246{,}100 - 240{,}833.3}{11} = \frac{5{,}266.7}{11} = 478.8$$

$$b = \frac{109{,}380 - (766 \times 1{,}700)/12}{171.67} = \frac{109{,}380 - 108{,}516.7}{171.67}$$

$$= \frac{863.3}{171.67} = 5.029$$

$$s_{y \cdot x}^2 = \tfrac{11}{10}\{478.8 - [(5.029)^2 \times 15.61]\} = 92.5$$

$$\bar{Y}_x = 141.67 + 5.029(X - 63.83)$$

Figure 11-2 shows the original data with this estimated regression line drawn. Figure 11-3 shows the original data with the estimated regression line and indicates the vertical deviations of the observations from the line.

**Least squares.** The line drawn from equation (3) has the property that the sum of squares of vertical deviations of observations from this line is

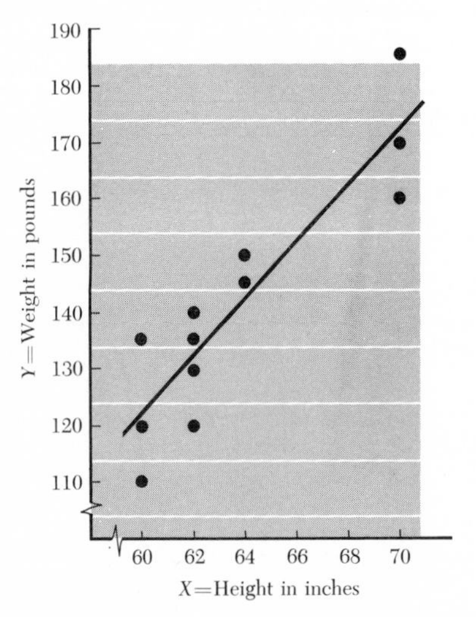


**Figure 11-2** Weights plotted against heights with "least-square" estimated regression line for the same 12 men.

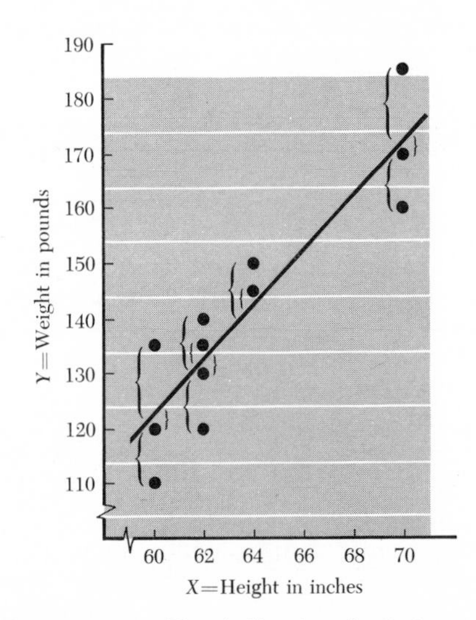


**Figure 11-3** Plot indicating deviations of weights from "least-square" estimated regression line for the same 12 men.

smaller than the corresponding sum of squares of deviations from any other line. This is called a *least-square property.*

The principle of least squares is frequently used as a justification for using formula (3) to connect $X$ and $Y$ values. This is basically a different principle from those developed here, and the principle of least squares is often used in problems unrelated to statistical inference. It happens that in many cases, as in this one, the least-squares principle and the principle of having an efficient unbiased estimate lead to the same formulas.

An equivalent formula for $b$ as given in formula (2) above is

$$b = \frac{\Sigma(X_i - \bar{X})(Y_i - \bar{Y})}{\Sigma(X_i - \bar{X})^2} \tag{2'}$$

An equivalent computation of the sum of squares in formula (4) above is

$$\Sigma y_i'^2 = \Sigma(Y_i - \bar{Y})^2 - \frac{[\Sigma(X_i - \bar{X})(Y_i - \bar{Y})]^2}{\Sigma(X_i - \bar{X})^2} \tag{4'}$$

where

$$y_i' = Y_i - [\bar{Y} + b(X_i - \bar{X})]$$

An alternative computing formula for the variance of residuals can then be written as

$$s_{y\cdot x}^2 = \frac{\Sigma y_i'^2}{N - 2} \tag{5'}$$

## ESTIMATION AND TESTS IN NORMAL POPULATIONS 11-4

In Sec. 11-3 we made no assumption about the distribution of $Y$ for given $X$, except that $\sigma_{y\cdot x}^2$ was constant for all $X$ and the regression was linear. If, in addition, we can assume that the distribution of $Y$ for any given $X$ is a normal distribution, then it is possible to indicate by confidence intervals estimates of $A$, $B$, $\sigma_{y\cdot x}^2$, and $\mu_{y\cdot x}$.

If the above assumptions are satisfied, then the sampling distribution of

$$t = \frac{(\bar{Y} - A)\sqrt{N}}{s_{y\cdot x}} \tag{6}$$

is a $t$ distribution with $N - 2$ degrees of freedom, the sampling distribution of

$$t = \frac{(b - B)s_x\sqrt{N - 1}}{s_{y\cdot x}} \tag{7}$$

is a $t$ distribution with $N - 2$ degrees of freedom, and the sampling distribution of

$$\chi^2/df = \frac{s_{y\cdot x}^2}{\sigma_{y\cdot x}^2} \tag{8}$$

is a $\chi^2/df$ distribution with $N - 2$ degrees of freedom. These distributions may be used in the same manner as in Chaps. 6 and 7 to give confidence limits for or to test hypotheses about $A$ or $B$ or $\sigma^2_{y\cdot x}$.

Thus $100(1 - \alpha)$ per cent confidence limits for $A$, $B$, and $\sigma^2_{y\cdot x}$ are obtained from equations (6), (7), and (8) above as

$$\bar{Y} + t_{\frac{1}{2}\alpha} \frac{s_{y\cdot x}}{\sqrt{N}} < A < \bar{Y} + t_{1-\frac{1}{2}\alpha} \frac{s_{y\cdot x}}{\sqrt{N}} \tag{6'}$$

$$b + t_{\frac{1}{2}\alpha} \frac{s_{y\cdot x}}{s_x \sqrt{N - 1}} < B < b + t_{1-\frac{1}{2}\alpha} \frac{s_{y\cdot x}}{s_x \sqrt{N - 1}} \tag{7'}$$

where the $t$ values are from Table A-5 for $N - 2$ degrees of freedom, and

$$\frac{s^2_{y\cdot x}}{\chi^2_{1-\frac{1}{2}\alpha}/df} < \sigma^2_{y\cdot x} < \frac{s^2_{y\cdot x}}{\chi_{\frac{1}{2}\alpha}{}^2/df} \tag{8'}$$

where the $\chi^2/df$ percentiles are obtained from Table A-6*b* for $N - 2$ degrees of freedom. For our example above 90 per cent confidence limits for these parameters are

$$141.67 - 1.81 \frac{9.61}{3.46} < A < 141.67 + 1.81 \frac{9.61}{3.46}$$

$$136.64 < A < 146.70$$

and for $B$

$$5.029 - 1.81 \frac{9.61}{13.1} < B < 5.029 + 1.81 \frac{9.61}{13.1}$$

$$3.70 < B < 6.35$$

Furthermore,

$$\frac{92.4}{1.83} < \sigma^2_{y\cdot x} < \frac{92.4}{.394}$$

$$50.5 < \sigma^2_{y\cdot x} < 235.$$

**Estimation of $\mu_{y\cdot x}$.** If the above assumptions hold, then $100(1 - \alpha)$ per cent confidence limits for the mean of the $Y$ values of all individuals having a particular $X$ value are

$$\bar{Y}_x + t_{\frac{1}{2}\alpha} s_{y\cdot x} \sqrt{\frac{1}{N} + \frac{(X - \bar{X})^2}{(N - 1)s_x{}^2}} < \mu_{y\cdot x} < \bar{Y}_x + t_{1-\frac{1}{2}\alpha} s_{y\cdot x} \sqrt{\frac{1}{N} + \frac{(X - \bar{X})^2}{(N - 1)s_x{}^2}} \tag{9}$$

where the $t$ values are read from Table A-5 for $N - 2$ degrees of freedom. This is an estimate of the *mean Y only*, not an estimate of some individual's $Y$ score from his $X$ score.

In the example we have for $X = 65$, say,

$$\bar{Y}_{65} = 141.67 + 5.029(65 - 63.83) = 147.55$$

and 95 per cent confidence limits for $\mu_{y\cdot 65}$ are

$$147.55 - (2.23 \times 9.61)\sqrt{\frac{1}{12} + \frac{(65 - 63.83)^2}{11 \times 15.61}} < \mu_{y\cdot 65}$$

$$< 147.55 + (2.23 \times 9.61)\sqrt{\frac{1}{12} + \frac{(65 - 63.83)^2}{11 \times 15.61}}$$

$$147.55 - 21.4\sqrt{.0913} < \mu_{y\cdot 65} < 147.55 + 21.4\sqrt{.0913}$$

$$141.09 < \mu_{y\cdot 65} < 154.01$$

We are 95 per cent confident that the mean weight of all individuals 65 inches tall is between 141.09 and 154.01. If we repeat this experiment with another sample we should obtain another set of limits instead of 141.09 and 154.01. In the long run 95 per cent of such intervals would cover $\mu_{y\cdot 65}$. This is the interpretation we use in our statements of level of confidence. We do not know that $\mu_{y\cdot 65}$ is between 141.09 and 154.01, but we do give a level of our belief that it is between them.

**Estimation of an individual's $Y$ value.** Interval (9) is an estimate of the mean $Y$ for a particular group of individuals. The interval given below can be used to estimate (or predict) the $Y$ value for a single observed $X$ value. The confidence level is correct for a single prediction. The confidence level is not correct for repeated predictions using the same sample.

The $100(1 - \alpha)$ per cent prediction interval for an individual's $Y$ score in terms of his $X$ score is

$$\bar{Y}_x + t_{\frac{1}{2}\alpha} s_{y\cdot x}\sqrt{1 + \frac{1}{N} + \frac{(X - \bar{X})^2}{(N - 1)s_x^2}} < Y$$

$$< \bar{Y}_x + t_{1-\frac{1}{2}\alpha} s_{y\cdot x}\sqrt{1 + \frac{1}{N} + \frac{(X - \bar{X})^2}{(N - 1)s_x^2}}$$

**Tolerance interval.** A formula which gives an interval having chance $\gamma$ of covering a proportion $P$ of the population is

$$\bar{Y}_x \pm k s_{y\cdot x}$$

where

$$k = z_{\frac{1}{2}(1+P)}\left(1 + \frac{1}{2N'} - \frac{2z^2_{\frac{1}{2}(1+P)} - 3}{24N'^2}\right)\sqrt{F_{1-\gamma}(\infty, N - 2)}$$

and

$$N' = \frac{1}{\dfrac{1}{N} + \dfrac{(X - \bar{X})^2}{(N - 1)s_x^2}}$$

This is an approximate formula which is good when $N' \geq 2$ and $N \geq 3$. The derivation of the formula assumes that $\sigma^2_{y \cdot x}$ is constant, the regression of $Y$ on $X$ is linear in the range of $X$ considered, and the distributions of $Y$ are normal for each $X$.

**Test for independence.** A hypothesis frequently tested in regression analysis is that a variable $Y$ is independent of a variable $X$. One criterion of independence is that the mean $Y$ is the same for each value of $X$, which, in the case of linear regression, means that $B = 0$.

1. We shall test the hypothesis $B = 0$, and if this hypothesis is rejected we shall say that there is sufficient reason to believe, at the specified level of significance, that $Y$ is dependent upon $X$.
2. $H: B = 0$, alternatives $B < 0$ or $B > 0$.
3. Choose $\alpha$.
4. As a test statistic, use $t = \dfrac{(b - 0)s_x \sqrt{N - 1}}{s_{y \cdot x}}$.
5. If the distribution of $Y$ for each $X$ is normal with the same variance and with the same mean ($\mu_{y \cdot x} = A$), then the sampling distribution of this statistic is a $t$ distribution with $N - 2$ degrees of freedom.
6. Reject if $t < t_{\frac{1}{2}\alpha}(N - 2)$ or $t > t_{1-\frac{1}{2}\alpha}(N - 2)$.
7. Perform the computation for $t$ and compare with the critical region.
8. Accept or reject the hypothesis.
9. Conclude the independence or dependence of $Y$ and $X$.

For our numerical example the test proceeds as follows:

1. We test the hypothesis that $Y$ is independent of $X$.
2. $H: B = 0$, alternatives $B < 0$ or $B > 0$.
3. Choose $\alpha = .05$.
4. $t = \dfrac{5.029 \times 13.1}{9.61} = 6.9$
5. If the distribution of $Y$ for each $X$ is normal with the same variance and with the same mean, then the sampling distribution of this statistic is a $t$ distribution with 10 degrees of freedom.
6. The critical region is $t < -2.23$ or $t > 2.23$.
7. Here $t = 6.9$, which is larger than 2.23.
8. We have sufficient reason at the 5 per cent level of significance to reject the hypothesis of $B = 0$.
9. We conclude that $Y$ is dependent on $X$.

## 11-5 TEST FOR LINEARITY OF REGRESSION

To test the hypothesis of linearity of regression we use an analysis-of-variance technique. Instead of the hypothesis $\mu_1 = \mu_2 = \cdots = \mu_k$,

**Table 11-1**

| | Sum of squares | Degrees of freedom | Mean square |
|---|---|---|---|
| *Total* | $\sum\sum Y_{ij}^2 - \frac{T_{Y++}^2}{N}$ | $N - 1$ | |
| Within groups | $\sum\sum Y_{ij}^2 - \sum \frac{T_{Y_{i+}}^2}{n_i}$ | $N - k$ | $s_p^2$ |
| Regression | $b^2 \left[\sum\sum X_{ij}^2 - \frac{(\Sigma\Sigma X_{ij})^2}{N}\right]$ | 1 | |
| About regression | Difference | $k - 2$ | |
| *Subtotal* | $\sum \frac{T_{Y_{i+}}^2}{n_i} - \frac{T_{Y++}^2}{N}$ | $k - 1$ | |

we test whether or not the means for the groups are located on a straight line. We shall compare the variance within groups with the variance of the deviations of the group means from the estimated regression line. The appropriate mean squares for the $F$ ratio are obtained from Table 11-1. The sum of squares of the deviations of the group-mean $Y$ values from the estimated regression line is obtained by subtracting the "Regression" sum of squares from the "Total."

The test for linearity is the ratio of the mean square about regression to the mean square within groups, which, if the assumptions of Sec. 11-4 concerning normality are true and if the hypothesis is true, has an $F(k - 2, N - k)$ sampling distribution. The analysis for this example is shown in Table 11-2. $F = 30.3/108.1 = .28$ is compared with $F_{.95}(2,8) = 4.46$ and declared not significant. Therefore we accept the hypothesis that the regression curve is a straight line.

The sums of squares in Table 11-2 can be extracted from the previous

**Table 11-2**

| | Sum of squares | Degrees of freedom | Mean square | F ratio |
|---|---|---|---|---|
| *Total* | 5,266.7 | 11 | | |
| Within groups | 864.6 | 8 | 108.1 | |
| Regression | 4,341.5 | 1 | | |
| About regression | 60.6 | 2 | 30.3 | .28 |
| *Subtotal* | 4,402.1 | 3 | | |

computations but are indicated completely for ease in reference:

$$\frac{T_{Y++}^2}{N} = 240{,}833.3$$

$$\sum \frac{T_{Y_{i+}}^2}{n_i} = \frac{(365)^2}{3} + \frac{(525)^2}{4} + \frac{(295)^2}{2} + \frac{(515)^2}{3} = 245{,}235.41$$

$$\Sigma\Sigma Y_{ij}^2 = 246{,}100$$

$$\sum\sum Y_{ij}^2 - \frac{T_{Y++}^2}{N} = 5{,}266.7$$

$$\sum \left(\frac{T_{Y_{i+}}^2}{n_i}\right) - \frac{T_{Y++}^2}{N} = 4{,}402.1$$

$$b^2 \left[\sum\sum X_{ij}^2 - \frac{(\Sigma\Sigma X_{ij})^2}{N}\right] = (5.029)^2 \times 171.67 = 4{,}341.5$$

## 11-6 CORRELATION PROBLEMS

In a correlation problem we sample from a population, observing two measurements on each individual in the sample. This contrasts with a purely regression problem, where the sample is chosen with preassigned $X$ values. A large part of the classical study of this subject is based upon the assumption that the distribution of values $(X,Y)$ is a *two-variable normal* distribution. In appearance this distribution surface is bellshaped. The distribution of $Y$ values for any fixed $X$ is normal, and the distribution of $X$ values for any fixed $Y$ is also normal. The regression curve of $Y$ on $X$ and the regression curve of $X$ on $Y$ are both straight lines with *homoscedasticity* (constant variance) for $Y$ for each value of $X$, and for $X$ for each value of $Y$. The regression lines intersect at a point which has as its $X$ coordinate $\mu_x$, the mean of the $X$ measurements of all individuals in the population, and as its $Y$ coordinate $\mu_y$, the mean of the $Y$ values of all individuals in the population. Figure 11-4 illustrates such a distribution surface.

We shall write the equation of the regression line of $Y$ on $X$ in the form

$$\frac{\mu_{y\cdot x} - \mu_y}{\sigma_y} = \rho \frac{X - \mu_x}{\sigma_x} \tag{10}$$

Here $\sigma_y$ and $\sigma_x$ are the standard deviations of all the $Y$ and $X$ measurements, respectively, in the population. $\rho$ is defined as the population *correlation coefficient* and is equal to

$$\rho = \sqrt{BB'}$$

where $B$ = regression coefficient from regression line of $Y$ on $X$
$B'$ = regression coefficient from regression line of $X$ on $Y$

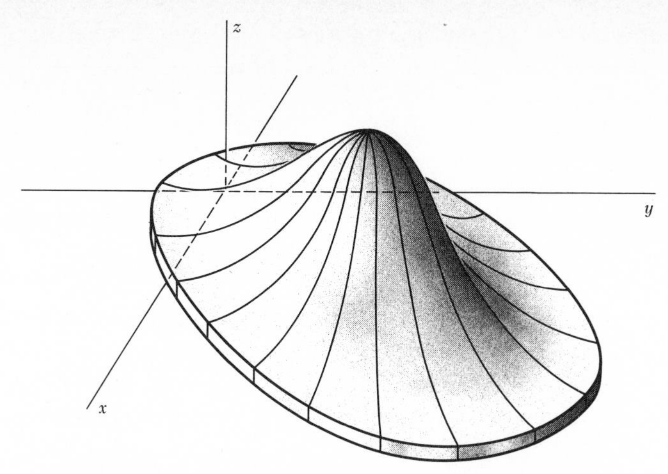


**Figure 11-4** Possible limiting two-dimensional "histogram" for weights and heights of a very large number of men.

The equation of the regression line of $X$ on $Y$ is

$$\frac{\mu_{x\cdot y} - \mu_x}{\sigma_x} = \rho \frac{Y - \mu_y}{\sigma_y} \tag{11}$$

Equation (10) is satisfied by the values $X = \mu_x$ and $\mu_{y\cdot x} = \mu_y$, and the line goes through the point which has the mean $X$ and the mean $Y$ for its coordinates. Similarly, the regression of $X$ on $Y$ goes through the same point. The lines coincide only if $\rho = +1$ or if $\rho = -1$.

The population values are estimated by sample values as follows. For $\mu_x$ use $\bar{X}$, for $\mu_y$ use $\bar{Y}$, for $\sigma_x$ use $s_x$, for $\sigma_y$ use $s_y$, and for $\rho$ use the sample correlation coefficient $r$ defined by

$$r = b\frac{s_x}{s_y} \qquad \text{or} \qquad r = \sqrt{bb'}$$

where $b$ is the estimate of $B$ defined on page 195 and $b'$ is the estimate of $B'$ found by interchanging the $X$ and $Y$ values in the formula for $b$. Figure 11-6 shows a scatter diagram with the two estimated regression lines drawn.

Equations (10) and (11) with the sample values substituted for the parameters are the same as those obtained by estimating the two regression lines separately by the methods of Sec. 11-3. One advantage of a correlation situation is that it allows us to estimate both equations from one sample. A serious disadvantage is the rare occurrence of populations which have bivariate normal distributions, that is, populations having both the distribution of $Y$ values for given $X$ and the distribution of $X$ values for given $Y$ normal. Another disadvantage lies in the sampling procedure, which requires that neither variable be controlled.

**Test for independence.** If the assumptions of normality are satisfied it is possible to use the observed value of $r$ to test for independence. If the two variables are independent the regression curves are horizontal and vertical straight lines. This implies that the parameter $\rho$ is equal to zero. If $r$, which is an estimate of $\rho$, is close to zero we shall say that there is not sufficient reason to doubt the independence, while if $r$ is far from zero we shall reject the hypothesis that the two variables are independent. Table A-30*a* gives several percentiles of the sampling distribution of $r$ under the assumption that $X$ and $Y$ have independent normal distributions. For example, if the observed $r$ in a sample of 50 observations is larger than .279 or less than $-.279$ we should reject the hypothesis that $\rho = 0$ at the 5 per cent level of significance. For regression problems the hypothesis of $B = 0$ can be tested by the $t$ test as illustrated in Sec. 11-4. This can also be done in correlation problems.

**Confidence interval for $\rho$.** Table A-30*a* should be used only to test the hypothesis that $\rho = 0$, and should not be used to test either the hypothesis that $\rho$ equals some number not zero or the hypothesis that two correlation coefficients are equal. For these hypotheses we can use the variable $\mathbf{z} = .5 \ln \frac{1+r}{1-r}$, which has a nearly normal sampling distribution with mean approximately $.5 \ln \frac{1+\rho}{1-\rho}$ and standard deviation approximately $1/\sqrt{N-3}$. In many practical problems it is more informative to deal with the regression coefficients in the manner previously studied. Values of $.5 \ln \frac{1+r}{1-r}$ are given in Table A-30*b*.

If a two-sided test with $\alpha = .05$ or if .95 confidence limits are desired the chart in Table A-27 may be used in place of this approximation.

EXAMPLE 11-1. Suppose for a sample of $N = 25$ observations we observe $r = .2$. To use the approximation we refer to Table A-30*b* to find $\mathbf{z} = .20273$. The standard deviation is $1/\sqrt{N-3} = 1/\sqrt{22} = .213$, so we may say with 95 per cent confidence that $.5 \ln \frac{1+\rho}{1-\rho}$ is between $.203 - (1.96 \times .213) = -.214$ and $.203 + (1.96 \times .213) = .620$. Using the table in reverse, we find corresponding to $-.214$ and .620 the limits $-.211$ and .551 for $\rho$. To solve the same problem by use of Table A-27 we enter the chart with $r = .2$ and read the heights of the two $N = 25$ curves above $r = .2$ on the horizontal scale as $\rho_1 = -.21$ and $\rho_2 = .54$. These two numbers are 95 per cent confidence limits for the population parameter $\rho$ and differ little from those obtained from the approximation. A test of any hypothetical value of $\rho$ can be made by either method. In this example any hypothetical $\rho$ between $-.21$ and .54 would be accepted at the 5 per cent level of significance.

Pairs of measurements $(X_1,Y_1)$, $(X_2,Y_2)$, $(X_3,Y_3)$, . . . , $(X_N,Y_N)$ have been obtained and we wish to describe the interrelationship as described by straight lines. As an example, consider the heights and weights of 356 men. The listing of the original data will not be given here, but the scatter plot is given in Fig. 11-5, and the frequencies as grouped to the nearest 2 inches and to the nearest 10 pounds are given in Table 11-3. Computations for problems of this size are usually done on a desk calculator or, if one is available, on an electronic digital computer. The linear regression of weight $Y$ on height $X$ is given by

$$\bar{Y}_x = \bar{Y} + b(X - \bar{X})$$

and that for height on weight is given by

$$\bar{X}_y = \bar{X} + b'(Y - \bar{Y})$$

As indicated in the previous section, the correlation coefficient $r$ may be obtained from $r = \sqrt{bb'}$ or from the direct computational form

$$r = \frac{\Sigma XY - \Sigma X \Sigma Y/N}{(N-1)s_x s_y}$$

For the data of Fig. 11-5 basic quantities were computed as follows:

$$\Sigma X = 23{,}814 \qquad \Sigma Y = 60{,}730$$
$$\Sigma X^2 = 1{,}595{,}628 \qquad \Sigma Y^2 = 10{,}522{,}500$$
$$\Sigma XY = 4{,}075{,}360$$

**Table 11-3** Height-weight frequency distribution

| $Y$ VARIABLE (WEIGHT, POUNDS) | $X$ VARIABLE (HEIGHT, INCHES) 60 | 62 | 64 | 66 | 68 | 70 | 72 | 74 |
|---|---|---|---|---|---|---|---|---|
| 230 | | | | | 1 | | 1 | 3 |
| 220 | | | | | | 4 | 4 | 3 |
| 210 | | | | | 5 | 1 | 2 | 1 |
| 200 | | 2 | 1 | 3 | 9 | 1 | 7 | 1 |
| 190 | | 1 | 3 | 8 | 16 | 3 | 5 | |
| 180 | | 1 | 5 | 8 | 15 | 12 | 1 | |
| 170 | | 2 | 8 | 18 | 26 | 8 | 1 | |
| 160 | | | 19 | 40 | 20 | 4 | | |
| 150 | | 5 | 15 | 26 | 9 | 2 | | |
| 140 | 1 | 4 | 6 | 5 | 1 | | | |
| 130 | 2 | 3 | 1 | 1 | | | | |
| 120 | 1 | 1 | | | | | | |

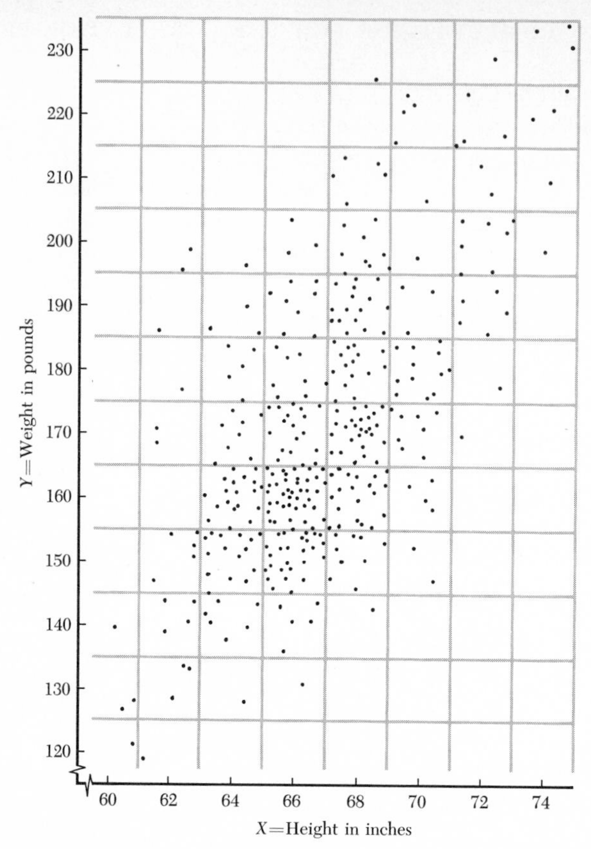


**Figure 11-5** Plot or scattergram of weights and heights of men. ($N = 356$)

and

$$\bar{X} = 66.89 \qquad \bar{Y} = 170.59$$
$$s_x = 2.72 \qquad s_y = 21.40$$
$$b' = .0795 \qquad b = 4.913$$

The two regression lines are

$$\bar{Y}_x = 170.59 + 4.913(X - 66.89)$$

and

$$\bar{X}_y = 66.89 + .0795(Y - 170.59)$$

These two lines are depicted in Fig. 11-6.

## 11-8 REGRESSION IN TWO POPULATIONS

The concepts described previously in this chapter can be extended immediately to two populations. We assume that the regression curves are

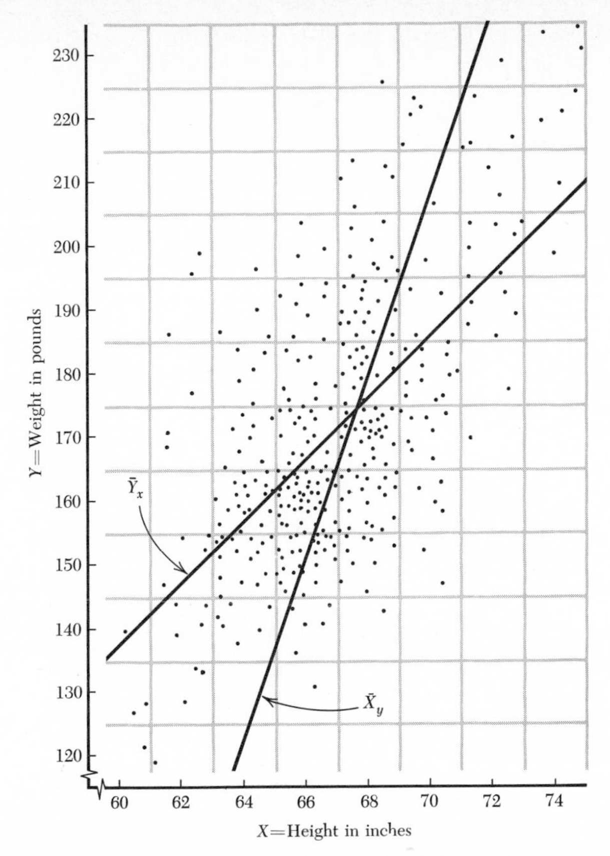


**Figure 11-6** Scattergram of weights and heights of the same men with "least-square" lines drawn for estimating mean weight from height, $\bar{Y}_x$, and estimating mean height from weight, $\bar{X}_y$.

straight and are given by the linear formulas

$$\mu_{y_1 \cdot x} = A_1 + B_1(X - \bar{X}_1) \qquad \text{and} \qquad \mu_{y_2 \cdot x} = A_2 + B_2(X - \bar{X}_2)$$

and we compare the slopes $B_1$ and $B_2$ or compare the mean values for a specified $X_0$.

**Comparison of slopes.** Represent the independent and dependent observations made in the random sample of size $n_1$ from the first population by

$$(X_{11},Y_{11}), (X_{12},Y_{12}), (X_{13},Y_{13}), \ldots, (X_{1n_1},Y_{1n_1})$$

and the observations in the random sample of size $n_2$ from the second population by

$$(X_{21},Y_{21}), (X_{22},Y_{22}), (X_{23},Y_{23}), \ldots, (X_{2n_2},Y_{2n_2})$$

Unbiased estimates $b_1$ of $B_1$ and $b_2$ of $B_2$ are given by appropriate use of formula (2), Sec. 11-3, as

$$b_1 = \frac{\Sigma X_{1i}Y_{1i} - \Sigma X_{1i}\Sigma Y_{1i}/N}{\Sigma X_{1i}^2 - (\Sigma X_{1i})^2/N}$$

and

$$b_2 = \frac{\Sigma X_{2i}Y_{2i} - \Sigma X_{2i}\Sigma Y_{2i}/N}{\Sigma X_{2i}^2 - (\Sigma X_{2i})^2/N}$$

Define $s_{x_1}{}^2$ and $s_{y_1}{}^2$ as the variances of the $X$'s and $Y$'s, respectively, in the first sample and $s_{x_2}{}^2$ and $s_{y_2}{}^2$ as the corresponding variances for the second sample. Then, if the regressions are linear,

$$s^2_{y_1 \cdot x} = \frac{n_1 - 1}{n_1 - 2}(s_{y_1}{}^2 - b_1{}^2 s_{x_1}{}^2)$$

and

$$s^2_{y_2 \cdot x} = \frac{n_2 - 1}{n_2 - 2}(s_{y_2}{}^2 - b_2{}^2 s_{x_2}{}^2)$$

are unbiased estimates of the variances about the population regression lines. If we assume that these population variances are equal, then the pooled variance

$$s^2_{y \cdot x \cdot p} = \frac{(n_1 - 2)s^2_{y_1 \cdot x} + (n_2 - 2)s^2_{y_2 \cdot x}}{n_1 + n_2 - 4} \tag{1}$$

is an unbiased estimate of this common variance.

The mean and variance of $b_1 - b_2$ are given by

$$\text{mean } (b_1 - b_2) = B_1 - B_2$$

$$\text{var } (b_1 - b_2) = \frac{\sigma^2_{y_1 \cdot x}}{(n_1 - 1)s_{x_1}{}^2} + \frac{\sigma^2_{y_2 \cdot x}}{(n_2 - 1)s_{x_2}{}^2}$$

If the two population variances are equal but are unknown, the variance of $b_1 - b_2$ is estimated by

$$s^2_{y \cdot x \cdot p}\left(\frac{1}{(n_1 - 1)s_{x_1}{}^2} + \frac{1}{(n_2 - 1)s_{x_2}{}^2}\right) \tag{2}$$

If the sample sizes $n_1$ and $n_2$ are large, or if the $Y$ values in the two populations are approximately normally distributed for each $X$, then $b_1 - b_2$ is approximately normally distributed, and a $100(1 - \alpha)$ per cent-level confidence-interval estimate for $b_1 - b_2$ is

$$b_1 - b_2 + z_{\frac{1}{2}\alpha}\sqrt{\frac{\sigma^2_{y_1 \cdot x}}{(n_1 - 1)s_{x_1}{}^2} + \frac{\sigma^2_{y_2 \cdot x}}{(n_2 - 1)s_{x_2}{}^2}} < B_1 - B_2$$

$$< b_1 - b_2 + z_{1-\frac{1}{2}\alpha}\sqrt{\frac{\sigma^2_{y_1 \cdot x}}{(n_1 - 1)s_{x_1}{}^2} + \frac{\sigma^2_{y_2 \cdot x}}{(n_2 - 1)s_{x_2}{}^2}} \tag{3}$$

If for each $X$ the $Y$ populations are normal with equal but unknown variance, then substitution of $s_{y \cdot x \cdot p}$ for each $\sigma_{y \cdot x}$ and $t_{\frac{1}{2}\alpha}$ and $t_{1-\frac{1}{2}\alpha}$ for $z_{\frac{1}{2}\alpha}$ and $z_{1-\frac{1}{2}\alpha}$ also provides a confidence interval with level $1 - \alpha$.

To test the hypothesis $H: B_1 - B_2 = 0$, given equal variances as above, we can define

$$z = \frac{b_1 - b_2}{\sigma_{y \cdot x}\sqrt{\dfrac{1}{(n_1 - 1)s_{x_1}{}^2} + \dfrac{1}{(n_2 - 1)s_{x_2}{}^2}}} \tag{4}$$

or, if $\sigma_{y \cdot x}$ is unknown,

$$t = \frac{b_1 - b_2}{s_{y \cdot x \cdot p}\sqrt{\dfrac{1}{(n_1 - 1)s_{x_1}{}^2} + \dfrac{1}{(n_2 - 1)s_{x_2}{}^2}}} \tag{5}$$

and reject if $z$ or $t$ is significantly different from zero ($df = n_1 + n_2 - 4$).

As an example, suppose we have two samples, I and II, with the observations shown in Table 11-4 and plotted in Fig. 11-7. The basic computations for the two samples are obtained as explained in Sec. 11-2 and listed below:

| | $n$ | $\bar{X}$ | $\bar{Y}$ | $s_x{}^2$ | $s_y{}^2$ | $b$ | $s_{y \cdot x}^2$ |
|---|---|---|---|---|---|---|---|
| *Sample* I | 4 | 3.75 | 11.75 | 2.75/3 | 4.75/3 | 1 | 2/2 |
| *Sample* II | 4 | 1.75 | 6.50 | 2.75/3 | 5/3 | .545 | 4.182/2 |

The value of $b_1 - b_2 = .455$ is an unbiased estimate of $B_1 - B_2$, and the variance of this estimate is estimated by

$$\frac{2 + 4.182}{4 + 4 - 4}\left(\frac{1}{2.75} + \frac{1}{2.75}\right) = 1.124$$

**Table 11-4** Two samples with two variables

| | SAMPLE I | | SAMPLE II | |
|---|---|---|---|---|
| | $X$ | $Y$ | $X$ | $Y$ |
| | 4 | 12 | 1 | 6 |
| | 3 | 12 | 2 | 5 |
| | 3 | 10 | 3 | 8 |
| | 5 | 13 | 1 | 7 |
| *Total* | 15 | 47 | 7 | 26 |

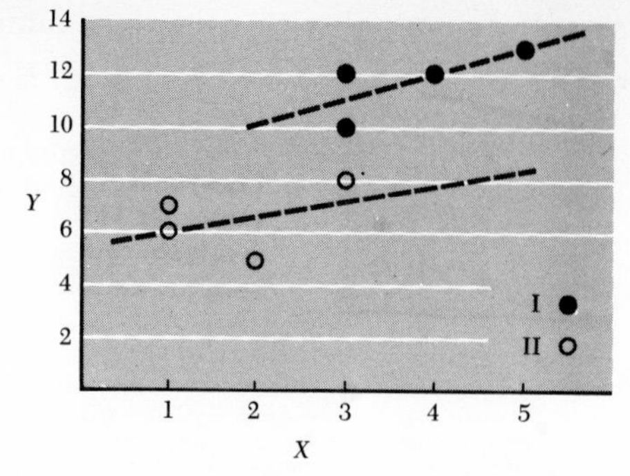


**Figure 11-7** Plot of $X$ and $Y$ values for two groups of four individuals each with least-square lines drawn for each group.

Here we have used formulas (1) and (2) above. We now use formula (5) to obtain the 95 per cent confidence interval

$$.455 - 2.776\sqrt{1.124} < B_1 - B_2 < .455 + 2.776\sqrt{1.124}$$

$$-2.5 < B_1 - B_2 < 3.4$$

Also,

$$t = \frac{.455}{\sqrt{1.124}} = .43$$

with 4 degrees of freedom; thus a hypothesis $H: B_1 - B_2 = 0$ would not be rejected. This agrees with the visual impression in Fig. 11-7. Of course, the sample sizes are so small that we have little support for equality of $B_1$ and $B_2$.

**Comparison of regression lines.** If a value, say, $X_0$, of the independent variable is specified, the heights of the regression lines are estimated by

$$\bar{Y}_1 + b_1(X_0 - \bar{X}_1) \qquad \text{and} \qquad \bar{Y}_2 + b_2(X_0 - \bar{X}_2)$$

The difference $d = \bar{Y}_1 + b_1(X_0 - \bar{X}_1) - \bar{Y}_2 - b_2(X_0 - \bar{X}_2)$ is an unbiased estimate of $\mu_{y_1} - \mu_{y_2}$ at $X_0$ and has variance

$$\text{var } d = \sigma^2_{y_1 \cdot x}\left[\frac{1}{n_1} + \frac{(X_0 - \bar{X}_1)^2}{(n_1 - 1)s_{x_1}{}^2}\right] + \sigma^2_{y_2 \cdot x}\left[\frac{1}{n_2} + \frac{(X_0 - \bar{X}_2)^2}{(n_2 - 1)s_{x_2}{}^2}\right] \tag{6}$$

Also, $d$ is approximately normally distributed, and thus if $\sigma_{y_1 \cdot x}$ and $\sigma_{y_2 \cdot x}$ are known, a confidence-interval estimate of the difference in means can be given. If the variances are equal but unknown and the $Y$ distributions are normal for each $X$, then use of $s_{y \cdot x \cdot p}$ and the $t$ distribution provides a confidence interval.

**Analysis-of-variance table.** The data relating to the slopes for the two populations can be presented in an analysis-of-variance table including the sum of squares for the contrast $b_1 - b_2$ with 1 degree of freedom and $s^2_{y \cdot x \cdot p}$ with $n_1 + n_2 - 4$ degrees of freedom. Sums of squares and cross products are computed for each of the two groups separately and for both groups together and are entered in the shaded area of the top three lines in Table 11-5. The sums of the entries for the two groups are entered in the

**Table 11-5**

| | $\Sigma(X - \bar{X})^2$ | $\Sigma(X - \bar{X})(Y - \bar{Y})$ | $\Sigma(Y = \bar{Y})^2$ | $\Sigma y'^2$ | |
|---|---|---|---|---|---|
| *Total* | 13.50 | 25.25 | 64.875 | 17.648 | |
| Group 1 | 2.75 | 2.75 | 4.750 | 2.000 | |
| Group 2 | 2.75 | 1.50 | 5.000 | 4.182 | |
| | | | | | 6.182 |
| *Within groups* | 5.50 | 4.25 | 9.750 | 6.466 | |

bottom line of the table. Next, the four entries in the column headed $\Sigma y'^2$ are entered by use of formula (4′) in Sec. 11-3. For example,

$$64.875 - \frac{(25.25)^2}{13.50} = 17.648$$

The analysis-of-variance table is then constructed from these values, which are entered in the shaded areas of Table 11-6. The remainder is computed from $17.648 - 6.466 = 11.182$, and the sum of squares for $b_1 - b_2$ is computed as $6.466 - 6.182 = .284$

The sum of squares for $b_1 - b_2$, .284 in this example, can be obtained from the formula

$$\frac{(b_1 - b_2)^2}{\dfrac{1}{(n_1 - 1)s_{x_1}{}^2} + \dfrac{1}{(n_2 - 1)s_{x_2}{}^2}}$$

Also $s^2_{y \cdot x \cdot p} = 1.545$. Thus the $F$ test is equivalent to the $t$ test of the hypothesis $H: B_1 = B_2$ as presented above. The entry labeled "Remainder" will be used in Chap. 12.

**Table 11-6** Analysis of variance

| | *Sum of squares* | *Degrees of freedom* | *Mean square* | *F ratio* |
|---|---|---|---|---|
| *Total* | 17.648 | | | |
| Remainder | 11.182 | 1 | | |
| Both groups | 6.182 | 4 | 1.545 | |
| $b_1 - b_2$ | .284 | 1 | .284 | $F = .184$ |
| *Within groups* | 6.466 | 5 | | |

## 11-9 *MULTIPLE REGRESSION

It is sometimes desirable to describe the joint relationship of a single $Y$ variable to several $X$ variables. For example, weight gain may depend on original weight, amount of food eaten, and perhaps several other variables. A measure of the extent of atherosclerosis in experimental animals may depend on such factors as the initial weight of the animals, the amount of food intake, and the cholesterol dosage administered to each animal. If the $Y$ variable is well described by the other variables we will want to know the extent of this dependence and the actual regression equation.

As in the example for a single $X$ in Sec. 11-3, where $A$ and $B$ were estimated in the equation $\bar{Y}_x = A + B(X - \bar{X})$, it is possible to compute a set of coefficients $b_i$ in the expression

$$\bar{Y}_x = b_0 + b_1X_1 + b_2X_2 + \cdots + b_kX_k$$

The coefficients are determined to provide the minimum sum of squares of differences between the observed $Y$'s and this linear combination of the $X$ values.

Generally, if we let the observations on the first individual be represented by

$$Y_1, X_{11}, X_{21}, X_{31}, \ldots, X_{k1}$$

those on the second individual by

$$Y_2, X_{12}, X_{22}, X_{32}, \ldots, X_{k2}$$

and so on to the $n$th, or last, individual,

$$Y_n, X_{1n}, X_{2n}, X_{3n}, \ldots, X_{kn}$$

then the required solution for $b_0, b_1, \ldots, b_n$ is obtained by solving for these coefficients in the following set of equations:

$$\begin{array}{l} nb_0 + b_1\Sigma X_{1i} + b_2\Sigma X_{2i} + \cdots + b_k\Sigma X_{ki} = \Sigma Y_i \\ b_0\Sigma X_{1i} + b_1\Sigma X_{1i}^2 + b_2\Sigma X_{1i}X_{2i} + \cdots + b_k\Sigma X_{1i}X_{ki} = \Sigma X_{1i}Y_i \\ b_0\Sigma X_{2i} + b_1\Sigma X_{2i}X_{1i} + b_2\Sigma X_{2i}^2 + \cdots + b_k\Sigma X_{2i}X_{ki} = \Sigma X_{2i}Y_i \\ \cdots\cdots\cdots\cdots\cdots\cdots\cdots\cdots\cdots\cdots \\ b_0\Sigma X_{ki} + b_1\Sigma X_{ki}X_{1i} + b_2\Sigma X_{ki}X_{2i} + \cdots + b_k\Sigma X_{ki}^2 = \Sigma X_{ki}Y_i \end{array}$$

The computations, although not theoretically complex, are difficult even for $k$ as small as 4 or 5. The arithmetic is not only tedious, but it is hazardous, since it is difficult to detect arithmetic errors or to be sure that errors resulting from the use of a limited number of significant digits are not building up and destroying the accuracy of the result. Today most university or research computing facilities have computer programs for carrying out the computations for a multiple-regression model.

The material discussed earlier in this chapter applies to the case where $k = 1$. If there are two $X$ variables ($k = 2$), the geometric model is in three dimensions and consequently is still at a level where it can be visualized. The least-squares formula provides a best-fitting *plane* to the data. The equations above reduce to three, giving values of $b_0$, $b_1$, and $b_2$. In the example below the $Y$ variable is weight, the $X_1$ variable is height, and the $X_2$ variable is age. The values given for $Y$ and $X_1$ are those chosen to illustrate the case with a single variable earlier in the chapter, and Fig. 11-1 shows the results of projecting the points in three dimensions onto one of the coordinate planes, the $Y,X_1$ plane. The observed values are recorded in the first three columns of Table 11-5. The calculations giving the least-squares equations and their solution are as follows:

$$\begin{array}{llll} n = 12 & \Sigma X_{1i} = 766 & \Sigma X_{2i} = 350 & \Sigma Y_i = 1{,}700 \\ & \Sigma X_{1i}^2 = 49{,}068 & \Sigma X_{1i}X_{2i} = 22{,}360 & \Sigma X_{1i}Y_i = 109{,}380 \\ & & \Sigma X_{2i}^2 = 10{,}900 & \Sigma X_{2i}Y_i = 49{,}400 \end{array}$$

$$\Sigma Y_i^2 = 246{,}100 \qquad \bar{Y} = 141.667 \qquad \bar{X}_1 = 63.833 \qquad \bar{X}_2 = 29.167$$

The three equations to be solved for $b_0$, $b_1$, and $b_2$ are

$$\begin{aligned} 12b_0 + 766b_1 + 350b_2 &= 1{,}700 \\ 766b_0 + 49{,}068b_1 + 22{,}360b_2 &= 109{,}380 \\ 350b_0 + 22{,}360b_1 + 10{,}900b_2 &= 49{,}400 \end{aligned}$$

Optimum computing techniques will not be discussed here. We shall solve the equations by dividing each equation by the coefficient of $b_0$ and subtracting to get two equations in $b_1$ and $b_2$. These in turn can be divided by the coefficients of $b_1$ and subtracted to give a single equation in $b_2$. The solution for $b_2$ is substituted in the equation containing both $b_1$ and $b_2$, and finally these two values are substituted in one of the original equations,

**Table 11-5**

| OBSERVED VALUES | | | | |
|---|---|---|---|---|
| $Y_i$ | $X_{1i}$ | $X_{2i}$ | *Est.* $Y = b_0 + b_1X_1 + b_2X_2$ | *Obs.* $Y_i$ −*est.* $Y_i$ |
| 110 | 60 | 40 | 117.897 | −7.897 |
| 135 | 60 | 20 | 125.887 | 9.113 |
| 120 | 60 | 30 | 121.892 | −1.892 |
| 120 | 62 | 20 | 136.030 | −16.030 |
| 140 | 62 | 30 | 132.035 | 7.965 |
| 130 | 62 | 40 | 128.040 | 1.960 |
| 135 | 62 | 20 | 136.030 | −1.030 |
| 150 | 64 | 30 | 142.179 | 7.821 |
| 145 | 64 | 30 | 142.179 | 2.821 |
| 170 | 70 | 20 | 176.604 | −6.604 |
| 185 | 70 | 30 | 172.609 | 12.391 |
| 160 | 70 | 40 | 168.614 | −8.614 |
| | | | | .004 *Total* |

which is then solved for $b_0$. For the example, we divide by the coefficients of $b_0$ to get

$$\begin{aligned} b_0 + 63.83333b_1 + 29.16667b_2 &= 141.66667 \\ b_0 + 64.05744b_1 + 29.19060b_2 &= 142.79373 \\ b_0 + 63.88571b_1 + 31.14286b_2 &= 141.14286 \end{aligned}$$

Then, subtracting the second and third equation from the first, we obtain

$$\begin{aligned} -.22411b_1 - .02393b_2 &= -1.12706 \\ -.05238b_1 - 1.97619b_2 &= .523810 \end{aligned}$$

Dividing by the coefficients of $b_1$ gives

$$\begin{aligned} b_1 + .10678b_2 &= 5.02905 \\ b_1 + 37.72795b_2 &= 10.00019 \end{aligned}$$

Subtracting gives

$$-37.62117b_2 = 15.02924 \qquad \text{or} \qquad b_2 = -.39949$$

Substituting this value gives

$$b_1 = 5.02905 - .10678b_2 = 5.02905 - .10678(-.39949) = 5.07171$$

Finally,

$$\begin{aligned} b_0 &= 141.6667 - 63.83333b_1 - 29.16667b_2 \\ &= 141.66667 - 63.83333(5.07171) - 29.16667(-.39949) = -170.426 \end{aligned}$$

Thus the estimated regression formula is

$$\bar{Y}_x = -170.426 + 5.07171X_1 - .39949X_2$$

For given $X_{1i}$ and $X_{2i}$ we may now use the formula to give estimated values of the dependent variable, say, $Y'_i$, which would be on the regression plane. The differences between the observed $Y_i$ and the regression estimates $Y'_i$ would measure the variability $\sigma_{y \cdot x}$ away from the plane. For example, if $X_1 = 60$ and $X_2 = 40$, then

$$Y' = -170.426 + 5.07171 \times 60 - .39949 \times 40 = 117.897$$

The differences for all the points are also given in Table 11-5.

Analogous to the case for $k = 1$, the differences between the observed $Y_i$ and the estimated $Y'_i$ provide indices of the goodness of fit of the plane. Thus a variance of $Y$'s given the values of $X_1$ and $X_2$ is estimated by

$$s^2_{y \cdot x_1 x_2} = \frac{\Sigma(Y_i - Y'_i)^2}{N - 1 - k}$$

which for the example is

$$s^2_{y \cdot x_1 x_2} = \frac{814.7704}{12 - 1 - 2} = 90.53$$

Also, the ratio of $s^2_{y \cdot x_1 x_2}$ to $s_y^2$ is used to define a multiple-correlation coefficient by

$$R^2 = 1 - \frac{s^2_{y \cdot x_1 x_2}(N - 1 - k)}{s_y^2(N - 1)} = 1 - \frac{814.7704}{5266.6667} = .8453$$

As in the case for $k = 1$, the multiple-correlation coefficient cannot be greater than unity (it will equal unity if the fit is exact, that is, if the points are on a plane). As defined, it cannot be negative. Also, if the fitted plane is horizontal, that is, if there is a random scatter of $Y$ values without relationship to the $X$ values, then $R$ will be close to zero. Consequently, as in the case for $k = 1$, a critical region consisting of large values of $R$ can be used to test for significance.

## OTHER ESTIMATES OF CORRELATION 11-10

Suppose there is a linear relationship between two variables and a number of observations are made on those variables. Let us plot the results on a graph and observe, if there is a positive relationship, that if we divide the data in half with a horizontal line and then in half by a vertical line, there will be more observations in the upper right and lower left quadrants. There will be more in the upper left and lower right if there is an inverse relationship.

If we designate by $n_1$, $n_2$, $n_3$, $n_4$ the number of points in each quadrant (for example, see Fig. 11-8),

| $n_2$ | $n_1$ |
|---|---|
| $n_3$ | $n_4$ |

we first compute

$$\frac{n_1 + n_3}{n_1 + n_2 + n_3 + n_4} \quad \text{or} \quad \frac{n_2 + n_4}{n_1 + n_2 + n_3 + n_4}$$

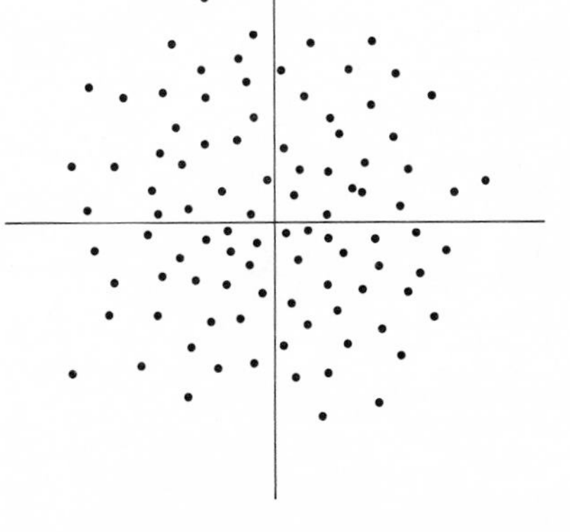

**Figure 11-8** Scatter diagram with median lines drawn.

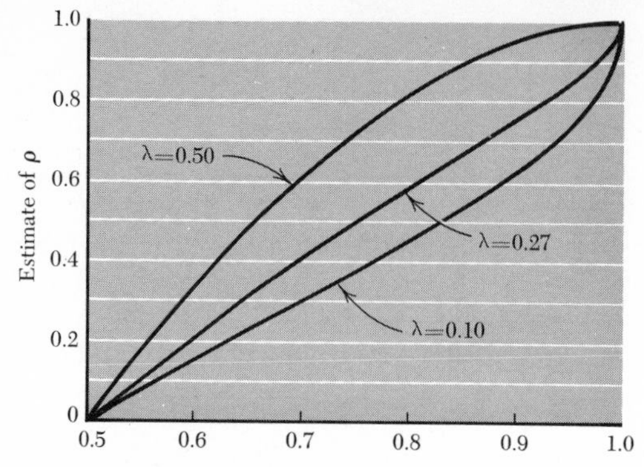


**Figure 11-9** Chart for the tetrachoric $r$ estimate of $\rho$.

whichever is larger, and refer to Fig. 11-9. Enter Fig. 11-9 at the bottom; read up to the curve marked $\lambda = .50$ and left to the scale marked "Estimate of $\rho$." The correlation is positive if the first ratio is used and negative if the second is used. The curve for $\lambda = .50$ is used, since the vertical line divides the observations into two equal groups.

The use of the graph for obtaining the estimate of $\rho$ avoids the need of substitution in a formula. If the observations are from a normal bivariate population, this estimate of $\rho$, called *tetrachoric* $r$, has efficiency of .40 compared with the efficient estimate $r$.

A similar estimate with efficiency equal to .52 can be obtained as follows:

1. Retain the observations whose $X$ is among the top or bottom 27 per cent and discard the remainder.

2. Separate these observations into upper and lower halves with respect to the $Y$ value.

Estimate $\rho$ as before, except that the line $\lambda = .27$ will be used in place of the line $\lambda = .50$. Any value of $\lambda$ from zero to .50 may be used, but the estimate of highest efficiency is obtained by using $\lambda = .27$. If the observations are plotted on a graph this estimate can be obtained by drawing lines as indicated in Fig. 11-10 and counting the dots in the four corner sections.

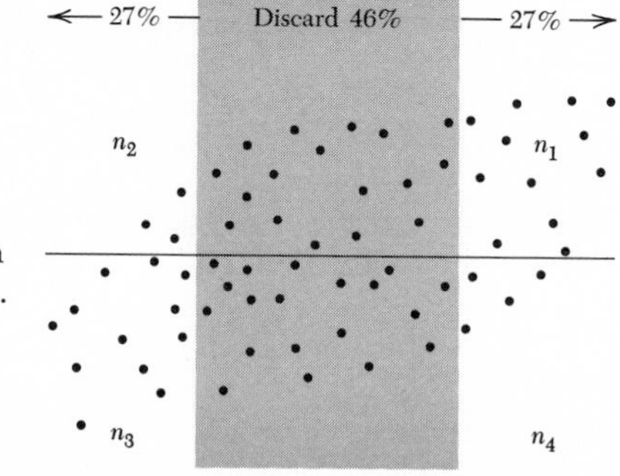


**Figure 11-10** Scatter diagram showing largest and smallest 27%.

## GLOSSARY

correlation
homoscedasticity
independence
least squares
regression
scatter diagram
standard error of estimate
tetrachoric $r$
variance of residuals

## DISCUSSION QUESTIONS

**1.** What are some methods of graphically representing data for two variables?

**2.** Suppose a school class has examinations at the beginning and the end of the school year. What is meant by "regression of final grades on beginning grades"? What is meant by "regression of beginning grades on final grades"? Which of these would be more useful in practice? Might these regressions coincide?

**3.** What estimate might be referred to in the expression "standard error of estimate"? Is the standard error of estimate a parameter or a statistic?

**4.** What is the meaning of a point on a sample regression line?

**5.** Is the regression in the population always a straight line? If not, give an example of a population where it is not.

**6.** What assumptions are made about the population for the applications of the procedures in this chapter?

**7.** Describe a sampling experiment to illustrate correlation.

**8.** Explain the difference between regression and correlation problems. Can a correlation problem also be a regression problem? Can a regression problem also be a correlation problem?

**9.** Label the following examples as regression- or correlation-type problems. Discuss methods of obtaining a sample and procedures of analysis in each case. Explain the interpretation of possible results. It is desired to study:

(*a*) The connection between IQ and weight of fifteen-year-old girls

(*b*) The connection between the velocity of the Willamette River and its depth at various points

(*c*) The connection between the amount of winter snow and the barley yield for some locality

(*d*) The connection between tensile strength and hardness of aluminum

(*e*) The connection between milk production and number of hours of light (either sunlight or artificial)

(*f*) The connection between the size of brains and success in life

(*g*) The effect of an antihistamine drug upon length of time it takes to recover from colds

(*h*) The effect upon the size of fish of dumping waste material from mills into a river

(*i*) The relation between age and weight of trout

(*j*) The relationship between a student's financial status and the amount of dating

(*k*) The connection between city size and amount of crime

(*l*) The relationship between the degree of body consciousness of an individual and the elapsed time between hospital discharge and the recurrence of active tuberculosis

## CLASS EXERCISES

**1.** Collect the heights $X$ and weights $Y$ of boys in a class and plot the points. Find the least-squares line and draw it on the same graph.

**2.** Repeat Exercise 1 for the girls in a class.

## PROBLEMS

*Section* 11-3

**1.** (*a*) Plot the six points having the coordinates $(X,Y)$ as follows: (0,0), (0,1), (1,1), (1,3), (2,2), (2,3). Draw a "close-fitting" line by eye. If the equation is $Y = A + BX$, then $A$ is the value of $Y$ where the line crosses the $Y$ axis. Read this value from the graph for the line drawn. Correspondingly, the value of $B$ is the slope of the line, the increase in $Y$ occurring when $X$ changes 1 unit. Read this value from the graph for the line drawn.

(*b*) Compute the values of $\bar{X}$ and $\bar{Y}$ and of $b$ using these data. Draw the line having equation $\bar{Y}_x = \bar{Y} + b(X - \bar{X})$. Read the values of the $Y$ intercept and slope for this least-squares line. Do these values agree with those obtained in part (*a*)?

(*c*) Measure the vertical deviations of the points from each of the lines drawn. Use the formulas to check the measurements. Compute the sum of squares of deviations from each line. Note which is larger.

*Section* 11-4*

**2.** (*a*) Plot the height $X$ and weight $Y$ of each individual in Table 2-2*a*. (*b*) The same data give,

$N = 100$ $\quad \Sigma X_i = 6{,}826$ $\quad \Sigma Y_i = 16{,}440$

$\Sigma X_i^2 = 466{,}540$ $\quad \Sigma Y_i^2 = 2{,}766{,}596$ $\quad \Sigma X_i Y_i = 1{,}124{,}828$

Compute the values of $\bar{X}$, $\bar{Y}$, $s_x^2$, $s_y^2$, $b$, and $s_{y \cdot x}^2$ and give the equation of the least-squares line which is used to estimate $\mu_{y \cdot x}$. Draw the line on the graph. (*c*) Assuming normality, give confidence interval estimates of $A$, $B$, $\sigma_{y \cdot x}^2$ and $\mu_{y \cdot 65}$.

**3.** (*a*) Plot the height $X$ and the weight $Y$ of each individual in Table 2-2*b*. (*b*) Compute $\Sigma X_i$, $\Sigma Y_i$, $\Sigma X_i^2$, $\Sigma Y_i^2$, and $\Sigma X_i Y_i$. (*c*) From these values find the equation of the least-squares line and draw the line on the graph. (*d*) Assuming normality, give confidence-interval estimates of $A$, $B$, $\sigma_{y \cdot x}^2$, and $\mu_{y \cdot 65}$.

*Section* 11-5

**4.** It is believed that the mean yield of wheat per acre plotted against the number of pounds per acre of a nitrogen fertilizer would be approximately a straight line for the amount of fertilizer applied. A sample of 82 acres of wheat has been treated with varying amounts of fertilizer and the yields recorded. Find $\bar{X}$, $\bar{Y}$, $s_x$, $s_y$, $b$, $s_{y \cdot x}$. Write the equation of the estimated regression line for yield on fertilizer. Test for linearity of regression.

| | FERTILIZER, POUNDS PER ACRE | | | | | |
|---|---|---|---|---|---|---|
| YIELD | 50 | 60 | 70 | 80 | 90 | 100 |
| 31–35 | | | | 2 | 6 | 3 |
| 26–30 | | | 5 | 12 | 7 | 2 |
| 21–25 | | 4 | 8 | 8 | 6 | |
| 16–20 | | 2 | 7 | | 1 | |
| 11–15 | 1 | 3 | | | | |
| 6–10 | 3 | 1 | | | | |
| 1–5 | 1 | | | | | |
| *Total* | 5 | 10 | 20 | 22 | 20 | 5 |

* Problems 2, 3, 9, 10, 11, 12 could also be done with a subset of the data, such as the first 10 individuals, the first 20, etc.

*Section* 11-6

**5.** In the tabulation are quantities $X$, the body weight in kilograms, and $Y$, the blood volume in cubic centimeters, for 30 goats.

| $X$ | $Y$ | $X$ | $Y$ | $X$ | $Y$ |
|---|---|---|---|---|---|
| 34 | 2,370 | 17 | 1,100 | 66 | 4,230 |
| 28 | 2,100 | 48 | 3,550 | 34 | 2,440 |
| 19 | 1,120 | 38 | 2,980 | 16 | 1,050 |
| 41 | 2,810 | 30 | 2,020 | 30 | 2,360 |
| 21 | 1,500 | 26 | 1,710 | 35 | 2,410 |
| 20 | 1,660 | 19 | 1,240 | 38 | 2,900 |
| 21 | 1,480 | 60 | 3,990 | 21 | 1,580 |
| 39 | 2,450 | 45 | 2,940 | 52 | 3,600 |
| 37 | 2,560 | 18 | 1,070 | 28 | 1,850 |
| 23 | 1,550 | 40 | 2,300 | 45 | 3,010 |

(*a*) Make a scatter diagram for the data. (*b*) Find $\bar{X}$, $\bar{Y}$, $s_x$, $s_y$, $r$, $s_{y \cdot x}$ and the slopes of the two regression lines. (*c*) Draw the estimated regression lines on the scatter diagram. (*d*) Estimate the mean $Y$ for $X = 50$. How close to the true mean $Y$ for $X = 50$ can you state such an estimate to be with 95 per cent confidence?

**6.** Measurements were taken of the ability of rats to run a maze before ($X$) and after ($Y$) a stimulus. In a sample of 300 rats it was found that $\bar{X} = 16$, $\bar{Y} = 12.8$, $s_x^2 = 4$, $s_y^2 = 3.7$, $r = .4$. (*a*) Draw the estimated regression line of $Y$ on $X$. (*b*) Estimate the mean $Y$ for $X = 17$. (*c*) What are 95 per cent confidence limits for this estimate? (*d*) If a rat has $X = 17$, estimate by a 90 per cent interval his $Y$ value.

**7.** The following data are measurements of students' ability, first by an IQ examination and second by an achievement test in a general science course. Let $X$ be the result of the IQ test and $Y$ the result of the achievement test. Make a scatter diagram to represent the observations. Collect the data into a frequency table, using intervals of 7 for the $X$ measurements and intervals of 5 for the $Y$ measurements. Use values of $X = 100$ and $Y = 50$ as mid-points of intervals. The interval corresponding to $X = 100$ extends from 96.5 to 103.5, and the interval for $Y = 50$ extends from 47.5 to 52.5, and so on. (*a*) Find $\bar{X}$, $\bar{Y}$, $s_x$, $s_y$, $r$ and the slopes of the two estimated regression lines. (*b*) Test the hypothesis: $\rho = 0$. (*c*) Estimate $\mu_{y \cdot 90}$ and $\mu_{x \cdot 60}$ by 95 per cent confidence intervals.

| *IQ* | *Achievement* | *IQ* | *Achievement* | *IQ* | *Achievement* |
|---|---|---|---|---|---|
| 100 | 49 | 105 | 47 | 92 | 31 |
| 117 | 47 | 95 | 46 | 125 | 53 |
| 98 | 69 | 126 | 67 | 120 | 64 |
| 87 | 47 | 111 | 66 | 107 | 43 |
| 106 | 45 | 121 | 59 | 121 | 75 |
| 134 | 55 | 106 | 49 | 90 | 40 |
| 77 | 72 | 134 | 78 | 132 | 80 |
| 107 | 59 | 125 | 39 | 116 | 55 |
| 125 | 27 | 140 | 66 | 137 | 73 |
| 105 | 50 | 137 | 69 | 113 | 48 |
| 89 | 72 | 142 | 68 | 110 | 41 |
| 96 | 45 | 130 | 71 | 114 | 29 |

Note: Data continues on next page.

| *IQ* | *Achievement* | *IQ* | *Achievement* | *IQ* | *Achievement* |
|---|---|---|---|---|---|
| 122 | 66 | 120 | 59 | 112 | 61 |
| 130 | 63 | 97 | 50 | 96 | 38 |
| 116 | 43 | 100 | 47 | 105 | 76 |
| 101 | 44 | 80 | 51 | 117 | 66 |
| 92 | 50 | 104 | 45 | 109 | 66 |
| 120 | 60 | 95 | 46 | 122 | 75 |
| 80 | 31 | 117 | 68 | 107 | 59 |
| 117 | 55 | 101 | 47 | 80 | 54 |
| 93 | 50 | 108 | 45 | 123 | 57 |
| 90 | 51 | 121 | 55 | 116 | 42 |
| 106 | 63 | 106 | 73 | 102 | 36 |
| 126 | 60 | 114 | 60 | 119 | 68 |
| 132 | 71 | 70 | 30 | 93 | 34 |
| 84 | 58 | 121 | 50 | 102 | 32 |
| 72 | 34 | 128 | 72 | 100 | 61 |
| 77 | 26 | 106 | 46 | 91 | 26 |
| 90 | 39 | 126 | 67 | 104 | 60 |
| 111 | 75 | 114 | 62 | 114 | 70 |
| 125 | 67 | 109 | 44 | 79 | 58 |
| 100 | 68 | 94 | 43 | 110 | 61 |
| 136 | 80 | 115 | 50 | 92 | 27 |
| 106 | 50 | | | | |

**8.** The data in the table give $X$, blood hemoglobin as the percentage of normal, and $Y$, red blood cells in millions per cubic millimeter, for dogs.

| $X$ | $Y$ | $X$ | $Y$ | $X$ | $Y$ |
|---|---|---|---|---|---|
| 93 | 7.3 | 70 | 5.8 | 93 | 7.4 |
| 96 | 6.5 | 92 | 7.4 | 118 | 8.4 |
| 108 | 7.7 | 94 | 6.8 | 98 | 7.0 |
| 86 | 5.4 | 112 | 7.5 | 62 | 6.0 |
| 92 | 6.7 | 81 | 6.4 | 86 | 6.7 |
| 80 | 5.1 | 102 | 6.6 | 98 | 7.5 |
| 96 | 7.0 | 92 | 6.8 | 110 | 7.7 |
| 117 | 8.5 | 94 | 7.8 | 61 | 4.6 |
| 95 | 7.8 | 80 | 6.4 | 96 | 6.6 |
| 94 | 6.7 | 90 | 6.6 | 101 | 6.8 |
| 96 | 6.7 | 111 | 7.8 | 79 | 6.3 |
| 111 | 8.6 | 98 | 6.3 | 109 | 7.5 |
| 104 | 7.5 | 96 | 7.8 | 101 | 6.9 |
| 84 | 7.2 | 99 | 7.5 | 100 | 7.5 |
| 80 | 6.5 | 100 | 7.2 | 80 | 6.0 |

(*a*) Make a scatter diagram for the data. (*b*) Find $\bar{X}$, $\bar{Y}$, $s_x$, $s_y$, $r$, $s_{y \cdot x}$, and the slopes of the two regression lines. (*c*) Plot the regression lines on the scatter diagram. (*d*) Compute a tolerance interval which has chance .90 of covering 50 per cent of the population.

**9.** (*a*) Plot the age $X$ and systolic blood pressure $Y$ of each individual in Table 2-2*a*. (*b*) The same data give

$$N = 100 \qquad \Sigma X_i = 4{,}421 \qquad \Sigma Y_i = 12{,}130$$
$$\Sigma X_i^2 = 208{,}349 \qquad \Sigma Y_i^2 = 1{,}498{,}976 \qquad \Sigma X_i Y_i = 542{,}735$$

Compute the values of $\bar{X}$, $\bar{Y}$, $s_x^2$, $s_y^2$, $b$, $s_{y \cdot x}^2$ and give the equation of the least-squares line used to estimate $\mu_{y \cdot x}$. Draw the line on the graph. (*c*) Assuming normality, give confidence-interval estimates of $A$, $B$, $\sigma_{y \cdot x}^2$, and $\mu_{y \cdot x}$.

**10.** (*a*) Plot the age $X$ and systolic blood pressure $Y$ of each individual in Table 2-2*b*. (*b*) Compute $\Sigma X_i$, $\Sigma Y_i$, $\Sigma X_i^2$, $\Sigma Y_i^2$, and $\Sigma X_i Y_i$. (*c*) From these values find the equation of the least-squares line and draw the line on the graph. (*d*) Assuming normality, give confidence-interval estimates of $A$, $B$, $\sigma_{y \cdot x}^2$, and $\mu_{y \cdot 50}$.

**11.** (*a*) Plot the weight $X$ and systolic blood pressure $Y$ of each individual in Table 2-2*a*. (*b*) The same data give

$$N = 100 \qquad \Sigma X_i = 16{,}440 \qquad \Sigma Y_i = 12{,}130$$
$$\Sigma X_i^2 = 2{,}766{,}596 \qquad \Sigma Y_i^2 = 1{,}498{,}976 \qquad \Sigma X_i Y_i = 2{,}005{,}093$$

Compute the values of $\bar{X}$, $\bar{Y}$, $s_x^2$, $s_y^2$, $b$, $s_{y \cdot x}^2$ and give the equation of the least-squares line used to estimate $\mu_{y \cdot x}$. Draw the line on the graph.

**12.** (*a*) Plot the weight $X$ and the systolic blood pressure $Y$ of each individual in Table 2-2*b*. (*b*) Compute the values of $\Sigma X_i$, $\Sigma Y_i$, $\Sigma X_i^2$, $\Sigma Y_i^2$, and $\Sigma X_i Y_i$. (*c*) From these values find the equation of the least-squares line and draw the line on the graph. (*d*) Assuming normality, give confidence interval estimates of $A$, $B$, $\sigma_{y \cdot x}^2$, and $\mu_{y \cdot 50}$.

*Section* 11-8

**13.** (*a*) Compute $\Sigma X_i$, $\Sigma Y_i$, $\Sigma X_i^2$, $\Sigma Y_i^2$, and $\Sigma X_i Y_i$ for each of the three sets of data below:

| A | | B | | C | |
|---|---|---|---|---|---|
| 3 | 10 | 4 | 12 | 1 | 6 |
| 2 | 8 | 3 | 12 | 2 | 5 |
| 1 | 8 | 3 | 10 | 3 | 8 |
| 2 | 11 | 5 | 13 | 1 | 7 |

(*b*) From these find the three least-squares lines. Indicate estimates $b_1$, $b_2$, $b_3$ of the slopes $B_1$, $B_2$, and $B_3$. (*c*) Test the hypotheses $H$: $B_1 = B_2$, $H$: $B_1 = B_3$, and $H$: $B_2 = B_3$. (*d*) Plot the points and draw the least-squares lines. Note whether they appear parallel.

*Section* 11-9

**14.** The data in Table 2-2*a* with systolic blood pressure $Y$, age $X_1$, and weight $X_2$, give the following:

$$N = 100 \qquad \Sigma Y_i = 12{,}130 \qquad \Sigma X_{1_i} = 4{,}421 \qquad \Sigma X_{2_i} = 16{,}440$$
$$\Sigma Y_i^2 = 1{,}498{,}976 \qquad \Sigma X_{1_i}^2 = 208{,}349 \qquad \Sigma X_{2_i}^2 = 2{,}766{,}596$$
$$\Sigma Y_i X_{1_i} = 542{,}735 \qquad \Sigma Y_i X_{2_i} = 2{,}005{,}093 \qquad \Sigma X_{1_i} X_{2_i} = 726{,}526$$

(*a*) Write the three equations to be solved to give the best-fitting plane of $Y$ on $X_1$ and $X_2$. (*b*) If a calculator is available, solve these equations and write the formula for $\bar{Y}_x$.

CHAPTER TWELVE

# Analysis of covariance

Procedures for estimating a difference in two means or for testing this equality were presented in Chap. 8. In Chap. 10 these procedures were extended to the study of several means through multiple comparisons and testing for differences by using the $F$ test in an analysis-of-variance table. The possible effect of a second categorical variable and adjustment by pairing were presented in Chap. 8, and in Chap. 10 the use of a second variable of classification was introduced in the two-way analysis-of-variance tables. In Chap. 11 the second variable was in the form of a measurement variable. Analysis was presented for the case where the strata, or conditional, means fall on a regression curve which is a straight line. Under the assumption of a linear relationship, (1) the slope was estimated, (2) the "amount" of dependence on the second variable was analyzed by comparison to the within-groups variability in an analysis-of-variance table (Table 11-6), and (3) two slopes were compared.

In this chapter we shall study differences in means which can be based jointly on one categorical variable and one measurement variable. Comparisons will be made between two or more populations, as in Chap. 10, with "adjustments" for the measurement variable. Two equivalent statements of the problem are:

1. Comparison of levels of regression lines, with the assumption (possibly after testing) that the slopes of the lines are equal

2. Comparison of column main effects, after adjustment by a fitted regression line on another variable

In the two-variables-of-classification problem treated by analysis-of-variance methods we tested for the effect of one variable separately from the effect of the second variable. In that case the second variable was represented by several categories. If the second variable represents an actual measurement or score for each individual (rather than a category), we can again test the effect of the first variable, separately from the effect of the second variable. The method of analysis to be presented is called the *analysis of covariance*. The second variable is often referred to as a *control variable*.

For instance, if we wish to compare the effects of different feeds on the weight of hogs, a measurement of the weight of each hog before the experiment would be valuable as a control. If we wish to say that food mixture $A$ is the best, we should be able to state that the extra weight of the group fed by mixture $A$ was not largely explained by the original weights. Even if the original weights are comparable, it is often impossible to have individuals in the experiment maintain equal amounts of intake of the particular diets. The related variable of total intake could be measured and taken into account in comparing the diets.

In an experiment designed to study the results of a program to increase the spelling ability of four classes of students we measure the spelling ability $Y$ of each student at the end of the program and introduce the original spelling ability $X$ for each student as a control variable. We may study the differences in the effectiveness for the four classes with the use of the $Y$ variable, "controlled," or "adjusted," for the $X$ variable.

The analysis-of-variance procedure for difference in means was based on the separation of a total sum of squares into several portions. If the mean square for means was significantly large we rejected the hypothesis of equal means. The analysis-of-covariance procedure also leads to a test for difference in means by separation of a sum of squares into several portions. In this case we test for a difference in means of *residuals*. The residuals are the differences of the actual observations and a regression quantity based on the associated second variable. The definition of the test procedure is developed in the following sections.

This chapter deals with a single-variable-of-classification problem where we have samples from a fixed number of populations.

## STATEMENT OF THE PROBLEM 12-1

For a single-variable-of-classification problem there are $k$ populations. Random samples of sizes $n_i$ are chosen from the respective populations. The sample from the $i$th population will be in the form $(X_{i1},Y_{i1})$, $(X_{i2},Y_{i2})$, . . . , $(X_{in_i},Y_{in_i})$, where the first subscript indicates the population and the second subscript indicates the particular individual from that population.

Data in this form are illustrated in Table 12-1 with $k = 3, n_1 = n_2 = n_3 = 4$, and a numerical example is given in Table 12-2.

We shall deal here with the general principle behind the analysis, and later sections give the details of the numerical analysis. Suppose the observations are plotted on $X$ and $Y$ coordinates about the grand mean. That is, we plot the deviations from $\bar{X}..$ and $\bar{Y}..$ as in Fig. 12-1, where the points representing the pairs of values $(X,Y)$ in the first treatment are designated by $A$, those in the second by $B$, and those in the third by $C$. From the methods of Chap. 11, we may compute the regression line of $Y$ on $X$ for these points. Designate this regression coefficient by $b_t$ and the variance of deviations from this regression line by $(s^2_{y \cdot x})_t$, where $t$ represents *total*. Differences between the samples, as well as differences within each sample, will affect the size of $(s^2_{y \cdot x})_t$.

We shall partition the sum of squares corresponding to this variance into two parts: (1) the portion of the sum of squares representing deviations from regression lines (with common slope) *within* each classification and (2) the portion of the sum of squares representing the remaining sum of squares of deviations which result from group differences. To represent this partition we next consider the three groups of values plotted about their own means (we plot deviations from $\bar{X}_i.$ and $\bar{Y}_i.$) as in Fig. 12-2, and then we superimpose these figures into a single figure as in Fig. 12-3. We compute a regression line and variance about the regression line for the points, as plotted in Fig. 12-3. Denote this within-group regression coefficient by $b_w$ and the variance by $(s^2_{y \cdot x})_w$.

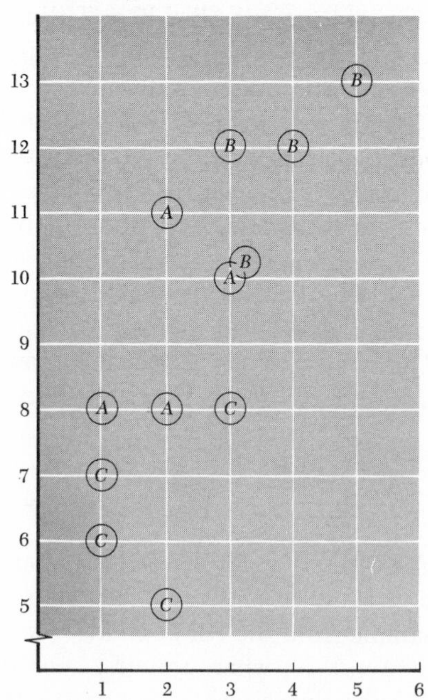


**Figure 12-1** Plot showing $X$ and $Y$ values for observations on four individuals from each of three populations.

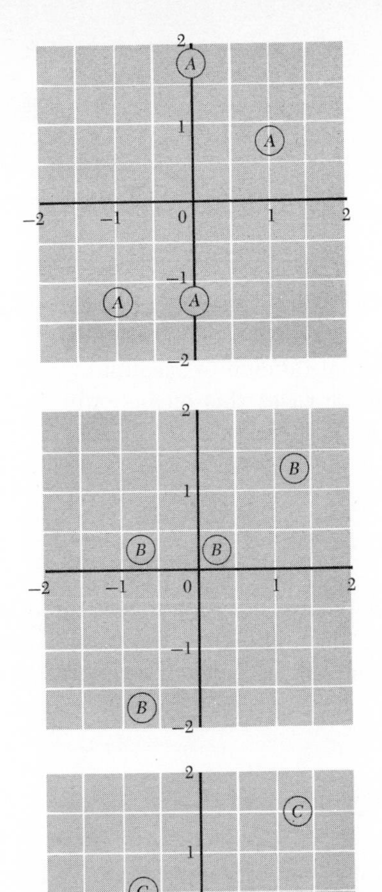


**Figure 12-2** Plot showing same points as Fig. 12-1 but scaled from the individual group means.

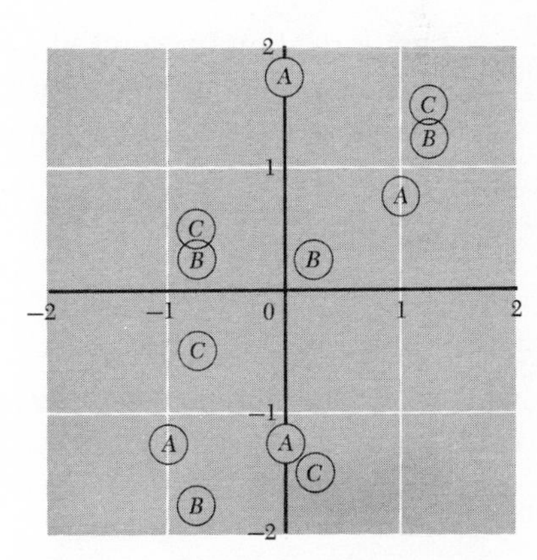


**Figure 12-3** Plot showing the same points scaled from a common location for the group means.

A regression line gives the mean $Y$ for different values of $X$. If in each of the three populations the regression lines have the same slope, then $b_w$ is an estimate of this slope and $(s^2_{y \cdot x})_w$ is an estimate of the variance about the regression line in each of the populations.

We expect the variation about the regression line in Fig. 12-3 to be less than in Fig. 12-1, since the means were made to coincide. However, if the dispersion $(s^2_{y \cdot x})_w$ for Fig. 12-3 is significantly smaller than the dispersion $(s^2_{y \cdot x})_t$ for Fig. 12-1, we conclude that the procedure of bringing all the means together has had a significant effect on the data and state that there is a difference in the $Y$ means that is not explained by the $X$ measurements.

Of course, it is possible that this significant reduction in sum of squares

**Table 12-1**

| | A | | B | | C | | | | |
|---|---|---|---|---|---|---|---|---|---|
| | $X_{11}$ | $Y_{11}$ | $X_{21}$ | $Y_{21}$ | $X_{31}$ | $Y_{31}$ | | | |
| | $X_{12}$ | $Y_{12}$ | $X_{22}$ | $Y_{22}$ | $X_{32}$ | $Y_{32}$ | | | |
| | $X_{13}$ | $Y_{13}$ | $X_{23}$ | $Y_{23}$ | $X_{33}$ | $Y_{33}$ | | | |
| | $X_{14}$ | $Y_{14}$ | $X_{24}$ | $Y_{24}$ | $X_{34}$ | $Y_{34}$ | | | |
| *Total* | $T_{X_{1+}}$ | $T_{Y_{1+}}$ | $T_{X_{2+}}$ | $T_{Y_{2+}}$ | $T_{X_{3+}}$ | $T_{Y_{3+}}$ | $T_{X_{++}}$ | $T_{Y_{++}}$ | *Grand total* |
| *Mean* | $\bar{X}_{1.}$ | $\bar{Y}_{1.}$ | $\bar{X}_{2.}$ | $\bar{Y}_{2.}$ | $\bar{X}_{3.}$ | $\bar{Y}_{3.}$ | $\bar{X}_{..}$ | $\bar{Y}_{..}$ | *Grand mean* |

of deviations might occur if the $X$ variable is not taken into account or if the regression coefficient is near zero, but in those cases this means only that the $X$ variable is not important or that the differences in $Y$ were so pronounced that this refinement in technique is not necessary.

## 12-2 COMPUTATION AND CONCLUSION

First a relation for sums of products similar to that for sums of squares will be shown, then altered into computing form, and then the computing form for the dispersion about a regression line will be given. The expression for sums of products is

$$\sum_i \sum_j (X_{ij} - \bar{X}_{..})(Y_{ij} - \bar{Y}_{..})$$
$$= n \sum_i (\bar{X}_{i.} - \bar{X}_{..})(\bar{Y}_{i.} - \bar{Y}_{..}) + \sum_i \sum_j (X_{ij} - \bar{X}_{i.})(Y_{ij} - \bar{Y}_{i.})$$

where $n$ is the number of observations in each of $k$ groups. Any of these sums of products divided by its degrees of freedom is called a *covariance*.

1. The computing form for the quantity on the left, called the *total sum of products*, is

$$\sum \sum X_{ij}Y_{ij} - \frac{T_{X_{++}}T_{Y_{++}}}{nk} \qquad \text{(total sum of products)}$$

2. The computing form for the first expression on the right above, called the *among-means* or *among-groups sum of products*, is

$$\frac{\sum_i T_{X_{i+}}T_{Y_{i+}}}{n} - \frac{T_{X_{++}}T_{Y_{++}}}{nk} \qquad \text{(among-means sum of products)}$$

3. The computing form for the second quantity on the right above, called the *within-groups sum of products*, is

$$\sum \sum X_{ij}Y_{ij} - \frac{\sum_i T_{X_{i+}}T_{Y_{i+}}}{n} \qquad \text{(within-groups sum of products)}$$

**Table 12-2**

| | A | | B | | C | | | |
|---|---|---|---|---|---|---|---|---|
| | X | Y | X | Y | X | Y | | |
| | 3 | 10 | 4 | 12 | 1 | 6 | | |
| | 2 | 8 | 3 | 12 | 2 | 5 | | |
| | 1 | 8 | 3 | 10 | 3 | 8 | | |
| | 2 | 11 | 5 | 13 | 1 | 7 | $T_{X_{++}}$ | $T_{Y_{++}}$ |
| *Total* | 8 | 37 | 15 | 47 | 7 | 26 | 30 | 110 |

The above computing forms can be verified algebraically in a manner similar to the derivation of the corresponding formulas for sums of squares. However, it can be seen from the computing forms that the first is the sum of the second and the third. Note that the computing forms for the sums of products are very similar to those for the sums of squares.

The formulas shown thus far will be applied to the data in Table 12-2, and then the computations for dispersion about a regression line will be shown. The sums of products are as follows:

Total:

$$(3 \times 10) + (2 \times 8) + \cdots + (1 \times 7) - \frac{30 \times 110}{12} = 302 - 275 = 27$$

Among means:

$$\frac{8 \times 37}{4} + \frac{15 \times 47}{4} + \frac{7 \times 26}{4} - \frac{30 \times 110}{12} = 295.75 - 275 = 20.75$$

Within groups:

$$27 - 20.75 = 6.25$$

The computations for sums of squares on $X$ and on $Y$ are made as for the analysis of variance. The results are entered in Table 12-3, where $x = X - \bar{X}$ and $y = Y - \bar{Y}$. The computations on $X$ and $Y$ are given in Table 12-4.

**Table 12-3**

| | *Degrees of freedom* | $\Sigma x^2$ | $\Sigma xy$ | $\Sigma y^2$ |
|---|---|---|---|---|
| *Total* | 11 | 17 | 27 | 71.67 |
| Among means | 2 | 9.5 | 20.75 | 55.17 |
| Within groups | 9 | 7.5 | 6.25 | 16.50 |

**Table 12-4**

| | $\Sigma x^2$ | $\Sigma y^2$ |
|---|---|---|
| *Total* | $92 - \frac{(30)^2}{12} = 17$ | $1{,}080 - \frac{(110)^2}{12} = 71.67$ |
| Means | $\frac{338}{4} - \frac{(30)^2}{12} = 9.5$ | $\frac{4{,}254}{4} - \frac{(110)^2}{12} = 55.17$ |
| Within | $17 - 9.5 = 7.5$ | $71.67 - 55.17 = 16.50$ |

To find the sum of squares about a regression line we compute the sum of squares of the residuals $Y - \bar{Y}_X$. Since

$$\bar{Y}_X = \bar{Y} + b(X - \bar{X})$$

the sum of squares of residuals is

$$\Sigma(Y - \bar{Y}_X)^2 = \Sigma[Y - \bar{Y} - b(X - \bar{X})]^2$$

It will be more convenient to use $y = Y - \bar{Y}$ and $x = X - \bar{X}$. This relation then appears as

$$\begin{aligned}\Sigma(Y - \bar{Y}_X)^2 &= \Sigma(y - bx)^2 = \Sigma(y^2 - 2bxy + b^2x^2) \\ &= \Sigma y^2 - 2b\Sigma xy + b^2\Sigma x^2\end{aligned}$$

In this notation the slope $b$ of the regression line is $b = \Sigma xy/\Sigma x^2$, and we substitute this into the above equation to obtain

$$\Sigma(Y - \bar{Y}_X)^2 = \Sigma y^2 - 2\frac{\Sigma xy}{\Sigma x^2}\Sigma xy + \left(\frac{\Sigma xy}{\Sigma x^2}\right)^2 \Sigma x^2 = \Sigma y^2 - \frac{(\Sigma xy)^2}{\Sigma x^2}$$

The three summations in the final form of this expression are given in Table 12-3. For the sum of squares about the regression line for "Total" we then have

$$71.67 - \frac{(27)^2}{17} = 28.79$$

and for "Within groups"

$$16.50 - \frac{(6.25)^2}{7.5} = 11.29$$

We now have estimates of the variances (dispersions) referred to at the beginning of this chapter, the sums of squares about the regression line with coefficient $b_t$ and the line with coefficient $b_w$, 28.79 and 11.29, respectively. The dispersion about the line with coefficient $b_t$ is the total dispersion of the $Y$ about the grand mean $Y_{..}$ minus the variance due to the regression. Even though all samples are from the same population, the means will not be exactly equal. Hence this second value will always be less than the first.

The reduction in sum of squares 28.79 − 11.29 = 17.50 is attributable to the dispersion of the means. When all groups are from the same population the ratio of this last quantity (17.50) to the estimate of variance obtained from within groups (11.29) has as a sampling distribution the $F$ distribution. The assumptions involved here will be discussed below. We can make a test of significance of difference in means, with each measurement adjusted for measurement $X$. The complete computation can be indicated in an analysis-of-covariance table. Note that the use of an estimated regression coefficient to adjust our values used 1 degree of freedom, as does the use of an estimated value of the mean. The regression coefficients $b_t$ and $b_w$ were used for "Total" and for "Within," respectively, and therefore the degrees of freedom have been decreased by 1 in the last two lines of the table. $y'$ is used to denote $Y - \bar{Y}_X$.

Here there is significance at the 5 per cent level for differences in means of the $Y$ values among the groups after the $Y$ values have been adjusted by the within-groups regression coefficient $b_w$. The test of significance of differences in the $Y$ means not making use of the $X$ values would give

$$F = \frac{55.17/2}{16.50/9} = \frac{27.58}{1.83} = 15.07 \qquad F_{.95}(2,9) = 4.26$$

The very large value for $F$ obtained here can be accounted for by differences in the $X$ values, since when the $Y$ values are adjusted for $X$, the value of $F$ is reduced to about the 2.5 per cent level.

As a more concrete example, suppose that the analysis of covariance in Table 12-5 reports the results of an experiment on three types of rations, $A$, $B$, $C$. Four animals were given each ration; $X$ is food intake in pounds per day and $Y$ is gain in ounces at the end of 1 week. The conclusions could be described as follows:

1. The three types of rations resulted in significantly different gains, $F = 15.07$, $F_{.95}(2,9) = 4.26$.

2. The three types of rations resulted in significantly different residual

**Table 12-5** Analysis of covariance

| | Degrees of freedom | $\Sigma x^2$ | $\Sigma xy$ | $\Sigma y^2$ | RESIDUALS Degrees of freedom | $\Sigma y'^2$ | Mean square |
|---|---|---|---|---|---|---|---|
| Among means | 2 | 9.5 | 20.75 | 55.17 | 2 | 17.50 | 8.75 |
| Within groups | 9 | 7.5 | 6.25 | 16.50 | 8 | 11.29 | 1.41 |
| *Total* | 11 | 17.0 | 27.00 | 71.67 | 10 | 28.79 | |

$$F = \frac{8.75}{1.41} = 6.2 \qquad F_{.95}(2,8) = 4.46$$

gains. The residual indicates a gain adjusted for the amount of food intake as given by the within-groups regression line

$$\bar{Y}_X = \bar{Y}_{i\cdot} + b_w(X - \bar{X}_{i\cdot})$$

For example, in group $A$ the regression line is

$$\bar{Y}_X = \frac{37}{4} + \frac{6.25}{7.5}\left(X - \frac{8}{4}\right) = 9.25 + .833(X - 2)$$

and the first residual is $10 - [9.25 + .833(3 - 2)] = 10 - 10.083 = -.083$. The $F$ ratio for this test is in Table 12-5; $F = 6.2$, $F_{.95}(2,8) = 4.46$. Therefore the residual gains are significantly different for the three rations.

It is evident that in analyses of this type the first test could be significant and the second test not significant. We could then infer that the differences in weight gains for the different rations were largely due to the differences in food intake.

## 12-3 HYPOTHESIS AND ASSUMPTIONS

The analysis-of-covariance test described above applies when it is known that the regression curves in the $k$ populations are parallel straight lines. The hypothesis states that the parallel lines coincide; that is, the populations have the same $Y$ means when the same $X$ is used for each population ($Y$ means are equal after adjustment for $X$ values). The population variances $\sigma^2_{y\cdot x}$ about the regression lines are assumed to be equal in each of the $k$ populations. Tests of the hypothesis that the regression curve is a straight line and of the hypothesis that the regression lines are parallel are given in Sec. 12-5.

For the percentiles of the computed $F$ statistic to be as given in Table A-7 we also assume that within each population and for each $X$ value the $Y$ values are approximately normally distributed.

## 12-4 EXTENSION TO OTHER PROBLEMS

The analysis-of-covariance procedure has been extended in several ways:

1. The procedure can be developed for problems having several variables of classification, such as the two-way table with single or multiple observations presented in Chap. 10 and the Latin-square designs discussed in Chap. 15.

2. The procedure can be developed for using several control variables to adjust the $Y$ observations.

3. Curves other than straight lines may be used as regression curves.

The first extension is fairly straightforward; the second, however, involves multiple-regression methods.

In the analysis of variance for single variable of classification we obtained two independent estimates of variance, (1) within groups and (2) among means. These were both estimates of the population $\sigma^2$ if the hypothesis is correct. In the analysis of covariance the population value for the variance about the regression line can be estimated in four independent ways. Comparisons of these estimates can be made to investigate departures from uniformity of various sorts. These estimates are obtained from sums of squares divided by their degrees of freedom. The four sums of squares will be denoted by $S_1$, $S_2$, $S_3$, $S_4$.

It will be convenient to denote the quantities in the covariance table by individual letters, as in Table 12-6. The definitions of the quantities to be computed are as follows:

$C_{xx1}$, $C_{xx2}$, . . . represent the computation $\Sigma X^2 - (\Sigma X)^2/n$ for groups 1, 2, . . . . For example, for the data in Table 12-2

$$C_{xx1} = (3)^2 + (2)^2 + (1)^2 + (2)^2 - \frac{(8)^2}{4} = 2$$

$C_{xy1}$, $C_{xy2}$, . . . represent the computation $\Sigma XY - \Sigma X \Sigma Y/n$ for groups 1, 2, . . . . In the example

$$C_{xy1} = (3 \times 10) + (2 \times 8) + (1 \times 8) + (2 \times 11) - \frac{8 \times 37}{4} = 2$$

$C_{yy1}$, $C_{yy2}$, . . . represent the computation $\Sigma Y^2 - (\Sigma Y)^2/n$ for groups 1, 2, . . . . In the example,

$$C_{yy1} = (10)^2 + (8)^2 + (8)^2 + (11)^2 - \frac{(37)^2}{4} = 6.75$$

The quantities in the column headed "$\Sigma y'^2$" are computed by the formula $\Sigma y^2 - (\Sigma xy)^2/\Sigma x^2$, using in every case the entries in the same line. For example, $C'_{yy1} = 6.75 - (2)^2/2 = 4.75$.

**Table 12-6**

| | $\Sigma x^2$ | $\Sigma xy$ | $\Sigma y^2$ | $\Sigma y'^2$ |
|---|---|---|---|---|
| Within each group: | | | | |
| 1 | $C_{xx1}$ | $C_{xy1}$ | $C_{yy1}$ | $C'_{yy1}$ |
| 2 | $C_{xx2}$ | $C_{xy2}$ | $C_{yy2}$ | $C'_{yy2}$ |
| . . . | . . . | . . . | . . . | . . . |
| $k$ | $C_{xxk}$ | $C_{xyk}$ | $C_{yyk}$ | $C'_{yyk}$ |
| *Among means* | $C_{xxm}$ | $C_{xym}$ | $C_{yym}$ | $C'_{yym}$ |
| Within groups | $C_{xxw}$ | $C_{xyw}$ | $C_{yyw}$ | $C'_{yyw}$ |
| *Total* | $C_{xxt}$ | $C_{xyt}$ | $C_{yyt}$ | $C'_{yyt}$ |

**Table 12-7**

| | $\Sigma x^2$ | $\Sigma xy$ | $\Sigma y^2$ | $\Sigma y'^2$ | *Slopes* |
|---|---|---|---|---|---|
| Within each group: | | | | | |
| 1 | 2.00 | 2.00 | 6.75 | 4.75 | 1.0 $= b_1$ |
| 2 | 2.75 | 2.75 | 4.75 | 2.00 | 1.0 $= b_2$ |
| 3 | 2.75 | 1.50 | 5.00 | 4.18 | .545 $= b_3$ |
| *Among means* | 9.50 | 20.75 | 55.17 | 9.85 | 2.18 $= b_m$ |
| Within groups | 7.50 | 6.25 | 16.50 | 11.29 | .833 $= b_w$ |
| *Total* | 17.00 | 27.00 | 71.67 | 28.79 | |

All entries in the last three lines are the same as entered in Table 12-5, except $C'_{yym}$, which is computed like the other $C'$ quantities. For example, $C'_{yym} = 55.17 - (20.75)^2/9.5 = 9.85$. In the analysis-of-covariance table the quantity appearing in this position was computed as the difference $C'_{yyt} - C'_{yyw}$.

The completed table for the example of this chapter is given in Table 12-7. The quantities $S_1$, $S_2$, $S_3$, $S_4$ are now defined as in Table 12-8 in terms of the $C'$ as follows:

$S_1$ = the sum of squares within each group from the regression line in each group, totaled for all groups.
$S_2$ = the variation among regression coefficients of the different groups.
$S_3$ = the sum of squares of deviations of the means from the regression line of the means
$S_4$ = the square of the difference between $b_w$ and $b_m$ multiplied by the appropriate factor to make it an estimate of variance

The quantities $b_w$ and $b_m$ are the regression coefficients obtained from the within and mean sums of squares, and $S_T = S_1 + S_2 + S_3 + S_4$.

For the example

$$S_1 = 4.75 + 2 + 4.18 = 10.93$$
$$S_2 = 11.29 - 10.93 = .36$$
$$S_3 = 9.85$$
$$S_4 = 28.79 - 11.29 - 9.85 = 7.65$$
$$S_T = 28.79$$

These four sums of squares, when divided by the appropriate degrees of freedom, are all estimates of the variance $\sigma^2_{y \cdot x}$ when the residuals are from the same population.

The four estimates of variance can be used to investigate the various ways in which the relationship of $X$ and $Y$ may deviate more than would be expected by sampling fluctuation when the groups are all from the same normal population.

**Table 12-8**

| | Definition of $S_i$ | Degrees of freedom | EXAMPLE $S_i$ | EXAMPLE Degrees of freedom |
|---|---|---|---|---|
| | $S_1 = \Sigma C'_{yyi}$ | $k(n-2)$ | 10.93 | 6 |
| | $S_2 = C'_{yyw} - S_1$ | $k-1$ | .36 | 2 |
| | $S_3 = C'_{yym}$ | $k-2$ | 9.85 | 1 |
| | $S_4 = C'_{yyt} - C'_{yyw} - C'_{yym}$ | 1 | 7.65 | 1 |
| *Total* | $S_T = C'_{yyt}$ | $nk-2$ | 28.79 | 10 |

1. With this notation the analysis of covariance for a test for difference in means as described in Table 12-5 is

$$F = \frac{\dfrac{S_3 + S_4}{k-1}}{\dfrac{S_1 + S_2}{k(n-1)-1}}$$

For example,

$$F = \frac{17.50/2}{11.29/8} = 6.20 \qquad \text{and} \qquad F_{.95}(2,8) = 4.46$$

2. The analysis-of-covariance procedure assumes a common regression line. A test of whether one regression line can be used for all the observations can be formed as follows:

$$F = \frac{\dfrac{S_2 + S_3 + S_4}{2(k-1)}}{\dfrac{S_1}{k(n-2)}}$$

For example,

$$F = \frac{17.86/4}{10.93/6} = 2.45 \qquad \text{and} \qquad F_{.95}(4,6) = 4.53$$

If this $F$ ratio is significant, that is, if a single regression line is not adequate, it may be of interest to see in what particular manner the data fail in this quality. These sums of squares can also be used for this purpose as follows:

*a.* As a test of whether the slopes of the regression lines within the groups are the same:

$$F = \frac{\dfrac{S_2}{k-1}}{\dfrac{S_1}{k(n-2)}}$$

For example,

$$F = \frac{.36/2}{10.93/6} = .10 \qquad \text{and} \qquad F_{.95}(2,6) = 5.14$$

If the regression slopes within groups are not significantly different, we may use the sums of squares

*b*. As a test of whether the regression for means is linear (assuming the slopes within the groups are the same):

$$F = \frac{\dfrac{S_3}{k-2}}{\dfrac{S_1 + S_2}{k(n-1)-1}}$$

For example,

$$F = \frac{9.85/1}{11.29/8} = 6.98 \qquad \text{and} \qquad F_{.95}(1,8) = 5.32$$

If this is accepted, then the sums of squares may be used

*c*. As a test of whether the regression coefficients $b_w$ and $b_m$ are the same (assuming the regression slopes within groups are the same and the regression for means is linear):

$$F = \frac{\dfrac{S_4}{1}}{\dfrac{S_1 + S_2}{k(n-1)-1}}$$

For example,

$$F = \frac{7.65/1}{11.29/8} = 5.42 \qquad \text{and} \qquad F_{.95}(1,8) = 5.32$$

For the numerical example, the conclusions may be stated as follows. The first $F$ ratio is the $F$ ratio computed in Table 12-5, and the interpretation is given there. The second $F$ ratio indicates that a single regression line may be used for all the observations. In this case the other $F$ ratios need not be computed. The numerical results are given to illustrate their use when such tests are appropriate.

## GLOSSARY

"adjusted" values
covariance
regression of means
regression within groups

## DISCUSSION QUESTIONS

**1.** State carefully what hypothesis is tested by the analysis-of-covariance technique. What assumptions are made before the $F$ test is valid?

**2.** Define each term in the Glossary.

**3.** What advantage does the analysis of covariance have over the analysis of variance? Is it easier to take samples for the analysis of covariance than for the analysis of variance?

**4.** It is desired to compare the IQ's of the students in three schools. It is suspected that IQ is related to the student's grade-point average. If samples of 12 students are taken from each school, should their IQ values be adjusted for their grade-point averages by an analysis of covariance, or should an analysis of variance of their IQ's be performed directly?

**5.** Describe a sampling experiment to illustrate the technique of analysis of covariance.

## PROBLEMS

**1.** Plot the points with coordinates below as was done for the example in the text in Figs. 12-1 to 12-3. Carry out the analysis of covariance and compare the conclusions with the visual impression from the plots.

| *A* | | *B* | | *C* | |
|---|---|---|---|---|---|
| $X$ | $Y$ | $X$ | $Y$ | $X$ | $Y$ |
| 1 | 2 | 4 | 1 | 0 | 0 |
| −2 | 1 | 2 | −2 | 4 | 2 |
| 3 | 5 | −3 | −3 | −7 | −7 |
| 5 | 7 | −3 | −6 | −3 | −4 |

**2.** Carry out an analysis of covariance with the data in the $A$ and $B$ subgroups in Prob. 1. Compare the results with an analysis as in Sec. 11-8.

**3.** Carry out an analysis of covariance with the data below and compare the results with an analysis as in Sec. 11-8 and Prob. 13 in Chap. 11.

| *A* | | *B* | |
|---|---|---|---|
| $X$ | $Y$ | $X$ | $Y$ |
| 3 | 10 | 4 | 12 |
| 2 | 8 | 3 | 12 |
| 1 | 8 | 3 | 10 |
| 2 | 11 | 5 | 13 |

**4.** As in Prob. 8, Chap. 10, stratify individuals from Table 2-2*a* (or Table 2-2*b*) as follows: men under 40 in group 1, men 40 to 49 in group 2, men 50 to 59 in group 3, and men 60 and over in group 4. Use only the first 10 men in each group from Table

2-2$a$ or the first 8 for Table 2-2$b$. Carry out an analysis of covariance for ($a$) systolic blood pressure (1952 readings), ($b$) diastolic blood pressure (1952 readings), and ($c$) blood cholesterol (1952 readings), with weight as an independent variable in each case.

**5.** Repeat Prob. 4 with height as an independent variable.

**6.** Stratify the data in Table 2-2$a$ (or Table 2-2$b$) into age-height strata, as in Prob. 12, Chap. 10, as follows: men less than 50 years of age and 66 inches or less in height in subgroup 1, men less than 50 years of age and over 66 inches in height in subgroup 2, men 50 years or over in age and 66 inches or less in height in subgroup 3, and men 50 years or over in age and over 66 inches in height in subgroup 4. Use only the first 10 men in each group from Table 2-2$a$ or the first 8 for Table 2-2$b$. Carry out an analysis of covariance for ($a$) systolic blood pressure (1952 readings), and ($b$) blood cholesterol (1952 readings), with weight as an independent variable in each case.

**7.** Complete the analysis of covariance in the table, comparing score $Y$ on the final examination in a statistics course for three different classes. The variable $X$ used to adjust this score is the grade-point average for each student. The data have been reduced for simplification to three classes of 16 students each.

| | |
|---|---|
| Sum of squares for $Y$ means | 504 |
| Total sum of squares for $Y$ | 13,813 |
| Sum of squares for $X$ means | .5864 |
| Total sum of squares for $X$ | 8.296 |
| Sum of products for means | 17.19 |
| Total sum of products | 157.88 |

**8.** An experiment on weight gain of rats resulted as shown in the table. $X$ indicates the quantity of food and $Y$ the gain in weight. Did the four rations, $A$, $B$, $C$, $D$, produce different gains among the rats? Are the gains affected materially by quantity of food?

| $A$ | | $B$ | | $C$ | | $D$ | |
|---|---|---|---|---|---|---|---|
| $X$ | $Y$ | $X$ | $Y$ | $X$ | $Y$ | $X$ | $Y$ |
| 96 | 98 | 109 | 64 | 179 | 71 | 127 | 72 |
| 108 | 102 | 125 | 86 | 132 | 84 | 100 | 54 |
| 94 | 102 | 85 | 51 | 163 | 71 | 151 | 109 |
| 128 | 108 | 82 | 72 | 143 | 62 | 116 | 93 |

**9.** Plot the points with coordinates given below. One of the assumptions behind the analysis of covariance as presented in this chapter is apparently not satisfied for these data. Which assumption? Carry out the analysis to see the effect on the results.

| $A$ | | $B$ | | $C$ | |
|---|---|---|---|---|---|
| $X$ | $Y$ | $X$ | $Y$ | $X$ | $Y$ |
| 3 | 1 | 0 | 1 | 0 | 5 |
| 2 | 4 | 3 | 0 | 2 | 7 |
| 1 | 5 | 2 | 3 | 1 | 8 |
| 0 | 2 | 1 | 4 | 3 | 4 |

CHAPTER THIRTEEN

# Enumeration statistics

Previously we have dealt with measurements of certain variables. There are many problems in which we are interested merely in counting the number of cases that fall in specified categories. For example, in tossing coins we count the number of heads in 20 tosses, in the study of genetics we count the number of progeny which inherit certain characteristics, in public-opinion polls we count the number of favorable votes in a sample, in acceptance sampling for quality control we count the number of defective items in a sample, in a study of performance we count the number passing or failing. We frequently change a problem involving measurements to one in which we count the number of occurrences by arbitrarily assigning certain measurements to each category. For example, we could measure the heights and count the number of cases between 60 inches and 65 inches, calling this the first category, and so on. We do this, of course, in constructing any frequency table. The difference is that we shall now forget the actual measurements and remember only the several categories and the number of observations falling into each.

One of our most useful tools will be a statistic called the *chi square*, which, although different in form from that in Chap. 8, has a $\chi^2$ distribution as given in Table A-6*a*. Other useful distributions are the *binomial* and *Poisson distributions*.

## 13-1 CHI SQUARE

We shall suppose that there are $k$ categories and that we have a random sample of $N$ observations such that each observation must fall in one and only one category. We then count the *observed frequency* in each category and denote these frequencies by $f_1, f_2, \ldots, f_k$, where $\Sigma f_i = N$. We shall be interested in situations in which there is some *theoretical frequency* $F_1, F_2, \ldots, F_k$ for each category, where $\Sigma F_i = N$. We shall ask whether the observations agree or disagree with the values $F_1, F_2, \ldots, F_k$; that is, we wish to test the hypothesis that states the values of the theoretical frequencies. The statistic we shall use is

$$\chi^2 = \sum_{i=1}^{k} \frac{(f_i - F_i)^2}{F_i}$$

The sampling distribution of this $\chi^2$ statistic is in approximately the form given in Table A-6*a*, a $\chi^2$ distribution with $k - 1$ degrees of freedom. Thus, for example, if there are 10 categories, we have 9 degrees of freedom, and the chance that $\chi^2$ will be less than 3.33 is .05, the chance that $\chi^2$ will be less than 16.92 is .95, and so on.

Note that the total number of observations does not enter the $\chi^2$ formula except as the total of all the $f_i$'s or the total of the $F_i$'s. In using the $\chi^2$ table we make reference only to the number of categories, and not to the total number of observations. However, for the approximation of the distribution to be close to that in Table A-6*a*, the sample size $N$ must be sufficiently large that none of the $F_i$'s is less than 1 and not more than 20 per cent of the $F_i$'s are less than 5. For 1 degree of freedom a correction for continuity should be applied. This is accomplished by reducing each different $f_i - F_i$ by $\frac{1}{2}$ before computing $\chi^2$.

## 13-2 SINGLE CLASSIFICATION

A single-classification problem is a problem in which the theoretical proportion of cases in each category is specified in advance.

**Table 13-1**

| | *Observed* | *Theoretical* |
|---|---|---|
| 1 | $f_1$ | $F_1$ |
| 2 | $f_2$ | $F_2$ |
| . . . | . . . | . . . |
| $k$ | $f_k$ | $F_k$ |
| *Total* | $N$ | $N$ |

**Table 13-2**

| | *Observed* | *Theoretical* |
|---|---|---|
| White | 1,981 | 2,423.25 |
| Red | 7,712 | 7,269.75 |
| *Total* | 9,693 | 9,693.00 |

EXAMPLE 13-1. In tossing a true coin we expect 50 per cent heads. The theoretical frequencies, therefore, in 140 tosses are 70 heads and 70 tails. If in 140 tosses we actually observe 60 heads and 80 tails, we compute

$$\chi^2 = \frac{(80 - 70 - .5)^2}{70} + \frac{(60 - 70 + .5)^2}{70} = \frac{90.25}{70} + \frac{90.25}{70} = 2.58$$

If the coin does have a 50 per cent chance of coming up heads, then $\chi^2$ has (from Table A-6*a*, with 1 degree of freedom) a 99 per cent chance of being less than 6.63 and a 95 per cent chance of being less than 3.84. The observed value $\chi^2 = 2.58$ is not large enough to cause us to reject the hypothesis (at the 5 per cent level of significance) that the proportion of heads is .5.

EXAMPLE 13-2. In the study of genetics it has been found that certain characteristics are inherited in a ratio 1:3. That is, in the long run, one-fourth of the progeny will have a given characteristic and three-fourths will not. In an experiment in genetics 1,981 fruit flies were found to have white eyes and 7,712 to have red eyes. If we wish to test the hypothesis that the ratio of the number of flies with white eyes to the number of flies with red eyes is 1:3, we have as theoretical frequencies one-fourth and three-fourths of the total, as in Table 13-2. This gives a $\chi^2$ of

$$\frac{(1{,}981 - 2{,}423.25 + .5)^2}{2{,}423.25} + \frac{(7{,}712 - 7{,}269.75 - .5)^2}{7{,}269.75} = 107.6$$

This is much larger than any number in Table A-6*a* for 1 degree of freedom, and the theoretical ratio 1:3 would be rejected.

EXAMPLE 13-3. As another example from genetics we could consider

**Table 13-3**

| | FREQUENCY | | | |
|---|---|---|---|---|
| | *Observed* | *Theoretical* | $f_i - F_i$ | $\frac{(f_i - F_i)^2}{F_i}$ |
| Round and yellow | 315 | 312.75 | 2.25 | .016 |
| Wrinkled and yellow | 101 | 104.25 | −3.25 | .101 |
| Round and green | 108 | 104.25 | 3.75 | .135 |
| Wrinkled and green | 32 | 34.75 | −2.75 | .218 |
| *Total* | 556 | 556.00 | | .470 |

the problem of crossing two types of peas. In one experiment Mendel counted the seeds of plants as shown in Table 13-3. The mendelian theory of inheritance states that the frequencies should be in the ratios 9:3:3:1. That is, $\frac{9}{16}$ should be round and yellow, and so on. The $\chi^2$ statistic here has $4 - 1 = 3$ degrees of freedom. The theoretical frequency of round and yellow peas is $\frac{9}{16} \times 556 = 312.75$, of wrinkled and yellow is $\frac{3}{16} \times 556 = 104.25$, and so on. Here $\chi^2 = .470$, the 5 per cent critical value is 7.81, so we do not have sufficient reason to reject the hypothesis.

## 13-3 TWO-WAY CLASSIFICATION: INDEPENDENCE

In this type of problem the observations are classified by two characteristics.

EXAMPLE 13-4. Hair and eye color might be the classifications, with the observed results as recorded in Table 13-4. If the theoretical frequencies for the six categories are given, the problem would be a one-variable-of-classification problem. The last example in Sec. 13-2 is not a two-way-classification example because the four theoretical proportions have been stated.

We shall examine a method of testing the hypothesis that the two characteristics are *independent*. Here the term "independent" has the same meaning as in Chap. 11: the distribution of one characteristic should be the same regardless of the other characteristic. For example, if eye color and hair color are independent, then the proportion of the blue-eyed people having light-colored hair should be the same as the proportion of the brown-eyed people having light-colored hair, etc.; these theoretical proportions are, of course, for the population, not for the sample. Our procedure will be to examine the proportions in the sample and to determine whether or not they are significantly different. If they are significantly different we reject the hypothesis that the two characteristics are independent, and if they are not significantly different we say that the sample does not contradict the hypothesis.

The statistic we use to compare the proportions is a $\chi^2$ statistic obtained in the following manner. First we record the number of observations that

**Table 13-4**

| EYE COLOR | HAIR COLOR *Light* | *Dark* | *Total* |
|---|---|---|---|
| *Blue* | 32 | 12 | 44 |
| *Brown* | 14 | 22 | 36 |
| *Other* | 6 | 9 | 15 |
| *Total* | 52 | 43 | 95 |

**Table 13-5**

| EYE COLOR | HAIR COLOR *Light* | *Dark* | *Total* |
|---|---|---|---|
| *Blue* | 32 (24.1*) | 12 (19.9*) | 44 |
| *Brown* | 14 (19.7*) | 22 (16.3*) | 36 |
| *Other* | 6 (8.2*) | 9 (6.8*) | 15 |
| *Total* | 52 | 43 | 95 |

fall into each category. In Table 13-4 there are 32 people having blue eyes and light hair, and so on. Next we look at each characteristic separately and record the total number of observations as before. Here there are 52 people with light hair, 43 with dark, 44 people with blue eyes, and so on. We find theoretical frequencies by using these "marginal" totals. We note that 52 out of 95 people had light hair. If the characteristics are independent, we should expect to find the same proportion of light-haired people among those who have blue eyes. Since we observe 44 people with blue eyes, we should expect to find that $\frac{52}{95} \times 44 = 24.1$ of these people have light hair. Also, we should expect $\frac{43}{95} \times 44 = 19.9$ of them to have dark hair. These two numbers, of course, add to 44, and the second could have been found by subtracting the first from 44. Similarly, the theoretical frequency of people with brown eyes and light hair is $\frac{52}{95} \times 36 = 19.7$. Table 13-5 shows both the observed and the theoretical frequencies. The theoretical frequencies are asterisked. The $\chi^2$ statistic is computed from the six categories in exactly the same fashion as before. The distribution of this statistic can be shown to be a $\chi^2$ distribution, given in Table A-6*a*, with, if $r$ denotes the number of rows and $c$ the number of columns, $(r - 1)(c - 1)$ degrees of freedom. Thus in this example there are $(3 - 1)(2 - 1) = 2$ degrees of freedom. Here

$$\chi^2 = \frac{(7.9)^2}{24.1} + \frac{(-7.9)^2}{19.9} + \frac{(-5.7)^2}{19.7} + \frac{(5.7)^2}{16.3} + \frac{(-2.2)^2}{8.2} + \frac{(2.2)^2}{6.8}$$
$$= 2.59 + 3.14 + 1.65 + 1.99 + .59 + .71 = 10.67$$

For 2 degrees of freedom $\chi_{.95}^2 = 5.99$. The hypothesis of independence is rejected at the 5 per cent level of significance. The $\chi^2$ approximation will be adequate if the minimum theoretical frequency is 2 or if no more than 20 per cent of the theoretical frequencies are less than 5 with a minimum of 1.

Two-way-classification tables of this type are frequently called *contingency tables.* If the totals for rows (or columns) are specified in advance, the test is called a test of *homogeneity.* We are actually testing that the various columns (or rows) have the same proportions of individuals in the various categories. The procedure is exactly the same as in the test of independence whether or not the marginal totals are specified.

**Two-by-two tables.** In general the test for independence can be carried out as above. In the case where there are only two rows and two columns the computing forms can be simplified. In the case of 1 degree of freedom the approximation of the discrete sampling distribution of the $\chi^2$ as computed to the continuous $\chi^2$ distribution in Table A-6*a* can be markedly improved by reducing the absolute value of each difference by .5 before it is squared. This modification is sometimes called a *continuity correction* or *Yates' correction*. In particular, for a 2 × 2 table we can include this correction and simplify the $\chi^2$ statistic formula. The minimum theoretical frequency for the 2 × 2 table should not be less than 5.

**Table 13-6**

| | I | II | *Total* |
|---|---|---|---|
| 1 | $a$ | $b$ | $a + b$ |
| 2 | $c$ | $d$ | $c + d$ |
| *Total* | $a + c$ | $b + d$ | $a + b + c + d = N$ |

Suppose we have two rows and two columns and have observed frequencies $a$, $b$, $c$, $d$, as in Table 13-6. Then the $\chi^2$ statistic for the test of independence, corrected as above, can be written in the form

$$\chi^2 = \frac{(|ad - bc| - \frac{1}{2}N)^2 N}{(a + b)(a + c)(b + d)(c + d)}$$

EXAMPLE 13-5. Suppose that in a public-opinion survey answers are tabulated to the following questions:

1. Do you drink beer?
2. Are you in favor of local option on the sale of liquor?

| | QUESTION 1 | | |
|---|---|---|---|
| QUESTION 2 | *Yes* | *No* | *Total* |
| *Yes* | 37 | 20 | 57 |
| *No* | 15 | 6 | 21 |
| *Total* | 52 | 26 | 78 |

$$\chi^2 = \frac{(|222 - 300| - 39)^2 \times 78}{52 \times 26 \times 57 \times 21} = \frac{(78 - 39)^2 \times 78}{52 \times 26 \times 57 \times 21}$$

$$= \frac{118{,}638}{1{,}618{,}344} = .073 \qquad \chi^2_{.95}(1) = 3.84$$

and the hypothesis of independence is accepted; we do not have sufficient reason to say that opinion on local option is dependent on whether or not an individual drinks beer.

EXAMPLE 13-6. *Fisher's exact test.* Suppose that a number of patients were treated for cancer with results as in the following table:

| TOXICITY PRESENT | TUMOR REGRESSION *Yes* | *No* | *Total* |
|---|---|---|---|
| *Yes* | $5 = X$ | 2 | $7 = S_2$ |
| *No* | 1 | 7 | 8 |
| *Total* | $6 = S_1$ | 9 | $15 = N$ |

The observed frequencies are too small for the sampling distribution of the $\chi^2$ statistic to be adequately approximated by Table A-6*a*. We may use Table A-9*e* to provide an "exact" test of independence for this $2 \times 2$ table.

The notation used in Table A-9*e* follows the example and is made to depend on the particular configuration of numbers observed. Here $N$ is the total number of observations (15 in the example), $S_1$ is the smallest marginal total (6), $S_2$ is the next smallest marginal total (7), and $X$ is the frequency in the cell corresponding to the two smallest marginal totals (5).

Table A-9*e* is entered with these values and the following line is obtained:

| $N$ | $S_1$ | $S_2$ | $X$ | PROBABILITY *Obs.* | *Other* | *Total* |
|---|---|---|---|---|---|---|
| 15 | 6 | 7 | 5 | .035 | .006 | .041 |

These values indicate that for the given marginal totals:

1. The one-tail probability of equaling or exceeding the deviation from equal proportions in the same direction of the observed difference is .035.

2. The probability of as large or larger deviation from equal proportions in the opposite direction is .006.

3. The probability of as large or larger deviation in either direction is the total of these numbers, $.035 + .006 = .041$. This total could be interpreted as the probability of having a value of $\chi^2$ as large as or larger than the one observed.

In this example the choice of a 5 per cent level of significance would lead to the rejection of the hypothesis of equal proportions.

## TEST FOR GOODNESS OF FIT 13-4

The term "goodness of fit" refers to the comparison of some observed sample distribution with a theoretical frequency distribution. Actually, all the tests for which we are using $\chi^2$ are problems of this type. In this section we shall make the comparison of a sample distribution with a normal distribution. That is, we shall test the hypothesis that the universe has a normal distribution. The general technique is the same for any theoretical distribution except for counting the degrees of freedom.

Suppose we have a sample of $N$ observations which have mean $\bar{X}$ and variance $s^2$. The normal curve fitted to these data has the equation

$$Y = \frac{N}{s\sqrt{2\pi}} e^{-\frac{1}{2}\left(\frac{X-\bar{X}}{s}\right)^2}$$

From Table A-4 and the techniques explained in Chap. 5 we can find the area under this curve between any two points. Corresponding to categories, we divide the range of the variable $X$ into a number of intervals. The choice of intervals is arbitrary, except that the smallest theoretical frequency should be 1 or larger. The *theoretical frequency* is the area under the normal curve in the interval, while the *observed frequency* is the actual number of observations which fall in the interval. The $\chi^2$ statistic is then computed in the same fashion as before. The number of degrees of freedom for this example is $k - 3$, where $k$ is the number of categories. In general, in measuring the fit of any frequency curve the number of degrees of freedom decreases one for each parameter estimated from the sample. Here we estimated the mean and standard deviation and thus lost 2 degrees of freedom,

$$(k - 1) - 2 = k - 3$$

EXAMPLE 13-7. Suppose we have a sample of 100 observations as shown in Table 13-7. Since the observed value of 1.99 is smaller than 9.49, we accept, at the 5 per cent level of significance, the hypothesis that the distribution is normal.

**Table 13-7**

| *Mid-point* | *Frequencies* | $\frac{X - \bar{X}}{s}$ *at end points* | *Interval* | *Theoretical frequency* | *Observed frequency* | $\frac{(f_i - F_i)^2}{F_i}$ |
|---|---|---|---|---|---|---|
| 145 | 8 | +1.60 | Above 140 | 5.5 | 8 | 1.14 |
| 135 | 9 | +1.00 | 130–140 | 10.4 | 9 | .19 |
| 125 | 16 | + .40 | 120–130 | 18.6 | 16 | .36 |
| 115 | 23 | − .20 | 110–120 | 23.5 | 23 | .01 |
| 105 | 21 | − .80 | 100–110 | 20.9 | 21 | .00 |
| 95 | 15 | −1.40 | 90–100 | 13.1 | 15 | .29 |
| 85 | 8 | | Below 90 | 8.1 | 8 | .00 |
| *Total* | 100 | | | 100.1 | 100 | 1.99 |

$\bar{X} = 113.3$ $s = 16.65$ $\chi_{.95}^2(4) = 9.49$

In Chaps. 5 and 7 we examined dichotomous populations, those in which a proportion $p$ of the individuals have a certain characteristic and a proportion $1 - p$ of the individuals do not have it. We saw that we could discuss this case as a special case of an arithmetic mean by assigning the score 1 to those individuals having the characteristic and the score 0 to the rest. For large samples we drew inferences under the assumption that the sampling distribution of the proportion $\bar{X}$ of the sample having the characteristic was normal. Let us now discuss the problem more exactly for small samples.

As usual, we assume that we have a random sample of $N$ individuals from the population. The assumption of randomness is important, and the analysis depends on it. Although this statement has been made before, it is emphasized here because of the wide use of stratified and quota sampling designs in public-opinion polls and other samples of human populations. Frequently these designs produce more precise results than those obtained from random sampling, but their analysis is different from that presented here.

Note that $p$ is a parameter. Since the use of $p$ is widely accepted as the population proportion, we are using it instead of a Greek letter.

The best estimate of $p$ is the proportion, say, $\bar{X} = X/N$, observed in the sample. Here $X$ is the number in the sample having the characteristic and $N$ is the size of the sample. $X/N$ is a statistic which can only have as its value one of the sequence

$$0, \frac{1}{N}, \frac{2}{N}, \frac{3}{N}, \ldots, \frac{N-1}{N}, 1$$

For random samples the sampling distribution of $X/N$ is given by

$$\frac{1 \times 2 \times 3 \cdots (N-2)(N-1)N}{(1 \times 2 \times 3 \cdots X)[1 \times 2 \times 3 \cdots (N-X)]} p^X(1-p)^{N-X}$$

for $X = 0, 1, 2, 3, \ldots, N$. A distribution where the frequencies are given by this formula is called a *binomial distribution*. The formula indicates the chance, or relative frequency, that the statistic $X/N$ will appear as a particular one of the values which it can take on. This is indicated in Fig. 13-1 in tabular and graphical form. The reasoning leading to this formula is mathematical in nature and is outlined in Chap. 20.

We have drawn a histogram here, although actually the variable $X/N$ can only be one of a discrete series of values. This somewhat inexact method of depicting the distribution is used so that we may continue to interpret the area as frequencies.

The computation involved in finding the values of these theoretical frequencies is rather formidable even for small values of $N$, and for large values of $N$ it is prohibitive. Table A-29$a$ lists values for sample sizes less

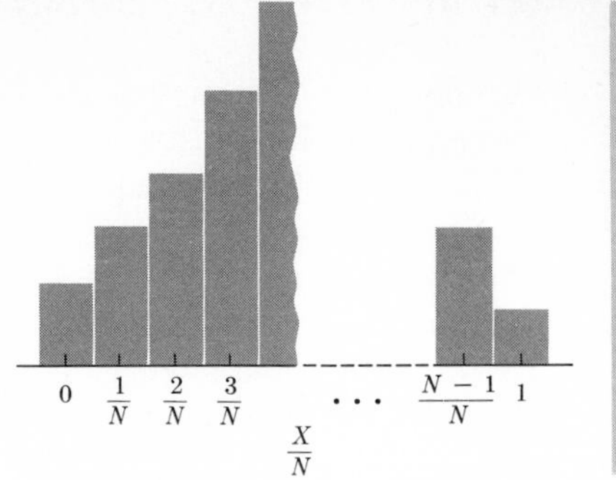


| $X$ | Theoretical frequency |
|---|---|
| 0 | $(1-p)^N$ |
| 1 | $Np(1-p)^{N-1}$ |
| 2 | $N\frac{N-1}{2}p^2(1-p)^{N-2}$ |
| 3 | $N\frac{N-1}{2}\frac{N-2}{3}p^3(1-p)^{N-3}$ |
| ⋮ | ⋮ |
| $N$ | $p^N$ |

**Figure 13-1** Binomial frequency formula and histogram.

than or equal to 10. As pointed out in Chap. 5, it can be shown that if $Np$ and $N(1-p)$ are fairly large, say, each larger than 5, the statistic

$$\frac{\bar{X}-p}{\sqrt{p(1-p)/N}}$$

has an approximately normal sampling distribution with mean 0 and unit variance. A verification of this approximation can be shown by sampling experiments. The mean and variance of the observed $\bar{X}$ distribution should be approximately $p$ and $p(1-p)/N$, and the cumulative should be approximately a straight line on normal-probability paper (see, for example, the experiment described in Chap. 5).

The chance of observing a value of $\bar{X}$ between $\bar{X} = a$ and $\bar{X} = b$ with $b > a$ is approximately the area under the unit normal curve between

$$z = \frac{a - 1/(2N) - p}{\sqrt{p(1-p)/N}} \qquad \text{and} \qquad z = \frac{b + 1/(2N) - p}{\sqrt{p(1-p)/N}}$$

The term $1/(2N)$ is included so that the area under the continuous curve will better approximate the sum of the probabilities of $\bar{X}$ falling at the points $a$, $(a + 1/N)$, . . . , $b$. That is, we wish the entire area corresponding to the rectangles to be centered at these points.

For large samples, [$Np > 5$, and $N(1-p) > 5$], $100(1-\alpha)$ per cent confidence limits for estimating $p$ are

$$\frac{N}{N + z_{1-\frac{1}{2}\alpha}^2}\left\{\bar{X} \mp \frac{1}{2N} + \frac{z_{1-\frac{1}{2}\alpha}^2}{2N} \mp z_{1-\frac{1}{2}\alpha}\sqrt{\frac{[\bar{X} \pm 1/(2N)][1 - \bar{X} \mp 1/(2N)]}{N} + \frac{z_{1-\frac{1}{2}\alpha}^2}{4N^2}}\right\}$$

where $z_{1-\frac{1}{2}\alpha}$ is read from Table A-4. If $N$ is very large these limits are approximately

$$\bar{X} + z_{\frac{1}{2}\alpha}\sqrt{\frac{\bar{X}(1-\bar{X})}{N}} < p < \bar{X} + z_{1-\frac{1}{2}\alpha}\sqrt{\frac{\bar{X}(1-\bar{X})}{N}}$$

as given in Sec. 7-2.

Similarly, for large samples we can test the hypothesis that $p$ has some specified value, say, $p = p_0$, at any desired level of significance by comparing the value of

$$z = \frac{\bar{X} - p_0 \pm 1/(2N)}{\sqrt{p_0(1 - p_0)/N}}$$

with values from Table A-4. If $z < z_{\frac{1}{2}\alpha}$ or if $z > z_{1-\frac{1}{2}\alpha}$, we should reject the hypothesis at the $\alpha$ level of significance. We use the plus sign with the factor $1/(2N)$ when comparing with $z_{\frac{1}{2}\alpha}$ and the minus sign when comparing with $z_{1-\frac{1}{2}\alpha}$.

In Table A-9 there are graphs giving confidence limits for $p$ corresponding to the observed $\bar{X}$ for several different sample sizes. Four graphs are given for 80 per cent, 90 per cent, 95 per cent, and 99 per cent confidence limits. The observed $\bar{X}$ is read on the horizontal scale, and values of $p$ are read from the vertical. Various belt-shaped curves, two for each $N$, indicate the confidence interval. The confidence limits for $p$ are read above the observed $\bar{X} = X/N$; those in Fig. 13-2 show 80 per cent confidence limits for $N = 10$. Suppose that for 10 items $\bar{X} = .6$. The confidence limits are .35 and .82. For larger $N$ the curves are closer together, and we obtain shorter confidence intervals.

We can test the hypothesis that $p = p_0$ at the $\alpha$ level of significance by observing whether or not $p_0$ is in the $100(1 - \alpha)$ per cent confidence interval corresponding to the observed $\bar{X}$. If it is, we accept the hypothesis $p = p_0$; if it is not, we reject the hypothesis.

Example 13-8. In a random sample of 100 articles there were 40 having a certain characteristic. Find 90 per cent confidence limits for the proportion of the population having the characteristic. We use the 90 per cent graph for $N = 100$. Above $\bar{X} = \frac{40}{100} = .40$ we read the limits as .32 and .48. Thus we are 90 per cent confident that $p$ is between .32 and .48.

Example 13-9. The binomial distribution can also be used to state the chance of rejecting at least one of $N$ true statistical hypotheses tested independently at the $\alpha$ level of significance. The chance of rejecting exactly

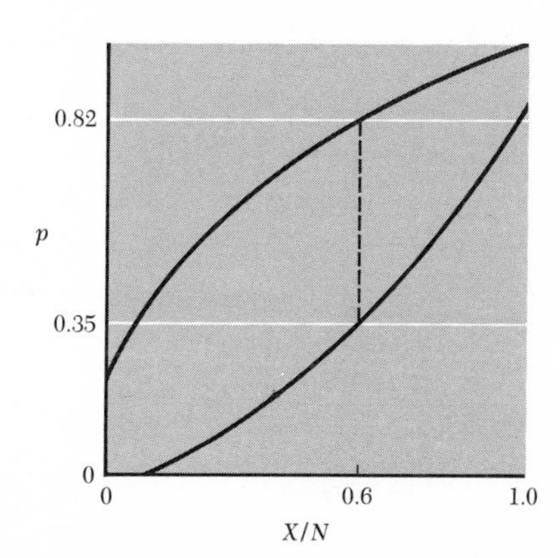


**Figure 13-2** Graph illustrating method of reading confidence limits for population proportion $p$.

none is $(1 - \alpha)^N$, and the chance of rejecting at least one is $1 - (1 - \alpha)^N$. Values of $(1 - \alpha)^N$ can be read from Table A-29. We read from the table for $p = \alpha$ and $X = 0$. For example, if $N = 5$ independent tests are performed at the $\alpha = .01$ level of significance, we read in the table under $p = .01$, $N = 5$, $X = 0$ that $(1 - \alpha)^N = .9510$, and so if all hypotheses are true, the chance of rejecting at least one is $1 - .9510 = .0490$.

**Poisson distribution.** If $N$ is large and $Np$ is fairly small, say, smaller than 5, the area under the normal curve will not be a good approximation to the area under the binomial-distribution histogram. In this case the value of

$$\frac{1 \times 2 \times 3 \cdots N}{(1 \times 2 \times 3 \cdots X)[1 \times 2 \times 3 \cdots (N - X)]} p^X(1 - p)^{N-X}$$

is given approximately by

$$e^{-Np} \frac{(Np)^X}{1 \times 2 \times 3 \cdots X} \qquad X = 0, 1, 2, 3, \cdots$$

A distribution where the frequencies are given by this formula is called a *Poisson distribution*. The only parameter in this formula is $Np$, which can be treated as a single quantity and is denoted by $\lambda$. Sometimes this value is known where $N$ alone is not. It happens that both the mean and variance of this distribution are equal to $\lambda$.

Table A-15 gives values of the cumulative Poisson distribution for various values of $\lambda$ and $X$. Values of $\lambda$ are recorded on the left of the page, and values of $X$ are across the top. The numbers in the table are the cumulated relative frequency (or probability) for values less than or equal to $X$. For example, if $\lambda = 1.2$ we read that 66.3 per cent of the time $X$ will have the values 0 or 1, 87.9 per cent of the time $X$ will have the values 0 or 1 or 2, and so on. These tabled numbers are actually $\sum_{k=0}^{k=X} e^{-\lambda} \frac{\lambda^k}{1 \times 2 \times 3 \cdots k}$, and this formula could be used for computation.

Further material on the binomial and Poisson distributions is in Chap. 20.

**Goodness of fit.** The general procedure of Sec. 13-4 using $\chi^2$ can be used to compare any hypothetical distribution with an observed distribution. However, for the binomial and Poisson distributions a comparison of the variances of the observed and theoretical distributions is recommended as an alternative.

For the binomial we compute

$$\chi^2 = \frac{N\Sigma(\bar{X}_i - p)^2}{p(1 - p)}$$

if $p$ is known. Here $\bar{X}_1, \bar{X}_2, \ldots, \bar{X}_n$ are the successive proportions observed, in samples of size $N$. This statistic has an approximate $\chi^2$ distribution with $n$ degrees of freedom when the hypothesis is correct. If $p$ is estimated by $\bar{X} = \Sigma \bar{X}_i/n$ and substituted in this formula for $p$, the degrees of freedom are $n - 1$.

In the Poisson case we compute

$$\chi^2 = \frac{\Sigma(X_i - \lambda)^2}{\lambda}$$

which analogously has an approximate $\chi^2$ distribution with $n$ degrees of freedom, or if $\bar{X} = \Sigma X_i/n$ obtained from the sample is substituted for $\lambda$ in this formula, it has $n - 1$ degrees of freedom.

## DIFFERENCE IN PROPORTIONS 13-6

Suppose we have two populations in which proportions $p_1$ and $p_2$, respectively, have some characteristic. We take random samples of size $N_1$ and $N_2$ from the populations and observe proportions $\bar{X}_1$ and $\bar{X}_2$. If $N_1$ and $N_2$ are large, say, such that $N_1p_1$, $N_2p_2$, $N_1(1 - p_1)$, $N_2(1 - p_2)$ are all larger than 5, we can estimate the difference between $p_1$ and $p_2$ by the following approximate confidence interval:

$$\bar{X}_1 - \bar{X}_2 + z_{\frac{1}{2}\alpha} \sqrt{\frac{\bar{X}_1(1 - \bar{X}_1)}{N_1} + \frac{\bar{X}_2(1 - \bar{X}_2)}{N_2}} < p_1 - p_2$$
$$< \bar{X}_1 - \bar{X}_2 + z_{1-\frac{1}{2}\alpha} \sqrt{\frac{\bar{X}_1(1 - \bar{X}_1)}{N_1} + \frac{\bar{X}_2(1 - \bar{X}_2)}{N_2}}$$

EXAMPLE 13-10. Suppose that in a survey of 400 people from one city 188 preferred a brand $A$ soap to all others and in a sample of 500 people from another city 210 preferred the same product.

$$\bar{X}_1 = \tfrac{188}{400} = .47 \qquad 1 - \bar{X}_1 = .53 \qquad \bar{X}_2 = \tfrac{210}{500} = .42 \qquad 1 - \bar{X}_2 = .58$$

The 95 per cent confidence limits for $p_1 - p_2$ are

$$.47 - .42 - 1.96 \sqrt{\frac{.47 \times .53}{400} + \frac{.42 \times .58}{500}} < p_1 - p_2$$
$$< .47 - .42 + 1.96 \sqrt{\frac{.47 \times .53}{400} + \frac{.42 \times .58}{500}}$$

or

$$.05 - .065 < p_1 - p_2 < .05 + .065$$
$$-.015 < p_1 - p_2 < .115$$

The hypothesis $H: p_1 = p_2$ can be tested by use of the statistic

$$z = \frac{\bar{X}_1 - \bar{X}_2}{\sqrt{\bar{p}(1 - \bar{p})(1/N_1 + 1/N_2)}}$$

where $\bar{p} = (N_1\bar{X}_1 + N_2\bar{X}_2)/(N_1 + N_2)$ is an estimate of the population proportion, equal in the two populations by hypothesis. For moderately large values of $N_1$ and $N_2$ the sampling distribution of $z$ is approximately normal with variance 1 and, if the hypothesis is true, has mean 0, so that Table A-4 can be used to define a critical region.

The use of the $z$ statistic above is equivalent to the $\chi^2$ test in Sec. 13-3, except it does not have the continuity correction. If the values of $N_1$ and $N_2$ are small, the distribution of the $\chi^2$ with continuity correction is somewhat better approximated by Table A-6*a*.

The rule of rejecting the hypothesis $p_1 - p_2 = 0$ if the above confidence limits for $p_1 - p_2$ do not include zero is approximately equivalent to the described test but has slightly less power. Thus, for example, approximate 95 per cent confidence limits for the difference in population proportions is $-.015 < p_1 - p_2 < .115$, and the value $p_1 - p_2 = 0$ would not be rejected.

**Correlated proportions.** A group of individuals are asked a question on public affairs which they are to answer yes or no. After a certain period of time or after a propaganda lecture they are asked the same question again. If we wish to test the hypothesis that the answers to the second question are independent of the answers to the first, these data can be analyzed as a contingency table. However, this analysis would not answer the question of whether the passage of time or the propaganda lecture was effective in changing the proportion saying yes.

| | BEFORE | | |
|---|---|---|---|
| AFTER | *Yes* | *No* | *Total* |
| *Yes* | 30 | 15 | 45 |
| *No* | 9 | 51 | 60 |
| *Total* | 39 | 66 | 105 |

Example 13-11. The data are recorded in the table. The proportion saying yes originally is $\frac{39}{105}$ and those saying yes finally is $\frac{45}{105}$. This case of difference of proportions differs from the case discussed above, since these proportions are not independent (both include the 30 individuals saying yes both times). The standard deviation of this difference can be estimated, with the notation of Table 13-6, by

$$\sqrt{\frac{b + c - (b - c)^2/N}{N(N - 1)}} = \sqrt{\frac{15 + 9 - (15 - 9)^2/105}{105 \times 104}} = .0466$$

An approximate 95 per cent confidence interval for the increase in the population answering yes is $(\frac{45}{105} - \frac{39}{105}) \pm (1.96 \times .0466)$, giving the interval $-.0342$ to $.1484$.

A test of the null hypothesis of no change in opinion can also be made

**Table 13-8**

| SECOND CLASSIFICATION | FIRST CLASSIFICATION 1 | 2 | . . . | $k$ |
|---|---|---|---|---|
| 1 | $n_{11}$ | $n_{21}$ | . . . | $n_{k1}$ |
| 2 | $n_{12}$ | $n_{22}$ | . . . | $n_{k2}$ |
| . . . | . . . | . . . | . . . | . . . |
| $k$ | $n_{1k}$ | $n_{2k}$ | . . . | $n_{kk}$ |

by testing the hypothesis that of those who changed their minds an equal number changed from yes to no and from no to yes. In this example we could test whether or not the observed binomial $\bar{X} = \frac{9}{24}$ is significantly different from $p = .50$ as in Sec. 13-5.

This test is in reality a test of symmetry and can be applied to tables with more than two categories. Suppose the data are recorded as in Table 13-8, where $n_{ij}$ represents the observed frequency in the $i$th column and $j$th row. We wish to test the hypothesis that the population frequencies are the same in symmetrically located cells, or that for all $i$'s and $j$'s the frequency in the $ij$ cell is equal to the frequency in the $ji$ cell. This hypothesis can be tested by computing the sum

$$\sum_{i<j} \frac{(n_{ij} - n_{ji})^2}{n_{ij} + n_{ji}}$$

The summation extends over all cells in the table where $i$ is less than $j$. When the hypothesis is true and the total frequency is large, this statistic has approximately a $\chi^2$ distribution with $k(k - 1)/2$ degrees of freedom.

## BINOMIAL-PROBABILITY PAPER 13-7

There is a type of graph paper, called *binomial-probability paper*, which frequently can be used in analyzing enumeration data. Figure 13-3 shows a sheet of this paper. The scales are marked off for observations, but the units of length change as the distance from the origin changes. Actually, the distance from the origin is the square root of the coordinate. We shall give here only three examples of the use of this paper, without going into detail about why it can be used in the stated manner.

A straight line through the origin passes through points whose coordinates are proportional. For example, a line through the origin and (80,20) also passes through (160,40), (40,10), (20,5), and so on. A line through the origin on this paper is called a *split* and is referred to in terms of the coordinates of the point where the line cuts the quarter circle marked on the graph.

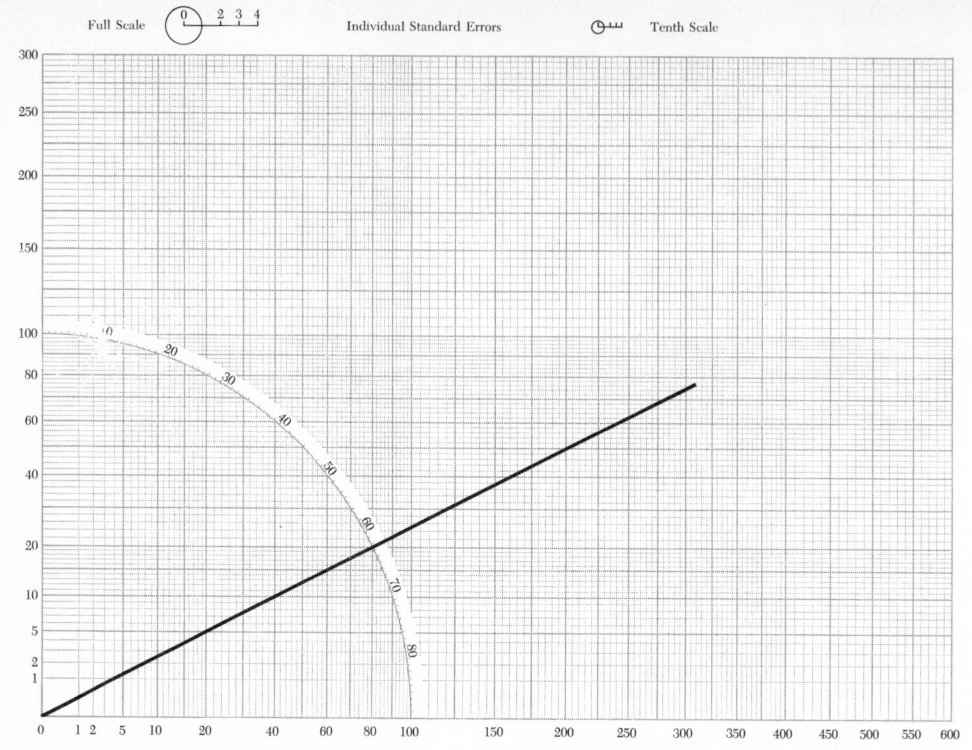


**Figure 13-3** Binomial-probability paper with 80-20 split drawn.

Note that these coordinates add up to 100. Thus the line on Fig. 13-3 is the 80:20 split.

*A paired count* refers to the numbers in the sample observed as having and as not having some characteristic. Thus if 80 in a sample of $N = 100$ have a particular characteristic, we say the paired count is (80,20). A paired count is plotted as a right triangle, with the vertex plotted in the usual manner and then two sides extended 1 unit parallel to the horizontal and vertical axes, respectively. For example, (5,7) is plotted with (5,7) as the right-angle vertex and with (6,7), (5,8) as the other two vertices. If one or both of the coordinates is larger than 100, the addition of 1 will not show on the paper, and the triangle appears as a short line (1 unit long) or as a point.

The distances from any split to the two acute angles of a plotted point are called the *short distance* and the *long distance*, respectively. These distances can be interpreted by referring to the scale at the top of the paper marked "Full scale, individual standard errors." The units on this scale refer to a normal-probability scale, and thus a deviation of 2 full-scale units would be considered as just significant at the 5 per cent level for a two-sided test. The probabilities corresponding to the long and short distances form a *significance zone,* and test results are significant at some level between them. For example, a zone of (15%, .1%) would not indicate significance at the 5 per cent level, while the zone (4%, .1%) would. The use of this zone instead of a single level takes into consideration the possible effect of another observation on our conclusion.

EXAMPLE 13-12. Suppose we wish to test the hypothesis that 50 per cent of a population has some particular characteristic. If we observe that 40 out of a sample of 100 have the characteristic, is there reason to reject the hypothesis at the 5 per cent level of significance?

*Answer.* Draw the 50:50 split (corresponding to the theoretical proportions of 50:50). Plot the paired count (40,60) and measure the long and short distances to be 2 and 2.1 (see Fig. 13-4). This gives a significance zone of (4.6%,3.6%), and we thus have reason to doubt the 50:50 hypothesis at the 5 per cent level of significance.

EXAMPLE 13-13. Form 95 per cent confidence limits for the population proportion if we obtain 321 individuals having the characteristic out of a sample of 500.

*Answer.* Draw splits which would be accepted at the 5 per cent level of significance, that is, within 2 full-scale units of (321,179). The proportions of these two splits are the desired limits, .60 and .69 for this problem.

EXAMPLE 13-14. Suppose the proportion of a population having a

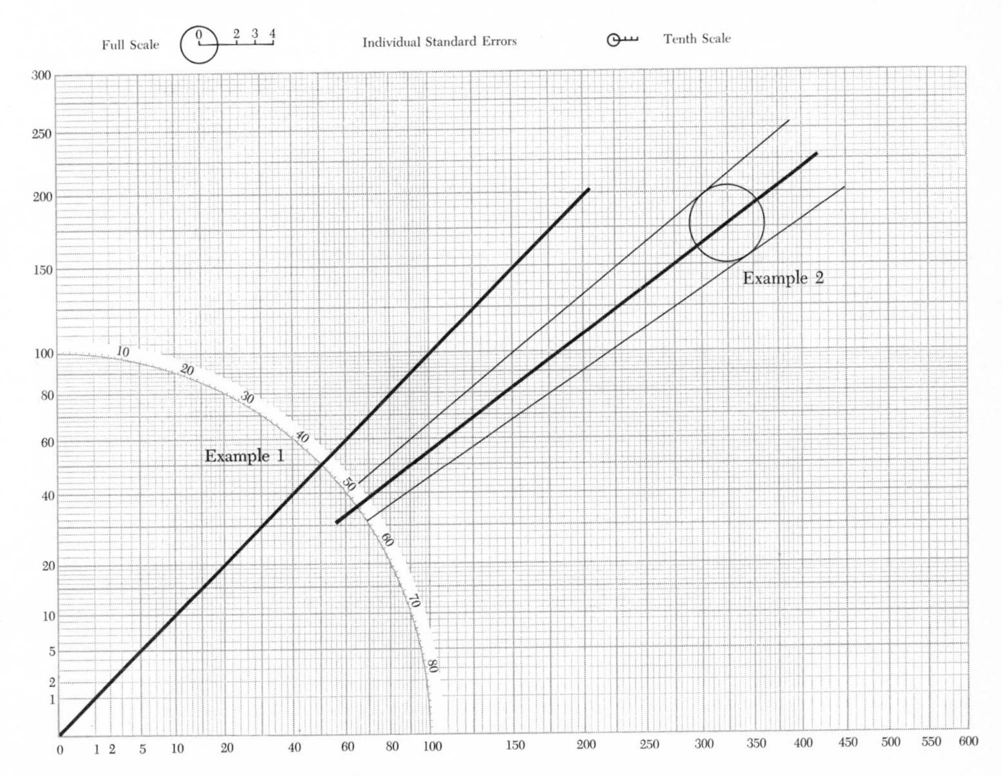


**Figure 13-4** Binomial-probability paper with examples of its use.

certain characteristic is either .50 or .25. A two-sided test of the hypothesis $p = .50$ is made at the 5 per cent level of significance. How large a sample should be taken to have a .005 chance of rejecting this hypothesis if $p = .25$?

*Answer.* Draw the 50:50 split and two parallel lines 2 full-scale units on each side of it. Points inside these parallel lines are acceptable for the 50:50 hypothesis at the 5 per cent level. Draw the 25:75 split and two lines 2.6 full-scale units on either side of it. Points inside this band are acceptable for $p = .25$ at the 1 per cent level of significance. The inside lines intersect at (32,51), and it may be noted that if a sample has size larger than $32 + 51 = 83$, then it is impossible to accept both $p = .50$ and $p = .25$ at the desired levels of significance, and the test must therefore indicate which is to be accepted. In a practical case it may happen that both values are rejected; that is, we may reject both .50 and .25.

Two things should be emphasized. The paper provides a quick method of obtaining (approximately) the same numbers arrived at by the other techniques. Also, we have only touched upon the uses of this type of graph paper. Rough approximations to the results of $t$ tests, analysis of variance, tolerance limits, two-sample tests, goodness-of-fit tests, and so on, can also be read from the paper. For these and further applications see the references at the end of the book.

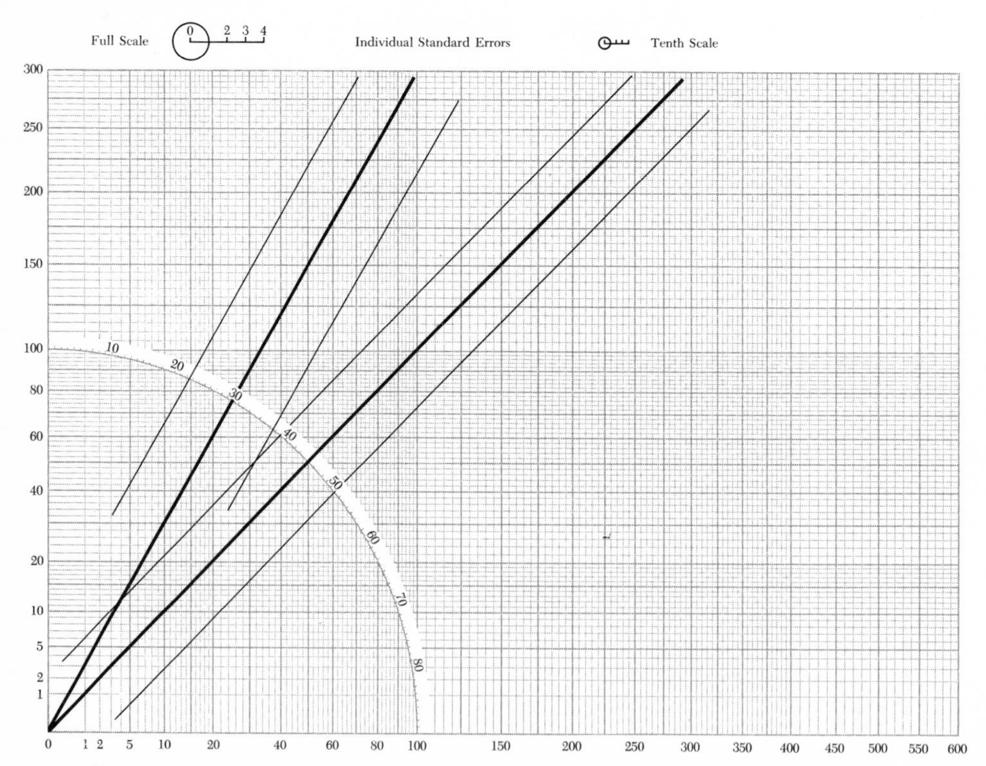


**Figure 13-5** Binomial-probability paper with examples of its use.

Control charts may be constructed for proportions in a manner similar to the construction of charts for the mean and range given in Chap. 9. Suppose $p$ is the proportion of the population having the characteristic. We have noted that the proportion $\bar{X}$ of a sample which has the characteristic has a sampling distribution which is approximately normal for large sample sizes. We stated the formula for the variance of $\bar{X}$ as

$$\sigma_{\bar{X}}^2 = \frac{p(1-p)}{N}$$

If some standard percentage $p$ is given and we wish to construct control lines, it is customary to draw lines $3\sqrt{p(1-p)/N}$ units above and below $p$. Usually the data are recorded in tabular form as well as on a graph.

EXAMPLE 13-15. Table 13-9 shows data for $p = .2$ and $N = 100$, which are pictured on the graph in Fig. 13-6. In the first sample there were

**Table 13-9**

| *Sample no.* | *Sample size* | *Observed* |
|---|---|---|
| 1 | 100 | .12 |
| 2 | 100 | .19 |
| 3 | 100 | .16 |
| 4 | 100 | .34 |
| 5 | 100 | .30 |
| 6 | 100 | .10 |
| 7 | 100 | .27 |

12 items having the characteristic; thus $\bar{X} = \frac{12}{100} = .12$. The control limits are $.2 \pm 3\sqrt{.2 \times .8/100}$, or $.2 \pm .12$.

If the sample size changes, the upper and lower control limits also change, and they are frequently recorded as in Table 13-10. The factor $3\sqrt{p(1-p)}$ is constant for all $N$, and this can be divided by the various $\sqrt{N}$. Again we use $p = .2$. Figure 13-7 shows the control lines and observations for the results in this table.

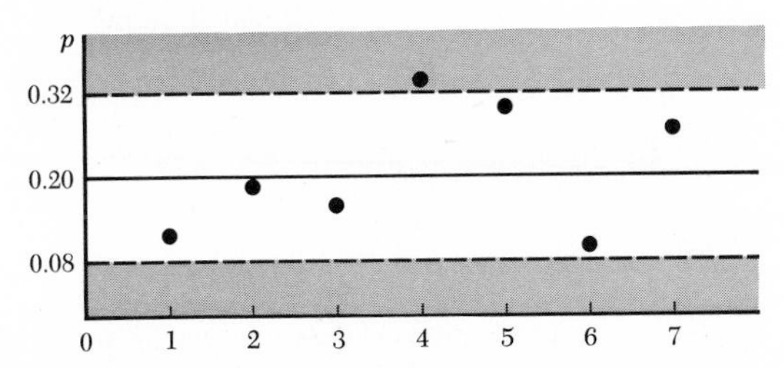


**Figure 13-6** Control chart for proportions.

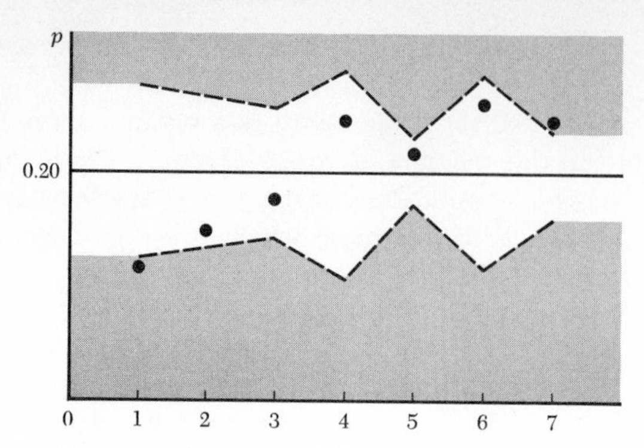


**Figure 13-7** Control chart for proportions. Unequal sample sizes.

**Table 13-10**

| *Sample no.* | *Sample size* | *Observed* $\bar{X}$ | $\frac{3\sqrt{.2 \times .8}}{\sqrt{N}}$ | *Upper control limit* | *Lower control limit* |
|---|---|---|---|---|---|
| 1 | 250 | .120 | .076 | .276 | .124 |
| 2 | 320 | .150 | .067 | .267 | .133 |
| 3 | 450 | .177 | .057 | .257 | .143 |
| 4 | 175 | .246 | .091 | .291 | .109 |
| 5 | 1,840 | .217 | .028 | .228 | .172 |
| 6 | 200 | .260 | .085 | .285 | .115 |
| 7 | 900 | .244 | .040 | .240 | .160 |

## GLOSSARY

binomial distribution
binomial-probability paper
$\chi^2$ distribution
$\chi^2$ statistic
contingency table
continuity correction
goodness of fit
homogeneity of proportions
independence
Poisson distribution
proportion
single classification
two-way classification
Yates' correction

## DISCUSSION QUESTIONS

1. What is the difference between a $\chi^2$ distribution and a $\chi^2/df$ distribution?

2. What is the difference between a single-classification and a two-way-classification experiment?

3. Is it always necessary to specify the hypothetical proportions to perform a $\chi^2$ test?

4. What does "independence" mean?

5. Verify that for large samples a $\chi^2$ test for a single-classification two-category table is exactly equivalent to the test using proportions (i.e., show that the critical regions coincide).

6. What connection is there between control charts and tests of hypotheses?

7. Define each term in the Glossary.

8. Has any assumption about the population been made before application of the $\chi^2$ test in a single-classification experiment?

9. A man took a survey by selecting a random sample of 100 names from a telephone book and interviewing those people. He found that 55 stated that they had voted in the last election, 37 stated that they had not voted, and the remainder refused to answer or could not be contacted. What conclusions can he draw? From what population did he sample?

## CLASS EXERCISES

1. For Class Exercise 1, Chap. 5, verify that the mean proportion of observed red beads is $\frac{2}{3}$ and that the variance of the proportion of observed red beads is

$$\frac{p(1-p)}{N} = .011$$

2. Draw samples of 25 beads from a box containing 200 red beads, 400 blue beads, and 400 white beads. Compute the value of $\chi^2$ for each sample and collect the observed values into a distribution. Compare the 95th percentile of the observed distribution with the theoretical value from Table A-6*a*.

3. In Class Exercise 1, Chap. 6, the number of 90 per cent confidence intervals for $\mu$ which actually covered $\mu$ were counted. Test the hypothesis that the theoretical proportion of intervals which cover $\mu$ is 90 per cent.

4. Work Exercise 3 for the data in Class Exercise 1, Chap. 7.

5. Work Exercise 3 for the data in Class Exercise 4, Chap. 6.

6. Draw samples of size 5 from a box of 500 white beads and 500 red beads and verify that the frequencies of red beads in the samples are as given by the binomial-distribution formula (verify that 1 time out of 32 you get no red beads out of 5, etc.). After a number of samples have been drawn, test the agreement with the theoretical frequencies by a $\chi^2$ test.

7. From a box of 100 red beads and 900 white beads draw samples of size 10 and observe the number of red beads in each sample. Test the hypothesis that the Poisson-distribution formula gives the frequencies with which various proportions of red beads occur in these samples.

8. From a two-category population draw samples of size 50. An example of such a population is a number of tags, some white and some colored, with each tag bearing a plus or a minus sign. Record the results of the sample in a table. Compute the

| | + | − |
|---|---|---|
| *White* | | |
| *Colored* | | |

$\chi^2$ statistic to test for independence. Tabulate a number of $\chi^2$ values found in this manner, compute the 95th percentile of the observed values, and compare it with $\chi_{.95}^2(1)$ in Table A-6*a*.

9. Suppose a variable takes on the values 0, 1, 2, 3, 4 with the frequencies shown

in the tabulation. Each student draws 5 samples of 25 from the random-number

| $X$ | 0 | 1 | 2 | 3 | 4 |
|---|---|---|---|---|---|
| $f$ | .1 | .2 | .4 | .2 | .1 |

table (Table A-1), assigning to the numbers drawn values $X$ so that the frequencies in the tabulation hold in the population (for example, assign to 0 the score 0, to 1 and 5 the score 1, to 2 and 6 and 7 and 8 the score 2, to 3 and 9 the score 3, and to 4 assign the score 4). Compute $\chi^2$ for each experiment and then collect the results into a single distribution of $\chi^2$. Compare the 90th percentile of the observed distribution with $\chi_{.90}^2(4)$ from Table A-6*a*.

## PROBLEMS

*Section* 13-2

**1.** In a botany experiment the results of crossing two hybrids of a species of flower gave observed frequencies of descriptive categories of 120, 48, 36, 13. Do these results disagree with theoretical frequencies which specify a 9:3:3:1 ratio?

**2.** Two hundred throws of a die resulted in the following observations:

| *No. of spots* | 1 | 2 | 3 | 4 | 5 | 6 |
|---|---|---|---|---|---|---|
| *Frequency* | 30 | 27 | 29 | 31 | 40 | 43 |

Is this reason to believe that the die is not balanced correctly?

**3.** From a $\chi^2$ test, how many occurrences of heads in 100 tosses of a coin would lead to rejection at the 1 per cent level of significance of the hypothesis that it was a true coin?

**4.** Of the 400 winning numbers in a sweepstakes drawing, 176 were odd numbers and 224 were even numbers. Is this significantly different from 50:50?

**5.** In a sampling experiment 100 samples of size 25 each were drawn from Table A-2. It was observed that 10 sample means were less than $-.20$, that 65 means were between $-.20$ and .20, and that the remaining 25 means were greater than .20. Use a $\chi^2$ statistic to test the hypothesis that random means were drawn from a normal population with $\mu = 0$ and $\sigma = .2$.

*Section* 13-3

**6.** Analyze the tabulated data on the number of children in families. State assumptions, hypotheses, and conclusions.

| FAMILY INCOME, DOLLARS | NUMBER OF CHILDREN 0 | 1 | 2 | *More than* 2 |
|---|---|---|---|---|
| *Under* 2,000 | 15 | 27 | 50 | 43 |
| 2,000–5,000 | 25 | 37 | 12 | 8 |
| *Over* 5,000 | 8 | 13 | 9 | 10 |

**7.** Celery seed is treated chemically in an attempt to reduce the incidence of blight. Of 100 plants from treated seeds there were 20 which showed blight. Of 100 plants from untreated seeds 46 showed blight. Is this difference significant?

**8.** The grade distribution of three instructors who taught the same course for a period of 2 years is given in the table. Did the instructors give significantly different percentages of the five grades?

| | GRADE | | | | |
|---|---|---|---|---|---|
| INSTRUCTOR | *A* | *B* | *C* | *D* | *F* |
| *Smith* | 21 | 39 | 128 | 19 | 23 |
| *Jones* | 35 | 52 | 212 | 24 | 29 |
| *Brown* | 15 | 42 | 178 | 20 | 17 |

**9.** A certain vitamin was thought to increase energy. In an experiment 100 men were given the vitamin. As a control, 100 men were given placebos but were told that they were being given the vitamin. Their reactions are given in the table. Analyze the data.

| | *More energy* | *Less energy* | *No change* |
|---|---|---|---|
| *Control group* | 20 | 10 | 70 |
| *Treated group* | 36 | 8 | 56 |

**10.** A baseball team played 140 games. Of the 70 home games it won 45 and of the 70 away from home it won 35. What hypothesis can be tested with a $\chi^2$ test? What assumptions are needed to make the $\chi^2$ analysis appropriate? Compute the $\chi^2$.

**11.** A manufacturer of brand *A* automobiles believed that his cars would last longer than those of certain competitors. A random sample of automobiles from a certain city showed the data given in the table. As constructed this experiment cannot prove or disprove the manufacturer's belief. Why not?

| | AGE OF CAR | | |
|---|---|---|---|
| CAR | *Over* 10 *years* | 5–10 *years* | *Less than* 5 *years* |
| *A* | 33 | 127 | 256 |
| *B* | 42 | 205 | 409 |
| *C* | 10 | 55 | 100 |

Later another survey was made of cars manufactured exactly 10 years ago. The results are as in the second table. What do these figures prove? State carefully the populations sampled from and what interpretations can and cannot be made from the data.

| | NO. OF YEARS CAR RAN WITHOUT MAJOR OVERHAUL | | | | | |
|---|---|---|---|---|---|---|
| CAR | 0–1 | 2–3 | 4–5 | 6–7 | 8–9 | *No overhaul yet* |
| *A* | 15 | 37 | 55 | 108 | 118 | 60 |
| *B* | 5 | 13 | 180 | 205 | 386 | 400 |
| *C* | 55 | 160 | 300 | 106 | 55 | 40 |

**12.** Samples of frozen food taken in New York in 1958, 1959, and 1960 showed the results in the table. Is there evidence of varying level over the years?

| | 1958 | 1959 | 1960 |
|---|---|---|---|
| *No. examined* | 240 | 278 | 144 |
| *No. satisfactory on several criteria* | 153 | 201 | 90 |

**13.** Four interviewers determined whether subjects had a cough symptom, with the results in the following table. Is there evidence that the interviewers do not have random samples from the same population? Indicate other reasons for the proportions to be unequal.

| | *A* | *B* | *C* | *D* |
|---|---|---|---|---|
| *No. interviewed* | 261 | 250 | 300 | 182 |
| *No. with cough* | 61 | 70 | 121 | 60 |

**14.** Analyze the data showing the percentages of a sample of 137 women according to number of pregnancies and number of additional children wanted.

| NO. OF ADDITIONAL CHILDREN WANTED | NO. OF PREGNANCIES 0 | 1–3 | 4+ | *Total* |
|---|---|---|---|---|
| 0 | 50% | 59.7% | 88.7% | 72.3% |
| 1–3 | 50 | 40.3 | 11.3 | 27.7 |
| *Number* | 8 | 67 | 62 | 137 |

**15.** Analyze the data giving the percentage distribution of hospitalizations for several age groups.

| TIMES HOSPITALIZED | AGE *Under* 82 | 82–83 | 84–85 | 86–87 | 88–89 | 90+ | *Total* |
|---|---|---|---|---|---|---|---|
| 0 | 80.7% | 71.2% | 76.6% | 66.2% | 66.5% | 65.4% | 70.9% |
| 1 | 12.9 | 23.0 | 19.6 | 27.3 | 25.0 | 28.0 | 23.1 |
| 2+ | 6.4 | 5.8 | 3.8 | 6.5 | 8.5 | 6.6 | 6.0 |
| *Number* | 171 | 344 | 470 | 458 | 284 | 182 | 1,909 |

**16.** Ten of 50 patients with a specific disease treated by a standard method died, while only 5 out of 50 treated by a new method died. Test for equality of proportions using a $\chi^2$ test.

**17.** Analyze the data giving mortality rates among vaccinated and nonvaccinated pheasants.

| | *Living* | *Died* | *Total* |
|---|---|---|---|
| *Vaccinated* | 399 (73.9%) | 141 (26.1%) | 540 |
| *Nonvaccinated* | 159 (55.2%) | 129 (44.8%) | 288 |
| *Total* | 558 (67.4%) | 270 (32.6%) | 828 |

**18.** A sample of five treated and five untreated sick animals resulted in four and two surviving, respectively. Analyze using Table A-9*e*.

*Section* 13-4

**19.** In Table 4-3 there are two distributions of sample means, one for samples of size 10 and one for samples of size 40. (*a*) Use a $\chi^2$ test to test the hypothesis that each set of means is from a normal population. (*b*) Test the hypothesis that the sample means for $N = 10$ are from a normal population with mean 0 and variance .1 by use of the $\chi^2$ statistic. (*c*) Test the hypothesis that the sample means for $N = 40$ come from a normal population with mean 0 and variance .025 by use of a $\chi^2$ test.

*Section* 13-5

**20.** A student newspaper polled 20 students and found that 14 of them favored a change from a term system to the semester plan. Read 90 per cent confidence limits for the proportion of students in the student body who favor this change.

**21.** An experiment is performed whereby one man looks at a playing card and another guesses whether the card is red or black. $N = 10$ cards in all are to be inspected. From Table A-29*a* determine a critical region to test the hypothesis $H$: $p = .5$ with $\alpha = .05$.

**22.** In a Poisson distribution with mean 2 find the probability that the variable will be equal to 0, less than or equal to 2, greater than 3.

**23.** In a random sample of 250 individuals 100 were wearing glasses. (*a*) From Table A-9 read 90 per cent confidence limits for $p$. (*b*) Using the normal approximation, find 90 per cent confidence limits for $p$. (*c*) Compare these limits.

**24.** Over a period of several years there had been an average of 12 accidents per year in a certain city. This year there were accidents in the 12 months as follows: 0, 0, 2, 2, 1, 1, 2, 0, 1, 0, 1, 2. Do these data agree with a theory that the number of accidents per month follows a Poisson distribution with $Np = 1$? (Do a $\chi^2$ test and also the variance test on page 249.)

*Section* 13-6

**25.** A class experiment consists of drawing 10 samples each of size $N$ from Table A-2 (where $\mu = 0$ and $\sigma = 1$) and constructing for each a confidence interval $\bar{X} \pm 1.645/\sqrt{N}$ for the population mean $\mu$. From Table A-29*a* read the probability that exactly 9 of the 10 intervals will include $\mu$. If the experiment is to take 100 samples, use Table A-4 to find the probability that the proportion of the 100 confidence intervals which cover $\mu$ will be between .87 and .93.

**26.** From the data in Prob. 17 determine 95 per cent confidence limits for the probability of living for vaccinated pheasants and for nonvaccinated pheasants. Would the analysis in Prob. 17 be equivalent to rejection if these confidence intervals do not overlap?

*Section* 13-6

**27.** Use a difference in proportions test to analyze the data in Prob. 17 using a 5 per cent level of significance test.

**28.** Using the data in Prob. 14, test the hypothesis that equal proportions of women having 4 or more pregnancies and of women having 1 to 3 pregnancies want no further children.

**29.** Use a difference in proportions test to analyze the data in Prob. 7.

**30.** Use a difference in proportions test to analyze the data in Prob. 10.

*Section* 13-7

**31.** Use binomial-probability paper to analyze the data in Prob. 4.

**32.** Use binomial-probability paper to determine whether 120 successes in 217 trials agrees with a 9 out of 16 theory.

**33.** Use binomial-probability paper to analyze the data in Prob. 3.

**34.** Use binomial-probability paper to analyze the data in Prob. 7.

**35.** Use binomial-probability paper to give 90 per cent confidence limits for $p$ with the data in Prob. 23.

*Section* 13-8

**36.** The data in the table record the number of calls and sales made by a life-insurance salesman. Fill in the table and draw a control chart showing control lines. Inspect this chart for significant changes in sale conditions. For $p$ use the proportion of sales during the entire year.

| | *Calls* | *Sales* | $\bar{X}$ | $\frac{3\sqrt{p(1-p)}}{\sqrt{N}}$ | *Upper control limit* | *Lower control limit* |
|---|---|---|---|---|---|---|
| January | 95 | 7 | | | | |
| February | 100 | 15 | | | | |
| March | 110 | 14 | | | | |
| April | 90 | 8 | | | | |
| May | 97 | 12 | | | | |
| June | 105 | 6 | | | | |
| July | 100 | 7 | | | | |
| August | 108 | 12 | | | | |
| September | 120 | 15 | | | | |
| October | 98 | 12 | | | | |
| November | 100 | 17 | | | | |
| December | 85 | 9 | | | | |
| *Total* | | | | | | |

**37.** The data in the table are the results of random samples of articles from five manufacturers. Indicate for each of the hypotheses tested the possible errors which could be made and the consequences of making each of them. (*a*) Test the hypothesis that the defective rate of the product made by manufacturer $A$ is 3 per cent. In case of rejection of the hypothesis, find the 95 per cent confidence interval for $p$. (*b*) Test the hypothesis that the defective rates of the products made by the manufacturers $A$ and $B$ are the same. In case of rejection of the hypothesis, find a 99 per cent confidence interval for the difference between the two defective rates. (*c*) Test the hypothesis that the defective rates of the products made by the five manufacturers are the same. In case of rejection of the hypothesis, what would you do next?

| | *A* | *B* | *C* | *D* | *E* |
|---|---|---|---|---|---|
| *Defective* | 41 | 15 | 21 | 13 | 19 |
| *Nondefective* | 259 | 301 | 401 | 304 | 405 |

**38.** An $\bar{X}$ chart with 95 per cent control limits was used in controlling the mean. Of the 500 points plotted on the chart, 14 points fell below the lower limit and 16 points fell above the upper limit. Use the $\chi^2$ test to determine whether or not this indicates that the population mean has been changing during the period the chart was used.

CHAPTER FOURTEEN

# Probability of accepting a false hypothesis

We have dealt with various techniques for testing a hypothesis. If the hypothesis were true, the chance, or probability, of rejecting it was some preassigned number $\alpha$. That is, the chances of rejecting true hypotheses have been controlled. Our method of testing any one hypothesis was not unique. It involved a choice of a test statistic and a choice of a critical region. While some of our choices seemed obvious, others did not. In this chapter we shall examine more carefully the advantages of some particular tests.

We shall also delve more deeply into the problem of evaluating the chance $\beta$ of failing to detect the falsity of a hypothesis. This topic deserves careful attention and thought, especially from those who must determine whether or not an experiment is to be performed at all. If it has already been decided to perform the experiment, the size of $\beta$ is of interest but is not particularly vital. However, if it is important to discover a deviation from the hypothesis of a certain magnitude and if the time and money available give, say, only a 15 per cent chance of recognizing such a difference if it is present, there seems little point in performing the experiment. Actually, with $\beta = .85$ the null hypothesis is very likely to be accepted even if these important deviations are present. The researcher must resist the temptation to believe the hypothesis to be true when it is accepted in such circumstances. With $\beta$ so large, even repeated tests constitute little support for the hypothesis. In such a case it may be better to use the resources on some problem that is either less expensive or where there can be more assurance of arriving at a correct conclusion.

## 14-1 ALTERNATIVES AND TWO TYPES OF ERROR

A *test* of a statistical hypothesis consists in choosing a test statistic and selecting a critical region. When the hypothesis is true, the test specifies that the chance of rejecting the hypothesis is some preassigned level of significance $\alpha$. It would be an error to reject the hypothesis when it is true, and this type of error is called an $\alpha$ error. If the hypothesis is not true, accepting it is also a mistake, and this type of error is called a $\beta$ error. Of course, only one of these two errors is possible in a single problem, but there is no way to determine which error, if either, may occur. Our method of testing has controlled the chance of making the $\alpha$ error. Let us now examine the chance of accepting a false hypothesis ($\beta$ error).

Declaring a hypothesis false should imply that we have some idea of what alternative situations might exist. For example, if for a normal distribution with $\sigma = 1$ we are testing the hypothesis that the mean $\mu = 5$, possible alternatives might be $\mu = 6$, $\mu = 3$, and so on. In fact, alternatives might include any number other than 5 for the value of $\mu$. If it happens that $\mu = 6$, then there is a certain probability of rejecting the hypothesis $\mu = 5$. If $\mu = 3$, there is some other probability of rejecting the hypothesis $\mu = 5$. This probability of rejecting a hypothesis is called the *power* of the test.

The power depends upon which alternative is actually true. In practice we do not know which alternative is true and are interested in the power of the test for several possible alternatives. If the hypothesis is not true we would like the chance of rejecting it to be as large as possible and thus would like the power to be large. Suppose that the chance of rejecting a certain hypothesis when it is true is equal to $\alpha$ for two different tests. We would then use the test which is more powerful.

These two types of error are sometimes called *producers'* and *consumers' risks*. To illustrate this usage, suppose a hypothesis states that a batch of manufactured articles meets some standard and a statistical test is to be applied. The risk of rejecting a true hypothesis would be a producers' risk, since if the hypothesis is rejected the articles are not sold and the producer takes an unjustified loss. Conversely, the chance that below-standard articles will be accepted for sale is a chance of loss for the consumer.

## 14-2 POWER FUNCTION FOR TESTING THE HYPOTHESIS $\mu = \mu_0$

**Choice of critical region.** In testing the hypothesis $\mu = 5$ for a normally distributed population with $\sigma = 1$ a random sample of $N$ items is taken from the population, and the mean $\bar{X}$ of the sample is computed. The sampling distribution of the statistic $\bar{X}$ is known to be a normal distribution with the same mean as the mean of the population and with a standard deviation $\sigma_{\bar{X}}$ equal to $\sigma/\sqrt{N}$. Thus $(\bar{X} - \mu)/(\sigma/\sqrt{N})$ has a normal distribution with mean 0 and unit standard deviation, and the table of this probability distribution may be used to determine critical regions.

One test of the hypothesis $\mu = 5$ at the 5 per cent level of significance ($\alpha = .05$) would be to reject it if $(\bar{X} - 5)/(1/\sqrt{N}) > 1.645$. A second test would be to reject it if $(\bar{X} - 5)/(1/\sqrt{N}) < -1.645$. A third test would be to reject it if either $(\bar{X} - 5)/(1/\sqrt{N}) > 1.960$ or $(\bar{X} - 5)/(1/\sqrt{N}) < -1.960$. If the hypothesis is true any one of these three tests would reject it 5 per cent of the time. In each case we have selected, using the sampling distribution of $\bar{X}$, regions where $\bar{X}$ would fall 5 per cent of the time by chance if $\mu = 5$. There are many other such regions, but these are perhaps three of the more useful.

Suppose we have a sample of 100 individuals ($N = 100$). Then the three critical regions can be described as values of $\bar{X}$ which satisfy the following relations:

(*a*) $\dfrac{\bar{X} - 5}{1/\sqrt{100}} > 1.645 \qquad (\bar{X} > 5 + .1645 \text{ or } \bar{X} > 5.1645)$

(*b*) $\dfrac{\bar{X} - 5}{1/\sqrt{100}} < -1.645 \qquad (\bar{X} < 4.8355)$

(*c*) Either $\dfrac{\bar{X} - 5}{1/\sqrt{100}} < -1.960$ or $\dfrac{\bar{X} - 5}{1/\sqrt{100}} > 1.960$, which can be written as either $\bar{X} > 5.1960$ or $\bar{X} < 4.8040$.

Figure 14-1 indicates the rejection regions for these three tests. Any one of these regions has 5 per cent of the area of a normal curve with mean 5 and $\sigma = 1/\sqrt{N} = \frac{1}{10}$. Thus if $\mu$ does equal 5, the probability of rejecting the hypothesis $\mu = 5$ is $\alpha = .05$ no matter which of the three tests we might decide to use. Let us find the power of each of these critical regions against certain alternatives.

First let us consider the operation of these tests in the event that the population mean $\mu$ is 5.2. Then the sampling distribution of $\dfrac{\bar{X} - 5.2}{1/\sqrt{100}}$ is a normal distribution with mean 0 and unit standard deviation. We shall find the area under this curve corresponding to each of the three regions (*a*), (*b*), (*c*) above. These areas will be the chances of rejecting the hypothesis $\mu = 5$ if actually $\mu = 5.2$ for the three tests. To find these areas we find $z$ scores for the end points of the critical regions:

(*a*) For $\bar{X} = 5.1645$, $z = \dfrac{5.1645 - 5.2}{1/\sqrt{100}} = -.355.$

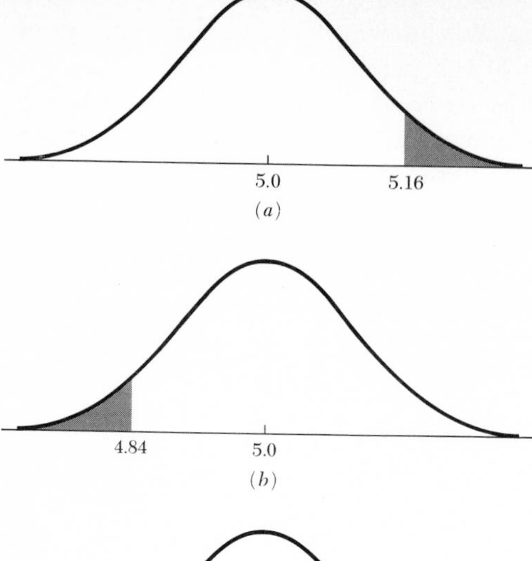


**Figure 14-1** Three possible critical regions for testing $H: \mu = 5.0$ given $\sigma = 1.0$ and $N = 100$. The shaded area in each case is 5 per cent of the total area under the sampling distribution curve.

$$(b) \text{ For } \bar{X} = 4.8355,\ z = \frac{4.8355 - 5.2}{1/\sqrt{100}} = -3.645.$$

$$(c) \text{ For } \bar{X} = 5.1960,\ z = \frac{5.1960 - 5.2}{1/\sqrt{100}} = -.040, \text{ and,}$$

$$\text{for } \bar{X} = 4.8040,\ z = \frac{4.8040 - 5.2}{1/\sqrt{100}} = -3.960.$$

The areas under the unit normal curve for these regions can be read from Table A-4 as (*a*) .639, (*b*) .000, and (*c*) .516 + .000 = .516.

If the true value of $\mu$ is 5.2, then with critical region (*a*) we would have a chance .639 of rejecting the hypothesis $\mu = 5$. With region (*b*) we would have a chance .000, while with region (*c*) the chance is .516. Thus for this alternative $\mu = 5.2$ region (*a*) is the most powerful of the three. Figure 14-2 illustrates the chance of falling in region (*a*) when $\mu = 5$ and when $\mu = 5.2$. Figure 14-3 illustrates the chance of falling in region (*b*) when $\mu = 5$ and when $\mu = 5.2$. Figure 14-4 illustrates the chance of falling in region (*c*) when $\mu = 5$ and when $\mu = 5.2$.

Next let us consider the operation of these tests in the event that the population mean $\mu = 4.8$. Then the sampling distribution of $\dfrac{\bar{X} - 4.8}{1/\sqrt{100}}$ is a normal distribution with mean 0 and unit standard deviation. We shall

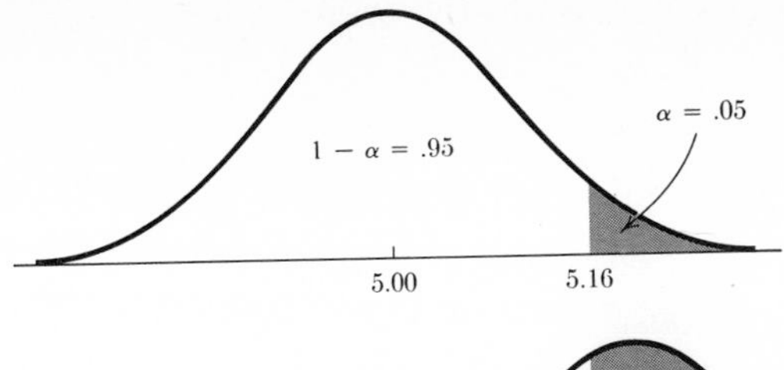


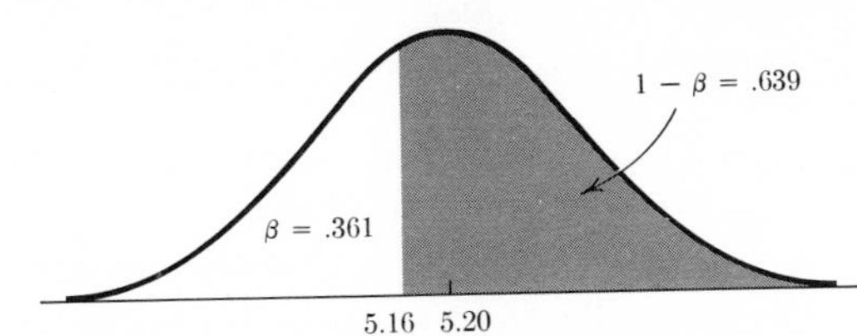


**Figure 14-2** Normal sampling distribution of $\bar{X}$ (1) if $\mu = 5.0$, $\sigma = 1.0$, $N = 100$ and (2) if $\mu = 5.20$, $\sigma = 1.0$, $N = 100$. If the $H: \mu = 5.0$ is rejected if $\bar{X} > 5.16$, then the shaded areas represent $\alpha$ and $1 - \beta$, respectively.

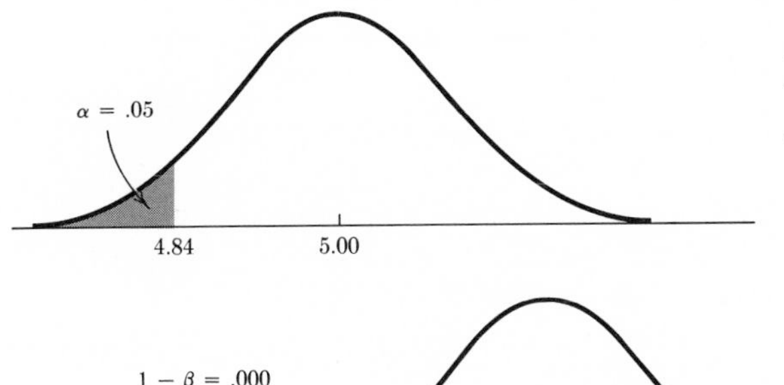


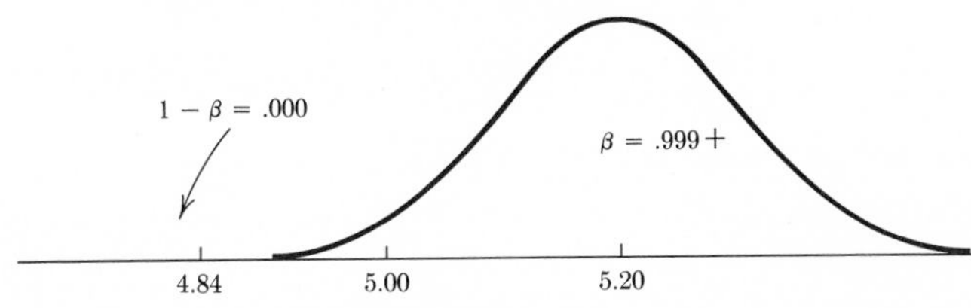


**Figure 14-3** The same curves as in Fig. 14-2 but with $\bar{X} < 4.84$ used as rejection region.

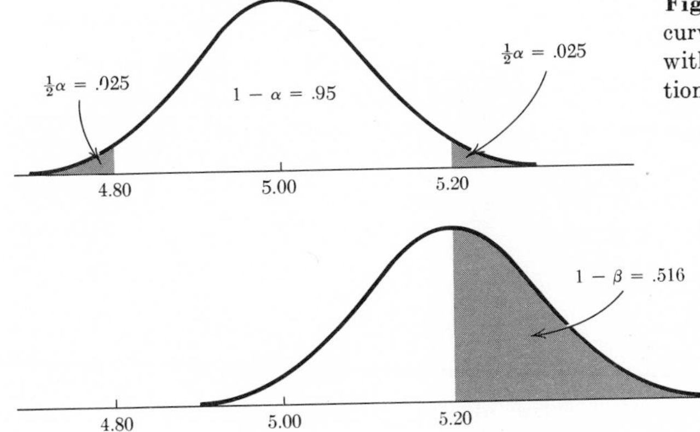


**Figure 14-4** The same curves as in Fig. 14-2 but with a two-sided rejection region.

find the areas under this curve corresponding to the three regions (*a*), (*b*), (*c*). The end points of the intervals are:

(*a*) For $\bar{X} = 5.1645$, $z = \dfrac{5.1645 - 4.8}{1/\sqrt{100}} = 3.645.$

(*b*) For $\bar{X} = 4.8355$, $z = \dfrac{4.8355 - 4.8}{1/\sqrt{100}} = .355.$

(*c*) For $\bar{X} = 5.1960$, $z = 3.960$, and, for $\bar{X} = 4.8040$, $z = .040$.

The areas under the unit normal curve for these regions are (*a*) .000, (*b*) .639, and (*c*) .516. Thus if $\mu = 4.8$, using region (*a*) would give us no chance of rejecting the hypothesis $\mu = 5$, while (*b*) gives a chance of .639 and (*c*) a chance of .516. Region (*b*) is the most powerful of the three against the alternative $\mu = 4.8$.

We can generalize these results by saying that region (*a*) is the best region if the true value of $\mu$ is greater than 5, while region (*b*) is the most powerful if the true value of $\mu$ is less than 5. Of course, if $\mu$ is less than 5 region (*a*) is very poor, and similarly, if $\mu$ is greater than 5 region (*b*) is very poor. Region (*c*), while not as uniformly powerful as (*a*) or (*b*), guards against both alternatives. If we know, or are only worried about the possibility, that $\mu > 5$, then we would use region (*a*). Similarly, we would use region (*b*) to test the hypothesis $\mu = 5$ against alternatives $\mu < 5$, while if we must consider the possibility that $\mu$ may be either less than or greater than 5, we would use region (*c*).

We have computed the power for the three points $\mu = 4.8$, $\mu = 5$ (here the probability of rejecting is $\alpha = .05$), and $\mu = 5.2$. We have computed similar values of the power function for other values of $\mu$, and these are shown by means of the three curves in Fig. 14-5. These curves correspond to a level of significance $\alpha = .05$. The hypothesis is now $\mu = \mu_0$, and the regions are now given in terms of $\mu_0$ and $\sigma$.

The three critical regions are as follows:

(*a*) $z = \dfrac{\bar{X} - \mu_0}{\sigma/\sqrt{N}}$, reject if $z > 1.645$.

(*b*) $z = \dfrac{\bar{X} - \mu_0}{\sigma/\sqrt{N}}$, reject if $z < -1.645$.

(*c*) $z = \dfrac{\bar{X} - \mu_0}{\sigma/\sqrt{N}}$, reject if $z < -1.960$ or if $z > +1.960$.

The alternative values of $\mu$ are read along the horizontal axis in units of $\sigma/\sqrt{N}$. The zero of the horizontal scale is $\mu = \mu_0$. Thus for the example above $\mu_0 = 5$, and the unit along the horizontal axis would be $\sigma/\sqrt{N} = 1/\sqrt{100} = .1$. An alternative of $\mu = 4.8$ would correspond to $(4.8 - 5)/.1 = -2$. Note that above $\mu_0 - 2\sigma/\sqrt{N}$ the heights of the three

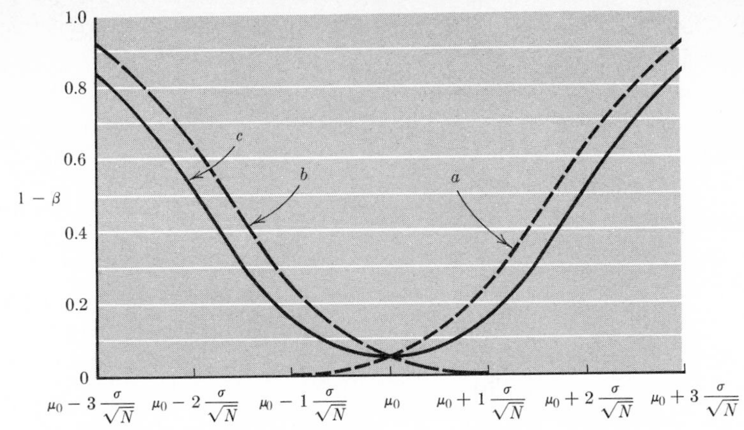


**Figure 14-5** Power curves corresponding to three possible rejection regions of size $\alpha = .05$.

curves are, respectively, (*a*) .000, (*b*) .639, (*c*) .516, as we found before. Similarly, above $\mu_0 + 2\sigma/\sqrt{N}$ the heights correspond to the numbers we obtained. The horizontal scale is read with the number $(\mu - \mu_0)/(\sigma/\sqrt{N})$. The height of the curve in each case tells how frequently we would reject the hypothesis $\mu = \mu_0$ for various possible values of $\mu$. We see that with region (*a*) the power is large if $\mu > \mu_0$ but very small if $\mu < \mu_0$. The reverse is true for (*b*), while (*c*) is fairly powerful in either case.

Here we assumed a normal universe. However, the procedure is valid if the sample is large enough so that $\bar{X}$ can be assumed to have an approximately normal sampling distribution.

There are infinitely many possible regions such that $\alpha = .05$. It can be proved that (*a*) has a power curve higher than that for any other region if $\mu > \mu_0$, while (*b*) is more powerful than any other region if $\mu < \mu_0$. If it is desired to have a test such that the power curve goes up on both sides of $\mu_0$, then region (*c*) is the *best* region that can be found.

The figures in Table A-12*a* show power curves for one- and two-sided tests for several values of $\alpha$. The deviation of the alternative mean $\mu$ from the hypothetical mean $\mu_0$ in units of $\sigma/\sqrt{N}$ is

$$d = \frac{\mu - \mu_0}{\sigma/\sqrt{N}}$$

and is given on the horizontal axis. The power $1 - \beta$ is on the vertical axis. The values of $1 - \beta$ have been plotted on a normal-probability scale to straighten the curves. The power curves on this paper for the one-sided test are exactly straight.

**Determination of power.** Thus far we have used the power curves merely to tell us which test would have the best chance of rejecting a false hypothesis. They can also be used to find the power of the adopted test against a *specified* alternative.

In the example above the hypothesis was $\mu = 5$, and any of the regions gives a chance of .05 of rejecting this hypothesis if it is true. What is the chance of rejecting this hypothesis if $\mu = 5.1$? We read

$$d = \frac{\mu - \mu_0}{\sigma/\sqrt{N}} = \frac{5.1 - 5}{1/\sqrt{100}} = 1$$

on the horizontal scale and the heights of the curves as (*a*) .26, (*b*) .004, (*c*) .170. If we used region (*a*) as a critical region, the chance of rejecting the hypothesis $\mu = 5$ if $\mu = 5.1$ is .26. With region (*b*) the chance is .004. With region (*c*) the chance is .170.

For four selected values of $\alpha$ Table A-12*d* gives values of $d' = (\mu - \mu_0)/\sigma$ which would be detected with power $1 - \beta$ for a number of sample sizes. For example, with a one-sided test with $\alpha = .05$ and $N = 100$ we read that if $d' = .16$, then there is a $1 - \beta = .50$ chance of recognizing that the hypothesis $H\colon \mu = \mu_0$ is false.

**Sample size.** The curves can also be used to help in designing experiments by indicating how large a sample may be needed. For example, suppose we have a normal distribution with $\sigma = 1$ and we wish to test the hypothesis $\mu = 5$ with $\alpha = .05$. We do not know whether the true $\mu$ is larger or smaller than 5, and so we shall use region (*c*) as a critical region. Suppose we wish to guard against accepting the hypothesis if $\mu$ is more than .3 unit larger or more than .3 unit smaller than 5 with power .80. That is, if $\mu \leq 4.7$ or if $\mu \geq 5.3$, we wish the chance of rejecting the hypothesis $\mu = 5$ to be at least .80. How large a sample is needed? We move across from .80 on the vertical scale until we reach the (*c*) curve. Below the intersection we read $2.8(\sigma/\sqrt{N})$ (approximately). We wish this to correspond to .3 unit, so we have

$$.3 = 2.8\frac{\sigma}{\sqrt{N}} \text{ or } .3 = 2.8\frac{1}{\sqrt{N}} \text{ or } .3\sqrt{N} = 2.8 \text{ or } .09N = 7.84$$

and thus

$$N = \frac{7.84}{.09} = 87.1 \text{ or } 88$$

Table A-12*c* gives the necessary sample size for selected values of $\alpha$, $\beta$, and $d' = (\mu - \mu_0)/\sigma$. For example, we read that a sample size $N = 19$ is required for a $1 - \beta = .70$ probability of obtaining a significant result with a one-sided test at the $\alpha = .05$ level when the true mean is $.5\sigma$ distant from the hypothetical value.

**Two-sample case.** A method of testing the hypothesis that two populations have equal means, where the value of the common variance is known, is discussed in Sec. 8-1. We compute

$$z = \frac{\bar{X}_1 - \bar{X}_2}{\sigma \sqrt{(1/N_1) + (1/N_2)}}$$

and accept the hypothesis if the observed $z$ is not significantly far from zero. The mean of the sampling distribution of $\bar{X}_1 - \bar{X}_2$ is equal to the difference in the means of the two populations, $\mu_1 - \mu_2$, and the hypothesis tested is $\mu_1 = \mu_2$ (that is, $\mu_1 - \mu_2 = 0$). If in fact the means are unequal and we use

$$d = \frac{\mu_1 - \mu_2}{\sigma \sqrt{(1/N_1) + (1/N_2)}}$$

the power curves are exactly the same as above and we may use Table $A$-12$a$. Suppose, for example, we take samples of sizes $N_1 = 20$, $N_2 = 20$ from two populations with $\sigma_1 = \sigma_2 = 10$ and prepare to test the hypothesis $\mu_1 = \mu_2$ at the 5 per cent level of significance. If in fact $\mu_1 - \mu_2 = 4$, that is, $d = 4/(10\sqrt{.10}) = 1.26$, we read from Table A-12$a$ that $1 - \beta = .24$. The chance of not recognizing that these populations do not have equal means is $\beta = .76$.

Analogous to the single-population case, Table A-12$d$ gives the magnitude of $d' = (\mu_1 - \mu_2)/\sigma$ between two population means which can be detected with probability $1 - \beta$ by means of an $\alpha$-level test with a sample of size $N$. Alternatively, Table A-12$c$ gives the necessary sample size for specified chance $1 - \beta$ of finding significance at the $\alpha$ level.

For example, if the one-sided hypothesis $H\colon \mu_1 \leq \mu_2$ is to be tested at an $\alpha = .05$ level and we wish to have power $1 - \beta = .9$ against an alternative where $\mu_1 - \mu_2$ is one standard deviation unit, i.e., $d' = 1/\sqrt{2} = .7$, then we read from Table A-12$c$ that samples of sizes $N_1 = N_2 = N = 18$ are required. Table A-12$c$ indicates that for the same $\alpha = .05$ samples of size $N = 75$ would have chance $1 - \beta = .9$ of detecting $\mu_1 - \mu_2 = .5\sigma$, samples of size $N = 857$ would have chance $1 - \beta = .9$ of detecting $\mu_1 - \mu_2 = .14\sigma$, and so on.

**Case where $\sigma$ is unknown.** Suppose we have a normally distributed population with unknown mean $\mu$ and unknown variance $\sigma^2$. To test the hypothesis that $\mu = \mu_0$ we have used the statistic

$$t = \frac{\bar{X} - \mu_0}{s/\sqrt{N}}$$

which, if the hypothesis is true, has a $t$ distribution with $N - 1$ degrees of freedom.

In this case the chance of rejecting $\mu_0$ when in fact some alternative is true cannot be computed quite so simply as in the case of $\sigma$ known. Two

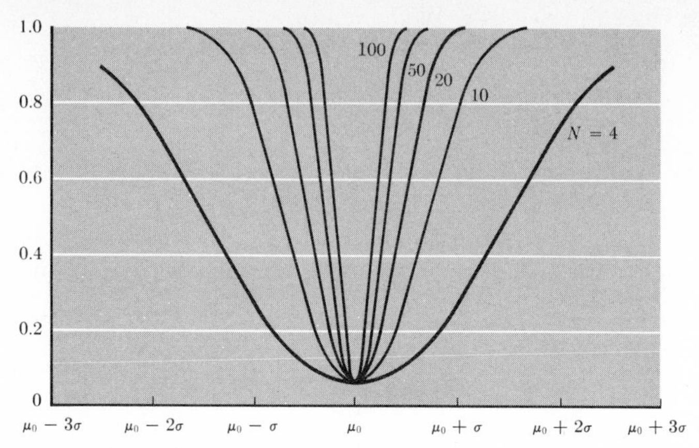


**Figure 14-6** Power of 5 per cent level test of $H:\mu = \mu_0$ using the two-sided $t$ test.

factors complicate the problem. First, the critical region, in terms of a number of standard deviations, depends on the number of degrees of freedom. For 10 degrees of freedom and $\alpha = .05$ we read from Table A-5 that $t = 2.23$ units is significant. For a large number of degrees of freedom $t = z$, and we find that $z = 1.96$ units is significant. Second, the value of $s$ in the denominator of $t$ varies from one sample to another. For large samples, say, $N > 30$, these factors could be neglected with little error and the preceding results used. However, for smaller sample sizes or for more precise results for larger sample sizes Table A-12*b* may be used. This table gives the relationships among:

(*a*) $d = \dfrac{\mu - \mu_0}{\sigma/\sqrt{N}}$ for the one sample case

$d = \dfrac{\mu_1 - \mu_2}{\sigma\sqrt{(1/N_1) + (1/N_2)}}$ for the two-sample case

(*b*) The level of significance $\alpha$

(*c*) The degrees of freedom for the one-sample case, $N - 1$, and for the two-sample case, $N_1 + N_2 - 2$

(*d*) The power of the test, $1 - \beta$

The values listed are for a one-sided test at the $\alpha$ level of significance or, if $d$ is not near zero, for a two-sided test at the $2\alpha$ level of significance.

For the two-sided test with $\alpha = .01$ or $.05$ the graphs in Table A-13 may be used with $\nu_1 = 1$, since a two-sided $t$ test is equivalent to the analysis-of-variance test with 1 degree of freedom for the numerator. In fact, the values in the $F$ table for $F(1,N - 1)$ are the squares of the values in the $t$ table for $N - 1$ degrees of freedom.

Figure 14-6 shows power curves for $\alpha = .05$ for a two-sided test. The shape of the curves can be seen from this figure. The figures in Table A-12*a*, for the test where $\sigma$ is known, have been plotted on a normal-probability scale and show only the portion for $\mu > \mu_0$.

An approximate formula for the numbers in the Table A-12*b* giving the value $d$ for a test at the $\alpha$ level of significance with power $1 - \beta$ is

$$d = (z_{1-\alpha} + z_{1-\beta}) \left[1 + \frac{1.21(z_{1-\alpha} - 1.06)}{df}\right]$$

where $z_{1-\alpha}$ and $z_{1-\beta}$ are percentiles of the standard normal distribution in Table A-4. This formula gives accuracy within one-half of 1 per cent for the range of $\alpha$ and $\beta$ considered in the table if $df > 9$. The values of $z_{1-\alpha} + z_{1-\beta}$ are listed in the $\infty$ line in Table A-12*b* and are the values of $d$ for the case where $\sigma$ is known.

For example, consider a two-sided test of the hypothesis $\mu = 50$ with $\alpha = .10$ and $N = 25$. The critical region is $t < -2.06$ and $t > 2.06$, where $t = (\bar{X} - 50)/\sqrt{s^2/25}$. Suppose that in fact $\mu = 52$ and $\sigma$ is approximately 10. Then we compute $d = (52 - 50)/\sqrt{\frac{100}{25}}$. In Table A-12*b* for $\alpha = .10$ (two-sided) we read across from $df = 24$ and find that $d = 1$ is between values for $1 - \beta = .2$ and $1 - \beta = .3$, showing that the chance of making the mistake of accepting $\mu = 50$ if in fact $\mu = 52$ is between 70 and 80 per cent.

Table A-12*c* giving required sample sizes when $\sigma$ is known can, with the changes noted in the table, be used when a $t$ test is anticipated. We see for the above example that to have 70 per cent power ($1 - \beta = .70$) against an alternative $d' = (52 - 50)/10 = .2$ requires a sample of 120 individuals.

If the sample size is to be fairly large, the approximate formula

$$N = [(z_{1-\alpha} + z_{1-\beta})/d']^2$$

can be used. In the above example we find $N = (2.169/.2)^2 = 118$, which is nearly equal to the value 120 obtained from the table.

An estimate of $\sigma$ is needed to determine power, since a deviation from a hypothesis must be quantified in order to use the tables.

**Paired data.** In Sec. 8-5 is a discussion of a technique for collecting and analyzing data in pairs in order to compare two populations. The chance with this design of accepting a hypothesis $\mu_1 - \mu_2 = 0$ when the means are actually not equal can be read from Table A-12*b*, with $d = (\mu_1 - \mu_2)/\sqrt{(\sigma_1^2 + \sigma_2^2)/N}$ and $N - 1$ degrees of freedom, where $N$ is the number of pairs of observations. For example, suppose we are testing the null hypothesis at the 5 per cent level of significance with 20 pairs of observations. If in fact $\mu_1 - \mu_2 = 4$ and $\sigma_1 = \sigma_2 = 10$, then $d = 4/\sqrt{10} = 1.26$ and in Table *A*-12*b* we see that for $df = 19$ this corresponds approximately to $1 - \beta = .22$. The linear interpolation for this result can be performed as follows:

| DEGREES OF FREEDOM | POWER $(1 - \beta)$ .2 | .22 | .3 |
|---|---|---|---|
| 16 | 1.19 | | 1.53 |
| 19 | 1.18 | 1.26 | 1.52 |
| 24 | 1.16 | | 1.50 |

$$\frac{1.26 - 1.18}{1.52 - 1.18} = \frac{8}{34} = .2 \text{ (approx.)}$$

Values are interpolated for 19 degrees of freedom for both $1 - \beta = .2$ and $1 - \beta = .3$, to give 1.18 and 1.52. It is then determined that 1.26 is about two-tenths of the way between 1.18 and 1.52, and thus $1 - \beta$ is read two-tenths of the way between $1 - \beta = .2$ and $1 - \beta = .3$, to give $1 - \beta = .22$.

Note that this example involves the same parameters and sample size as the preceding examples and that the paired design has a slightly smaller power $1 - \beta$. This shows that unnecessary pairing has increased our chance of making an error, mainly because of having $N - 1$ instead of $2N - 2$ degrees of freedom. The increase in this case is not great: .76 for $\sigma$ known, to .77 for $\sigma_1$, $\sigma_2$ not known but equal, to .78 for paired data with $\sigma_1 = \sigma_2$.

This discussion applies only to the case where independent measurements are taken from two populations. It does not apply to the case where a pair of measurements are made on each individual. For this case refer to Example 8-7. The appropriate $d'$ for use with Table A-12 is

$$d' = \frac{\mu_1 - \mu_2}{\sqrt{\sigma_1^2 + \sigma_2^2 - 2\rho\sigma_1\sigma_2}}$$

where $\rho$ is the correlation coefficient between the measurements.

## 14-3 OPERATING-CHARACTERISTIC CURVES

Curves giving the chance of accepting a hypothesis are called *operating-characteristic curves.* They are related to power curves in that the height of an operating-characteristic curve is 1 minus the height of the power curve for the same alternative. Figure 14-7 shows operating-characteristic curves for the three tests of the hypothesis that $\mu = \mu_0$ in a normal distribution (Sec. 14-2).

## 14-4 POWER FUNCTION OF TEST FOR PROPORTIONS

In this section we shall examine the problem of testing the hypothesis that a given proportion $p_0$ of a population has some attribute. Alternatives

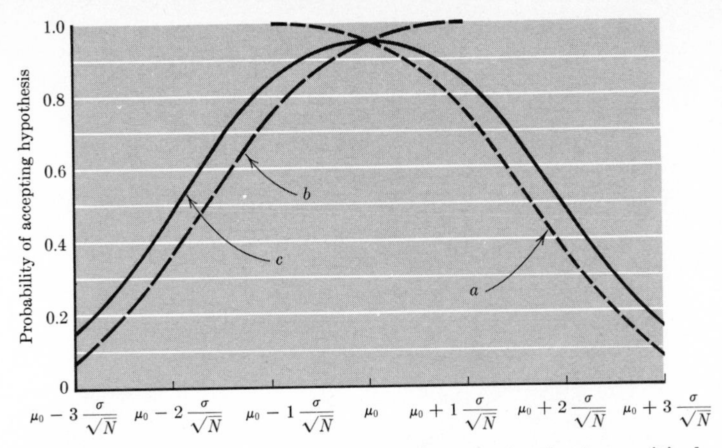


**Figure 14-7** Operating-characteristic ($\beta$) curves for the three critical regions discussed.

will be that the true proportion $p$ of the population having the attribute is some value other than $p_0$. Thus we shall test the hypothesis $p = p_0$.

It has been pointed out that if $N$ is large ($Np(1 - p) > 5$) the sampling distribution of the observed proportion $\bar{X}$ is approximately normal. In this case we can proceed as we did in Sec. 14-2 to test the hypothesis that the mean has a given value, say, $H: p = p_0$, with alternative $p = p_1 > p_0$. If the hypothesis is true, then $\sigma = \sqrt{p_0(1 - p_0)}$ and a one-sided test with level of significance $\alpha$ is to reject if

$$\bar{X} > p_0 + z_{1-\alpha}\sqrt{\frac{p_0\,(1 - p_0)}{N}}$$

where $z_{1-\alpha}$ is the $100(1 - \alpha)$ percentile of the normal distribution read from Table A-4.

To determine the value of $\beta$ if in fact the alternative $p = p_1$ is true we standardize this critical region to read

$$\frac{\bar{X} - p_1}{\sqrt{p_1(1 - p_1)/N}} > z_1$$

where

$$z_1 = \frac{p_0 + z_{1-\alpha}\sqrt{p_0(1 - p_0)/N} - p_1}{\sqrt{p_1(1 - p_1)/N}}$$

For this value of $z_1$ we can read $1 - \beta$ as the cumulative area in Table A-4.

As an example, consider the hypothesis $H: p \leq .2$ to be tested with $\alpha = .05$ level of significance using a random sample of size $N = 400$. The

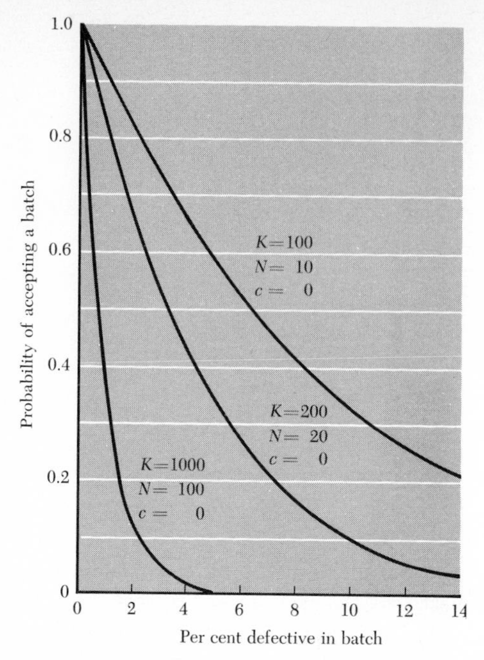


**Figure 14-8** Operating-characteristic curves; batch size $K$; sample of size $N$.

one-sided rejection region is

$$\bar{X} > .2 + 1.645\sqrt{(.2 \times .8)/400} = .233$$

If, for example, $p = p_1 = .25$, we compute

$$z_1 = \frac{.233 - .25}{\sqrt{(.25 \times .75)/400}} = -.78$$

Table A-4 gives power $1 - \beta = .21$, and the chance of the second type of error is $\beta = .79$. This procedure is the same as in Sec. 14-2, except that $\sigma$ depends on $p$. Some increase in the accuracy from Table A-4 can be achieved by using the continuity correction in Sec. 13-5. For small values of $Np(1 - p)$ the sampling distribution of $\bar{X}$ differs considerably from normal and Table A-4 should not be used. Table 17-2 gives for $p_0 = .50$ and a number of values of $\alpha$ the necessary sample sizes to have specified power $1 - \beta = .95$ against certain alternatives. The exact distribution is discussed in Chap. 20.

**Acceptance sampling.** Suppose we are to inspect batches of manufactured articles, each batch having $K$ articles. Our inspection will consist in choosing a random sample of $N$ articles, inspecting them, and rejecting the entire batch if we find more than $c$ defective articles among the $N$. The curves in Fig. 14-8 are the operating-characteristic curves for this procedure. Obviously, there are many possible sets of numbers $K, N, c$, and a statistician would need to have either many such sets of curves or else the mathematical training needed to draw them.

These curves tell us how frequently we would accept batches having certain percentages of defective items. Thus in batches of 100 items the plan of inspecting 10 articles and rejecting the batch if we observe one or more defective items will result in accepting batches having 2 per cent defective items 82 per cent of the time.

## POWER OF THE ANALYSIS-OF-VARIANCE TESTS 14-5

The analysis-of-variance technique is used to compare means of several populations. It is assumed that each of the populations has a normal distribution with a common value $\sigma^2$ for the variance. The means of the populations are $\mu_1, \ldots, \mu_k$, and the usual hypothesis to be tested is $\mu_1 = \mu_2 = \cdots = \mu_k$. Alternatives to this hypothesis would specify values for the $\mu_1, \ldots, \mu_k$ not all the same. We can measure the dispersion of the $\mu_i$'s by using the variance of these quantities,

$$\frac{\sum_{i=1}^{k} (\mu_i - \bar{\mu})^2}{k}$$

where $\bar{\mu} = \sum_{i=1}^{k} \mu_i/k$.

It is convenient to divide this quantity by $\sigma^2/n$, and call it $\phi^2$.

$$\phi^2 = \frac{\sum_{i=1}^{k} (\mu_i - \bar{\mu})^2/k}{\sigma^2/n}$$

where $n$ is the number of observations from each population.

This number $\phi^2$ can be used to measure an alternative. The probability of rejecting the hypothesis $\mu_1 = \mu_2 = \cdots = \mu_k$ when actually they have specified unequal values can be obtained in terms of $\phi^2$.

In Table A-13 are graphs giving the value of $1 - \beta$ on the vertical scale related to $\phi$ on the horizontal. The graphs are for two levels of significance, $\alpha = .01$ and $.05$, for eight values of $\nu_1$, the number of degrees of freedom for the numerator, and several values of $\nu_2$, the number of degrees of freedom in the denominator of the $F$ ratio. Note that there is a different curve for each set of values $\nu_1$, $\nu_2$, and $\alpha$. Consider the curve on the third page of Table A-13 for $\alpha = .05$, $\nu_1 = 3$, and $\nu_2 = 12$. This would be used, for example, in testing the hypothesis that four ($\nu_1 = 4 - 1 = 3$) populations have equal means with samples of size $n = 4$ from each population ($\nu_2 = 16 - 4 = 12$). Reading above $\phi = 2$, we see that the chance of recognizing that the four populations do not have equal means when actually $\phi^2 = 4$ is $1 - \beta = .82$. Thus the power of the test against any set of values $\mu_1, \mu_2, \mu_3, \mu_4$ giving $\phi^2 = 4$ is .82. The alternatives are in terms of $\phi^2$ alone, not distinguishing among different sets of means which give the same value to $\phi^2$. For example, with $\sigma^2 = 1$ the four population means could be 50, 50, 52, 52 or 50, 50, 50, 52.31 and give $\phi^2 = 4$.

EXAMPLE 14-1. Suppose we are to test the hypothesis that the means of five populations are equal, using the analysis-of-variance technique at the 1 per cent level of significance. Suppose further that we know that $\sigma^2$ is approximately 100. We shall take three observations from each population. Our $F$ ratio will have $\nu_1 = 5 - 1 = 4$ and $\nu_2 = (5 \times 3) - 5 = 10$. We can read from the graph that we have a chance of .7 of detecting a $\phi = 2.34$ or $\phi^2 = 5.48$. Many different sets of $\mu_1, \ldots, \mu_5$ will give $\phi^2 = 5.48$. One approximate example is

$$\begin{aligned} \mu_1 &= \bar{\mu} - 27 \\ \mu_2 &= \bar{\mu} + 6.75 \\ \mu_3 &= \bar{\mu} + 6.75 \\ \mu_4 &= \bar{\mu} + 6.75 \\ \mu_5 &= \bar{\mu} + 6.75 \end{aligned} \qquad \phi^2 = \frac{\dfrac{(-27)^2 + 4(6.75)^2}{5}}{\frac{100}{3}} = 5.47$$

Here we would reject the hypothesis when such alternatives were true about 7 out of 10 times.

EXAMPLE 14-2. A different problem which can be solved in a similar manner is as follows. Suppose $\mu_1, \mu_2, \mu_3$ are 10 units larger than $\mu_4$ and $\mu_5, \mu_6, \mu_7$ are 10 units smaller than $\mu_4$. Again we suppose a normal distribution with $\sigma^2$ known, say, $\sigma^2 = 100$. How many items should we take from each of the seven populations so that the analysis of variance with $\alpha = .05$ will have chance .7 of detecting this difference? Let $n$ be the number taken from each population. The degrees of freedom for $F$ are $\nu_1 = 6$ and $\nu_2 = 7n - 7$. We compute $\phi^2$:

$$\phi^2 = \frac{\dfrac{(-10)^2 \times 3 + (10)^2 \times 3}{7}}{100/n} = \frac{6}{7} n = .86n$$

Now we read from the $\alpha = .05$ graph for $\nu_1 = 6$ and $\nu_2 = 7n - 7$ corresponding to several values of $n$, each time comparing the tabled value with $.86n$ until approximate agreement is obtained. Reference was made to the graphs using $n = 2, 3, 4, 5$, and the following results were obtained:

| $n$ | $\nu_2$ | $.86n$ | FROM TABLE A-13 $\phi$ | $\phi^2$ |
|---|---|---|---|---|
| 2 | 7 | 1.72 | 1.88 | 3.5 |
| 3 | 14 | 2.58 | 1.55 | 2.4 |
| 4 | 21 | 3.44 | 1.48 | 2.2 |
| 5 | 28 | 4.30 | 1.42 | 2.0 |

For $n = 2$ the tabular $\phi^2$ is greater than $.86n$ and for $n = 3, 4, 5$ it is less than $.86n$. This indicates a solution between $n = 2$ and $n = 3$. We conclude that three items should be taken from each population.

**Two-way classification.** In dealing with a two-way classification (without replication) the procedure is the same, except that

$$\nu_2 = (n - 1)(k - 1)$$

($k$ columns and $n$ rows). $\phi^2$ is computed from the same formulas as above, and the examples are exactly the same.

It is also possible to use the analysis-of-variance procedure to test for differences in row means, in which case $\nu_1 = n - 1$, $\nu_2 = (n - 1)(k - 1)$, and

$$\phi^2 = \frac{\sum_{i=1}^{n} (\mu'_i - \bar{\mu}')^2/n}{\sigma^2/k}$$

where the $\mu'_i$ are the population means for rows and

$$\bar{\mu}' = \frac{\sum_{i=1}^{n} \mu'_i}{n}$$

## GLOSSARY

acceptance sampling
alternative
operating-characteristic curve
power function

## DISCUSSION QUESTIONS

1. What are the two types of error? Why can only one of these errors be made in a particular problem?

2. What is the power of a test?

3. How can there be different tests of the same hypothesis? Will different tests of the same hypothesis have the same power? Can they have the same level of significance?

4. In what way is a test useful if you do not have any information about its power function? What additional information does the power function give?

5. What is the effect on the power function of increasing the sample size?

6. Which is more useful—a power curve or an operating-characteristic curve?

7. Define each term in the Glossary.

8. Explain first in your own words and second by means of an example the use and interpretation of each of the graphs in this chapter.

9. In what way does a power curve help you to design an experiment? In what way is it dangerous to ignore the power function? Can you give a practical example where you are not particularly interested in the power of a test?

## CLASS EXERCISES

1. Suppose you wish to test the hypothesis that a population is normally distributed with zero mean and unit variance. As a test you agree to reject if $\bar{X} > 1.96/\sqrt{10}$. From the samples of size 5 drawn from Table A-23 for the Class Exercise in Chap. 3, estimate the chance of rejecting this hypothesis if actually the mean is 2. Use this result to test the hypothesis that the chance of rejecting is that given by Table A-12*a*.

2. Put 4 red beads and 96 white beads in a box and take samples of size $N = 10$ without replacement (take all 10 at once, record the number of red beads, and then replace the 10). From a number of such trials test the hypothesis that the value in Fig. 14-8 is correct. Repeat this experiment with 10 red beads and 90 white beads. Repeat with 14 red beads and 86 white beads.

3. In Class Exercise 1, Chap. 10, an estimate was obtained of the chance that a particular false hypothesis would be detected by an analysis-of-variance-designed experiment. Compare the observed result with that in Table A-13. Make the same comparison for Class Exercise 2, Chap. 10.

## PROBLEMS

*Section* 14-1

1. A hypothesis that a new drug is equally as good as an old drug was tested and rejected at the $\alpha = .01$ level of significance. Either an $\alpha$ error or a $\beta$ error may have been made. If so, which? Is it true that the hypothesis has a 99 per cent chance of being false?

2. In general if an alternative to a hypothesis is far removed, the power of the test is greater (if $H\colon \mu = 50$, $\sigma = 10$ is tested and the true value of $\mu$ is 80 or 90, then $\beta$ is small). Can we conclude that to make a test more acceptable we should select alternatives far removed from the hypothesis?

*Section* 14-2

3. The power curves drawn in Fig. 14-5 were for the 5 per cent level of significance. Draw the similar curves for the same hypothesis for the 1 per cent level of significance.

4. In a normal distribution $\sigma = 5$. It is desired to test the hypothesis $\mu = 12$ with $\alpha = .05$. With a sample of 64 cases what is the chance of rejecting the hypothesis if $\mu = 14$? If $\mu = 9$? How large a sample (instead of 64) should be used if it is desired to have a chance of .60 of rejecting the hypothesis if $\mu \geq 14$ or if $\mu \leq 10$? Be careful here to use region (*c*).

5. In testing the hypothesis $\mu = 50$ in a normal distribution with what chance would you accept the hypothesis if the true mean is $.5\sigma$ above 50? ($\alpha = .05$, $N = 10$.)

6. We wish to test the hypothesis that $\mu = 50$, and we know $\sigma$ is between 3 and 8. At the 5 per cent level of significance how large a sample must be taken so that the probability of rejecting $\mu = 50$ is at least .80 if $\mu$ is more than 3 units away from 50?

7. If we wish to select a school for a special experiment, and wish to be at least 90 per cent sure of accepting a school for the purpose if the mean IQ of its students is 105 or above, and wish to be at least 90 per cent sure of rejecting the school for the purpose if the mean IQ is 100 or less, how large a sample is needed? (Assume $\sigma = 16$.)

8. A two-sided $t$ test is used to test at the 5 per cent level of significance that the mean height of a population is 66 inches. A random sample of 10 individuals is

taken. Using 3 inches as an approximate value of $\sigma$, find the chance that the hypothesis will be rejected if the sample is taken from a population of mean height 67 inches. Of mean height 64.5 inches.

**9.** Two groups of 20 animals each are to be compared. If the standard deviations of the two populations are taken to be 10 pounds each, what is the chance that a $t$ test at the 1 per cent level of significance will recognize a difference of 1 pound in the two means?

**10.** A one-sided $t$ test of the hypothesis that the means of two populations are equal is to be performed at the 5 per cent level of significance on 20 pairs of individuals, one member of each pair from each population. If the standard deviations of the populations are taken to be 10 and 20 pounds, respectively, what is the chance that the $t$ test will detect a difference of 5 pounds in the population means?

**11.** Suppose a mean response changes linearly with height in inches with $\sigma_{y \cdot x} = .15$ units. For the population available the standard deviations of response for all heights is $\sigma_y = .30$ units. It is thought that a drug would increase the mean response by approximately .2 units. Two experiments are suggested as follows: (1) random samples of 100 treated and 100 controls are chosen and measured and (2) random samples of 25 treated and 25 controls are chosen at each of two selected heights (100 in all). (*a*) In each case a test at the 5 per cent level of significance is to be carried out. Which of the two experiments has a better chance of recognizing an effect of the drug? (*b*) What size samples at each of two selected heights would make the two methods have equal power?

**12.** Lung-capacity determinations vary from day to day about individual means with $\sigma = .3$ liters. How many people are required to warrant a 90 per cent chance of recognizing a shift in mean of .6 liters with an $\alpha = .05$ level of significance?

**13.** A certain lung-capacity measurement varies from day to day about individual means with $\sigma = .3$ liters. With aging from 30 to 50 years the measurements decrease .02 liters per year for nonsmokers and .04 liters per year for smokers. How many thirty-year-old smokers and nonsmokers should be followed for 20 years for there to be a $1 - \beta = .90$ chance of recognizing this difference with an $\alpha = .05$ level test? How many should be followed for 5 years for the same criterion?

*Section* 14-4

**14.** Approximately 20 per cent of people treated with a particular drug develop an allergy to it. A buffered solution is to be tested in the hope that this percentage would decrease.

(*a*) If a sample of 100 people is to be used what chance is there that a 5 per cent level-of-significance test will reject the hypothesis of no change if in fact there is no effect from the buffer?

(*b*) If a sample of 100 people is to be used what chance is there that a 5 per cent level-of-significance test will reject the hypothesis of no change if in fact the percentage is reduced to 10 per cent?

(*c*) Repeat (*a*) and (*b*) for a test at the 1 per cent level.

(*d*) Repeat (*a*) and (*b*) and (*c*) with a sample of size 300.

(*e*) How large a sample is required for a 90 per cent chance of obtaining significance at the 5 per cent level of significance if in fact the percentage is reduced to 10 per cent?

**15.** The death rate from a certain disease is approximately 1 per 100,000 people per year. How large a vaccinated sample would be required to test the hypothesis of no effect with $\alpha = .05$ so that $\beta = .05$ if a vaccine reduced the rate to 1 per 200,000 people per year?

*Section* 14-5

**16.** Effects of two treatments are to be measured with an analysis of variance (or, equivalently, with a two-sample $t$-test analysis). Individuals are known to vary with $\sigma = 10$ pounds.

(*a*) If two samples each of size 20 are used what is the chance that a 1 per cent level-of-significance test will reject the no-difference hypothesis if there is in fact no difference in the treatment effects?

(*b*) If two samples each of size 20 are used what is the chance that a 1 per cent level-of-significance test will reject the no-difference hypothesis if in fact one treatment results in a population mean 5 pounds larger than that resulting from the other treatment?

(*c*) Repeat (*a*) and (*b*) for a test at the 5 per cent level.

(*d*) Repeat (*a*), (*b*), and (*c*) if two samples of size 40 each are to be used.

(*e*) What size samples should be used so that a 1 per cent level-of-significance test has a 90 per cent chance of rejecting the no-difference hypothesis if in fact there is a 5-pound difference between the mean results?

**17.** Four schools are to be compared on the basis of the amount of knowledge the students have of current affairs. A standardized test is given to random samples of 20 students at each school. It is desired to test the hypothesis, at the 5 per cent level of significance, that the mean knowledge of current affairs, as measured by this test, is the same for all four schools. Past experience with the test has indicated that $\sigma^2$ is approximately 100. What chance would the analysis of variance have of detecting the fact that two of the school means are 10 units higher than the other two?

**18.** Suppose we have a single-variable-of-classification experiment with five categories and $n$ items in each category. We wish to test the hypothesis that the means are equal at the 5 per cent level of significance. Suppose the standard deviation $\sigma$ is thought to be about 10 per cent of $\bar{\mu}$. Suppose that four means are equal and the other is 10 per cent higher than the four. How large a sample is needed for a chance of .7 of detecting this difference?

CHAPTER FIFTEEN

# More analysis of variance

In this chapter we shall extend the ideas of Chap. 10 to provide analyses appropriate to ordered variables of classification and to some special experiment designs, and we shall discuss several topics related to the assumptions underlying the analysis of variance. The use of the analysis-of-variance table to display the sums of squares, mean squares, and $F$ ratios was demonstrated for several simple problems concerning the means of samples in several categories in Chap. 10. The sums of squares attributable to regression were studied in Chap. 11, and the combinations of these two problems, analysis of covariance, was presented in Chap. 12. These same problems can be approached on the basis of forming special contrasts related to the effects we wish to study. These will be introduced in Sec. 15-1 and applied to the problem of bioassay in Sec. 15-2. Later sections cover the components of variance assumptions for analysis of variance and other topics.

The plan of an experiment will determine the choice of sums of squares to display and analyze. For the one-variable analysis-of-variance design we display the sum of squares for means and use this in our test of equality of means as an index of general inequality of means. For the two-variable analysis-of-variance design with replications we display three sums of squares as indices of differences in row means, column means, and interactions. In this chapter we shall break down these sums of squares to provide indices of more specific differences (contrasts) among the means.

The number and type of contrasts that can be investigated depends, of course, on the design of the experiment. For example, we can separate any interaction effects (measured by $\sigma_I^2$ in Table 10-18) from the within-cell variance $\sigma^2$ only if there are replications in the two-variable design.

Similarly, in a regression problem an experiment so designed as to use only two values of the independent variable allows an estimate of slope for the linear model but does not permit inspection of the data for a curvature in the regression. A design using three values of the independent variable does allow such inspection, and we will note that in certain designs the linear and curvature effects can be displayed in an analysis-of-variance-table format.

## 15-1 ORTHOGONAL CONTRASTS

**Orthogonal contrasts in a factorial experiment.** A *contrast* is a linear combination of the observations or the parameters with the coefficients adding to zero. Suppose the observations are as given in Table 15-1. As illustrated in Chap. 10 a contrast on means can be written (using as an example the difference in the two column means of Table 15-1). This is

$$\bar{X}_1 - \bar{X}_2 = 10 - 6$$

where

$$\bar{X}_1 = \tfrac{1}{6}(6 + 8 + 7 + 11 + 13 + 15) = 10$$
$$\bar{X}_2 = \tfrac{1}{6}(5 + 4 + 6 + 7 + 9 + 5) = 6$$

If these 12 observations are designated as in Table 15-2, we can see that

$$\bar{X}_1 - \bar{X}_2 = \tfrac{1}{6}(X_1 + X_2 + X_3 + X_7 + X_8 + X_9) - \tfrac{1}{6}(X_4 + X_5 + X_6 + X_{10} + X_{11} + X_{12})$$

**Table 15-1**

| | | COLUMN 1 | COLUMN 2 | *Total* | *Mean* |
|---|---|---|---|---|---|
| ROW | 1 | 6<br>8<br>7 | 5<br>4<br>6 | 36 | 6 |
| ROW | 2 | 11<br>13<br>15 | 7<br>9<br>5 | 60 | 10 |
| | *Total* | 60 | 36 | 96 | |
| | *Mean* | 10 | 6 | | 8 |

**Table 15-2**

| | |
|---|---|
| $X_1$ | $X_4$ |
| $X_2$ | $X_5$ |
| $X_3$ | $X_6$ |
| | |
| $X_7$ | $X_{10}$ |
| $X_8$ | $X_{11}$ |
| $X_9$ | $X_{12}$ |

These terms may be rearranged in the order of the subscripts for a linear contrast of the 12 observations:

$$\bar{X}_1 - \bar{X}_2 = \tfrac{1}{6}(X_1 + X_2 + X_3 - X_4 - X_5 - X_6 + X_7 + X_8 + X_9 - X_{10} - X_{11} - X_{12})$$

More generally we define a linear contrast as

$$\Sigma a_i X_i = a_1 X_1 + a_2 X_2 + \cdots + a_N X_N \qquad \text{with } \Sigma a_i = 0$$

and have illustrated that the difference in column means can be written in this form by taking

$$a_1 = a_2 = a_3 = a_7 = a_8 = a_9 = \tfrac{1}{6}$$

and

$$a_4 = a_5 = a_6 = a_{10} = a_{11} = a_{12} = -\tfrac{1}{6}$$

The difference in row means and the interaction contrast can also be written in this form.

Any mean square in the analysis-of-variance table which has a single degree of freedom can be obtained from the square of a contrast divided by the sum of squares of the contrast coefficients. To illustrate this fact let us first construct the usual analysis-of-variance table for the data in Table 15-1; this is given in Table 15-3. The contrast for difference in column means was computed to be 4. If we square this number and divide by the sum of squares of the coefficients ($12(\frac{1}{6})^2 = \frac{1}{3}$) we obtain $4^2/\frac{1}{3} = 48$, as reported in the analysis-of-variance table. Since the multiplication of a contrast by a constant does not change its *mean square* $(\Sigma a_i X_i)^2/\Sigma a_i^2$, we

**Table 15-3** Analysis of variance

| | *Sum of squares* | *Degrees of freedom* | *Mean square* |
|---|---|---|---|
| Rows | 48 | 1 | 48 |
| Columns | 48 | 1 | 48 |
| Interaction | 12 | 1 | 12 |
| Residual | 20 | 8 | 2.5 |
| *Total* | 128 | 11 | |

could equally well refer to the contrast for columns as

$$X_1 + X_2 + X_3 - X_4 - X_5 - X_6 + X_7 + X_8 + X_9 - X_{10} - X_{11} - X_{12}$$

and find the same mean square as $(24)^2/12 = 48$.

For the particular example above, three replications in a $2 \times 2$ design, this mean square could also be obtained (with appropriate modification) from the difference in the two means. The illustration with individual coefficients for each observation will facilitate transition to other designs.

We now define orthogonal contrasts. Any two contrasts

$$a_1\mu_1 + a_2\mu_2 + \cdots + a_k\mu_k$$

and

$$b_1\mu_1 + b_2\mu_2 + \cdots + b_k\mu_k$$

with

$$\Sigma a_i = 0 \qquad \text{and} \qquad \Sigma b_i = 0$$

are called *orthogonal* if, in addition,

$$a_1b_1 + a_2b_2 + \cdots + a_kb_k = 0$$

It can be shown mathematically that any sum of squares with $\nu$ degrees of freedom can be written as the sum of $\nu$ separate terms, each of which is a mean-square-contrast orthogonal to the rest. If this is done the analysis is said to be reduced to single degrees of freedom.

For example, an analysis of variance of the data in Table 15-2, with the four groups considered as four categories of a single variable of classification (that is, with the row and column variable arrangement ignored), would appear as in Table 15-4. It can be noted that the row, column, and interaction mean squares in Table 15-3 add to the between-means sum of squares in Table 15-4.

There are a number of special cases in which it is useful to display a set of orthogonal contrasts making up a particular sum of squares. The sums of squares for the main effects and interaction effects in a factorial experiment are the squares of orthogonal contrasts or the sums of squares of orthogonal contrasts. The measures of linear and higher-order effects in regression problems can also be studied as orthogonal contrasts.

As an illustration we display the three orthogonal contrasts corre-

**Table 15-4** Analysis of variance

| | *Sum of squares* | *Degrees of freedom* | *Mean square* |
|---|---|---|---|
| Between means | 108 | 3 | 36.00 |
| Within groups | 20 | 8 | 2.50 |
| *Total* | 128 | 11 | |

**Table 15-5** Three orthogonal contrasts

| | OBSERVATIONS $X_i$ | | | | | | | | | | | | *Con-* | | *Mean* |
|---|---|---|---|---|---|---|---|---|---|---|---|---|---|---|---|
| | 6 | 8 | 7 | 5 | 4 | 6 | 11 | 13 | 15 | 7 | 9 | 5 | *trast* | $\Sigma a_i^2$ | *square* |
| | *Contrast coefficients* | | | | | | | | | | | | | | |
| *Column* $a_i$ | 1 | 1 | 1 | −1 | −1 | −1 | 1 | 1 | 1 | −1 | −1 | −1 | 24 | 12 | 48 |
| *Row* $b_i$ | 1 | 1 | 1 | 1 | 1 | 1 | −1 | −1 | −1 | −1 | −1 | −1 | −24 | 12 | 48 |
| *Interaction* $c_i$ | 1 | 1 | 1 | −1 | −1 | −1 | −1 | −1 | −1 | 1 | 1 | 1 | −12 | 12 | 12 |

sponding to the first three rows of the analysis of variance in Table 15-3. We can further simplify the writing of these contrasts by indicating only the coefficients, as in Table 15-5. The column contrast discussed above appears in the table, along with the contrast for row and interaction. We can easily verify orthogonality by computing $\Sigma a_i b_i$, $\Sigma a_i c_i$, and $\Sigma b_i c_i$ to be zero.

**Orthogonal contrasts for unequal cell frequencies.** Consider an experiment with fewer replications for one column or one row, as in Table 15-6, where for illustrative purposes we have removed two observations

**Table 15-6** Ten observations in a 2 × 2 design

| ROW | COLUMN 1 | COLUMN 2 | *Total* | *Mean* |
|---|---|---|---|---|
| 1 | 6, 8, 7 | 5, 4 | 30 | 6 |
| 2 | 11, 13, 15 | 7, 9 | 55 | 11 |
| *Total* | 60 | 25 | 85 | |
| *Mean* | 10 | 6.25 | | 8.5 |

from Table 15-1. Three orthogonal contrasts associated with row, column, and interaction effects are given in Table 15-7 for the data in Table 15-6. This is a representation for this special set of cell frequencies.

**Table 15-7** Three orthogonal contrasts

| | OBSERVATIONS $X_i$ | | | | | | | | | | *Con-* | | *Mean* |
|---|---|---|---|---|---|---|---|---|---|---|---|---|---|
| | 6 | 8 | 7 | 5 | 4 | 11 | 13 | 15 | 7 | 9 | *trast* | $\Sigma a_i^2$ | *square* |
| | *Contrast coefficients* | | | | | | | | | | | | |
| *Column* | 2 | 2 | 2 | −3 | −3 | 2 | 2 | 2 | −3 | −3 | 45 | 60 | 33.75 |
| *Row* | 1 | 1 | 1 | 1 | 1 | −1 | −1 | −1 | −1 | −1 | −25 | 10 | 62.50 |
| *Interaction* | 2 | 2 | 2 | −3 | −3 | −2 | −2 | −2 | 3 | 3 | −15 | 60 | 3.75 |

**Table 15-8** Analysis of variance

| | *Sum of squares* | *Degrees of freedom* | *Mean square* |
|---|---|---|---|
| Column | 33.75 | 1 | 33.75 |
| Row | 62.50 | 1 | 62.50 |
| Interaction | 3.75 | 1 | 3.75 |
| Remainder | 12.50 | 6 | 2.08 |
| *Total* | 112.50 | 9 | |

Orthogonal contrasts cannot be obtained for all possible cell frequencies. However, if the cell frequencies are proportional ($n_{11}/n_{12} = n_{21}/n_{22}$) orthogonal contrasts can be constructed for row, column, and interaction effects. General rules for determining orthogonal-contrast coefficients will not be given beyond a note that the contrast coefficients must sum to zero and satisfy the rule of $\Sigma a_i b_i = 0$.

For experiments with unstructured inequality of the numbers of replications it may not be possible to construct orthogonal contrasts for the comparisons of interest. An alternative approach, with contrasts that are not necessarily orthogonal, was given in Sec. 10-3. With the orthogonal contrasts in Table 15-7 an analysis-of-variance table for the data in Table 15-6 is given in Table 15-8. Tests of significance can be made in the usual fashion by the use of $F$ ratios and Table A-7. We note that row and column effects are significant at the $\alpha = .01$ level, while the interaction effect is relatively small.

**Contrasts for means and sums.** When the number of replications are equal, the specifications of contrasts can be simplified by applying the coefficients to the means or to the sums of the observations in each cell. This corresponds to the example of Sec. 10-6.

For the $2 \times 2$ experiment with $n$ observations per cell we can write a table of contrasts as in Table 15-9 or 15-10. The data from Table 15-1 is used in each to illustrate the results. In Table 15-9 the mean square is obtained by dividing the squared contrast by $n\Sigma a_i^2$, while in Table 15-10 it is obtained by dividing by $\Sigma a_i^2/n$. Note that the mean squares are the same in each case.

**Table 15-9** Contrasts for sums

| | CELL SUMS $\Sigma X$ | | | | *Con-* | | *Mean* |
|---|---|---|---|---|---|---|---|
| | 21 | 15 | 39 | 21 | *trast* | $\Sigma a_i^2$ | *square* |
| | *Contrast coefficients* | | | | | | |
| *Column* | 1 | −1 | 1 | −1 | 24 | 4 | 48 |
| *Row* | 1 | 1 | −1 | −1 | −24 | 4 | 48 |
| *Interaction* | 1 | −1 | −1 | 1 | −12 | 4 | 12 |

**Table 15-10** Contrasts for means

| | CELL MEANS $\bar{X}$ 7 | 5 | 13 | 7 | *Con-trast* | $\Sigma a_i^2$ | *Mean square* |
|---|---|---|---|---|---|---|---|
| | *Contrast coefficients* | | | | | | |
| *Column* | 1 | −1 | 1 | −1 | 8 | 4 | 48 |
| *Row* | 1 | 1 | −1 | −1 | −8 | 4 | 48 |
| *Interaction* | 1 | −1 | −1 | 1 | −4 | 4 | 12 |

**Contrasts for other designs.** Other sets of orthogonal contrasts can be constructed and may be appropriate for other analyses or other experimental designs. The discussion above illustrates the use of orthogonal contrasts for the factorial row and column effects. The coefficients in Table 15-11 illustrate successive comparisons of each cell mean with means of

**Table 15-11** Successive contrasts

| | CELL MEANS $\bar{X}_{11}$ | $\bar{X}_{21}$ | $\bar{X}_{12}$ | $\bar{X}_{22}$ | $\Sigma a_i^2$ |
|---|---|---|---|---|---|
| | *Contrast coefficients* | | | | |
| *Diff. for cell* 2 *and cell* 1 | 1 | −1 | 0 | 0 | 2 |
| *Diff. for cell* 3 *and cells* 1 *and* 2 | 1 | 1 | −2 | 0 | 6 |
| *Diff. for cell* 4 *and cells* 1, 2, 3 | 1 | 1 | 1 | −3 | 12 |

previous cells. The coefficients in Table 15-12 compare column means separately for each row. The various sets of contrasts will be constructed to correspond to various experimental designs. Suppose cells 11 and 21 represent two alternative forms of drug therapy, cell 12 represents a placebo therapy, and cell 22 represents no therapy. The first contrast in Table 15-11 compares the two active drugs. The second contrast compares the placebo therapy with the two active drugs, and the third contrast compares no therapy with the three other therapies.

In Table 15-12 the rows could represent a basic variable under study and the column classification a "nesting," or subdivision into two classes

**Table 15-12** Contrasts for columns nested in rows

| | CELL MEANS $\bar{X}_{11}$ | $\bar{X}_{21}$ | $\bar{X}_{12}$ | $\bar{X}_{22}$ | $\Sigma a_i^2$ |
|---|---|---|---|---|---|
| | *Contrast coefficients* | | | | |
| *Col. diff. in row* 1 | 1 | −1 | 0 | 0 | 2 |
| *Col. diff. in row* 2 | 0 | 0 | 1 | −1 | 2 |
| *Row diff.* | 1 | 1 | −1 | −1 | 4 |

**Table 15-13** Contrasts for rows nested in columns

| | CELL MEANS $\bar{X}_{11}$ | $\bar{X}_{21}$ | $\bar{X}_{12}$ | $\bar{X}_{22}$ | $\Sigma a_i^2$ |
|---|---|---|---|---|---|
| | *Contrast coefficients* | | | | |
| *Row diff. in col.* 1 | 1 | 0 | −1 | 0 | 2 |
| *Row diff. in col.* 2 | 0 | 1 | 0 | −1 | 2 |
| *Col. diff.* | 1 | −1 | 1 | −1 | 4 |

separately for each row. The columns then do not represent a second variable of classification crossed with the first. Applications of these designs may occur in agriculture field studies, in subsampling by census districts, and so on.

In Table 15-13 the contrasts reflect nesting of rows in columns.

**Orthogonal contrasts in a 2 × 3 factorial experiment.** Table 15-14 presents cell totals for data from a 2 × 3 factorial experiment with

**Table 15-14** Data from a 2 × 3 experiment

| | | | | *Total* |
|---|---|---|---|---|
| | $T_{11+} = 16$ | $T_{21+} = 7$ | $T_{31+} = 15$ | 38 |
| | $T_{12+} = 25$ | $T_{22+} = 20$ | $T_{32+} = 25$ | 70 |
| *Total* | 41 | 27 | 40 | 108 |

three replicates. These are the data analyzed in Sec. 10-5 and presented originally in Table 10-15. The conventional analysis of variance was presented in Table 10-17 and is repeated here as Table 15-15.

The first contrast of Table 15-16 corresponds to the mean square for row means. The second and third lines of Table 15-16 display coefficients of contrasts, orthogonal to each other and to the first, which have mean squares adding to the column sum of squares. Other contrasts could have been used

**Table 15-15** Analysis of variance, 2 × 3 experiment

| | *Sum of squares* | *Degrees of freedom* | *Mean square* |
|---|---|---|---|
| Row mean | 56.89 | 1 | 56.89 |
| Column mean | 20.33 | 2 | 10.17 |
| Interaction | 1.45 | 2 | .72 |
| *Subtotal* | 78.67 | 5 | |
| Within groups | 17.33 | 12 | 1.44 |
| *Total* | 96.00 | 17 | |

**Table 15-16** Contrasts in a 2 × 3 experiment

| | CELL SUMS | | | | | | Con- | | Mean |
|---|---|---|---|---|---|---|---|---|---|
| | 16 | 7 | 15 | 25 | 20 | 25 | trast | $\Sigma a_i^2$ | square |
| | *Contrast coefficients* | | | | | | | | |
| Row | 1 | 1 | 1 | −1 | −1 | −1 | −32 | 6 | 56.89 |
| Col. 1 | 1 | 0 | −1 | 1 | 0 | −1 | 1 | 4 | .08 |
| Col. 2 | 1 | −2 | 1 | 1 | −2 | 1 | 27 | 12 | 20.25 |
| Interaction 1 | 1 | 0 | −1 | −1 | 0 | 1 | 1 | 4 | .08 |
| Interaction 2 | 1 | −2 | 1 | −1 | 2 | −1 | 7 | 12 | 1.37 |

here. The last two contrasts corresponding to interaction have mean squares adding to the interaction term in Table 15-15.

An expanded form of Table 15-15 then appears as Table 15-17.

**Table 15-17** Analysis of variance; contrasts in a 2 × 3 experiment

| | | Sum of squares | | Degrees of freedom | Mean square |
|---|---|---|---|---|---|
| Row | | | 56.89 | 1 | 56.89 |
| Column | (1) | .08 | | | |
| | (2) | 20.25 | | | |
| | *Total* | | 20.33 | 2 | 10.17 |
| Interaction | (1) | .08 | | | |
| | (2) | 1.37 | | | |
| | *Total* | | 1.45 | 2 | .72 |
| *Subtotal* | | | 78.67 | 5 | |
| Within groups | | | 17.33 | 12 | 1.44 |
| *Total* | | | 96.00 | 17 | |

**Orthogonal contrasts in regression.** Suppose that the observations in a 2 × 2 table represent data for two regression lines, as in Table 15-18, where the data are responses observed for two amounts of two chemicals. The observations are plotted in Fig. 15-1, and the regression lines are drawn.

**Table 15-18** Data for two regressions

| | | AMOUNT 1 | AMOUNT 2 |
|---|---|---|---|
| CHEMICAL | 1 | 6 | 9 |
| | | 5 | 8 |
| | 2 | 4 | 6 |
| | | 3 | 5 |

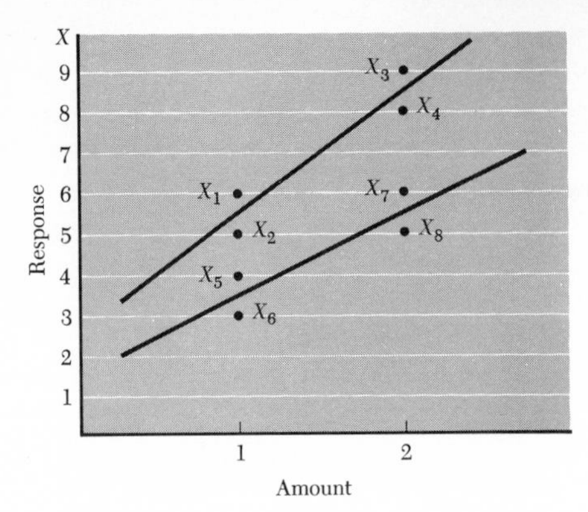


**Figure 15-1** Plot of responses at two "dosage" levels for two chemicals.

If we interpret the data in Table 15-18 as a design with four categories, the analysis-of-variance table constructed as in Chap. 10 would appear as in Table 15-19.

**Table 15-19** Analysis of variance

| | *Sum of squares* | *Degrees of freedom* | *Mean square* | *F ratio* |
|---|---|---|---|---|
| Between means | 25.5 | 3 | 8.5 | 17 |
| Within groups | 2.0 | 4 | .5 | |
| *Total* | 27.5 | | | |

We now display a set of three orthogonal contrasts in Table 15-20. The first contrast measures the mean difference in the two chemicals, and the other two contrasts are orthogonal to the first and have been chosen to measure effects of interest in the model. The second provides an estimate of the mean slope, which could be labeled as a "common slope" if the separate slopes are equal, and the third measures the difference in the two slopes.

The between-means sum of squares with 3 degrees of freedom is equal to the sum of mean squares for these three orthogonal contrasts, and an analysis of variance can be presented as in Table 15-21. This is another

**Table 15-20** Three contrasts for regression

| | CHEMICAL 1 | | | | CHEMICAL 2 | | | | | | |
|---|---|---|---|---|---|---|---|---|---|---|---|
| | $X_1$ | $X_2$ | $X_3$ | $X_4$ | $X_5$ | $X_6$ | $X_7$ | $X_8$ | *Contrast* | $\Sigma a_i^2$ | *Mean square* |
| Data | 6 | 5 | 9 | 8 | 4 | 3 | 6 | 5 | | | |
| | | | | | *Coefficients* | | | | | | |
| Chemicals | −1 | −1 | −1 | −1 | 1 | 1 | 1 | 1 | −10 | 8 | 12.5 |
| Common slope | −1 | −1 | 1 | 1 | −1 | −1 | 1 | 1 | 10 | 8 | 12.5 |
| Diff. in slopes | −1 | −1 | 1 | 1 | 1 | 1 | −1 | −1 | 2 | 8 | .5 |

**Table 15-21** Analysis of variance for regression

| | *Sum of squares* | *Degrees of freedom* | *Mean squares* |
|---|---|---|---|
| Chemicals | 12.5 | 1 | 12.5 |
| Slope | 12.5 | 1 | 12.5 |
| Diff. in slopes | .5 | 1 | .5 |
| Within groups | 2.0 | 4 | .5 |
| *Total* | 27.5 | 7 | |

example of the general theorem that the sum of squares for any $\nu$-degree-of-freedom portion of an analysis-of-variance table can be written as the sum of $\nu$ mean squares obtained from $\nu$ orthogonal contrasts.

The computation for Table 15-21 proceeds much like that of the preceding example. The first three mean squares are available from Table 15-21, and they add to the value 25.5, as given in Table 15-19 for the between-means sum of squares. The total sum of squares is computed in the usual manner, and the sum of squares within groups is obtained by subtraction, $27.5 - 12.5 - 12.5 - 0.5 = 2$ or, alternatively, as the pooled cell variance. The difference in slopes is directly analogous to the interaction discussed earlier. When the slopes are not the same, the level of concentrations must be taken into account when comparing responses.

**Simple nonlinearity.** The two-point design presented above gives estimates of linear slopes but does not allow a measurement of curvature. Data from a three-point design with equally spaced concentration of each of two chemicals are presented in Fig. 15-2. The values of $X_3 - X_2$ and $X_2 - X_1$ both measure linear slopes, while the difference

$$X_3 - X_2 - (X_2 - X_1) = X_3 - 2X_2 + X_1$$

will be zero if the three points are on a single line and not zero if the points do not lie on a single line. Thus this contrast provides a measure of curvature, or deviation from "straightness."

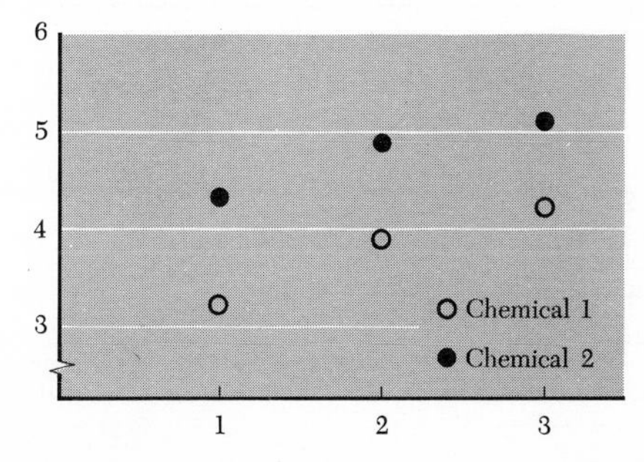


**Figure 15-2** Plot for six-point regression problem.

**Table 15-22** Contrasts for a six-point regression problem

| | CHEMICAL 1 $X_1$ | $X_2$ | $X_3$ | CHEMICAL 2 $X_4$ | $X_5$ | $X_6$ |
|---|---|---|---|---|---|---|
| | *Contrast coefficients* | | | | | |
| Diff. in chemicals | −1 | −1 | −1 | 1 | 1 | 1 |
| Common slope | −1 | 0 | 1 | −1 | 0 | 1 |
| Diff. in slope | −1 | 0 | 1 | 1 | 0 | −1 |
| Common nonlinearity | 1 | −2 | 1 | 1 | −2 | 1 |
| Reverse nonlinearity | 1 | −2 | 1 | −1 | 2 | −1 |

Table 15-22 presents one set of five orthogonal contrasts for this design. Another set of possibly equal interest could be written by replacing the last two by

| | | | | | |
|---|---|---|---|---|---|
| 1 | −2 | 1 | 0 | 0 | 0 |
| 0 | 0 | 0 | 1 | −2 | 1 |

These two contrasts measure the nonlinearity for the two chemicals separately. If the model specifies linearity with possibly different slopes, an analysis-of-variance table would display the first three contrasts separately and use the last two to estimate $\sigma^2$, as in Table 15-23.

For the six observations 3.3, 3.9, 4.2, 4.4, 4.9, 5.1 from an experiment of the type described in Fig. 15-2 and Table 15-22 we find mean squares as given in Table 15-23. The residual sum of squares can be obtained by subtraction, or equivalently by adding the mean squares for the remaining two orthogonal contrasts. The analysis could proceed as in Sec. 10-5. The $F$ ratio for difference in slopes, $F = .01/.015 = .67$, is not significant, the $F$ ratio for common slope, $F = 42.7$, is not significant at the $\alpha = .01$ level, while that for difference in chemicals, $F = 100$, is significant when compared with $F_{.99}(1,2) = 98.5$. Following the procedure of pooling sums of squares, if the

**Table 15-23** Analysis of variance for a 2 × 3 regression design

| | *Sum of squares* | | *Degrees of freedom* | *Mean square* |
|---|---|---|---|---|
| Diff. in chemicals | $\dfrac{(-X_1 - X_2 - X_3 + X_4 + X_5 + X_6)^2}{6}$ | = 1.50 | 1 | 1.50 |
| Common slope | $\dfrac{(-X_1 + X_3 - X_4 + X_6)^2}{4}$ | = .64 | 1 | .64 |
| Diff. in slope | $\dfrac{(-X_1 + X_3 + X_4 - X_6)^2}{4}$ | = .01 | 1 | .01 |
| Residual | By subtraction | = .03 | 2 | .015 |
| *Total* | $\Sigma(X_i - \bar{X})^2$ | = 2.18 | 5 | |

ratio is less than $2F_{.50}(1,2) = 1.33$, we would use $(.01 + .03)/3 = .0133$ for the denominator mean square and compare with $F_{.99}(1,3) = 34.1$ to see that both common slope and difference in chemicals are significant.

**Slope for several points.** The hypothesis that the means $\mu_1, \mu_2, \ldots, \mu_k$ for $k$ categories are linearly related to the variable which identifies the $k$ categories can be tested with a contrast, as was done above for $k = 3$. We assume that the means of the successive categories can be written in terms of a constant quantity $\mu$ and a common increment (slope) $d$ as

$$\begin{aligned} \mu_1 &= \mu \\ \mu_2 &= \mu + d \\ \mu_3 &= \mu + 2d \\ &\cdots\cdots \\ \mu_k &= \mu + (k-1)d \end{aligned}$$

For an analysis of variance we imposed the condition that the coefficients in a contrast add to zero. We restate the above formulation with equally spaced coefficients which sum to zero.

For $k = 2$: $\mu_1 = -\frac{1}{2}d \quad \mu_2 = \frac{1}{2}d$

For $k = 3$: $\mu_1 = -d \quad \mu_2 = 0 \quad \mu_3 = d$

For $k = 4$: $\mu_1 = -\frac{3}{2}d \quad \mu_2 = -\frac{1}{2}d \quad \mu_3 = \frac{1}{2}d \quad \mu_4 = \frac{3}{2}d$

Each of these contrasts representing slope is part of the sum of squares for difference among means given in Table 10-3. They have been chosen to be orthogonal to other contrasts of interest to be introduced later. The expected mean square of each contrast in the analysis-of-variance table is given by $\sigma^2 + d^2$. For even values of $k$ the fractional coefficients can be avoided by setting $d = 2d'$.

For $k = 2$: $\mu_1 = -d' \quad \mu_2 = d'$

For $k = 4$: $\mu_1 = -3d' \quad \mu_2 = -d' \quad \mu_3 = d' \quad \mu_4 = 3d'$

An illustration of testing for zero slopes is provided by an experiment with measurements as given in Table 15-24 from a $3 \times 4$ factorial experiment with three different weight classes of animals and four different time

**Table 15-24**

| WEIGHT | TIME PERIODS | | | |
|---|---|---|---|---|
| CLASSES | 1 | 2 | 3 | 4 |
| 1 | 1.856 | − .249 | −2.841 | −6.352 |
| 2 | 4.181 | − .552 | −5.056 | −6.950 |
| 3 | 1.860 | −2.246 | −4.518 | −6.129 |

**Table 15-25** Analysis of variance

| | *Sum of squares* | *Degrees of freedom* | *Mean square* |
|---|---|---|---|
| Time | 140.382 | 3 | 46.794 |
| Weight | 1.630 | 2 | .815 |
| Residual | 7.311 | 6 | 1.218 |
| *Total* | 149.323 | 11 | |

periods. The time periods and weight classes are to be considered ordered and equally spaced.

If we proceed with the analysis in the usual fashion we obtain the analysis of variance Table 15-25.

This analysis does not take account of the ordered categories as designed for the experiment. The single-degree-of-freedom terms with linearity assumed can be obtained by computing two orthogonal contrasts.

The coefficients for the linear effect for $k = 4$ time periods are given in Table 15-26 where, for convenience, the coefficients are arranged in the same pattern as the original 12 data values rather than in a single line. Applying these coefficients to the data in Table 15-24 gives $-91.352$, and since $\Sigma a_i^2 = 60$, the mean square is $\frac{8{,}345.19}{60} = 139.086$.

**Table 15-26** Coefficients for time period in a 3 $\times$ 4 experiment

| | | | |
|---|---|---|---|
| −3 | −1 | 1 | 3 |
| −3 | −1 | 1 | 3 |
| −3 | −1 | 1 | 3 |

The coefficients for linear effect for $k = 3$ weights are given in Table 15-27, giving $\Sigma a_i X_i = -3.447$ and mean square equal to $\frac{11.8818}{8} = 1.485$. The analysis-of-variance table is given in Table 15-28.

In this case the experiment was designed to study the effect of the variables time and weight on the measurements of response. We shall examine as main effects the slope of the linear relationship of response on time and weight. The standard analysis as in Table 15-25 has less power for

**Table 15-27** Coefficients for weights in a 3 $\times$ 4 experiment

| | | | |
|---|---|---|---|
| −1 | −1 | −1 | −1 |
| 0 | 0 | 0 | 0 |
| 1 | 1 | 1 | 1 |

**Table 15-28** Analysis of variance

| | Sum of squares | Degrees of freedom | Mean square |
|---|---|---|---|
| Linear time | 139.086 | 1 | 139.086 |
| Linear weight | 1.485 | 1 | 1.485 |
| Remainder | 8.752 | 9 | .972 |
| *Total* | 149.323 | 11 | |

discovering these effects, since the several degrees of freedom associated with time and weight are for general differences in categories whatever their order.

In an experiment of the type illustrated above it may also be desirable to study the interaction of the two linear main effects. Another contrast orthogonal to the two main effects appropriate to this purpose can be constructed for this case from the products of the coefficients of the two linear

**Table 15-29** Coefficients for linear time and linear weight interaction in a 3 × 4 experiment

| | | | |
|---|---|---|---|
| 3 | 1 | −1 | −3 |
| 0 | 0 | 0 | 0 |
| −3 | −1 | 1 | 3 |

effects. The coefficients are given in Table 15-29. The analysis-of-variance table, including a mean square for this contrast, is given in Table 15-30.

**Table 15-30** Analysis of variance

| | Sum of squares | Degrees of freedom | Mean square |
|---|---|---|---|
| Linear time | 139.086 | 1 | 139.086 |
| Linear weight | 1.485 | 1 | 1.485 |
| Interaction | .024 | 1 | .024 |
| Remainder | 8.728 | 8 | 1.091 |
| *Total* | 149.323 | 11 | |

In some experimental situations it may be desirable to estimate quadratic or higher-degree polynomial effects, either for direct information about these effects or to verify the reliability of an assumption that the assumed model is linear. For equal replications and equal spacing this can be done with the orthogonal contrasts described below.

**Ordered effects among categories.** In Sec. 10-4 determination of column and row effects were obtained by assuming the additive model

$\mu_{ij} = \mu + r_j + c_i$. This provides mean squares which estimate the $r_j$ and $c_i$ separately and independently. Often the experiment is designed to test more specifically for certain differences among means than in Secs. 10-4 and 10-5. For example, if categories for columns, rows, or both correspond to equally spaced levels of a continuous variable, we may wish to test for polynomial differences among effects.

For two groups only slope can be estimated:

| OBSERVATIONS $X_1$ | $X_2$ | $\Sigma a_i^2$ |
|---|---|---|
| *Coefficients* | | |
| −1 | 1 | 2 |

For three groups both slope (linear) and quadratic effects can be estimated:

| DEGREE | OBSERVATIONS $X_1$ | $X_2$ | $X_3$ | $\Sigma a_i^2$ |
|---|---|---|---|---|
| | *Coefficients* | | | |
| 1 | −1 | 0 | 1 | 2 |
| 2 | 1 | −2 | 1 | 6 |

For four groups linear, quadratic, and cubic effects can be estimated. It is not usual to estimate all possible higher-order effects, but instead the first two or three effects are estimated and the higher-order contrasts are left in the remainder term. Coefficients for the first three contrasts are given for $k = 4, 5, 6$ in Table 15-31.

**Table 15-31** Coefficients for orthogonal contrasts for $k = 4, 5, 6$

| | DEGREE | OBSERVATIONS $X_1$ | $X_2$ | $X_3$ | $X_4$ | $X_5$ | $X_6$ | $\Sigma a_i^2$ |
|---|---|---|---|---|---|---|---|---|
| | | *Coefficients* | | | | | | |
| $k = 4$ | 1 | −3 | −1 | 1 | 3 | | | 20 |
| | 2 | 1 | −1 | −1 | 1 | | | 4 |
| | 3 | −1 | 3 | −3 | 1 | | | 20 |
| $k = 5$ | 1 | −2 | −1 | 0 | 1 | 2 | | 10 |
| | 2 | 2 | −1 | −2 | −1 | 2 | | 14 |
| | 3 | −1 | 2 | 0 | −2 | 1 | | 10 |
| $k = 6$ | 1 | −5 | −3 | −1 | 1 | 3 | 5 | 70 |
| | 2 | 5 | −1 | −4 | −4 | −1 | 5 | 84 |
| | 3 | −5 | 7 | 4 | −4 | −7 | 5 | 180 |

Problems discussed in previous sections and chapters related to the comparison of means (difference in means, regression, etc.) for specified known values of a stimulus variable. For example, if two concentrations of two mixtures are used in an experiment, with each mixture tested at each concentration, then we may analyze the differences in response for the two concentrations as well as for the two mixtures in the $2 \times 2$ design discussed in Sec. 15-1. Analysis for three concentrations of the two mixtures can be based on a six-point, or $2 \times 3$, design.

Here the $X$-variable concentration is specified or can be determined, and the distribution of the response $Y$ variable is studied. The means and variances of the $Y$-variable distributions are estimated and contrasted. An alternative problem exists when the values of $X$ cannot be determined directly but are to be analyzed by a study of the observed responses. This technique is widely used with a chemical stimulus and with the responses of biological systems (animals or plants) and in this situation is called *bioassay*. Thus the chemical is quantified, or assayed, by the biological response. It is an important and complex problem, and only the simplest examples are discussed below. Typically the chemical is a drug, such as a vitamin, and a bioassay may be required when a chemical assay is not available.

In the following examples we shall make the assumption, standard in bioassay, that the biological response being studied has a linear regression on the logarithm of the amount of the mixture. It is also customary to select concentrations at equal intervals for this regression for convenience both in analysis and in performing the experiment. On a logarithmic scale this is accomplished by successive dilutions of the original mixture. For example, a 1:1 dilution will provide concentration points differing by log 2 on a log scale; a second 1:1 dilution will differ by 2 log 2 from the original, and so on. An analysis by regression will give the results in terms of the concentration of the original mixture. We can estimate the mean biological response for specified aliquots of the substance and can estimate the slope of the mean response to log dosage. If we have at least three concentrations, we can also test the linearity assumption.

The linearity assumption seldom, if ever, holds for all possible concentrations of the chemical. Especially in dealing with new substances, the experiment should be set up to allow this to be tested. An important problem in biochemistry or pharmacology is the study of biological responses to various drugs to provide some knowledge of the validity of such assumptions.

A second common problem in the use of regressions on log doses is the comparison of an unknown quantity of the chemical with a "standard." In practice a relatively large quantity of the standard can be set aside for comparison with new substances as the occasion arises. This is somewhat analogous to keeping a standard yardstick and calibrating new measurement tools to it.

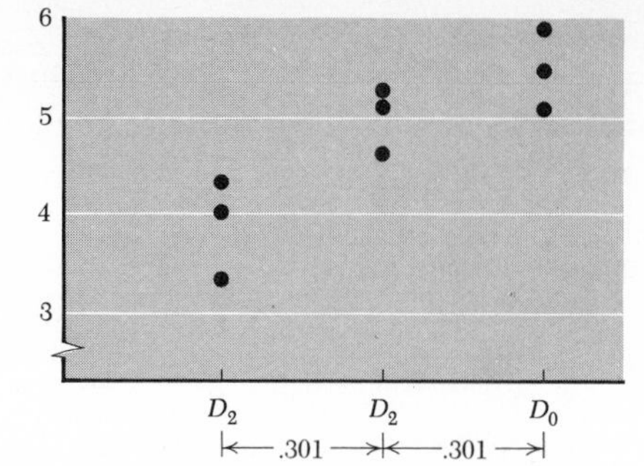


**Figure 15-3** Drug dosage on log scale equally spaced for 1:1 dilutions.

To use the linear regression analysis as in Chap. 11 we assume the distribution of the response $Y$ for a given stimulus to be normal with constant variance. It is common for the biological response as directly measured to require transformation to satisfy this assumption as well as to give a linear regression. Frequently the transformation found to be appropriate is the logarithm of the measured response.

**Estimation of the effect of a given dose.** As noted, we assume a response $Y$ which has a linear regression on the logarithm of the amount of the chemical. By taking different amounts of a substance containing an unknown percentage of the chemical we may set up an experiment illustrated in Fig. 15-3. These different amounts are obtained by *dilution*, that is, mixing a part of the substance with a part of neutral material to reduce the concentration. In Fig. 15-3 two steps of this type of dilution are shown. $D_0$ is the original concentration, $D_1$ is a 1:1 dilution of $D_0$, which is then one-half as concentrated, and $D_2$ is a 1:1 dilution of $D_1$, which is one-fourth as concentrated as $D_0$. A logarithm scale was used, and the relative locations of the dosages are plotted from the amount of dilution chosen. Thus in Fig. 15-3 the $D_1$ value was obtained as a 1:1 dilution of the $D_0$ and is a distance equal to $\log_{10} 2 = .301$ units to the left of $D_0$; correspondingly, $D_2$ is .301 units to the left of $D_1$.

The equation of a least-squares line can be obtained from the data obtained in the dilution experiment with the results in terms of the unknown concentration of the chemical in the original substance. As in Sec. 11-4, estimates may be made of the mean response for any given dilution. Thus "prescriptions" based on the analyzed response curve may be made of quantities of the original substance. Given a desired value of the mean $Y$, say, $Y_0$, then the estimated regression equation from Sec. 11-3,

$$Y_0 = \bar{Y} + b(X - \bar{X})$$

can be solved for the required $X$ to give the estimate

$$X = \bar{X} + (Y_0 - \bar{Y})/b$$

As usual, we should concern ourselves with the sampling distribution of the statistic. If the linear regressions and normality assumptions are satisfied, the statistic

$$t = \frac{[Y_0 - \bar{Y} - b(X - \bar{X})]/\sqrt{1/N + (X - \bar{X})^2/\Sigma(X_i - \bar{X})^2}}{s_{y \cdot x}}$$

where $s_{y \cdot x}$ is as given in Sec. 11-3, has a $t$ distribution with $N - 2$ degrees of freedom. With a $t$ value from Table A-5 and solutions of this equation we can obtain confidence limits for the unknown concentration $X$. Note that $X$ occurs in both the numerator and the denominator. The values of $X$ which satisfy this equation may be obtained by trial and error or preferably by squaring both sides and solving the resulting quadratic equation.

As an example, suppose the data for chemical 1 in Table 15-18 is for an unknown concentration in column 2 and a 1:1 dilution of that concentration in column 1. Using $D$ for log concentration, we can write the log concentration in column 1 as

$$D_1 = D_0 - \log_{10} 2 = D_0 - .301$$

We find then

$$\bar{D} = D_0 - .150$$
$$\Sigma(D_i - \bar{D})^2 = 2(.0225) + 2(.0225) = .09$$

Correspondingly, for the dependent variable now labeled $D$ we have

$$\bar{Y} = 7 \qquad b = 3/.301 = 10 \qquad s_{y \cdot x}^2 = \tfrac{1}{2} \times 1 = .5$$

If, for example, we wished to estimate with a 90 per cent confidence interval the value of $D$ which results in a mean $Y_0 = 8$, we would set up the equation

$$2.92 = \frac{[1. - 10(D - D_0 + .150)]/\sqrt{\tfrac{1}{4} + (D - D_0 + .150)^2/.09}}{\sqrt{.5}}$$

which reduces to

$$D_0 - .153 < D < D_0 + .233$$

Limits on the required relative concentration of the original mixture are antilogarithms of these limits, which in this example are .70 and 1.71 times the original concentration of the second column.

It should be noted that for certain combinations of values of $t$, $N$, and $s_{y \cdot x}$ the above equation may not have solutions or they may not correspond to useful limits. This is more likely to occur with large $t$ or small $N$. For example, using $t = 4.303$, corresponding to 95 per cent confidence in the problem above causes difficulty.

**Comparison with a standard.** When a new substance containing a different concentration of the chemical is prepared for use, the prescriptions

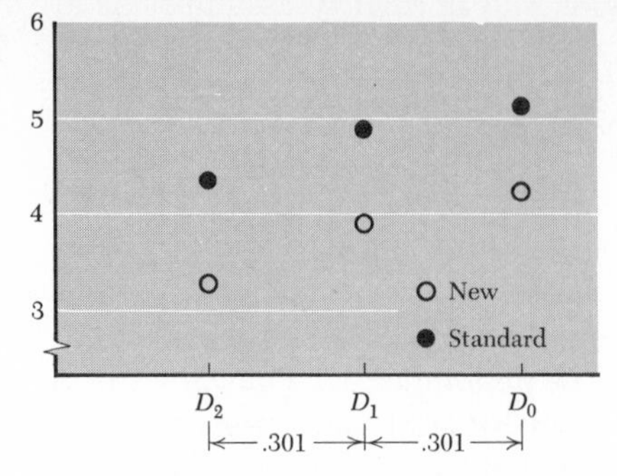


**Figure 15-4** Responses to standard and new drugs.

based on the old substance can be used only if the concentration or strength of the new preparation relative to the old one can be determined. Figure 15-4 shows observed response to dilutions of new (unknown) and old (standard) substances. Again, the horizontal scales are in relative rather than absolute units. However, under the assumptions, if the two preparations differ only in concentration, the lines will be parallel.

We shall consider the case where the same dilutions are used for the unknown and standard solutions, with an equal number $N$ of observations for each. Thus the concentration variable has equal sample variances $s_1^2 = s_2^2$ and means which differ by the logarithm of the ratio of concentrations. For this we will write $M = \bar{X}_1 - \bar{X}_2$, where the antilogarithm of $M$ is the relative concentration. Define $b$ to be the estimated common slope and $s_{y \cdot x \cdot p}^2$ the within-groups variance. Then, as in the example above for a single substance, the statistic

$$t = \frac{(\bar{Y}_1 - \bar{Y}_2 + bM)/\sqrt{2/N + 2M^2/[(N-1)s_1^2]}}{s_{y \cdot x \cdot p}}$$

has a $t$ distribution with $2N - 4$ degrees of freedom. A slight modification of assumption and analysis would result in a $t$ with $2N - 3$ degrees of freedom. The additional degree of freedom would come from the mean square of the difference in slope contrast, as displayed, for example, in Table 15-21. Again this could be solved to give confidence limits for the unknown relative potency.

## 15-3 COMPONENTS-OF-VARIANCE MODEL: MODEL II

**Single variable of classification.** In Sec. 10-1 the expression "single-variable components of variance" was introduced for the model in which $k$ categories are chosen from a parent population of categories and we wish to make inferences to the whole population of categories. A concept equivalent to the selection of categories is the selection of the means $\mu_1, \mu_2, \ldots,$ $\mu_k$ of the chosen categories from a population of means. Let $\mu$ and $\sigma_m^2$ be the mean and variance of the means of the parent population of categories. A category is itself a population, and from each one selected we shall draw a sample of size $n_i$. We assume that all categories have the same variance $\sigma^2$ and wish to estimate the quantities $\mu$, $\sigma_m{}^2$, and $\sigma^2$. We shall also test the hypothesis that $\sigma_m{}^2 = 0$. Although the formula for $\sigma_m{}^2$ is the same as used in the fixed-constants model, where it was a parameter based only on the categories from which samples were taken, here it involves all category means, including the means of those categories from which no samples were taken.

As an example, suppose we select at random five men ($k = 5$) from a population of men and measure each man's height three times ($n_i = 3$). The categories selected are the five men. The samples drawn from the selected categories are the three measurements on each. Then $\mu$ is the mean and $\sigma_m{}^2$ is the variance of the heights of the population of men. The set of three measurements for each man can be considered to be a sample of size 3 from the population of all possible measurements of the height of the man. The variance of each of these populations of measurements is $\sigma^2$. The three measurements will vary with the degree of precision of the measuring device or with slight variations in the position or posture of the person being measured. For the measurements described $\sigma^2$ would be quite small, perhaps .01 square inch, and $\sigma_m{}^2$ would be comparatively large, perhaps about 9 square inches. It is clear in this example that taking many measurements on each person is of little value, since most of the variation is between people rather than between repeated measurements of (within) each person.

*Test of hypothesis of equal means.* Since we know that people do vary in height, we know that $\sigma_m{}^2$ is not zero, so in this case we would not wish to test the hypothesis $\sigma_m{}^2 = 0$. However, had we not known it, the test procedure outlined in Sec. 10-2 would still be appropriate here. To test the hypothesis that all category means are equal, we divide $s_M{}^2$ by $s_p{}^2$ and reject the hypothesis if the ratio is larger than $F_{1-\alpha}(k - 1, N - k)$.

*Estimation of $\sigma^2$, $\sigma_m{}^2$, and $\mu$.* Table 15-32 contains an example of data collected as described above. The computations are carried out in exactly the same way as for the fixed-constants model and are recorded in Table 15-33. Note that the first, second, and third measurements do not form a second variable of classification, since the order of measurement should have no influence on the result. Note also that in this example, as in all other

**Table 15-32** Example for one-variable components-of-variance model

| MEASUREMENT | MAN 1 | 2 | 3 | 4 | 5 | |
|---|---|---|---|---|---|---|
| 1 | 65.4 | 68.7 | 63.1 | 70.3 | 66.0 | |
| 2 | 65.3 | 68.7 | 63.2 | 70.2 | 66.1 | |
| 3 | 65.5 | 68.9 | 63.3 | 70.1 | 65.8 | |
| $T_{i+}$ | 196.2 | 206.3 | 189.6 | 210.6 | 197.9 | $T_{++} = 1{,}000.6$ |

**Table 15-33** Components-of-variance analysis

| | Sum of squares | Degrees of freedom | Mean square | Estimate of |
|---|---|---|---|---|
| Men | 93.00 | 4 | $s_M^2 = 23.25$ | $\sigma^2 + 3\sigma_m^2$ |
| Within | .13 | 10 | $s_p^2 = .013$ | $\sigma^2$ |
| *Total* | 93.13 | 14 | | |

problems of this chapter, we are assuming that the repeated measurements are independent determinations.

As in the fixed-constants model, $s_p^2 = .013$ is an unbiased estimate of $\sigma^2$, and $(s_M^2 - s_p^2)/n = \dfrac{23.24}{3} = 7.75$ is an unbiased estimate of $\sigma_m^2$. An unbiased estimate of $\mu$ is $\bar{X} = \Sigma\bar{X}_{i.}/k = 66.71$.

Confidence intervals for the parameters depend on further assumptions concerning the populations. If measurement errors are normally distributed, then $s_p^2/\sigma^2$ has a $\chi^2/df$ distribution with $N - k$ degrees of freedom. Thus Table A-6*b* can be used to estimate $\sigma^2$ by confidence intervals. For the example, $N - k = 10$, and a 95 per cent confidence-interval estimate of $\sigma^2$ is $.325 < .013/\sigma^2 < 2.05$ or $.0063 < \sigma^2 < .40$.

If, in addition, the category means are normally distributed, then

$$F = \frac{s_M^2/(\sigma^2 + n\sigma_m^2)}{s_p^2/\sigma^2}$$

has an $F(4,10)$ distribution and Table A-7 can be used to give confidence limits for $\sigma_m^2/\sigma^2$. This is done by setting

$$F_{\frac{1}{2}\alpha}(k - 1, N - k) < \frac{s_M^2/(\sigma^2 + n\sigma_m^2)}{s_p^2/\sigma^2} < F_{1-\frac{1}{2}\alpha}(k - 1, N - k)$$

and rearranging terms to find the limits

$$\frac{s_M^2}{ns_p^2F_{1-\frac{1}{2}\alpha}(k - 1, N - k)} - \frac{1}{n} < \frac{\sigma_m^2}{\sigma^2} < \frac{s_M^2}{ns_p^2F_{\frac{1}{2}\alpha}(k - 1, N - k)} - \frac{1}{n}$$

If the term on the left of the inequality is negative, it is usually replaced by zero, since $\sigma_m^2$ is never negative. This same rule may be followed in computing the estimate of $\sigma_m^2$ from $(s_M^2 - s_p^2)/n$, but the resulting estimate will not then be unbiased.

The sampling distribution of $\bar{X} = \Sigma\bar{X}_{i.}/k$ is approximately normal if the sample sizes taken from the $k$ populations are large and exactly normal for any size sample if the measuring errors as well as the population of means are normal. Here the mean of $\bar{X}$ is $\mu$, and the variance of $\bar{X}$ is $\sigma_m^2/k + (\sigma^2/k^2)\,\Sigma(1/n_i)$, where $n_i$ is the sample size of the $i$th category. For this example $k = 5$ and all $n_i$'s are equal to 3, so the variance of $\bar{X}$ is $(3\sigma_m^2 + \sigma^2)/15$. An estimate of this variance is $s_M^2/15$. If the populations are normal the sampling distribution statistic $t = (\bar{X} - \mu)/\sqrt{s_M^2/15}$ is given by Table A-5 for $df = k - 1 = 4$, allowing the computation of confidence intervals for $\mu$.

In the example $\bar{X} = 66.71$, $s_M^2 = 23.25$, and approximate 95 per cent confidence limits for $\mu$ are $66.71 \pm 2.776\sqrt{23.25/15}$ or $63.25 < \mu < 70.17$.

If the $n_i$'s are not equal, $s_M^2$ is an unbiased estimate of $\sigma^2 + n'\sigma_m^2$, where $n' = (N - \Sigma n_i^2/N)/(k - 1)$. In this case the estimate of the variance of $\bar{X}$ is $\dfrac{s_M^2 - s_p^2}{n'k} + \dfrac{s_p^2\,\Sigma(1/n_i)}{k^2}$. Note that if the $n_i$'s are not equal, $\bar{X} = \Sigma\bar{X}_{i\cdot}/k$ is not the direct mean of all $N$ observations. See also Sec. 10-7.

**Two variables of classification.** In the single-variable case the analysis-of-variance table is the same for the components model and for the fixed-constants model, and the difference arises in the interpretation of $\sigma_m^2$. In the two-variable case where the $c$ columns and $r$ rows each consist of samples from larger groups of categories, the table differs from the fixed-constants model only in the column headed "Estimate of." This column, shown in Table 15-34, gives the expression which the mean squares estimate. The remainder of the table is the same as Table 10-18 and is not repeated here. As in Sec. 10-5, $n$ observations come from each row and column category combination. Inspection of formulas in Table 15-34 indicates that if $\sigma_I^2 = 0$ is rejected further tests for $\sigma_c^2$ and $\sigma_r^2$ should be carried out with the "Interaction" mean square as the denominator. Estimation of $\sigma_I^2$, $\sigma_c^2$, $\sigma_r^2$, the mean of row populations, and so on can be carried out in a manner similar to that shown for estimating $\sigma_m^2$. This will not be discussed here.

**Table 15-34** Expected mean squares

| | *Components of variance* | *Fixed components* |
|---|---|---|
| Column means | $\sigma^2 + n\sigma_I^2 + rn\sigma_c^2$ | $\sigma^2 + rn\sigma_c^2$ |
| Row means | $\sigma^2 + n\sigma_I^2 + cn\sigma_r^2$ | $\sigma^2 + cn\sigma_r^2$ |
| Interaction | $\sigma^2 + n\sigma_I^2$ | $\sigma^2 + n\sigma_I^2$ |
| *Subtotal* | | |
| Within | $\sigma^2$ | $\sigma^2$ |
| *Total* | | |

## VARIOUS TOPICS RELATED TO THE ANALYSIS OF VARIANCE 15-4

The procedure outlined for testing hypotheses includes the agreement to say "We reject the hypothesis" if a significant result is obtained. In testing for interaction it can be noted that some possible factors causing a significant result are as follows:

1. There is no interaction, but we have obtained a value which we have declared significant. This will occur with chance equal to the level of significance when there is no interaction in the populations.
2. The two variables are interacting, and we have correctly recognized this fact.
3. An uncontrolled and unmeasured variable may be of sufficient importance to appear as an interaction effect.
4. The items in the subgroups are not randomly drawn.

We have emphasized particularly the first and second interpretation. The third factor is of concern in any experiment, whether the model contains interaction effects or not. Researchers should always attempt to randomize the placing of the experimental units among the various categories so that any unnoticed extraneous variable is scattered randomly over the categories. Table A-1 can be used as an aid to randomization.

The fourth item supposedly has no place here, since it has been assumed at all times that the observations are randomly drawn. Unfortunately, however, these procedures are sometimes applied without allowance for the fact that the observations are on individuals selected in intact groups, such as classes of school children, family groups, or groups selected subjectively by experimenters.

Similar comments can be made in any of the analysis-of-variance problems about the occurrence of a significantly small $F$ ratio. Although there are many possible causes for this result, one important cause may be nonrandom choice of individuals or groups.

The analysis of variance has sometimes been misused on measurements made on the same individuals over a period of time representing growth curves or on measurements on the same person recorded in different categories. Special care must be taken not to confuse the variance of repeated measurements on an individual and the variance of measurements made on different randomly selected individuals.

Similarly, measurements may be made on the same person under several environmental conditions. Correlations beyond an additive effect may exist and change the interpretation of the analysis of variance. Differences of paired measurements or other derived scores may be helpful. Alternatively, the experimenter may choose a sample (that is, design his experiment) in such a way that dependency is avoided. For example, if a study is to be made of weights of animals at 1 month, 6 months, and 12 months, a choice of different animals at each age level will avoid dependency. However, when different animals are chosen, less information is gained about individual growth patterns; so the experimenter may wish to use the same animals throughout the experiment but different techniques to allow for this dependency. Chapters 11 and 12 introduce some relatively simple alternative techniques.

This discussion is only an introduction to the topic of analysis of variance. The analysis extends readily to factorial designs where four or more variables of classification are included in the experiment. Such cases are used frequently in agriculture, industry, and psychology. For example, an experiment involving five high schools, four grades in the high schools, two sexes of students, and three income categories for parents has four variables of classification. In this case there are four main effects, six two-category interactions, four three-category interactions, and one four-category interaction. A factorial design for this situation replicated three times would require three students from each of the $5 \times 4 \times 2 \times 3 = 120$ categories, or 360 students in all. The analysis may be undertaken to estimate or to test hypotheses about the main effects or any of the interaction effects.

A number of modifications of the factorial and of randomized-block experiments are available for situations where measurements are made only for a portion of the categories of each variable of classification. The Latin-square and the Greco-Latin-square designs (discussed in Sec. 15-6) are two examples of such designs, and there are many others.

**Estimation of means from analysis-of-variance experiments.** The confidence-interval estimate for the population mean $\mu$ for a sample of $N$ observations is $\bar{X} + t_{\frac{1}{2}\alpha}s/\sqrt{N}$ and $\bar{X} + t_{1-\frac{1}{2}\alpha}s/\sqrt{N}$, where the values of $t$ are obtained from Table A-5 for the number of degrees of freedom for $s$ and the level of confidence $1 - \alpha$ desired. The analysis-of-variance computations include an estimate of the population $\sigma^2$ as the residual mean square or the within-groups mean square. This estimate may be used to compute confidence-interval estimates for the mean for any group in an analysis-of-variance experiment. For example, the 95 per cent confidence-interval estimate for the mean $\mu_{11}$ of the population corresponding to the first cell of Table 10-15 is computed as follows:

$$\bar{X}_{11\cdot} = \frac{4 + 7 + 5}{3} = 5.33$$

Residual mean square = 1.34, and $\sqrt{1.34} = 1.16$. The two percentiles of $t$ are $t_{.025}(14) = -2.14$ and $t_{.975}(14) = 2.14$. The confidence-interval estimate is

$$5.33 - \frac{2.14 \times 1.16}{\sqrt{3}} < \mu_{11} < 5.33 + \frac{2.14 \times 1.16}{\sqrt{3}}$$

$$3.90 < \mu_{11} < 6.76$$

In Sec. 15-7 we shall explore the possibility of simultaneously estimating several of the cell means with given probability that all confidence interval estimates are correct.

**Equal sample sizes.** The computations for the analysis of variance and the computations for the test for homogeneity of variances are simplified if all samples are of the same size. An additional reason for selecting samples of equal size is an expected improvement by decreasing the variance of the difference of two means. As an illustration of this effect, consider the denominator for the $t$ test of difference between two means:

$$t = \frac{\bar{X}_1 - \bar{X}_2}{s_p \sqrt{(1/N_1) + (1/N_2)}}$$

If a total of 100 observations is made, and in one case each group has 50 observations and in the other case one group has 95 and the other group 5, we obtain

$$t_1 = \frac{\bar{X}_1 - \bar{X}_2}{s_p \sqrt{\frac{1}{50} + \frac{1}{50}}} \qquad \text{or} \qquad t_2 = \frac{\bar{X}_1 - \bar{X}_2}{s_p \sqrt{\frac{1}{95} + \frac{1}{5}}}$$

Computing the square root in the denominator, we obtain

$$t_1 = \frac{\bar{X}_1 - \bar{X}_2}{s_p(.200)} \qquad \text{or} \qquad t_2 = \frac{\bar{X}_1 - \bar{X}_2}{s_p(.459)}$$

and see that equal sample sizes will give a much smaller denominator and thus a large $t$. Similar effects hold for the analysis of variance. Whether the samples are of equal size or not has no effect on the level of significance $\alpha$, the chance of rejecting when true the hypothesis of equal means. However, in cases of inequality of means we shall have somewhat greater power for discovering this inequality if the samples are of equal size. Note also that the equality of sample sizes gives a much shorter confidence interval for the difference in means.

## 15-5 TESTS FOR HOMOGENEITY OF VARIANCES

Since the test of the hypothesis $\mu_1 = \mu_2 = \cdots = \mu_k$ made in the analysis-of-variance $F$ test assumes homogeneity of variances, we may wish to test the hypothesis

$$\sigma_1^2 = \sigma_2^2 = \cdots = \sigma_k^2$$

Two frequently used tests are given below. They have lower power then is desirable for most applications.

**Bartlett's test.** This test requires computation with natural logarithms, which may be found in Table A-14$a$. The hypothesis to be tested is that the variances of $k$ normally distributed populations are equal. The samples are of size $n_i$, where $\Sigma n_i = N$. We denote the variance of the $i$th sample by $s_i^2$. Let

$$M = (N - k) \ln s_p^2 - \Sigma[(n_i - 1) \ln s_i^2]$$

$$s_p^2 = \frac{\Sigma(n_i - 1)s_i^2}{N - k}$$

$$A = \frac{1}{3(k - 1)} \left[\sum \left(\frac{1}{n_i - 1}\right) - \frac{1}{N - k}\right]$$

$$\nu_1 = k - 1$$

$$\nu_2 = \frac{k + 1}{A^2}$$

$$b = \frac{\nu_2}{1 - A + (2/\nu_2)}$$

Bartlett's test is based on the statistic $M$ rejecting if it is significantly large. $F = \nu_2 M/[\nu_1(b - M)]$ gives us a modification of Bartlett's test statistic which has a sampling distribution better approximated by the $F(\nu_1,\nu_2)$ distribution. The values of $\nu_2$ will usually not be an integer, and it may be necessary to interpolate in the $F$ table. Good accuracy can be obtained by the method of interpolating on the reciprocals of the degrees of freedom, as in the example below. Usually, however, the observed value will differ sufficiently from the table value that interpolation is not necessary.

Suppose, for example, that we have four samples of sizes 3, 3, 3, 4 with variances 6.33, 1.33, 4.33, 4.33, respectively. Substituting these numbers, we obtain

$$\begin{aligned} M &= 9 \ln 4.11 - 2 \ln 6.33 - 2 \ln 1.33 - 2 \ln 4.33 - 3 \ln 4.33 \\ &= (9 \times 1.413) - (2 \times 1.846) - (2 \times 0.284) - (2 \times 1.466) - (3 \times 1.466) \\ &= 1.127 \end{aligned}$$

$$A = \frac{1}{3 \times 3}\left(\frac{1}{2} + \frac{1}{2} + \frac{1}{2} + \frac{1}{3} - \frac{1}{9}\right) = .1913$$

$$\nu_1 = 3$$

$$\nu_2 = \frac{5}{(.1913)^2} = 136.6$$

$$b = \frac{136.6}{1 - .1913 + 2/136.6} = 165.9$$

For $\alpha = .05$ we read

$$F_{.95}(3,120) = 2.68$$
$$F_{.95}(3,\infty) = 2.60$$

Interpolating by means of reciprocals, we proceed as follows:

Corresponding to $1/\infty = .0000$, we have 2.60

Corresponding to $\frac{1}{120} = .0083$, we have 2.68.

We are interested in $1/136.6 = .0073$, which is $\frac{73}{83} = .88$ of the way between .0000 and .0083. We take a value .88 of the way between 2.60 and 2.68, or $2.60 + (.88 \times .08) = 2.67$.

Thus if the observed $F$ is larger than 2.67 we shall reject at the 5 per cent level of significance the hypothesis that the four populations have equal variances. The observed value is

$$F = \frac{136.6 \times 1.127}{3(165.9 - 1.127)} = .31$$

Since this value is less than 2.67, we accept the hypothesis. Note that the interpolation was not necessary, since $F(3,136.6)$ is between 2.60 and 2.68, and that is sufficient information in this case.

**Cochran's test.** One type of deviation from homogeneous variance which is serious in terms of invalidating the analysis-of-variance test for means occurs when one variance is very much larger than the remainder of the variances. The following test statistic is designed especially for this situation and is referred to as Cochran's test:

$$C = \frac{\text{largest } s_i^2}{\Sigma s_i^2}$$

The 95th and 99th percentiles of the sampling distribution of $C$ are in Table A-17. They can be used for a test at the 5 per cent and 1 per cent levels of significance. The hypothesis of equal variances is rejected if the computed value of the above statistic exceeds the value in the table. Critical values are indicated only for the case where the sample variances have an equal number of degrees of freedom.

## 15-6 LATIN SQUARE

When several variables of classification must be included in an experiment and each variable has several categories, the conduct of a factorial experiment may require a very great number of test subjects. For example, $n$ replicates of a factorial design with three variables, each of which has four categories, will require $4 \times 4 \times 4 \times n = 64n$ subjects. There is an extensive literature on the design of experiments for carrying out only a fraction of the full factorial experiment and selecting the fraction in a balanced way so that the main effects or interactions of interest in the particular experiment are estimated by orthogonal contrasts or with equal standard errors. An example of this type of design is the *Latin square*. This design may be used when each variable of classification has the same number of categories, and effects are additive with zero interactions. For the example above a Latin square would require $16n$ subjects in place of $64n$ subjects. For illustration the design is presented with $n = 1$ for a $4 \times 4$ Latin square, and 16 subjects will be required. This is called a $4 \times 4$ Latin square.

**Table 15-35**

| | I | II | III | IV |
|---|---|---|---|---|
| 1 | *A* | *B* | *C* | *D* |
| 2 | *D* | *A* | *B* | *C* |
| 3 | *C* | *D* | *A* | *B* |
| 4 | *B* | *C* | *D* | *A* |

To construct this design we lay out a plan showing two variables of classification. Since there are the same number of categories for each, we may superimpose a third variable of classification, represented by *A*, *B*, *C*, *D*, in such a way that each category appears once and only once in each row and column. It is easily seen that the design presented in Table 15-35 is not the only one available for four rows and columns; there are 576 in all. The experimenter may wish to pick one at random from all the possibilities.

In Table 15-35 the roman numerals may represent four counties in a state, the arabic numbers four types of soil preparation, and the letters four types of fertilizers. The observations to be analyzed may be crop yield. The experiment can now be performed on 16 plots of ground. The computing techniques are illustrated in the following discussion.

**Computational example.** Consider the data in Table 15-36. We shall defer a complete description until later. For the moment consider three variables of classification with the following categories:

| | |
|---|---|
| Roman | I, II, III |
| Arabic | 1, 2, 3 |
| Latin | *A*, *B*, *C* |

**Table 15-36** 3 × 3 Latin square

| | I | II | III | *Arabic totals* |
|---|---|---|---|---|
| 1 | .194 *A* | .730 *B* | 1.187 *C* | 2.111 |
| 2 | .758 *C* | .311 *A* | .589 *B* | 1.658 |
| 3 | .369 *B* | .558 *C* | .311 *A* | 1.238 |
| *Roman totals* | 1.321 | 1.599 | 2.087 | 5.007 |

| *Latin totals* | |
|---|---|
| *A* | .816 |
| *B* | 1.688 |
| *C* | 2.503 |

The sums of squares are computed as follows:

$$\text{Roman: } \frac{(1.321)^2}{3} + \frac{(1.599)^2}{3} + \frac{(2.087)^2}{3} - \frac{(5.007)^2}{9} = .1002$$

$$\text{Arabic: } \frac{(2.111)^2}{3} + \frac{(1.658)^2}{3} + \frac{(1.238)^2}{3} - \frac{(5.007)^2}{9} = .1271$$

$$\text{Latin: } \frac{(.816)^2}{3} + \frac{(1.688)^2}{3} + \frac{(2.503)^2}{3} - \frac{(5.007)^2}{9} = .4745$$

$$\text{Total: } (.194)^2 + (.730)^2 + (1.187)^2 + (.758)^2 + (.311)^2 + (.589)^2 + (.369)^2 + (.558)^2 + (.311)^2 - \frac{(5.007)^2}{9} = .7564$$

The analysis-of-variance table is given in Table 15-37.

**Table 15-37**

| | *Sum of squares* | *Degrees of freedom* | *Mean square* |
|---|---|---|---|
| Roman | .1002 | 2 | .0501 |
| Arabic | .1271 | 2 | .0635 |
| Latin | .4745 | 2 | .2372 |
| Remainder | .0546 | 2 | .0273 |
| *Total* | .7564 | 8 | |

The residual sum of squares is obtained by subtraction.

As an illustration of the analysis for ordered variables of classification, let us assume that the variables for the data in Table 15-36 to be as follows:

| | |
|---|---|
| Roman | Three equally spaced concentrations of poison as extracted from the scorpion fish |
| Arabic | Three equally spaced body weights for the animals tested |
| Latin | Three equally spaced times of storage of the poison before it is administered to the animals |

The recorded observation is the lethal amount of drug. The data presented come from an actual experiment, but some modifications have been made to better illustrate the experiment.

If the categories were not ordered we could proceed similarly to the analysis of Sec. 10-4, forming $F$ ratios for any of the three main effects compared to the remainder mean square.

To take account of the ordered categories in the three variables we may employ the method of Sec. 15-1. Linear effects for each of the three variables of classification can be found by using coefficients as in Table 15-38.

**Table 15-38** Coefficients for linear effects in Latin-square experiment

| *Roman* | | | *Arabic* | | | *Latin* | | |
|---|---|---|---|---|---|---|---|---|
| −1 | 0 | 1 | 1 | 1 | 1 | −1 | 0 | 1 |
| −1 | 0 | 1 | 0 | 0 | 0 | 1 | −1 | 0 |
| −1 | 0 | 1 | −1 | −1 | −1 | 0 | 1 | −1 |

The resulting analysis-of-variance table is given in Table 15-39.

**Table 15-39** Linear effects in a Latin-square experiment

| | *Sum of squares* | *Degrees of freedom* | *Mean square* |
|---|---|---|---|
| Linear concentration | .0978 | 1 | .0978 |
| Linear body weight | .1270 | 1 | .1270 |
| Linear storage | .4743 | 1 | .4743 |
| Residual | .0573 | 5 | .0115 |
| *Total* | .7564 | 8 | |

If we hypothesize only linear effects, we can test their significance with $F$ ratios, for example, a test of $H$: $b_{\text{conc.}} = 0$ by $F = .0978/.0115 = 8.50$ compared to $F_{.95}(1,5) = 6.61$ for a significant slope attributable to concentration.

The usual assumptions that distribution of the observations is normal and that the effects are additive are necessary for the tests of significance to be valid.

Generalizations of the Latin square are also used. For example, suppose four treatments of a different type, say, $\alpha$, $\beta$, $\gamma$, $\delta$, were superimposed on the Latin square. We would have what is called a *Greco-Latin square.* This superimposition may be performed so that each treatment of the first type appears with each treatment of the second type. Each treatment of each type appears just once in each row and column. A third type of

| | | | |
|---|---|---|---|
| $A\alpha$ | $B\beta$ | $C\gamma$ | $D\delta$ |
| $B\gamma$ | $A\delta$ | $D\alpha$ | $C\beta$ |
| $C\delta$ | $D\gamma$ | $A\beta$ | $B\alpha$ |
| $D\beta$ | $C\alpha$ | $B\delta$ | $A\gamma$ |

treatment may be superimposed and analyzed in a similar manner when there are five or more classifications for each variable.

## MULTIPLE ORTHOGONAL ESTIMATES 15-7

We have noted in Sec. 10-3, (1) the possibility of estimating a *single* linear function of the cell means with its standard error leading to a confidence interval, and (2) the possibility of simultaneously estimating all linear functions which were in the form of contrasts by use of the $q$- or $F$-distribution tables. It is also possible, as described below, to make probability statements relating to all of a specified set of orthogonal linear functions. Usually confidence intervals in this case are intermediate in length between those for a *single* contrast and those used when *all possible* contrasts may be inspected.

Let

$$\Sigma c_{1j}\bar{X}_j \qquad \Sigma c_{2j}\bar{X}_j \qquad \cdots \qquad \Sigma c_{kj}\bar{X}_j$$

be orthogonal functions of the means of random samples, each of size $n$, from several ($k$ or more) normally distributed populations with means $\mu_i$ and variances $\sigma^2$. Let $s_p{}^2$ be the within-groups estimate of $\sigma^2$ with $(n-1)k$ degrees of freedom. If $h_\alpha$ is read from Table A-18*b*, then there is probability $\alpha$ that *all* the $k$ functions satisfy the following inequality:

$$-s_p h_\alpha \sqrt{\frac{\Sigma c_{1j}{}^2}{n}} < \Sigma c_{1j}\bar{X}_j - \Sigma c_{1j}\mu_j < s_p h_\alpha \sqrt{\frac{\Sigma c_{1j}{}^2}{n}}$$

$$-s_p h_\alpha \sqrt{\frac{\Sigma c_{2j}{}^2}{n}} < \Sigma c_{2j}\bar{X}_j - \Sigma c_{2j}\mu_j < s_p h_\alpha \sqrt{\frac{\Sigma c_{2j}{}^2}{n}}$$

$$\cdots\cdots\cdots\cdots\cdots\cdots\cdots\cdots\cdots$$

$$-s_p h_\alpha \sqrt{\frac{\Sigma c_{kj}{}^2}{n}} < \Sigma c_{kj}\bar{X}_j - \Sigma c_{kj}\mu_j < s_p h_\alpha \sqrt{\frac{\Sigma c_{kj}{}^2}{n}}$$

Thus simultaneous confidence intervals can be obtained for the $k$ indicated orthogonal functions.

It should be recognized that exploitation of this theorem depends on the investigator's being interested only in a particular set of orthogonal functions. In a two-variable analysis-of-variance model with equal number of observations per cell the over-all mean, the row effects, the column effects, and the interactions do form an orthogonal set, and the theorem may be used. Consider the $2 \times 2$ example in Table 10-20 with data copied in Table 15-40. The three functions represented in the analysis-of-variance table, the row, column, and interaction contrasts, can be written as follows:

Row effect: $$\frac{\bar{X}_{11} + \bar{X}_{21} - \bar{X}_{12} - \bar{X}_{22}}{2}$$

Column effect: $$\frac{\bar{X}_{11} - \bar{X}_{21} + \bar{X}_{12} - \bar{X}_{22}}{2}$$

Interaction: $$\frac{\bar{X}_{11} - \bar{X}_{21} - \bar{X}_{12} + \bar{X}_{22}}{2}$$

The three functions (or effects) are orthogonal, and

$$\Sigma c_{1j}^2 = \Sigma c_{2j}^2 = \Sigma c_{3j}^2 = 1$$

Thus if we want an over-all probability equal to .95 we read $k = 3$, $\nu = 8$, the value $h_{.95} = 2.96$ from Table A-18*b* and use $s_p = \sqrt{2.50} = 1.58$ from Table 10-21.

**Table 15-40** Data from Table 10-20 with means

| | DATA *A* | DATA *B* | MEANS *A* | MEANS *B* | *Row* |
|---|---|---|---|---|---|
| MEN | 5 | 5 | $\bar{X}_{11\cdot} = 7.33$ | $\bar{X}_{21\cdot} = 5.33$ | $\bar{X}_{\cdot 1\cdot} = 6.33$ |
| | 8 | 5 | | | |
| | 9 | 6 | | | |
| WOMEN | 11 | 7 | $\bar{X}_{12\cdot} = 13.33$ | $\bar{X}_{22\cdot} = 7.00$ | $\bar{X}_{\cdot 2\cdot} = 10.17$ |
| | 14 | 8 | | | |
| | 15 | 6 | | | |
| *Column* | | | $\bar{X}_{1\cdot\cdot} = 10.33$ | $\bar{X}_{2\cdot\cdot} = 6.17$ | $\bar{X}_{\cdot\cdot\cdot} = 8.25$ |

With the notation of Sec. 10-5 we set up the intervals in Table 15-41 in the same format as in Table 10-22 to compare the interval estimates obtained. Restriction to three orthogonal functions allows the reduction of $\pm 4.13$ to $\pm 2.70$ for the confidence intervals.

**Table 15-41** Orthogonal comparisons

| | $\bar{X}_{11\cdot} = \frac{22}{3}$ | $\bar{X}_{21\cdot} = \frac{16}{3}$ | $\bar{X}_{12\cdot} = \frac{40}{3}$ | $\bar{X}_{22\cdot} = \frac{21}{3}$ | $\Sigma c_{ij}\bar{X}_{ij\cdot} \pm h_\alpha s_p \sqrt{\frac{\Sigma c_{ij}^2}{n}}$ | *Estimate of* |
|---|---|---|---|---|---|---|
| | *Coefficients* | | | | | |
| 1 | $\frac{1}{2}$ | $\frac{1}{2}$ | $-\frac{1}{2}$ | $-\frac{1}{2}$ | $-3.83 \pm 2.70$ | $r_1 - r_2$, difference for men and women |
| 2 | $\frac{1}{2}$ | $-\frac{1}{2}$ | $\frac{1}{2}$ | $-\frac{1}{2}$ | $4.17 \pm 2.70$ | $c_1 - c_2$, difference for method |
| 3 | $\frac{1}{2}$ | $-\frac{1}{2}$ | $-\frac{1}{2}$ | $\frac{1}{2}$ | $-2.17 \pm 2.70$ | $2I_{11}$, interaction effect |

**Estimation of several means.** One application of the $h$ distribution is the simultaneous estimation of several mean values. For example, using the data in Table 15-40, we would estimate the four cell means by noting

**Table 15-42** Estimation of cell means; confidence level .95

| | $\bar{X}_{11\cdot} = \frac{22}{3}$ | $\bar{X}_{21\cdot} = \frac{16}{3}$ | $\bar{X}_{12\cdot} = \frac{40}{3}$ | $\bar{X}_{22\cdot} = \frac{21}{3}$ | $\Sigma c_{ij}\bar{X}_{ij\cdot} \pm h_\alpha s_p \sqrt{\frac{\Sigma c_{ij}^2}{n}}$ | *Estimate of mean of* |
|---|---|---|---|---|---|---|
| | | *Coefficients* | | | | |
| 1 | 1 | 0 | 0 | 0 | 7.33 ± 2.86 | Cell 11 |
| 2 | 0 | 1 | 0 | 0 | 5.33 ± 2.86 | Cell 21 |
| 3 | 0 | 0 | 1 | 0 | 13.33 ± 2.86 | Cell 12 |
| 4 | 0 | 0 | 0 | 1 | 7.00 ± 2.86 | Cell 22 |

that each is a linear function with coefficients given in Table 15-42. We are 95 per cent confident that all four interval estimates cover the corresponding cell mean. For this case the technique is appropriate for unequal cell frequencies.

## GLOSSARY

Bartlett's test
bioassay
Cochran's test
components-of-variance
contrast
contrast mean square
Greco-Latin square
*h* distribution
homogeneity of variances
Latin square
linear contrast
linear effect
linearity
multiple orthogonal contrast
nesting
orthogonal contrasts
quadratic effect

## DISCUSSION QUESTIONS

**1.** Define each term in the Glossary.

**2.** Give an example of a 2 × 3 experiment from a field of application and discuss whether or not the main effects would be of interest if the interaction effects were sizable.

**3.** What assumptions on a 3 × 3 × 3 factorial design justify a Latin-square design?

**4.** Construct a 5 × 5 Latin square.

**5.** If a Latin-square-design experiment is used in agriculture, and the rows and columns of data actually correspond to rows and columns of plots in a field, is it reasonable to assume that there is no interaction? Can you think of a possible cause of interaction in such an experiment?

## PROBLEMS

*Section* 15-1

**1.** Two fertilizers are used together in either 1 or 2 pounds per plot. A 2 × 2 factorial experiment replicated four times was carried out, with the resulting yields as given:

| | | FERTILIZER 1, POUNDS | |
|---|---|---|---|
| | | 1 | 2 |
| FERTILIZER 2, POUNDS | 1 | 17 | 13 |
| | | 16 | 13 |
| | | 15 | 14 |
| | | 18 | 12 |
| | 2 | 21 | 14 |
| | | 20 | 16 |
| | | 19 | 16 |
| | | 18 | 14 |

(*a*) Display orthogonal contrasts for row, column, and interaction effects. (*b*) Present the mean squares of these contrasts in an analysis-of-variance table and draw conclusions about the population means. (*c*) Verify that the three mean squares add to the between-means sum of squares for the four samples.

**2.** For the data given below: (*a*) Display the orthogonal contrasts to estimate the row, column, and interaction effects. (*b*) Present the mean squares in an analysis-of-variance table and draw conclusions. (*c*) Verify that the mean squares of these contrasts add to the between-means sum of squares for the four-cell table.

| | | TREATMENT | |
|---|---|---|---|
| | | I | II |
| TYPE | 1 | 3.5 | 4.6 |
| | | 4.0 | 4.8 |
| | 2 | 2.9 | 5.0 |
| | | 3.1 | 5.4 |

**3.** Three observations at each of four equally spaced drug levels resulted in the following values:

| | | STIMULUS LEVEL | | | |
|---|---|---|---|---|---|
| | | 1 | 2 | 3 | 4 |
| RESPONSE | 1 | 5 | 6 | 8 | 6 |
| | 2 | 2 | 4 | 4 | 4 |
| | 3 | 4 | 6 | 7 | 3 |

(*a*) Display three orthogonal contrasts measuring linear, quadratic, and cubic effects. (*b*) Verify that the three mean squares add to the between-means sum of squares for the four groups. (*c*) Display these in an analysis-of-variance table. (*d*) Plot the points with the drug level on the horizontal axis and mark the mean response for each group. (*e*) Do the plotted means seem to fall on a line? On a parabola? Does your visual impression agree with the analysis?

**4.** The following data from Table 10-15 are from a 2 × 3 factorial experiment. Find five orthogonal contrasts measuring row difference, column differences, and two interactions. Verify that the mean squares of the contrasts add to the between-means sum of squares (78.67 in Table 10-16). Verify that the column and interaction mean squares add separately to the sum-of-squares terms in Table 10-17.

| | *A* | *B* | *C* |
|---|---|---|---|
| | 4 | 2 | 5 |
| *a* | 7 | 3 | 6 |
| | 5 | 2 | 4 |
| | 9 | 8 | 10 |
| *b* | 8 | 7 | 8 |
| | 8 | 5 | 7 |

**5.** The following data give improvement scores after treatment by one of two types of radiation or by placebo. Give orthogonal contrasts for the difference in the two types of radiation and for the difference between radiation and placebo. Present the results in an analysis-of-variance table and indicate the conclusions drawn.

| RADIATION | | NO RADIATION |
|---|---|---|
| *x ray* | *γ ray* | *Placebo* |
| 4 | 5 | 6 |
| 9 | 10 | 3 |
| 6 | 9 | 9 |
| 12 | 8 | 2 |
| 5 | 11 | 1 |

**6.** The drained weight in ounces of frozen apricots was measured for various types of sirup and various concentrations of sirup. The original weight of the apricots was the same. Differences in drained weight would be attributable to differences in concentration or type of sirup.

| | SIRUP COMPOSITION | | | |
|---|---|---|---|---|
| CONCENTRATION OF SIRUP | *All sucrose* | $\frac{2}{3}$ *sucrose,* $\frac{1}{3}$ *corn sirup* | $\frac{1}{2}$ *sucrose,* $\frac{1}{2}$ *corn sirup* | *All corn sirup* |
| 30 | 28.80 | 28.21 | 29.28 | 29.12 |
| 40 | 29.12 | 28.64 | 29.12 | 30.24 |
| 50 | 29.76 | 30.40 | 29.12 | 28.32 |
| 60 | 30.56 | 29.44 | 28.96 | 29.60 |

(*a*) Assuming no interaction, give an analysis-of-variance table showing concentration effects and composition effects. (*b*) For the all-sucrose composition display the three orthogonal contrasts (over concentrations) measuring linear, quadratic, and cubic effects of concentration. Compute the mean square for each and display as part of an analysis-of-variance table. Would any of the three be considered significantly large at the 5 per cent level? Show that the sum of the three mean squares equals the between-means sum of squares for the all-sucrose observations.

**7.** The data given below are from a $2 \times 2 \times 2$ factorial design for three variables, each with two categories.

| | LAYER 1 | | LAYER 2 | |
|---|---|---|---|---|
| | *Col.* 1 | *Col.* 2 | *Col.* 1 | *Col.* 2 |
| *Row* 1 | $X_{111} = 3.0$ | $X_{211} = 2.0$ | $X_{112} = 2.1$ | $X_{212} = 1.5$ |
| *Row* 2 | $X_{121} = 4.1$ | $X_{221} = 3.5$ | $X_{122} = 3.9$ | $X_{222} = 2.0$ |

(*a*) Eight orthogonal linear expressions are as follows. Compute the observed values of the eight functions.

| | OBSERVATIONS | | | | | | | | |
|---|---|---|---|---|---|---|---|---|---|
| | $X_{111}$ | $X_{211}$ | $X_{121}$ | $X_{221}$ | $X_{112}$ | $X_{212}$ | $X_{122}$ | $X_{222}$ | $\Sigma a_i^2$ |
| | *Coefficients* | | | | | | | | |
| 1 | 1 | 1 | 1 | 1 | 1 | 1 | 1 | 1 | 8 |
| 2 | 1 | 1 | −1 | −1 | 1 | 1 | −1 | −1 | 8 |
| 3 | 1 | −1 | 1 | −1 | 1 | −1 | 1 | −1 | 8 |
| 4 | 1 | 1 | 1 | 1 | −1 | −1 | −1 | −1 | 8 |
| 5 | 1 | −1 | −1 | 1 | 1 | −1 | −1 | 1 | 8 |
| 6 | 1 | 1 | −1 | −1 | −1 | −1 | 1 | 1 | 8 |
| 7 | 1 | −1 | 1 | −1 | −1 | 1 | −1 | 1 | 8 |
| 8 | 1 | −1 | −1 | 1 | −1 | 1 | 1 | −1 | 8 |

(*b*) Verify that the eight linear functions, the seven contrasts and the total, form a mutually orthogonal set.

(*c*) As in Sec. 10-4, a no-interaction model giving the means of the eight populations could be written as follows, where $R$ is a row effect, $C$ a column effect, and $L$ the third-variable (layer) effect:

| | LAYER 1 | | LAYER 2 | |
|---|---|---|---|---|
| | *Col.* 1 | *Col.* 2 | *Col.* 1 | *Col.* 2 |
| *Row* 1 | $\mu + R + C + L$ | $\mu + R - C + L$ | $\mu + R + C - L$ | $\mu + R - C - L$ |
| *Row* 2 | $\mu - R + C + L$ | $\mu - R - C + L$ | $\mu - R + C - L$ | $\mu - R - C - L$ |

Verify by substitution that the first three contrasts (functions 2, 3, 4) give the parameters $8R$, $8C$, and $8L$ and the last four contrasts give the value zero if the sample values fit the model exactly. Verify that the first function gives the parameter $8\mu$ if the sample values fit the model exactly.

(*d*) If the model is correct (there is no interaction) the mean squares of the last four contrasts add to an unbiased estimate of $\sigma^2$ with 4 degrees of freedom. Estimate $\sigma^2$ by a 95 per cent confidence interval and estimate $R$, $C$, and $L$ by 95 per cent confidence intervals.

*Section* 15-2

**8.** Two successive 1:1 dilutions were made on a drug, and the responses of five animals with the original drug and each dilution were observed as given below. Test the hypothesis that the regression is a straight line on log dose. Assuming that this is true, estimate the necessary dosage which would give a mean log response equal to log 9 = .954.

| *Original dose* | 10 | 12 | 9 | 10 | 9 |
|---|---|---|---|---|---|
| *First dilution* (1:1) | 7 | 7 | 8 | 7 | 8 |
| *Second dilution* (1:3) | 6 | 7 | 6 | 5 | 6 |

**9.** (*a*) Fit a linear regression of log response on log dose using the data below:

| *Dilution* | Full strength | 1:1 | 1:3 | 1:7 | 1:15 |
|---|---|---|---|---|---|
| *Quantity, per cent* | 100 | 50 | 25 | $12\frac{1}{2}$ | $6\frac{1}{4}$ |
| *Response* | 10.8 | 8.5 | 5.3 | 3.1 | 2.3 |

(*b*) Estimate with a 95 per cent confidence interval the required dilution to result in a mean log response equal to log 7.5. = .875. What assumptions are made leading to this estimate? (*c*) Plot the points (dose, responses) on log-log graph paper (page 432) and decide on a visual basis whether the linearity assumption appears reasonable. (*d*) Test the hypothesis that the regression is linear.

*Section* 15-3

**10.** A measurement, in liters, of lung capacity was made on each of five men every day for a week, with observations as given:

| | | DAY | | | | | | |
|---|---|---|---|---|---|---|---|---|
| | | 1 | 2 | 3 | 4 | 5 | 6 | 7 |
| MAN | 1 | 6.3 | 6.1 | 5.5 | 6.2 | 6.2 | 5.8 | 5.8 |
| | 2 | 7.2 | 6.9 | 6.6 | 7.3 | 7.1 | 6.9 | 6.8 |
| | 3 | 6.6 | 6.4 | 6.0 | 6.6 | 6.7 | 6.5 | 6.3 |
| | 4 | 7.0 | 6.8 | 6.5 | 7.1 | 7.0 | 7.0 | 6.6 |
| | 5 | 6.0 | 5.7 | 5.6 | 6.0 | 5.9 | 5.8 | 5.7 |

(*a*) What assumptions need be made so that the single-variable-of-classification components-of-variance model is appropriate? (*b*) Analyze the data as a single variable model giving estimates of $\mu$, $\sigma_m^2$, and $\sigma^2$. (*c*) What different assumptions make a two variable model appropriate? Analyze the data as a two-variable components-of-variance model.

**11.** Measurements made on the size of certain objects have variability with $\sigma = 10$ cubic centimeters owing to random instrument errors. Thus $N$ repeated measurements of the same object would increase the precision $10/\sqrt{N}$. The various objects in a population have $\sigma_m$ approximately 100 cubic centimeters. (*a*) To estimate $\mu$, would it be better to take 100 objects and measure them each once or 10 objects and measure each 10 times? (*b*) If the cost of acquiring an object is \$10 and the cost to measure it is \$1, how many objects should be selected and measured such that at most \$100 is spent and the standard error of the estimate of $\mu$ is a minimum?

*Section* 15-6

**12.** Four fertilizers, *A*, *B*, *C*, *D*, were tested by arranging plants in a Latin-square design in a field. Thus rows and columns in the table are rows and columns in the field. The yields are as shown. Analyze the data for evidence at the 5 per cent level that the mean yields are not equal for the four fertilizers.

| | | | |
|---|---|---|---|
| *A* 17 | *B* 13 | *C* 19 | *D* 16 |
| *B* 13 | *C* 20 | *D* 14 | *A* 18 |
| *C* 21 | *D* 16 | *A* 15 | *B* 12 |
| *D* 14 | *A* 16 | *B* 14 | *C* 18 |

*Section* 15-7

**13.** Random samples of 25 students each were taken from five schools and given an examination on current events. The mean scores for the five samples were $\bar{X}_{1.} = 63$, $\bar{X}_{2.} = 72$, $\bar{X}_{3.} = 60$, $\bar{X}_{4.} = 80$, $\bar{X}_{5.} = 92$, with $s_p^2 = 26.01$. (*a*) Give a set of interval estimates of the five population means such that we can be 95 per cent confident that all are correct. (*b*) Give a 95 per cent confidence interval for the mean of the first school.

**14.** The following experiment was designed to determine the relative merits of four different feeds in regard to the gain in weight of pigs. Twenty pigs are divided at random into four lots of five pigs each. Each lot is given a different feed. The weight gain in pounds by each of the pigs during a fixed length of time is given in the table.

| | | | | | |
|---|---|---|---|---|---|
| *Feed A* | 133 | 144 | 135 | 149 | 143 |
| *Feed B* | 163 | 148 | 152 | 146 | 157 |
| *Feed C* | 210 | 233 | 220 | 226 | 229 |
| *Feed D* | 195 | 184 | 199 | 187 | 193 |

(*a*) Give confidence-interval estimates of the four cell means such that you are 99 per cent sure that all are correct. (*b*) Give a 99 per cent confidence-interval estimate of the single population mean for feed *A*.

**15.** With the data and contrasts in Prob. 1, use the $q$ distribution to simultaneously estimate the row, column, and interaction population effects with a set of confidence intervals. Repeat with the $h$ distribution and note which, the $q$ or $h$, results in shorter intervals.

**16.** Show that $b$ given by equation (2) of Sec. 11-3 and $\bar{Y}$ are orthogonal functions. Use the $h$ distribution to obtain simultaneous confidence limits for $\mu_{y \cdot \bar{x}}$, the mean $Y$ for $X = \bar{X}$, and $B$, the population regression slope. Apply this to the example in Sec. 11-3.

**17.** For the Latin square with three treatments in Table 15-36 the treatment sum of squares in Table 15-37 has 2 degrees of freedom. (*a*) Find two orthogonal contrasts of the nine observations which have mean squares adding to the between treatment sum of squares. (*b*) Find two other orthogonal contrasts which have mean squares adding to the between-row sum of squares. (*c*) Verify that these two sets of contrasts are mutually orthogonal. (*d*) Use the $h$ distribution to give a set of intervals for the two treatment differences such that you are 95 per cent confident that both are correct. (*e*) Use the $h$ distribution to give a set of intervals for the two treatment differences and for the two row differences such that you are 95 per cent confident that all four are correct.

CHAPTER SIXTEEN

# Questions of normality

For most of the methods provided thus far normality of the population being sampled is assumed. Few populations will satisfy the assumption strictly, and it is very difficult to state precisely the effect on various statistical procedures if the population differs somewhat from normal. Furthermore, since no single measure of nonnormality is generally accepted, it is impossible to give general rules which will apply in every case. Also, small deviations from normality appear to have a greater effect on some statistical tests and estimates than on others. The choice of methods to use when normality is not assured is something of an art at present.

We can approach the problem by discarding almost all assumptions on the form of the distribution. The nonparametric procedures of Chap. 17 will provide methods that are valid whatever the form of the population distributions. Possible approaches for a population which is nearly normal or for a normal population admixed (to a small amount) with some other population will be presented in this chapter.

If very little is known about the population the nonparametric methods may provide the best solution. If the form of the distribution is known and perhaps is such that it can be brought closer to normality, this may allow the use of statistical tests with greater power and provide shorter confidence-interval estimates.

**Specified nonnormality.** If the population distribution is of a known form that can be transformed to the normal form, the application of this transformation to the observations may be the most effective way to handle the problem. A discussion of several transformations is given in the next section.

**Approximate normality.** Sampling is frequently from a distribution which is approximately normal. If the deviations from normality are not large, the best procedure may be to proceed with the standard methods which assume normality and interpret the results with some degree of caution.

**Admixed normality.** Sampling may be from a population consisting for the most part of individuals described by a single normal distribution, but also containing in part individuals described by other admixed distributions. In our attempt to estimate or test hypotheses about the parameters of the single normal distribution the usual procedures may be grossly affected by the occurrence in the sample of observations from these admixed populations, particularly if the distributions are widely different.

If there is approximate normality or some type of admixed normality, one of the most serious consequences occurs when one or more observations in the sample are at the extremes of the sample, perhaps at some distance from the other observations. The mean may be greatly biased either positively or negatively, and the variance estimate will be too large. Section 16-3 provides a technique for detecting the presence of such extremes. Sections 16-3 and 16-4 provide a technique for estimates and tests on the mean which are largely unaffected by the presence of such extremes.

## TRANSFORMATIONS 16-2

Certain assumptions about the distributions of the populations are necessary for most statistical procedures. Examples of these assumptions are that one or more populations are normally distributed, that two or more populations have equal variances, or that a two-variable analysis-of-variance model has no interaction (also assumed in the Latin-square design). It sometimes happens that an appropriate change of scale, such as using the square root or logarithm of the observations, will more nearly satisfy some of the assumptions. The exact change of scale is, in general, difficult to determine, and success in finding a good transformation depends in part on experience in the particular field of application. For example, it is often found that size measurements on plants and animals are approximately normally distributed if logarithms are used.

A large number of observations is usually required before any definite statement can be made about the appropriate transformation. Without prior information the experimenter has little chance of proving or disproving normality or of finding a normalizing transformation from samples as small as 10 or 20.

**Means of several observations.** Since a mean is usually more nearly normally distributed than individual measurements, it is sometimes convenient to draw a series of small samples and record the means for use in the analysis in place of the single observations. An example was given in Sec. 9-7 for the case where $\sigma$ was known. For the cases where $\sigma$ is not known we may use the observed means with Table A-5 to estimate $\mu$ or with Table A-6*b* to estimate $\sigma^2/n$, where $n$ is the size of each sample. We can then estimate $\sigma^2$ by multiplying the estimate of $\sigma^2/n$ by $n$.

**Square-root transformation.** If the sample means are approximately proportional to the variances of the respective samples (or the squares of the ranges), replacing each measurement by its square root will often result in homogeneous variances. This type of transformation is appropriate for data having the Poisson distribution discussed in Chaps. 13 and 20. In this case the transformation $\sqrt{X} + \sqrt{X+1}$ will approximately stabilize the variance at $\sigma^2 = 1$ if the mean of the original observations is greater than .8.

**Logarithmic transformation.** If the means of the samples are proportional to the range or to the standard deviation of the respective samples, replacing each measurement by its logarithm will often result in the variances being more nearly equal. Actually, it happens in many applications that the logarithmic transformation also tends to normalize the distribution. Graph paper is available with a logarithmic scale on one axis and a normal-probability scale on the other (see page 435). This paper simplifies examination of data for normality of the transformed observations. An approximately straight cumulative-distribution polygon is an indication of normality of the transformed observations.

**Arc Sine transformation.** If data are collected in the form of proportions the variance of the observed proportion is $p(1-p)/N$, as noted in Chap. 5. The arc sine transformation of the observed proportions produces an approximately constant variance unless $Np < .8$ or $N(1-p) < .8$. The transformation can be obtained by using Table A-28. The tabled values are in radians, and the variance of the transformed measurements is approximately $1/(N+\frac{1}{2})$. The formula for the transformation is

$$\phi = \arcsin\sqrt{\frac{X}{N+1}} + \arcsin\sqrt{\frac{X+1}{N+1}}$$

In this case the transformed observations are also approximately normal. We may refer to Table A-4 and use

$$z = \sqrt{N+1}\left(2 \arcsin \sqrt{\frac{X+1}{N+1}} - 2 \arcsin \sqrt{p}\right)$$

to give approximate confidence limits for $p$ or to test a hypothesis $p = p_0$. The use of the arc sine transformation avoids the difficulty (encountered in Chaps. 7 and 13) that $\sigma^2$ is unknown if $p$ is unknown.

**Normalized scores.** In some cases it is desirable to force the observations into a normal distribution. If the sample sizes are not greater than 20, Table A-8*b*(5) can be used to obtain normalized scores for the ranks of observations. These scores are the mean positions of ordered observations from a normal population with mean 0 and unit variance.

**Use of *T* scores.** A method for assigning scores similar to normalized scores has been used frequently in educational and psychological studies to standardize the scores of examinations. Knowing that a student's score on an examination is 63 gives no information about the comparative standing of the student. Also, saying that a man is 70 inches tall does not in itself indicate whether he is exceptionally tall, or of medium height, or short. In no case does a single measurement give us a great deal of information. A single measurement selected from a known distribution tells us much more, for then we can say whether the item picked is exceptionally large or small. In fact, we can say what proportion of cases fall above or below it. Many times it is difficult to look at an entire distribution, and we should like some way of designating a single score (measurement) so that its value tells at a glance whether it is a comparatively large or small value. $Z$ scores, introduced in Chap. 3, perform this function to a certain extent. A $Z$ score of 700 is two standard deviations above the mean and thus probably a very high score. We still would not know exactly what proportion of the cases are above 700. If the distribution is normal, there are about 2.3 per cent of the scores above a $Z$ score of 700. In a normal distribution there are

50 per cent above a $Z$ score of 500
16 per cent above a $Z$ score of 600
7 per cent above a $Z$ score of 650
2.3 per cent above a $Z$ score of 700
.6 per cent above a $Z$ score of 750
.1 per cent above a $Z$ score of 800

Thus a $Z$ score in a *normal distribution* gives us considerable information. This indicates a need for a normalized $Z$ score. Such scores, called *T scores*, are defined in the following manner. Suppose a proportion $p$ of the observed cases falls below a particular value of the variable. We shall assign to

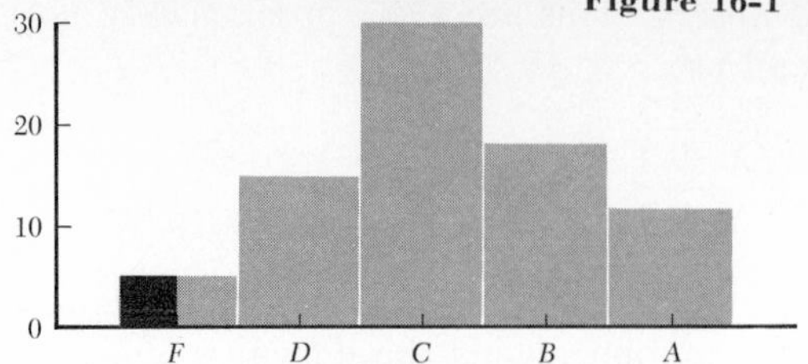


**Figure 16-1** Scaling from a histogram.

$(T - 500)/100$, that is, to $(T - \mu_T)/\sigma_T$, a value from the normal table (Table A-4) such that a proportion $p$ of the area of the normal curve lies to the left of that value. Thus if a value is larger than 90 per cent of the cases observed we shall say

$$\frac{T - 500}{100} = 1.28 \qquad \text{or} \qquad T - 500 = 100 \times 1.28$$

and the $T$ score is $T = 500 + 100 \times 1.28 = 630$.

Three examples illustrate this method of finding $T$ scores.

EXAMPLE 16-1. *$T$ score for grades.* Suppose the following table of grades and frequencies is given. What $T$ score would be assigned to each grade?

| *Grade* | F | D | C | B | A | |
|---|---|---|---|---|---|---|
| *Frequency* | 5 | 15 | 30 | 18 | 12 | 80 |

Below (to the left of) the center of the F rectangle in the histogram there are 2.5 units out of a total of 80 units of area. This is based on the premise that 2.5 of the 5 F grades were below the middle. Thus the proportion of cases below the center grade of F is $2.5/80 = .03125$, and this corresponds to $(T - 500)/100 = -1.9$. Thus the $T$ score for a grade of F is $T = 500 + 100(-1.9) = 310$. Below the D center there are 12.5 units of area (5 from the F and half of the 15 D's). The proportion of cases below a

**Table 16-1**

| *Grade* | *Freq.* | *Cum. freq. to mid-point* | *Cum. prop. to mid-point* | FROM TABLE A-4 $\frac{T-500}{100}$ | *T score* |
|---|---|---|---|---|---|
| A | 12 | 74 | .925 | +1.4 | 640 |
| B | 18 | 59 | .738 | + .6 | 560 |
| C | 30 | 35 | .438 | − .2 | 480 |
| D | 15 | 12.5 | .156 | −1.0 | 400 |
| F | 5 | 2.5 | .031 | −1.9 | 310 |
| | 80 | | | | |

**Table 16-2**

| *Rank* | *Freq.* | *Cum. freq. to mid-point* | *Cum. prop. to mid-point* | $\frac{T-500}{100}$ | $T$ |
|---|---|---|---|---|---|
| 1 | 1 | 4.5 | .9 | +1.3 | 630 |
| 2 | 1 | 3.5 | .7 | + .5 | 550 |
| 3 | 1 | 2.5 | .5 | 0 | 500 |
| 4 | 1 | 1.5 | .3 | − .5 | 450 |
| 5 | 1 | .5 | .1 | −1.3 | 370 |

D is $12.5/80 = .156$. Table A-4 gives approximately $(T - 500)/100 = -1.0$ and $T = 500 + 100(-1.0) = 400$. A form outlining the complete calculations for this problem is shown in Table 16-1.

EXAMPLE 16-2. *T scores for ranks.* Suppose that a group of scores have been assigned ranks; say, the highest score is assigned rank 1, the next highest rank 2, etc. These ranks can be thought of as grades with one score to each grade, and the $T$ score for each rank can be computed by the method illustrated above. For example, suppose that five scores have been given a rank order. Table 16-2 shows the computations to find the $T$ scores.

EXAMPLE 16-3. *T scores for data arranged in a frequency table.* $T$ scores for numerical data are obtained in exactly the same way as $T$ scores for grades. Table 16-3 shows the computation and the results.

$T$ scores can be read from normal-probability paper which has been marked with $\mu = 500$ and $\sigma = 100$ on the horizontal scale and has the corresponding cumulative line drawn. The line drawn on Fig. 5-8 can be used for this purpose if the $T$ scale marked on the horizontal is used.

To find $T$ scores with this paper find the cumulative proportion as in the examples above, and then read the score from the horizontal scale. Verify for Example 16-3 that, corresponding to $X = 17.5$, which has a cumulative proportion of .80, the point on the line is above $T = 590$. The use of this prepared paper omits the reference to Table A-4, the multiplication by 100, and the addition of 500.

**Table 16-3**

| $X$ | *Freq.* | *Cum. freq. to mid-point* | *Cum. prop. to mid-point* | $\frac{T-500}{100}$ | $T$ |
|---|---|---|---|---|---|
| 18.5 | 5 | 37.5 | .94 | +1.6 | 660 |
| 17.5 | 6 | 32 | .80 | + .8 | 580 |
| 16.5 | 8 | 25 | .63 | + .3 | 530 |
| 15.5 | 12 | 15 | .38 | − .3 | 470 |
| 14.5 | 7 | 5.5 | .14 | −1.1 | 390 |
| 13.5 | 2 | 1 | .03 | −1.9 | 310 |

## 16-3 DETECTION OF EXTREMES

Sometimes a statistical criterion is desired which will indicate whether or not the largest or smallest observation is significantly far removed from the main body of the data. Such extreme observations, or *outliers*, may occur because of *gross errors*, or *blunders*, or may be from a population other than the population from which the rest of the data come, or may result from the fact that the population under investigation contains a certain proportion of extreme cases and our sample happens to include one. Of course, individual observations of any of the three types above may occur in the main body of the sample and not be recognized by statistical techniques which test the extreme observations in the sample.

As an example of a possible gross error, consider the results of weighing an object four times. The four observations are 20.1, 20.0, 20.1, 19.1. It seems likely that the observation 19.1 is an incorrect reading or gross error.

An anthropologist may wish to decide whether a particular skeleton belongs to the same population as the remainder of a group of skeletons. We have an example of populations which contain a group of extreme measurements in income studies where a small portion of the population has very high incomes. In the first two examples the research worker may wish to eliminate extreme values from his sample before making statistical inferences about the populations. However, in the third example we may wish to treat an individual with high income in the sample as representative of the population on equal terms with any other member of the sample.

A statistic which can be used to detect outliers in either direction (too large or too small) is $w/\sigma$, the range divided by the population $\sigma$ if it is known. Percentiles of the sampling distribution of $w/\sigma$ if the observations come from a normally distributed population are given in Table A-8*b*(1). If a significantly large value is observed, we view this as evidence that the most extreme observation is from a different population.

An alternative set of test statistics and several percentiles of their sampling distributions are given in Table A-8*e*. These $r$ ratios compare the distance of one end observation from its neighbors with the range of all, or all but one or two, of the observations. For example, for the four observations ordered by size 326, 177, 176, 157 Table A-8*e* indicates the use of

$$r_{10} = \frac{X_2 - X_1}{X_4 - X_1} = \frac{177 - 326}{157 - 326} = .882$$

and since this is greater than the 95th percentile .765, we reject at the 5 per cent level the hypothesis that the observation 326 is from the same population as the other observations. Note that $X_1$ in Table A-8*e* can be either the largest or the smallest observation in the sample. Also, the $X_i$'s may be individual observations or the means of equal-size samples. The percentiles in Table A-8*e* are based on the assumption that the observations are from a normal population.

**Table 16-4** Rules for estimation in the presence of outliers

| Degree | Examples | $\alpha$ | $N \leq 10$ Mean | $N \leq 10$ Std. dev. | $10 < N < 20$ Mean | $10 < N < 20$ Std. dev. |
|---|---|---|---|---|---|---|
| Light | Up to {10% obs. shifted $1\sigma$; 5% obs. shifted $2\sigma$; 1% obs. shifted $10\sigma$} | .05 | $\bar{X}$ | Range | $\bar{X}$ | $s$ |
| Medium | Up to {30% obs. shifted $1\sigma$; 10% obs. shifted $3\sigma$; 5% obs. shifted $9\sigma$} | .10 | $\bar{X}$ | Range | Median | $s$ |
| Heavy | Over amounts for medium | .20 | Median | Range | Median | $s$ |

In place of tagging an individual observation from another population, we may be concerned with estimating the parameters of the basic distribution free from these contaminating effects. How might we process the data so that our estimates will come closer to the mean and standard deviation of this basic distribution? If very little is known about the distribution from which the extraneous observations may come, about the best we can do is to tag observations and remove them from estimates of $\mu$ and $\sigma$. If even a moderate amount is known about the distribution of extreme values in the population rules can be stated for samples of size less than 20 which minimize the effects on the estimates of $\mu$ and $\sigma$ of any outliers which may be present. Such rules are given in Table 16-4 and are used as follows. Decide from past experience or from knowledge of the measuring apparatus whether the outliers are causing light, medium, or heavy effects on the samples. The table gives examples of what is meant by these rough categories. The value $1 - \alpha$ indicates the percentiles of the $r$ distributions to be used in tagging outliers. The ratios are used repeatedly until no additional observations can be tagged. The mean and standard deviation are then estimated from the remaining observations as recommended in the table, the mean by $\bar{X}$ or the median and the standard deviation by the range or the sample standard deviation.

EXAMPLE 16-4. Suppose samples of size 5 are taken and it is expected that about 10 per cent of the observations will be shifted by three to four standard deviations. This falls in the medium classification which prescribes $\alpha = .10$ and estimation with $\bar{X}$ and the range for sample sizes under 10. The observations recorded in order of size are $X_1 = 23.2$, $X_2 = 23.4$, $X_3 = 23.5$, $X_4 = 24.1$, $X_5 = 25.5$. We note by inspection that $X_1 = 23.2$ would not be rejected. To test $X_5 = 25.5$ we compute

$$r_{10} = \frac{25.5 - 24.1}{25.5 - 23.2} = \frac{1.4}{2.3} = .609 \qquad (r_{10})_{.90} = .557$$

Therefore $X_5$ is rejected. We now proceed with $N = 4$ and test $X_4$ by computing

$$r_{10} = \frac{24.1 - 23.5}{24.1 - 23.2} = \frac{.6}{.9} = .667 \qquad (r_{10})_{.90} = .679$$

Therefore $X_4$ is not rejected. We estimate $\mu$ with

$$\bar{X} = \frac{23.2 + 23.4 + 23.5 + 24.1}{4} = 23.55$$

and estimate $\sigma$ by $.486w = .486(24.1 - 23.2) = .44$. The use of a large value of $\alpha$ will discover more outliers when they are present but, of course, will make the estimates of $\mu$ and $\sigma$ less precise if the samples do not contain outliers. This loss of precision is small, however, and usually is very small compared with the reduction in precision in the estimates if even occasional extreme values are not removed.

## 16-4 WINSORIZED ESTIMATES

If it is suspected that some individuals in the sample may have come from a population other than that of interest we may wish to estimate the mean in a manner to minimize the effect of these extraneous values. If the smallest and largest observations are given the value of their nearest neighbors, which is called *first-level Winsorization*, the mean computed on the modified sample will not have lost much efficiency if the extremes are actually valid (from the appropriate population). If the extremes are not valid we have obviously improved our estimate. This procedure can be extended for several observations at each extreme.

Referring to Example 16-4, we would replace the sample observations

| | | | | | | |
|---|---|---|---|---|---|---|
| | 23.2 | 23.4 | 23.5 | 24.1 | 25.5 | with sum 119.7 |
| by | 23.4 | 23.4 | 23.5 | 24.1 | 24.1 | with sum 118.5 |

and obtain the Winsorized estimate $\bar{X}_w = \frac{118.5}{5} = 23.70$.

The efficiency of this estimating procedure is known for samples with all observations from a single normal population. Table A-8*f* lists efficiencies for various levels of Winsorizing. We refer to the first level of Winsorizing when the extreme observations are made equal to their nearest neighbors. The second level refers to making two extremes at each end equal to their nearest neighbors, and so on.

For sample sizes up to 20 this procedure always has an efficiency of at least .65 and above .90 if $N > 5$ for first-level, $N > 10$ for second level, and $N > 15$ for third-level Winsorizing.

The efficiency of the $k$th-level Winsorized estimate of the mean is approximately $1 - k/[3(N - 2)]$.

An estimate may be desired which disregards the extremes, or we may wish to obtain an estimate of $\mu$ when missing observations are known to be

**Table 16-5** Efficiency of trimmed and Winsorized estimates of the mean

| | $k = 1$ | | $k = 2$ | | $k = 3$ | | $k = 4$ | |
|---|---|---|---|---|---|---|---|---|
| $N$ | *Trim.* | *Wins.* | *Trim.* | *Wins.* | *Trim.* | *Wins.* | *Trim.* | *Wins.* |
| 10 | .949 | .958 | .883 | .889 | .808 | .821 | .723 | .723 |
| 20 | .978 | .984 | .948 | .962 | .915 | .936 | .880 | .905 |

at the extreme. If we use the Winsorizing method when observations are missing (or to be omitted) in equal numbers at either extreme, the estimates are known to be 99.9 per cent efficient relative to the best possible linear estimate based on these same observations for samples from normal populations with sample sizes 20 or less.*

Winsorized estimates can be used if missing observations are known to be at the extreme and are missing almost symmetrically.

EXAMPLE 16-5. Consider the observations

— — 108 111 119 121 125 — — —

where dashes indicate missing observations. We can obtain the third-level Winsorized estimate of $\mu$ as

$$\frac{(4 \times 111) + 119 + 121 + (4 \times 125)}{10} = 118.4$$

It can be determined† that this estimate is 97 per cent efficient in comparison with the best linear estimate of $\mu$ (82 per cent efficient, however, in comparison with an estimate with all values present).

If symmetrically placed extreme observations are omitted, that is, trimmed rather than being Winsorized, the arithmetic mean of the remaining observations provides an estimate of somewhat smaller efficiency. Table A-8*b*(4) lists efficiencies for first-level trimming. Table 16-5 lists some efficiencies for higher levels of trimming.

## WINSORIZED $t$ 16-5

Confidence-interval estimates and tests of hypotheses for the case of disregarded missing extremes may be computed with the use of a $t$ ratio based on Winsorization. If observations at the extremes are replaced by the values of nearest neighbors as in the last section and the $t$ ratio is computed as in Chap. 6, the ratio multiplied by a constant depending only on $N$ has, to a high degree of accuracy, a $t$ distribution. The number of observations

* See W. J. Dixon, Simplified Estimation from Censored Normal Samples, *Ann. Math. Stat.*, vol. 31, pp. 385–391 (1960).

† See W. J. Dixon, and John W. Tukey, Approximate Behavior of the Distribution of Winsorized *t*, *Technometrics* vol. 10, pp. 83–98 (1968).

in a sample of size $N$ which are not changed in value is designated $h$; for example, for second-level Winsorizing $h = N - 4$. The $t$ ratio computed on a Winsorized sample is denoted by $t_w$. The quantity

$$t = \frac{h - 1}{N - 1} t_w$$

has approximately a $t$ distribution with $h - 1$ degrees of freedom (in place of $N - 1$). Then values from Table A-5 can be used to construct confidence intervals or to test hypotheses.

EXAMPLE 16-6. Given the observations

$$-36 \quad -30 \quad -24 \quad -17 \quad -3 \quad -3 \quad -1 \quad 5 \quad 12 \quad 19 \quad 24 \quad 63$$

find the 95 per cent confidence interval for $\mu$, based on first-level Winsorizing. We replace $-36$ by $-30$ and 63 by 24 to obtain

$$\Sigma X = -24 \qquad \Sigma X^2 = 4366 \qquad s_w^2 = 392.54$$

with the confidence interval

$$\bar{X}_w \pm \frac{N - 1}{h - 1} t_{.975}(h - 1) \frac{s_w}{\sqrt{N}}$$

which for this example is

$$-\frac{24}{12} \pm \frac{11}{9} 2.26 \sqrt{\frac{392.54}{12}} \qquad \text{or} \qquad -2 \pm 15.8$$

The use of the $t$ distribution here is approximate. The approximation is good for $N > 6$ and $\alpha$ in the range .01 to .20 in samples where fewer than half the observations have been Winsored away.

## GLOSSARY

admixed normality
arc sine transformation
logarithm transformation
log normal
outliers
square-root transformation
$T$ scores
trimmed estimates
Winsorized estimates
Winsorized $t$

## CLASS EXERCISES

1. Construct a sample of size 10 in which each observation has a 10 per cent chance of being drawn from a population with mean increased by 2. This may be accomplished by drawing observations from Table A-2. After each observation is selected, refer to Table A-1. If the digit observed in Table A-1 is 9, increase the observation selected from Table A-2 by 2. Test the resulting sample for outliers and compute the

mean of the remaining observations. Collect the results of all students and compute the mean and variance of the resulting distribution of means.

2. For samples formed as in Exericse 1 estimate the mean by Winsorizing the sample. Tabulate the resulting estimates and compute the mean and variance of the resulting distribution of estimates of $\mu$.

## DISCUSSION QUESTIONS

1. Give an example of an observational situation in which occasional extraneous observations might be introduced.

2. Give an example in which the discovery of an outlier may be important in itself rather than for its effect on the sample mean.

3. Describe an experiment in which the values of the extreme observations might not be known.

## PROBLEMS

*Section* 16-2

1. Plot the cumulative distributions from Probs. 5 to 26, Chap. 2, on arithmetic-normal and on log-normal paper to determine if the distribution is either normal or log normal.

2. Fish were weighed four at a time and the resulting measurements of 100 sets (400 fish) had a mean equal to 4.1 pounds and a variance $s^2 = 1.21$. Estimate the variance of weights of the individual fish and the mean of the individual fish by 95 per cent confidence intervals.

3. Table A-15 gives cumulative Poisson distributions. For $\lambda = 1$, $\lambda = 2$, and $\lambda = 3$ apply the square-root transformation and compute the means and variances of each transformed distribution. Compare each variance with 1.

4. Find $T$ scores for the ranks of seven scores.

5. Find $T$ scores for the following distribution:

| *Test score* | 95 | 90 | 85 | 80 | 75 | 70 | 65 | 60 |
|---|---|---|---|---|---|---|---|---|
| *Frequency* | 2 | 3 | 5 | 6 | 8 | 12 | 8 | 6 |

6. With the height data in Tables 2-2*a* and 2-2*b*, give for each inch of height a transformed value such that the transformed distribution is normal.

*Section* 16-3

7. With the data for gains in weight of pairs of rats fed raw or roasted peanuts, analyze each set and the set of ten differences for outliers.

| | PAIR | | | | | | | | | |
|---|---|---|---|---|---|---|---|---|---|---|
| | 1 | 2 | 3 | 4 | 5 | 6 | 7 | 8 | 9 | 10 |
| *Raw* | 61 | 60 | 56 | 63 | 56 | 63 | 59 | 56 | 44 | 61 |
| *Roasted* | 55 | 54 | 47 | 59 | 51 | 61 | 57 | 54 | 62 | 58 |

*Section* 16-4

**8.** The percentage of aluminum oxide found in synthetic granite glass was reported to be as follows:

| *Lab. no.* | 1 | 2 | 3 | 4 | 5 | 6 |
|---|---|---|---|---|---|---|
| *Per cent* | 16.20 | 15.85 | 16.78 | 16.39 | 16.12 | 16.12 |

| *Lab. no.* | 7 | 8 | 9 | 10 | 11 | 12 |
|---|---|---|---|---|---|---|
| *Per cent* | 16.27 | 16.34 | 15.91 | 15.98 | 16.14 | 15.79 |

Obtain 95 per cent confidence intervals for the mean based on first-level and second-level Winsorizing.

CHAPTER SEVENTEEN

# Nonparametric statistics

In previous chapters we usually assumed that the population had some known form. Frequently we assumed a normal distribution, and we estimated or tested hypotheses about the means and variances. Recent advances in statistics have been in the direction of attempts to find test statistics which would compare distributions without specifying their form. Since the comparison is between distributions and not between parameters, the procedures are frequently called *nonparametric statistics*. Probably the most widely used nonparametric technique is the $\chi^2$ test for independence and goodness of fit. In this chapter we shall consider certain other nonparametric techniques.

## THE SIGN TEST 17-1

In experimental investigations it is often desired to compare two materials or treatments under various sets of conditions. Pairs of observations (one observation for each of the two materials or treatments) are obtained for each of the separate sets of conditions. For example, in comparing the yield of two hybrid lines of corn, $A$ and $B$, we might have a few results from each of several experiments carried out under widely varying conditions. The experiments may have been performed on different soil types, with different fertilizers, and in different years, with consequent variations in seasonal effects such as rainfall, temperature, or amount of sunshine. It is supposed that both hybrids appeared equally often in each block of each experiment, so that the observed yields occur in pairs (one yield for each line) produced under quite similar conditions. The sign test is most useful, under the following circumstances:

1. There are independent pairs of observations on two things being compared.

2. Each of the two observations of a given pair was made under similar conditions.

3. The different pairs were observed under different conditions.

This last condition generally makes the $t$ test invalid, since this would usually mean that the differences have different variances. If this were not the case (that is, if all the pairs of observations were comparable), the $t$ test would ordinarily be employed unless there were other reasons, such as obvious nonnormality, for not using it.

Even when the $t$ test is the appropriate technique, the sign test may be used because of its extreme simplicity. We merely count the number of positive and negative differences and refer to a table of significance values. Frequently the question of significance may be settled at once by the sign test, without any need for calculations.

It should be pointed out that, strictly speaking, the methods of this section are applicable only to the case in which no ties in paired comparisons occur. In practice, however, even when ties would not occur if measurements were sufficiently precise, ties do occur because measurements are often made only to the nearest unit or tenth of a unit, for example. Such ties should be excluded, resulting in a reduction in $N$, the sample size.

**Procedure.** Let $A$ and $B$ represent two materials or treatments to be compared. Let $X$ and $Y$ represent measurements made on $A$ and $B$. Let the number of pairs of observations be $N$. The $N$ pairs of observations and their differences may be denoted by

$$(X_1,Y_1),\ (X_2,Y_2),\ \ldots\ ,\ (X_N,Y_N)$$

and

$$X_1 - Y_1,\ X_2 - Y_2,\ \ldots\ ,\ X_N - Y_N$$

The sign test is based on the signs of these differences. The letter $r$ will be used to denote the number of times the less frequent sign occurs. If some of the differences are zero they will be excluded and the sample size reduced.

Example 17-1. As an example of the type of data for which the sign test is appropriate, consider the data in Table 17-1, which gives the yields of two hybrid lines of corn obtained from several different experiments. In this example $N = 28$ and $r = 7$.

The null hypothesis here is that each difference has a probability distribution (which need not be the same for all differences) with median 0. This null hypothesis will be true, for instance, if each difference is sym-

**Table 17-1** Yields of two hybrid lines of corn

| | YIELD | | | | YIELD | | |
|---|---|---|---|---|---|---|---|
| *Experiment* | *A* | *B* | *Sign of* $X - Y$ | *Experiment* | *A* | *B* | *Sign of* $X - Y$ |
| 1 | 47.8 | 46.1 | + | 4 | 40.8 | 41.3 | − |
| | 48.6 | 50.1 | − | | 39.8 | 40.8 | − |
| | 47.6 | 48.2 | − | | 42.2 | 42.0 | + |
| | 43.0 | 48.6 | − | | 41.4 | 42.5 | − |
| | 42.1 | 43.4 | − | | | | |
| | 41.0 | 42.9 | − | 5 | 38.9 | 39.1 | − |
| | | | | | 39.0 | 39.4 | − |
| 2 | 28.9 | 38.6 | − | | 37.5 | 37.3 | + |
| | 29.0 | 31.1 | − | | | | |
| | 27.4 | 28.0 | − | 6 | 36.8 | 37.5 | − |
| | 28.1 | 27.5 | + | | 35.9 | 37.3 | − |
| | 28.0 | 28.7 | − | | 33.6 | 34.0 | − |
| | 28.3 | 28.8 | − | | | | |
| | 26.4 | 26.3 | + | 7 | 39.2 | 40.1 | − |
| | 26.8 | 26.1 | + | | 39.1 | 42.6 | − |
| 3 | 33.3 | 32.4 | + | | | | |
| | 30.6 | 31.7 | − | | | | |

metrically distributed about a mean of zero, although such symmetry is not necessary. We shall reject the null hypothesis when the numbers of positive and negative signs differ significantly from equality.

Table A-10*a* gives the critical values of $r$ for 1, 5, 10, and 25 per cent levels of significance. A value of $r$ less than or equal to the tabled value is significant at the given per cent level.

Table A-10*b* gives percentiles for the sampling distribution of the number of plus signs when the hypothesis is true. This is the binomial distribution with $p = \frac{1}{2}$. In general there are no values of $r$ which correspond exactly to the usual levels of significance, for example, $\alpha = .05$ or $\alpha = .01$. However, we can find levels of significance close to any desired level if the sample size is fairly large. In Example 17-1, where $N = 28$, we find from Table A-10*b* that a one-sided test of the hypothesis that the population consists of 50 per cent or more plus signs can be made at the $\alpha = .044$ level by rejecting if nine or fewer *plus* signs are observed in the sample. Similarly, a one-sided test of the hypothesis that the population consists of 50 per cent or fewer plus signs can be made at the $\alpha = .044$ level by rejecting if nine or fewer *minus* signs (more than 18 plus signs) are observed in the sample. A two-sided test at the $\alpha = 2 \times .044 = .088$ level of significance can be made by rejecting if nine or fewer plus signs or if nine or fewer minus signs are observed in the sample. This two-sided test criterion designates the critical region as values of $r$ less than or equal to 9. If the data in Table 17-1 are used to test with two-sided alternatives we shall reject the hypothesis at the $\alpha = .088$ level, since there are only $r = 7$ plus signs in the sample.

For small samples it is often impossible to find a critical region of a given size. For example, if we desire a two-sided test at the $\alpha = .10$ level with $N = 12$ observations, we see from Table A-10*b* that the closest choices are $\alpha = 2 \times .019 = .038$ (rejecting if $r \leq 2$) and $\alpha = 2 \times .073 = .146$ (rejecting it $r \leq 3$). Even if the hypothesis of the population having 50 per cent plus and 50 per cent minus signs is true, a sample of size 4, or even one of size 5, will have more than a 5 per cent chance of having all signs alike. Four signs alike in a sample of size 4 will occur with a chance .125, and five signs alike in a sample of size 5 will occur with chance .0625 if the hypothesis is true. Therefore it is necessary to have at least six pairs of observations if any value of $r$ is to cause rejection of the hypothesis at the 5 per cent level of significance.

**Power function and sample size.** Just as with any statistical test, the power of the sign test to recognize cases where the hypothesis is not true increases when a larger sample size is used. The probability of rejecting the hypothesis can be given for alternatives which specify that some proportion $p$ of the distribution of the differences are plus and a proportion $1 - p$ are negative. The hypothesis states that 50 per cent ($p = .50$) are plus and 50 per cent minus. Table 17-2 gives the minimum sample size $N$ needed for a probability of at least .95 of rejecting the hypothesis $p = .50$ when $p$ is actually the value given in the left-hand column.

For example, Table 17-2 shows that if the signs are actually distributed 45:55, then we must take samples of 1,080 pairs in order to give signifi-

**Table 17-2** Minimum values of $N$ necessary to find significant differences 95 per cent of the time for various true proportions of the hypothesis that $p = .50$. The asterisked values are approximate. The maximum error is about 5.

| | SAMPLE SIZE $N$ | | | |
|---|---|---|---|---|
| $p$ | $\alpha = .01$ | $\alpha = .05$ | $\alpha = .10$ | $\alpha = .25$ |
| .45(.55) | 1,777* | 1,297* | 1,080* | 780* |
| .40(.60) | 442* | 327 | 267* | 193* |
| .35(.65) | 193* | 143 | 118* | 86 |
| .30(.70) | 106* | 79 | 67 | 47 |
| .25(.75) | 66 | 49 | 42 | 32 |
| .20(.80) | 44 | 35 | 28 | 21 |
| .15(.85) | 32 | 23 | 18 | 14 |
| .10(.90) | 24 | 17 | 13 | 11 |
| .05(.95) | 15 | 12 | 11 | 6 |

cance 95 per cent of the time at the 10 per cent level; if a large number of samples of 1,080 each were drawn from a 45:55 distribution, then 95 per cent of those samples could be expected to indicate a significant departure (at the 10 per cent level) from a 50:50 distribution.

Of course, in practice we would not do any testing if we knew in advance the expected distribution of signs (that it was 45:55, for example). The practical significance of Table 17-2 is of the following nature. In comparing two materials we are interested in determining whether they are of about equal or of different value. Before the investigation is begun a decision must be made as to how different the materials must be to be classed as different. Expressed in another way, how large a difference may be tolerated in the statement "The two materials are of about equal value"? This decision, together with Table 17-2, determines the sample size. If we are interested in detecting a difference so small that the signs may be distributed 45:55, we must be prepared to take a very large sample. If, however, we are interested only in detecting larger differences (for example, differences represented by a 30:70 distribution of signs), a smaller size sample will suffice.

**Modifications of the sign test.** If the same type or unit of measurement is made for every pair, that is, if the measurements are comparable for all pairs of observations, then the sign test can be used to answer questions of the following kind:

1. Is material $A$ better than $B$ by $P$ per cent?
2. Is material $A$ better than $B$ by $Q$ units?

The first question would be tested by increasing the measurement on $B$ by $P$ per cent and comparing the results with the measurements on $A$. Thus let

$$(X_1,Y_1),\ (X_2,Y_2),\ (X_3,Y_3),\ \ldots,\ (X_N,Y_N)$$

be pairs of measurements on $A$ and $B$, and suppose we wish to test the hypothesis that the measurements $X$ on $A$ were 5 per cent higher than the measurements $Y$ on $B$. The sign test would simply be applied to the signs of the differences

$$(X_1 - 1.05Y_1),\ (X_2 - 1.05Y_2),\ (X_3 - 1.05Y_3),\ \ldots,\ (X_N - 1.05Y_N)$$

In the case of the second question the sign test would be applied to the differences

$$[X_1 - (Y_1 + Q)],\ [X_2 - (Y_2 + Q)],\ [X_3 - (Y_3 + Q)],\ \ldots,\ [X_N - (Y_N + Q)]$$

In either case, if the resulting distribution of signs is not significantly different from 50:50, the data are not inconsistent with a positive answer to the question. Usually there will be a range of values of $P$ (or $Q$) which will produce a nonsignificant distribution of signs. If such a range is deter-

**Table 17-3** Power efficiency of sign test for normal populations

| $N$ | $\alpha$ | DIFFERENCE $\delta$ *Near* 0 | .5 | 1 | 1.5 | 2 |
|---|---|---|---|---|---|---|
| 5 | .0625 | .96 | .96 | .95 | .93 | .91 |
| 10 | .0020 | .94 | .92 | .90 | .87 | .84 |
| 10 | .0215 | .85 | .84 | .82 | .80 | .77 |
| 10 | .1094 | .77 | .76 | .74 | .72 | |
| 20 | .0118 | .76 | .75 | .73 | .70 | |
| 20 | .0414 | .73 | .72 | .70 | .68 | |
| 20 | .1153 | .70 | .69 | .67 | .65 | |
| . . . | . . . | . . . | | | | |
| $\infty$ | $\alpha$ | .637 | | | | |

mined at the 5 per cent level of significance, for example, it will be a 95 per cent confidence interval for $P$ (or $Q$).

The efficiency of the sign test compared with the $t$ test when the observations are from normal populations can be assessed by finding the sample size $N_t$ for which a $t$ test will have the same power as a sign test based on a sample of size $N$. The ratio $N_t/N$ is called the *power efficiency* of the sign test relative to the $t$ test. For the sign test the power efficiency decreases (1) with increasing $\alpha$, (2) with increasing sample size, and (3) with increasing difference in the population means. Table 17-3 contains power efficiencies for the sign test used for two normal populations with means $\mu_x$ and $\mu_y$, and with equal variances $\sigma^2$. The quantity $\delta$ in the table is the difference in population means divided by the standard deviation of a difference of two observations, that is, $\delta = |\mu_1 - \mu_2|/(\sqrt{2}\sigma)$. It can be seen from this table that the sign test used with a sample of size $N = 20$ at the $\alpha = .0414$ level of significance has approximately the same power as a $t$ test used with a sample of size $N_t = .70N = 14$ when $\delta = 1$. If $|\mu_1 - \mu_2|/(\sqrt{2}\sigma)$ is small and $N$ is large, the $t$ test needs approximately 64 per cent as many observations to have the same power as the sign test.

## 17-2 EXTENSIONS OF THE SIGN TEST

Table A-26 lists the level of significance for some of the values in the sign test and includes a number of additional tests for other significance levels. The sign test can be described as follows. Suppose, for example, we arrange a sample of four differences in order of size, $d_1 < d_2 < d_3 < d_4$, and agree to reject the hypothesis $\mu = 0$ if $d_4$ is less than zero. Table A-26 gives the level of significance (chance that $d_4$ will be less than zero if the hypothesis is true) as $\alpha = 6.2$ per cent. If we agree to reject if either $d_4 < 0$ or $d_1 > 0$, then $\alpha = 12.5$ per cent. Similarly, if we wish to compare $X_i$ and $Y_i + \mu_0$, as indicated in the previous section, we might reject the hypothesis $\mu_x - \mu_y = \mu_0$ if the largest of the differences $X_i - Y_i$ in a sample of four pairs is less than $\mu_0$. Table A-26 gives $\alpha = 6.2$ per cent for this test.

The additional tests included in Table A-26 are based on slightly stricter assumptions than those used for the sign test. The assumptions are such that the tests are valid if they are based on $N$ observations independently drawn from $N$ symmetrical populations each having the same median $\mu$.

EXAMPLE 17-2. Consider the following six observations arranged in order of size: $-2.5$, $-1.5$, $-1.3$, 0.1, 0.3, 0.8. Test the hypothesis $\mu = .5$ at the 9.4 per cent level of significance. We see from Table A-26 that we are to compute

1. The maximum of .3 and $\frac{1}{2}(.1 + .8)$ which is .45.
2. The minimum of $-1.5$ and $\frac{1}{2}(-2.5 - 1.3)$ which is $-1.9$.

If either the first of these is less than .5 or the second is greater than .5, we reject the hypothesis. Since $.45 < .5$, we reject the hypothesis $\mu = .5$ at the 9.4 per cent level of significance.

Power efficiencies for the tests in Table A-26 (which include several cases of the sign test) would be similar to those given in Table 17-3 for the sign test.

**The signed-rank test.** Another similar modification of the sign test is the *Wilcoxon signed-rank test.* This is used to test the hypothesis that observations have come from symmetrical populations with a common specified median, say, $\mu_0$. The signed-rank statistic $T$ is computed as follows:

1. Subtract $\mu_0$ from each observation.
2. Rank the resulting differences in order of size, disregarding sign.
3. Restore the sign of the original difference to the corresponding rank.
4. Obtain $T$, the sum of the positive ranks.

Table A-19 lists percentiles of the sampling distribution of the $T$ statistic.

**Table 17-4**

| $X$ | $X - \mu_0$ | *Rank* $\lvert X - \mu_0 \rvert$ | *Signed rank* | |
|---|---|---|---|---|
| 2.55 | .55 | 3 | | 3 |
| 4.62 | 2.62 | 8 | | 8 |
| 2.93 | .93 | 4 | | 4 |
| 2.46 | .46 | 2 | | 2 |
| 1.95 | $-$ .05 | 1 | $-1$ | |
| 4.55 | 2.55 | 7 | | 7 |
| 3.11 | 1.11 | 6 | | 6 |
| .90 | $-1.10$ | 5 | $-5$ | |
| | | | | $30 = T$ |

An example of the use of this statistic to test the hypothesis that the median of a population is $\mu_0 = 2$ is given in Table 17-4. A sample of size 8 was chosen from the population. From Table A-19 we see that to have a level of significance close to .05 we could use for a critical region values of $T$ less than or equal to 4 and values of $T$ greater than or equal to 32. The level of significance for this critical region is $\alpha = 2 \times .027 = .054$. The data in Table 17-4 are not evidence at the .054 level that the hypothesis is incorrect.

## 17-3 RUNS

In the previous sections on the sign test we compared the total number of positive differences observed with the total number of negative differences observed. It is reasonable also to ask whether the positive differences are scattered in a fairly random manner among the negative ones when the observations are in the order observed, that is, in order of time. For example, if we observed the sequence

$$+ + + + - - - - -$$

we might want to consider it as different from the sequence

$$- + - + - + - + -$$

Each sequence has four positive differences and five negative differences. However, in the first case the positive differences all occurred together. Any set of differences having the same sign which occur in a row we shall call a *run*. We shall understand that a run includes as many elements as possible; thus the sequence $-, +, +, +, -$ has three runs, two with a single negative and one with three positives. The sequence $+, +, +, +, -, -, -, -, -$ has two runs, one of length 4 and one of length 5. The sequence of $n$ differences may have only one run (all positive or all negative) and may have as many as $n$ runs if the signs alternate. We may wish to reject the hypothesis of random arrangement of plus and minus signs if there are too few or too many runs.

Runs can be used to test whether or not two random samples come from populations having the same frequency distribution. We arrange the observations of the two samples together in one series according to size. We then count the number of runs of items from each of the samples. If there are fewer runs than would be expected by chance if both populations are the same, we reject the hypothesis that the two populations have the same distribution.

The theory of runs may also be used to test the hypothesis that observations have been drawn at random from a single population. To do this, find the median of the sample and denote observations below the median by a minus sign and observations above the median by a plus sign. If the number of runs of plus and minus signs is larger or smaller than might be expected by chance, we reject the hypothesis.

Let $N_1$ be the number of occurrences of one type (positive differences, observations below median) and $N_2$ be the number of occurrences of the other type (negative differences, observations above median). Let $u$ equal the total number of runs among the $N_1 + N_2$ observations. Table A-11 gives the sampling distribution of $u$ for values of $N_1$ and $N_2$ less than or equal to 10 and a number of percentiles of the distributions for larger sample sizes.

EXAMPLE 17-3. Suppose steel rods are manufactured and the diameters of the rods are measured. In the first 40 rods measured there is a total of 16 runs above and below the median. Let us test the hypothesis that the machinery is turning out rods whose diameters vary randomly. Here $N_1 = N_2 = 20$ (since half the observations are above the median and half are below the median). The observed number 16 is not less than or equal to $u_{.025} = 14$ and is not greater than or equal to $u_{.975} = 27$, so there is not sufficient evidence to reject the hypothesis at the 5 per cent level of significance.

EXAMPLE 17-4. Suppose 10 plots of wheat are treated with fertilizer $A$ and 10 plots are treated with fertilizer $B$. The yields are recorded in the tabulation, and then all 20 yields are arranged in order of size, with the

| | | | | | | | | | | |
|---|---|---|---|---|---|---|---|---|---|---|
| $A$ | 26.3 | 28.6 | 25.4 | 29.2 | 27.6 | 25.6 | 26.4 | 27.7 | 28.2 | 29.0 |
| $B$ | 28.5 | 30.0 | 28.8 | 25.3 | 28.4 | 26.5 | 27.2 | 29.3 | 26.2 | 27.5 |

observations in sample $A$ in boldface type:

25.3 **25.4** **25.6** 26.2 **26.3** **26.4** 26.5 27.2 27.5 **27.6**
**27.7** **28.2** 28.4 28.5 **28.6** 28.8 **29.0** **29.2** 29.3 30.0

There are 11 runs, which, for $N_1 = N_2 = 10$, corresponds to the 58.6 percentile of the distribution of $u$. Since this is between 2.5 and 97.5, there is no reason to reject at the 5 per cent level of significance the hypothesis that the two populations have the same distribution. Of course, it is unlikely that anyone would be interested in rejecting the hypothesis of no difference in fertilizers if there were too many runs. In this case too many runs would indicate a lack of randomness in the selection of the sample. Therefore we might use a one-sided test and reject only if the percentile rank of the observed $u$ were less than .05, that is, if there were 6 or fewer runs.

**Normal approximation.** If $N_1$ and $N_2$ are both larger than 10, the cumulative sampling distribution of $u$ can be approximated from the normal tables with $z = \dfrac{u - \mu_u + \frac{1}{2}}{\sigma_u}$, where

$$\mu_u = \frac{2N_1N_2}{N_1 + N_2} + 1 \qquad \text{and} \qquad \sigma_u{}^2 = \frac{2N_1N_2(2N_1N_2 - N_1 - N_2)}{(N_1 + N_2)^2(N_1 + N_2 - 1)}$$

In Example 17-4 $N_1 = N_2 = 10$, $\mu_u = 11$, and $\sigma_u = \sqrt{4.73} = 2.17$. The observed value of $u = 11$ gives a $z$ score of $.5/2.17 = .23$, or approximately the 59th percentile.

## 17-4 THE RANK-SUM TEST

In Example 17-4 we counted the number of runs in the arrangement of 20 observations, 10 from each of two populations. Another statistic which may be used to compare the two samples is the *rank-sum statistic* $T'$, defined as follows. Arrange the two samples together in order of size and assign rank scores to the individual observations, score 1 to the smallest, score 2 to the second smallest, and so on. Then $T'$ is the sum of ranks of the observations in the smaller of the two samples. If the samples are of the same size we may choose either sample. Note that if $T'$ is the sum of the $N_1$ ranks for samples of sizes $N_1$ and $N_2$, respectively, then $T'$ can be as small as $1 + 2 + 3 + \cdots + N_1 = N_1(N_1 + 1)/2$ or as large as $N_1(2N_2 + N_1 + 1)/2$.

In Table A-20 we have recorded some of the percentiles of the sampling distribution of $T'$ in the case where the two samples are from populations having identical distributions. We reject the hypothesis that we have random samples from identically distributed populations if $T'$ is significantly large or significantly small.

For example, if $N_1 = N_2 = 10$ we see from Table A-20 that the chance that $T'$ is less than or equal to 79 is .026 and the chance that $T'$ is greater than or equal to 131 is .026. Thus values of $T' \leq 79$ and $T' \geq 131$ form a 5.2 per cent critical region for the above hypothesis. In Example 17-4 the ranks of the $A$ sample are 2, 3, 5, 6, 10, 11, 12, 15, 17, 18, and so $T' = 99$; since this is not in the critical region, we accept the hypothesis that we have random samples from identically distributed populations.

In the case of ties we replace the observation by the mean of the ranks for which it is tied.

**Normal approximation.** For $N_1$ and $N_2$ both larger than 10 we use the fact that the sampling distribution of $T'$ is approximately normal with mean and variance

$$\mu_{T'} = \frac{N_1(N_1 + N_2 + 1)}{2} \qquad \text{and} \qquad \sigma_{T'}{}^2 = \frac{N_1N_2(N_1 + N_2 + 1)}{12}$$

and obtain the approximate chance that $T'$ will be less than or equal to $T'_0$ by finding the area to the left of $z = (T'_0 - \mu_{T'} + \frac{1}{2})/\sigma_{T'}$ from Table A-4. For the preceding example, where $N_1 = N_2 = 10$, these formulas give $\mu_{T'} = 105$ and $\sigma_{T'} = \sqrt{175} = 13.2$. The observed value of $T' = 99$ gives a $z$ score of $z = -5.5/13.2 = -.42$, and this, if compared with $z_{.025} = -1.96$ and $z_{.975} = 1.96$, is seen to be not significant at the 5 per cent level. Since the exact distribution of $T'$ is given for $N_1 = N_2 = 10$, we can compare the normal approximation with the exact chance that $T'$ will be less than or equal to 99. The exact chance is .342. The normal approximation read from Table A-4 for $z = -.42$ gives the approximate chance .337. Also note that for $T'_0 = 79$, $z = (79 - 105 + \frac{1}{2})/13.2 = -1.93$, which gives the approximate chance .027 corresponding to the exact chance .026.

The rank-sum test requires approximately 5 per cent more observations than a $t$ test to provide the same power as a $t$ test for shifts in means of two normally distributed populations. For nonnormal populations the rank-sum test may be more powerful than the $t$ test. In some cases the rank-sum test requires only 80 per cent as many observations for equal power. It should be noted that for nonnormal populations Table A-5 does not apply to the distribution of $t$, whereas the distribution of $T'$ in Table A-20 may be used whether or not the populations are normal.

**Rank-sum test for several samples.** Ranks can be used to test the hypothesis that $k$ samples of sizes $n_1, n_2, \ldots, n_k$ are randomly drawn from $k$ identically distributed populations. We arrange the $N = \Sigma n_i$ observations together in order of size and assign ranks as was done for the two-sample rank-sum test. Let $R_i$ be the sum of ranks of the $i$th sample, and let

$$H = \frac{12}{N(N+1)} \sum \frac{R_i^2}{n_i} - 3(N+1)$$

If the hypothesis is true and the $n_i$'s are not small, the sampling distribution of the statistic $H$ is approximately $\chi^2$ with $k - 1$ degrees of freedom. If all $n_i$'s are greater than 5, the 95th and 99th percentiles in Table A-6$a$ are reasonably accurate. In the case of ties we replace the observation by the mean of the ranks for which it is tied. The statistic $H$ is essentially the variance of the sample rank sums $R_i$. If a significantly large value of $H$ is observed the hypothesis is rejected.

## ESTIMATION OF A CUMULATIVE FREQUENCY DISTRIBUTION 17-5

Frequently in the earlier chapters of the book we approximated a distribution by sampling. It is possible to predict by means of confidence intervals how close the cumulative distribution of a sample can be expected to be to the cumulative distribution of the population. Draw a cumulative-

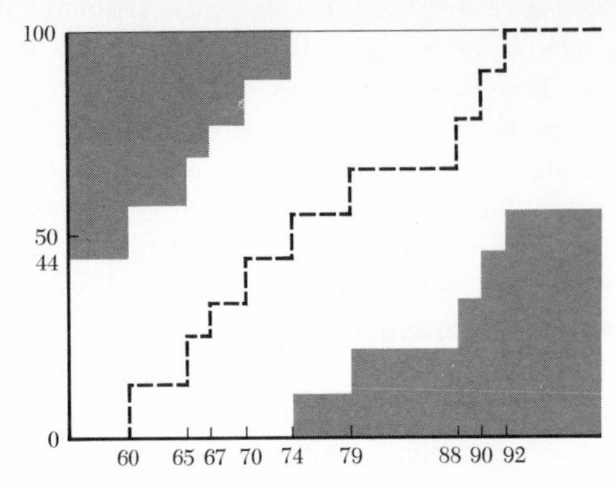


**Figure 17-1** Cumulative sample percentages with confidence interval for population cumulative.

percentage distribution for observations in the sample (dotted line in Fig. 17-1). Above and below this distribution draw two parallel polygons at a distance $100d_{1-\alpha}$. The width of a band giving confidence $1 - \alpha$ for the statement "The cumulative frequency distribution of the population is in this band" is $200d_{1-\alpha}$. Values of $d_{1-\alpha}$ can be read from Table A-21. The sample size is on the left, and five levels of confidence are referred to at the top. For example, if $N = 50$ and we wish to be 90 per cent confident that the band covers the population cumulative distribution, then the band should be of width $2 \times .17 = .34$, or 34 percentage points.

Actually, Table A-21 is a table of percentiles of the frequency distribution of the maximum deviation of a sample cumulative distribution from the population cumulative distribution. Thus in 90 per cent of samples of size 50 the maximum deviation of the sample distribution from the population distribution will be less than .17.

EXAMPLE 17-5. Suppose the observations of a sample of size $N = 9$ are 88, 92, 65, 74, 67, 79, 90, 70, 60. For $1 - \alpha = .95$ and for $N = 9$, we read from Table A-21 (interpolating between 5 and 10) that $d_{.95} = .44$. Figure 17-1 shows the observed distribution by dotted lines and the confidence band, 44 percentage points above and below the observed distribution, by solid lines. We are 95 per cent confident that the population cumulative distribution is entirely inside this band.

**Determination of sample size.** Table A-21 can also be used to determine the required sample size for 95 per cent confidence of containing the population cumulative distribution within some interval of given width. Suppose, for example, that we wish to be 95 per cent sure of containing the distribution within an interval of width 5 percentage points. Here we require $d_{.95} = .05/2$. Thus $1.36/\sqrt{N} = .025$, or

$$N = \left(\frac{1.36}{.025}\right)^2 = (54.4)^2 = 2{,}960$$

For 99 per cent confidence the sample must be of size

$$N = \left(\frac{1.63}{.025}\right)^2 = 4{,}251$$

For 99 per cent confidence that the sample distribution will be within 1 percentage point of the population distribution we would need a sample of size $N = (1.63/.01)^2 = 26{,}569$.

This procedure can also be used as a test for goodness of fit. If we have a hypothetical cumulative curve that is completely inside the 95 per cent–confidence strip for the sample drawn we would accept the hypothetical curve at the 5 per cent level of significance; if the hypothetical curve is outside the strip at one or more points we would reject this curve as describing the population.

A similar procedure can be used to test the hypothesis that two samples are from populations having the same distribution. If the maximum vertical difference between the two sample cumulative distributions is greater than $1.36\sqrt{(1/N_1) + (1/N_2)}$, we would reject the hypothesis at the 5 per cent level of significance.

No complete investigation has been made of the power function of these tests. However, preliminary investigations indicate that in many examples it is as good as or better than corresponding $\chi^2$ tests.

## CHEBYSHEV'S INEQUALITY 17-6

The following inequality is not strictly nonparametric, since it does involve $\mu$ and $\sigma^2$. However, it is distribution-free in the sense that it holds for any distribution. *Chebyshev's inequality* states that the amount of area under any distribution curve which is farther away from the mean than $k$ standard deviation units is less than $1/k^2$. Thus, for example, there is always less than $(\frac{1}{2})^2 = .25$ unit of area farther than two standard deviations from the mean. In case the form of the distribution is known, more exact statements can be made. In a normal distribution there is approximately .046 unit of area farther than two standard deviations from the mean.

**Table 17-5** Tail areas beyond $\mu - k\sigma$ and $\mu + k\sigma$ for several distributions

| | $t$ DISTRIBUTION | | | $\chi^2$ | | | NORMAL | CHEBYSHEV MAX. TAIL AREA | |
|---|---|---|---|---|---|---|---|---|---|
| $k$ | $\nu = 3$ | $\nu = 4$ | $\nu = 10$ | $\nu = 1$ | $\nu = 2$ | $\nu = 10$ | | *Any* | *Symmetric* |
| 1 | .182 | .232 | .290 | .121 | .135 | .297* | .317 | 1.000 | .444 |
| 2 | .020 | .048 | .050 | .050 | .050 | .041* | .046 | .250 | .111 |
| 3 | .015 | .013 | .007 | .022 | .018 | .009 | .003 | .111 | .049 |
| 4 | .0064 | .0046 | .0011 | .0099 | .0067 | .0019 | .00006 | .063 | .028 |

For symmetrical distributions with a single mode the maximum area farther than $k\sigma$ from the mean is $\frac{4}{9}/k^2$.

A comparison of the actual and Chebyshev-maximum area or probability in the tails beyond $\mu - k\sigma$ and $\mu + k\sigma$ is given for a number of distributions in Table 17-5. If the number $\nu$ of degrees of freedom is greater than 2, the $t$ distribution has mean 0 and variance $\nu/(\nu - 2)$. The $\chi^2$ distribution with $\nu$ degrees of freedom has mean $\nu$ and variance $2\nu$. For small values of $\nu$ the $\chi^2$ distribution is so skewed that the mean is less than one standard deviation above zero, and consequently, for this distribution only the values marked with an asterisk actually represent two-tail areas.

**Law of large numbers.** Chebyshev's inequality can be applied to the sampling distribution of the mean. This distribution has standard deviation $\sigma/\sqrt{N}$. Therefore the chance that a sample mean from any distribution having variance $\sigma^2$ will fall farther than $k\sigma/\sqrt{N}$ from the population mean is less than $1/k^2$. By letting $d = k\sigma/\sqrt{N}$ we can say that the chance of the sample mean falling farther than $d$ units from the population mean is less than $\sigma^2/(d^2N)$. If $N$ is increased this maximum chance approaches zero. This result is called the *law of large numbers.*

EXAMPLE 17-6. How large should $N$ be for 95 per cent confidence that $\bar{X}$ will not fall farther than $\sigma/2$ from the mean? Here $d = .5\sigma$, and the chance that $\bar{X}$ is farther than $\sigma/2$ from the mean is less than

$$\frac{\sigma^2}{.25\sigma^2 N} = \frac{1}{.25N}$$

This should be equal to .05, so we set $1/(.25N) = .05$ and obtain

$$N = \frac{1}{.25 \times .05} = 80$$

Thus the sample size should be at least 80.

Of course, knowing that the sampling distribution of $\bar{X}$ is approximately normal would allow a sample size of only $N = 16$ to be used.

## 17-7 NONPARAMETRIC TOLERANCE LIMITS

A method for constructing intervals which had a specified chance of covering a certain proportion of a normal population was given in Chap. 9. Such intervals were called *tolerance intervals.* It is possible to form tolerance intervals which are valid for any population. Generally they are somewhat less sensitive (longer) than those for a particular distribution, and it is usually necessary to take larger samples in order to obtain the same precision.

The range between the smallest and largest observations of a random sample can be expected to include a certain proportion of a population. Table A-25*b* provides probabilities $\gamma$ of including a proportion $P$ for a number of sample sizes. For example, with a sample of size $N = 100$ there is $\gamma = .90$ probability of including $P = 96$ per cent of the population between the sample extremes. The interval is called a tolerance interval and $\gamma$ is the chance the assertion is correct. Thus $\gamma$, called the *tolerance level*, corresponds to a level of confidence, except that the location of a portion of the population is being estimated rather than the value of a parameter. Table A-25*c* gives information as in Table A-25*b*, except that the table is entered with the proportion desired. For example, with a sample of $N = 20$ the probability of including $P = 70$ per cent of the population is $\gamma = .992$. Required sample sizes can be read from either of these tables and are also presented in Table A-25*d*. We read, for example, that for a $\gamma = 90$ per cent chance of including $P = 99$ per cent of the population between the sample extremes a sample of size $N = 388$ is required.

In quality inspection it is sometimes desired to guard against or to detect individuals at only one extreme of the observational scale. Table A-25*e* gives sample sizes corresponding to the probability $\gamma$ of finding a proportion $P$ of the population at a specified end of the distribution; thus the single extreme provides one-sided tolerance limits. For example, for probability $\gamma = .50$ that $P = 99$ per cent of the population is larger than the smallest sample observation a sample of size $N = 69$ is required.

## CONFIDENCE LIMITS FOR THE MEDIAN 17-8

A confidence interval for the median of the population can be obtained from the observations of a continuous variable of a random sample from any population. We arrange the observations in order $X_1 < X_2 < \cdots < X_N$ and state that the population median is between two of these observations. For example, for a sample of size 20 we might state that the median is between $X_1$ and $X_{20}$ or between $X_2$ and $X_{19}$. As we progress toward the central values of the sample, we are less confident that our statement is true. Table A-25*a* gives the largest values of $k$ for confidence of more than .95 (or more than .99) that the population median is between $X_k$ and $X_{N+1-k}$. The values of $\alpha$ listed are the exact chances of the statement being false. For example, if $N = 20$, we see from Table A-25*a* that the sixth and fifteenth observations are confidence limits of more than 95 per cent for the population median. For $N = 40$ we can be 99.4 per cent confident that the median is between the 12th and 29th observations and 96.2 per cent confident that the median is between the 14th and 27th observations.

## RANK CORRELATION 17-9

Suppose we have a sample of individuals and we make two measurements on each. We have then $N$ pairs of observations $(X_1,Y_1)$, $(X_2,Y_2)$,

. . . , $(X_N, Y_N)$. We arrange the observed $X$ values in order of size and assign a rank to each value; the largest value is assigned a rank of 1, the second largest a rank of 2, and so on. We then do the same for the $Y$ values. Next we subtract each $X_i$ rank from its paired $Y_i$ rank, and denote the difference by $d_i$. The statistic

$$r_s = 1 - \frac{6(\Sigma d_i^2)}{N(N^2 - 1)}$$

is called the *rank-correlation coefficient.* It can be used to test the hypothesis that the two variables $X$ and $Y$ are independent. It has the advantage that no assumptions are made about the distributions of $X$ or $Y$. $r_s$ is similar to the correlation coefficient in that its values range from $-1$ to 1. A value of 1 indicates perfect agreement, while a value of $-1$ indicates exactly opposite ranking. Table A-30*c* gives the distribution of $\Sigma d_i^2$ for samples of size less than or equal to 10. These values are such that $\Sigma d_i^2$ computed for a sample drawn from a population of independent variables $X$ and $Y$ will be less than or equal to the tabulated value with probability $\alpha$ as given in the table. The values of the rank-correlation coefficient $r_s$ are given for each $\Sigma d_i^2$, so that Table A-30*c* also records the sampling distribution of $r_s$. For example, in a sample of $N = 10$ observations the probability that $\Sigma d_i^2$ will be less than or equal to 40 is .007, and this is also the probability that $r_s$ will be greater than or equal to .76.

The tables may be used for two-sided tests. For example, if, with $N = 10$, we agree to reject the independence hypothesis if $\Sigma d_i^2$ is less than or equal to 40 or greater than 288, the test has level of significance equal to $.007 + (1 - .993) = .014$. This test is equivalent to rejecting the hypothesis if $r_s$ is less than $-.75$ or greater than or equal to .76.

For values of $N$ larger than those in the table approximate percentiles of the sampling distribution of $r_s$ can be read from Table A-27 ($\rho = 0$) or from Table A-30*a*. For example, with $N = 12$, rejection for $r_s$ greater than .497 provides a one-sided test with level of significance $\alpha = .05$. Correspondingly, rejection for $r_s$ greater than .497 or less than $-.497$ provides a two-sided test, with $\alpha = .10$.

**Table 17-6**

| | *First judge* | *Second judge* | $d_i$ |
|---|---|---|---|
| *A* | 2 | 3 | $-1$ |
| *B* | 1 | 4 | $-3$ |
| *C* | 4 | 2 | 2 |
| *D* | 5 | 5 | 0 |
| *E* | 3 | 1 | 2 |
| *F* | 7 | 6 | 1 |
| *G* | 6 | 7 | $-1$ |

$$\Sigma d_i^2 = 20$$

$$r_s = 1 - \frac{6 \times 20}{7 \times 48} = .643$$

In comparing rankings we are frequently interested in a one-sided test, rejecting the hypothesis of independence only when there is sufficiently close agreement in the rankings.

EXAMPLE 17-7. Suppose two judges rank seven contestants in a beauty contest as in Table 17-6. We do not reject the hypothesis of independence in the ranking at the 5 per cent level of significance, since $\Sigma d_i^2 = 20$ is not less than or equal to 16, the tabular value corresponding to the first available level of significance, less than .05.

## MEDIAN TESTS 17-10

The tests to be described in this section are particularly useful in situations where percentiles of a group of observations can be determined easily. In fact, if relative dimensions can be assigned, we may even avoid the actual measurement of more than a few of the individuals.

**Median test for two samples.** The number of cases in two samples of sizes $N_1$ and $N_2$ falling above and below the median of the combined $N = N_1 + N_2$ observations can be used to test the hypothesis that the samples are randomly drawn from two identically distributed populations.

EXAMPLE 17-8. Table 17-7 gives the results of two samples of sizes $N_1 = N_2 = 15$ with the number of individuals falling above and below the median for each sample. The table is analyzed as a contingency table (Sec. 13-3), and the hypothesis is rejected if the observed $\chi^2$ is larger than the critical value read from Table A-6*a* for 1 degree of freedom. For this example the value of $\chi^2$ corrected for continuity is

$$\chi^2 = \frac{(|6^2 - 9^2| - 15)^2 \times 30}{15 \times 15 \times 15 \times 15} = .53$$

The critical region for $\alpha = .05$ is $\chi^2 > 3.84$, so we have no reason to reject the hypothesis at this level.

**Median test for *k* samples.** The hypothesis that $k$ samples are randomly drawn from populations having identical distributions can be tested in a manner similar to that of the preceding section. For this case the data are arranged in a table showing the numbers of each sample above

**Table 17-7**

| | *Sample* I | *Sample* II | *Total* |
|---|---|---|---|
| Above median | 6 | 9 | 15 |
| Below median | 9 | 6 | 15 |
| *Total* | 15 | 15 | 30 |

$\chi^2 = .53$

$\chi_{.95}^2\,(1) = 3.84$

and below the median of the combined set. The table is then analyzed as a contingency table to obtain a $\chi^2$ statistic having $k - 1$ degrees of freedom. The hypothesis is rejected if the observed $\chi^2$ is significantly large.

**Extension of the median test.** The median test may be extended by using any fixed number of percentiles of the grouped data instead of the median only. The numbers in each sample falling between those percentiles are recorded in a table, and the data are then analyzed as a contingency table. The hypothesis that the $k$ samples are randomly drawn from populations having identical distributions is rejected if the observed $\chi^2$ is significantly large. For this problem the number of degrees of freedom is $(k - 1)(r - 1)$, where $r$ is the number of categories used.

EXAMPLE 17-9. Table 17-8 records for three samples ($k = 3$) of size 20 each the number of observations above $P_{75}$, between $P_{50}$ and $P_{75}$, between $P_{25}$ and $P_{50}$, and below $P_{25}$. The $\chi^2$ statistic has $2 \times 3 = 6$ degrees of freedom. The theoretical frequencies for the analysis of this example are all equal to $15 \times 20/60 = 5$.

**Table 17-8**

| | *Sample* I | *Sample* II | *Sample* III | *Total* |
|---|---|---|---|---|
| Above $P_{75}$ | 5 | 7 | 3 | 15 |
| Between $P_{50}$ and $P_{75}$ | 3 | 3 | 9 | 15 |
| Between $P_{25}$ and $P_{50}$ | 4 | 7 | 4 | 15 |
| Below $P_{25}$ | 8 | 3 | 4 | 15 |
| *Total* | 20 | 20 | 20 | 60 |

The value of $\chi^2$ is

$$\chi^2 = \frac{(5 - 5)^2}{5} + \frac{(7 - 5)^2}{5} + \cdots + \frac{(4 - 5)^2}{5} = \frac{52}{5} = 10.4$$

which, since $\chi_{.95}^2(6) = 11.07$, is not cause for rejection of the hypothesis at the 5 per cent level of significance.

## 17-11 CORNER TEST FOR ASSOCIATION

The technique described in this section uses a very easily computed nonparametric statistic to test the hypothesis that two continuous variables are independent. It is especially useful if the data are plotted on a scatter diagram, as in Fig. 17-2, or if the data can be easily ranked.

EXAMPLE 17-10. Figure 17-2 shows a scatter diagram of scores $X$ and $Y$ made by $N = 50$ people on two types of aptitude tests. The corner-test statistic $S$ is obtained as follows. Label the quadrants with plus and minus

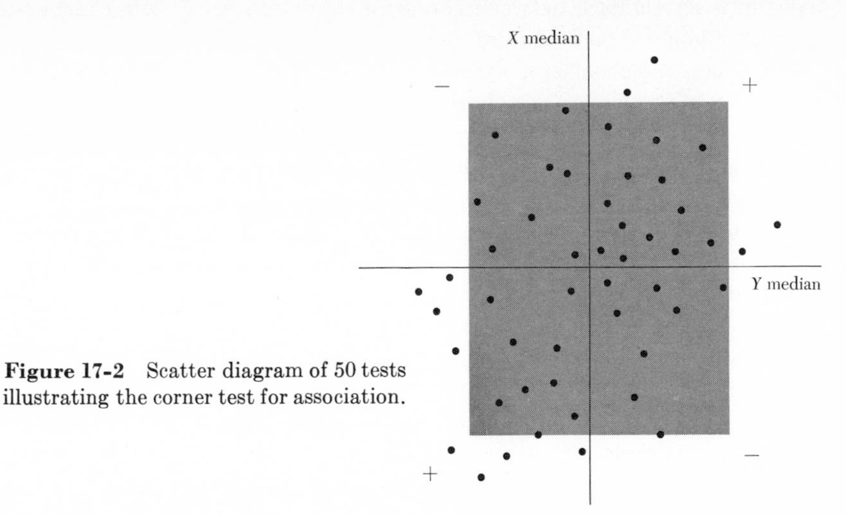

**Figure 17-2** Scatter diagram of 50 tests illustrating the corner test for association.

signs, as in the figure. Locate the median values of $X$ and $Y$. Starting at the extreme right and moving toward the center, count the number of points encountered before a point is found across the $Y$ median line and give this total the sign of the quadrant in which the extreme points occur. In the figure this is $+2$. Proceed similarly from the top, left, and bottom, obtaining 2, 5, and $4\frac{1}{2}$ (note that the point at the lower left is counted twice and that the $\frac{1}{2}$ comes from a tie). The statistic $S$ then is the sum of these four numbers:

$$S = 2 + 2 + 5 + 4\tfrac{1}{2} = 13\tfrac{1}{2}$$

Table A-30*d* gives the sampling distribution of $S$ in the case where $X$ and $Y$ are independent. We see that $S = 13\frac{1}{2}$ is a significant departure from independence at the $\alpha = .05$ level of significance.

If the sample size $N$ is odd the two median points can be replaced by a single point having as its coordinates the nonmedian values of the two. For example, if the individual (5,7) is median for $X$ (median $X$ value equals 5) and (12,6) is median for $Y$, these points are replaced by (12,7). The test may then be applied to the sample thus reduced by one observation.

## TESTS OF RANDOMNESS 17-12

In the test and estimation techniques we have discussed, the method of drawing the sample from the population was in most cases specified to be random. If some characteristic of the population is known the sample may be checked for the corresponding characteristic in order to test the hypothesis that it was randomly drawn. Actually, a test of any hypothesis can be considered to be a test of randomness if the hypothesis is known to be correct. However, most of the test techniques were designed for specific types of discrepancies between the sample and the population rather than tests for a general lack of randomness.

The control chart is a widely used technique for examining a series of observations to determine whether or not it could be considered to be a random series. In particular, it is effective in detecting extreme values. Other techniques are given below.

EXAMPLE 17-11. If a population of people is known to contain 45 per cent men and 55 per cent women, then the proportion $\bar{X}$ of men in a sample can be used to test the hypothesis that the sample was drawn at random. Either the binomial distribution or its normal approximation with $z = (\bar{X} - .45)/\sqrt{.45 \times .55/N}$ is used and the hypothesis rejected if $z$ is very large, either positive or negative. In this case it may also be considered reasonable to reject the hypothesis if $z$ is close to zero, since it may be argued that the agreement is so close that it would occur rarely in a random sample.

If items are being manufactured in sequence, the number of runs above and below the median can be used to test for randomness, as discussed in Sec. 17-3.

If observations are chosen in a sequence $X_1, X_2, \ldots, X_N$, the *mean-square successive difference*

$$\eta = \frac{\sum_{i=1}^{N-1} (X_{i+1} - X_i)^2}{\Sigma(X_i - \bar{X})^2}$$

can be used to test for randomness. If $N$ is moderately large, say, greater than 20, and the population is normal, then

$$z = \frac{1 - \eta/2}{\sqrt{\dfrac{N-2}{(N-1)(N+1)}}}$$

is approximately normally distributed with mean 0 and unit standard deviation. Long trends are associated with high positive values of $z$ and short oscillations with high negative values.

## 17-13 REMARKS ON NONPARAMETRIC METHODS

When should nonparametric tests be used instead of the more classical tests? If a nonparametric test is to be used, which one should be used? For example, when should the sign test be used as an alternative to the $t$ test, or when should the $H$ statistic be used as an alternative to the one variable analysis of variance? As alternatives to the $t$ test for two samples, when should the rank-sum statistic be used and when should the median test be used?

If the basic assumptions on the population are incorrect any inferences based on these assumptions are of dubious validity. Specifically, the numbers read from the percentile tables for these statistics will not provide the appropriate level of significance in testing hypotheses or the appropriate level of confidence in estimation. It is difficult to say how much the levels change if the assumptions are not met. However, for a moderate departure from normality the $t$ and $F$ tables are sufficiently close for most practical purposes. If it is believed that the assumptions are not satisfied even moderately well, the levels of significance and confidence will be better satisfied with nonparametric methods.

The level of significance for a nonparametric method is not affected by the population distribution. Therefore the experimenter can be sure that he knows the chance of rejecting the null hypothesis when it is true. However, we know that some tests are better than others in detecting certain false hypotheses. If the necessary assumptions are correct the classical tests are in most cases the most powerful ones and are to be preferred. Some comparisons of classical and nonparametric tests have been made in cases where the classical methods are known to be best. Some of these were noted previously. For example, for normal populations the sign test requires 4 to 50 per cent more observations than the $t$ test, and the rank-sum test requires approximately 5 per cent more observations than the $t$ test to be equally powerful. Of course, these figures apply only if the populations are normal. The comparison for nonnormal populations may even be reversed.

Comparisons among the various nonparametric tests have been made for samples of several sizes from two normal populations in an attempt to assess their relative efficiency. Some results for samples of size 5 are given in Table 17-9 for the case where the normal populations have equal variances and a difference in means $\delta = |\mu_1 - \mu_2|/\sigma$. The power efficiencies are relative to the $t$ test.

Further comparisons for samples of size 10 and 20 have been made by sampling experiments. The tests listed in order of their observed power to reject the hypothesis $\delta = 0$ for samples of size 5, 10, and 20 are (1) rank sum, (2) maximum absolute deviation, (3) median, and (4) run. The power of the rank-sum test is little less than the power of $t$. The power efficiency of the run test is less than .50. The rank-sum test is almost completely

**Table 17-9** Power efficiency of several nonparametric tests for $\alpha = .025$ and $N_1 = N_2 = 5$

| | DIFFERENCE $\delta$ | | | | |
|---|---|---|---|---|---|
| | *Near* 0 | 1 | 2 | 3 | 4 |
| Rank-sum test | .96 | .95 | .93 | .91 | .89 |
| Maximum absolute deviation | .81 | .80 | .78 | .76 | .74 |
| Median | .70 | .70 | .71 | .73 | .73 |

insensitive to differences in standard deviations of two normal populations with the same mean. The median test is insensitive to changes in standard deviation. The maximum absolute deviation has some power against this alternative, but not so much as the run test, which has close to 50 per cent power efficiency compared with the $F$ test in detecting differences in standard deviations.

The various nonparametric statistics may be used to estimate parameters. As noted, the sign-test statistic may be used to estimate the population median. Distribution-free estimation can also be based on the material of Secs. 17-5, 17-7, and 17-8.

Following are some additional comments on the use of nonparametric methods:

1. In some cases nonparametric methods of analysis permit an easier method of collecting data. For example, no actual measurements may be needed to use the sign test because we need to know only whether one observation is "better" or "worse" than another.

2. Nonparametric statistics for small samples are usually easier to compute and apply than the classical techniques.

3. In some cases data may be collected from several populations about which very little is assumed. It may be possible to use a nonparametric method of analysis.

4. Sometimes the data occur naturally in the form of ranks. We may use the nonparametric methods directly and need not make assumptions about the population distribution.

5. The classical methods may be easier to apply in cases where there is no automatic sorting machinery to rank data for large samples.

6. If the form of the distribution of the population is known to agree with the assumptions of the classical methods or can be made to agree by means of transformations, the nonparametric test may be wasteful of information.

7. Some tests detect differences in population means but not other differences. For example, the rank-sum tests and $t$ tests are designed to detect shifts in means but not in variances. Similarly, an $F$ test for variances will detect a difference in variances but is ineffectual in discovering differences in means.

8. In any problem the cost of collecting and analyzing data must be considered in the choice of a statistical method. Different amounts of effort or cost are required to collect data in the appropriate form for different techniques of analysis.

## GLOSSARY

Chebyshev's inequality
confidence limits for
  cumulative-distribution functions
law of large numbers
nonparametric inference
rank correlation
runs
sign test
tolerance limits

## DISCUSSION QUESTIONS

**1.** What situations favor the use of the sign test instead of a $t$ test?

**2.** What concept is considered in the study of runs which is not considered in a sign test or in a $t$ test?

**3.** What concept is considered in the use of a $t$ test which is not considered in the sign test or in a run test?

**4.** Describe favorable and unfavorable aspects in the application of Chebyshev's inequality.

**5.** In what type of situation is rank correlation a better measure than the correlation coefficient?

**6.** Define each term in the Glossary.

**7.** How many samples would be needed to have a 99 per cent confidence belt for the cumulative distribution of $\bar{X}$ which would be 1 percentage point wide? The cumulative distribution which would be .1 percentage point wide?

**8.** What is the difference between confidence limits and tolerance limits?

## CLASS EXERCISES

**1.** For each of the samples of size $N = 5$ drawn from Table A-2 for the Class Exercise in Chap. 3, count the number of runs of elements above and below the median. Collect the results of all the students and test the hypothesis that the values in Table A-11 are correct.

**2.** Make up some distributions of 10 items and verify that Chebyshev's inequality holds for each. These distributions are to be considered as populations and the variance computed by dividing by 10.

**3.** Take a shuffled deck of playing cards, count the number of runs of red and black cards and decide on this basis whether or not the pack of cards was thoroughly shuffled.

**4.** Draw on a sheet of tracing paper a cumulative-percentage normal curve with mean 0 and unit variance. Above and below this curve draw two parallel curves, one 41 percentage points above and the other 41 percentage points below the original curve. From a normal population having $\mu = 0$ and $\sigma = 1$ draw a sample of $N = 10$ observations. Draw the cumulative distribution of this sample and see whether or not it falls in the strip. This can be done easily by drawing the distribution on graph paper and superimposing the curves.

Each student should repeat this 10 times, and the total number of samples which have distributions completely in the strip should be counted. Approximately 95 per cent of the distributions should be entirely within the strip. The normal numbers (Table A-2) can be used as a population, or a population of numbered tags can be used.

As an alternative exercise the uniform population may be used instead of a normal population. In a uniform distribution every number should have an equal chance of being chosen in the sample. Samples can be drawn from a table of random numbers (Table A-1), or a population of tags can be constructed as described for set 6 in Table A-22.

**5.** In samples of size 10 the chance of including 66 per cent of the population between $X_1$ and $X_{10}$ is .90 (obtained from Table A-25*b*). From the samples drawn in Exercise 4 (or other samples from Table A-2) count the number of cases in which $X_1$ and $X_{10}$ include 66 per cent of the area under the normal curve.

## PROBLEMS

*Section* 17-1

**1.** Use the sign test to analyze the data in the table giving the gains of 10 pairs of rats, half of which received their protein from raw peanuts, while the other half received their protein from roasted peanuts. Test to see whether or not roasting the peanuts had any effect on their protein value. Compare the results with the results of a $t$ test and explain the difference.

| *Raw* | 61 | 60 | 56 | 63 | 56 | 63 | 59 | 56 | 44 | 61 |
|---|---|---|---|---|---|---|---|---|---|---|
| *Roasted* | 55 | 54 | 47 | 59 | 51 | 61 | 57 | 54 | 62 | 58 |

**2.** A teataster claims that he can distinguish between two brands of tea by taste. In an experiment he correctly identified the brands 14 times out of 20 trials in the following order,

$$+ + - - + + + + + + - - - + + + + + + -$$

where + stands for a correct identification. Analyze the data by using the sign test.

**3.** An experiment used mice from two genetic strains and three environments differing in temperature. Three pairs of brothers were taken from each of the six cells formed by these strata. In each pair one mouse received diet I and the other diet II. The weight gains in 2 weeks are as recorded:

| | | STRAIN 1 | | STRAIN 2 | |
|---|---|---|---|---|---|
| | | I | II | I | II |
| TEMPERATURE | 1 | 15 | 16 | 14 | 15 |
| | | 14 | 15 | 14 | 16 |
| | | 16 | 17 | 15 | 18 |
| | 2 | 19 | 18 | 17 | 16 |
| | | 20 | 18 | 16 | 14 |
| | | 22 | 19 | 17 | 15 |
| | 3 | 16 | 15 | 13 | 14 |
| | | 16 | 14 | 12 | 11 |
| | | 15 | 16 | 13 | 12 |

(*a*) For each genetic strain analyze the data by use of the sign test. (*b*) For each temperature environment analyze the data by use of the sign test. (*c*) For all cells analyze the data by use of the sign test.

For each analysis state the hypothesis and the conclusion.

**4.** Measurements on blood pressure before and after taking a drug showed, for a total of 20 men, 10 increases, 5 decreases, and 5 no change. Analyze by means of a sign test.

**5.** Two types of fertilizer used on six pairs of plants at three experiment stations had resulting measurements as follows:

| STATION | | TYPE 1 | TYPE 2 |
|---|---|---|---|
| | 1 | 7.1 | 7.5 |
| | | 8.2 | 8.0 |
| | | 6.0 | 6.2 |
| | | 7.1 | 7.0 |
| | | 8.1 | 7.8 |
| | | 8.1 | 6.0 |
| | 2 | 6.9 | 7.5 |
| | | 7.8 | 7.7 |
| | | 9.2 | 8.1 |
| | | 9.4 | 7.6 |
| | | 9.4 | 7.8 |
| | | 9.8 | 9.0 |
| | 3 | 5.1 | 7.0 |
| | | 5.2 | 7.1 |
| | | 5.8 | 6.8 |
| | | 5.6 | 6.4 |
| | | 6.2 | 6.0 |
| | | 6.4 | 6.6 |

Analyze with a sign test.

**6.** (*a*) Pair the numbers in column 1 with the numbers in column 2 of Table A-1 and perform a sign test with the first 50 pairs of numbers. What hypothesis is being tested? (*b*) Do the same for the first two columns of Table A-2. (*c*) Pair the numbers in the first column of Table A-2 with those in the first column of Table A-23 and perform a sign test. (*d*) Do the same with the first column of Table A-23 and the first column of Table A-24.

*Section* 17-2

**7.** Analyze the data in Prob. 1 with a signed-rank test.

*Section* 17-3

**8.** Consider the data in Prob. 1 to be two random samples and use the number of runs to test the hypothesis that raw or roasted peanuts resulted in the same distribution of weight gains.

**9.** Each day a sample of 10 production items was taken and the mean weight computed. Following are the first 20 daily means:

| | | | | | | | | | |
|---|---|---|---|---|---|---|---|---|---|
| 13.0 | 12.8 | 12.9 | 13.0 | 13.1 | 12.9 | 12.6 | 12.6 | 12.7 | 12.9 |
| 13.1 | 13.1 | 13.2 | 13.3 | 13.2 | 13.1 | 12.9 | 13.2 | 13.3 | 13.2 |

Are the number of runs below and above the median significant at the 5 per cent level?

**10.** In the manufacture of automobile gears the following data were obtained for the daily number defective for a production of 100 parts per day:

| | | | | | | | |
|---|---|---|---|---|---|---|---|
| 22 | 25 | 15 | 26 | 31 | 22 | 17 | 26 |
| 23 | 20 | 28 | 32 | 43 | 18 | 16 | 36 |
| 21 | 16 | 29 | 26 | 18 | 24 | 28 | 42 |
| 17 | 14 | 26 | 33 | 26 | 24 | 32 | 36 |
| 38 | 26 | 25 | 30 | 21 | 16 | 18 | 34 |

Is the number of runs above or below the median significant at the 5 per cent level?

*Section* 17-4

**11.** Five samples of each of two types of paint are scored as follows:

| | | | | | |
|---|---|---|---|---|---|
| *Paint* I | 85 | 87 | 92 | 80 | 84 |
| *Paint* II | 89 | 89 | 90 | 84 | 88 |

Analyze this with the rank-sum test.

**12.** Analyze the data in Prob. 9, Chap. 10, by means of the $H$ statistic.

*Section* 17-5

**13.** For what size random sample will the probability be .99 that the sample cumulative distribution agrees within $\pm.002$ with the population cumulative distribution?

**14.** Using the 40 observations in Prob. 10 as a random sample, draw the cumulative distribution and give a 95 per cent confidence band for the population cumulative distribution.

**15.** Using the data in Prob. 9 as a random sample, draw a cumulative distribution and give a 90 per cent confidence band for the population cumulative distribution.

**16.** Consider the first 20 observations in Prob. 10 as one random sample and the second 20 as another. Draw the two cumulative-distribution curves. Is the maximum vertical distance between them significant at the 5 per cent level?

*Section* 17-6

**17.** If it is known that a distribution is symmetrical and has only one mode, how large a sample is needed for the chance that $\bar{X}$ will be within $\sigma/10$ of the population mean $\mu$ to be at least .99? Do this by two methods, one with Chebyshev's inequality and the other with the assumption that the distribution of $\bar{X}$ is normal.

*Section* 17-7

**18.** How large a sample is needed for a 90 per cent chance of including 95 per cent of the population between the extremes?

**19.** With the data in Prob. 10, what chance is there that the range includes 90 per cent of the population?

*Section* 17-8

**20.** Estimate with a 95 per cent confidence interval the population median from the raw-peanut data in Prob. 1.

**21.** Estimate with a 95 per cent confidence interval the population median from the data on gears in Prob. 10.

*Section* 17-9

**22.** A board of judges at a county fair ranked a group of 10 pigs. A 4-H club member also ranked the 10 pigs. The results were as in the table. Is there a lack of independence in these rankings?

| *Judges* | 9 | 4 | 3 | 7 | 2 | 1 | 5 | 8 | 10 | 6 |
|---|---|---|---|---|---|---|---|---|---|---|
| 4-H *member* | 7 | 6 | 4 | 9 | 2 | 3 | 8 | 5 | 10 | 1 |

**23.** Analyze the data in Prob. 5, Chap. 11, by means of the rank-correlation coefficient.

*Section* 17-10

**24.** Analyze the data in Prob. 1 with the median test.

**25.** Analyze the data in Prob. 15, Chap. 8, with the median test.

**26.** Analyze the data in Prob. 1, Chap. 10, with the median test by dividing the data into approximately four equal parts as in Example 17-9.

CHAPTER EIGHTEEN

# Sequential analysis

If in testing a hypothesis we wish to decide in advance what risks we are willing to take of rejecting a true hypothesis or of accepting a particular false hypothesis, our sample size is determined. This implies, however, that we are going to take a sample of this size no matter what results we obtain from our first few observations. It would seem reasonable to request a procedure which will not require more observations than are necessary to make a decision. *Sequential analysis* is a procedure which leads to a statistical inference and in which the number of observations to be made is not determined before the experiment is begun. The procedure indicates when sufficient observations have been gathered to make our decisions with the risks we have chosen. On the average, fewer observations will be required by this procedure, and its use will not increase the risks $\alpha$ and $\beta$. For some problems only half the number of observations will be required on the average for the sequential procedure in comparison with the number required if the sample size is fixed in advance.

## 18-1 SEQUENTIAL TEST

In this procedure of testing a hypothesis observations are taken one at a time, and after every observation we decide to do one of the following three things:

1. Accept the hypothesis
2. Reject the hypothesis
3. Make an additional observation

In order to determine which one of these three possible actions to take we must determine the critical region for each sample size. To do this we compute $p_{0m}$, the probability that the $m$ observations collected thus far would occur if our hypothesis $H_0$ were true, and $p_{1m}$, the probability that these observations would occur if some alternative statement $H_1$ were true. To find $p_{0m}$ we assume that sampling is from the population stated in $H_0$ and compute the probability that such a result would occur. Similarly, to find $p_{1m}$ we assume that sampling is from the population stated in $H_1$ and again compute the probability that such a result would occur.

When $p_{0m}$ is much larger than $p_{1m}$ we shall accept $H_0$; when $p_{1m}$ is much larger than $p_{0m}$ we shall accept $H_1$. If there is not much difference between $p_{1m}$ and $p_{0m}$ we shall take another observation. We shall compare the two by the use of the ratio $p_{1m}/p_{0m}$. It can be shown mathematically that for the given risks $\alpha$ of rejecting $H_0$ when it is true and $\beta$ of accepting $H_0$ when $H_1$ is true the characterization of the sequential test may be written as follows:

1. If $\dfrac{p_{1m}}{p_{0m}} \leq \dfrac{\beta}{1-\alpha}$, accept $H_0$.
2. If $\dfrac{p_{1m}}{p_{0m}} \geq \dfrac{1-\beta}{\alpha}$, accept $H_1$.
3. If $\dfrac{\beta}{1-\alpha} < \dfrac{p_{1m}}{p_{0m}} < \dfrac{1-\beta}{\alpha}$, take another observation.

This procedure is continued until either condition 1 or condition 2 is satisfied.

The mathematical development of these inequalities involves certain approximations, but an experimenter following the above procedure can be sure that the sum of the risks $\alpha$ and $\beta$ will not be increased and the increase in the number of observations caused by a possible decrease in these risks will be slight (say, a maximum of 3 when 30 observations are required and 5 when 100 are required when $\alpha = \beta = .05$).

## TEST FOR PROPORTION 18-2

The procedure to be developed here is applicable to any problem concerning a test of proportion. Such types of problems were discussed in Chaps. 7, 13, and 14. The sequential test will be illustrated by an application to acceptance inspection of a large lot of items, each of which is either good or bad. Suppose we wish to reject the lot only 1 per cent of the time if the proportion of defective parts is .10 and accept the lot only 5 per cent of the time if the proportion of defectives is .20. Of course, this also implies that if the proportion were less than .10 we would be rejecting even less frequently than 1 per cent of the time, and if the proportion were greater than .20 we would be accepting less frequently than 5 per cent.

$H_0: p = p_0 = .10$
$H_1: p = p_1 = .20$
$\alpha = .01$
$\beta = .05$

Suppose we select items at random from the entire lot. If we assume that $p_1$ is the proportion of defectives in the whole lot, the probability that we get, say, $d_m$ defectives and $g_m$ good items in some particular order among the first $m$ observations is

$$p_{1m} = p_1^{d_m}(1 - p_1)^{g_m} \qquad \text{with } d_m + g_m = m \tag{1}$$

If $p_0$ is the proportion of defectives in the lot, the probability that we get $d_m$ defectives and $g_m$ good items among the first $m$ observations is

$$p_{0m} = p_0^{d_m}(1 - p_0)^{g_m} \tag{2}$$

For example, if we obtain one good, one bad, and one good item for the first three observations, then, under the assumption that $p_1$ is the proportion of times we shall get a defective item, the probability of this result is

$$(1 - p_1)p_1(1 - p_1) = p_1^1(1 - p_1)^2$$

and if the assumed proportion is $p_0$, the probability is

$$(1 - p_0)p_0(1 - p_0) = p_0^1(1 - p_0)^2$$

The ratio $p_{1m}/p_{0m}$ after obtaining one good observation is

$$\frac{1 - p_1}{1 - p_0} = \frac{1 - .2}{1 - .1} = .889$$

and after obtaining one good and one bad is

$$\frac{(1 - p_1)p_1}{(1 - p_0)p_0} = \frac{.8}{.9}\frac{.2}{.1} = 1.778$$

and after obtaining one good, one bad, and one good is

$$\frac{p_1^1(1 - p_1)^2}{p_0^1(1 - p_0)^2} = \left(\frac{p_1}{p_0}\right)^1\left(\frac{1 - p_1}{1 - p_0}\right)^2 = \frac{.2}{.1}\left(\frac{.8}{.9}\right)^2 = 1.58$$

Successive observations will continue to change this value, and if it at any time exceeds

$$\frac{1 - \beta}{\alpha} = \frac{1 - .05}{.01} = 95$$

or is less than

$$\frac{\beta}{1-\alpha} = \frac{.05}{.99} = .0505$$

we shall stop sampling and accept either $H_1$: $p = .2$ or $H_0$: $p = .1$. If $H_1$ is accepted we reject the lot, and if $H_0$ is accepted we accept the lot.

The quantities actually observed are the number of defective and good parts. The ratio

$$\frac{p_{1m}}{p_{0m}} = \left(\frac{p_1}{p_0}\right)^{d_m} \left(\frac{1-p_1}{1-p_0}\right)^{g_m} \tag{3}$$

can be transformed along with the values $(1-\beta)/\alpha$ and $\beta/(1-\alpha)$ to natural logarithms:

$$\ln \frac{p_{1m}}{p_{0m}} = d_m \ln \frac{p_1}{p_0} + g_m \ln \frac{1-p_1}{1-p_0} \tag{4}$$

The critical values may be obtained by setting

$$d_m \ln \frac{p_1}{p_0} + g_m \ln \frac{1-p_1}{1-p_0} = \ln \frac{1-\beta}{\alpha}$$

and

$$d_m \ln \frac{p_1}{p_0} + g_m \ln \frac{1-p_1}{1-p_0} = \ln \frac{\beta}{1-\alpha}$$

For our problem we write

$$d_m \ln \frac{.2}{.1} + g_m \ln \frac{.8}{.9} = \ln 95$$

and

$$d_m \ln \frac{.2}{.1} + g_m \ln \frac{.8}{.9} = \ln .0505$$

We substitute the values of the logarithms obtained from Table A-14 to obtain

$$.693d_m - .118g_m = 4.554$$
$$.693d_m - .118g_m = -2.986$$

These equations can be represented by lines on a graph. To draw a line we find two points and draw the line through them. Actually, for check purposes we shall determine three points:

1. For $g_m = 0$, $d_m = 6.57$ for the first line, and $d_m = -4.31$ for the second.

2. For $g_m = 20$, $d_m = 9.98$ for the first line, and $d_m = -.77$ for the second.

3. For $g_m = 40$, $d_m = 13.38$ for the first line, and $d_m = 2.50$ for the second.

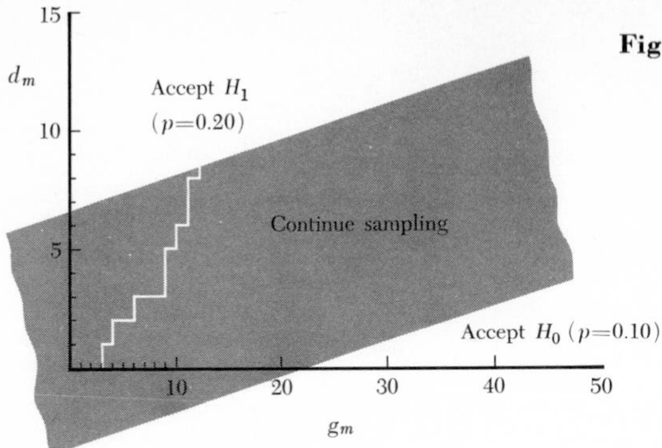


**Figure 18-1** Critical region for a sequential test of proportions.

The chart in Fig. 18-1 can be constructed in advance so that the data can be rapidly recorded as they are observed. The result of each observation can then be recorded easily by a line drawn 1 unit to the right if the item is good and a line drawn 1 unit up if the item is defective. Suppose we represent by $d$ the defective item and by $g$ the good items and obtain the following results, which are placed on the graph:

$g\,g\,g\,d\,g\,d\,g\,g\,d\,g\,g\,g\,d\,d\,g\,d\,g\,d\,d\,g\,d$

The sampling is stopped at this point and the lot rejected.

**Operating characteristic.** We have defined an *operating-characteristic function* as the function which gives the probability that the lot will be accepted when $p$ is the proportion of defectives in the whole lot. For this test we know four values and can sketch the curve roughly.

An additional point between $p_0$ and $p_1$ can be plotted for the proportion

$$p' = \frac{\ln \dfrac{1 - p_1}{1 - p_0}}{\ln \dfrac{1 - p_1}{1 - p_0} - \ln \dfrac{p_1}{p_0}}$$

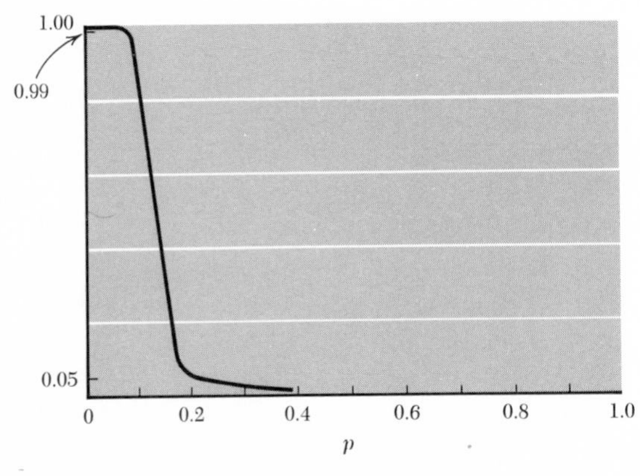


**Figure 18-2** Operating-characteristic curve for a sequential test.

The height at that point is

$$\frac{\ln \dfrac{1-\beta}{\alpha}}{\ln \dfrac{1-\beta}{\alpha} - \ln \dfrac{\beta}{1-\alpha}}$$

Of course, if the true proportion of defective items is less than .10, then the lot will be accepted more than 99 per cent of the time and lots that have more than a .20 proportion defective will be accepted less than 5 per cent of the time.

| $p$ | *Operating-characteristic function* |
|---|---|
| 0 | 1 |
| $p_0 = .1$ | $1 - \alpha = .99$ |
| $p' = .146$ | .604 |
| $p_1 = .2$ | $\beta = .05$ |
| 1 | 0 |

**Average sample number.** The number of trials required to reach a decision depends on the sample. The mean of the sampling distribution of this statistic is called the *average sample number*. It is easy to see that if the proportion of defectives is much less than $p_0$ or much greater than $p_1$ the sampling will terminate earlier than if the proportion is close to or between $p_0$ and $p_1$. If all the items are defective we can see from our chart for recording observations that the sampling will stop with 7 observations; if all items are good it will stop with 26. The average number of observations necessary for other proportions can be determined mathematically.

A curve adequate for most purposes can be drawn from five points. The numerical results in the following expressions were obtained by substituting the quantities $p_0 = .10$, $p_1 = .20$, $\alpha = 1$ per cent, and $\beta = 5$ per cent. If the true proportion is $p_0$, the average sample number is

$$\frac{(1-\alpha)\ln \dfrac{\beta}{1-\alpha} + \alpha \ln \dfrac{1-\beta}{\alpha}}{p_0 \ln \dfrac{p_1}{p_0} + (1-p_0)\ln \dfrac{1-p_1}{1-p_0}} = \frac{.99(-2.986) + .01(4.554)}{.10(.693) + .90(-.118)} = 79$$

and at $p_1$ it is

$$\frac{\beta \ln \dfrac{\beta}{1-\alpha} + (1-\beta)\ln \dfrac{1-\beta}{\alpha}}{p_1 \ln \dfrac{p_1}{p_0} + (1-p_1)\ln \dfrac{1-p_1}{1-p_0}} = \frac{.05(-2.986) + .95(4.554)}{.20(.693) + .80(-.118)} = 95$$

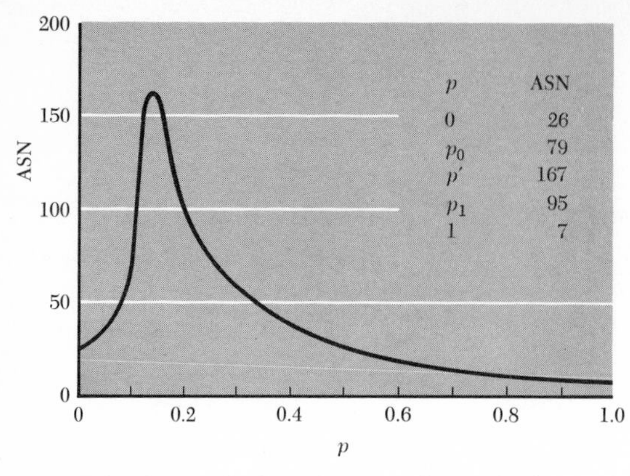


**Figure 18-3** Average sample number curve for a sequential test.

The formula for the number of observations when all items obtained are defective is

$$\frac{\ln \dfrac{1-\beta}{\alpha}}{\ln \dfrac{p_1}{p_0}} = 7$$

and for none defective is

$$\frac{\ln \dfrac{\beta}{1-\alpha}}{\ln \dfrac{1-p_1}{1-p_0}} = 26$$

It is also possible to obtain the average sample number for the proportion $p'$ between $p_0$ and $p_1$. The formula for this is

$$\frac{\ln \dfrac{\beta}{1-\alpha} \ln \dfrac{1-\beta}{\alpha}}{\ln \dfrac{p_1}{p_0} \ln \dfrac{1-p_1}{1-p_0}} = \frac{4.554(-2.986)}{.693(-.118)} = 167$$

The maximum value for the curve occurs for a point which is generally very near $p'$ (see Fig. 18-3).

**Observations in groups.** If we wish to make several observations at a time and then record results as above, the effect on the sequential procedure is an increase in the average sample number by an amount up to the number of items in each group. For example, if we examine five at a time, the number is increased not more than 5.

A test of the hypothesis $H_0$: $\mu = \mu_0$ against the alternative $H_1$: $\mu = \mu_1$ will be developed for measurements from a *normal* population. For example, if yarn has been given an additional treatment to increase breaking strength, we would wish to investigate whether we should reject the hypothesis $\mu = \mu_0$ (standard breaking strength) in favor of a larger value $\mu_1$. As another example, assume a change has been made in the teaching process. We may wish to see whether we should reject the hypothesis $\mu = \mu_0$ (previous performance level) in favor of $\mu = \mu_1$ (some higher performance level).

The test procedure is developed in the same manner as the test on proportions. The measurements on the successive items put to test are $X_1, X_2, X_3, \ldots$. The ratio of the probability that we would get $X_1$ if $\mu = \mu_1$ to the probability that we would get $X_1$ if $\mu = \mu_0$ is

$$\frac{[1/(\sqrt{2\pi}\,\sigma)]e^{-\frac{1}{2\sigma^2}(X_1-\mu_1)^2}}{[1/(\sqrt{2\pi}\,\sigma)]e^{-\frac{1}{2\sigma^2}(X_1-\mu_0)^2}} = e^{-\frac{1}{2\sigma^2}(X_1-\mu_1)^2+\frac{1}{2\sigma^2}(X_1-\mu_0)^2}$$

$$= e^{-\frac{1}{2\sigma^2}(X_1^2-2X_1\mu_1+\mu_1^2-X_1^2+2X_1\mu_0-\mu_0^2)}$$

$$= e^{\frac{\mu_1-\mu_0}{\sigma^2}X_1+\frac{\mu_0^2-\mu_1^2}{2\sigma^2}} = R_{X_1}$$

The probability ratio for $X_2$ is the same, except that $X_2$ appears in place of $X_1$, and after $m$ observations we have

$$\frac{p_{1m}}{p_{0m}} = R_{X_1}R_{X_2}R_{X_3} \cdots R_{X_m}$$

Using logarithms as before, we see that sampling will continue as long as

$$\ln\frac{\beta}{1-\alpha} < \ln\frac{p_{1m}}{p_{0m}} < \ln\frac{1-\beta}{\alpha}$$

that is, as long as

$$\ln\frac{\beta}{1-\alpha} < \ln R_{X_1} + \ln R_{X_2} + \cdots + \ln R_{X_m} < \ln\frac{1-\beta}{\alpha}$$

Now notice that

$$\ln R_{X_1} = \ln e^{\frac{\mu_1-\mu_0}{\sigma^2}X_1+\frac{\mu_0^2-\mu_1^2}{2\sigma^2}}$$

which, by the definition of natural logarithms, gives

$$\ln R_{X_1} = \frac{\mu_1-\mu_0}{\sigma^2}X_1 + \frac{\mu_0^2-\mu_1^2}{2\sigma^2}$$

and

$$\sum \ln R_{X_i} = \frac{\mu_1 - \mu_0}{\sigma^2} \sum X_i + \frac{\mu_0{}^2 - \mu_1{}^2}{2\sigma^2} m$$

If we set this expression equal first to $\ln \dfrac{\beta}{1 - \alpha}$ and then to $\ln \dfrac{1 - \beta}{\alpha}$, we can obtain the critical lines for our test.

To illustrate this procedure, consider the problem of determining the passing or failure of students by a sequence of examination problems or projects which are of approximately equal difficulty and on which students in the past have obtained scores which for each student are normally distributed with $\sigma = 10$. Suppose an average of 60 points is a passing mark and we wish to have a chance of only 1 per cent of passing a student whose examinations are from a population with mean 50 and a chance of only 1 per cent of failing a student whose examinations are from a population with mean 70. Let us determine a chart or table for recording scores and see whether the plan is feasible in terms of the average number of problems or projects required to reach a decision. Here we have

$$\begin{aligned} \mu_0 &= 50 \qquad & \alpha &= .01 \qquad & \sigma &= 10 \\ \mu_1 &= 70 & \beta &= .01 \end{aligned}$$

We substitute these values in the expression above to obtain

$$\frac{20}{(10)^2} \sum X_i + \frac{(50)^2 - (70)^2}{2(10)^2} m = \ln \frac{\beta}{1 - \alpha}$$

$$\frac{20}{(10)^2} \sum X_i + \frac{(50)^2 - (70)^2}{2(10)^2} m = \ln \frac{1 - \beta}{\alpha}$$

which simplifies to

$$\begin{aligned} .2\Sigma X_i - 12m &= +4.595 \\ .2\Sigma X_i - 12m &= -4.595 \end{aligned}$$

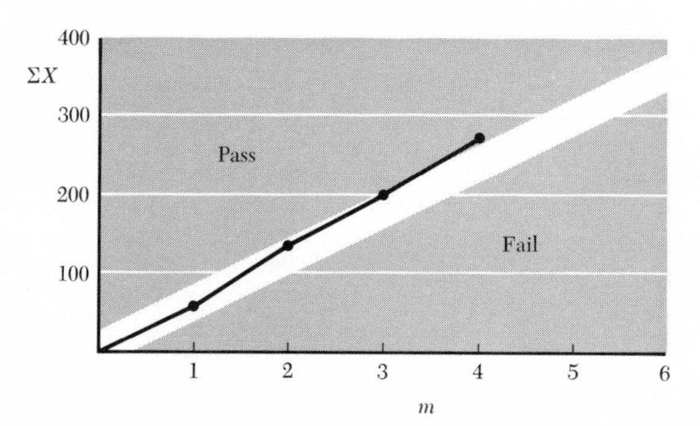


**Figure 18-4** Critical region for a sequential test of mean value.

**Table 18-1** The columns headed $X$ and $\Sigma X$ indicate the recording of any particular set of scores

| *No. of tests, m* | *Passing score* | $X$ | $\Sigma X$ | *Failing score* |
|---|---|---|---|---|
| 1 | 83 | 60 | 60 | 37 |
| 2 | 143 | 75 | 135 | 97 |
| 3 | 203 | 65 | 200 | 157 |
| 4 | 263 | 70 | 270 | 217 |
| 5 | 323 | | Pass | 277 |
| 6 | 383 | | | 337 |
| 7 | 443 | | | 397 |

The solution for $\Sigma X_i$ is

$$\Sigma X_i = 60m - 23$$
$$\Sigma X_i = 60m + 23$$

We can draw a graph as before or form a table stating the critical values. These can be easily obtained by substituting $m = 1, 2, 3, \ldots$ in succession in the two equations just above. Frequently the table of values of $\Sigma X$ will be more useful, since it may be difficult to plot the points on the graph with sufficient precision to determine whether or not it is between the lines. A student's successive scores on examinations of 60, 75, 65, 70 leading to a decision of passing are plotted on the graph and inserted in Table 18-1.

The operating-characteristic curve for this test is very similar to that drawn for the test on proportions. It is difficult to obtain specific values, but it is true, of course, that individuals with true mean above 70 will be failed less than 1 per cent of the time and those with true mean scores below 50 will be passed less than 1 per cent of the time by this procedure.

**Average sample number.** The number of tests we can expect to make before a decision is reached is available for the true situations $\mu = \mu_0$, $\mu = \mu_1$, and $\mu = (\mu_0 + \mu_1)/2$. We designate the $Y$-axis intercepts of the two lines in our graph by $h_0$ and $h_1$, which are found by setting $m = 0$ and solving the equations in the last section for $\Sigma X_i$. Thus

$$h_0 = \frac{\sigma^2}{\mu_1 - \mu_0} \ln \frac{1 - \beta}{\alpha} \qquad \text{and} \qquad h_1 = \frac{\sigma^2}{\mu_1 - \mu_0} \ln \frac{\beta}{1 - \alpha}$$

For $\mu = \mu_0$ the average sample number is

$$2\,\frac{h_1 + (1 - \alpha)(h_0 - h_1)}{\mu_0 - \mu_1}$$

for $\mu = \mu_1$ it is

$$2\,\frac{h_1 + \beta(h_0 - h_1)}{\mu_1 - \mu_0}$$

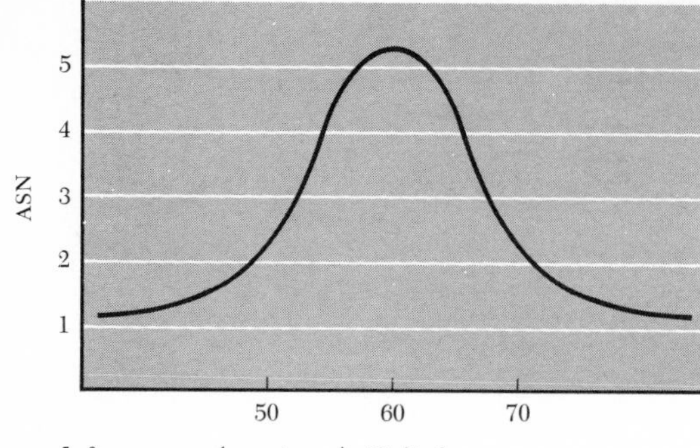


**Figure 18-5** Average sample number curve for a sequential test of mean value.

and for $\mu = (\mu_0 + \mu_1)/2$ it is

$$-\frac{h_0 h_1}{\sigma^2}$$

For our example, $h_0 = -23$, $h_1 = 23$, $\mu_0 = 50$, $\mu_1 = 70$, $\sigma = 10$, and we obtain the numbers 2.25 at $\mu_0$ and $\mu_1$ and 5.3 at $(\mu_0 + \mu_1)/2$. These values are plotted in Fig. 18-5. If we follow the procedures of Chap. 14 we find the number of examinations required if given all at once to be 5.4 or 6.

## 18-4 TEST FOR STANDARD DEVIATION

Suppose we wish to investigate the effect on variability in breaking strength of a material produced on machinery operating at a greater rate of speed than formerly. Or suppose we investigate whether the variability in the length of time that white rats can withstand a loud noise is greater for unrelated or related rats. We shall test the hypothesis $H_0$: $\sigma^2 = \sigma_0{}^2$ against the alternative $H_1$: $\sigma^2 = \sigma_1{}^2$.

We shall assume that the observations are normally distributed. For the sequential test to be presented in this section it is necessary to know the mean value $\mu$. The probability-ratio inequality for $X_1$ is given first. A second observation will be taken if

$$\frac{\beta}{1-\alpha} < \frac{\sigma_0}{\sigma_1} e^{-\frac{1}{2\sigma_1{}^2}(X_1-\mu)^2 + \frac{1}{2\sigma_0{}^2}(X_1-\mu)^2} < \frac{1-\beta}{\alpha}$$

The central portion of the above inequality is obtained from the probability ratio

$$\frac{p_{11}}{p_{01}} = \frac{[1/(\sqrt{2\pi}\,\sigma_1)]\, e^{-\frac{1}{2\sigma_1{}^2}(X_1-\mu)^2}}{[1/(\sqrt{2\pi}\,\sigma_0)]\, e^{-\frac{1}{2\sigma_0{}^2}(X_1-\mu)^2}}$$

By taking logarithms in this inequality we obtain

$$\ln \frac{\beta}{1-\alpha} < \ln \frac{\sigma_0}{\sigma_1} + \frac{1}{2}(X_1 - \mu)^2 \left(\frac{1}{\sigma_0^2} - \frac{1}{\sigma_1^2}\right) < \ln \frac{1-\beta}{\alpha}$$

The successive ratios for $X_2$, $X_3$, . . . will have the same form and will be multiplied into the probability ratio above or added into the central portion where the logarithms have been taken, so that we obtain after $m$ observations the following inequality, which determines whether sampling should continue:

$$\ln \frac{\beta}{1-\alpha} < m \ln \frac{\sigma_0}{\sigma_1} + \frac{1}{2}\left(\frac{1}{\sigma_0^2} - \frac{1}{\sigma_1^2}\right) \sum (X_i - \mu)^2 < \ln \frac{1-\beta}{\alpha}$$

As before, we set the central portion of this inequality equal to the two extremes and obtain the critical lines:

$$\sum (X_i - \mu)^2 = 2 \frac{\ln \dfrac{\beta}{1-\alpha}}{(1/\sigma_0^2) - (1/\sigma_1^2)} + m \frac{2 \ln \dfrac{\sigma_1}{\sigma_0}}{(1/\sigma_0^2) - (1/\sigma_1^2)}$$

$$\sum (X_i - \mu)^2 = 2 \frac{\ln \dfrac{1-\beta}{\alpha}}{(1/\sigma_0^2) - (1/\sigma_1^2)} + m \frac{2 \ln \dfrac{\sigma_1}{\sigma_0}}{(1/\sigma_0^2) - (1/\sigma_1^2)}$$

These can be written as

$$\Sigma(X_i - \mu)^2 = h_0 + mD$$
$$\Sigma(X_i - \mu)^2 = h_1 + mD$$

where $h_0 = 2 \dfrac{\ln \dfrac{\beta}{1-\alpha}}{(1/\sigma_0^2) - (1/\sigma_1^2)}$

$$h_1 = 2 \frac{\ln \dfrac{1-\beta}{\alpha}}{(1/\sigma_0^2) - (1/\sigma_1^2)}$$

$$D = \frac{\ln \dfrac{\sigma_1^2}{\sigma_0^2}}{(1/\sigma_0^2) - (1/\sigma_1^2)}$$

Testing now proceeds in the same fashion as with the sequential test for means.

We can write the average sample number when $H_0$ is true as

$$\frac{(1-\alpha)h_0 + \alpha h_1}{\sigma_0^2 - D}$$

and when $H_1$ is true as

$$\frac{\beta h_0 + (1 - \beta)h_1}{\sigma_0^2 - D}$$

The largest number occurs when $\sigma^2$ is larger than $\sigma_0^2$ and smaller than $\sigma_1^2$. The number for $\sigma^2 = D$ which lies between $\sigma_0^2$ and $\sigma_1^2$ is

$$\frac{-h_0 h_1}{2D^2}$$

**Mean unknown.** If $\mu$ is not known the same sequential table or graph may be used by computing at each stage $\Sigma(X_i - \bar{X})^2$ in place of $\Sigma(X_i - \mu)^2$ and comparing with the acceptance and rejection values for $m - 1$. This is a correct procedure but is perhaps less useful than the preceding tests presented, since the value of $\bar{X}$ will also change after every observation, thus increasing the amount of computation necessary.

## 18-5 DISCUSSION OF SEQUENTIAL TESTS

Much research is done in a sequential fashion. Each step in the research process may be better planned on the basis of the findings of earlier steps. Experimental material is then not wasted on experiments remote from the area of interest or performed in a nonoptimum way. Each step of this general strategy may also benefit from sequential study of each separate case in a given experiment. Each case is evaluated in turn, and additional testing is discontinued when sufficient data have been gathered. Experiments that can profit from sequential testing have some of the following properties:

1. Each individual may be tested separately.
2. Each test is expensive, increasing the importance of minimum sample size (for example, tests on baboons).
3. The response time is short. Since the experiment will be many times the length of a single test, the increased time required must be acceptable. If survival following therapy is months or years, sequential procedures would not be feasible.
4. Test conditions do not permit testing of more than one individual at a time. For example, only one test laboratory or station may be available, or measurement staff or instruments may be limited.
5. Test cases are available only at wide intervals in time, as, for example, in cases of rare diseases.

Sequential tests may have limited advantage if conditions such as these do not apply. For example, the efficiency of the sequential procedure may be unimportant under the following circumstances:

1. Test materials are inexpensive, so that minimum sample size is not important.

2. Long response times extend the total experimental time greatly if tests are done sequentially. Staff and equipment may be required over a lengthy time period.

3. A decision must be reached in a specified short time.

4. A loss in uniformity of tests occurs if the tests are not all done simultaneously, as, for example, from seasonal variations in response or decay in drug strength.

In the test for means it is assumed that $\sigma$ is known. Sequential procedures exist for testing the mean $\mu$ when $\sigma$ is not known, but they are not so simple to use as those presented in this chapter and are therefore not given. The sequential procedures presented in this chapter are for one-sided alternatives, the proportion $p_0$ against an alternative $p_1$, the mean $\mu_0$ against an alternative $\mu_1$, the variance $\sigma_0^2$ against an alternative $\sigma_1^2$. Particularly in the case of the mean $\mu$ we are often interested in testing with two-sided alternatives, the mean $\mu$ against alternatives $\mu_0 \pm d$; that is, we wish to reject $\mu_0$ for significant deviations in either direction. Sequential procedures for testing this hypothesis have been developed but are not given here.

## GLOSSARY

average sample number

probability ratio

sequential test

## DISCUSSION QUESTION

1. Discuss the change in the positions of the lines and the change in the method of recording necessary if the sequential procedure for testing for average proportion were set up in terms of the number of defective items and the total number tested in place of the number of defective and number of good items.

2. Explain why the graph in Fig. 18-5 does not drop below 1 at either end.

3. What is the minimum number of observations necessary to make a decision in testing for mean as described in Sec. 18-3?

## CLASS EXERCISES

1. Construct a chart for recording observations for testing $p_0 = .5$ against $p_1 = .9$ with $\alpha = .05$ and $\beta = .05$. Sketch the operating-characteristic function for this test. Sketch the average-sample-number curve for this test and determine experimentally the number of observations necessary to reach a decision when the true proportion is .50. This can be done by drawing red and white beads from a box or by tossing a coin. With two colors of beads, one color could represent "good" items and the other color could represent "bad" items. Draw beads one at a time, replacing each before the next is drawn. Record the number of beads drawn before a decision to accept or reject is reached. If each student determines several such results, a distribution of the required number can be formed and the mean obtained from this distribution.

2. Construct a table for the sequential test for $\mu_0 = -.5$ against the alternative $\mu_1 = .5$ with $\sigma = 1$ and $\alpha = .05$ and $\beta = .05$. Using the random normal numbers in Table A-2, sample from that population until a decision is reached and record whether $\mu_0$ or $\mu_1$ is accepted and the number of observations required to reach a decision. Perform this experiment 10 times and collect the results of all the students. Compare the results with those to be expected theoretically, that is, $\mu_0$ accepted as frequently as $\mu_1$ and the average sample size 8.67. Note that the true mean is neither $\mu_0$ nor $\mu_1$.

## PROBLEMS

*Section* 18-2

1. A hypothesis $H_0$: $p \leq .10$ is to be tested at the $\alpha = .05$ level of significance, and it is desired to have $\beta = .10$ if $H_1$: $p = .30$ is true. Find the equations defining the appropriate sequential test. Draw the sampling-plan chart corresponding to Fig. 18-1. Sketch the operating-characteristic curve and the average-sample-size curve.

2. A hypothesis $H_0$: $p \leq .30$ is to be tested with $\alpha = .10$ such that $\beta = .10$ if $p$ is .5 or greater. Find the equations defining the appropriate sequential test. Draw the sampling-plan chart corresponding to Fig. 18-1. Sketch the operating-characteristic curve and the average-sample-size curve.

3. Construct the sampling-plan chart to test $H_0$: $p = .2$ with $\alpha = .10$ against $H_1$: $p = .5$ with $\beta = .10$. Using the series of single-digit numbers in the first column in Table A-1, record a success if the number is 0 or 1 and a failure if the number is 2, 3, 4, 5, 6, 7, 8, or 9 and carry the experiment to a conclusion. Record whether the hypothesis was accepted or rejected and the number of trials needed. Repeat this with each of the next 19 columns in Table A-1. This experiment will give you a check on the correctness of some statements in this chapter. Do the results observed agree with those statements?

4. Prepare a sheet of graph paper on which an inspector may record his tests of a batch of fuses and reject batches only 1 per cent of the time if there are 4 per cent defective items and accept batches only 5 per cent of the time if there are 8 per cent defective items.

5. The following data were drawn one observation at a time in the order recorded; $d$ denotes a defective item and $g$ denotes a good item. The experiment was performed to test the hypothesis that the proportion $p$ of defective items in the population was .1, $H$: $p = .1$, with $\alpha = .05$, and so that if $p = .3$ the chance of rejecting is .90 ($\beta = .10$). With the assumption that $\bar{X}$ is approximately normal and the techniques of Chaps. 13 and 14 with $\alpha = .05$, is $\beta = .10$? Do you accept or reject $H$: $p = .1$? Analyze the data using sequential analysis and see whether or not you would accept or reject before you had used the entire sample.

*ggdgdgggdd gggggdgggg gggdggggdg*

*Section* 18-3

6. Recompute the sequential test on means given in this chapter with $\sigma = 20$ and $\alpha = \beta = .05$. With $\sigma = 20$ and $\alpha = \beta = .01$. Note the changes in the positions of the lines and the change in the average sample size for $\mu = (\mu_1 + \mu_2)/2$.

*Section* 19-4

7. For $\alpha = \beta = .05$, construct a table of acceptance and rejection values for a sequential test for $H_0$: $\sigma = 10$ against $H_1$: $\sigma = 15$. Find the average sample size corresponding to $\sigma^2 = D$.

CHAPTER NINETEEN

# Sensitivity experiments

Experimental investigations often deal with continuous variables which cannot be measured in practice. For example, in testing the sensitivity of explosives to shock a common procedure is to drop a weight on specimens of the same explosive mixture from various heights. There are heights at which some specimens will explode and others will not, and it is assumed that those which do not explode would have exploded were the weight dropped from a sufficiently greater height. It is supposed, therefore, that there is a critical height associated with each specimen and that the specimen will explode when the weight is dropped from a greater height and will not explode when the weight is dropped from a lesser height. The population of specimens is thus characterized by a continuous variable—the critical height—which cannot be measured. All we can do is select some height arbitrarily and determine whether the critical height for a given specimen is less than or greater than the selected height.

This situation arises in many fields of research. In testing insecticides, for example, a critical dose is associated with each insect, but it cannot be measured. We can only try some dose and observe whether or not the insect is killed, that is, whether the critical dose for the insect is less than or greater than the chosen dose. The same difficulty arises in pharmaceutical research dealing with germicides, anesthetics, and other drugs, in testing strength of materials, in psychophysical research dealing with threshold stimuli, and in other areas of biological and medical research.

In true sensitivity experiments it is not possible to make more than one observation on a given specimen. Once a test has been made, the specimen is altered (the explosive is packed, the insect is weakened), so that a bona fide result cannot be obtained from a second test. A common procedure in experiments of this kind is to divide the sample of specimens into several groups (usually but not necessarily of the same size) and to test one group at a chosen level, a second group at a second level, and so on. The data then consist of the numbers responding at each level. Several methods are given for analyzing such data (variously called *sensitivity data, all-or-none data,* or *quantal-response data*).

We assume that the distribution of initial or threshold values, or a known transformation of these values, is normal. It is often the case in dosage-mortality experiments and in experiments on explosives that the logarithm of the dosage concentration or of the height is reasonably normally distributed. But in other areas of research, and sometimes in these areas as well, other transformations may be more appropriate.

If we have no idea of the shape of our distribution function, then the data of the experiment itself must be used to provide this information. The common procedure here is to plot the percentage affected at each level on arithmetic-probability paper against various functions of the variate in question. Usually it soon becomes apparent what sort of function will force the percentages to lie sensibly along a straight line. There are, of course, infinitely many functions to choose from; the chosen function should be as simple as possible consistent with whatever knowledge is available concerning the nature of the material at hand.

## 19-1 GRAPHICAL SOLUTION

If no special method is used for obtaining the data, that is, if a certain number of tests are made at various levels of intensity, a graphical analysis of the data may be made by use of the method discussed in Chap. 5. The percentages responding at each level are plotted against that level on normal-probability paper (or on log-normal-probability paper if logarithms are thought to be normally distributed). Then a straight line is drawn which passes as closely as possible through the points. The levels at which all specimens respond or at which no specimens respond give little information concerning the distribution, and in fact, since 0 and 100 per cent do not show on the graph, they give none at all for the graphical analysis.

EXAMPLE 19-1. Consider the data in Table 19-1 for tests on 20 specimens at each of seven levels, with the numbers responding given. These data were plotted on log-normal-probability paper in Fig. 19-1, and the line shown was drawn by eye (the subjective fitting by eye may vary considerably from one person to another and opens the technique to criticism, particularly if the data are such that there is not reasonable agreement on the position of the line). From this line the median is estimated to be at $P_{50} = 38$, and, for example, the values $P_{95} = 260$ and $P_{05} = 5.3$ can be read.

The line seems to be a reasonable fit to the points, implying that the logarithms of the original data are approximately normally distributed. If desired, the parameters of the logarithm distribution can be estimated. The

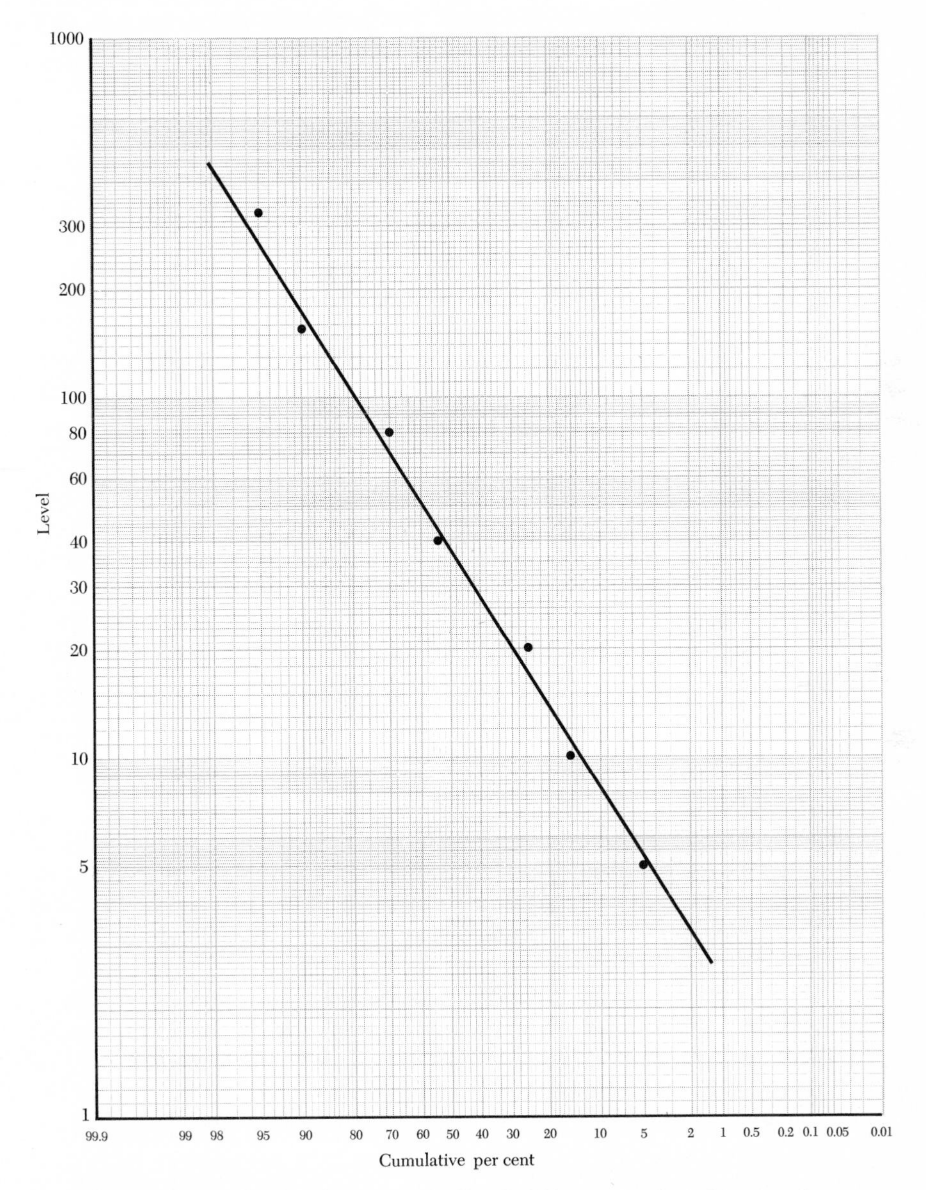


**Figure 19-1** Cumulative on normal probability scale plotted against logarithm of stimulus level.

**Table 19-1**

| Level | *No. exposed* | *No. responding* | *Per cent responding* |
|---|---|---|---|
| 5 | 20 | 1 | 5 |
| 10 | 20 | 3 | 15 |
| 20 | 20 | 5 | 25 |
| 40 | 20 | 11 | 55 |
| 80 | 20 | 14 | 70 |
| 160 | 20 | 18 | 90 |
| 320 | 20 | 20 | 100 |

mean is estimated as log 38 = 1.580, and since $P_{95}$ and $P_{05}$ differ by 3.29 standard deviations, the standard deviation is estimated as

$$\frac{\log 260 - \log 5.3}{3.29} = \frac{2.415 - .724}{3.29} = \frac{1.691}{3.29} = .514$$

Computing procedures exist (see the references at the end of the book) for obtaining the line which fits best, but they are fairly complex and will not be presented here.

If the data are obtained in a variety of special ways the analysis is very much simplified and in many cases results in an increased precision for a given number of observations. The special method described in the next section is the best method available if it is desired to obtain general information about the whole distribution (both the mean and standard deviation). For small samples the estimate of standard deviation is not reliable. A modified procedure to estimate the median only is described in Sec. 19-5.

## 19-2 THE "UP-AND-DOWN" METHOD

This technique for obtaining sensitivity data has been developed and used in explosives research. The method may be employed in any sensitivity experiment, but we shall discuss it in terms of explosives to avoid general terminology. The technique is to choose some initial height $h_0$, and a succession of heights $h_1, h_2, h_3, \ldots$ above $h_0$, and a succession $h_{-1}, h_{-2}, h_{-3}, \ldots$ below $h_0$. The first specimen is tested by dropping the weight from height $h_0$. If the specimen explodes, the second specimen will be tested at $h_{-1}$; otherwise it will be tested at $h_1$. In general any specimen will be tested at the level immediately below or immediately above the level of the previous test, according to whether there was or was not an explosion on the previous test.

EXAMPLE 19-2. The result of an up-and-down experiment is depicted in Fig. 19-2, where X represents explosions and O nonexplosions. The

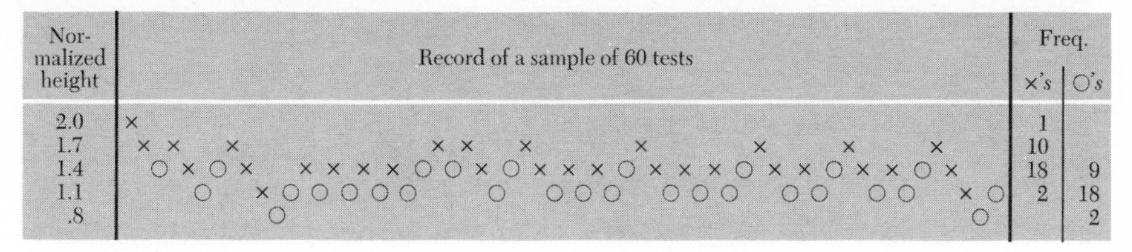


**Figure 19-2** Results of 60 tests in an up-down experiment.

first test is on the left at the highest level; this was a success (an explosion), so the second test was made at the next lower level and was also a success; the third test was therefore made at the level below that of the second; since it was a failure, the fourth test was made at the level above that of the third test.

In this notation an × indicates a success and also that the following test is to be made at the next lower level. An O indicates a failure and hence that the following test is to be made at the next higher level.

The primary advantage of this method is that it automatically concentrates testing near the mean. We shall see later that this increases the accuracy with which the mean can be estimated. In other words, for a given accuracy the up-and-down method will require fewer tests than the ordinary method of testing groups of equal size at preassigned heights. The saving in the number of observations may be of the order of 30 to 40 per cent.

The method has one obvious disadvantage in certain kinds of experiments in that it requires that each specimen be tested separately and sequentially. This is not important in explosives experiments, because each test must be made separately anyway. But in tests of insecticides, for example, a large group of insects can sometimes be exposed as easily as a single one, and in large experiments of this kind any advantage of the up-and-down method might well be outweighed by this requirement of single tests. Even here, if expensive laboratory animals were being used the advantage in economy of tests might offset the trouble of making single tests. Also, several sequences of tests may be run simultaneously and the results combined as described in Sec. 19-5.

The statistical analysis of data of this type can be quite simple, provided that the experiment satisfies certain conditions. These will be discussed here, and the actual analysis will be given in the following section.

While the up-and-down method is particularly effective for estimating the mean, it is not a good method for estimating small or large percentage points (for example, the height at which 99 per cent of specimens explode) unless normality of the distribution throughout a wide range is assured. In fact, no method which uses the normal distribution can be relied on to estimate extreme percentage points because such estimates and percentiles depend critically on the assumption of normality. In most experimental research it is possible to find simple transformations which make the variate approximately normal in the region of the mean, but to make it normal in the tails is quite another matter. Nothing short of an extensive exploration of the distribution, involving perhaps thousands of observations, will suffice in that case.

The information obtained by the up-and-down method is approximately equivalent to that which would be obtained with a sample only half as large if the minimum exploding height could be measured. This fact can be used to assist in planning the size of the experiment (see Sec. 19-4).

A further condition is necessary if the statistical analysis is to be simple. We must be able to estimate roughly in advance the standard deviation of the normally distributed variate. The interval between testing levels should be approximately equal to the standard deviation. This condition is not severe, since it will be well enough satisfied if the interval actually used is less than twice the standard deviation. Furthermore, researchers who perform these experiments repeatedly on essentially similar materials can usually make very good preliminary estimates. This is the case in explosives research or biological assay, for example. This circumstance of repeated experiments is precisely the one in which a simple analysis is most desirable.

## 19-3 STATISTICAL ANALYSIS FOR $N$ LARGE

The method of analysis given in this section is applicable when all the conditions described in the preceding section are fulfilled. More complex analyses, not included here, are required when the levels are not equally spaced or when the distance between levels exceeds twice the standard deviation.

We again return to the explosives experiment in describing the technique. Suppose it is known for the given type of explosive that the logarithms of the critical heights are normally distributed. If $h$ represents the height, $y = \ln h$ will then be the normally distributed variate. We shall call $y$ the normalized height and represent the mean and variance of its distribution by $\mu$ and $\sigma^2$. The experiment is performed by choosing an initial height for the first test, say, $h_0$. This should be chosen near the anticipated mean. The other testing levels are determined such that the values of the normalized height $y$ are equally spaced. If $d$ is the preliminary estimate of $\sigma$, and if $y_0 = \ln h_0$, then the actual testing heights are obtained by putting $\ln h = y_0 \pm d,\ y_0 \pm 2d,\ y_0 \pm 3d,\ \ldots$ and solving for $h$. The heights will then be so spaced that the transformed variate is equally spaced, with the spacing equal to its anticipated standard deviation. All computations are in terms of $y$.

In any experiment the total number of successes will be approximately equal to the total number of failures. In fact, the number of failures at any level cannot differ by more than 1 from the number of successes at the next higher level. For estimating $\mu$ and $\sigma$ only the successes or only the failures are used, depending on which has the smaller total. In the example shown in Fig. 19-2 there are fewer failures than successes, so the failures would be used. We shall let $n$ denote the smaller total and let $n_0, n_1, n_2, \ldots, n_k$ denote the frequencies at each level for this less frequent event, where $n_0$ corresponds to the lowest level and $n_k$ the highest level on which the event occurs. We have then $\Sigma n_i = n$.

The estimates of $\mu$ and $\sigma$ are based on the mean and variance, $\bar{y}$ and $s_y^2$, of the $y$ values, with the observed frequencies $n_i$. In this notation the estimate of $\mu$, say, $\bar{X}$, is

$$\bar{X} = \bar{y} \pm \tfrac{1}{2}d \tag{1}$$

The plus sign is used when the analysis is based on the O's and the minus sign when it is based on the X's. The estimate of $\sigma$ is

$$s = 1.620d\left(\frac{s_y^2}{d^2} + .029\right) \tag{2}$$

This is a curious estimate in that, while it is a linear function of $s_y^2$, it gives an estimate of the standard deviation, not the square of the standard deviation. The formula is an approximate one which can be used when $s_y^2/d^2$ is larger than .3 but which becomes inaccurate for $s_y^2/d^2$ much less than .3. In the latter instance the formula should not be used to give an estimate of $\sigma$.

Example 19-2 is used to illustrate the use of the formulas. Here the $y$ values used were 2.0, 1.7, 1.4, 1.1, .8, and $d = .3$. Among 60 tests there were 31 explosions and 29 failures, and hence the failures are used to estimate the parameters. The failures occurred on three levels (.8, 1.1, 1.4) with frequencies $n_0 = 2$, $n_1 = 18$, $n_2 = 9$ as given in Table 19-2. The estimated mean is $\bar{X} = 34/29 + .3/2 = 1.32$. We next compute

$$s_y^2 = \frac{40.70 - (34)^2/29}{28} = .030$$

and enter this value in the formula for the estimated standard deviation to obtain

$$s = 1.620 \times .3\left(\frac{.030}{.09} + .029\right) = .176$$

**Table 19-2**

| $y_i$ | $n_i$ | $y_i n_i$ | $y_i^2 n_i$ |
|---|---|---|---|
| 1.4 | 9 | 12.6 | 17.64 |
| 1.1 | 18 | 19.8 | 21.78 |
| .8 | 2 | 1.6 | 1.28 |
| *Total* | 29 | 34.0 | 40.70 |

The data for the example were generated by reading values from the table of random normal deviates, Table A-2. Each value was multiplied by .2 and then a constant 1.3 was added, so the $X$ values were a random sample from a normal population with $\mu = 1.3$ and $\sigma = .2$. The mean and standard deviation of the 60 transformed observations from Table A-2 were 1.312 and .158, which are close, respectively, to both $\mu$ and $\sigma$ as well as to the $\bar{X}$ and $s$ estimates obtained by the up-and-down method.

Percentage points can be estimated by $\bar{X} + z_\alpha s$, where $z_\alpha$ is chosen from tables of the normal deviate to give the desired percentage. Thus in the example $P_{05}$ is estimated by $1.32 - (1.645 \times .176) = 1.03$. If the $y$ values had been natural logarithms of actual heights in feet in an explosives experiment, the antilogarithms of estimated percentage points would be estimates of the corresponding points for the distribution of $h$. Thus the median (not mean) value of $h$ can be estimated by antiln $1.32 = 3.74$ feet and the 5 per cent height by antiln $1.03 = 2.80$ feet. The antilogarithm of $s$ does not estimate the standard deviation for $h$, however, and any computation which involves the standard deviation (estimates of percentage points, confidence limits) must be done in terms of the normalized height and only the final result transformed to actual heights.

## 19-4 CONFIDENCE INTERVALS

For a random sample the standard error of a sample mean $\bar{X}$ is given by $\sigma_{\bar{X}} = \sigma/\sqrt{N}$, where $\sigma$ is the population standard deviation and $N$ is the sample size. For data collected by the up-and-down method, $n$ will be approximately $N/2$. The standard error of $\bar{X}$ is

$$\sigma_{\bar{X}} = \frac{G\sigma}{\sqrt{n}} \tag{3}$$

where $G$ depends on the ratio $d/\sigma$ and on the position of the mean relative to the testing levels. $G$ is plotted in Fig. 19-3 as a function of $d/\sigma$. The position of the mean relative to the testing levels does not affect $G$ unless the interval $d$ is large; the solid branch of the curve gives the value of $G$ when the mean falls on one of the testing levels, while the dashed branch gives the value when the mean falls midway between two levels. Curves for other positions of the mean would fall between the two branches.

In practice $\sigma$ is not known, and $s$ must be used in equation (3) to obtain an estimate, say, $s_{\bar{X}}$, of $\sigma_{\bar{X}}$. In the illustrative example, with $s = .176$ we have $d/s = 1.7$, so that $G$ is about 1.12. The estimate of $\sigma_{\bar{X}}$ is therefore

$$s_{\bar{X}} = \frac{.176 \times 1.12}{\sqrt{29}} = .037$$

A confidence interval for $\mu$ may now be estimated by $\bar{X} \pm z_\alpha s_{\bar{X}}$. Thus a

95 per cent confidence interval is

$$1.32 \pm (1.96 \times .037) \qquad \text{or} \qquad 1.25 \text{ to } 1.39$$

For moderate values of $n$ it might be more accurate to use $t$ values in place of $z_\alpha$, but it is likely that this is a minor matter relative to the error caused by the approximations already used. Again with the assumption that the confidence interval refers to the logarithm of an actual height, it gives rise to an asymmetric 95 per cent confidence interval (3.5 to 4 feet) for the median height.

The standard error of the sample standard deviation, say, $\sigma_s$, is estimated in the same manner as the standard error of the mean. We shall write

$$\sigma_s = \frac{H\sigma}{\sqrt{n}} \tag{4}$$

$H$ is plotted in Fig. 19-3, where the solid branch gives the value of $H$ when the mean falls on a level, while the dashed branch gives the value when the mean is midway between two levels. When $d/\sigma$ is less than 2 there will be little error introduced by interpolating linearly between the two branches for other positions of the mean. Thus if the mean falls $d/4$ from a testing level we may use the value of $G$ midway between the two branches. For the illustrative example with $d/s = 1.7$ we find $H$ to be about 1.24, so that the estimate of $\sigma_s$ is

$$s_s = \frac{1.24 \times .176}{\sqrt{29}} = .041$$

With the assumption of normality the up-and-down method gives an estimate of a percentage point as $\bar{X} + z_\alpha s$, where $z_\alpha$ is read from Table A-4.

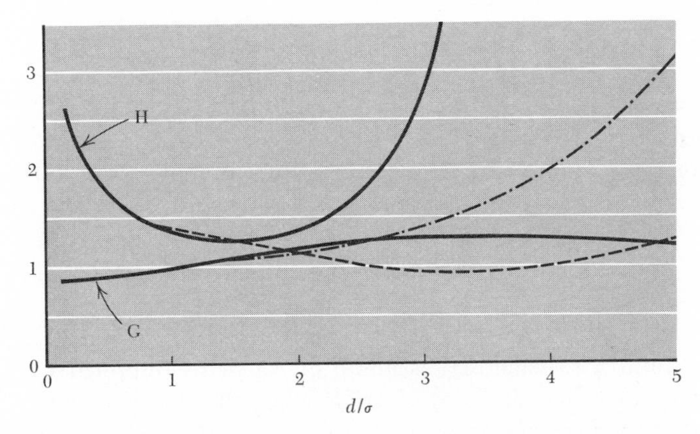


**Figure 19-3** Values of $G$ and $H$ (used in up-down testing) plotted against $d/\sigma$.

This estimate has standard error equal to $\sqrt{\sigma_{\bar{X}}^2 + z_\alpha^2\sigma_s^2}$, and this is estimated by $\sqrt{s_{\bar{X}}^2 + z_\alpha^2 s_s^2}$. Thus in the example $P_{05}$ was estimated as $\bar{X} - 1.645s = 1.03$ and an approximate 95 per cent confidence interval, for its true value would be

$$1.03 \pm 1.96\sqrt{(.037)^2 + (1.645)^2(.041)^2} \qquad \text{or} \qquad .88 < P_{05} < 1.18$$

**Choice of testing interval.** The curves in Fig. 19-3 have been extended beyond $d = 2\sigma$ in order to show what happens to the measures of precision for larger intervals. Curve $G$ shows that the precision of the mean steadily decreases as $d$ increases. The two branches of $H$ show that there is an optimum spacing for estimating the standard deviation depending on the position of the mean relative to the testing levels. Since the mean is usually unknown, this information is of little practical value.

Curve $G$ indicates that the interval should be quite small for maximum precision in the mean, but in practice this is not true, for several reasons. In the first place, the curves are for expected values, and essentially infinite sample sizes are assumed; in fact, very large samples are required to get good estimates of the mean for a very small interval. The estimate may be biased appreciably toward the initial testing level unless the sample is large. In the second place, a small interval may result in wasted observations unless a good choice for the initial level is made; if a poor choice is made many observations must be spent getting from the level to the region of the mean. And finally, since $\sigma$ is usually unknown, the precision of the mean must actually be measured by $s$, and the accuracy of $s$ becomes poor for very small intervals, as shown by curve $H$.

All these considerations indicate that the interval should be within the range of about $.5\sigma$ to $2\sigma$, and experiments with the method support this conclusion.

## 19-5 STATISTICAL ANALYSIS FOR $N$ SMALL

In analyses of measured response the usual $t$ test and analysis-of-variance procedures are widely used for (1) comparing mean response between two or more groups and (2) separating the components of variation due to the design variables of an experiment by the use of factorial designs, Latin squares, and so on. The same analyses are desirable when the basic measurement is an all-or-none response. Such comparisons or analyses can more easily be made from shorter series of trials.

The method given in Sec. 19-3 provides estimates of $\mu$ and $\sigma$ for a long series of trials, and a satisfactory estimate of $\mu$ can be obtained by that method for series of as few as 10 or 15 trials when the starting level is not too far from the mean. However, even shorter series are desirable when the elapsed time for observing the response is long or when several series are to be tested concurrently, with each series at some different level of an associated variable. For very short series the estimate of $\mu$ is somewhat dependent on the starting level, and if a predetermined number of tests are performed the standard error of the mean will also depend on the starting level.

This section presents a modified up-and-down estimate of $\mu$ which is almost independent of the choice of starting level and has a smaller and uniform standard error. This is accomplished by allowing the sample size to vary slightly, depending on the outcome of the first few trials, and by using the list of estimates provided in Table 19-3 for each particular sequence of results.

The trials are performed in the same up-and-down manner described in the first sections of this chapter, except that the testing continues for a total number of tests $N'$ performed in each series, which is determined by choosing a "nominal" sample size $N$. This nominal $N$ is the total number of trials reduced by one less than the number of like responses at the beginning of the series.

Example 19-3. A test-series shown in Fig. 19-4 with the outcome O, X, X, O, X, O has $N' = 6$. Since the response changed after the first trial, $N$ is also 6. For the series O, O, O, X, X, O, X, O, where there are three like responses at the beginning, we have $N' = 8$ and $N = 6$.

We obtain several estimates of $P_{50}$ with equal standard errors by continuing testing in each series so that each series is of the same nominal sample size. If, for example, we wish the standard error of $P_{50}$ to be $.56\sigma$, we see from Table 19-3 that a nominal sample of size 6 is required. Thus four additional observations will be needed after the first reversal of response. For example, a sequence starting O, X, or O, O, X, or O, O, O, X, and so on, or X, O, or X, X, O, and so on would be followed by four more observations.

The resulting configuration of responses and nonresponses for each series is referred to in Table 19-3, and we compute

$$X_f + kd$$

where $X_f$ = the last dose administered
$k$ = the tabular value
$d$ = the interval between dose levels

Table 19-3 lists all solutions for all $N'$ and for $N \leq 6$. If the series begins with more than four like responses, that is, $N' - N > 3$, the entry in the

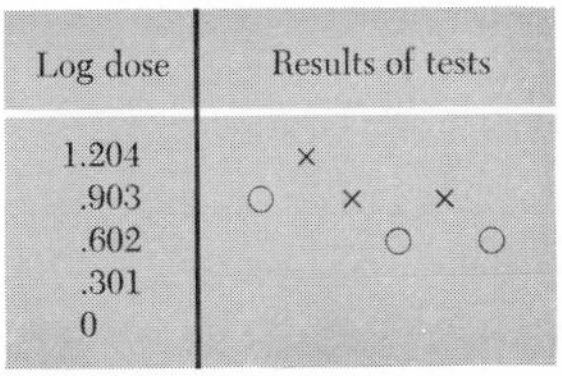

| Log dose | Results of tests | | | | | |
|---|---|---|---|---|---|---|
| 1.204 | | × | | | | |
| .903 | ○ | | × | | × | |
| .602 | | | | ○ | | ○ |
| .301 | | | | | | |
| 0 | | | | | | |

**Figure 19-4** Results of six tests in an up-down experiment.

**Table 19-3** Values of $k$ for estimating $P_{50}$ from up-and-down sequence of trials of nominal length $N$. The estimate of $P_{50}$ is $X_f + kd$, where $X_f$ is the final test level and $d$ is the interval between dose levels. If the table is entered from the foot, the sign of $k$ is to be reversed

| Sample size $N$ | *Second part of series* | $k$ for test series whose first part is: O | OO | OOO | OOOO | | Standard error of $P_{50}$ |
|---|---|---|---|---|---|---|---|
| 2 | × | −.500 | −.388 | −.378 | −.377 | O | .88σ |
| 3 | × O | .842 | .890 | .894 | .894 | O × | .76σ |
| | × × | −.178 | .000 | .026 | .028 | O O | |
| 4 | × O O | .299 | .314 | .315 | .315 | O × × | .67σ |
| | × O × | −.500 | −.439 | −.432 | −.432 | O × O | |
| | × × O | 1.000 | 1.122 | 1.139 | 1.140 | O O × | |
| | × × × | .194 | .449 | .500 | .506 | O O O | |
| 5 | × O O O | −.157 | −.154 | −.154 | −.154 | O × × × | .61σ |
| | × O O × | −.878 | −.861 | −.860 | −.860 | O × × O | |
| | × O × O | .701 | .737 | .741 | .741 | O × O × | |
| | × O × × | .084 | .169 | .181 | .182 | O × O O | |
| | × × O O | .305 | .372 | .380 | .381 | O O × × | |
| | × × O × | −.305 | −.169 | −.144 | −.142 | O O × O | |
| | × × × O | 1.288 | 1.500 | 1.544 | 1.549 | O O O × | |
| | × × × × | .555 | .897 | .985 | 1.000$^{+1}$ | O O O O | |
| 6 | × O O O O | −.547 | −.547 | −.547 | −.547 | O × × × × | .56σ |
| | × O O O × | −1.250 | −1.247 | −1.246 | −1.246 | O × × × O | |
| | × O O × O | .372 | .380 | .381 | .381 | O × × O × | |
| | × O O × × | −.169 | −.144 | −.142 | −.142 | O × × O O | |
| | × O × O O | .022 | .039 | .040 | .040 | O × O × × | |
| | × O × O × | −.500 | −.458 | −.453 | −.453 | O × O × O | |
| | × O × × O | 1.169 | 1.237 | 1.247 | 1.248 | O × O O × | |
| | × O × × × | .611 | .732 | .756 | .758 | O × O O O | |
| | × × O O O | −.296 | −.266 | −.263 | −.263 | O O × × × | |
| | × × O O × | −.831 | −.763 | −.753 | −.752 | O O × × O | |
| | × × O × O | .831 | .935 | .952 | .954 | O O × O × | |
| | × × O × × | .296 | .463 | .500 | .504$^{+1}$ | O O × O O | |
| | × × × O O | .500 | .648 | .678 | .681 | O O O × × | |
| | × × × O × | −.043 | .187 | .244 | .252$^{+1}$ | O O O × O | |
| | × × × × O | 1.603 | 1.917 | 2.000 | 2.014$^{+1}$ | O O O O × | |
| | × × × × × | .893 | 1.329 | 1.465 | 1.496$^{+1}$ | O O O O O | |
| | | $-k$ for series whose first part is: × | × × | × × × | × × × × | *Second part of series* | |

final column of Table 19-3 may be used (except for five tabular entries where an additional increment in the third decimal place is indicated). The estimate for Example 19-3 is $P_{50} = .602 + (.831 \times .301) = .852$.

For $N$ greater than 6, $P_{50}$ may be estimated by computing for the last $N$ trials the mean of the test levels corrected by a factor which is dependent on the constants $A$ and $C$ of Table 19-4. The estimate is

$$\frac{\Sigma X_i}{N} + \frac{d}{N}(A + C)$$

where the $X_i$'s are the test levels and $A$ and $C$ are obtained from Table 19-4. In Table 19-4 $n_\bigcirc$ refers to the number of O's and $n_\times$ to the number of X's in the final $N$ trials. The standard error of this estimated mean is approximately $\sigma\sqrt{2/N}$. The additional adjustment $C$, which improves this estimate, particularly for the smaller sample sizes, is based on the initial trials. This adjustment has little effect except in a few cases with small probability of occurrence, that is, where there are great differences of the number of X's and O's in the final $N$ trials and where the series starts with a compensating run of O's or X's, respectively. If the use of this adjustment has any appreciable effect on the estimate it is advisable to investigate the possible disagreement of the experimental situation with the assumptions of this model; for example, the assumption that the interval may be very much smaller than $\sigma$ or that the sampling is not all from the same population.

The estimates are given in Table 19-3 for each possible configuration of responses, with the assumption that the proportion of successes is given by a normal cumulative distribution. For estimates as given in Table 19-3 it is assumed that $d = \sigma$. Fortunately, estimates for this design with $N \geq 3$ prove to have standard errors which depend very little on the actual value of $\sigma$ and, in addition, are almost independent of the starting level and of $\mu$. This is approximately true even when the spacing $d$ differs from $\sigma$.

**Table 19-4** Values of $A$ and $C$ for approximate estimate of $P_{50}$ for $N > 6$. The estimate is $\Sigma X_i/N + d(A + C)/N$, where the $X_i$'s are the test levels of the final $N$ trials with $n_\bigcirc$ nonresponses and $n_\times$ responses and $d$ is the interval between dose levels. $C = 0$ for a series whose first part is a single O or X

| | | $C$ FOR TEST SERIES WHOSE FIRST PART IS | | | |
|---|---|---|---|---|---|
| $n_\bigcirc - n_\times$ | $A$ | OO | OOO | OOOO | OOOOO |
| 5 | 10.8 | 0 | 0 | 0 | 0 |
| 4 | 7.72 | 0 | 0 | 0 | 0 |
| 3 | 5.22 | .03 | .03 | .03 | .03 |
| 2 | 3.20 | .10 | .10 | .10 | .10 |
| 1 | 1.53 | .16 | .17 | .17 | .17 |
| 0 | 0 | .44 | .48 | .48 | .48 |
| −1 | −1.55 | .55 | .65 | .65 | .65 |
| −2 | −3.30 | 1.14 | 1.36 | 1.38 | 1.38 |
| −3 | −5.22 | 1.77 | 2.16 | 2.22 | 2.22 |
| −4 | −7.55 | 2.48 | 3.36 | 3.52 | 3.56 |
| −5 | −10.3 | 3.5 | 4.8 | 5.2 | 5.3 |
| $n_\times - n_\bigcirc$ | $-A$ | XX | XXX | XXXX | XXXXX |
| | | $-C$ FOR TEST SERIES WHOSE FIRST PART IS | | | |

The estimates of Sec. 19-2 do not make full use of the initial series of trials and the occurrence of an unequal number of X's and O's. This is not important in a long sequence, but it does make a sizable increase in standard error for small samples.

Much has been written concerning stochastic estimation methods which involve reduction of interval size during the seriês of trials. There appears to be little to gain from any change in interval size for the short trials described here. A primary drawback to early reduction in interval size is the possibility of inappropriate reduction in interval size before the region of $P_{50}$ is reached. In this case a poor estimate will result.

EXAMPLE 19-4. *Use of several up-and-down series.* An experiment on the effects of a drug (Ryanodine) will illustrate the use of the method described in this section. Several series of tests can be in progress at the same time, so that additional animals can be treated without waiting for the outcome of each separate trial. Also, this example illustrates the use of different sets of test levels for different series. A common log dose interval, $d = 1$, is used for log dose for all trials. The end point is considered to be the time lapse from drug injection to the last visible movement. Four cutoff points with equal spacing in log time, 64, 96, 144, and 216 seconds, are chosen for observing the status of the animal. Dosage is computed by body weight, and an experiment with body weight as a second variable (see Table 19-5) is used to obtain information on the validity of the body-weight basis for dosage.

A nominal sample size of 5 is chosen to provide a standard error for $P_{50}$ of $.61\sigma$. For example, in the 64-second 18-to-20-weight cell the first three O's determine the total number of observations in the cell to be seven and Table 19-3 gave $k = -.144$. The final test level was $X_f = 2$, and the corresponding estimate is $\bar{X}_f + kd = 2 + (-.144 \times 1) + 1.856$.

**Table 19-5** Analysis of drug experiment

| WEIGHT, GRAMS | | TIME TO CUTOFF POINT, SEC 64 | 96 | 144 | 216 |
|---|---|---|---|---|---|
| 18–20 | | | | | |
| | *Tests* | OOOXXOX | OOOOXXOX | OOXXOO | XXOXXO |
| | $X_f$ | 2.000 | − .107 | −3.213 | −7.213 |
| | $k$ | −.144 | − .142 | .372 | .861 |
| | *Est.* | 1.856 | − .249 | −2.841 | −6.352 |
| 21–23 | | | | | |
| | *Tests* | OOOXOXX | OXXXX | XOXXX | XXOXOX |
| | $X_f$ | 4.000 | −1.107 | −5.213 | −6.213 |
| | $k$ | .181 | .555 | .157 | − .737 |
| | *Est.* | 4.181 | − .552 | −5.056 | −6.950 |
| 24–26 | | | | | |
| | *Tests* | XXXOXXO | XXOXXO | OXXOX | OXOXX |
| | $X_f$ | 1.000 | −3.107 | −4.213 | −6.213 |
| | $k$ | .860 | .861 | − .305 | .084 |
| | *Est.* | 1.860 | −2.246 | −4.518 | −6.129 |

**Table 19-6** Estimated $P_{50}$ dosages (log units) for reduced drug experiment

| WEIGHT, GRAMS | TIME, SECONDS 64 | 96 |
|---|---|---|
| 18–20 | 1.856 | −.249 |
| 21–23 | 4.181 | −.552 |

In the 12 experiments which form the basis of this analysis 13 additional animals beyond the minimum 12 × 5 were used.

To simplify the presentation of the analysis let us first assume that only two cutoff time points, 64 and 96, were used and that only the two lower weight classes were included. The results of this experiment are presented in Table 19-6. The analysis-of-variance table can be computed in the manner of Sec. 10-4. The results are given in Table 19-7. The analysis

**Table 19-7** Analysis of variance for reduced drug experiment

| | *Sum of squares* | *Degrees of freedom* | *Mean square* |
|---|---|---|---|
| Time | 11.690 | 1 | 11.690 |
| Weight | 1.022 | 1 | 1.022 |
| Remainder | 1.727 | 1 | 1.727 |
| *Total* | 14.439 | 3 | |

of this table could then proceed as described in Chap. 10. However, $F$ ratios with a single degree of freedom for the denominator must be very large to be declared significant. The original experiment has the advantage of more detailed coverage of time and weight variables and also provides, with more degrees of freedom, a better estimate of remainder mean square. The estimates from the original experiment are repeated in Table 19-8.

Both variables of classification are ordered and weight classes are equally spaced. Time is equally spaced in log time. Analysis will be performed by the methods of Sec. 15-1. We shall obtain only the linear component for each of the two variables of classification. The coefficients for

**Table 19-8** Data from drug experiment

| WEIGHT, GRAMS | TIME TO CUTOFF, SECONDS 64 | 96 | 144 | 216 |
|---|---|---|---|---|
| 18–20 | 1.856 | −.249 | −2.841 | −6.352 |
| 21–23 | 4.181 | −.552 | −5.056 | −6.950 |
| 24–26 | 1.860 | −2.246 | −4.518 | −6.129 |

**Table 19-9** Coefficients for linear time effect

| | | | |
|---|---|---|---|
| −3 | −1 | 1 | 3 |
| −3 | −1 | 1 | 3 |
| −3 | −1 | 1 | 3 |

**Table 19-10** Coefficients for linear weight effect

| | | | |
|---|---|---|---|
| −1 | −1 | −1 | −1 |
| 0 | 0 | 0 | 0 |
| 1 | 1 | 1 | 1 |

the linear contrast on time are given in Table 19-9, and the coefficients for linear weight effect are given in Table 19-10. The sum of squares for linear time effect is computed as follows:

$$\frac{(-3 \times 1.856) - 1(-.249) + \cdots + 3(-6.129)}{60} = 139.086$$

the denominator is computed as

$$(-3)^2 + (-1)^2 + \cdots + 3^2 = 60$$

We proceed in a similar fashion for the linear weight effect. The sum of squares for these two effects can be subtracted from the total sum of squares to provide the remainder sum of squares. These results are presented in Table 19-11.

An $F$ test for time effect is significant at the $\alpha = .0005$ level of significance, and the weight effect is not significant. We can obtain additional results from the analysis-of-variance tables. Since tests carried out at nominal sample size 5 are expected to provide estimates of $P_{50}$ with standard error $.61\sigma$ (see Table 19-3), we estimate $\sigma$ as $\sqrt{.972}/.61 = 1.61$.

We can indicate in a linear form the relationship between log dose and log time. The estimate of the slope of the dose-response line for changes in cutoff time is $-\sqrt{139.086}/3 = -3.931$. The minus sign results from the fact that the required dose decreases with increasing time. Division by 3 is required because the experiment ranges across three time intervals above 64 seconds.

The mean $P_{50}$ log dose in this experiment is $-2.250$. The mean log time is 2.032. Therefore the regression line is

$$\text{Log dose} = -2.250 - 3.931 \times (\text{log time} - 2.032)$$

**Table 19-11** Analysis of variance for drug experiment

| | *Sum of squares* | *Degrees of freedom* | *Mean square* | *F ratio* |
|---|---|---|---|---|
| Time (linear) | 139.086 | 1 | 139.086 | 143. |
| Weight (linear) | 1.485 | 1 | 1.485 | 1.5 |
| Remainder | 8.752 | 9 | .972 | |
| *Total* | 149.323 | 11 | | |

This may be written in the form

$$\text{Dose} = 311e^{-51\ \text{time}}$$

which relates median dose for lethal effect to time in seconds to cutoff point.

## GLOSSARY

all-or-none data
critical height or dose
nominal sample size
quantal response
sensitivity data
threshold
up-and-down method

## DISCUSSION QUESTIONS

**1.** Describe an experiment involving materials or subjects where the observations are all-or-none, or quantal, responses. In what way is the material or subject altered so that repeated tests may not be made on the same individual even though no reaction or response occurs?

**2.** Discuss the reason that the curves in Fig. 19-3 branch into two curves when $d/\sigma$ becomes large.

**3.** If the observations are such that the logarithms of the observations are normally distributed, the graphical solution discussed at the beginning of this chapter may be effected by using logarithmic-probability paper. Logarithmic-probability paper differs from normal-probability paper in that the scale of measurements is marked so that in effect the logarithms of the measurements are plotted. Discuss how the estimates of the mean and standard deviation may be obtained from this paper.

## CLASS EXERCISES

**1.** Make a record of 50 tests, using as observations the random normal numbers in Table A-2 by taking the heights $-3$, $-2$, $-1$, 0, 1, 2, 3 as various levels and recording a $\times$ if the number is below the level considered at the moment and a $\bigcirc$ if it is above. Change level as indicated by the up-and-down method. Compute $\bar{X}$, $s$, $P_{05}$, and $P_{95}$. Collect the results of all the students into frequency distributions and comment on the results.

**2.** Draw a cumulative polygon on normal-probability paper for the same 50 observations used for Exercise 1, with $-2$, $-1$, 0, 1, 2 as interval limits. Draw a best-fitting line by eye and estimate the mean, the standard deviation, $P_{05}$, and $P_{95}$ as discussed in Secs. 5-5 and 19-1. Collect the results of all the students into frequency distributions and comment on the results. (Note that, in contrast to Prob. 1 below, different answers are expected because of sampling variation. In Prob. 1 the variation observed arises from individual preference in placing the line.)

**3.** Sampling verification of some of the results of this chapter can be obtained by comparing random normal numbers with a set of levels chosen in the manner of Sec. 19-5. Choose the nominal sample size $N = 4$ and $d = 1$ for normal numbers with $\sigma = 1$. Generate 20 series, estimate $P_{50}$ for each, and compute the standard deviation of these estimates. Compare this result with the value .67 given in Table 19-3 for $N = 4$. Count the total number of trials required and compute $N'/N$. Repeat this experiment for different starting levels $-2$, 0, 2, 4.

## PROBLEMS

*Section* 19-1

**1.** Plot the results in the table below on normal-probability and log-normal-probability paper. Using the plot which seems to fit best, draw a straight line by eye and estimate the mean and standard deviation. Compare results with other members of the class.

| *Level* | *No. exposed* | *No. affected* |
|---|---|---|
| 2 | 100 | 5 |
| 5 | 100 | 30 |
| 10 | 100 | 60 |
| 20 | 100 | 70 |
| 30 | 100 | 80 |
| 40 | 100 | 95 |

*Section* 19-3

**2.** The tabulated data were obtained by the up-and-down technique.

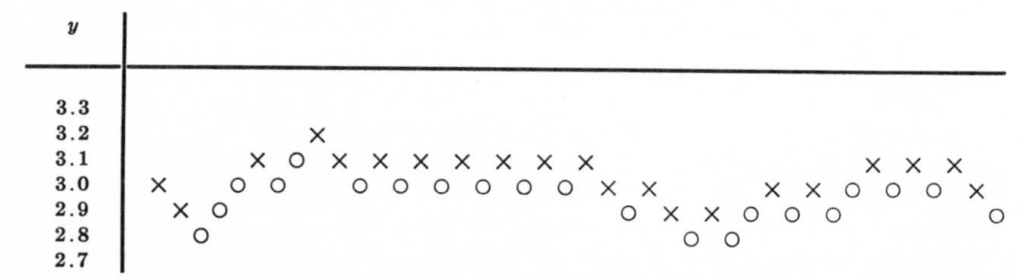


Analyze the data by the techniques discussed in this chapter as follows: (*a*) Estimate the mean. (*b*) Estimate the standard deviation. (*c*) Give 95 per cent confidence limits for the mean. (*d*) Give 90 per cent confidence limits for $P_{20}$. (*e*) If the $y$ values are natural logarithms of dosages of a poisonous drug, estimate the dosage which kills 20 per cent of the individuals. Estimate, by a 95 per cent confidence interval, the dosage which kills 50 per cent of the individuals.

*Section* 19-5

**3.** (*a*) From Table 19-3 find the estimate of $P_{50}$ for the series O, O, O, X, X, O, X, O with final test .602 and interval .301. (*b*) If these levels represent logarithms to base 10, find the estimate of $P_{50}$ in the original units.

CHAPTER TWENTY

# Probability

The concept of probability may be discussed in a general way as a theory based on an equally likely set of events, or as relative frequencies, or as a subjective determination of fair odds.

In an intuitive sense *probability* is usually related to a set of events which are *equally likely*. For example, if we select one person from a *well-mixed* group of 50 people we may consider any of the 50 people to be equally likely to be chosen. If we draw a card from a *well-shuffled* deck of ordinary playing cards we expect to have an equal chance of selecting any particular one of the 52 cards in the deck. In drawing a ball from a box containing seven *well-mixed* balls we expect that any one of the balls is equally likely to be chosen.

A comparable situation occurs when a table of random numbers is used to select people for a public-opinion poll or to select a few plants for harvesting to estimate a total yield.

We may think of probability as the *relative frequency* of particular events in a very long sequence of trials. In tossing a coin, for example, we generally expect heads or tails to be equally likely. This is based on the fact that a coin has two sides, and when a fair (or true) coin is tossed repeatedly, it will fall heads with approximately the same frequency as it will fall tails. However, a *particular* coin in a long series of tosses may come up heads considerably less often than tails, so that heads and tails would not be equally likely. Therefore we may think of the probability of heads for a particular coin as the ratio of the number of heads to the total number of tosses in a sequence of tosses. Similarly, we may define the probability that a particular item in a group of manufactured articles will be defective as the ratio of the number of defective items to the total number of articles manufactured. The probability of an explosion when a bomb is dropped may be defined as the ratio of the number of explosions to the total number of bombs. We can connect this concept of repeated trials to the equally likely concept by considering a long sequence of draws from a deck of cards (replacing the card drawn each time and reshuffling the deck) or a long sequence of draws from a box of balls (replacing and remixing each time).

A third reference to probability may be described as *subjective*. A man may feel that enough is known about a situation to enable him to assess the "odds" that some hypothesis might be true. He may indicate the extent of his belief that an event will occur or that a particular hypothesis is true by stating the odds he would consider a fair bet. That is, 1:10 odds indicate a different level of belief or information about a situation from 1:2 odds. Odds of 1:10 indicate an assessment of probability as $\frac{1}{11}$, and odds of 1:2 indicate $\frac{1}{3}$. For example, a gambler may feel that today is his lucky day and be willing to bet on odds different from those he would accept under other circumstances. Even though no bet is involved, the degree of belief in a hypothesis may be indicated, for example, by the statement that the odds against life on the moon are 1:1,000,000. Presumably the odds would change with the addition of any further knowledge about the moon.

This last concept is not interpreted on a frequency basis, and, of course, the theorems derived from a relative-frequency or equally-likely basis should not be used in this case.

We shall approach probability from the first viewpoint discussed above, taking a finite set of equally likely possibilities, writing down the relationships which seem reasonable, and then formalizing the theory by stating these "reasonable" results as axioms or as theorems which follow from the axioms.

## 20-1 TWO-WAY FREQUENCY TABLE

EXAMPLE 20-1. Suppose we have a box containing 100 balls of the following description:

10 red balls marked with the digit 1
20 red balls marked with the digit 2
10 red balls marked with the digit 3
20 white balls marked with the digit 1
10 white balls marked with the digit 2
30 white balls marked with the digit 3

**Table 20-1** Two-way frequency table

| | | MARKING | | |
|---|---|---|---|---|
| | 1 | 2 | 3 | *Total for each color* |
| COLOR Red | 10 | 20 | 10 | 40 |
| COLOR White | 20 | 10 | 30 | 60 |
| *Total for each marking* | 30 | 30 | 40 | 100 *Total no. of balls* |

A tabular presentation of the composition of the box of balls is given in Table 20-1.

Instead of using *color* and *number* to describe a ball, we can consider classifications such as *sex* and *hair color* for people or *quality* and *type of process* for manufactured parts. For test animals we could consider one classification *infected-uninfected* and the other classification *inoculated-uninoculated.*

Now, returning to the example of 100 balls described above, we shall agree to consider that if a ball is drawn from the box, each ball is equally likely to be chosen. Furthermore we shall interpret the fraction $\frac{1}{100}$ as the probability of choosing any particular ball that is specified. Since there are a number of balls that have the same description—10 which are red and marked 3, 40 which are red, 30 which are marked 2, and so on—we shall define the probability of drawing a red ball marked 3 to be $\frac{10}{100}$, the probability of drawing a red ball to be $\frac{40}{100}$, the probability of drawing a ball marked 2 to be $\frac{30}{100}$, and so on. That is, the probability of choosing one of these 100 equally likely balls having a certain characteristic will be defined as the number of balls in the group having the characteristic divided by the total number of balls in the box.

We shall use the notation $P(A)$ to represent the probability that an object drawn will have characteristic $A$. We shall also interpret $A$ as the *event* of drawing an object having a certain characteristic. In this notation

$P(\text{red ball marked 2}) = \frac{20}{100}$
$P(\text{red ball}) = P(\text{red ball marked 1, 2, or 3}) = \frac{40}{100}$
$P(\text{red ball marked 2 or white ball marked 3}) = \frac{50}{100}$
Etc.

The conditions of the drawing are often added to the above notation and must, of course, be stated if there is any doubt or if we wish to specify different conditions. For example, we may wish to write the $P(\text{red ball})$ of the above example as

$$P(\text{red ball} \mid \text{set of 100 equally likely balls of which 40 are red}) = \frac{40}{100}$$

The conditions of the drawing are stated following the vertical bar, which is often read "given." If a drawing is to be made from some part of the originally specified group of objects, this condition should always be stated.

Thus

$$P(\text{red ball} \mid \text{ball marked 1}) = \tfrac{10}{30}$$

which is obtained by following the same rules for computing a probability, that is, by computing the ratio of the number of objects having the characteristic to the total number of balls available for the drawing. In this context this probability is referred to as a *conditional probability*. Another example is

$$P(\text{ball marked 2} \mid \text{white ball}) = \tfrac{10}{60}$$

In dealing either with two possible events or with two descriptive characteristics we find it very useful to consider the use of the words *and* and *or* in the following statements:

1. The first event *and* the second event both occur.
2. One event *or* the other event occurs.

Thus we speak of a ball that is red *and* marked 1, or a ball marked 1 *or* marked 2, or a ball that is red *or* marked 1. Note that the word "or" is used in the sense of including *either or both*. From Table 20-1 we see that we can write

$$P(\text{red } \textit{and} \text{ marked 1}) = \frac{10}{100}$$

$$P(\text{marked 1 } \textit{or} \text{ 2}) = \frac{30 + 30}{100}$$

Care must be taken in computing probabilities of the type $P(A \text{ } or \text{ } B)$ to avoid counting any ball more than once. For example, we write

$$P(\text{red } \textit{or} \text{ marked 1}) = \frac{10 + 20 + 10 + 20}{100} = \frac{60}{100}$$

or we can actually count some of the balls twice and then subtract to correct the total as follows:

$$P(\text{red } \textit{or} \text{ marked 1}) = \frac{40 + 30 - 10}{100} = \frac{60}{100}$$

The marginal total for red is 40, and the marginal total for balls marked 1 is 30. The sum of these two marginal totals will include in both cases the 10 red balls marked 1. This can be restated as

$$\begin{aligned} P(\text{red } \textit{or} \text{ marked 1}) &= \tfrac{40}{100} + \tfrac{30}{100} - \tfrac{10}{100} \\ &= P(\text{red}) + P(\text{marked 1}) - P(\text{red } \textit{and} \text{ marked 1}) \end{aligned}$$

Events can also be classified as *complementary, exhaustive, independent,* and *exclusive.*

In the above example a ball is either red or white, and we refer to drawing a red ball as *complementary* to drawing a white ball. The event complementary to drawing a ball marked 1 is drawing a ball marked either 2 or 3.

An *exhaustive* set of events includes every event that could possibly happen. In our example drawing a red ball and drawing a white ball are two events which include every possibility. The probability that some event of a set of exhaustive events will occur is 1.

If the conditional probability of event $A$, given another event $B$, is equal to the probability of $A$ without that condition, then events $A$ and $B$ are called *independent*. The condition for independence can be stated with the symbols

$$P(A \mid B) = P(A)$$

In our example $P(\text{red} \mid \text{marked } 1) = \frac{10}{30}$, which is not equal to $P(\text{red}) = \frac{40}{100}$; so these events are *dependent*.

*Exclusive* events cannot occur simultaneously; for example, a single ball cannot be both red and white, so red and white are exclusive characteristics.

## AXIOMS AND THEOREMS OF PROBABILITY 20-2

Table 20-1 can be generalized by introducing letters for the frequencies as in Table 20-2. With this notation we can write

$$P(\text{marked } 1) = \frac{n_{11} + n_{21}}{N}$$

$$P(\text{red} \mid \text{marked } 1) = \frac{n_{11}}{n_{11} + n_{21}}$$

$$\begin{aligned} P(\text{red } \textit{and} \text{ marked } 1) &= \frac{n_{11}}{N} = \frac{n_{11} + n_{21}}{N} \frac{n_{11}}{n_{11} + n_{21}} \\ &= P(\text{marked } 1) \times P(\text{red} \mid \text{marked } 1) \\ P(\text{red } \textit{or} \text{ marked } 1) &= \frac{(n_{11} + n_{12} + n_{13}) + (n_{11} + n_{21}) - n_{11}}{N} \\ &= P(\text{red}) + P(\text{marked } 1) - P(\text{red } \textit{and} \text{ marked } 1) \end{aligned}$$

Similar formulas may be written for the white balls and for the balls marked 2 and 3.

**Table 20-2**

| | MARKING 1 | 2 | 3 | *Total for each color* |
|---|---|---|---|---|
| COLOR Red | $n_{11}$ | $n_{12}$ | $n_{13}$ | $n_{11} + n_{12} + n_{13}$ |
| COLOR White | $n_{21}$ | $n_{22}$ | $n_{23}$ | $n_{21} + n_{22} + n_{23}$ |
| *Total for each marking* | $n_{11} + n_{21}$ | $n_{12} + n_{22}$ | $n_{13} + n_{23}$ | $n_{11} + n_{12} + n_{13} + n_{21} + n_{22} + n_{23} = N$ |

In general, we may note from this table that if $A$ and $B$ are two events, we always have the following relationships among the probabilities:

AXIOM 1. $0 \leq P(A) \leq 1$.

AXIOM 2. $P(A \text{ and } B) = P(A)P(B \mid A) = P(B)P(A \mid B)$.

AXIOM 3. $P(A \text{ or } B) = P(A) + P(B) - P(A \text{ and } B)$.

We shall consider these three relationships as axioms for our probability calculations. If the possible events were not assumed to be equally likely, as was assumed in the tables above, a more elaborate development would be necessary.

Various theorems may be deduced from the above axioms. For example, using $A_1, A_2, \ldots, A_k$ to represent events, we have the following:

THEOREM 1

*Two events:*

$$P(A_1 \text{ or } A_2) = P(A_1) + P(A_2) - P(A_1 \text{ and } A_2)$$

*Three events:*

$$\begin{aligned} P(A_1 \text{ or } A_2 \text{ or } A_3) = {} & P(A_1) + P(A_2) + P(A_3) - P(A_1 \text{ and } A_2) \\ & - P(A_1 \text{ and } A_3) - P(A_2 \text{ and } A_3) \\ & + P(A_1 \text{ and } A_2 \text{ and } A_3) \end{aligned}$$

*Four events:*

$$\begin{aligned} P(A_1 \text{ or } A_2 \text{ or } A_3 \text{ or } A_4) = {} & P(A_1) + P(A_2) + P(A_3) + P(A_4) \\ & - P(A_1 \text{ and } A_2) - P(A_1 \text{ and } A_3) \\ & \qquad - P(A_1 \text{ and } A_4) \\ & - P(A_2 \text{ and } A_3) - P(A_2 \text{ and } A_4) \\ & \qquad - P(A_3 \text{ and } A_4) \\ & + P(A_1 \text{ and } A_2 \text{ and } A_3) \\ & \qquad + P(A_1 \text{ and } A_3 \text{ and } A_4) \\ & + P(A_1 \text{ and } A_2 \text{ and } A_4) \\ & \qquad + P(A_2 \text{ and } A_3 \text{ and } A_4) \\ & - P(A_1 \text{ and } A_2 \text{ and } A_3 \text{ and } A_4) \end{aligned}$$

Similar relationships hold for more than four events.

THEOREM 2. *If $A_1$ and $A_2$ are mutually exclusive, that is, if*

$$P(A_1 \text{ and } A_2) = 0$$

*then* $P(A_1 \text{ or } A_2) = P(A_1) + P(A_2)$.

This is merely a special case of Theorem 1.

Theorem 3. *If* $P(A_1 \mid A_2) = P(A_1)$ *and* $P(A_1) \neq 0$, *then*

$$P(A_2 \mid A_1) = P(A_2)$$

This follows directly from Axiom 2.

Definition. *If* $P(A_1 \text{ or } A_2 \text{ or } \cdots \text{ or } A_k) = 1$, *then* $A_1, A_2, \ldots, A_k$ *are an exhaustive set of events.*

Definition. *If* $A_1$ *represents the occurrence of an event and* $\tilde{A}_1$, *read "A tilde," represents the nonoccurrence of the same event, then* $A_1$ *and* $\tilde{A}_1$ *are called complementary events.*

It follows, of course, that

$$P(A_1) + P(\tilde{A}_1) = 1$$

## COUNTING TECHNIQUES: PERMUTATIONS AND COMBINATIONS 20-3

The consideration of probability problems will often require the enumeration of the possible ways events can occur. This section will present several methods to aid in this enumeration.

Definition. *If* $N$ *is a positive integer, the product*

$$1 \times 2 \times 3 \cdots (N - 1) \times N$$

*is defined by the symbol* $N!$ *and is read "N factorial."*

It will be convenient to define $0!$ to be 1.

Definition. $0! = 1$.

Theorem 1. *The number of different ways* $N$ *distinct objects can be arranged in line is* $N!$.

An arrangement of $N$ objects in a line is called a permutation of the $N$ objects. The total number of permutations, then, of $N$ objects is $N!$.

Theorem 2. *The number of ways of selecting and arranging* $r$ *objects taken from* $N$ *distinct objects is* $P(N,r) = \dfrac{N!}{(N - r)!}$.

Definition. *The total number of possible selections of* $r$ *objects from* $N$ *distinct objects is called the number of combinations of* $N$ *objects taken* $r$ *at a time and will be denoted by* $C(N,r)$.

Other notations used are $\binom{N}{r}$ or ${}_NC_r$ or ${}^NC_r$.

THEOREM 3. $C(N,r) = \dfrac{N!}{r!(N-r)!}$.

THEOREM 4. $C(N,r) = C(N,N-r)$.

THEOREM 5. $C(N,N) = C(N,0) = 1$.

DEFINITION. *If* $r < 0$ *or if* $r > N$, *then* $C(N,r) = 0$.

THEOREM 6

$$
\begin{aligned}
(a+b)^N &= \sum_{r=0}^{N} C(N,r)a^{N-r}b^r \\
&= C(N,0)a^N + C(N,1)a^{N-1}b + C(N,2)a^{N-2}b^2 \\
&\quad + \cdots + C(N,N-1)ab^{N-1} + C(N,N)b^N
\end{aligned}
$$

The values of $C(N,r)$ are often called *binomial coefficients*, since they occur as coefficients in the relationship shown in Theorem 6. Values of $C(N,r)$ for $N \leq 10$ are given in Table A-29*b*.

Setting $a = b = 1$ in Theorem 6 gives

THEOREM 7. $\displaystyle\sum_{r=0}^{N} C(N,r) = 2^N$.

The use of the above axioms, theorems, and definitions will be illustrated by several examples.

EXAMPLE 20-2. We have a box of 50 light bulbs, of which five are defective. What is the probability, or chance, of finding exactly two defective bulbs in a sample of size 4 chosen from the box? We consider any one of the $C(50,4)$ possible samples as being equally likely and define the probability as

$$
\begin{aligned}
&P(\text{exactly 2 defectives in sample of 4} \mid \text{box of 50 with 5 defectives}) \\
&\qquad = \frac{\text{number of possible samples of 4 bulbs with 2 defectives}}{C(50,4)}
\end{aligned}
$$

To count the number of possible samples having two defectives, we note that the two defectives must come from the five defectives in the box [this can be done in $C(5,2)$ ways] and the two nondefectives from the 45 nondefectives in the box [this can be done in $C(45,2)$ ways].

Note that the number of ways two independent events can occur together is the product of the numbers of ways each event can occur separately. Similarly, note that the number of ways that some one of several exclusive events can occur is the sum of the number of ways each can occur. In this case the two events which must occur are the selection of two light bulbs from the five which are defective and the selection of two light bulbs from the 45 which are not defective. Thus the number of ways we can select a sample of size 4 having two defectives is equal to the product of $C(5,2)$ and $C(45,2)$; the desired probability is

$$\frac{C(5,2)C(45,2)}{C(50,4)} = \frac{\dfrac{5!}{2!3!}\,\dfrac{45!}{2!43!}}{\dfrac{50!}{4!46!}}$$

$$= \frac{5 \times 4}{2}\,\frac{45 \times 44}{2}\,\frac{4 \times 3 \times 2 \times 1}{50 \times 49 \times 48 \times 47} = .0430$$

In a similar manner we can compute the probabilities of no defectives, of exactly one defective, of exactly three defectives, of exactly four defectives in a sample of size 4. The results are displayed in Table 20-3 [the notation $f(x)$ and $F(x)$ in the column headings is used later in this chapter].

The above discussion is *mathematical* in nature in that we are given complete information about the population sampled, that is, 50 items of which five are defective. A *statistical* question occurs when we do not know how many of the 50 items are defective but wish to infer from the sample the proportion defective in the population. For example, as in Chap. 7, we could use a rule of behavior (decision rule) which will declare the proportion of defectives in the population too high if any of the five in the sample are defective. If this occurs, we may decide to inspect all 50 items. From Table 20-3 we see that 64.7 per cent of the boxes containing five defective items would not be examined further if this decision rule were followed.

EXAMPLE 20-3. Suppose that an animal with a particular disease has a 40 per cent chance of recovering if treated with a certain drug (and, of

**Table 20-3** Probability of $x$ defectives in a sample of size 4 from a box containing 45 good and 5 defective items. The total of $P(x \text{ defectives})$ for all the values of $x$ is 1

| *Measurement* $x$ | *Frequency* $P(x \text{ defectives}) = f(x)$ | *Cumulative frequency* $P(x \text{ or fewer defectives}) = F(x)$ |
|---|---|---|
| 4 | $\frac{C(5,4)C(45,0)}{C(50,4)} = .00002$ | 1.00000 |
| 3 | $\frac{C(5,3)C(45,1)}{C(50,4)} = .00195$ | .99998 |
| 2 | $\frac{C(5,2)C(45,2)}{C(50,4)} = .04299$ | .99803 |
| 1 | $\frac{C(5,1)C(45,3)}{C(50,4)} = .30808$ | .95504 |
| 0 | $\frac{C(5,0)C(45,4)}{C(50,4)} = .64696$ | .64696 |

course, a 60 per cent chance of not recovering). One interpretation of this statement is that if all of a very large number of animals with the disease were treated with the drug, 40 per cent of the animals would recover and we state a probability of 40 per cent on the basis of our "random choice" of an animal from this large number of animals. The interpretation of this example is the same as in Example 20-2, except that the number of items from which the sample is chosen is very large or infinite. An alternative interpretation is that there is a 40 per cent chance of recovery for any animal when given the drug; that is, the probability of recovery is inherent in the animal itself rather than representing the chance of selection of an animal which will recover when given the drug. This second interpretation is analogous to the occurrence of heads in the toss of a coin where the chance of heads is 40 per cent.

Suppose we want to know the probability that exactly two out of five treated animals will recover. We take .4 as the probability of recovery of the animal chosen first, .4 as the probability of recovery of the animal chosen second (it is independent of the first choice, as in tossing coins), and so on. The probability of recovery of any one of the five animals is .4 (independent of the recovery or nonrecovery of the other choices).

We shall now attack the problem with a procedure useful for a wide variety of probability problems. We write the event of two recoveries resulting when five animals are treated as a series of exclusive events as follows:

"2 recoveries out of 5 treated" is the same as

"Animals 1 *and* 2 recover *and* animals 3, 4, 5 do not recover"
*or* "Animals 1 *and* 3 recover *and* animals 2, 4, 5 do not recover"
*or* "Animals 1 *and* 4 recover *and* animals 2, 3, 5 do not recover"
*or* "Animals 1 *and* 5 recover *and* animals 2, 3, 4 do not recover"
*or* "Animals 2 *and* 3 recover *and* animals 1, 4, 5 do not recover"
*or* "Animals 2 *and* 4 recover *and* animals 1, 3, 5 do not recover"
*or* "Animals 2 *and* 5 recover *and* animals 1, 3, 4 do not recover"
*or* "Animals 3 *and* 4 recover *and* animals 1, 2, 5 do not recover"
*or* "Animals 3 *and* 5 recover *and* animals 1, 2, 4 do not recover"
*or* "Animals 4 *and* 5 recover *and* animals 1, 2, 3 do not recover"

There are $C(5,2)$ exclusive events combined in the either/or statements listed above, and the probability of the combined event is the sum of the probabilities of the individual events. Since the chances for recovery for individual animals are independent, the probability that the first and second animals recover and the remaining three do not is

$$P(\text{1st recovers}) \times P(\text{2d recovers}) \times P(\text{3d does not recover}) \times P(\text{4th does not recover}) \times P(\text{5th does not recover}) = .4 \times .4 \times .6 \times .6 \times .6 = (.4)^2(.6)^3$$

**Table 20-4** Probability of exactly $x$ recoveries out of five independent trials where the chance of any individual recovering is .4

| $x$ | $P$(*exactly x recoveries*) $= f(x)$ | $P$(*x or fewer recoveries*) $= F(x)$ |
|---|---|---|
| 5 | $C(5,5)(.4)^5 = .01024$ | 1.00000 |
| 4 | $C(5,4)(.4)^4(.6) = .07680$ | .98976 |
| 3 | $C(5,3)(.4)^3(.6)^2 = .23040$ | .91296 |
| 2 | $C(5,2)(.4)^2(.6)^3 = .34560$ | .68256 |
| 1 | $C(5,1)(.4)(.6)^4 = .25920$ | .33696 |
| 0 | $C(5,0)(.6)^5 = .07776$ | .07776 |

The chance of any of the $C(5,2) = 10$ events listed above will be the same, since the only change in writing down the probability is an interchange in the order in which the factors are written. Therefore we have

$$P(\text{exactly 2 recover from 5}) = C(5,2)(.4)^2(.6)^3 = .3456$$

Similarly, we can compute the probability of exactly three recoveries out of five, etc. These results are presented in Table 20-4.

Again, the above discussion is mathematical. The statistical question would occur here if we wished to estimate the chance of recovery from the observations made on the five animals.

EXAMPLE 20-4. $Y$ claims that he can tell whether or not there is water in a closed container by holding a forked stick above the container. To test his talent it is suggested that the following experiment be performed. Six sets of three containers each are presented to $Y$. One container of each set of three is filled with water, the other two are empty, and $Y$ is so informed. He attempts to choose the one container in each set which is filled with water. If he has no talent, he has a chance of $\frac{1}{3}$ of correctly choosing the container with water in any one set, independently of his other choices. If he has some ability he should do better than this. In other words, he would have an independent chance greater than $\frac{1}{3}$ of correctly choosing the container with water in any one set of three containers. If we specify the chance $p$ that $Y$ will correctly choose the container with water, this example is the same as Example 20-3, and we can compute the probability of his picking exactly $x$ correctly in six trials as in that example. For example,

$$P(\text{4 correct choices in 6 trials with } p = \tfrac{1}{3}) = C(6,4)(\tfrac{1}{3})^4(\tfrac{2}{3})^2 = .0823$$

The results of this computation are given in Table 20-5 for four values of $p$.

A possible assessment of the ability of $Y$ may be made by agreeing to say there is evidence that he has some ability if he is correct in five or more of the six trials. If he is guessing, we would then have probability of concluding he has some ability (when he actually possesses no unusual ability) equal to $.0179 = .0014 + .0165$. Drawing an inference about his talent from a sample of his work is a *statistical* problem. As in Chaps. 6 and 14,

**Table 20-5** Probability of correctly picking exactly $x$ containers of water in six trials with constant probability of success $p$ in each trial

| $x$ | $p = \frac{1}{3}$ | $p = \frac{1}{2}$ | $p = \frac{4}{5}$ | $p = 1$ |
|---|---|---|---|---|
| 6 | .0014 | .0156 | .2621 | 1 |
| 5 | .0165 | .0938 | .3932 | 0 |
| 4 | .0823 | .2344 | .2458 | 0 |
| 3 | .2195 | .3125 | .0819 | 0 |
| 2 | .3292 | .2344 | .0154 | 0 |
| 1 | .2634 | .0938 | .0015 | 0 |
| 0 | .0878 | .0156 | .0001 | 0 |

the probability of our concluding that $Y$ has some talent when he actually has none is called the *level of significance* of the experiment.

The $P$(5 *or* 6 correct choices in 6 trials) is graphed in Fig. 20-1. This curve is a power curve for the test of $Y$'s ability. The height of the curve gives the probability that we will decide to say "$Y$ is not guessing" if actually he has probability $p$ of selecting the right container. If he has a 50 per cent chance of being correct there is a probability of .11 that we will conclude that he is not guessing; if he has an 80 per cent chance there is a probability of .66 that we will conclude that he is not guessing. The curve in Fig. 20-1 is the power curve of the test of the hypothesis $H: p = \frac{1}{3}$, where the critical region of the test is $x = 5, 6$.

Other experiments could be designed to test $Y$. For example, the containers could be presented two at a time instead of three at a time, or they could be presented singly, with a coin tossed to determine whether or not each container was empty. As noted in earlier chapters the specific way the data are collected is referred to as the *design of the experiment*, the statement concerning the population (for example, $p = \frac{1}{3}$ to specify that $Y$ is guessing) is called a *statistical hypothesis*, and the rule which indicates

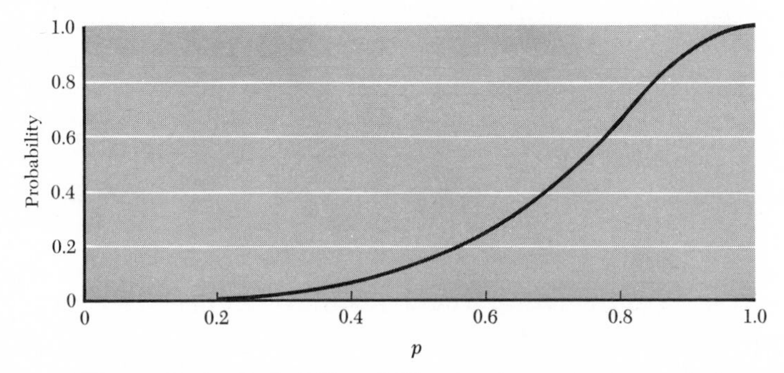


**Figure 20-1** Probability of $Y$ having five *or* six correct choices out of six trials with probability $p$ for each trial.

the experimenter's procedure (for example, rejection of the hypothesis if five or six successes result) is a *statistical test* of the hypothesis.

EXAMPLE 20-5. Yellow guinea pigs crossed with yellow produce yellow guinea pigs, and white crossed with white produce white. Yellow guinea pigs crossed with white always produce cream-colored offspring. Cream-colored guinea pigs when crossed produce in the long run 25 per cent yellow offspring, 50 per cent cream offspring, and 25 per cent white offspring. This type of inheritance may be explained with the use of the genetic formulas $AA$, $Aa$, $aa$, which represent yellow, cream, and white, respectively. When two animals are crossed, the offspring receive one factor represented by $A$ or $a$ from each parent. Therefore if both parents are $AA$ all offspring will be $AA$, and if both parents are $aa$ all offspring will be $aa$.

If both parents are $Aa$, then their offspring may be $AA$, $Aa$, or $aa$. Since the cream guinea pigs are of the type $Aa$, all three types will result, and we can compute the probabilities as follows, assuming that either factor has an equal chance of being transmitted by each parent:

$$\begin{aligned} P(\text{offspring } AA) &= P(\text{receiving } A \text{ from father } \textit{and } A \text{ from mother}) \\ &= \tfrac{1}{2} \times \tfrac{1}{2} = .25 \\ P(\text{offspring } Aa) &= P\,[(\text{receiving } A \text{ from father } \textit{and } a \text{ from mother}) \textit{ or} \\ &\qquad (\text{receiving } a \text{ from father } \textit{and } A \text{ from mother})] \\ &= (\tfrac{1}{2} \times \tfrac{1}{2}) + (\tfrac{1}{2} \times \tfrac{1}{2}) = .50 \\ P(\text{offspring } aa) &= P(\text{receiving } a \text{ from father } \textit{and } a \text{ from mother})] \\ &= \tfrac{1}{2} \times \tfrac{1}{2} = .25 \end{aligned}$$

With the above assumptions and reasoning we can compute various probabilities. In a group of 10 offspring from type $Aa$ parents the probability that exactly three will be cream is $C(10,3)(.50)^{10}$. The probability that exactly four will be yellow is $C(10,4)(.25)^4(.75)^6$. The probability that exactly three will be cream and exactly four will be yellow is $C(10,3)(.50)^{10}$ $C(7,4)(.50)^4(.50)^3$.

If one parent is $AA$ and the other $Aa$, the probability that an offspring will be $AA$ is .50 and that an offspring will be $Aa$ is .50. Thus the probability that exactly four of 10 offspring from such parents will be $Aa$ is $C(10,4)(\frac{1}{2})^{10}$.

EXAMPLE 20-6. We plan to plant varieties $A$ and $B$ of corn on pairs of equal-size plots at six localities in the state. Since any pair of plots at the six places has uniform environmental conditions, we may suppose that the probability is $\frac{1}{2}$ that type $A$ will produce a larger yield than type $B$ at any one location if the varieties are equal in yield. When the yield of type $A$ exceeds that of type $B$ at any location we record a plus, and otherwise a minus. We suppose that the probability of a tie is 0. We can now compute the probabilities: $P$(0 plus values), $P$(1 plus value), $P$(2 plus values), and so on. In fact, a similar computation was done in Example 20-4. The results are given in Table 20-6. This situation will be recognized as that described in Sec. 17-1, in which the sign test is used. There we denoted by $r$ the number

**Table 20-6** Probability of $x$ plus values and $6 - x$ minus values in six independent trials each having probability $\frac{1}{2}$. For example, $P(1 \text{ plus value}) = C(6,1)(\frac{1}{2})^6 = .093750$

| $x$ | $6-x$ | $P$ | $r$ |
|---|---|---|---|
| 6 | 0 | .015625 | 0 |
| 5 | 1 | .093750 | 1 |
| 4 | 2 | .234375 | 2 |
| 3 | 3 | .312500 | 3 |
| 2 | 4 | .234375 | 2 |
| 1 | 5 | .093750 | 1 |
| 0 | 6 | .015625 | 0 |

of times the less frequent sign occurs. From Table 20-6 for $N = 6$ we can read possible values of $r$ with the probabilities as recorded in Table 20-7.

**Table 20-7** Values of the smallest number of plus or minus signs with the probabilities for each

| $r$ | $P(r) = f(r)$ | $P(r \text{ or less}) = F(r)$ |
|---|---|---|
| 3 | .31250 | 1.00000 |
| 2 | .46875 | .68750 |
| 1 | .18750 | .21875 |
| 0 | .03125 | .03125 |

Note that Table 20-7 contains the complete list of cumulative probabilities $P(r \text{ or less})$ for $N = 6$. Therefore we cannot find a value of $r$ to correspond to any arbitrarily chosen value of $P(r \text{ or less})$ unless the probability is listed. We can, however, state for the arbitrarily chosen value .01 that no integer $r$ satisfies the relationship $P(r \text{ or less}) \leq .01$. The number $r = 0$ can be used as the largest integer for .05 and .10, since $P(0 \text{ or less}) \leq .05$ and $P(0 \text{ or less}) \leq .10$. The number $r = 1$ is the largest integer having the property that $P(r \text{ or less}) \leq .25$.

EXAMPLE 20-7. Suppose for data of the type described in Example 20-6 we write the plus and minus signs in the order observed and compute for the given number of plus and minus signs the probability of any single arrangement. For example, if there are three plus and three minus signs the possible arrangements are given in Table 20-8. Following each arrangement is a

**Table 20-8** All possible arrangements of three plus and three minus signs

| | | | |
|---|---|---|---|
| + + + − − − 2 | + − + − + − 6 | − + + + − − 3 | − + − − + + 4 |
| + + − + − − 4 | + − + − − + 5 | − + + − + − 5 | − − + + + − 3 |
| + + − − + − 4 | + − − + + − 4 | − + + − − + 4 | − − + + − + 4 |
| + + − − − + 3 | + − − + − + 5 | − + − + + − 5 | − − + − + + 4 |
| + − + + − − 4 | + − − − + + 3 | − + − + − + 6 | − − − + + + 2 |

**Table 20-9** Number of runs with their probabilities in arranging three plus signs and three minus signs at random

| $u$ | $P(u) = f(u)$ | $P(u \text{ or fewer}) = F(u)$ |
|---|---|---|
| 6 | $\frac{2}{20}$ | 1.0 |
| 5 | $\frac{4}{20}$ | .9 |
| 4 | $\frac{8}{20}$ | .7 |
| 3 | $\frac{4}{20}$ | .3 |
| 2 | $\frac{2}{20}$ | .1 |

number $u$, which is the number of runs of plus and minus signs in that sequence. If we take each arrangement as equally likely we can compute the probability for any value of $u$ as in Table 20-9.

The total number of arrangements is the number of ways three of the six places can be chosen for the three plus signs, $C(6,3)$. In general, if there are $N_1$ plus signs and $N_2$ minus signs, then there are $C(N_1 + N_2, N_1)$ possible arrangements. It is more difficult to count the arrangements where $u$ has a specified value. It has been shown that if $u$ is even the number of arrangements is

$$2C(N_1 - 1, \tfrac{1}{2}u - 1)C(N_2 - 1, \tfrac{1}{2}u - 1)$$

and if $u$ is odd the number of arrangements is

$$C(N_1 - 1, \tfrac{1}{2}u - \tfrac{1}{2})C(N_2 - 1, \tfrac{1}{2}u - \tfrac{3}{2}) + C(N_1 - 1, \tfrac{1}{2}u - \tfrac{3}{2})C(N_2 - 1, \tfrac{1}{2}u - \tfrac{1}{2})$$

Agreement of these formulas with the distribution computed in Table 20-9 can be shown by setting $N_1 = N_2 = 3$. For example, for $u = 2$ the number of arrangements is $2C(2,0) = 2$, and for $u = 3$ the number of arrangements is $C(2,1)C(2,0) + C(2,0)C(2,1) = 2 + 2 = 4$, and so on.

The application of this theory is given in Sec. 17-3, where the use of Table A-11 is described.

It should be noted that we are computing the probability that a certain number of runs results when we are given $N_1$ and $N_2$. For some problems we may wish to allow $N_1$ and $N_2$ to vary also, that is, to allow the number as well as the arrangement of plus and minus values to be a chance variable. This will lead to different probabilities, and Table A-11 will not be appropriate. For example, if we toss a coin three times the probability that there are exactly two runs ($u = 2$) is the probability that one of the following exclusive sequences of heads and tails occurs: HHT, THH, TTH, HTT. Since the probability of any one of the sequences is $\frac{1}{8}$, the probability that two runs will occur is .5:

$$\begin{aligned} P(\text{two runs}) &= P(\text{HHT } or \text{ THH } or \text{ TTH } or \text{ HTT}) \\ &= P(\text{HHT}) + P(\text{THH}) + P(\text{TTH}) + P(\text{HTT}) \\ &= \tfrac{1}{8} + \tfrac{1}{8} + \tfrac{1}{8} + \tfrac{1}{8} = \tfrac{1}{2} \end{aligned}$$

**Table 20-10** Possible arrangements of three $x$ and three $y$ values with the sum of the ranks of the $x$ values

| | | | | | | | |
|---|---|---|---|---|---|---|---|
| $xxxyyy$ | 6 | $xyxyxy$ | 9 | $xyyxyx$ | 11 | $yyxxxy$ | 12 |
| $xxyxyy$ | 7 | $yxxxyy$ | 9 | $yxxyyx$ | 11 | $yxyyxx$ | 13 |
| $xxyyxy$ | 8 | $xyxyyx$ | 10 | $yxyxxy$ | 11 | $yyxxyx$ | 13 |
| $xyxxyy$ | 8 | $xyyxxy$ | 10 | $yxyxyx$ | 12 | $yyxyxx$ | 14 |
| $xxyyyx$ | 9 | $yxxyxy$ | 10 | $xyyyxx$ | 12 | $yyyxxx$ | 15 |

EXAMPLE 20-8. Suppose we denote by $x_1$, $x_2$, $x_3$ and $y_1$, $y_2$, $y_3$ the yields of three plots of each of two varieties of corn. When the yields of the six plots are known, they are arranged in order of size and ranks are assigned: 1 to the largest, 2 to the next largest, and so on, to 6 for the smallest. Let $T'$ be the sum of the ranks assigned to $x_1$, $x_2$, $x_3$, the yields of the first variety. The sum of all six ranks is 21; therefore the sum of the ranks assigned to $y_1$, $y_2$, $y_3$ will be 21 minus the sum assigned to the first variety. If the varieties are equally good in terms of measured yield we shall consider any of the 20 possible arrangements of the ranks of the plot yields as equally likely. Table 20-10 lists the possible arrangements, with the corresponding values of $T'$, and Table 20-11 gives the distribution of $T'$. The statistic $T'$ is called the *rank-sum statistic*. Table A-20 gives the complete distributions of the rank-sum statistic $T'$ for two samples of size $N_1$ and $N_2$, respectively, from the same population for all sample sizes less than or equal to 10 and a normal approximation for larger sample sizes.

## 20-4 DISCRETE FREQUENCY DISTRIBUTIONS

The examples in the preceding section show the recording of various probabilities in a frequency (probability) table. Tables 20-3 to 20-7, 20-9, and 20-11 contain particular examples in which are given the values $x$ of

**Table 20-11** Distribution of $T'$ for two samples of three observations each

| $T'$ | $P(T') = f(T')$ | $P(T'$ *or fewer*$) = F(T')$ |
|---|---|---|
| 15 | .05 | 1.00 |
| 14 | .05 | .95 |
| 13 | .10 | .90 |
| 12 | .15 | .80 |
| 11 | .15 | .65 |
| 10 | .15 | .50 |
| 9 | .15 | .35 |
| 8 | .10 | .20 |
| 7 | .05 | .10 |
| 6 | .05 | .05 |

the observable variable and the chance $f(x)$ that these values occur. The notation for a general table of this sort is given in Table 20-12. For a given $x$, $f(x)$ gives the probability that $x$ will occur, $F(x)$ gives the probability that $x$ or any smaller $x$ occurs. In symbols,

$$P(x) = f(x)$$
$$P(x \text{ or fewer}) = F(x)$$

The group of individual probabilities $f(x)$ is referred to as a *density distribution* (or *frequency function* or *elementary probability law*). The group of cumulative probabilities $F(x)$ is referred to as a *cumulative distribution* or sometimes merely *distribution*. A variable which has associated with it a distribution given either in the density or in the cumulative form is called a *random variable* (or sometimes a *stochastic variable*). In Table 20-12 the random variable $x$ takes on the values 0, 1, 2, 3, . . . with probabilities as given by the corresponding $f(0)$, $f(1)$, $f(2)$, $f(3)$, . . . . In Table 20-12 the random variable is illustrated as taking on only these values, but in any specific problem other values may occur. For example, the distribution given in Table 20-4 is for a random variable $x$ which takes on the values 0, 1, 2, 3, 4, 5, and the random variable representing the proportion of recoveries takes on the values 0, .2, .4, .6, .8, 1.

The density function is never negative and the sum of all probabilities is equal to unity.

Since the cumulative distribution gives the probability that the random variable is equal to or less than a particular value, it is usually used only where the random variable has an obvious ranking, such as size, age, or number, and not for random variables which represent categories such as color, sex, or variety.

Distribution functions may be represented by tables, as in the previous examples. In many cases, however, it is possible and convenient to write the density and cumulative distributions in the form of an equation. When a distribution can be presented in formula form, we can often find formulas for the mean and for the variance as well. Formulas will be presented for the distributions which were developed in Examples 20-2 and 20-3.

If, as in Example 20-2, we have a box containing $K$ objects of which $d$ are defective, the probability of exactly $x$ defectives in a sample of $N$ can be

**Table 20-12** General notation for a frequency and cumulative distribution

| $x$ | $f(x)$ | $F(x)$ |
|---|---|---|
| . . . | . . . | . . . |
| 3 | $f(3)$ | $f(0) + f(1) + f(2) + f(3)$ |
| 2 | $f(2)$ | $f(0) + f(1) + f(2)$ |
| 1 | $f(1)$ | $f(0) + f(1)$ |
| 0 | $f(0)$ | $f(0)$ |

written as

$$f(x) = \frac{C(d,x)C(K - d,N - x)}{C(K,N)}$$

for $x = 0, 1, \ldots, s$, where $s$ is the smaller of $N$ and $d$. The probability of $x$ or fewer defectives can be written as

$$F(x) = \frac{\sum_{i=0}^{x} C(d,i)C(K - d,N - i)}{C(K,N)}$$

For example, from Table 20-3 the probability of one or fewer defectives is represented by

$$\begin{aligned} F(1) &= \frac{\sum_{i=0}^{1} C(5,i)C(45,4 - i)}{C(50,4)} \\ &= \frac{C(5,0)C(45,4)}{C(50,4)} + \frac{C(5,1)C(45,3)}{C(50,4)} \\ &= .64696 + .30808 = .95504 \end{aligned}$$

This distribution is called a *hypergeometric distribution*. The variable here is discrete (as it has been in every case so far in this chapter), taking on values $0, 1, 2, \ldots, s$, and the distribution depends on three parameters, $K$, $d$, $N$. However, when this distribution is used in a problem of statistical inference, $d$ is usually the only unknown parameter, since $K$, the total size of the population, and $N$, the size of the sample, would be known.

If, as in Example 20-3, there are repeated independent trials with constant probability $p$ of success, the probability of exactly $x$ successes in $N$ trials can be written as

$$f(x) = C(N,x)p^x(1 - p)^{N-x}$$

for $x = 0, 1, \ldots, N$, and the probability of $x$ or fewer successes in $N$ trials is

$$F(x) = \sum_{i=0}^{x} C(N,i)p^i(1 - p)^{N-i}$$

This distribution is called a *binomial distribution*. The random variable is discrete, taking on the values $0, 1, \ldots, N$. The distribution depends on two parameters, $p$ and $N$. The sample size $N$ is usually known, with $p$ the only unknown parameter.

There are, of course, many discrete density distributions which can be expressed in formula form. In addition to the hypergeometric and binomial distributions discussed above, the Poisson distribution, to be presented in Sec. 20-6, is among the most useful. In general, a discrete density distribution is zero except at a countable number of points. In the examples described above the points where $f(x)$ is not zero are finite in number. The Poisson distribution has an infinite number of points where $f(x)$ is not zero.

We noted in Chaps. 2 and 3 that we could describe a distribution in part by computing such measurements as the arithmetic mean $\mu$, the variance $\sigma^2$, the standard deviation $\sigma$, and the various percentiles. This can be done for the probability distributions which we are now studying. Thus from Tables 20-3 and 20-4, reproduced below in Table 20-13, we compute $\mu$ and $\sigma^2$. Note that since the entire population is present, we divide by $N$ in computing $\sigma^2$. The values of $f(x)$ correspond to $f_i/N$ in the formulas for $\mu$ and $\sigma^2$, so

$$\mu = \Sigma x_i f(x_i) \qquad \text{and} \qquad \sigma^2 = \Sigma(x_i - \mu)^2 f(x_i) = \Sigma x_i^2 f(x_i) - \mu^2$$

As noted above, in cases where $f(x)$ is given by formula, the values of $\mu$ and $\sigma^2$ can also be given by formula, sometimes by a fairly simple formula. As an example, consider the binomial distribution. The computation is arranged in Table 20-14. To avoid extra difficulty in computing $\sigma^2$ we shall use a trick. Since $x^2 = x(x-1) + x$, it follows that $\Sigma x^2 f(x) = \Sigma x(x-1)f(x) + \Sigma x f(x)$. We shall obtain $\Sigma x^2 f(x)$ in this indirect way. The trick allows cancellation of the factors $x$ and $x - 1$ in the numerator, with two factors of $x!$ to give $(x-2)!$ in the denominator. From Table 20-14 we find $\Sigma x f(x) = Np$ and $\Sigma x(x-1)f(x) = N(N-1)p^2$, so that $\Sigma x^2 f(x) = N(N-1)p^2 + Np$. We can now compute $\mu = Np$ and $\sigma^2 = N(N-1)p^2 + Np - (Np)^2 = Np(1-p)$. Table 20-13 shows the numerical computation of the mean and variance of the binomial distribution of Table 20-4. For that example $N = 5$ and $p = .4$, so that the formulas give $\mu = 5 \times .4 = 2$ and $\sigma^2 = 5 \times .4 \times .6 = 1.2$, which, of course, agrees with the numerical computation.

The final formulas obtained for the mean and variance of the binomial

**Table 20-13** Computation of $\mu$ and $\sigma^2$ for examples of the hypergeometric and binomial distributions of Tables 20-3 and 20-4

FROM TABLE 20-4

| $x$ | $f(x)$ | $xf(x)$ | $x^2f(x)$ |
|---|---|---|---|
| 5 | .01024 | .05120 | .25600 |
| 4 | .07680 | .30720 | 1.22880 |
| 3 | .23040 | .69120 | 2.07360 |
| 2 | .34560 | .69120 | 1.38240 |
| 1 | .25920 | .25920 | .25920 |
| 0 | .07776 | 0 | 0 |
| *Total* | 1.00000 | 2.00000 | 5.20000 |

$\mu = 2.00000$
$\sigma^2 = 5.20000 - (2.00000)^2 = 1.2$
$\sigma = 1.09544$

FROM TABLE 20-3

| $x$ | $f(x)$ | $xf(x)$ | $x^2f(x)$ |
|---|---|---|---|
| 4 | .00002 | .00008 | .00032 |
| 3 | .00195 | .00585 | .01755 |
| 2 | .04299 | .08598 | .17196 |
| 1 | .30808 | .30808 | .30808 |
| 0 | .64696 | 0 | 0 |
| *Total* | 1.00000 | .39999 | .49791 |

$\mu = .39999$
$\sigma^2 = .49791 - (.39999)^2 = .33791$
$\sigma = .58130$

**Table 20-14** Computation of mean and variance for binomial distribution

| $x$ | $f(x)$ | $xf(x)$ | $x(x-1)f(x)$ |
|---|---|---|---|
| $N$ | $p^N$ | $Np^N$ | $N(N-1)p^N$ |
| $N-1$ | $Np^{N-1}(1-p)$ | $N(N-1)p^{N-1}(1-p)$ | $N(N-1)(N-2)p^{N-1}(1-p)$ |
| . . . | . . . | . . . | . . . |
| 3 | $\frac{N(N-1)(N-2)}{1 \times 2 \times 3} p^3(1-p)^{N-3}$ | $\frac{N(N-1)(N-2)}{1 \times 2} p^3(1-p)^{N-3}$ | $\frac{N(N-1)(N-2)}{1} p^3(1-p)^{N-3}$ |
| 2 | $\frac{N(N-1)}{1 \times 2} p^2(1-p)^{N-2}$ | $\frac{N(N-1)}{1} p^2(1-p)^{N-2}$ | $N(N-1)p^2(1-p)^{N-2}$ |
| 1 | $Np(1-p)^{N-1}$ | $Np(1-p)^{N-1}$ | 0 |
| 0 | $(1-p)^N$ | 0 | 0 |

The sum of the $f(x)$ column is obtained directly from Theorem 6 to be 1.

The sum of the $xf(x)$ column is obtained by factoring out $Np$ from each term and then noting that the terms remaining are the same as those in the $f(x)$ column, except that $N$ is replaced by $N-1$, and therefore, using Theorem 6, the column sum is

$$Np[p + (1-p)]^{N-1} = Np$$

The sum of the $x(x-1)f(x)$ column is obtained by factoring out $N(N-1)p^2$ from each term and noting that the terms remaining are the same as in the $f(x)$ column, except that $N$ is replaced by $N-2$, and therefore, using Theorem 6, the column sum is

$$N(N-1)p^2[p + (1-p)]^{N-2} = N(N-1)p^2$$

turned out to be quite simple. The algebraic derivation of these values would have been much more difficult, however, without the use of the trick of writing $x^2 = x(x-1) + x$. Finding the mean and variance of any distribution given as a formula is a mathematical problem which is sometimes quite easy and sometimes of considerable difficulty.

Referring again to the binomial distribution, let us consider the random variable in the form of the proportion of successes

$$\frac{x}{N} = 0, \frac{1}{N}, \frac{2}{N}, \frac{3}{N}, \ldots, \frac{N-1}{N}, 1$$

in place of $x = 0, 1, 2, 3, \ldots, N-1, N$. This change would result in a distribution with mean $\mu = Np/N = p$ and variance

$$\sigma^2 = \frac{Np(1-p)}{N^2} = \frac{p(1-p)}{N}$$

These results can, of course, be obtained directly by the use of the rule that dividing each observation in a distribution by a constant divides the mean by that constant and divides the variance by the square of that constant.

Formulas can also be computed for the mean and variance of the hypergeometric distribution with somewhat more difficulty. They are

$$\mu = \frac{Nd}{K} \quad \text{and} \quad \sigma^2 = \frac{Nd}{K}\left(1 - \frac{d}{K}\right)\frac{K-N}{K-1}$$

The numerical computation in Table 20-13 with the distribution from Table 20-3 is for $K = 50$, $d = 5$, $N = 4$ (thus $s = 4$); therefore we find

$$\mu = \frac{4 \times 5}{50} = .4 \qquad \text{and} \qquad \sigma^2 = 4 \times \frac{5}{50} \times \frac{45}{50} \times \frac{46}{49} = .33796$$

These agree with the numerical computation, except for rounding errors caused by recording the probabilities to five places only.

Formulas for the mean and variance of the rank-sum statistic $T'$ and the run statistic $u$ are given in Chap. 17.

## MEAN AND VARIANCE FORMULAS 20-5

If independent random variables are added, subtracted, or combined in any linear expression to form a new random variable, the mean and variance of the distribution of the new variable are given by relatively simple formulas.

THEOREM. *Let $X_1$ and $X_2$ be independent and from populations having means $\mu_1$ and $\mu_2$ and variances $\sigma_1^2$ and $\sigma_2^2$, respectively. Let $a_1$ and $a_2$ be any two constants, and define $z = a_1X_1 + a_2X_2$. Then the distribution of $z$ has mean equal to $a_1\mu_1 + a_2\mu_2$ and variance equal to $a_1^2\sigma_1^2 + a_1^2\sigma_2^2$.*

EXAMPLE 20-9. As an illustration, refer to Example 4-1. We can consider $X_1$ to be the weight of the first rabbit and $X_2$ the weight of the second rabbit. As given in Example 4-1, $\mu_1 = \mu_2 = 14$ and $\sigma_1^2 = \sigma_2^2 = \frac{11}{3}$. Now, if we choose $a_1 = .5$ and $a_2 = .5$, then $z$ is the mean of the sample consisting of $X_1$ and $X_2$. By the above formulas, the mean of $z$ is given by

$$\text{mean } z = (.5 \times 14) + (.5 \times 14) = 14$$

and the variance of $z$ is given by

$$\begin{aligned} \text{var } z &= (.5)^2\sigma_1^2 + (.5)^2\sigma_2^2 \\ &= (.25 \times \tfrac{11}{3}) + (.25 \times \tfrac{11}{3}) = \tfrac{11}{6} \end{aligned}$$

which agrees with the results in Example 4-1.

These formulas generalize to $N$ independent random variables, so that if

$$z = \Sigma a_i X_i$$

then for the distribution of $z$

$$\text{mean } z = \Sigma a_i \mu_i \qquad \text{and} \qquad \text{var } z = \Sigma a_i^2 \sigma_i^2$$

Many formulas presented in this text come from the above formulas.

**The arithmetic mean.** Let $X_1, \ldots, X_N$ be randomly chosen, with replacement, from a population with mean $\mu$ and variance $\sigma^2$. If $a_i = 1/N$, then $z = \bar{X}$, and since $\mu_i = \mu$ and $\sigma_i^2 = \sigma^2$, we get

$$\text{mean } z = \Sigma a_i \mu_i = \sum \frac{1}{N} = \mu \qquad \text{and}$$

$$\text{var } z = \Sigma a_i^2 \sigma_i^2 = \sum \frac{1}{N^2} \sigma^2 = \frac{1}{N^2} \sigma^2 N = \frac{\sigma^2}{N}$$

**Difference of two means.** If $X_{1i}$ and $X_{2j}$ are independent and $z = \bar{X}_{1\cdot} - \bar{X}_{2\cdot}$, then the results above give us

$$\text{mean } z = \mu_1 - \mu_2 \qquad \text{and} \qquad \text{var } z = (1)^2 \frac{\sigma_1^2}{N_1} + (-1)^2 \frac{\sigma_2^2}{N_2} = \frac{\sigma_1^2}{N_1} + \frac{\sigma_2^2}{N_2}$$

**Regression coefficient.** Let $X_i$ be constants and $y_i$ be independent random variables with mean $y_i = a + bX_i$ and var $y_i = \sigma^2$. Let

$$a_i = \frac{X_i - \bar{X}}{\Sigma(X_i - \bar{X})^2}$$

and define

$$b' = \frac{\Sigma y_i (X_i - \bar{X})}{\Sigma(X_i - \bar{X})^2}$$

After some manipulations and use of the theorem at the beginning of this section, it can be shown that $b'$ is an unbiased estimate of $b$ with variance given by

$$\text{var } b' = \frac{\sigma^2}{\Sigma(X_i - \bar{X})^2}$$

In this example the $X$'s play the role of the constants, the $y_i$ are the random variables, and $z = b'$.

For the formula for the variance to apply, the random variables must be independent (or at least have zero correlation). In random sampling without replacement the variables are correlated, and the formula for the variance does not apply. For dependent variables the formula for the mean is the same as given, but the formula for the variance is

$$\text{var } z = \Sigma a_i^2 \sigma_i^2 + \Sigma\Sigma a_i a_j \sigma_i \sigma_j \rho_{ij}$$

The second summation includes all values of $i$ and $j$ except $i = j$, and the $\rho_{ij}$ are the population correlation coefficients (see Chap. 11) between $X_i$ and $X_j$.

The Poisson distribution has very wide application. It arises from two different approaches, which we now discuss.

**The Poisson distribution as a limit of the binomial distribution.** Consider the following modifications in the way the binomial distribution may be written:

$$\begin{aligned} f(x) &= C(N,x)p^x(1-p)^{N-x} \qquad x = 0, 1, 2, \ldots, N \\ &= \frac{N!}{x!(N-x)!}p^x(1-p)^{N-x} \\ &= N(N-1)(N-2)\cdots(N-x+1)\frac{p^x(1-p)^{N-x}}{x!} \end{aligned}$$

Let $p = \lambda/N$ (that is, define $\lambda = Np$). We then have

$$f(x) = \frac{N}{N}\frac{N-1}{N}\frac{N-2}{N}\cdots\frac{N-x+1}{N}\frac{\lambda^x(1-\lambda/N)^N(1-\lambda/N)^{-x}}{x!}$$

For large $N$ and for fixed small values of $x$ the terms

$$\frac{N}{N}, \frac{N-1}{N}, \ldots, \frac{N-x+1}{N}, \left(1-\frac{\lambda}{N}\right)^{-x}$$

are all close to unity, so $f(x)$ is approximately

$$f(x) \doteq \frac{\lambda^x}{x!}\left(1-\frac{\lambda}{N}\right)^N$$

where we have used the dot over the equal sign to represent "approximately equal." It can be shown that the value of $(1-\lambda/N)^N$ can be approximated as a power of $e$, where $e$ is the limit as $N$ becomes infinite of $(1+1/N)^N \doteq 2.7183\cdots$. For our problem the factor $(1-\lambda/N)^N$ is approximated by $e^{-\lambda}$, so

$$f(x) \doteq \frac{\lambda^x}{x!}e^{-\lambda} \qquad x = 0, 1, 2, \ldots$$

This formula, defined for zero and all positive integral values $x$, is called the *Poisson distribution*. It will give approximately the same probabilities as the binomial distribution if $N$ is large and $Np$ is fairly small ($Np < 5$ is the requirement usually stated). The cumulative Poisson distribution is given for several values of $\lambda$ in Table A-15.

One computation will be presented to indicate how closely the Poisson distribution approximates the binomial distribution. The binomial with $N = 40$ and $p = .02$ is

$$f(x) = C(40,x)(.02)^x(.98)^{40-x}$$

**Table 20-15** Comparison of the binomial probabilities for $N = 40$, $p = .02$ and the Poisson probabilities for $\lambda = .8$

| $x$ | $C(40,x)(.02)^x(.98)^{40-x}$ | $\frac{(.8)^x e^{-.8}}{x!}$ |
|---|---|---|
| *5 or over* | .001 | .001 |
| 4 | .007 | .008 |
| 3 | .037 | .038 |
| 2 | .145 | .144 |
| 1 | .364 | .360 |
| 0 | .446 | .449 |
| *Total* | 1.000 | 1.000 |

The approximating Poisson distribution has $\lambda = Np = .8$. Individual probabilities can be computed from the binomial formula and for the Poisson distribution can be read from Table A-15. Table 20-15 lists these probabilities for comparison.

**The Poisson distribution derived from a set of axioms.** The Poisson distribution is derived above as an approximation to the binomial distribution for a certain range of values for $N$ and $p$. The binomial distribution was originally derived to describe probabilities for a fixed number of independent trials having a constant probability of success. The Poisson distribution also can be shown to arise from a different axiomatic approach.

We consider a fixed unit of time, $T$ (perhaps a second, day, or week), in which certain events may occur. We shall assume that events occur independently and that for short periods of time $\Delta t$ the probability of one event is proportional to the length of time $\Delta t$, that is, is equal to $c\,\Delta t/T$, where $c$ is constant during the time period $T$. We shall assume that the probability of two or more events in time $\Delta t$ is so small that it may be neglected. With these assumptions it can be shown, with tools studied in the calculus, that with $\lambda = ct/T$,

$$f(x) = \frac{\lambda^x e^{-\lambda}}{x!}$$

That is, the probability that $x$ events will occur before time $t$ is given by the Poisson distribution.

If we are observing real events, and the conditions under which they are being observed seem to correspond to the above assumption, that is, the events are reasonably independent and the probability that an event will occur is proportional to the length of the time interval, then the frequency of observed events may follow the Poisson distribution. It should be noted that although the problem was stated in terms of time intervals, it would be just as reasonable to use units of length, area, or volume.

The assumptions leading to the Poisson distribution may apply in the following situations:

1. Let $x$ be the number of telephone calls originated in an exchange between, say, 3 and $3 + t$ o'clock. We have noted in the past that calls are originated at the rate of 120 per hour between 2 and 4 o'clock. If we then use $\frac{120}{60} = 2$ per minute as our value of $c/T$, we would obtain $f(x) = (2t)^x e^{-2t}/x!$ as the probability of $x = 0, 1, 2, 3, \ldots$ calls per $t$ minutes.

2. Let $x$ be the number of boxes of cereal sold on Monday in a grocery store. Here $c/T$ is the average number sold on Monday, and $t$ is 1.

3. Let $x$ be the number of tire failures per week for a fleet of delivery trucks.

4. Let $x$ be the number of atoms disintegrating per second in a certain quantity of radioactive material.

5. Let $x$ be the number of defects in the enamel of manufactured pans.

6. Let $x$ be the number of typing errors per page.

7. Let $x$ be the number of electrons emitted by a heated cathode in a given time interval.

8. Let $x$ be the number of molecules of a gas which are in a subregion $v$ of a container of volume $V$. This is an example for the limiting case of the binomial $C(n,x)(v/V)^x(1 - v/V)^{n-x}$ for $n$ very large.

Formulas for the mean and variance of the Poisson distribution are

$$\mu = \lambda \qquad \text{and} \qquad \sigma^2 = \lambda$$

These results can be obtained by setting $\lambda = Np$ in the formulas for the mean and variance of the binomial and considering $N$ to be very large. We see that the mean is equal to the variance for the Poisson distribution.

## CONTINUOUS DENSITY FUNCTIONS 20-7

In the preceding examples we have considered only discrete distributions, with the random variable taking on only a countable number of possible values. Following are examples in which the random variable can take on any one of a continuous range of values.

EXAMPLE 20-10. Suppose a pointer is spun on a pivot and the angle the pointer makes with a prechosen direction is measured when it comes to rest. We may wish to consider that any angle between 0 and 360° is equally likely to occur. Therefore the probability that the observed angle $\theta$ is between two particular angles, say, $a$ and $b$, is equal to a quantity proportional to the difference $b - a$. Since $\theta$ must be between 0 and 360°, we say

$$P(a \leq \theta \leq b) = (b - a)/360$$

where, of course, $a$ and $b$ are both between 0 and 360 and $a$ is less than $b$. We could express the same result by defining a density function $f(\theta) = \frac{1}{360}$ and

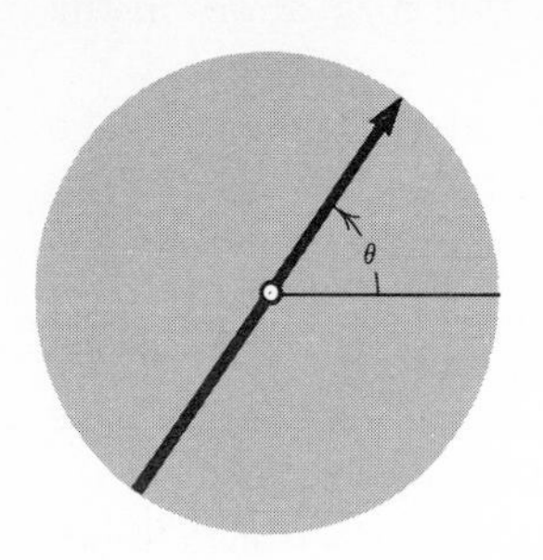


**Figure 20-2** Spinner generating a uniform or rectangular distribution.

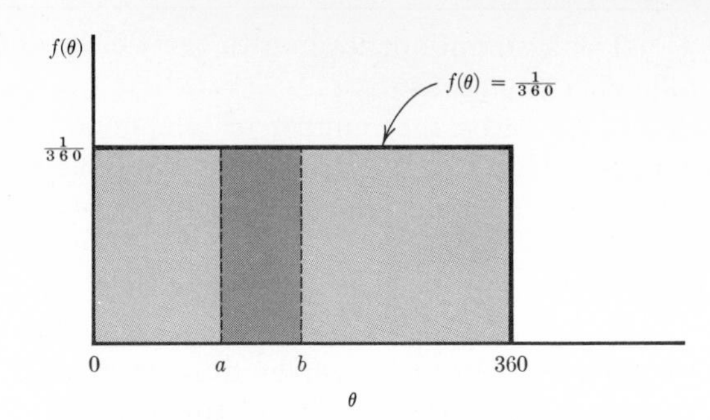


**Figure 20-3** Uniform distribution generated by spinner.

obtain the probability as the area under the density function $f(\theta)$ between $a$ and $b$. The graph of the density function is drawn in Fig. 20-3.

The density function of this example is called a *uniform*, or *rectangular*, distribution. To represent equal probabilities for all values of $x$ ranging from $d$ to $d + r$ we can set $f(x) = 1/r$ for $d \leq x \leq d + r$. The mean and variance for this distribution are

$$\mu = d + \frac{r}{2} \qquad \text{and} \qquad \sigma^2 = \frac{r^2}{12}$$

EXAMPLE 20-11. Suppose that in shooting at a target the lateral errors from the aiming point are such that all hit within 6 inches of the aiming point and the numbers missing the center by $x$ inches decrease in proportion to $x$ (this assumption is not very realistic, but it serves as an example). We suppose the density function to be $f(x) = \frac{1}{3}(1 - x/6)$. The density function is graphed in Fig. 20-4. The probability that the lateral error will be less than $t$ inches is the area of the shaded portion of Fig. 20-4.

In general we can state that a density function is nonnegative for all possible values of the variable, since this function is used to represent probabilities, and further that the total area under the density function is

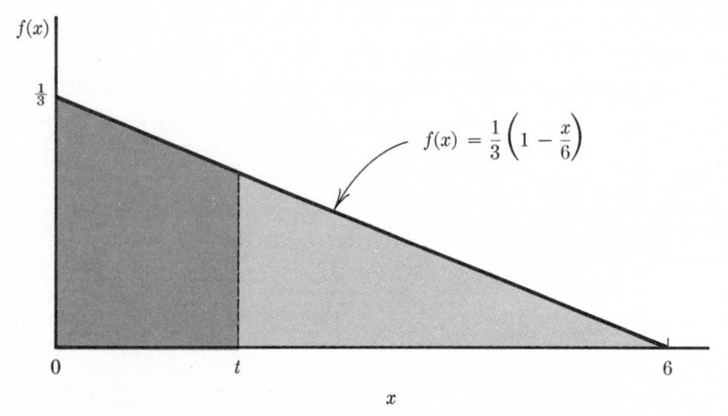


**Figure 20-4** Triangular distribution.

equal to 1, since we wish the total probability to be 1. In fact, any continuous function which is nonnegative and which has unit total area can be called a continuous density function. These requirements are the same as those stated for the discrete distribution functions, except that area replaces the sum of individual probabilities.

## EXPECTATION 20-8

Let $x$ be a random variable with density function $f(x)$. The mean of the distribution for discrete variables is $\Sigma x f(x)$. For either the discrete or the continuous case the mean is called the *expected value of x.*

It should be noted that the expected value of $x$ is the arithmetic mean and, of course, is "typical" only in this sense. The probability that a value of $x$ will be less than the expected value is equal to $\frac{1}{2}$ only for very special types of distributions. Also, the expected value is the most probable value only for certain types of distributions, and in certain cases the expected value cannot occur. For example, in tossing a coin the expected number of heads may be taken as .5 (see page 245 for the 0 and 1 convention), and this, of course, will never occur. The confusion does not seem to arise when we use the word "mean," as in the mean family size of 2.6, where we do not anticipate finding a family that has $\frac{6}{10}$ of a child. The word "expectation" also occurs in actuarial tables, where a distribution of length of life is given and its mean is labeled as "expected life" or "expected number of years to death."

EXAMPLE 20-12. An example of a life table is given in Table 20-16. The basic elements are the age intervals and the probability of death ${}_nq_x$ during the age interval following each age $x$. This is augmented by an arbitrary number, $l_0 = 100{,}000$ in this case, of individuals forming a cohort, which starts at age $x = 0$ and is depleted over time by the rates ${}_nq_x$. Thus it is assumed that the number of deaths in the age interval starting at $x$ is ${}_nd_x = {}_nq_x \cdot l_x$, and the values of ${}_nd_x$ are also given in the table. The "age" and "death" columns can be interpreted as a frequency table of age at death, and the mean can be computed starting with the cohort at any age. Typically these means are given as in Table 20-16. Note that the mean length of life for a newborn boy is given as $e_0 = 66.9$ years. For a 10-year-old boy ($x = 10$) the mean length of life remaining is $e_x = e_{10} = 59.2$ and for the 10-year-old the mean age at death is

$$x + e_x = 10 + e_{10} = 10 + 59.2 = 69.2$$

In gambling games the expected value of a win is sometimes referred to as a "fair" amount to pay to play the game. This is correct in the sense that if one plays the game over a long period of time (paying that amount each time), the amounts he may win or lose will average out. However,

**Table 20-16** Abridged life table for United States male population, 1964

| PERIOD OF LIFE BETWEEN TWO EXACT AGES, YEARS ($x$ TO $x + n$) | *Proportion dying during age interval* (${}_nq_x$) | *No. living at beginning of age interval* ($l_x$) | *No. dying during age interval* (${}_nd_x$) | *Mean no. of years of life remaining at beginning of age interval* ($e_x$) |
|---|---|---|---|---|
| 0–1 | .0278 | 100,000 | 2,779 | 66.9 |
| 1–5 | .0042 | 97,221 | 407 | 67.8 |
| 5–10 | .0026 | 96,814 | 249 | 64.0 |
| 10–15 | .0026 | 96,565 | 254 | 59.2 |
| 15–20 | .0066 | 96,311 | 637 | 54.3 |
| 20–25 | .0092 | 95,674 | 877 | 49.7 |
| 25–30 | .0090 | 94,797 | 850 | 45.1 |
| 30–35 | .0107 | 93,947 | 1,010 | 40.5 |
| 35–40 | .0149 | 92,937 | 1,386 | 35.9 |
| 40–45 | .0231 | 91,551 | 2,111 | 31.4 |
| 45–50 | .0366 | 89,440 | 3,270 | 27.1 |
| 50–55 | .0586 | 86,170 | 5,045 | 23.0 |
| 55–60 | .0897 | 81,125 | 7,280 | 19.3 |
| 60–65 | .1309 | 73,845 | 9,668 | 15.9 |
| 65–70 | .1924 | 64,177 | 12,347 | 12.9 |
| 70–75 | .2579 | 51,830 | 13,369 | 10.4 |
| 75–80 | .3454 | 38,461 | 13,283 | 8.2 |
| 80–85 | .4699 | 25,178 | 11,830 | 6.2 |
| 85 and over | 1.0000 | 13,348 | 13,348 | 4.5 |

many people are willing to pay more than the expected winning to play for stakes which are very high.

EXAMPLE 20-13. Suppose a lottery has 1 million tickets at $1 per ticket, with a first prize of $100,000 and three prizes of $50,000 each. A person buying one ticket has expected winnings of

$$\left(100{,}000 \times \frac{1}{1{,}000{,}000}\right) + \left(50{,}000 \times \frac{3}{1{,}000{,}000}\right) + \left(0 \times \frac{999{,}996}{1{,}000{,}000}\right) = .25$$

so a fair price would be 25 cents. Many people, however, are willing to pay $1 for tickets to such a lottery. The explanation has been advanced that for these people the loss of $1 is of small consequence but the gain of $100,000 or even $50,000 is such a fortune that they are willing to play, even against such odds. Apparently each dollar does not have the same value. The value will depend on whether it is the first, second, one-thousandth, one-millionth.

## GLOSSARY

binomial distribution
combinations
conditional probability
equally likely
expectation
expected length of life
hypergeometric
independent events
life table
permutations
Poisson distribution
probability
random variable
stochastic variable
uniform distribution

## PROBLEMS

*Section* 20-3

**1.** Before buying a batch of 100 fuses a man tests four. If the batch contains 10 defective fuses, what is the probability that he will find none in the sample? That he will find four defective fuses in the sample?

**2.** An aircraft factory plans to purchase small parts in batches of 40. The factory will inspect three from each batch. If a batch has two defective parts, what is the probability that exactly one will be found in the sample? That none will be found? That exactly two will be found? That two or fewer will be found?

**3.** A shipment of electrical switches is packed in boxes of 100 each. A box is inspected by examining 20 switches and rejecting the box if any of the 20 switches is defective. What is the probability of rejecting a box containing two defective switches?

**4.** An experiment is conducted in which $A$ looks at five cards (marked 1, 2, 3, 4, 5), concentrating on two particular cards, and $B$ attempts to "read his mind" and name the two cards. If $B$ actually has no talent for mind reading, what is the probability that he will guess the two cards correctly?

**5.** In Fig. 14-8 curves are drawn for the probability that a sample of size $N$ from a batch of $K$ items, with a certain number of defectives present will contain no defectives. For example, if $K = 100$, $N = 10$, with 2 per cent defectives (two defective items among 100), then the probability

$$P(0 \text{ defectives in a sample of } 10) = \frac{C(98,10)}{C(100,10)} = \frac{98!}{10!88!}\,\frac{10!90!}{100!} = \frac{90 \times 89}{100 \times 99} = .809$$

which agrees with the curve in Fig. 14-8. Verify the point on the curve for $K = 200$, $N = 20$, above 2 per cent of defective. Given a sample of $N = 10$ for a batch of $K = 100$ items having 0, 2, 4, 6, 8, or 10 per cent defectives, compute the probabilities $P(0 \textit{ or } 1 \text{ defective})$ for each case and draw an operating-characteristic curve comparable with those in Fig. 14-8. Compare this curve and the corresponding curve of Fig. 14-8 to decide whether it seems better to reject the batch if the sample contains one or more defectives or to reject the batch if the sample contains two or more defectives.

**6.** A process yields ball bearings of which, in the long run, 5 per cent are defective. What is the probability that in a run of 10 ball bearings there would be no defectives, 1, . . . , 10 defectives?

**7.** Ten per cent of people contracting a certain disease do not recover. What is the probability that all of three people contracting the disease do not recover? If a new drug is given to 20 people with the disease and all recover, can we conclude that the drug is effective? As an aid in making a decision, compute the probability that the observed result would occur if the drug had no effect at all.

**8.** If we assume that boy and girl births are equally likely, compute the probability that exactly 40 out of 100 babies born will be boys.

**9.** With the same assumption as in Prob. 8, how frequently will families of five children (no multiple births) have (*a*) no girls, (*b*) one girl, (*c*) two girls, (*d*) three girls, (*e*) four girls, (*f*) five girls?

Note that if actual observation disagrees too markedly with the assumptions used in computing the probabilities, we may be forced to discard such assumptions. The probability that a baby will be a boy may actually differ from $\frac{1}{2}$. Furthermore, the probability that a baby will be a boy may be different for different families and even at different times for the same family.

**10.** Assume that a man walking along a north-south street tosses a coin at each corner to decide whether to go north or south. What is the probability that in walking 10 blocks he is again at his starting point?

Note that this problem may also be worded in terms of the motion of a particle, where the change of direction is caused by collision rather than the coin toss. The probability of a change in direction will be the probability of a collision. The more practical problem of the motion of a particle in three dimensions involves more mathematics.

**11.** Actuarial tables indicate that in a large group of 10-year-old children 85 per cent will survive at least 40 years. What is the probability that of three children each age 10 exactly one will survive 40 years?

**12.** Consider an experiment which requires $Y$ to choose between two containers, one empty and one with water, for six trials. Find the resulting level of significance if the critical region is, as before, five or six correct.

**13.** $A$ cuts a well-shuffled deck of playing cards (52 cards, 26 red and 26 black) and observes the color of the cut card. As in Prob. 4, $B$ attempts to read his mind and writes down red or black. The cards are then reshuffled and the trial is repeated. If $B$ has no talent and is guessing, what is the probability that in 10 such trials he will call exactly eight correctly? At least eight correctly?

**14.** Suppose that in Prob. 13 $B$ has sufficient ability to enable him to recognize the correct color 80 per cent of the time. What is the probability that he will call at least eight out of the 10 correctly?

**15.** In recording the performance of $B$ in Prob. 13 we may record the number of red cards called correctly and the number of black cards called correctly as in the accompanying table. Assume that $B$ has no talent and verify that the probability of obtaining exactly the recorded results is

$$P(\text{6 red, of which 4 are called correctly, } \textit{and} \text{ 4 black all called incorrectly})$$
$$= C(10,6)(\tfrac{1}{2})^{10}C(6,4)(\tfrac{1}{2})^{6}C(4,4)(\tfrac{1}{2})^{4}$$
$$= \frac{10!}{6!4!}\frac{6!}{4!2!}\left(\frac{1}{2}\right)^{20} = .0030$$

Note that this is not the same as the probability that $B$ calls exactly four correctly, independently of whether the correct cards are red or black. This latter probability is $C(10,4)(\frac{1}{2})^{10} = .2051$.

| | *Correct* | *Incorrect* | *Total* |
|---|---|---|---|
| *Red* | 4 | 2 | 6 |
| *Black* | 0 | 4 | 4 |
| *Total* | 4 | 6 | 10 |

**16.** For a cross of two cream-colored guinea pigs compute the probability that all of three offspring will be yellow. That five offspring will consist of three yellow and two cream. That all of five offspring will be cream. That five offspring will consist of two yellow, two cream, and one white. For a cross of cream guinea pigs with yellow compute the probabilities of the same results as above.

**17.** The factor $A$ is called *dominant* and the factor $a$ is called *recessive* if individuals of type $Aa$ have the same observable characteristic as $AA$ but this differs from those of type $aa$. Mendel observed that in crossing pea plants from a tall race (no dwarfness in their ancestry so far as he knew) with plants from a dwarf race all the resulting plants were tall. He then crossed these resulting plants (hybrids) with each other and observed that about three-fourths were tall and about one-fourth were dwarf. Explain in a manner similar to that of Example 20-5 the occurrence of these proportions.

**18.** For the cross of the hybrids of Prob. 17 compute the probability that of four offspring two will be tall and two will be dwarf.

**19.** Suppose we planned to plant the two varieties of corn of Example 20-6 in 10 locations instead of 6. Write down the distribution of $x$ as in Table 20-6 and verify the entries in Table A-10 for $N = 10$.

**20.** The distribution in Table 20-7 was constructed for $r$, the number of times the less frequent sign occurs. Consider $x$, the number of plus signs, and determine the largest $x$ which satisfies the relationship $P(x \text{ or less}) \leq .01, .05, .10, .25$, for $N = 6$.

**21.** The median $M$ of a population is defined as that $M$ for which $P(x \leq M) = \frac{1}{2}$. If the population is sufficiently large that this probability remains equal to $\frac{1}{2}$ as successive observations are taken from this population or if each of the observations is returned to the population before another is taken, we may use the same probability that each successive observation is less than the median. Compute the probability that of 10 observations the second largest observation will lie on one side of the median and the second smallest will lie on the other side. (*Hint:* This will occur unless there are exactly 0, 1, 9, or 10 observations smaller than the median.) Find the probability that the largest and the smallest observations will be on opposite sides of the median.

**22.** Write down all possible arrangements of $N_1 = 4$ plus and $N_2 = 3$ minus signs and verify the counting formulas for the number of runs in this case. Write down in a table the probabilities of 2, 3, 4, 5, 6, 7 runs.

**23.** Suppose there is a 50 per cent chance for a plus sign to occur and a 50 per cent chance for a minus sign to occur in each of four trials. What is the probability that one run will result? Two runs? Three runs? Four runs? What is the probability that two runs will result if we know that two plus and two minus signs will result?

**24.** Write down all possible arrangements of three $x$ values and four $y$ values and verify the distribution of $T'$ as given for $N_1 = 3$ and $N_2 = 4$ in Table A-20.

*Section* 20-4

**25.** Which of the following functions are density distributions?

(*a*) $f(x) = 0$, except $f(0) = .4$ and $f(1) = .6$

(*b*) $f(x) = \dfrac{e^{-x}}{3}$ for $x = 0, 1, 2, f(3) = \dfrac{2}{3} - \dfrac{1}{3e} - \dfrac{1}{3e^2}, f(x) = 0$ for other values of $x$

(*c*) $f(x) = 0$, except $f(0) = p^3$, $f(1) = 3p^2(1 - p)$, $f(2) = 3p(1 - p)^2$, and $f(3) = (1 - p)^3$

**26.** Write down formulas for $f(x)$ and $F(x)$ in the following instances, specifying the values of $x$ for which the formula applies:

(*a*) $x$ is the number of defectives in a sample of 20 chosen from a batch of 200 articles of which 5 per cent are defective.

(*b*) $x$ is the number of baby boys in a series of 30 births, with the assumption that the chance for a boy at each birth is 50 per cent.

(*c*) $x$ is the number of heads in 12 tosses of a fair coin.

(*d*) $x$ is the number of shots among 50 fired which hit the center of a target, with the assumption that in the long run the frequency of such successes is 10 per cent.

(*e*) $x$ is the number of patients in a group of 35 recovering from a disease if the long-run frequency of recovery is 75 per cent.

**27.** Compute the mean and variance of the distribution in Table 20-9. Compare the results with those given by the formulas for the mean and variance given in Sec. 17-3. (*Note:* $N_1 = N_2 = 3$.)

**28.** Compute the mean and variance of the distribution in Table 20-6 and compare the results with those given by the formulas for the binomial.

**29.** Compute the mean and variance of the distribution in Table 20-11. Compare the results with those given by the formulas for the mean and variance in Sec. 17-4.

*Section* 20-5

**30.** Verify that the general formulation in Sec. 20-5 gives $\sigma/\sqrt{3}$ for the standard deviation of the mean of a random sample, with replacement, of size 3 from a population with standard deviation $\sigma$.

**31.** In a random sample, without replacement, of size $n$ from a population of size $N_p$ the correlation coefficient between any two observations is $-(n - 1)/(N - 1)$. Verify this in the special case of the data on $N = 6$ rabbits on page 44 and the 15 possible samples of size $n = 2$. (The weights of the two rabbits are the $X$ and $Y$ values in the formula on page 205 for the correlation coefficient.)

*Section* 20-6

**32.** Suppose 10,000 bacteria are moving independently and randomly in a volume of 20,000 cubic centimeters. What is the probability that a selected cubic centimeter would contain no bacteria? (Note that the binomial distribution gives an exact result and the Poisson distribution gives a very good approximation.)

**33.** Suppose there is an average of one bacteria per 2 cubic centimeters of water in a swimming pool. What is the probability that a selected cubic centimeter would have no bacteria? What is the probability that none of five independently selected cubic centimeters of water would have any bacteria?

*Section* 20-7

**34.** What are $\mu$ and $\sigma$ for the density in Fig. 20-3?

**35.** Use the density in Fig. 20-3 to find and draw the corresponding cumulative distribution.

**36.** In the density in Fig. 20-3 what is the probability that a reading between 20° and 40° would be observed in a single spin? What is the probability of exactly zero readings between 20° and 40° in 10 spins?

**37.** Draw the cumulative distribution for the density in Fig. 20-4. (*Hint:* The shaded area can be obtained by subtracting the area of the unshaded triangle from 1.) Read from the cumulative curve the values of $P_{05}$, $P_{50}$, $P_{90}$.

*Section* 20-8

**38.** Verify in Table 20-16 the value of $e_{70}$.

**39.** It was noted that people with a certain disease had 60 per cent mortality the first year, the survivors had 50 per cent mortality the second year, those surviving 2 years had 40 per cent mortality the third year, and the survivors had 30 per cent mortality in each of the following 10 years. Construct a life table for 100 people who contract the disease and find the expected length of life for a survivor of 2 years.

**40.** Mosquitoes have a risk of death from natural enemies of .6 per time period. Construct a life table for 100,000 newly hatched mosquitoes.

# General comments

**Misuses of statistics.** In general we recommend a positive approach to the study of statistics, especially in a first course. We find that an initial stress on misuses so overwhelms some beginners (who, perhaps because of rumors about the difficulties of the subject matter, come already frightened or resentful) that they hesitate to use the techniques in any real way. However, if caution is used there are definite and important lessons to be learned from examples of misuses of statistics as well as from examples of appropriate uses. A variety of examples may be drawn from Huff (40), Hill (37), and Mainland (46). Wallis and Roberts (73) gives an especially good collection of examples of uses and misuses of statistics from many fields of applications.

**Objectives.** In the discussion of applications it is well to pose such question as: Is a population defined? What is it about the population that we wish to study? Have appropriate characteristics on the individual been specified? Do these data bear on the problem under investigation? Do we know how to measure these characteristics? Do the individuals in the sample come from the population of interest? Do we know the manner of sampling individuals from the population? (This last point may be particularly important when the statistical problem is first discussed after the data have been obtained. If the method of collection is such that no valid inference can be made, one should say so.) A discussion on these questions is most effective if centered on experiences familiar to the members of the group. At a beginning level students generally consider assigned reading on these points "too philosophical" but will participate actively in classroom discussions.

**Motivation.** One of the best ways to increase interest is for the reader to supply sample examples or problems, first from the literature and then original examples, which illustrate measurement, sampling, inference, etc. Newspapers, magazines, and various journals provide current sources of real data. Since these sources are readily available, the student is more easily convinced of the wide usage of statistical concepts in our modern industrialized world.

**Errors.** The major emphasis in this text has been on sampling-errors, that is, variation which occurs randomly from one sample to another and which, given the size of the sample and the method of sampling, cannot be controlled further. Investigators should be aware that other sources of variability exist and may contribute considerably to the total error. Some of these are

1. Measurement error due to precision of the instrument
2. Measurement error due to careless or unskilled usage
3. Copying errors
4. Use of incorrect (printed or copied incorrectly) formulas or completely inappropriate formulas
5. Errors in arithmetic

**Arithmetic, mathematics and statistics.** Mathematics, especially probability, is essential in the development of statistical theory. However, for applicational concepts in statistics, the deductive reasoning and analytical manipulation of mathematics is relatively unimportant. But, since numerical values are continually used and interpreted, techniques of arithmetic calculations should be reviewed if necessary. The student who feels there will be difficulty in keeping up with normal class progress may wish some type of arithmetic review before starting this course. Walker (71), and Clark and Tarter (9) are good self-help texts.

**Analysis time.** In a total investigation a relatively small amount of time will be spent on arithmetic compared to time spent in planning, collection of data, and verbal interpretation. The advent of computers has dramatically changed the kinds and extent of problems which may be profitably analyzed. In actual practice computer support will likely be available for problems which require extensive computation. It seems appropriate, therefore, to place increasing emphasis on advance planning, experimental design, data collection, etc. Extensive computational skill is of much less importance than the understanding of statistical concepts and the ability to interpret results.

**Costs.** Sources of cost, such as planning, sampling, measuring, data processing, interpretation, must also be taken into account. These points should be discussed in connection with the choice of sample size.

**Sampling experiments.** The investigation of sampling distributions is effectively aided by the use of uniform random numbers, random normal numbers, or random numbers from other distributions. The concept is directly illustrated and the student is able to observe sampling distributions for a much wider variety of problems than his mathematical ability would provide by an analytical approach. This is true whatever his level of mathematical competence may be. This technique, often called simulation or a "Monte Carlo" method, is greatly simplified if computers are available so that automatic "random" number generators may be used.

**Graphical aids.** The following pages contain full-page reproductions of graph paper which may be of particular use to statisticians.

1. *Arithmetic paper.* The paper illustrated was ruled with five lines per inch each way and is of use for general plotting. The particular page is reproduced from Keuffel & Esser, 358-2.

2. *Semilogarithmic paper.* More accurately the paper is arithmetic logarithmic. The particular page has two cycles of the logarithmic scale (K & E, 359-63). Paper is available with fewer or more cycles.

3. *Log-log paper.* Both scales are logarithmic. The page included has two cycles in each direction (K & E, 359-110). Paper is available with many combinations of the numbers of cycles $1 \times 1$, $1 \times 2$, $2 \times 3$, etc.

4. *Binomial paper.* This paper has a square-root scale in both directions (Codex Book Co., 32,298). Its use is described in Chap. 13.

5. *Arithmetic normal paper.* This paper has an arithmetic scale in one direction and a cumulative normal scale in the other direction. This scale produces a straight line for any cumulative normal distribution. Use of this paper is described in Chaps. 5, 16, and 19. The cumulative probabilities range from .0001 to .9999 (K & E, 358-23). Other ranges are also available, for example, .01 to .99 and .000001 to .999999.

6. *Log-normal paper.* This paper differs from arithmetic normal paper by use of a two-cycle logarithmic scale in place of the arithmetic scale and the probability scale is placed lengthwise on the paper (Codex, 3228). A three-cycle paper with probability scale range of .02 to .98 is available (K & E, 358-22).

7. *Reciprocal normal paper.* This paper is the same as arithmetic normal paper except for the use of a reciprocal scale in place of the arithmetic scale (Team-316).

8. *Half-normal paper.* This paper has an arithmetic scale and a cumulative probability scale. The cumulative distribution which appears as a straight line on this paper is a cumulative half-normal distribution. This is the distribution of normal quantities with sign disregarded. An equivalent distribution is the chi-distribution with a single degree of freedom, that is the distribution of $y = \sqrt{\chi^2(1)}$.

**Figure 1** Arithmetic paper.

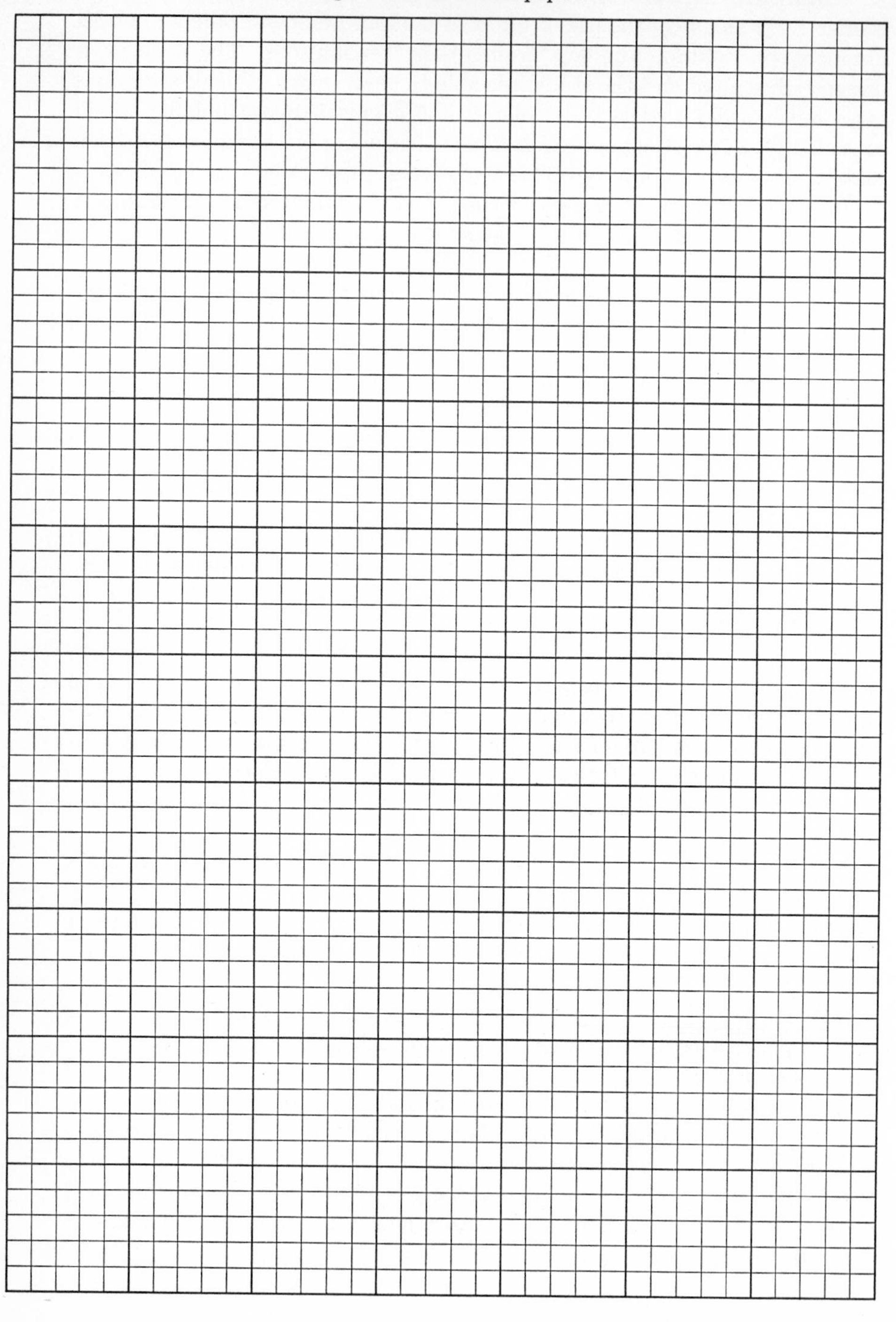

**Figure 2** Semilogarithmic paper.

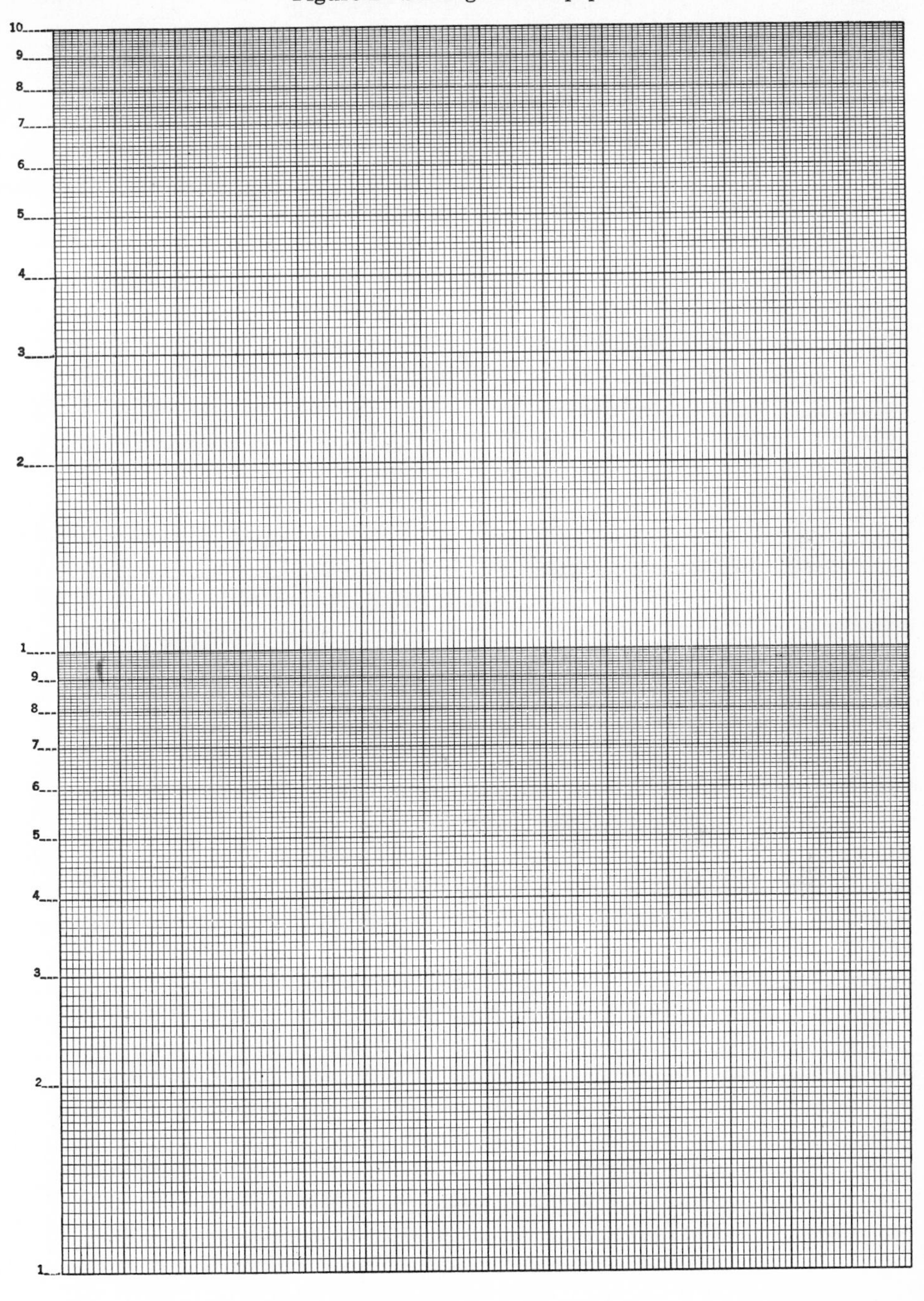

**Figure 3** Log-log paper.

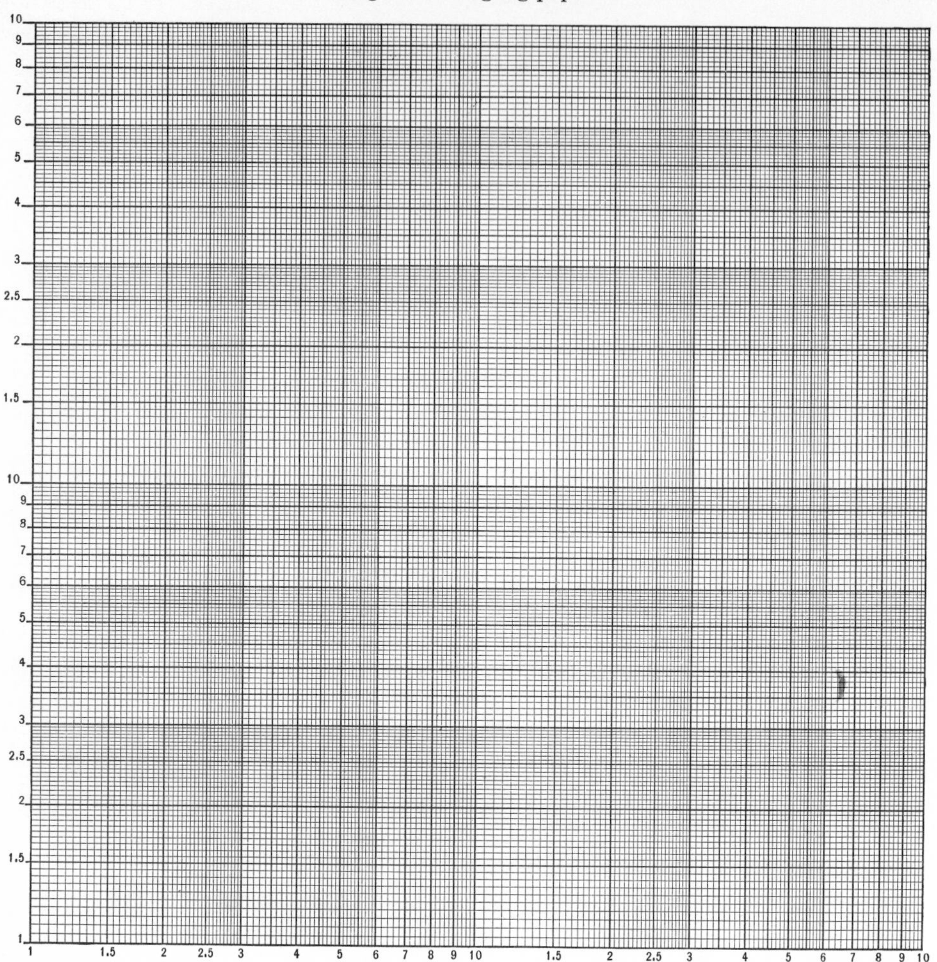

**Figure 4** Binomial paper.

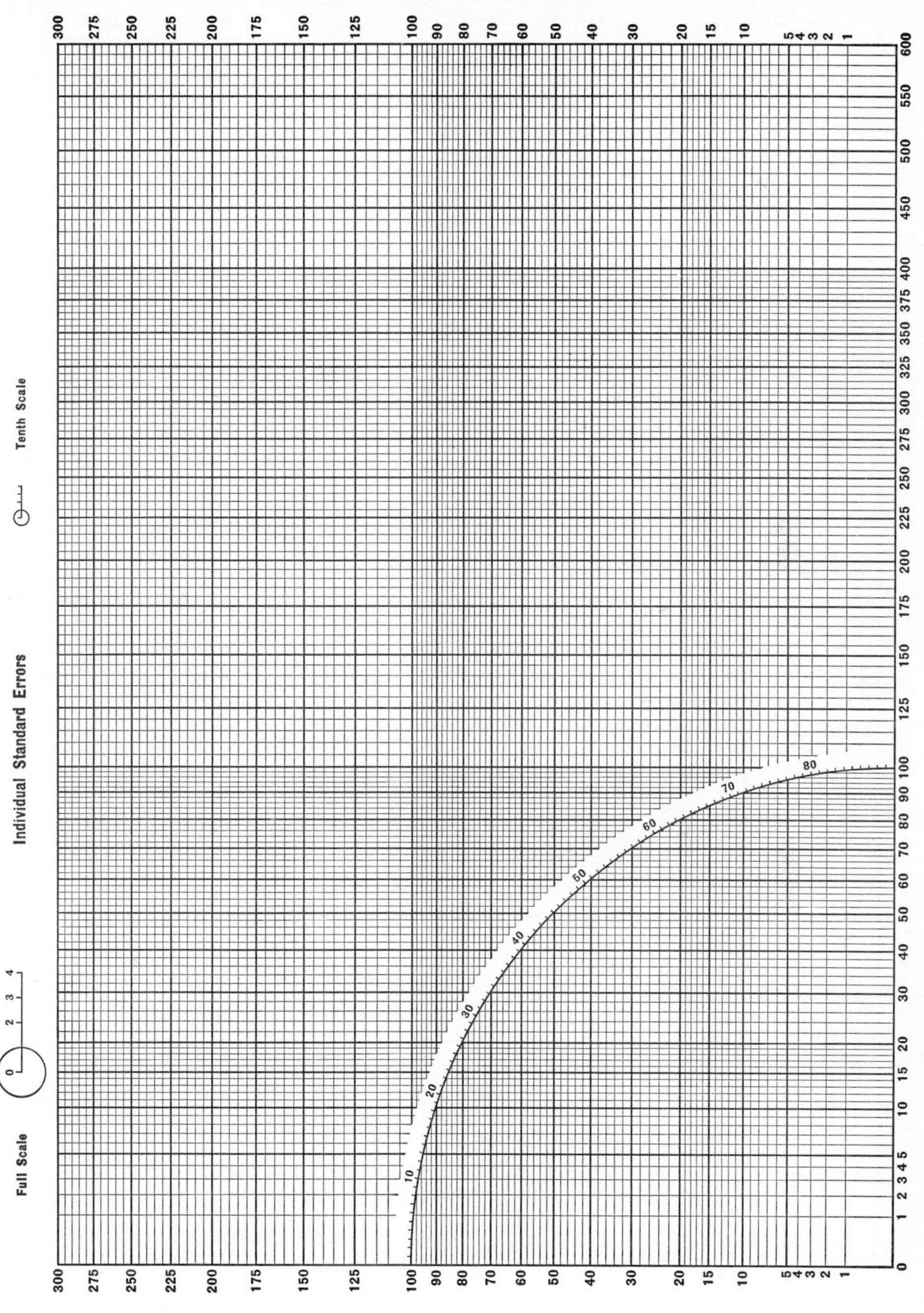

**Figure 5** Arithmetic normal paper.

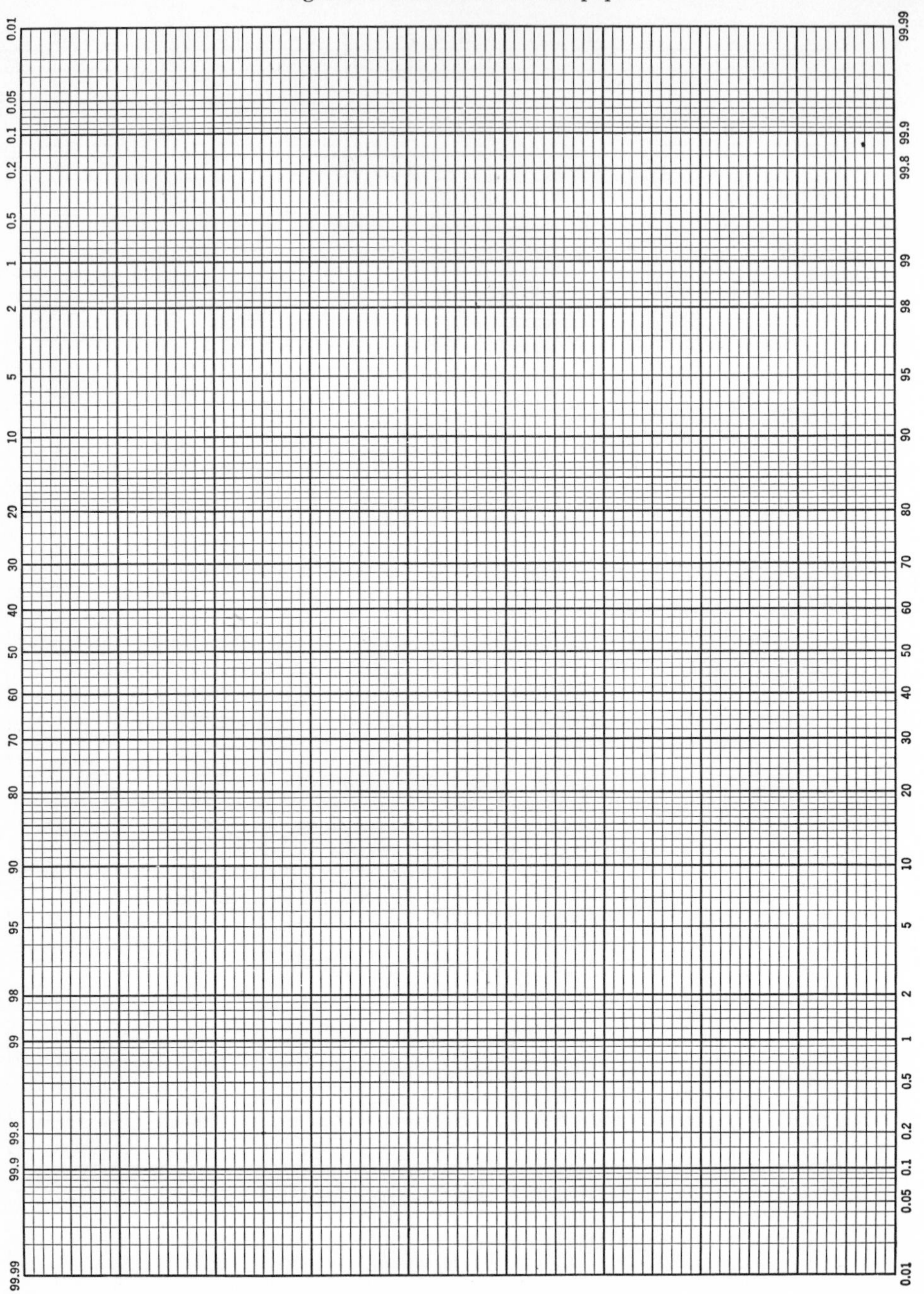

**Figure 6** Log-normal paper.

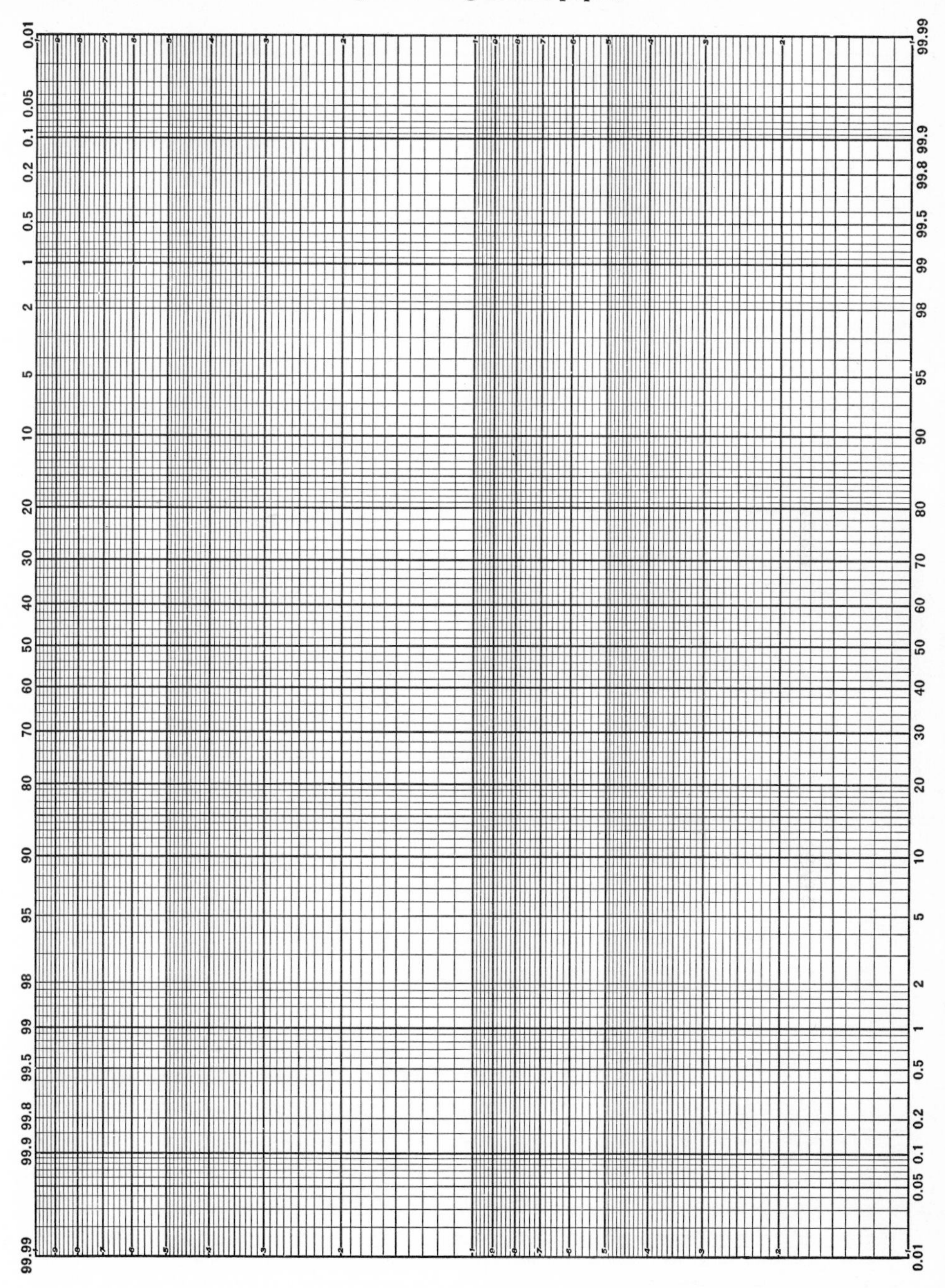

**Figure 7** Reciprocal normal paper.

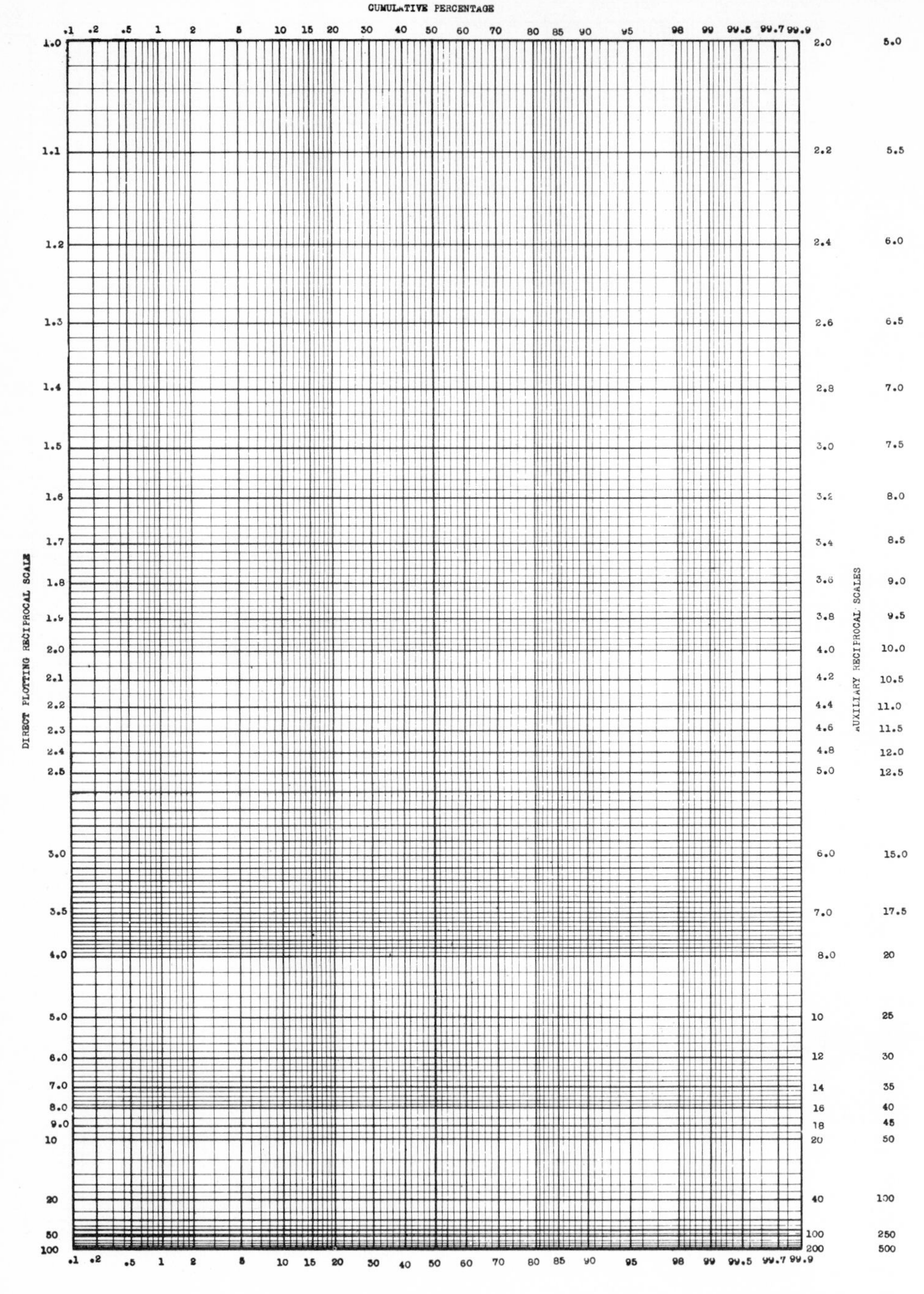

**Figure 8** Half-normal paper.

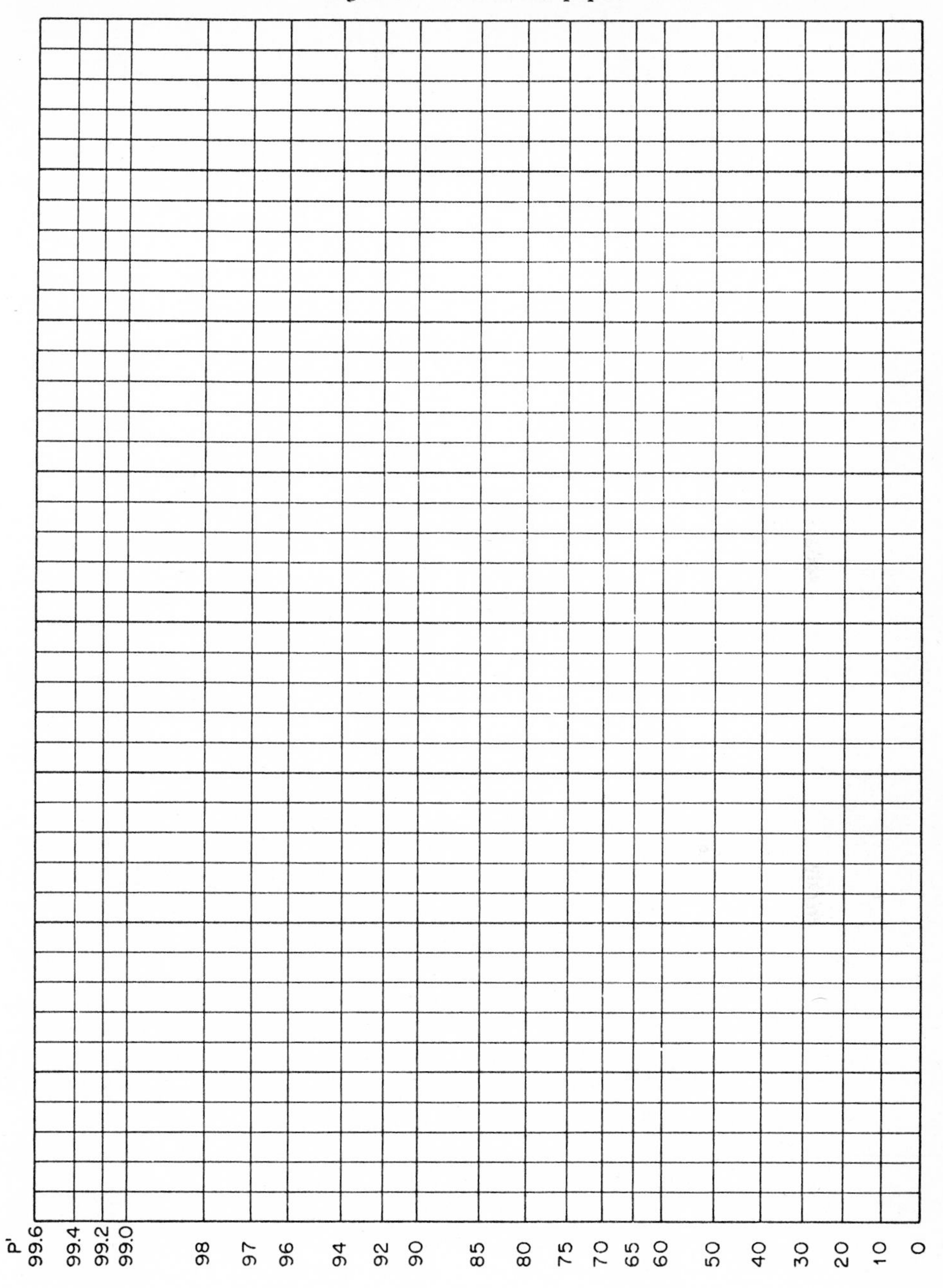

# References

The list of references following this section includes general statistical texts, books which stress certain fields of applications, and articles in the statistical journals providing extended reading on subjects introduced in this text. As a partial guide to these references, some comments are given by specific chapters.

**Chapters 1–5.** These chapters present fundamental definitions. Texts presenting statistics in a more mathematical context include

| | |
|---|---|
| Anderson and Bancroft (1) | Li (45) |
| Bennett and Franklin (4) | Mood and Graybill (49) |
| Feller (28) | Neyman (56) |
| Hodges and Lehmann (38) | Rao (62) |
| Hoel (39) | Scheffé (64) |
| Kempthorne (41) | Wilks (75) |

**Chapters 6–8.** The topics of tests and estimates are illustrated with examples from many fields of application. References stressing various fields of application are

Biology: Bliss (5), Finney (30), Fisher (31), (33), Li (44)
Engineering: Bowker and Goode (6), Brownlee (7)
Social Sciences: Cochran, Mosteller, and Tukey (13), Cochran (12), Deming (17), Hansen, Hurwitz, and Madow (35), Mosteller, Rourke, and Thomas (52)
Business: Croxton and Cowden (15), Sprowls (66)
Public Health: Dunn (23)
Psychology: Edwards (24), Moses (51)
Bacteriology: Eisenhart and Wilson (26), Stearman (67)
Medicine: Hill (37), Mainland (46)
Economics: Huff (40), Moroney (50)
Agriculture: Snedecor (65)
Chemistry: Youden (77)
Education: Walker and Lev (72)

**Chapter 9**

Systematic statistics, inefficient statistics, percentiles: Dixon (18). See also references in Appendix Table A-8.

**Chapter 10**

Assumptions of analysis of variance: Cochran (10), Eisenhart (27)
Surveys: Cochran (12)
Variance components: Crump (16)
Pooling of sums of squares: Paull (60)
Individual comparisons: Sheffé (64), Tukey (69)

**Chapter 11**

Interpretation of regression: Eisenhart (25), Winsor (76)
Rank correlation: Kendall (42)

**Chapter 13**

Discussion of the chi-square test: Cochran (14)
Binomial probability paper: Mosteller and Tukey (53)
Applications of acceptance sampling: Bowker and Goode (6)

**Chapter 15**

Transformation: Bartlett (3), Cochran (10)
Designs: Cochran and Cox (11)
Table of Latin squares: Fisher and Yates (32)

**Chapter 16**

Outliers: Dixon (18)
Winsorizing: Dixon and Tukey (22)

**Chapter 17**

Sign test: Dixon and Mood (19)
Treatment of ties: Kruskal and Wallis (43)
Maximum deviation statistic: Massey (47)
Runs: Swed and Eisenhart (68)
Rank sum test: Wilcoxon (74)
Rank correlation: Kendall (42)
Corner test: Olmsted and Tukey (58)

**Chapter 18**

General theory of sequential analysis: Wald (70)

**Chapter 19**

Up-and-down (staircase) method: Dixon and Mood (20), Dixon (21). See also Brownlee, Hodges, and Rosenblatt (8). An alternative sequential variable-level method for estimating $P_{50}$ is the Robbins-Munro method discussed in H. Robbins and S. Munro, "A Stochastic Approximation Method," *Annals of Mathematical Statistics*, vol. 22 (1951), pp. 400–407. This method does not require the assumption of normality, but uses continuously modified step sizes. Greater efficiency is possible in long series but in general no gains in efficiency occur for short series.

Other analyses of sensitivity data: Bliss (5), Finney (29)

**Tables**

Statistical tables: Fisher and Yates (32), Hald (34), Owen (59), Biometrika (61).
Tables of the binomial distribution (36), (55) and Romig (63)
Normal distributions: National Bureau of Standards (54)
Poisson: Molina (48)
Arithmetical tables: Barlow (2)


1. Anderson, R. L., Bancroft, T. A., *Statistical Theory in Research*. McGraw-Hill Book Company, New York, 1952.

2. Barlow, P., *Tables of Squares,* 4th ed. Chemical Publishing Company, Inc., New York, 1960.
3. Bartlett, M. S., "The use of transformations," *Biometrics,* vol. 3 (1947), p. 39.
4. Bennett, C. A., Franklin, N. L., *Statistical Analysis in Chemistry and the Chemical Industry.* John Wiley & Sons, Inc., New York, 1954.
5. Bliss, C. I., *Statistics in Biology.* McGraw-Hill Book Company, New York, 1967.
6. Bowker, A. H., Goode, H. P., *Sampling Inspection by Variables.* McGraw-Hill Book Company, New York, 1952.
7. Brownlee, K. A., *Statistical Theory and Methodology in Science and Engineering,* 2d ed. John Wiley & Sons, Inc., New York, 1965.
8. Brownlee, K. A., Hodges, J. L., Rosenblatt, M., "The up-and-down method with small samples," *Journal of the American Statistical Association,* vol. 48 (1953), p. 262.
9. Clark, V. A., Tartar, M. E., *Preparation for Basic Statistics.* McGraw-Hill Book Company, New York, 1968.
10. Cochran, W. G., "Some consequences when the assumptions for the analysis of variance are not satisfied," *Biometrics,* vol. 3 (1947), p. 22.
11. Cochran, W. G., Cox, G. M., *Experimental Designs,* 2d ed. John Wiley & Sons, Inc., New York, 1957.
12. Cochran, W. G., *Sampling Techniques,* 2d ed. John Wiley & Sons, Inc., New York, 1963.
13. Cochran, W. G., Mosteller, F., Tukey, J. W., *Statistical Problems of the Kinsey Report.* The American Statistical Association, Washington, 1954.
14. Cochran, W. G., "Some methods for strengthening the common chi-square tests," *Biometrics,* vol. 10 (1954), p. 417.
15. Croxton, F. E., Cowden, D. J., *Applied General Statistics,* 3d ed., Prentice-Hall, Inc., Englewood Cliffs, N.J., 1967.
16. Crump, S. L., "The present status of variance component analysis," *Biometrics,* vol. 7 (1951), p. 1.
17. Deming, W., *Some Theory of Sampling.* John Wiley & Sons, Inc., New York, 1950.
18. Dixon, W. J., "Analysis of extreme values," *Annals of Mathematical Statistics,* vol. 21 (1950), p. 488.
19. Dixon, W. J., Mood, A. M., "The statistical sign test," *Journal of the American Statistical Association,* vol. 41 (1946), p. 557.
20. Dixon, W. J., Mood, A. M., "A method for obtaining and analyzing sensitivity data," *Journal of the American Statistical Association,* vol. 43 (1948), p. 109.
21. Dixon, W. J., "The up-and-down method for small samples," *Journal of the American Statistical Association,* vol. 60 (1965), p. 967.
22. Dixon, W. J., Tukey, J. W., "Approximate behavior of the distribution of Winsorized *t*," *Technometrics,* vol. 10 (1968).
23. Dunn, O. J., *Basic Statistics.* John Wiley & Sons, Inc., New York, 1964.
24. Edwards, A. L., *Experimental Design in Psychological Research.* Holt, Rinehart and Winston, Inc., New York, 1950.
25. Eisenhart, C., "The interpretation of certain regression methods and their use in biological and industrial research," *Annals of Mathematical Statistics,* vol. 10 (1939), p. 162.
26. Eisenhart, C., Wilson, P. W., "Statistical methods and control in bacteriology," *Bacteriological Reviews,* vol. 7 (1943), p. 57.
27. Eisenhart, C., "The assumptions underlying the analysis of variance," *Biometrics,* vol. 3 (1947), p. 1.
28. Feller, W., *An Introduction to Probability Theory and Its Applications,* vol. 1, 2d ed. John Wiley & Sons, Inc., New York, 1957.
29. Finney, D. J., *Experimental Design and Its Statistical Basis.* The University of Chicago Press, Chicago, 1955.

30. Finney, D. J., *Statistical Method in Biological Assay,* 2d ed. Charles Griffin & Co., Ltd., London, 1964.
31. Fisher, R. A., *Statistical Methods for Research Workers,* 12th ed. Oliver & Boyd Ltd., Edinburgh and London, 1954.
32. Fisher, R. A., Yates, F., *Statistical Tables for Biological, Agricultural and Medical Research,* 6th ed. Oliver and Boyd Ltd., Edinburgh and London, 1963.
33. Fisher, R. A., *The Design of Experiments,* 8th ed. Oliver and Boyd Ltd., Edinburgh and London, 1966.
34. Hald, A., *Statistical Tables and Formulas.* John Wiley & Sons, Inc., New York, 1952.
35. Hansen, M. H., Hurwitz, W. N., Madow, W. G., *Sample Survey Methods and Theory.* John Wiley & Sons, Inc., New York, 1953.
36. Harvard University, *Tables of the Cumulative Binomial Probability Distribution.* Annals of the Computation Laboratory, Cambridge, Mass., vol. 35.
37. Hill, A. B., *Principles of Medical Statistics,* 8th ed. Oxford University Press, Fair Lawn, N.J., 1966.
38. Hodges, J., Lehman, E., *Elements of Finite Probability.* Holden-Day, Inc., San Francisco, 1965.
39. Hoel, P. G., *Introduction to Mathematical Statistics,* 3d ed. John Wiley & Sons, Inc., New York, 1962.
40. Huff, D., *How to Lie with Statistics.* W. W. Norton & Company, Inc., New York, 1954.
41. Kempthorne, O., *The Design and Analysis of Experiments.* John Wiley & Sons, Inc., New York, 1952.
42. Kendall, M. G., *Rank and Correlation Methods.* Charles Griffin & Co., Ltd., London, 1948.
43. Kruskal, W. H., Wallis, W. A., "Use of ranks in one-criterion variance analysis," *Journal of the American Statistical Association,* vol. 47 (1952) p. 583.
44. Li, C. C., *Introduction to Experimental Statistics.* McGraw-Hill Book Company, New York, 1964.
45. Li, J. C. R., *Introduction to Statistical Inference.* Edwards Brothers, Inc., Ann Arbor, Mich., 1957.
46. Mainland, D., *Elementary Medical Statistics,* 2d ed. W. B. Saunders Company, Philadelphia, 1963.
47. Massey, F. J., "The Kolmogorov-Smirnov test for goodness of fit," *Journal of the American Statistical Association,* vol. 46 (1951), p. 68.
48. Molina, E. C., *Poisson's Exponential Binomial Limit.* D. Van Nostrand Company, Inc., New York, 1942.
49. Mood, A., Graybill, F., *Introduction to the Theory of Statistics.* McGraw-Hill Book Company, New York, 1963.
50. Moroney, M. J., *Facts from Figures.* Penguin Books, Inc., Baltimore, 1953.
51. Moses, L., "Non-parametric statistics for psychological research," *Psychological Bulletin,* vol. 49 (1952), p. 122.
52. Mosteller, F., Rourke, R. E. K., Thomas, G. B., *Probability and Statistics.* Addison-Wesley Publishing Company, Inc., Reading, Mass., 1961.
53. Mosteller, F., Tukey, J. W., "The uses and usefulness of binomial probability paper," *Journal of the American Statistical Association,* vol. 44 (1949), p. 174.
54. National Bureau of Standards, *Tables of Probability Functions,* vols. I, II. Mathematical Tables 8, Washington, 1941.
55. National Bureau of Standards, *Tables of the Binomial Probability Distribution.* Applied Mathematics Series, 6, Washington, 1949.
56. Neyman, J., *First Course in Probability and Statistics.* Holt, Rinehart and Winston, Inc., New York, 1950.
57. Neyman, J., *Lectures and Conferences on Mathematical Statistics and Probability,* 2d ed. Graduate School of the U.S. Department of Agriculture, Washington, 1952.

58. Olmstead, P. S., Tukey, J. W., "A corner test for association," *Annals of Mathematical Statistics*, vol. 18 (1947) p. 495.
59. Owen, D. B., *Handbook of Statistical Tables*, Addison-Wesley Publishing Company, Inc., Reading, Mass., 1962.
60. Paull, A. E., "On a preliminary test for pooling mean squares in the analysis of variance," *Annals of Mathematical Statistics*, vol. 21 (1950), p. 539.
61. Pearson, E. S., Hartley, H. O., *Biometrika Tables for Statisticians*, vol. 1. Cambridge University Press, New York, 1954.
62. Rao, C. R., *Linear Statistical Inference and Its Applications*. John Wiley & Sons, Inc., New York, 1965.
63. Romig, H. G., 50–100 *Binomial Tables*. John Wiley & Sons, Inc., New York, 1953.
64. Scheffé, H., *The Analysis of Variance*. John Wiley & Sons, Inc., New York, 1959.
65. Snedecor, G. W., *Statistical Methods*, 5th ed. Iowa State College Press, Ames, Iowa, 1956.
66. Sprowls, R. C., *Elementary Statistics—For Students of Social Science and Business*. McGraw-Hill Book Company, New York, 1955.
67. Stearman, R. L., "Statistical concepts in microbiology," *Bacteriological Reviews*, vol. 19 (1955), p. 160.
68. Swed, F., Eisenhart, C., "Tables for testing randomness of grouping in a sequence of alternatives," *Annals of Mathematical Statistics*, vol. 14 (1943), p. 66.
69. Tukey, J. W., "Comparing individual means in the analysis of variance," *Biometrics*, vol. 5 (1949), p. 99.
70. Wald, A., *Sequential Analysis*. John Wiley & Sons, Inc., New York, 1947.
71. Walker, H. M., *Mathematics Essential for Elementary Statistics*, 2d ed. Holt, Rinehart and Winston, Inc., New York, 1951.
72. Walker, H. M., Lev, J., *Statistical Inference*. Holt, Rinehart and Winston, Inc., New York, 1953.
73. Wallis, W. A., Roberts, H. V., *Statistics: A New Approach*. The Free Press of Glencoe, New York, 1956.
74. Wilcoxon, F., "Individual comparisons by ranking methods," *Biometrics*, vol. 1 (1945), p. 80.
75. Wilks, S. S., *Elementary Statistical Analysis*, 2d ed. John Wiley & Sons, Inc., New York, 1962.
76. Winsor, C. P., "Which regression?" *Biometrics*, vol. 2 (1946), p. 101.
77. Youden, W. J., *Statistical Methods for Chemists*. John Wiley & Sons, Inc., New York, 1951.

# List of tables

## A-1 RANDOM NUMBERS

| | | | | |
|---|---|---|---|---|
| 10 09 73 25 33 | 76 52 01 35 86 | 34 67 35 48 76 | 80 95 90 91 17 | 39 29 27 49 45 |
| 37 54 20 48 05 | 64 89 47 42 96 | 24 80 52 40 37 | 20 63 61 04 02 | 00 82 29 16 65 |
| 08 42 26 89 53 | 19 64 50 93 03 | 23 20 90 25 60 | 15 95 33 47 64 | 35 08 03 36 06 |
| 99 01 90 25 29 | 09 37 67 07 15 | 38 31 13 11 65 | 88 67 67 43 97 | 04 43 62 76 59 |
| 12 80 79 99 70 | 80 15 73 61 47 | 64 03 23 66 53 | 98 95 11 68 77 | 12 17 17 68 33 |
| 66 06 57 47 17 | 34 07 27 68 50 | 36 69 73 61 70 | 65 81 33 98 85 | 11 19 92 91 70 |
| 31 06 01 08 05 | 45 57 18 24 06 | 35 30 34 26 14 | 86 79 90 74 39 | 23 40 30 97 32 |
| 85 26 97 76 02 | 02 05 16 56 92 | 68 66 57 48 18 | 73 05 38 52 47 | 18 62 38 85 79 |
| 63 57 33 21 35 | 05 32 54 70 48 | 90 55 35 75 48 | 28 46 82 87 09 | 83 49 12 56 24 |
| 73 79 64 57 53 | 03 52 96 47 78 | 35 80 83 42 82 | 60 93 52 03 44 | 35 27 38 84 35 |
| 98 52 01 77 67 | 14 90 56 86 07 | 22 10 94 05 58 | 60 97 09 34 33 | 50 50 07 39 98 |
| 11 80 50 54 31 | 39 80 82 77 32 | 50 72 56 82 48 | 29 40 52 42 01 | 52 77 56 78 51 |
| 83 45 29 96 34 | 06 28 89 80 83 | 13 74 67 00 78 | 18 47 54 06 10 | 68 71 17 78 17 |
| 88 68 54 02 00 | 86 50 75 84 01 | 36 76 66 79 51 | 90 36 47 64 93 | 29 60 91 10 62 |
| 99 59 46 73 48 | 87 51 76 49 69 | 91 82 60 89 28 | 93 78 56 13 68 | 23 47 83 41 13 |
| 65 48 11 76 74 | 17 46 85 09 50 | 58 04 77 69 74 | 73 03 95 71 86 | 40 21 81 65 44 |
| 80 12 43 56 35 | 17 72 70 80 15 | 45 31 82 23 74 | 21 11 57 82 53 | 14 38 55 37 63 |
| 74 35 09 98 17 | 77 40 27 72 14 | 43 23 60 02 10 | 45 52 16 42 37 | 96 28 60 26 55 |
| 69 91 62 68 03 | 66 25 22 91 48 | 36 93 68 72 03 | 76 62 11 39 90 | 94 40 05 64 18 |
| 09 89 32 05 05 | 14 22 56 85 14 | 46 42 75 67 88 | 96 29 77 88 22 | 54 38 21 45 98 |
| 91 49 91 45 23 | 68 47 92 76 86 | 46 16 28 35 54 | 94 75 08 99 23 | 37 08 92 00 48 |
| 80 33 69 45 98 | 26 94 03 68 58 | 70 29 73 41 35 | 53 14 03 33 40 | 42 05 08 23 41 |
| 44 10 48 19 49 | 85 15 74 79 54 | 32 97 92 65 75 | 57 60 04 08 81 | 22 22 20 64 13 |
| 12 55 07 37 42 | 11 10 00 20 40 | 12 86 07 46 97 | 96 64 48 94 39 | 28 70 72 58 15 |
| 63 60 64 93 29 | 16 50 53 44 84 | 40 21 95 25 63 | 43 65 17 70 82 | 07 20 73 17 90 |
| 61 19 69 04 46 | 26 45 74 77 74 | 51 92 43 37 29 | 65 39 45 95 93 | 42 58 26 05 27 |
| 15 47 44 52 66 | 95 27 07 99 53 | 59 36 78 38 48 | 82 39 61 01 18 | 33 21 15 94 66 |
| 94 55 72 85 73 | 67 89 75 43 87 | 54 62 24 44 31 | 91 19 04 25 92 | 92 92 74 59 73 |
| 42 48 11 62 13 | 97 34 40 87 21 | 16 86 84 87 67 | 03 07 11 20 59 | 25 70 14 66 70 |
| 23 52 37 83 17 | 73 20 88 98 37 | 68 93 59 14 16 | 26 25 22 96 63 | 05 52 28 25 62 |
| 04 49 35 24 94 | 75 24 63 38 24 | 45 86 25 10 25 | 61 96 27 93 35 | 65 33 71 24 72 |
| 00 54 99 76 54 | 64 05 18 81 59 | 96 11 96 38 96 | 54 69 28 23 91 | 23 28 72 95 29 |
| 35 96 31 53 07 | 26 89 80 93 54 | 33 35 13 54 62 | 77 97 45 00 24 | 90 10 33 93 33 |
| 59 80 80 83 91 | 45 42 72 68 42 | 83 60 94 97 00 | 13 02 12 48 92 | 78 56 52 01 06 |
| 46 05 88 52 36 | 01 39 09 22 86 | 77 28 14 40 77 | 93 91 08 36 47 | 70 61 74 29 41 |
| 32 17 90 05 97 | 87 37 92 52 41 | 05 56 70 70 07 | 86 74 31 71 57 | 85 39 41 18 38 |
| 69 23 46 14 06 | 20 11 74 52 04 | 15 95 66 00 00 | 18 74 39 24 23 | 97 11 89 63 38 |
| 19 56 54 14 30 | 01 75 87 53 79 | 40 41 92 15 85 | 66 67 43 68 06 | 84 96 28 52 07 |
| 45 15 51 49 38 | 19 47 60 72 46 | 43 66 79 45 43 | 59 04 79 00 33 | 20 82 66 95 41 |
| 94 86 43 19 94 | 36 16 81 08 51 | 34 88 88 15 53 | 01 54 03 54 56 | 05 01 45 11 76 |
| 98 08 62 48 26 | 45 24 02 84 04 | 44 99 90 88 96 | 39 09 47 34 07 | 35 44 13 18 80 |
| 33 18 51 62 32 | 41 94 15 09 49 | 89 43 54 85 81 | 88 69 54 19 94 | 37 54 87 30 43 |
| 80 95 10 04 06 | 96 38 27 07 74 | 20 15 12 33 87 | 25 01 62 52 98 | 94 62 46 11 71 |
| 79 75 24 91 40 | 71 96 12 82 96 | 69 86 10 25 91 | 74 85 22 05 39 | 00 38 75 95 79 |
| 18 63 33 25 37 | 98 14 50 65 71 | 31 01 02 46 74 | 05 45 56 14 27 | 77 93 89 19 36 |
| 74 02 94 39 02 | 77 55 73 22 70 | 97 79 01 71 19 | 52 52 75 80 21 | 80 81 45 17 48 |
| 54 17 84 56 11 | 80 99 33 71 43 | 05 33 51 29 69 | 56 12 71 92 55 | 36 04 09 03 24 |
| 11 66 44 98 83 | 52 07 98 48 27 | 59 38 17 15 39 | 09 97 33 34 40 | 88 46 12 33 56 |
| 48 32 47 79 28 | 31 24 96 47 10 | 02 29 53 68 70 | 32 30 75 75 46 | 15 02 00 99 94 |
| 69 07 49 41 38 | 87 63 79 19 76 | 35 58 40 44 01 | 10 51 82 16 15 | 01 84 87 69 38 |

From tables of the RAND Corporation, by permission.

| | | | | |
|---|---|---|---|---|
| 09 18 82 00 97 | 32 82 53 95 27 | 04 22 08 63 04 | 83 38 98 73 74 | 64 27 85 80 44 |
| 90 04 58 54 97 | 51 98 15 06 54 | 94 93 88 19 97 | 91 87 07 61 50 | 68 47 66 46 59 |
| 73 18 95 02 07 | 47 67 72 62 69 | 62 29 06 44 64 | 27 12 46 70 18 | 41 36 18 27 60 |
| 75 76 87 64 90 | 20 97 18 17 49 | 90 42 91 22 72 | 95 37 50 58 71 | 93 82 34 31 78 |
| 54 01 64 40 56 | 66 28 13 10 03 | 00 68 22 73 98 | 20 71 45 32 95 | 07 70 61 78 13 |
| | | | | |
| 08 35 86 99 10 | 78 54 24 27 85 | 13 66 15 88 73 | 04 61 89 75 53 | 31 22 30 84 20 |
| 28 30 60 32 64 | 81 33 31 05 91 | 40 51 00 78 93 | 32 60 46 04 75 | 94 11 90 18 40 |
| 53 84 08 62 33 | 81 59 41 36 28 | 51 21 59 02 90 | 28 46 66 87 95 | 77 76 22 07 91 |
| 91 75 75 37 41 | 61 61 36 22 69 | 50 26 39 02 12 | 55 78 17 65 14 | 83 48 34 70 55 |
| 89 41 59 26 94 | 00 39 75 83 91 | 12 60 71 76 46 | 48 94 97 23 06 | 94 54 13 74 08 |
| | | | | |
| 77 51 30 38 20 | 86 83 42 99 01 | 68 41 48 27 74 | 51 90 81 39 80 | 72 89 35 55 07 |
| 19 50 23 71 74 | 69 97 92 02 88 | 55 21 02 97 73 | 74 28 77 52 51 | 65 34 46 74 15 |
| 21 81 85 93 13 | 93 27 88 17 57 | 05 68 67 31 56 | 07 08 28 50 46 | 31 85 33 84 52 |
| 51 47 46 64 99 | 68 10 72 36 21 | 94 04 99 13 45 | 42 83 60 91 91 | 08 00 74 54 49 |
| 99 55 96 83 31 | 62 53 52 41 70 | 69 77 71 28 30 | 74 81 97 81 42 | 43 86 07 28 34 |
| | | | | |
| 33 71 34 80 07 | 93 58 47 28 69 | 51 92 66 47 21 | 58 30 32 98 22 | 93 17 49 39 72 |
| 85 27 48 68 93 | 11 30 32 92 70 | 28 83 43 41 37 | 73 51 59 04 00 | 71 14 84 36 43 |
| 84 13 38 96 40 | 44 03 55 21 66 | 73 85 27 00 91 | 61 22 26 05 61 | 62 32 71 84 23 |
| 56 73 21 62 34 | 17 39 59 61 31 | 10 12 39 16 22 | 85 49 65 75 60 | 81 60 41 88 80 |
| 65 13 85 68 06 | 87 64 88 52 61 | 34 31 36 58 61 | 45 87 52 10 69 | 85 64 44 72 77 |
| | | | | |
| 38 00 10 21 76 | 81 71 91 17 11 | 71 60 29 29 37 | 74 21 96 40 49 | 65 58 44 96 98 |
| 37 40 29 63 97 | 01 30 47 75 86 | 56 27 11 00 86 | 47 32 46 26 05 | 40 03 03 74 38 |
| 97 12 54 03 48 | 87 08 33 14 17 | 21 81 53 92 50 | 75 23 76 20 47 | 15 50 12 95 78 |
| 21 82 64 11 34 | 47 14 33 40 72 | 64 63 88 59 02 | 49 13 90 64 41 | 03 85 65 45 52 |
| 73 13 54 27 42 | 95 71 90 90 35 | 85 79 47 42 96 | 08 78 98 81 56 | 64 69 11 92 02 |
| | | | | |
| 07 63 87 79 29 | 03 06 11 80 72 | 96 20 74 41 56 | 23 82 19 95 38 | 04 71 36 69 94 |
| 60 52 88 34 41 | 07 95 41 98 14 | 59 17 52 06 95 | 05 53 35 21 39 | 61 21 20 64 55 |
| 83 59 63 56 55 | 06 95 89 29 83 | 05 12 80 97 19 | 77 43 35 37 83 | 92 30 15 04 98 |
| 10 85 06 27 46 | 99 59 91 05 07 | 13 49 90 63 19 | 53 07 57 18 39 | 06 41 01 93 62 |
| 39 82 09 89 52 | 43 62 26 31 47 | 64 42 18 08 14 | 43 80 00 93 51 | 31 02 47 31 67 |
| | | | | |
| 59 58 00 64 78 | 75 56 97 88 00 | 88 83 55 44 86 | 23 76 80 61 56 | 04 11 10 84 08 |
| 38 50 80 73 41 | 23 79 34 87 63 | 90 82 29 70 22 | 17 71 90 42 07 | 95 95 44 99 53 |
| 30 69 27 06 68 | 94 68 81 61 27 | 56 19 68 00 91 | 82 06 76 34 00 | 05 46 26 92 00 |
| 65 44 39 56 59 | 18 28 82 74 37 | 49 63 22 40 41 | 08 33 76 56 76 | 96 29 99 08 36 |
| 27 26 75 02 64 | 13 19 27 22 94 | 07 47 74 46 06 | 17 98 54 89 11 | 97 34 13 03 58 |
| | | | | |
| 91 30 70 69 91 | 19 07 22 42 10 | 36 69 95 37 28 | 28 82 53 57 93 | 28 97 66 62 52 |
| 68 43 49 46 88 | 84 47 31 36 22 | 62 12 69 84 08 | 12 84 38 25 90 | 09 81 59 31 46 |
| 48 90 81 58 77 | 54 74 52 45 91 | 35 70 00 47 54 | 83 82 45 26 92 | 54 13 05 51 60 |
| 06 91 34 51 97 | 42 67 27 86 01 | 11 88 30 95 28 | 63 01 19 89 01 | 14 97 44 03 44 |
| 10 45 51 60 19 | 14 21 03 37 12 | 91 34 23 78 21 | 88 32 58 08 51 | 43 66 77 08 83 |
| | | | | |
| 12 88 39 73 43 | 65 02 76 11 84 | 04 28 50 13 92 | 17 97 41 50 77 | 90 71 22 67 69 |
| 21 77 83 09 76 | 38 80 73 69 61 | 31 64 94 20 96 | 63 28 10 20 23 | 08 81 64 74 49 |
| 19 52 35 95 15 | 65 12 25 96 59 | 86 28 36 82 58 | 69 57 21 37 98 | 16 43 59 15 29 |
| 67 24 55 26 70 | 35 58 31 65 63 | 79 24 68 66 86 | 76 46 33 42 22 | 26 65 59 08 02 |
| 60 58 44 73 77 | 07 50 03 79 92 | 45 13 42 65 29 | 26 76 08 36 37 | 41 32 64 43 44 |
| | | | | |
| 53 85 34 13 77 | 36 06 69 48 50 | 58 83 87 38 59 | 49 36 47 33 31 | 96 24 04 36 42 |
| 24 63 73 87 36 | 74 38 48 93 42 | 52 62 30 79 92 | 12 36 91 86 01 | 03 74 28 38 73 |
| 83 08 01 24 51 | 38 99 22 28 15 | 07 75 95 17 77 | 97 37 72 75 85 | 51 97 23 78 67 |
| 16 44 42 43 34 | 36 15 19 90 73 | 27 49 37 09 39 | 85 13 03 25 52 | 54 84 65 47 59 |
| 60 79 01 81 57 | 57 17 86 57 62 | 11 16 17 85 76 | 45 81 95 29 79 | 65 13 00 48 60 |

## A-1 RANDOM NUMBERS (CONTINUED)

| | | | | |
|---|---|---|---|---|
| 03 99 11 04 61 | 93 71 61 68 94 | 66 08 32 46 53 | 84 60 95 82 32 | 88 61 81 91 61 |
| 38 55 59 55 54 | 32 88 65 97 80 | 08 35 56 08 60 | 29 73 54 77 62 | 71 29 92 38 53 |
| 17 54 67 37 04 | 92 05 24 62 15 | 55 12 12 92 81 | 59 07 60 79 36 | 27 95 45 89 09 |
| 32 64 35 28 61 | 95 81 90 68 31 | 00 91 19 89 36 | 76 35 59 37 79 | 80 86 30 05 14 |
| 69 57 26 87 77 | 39 51 03 59 05 | 14 06 04 06 19 | 29 54 96 96 16 | 33 56 46 07 80 |
| | | | | |
| 24 12 26 65 91 | 27 69 90 64 94 | 14 84 54 66 72 | 61 95 87 71 00 | 90 89 97 57 54 |
| 61 19 63 02 31 | 92 96 26 17 73 | 41 83 95 53 82 | 17 26 77 09 43 | 78 03 87 02 67 |
| 30 53 22 17 04 | 10 27 41 22 02 | 39 68 52 33 09 | 10 06 16 88 29 | 55 98 66 64 85 |
| 03 78 89 75 99 | 75 86 72 07 17 | 74 41 65 31 66 | 35 20 83 33 74 | 87 53 90 88 23 |
| 48 22 86 33 79 | 85 78 34 76 19 | 53 15 26 74 33 | 35 66 35 29 72 | 16 81 86 03 11 |
| | | | | |
| 60 36 59 46 53 | 35 07 53 39 49 | 42 61 42 92 97 | 01 91 82 83 16 | 98 95 37 32 31 |
| 83 79 94 24 02 | 56 62 33 44 42 | 34 99 44 13 74 | 70 07 11 47 36 | 09 95 81 80 65 |
| 32 96 00 74 05 | 36 40 98 32 32 | 99 38 54 16 00 | 11 13 30 75 86 | 15 91 70 62 53 |
| 19 32 25 38 45 | 57 62 05 26 06 | 66 49 76 86 46 | 78 13 86 65 59 | 19 64 09 94 13 |
| 11 22 09 47 47 | 07 39 93 74 08 | 48 50 92 39 29 | 27 48 24 54 76 | 85 24 43 51 59 |
| | | | | |
| 31 75 15 72 60 | 68 98 00 53 39 | 15 47 04 83 55 | 88 65 12 25 96 | 03 15 21 91 21 |
| 88 49 29 93 82 | 14 45 40 45 04 | 20 09 49 89 77 | 74 84 39 34 13 | 22 10 97 85 08 |
| 30 93 44 77 44 | 07 48 18 38 28 | 73 78 80 65 33 | 28 59 72 04 05 | 94 20 52 03 80 |
| 22 88 84 88 93 | 27 49 99 87 48 | 60 53 04 51 28 | 74 02 28 46 17 | 82 03 71 02 68 |
| 78 21 21 69 93 | 35 90 29 13 86 | 44 37 21 54 86 | 65 74 11 40 14 | 87 48 13 72 20 |
| | | | | |
| 41 84 98 45 47 | 46 85 05 23 26 | 34 67 75 83 00 | 74 91 06 43 45 | 19 32 58 15 49 |
| 46 35 23 30 49 | 69 24 89 34 60 | 45 30 50 75 21 | 61 31 83 18 55 | 14 41 37 09 51 |
| 11 08 79 62 94 | 14 01 33 17 92 | 59 74 76 72 77 | 76 50 33 45 13 | 39 66 37 75 44 |
| 52 70 10 83 37 | 56 30 38 73 15 | 16 52 06 96 76 | 11 65 49 98 93 | 02 18 16 81 61 |
| 57 27 53 68 98 | 81 30 44 85 85 | 68 65 22 73 76 | 92 85 25 58 66 | 88 44 80 35 84 |
| | | | | |
| 20 85 77 31 56 | 70 28 42 43 26 | 79 37 59 52 20 | 01 15 96 32 67 | 10 62 24 83 91 |
| 15 63 38 49 24 | 90 41 59 36 14 | 33 52 12 66 65 | 55 82 34 76 41 | 86 22 53 17 04 |
| 92 69 44 82 97 | 39 90 40 21 15 | 59 58 94 90 67 | 66 82 14 15 75 | 49 76 70 40 37 |
| 77 61 31 90 19 | 88 15 20 00 80 | 20 55 49 14 09 | 96 27 74 82 57 | 50 81 69 76 16 |
| 38 68 83 24 86 | 45 13 46 35 45 | 59 40 47 20 59 | 43 94 75 16 80 | 43 85 25 96 93 |
| | | | | |
| 25 16 30 18 89 | 70 01 41 50 21 | 41 29 06 73 12 | 71 85 71 59 57 | 68 97 11 14 03 |
| 65 25 10 76 29 | 37 23 93 32 95 | 05 87 00 11 19 | 92 78 42 63 40 | 18 47 76 56 22 |
| 36 81 54 36 25 | 18 63 73 75 09 | 82 44 49 90 05 | 04 92 17 37 01 | 14 70 79 39 97 |
| 64 39 71 16 92 | 05 32 78 21 62 | 20 24 78 17 59 | 45 19 72 53 32 | 83 74 52 25 67 |
| 04 51 52 56 24 | 95 09 66 79 46 | 48 46 08 55 58 | 15 19 11 87 82 | 16 93 03 33 61 |
| | | | | |
| 83 76 16 08 73 | 43 25 38 41 45 | 60 83 32 59 83 | 01 29 14 13 49 | 20 36 80 71 26 |
| 14 38 70 63 45 | 80 85 40 92 79 | 43 52 90 63 18 | 38 38 47 47 61 | 41 19 63 74 80 |
| 51 32 19 22 46 | 80 08 87 70 74 | 88 72 25 67 36 | 66 16 44 94 31 | 66 91 93 16 78 |
| 72 47 20 00 08 | 80 89 01 80 02 | 94 81 33 19 00 | 54 15 58 34 36 | 35 35 25 41 31 |
| 05 46 65 53 06 | 93 12 81 84 64 | 74 45 79 05 61 | 72 84 81 18 34 | 79 98 26 84 16 |
| | | | | |
| 39 52 87 24 84 | 82 47 42 55 93 | 48 54 53 52 47 | 18 61 91 36 74 | 18 61 11 92 41 |
| 81 61 61 87 11 | 53 34 24 42 76 | 75 12 21 17 24 | 74 62 77 37 07 | 58 31 91 59 97 |
| 07 58 61 61 20 | 82 64 12 28 20 | 92 90 41 31 41 | 32 39 21 97 63 | 61 19 96 79 40 |
| 90 76 70 42 35 | 13 57 41 72 00 | 69 90 26 37 42 | 78 46 42 25 01 | 18 62 79 08 72 |
| 40 18 82 81 93 | 29 59 38 86 27 | 94 97 21 15 98 | 62 09 53 67 87 | 00 44 15 89 97 |
| | | | | |
| 34 41 48 21 57 | 86 88 75 50 87 | 19 15 20 00 23 | 12 30 28 07 83 | 32 62 46 86 91 |
| 63 43 97 53 63 | 44 98 91 68 22 | 36 02 40 08 67 | 76 37 84 16 05 | 65 96 17 34 88 |
| 67 04 90 90 70 | 93 39 94 55 47 | 94 45 87 42 84 | 05 04 14 98 07 | 20 28 83 40 60 |
| 79 49 50 41 46 | 52 16 29 02 86 | 54 15 83 42 43 | 46 97 83 54 82 | 59 36 29 59 38 |
| 91 70 43 05 52 | 04 73 72 10 31 | 75 05 19 30 29 | 47 66 56 43 82 | 99 78 29 34 78 |

| | | | | |
|---|---|---|---|---|
| 94 01 54 68 74 | 32 44 44 82 77 | 59 82 09 61 63 | 64 65 42 58 43 | 41 14 54 28 20 |
| 74 10 88 82 22 | 88 57 07 40 15 | 25 70 49 10 35 | 01 75 51 47 50 | 48 96 83 86 03 |
| 62 88 08 78 73 | 95 16 05 92 21 | 22 30 49 03 14 | 72 87 71 73 34 | 39 28 30 41 49 |
| 11 74 81 21 02 | 80 58 04 18 67 | 17 71 05 96 21 | 06 55 40 78 50 | 73 95 07 95 52 |
| 17 94 40 56 00 | 60 47 80 33 43 | 25 85 25 89 05 | 57 21 63 96 18 | 49 85 69 93 26 |
| | | | | |
| 66 06 74 27 92 | 95 04 35 26 80 | 46 78 05 64 87 | 09 97 15 94 81 | 37 00 62 21 86 |
| 54 24 49 10 30 | 45 54 77 08 18 | 59 84 99 61 69 | 61 45 92 16 47 | 87 41 71 71 98 |
| 30 94 55 75 89 | 31 73 25 72 60 | 47 67 00 76 54 | 46 37 62 53 66 | 94 74 64 95 80 |
| 69 17 03 74 03 | 86 99 59 03 07 | 94 30 47 18 03 | 26 82 50 55 11 | 12 45 99 13 14 |
| 08 34 58 89 75 | 35 84 18 57 71 | 08 10 55 99 87 | 87 11 22 14 76 | 14 71 37 11 81 |
| | | | | |
| 27 76 74 35 84 | 85 30 18 89 77 | 29 49 06 97 14 | 73 03 54 12 07 | 74 69 90 93 10 |
| 13 02 51 43 38 | 54 06 61 52 43 | 47 72 46 67 33 | 47 43 14 39 05 | 31 04 85 66 99 |
| 80 21 73 62 92 | 98 52 52 43 35 | 24 43 22 48 96 | 43 27 75 88 74 | 11 46 61 60 82 |
| 10 87 56 20 04 | 90 39 16 11 05 | 57 41 10 63 68 | 53 85 63 07 43 | 08 67 08 47 41 |
| 54 12 75 73 26 | 26 62 91 90 87 | 24 47 28 87 79 | 30 54 02 78 86 | 61 73 27 54 54 |
| | | | | |
| 60 31 14 28 24 | 37 30 14 26 78 | 45 99 04 32 42 | 17 37 45 20 03 | 70 70 77 02 14 |
| 49 73 97 14 84 | 92 00 39 80 86 | 76 66 87 32 09 | 59 20 21 19 73 | 02 90 23 32 50 |
| 78 62 65 15 94 | 16 45 39 46 14 | 39 01 49 70 66 | 83 01 20 98 32 | 25 57 17 76 28 |
| 66 69 21 39 86 | 99 83 70 05 82 | 81 23 24 49 87 | 09 50 49 64 12 | 90 19 37 95 68 |
| 44 07 12 80 91 | 07 36 29 77 03 | 76 44 74 25 37 | 98 52 49 78 31 | 65 70 40 95 14 |
| | | | | |
| 41 46 88 51 49 | 49 55 41 79 94 | 14 92 43 96 50 | 95 29 40 05 56 | 70 48 10 69 05 |
| 94 55 93 75 59 | 49 67 85 31 19 | 70 31 20 56 82 | 66 98 63 40 99 | 74 47 42 07 40 |
| 41 61 57 03 60 | 64 11 45 86 60 | 90 85 06 46 18 | 80 62 05 17 90 | 11 43 63 80 72 |
| 50 27 39 31 13 | 41 79 48 68 61 | 24 78 18 96 83 | 55 41 18 56 67 | 77 53 59 98 92 |
| 41 39 68 05 04 | 90 67 00 82 89 | 40 90 20 50 69 | 95 08 30 67 83 | 28 10 25 78 16 |
| | | | | |
| 25 80 72 42 60 | 71 52 97 89 20 | 72 68 20 73 85 | 90 72 65 71 66 | 98 88 40 85 83 |
| 06 17 09 79 65 | 88 30 29 80 41 | 21 44 34 18 08 | 68 98 48 36 20 | 89 74 79 88 82 |
| 60 80 85 44 44 | 74 41 28 11 05 | 01 17 62 88 38 | 36 42 11 64 89 | 18 05 95 10 61 |
| 80 94 04 48 93 | 10 40 83 62 22 | 80 58 27 19 44 | 92 63 84 03 33 | 67 05 41 60 67 |
| 19 51 69 01 20 | 46 75 97 16 43 | 13 17 75 52 92 | 21 03 68 28 08 | 77 50 19 74 27 |
| | | | | |
| 49 38 65 44 80 | 28 60 42 35 54 | 21 78 54 11 01 | 91 17 81 01 74 | 29 42 09 04 38 |
| 06 31 28 89 40 | 15 99 56 93 21 | 47 45 86 48 09 | 98 18 98 18 51 | 29 65 18 42 15 |
| 60 94 20 03 07 | 11 89 79 26 74 | 40 40 56 80 32 | 96 71 75 42 44 | 10 70 14 13 93 |
| 92 32 99 89 32 | 78 28 44 63 47 | 71 20 99 20 61 | 39 44 89 31 36 | 25 72 20 85 64 |
| 77 93 66 35 74 | 31 38 45 19 24 | 85 56 12 96 71 | 58 13 71 78 20 | 22 75 13 65 18 |
| | | | | |
| 38 10 17 77 56 | 11 65 71 38 97 | 95 88 95 70 67 | 47 64 81 38 85 | 70 66 99 34 06 |
| 39 64 16 94 57 | 91 33 92 25 02 | 92 61 38 97 19 | 11 94 75 62 03 | 19 32 42 05 04 |
| 84 05 44 04 55 | 99 39 66 36 80 | 67 66 76 06 31 | 69 18 19 68 45 | 38 52 51 16 00 |
| 47 46 80 35 77 | 57 64 96 32 66 | 24 70 07 15 94 | 14 00 42 31 53 | 69 24 90 57 47 |
| 43 32 13 13 70 | 28 97 72 38 96 | 76 47 96 85 62 | 62 34 20 75 89 | 08 89 90 59 85 |
| | | | | |
| 64 28 16 18 26 | 18 55 56 49 37 | 13 17 33 33 65 | 78 85 11 64 99 | 87 06 41 30 75 |
| 66 84 77 04 95 | 32 35 00 29 85 | 86 71 63 87 46 | 26 31 37 74 63 | 55 38 77 26 81 |
| 72 46 13 32 30 | 21 52 95 34 24 | 92 58 10 22 62 | 78 43 86 62 76 | 18 39 67 35 38 |
| 21 03 29 10 50 | 13 05 81 62 18 | 12 47 05 65 00 | 15 29 27 61 39 | 59 52 65 21 13 |
| 95 36 26 70 11 | 06 65 11 61 36 | 01 01 60 08 57 | 55 01 85 63 74 | 35 82 47 17 08 |
| | | | | |
| 49 71 29 73 80 | 10 40 45 54 52 | 34 03 06 07 26 | 75 21 11 02 71 | 36 63 36 84 24 |
| 58 27 56 17 64 | 97 58 65 47 16 | 50 25 94 63 45 | 87 19 54 60 92 | 26 78 76 09 39 |
| 89 51 41 17 88 | 68 22 42 34 17 | 73 95 97 61 45 | 30 34 24 02 77 | 11 04 97 20 49 |
| 15 47 25 06 69 | 48 13 93 67 32 | 46 87 43 70 88 | 73 46 50 98 19 | 58 86 93 52 20 |
| 12 12 08 61 24 | 51 24 74 43 02 | 60 88 35 21 09 | 21 43 73 67 86 | 49 22 67 78 37 |

## A-1 RANDOM NUMBERS (CONTINUED)

| | | | | |
|---|---|---|---|---|
| 19 61 27 84 30 | 11 66 19 47 70 | 77 60 36 56 69 | 86 86 81 26 65 | 30 01 27 59 89 |
| 39 14 17 74 00 | 28 00 06 42 38 | 73 25 87 17 94 | 31 34 02 62 56 | 66 45 33 70 16 |
| 64 75 68 04 57 | 08 74 71 28 36 | 03 46 95 06 78 | 03 27 44 34 23 | 66 67 78 25 56 |
| 92 90 15 18 78 | 56 44 12 29 98 | 29 71 83 84 47 | 06 45 32 53 11 | 07 56 55 37 71 |
| 03 55 19 00 70 | 09 48 39 40 50 | 45 93 81 81 35 | 36 90 84 33 21 | 11 07 35 18 03 |
| 98 88 46 62 09 | 06 83 05 36 56 | 14 66 35 63 46 | 71 43 00 49 09 | 19 81 80 57 07 |
| 27 36 98 68 82 | 53 47 30 75 41 | 53 63 37 08 63 | 03 74 81 28 22 | 19 36 04 90 88 |
| 59 06 67 59 74 | 63 33 52 04 83 | 43 51 43 74 81 | 58 27 82 69 67 | 49 32 54 39 51 |
| 91 64 79 37 83 | 64 16 94 90 22 | 98 58 80 94 95 | 49 82 95 90 68 | 38 83 10 48 38 |
| 83 60 59 24 19 | 39 54 20 77 72 | 71 56 87 56 73 | 35 18 58 97 59 | 44 90 17 42 91 |
| 24 89 58 85 30 | 70 77 43 54 39 | 46 75 87 04 72 | 70 20 79 26 75 | 91 62 36 12 75 |
| 35 72 02 65 56 | 95 59 62 00 94 | 73 75 08 57 88 | 34 26 40 17 03 | 46 83 36 52 48 |
| 14 14 15 34 10 | 38 64 90 63 43 | 57 25 66 13 42 | 72 70 97 53 18 | 90 37 93 75 62 |
| 27 41 67 56 70 | 92 17 67 25 35 | 93 11 95 60 77 | 06 88 61 82 44 | 92 34 43 13 74 |
| 82 07 10 74 29 | 81 00 74 77 49 | 40 74 45 69 74 | 23 33 68 88 21 | 53 84 11 05 36 |
| 21 44 58 27 93 | 24 83 19 32 41 | 14 19 97 62 68 | 70 88 36 80 02 | 03 82 91 74 43 |
| 72 51 37 64 00 | 52 22 59 23 48 | 62 30 89 84 81 | 29 74 43 31 65 | 33 14 16 10 20 |
| 71 47 94 50 27 | 76 16 05 74 11 | 13 78 01 36 32 | 52 30 87 77 62 | 88 87 43 36 97 |
| 83 21 05 14 66 | 09 08 85 03 95 | 26 74 30 53 06 | 21 70 67 00 01 | 99 43 98 07 67 |
| 68 74 99 51 48 | 94 89 77 86 36 | 96 75 00 90 24 | 94 53 89 11 43 | 96 69 36 18 86 |
| 05 18 47 57 63 | 47 07 58 81 58 | 05 31 35 34 39 | 14 90 80 88 30 | 60 09 62 15 51 |
| 13 65 16 25 46 | 96 89 22 52 40 | 47 51 15 84 83 | 87 34 27 88 18 | 07 85 53 92 69 |
| 00 56 62 12 20 | 00 29 22 40 69 | 25 07 22 95 19 | 52 54 85 40 91 | 21 28 22 12 96 |
| 50 95 81 76 95 | 58 07 26 89 90 | 60 32 99 59 55 | 71 58 66 34 17 | 35 94 76 78 07 |
| 57 62 16 45 47 | 46 85 03 79 81 | 38 52 70 90 37 | 64 75 60 33 24 | 04 98 68 36 66 |
| 09 28 22 58 44 | 79 13 97 84 35 | 35 42 84 35 61 | 69 79 96 33 14 | 12 99 19 35 16 |
| 23 39 49 42 06 | 93 43 23 78 36 | 94 91 92 68 46 | 02 55 57 44 10 | 94 91 54 81 99 |
| 05 28 03 74 70 | 93 62 20 43 45 | 15 09 21 95 10 | 18 09 41 66 13 | 78 23 45 00 01 |
| 95 49 19 79 76 | 38 30 63 21 92 | 82 63 95 46 24 | 72 43 49 26 06 | 23 19 17 46 93 |
| 78 52 10 01 04 | 18 24 87 55 83 | 90 32 65 07 85 | 54 03 46 62 51 | 35 77 41 46 92 |
| 96 34 54 45 79 | 85 93 24 40 53 | 75 70 42 08 40 | 86 58 38 39 44 | 52 45 67 37 66 |
| 77 96 33 11 51 | 32 36 49 16 91 | 47 35 74 03 38 | 23 43 52 40 65 | 08 45 89 53 66 |
| 07 52 01 12 94 | 23 23 80 17 48 | 41 69 06 73 28 | 54 81 43 77 77 | 10 05 74 23 32 |
| 38 42 30 23 09 | 70 70 38 57 36 | 46 14 81 42 58 | 29 23 61 21 52 | 05 08 86 58 25 |
| 02 46 36 55 33 | 21 19 96 05 55 | 33 92 80 18 17 | 07 39 68 92 15 | 30 72 22 21 02 |
| 15 88 09 22 61 | 17 29 28 81 90 | 61 78 14 88 98 | 92 52 52 12 83 | 88 58 16 00 98 |
| 71 92 60 08 19 | 59 14 40 02 24 | 30 57 09 01 94 | 18 32 90 69 99 | 26 85 71 92 38 |
| 64 42 52 81 08 | 16 55 41 60 16 | 00 04 28 32 29 | 10 33 33 61 68 | 65 61 79 48 34 |
| 79 78 22 39 24 | 49 44 03 04 32 | 81 07 73 15 43 | 95 21 66 48 65 | 13 65 85 10 81 |
| 35 33 77 45 38 | 44 55 36 46 72 | 90 96 04 18 49 | 93 86 54 46 08 | 93 17 63 48 51 |
| 05 24 92 93 29 | 19 71 59 40 82 | 14 73 88 66 67 | 43 70 86 63 54 | 93 69 22 55 27 |
| 56 46 39 93 80 | 38 79 38 57 74 | 19 05 61 39 39 | 46 06 22 76 47 | 66 14 66 32 10 |
| 96 29 63 31 21 | 54 19 63 41 08 | 75 81 48 59 86 | 71 17 11 51 02 | 28 99 26 31 65 |
| 98 38 03 62 69 | 60 01 40 72 01 | 62 44 84 63 85 | 42 17 58 83 50 | 46 18 24 91 26 |
| 52 56 76 43 50 | 16 31 55 39 69 | 80 39 58 11 14 | 54 35 86 45 78 | 47 26 91 57 47 |
| 78 49 89 08 30 | 25 95 59 92 36 | 43 28 69 10 64 | 99 96 99 51 44 | 64 42 47 73 77 |
| 49 55 32 42 41 | 08 15 08 95 35 | 08 70 39 10 41 | 77 32 38 10 79 | 45 12 79 36 86 |
| 32 15 10 70 75 | 83 15 51 02 52 | 73 10 08 86 18 | 23 89 18 74 18 | 45 41 72 02 68 |
| 11 31 45 03 63 | 26 86 02 77 99 | 49 41 68 35 34 | 19 18 70 80 59 | 76 67 70 21 10 |
| 12 36 47 12 10 | 87 05 25 02 41 | 90 78 59 78 89 | 81 39 95 81 30 | 64 43 90 56 14 |

| 01 | 02 | 03 | 04 | 05 | 06 | 07 | 08 | 09 | 10 |
|---|---|---|---|---|---|---|---|---|---|
| 0.464 | 0.137 | 2.455 | −0.323 | −0.068 | 0.296 | −0.288 | 1.298 | 0.241 | −0.957 |
| 0.060 | −2.526 | −0.531 | −0.194 | 0.543 | −1.558 | 0.187 | −1.190 | 0.022 | 0.525 |
| 1.486 | −0.354 | −0.634 | 0.697 | 0.926 | 1.375 | 0.785 | −0.963 | −0.853 | −1.865 |
| 1.022 | −0.472 | 1.279 | 3.521 | 0.571 | −1.851 | 0.194 | 1.192 | −0.501 | −0.273 |
| 1.394 | −0.555 | 0.046 | 0.321 | 2.945 | 1.974 | −0.258 | 0.412 | 0.439 | −0.035 |
| 0.906 | −0.513 | −0.525 | 0.595 | 0.881 | −0.934 | 1.579 | 0.161 | −1.885 | 0.371 |
| 1.179 | −1.055 | 0.007 | 0.769 | 0.971 | 0.712 | 1.090 | −0.631 | −0.255 | −0.702 |
| −1.501 | −0.488 | −0.162 | −0.136 | 1.033 | 0.203 | 0.448 | 0.748 | −0.423 | −0.432 |
| −0.690 | 0.756 | −1.618 | −0.345 | −0.511 | −2.051 | −0.457 | −0.218 | 0.857 | −0.465 |
| 1.372 | 0.225 | 0.378 | 0.761 | 0.181 | −0.736 | 0.960 | −1.530 | −0.260 | 0.120 |
| −0.482 | 1.678 | −0.057 | −1.229 | −0.486 | 0.856 | −0.491 | −1.983 | −2.830 | −0.238 |
| −1.376 | −0.150 | 1.356 | −0.561 | −0.256 | −0.212 | 0.219 | 0.779 | 0.953 | −0.869 |
| −1.010 | 0.598 | −0.918 | 1.598 | 0.065 | 0.415 | −0.169 | 0.313 | −0.973 | −1.016 |
| −0.005 | −0.899 | 0.012 | −0.725 | 1.147 | −0.121 | 1.096 | 0.481 | −1.691 | 0.417 |
| 1.393 | −1.163 | −0.911 | 1.231 | −0.199 | −0.246 | 1.239 | −2.574 | −0.558 | 0.056 |
| −1.787 | −0.261 | 1.237 | 1.046 | −0.508 | −1.630 | −0.146 | −0.392 | −0.627 | 0.561 |
| −0.105 | −0.357 | −1.384 | 0.360 | −0.992 | −0.116 | −1.698 | −2.832 | −1.108 | −2.357 |
| −1.339 | 1.827 | −0.959 | 0.424 | 0.969 | −1.141 | −1.041 | 0.362 | −1.726 | 1.956 |
| 1.041 | 0.535 | 0.731 | 1.377 | 0.983 | −1.330 | 1.620 | −1.040 | 0.524 | −0.281 |
| 0.279 | −2.056 | 0.717 | −0.873 | −1.096 | −1.396 | 1.047 | 0.089 | −0.573 | 0.932 |
| −1.805 | −2.008 | −1.633 | 0.542 | 0.250 | −0.166 | 0.032 | 0.079 | 0.471 | −1.029 |
| −1.186 | 1.180 | 1.114 | 0.882 | 1.265 | −0.202 | 0.151 | −0.376 | −0.310 | 0.479 |
| 0.658 | −1.141 | 1.151 | −1.210 | −0.927 | 0.425 | 0.290 | −0.902 | 0.610 | 2.709 |
| −0.439 | 0.358 | −1.939 | 0.891 | −0.227 | 0.602 | 0.873 | −0.437 | −0.220 | −0.057 |
| −1.399 | −0.230 | 0.385 | −0.649 | −0.577 | 0.237 | −0.289 | 0.513 | 0.738 | −0.300 |
| 0.199 | 0.208 | −1.083 | −0.219 | −0.291 | 1.221 | 1.119 | 0.004 | −2.015 | −0.594 |
| 0.159 | 0.272 | −0.313 | 0.084 | −2.828 | −0.439 | −0.792 | −1.275 | −0.623 | −1.047 |
| 2.273 | 0.606 | 0.606 | −0.747 | 0.247 | 1.291 | 0.063 | −1.793 | −0.699 | −1.347 |
| 0.041 | −0.307 | 0.121 | 0.790 | −0.584 | 0.541 | 0.484 | −0.986 | 0.481 | 0.996 |
| −1.132 | −2.098 | 0.921 | 0.145 | 0.446 | −1.661 | 1.045 | −1.363 | −0.586 | −1.023 |
| 0.768 | 0.079 | −1.473 | 0.034 | −2.127 | 0.665 | 0.084 | −0.880 | −0.579 | 0.551 |
| 0.375 | −1.658 | −0.851 | 0.234 | −0.656 | 0.340 | −0.086 | −0.158 | −0.120 | 0.418 |
| −0.513 | −0.344 | 0.210 | −0.736 | 1.041 | 0.008 | 0.427 | −0.831 | 0.191 | 0.074 |
| 0.292 | −0.521 | 1.266 | −1.206 | −0.899 | 0.110 | −0.528 | −0.813 | 0.071 | 0.524 |
| 1.026 | 2.990 | −0.574 | −0.491 | −1.114 | 1.297 | −1.433 | −1.345 | −3.001 | 0.479 |
| −1.334 | 1.278 | −0.568 | −0.109 | −0.515 | −0.566 | 2.923 | 0.500 | 0.359 | 0.326 |
| −0.287 | −0.144 | −0.254 | 0.574 | −0.451 | −1.181 | −1.190 | −0.318 | −0.094 | 1.114 |
| 0.161 | −0.886 | −0.921 | −0.509 | 1.410 | −0.518 | 0.192 | −0.432 | 1.501 | 1.068 |
| −1.346 | 0.193 | −1.202 | 0.394 | −1.045 | 0.843 | 0.942 | 1.045 | 0.031 | 0.772 |
| 1.250 | −0.199 | −0.288 | 1.810 | 1.378 | 0.584 | 1.216 | 0.733 | 0.402 | 0.226 |
| 0.630 | −0.537 | 0.782 | 0.060 | 0.499 | −0.431 | 1.705 | 1.164 | 0.884 | −0.298 |
| 0.375 | −1.941 | 0.247 | −0.491 | 0.665 | −0.135 | −0.145 | −0.498 | 0.457 | 1.064 |
| −1.420 | 0.489 | −1.711 | −1.186 | 0.754 | −0.732 | −0.066 | 1.006 | −0.798 | 0.162 |
| −0.151 | −0.243 | −0.430 | −0.762 | 0.298 | 1.049 | 1.810 | 2.885 | −0.768 | −0.129 |
| −0.309 | 0.531 | 0.416 | −1.541 | 1.456 | 2.040 | −0.124 | 0.196 | 0.023 | −1.204 |
| 0.424 | −0.444 | 0.593 | 0.993 | −0.106 | 0.116 | 0.484 | −1.272 | 1.066 | 1.097 |
| 0.593 | 0.658 | −1.127 | −1.407 | −1.579 | −1.616 | 1.458 | 1.262 | 0.736 | −0.916 |
| 0.862 | −0.885 | −0.142 | −0.504 | 0.532 | 1.381 | 0.022 | −0.281 | −0.342 | 1.222 |
| 0.235 | −0.628 | −0.023 | −0.463 | −0.899 | −0.394 | −0.538 | 1.707 | −0.188 | −1.153 |
| −0.853 | 0.402 | 0.777 | 0.833 | 0.410 | −0.349 | −1.094 | 0.580 | 1.395 | 1.298 |

From tables of the RAND Corporation, by permission.

## A-2 RANDOM NORMAL NUMBERS, $\mu = 0$ AND $\sigma = 1$ (CONTINUED)

| 11 | 12 | 13 | 14 | 15 | 16 | 17 | 18 | 19 | 20 |
|---|---|---|---|---|---|---|---|---|---|
| −1.329 | −0.238 | −0.838 | −0.988 | −0.445 | 0.964 | −0.266 | −0.322 | −1.726 | 2.252 |
| 1.284 | −0.229 | 1.058 | 0.090 | 0.050 | 0.523 | 0.016 | 0.277 | 1.639 | 0.554 |
| 0.619 | 0.628 | 0.005 | 0.973 | −0.058 | 0.150 | −0.635 | −0.917 | 0.313 | −1.203 |
| 0.699 | −0.269 | 0.722 | −0.994 | −0.807 | −1.203 | 1.163 | 1.244 | 1.306 | −1.210 |
| 0.101 | 0.202 | −0.150 | 0.731 | 0.420 | 0.116 | −0.496 | −0.037 | −2.466 | 0.794 |
| −1.381 | 0.301 | 0.522 | 0.233 | 0.791 | −1.017 | −0.182 | 0.926 | −1.096 | 1.001 |
| −0.574 | 1.366 | −1.843 | 0.746 | 0.890 | 0.824 | −1.249 | −0.806 | −0.240 | 0.217 |
| 0.096 | 0.210 | 1.091 | 0.990 | 0.900 | −0.837 | −1.097 | −1.238 | 0.030 | −0.311 |
| 1.389 | −0.236 | 0.094 | 3.282 | 0.295 | −0.416 | 0.313 | 0.720 | 0.007 | 0.354 |
| 1.249 | 0.706 | 1.453 | 0.366 | −2.654 | −1.400 | 0.212 | 0.307 | −1.145 | 0.639 |
| 0.756 | −0.397 | −1.772 | −0.257 | 1.120 | 1.188 | −0.527 | 0.709 | 0.479 | 0.317 |
| −0.860 | 0.412 | −0.327 | 0.178 | 0.524 | −0.672 | −0.831 | 0.758 | 0.131 | 0.771 |
| −0.778 | −0.979 | 0.236 | −1.033 | 1.497 | −0.661 | 0.906 | 1.169 | −1.582 | 1.303 |
| 0.037 | 0.062 | 0.426 | 1.220 | 0.471 | 0.784 | −0.719 | 0.465 | 1.559 | −1.326 |
| 2.619 | −0.440 | 0.477 | 1.063 | 0.320 | 1.406 | −0.701 | −0.128 | 0.518 | −0.676 |
| −0.420 | −0.287 | −0.050 | −0.481 | 1.521 | −1.367 | 0.609 | 0.292 | 0.048 | 0.592 |
| 1.048 | 0.220 | 1.121 | −1.789 | −1.211 | −0.871 | −0.740 | 0.513 | −0.558 | −0.395 |
| 1.000 | −0.638 | 1.261 | 0.510 | −0.150 | 0.034 | 0.054 | −0.055 | 0.639 | −0.825 |
| 0.170 | −1.131 | −0.985 | 0.102 | −0.939 | −1.457 | 1.766 | 1.087 | −1.275 | 2.362 |
| 0.389 | −0.435 | 0.171 | 0.891 | 1.158 | 1.041 | 1.048 | −0.324 | −0.404 | 1.060 |
| −0.305 | 0.838 | −2.019 | −0.540 | 0.905 | 1.195 | −1.190 | 0.106 | 0.571 | 0.298 |
| −0.321 | −0.039 | 1.799 | −1.032 | −2.225 | −0.148 | 0.758 | −0.862 | 0.158 | −0.726 |
| 1.900 | 1.572 | −0.244 | −1.721 | 1.130 | 0.495 | −0.484 | 0.014 | −0.778 | −1.483 |
| −0.778 | −0.288 | −0.224 | −1.324 | −0.072 | 0.890 | −0.410 | 0.752 | 0.376 | −0.224 |
| 0.617 | −1.718 | −0.183 | −0.100 | 1.719 | 0.696 | −1.339 | −0.614 | 1.071 | −0.386 |
| −1.430 | −0.953 | 0.770 | −0.007 | −1.872 | 1.075 | −0.913 | −1.168 | 1.775 | 0.238 |
| 0.267 | −0.048 | 0.972 | 0.734 | −1.408 | −1.955 | −0.848 | 2.002 | 0.232 | −1.273 |
| 0.978 | −0.520 | −0.368 | 1.690 | −1.479 | 0.985 | 1.475 | −0.098 | −1.633 | 2.399 |
| −1.235 | −1.168 | 0.325 | 1.421 | 2.652 | −0.486 | −1.253 | 0.270 | −1.103 | 0.118 |
| −0.258 | 0.638 | 2.309 | 0.741 | −0.161 | −0.679 | 0.336 | 1.973 | 0.370 | −2.277 |
| 0.243 | 0.629 | −1.516 | −0.157 | 0.693 | 1.710 | 0.800 | −0.265 | 1.218 | 0.655 |
| −0.292 | −1.455 | −1.451 | 1.492 | −0.713 | 0.821 | −0.031 | −0.780 | 1.330 | 0.977 |
| −0.505 | 0.389 | 0.544 | −0.042 | 1.615 | −1.440 | −0.989 | −0.580 | 0.156 | 0.052 |
| 0.397 | −0.287 | 1.712 | 0.289 | −0.904 | 0.259 | −0.600 | −1.635 | −0.009 | −0.799 |
| −0.605 | −0.470 | 0.007 | 0.721 | −1.117 | 0.635 | 0.592 | −1.362 | −1.441 | 0.672 |
| 1.360 | 0.182 | −1.476 | −0.599 | −0.875 | 0.292 | −0.700 | 0.058 | −0.340 | −0.639 |
| 0.480 | −0.699 | 1.615 | −0.225 | 1.014 | −1.370 | −1.097 | 0.294 | 0.309 | −1.389 |
| −0.027 | −0.487 | −1.000 | −0.015 | 0.119 | −1.990 | −0.687 | −1.964 | −0.366 | 1.759 |
| −1.482 | −0.815 | −0.121 | 1.884 | −0.185 | 0.601 | 0.793 | 0.430 | −1.181 | 0.426 |
| −1.256 | −0.567 | −0.994 | 1.011 | −1.071 | −0.623 | −0.420 | −0.309 | 1.362 | 0.863 |
| −1.132 | 2.039 | 1.934 | −0.222 | 0.386 | 1.100 | 0.284 | 1.597 | −1.718 | −0.560 |
| −0.780 | −0.239 | −0.497 | −0.434 | −0.284 | −0.241 | −0.333 | 1.348 | −0.478 | −0.169 |
| −0.859 | −0.215 | 0.241 | 1.471 | 0.389 | −0.952 | 0.245 | 0.781 | 1.093 | −0.240 |
| 0.447 | 1.479 | 0.067 | 0.426 | −0.370 | −0.675 | −0.972 | 0.225 | 0.815 | 0.389 |
| 0.269 | 0.735 | −0.066 | −0.271 | −1.439 | 1.036 | −0.306 | −1.439 | −0.122 | −0.336 |
| 0.097 | −1.883 | −0.218 | 0.202 | −0.357 | 0.019 | 1.631 | 1.400 | 0.223 | −0.793 |
| −0.686 | 1.596 | −0.286 | 0.722 | 0.655 | −0.275 | 1.245 | −1.504 | 0.066 | −1.280 |
| 0.957 | 0.057 | −1.153 | 0.701 | −0.280 | 1.747 | −0.745 | 1.338 | −1.421 | 0.386 |
| −0.976 | −1.789 | −0.696 | −1.799 | −0.354 | 0.071 | 2.355 | 0.135 | −0.598 | 1.883 |
| 0.274 | 0.226 | −0.909 | −0.572 | 0.181 | 1.115 | 0.496 | 0.453 | −1.218 | −0.115 |

| 21 | 22 | 23 | 24 | 25 | 26 | 27 | 28 | 29 | 30 |
|---|---|---|---|---|---|---|---|---|---|
| −1.752 | −0.329 | −1.256 | 0.318 | 1.531 | 0.349 | −0.958 | −0.059 | 0.415 | −1.084 |
| −0.291 | 0.085 | 1.701 | −1.087 | −0.443 | −0.292 | 0.248 | −0.539 | −1.382 | 0.318 |
| −0.933 | 0.130 | 0.634 | 0.899 | 1.409 | −0.883 | −0.095 | 0.229 | 0.129 | 0.367 |
| −0.450 | −0.244 | 0.072 | 1.028 | 1.730 | −0.056 | −1.488 | −0.078 | −2.361 | −0.992 |
| 0.512 | −0.882 | 0.490 | −1.304 | −0.266 | 0.757 | −0.361 | 0.194 | −1.078 | 0.529 |
| −0.702 | 0.472 | 0.429 | −0.664 | −0.592 | 1.443 | −1.515 | −1.209 | −1.043 | 0.278 |
| 0.284 | 0.039 | −0.518 | 1.351 | 1.473 | 0.889 | 0.300 | 0.339 | −0.206 | 1.392 |
| −0.509 | 1.420 | −0.782 | −0.429 | −1.266 | 0.627 | −1.165 | 0.819 | −0.261 | 0.409 |
| −1.776 | −1.033 | 1.977 | 0.014 | 0.702 | −0.435 | −0.816 | 1.131 | 0.656 | 0.061 |
| −0.044 | 1.807 | 0.342 | −2.510 | 1.071 | −1.220 | −0.060 | −0.764 | 0.079 | −0.964 |
| 0.263 | −0.578 | 1.612 | −0.148 | −0.383 | −1.007 | −0.414 | 0.638 | −0.186 | 0.507 |
| 0.986 | 0.439 | −0.192 | −0.132 | 0.167 | 0.883 | −0.400 | −1.440 | −0.385 | −1.414 |
| −0.441 | −0.852 | −1.446 | −0.605 | −0.348 | 1.018 | 0.963 | −0.004 | 2.504 | −0.847 |
| −0.866 | 0.489 | 0.097 | 0.379 | 0.192 | −0.842 | 0.065 | 1.420 | 0.426 | −1.191 |
| −1.215 | 0.675 | 1.621 | 0.394 | −1.447 | 2.199 | −0.321 | −0.540 | −0.037 | 0.185 |
| −0.475 | −1.210 | 0.183 | 0.526 | 0.495 | 1.297 | −1.613 | 1.241 | −1.016 | −0.090 |
| 1.200 | 0.131 | 2.502 | 0.344 | −1.060 | −0.909 | −1.695 | −0.666 | −0.838 | −0.866 |
| −0.498 | −1.202 | −0.057 | −1.354 | −1.441 | −1.590 | 0.987 | 0.441 | 0.637 | −1.116 |
| −0.743 | 0.894 | −0.028 | 1.119 | −0.598 | 0.279 | 2.241 | 0.830 | 0.267 | −0.156 |
| 0.779 | −0.780 | −0.954 | 0.705 | −0.361 | −0.734 | 1.365 | 1.297 | −0.142 | −1.387 |
| −0.206 | −0.195 | 1.017 | −1.167 | −0.079 | −0.452 | 0.058 | −1.068 | −0.394 | −0.406 |
| −0.092 | −0.927 | −0.439 | 0.256 | 0.503 | 0.338 | 1.511 | −0.465 | −0.118 | −0.454 |
| −1.222 | −1.582 | 1.786 | −0.517 | −1.080 | −0.409 | −0.474 | −1.890 | 0.247 | 0.575 |
| 0.068 | 0.075 | −1.383 | −0.084 | 0.159 | 1.276 | 1.141 | 0.186 | −0.973 | −0.266 |
| 0.183 | 1.600 | −0.335 | 1.553 | 0.889 | 0.896 | −0.035 | 0.461 | 0.486 | 1.246 |
| −0.811 | −2.904 | 0.618 | 0.588 | 0.533 | 0.803 | −0.696 | 0.690 | 0.820 | 0.557 |
| −1.010 | 1.149 | 1.033 | 0.336 | 1.306 | 0.835 | 1.523 | 0.296 | −0.426 | 0.004 |
| 1.453 | 1.210 | −0.043 | 0.220 | −0.256 | −1.161 | −2.030 | −0.046 | 0.243 | 1.082 |
| 0.759 | −0.838 | −0.877 | −0.177 | 1.183 | −0.218 | −3.154 | −0.963 | −0.822 | −1.114 |
| 0.287 | 0.278 | −0.454 | 0.897 | −0.122 | 0.013 | 0.346 | 0.921 | 0.238 | −0.586 |
| −0.669 | 0.035 | −2.077 | 1.077 | 0.525 | −0.154 | −1.036 | 0.015 | −0.220 | 0.882 |
| 0.392 | 0.106 | −1.430 | −0.204 | −0.326 | 0.825 | −0.432 | −0.094 | −1.566 | 0.679 |
| −0.337 | 0.199 | −0.160 | 0.625 | −0.891 | −1.464 | −0.318 | 1.297 | 0.932 | −0.032 |
| 0.369 | −1.990 | −1.190 | 0.666 | −1.614 | 0.082 | 0.922 | −0.139 | −0.833 | 0.091 |
| −1.694 | 0.710 | −0.655 | −0.546 | 1.654 | 0.134 | 0.466 | 0.033 | −0.039 | 0.838 |
| 0.985 | 0.340 | 0.276 | 0.911 | −0.170 | −0.551 | 1.000 | −0.838 | 0.275 | −0.304 |
| −1.063 | −0.594 | −1.526 | −0.787 | 0.873 | −0.405 | −1.324 | 0.162 | −0.163 | −2.716 |
| 0.033 | −1.527 | 1.422 | 0.308 | 0.845 | −0.151 | 0.741 | 0.064 | 1.212 | 0.823 |
| 0.597 | 0.362 | −3.760 | 1.159 | 0.874 | −0.794 | −0.915 | 1.215 | 1.627 | −1.248 |
| −1.601 | −0.570 | 0.133 | −0.660 | 1.485 | 0.682 | −0.898 | 0.686 | 0.658 | 0.346 |
| −0.266 | −1.309 | 0.597 | 0.989 | 0.934 | 1.079 | −0.656 | −0.999 | −0.036 | −0.537 |
| 0.901 | 1.531 | −0.889 | −1.019 | 0.084 | 1.531 | −0.144 | −1.920 | 0.678 | −0.402 |
| −1.433 | −1.008 | −0.990 | 0.090 | 0.940 | 0.207 | −0.745 | 0.638 | 1.469 | 1.214 |
| 1.327 | 0.763 | −1.724 | −0.709 | −1.100 | −1.346 | −0.946 | −0.157 | 0.522 | −1.264 |
| −0.248 | 0.788 | −0.577 | 0.122 | −0.536 | 0.293 | 1.207 | −2.243 | 1.642 | 1.353 |
| −0.401 | −0.679 | 0.921 | 0.476 | 1.121 | −0.864 | 0.128 | −0.551 | −0.872 | 1.511 |
| 0.344 | −0.324 | 0.686 | −1.487 | −0.126 | 0.803 | −0.961 | 0.183 | −0.358 | −0.184 |
| 0.441 | −0.372 | −1.336 | 0.062 | 1.506 | −0.315 | −0.112 | −0.452 | 1.594 | −0.264 |
| 0.824 | 0.040 | −1.734 | 0.251 | 0.054 | −0.379 | 1.298 | −0.126 | 0.104 | −0.529 |
| 1.385 | 1.320 | −0.509 | −0.381 | −1.671 | −0.524 | −0.805 | 1.348 | 0.676 | 0.799 |

## A-2 RANDOM NORMAL NUMBERS, $\mu = 0$ AND $\sigma = 1$ (CONTINUED)

| 31 | 32 | 33 | 34 | 35 | 36 | 37 | 38 | 39 | 40 |
|---|---|---|---|---|---|---|---|---|---|
| 1.556 | 0.119 | −0.078 | 0.164 | −0.455 | 0.077 | −0.043 | −0.299 | 0.249 | −0.182 |
| 0.647 | 1.029 | 1.186 | 0.887 | 1.204 | −0.657 | 0.644 | −0.410 | −0.652 | −0.165 |
| 0.329 | 0.407 | 1.169 | −2.072 | 1.661 | 0.891 | 0.233 | −1.628 | −0.762 | −0.717 |
| −1.188 | 1.171 | −1.170 | −0.291 | 0.863 | −0.045 | −0.205 | 0.574 | −0.926 | 1.407 |
| −0.917 | −0.616 | −1.589 | 1.184 | 0.266 | 0.559 | −1.833 | −0.572 | −0.648 | −1.090 |
| 0.414 | 0.469 | −0.182 | 0.397 | 1.649 | 1.198 | 0.067 | −1.526 | −0.081 | −0.192 |
| 0.107 | −0.187 | 1.343 | 0.472 | −0.112 | 1.182 | 0.548 | 2.748 | 0.249 | 0.154 |
| −0.497 | 1.907 | 0.191 | 0.136 | −0.475 | 0.458 | 0.183 | −1.640 | −0.058 | 1.278 |
| 0.501 | 0.083 | −0.321 | 1.133 | 1.126 | −0.299 | 1.299 | 1.617 | 1.581 | 2.455 |
| −1.382 | −0.738 | 1.225 | 1.564 | −0.363 | −0.548 | 1.070 | 0.390 | −1.398 | 0.524 |
| −0.590 | 0.699 | −0.162 | −0.011 | 1.049 | −0.689 | 1.225 | 0.339 | −0.539 | −0.445 |
| −1.125 | 1.111 | −1.065 | 0.534 | 0.102 | 0.425 | −1.026 | 0.695 | −0.057 | 0.795 |
| 0.849 | 0.169 | −0.351 | 0.584 | 2.177 | 0.009 | −0.696 | −0.426 | −0.692 | −1.638 |
| −1.233 | −0.585 | 0.306 | 0.773 | 1.304 | −1.304 | 0.282 | −1.705 | 0.187 | −0.880 |
| 0.104 | −0.468 | 0.185 | 0.498 | −0.624 | −0.322 | −0.875 | 1.478 | −0.691 | −0.281 |
| 0.261 | −1.883 | −0.181 | 1.675 | −0.324 | −1.029 | −0.185 | 0.004 | −0.101 | −1.187 |
| −0.007 | 1.280 | 0.568 | −1.270 | 1.405 | 1.731 | 2.072 | 1.686 | 0.728 | −0.417 |
| 0.794 | −0.111 | 0.040 | −0.536 | −0.976 | 2.192 | 1.609 | −0.190 | −0.279 | −1.611 |
| 0.431 | −2.300 | −1.081 | −1.370 | 2.943 | 0.653 | −2.523 | 0.756 | 0.886 | −0.983 |
| −0.149 | 1.294 | −0.580 | 0.482 | −1.449 | −1.067 | 1.996 | −0.274 | 0.721 | 0.490 |
| −0.216 | −1.647 | 1.043 | 0.481 | −0.011 | −0.587 | −0.916 | −1.016 | −1.040 | −1.117 |
| 1.604 | −0.851 | −0.317 | −0.686 | −0.008 | 1.939 | 0.078 | −0.465 | 0.533 | 0.652 |
| −0.212 | 0.005 | 0.535 | 0.837 | 0.362 | 1.103 | 0.219 | 0.488 | 1.332 | −0.200 |
| 0.007 | −0.076 | 1.484 | 0.455 | −0.207 | −0.554 | 1.120 | 0.913 | −0.681 | 1.751 |
| −0.217 | 0.937 | 0.860 | 0.323 | 1.321 | −0.492 | −1.386 | −0.003 | −0.230 | 0.539 |
| −0.649 | 0.300 | −0.698 | 0.900 | 0.569 | 0.842 | 0.804 | 1.025 | 0.603 | −1.546 |
| −1.541 | 0.193 | 2.047 | −0.552 | 1.190 | −0.087 | 2.062 | −2.173 | −0.791 | −0.520 |
| 0.274 | −0.530 | 0.112 | 0.385 | 0.656 | 0.436 | 0.882 | 0.312 | −2.265 | −0.218 |
| 0.876 | −1.498 | −0.128 | −0.387 | −1.259 | −0.856 | −0.353 | 0.714 | 0.863 | 1.169 |
| −0.859 | −1.083 | 1.288 | −0.078 | −0.081 | 0.210 | 0.572 | 1.194 | −1.118 | −1.543 |
| −0.015 | −0.567 | 0.113 | 2.127 | −0.719 | 3.256 | −0.721 | −0.663 | −0.779 | −0.930 |
| −1.529 | −0.231 | 1.223 | 0.300 | −0.995 | −0.651 | 0.505 | 0.138 | −0.064 | 1.341 |
| 0.278 | −0.058 | −2.740 | −0.296 | −1.180 | 0.574 | 1.452 | 0.846 | −0.243 | −1.208 |
| 1.428 | 0.322 | 2.302 | −0.852 | 0.782 | −1.322 | −0.092 | −0.546 | 0.560 | −1.430 |
| 0.770 | −1.874 | 0.347 | 0.994 | −0.485 | −1.179 | 0.048 | −1.324 | 1.061 | 0.449 |
| −0.303 | −0.629 | 0.764 | 0.013 | −1.192 | −0.475 | −1.085 | −0.880 | 1.738 | −1.225 |
| −0.263 | −2.105 | 0.509 | −0.645 | 1.362 | 0.504 | −0.755 | 1.274 | 1.448 | 0.604 |
| 0.997 | −1.187 | −0.242 | 0.121 | 2.510 | −1.935 | 0.350 | 0.073 | 0.458 | −0.446 |
| −0.063 | −0.475 | −1.802 | −0.476 | 0.193 | −1.199 | 0.339 | 0.364 | −0.684 | 1.353 |
| −0.168 | 1.904 | −0.485 | −0.032 | −0.554 | 0.056 | −0.710 | −0.778 | 0.722 | −0.024 |
| 0.366 | −0.491 | 0.301 | −0.008 | −0.894 | −0.945 | 0.384 | −1.748 | −1.118 | 0.394 |
| 0.436 | −0.464 | 0.539 | 0.942 | −0.458 | 0.445 | −1.883 | 1.228 | 1.113 | −0.218 |
| 0.597 | −1.471 | −0.434 | 0.705 | −0.788 | 0.575 | 0.086 | 0.504 | 1.445 | −0.513 |
| −0.805 | −0.624 | 1.344 | 0.649 | −1.124 | 0.680 | −0.986 | 1.845 | −1.152 | −0.393 |
| 1.681 | −1.910 | 0.440 | 0.067 | −1.502 | −0.755 | −0.989 | −0.054 | −2.320 | 0.474 |
| −0.007 | −0.459 | 1.940 | 0.220 | −1.259 | −1.729 | 0.137 | −0.520 | −0.412 | 2.847 |
| 0.209 | −0.633 | 0.299 | 0.174 | 1.975 | −0.271 | 0.119 | −0.199 | 0.007 | 2.315 |
| 1.254 | 1.672 | −1.186 | −1.310 | 0.474 | 0.878 | −0.725 | −0.191 | 0.642 | −1.212 |
| −1.016 | −0.697 | 0.017 | −0.263 | −0.047 | −1.294 | −0.339 | 2.257 | −0.078 | −0.049 |
| −1.169 | −0.355 | 1.086 | −0.199 | 0.031 | 0.396 | −0.143 | 1.572 | 0.276 | 0.027 |

| 41 | 42 | 43 | 44 | 45 | 46 | 47 | 48 | 49 | 50 |
|---|---|---|---|---|---|---|---|---|---|
| −0.856 | −0.063 | 0.787 | −2.052 | −1.192 | −0.831 | 1.623 | 1.135 | 0.759 | −0.189 |
| −0.276 | −1.110 | 0.752 | −1.378 | −0.583 | 0.360 | 0.365 | 1.587 | 0.621 | 1.344 |
| 0.379 | −0.440 | 0.858 | 1.453 | −1.356 | 0.503 | −1.134 | 1.950 | −1.816 | −0.283 |
| 1.468 | 0.131 | 0.047 | 0.355 | 0.162 | −1.491 | −0.739 | −1.182 | −0.533 | −0.497 |
| −1.805 | −0.772 | 1.286 | −0.636 | −1.312 | −1.045 | 1.559 | −0.871 | −0.102 | −0.123 |
| 2.285 | 0.554 | 0.418 | −0.577 | −1.489 | −1.255 | 0.092 | −0.597 | −1.051 | −0.980 |
| −0.602 | 0.399 | 1.121 | −1.026 | 0.087 | 1.018 | −1.437 | 0.661 | 0.091 | −0.637 |
| 0.229 | −0.584 | 0.705 | 0.124 | 0.341 | 1.320 | −0.824 | −1.541 | −0.163 | 2.329 |
| 1.382 | −1.454 | 1.537 | −1.299 | 0.363 | −0.356 | −0.025 | 0.294 | 2.194 | −0.395 |
| 0.978 | 0.109 | 1.434 | −1.094 | −0.265 | −0.857 | −1.421 | −1.773 | 0.570 | −0.053 |
| −0.678 | −2.335 | 1.202 | −1.697 | 0.547 | −0.201 | −0.373 | −1.363 | −0.081 | 0.958 |
| −0.366 | −1.084 | −0.626 | 0.798 | 1.706 | −1.160 | −0.838 | 1.462 | 0.636 | 0.570 |
| −1.074 | −1.379 | 0.086 | −0.331 | −0.288 | −0.309 | −1.527 | −0.408 | 0.183 | 0.856 |
| −0.600 | −0.096 | 0.696 | 0.446 | 1.417 | −2.140 | 0.599 | −0.157 | 1.485 | 1.387 |
| 0.918 | 1.163 | −1.445 | 0.759 | 0.878 | −1.781 | −0.056 | −2.141 | −0.234 | 0.975 |
| −0.791 | −0.528 | 0.946 | 1.673 | −0.680 | −0.784 | 1.494 | −0.086 | −1.071 | −1.196 |
| 0.598 | −0.352 | 0.719 | −0.341 | 0.056 | −1.041 | 1.429 | 0.235 | 0.314 | −1.693 |
| 0.567 | −1.156 | −0.125 | −0.534 | 0.711 | −0.511 | 0.187 | −0.644 | −1.090 | −1,281 |
| 0.963 | 0.052 | 0.037 | 0.637 | −1.335 | 0.055 | 0.010 | −0.860 | −0.621 | 0.713 |
| 0.489 | −0.209 | 1.659 | 0.054 | 1.635 | 0.169 | 0.794 | −1.550 | 1.845 | −0.388 |
| −1.627 | −0.017 | 0.699 | 0.661 | −0.073 | 0.188 | 1.183 | −1.054 | −1.615 | −0.765 |
| −1.096 | 1.215 | 0.320 | 0.738 | 1.865 | −1.169 | −0.667 | −0.674 | −0.062 | 1.378 |
| −2.532 | 1.031 | −0.799 | 1.665 | −2.756 | −0.151 | −0.704 | 0.602 | −0.672 | 1.264 |
| 0.024 | −1.183 | −0.927 | −0.629 | 0.204 | −0.825 | 0.496 | 2.543 | 0.262 | −0.785 |
| 0.192 | 0.125 | 0.373 | −0.931 | −0.079 | 0.186 | −0.306 | 0.621 | −0.292 | 1.131 |
| −1.324 | −1.229 | −0.648 | −0.430 | 0.811 | 0.868 | 0.787 | 1.845 | −0.374 | −0.651 |
| −0.726 | −0.746 | 1.572 | −1.420 | 1.509 | −0.361 | −0.310 | −3.117 | 1.637 | 0.642 |
| −1.618 | 1.082 | −0.319 | 0.300 | 1.524 | −0.418 | −1.712 | 0.358 | −1.032 | 0.537 |
| 1.695 | 0.843 | 2.049 | 0.388 | −0.297 | 1.077 | −0.462 | 0.655 | 0.940 | −0.354 |
| 0.790 | 0.605 | −3.077 | 1.009 | −0.906 | −1.004 | 0.693 | −1.098 | 1.300 | 0.549 |
| 1.792 | −0.895 | −0.136 | −1.765 | 1.077 | 0.418 | −0.150 | 0.808 | 0.697 | 0.435 |
| 0.771 | −0.741 | −0.492 | −0.770 | −0.458 | −0.021 | 1.385 | −1.225 | −0.066 | −1.471 |
| −1.438 | 0.423 | −1.211 | 0.723 | −0.731 | 0.883 | −2.109 | −2.455 | −0.210 | 1.644 |
| −0.294 | 1.266 | −1.994 | −0.730 | 0.545 | 0.397 | 1.069 | −0.383 | −0.097 | −0.985 |
| −1.966 | 0.909 | 0.400 | 0.685 | −0.800 | 1.759 | 0.268 | 1.387 | −0.414 | 1.615 |
| −0.999 | 1.587 | 1.423 | 0.937 | −0.943 | 0.090 | 1.185 | −1.204 | 0.300 | −1.354 |
| 0.581 | 0.481 | −2.400 | 0.000 | 0.231 | 0.079 | −2.842 | −0.846 | −0.508 | −0.516 |
| 0.370 | −1.452 | −0.580 | −1.462 | −0.972 | 1.116 | −0.994 | 0.374 | −3.336 | −0.058 |
| 0.834 | −1.227 | −0.709 | −1.039 | −0.014 | −0.383 | −0.512 | −0.347 | 0.881 | −0.638 |
| −0.376 | −0.813 | 0.660 | −1.029 | −0.137 | 0.371 | 0.376 | 0.968 | 1.338 | −0.786 |
| −1.621 | 0.815 | −0.544 | −0.376 | −0.852 | 0.436 | 1.562 | 0.815 | −1.048 | 0.188 |
| 0.163 | −0.161 | 2.501 | −0.265 | −0.285 | 1.934 | 1.070 | 0.215 | −0.876 | 0.073 |
| 1.786 | −0.538 | −0.437 | 0.324 | 0.105 | −0.421 | −0.410 | −0.947 | 0.700 | −1.006 |
| 2.140 | 1.218 | −0.351 | −0.068 | 0.254 | 0.448 | −1.461 | 0.784 | 0.317 | 1.013 |
| 0.064 | 0.410 | 0.368 | 0.419 | −0.982 | 1.371 | 0.100 | −0.505 | 0.856 | 0.890 |
| 0.789 | −0.131 | 1.330 | 0.506 | −0.645 | −1.414 | 2.426 | 1.389 | −0.169 | −0.194 |
| −0.011 | −0.372 | −0.699 | 2.382 | −1.395 | −0.467 | 1.256 | −0.585 | −1.359 | −1.804 |
| −0.463 | 0.003 | −1.470 | 1.493 | 0.960 | 0.364 | −1.267 | −0.007 | 0.616 | 0.624 |
| −1.210 | −0.669 | 0.009 | 1.284 | −0.617 | 0.355 | −0.589 | −0.243 | −0.015 | −0.712 |
| −1.157 | 0.481 | 0.560 | 1.287 | 1.129 | −0.126 | 0.006 | 1.532 | 1.328 | 0.980 |

## A-2 RANDOM NORMAL NUMBERS, $\mu = 0$ AND $\sigma = 1$ (CONTINUED)

| 51 | 52 | 53 | 54 | 55 | 56 | 57 | 58 | 59 | 60 |
|---|---|---|---|---|---|---|---|---|---|
| 0.240 | 1.774 | 0.210 | -1.471 | 1.167 | -1.114 | 0.182 | -0.485 | -0.318 | 1.156 |
| 0.627 | -0.758 | -0.930 | 1.641 | 0.162 | -0.874 | -0.235 | 0.203 | -0.724 | -0.155 |
| -0.594 | 0.098 | 0.158 | -0.722 | 1.385 | -0.985 | -1.707 | 0.175 | 0.449 | 0.654 |
| 1.082 | -0.753 | -1.944 | -1.964 | -2.131 | -2.796 | -1.286 | 0.807 | -0.122 | 0.527 |
| 0.060 | -0.014 | 1.577 | -0.814 | -0.633 | 0.275 | -0.087 | 0.517 | 0.474 | -1.432 |
| -0.013 | 0.402 | -0.086 | -0.394 | 0.292 | -2.862 | -1.660 | -1.658 | 1.610 | -2.205 |
| 1.586 | -0.833 | 1.444 | -0.615 | -1.157 | -0.220 | -0.517 | -1.668 | -2.036 | -0.850 |
| -0.405 | -1.315 | -1.355 | -1.331 | 1.394 | -0.381 | -0.729 | -0.447 | -0.906 | 0.622 |
| -0.329 | 1.701 | 0.427 | 0.627 | -0.271 | -0.971 | -1.010 | 1.182 | -0.143 | 0.844 |
| 0.992 | 0.708 | -0.115 | -1.630 | 0.596 | 0.499 | -0.862 | 0.508 | 0.474 | -0.974 |
| 0.296 | -0.390 | 2.047 | -0.363 | 0.724 | 0.788 | -0.089 | 0.930 | -0.497 | 0.058 |
| -2.069 | -1.422 | -0.948 | -1.742 | -1.173 | 0.215 | 0.661 | 0.842 | -0.984 | -0.577 |
| -0.211 | -1.727 | -0.277 | 1.592 | -0.707 | 0.327 | -0.527 | 0.912 | 0.571 | -0.525 |
| -0.467 | 1.848 | -0.263 | -0.862 | 0.706 | -0.533 | 0.626 | -0.200 | -2.221 | 0.368 |
| 1.284 | 0.412 | 1.512 | 0.328 | 0.203 | -1.231 | -1.480 | -0.400 | -0.491 | 0.913 |
| 0.821 | -1.503 | -1.066 | 1.624 | 1.345 | 0.440 | -1.416 | 0.301 | -0.355 | 0.106 |
| 1.056 | 1.224 | 0.281 | -0.098 | 1.868 | -0.395 | 0.610 | -1.173 | -1.449 | 1.171 |
| 1.090 | -0.790 | 0.882 | 1.687 | -0.009 | -2.053 | -0.030 | -0.421 | 1.253 | -0.081 |
| 0.574 | 0.129 | 1.203 | 0.280 | 1.438 | -2.052 | -0.443 | 0.522 | 0.468 | -1.211 |
| -0.531 | 2.155 | 0.334 | 0.898 | -1.114 | 0.243 | 1.026 | 0.391 | -0.011 | -0.024 |
| 0.896 | 0.181 | -0.941 | -0.511 | 0.648 | -0.710 | -0.181 | -1.417 | -0.585 | 0.087 |
| 0.042 | 0.579 | -0.316 | 0.394 | 1.133 | -0.305 | -0.683 | -1.318 | -0.050 | 0.993 |
| 2.328 | -0.243 | 0.534 | 0.241 | 0.275 | 0.060 | 0.727 | -1.459 | 0.174 | -1.072 |
| 0.486 | -0.558 | 0.426 | 0.728 | -0.360 | -0.068 | 0.058 | 1.471 | -0.051 | 0.337 |
| -0.304 | -0.309 | 0.646 | 0.309 | -1.320 | 0.311 | -1.407 | -0.011 | 0.387 | 0.128 |
| -2.319 | -0.129 | 0.866 | -0.424 | 0.236 | 0.419 | -1.359 | -1.088 | -0.045 | 1.096 |
| 1.098 | -0.875 | 0.659 | -1.086 | -0.424 | -1.462 | 0.743 | -0.787 | 1.472 | 1.677 |
| -0.038 | -0.118 | -1.285 | -0.545 | -0.140 | 1.244 | -1.104 | 0.146 | 0.058 | 1.245 |
| -0.207 | -0.746 | 1.681 | 0.137 | 0.104 | -0.491 | -0.935 | 0.671 | -0.448 | -0.129 |
| 0.333 | -1.386 | 1.840 | 1.089 | 0.837 | -1.642 | -0.273 | -0.798 | 0.067 | 0.334 |
| 1.190 | -0.547 | -1.016 | 0.540 | -0.993 | 0.443 | -0.190 | 1.019 | -1.021 | -1.276 |
| -1.416 | -0.749 | 0.325 | 0.846 | 2.417 | -0.479 | -0.655 | -1.326 | -1.952 | 1.234 |
| 0.622 | 0.661 | 0.028 | 1.302 | -0.032 | -0.157 | 1.470 | -0.766 | 0.697 | -0.303 |
| -1.134 | 0.499 | 0.538 | 0.564 | -2.392 | -1.398 | 0.010 | 1.874 | 1.386 | 0.000 |
| 0.725 | -0.242 | 0.281 | 1.355 | -0.036 | 0.204 | -0.345 | 0.395 | -0.753 | 1.645 |
| -0.210 | 0.611 | -0.219 | 0.450 | 0.308 | 0.993 | -0.146 | 0.225 | -1.496 | 0.246 |
| 0.219 | 0.302 | 0.000 | -0.437 | -2.127 | 0.883 | -0.599 | -1.516 | 0.826 | 1.242 |
| -1.098 | -0.252 | -2.480 | -0.973 | 0.712 | -1.430 | -0.167 | -1.237 | 0.750 | -0.763 |
| 0.144 | 0.489 | -0.637 | 1.990 | 0.411 | -0.563 | 0.027 | 1.278 | 2.105 | -1.130 |
| -1.738 | -1.295 | 0.431 | -0.503 | 2.327 | -0.007 | -1.293 | -1.206 | -0.066 | 1.370 |
| -0.487 | -0.097 | -1.361 | -0.340 | 0.204 | 0.938 | -0.148 | -1.099 | -0.252 | -0.384 |
| -0.636 | -0.626 | 1.967 | 1.677 | -0.331 | -0.440 | -1.440 | 1.281 | 1.070 | -1.167 |
| -1.464 | -1.493 | 0.945 | 0.180 | -0.672 | -0.035 | -0.293 | -0.905 | 0.196 | -1.122 |
| 0.561 | -0.375 | -0.657 | 1.304 | 0.833 | -1.159 | 1.501 | 1.265 | 0.438 | -0.437 |
| -0.525 | -0.017 | 1.815 | 0.789 | -1.908 | -0.353 | 1.383 | -1.208 | -1.135 | 1.082 |
| 0.980 | -0.111 | -0.804 | -1.078 | -1.930 | 0.171 | -1.318 | 2.377 | -0.303 | 1.062 |
| 0.501 | 0.835 | -0.518 | -1.034 | -1.493 | 0.712 | 0.421 | -1.165 | 0.782 | -1.484 |
| 1.081 | -1.176 | -0.542 | 0.321 | 0.688 | 0.670 | -0.771 | -0.090 | -0.611 | -0.813 |
| -0.148 | -1.203 | -1.553 | 1.244 | 0.826 | 0.077 | 0.128 | -0.772 | 1.683 | 0.318 |
| 0.096 | -0.286 | 0.362 | 0.888 | 0.551 | 1.782 | 0.335 | 2.083 | 0.350 | 0.260 |

| 61 | 62 | 63 | 64 | 65 | 66 | 67 | 68 | 69 | 70 |
|---|---|---|---|---|---|---|---|---|---|
| 0.052 | 1.504 | −1.350 | −1.124 | −0.521 | 0.515 | 0.839 | 0.778 | 0.438 | −0.550 |
| −0.315 | −0.865 | 0.851 | 0.127 | −0.379 | 1.640 | −0.441 | 0.717 | 0.670 | −0.301 |
| 0.938 | −0.055 | 0.947 | 1.275 | 1.557 | −1.484 | −1.137 | 0.398 | 1.333 | 1.988 |
| 0.497 | 0.502 | 0.385 | −0.467 | 2.468 | −1.810 | −1.438 | 0.283 | 1.740 | 0.420 |
| 2.308 | −0.399 | −1.798 | 0.018 | 0.780 | 1.030 | 0.806 | −0.408 | −0.547 | −0.280 |
| 1.815 | 0.101 | −0.561 | 0.236 | 0.166 | 0.227 | −0.309 | 0.056 | 0.610 | 0.732 |
| −0.421 | 0.432 | 0.586 | 1.059 | 0.278 | −1.672 | 1.859 | 1.433 | −0.919 | −1.770 |
| 0.008 | 0.555 | −1.310 | −1.440 | −0.142 | −0.295 | −0.630 | −0.911 | 0.133 | −0.308 |
| 1.191 | −0.114 | 1.039 | 1.083 | 0.185 | −0.492 | 0.419 | −0.433 | −1.019 | −2.260 |
| 1.299 | 1.918 | 0.318 | 1.348 | 0.935 | 1.250 | −0.175 | −0.828 | −0.336 | 0.726 |
| 0.012 | −0.739 | −1.181 | −0.645 | −0.736 | 1.801 | −0.209 | −0.389 | 0.867 | −0.555 |
| −0.586 | −0.044 | −0.983 | 0.332 | 0.371 | −0.072 | −1.212 | 1.047 | −1.930 | 0.812 |
| −0.122 | 1.515 | 0.338 | −1.040 | −0.008 | 0.467 | −0.600 | 0.923 | 1.126 | −0.752 |
| 0.879 | 0.516 | −0.920 | 2.121 | 0.674 | 1.481 | 0.660 | −0.986 | 1.644 | −2.159 |
| 0.435 | 1.149 | −0.065 | 1.391 | 0.707 | 0.548 | −0.490 | −1.139 | 0.249 | −0.933 |
| 0.645 | 0.878 | −0.904 | 0.896 | −1.284 | 0.237 | −0.378 | −0.510 | −1.123 | −0.129 |
| −0.514 | −1.017 | 0.529 | 0.973 | −1.202 | 0.005 | −0.644 | −0.167 | −0.664 | 0.167 |
| 0.242 | −0.427 | −0.727 | −1.150 | −1.092 | −0.736 | 0.925 | −0.050 | −0.200 | −0.770 |
| 0.443 | 0.445 | −1.287 | −1.463 | −0.650 | 0.412 | −2.714 | −0.903 | −0.341 | 0.957 |
| 0.273 | 0.203 | 0.423 | 1.423 | 0.508 | 1.058 | −0.828 | 0.143 | −1.059 | 0.345 |
| 0.255 | 1.036 | 1.471 | 0.476 | 0.592 | −0.658 | 0.677 | 0.155 | 1.068 | −0.759 |
| 0.858 | −0.370 | 0.522 | −1.890 | −0.389 | 0.609 | 1.210 | 0.489 | −0.006 | 0.834 |
| 0.097 | −1.709 | 1.790 | −0.929 | 0.405 | 0.024 | −0.036 | 0.580 | −0.642 | −1.121 |
| 0.520 | 0.889 | −0.540 | 0.266 | −0.354 | 0.524 | −0.788 | −0.497 | −0.973 | 1.481 |
| −0.311 | −1.772 | −0.496 | 1.275 | −0.904 | 0.147 | 1.497 | 0.657 | −0.469 | −0.783 |
| −0.604 | 0.857 | −0.695 | 0.397 | 0.296 | −0.285 | 0.191 | 0.158 | 1.672 | 1.190 |
| −0.001 | 0.287 | −0.868 | −0.013 | −1.576 | −0.168 | 0.047 | −0.159 | 0.086 | −1.077 |
| 1.160 | 0.989 | 0.205 | 0.937 | −0.099 | −1.281 | −0.276 | 0.845 | 0.752 | 0.663 |
| 1.579 | −0.303 | −1.174 | −0.960 | −0.470 | −0.556 | −0.689 | 1.535 | −0.711 | −0.743 |
| −0.615 | −0.154 | 0.008 | 1.353 | −0.381 | 1.137 | 0.022 | 0.175 | 0.586 | 2.941 |
| 1.578 | 1.529 | −0.294 | −1.301 | 0.614 | 0.099 | −0.700 | −0.003 | 1.052 | 1.643 |
| 0.626 | −0.447 | −1.261 | −2.029 | 0.182 | −1.176 | 0.083 | 1.868 | 0.872 | 0.965 |
| −0.493 | −0.020 | 0.920 | 1.473 | 1.873 | −0.289 | 0.410 | 0.394 | 0.881 | 0.054 |
| −0.217 | 0.342 | 1.423 | 0.364 | −0.119 | 0.509 | −2.266 | 0.189 | 0.149 | 1.041 |
| −0.792 | 0.347 | −1.367 | −0.632 | −1.238 | −0.136 | −0.352 | −0.157 | −1.163 | 1.305 |
| 0.568 | −0.226 | 0.391 | −0.074 | −0.312 | 0.400 | 1.583 | 0.481 | −1.048 | 0.759 |
| 0.051 | 0.549 | −2.192 | 1.257 | −1.460 | 0.363 | 0.127 | −1.020 | −1.192 | 0.449 |
| −0.891 | 0.490 | 0.279 | 0.372 | −0.578 | −0.836 | 2.285 | −0.448 | 0.720 | 0.510 |
| 0.622 | −0.126 | −0.637 | 1.255 | −0.354 | 0.032 | −1.076 | 0.352 | 0.103 | −0.496 |
| 0.623 | 0.819 | −0.489 | 0.354 | −0.943 | −0.694 | 0.248 | 0.092 | −0.673 | −1.428 |
| −1.208 | −1.038 | 0.140 | −0.762 | −0.854 | −0.249 | 2.431 | 0.067 | −0.317 | −0.874 |
| −0.487 | −2.117 | 0.195 | 2.154 | 1.041 | −1.314 | −0.785 | −0.414 | −0.695 | 2.310 |
| 0.522 | 0.314 | −1.003 | 0.134 | −1.748 | −0.107 | 0.459 | 1.550 | 1.118 | −1.004 |
| 0.838 | 0.613 | 0.227 | 0.308 | −0.757 | 0.912 | 2.272 | 0.556 | −0.041 | 0.008 |
| −1.534 | −0.407 | 1.202 | 1.251 | −0.891 | −1.588 | −2.380 | 0.059 | 0.682 | −0.878 |
| −0.099 | 2.391 | 1.067 | −2.060 | −0.464 | −0.103 | 3.486 | 1.121 | 0.632 | −1.626 |
| 0.070 | 1.465 | −0.080 | −0.526 | −1.090 | −1.002 | 0.132 | 1.504 | 0.050 | −0.393 |
| 0.115 | −0.601 | 1.751 | 1.956 | −0.196 | 0.400 | −0.522 | 0.571 | −0.101 | −2.160 |
| 0.252 | −0.329 | −0.586 | −0.118 | −0.242 | −0.521 | 0.818 | −0.167 | −0.469 | 0.430 |
| 0.017 | 0.185 | 0.377 | 1.883 | −0.443 | −0.039 | −1.244 | −0.820 | −1.171 | 0.104 |

## A-2 RANDOM NORMAL NUMBERS, $\mu = 0$ AND $\sigma = 1$ (CONTINUED)

| 71 | 72 | 73 | 74 | 75 | 76 | 77 | 78 | 79 | 80 |
|---|---|---|---|---|---|---|---|---|---|
| 2.988 | 0.423 | −1.261 | −1.893 | 0.187 | −0.412 | −0.228 | 0.002 | −0.384 | −1.032 |
| 0.760 | 0.995 | −0.256 | −0.505 | 0.750 | −0.654 | 0.647 | 0.613 | 0.086 | −0.118 |
| −0.650 | −0.927 | −1.071 | −0.796 | 1.130 | −1.042 | −0.181 | −1.020 | 1.648 | −1.327 |
| −0.394 | −0.452 | 0.893 | 1.410 | 1.133 | 0.319 | 0.537 | −0.789 | 0.078 | −0.062 |
| −1.168 | 1.902 | 0.206 | 0.303 | 1.413 | 2.012 | 0.278 | −0.566 | −0.900 | 0.200 |
| 1.343 | −0.377 | −0.131 | −0.585 | 0.053 | 0.137 | −1.371 | −0.175 | −0.878 | 0.118 |
| −0.733 | −1.921 | 0.471 | −1.394 | −0.885 | −0.523 | 0.553 | 0.344 | −0.775 | 1.545 |
| −0.172 | −0.575 | 0.066 | −0.310 | 1.795 | −1.148 | 0.772 | −1.063 | 0.818 | 0.302 |
| 1.457 | 0.862 | 1.677 | −0.507 | −1.691 | −0.034 | 0.270 | 0.075 | −0.554 | 1.420 |
| −0.087 | 0.744 | 1.829 | 1.203 | −0.436 | −0.618 | −0.200 | −1.134 | −1.352 | −0.098 |
| −0.092 | 1.043 | −0.255 | 0.189 | 0.270 | −1.034 | −0.571 | −0.336 | −0.742 | 2.141 |
| 0.441 | −0.379 | −1.757 | 0.608 | 0.527 | −0.338 | −1.995 | 0.573 | −0.034 | −0.056 |
| 0.073 | −0.250 | 0.531 | −0.695 | 1.402 | −0.462 | −0.938 | 1.130 | 1.453 | −0.106 |
| 0.637 | 0.276 | −0.013 | 1.968 | −0.205 | 0.486 | 0.727 | 1.416 | 0.963 | 1.349 |
| −0.792 | −1.778 | 1.284 | −0.452 | 0.602 | 0.668 | 0.516 | −0.210 | 0.040 | −0.103 |
| −1.223 | 1.561 | −2.099 | 1.419 | 0.223 | −0.482 | 1.098 | 0.513 | 0.418 | −1.686 |
| −0.407 | 1.587 | 0.335 | −2.475 | −0.284 | 1.567 | −0.248 | −0.759 | 1.792 | −2.319 |
| −0.462 | −0.193 | −0.012 | −1.208 | 2.151 | 1.336 | −1.968 | −1.767 | −0.374 | 0.783 |
| 1.457 | 0.883 | 1.001 | −0.169 | 0.836 | −1.236 | 1.632 | −0.142 | −0.222 | 0.340 |
| −1.918 | −1.246 | −0.209 | 0.780 | −0.330 | −2.953 | −0.447 | −0.094 | 1.344 | −0.196 |
| −0.126 | 1.094 | −1.206 | −1.426 | 1.474 | −1.080 | 0.000 | 0.764 | 1.476 | −0.016 |
| −0.306 | −0.847 | 0.639 | −0.262 | −0.427 | 0.391 | −1.298 | −1.013 | 2.024 | −0.539 |
| 0.477 | 1.595 | −0.762 | 0.424 | 0.799 | 0.312 | 1.151 | −1.095 | 1.199 | −0.765 |
| 0.369 | −0.709 | 1.283 | −0.007 | −1.440 | −0.782 | 0.061 | 1.427 | 1.656 | 0.974 |
| −0.579 | 0.606 | −0.866 | −0.715 | −0.301 | −0.180 | 0.188 | 0.668 | −1.091 | 1.476 |
| −0.418 | −0.588 | 0.919 | −0.083 | 1.084 | 0.944 | 0.253 | −1.833 | 1.305 | 0.171 |
| 0.128 | −0.834 | 0.009 | 0.742 | 0.539 | −0.948 | −1.055 | −0.689 | −0.338 | 1.091 |
| −0.291 | 0.235 | −0.971 | −1.696 | 1.119 | 0.272 | 0.635 | −0.792 | −1.355 | 1.291 |
| −1.024 | 1.212 | −1.100 | −0.348 | 1.741 | 0.035 | 1.268 | 0.192 | 0.729 | −0.467 |
| −0.378 | 1.026 | 0.093 | 0.468 | −0.967 | 0.675 | 0.807 | −2.109 | −1.214 | 0.559 |
| 1.232 | −0.815 | 0.608 | 1.429 | −0.748 | 0.201 | 0.400 | −1.230 | −0.398 | −0.674 |
| 1.793 | −0.581 | −1.076 | 0.512 | −0.442 | −1.488 | −0.580 | 0.172 | −0.891 | 0.311 |
| 0.766 | 0.310 | −0.070 | 0.624 | −0.389 | 1.035 | −0.101 | −0.926 | 0.816 | −1.048 |
| −0.606 | −1.224 | 1.465 | 0.012 | 1.061 | 0.491 | −1.023 | 1.948 | 0.866 | −0.737 |
| 0.106 | −2.715 | 0.363 | 0.343 | −0.159 | 2.672 | 1.119 | 0.731 | −1.012 | −0.889 |
| −0.060 | 0.444 | 1.596 | −0.630 | 0.362 | −0.306 | 1.163 | −0.974 | 0.486 | −0.373 |
| 2.081 | 1.161 | −1.167 | 0.021 | 0.053 | −0.094 | 0.381 | −0.628 | −2.581 | −1.243 |
| −1.727 | −1.266 | 0.088 | 0.936 | 0.368 | 0.648 | −0.799 | 1.115 | −0.968 | −2.588 |
| 0.091 | 1.364 | 1.677 | 0.644 | 1.505 | 0.440 | −0.329 | 0.498 | 0.869 | −0.965 |
| −1.114 | −0.239 | −0.409 | −0.334 | −0.605 | 0.501 | −1.921 | −0.470 | 2.354 | −0.660 |
| 0.189 | −0.547 | −1.758 | −0.295 | −0.279 | −0.515 | −1.053 | 0.553 | −0.297 | 0.496 |
| −0.065 | −0.023 | −0.267 | −0.247 | 1.318 | 0.904 | −0.712 | −1.152 | −0.543 | 0.176 |
| −1.742 | −0.599 | 0.430 | −0.615 | 1.165 | 0.084 | 2.017 | −1.207 | 2.614 | 1.490 |
| 0.732 | 0.188 | 2.343 | 0.526 | −0.812 | 0.389 | 1.036 | −0.023 | 0.229 | −2.262 |
| −1.490 | 0.014 | 0.167 | 1.422 | 0.015 | 0.069 | 0.133 | 0.897 | −1.678 | 0.323 |
| 1.507 | −0.571 | −0.724 | 1.741 | −0.152 | −0.147 | −0.158 | −0.076 | 0.652 | 0.447 |
| 0.513 | 0.168 | −0.076 | −0.171 | 0.428 | 0.205 | −0.865 | 0.107 | 1.023 | 0.077 |
| −0.834 | −1.121 | 1.441 | 0.492 | 0.559 | 1.724 | −1.659 | 0.245 | 1.354 | −0.041 |
| 0.258 | 1.880 | −0.536 | 1.246 | −0.188 | −0.746 | 1.097 | 0.258 | 1.547 | 1.238 |
| −0.818 | 0.273 | 0.159 | −0.765 | 0.526 | 1.281 | 1.154 | −0.687 | −0.793 | 0.795 |

| 81 | 82 | 83 | 84 | 85 | 86 | 87 | 88 | 89 | 90 |
|---|---|---|---|---|---|---|---|---|---|
| −0.713 | −0.541 | −0.571 | −0.807 | −1.560 | 1.000 | 0.140 | −0.549 | 0.887 | 2.237 |
| −0.117 | 0.530 | −1.599 | −1.602 | 0.412 | −1.450 | −1.217 | 1.074 | −1.021 | −0.424 |
| 1.187 | −1.523 | 1.437 | 0.051 | 1.237 | −0.798 | 1.616 | −0.823 | −1.207 | 1.258 |
| −0.182 | −0.186 | 0.517 | 1.438 | 0.831 | −1.319 | −0.539 | −0.192 | 0.150 | 2.127 |
| 1.964 | −0.629 | −0.944 | −0.028 | 0.948 | 1.005 | 0.242 | −0.432 | −0.329 | 0.113 |
| 0.230 | 1.523 | 1.658 | 0.753 | 0.724 | 0.183 | −0.147 | 0.505 | 0.448 | −0.053 |
| 0.839 | −0.849 | −0.145 | −1.843 | −1.276 | 0.481 | −0.142 | −0.534 | 0.403 | 0.370 |
| −0.801 | 0.343 | −1.822 | 0.447 | −0.931 | −0.824 | −0.484 | 0.864 | −1.069 | 0.860 |
| −0.124 | 0.727 | 1.654 | −0.182 | −1.381 | −1.146 | −0.572 | 0.159 | 0.186 | 1.221 |
| −0.088 | 0.032 | −0.564 | 0.654 | 1.141 | −0.056 | −0.343 | 0.067 | −0.267 | −0.219 |
| 0.912 | −1.114 | −1.035 | −1.070 | −0.297 | 1.195 | 0.030 | 0.022 | 0.406 | −0.414 |
| 1.397 | −0.473 | 0.433 | 0.023 | −1.204 | 1.254 | 0.551 | −1.012 | −0.789 | 0.906 |
| −0.652 | −0.029 | 0.064 | 0.511 | 1.117 | −0.465 | 0.523 | −0.083 | 0.386 | 0.259 |
| 1.236 | −0.457 | −1.354 | −0.898 | −0.270 | −1.837 | 1.641 | −0.657 | −0.753 | −1.686 |
| −0.498 | 1.302 | 0.816 | −0.936 | 1.404 | 0.555 | 2.450 | −0.789 | −0.120 | 0.505 |
| −0.005 | 2.174 | 1.893 | −1.361 | −0.991 | 0.508 | −0.823 | 0.918 | 0.524 | 0.488 |
| 0.115 | −1.373 | −0.900 | −1.010 | 0.624 | 0.946 | 0.312 | −1.384 | 0.224 | 2.343 |
| 0.167 | 0.254 | 1.219 | 1.153 | −0.510 | −0.007 | −0.285 | −0.631 | −0.356 | 0.254 |
| 0.976 | 1.158 | −0.469 | 1.099 | 0.509 | −1.324 | −0.102 | −0.296 | −0.907 | 0.449 |
| 0.653 | −0.366 | 0.450 | −2.653 | −0.592 | −0.510 | 0.983 | 0.023 | −0.881 | 0.876 |
| −0.150 | −0.088 | 0.457 | −0.448 | 0.605 | 0.668 | −0.613 | 0.261 | 0.023 | −0.050 |
| 0.060 | 0.276 | 0.229 | −1.527 | −0.316 | −0.834 | −1.652 | −0.387 | 0.632 | 0.895 |
| −0.678 | 0.547 | 0.243 | −2.183 | −0.368 | 1.158 | −0.996 | −0.705 | −0.314 | 1.464 |
| 2.139 | 0.395 | −0.376 | −0.175 | 0.406 | 0.309 | −1.021 | −0.460 | −0.217 | 0.307 |
| 0.091 | 1.793 | 0.822 | 0.054 | 0.573 | −0.729 | −0.517 | 0.589 | 1.927 | 0.940 |
| −0.003 | 0.344 | 1.242 | −1.105 | 0.234 | −1.222 | −0.474 | 1.831 | 0.124 | −0.840 |
| −0.965 | 0.268 | −1.543 | 0.690 | 0.917 | 2.017 | −0.297 | 1.087 | 0.371 | 1.495 |
| −0.076 | −0.495 | −0.103 | 0.646 | 2.427 | −2.172 | 0.660 | −1.541 | −0.852 | 0.583 |
| −0.365 | −3.305 | 0.805 | −0.418 | −1.201 | 0.623 | −0.223 | 0.109 | 0.205 | −0.663 |
| 0.578 | 0.145 | −1.438 | 1.122 | −1.406 | 1.172 | 0.272 | −2.245 | 1.207 | 1.227 |
| −0.398 | −0.304 | 0.529 | −0.514 | −0.681 | −0.366 | 0.338 | 0.801 | −0.301 | −0.790 |
| −0.951 | −1.483 | −0.613 | −0.171 | −0.459 | 1.231 | −1.232 | −0.497 | −0.779 | 0.247 |
| 1.025 | −0.039 | −0.721 | 0.813 | 1.203 | 0.245 | 0.402 | 1.541 | 0.691 | −1.420 |
| −0.958 | 0.791 | 0.948 | 0.222 | −0.704 | −0.375 | −0.246 | −0.682 | −0.871 | 0.056 |
| 1.097 | −1.428 | 1.402 | −1.425 | −0.877 | 0.536 | 0.988 | 2.529 | 0.768 | −1.321 |
| 0.377 | 2.240 | 0.854 | −1.158 | 0.066 | −1.222 | 0.821 | −1.602 | −0.760 | −0.871 |
| 1.729 | 0.073 | 1.022 | 0.891 | 0.659 | −1.040 | 0.251 | −0.710 | −1.734 | −0.038 |
| −1.329 | −0.381 | −0.515 | 1.484 | −0.430 | −0.466 | −0.167 | −0.788 | −0.660 | 0.003 |
| −0.132 | 0.391 | 2.205 | −1.165 | 0.200 | 0.415 | −0.765 | 0.239 | −1.182 | 1.135 |
| 0.336 | 0.657 | −0.805 | 0.150 | −0.938 | 1.057 | −1.090 | 1.604 | −0.598 | −0.760 |
| 0.124 | −1.812 | 1.750 | 0.270 | −0.114 | 0.517 | −0.226 | 0.127 | 0.129 | −0.751 |
| −0.036 | 0.365 | 0.766 | 0.877 | −0.804 | −0.140 | 0.182 | −0.483 | −0.376 | −0.564 |
| −0.609 | −0.019 | −0.992 | −1.193 | −0.516 | 0.517 | 1.677 | 0.839 | −1.134 | 0.675 |
| −0.894 | 0.318 | 0.607 | −0.865 | 0.526 | −0.971 | 1.365 | 0.319 | 1.804 | 1.740 |
| −0.357 | −0.802 | 0.635 | −0.491 | −1.110 | 0.785 | −0.042 | −1.042 | −0.572 | 0.243 |
| −0.258 | −0.383 | −1.013 | 0.001 | −1.673 | 0.561 | −1.054 | −0.106 | −0.760 | −1.009 |
| 2.245 | −0.431 | −0.496 | 0.796 | 0.193 | 1.202 | −0.429 | −0.217 | 0.333 | −0.643 |
| 1.956 | 0.477 | 0.812 | −0.117 | 0.606 | −0.330 | 0.425 | −0.232 | 0.802 | 0.656 |
| 1.358 | 0.139 | 0.199 | −0.475 | −0.120 | 0.184 | −0.020 | −1.326 | 0.517 | −1.708 |
| 0.656 | 1.081 | 0.180 | 0.145 | 0.376 | −1.363 | −0.491 | 0.352 | −1.477 | 1.280 |

# A-2 RANDOM NORMAL NUMBERS, $\mu = 0$ AND $\sigma = 1$ (CONTINUED)

| 91 | 92 | 93 | 94 | 95 | 96 | 97 | 98 | 99 | 100 |
|---|---|---|---|---|---|---|---|---|---|
| −0.181 | 0.583 | −1.478 | −0.181 | 0.281 | −0.559 | 1.985 | −1.122 | −1.106 | 1.441 |
| 1.549 | −1.183 | −2.089 | −1.997 | −0.343 | 1.275 | 0.676 | −0.212 | 1.252 | 0.163 |
| 0.978 | −1.067 | −2.640 | 0.134 | 0.328 | −0.052 | −0.030 | −0.273 | −0.570 | 1.026 |
| −0.596 | −0.420 | −0.318 | −0.057 | −0.695 | −1.148 | 0.333 | −0.531 | −2.037 | −1.587 |
| −0.440 | 0.032 | 0.163 | 1.029 | 0.079 | 1.148 | 0.762 | −1.961 | −0.674 | −0.486 |
| 0.443 | −1.100 | 0.728 | −2.397 | −0.543 | 0.872 | −0.568 | 0.980 | −0.174 | 0.728 |
| −2.401 | −1.375 | −1.332 | −2.177 | −2.064 | −0.245 | −0.039 | 0.585 | 1.344 | 1.386 |
| 0.311 | 0.322 | −0.158 | 0.359 | 0.103 | 0.371 | 0.735 | 0.011 | 2.091 | 0.490 |
| −1.209 | 0.241 | −1.488 | −0.667 | −1.772 | −0.197 | 0.741 | −1.303 | −1.149 | 2.251 |
| 0.575 | −1.227 | −1.674 | 1.400 | 0.289 | 0.005 | 0.185 | −1.072 | 0.431 | −1.096 |
| −0.190 | 0.272 | 1.216 | 0.227 | 1.358 | 0.215 | −2.306 | −1.301 | −0.597 | −1.401 |
| −0.817 | −0.769 | −0.470 | −0.633 | 0.187 | −0.517 | −0.888 | −1.712 | 1.774 | −0.162 |
| 0.265 | −0.676 | 0.244 | 1.897 | −0.629 | −0.206 | −1.419 | 1.049 | 0.266 | −0.438 |
| −0.221 | 0.678 | 2.149 | 1.486 | −1.361 | 1.402 | −0.028 | 0.493 | 0.744 | 0.195 |
| −0.436 | 0.358 | −0.602 | 0.107 | 0.085 | 0.573 | 0.529 | 1.577 | 0.239 | 1.898 |
| −0.010 | 0.475 | 0.655 | 0.659 | −0.029 | −0.029 | 0.126 | −1.335 | −1.261 | 2.036 |
| −0.244 | 1.654 | 1.335 | −0.610 | 0.617 | 0.642 | 0.371 | 0.241 | 0.001 | −1.799 |
| −0.932 | −1.275 | −1.134 | −1.246 | −1.508 | 0.949 | 1.743 | −0.271 | −1.333 | −1.875 |
| −0.199 | −1.285 | −0.387 | 0.191 | 0.726 | −0.151 | 0.064 | −0.803 | −0.062 | 0.780 |
| −0.251 | −0.431 | −0.831 | 0.036 | −0.464 | −1.089 | 0.284 | −0.451 | 1.693 | 1.004 |
| 1.074 | −1.323 | −1.659 | −0.186 | −0.612 | 1.612 | −2.159 | −1.210 | 0.596 | −1.421 |
| 1.518 | 2.101 | 0.397 | 0.516 | −1.169 | −1.821 | 1.346 | 2.435 | 1.165 | −0.428 |
| 0.935 | −0.206 | 1.117 | −0.241 | −0.963 | −0.099 | 0.412 | −1.344 | 0.411 | 0.583 |
| 1.360 | −0.380 | 0.031 | 1.066 | 0.893 | 0.431 | −0.081 | 0.099 | 0.500 | −2.441 |
| 0.115 | −0.211 | 1.471 | 0.332 | 0.750 | 0.652 | −0.812 | 1.383 | −0.355 | −0.638 |
| 0.082 | −0.309 | −0.355 | −0.402 | 0.774 | 0.150 | 0.015 | 2.539 | −0.756 | −1.049 |
| −1.492 | 0.259 | 0.323 | 0.697 | −0.509 | 0.968 | −0.053 | 1.033 | −0.220 | −2.322 |
| −0.203 | 0.548 | 1.494 | 1.185 | 0.083 | −1.196 | −0.749 | −1.105 | 1.324 | 0.689 |
| 1.857 | −0.167 | −1.531 | 1.551 | 0.848 | 0.120 | 0.415 | −0.317 | 1.446 | 1.002 |
| 0.669 | −1.017 | −2.437 | −0.558 | −0.657 | 0.940 | 0.985 | 0.483 | −0.361 | 0.095 |
| 0.128 | 1.463 | −0.436 | −0.239 | −1.443 | 0.732 | 0.168 | −0.144 | −0.392 | 0.989 |
| 1.879 | −2.456 | 0.029 | 0.429 | 0.618 | −1.683 | −2.262 | 0.034 | −0.002 | 1.914 |
| 0.680 | 0.252 | 0.130 | 1.658 | −1.023 | 0.407 | −0.235 | −0.224 | −0.434 | 0.253 |
| −0.631 | 0.225 | −0.951 | 1.072 | −0.285 | −1.731 | −0.427 | −1.446 | −0.873 | 0.619 |
| −1.273 | 0.723 | 0.201 | 0.505 | −0.370 | −0.421 | −0.015 | −0.463 | 0.288 | 1.734 |
| −0.643 | −1.485 | 0.403 | 0.003 | −0.243 | 0.000 | 0.964 | −0.703 | 0.844 | −0.686 |
| −0.435 | −2.162 | −0.169 | −1.311 | −1.639 | 0.193 | 2.692 | −1.994 | 0.326 | 0.562 |
| −1.706 | 0.119 | −1.566 | 0.637 | −1.948 | −1.068 | 0.935 | 0.738 | 0.650 | 0.491 |
| −0.498 | 1.640 | 0.384 | −0.945 | −1.272 | 0.945 | −1.013 | −0.913 | −0.469 | 2.250 |
| −0.065 | −0.005 | 0.618 | −0.523 | −0.055 | 1.071 | 0.758 | −0.736 | −0.959 | 0.598 |
| 0.190 | −1.020 | −1.104 | 0.936 | −0.029 | −1.004 | −0.657 | 1.270 | −0.060 | −0.809 |
| 0.879 | −0.642 | 1.155 | −0.523 | −0.757 | −1.027 | 0.985 | −1.222 | 1.078 | 0.163 |
| 0.559 | 1.094 | 1.587 | −0.384 | −1.701 | 0.418 | 0.327 | 0.669 | 0.019 | 0.782 |
| −0.261 | 1.234 | −0.505 | −0.664 | −0.446 | −0.747 | 0.427 | −0.369 | 0.089 | −1.302 |
| 3.136 | 1.120 | −0.591 | 2.515 | −2.853 | 1.375 | 2.421 | 0.672 | 1.817 | −0.067 |
| −1.307 | −0.586 | −0.311 | −0.026 | 1.633 | −1.340 | −1.209 | 0.110 | −0.126 | −0.288 |
| 1.455 | 1.099 | −1.225 | −0.817 | 0.667 | −0.212 | 0.684 | 0.349 | −1.161 | −2.432 |
| −0.443 | −0.415 | −0.660 | 0.098 | 0.435 | −0.846 | −0.375 | −0.410 | −1.747 | −0.790 |
| −0.326 | 0.798 | 0.349 | 0.524 | 0.690 | −0.520 | −0.522 | 0.602 | −0.193 | −0.535 |
| −1.027 | −1.459 | −0.840 | −1.637 | −0.462 | 0.607 | −0.760 | 1.342 | −1.916 | 0.424 |

Ordinates $Y$ at $\pm z$, and areas $A$ between $-z$ and $+z$, of the normal distribution

| $z$ | $X$ | $Y$ | $A$ | $1 - A$ | $z$ | $X$ | $Y$ | $A$ | $1 - A$ |
|---|---|---|---|---|---|---|---|---|---|
| 0 | $\mu$ | .399 | .0000 | 1.0000 | ±1.50 | $\mu \pm 1.50\sigma$ | .1295 | .8664 | .1336 |
| ± .05 | $\mu \pm .05\sigma$ | .398 | .0399 | .9601 | ±1.55 | $\mu \pm 1.55\sigma$ | .1200 | .8789 | .1211 |
| ± .10 | $\mu \pm .10\sigma$ | .397 | .0797 | .9203 | ±1.60 | $\mu \pm 1.60\sigma$ | .1109 | .8904 | .1096 |
| ± .15 | $\mu \pm .15\sigma$ | .394 | .1192 | .8808 | ±1.65 | $\mu \pm 1.65\sigma$ | .1023 | .9011 | .0989 |
| ± .20 | $\mu \pm .20\sigma$ | .391 | .1585 | .8415 | ±1.70 | $\mu \pm 1.70\sigma$ | .0940 | .9109 | .0891 |
| ± .25 | $\mu \pm .25\sigma$ | .387 | .1974 | .8026 | ±1.75 | $\mu \pm 1.75\sigma$ | .0863 | .9199 | .0801 |
| ± .30 | $\mu \pm .30\sigma$ | .381 | .2358 | .7642 | ±1.80 | $\mu \pm 1.80\sigma$ | .0790 | .9281 | .0719 |
| ± .35 | $\mu \pm .35\sigma$ | .375 | .2737 | .7263 | ±1.85 | $\mu \pm 1.85\sigma$ | .0721 | .9357 | .0643 |
| ± .40 | $\mu \pm .40\sigma$ | .368 | .3108 | .6892 | ±1.90 | $\mu \pm 1.90\sigma$ | .0656 | .9426 | .0574 |
| ± .45 | $\mu \pm .45\sigma$ | .361 | .3473 | .6527 | ±1.95 | $\mu \pm 1.95\sigma$ | .0596 | .9488 | .0512 |
| ± .50 | $\mu \pm .50\sigma$ | .352 | .3829 | .6171 | ±2.00 | $\mu \pm 2.00\sigma$ | .0540 | .9545 | .0455 |
| ± .55 | $\mu \pm .55\sigma$ | .343 | .4177 | .5823 | ±2.05 | $\mu \pm 2.05\sigma$ | .0488 | .9596 | .0404 |
| ± .60 | $\mu \pm .60\sigma$ | .333 | .4515 | .5485 | ±2.10 | $\mu \pm 2.10\sigma$ | .0440 | .9643 | .0357 |
| ± .65 | $\mu \pm .65\sigma$ | .323 | .4843 | .5157 | ±2.15 | $\mu \pm 2.15\sigma$ | .0396 | .9684 | .0316 |
| ± .70 | $\mu \pm .70\sigma$ | .312 | .5161 | .4839 | ±2.20 | $\mu \pm 2.20\sigma$ | .0355 | .9722 | .0278 |
| ± .75 | $\mu \pm .75\sigma$ | .301 | .5467 | .4533 | ±2.25 | $\mu \pm 2.25\sigma$ | .0317 | .9756 | .0244 |
| ± .80 | $\mu \pm .80\sigma$ | .290 | .5763 | .4237 | ±2.30 | $\mu \pm 2.30\sigma$ | .0283 | .9786 | .0214 |
| ± .85 | $\mu \pm .85\sigma$ | .278 | .6047 | .3953 | ±2.35 | $\mu \pm 2.35\sigma$ | .0252 | .9812 | .0188 |
| ± .90 | $\mu \pm .90\sigma$ | .266 | .6319 | .3681 | ±2.40 | $\mu \pm 2.40\sigma$ | .0224 | .9836 | .0164 |
| ± .95 | $\mu \pm .95\sigma$ | .254 | .6579 | .3421 | ±2.45 | $\mu \pm 2.45\sigma$ | .0198 | .9857 | .0143 |
| ±1.00 | $\mu \pm 1.00\sigma$ | .242 | .6827 | .3173 | ±2.50 | $\mu \pm 2.50\sigma$ | .0175 | .9876 | .0124 |
| ±1.05 | $\mu \pm 1.05\sigma$ | .230 | .7063 | .2937 | ±2.55 | $\mu \pm 2.55\sigma$ | .0154 | .9892 | .0108 |
| ±1.10 | $\mu \pm 1.10\sigma$ | .218 | .7287 | .2713 | ±2.60 | $\mu \pm 2.60\sigma$ | .0136 | .9907 | .0093 |
| ±1.15 | $\mu \pm 1.15\sigma$ | .206 | .7499 | .2501 | ±2.65 | $\mu \pm 2.65\sigma$ | .0119 | .9920 | .0080 |
| ±1.20 | $\mu \pm 1.20\sigma$ | .194 | .7699 | .2301 | ±2.70 | $\mu \pm 2.70\sigma$ | .0104 | .9931 | .0069 |
| ±1.25 | $\mu \pm 1.25\sigma$ | .183 | .7887 | .2113 | ±2.75 | $\mu \pm 2.75\sigma$ | .0091 | .9940 | .0060 |
| ±1.30 | $\mu \pm 1.30\sigma$ | .171 | .8064 | .1936 | ±2.80 | $\mu \pm 2.80\sigma$ | .0079 | .9949 | .0051 |
| ±1.35 | $\mu \pm 1.35\sigma$ | .160 | .8230 | .1770 | ±2.85 | $\mu \pm 2.85\sigma$ | .0069 | .9956 | .0044 |
| ±1.40 | $\mu \pm 1.40\sigma$ | .150 | .8385 | .1615 | ±2.90 | $\mu \pm 2.90\sigma$ | .0060 | .9963 | .0037 |
| ±1.45 | $\mu \pm 1.45\sigma$ | .139 | .8529 | .1471 | ±2.95 | $\mu \pm 2.95\sigma$ | .0051 | .9968 | .0032 |
| ±1.50 | $\mu \pm 1.50\sigma$ | .130 | .8664 | .1336 | ±3.00 | $\mu \pm 3.00\sigma$ | .0044 | .9973 | .0027 |
| | | | | | ±4.00 | $\mu \pm 4.00\sigma$ | .0001 | .99994 | .00006 |
| | | | | | ±5.00 | $\mu \pm 5.00\sigma$ | .000001 | .9999994 | .0000006 |

| $z$ | $X$ | $Y$ | $A$ | $1 - A$ | $z$ | $X$ | $Y$ | $A$ | $1 - A$ |
|---|---|---|---|---|---|---|---|---|---|
| ± .000 | $\mu$ | .3989 | .0000 | 1.0000 | ±1.036 | $\mu \pm 1.036\sigma$ | .2331 | .7000 | .3000 |
| ± .126 | $\mu \pm .126\sigma$ | .3958 | .1000 | .9000 | ±1.282 | $\mu \pm 1.282\sigma$ | .1755 | .8000 | .2000 |
| ± .253 | $\mu \pm .253\sigma$ | .3863 | .2000 | .8000 | ±1.645 | $\mu \pm 1.645\sigma$ | .1031 | .9000 | .1000 |
| ± .385 | $\mu \pm .385\sigma$ | .3704 | .3000 | .7000 | ±1.960 | $\mu \pm 1.960\sigma$ | .0584 | .9500 | .0500 |
| ± .524 | $\mu \pm .524\sigma$ | .3477 | .4000 | .6000 | ±2.576 | $\mu \pm 2.576\sigma$ | .0145 | .9900 | .0100 |
| ± .674 | $\mu \pm .674\sigma$ | .3178 | .5000 | .5000 | ±3.291 | $\mu \pm 3.291\sigma$ | .0018 | .9990 | .0010 |
| ± .842 | $\mu \pm .842\sigma$ | .2800 | .6000 | .4000 | ±3.891 | $\mu \pm 3.891\sigma$ | .0002 | .9999 | .0001 |

| $z$ | $X$ | Area | $z$ | $X$ | Area |
|---|---|---|---|---|---|
| −3.25 | $\mu - 3.25\sigma$ | .0006 | −1.00 | $\mu - 1.00\sigma$ | .1587 |
| −3.20 | $\mu - 3.20\sigma$ | .0007 | − .95 | $\mu - .95\sigma$ | .1711 |
| −3.15 | $\mu - 3.15\sigma$ | .0008 | − .90 | $\mu - .90\sigma$ | .1841 |
| −3.10 | $\mu - 3.10\sigma$ | .0010 | − .85 | $\mu - .85\sigma$ | .1977 |
| −3.05 | $\mu - 3.05\sigma$ | .0011 | − .80 | $\mu - .80\sigma$ | .2119 |
| −3.00 | $\mu - 3.00\sigma$ | .0013 | − .75 | $\mu - .75\sigma$ | .2266 |
| −2.95 | $\mu - 2.95\sigma$ | .0016 | − .70 | $\mu - .70\sigma$ | .2420 |
| −2.90 | $\mu - 2.90\sigma$ | .0019 | − .65 | $\mu - .65\sigma$ | .2578 |
| −2.85 | $\mu - 2.85\sigma$ | .0022 | − .60 | $\mu - .60\sigma$ | .2743 |
| −2.80 | $\mu - 2.80\sigma$ | .0026 | − .55 | $\mu - .55\sigma$ | .2912 |
| −2.75 | $\mu - 2.75\sigma$ | .0030 | − .50 | $\mu - .50\sigma$ | .3085 |
| −2.70 | $\mu - 2.70\sigma$ | .0035 | − .45 | $\mu - .45\sigma$ | .3264 |
| −2.65 | $\mu - 2.65\sigma$ | .0040 | − .40 | $\mu - .40\sigma$ | .3446 |
| −2.60 | $\mu - 2.60\sigma$ | .0047 | − .35 | $\mu - .35\sigma$ | .3632 |
| −2.55 | $\mu - 2.55\sigma$ | .0054 | − .30 | $\mu - .30\sigma$ | .3821 |
| −2.50 | $\mu - 2.50\sigma$ | .0062 | − .25 | $\mu - .25\sigma$ | .4013 |
| −2.45 | $\mu - 2.45\sigma$ | .0071 | − .20 | $\mu - .20\sigma$ | .4207 |
| −2.40 | $\mu - 2.40\sigma$ | .0082 | − .15 | $\mu - .15\sigma$ | .4404 |
| −2.35 | $\mu - 2.35\sigma$ | .0094 | − .10 | $\mu - .10\sigma$ | .4602 |
| −2.30 | $\mu - 2.30\sigma$ | .0107 | − .05 | $\mu - .05\sigma$ | .4801 |
| −2.25 | $\mu - 2.25\sigma$ | .0122 | | | |
| −2.20 | $\mu - 2.20\sigma$ | .0139 | | | |
| −2.15 | $\mu - 2.15\sigma$ | .0158 | .00 | $\mu$ | .5000 |
| −2.10 | $\mu - 2.10\sigma$ | .0179 | | | |
| −2.05 | $\mu - 2.05\sigma$ | .0202 | | | |
| −2.00 | $\mu - 2.00\sigma$ | .0228 | .05 | $\mu + .05\sigma$ | .5199 |
| −1.95 | $\mu - 1.95\sigma$ | .0256 | .10 | $\mu + .10\sigma$ | .5398 |
| −1.90 | $\mu - 1.90\sigma$ | .0287 | .15 | $\mu + .15\sigma$ | .5596 |
| −1.85 | $\mu - 1.85\sigma$ | .0322 | .20 | $\mu + .20\sigma$ | .5793 |
| −1.80 | $\mu - 1.80\sigma$ | .0359 | .25 | $\mu + .25\sigma$ | .5987 |
| −1.75 | $\mu - 1.75\sigma$ | .0401 | .30 | $\mu + .30\sigma$ | .6179 |
| −1.70 | $\mu - 1.70\sigma$ | .0446 | .35 | $\mu + .35\sigma$ | .6368 |
| −1.65 | $\mu - 1.65\sigma$ | .0495 | .40 | $\mu + .40\sigma$ | .6554 |
| −1.60 | $\mu - 1.60\sigma$ | .0548 | .45 | $\mu + .45\sigma$ | .6736 |
| −1.55 | $\mu - 1.55\sigma$ | .0606 | .50 | $\mu + .50\sigma$ | .6915 |
| −1.50 | $\mu - 1.50\sigma$ | .0668 | .55 | $\mu + .55\sigma$ | .7088 |
| −1.45 | $\mu - 1.45\sigma$ | .0735 | .60 | $\mu + .60\sigma$ | .7257 |
| −1.40 | $\mu - 1.40\sigma$ | .0808 | .65 | $\mu + .65\sigma$ | .7422 |
| −1.35 | $\mu - 1.35\sigma$ | .0885 | .70 | $\mu + .70\sigma$ | .7580 |
| −1.30 | $\mu - 1.30\sigma$ | .0968 | .75 | $\mu + .75\sigma$ | .7734 |
| −1.25 | $\mu - 1.25\sigma$ | .1056 | .80 | $\mu + .80\sigma$ | .7881 |
| −1.20 | $\mu - 1.20\sigma$ | .1151 | .85 | $\mu + .85\sigma$ | .8023 |
| −1.15 | $\mu - 1.15\sigma$ | .1251 | .90 | $\mu + .90\sigma$ | .8159 |
| −1.10 | $\mu - 1.10\sigma$ | .1357 | .95 | $\mu + .95\sigma$ | .8289 |
| −1.05 | $\mu - 1.05\sigma$ | .1469 | 1.00 | $\mu + 1.00\sigma$ | .8413 |

| $z$ | $X$ | Area | $z$ | $X$ | Area |
|---|---|---|---|---|---|
| 1.05 | $\mu + 1.05\sigma$ | .8531 | −4.265 | $\mu - 4.265\sigma$ | .00001 |
| 1.10 | $\mu + 1.10\sigma$ | .8643 | −3.719 | $\mu - 3.719\sigma$ | .0001 |
| 1.15 | $\mu + 1.15\sigma$ | .8749 | −3.090 | $\mu - 3.090\sigma$ | .001 |
| 1.20 | $\mu + 1.20\sigma$ | .8849 | −2.576 | $\mu - 2.576\sigma$ | .005 |
| 1.25 | $\mu + 1.25\sigma$ | .8944 | −2.326 | $\mu - 2.326\sigma$ | .01 |
| 1.30 | $\mu + 1.30\sigma$ | .9032 | −2.054 | $\mu - 2.054\sigma$ | .02 |
| 1.35 | $\mu + 1.35\sigma$ | .9115 | −1.960 | $\mu - 1.960\sigma$ | .025 |
| 1.40 | $\mu + 1.40\sigma$ | .9192 | −1.881 | $\mu - 1.881\sigma$ | .03 |
| 1.45 | $\mu + 1.45\sigma$ | .9265 | −1.751 | $\mu - 1.751\sigma$ | .04 |
| 1.50 | $\mu + 1.50\sigma$ | .9332 | −1.645 | $\mu - 1.645\sigma$ | .05 |
| 1.55 | $\mu + 1.55\sigma$ | .9394 | −1.555 | $\mu - 1.555\sigma$ | .06 |
| 1.60 | $\mu + 1.60\sigma$ | .9452 | −1.476 | $\mu - 1.476\sigma$ | .07 |
| 1.65 | $\mu + 1.65\sigma$ | .9505 | −1.405 | $\mu - 1.405\sigma$ | .08 |
| 1.70 | $\mu + 1.70\sigma$ | .9554 | −1.341 | $\mu - 1.341\sigma$ | .09 |
| 1.75 | $\mu + 1.75\sigma$ | .9599 | −1.282 | $\mu - 1.282\sigma$ | .10 |
| 1.80 | $\mu + 1.80\sigma$ | .9641 | −1.036 | $\mu - 1.036\sigma$ | .15 |
| 1.85 | $\mu + 1.85\sigma$ | .9678 | − .842 | $\mu - .842\sigma$ | .20 |
| 1.90 | $\mu + 1.90\sigma$ | .9713 | − .674 | $\mu - .674\sigma$ | .25 |
| 1.95 | $\mu + 1.95\sigma$ | .9744 | − .524 | $\mu - .524\sigma$ | .30 |
| 2.00 | $\mu + 2.00\sigma$ | .9772 | − .385 | $\mu - .385\sigma$ | .35 |
| 2.05 | $\mu + 2.05\sigma$ | .9798 | − .253 | $\mu - .253\sigma$ | .40 |
| 2.10 | $\mu + 2.10\sigma$ | .9821 | − .126 | $\mu - .126\sigma$ | .45 |
| 2.15 | $\mu + 2.15\sigma$ | .9842 | 0 | $\mu$ | .50 |
| 2.20 | $\mu + 2.20\sigma$ | .9861 | .126 | $\mu + .126\sigma$ | .55 |
| 2.25 | $\mu + 2.25\sigma$ | .9878 | .253 | $\mu + .253\sigma$ | .60 |
| 2.30 | $\mu + 2.30\sigma$ | .9893 | .385 | $\mu + .385\sigma$ | .65 |
| 2.35 | $\mu + 2.35\sigma$ | .9906 | .524 | $\mu + .524\sigma$ | .70 |
| 2.40 | $\mu + 2.40\sigma$ | .9918 | .674 | $\mu + .674\sigma$ | .75 |
| 2.45 | $\mu + 2.45\sigma$ | .9929 | .842 | $\mu + .842\sigma$ | .80 |
| 2.50 | $\mu + 2.50\sigma$ | .9938 | 1.036 | $\mu + 1.036\sigma$ | .85 |
| 2.55 | $\mu + 2.55\sigma$ | .9946 | 1.282 | $\mu + 1.282\sigma$ | .90 |
| 2.60 | $\mu + 2.60\sigma$ | .9953 | 1.341 | $\mu + 1.341\sigma$ | .91 |
| 2.65 | $\mu + 2.65\sigma$ | .9960 | 1.405 | $\mu + 1.405\sigma$ | .92 |
| 2.70 | $\mu + 2.70\sigma$ | .9965 | 1.476 | $\mu + 1.476\sigma$ | .93 |
| 2.75 | $\mu + 2.75\sigma$ | .9970 | 1.555 | $\mu + 1.555\sigma$ | .94 |
| 2.80 | $\mu + 2.80\sigma$ | .9974 | 1.645 | $\mu + 1.645\sigma$ | .95 |
| 2.85 | $\mu + 2.85\sigma$ | .9978 | 1.751 | $\mu + 1.751\sigma$ | .96 |
| 2.90 | $\mu + 2.90\sigma$ | .9981 | 1.881 | $\mu + 1.881\sigma$ | .97 |
| 2.95 | $\mu + 2.95\sigma$ | .9984 | 1.960 | $\mu + 1.960\sigma$ | .975 |
| 3.00 | $\mu + 3.00\sigma$ | .9987 | 2.054 | $\mu + 2.054\sigma$ | .98 |
| 3.05 | $\mu + 3.05\sigma$ | .9989 | 2.326 | $\mu + 2.326\sigma$ | .99 |
| 3.10 | $\mu + 3.10\sigma$ | .9990 | 2.576 | $\mu + 2.576\sigma$ | .995 |
| 3.15 | $\mu + 3.15\sigma$ | .9992 | 3.090 | $\mu + 3.090\sigma$ | .999 |
| 3.20 | $\mu + 3.20\sigma$ | .9993 | 3.719 | $\mu + 3.719\sigma$ | .9999 |
| 3.25 | $\mu + 3.25\sigma$ | .9994 | 4.265 | $\mu + 4.265\sigma$ | .99999 |

## A-5 PERCENTILES OF THE $t$ DISTRIBUTIONS

| df | $t_{.60}$ | $t_{.70}$ | $t_{.80}$ | $t_{.90}$ | $t_{.95}$ | $t_{.975}$ | $t_{.99}$ | $t_{.995}$ |
|---|---|---|---|---|---|---|---|---|
| 1 | .325 | .727 | 1.376 | 3.078 | 6.314 | 12.706 | 31.821 | 63.657 |
| 2 | .289 | .617 | 1.061 | 1.886 | 2.920 | 4.303 | 6.965 | 9.925 |
| 3 | .277 | .584 | .978 | 1.638 | 2.353 | 3.182 | 4.541 | 5.841 |
| 4 | .271 | .569 | .941 | 1.533 | 2.132 | 2.776 | 3.747 | 4.604 |
| 5 | .267 | .559 | .920 | 1.476 | 2.015 | 2.571 | 3.365 | 4.032 |
| 6 | .265 | .553 | .906 | 1.440 | 1.943 | 2.447 | 3.143 | 3.707 |
| 7 | .263 | .549 | .896 | 1.415 | 1.895 | 2.365 | 2.998 | 3.499 |
| 8 | .262 | .546 | .889 | 1.397 | 1.860 | 2.306 | 2.896 | 3.355 |
| 9 | .261 | .543 | .883 | 1.383 | 1.833 | 2.262 | 2.821 | 3.250 |
| 10 | .260 | .542 | .879 | 1.372 | 1.812 | 2.228 | 2.764 | 3.169 |
| 11 | .260 | .540 | .876 | 1.363 | 1.796 | 2.201 | 2.718 | 3.106 |
| 12 | .259 | .539 | .873 | 1.356 | 1.782 | 2.179 | 2.681 | 3.055 |
| 13 | .259 | .538 | .870 | 1.350 | 1.771 | 2.160 | 2.650 | 3.012 |
| 14 | .258 | .537 | .868 | 1.345 | 1.761 | 2.145 | 2.624 | 2.977 |
| 15 | .258 | .536 | .866 | 1.341 | 1.753 | 2.131 | 2.602 | 2.947 |
| 16 | .258 | .535 | .865 | 1.337 | 1.746 | 2.120 | 2.583 | 2.921 |
| 17 | .257 | .534 | .863 | 1.333 | 1.740 | 2.110 | 2.567 | 2.898 |
| 18 | .257 | .534 | .862 | 1.330 | 1.734 | 2.101 | 2.552 | 2.878 |
| 19 | .257 | .533 | .861 | 1.328 | 1.729 | 2.093 | 2.539 | 2.861 |
| 20 | .257 | .533 | .860 | 1.325 | 1.725 | 2.086 | 2.528 | 2.845 |
| 21 | .257 | .532 | .859 | 1.323 | 1.721 | 2.080 | 2.518 | 2.831 |
| 22 | .256 | .532 | .858 | 1.321 | 1.717 | 2.074 | 2.508 | 2.819 |
| 23 | .256 | .532 | .858 | 1.319 | 1.714 | 2.069 | 2.500 | 2.807 |
| 24 | .256 | .531 | .857 | 1.318 | 1.711 | 2.064 | 2.492 | 2.797 |
| 25 | .256 | .531 | .856 | 1.316 | 1.708 | 2.060 | 2.485 | 2.787 |
| 26 | .256 | .531 | .856 | 1.315 | 1.706 | 2.056 | 2.479 | 2.779 |
| 27 | .256 | .531 | .855 | 1.314 | 1.703 | 2.052 | 2.473 | 2.771 |
| 28 | .256 | .530 | .855 | 1.313 | 1.701 | 2.048 | 2.467 | 2.763 |
| 29 | .256 | .530 | .854 | 1.311 | 1.699 | 2.045 | 2.462 | 2.756 |
| 30 | .256 | .530 | .854 | 1.310 | 1.697 | 2.042 | 2.457 | 2.750 |
| 40 | .255 | .529 | .851 | 1.303 | 1.684 | 2.021 | 2.423 | 2.704 |
| 60 | .254 | .527 | .848 | 1.296 | 1.671 | 2.000 | 2.390 | 2.660 |
| 120 | .254 | .526 | .845 | 1.289 | 1.658 | 1.980 | 2.358 | 2.617 |
| $\infty$ | .253 | .524 | .842 | 1.282 | 1.645 | 1.960 | 2.326 | 2.576 |
| df | $-t_{.40}$ | $-t_{.30}$ | $-t_{.20}$ | $-t_{.10}$ | $-t_{.05}$ | $-t_{.025}$ | $-t_{.01}$ | $-t_{.005}$ |

When the table is read from the foot, the tabled values are to be prefixed with a negative sign. Interpolation should be performed using the reciprocals of the degrees of freedom.

Data are extracted from Table III of Fisher and Yates, *Statistical Tables*, with the permission of the authors and publishers, Oliver & Boyd, Ltd., Edinburgh and London.

| *df* | $P_{0.5}$ | $P_{01}$ | $P_{02.5}$ | $P_{05}$ | $P_{10}$ | $P_{90}$ | $P_{95}$ | $P_{97.5}$ | $P_{99}$ | $P_{99.5}$ |
|---|---|---|---|---|---|---|---|---|---|---|
| 1 | .000039 | .00016 | .00098 | .0039 | .0158 | 2.71 | 3.84 | 5.02 | 6.63 | 7.88 |
| 2 | .0100 | .0201 | .0506 | .1026 | .2107 | 4.61 | 5.99 | 7.38 | 9.21 | 10.60 |
| 3 | .0717 | .115 | .216 | .352 | .584 | 6.25 | 7.81 | 9.35 | 11.34 | 12.84 |
| 4 | .207 | .297 | .484 | .711 | 1.064 | 7.78 | 9.49 | 11.14 | 13.28 | 14.86 |
| 5 | .412 | .554 | .831 | 1.15 | 1.61 | 9.24 | 11.07 | 12.83 | 15.09 | 16.75 |
| 6 | .676 | .872 | 1.24 | 1.64 | 2.20 | 10.64 | 12.59 | 14.45 | 16.81 | 18.55 |
| 7 | .989 | 1.24 | 1.69 | 2.17 | 2.83 | 12.02 | 14.07 | 16.01 | 18.48 | 20.28 |
| 8 | 1.34 | 1.65 | 2.18 | 2.73 | 3.49 | 13.36 | 15.51 | 17.53 | 20.09 | 21.96 |
| 9 | 1.73 | 2.09 | 2.70 | 3.33 | 4.17 | 14.68 | 16.92 | 19.02 | 21.67 | 23.59 |
| 10 | 2.16 | 2.56 | 3.25 | 3.94 | 4.87 | 15.99 | 18.31 | 20.48 | 23.21 | 25.19 |
| 11 | 2.60 | 3.05 | 3.82 | 4.57 | 5.58 | 17.28 | 19.68 | 21.92 | 24.73 | 26.76 |
| 12 | 3.07 | 3.57 | 4.40 | 5.23 | 6.30 | 18.55 | 21.03 | 23.34 | 26.22 | 28.30 |
| 13 | 3.57 | 4.11 | 5.01 | 5.89 | 7.04 | 19.81 | 22.36 | 24.74 | 27.69 | 29.82 |
| 14 | 4.07 | 4.66 | 5.63 | 6.57 | 7.79 | 21.06 | 23.68 | 26.12 | 29.14 | 31.32 |
| 15 | 4.60 | 5.23 | 6.26 | 7.26 | 8.55 | 22.31 | 25.00 | 27.49 | 30.58 | 32.80 |
| 16 | 5.14 | 5.81 | 6.91 | 7.96 | 9.31 | 23.54 | 26.30 | 28.85 | 32.00 | 34.27 |
| 18 | 6.26 | 7.01 | 8.23 | 9.39 | 10.86 | 25.99 | 28.87 | 31.53 | 34.81 | 37.16 |
| 20 | 7.43 | 8.26 | 9.59 | 10.85 | 12.44 | 28.41 | 31.41 | 34.17 | 37.57 | 40.00 |
| 24 | 9.89 | 10.86 | 12.40 | 13.85 | 15.66 | 33.20 | 36.42 | 39.36 | 42.98 | 45.56 |
| 30 | 13.79 | 14.95 | 16.79 | 18.49 | 20.60 | 40.26 | 43.77 | 46.98 | 50.89 | 53.67 |
| 40 | 20.71 | 22.16 | 24.43 | 26.51 | 29.05 | 51.81 | 55.76 | 59.34 | 63.69 | 66.77 |
| 60 | 35.53 | 37.48 | 40.48 | 43.19 | 46.46 | 74.40 | 79.08 | 83.30 | 88.38 | 91.95 |
| 120 | 83.85 | 86.92 | 91.58 | 95.70 | 100.62 | 140.23 | 146.57 | 152.21 | 158.95 | 163.64 |

For large values of degrees of freedom the approximate formula

$$\chi_\alpha^2 = \nu\left(1 - \frac{2}{9\nu} + z_\alpha\sqrt{\frac{2}{9\nu}}\right)^3$$

where $z_\alpha$ is the normal deviate and $\nu$ is the number of degrees of freedom, may be used. For example $\chi^2_{.99} = 60[1 - .00370 + (2.326 \times .06086)]^3 = 60(1.1379)^3 = 88.4$ for the 99th percentile for 60 degrees of freedom. This same formula after division by $\nu$ can be used for percentiles of $\chi^2/df$ in Table A-6*b*.

## A-6b PERCENTILES OF THE $\chi^2/df$ DISTRIBUTIONS

| $df$ | $P_{0.05}$ | $P_{0.1}$ | $P_{0.5}$ | $P_{01}$ | $P_{02.5}$ | $P_{05}$ | $P_{10}$ | $P_{20}$ | $P_{30}$ | $P_{40}$ |
|---|---|---|---|---|---|---|---|---|---|---|
| 1 | $.0^{6}39$ | $.0^{5}157$ | $.0^{4}39$ | $.0^{3}16$ | $.0^{3}98$ | $.0^{2}39$ | .016 | .064 | .148 | .275 |
| 2 | .001 | .001 | .005 | .010 | .025 | .052 | .106 | .223 | .356 | .511 |
| 3 | .005 | .008 | .024 | .038 | .072 | .117 | .195 | .335 | .475 | .623 |
| 4 | .016 | .023 | .052 | .074 | .121 | .178 | .266 | .412 | .549 | .688 |
| 5 | .032 | .042 | .082 | .111 | .166 | .229 | .322 | .469 | .600 | .731 |
| 6 | .050 | .064 | .113 | .145 | .206 | .272 | .367 | .512 | .638 | .762 |
| 7 | .069 | .085 | .141 | .177 | .241 | .310 | .405 | .546 | .667 | .785 |
| 8 | .089 | .107 | .168 | .206 | .272 | .342 | .436 | .574 | .691 | .803 |
| 9 | .108 | .128 | .193 | .232 | .300 | .369 | .463 | .598 | .710 | .817 |
| 10 | .126 | .148 | .216 | .256 | .325 | .394 | .487 | .618 | .727 | .830 |
| 11 | .144 | .167 | .237 | .278 | .347 | .416 | .507 | .635 | .741 | .840 |
| 12 | .161 | .184 | .256 | .298 | .367 | .436 | .525 | .651 | .753 | .848 |
| 13 | .177 | .201 | .274 | .316 | .385 | .453 | .542 | .664 | .764 | .856 |
| 14 | .193 | .217 | .291 | .333 | .402 | .469 | .556 | .676 | .773 | .863 |
| 15 | .207 | .232 | .307 | .349 | .418 | .484 | .570 | .687 | .781 | .869 |
| 16 | .221 | .246 | .321 | .363 | .432 | .498 | .582 | .697 | .789 | .874 |
| 17 | .234 | .260 | .335 | .377 | .445 | .510 | .593 | .706 | .796 | .879 |
| 18 | .247 | .272 | .348 | .390 | .457 | .522 | .604 | .714 | .802 | .883 |
| 19 | .258 | .285 | .360 | .402 | .469 | .532 | .613 | .722 | .808 | .887 |
| 20 | .270 | .296 | .372 | .413 | .480 | .543 | .622 | .729 | .813 | .890 |
| 22 | .291 | .317 | .393 | .434 | .499 | .561 | .638 | .742 | .823 | .897 |
| 24 | .310 | .337 | .412 | .452 | .517 | .577 | .652 | .753 | .831 | .902 |
| 26 | .328 | .355 | .429 | .469 | .532 | .592 | .665 | .762 | .838 | .907 |
| 28 | .345 | .371 | .445 | .484 | .547 | .605 | .676 | .771 | .845 | .911 |
| 30 | .360 | .386 | .460 | .498 | .560 | .616 | .687 | .779 | .850 | .915 |
| 35 | .394 | .420 | .491 | .529 | .588 | .642 | .708 | .795 | .862 | .922 |
| 40 | .423 | .448 | .518 | .554 | .611 | .663 | .726 | .809 | .872 | .928 |
| 45 | .448 | .472 | .540 | .576 | .630 | .680 | .741 | .820 | .880 | .933 |
| 50 | .469 | .494 | .560 | .594 | .647 | .695 | .754 | .829 | .886 | .937 |
| 55 | .488 | .512 | .577 | .610 | .662 | .708 | .765 | .837 | .892 | .941 |
| 60 | .506 | .529 | .592 | .625 | .675 | .720 | .774 | .844 | .897 | .944 |
| 70 | .535 | .558 | .618 | .649 | .697 | .739 | .790 | .856 | .905 | .949 |
| 80 | .560 | .582 | .640 | .669 | .714 | .755 | .803 | .865 | .911 | .952 |
| 90 | .581 | .602 | .658 | .686 | .729 | .768 | .814 | .873 | .917 | .955 |
| 100 | .599 | .619 | .673 | .701 | .742 | .779 | .824 | .879 | .921 | .958 |
| 120 | .629 | .648 | .699 | .724 | .763 | .798 | .839 | .890 | .929 | .962 |
| 140 | .653 | .671 | .719 | .743 | .780 | .812 | .850 | .898 | .934 | .965 |
| 160 | .673 | .690 | .736 | .758 | .793 | .824 | .860 | .905 | .939 | .968 |
| 180 | .689 | .706 | .749 | .771 | .804 | .833 | .868 | .910 | .942 | .970 |
| 200 | .703 | .719 | .761 | .782 | .814 | .841 | .874 | .915 | .945 | .972 |
| 250 | .732 | .746 | .785 | .804 | .832 | .858 | .887 | .924 | .951 | .975 |
| 300 | .753 | .767 | .802 | .820 | .846 | .870 | .897 | .931 | .956 | .977 |
| 350 | .770 | .783 | .816 | .833 | .857 | .879 | .904 | .936 | .959 | .979 |
| 400 | .784 | .796 | .827 | .843 | .866 | .887 | .911 | .940 | .962 | .981 |
| 450 | .795 | .807 | .837 | .852 | .874 | .893 | .916 | .944 | .964 | .982 |
| 500 | .805 | .816 | .845 | .859 | .880 | .898 | .920 | .946 | .966 | .983 |
| 750 | .839 | .848 | .872 | .884 | .901 | .917 | .934 | .956 | .972 | .986 |
| 1,000 | .859 | .868 | .889 | .899 | .914 | .928 | .943 | .962 | .976 | .988 |
| 5,000 | .936 | .939 | .949 | .954 | .961 | .967 | .974 | .983 | .989 | .995 |
| $\infty$ | 1 | 1 | 1 | 1 | 1 | 1 | 1 | 1 | 1 | 1 |

Read $.0^{3}16$ as .00016, and so on.

| $P_{50}$ | $P_{60}$ | $P_{70}$ | $P_{80}$ | $P_{90}$ | $P_{95}$ | $P_{97.5}$ | $P_{99}$ | $P_{99.5}$ | $P_{99.9}$ | $P_{99.95}$ | $df$ |
|---|---|---|---|---|---|---|---|---|---|---|---|
| .455 | .708 | 1.07 | 1.64 | 2.71 | 3.84 | 5.02 | 6.64 | 7.88 | 10.83 | 12.12 | 1 |
| .693 | .916 | 1.20 | 1.61 | 2.30 | 3.00 | 3.69 | 4.61 | 5.30 | 6.91 | 7.60 | 2 |
| .789 | .982 | 1.22 | 1.55 | 2.08 | 2.60 | 3.12 | 3.78 | 4.28 | 5.42 | 5.91 | 3 |
| .839 | 1.011 | 1.22 | 1.50 | 1.94 | 2.37 | 2.79 | 3.32 | 3.72 | 4.62 | 5.00 | 4 |
| .870 | 1.03 | 1.21 | 1.46 | 1.85 | 2.21 | 2.57 | 3.02 | 3.35 | 4.10 | 4.42 | 5 |
| .891 | 1.04 | 1.21 | 1.43 | 1.77 | 2.10 | 2.41 | 2.80 | 3.09 | 3.74 | 4.02 | 6 |
| .907 | 1.04 | 1.20 | 1.40 | 1.72 | 2.01 | 2.29 | 2.64 | 2.90 | 3.47 | 3.72 | 7 |
| .918 | 1.04 | 1.19 | 1.38 | 1.67 | 1.94 | 2.19 | 2.51 | 2.74 | 3.27 | 3.48 | 8 |
| .927 | 1.05 | 1.18 | 1.36 | 1.63 | 1.88 | 2.11 | 2.41 | 2.62 | 3.10 | 3.30 | 9 |
| .934 | 1.05 | 1.18 | 1.34 | 1.60 | 1.83 | 2.05 | 2.32 | 2.52 | 2.96 | 3.14 | 10 |
| .940 | 1.05 | 1.17 | 1.33 | 1.57 | 1.79 | 1.99 | 2.25 | 2.43 | 2.84 | 3.01 | 11 |
| .945 | 1.05 | 1.17 | 1.32 | 1.55 | 1.75 | 1.94 | 2.18 | 2.36 | 2.74 | 2.90 | 12 |
| .949 | 1.05 | 1.16 | 1.31 | 1.52 | 1.72 | 1.90 | 2.13 | 2.29 | 2.66 | 2.81 | 13 |
| .953 | 1.05 | 1.16 | 1.30 | 1.50 | 1.69 | 1.87 | 2.08 | 2.24 | 2.58 | 2.72 | 14 |
| .956 | 1.05 | 1.15 | 1.29 | 1.49 | 1.67 | 1.83 | 2.04 | 2.19 | 2.51 | 2.65 | 15 |
| .959 | 1.05 | 1.15 | 1.28 | 1.47 | 1.64 | 1.80 | 2.00 | 2.14 | 2.45 | 2.58 | 16 |
| .961 | 1.05 | 1.15 | 1.27 | 1.46 | 1.62 | 1.78 | 1.97 | 2.10 | 2.40 | 2.52 | 17 |
| .963 | 1.05 | 1.14 | 1.26 | 1.44 | 1.60 | 1.75 | 1.93 | 2.06 | 2.35 | 2.47 | 18 |
| .965 | 1.05 | 1.14 | 1.26 | 1.43 | 1.59 | 1.73 | 1.90 | 2.03 | 2.31 | 2.42 | 19 |
| .967 | 1.05 | 1.14 | 1.25 | 1.42 | 1.57 | 1.71 | 1.88 | 2.00 | 2.27 | 2.37 | 20 |
| .970 | 1.05 | 1.13 | 1.24 | 1.40 | 1.54 | 1.67 | 1.83 | 1.95 | 2.19 | 2.30 | 22 |
| .972 | 1.05 | 1.13 | 1.23 | 1.38 | 1.52 | 1.64 | 1.79 | 1.90 | 2.13 | 2.23 | 24 |
| .974 | 1.05 | 1.12 | 1.22 | 1.37 | 1.50 | 1.61 | 1.76 | 1.86 | 2.08 | 2.17 | 26 |
| .976 | 1.04 | 1.12 | 1.22 | 1.35 | 1.48 | 1.59 | 1.72 | 1.82 | 2.03 | 2.12 | 28 |
| .978 | 1.04 | 1.12 | 1.21 | 1.34 | 1.46 | 1.57 | 1.70 | 1.79 | 1.99 | 2.07 | 30 |
| .981 | 1.04 | 1.11 | 1.19 | 1.32 | 1.42 | 1.52 | 1.64 | 1.72 | 1.90 | 1.98 | 35 |
| .983 | 1.04 | 1.10 | 1.18 | 1.30 | 1.39 | 1.48 | 1.59 | 1.67 | 1.84 | 1.90 | 40 |
| .985 | 1.04 | 1.10 | 1.17 | 1.28 | 1.37 | 1.45 | 1.55 | 1.63 | 1.78 | 1.84 | 45 |
| .987 | 1.04 | 1.09 | 1.16 | 1.26 | 1.35 | 1.43 | 1.52 | 1.59 | 1.73 | 1.79 | 50 |
| .988 | 1.04 | 1.09 | 1.16 | 1.25 | 1.33 | 1.41 | 1.50 | 1.56 | 1.69 | 1.75 | 55 |
| .989 | 1.04 | 1.09 | 1.15 | 1.24 | 1.32 | 1.39 | 1.47 | 1.53 | 1.66 | 1.71 | 60 |
| .990 | 1.03 | 1.08 | 1.14 | 1.22 | 1.29 | 1.36 | 1.43 | 1.49 | 1.60 | 1.65 | 70 |
| .992 | 1.03 | 1.08 | 1.13 | 1.21 | 1.27 | 1.33 | 1.40 | 1.45 | 1.56 | 1.60 | 80 |
| .993 | 1.03 | 1.07 | 1.12 | 1.20 | 1.26 | 1.31 | 1.38 | 1.43 | 1.52 | 1.56 | 90 |
| .993 | 1.03 | 1.07 | 1.12 | 1.18 | 1.24 | 1.30 | 1.36 | 1.40 | 1.49 | 1.53 | 100 |
| .994 | 1.03 | 1.06 | 1.11 | 1.17 | 1.22 | 1.27 | 1.32 | 1.36 | 1.45 | 1.48 | 120 |
| .995 | 1.03 | 1.06 | 1.10 | 1.16 | 1.20 | 1.25 | 1.30 | 1.33 | 1.41 | 1.44 | 140 |
| .996 | 1.02 | 1.06 | 1.09 | 1.15 | 1.19 | 1.23 | 1.28 | 1.31 | 1.38 | 1.41 | 160 |
| .996 | 1.02 | 1.05 | 1.09 | 1.14 | 1.18 | 1.22 | 1.26 | 1.29 | 1.36 | 1.38 | 180 |
| .997 | 1.02 | 1.05 | 1.08 | 1.13 | 1.17 | 1.21 | 1.25 | 1.28 | 1.34 | 1.36 | 200 |
| .997 | 1.02 | 1.04 | 1.07 | 1.12 | 1.15 | 1.18 | 1.22 | 1.25 | 1.30 | 1.32 | 250 |
| .998 | 1.02 | 1.04 | 1.07 | 1.11 | 1.14 | 1.17 | 1.20 | 1.22 | 1.27 | 1.29 | 300 |
| .998 | 1.02 | 1.04 | 1.06 | 1.10 | 1.13 | 1.15 | 1.18 | 1.21 | 1.25 | 1.27 | 350 |
| .998 | 1.02 | 1.04 | 1.06 | 1.09 | 1.12 | 1.14 | 1.17 | 1.19 | 1.24 | 1.25 | 400 |
| .999 | 1.02 | 1.03 | 1.06 | 1.09 | 1.11 | 1.13 | 1.16 | 1.18 | 1.22 | 1.23 | 450 |
| .999 | 1.01 | 1.03 | 1.05 | 1.08 | 1.11 | 1.13 | 1.15 | 1.17 | 1.21 | 1.22 | 500 |
| .999 | 1.01 | 1.03 | 1.04 | 1.07 | 1.09 | 1.10 | 1.12 | 1.14 | 1.17 | 1.18 | 750 |
| .999 | 1.01 | 1.02 | 1.04 | 1.06 | 1.07 | 1.09 | 1.11 | 1.12 | 1.14 | 1.15 | 1,000 |
| 1.00 | 1.00 | 1.01 | 1.02 | 1.02 | 1.03 | 1.04 | 1.05 | 1.05 | 1.06 | 1.07 | 5,000 |
| 1 | 1 | 1 | 1 | 1 | 1 | 1 | 1 | 1 | 1 | 1 | ∞ |

## A-6c NECESSARY SIZE OF A RANDOM SAMPLE FROM A NORMAL POPULATION TO HAVE PROBABILITY $1 - \alpha$ THAT $s^2$ WILL BE LESS THAN $k\sigma^2$ AND THAT $s$ WILL BE LESS THAN $k'\sigma$

| $k$ \ $1-\alpha$ | For $s^2 < k\sigma^2$ .95 | .975 | .99 | .995 | .999 | $k'$ \ $1-\alpha$ | For $s < k'\sigma$ .95 | .975 | .99 | .995 | .999 |
|---|---|---|---|---|---|---|---|---|---|---|---|
| 3 | 3 | 5 | 7 | 8 | 11 | 3 | | | 2 | 2 | 3 |
| 2.5 | 5 | 7 | 10 | 12 | 17 | 2.5 | | 2 | 3 | 3 | 4 |
| 2 | 9 | 12 | 18 | 21 | 31 | 2 | 2 | 3 | 4 | 5 | 7 |
| 1.50 | 27 | 39 | 56 | 69 | 100 | 1.5 | 6 | 9 | 12 | 15 | 22 |
| 1.25 | 97 | 139 | 197 | 243 | 352 | 1.25 | 22 | 32 | 45 | 56 | 81 |
| 1.10 | 565 | 807 | 1,142 | 1,404 | 2,026 | 1.10 | 134 | 193 | 274 | 338 | 488 |
| 1.05 | 2,211 | 3,151 | 4,448 | 5,462 | 7,872 | 1.05 | 538 | 769 | 1,089 | 1,338 | 1,931 |
| 1.01 | 54,310 | 77,180 | 108,760 | 133,444 | 192,050 | 1.01 | 13,509 | 19,207 | 27,078 | 33,224 | 47,830 |

## A-6d PERCENTILES OF $\sigma^2/s^2 = F(\infty, df)$

| $df$ | $P_{0.05}$ | $P_{0.1}$ | $P_{0.5}$ | $P_1$ | $P_{2.5}$ | $P_5$ | $P_{95}$ | $P_{97.5}$ | $P_{99}$ | $P_{99.5}$ | $P_{99.9}$ | $P_{99.95}$ |
|---|---|---|---|---|---|---|---|---|---|---|---|---|
| 4 | .200 | .217 | .269 | .301 | .359 | .422 | 5.624 | 8.264 | 13.477 | 19.305 | 44.053 | 62.500 |
| 5 | .226 | .244 | .299 | .331 | .390 | .452 | 4.367 | 6.017 | 9.025 | 12.136 | 23.810 | 31.646 |
| 6 | .249 | .267 | .323 | .357 | .415 | .476 | 3.670 | 4.850 | 6.882 | 8.873 | 15.748 | 20.040 |
| 7 | .269 | .288 | .345 | .379 | .437 | .498 | 3.230 | 4.143 | 5.650 | 7.077 | 11.710 | 14.430 |
| 8 | .287 | .306 | .364 | .398 | .456 | .516 | 2.927 | 3.670 | 4.859 | 5.952 | 9.337 | 11.261 |
| 9 | .303 | .323 | .382 | .415 | .473 | .532 | 2.707 | 3.333 | 4.310 | 5.187 | 7.806 | 9.259 |
| 10 | .318 | .338 | .397 | .431 | .488 | .546 | 2.538 | 3.080 | 3.909 | 4.638 | 6.761 | 7.905 |
| 12 | .345 | .365 | .424 | .458 | .514 | .571 | 2.296 | 2.725 | 3.360 | 3.903 | 5.420 | 6.203 |
| 14 | .367 | .388 | .447 | .480 | .536 | .591 | 2.130 | 2.487 | 3.004 | 3.435 | 4.604 | 5.192 |
| 16 | .387 | .408 | .467 | .500 | .555 | .608 | 2.010 | 2.316 | 2.753 | 3.111 | 4.058 | 4.525 |
| 18 | .405 | .425 | .484 | .517 | .571 | .624 | 1.917 | 2.187 | 2.566 | 2.873 | 3.670 | 4.055 |
| 20 | .421 | .441 | .500 | .532 | .585 | .637 | 1.843 | 2.085 | 2.421 | 2.690 | 3.377 | 3.705 |
| 22 | .436 | .456 | .514 | .546 | .598 | .649 | 1.783 | 2.003 | 2.306 | 2.545 | 3.151 | 3.435 |
| 24 | .449 | .469 | .527 | .558 | .610 | .659 | 1.733 | 1.935 | 2.211 | 2.428 | 2.968 | 3.221 |
| 26 | .461 | .481 | .538 | .570 | .620 | .669 | 1.691 | 1.878 | 2.131 | 2.330 | 2.819 | 3.045 |
| 28 | .472 | .492 | .549 | .580 | .630 | .677 | 1.654 | 1.829 | 2.064 | 2.247 | 2.695 | 2.899 |
| 30 | .483 | .502 | .559 | .589 | .639 | .685 | 1.622 | 1.787 | 2.006 | 2.176 | 2.589 | 2.777 |
| 35 | .506 | .525 | .581 | .610 | .658 | .703 | 1.558 | 1.702 | 1.891 | 2.036 | 2.383 | 2.539 |
| 40 | .526 | .545 | .599 | .628 | .674 | .717 | 1.509 | 1.637 | 1.805 | 1.932 | 2.233 | 2.366 |
| 45 | .543 | .562 | .615 | .643 | .688 | .730 | 1.470 | 1.586 | 1.737 | 1.851 | 2.118 | 2.235 |
| 50 | .558 | .577 | .629 | .657 | .700 | .741 | 1.438 | 1.545 | 1.683 | 1.786 | 2.026 | 2.131 |
| 60 | .584 | .602 | .653 | .679 | .720 | .759 | 1.389 | 1.482 | 1.601 | 1.689 | 1.890 | 1.977 |
| 70 | .606 | .623 | .672 | .697 | .737 | .773 | 1.353 | 1.436 | 1.540 | 1.618 | 1.793 | 1.868 |
| 80 | .624 | .641 | .688 | .712 | .750 | .785 | 1.325 | 1.400 | 1.494 | 1.563 | 1.720 | 1.786 |
| 90 | .639 | .656 | .702 | .725 | .762 | .795 | 1.302 | 1.371 | 1.457 | 1.520 | 1.662 | 1.722 |
| 100 | .653 | .669 | .713 | .736 | .772 | .804 | 1.283 | 1.347 | 1.427 | 1.485 | 1.615 | 1.669 |
| 120 | .676 | .691 | .733 | .755 | .788 | .819 | 1.254 | 1.310 | 1.381 | 1.431 | 1.543 | 1.590 |
| 140 | .694 | .709 | .749 | .770 | .802 | .830 | 1.232 | 1.283 | 1.346 | 1.391 | 1.491 | 1.532 |
| 160 | .710 | .724 | .763 | .782 | .812 | .840 | 1.214 | 1.261 | 1.319 | 1.360 | 1.450 | 1.487 |
| 180 | .723 | .737 | .774 | .793 | .822 | .848 | 1.200 | 1.244 | 1.297 | 1.334 | 1.417 | 1.451 |
| 200 | .734 | .748 | .784 | .802 | .830 | .855 | 1.188 | 1.229 | 1.279 | 1.314 | 1.390 | 1.422 |
| 300 | .775 | .787 | .818 | .834 | .857 | .879 | 1.150 | 1.182 | 1.220 | 1.246 | 1.305 | 1.328 |
| 400 | .801 | .811 | .839 | .853 | .875 | .894 | 1.128 | 1.154 | 1.186 | 1.209 | 1.257 | 1.276 |
| 500 | .819 | .829 | .854 | .867 | .887 | .904 | 1.113 | 1.136 | 1.164 | 1.184 | 1.225 | 1.242 |
| 600 | .833 | .842 | .866 | .878 | .896 | .912 | 1.103 | 1.124 | 1.149 | 1.166 | 1.203 | 1.218 |
| 800 | .853 | .861 | .882 | .893 | .909 | .923 | 1.088 | 1.106 | 1.127 | 1.142 | 1.173 | 1.185 |
| 1,000 | .867 | .874 | .894 | .903 | .918 | .930 | 1.078 | 1.094 | 1.112 | 1.125 | 1.153 | 1.164 |
| 2,000 | .903 | .909 | .923 | .930 | .941 | .950 | 1.054 | 1.065 | 1.078 | 1.086 | 1.105 | 1.112 |
| 5,000 | .937 | .941 | .950 | .955 | .962 | .968 | 1.034 | 1.040 | 1.048 | 1.053 | 1.065 | 1.069 |
| 10,000 | .955 | .958 | .965 | .968 | .973 | .977 | 1.024 | 1.028 | 1.034 | 1.037 | 1.045 | 1.048 |

| $df$ | $P_{0.05}$ | $P_{0.1}$ | $P_{0.5}$ | $P_{1}$ | $P_{2.5}$ | $P_{5}$ | $P_{95}$ | $P_{97.5}$ | $P_{99}$ | $P_{99.5}$ | $P_{99.9}$ | $P_{99.95}$ |
|---|---|---|---|---|---|---|---|---|---|---|---|---|
| 2 | .363 | .380 | .434 | .466 | .521 | .578 | 4.407 | 6.287 | 10.000 | 14.142 | 31.623 | 44.721 |
| 3 | .411 | .429 | .483 | .514 | .567 | .620 | 2.920 | 3.727 | 5.110 | 6.468 | 11.111 | 14.003 |
| 4 | .447 | .465 | .519 | .549 | .599 | .649 | 2.372 | 2.875 | 3.671 | 4.394 | 6.637 | 7.906 |
| 5 | .476 | .494 | .546 | .576 | .624 | .672 | 2.090 | 2.453 | 3.004 | 3.484 | 4.879 | 5.625 |
| 6 | .499 | .517 | .569 | .597 | .644 | .690 | 1.916 | 2.202 | 2.623 | 2.979 | 3.968 | 4.477 |
| 7 | .519 | .536 | .588 | .616 | .661 | .705 | 1.797 | 2.035 | 2.377 | 2.660 | 3.422 | 3.799 |
| 8 | .536 | .553 | .604 | .631 | .675 | .718 | 1.711 | 1.916 | 2.204 | 2.440 | 3.056 | 3 356 |
| 9 | .551 | .568 | .618 | .645 | .688 | .729 | 1.645 | 1.826 | 2.076 | 2.277 | 2.794 | 3.043 |
| 10 | .564 | .581 | .630 | .656 | .699 | .739 | 1.593 | 1.755 | 1.977 | 2.154 | 2.600 | 2.812 |
| 11 | .576 | .593 | .641 | .667 | .708 | .748 | 1.551 | 1.698 | 1.898 | 2.056 | 2.449 | 2.632 |
| 12 | .587 | .604 | .651 | .677 | .717 | .755 | 1.515 | 1.651 | 1.833 | 1.976 | 2.328 | 2.491 |
| 13 | .597 | .614 | .660 | .685 | .725 | .762 | 1.485 | 1.611 | 1.779 | 1.910 | 2.229 | 2.375 |
| 14 | .606 | .623 | .669 | .693 | .732 | .769 | 1.460 | 1.577 | 1.733 | 1.853 | 2.146 | 2.279 |
| 15 | .615 | .631 | .676 | .700 | .739 | .775 | 1.437 | 1.548 | 1.694 | 1.806 | 2.075 | 2.197 |
| 16 | .622 | .638 | .683 | .707 | .745 | .780 | 1.418 | 1.522 | 1.659 | 1.764 | 2.015 | 2.127 |
| 17 | .630 | .646 | .690 | .713 | .750 | .785 | 1.400 | 1.499 | 1.629 | 1.727 | 1.962 | 2.067 |
| 18 | .636 | .652 | .696 | .719 | .756 | .790 | 1.384 | 1.479 | 1.602 | 1.695 | 1.916 | 2.014 |
| 19 | .643 | .658 | .702 | .725 | .760 | .794 | 1.370 | 1.461 | 1.578 | 1.666 | 1.874 | 1.967 |
| 20 | .649 | .664 | .707 | .730 | .765 | .798 | 1.358 | 1.444 | 1.556 | 1.640 | 1.838 | 1.925 |
| 22 | .660 | .675 | .717 | .739 | .773 | .805 | 1.335 | 1.415 | 1.518 | 1.595 | 1.775 | 1.853 |
| 24 | .670 | .685 | .726 | .747 | .781 | .812 | 1.316 | 1.391 | 1.487 | 1.558 | 1.723 | 1.795 |
| 26 | .679 | .694 | .734 | .755 | .788 | .818 | 1.300 | 1.370 | 1.460 | 1.526 | 1.679 | 1.745 |
| 28 | .687 | .702 | .741 | .762 | .794 | .823 | 1.286 | 1.352 | 1.437 | 1.499 | 1.642 | 1.703 |
| 30 | .695 | .709 | .748 | .768 | .799 | .828 | 1.274 | 1.337 | 1.416 | 1.475 | 1.609 | 1.666 |
| 35 | .711 | .725 | .762 | .781 | .811 | .838 | 1.248 | 1.304 | 1.375 | 1.427 | 1.544 | 1.593 |
| 40 | .725 | .738 | .774 | .792 | .821 | .847 | 1.228 | 1.280 | 1.343 | 1.390 | 1.494 | 1.538 |
| 45 | .737 | .750 | .784 | .802 | .829 | .854 | 1.212 | 1.259 | 1.318 | 1.361 | 1.455 | 1.495 |
| 50 | .747 | .760 | .793 | .810 | .837 | .861 | 1.199 | 1.243 | 1.297 | 1.337 | 1.423 | 1.460 |
| 55 | .756 | .768 | .801 | .818 | .843 | .866 | 1.188 | 1.229 | 1.280 | 1.316 | 1.397 | 1.431 |
| 60 | .764 | .776 | .808 | .824 | .849 | .871 | 1.179 | 1.217 | 1.265 | 1.299 | 1.375 | 1.406 |
| 65 | .772 | .783 | .814 | .830 | .854 | .875 | 1.170 | 1.207 | 1.252 | 1.285 | 1.356 | 1.385 |
| 70 | .778 | .789 | .820 | .835 | .858 | .879 | 1.163 | 1.198 | 1.241 | 1.272 | 1.339 | 1.367 |
| 80 | .790 | .801 | .829 | .844 | .866 | .886 | 1.151 | 1.183 | 1.222 | 1.250 | 1.311 | 1.336 |
| 90 | .800 | .810 | .838 | .852 | .873 | .892 | 1.141 | 1.171 | 1.207 | 1.233 | 1.289 | 1.312 |
| 100 | .808 | .818 | .845 | .858 | .879 | .897 | 1.133 | 1.161 | 1.195 | 1.219 | 1.271 | 1.292 |
| 120 | .822 | .831 | .856 | .869 | .888 | .905 | 1.120 | 1.145 | 1.175 | 1.196 | 1.242 | 1.261 |
| 140 | .833 | .842 | .866 | .877 | .895 | .911 | 1.110 | 1.133 | 1.160 | 1.179 | 1.221 | 1.238 |
| 160 | .842 | .851 | .873 | .884 | .901 | .916 | 1.102 | 1.123 | 1.148 | 1.166 | 1.204 | 1.219 |
| 180 | .850 | .858 | .880 | .890 | .906 | .921 | 1.096 | 1.115 | 1.139 | 1.155 | 1.190 | 1.205 |
| 200 | .857 | .865 | .885 | .895 | .911 | .925 | 1.090 | 1.109 | 1.131 | 1.146 | 1.179 | 1.192 |
| 300 | .880 | .887 | .904 | .913 | .926 | .937 | 1.072 | 1.087 | 1.104 | 1.116 | 1.142 | 1.152 |
| 400 | .895 | .901 | .916 | .924 | .935 | .945 | 1.062 | 1.074 | 1.089 | 1.099 | 1.121 | 1.130 |
| 500 | .905 | .910 | .924 | .931 | .942 | .951 | 1.055 | 1.066 | 1.079 | 1.088 | 1.107 | 1.115 |
| 600 | .912 | .917 | .930 | .937 | .946 | .955 | 1.050 | 1.060 | 1.072 | 1.080 | 1.097 | 1.104 |
| 800 | .923 | .928 | .939 | .945 | .953 | .961 | 1.043 | 1.052 | 1.062 | 1.068 | 1.083 | 1.089 |
| 1,000 | .931 | .935 | .945 | .950 | .958 | .965 | 1.038 | 1.046 | 1.055 | 1.061 | 1.074 | 1.079 |
| 2,000 | .950 | .953 | .961 | .964 | .970 | .975 | 1.027 | 1.032 | 1.038 | 1.042 | 1.051 | 1.055 |
| 5,000 | .968 | .970 | .975 | .977 | .981 | .984 | 1.017 | 1.020 | 1.024 | 1.026 | 1.032 | 1.034 |
| 10,000 | .977 | .979 | .982 | .984 | .986 | .989 | 1.012 | 1.014 | 1.017 | 1.019 | 1.022 | 1.024 |

## A-7a $F$ DISTRIBUTION, UPPER 5 PER CENT POINTS ($F_{.95}$)

DEGREES OF FREEDOM FOR NUMERATOR (columns); DEGREES OF FREEDOM FOR DENOMINATOR (rows)

| | 1 | 2 | 3 | 4 | 5 | 6 | 7 | 8 | 9 | 10 | 12 | 15 | 20 | 24 | 30 | 40 | 60 | 120 | ∞ |
|---|---|---|---|---|---|---|---|---|---|---|---|---|---|---|---|---|---|---|---|
| 1 | 161 | 200 | 216 | 225 | 230 | 234 | 237 | 239 | 241 | 242 | 244 | 246 | 248 | 249 | 250 | 251 | 252 | 253 | 254 |
| 2 | 18.5 | 19.0 | 19.2 | 19.2 | 19.3 | 19.3 | 19.4 | 19.4 | 19.4 | 19.4 | 19.4 | 19.4 | 19.4 | 19.5 | 19.5 | 19.5 | 19.5 | 19.5 | 19.5 |
| 3 | 10.1 | 9.55 | 9.28 | 9.12 | 9.01 | 8.94 | 8.89 | 8.85 | 8.81 | 8.79 | 8.74 | 8.70 | 8.66 | 8.64 | 8.62 | 8.59 | 8.57 | 8.55 | 8.53 |
| 4 | 7.71 | 6.94 | 6.59 | 6.39 | 6.26 | 6.16 | 6.09 | 6.04 | 6.00 | 5.96 | 5.91 | 5.86 | 5.80 | 5.77 | 5.75 | 5.72 | 5.69 | 5.66 | 5.63 |
| 5 | 6.61 | 5.79 | 5.41 | 5.19 | 5.05 | 4.95 | 4.88 | 4.82 | 4.77 | 4.74 | 4.68 | 4.62 | 4.56 | 4.53 | 4.50 | 4.46 | 4.43 | 4.40 | 4.37 |
| 6 | 5.99 | 5.14 | 4.76 | 4.53 | 4.39 | 4.28 | 4.21 | 4.15 | 4.10 | 4.06 | 4.00 | 3.94 | 3.87 | 3.84 | 3.81 | 3.77 | 3.74 | 3.70 | 3.67 |
| 7 | 5.59 | 4.74 | 4.35 | 4.12 | 3.97 | 3.87 | 3.79 | 3.73 | 3.68 | 3.64 | 3.57 | 3.51 | 3.44 | 3.41 | 3.38 | 3.34 | 3.30 | 3.27 | 3.23 |
| 8 | 5.32 | 4.46 | 4.07 | 3.84 | 3.69 | 3.58 | 3.50 | 3.44 | 3.39 | 3.35 | 3.28 | 3.22 | 3.15 | 3.12 | 3.08 | 3.04 | 3.01 | 2.97 | 2.93 |
| 9 | 5.12 | 4.26 | 3.86 | 3.63 | 3.48 | 3.37 | 3.29 | 3.23 | 3.18 | 3.14 | 3.07 | 3.01 | 2.94 | 2.90 | 2.86 | 2.83 | 2.79 | 2.75 | 2.71 |
| 10 | 4.96 | 4.10 | 3.71 | 3.48 | 3.33 | 3.22 | 3.14 | 3.07 | 3.02 | 2.98 | 2.91 | 2.85 | 2.77 | 2.74 | 2.70 | 2.66 | 2.62 | 2.58 | 2.54 |
| 11 | 4.84 | 3.98 | 3.59 | 3.36 | 3.20 | 3.09 | 3.01 | 2.95 | 2.90 | 2.85 | 2.79 | 2.72 | 2.65 | 2.61 | 2.57 | 2.53 | 2.49 | 2.45 | 2.40 |
| 12 | 4.75 | 3.89 | 3.49 | 3.26 | 3.11 | 3.00 | 2.91 | 2.85 | 2.80 | 2.75 | 2.69 | 2.62 | 2.54 | 2.51 | 2.47 | 2.43 | 2.38 | 2.34 | 2.30 |
| 13 | 4.67 | 3.81 | 3.41 | 3.18 | 3.03 | 2.92 | 2.83 | 2.77 | 2.71 | 2.67 | 2.60 | 2.53 | 2.46 | 2.42 | 2.38 | 2.34 | 2.30 | 2.25 | 2.21 |
| 14 | 4.60 | 3.74 | 3.34 | 3.11 | 2.96 | 2.85 | 2.76 | 2.70 | 2.65 | 2.60 | 2.53 | 2.46 | 2.39 | 2.35 | 2.31 | 2.27 | 2.22 | 2.18 | 2.13 |
| 15 | 4.54 | 3.68 | 3.29 | 3.06 | 2.90 | 2.79 | 2.71 | 2.64 | 2.59 | 2.54 | 2.48 | 2.40 | 2.33 | 2.29 | 2.25 | 2.20 | 2.16 | 2.11 | 2.07 |
| 16 | 4.49 | 3.63 | 3.24 | 3.01 | 2.85 | 2.74 | 2.66 | 2.59 | 2.54 | 2.49 | 2.42 | 2.35 | 2.28 | 2.24 | 2.19 | 2.15 | 2.11 | 2.06 | 2.01 |
| 17 | 4.45 | 3.59 | 3.20 | 2.96 | 2.81 | 2.70 | 2.61 | 2.55 | 2.49 | 2.45 | 2.38 | 2.31 | 2.23 | 2.19 | 2.15 | 2.10 | 2.06 | 2.01 | 1.96 |
| 18 | 4.41 | 3.55 | 3.16 | 2.93 | 2.77 | 2.66 | 2.58 | 2.51 | 2.46 | 2.41 | 2.34 | 2.27 | 2.19 | 2.15 | 2.11 | 2.06 | 2.02 | 1.97 | 1.92 |
| 19 | 4.38 | 3.52 | 3.13 | 2.90 | 2.74 | 2.63 | 2.54 | 2.48 | 2.42 | 2.38 | 2.31 | 2.23 | 2.16 | 2.11 | 2.07 | 2.03 | 1.98 | 1.93 | 1.88 |
| 20 | 4.35 | 3.49 | 3.10 | 2.87 | 2.71 | 2.60 | 2.51 | 2.45 | 2.39 | 2.35 | 2.28 | 2.20 | 2.12 | 2.08 | 2.04 | 1.99 | 1.95 | 1.90 | 1.84 |
| 21 | 4.32 | 3.47 | 3.07 | 2.84 | 2.68 | 2.57 | 2.49 | 2.42 | 2.37 | 2.32 | 2.25 | 2.18 | 2.10 | 2.05 | 2.01 | 1.96 | 1.92 | 1.87 | 1.81 |
| 22 | 4.30 | 3.44 | 3.05 | 2.82 | 2.66 | 2.55 | 2.46 | 2.40 | 2.34 | 2.30 | 2.23 | 2.15 | 2.07 | 2.03 | 1.98 | 1.94 | 1.89 | 1.84 | 1.78 |
| 23 | 4.28 | 3.42 | 3.03 | 2.80 | 2.64 | 2.53 | 2.44 | 2.37 | 2.32 | 2.27 | 2.20 | 2.13 | 2.05 | 2.01 | 1.96 | 1.91 | 1.86 | 1.81 | 1.76 |
| 24 | 4.26 | 3.40 | 3.01 | 2.78 | 2.62 | 2.51 | 2.42 | 2.36 | 2.30 | 2.25 | 2.18 | 2.11 | 2.03 | 1.98 | 1.94 | 1.89 | 1.84 | 1.79 | 1.73 |
| 25 | 4.24 | 3.39 | 2.99 | 2.76 | 2.60 | 2.49 | 2.40 | 2.34 | 2.28 | 2.24 | 2.16 | 2.09 | 2.01 | 1.96 | 1.92 | 1.87 | 1.82 | 1.77 | 1.71 |
| 30 | 4.17 | 3.32 | 2.92 | 2.69 | 2.53 | 2.42 | 2.33 | 2.27 | 2.21 | 2.16 | 2.09 | 2.01 | 1.93 | 1.89 | 1.84 | 1.79 | 1.74 | 1.68 | 1.62 |
| 40 | 4.08 | 3.23 | 2.84 | 2.61 | 2.45 | 2.34 | 2.25 | 2.18 | 2.12 | 2.08 | 2.00 | 1.92 | 1.84 | 1.79 | 1.74 | 1.69 | 1.64 | 1.58 | 1.51 |
| 60 | 4.00 | 3.15 | 2.76 | 2.53 | 2.37 | 2.25 | 2.17 | 2.10 | 2.04 | 1.99 | 1.92 | 1.84 | 1.75 | 1.70 | 1.65 | 1.59 | 1.53 | 1.47 | 1.39 |
| 120 | 3.92 | 3.07 | 2.68 | 2.45 | 2.29 | 2.18 | 2.09 | 2.02 | 1.96 | 1.91 | 1.83 | 1.75 | 1.66 | 1.61 | 1.55 | 1.50 | 1.43 | 1.35 | 1.25 |
| ∞ | 3.84 | 3.00 | 2.60 | 2.37 | 2.21 | 2.10 | 2.01 | 1.94 | 1.88 | 1.83 | 1.75 | 1.67 | 1.57 | 1.52 | 1.46 | 1.39 | 1.32 | 1.22 | 1.00 |

Interpolation should be performed using reciprocals of the degrees of freedom.

By permission of Prof. E. S. Pearson from M. Merrington, C. M. Thompson, "Tables of percentage points of the inverted beta ($F$) distribution," *Biometrika*, vol. 33 (1943), p. 73.

DEGREES OF FREEDOM FOR NUMERATOR

| Degrees of freedom for denominator | 1 | 2 | 3 | 4 | 5 | 6 | 7 | 8 | 9 | 10 | 12 | 15 | 20 | 24 | 30 | 40 | 60 | 120 | ∞ |
|---|---|---|---|---|---|---|---|---|---|---|---|---|---|---|---|---|---|---|---|
| 1 | 4,052 | 5,000 | 5,403 | 5,625 | 5,764 | 5,859 | 5,928 | 5,982 | 6,023 | 6,056 | 6,106 | 6,157 | 6,209 | 6,235 | 6,261 | 6,287 | 6,313 | 6,339 | 6,366 |
| 2 | 98.5 | 99.0 | 99.2 | 99.2 | 99.3 | 99.3 | 99.4 | 99.4 | 99.4 | 99.4 | 99.4 | 99.4 | 99.4 | 99.5 | 99.5 | 99.5 | 99.5 | 99.5 | 99.5 |
| 3 | 34.1 | 30.8 | 29.5 | 28.7 | 28.2 | 27.9 | 27.7 | 27.5 | 27.3 | 27.2 | 27.1 | 26.9 | 26.7 | 26.6 | 26.5 | 26.4 | 26.3 | 26.2 | 26.1 |
| 4 | 21.2 | 18.0 | 16.7 | 16.0 | 15.5 | 15.2 | 15.0 | 14.8 | 14.7 | 14.5 | 14.4 | 14.2 | 14.0 | 13.9 | 13.8 | 13.7 | 13.7 | 13.6 | 13.5 |
| 5 | 16.3 | 13.3 | 12.1 | 11.4 | 11.0 | 10.7 | 10.5 | 10.3 | 10.2 | 10.1 | 9.89 | 9.72 | 9.55 | 9.47 | 9.38 | 9.29 | 9.20 | 9.11 | 9.02 |
| 6 | 13.7 | 10.9 | 9.78 | 9.15 | 8.75 | 8.47 | 8.26 | 8.10 | 7.98 | 7.87 | 7.72 | 7.56 | 7.40 | 7.31 | 7.23 | 7.14 | 7.06 | 6.97 | 6.88 |
| 7 | 12.2 | 9.55 | 8.45 | 7.85 | 7.46 | 7.19 | 6.99 | 6.84 | 6.72 | 6.62 | 6.47 | 6.31 | 6.16 | 6.07 | 5.99 | 5.91 | 5.82 | 5.74 | 5.65 |
| 8 | 11.3 | 8.65 | 7.59 | 7.01 | 6.63 | 6.37 | 6.18 | 6.03 | 5.91 | 5.81 | 5.67 | 5.52 | 5.36 | 5.28 | 5.20 | 5.12 | 5.03 | 4.95 | 4.86 |
| 9 | 10.6 | 8.02 | 6.99 | 6.42 | 6.06 | 5.80 | 5.61 | 5.47 | 5.35 | 5.26 | 5.11 | 4.96 | 4.81 | 4.73 | 4.65 | 4.57 | 4.48 | 4.40 | 4.31 |
| 10 | 10.0 | 7.56 | 6.55 | 5.99 | 5.64 | 5.39 | 5.20 | 5.06 | 4.94 | 4.85 | 4.71 | 4.56 | 4.41 | 4.33 | 4.25 | 4.17 | 4.08 | 4.00 | 3.91 |
| 11 | 9.65 | 7.21 | 6.22 | 5.67 | 5.32 | 5.07 | 4.89 | 4.74 | 4.63 | 4.54 | 4.40 | 4.25 | 4.10 | 4.02 | 3.94 | 3.86 | 3.78 | 3.69 | 3.60 |
| 12 | 9.33 | 6.93 | 5.95 | 5.41 | 5.06 | 4.82 | 4.64 | 4.50 | 4.39 | 4.30 | 4.16 | 4.01 | 3.86 | 3.78 | 3.70 | 3.62 | 3.54 | 3.45 | 3.36 |
| 13 | 9.07 | 6.70 | 5.74 | 5.21 | 4.86 | 4.62 | 4.44 | 4.30 | 4.19 | 4.10 | 3.96 | 3.82 | 3.66 | 3.59 | 3.51 | 3.43 | 3.34 | 3.25 | 3.17 |
| 14 | 8.86 | 6.51 | 5.56 | 5.04 | 4.70 | 4.46 | 4.28 | 4.14 | 4.03 | 3.94 | 3.80 | 3.66 | 3.51 | 3.43 | 3.35 | 3.27 | 3.18 | 3.09 | 3.00 |
| 15 | 8.68 | 6.36 | 5.42 | 4.89 | 4.56 | 4.32 | 4.14 | 4.00 | 3.89 | 3.80 | 3.67 | 3.52 | 3.37 | 3.29 | 3.21 | 3.13 | 3.05 | 2.96 | 2.87 |
| 16 | 8.53 | 6.23 | 5.29 | 4.77 | 4.44 | 4.20 | 4.03 | 3.89 | 3.78 | 3.69 | 3.55 | 3.41 | 3.26 | 3.18 | 3.10 | 3.02 | 2.93 | 2.84 | 2.75 |
| 17 | 8.40 | 6.11 | 5.19 | 4.67 | 4.34 | 4.10 | 3.93 | 3.79 | 3.68 | 3.59 | 3.46 | 3.31 | 3.16 | 3.08 | 3.00 | 2.92 | 2.83 | 2.75 | 2.65 |
| 18 | 8.29 | 6.01 | 5.09 | 4.58 | 4.25 | 4.01 | 3.84 | 3.71 | 3.60 | 3.51 | 3.37 | 3.23 | 3.08 | 3.00 | 2.92 | 2.84 | 2.75 | 2.66 | 2.57 |
| 19 | 8.19 | 5.93 | 5.01 | 4.50 | 4.17 | 3.94 | 3.77 | 3.63 | 3.52 | 3.43 | 3.30 | 3.15 | 3.00 | 2.92 | 2.84 | 2.76 | 2.67 | 2.58 | 2.49 |
| 20 | 8.10 | 5.85 | 4.94 | 4.43 | 4.10 | 3.87 | 3.70 | 3.56 | 3.46 | 3.37 | 3.23 | 3.09 | 2.94 | 2.86 | 2.78 | 2.69 | 2.61 | 2.52 | 2.42 |
| 21 | 8.02 | 5.78 | 4.87 | 4.37 | 4.04 | 3.81 | 3.64 | 3.51 | 3.40 | 3.31 | 3.17 | 3.03 | 2.88 | 2.80 | 2.72 | 2.64 | 2.55 | 2.46 | 2.36 |
| 22 | 7.95 | 5.72 | 4.82 | 4.31 | 3.99 | 3.76 | 3.59 | 3.45 | 3.35 | 3.26 | 3.12 | 2.98 | 2.83 | 2.75 | 2.67 | 2.58 | 2.50 | 2.40 | 2.31 |
| 23 | 7.88 | 5.66 | 4.76 | 4.26 | 3.94 | 3.71 | 3.54 | 3.41 | 3.30 | 3.21 | 3.07 | 2.93 | 2.78 | 2.70 | 2.62 | 2.54 | 2.45 | 2.35 | 2.26 |
| 24 | 7.82 | 5.61 | 4.72 | 4.22 | 3.90 | 3.67 | 3.50 | 3.36 | 3.26 | 3.17 | 3.03 | 2.89 | 2.74 | 2.66 | 2.58 | 2.49 | 2.40 | 2.31 | 2.21 |
| 25 | 7.77 | 5.57 | 4.68 | 4.18 | 3.86 | 3.63 | 3.46 | 3.32 | 3.22 | 3.13 | 2.99 | 2.85 | 2.70 | 2.62 | 2.53 | 2.45 | 2.36 | 2.27 | 2.17 |
| 30 | 7.56 | 5.39 | 4.51 | 4.02 | 3.70 | 3.47 | 3.30 | 3.17 | 3.07 | 2.98 | 2.84 | 2.70 | 2.55 | 2.47 | 2.39 | 2.30 | 2.21 | 2.11 | 2.01 |
| 40 | 7.31 | 5.18 | 4.31 | 3.83 | 3.51 | 3.29 | 3.12 | 2.99 | 2.89 | 2.80 | 2.66 | 2.52 | 2.37 | 2.29 | 2.20 | 2.11 | 2.02 | 1.92 | 1.80 |
| 60 | 7.08 | 4.98 | 4.13 | 3.65 | 3.34 | 3.12 | 2.95 | 2.82 | 2.72 | 2.63 | 2.50 | 2.35 | 2.20 | 2.12 | 2.03 | 1.94 | 1.84 | 1.73 | 1.60 |
| 120 | 6.85 | 4.79 | 3.95 | 3.48 | 3.17 | 2.96 | 2.79 | 2.66 | 2.56 | 2.47 | 2.34 | 2.19 | 2.03 | 1.95 | 1.86 | 1.76 | 1.66 | 1.53 | 1.38 |
| ∞ | 6.63 | 4.61 | 3.78 | 3.32 | 3.02 | 2.80 | 2.64 | 2.51 | 2.41 | 2.32 | 2.18 | 2.04 | 1.88 | 1.79 | 1.70 | 1.59 | 1.47 | 1.32 | 1.00 |

Interpolation should be performed using reciprocals of the degrees of freedom.

By permission of Prof. E. S. Pearson from M. Merrington, C. M. Thompson, "Tables of percentage points of the inverted beta ($F$) distribution," *Biometrika*, vol. 33 (1943), p. 73.

## A-7c PERCENTILES OF THE $F(\nu_1, \nu_2)$ DISTRIBUTION

| $\nu_2$, DEGREES OF FREEDOM FOR DENOMINATOR | | $\nu_1$, DEGREES OF FREEDOM FOR NUMERATOR | | | | | | | | | | | | |
|---|---|---|---|---|---|---|---|---|---|---|---|---|---|---|
| | *Cum. prop.* | 1 | 2 | 3 | 4 | 5 | 6 | 7 | 8 | 9 | 10 | 11 | 12 | *Cum. prop.* |
| 1 | .0005 | $.0^{6}62$ | $.0^{3}50$ | $.0^{2}38$ | $.0^{2}94$ | .016 | .022 | .027 | .032 | .036 | .039 | .042 | .045 | .0005 |
| | .001 | $.0^{5}25$ | $.0^{2}10$ | $.0^{2}60$ | .013 | .021 | .028 | .034 | .039 | .044 | .048 | .051 | .054 | .001 |
| | .005 | $.0^{4}62$ | $.0^{2}51$ | .018 | .032 | .044 | .054 | .062 | .068 | .073 | .078 | .082 | .085 | .005 |
| | .010 | $.0^{3}25$ | .010 | .029 | .047 | .062 | .073 | .082 | .089 | .095 | .100 | .104 | .107 | .010 |
| | .025 | $.0^{2}15$ | .026 | .057 | .082 | .100 | .113 | .124 | .132 | .139 | .144 | .149 | .153 | .025 |
| | .05 | $.0^{2}62$ | .054 | .099 | .130 | .151 | .167 | .179 | .188 | .195 | .201 | .207 | .211 | .05 |
| | .10 | .025 | .117 | .181 | .220 | .246 | .265 | .279 | .289 | .298 | .304 | .310 | .315 | .10 |
| | .25 | .172 | .389 | .494 | .553 | .591 | .617 | .637 | .650 | .661 | .670 | .680 | .684 | .25 |
| | .50 | 1.00 | 1.50 | 1.71 | 1.82 | 1.89 | 1.94 | 1.98 | 2.00 | 2.03 | 2.04 | 2.05 | 2.07 | .50 |
| | .75 | 5.83 | 7.50 | 8.20 | 8.58 | 8.82 | 8.98 | 9.10 | 9.19 | 9.26 | 9.32 | 9.36 | 9.41 | .75 |
| | .90 | 39.9 | 49.5 | 53.6 | 55.8 | 57.2 | 58.2 | 58.9 | 59.4 | 59.9 | 60.2 | 60.5 | 60.7 | .90 |
| | .95 | 161 | 200 | 216 | 225 | 230 | 234 | 237 | 239 | 241 | 242 | 243 | 244 | .95 |
| | .975 | 648 | 800 | 864 | 900 | 922 | 937 | 948 | 957 | 963 | 969 | 973 | 977 | .975 |
| | .99 | $405^{1}$ | $500^{1}$ | $540^{1}$ | $562^{1}$ | $576^{1}$ | $586^{1}$ | $593^{1}$ | $598^{1}$ | $602^{1}$ | $606^{1}$ | $608^{1}$ | $611^{1}$ | .99 |
| | .995 | $162^{2}$ | $200^{2}$ | $216^{2}$ | $225^{2}$ | $231^{2}$ | $234^{2}$ | $237^{2}$ | $239^{2}$ | $241^{2}$ | $242^{2}$ | $243^{2}$ | $244^{2}$ | .995 |
| | .999 | $406^{3}$ | $500^{3}$ | $540^{3}$ | $562^{3}$ | $576^{3}$ | $586^{3}$ | $593^{3}$ | $598^{3}$ | $602^{3}$ | $606^{3}$ | $609^{3}$ | $611^{3}$ | .999 |
| | .9995 | $162^{4}$ | $200^{4}$ | $216^{4}$ | $225^{4}$ | $231^{4}$ | $234^{4}$ | $237^{4}$ | $239^{4}$ | $241^{4}$ | $242^{4}$ | $243^{4}$ | $244^{4}$ | .9995 |
| 2 | .0005 | $.0^{6}50$ | $.0^{3}50$ | $.0^{2}42$ | .011 | .020 | .029 | .037 | .044 | .050 | .056 | .061 | .065 | .0005 |
| | .001 | $.0^{5}20$ | $.0^{2}10$ | $.0^{2}68$ | .016 | .027 | .037 | .046 | .054 | .061 | .067 | .072 | .077 | .001 |
| | .005 | $.0^{4}50$ | $.0^{2}50$ | .020 | .038 | .055 | .069 | .081 | .091 | .099 | .106 | .112 | .118 | .005 |
| | .01 | $.0^{3}20$ | .010 | .032 | .056 | .075 | .092 | .105 | .116 | .125 | .132 | .139 | .144 | .01 |
| | .025 | $.0^{2}13$ | .026 | .062 | .094 | .119 | .138 | .153 | .165 | .175 | .183 | .190 | .196 | .025 |
| | .05 | $.0^{2}50$ | .053 | .105 | .144 | .173 | .194 | .211 | .224 | .235 | .244 | .251 | .257 | .05 |
| | .10 | .020 | .111 | .183 | .231 | .265 | .289 | .307 | .321 | .333 | .342 | .350 | .356 | .10 |
| | .25 | .133 | .333 | .439 | .500 | .540 | .568 | .588 | .604 | .616 | .626 | .633 | .641 | .25 |
| | .50 | .667 | 1.00 | 1.13 | 1.21 | 1.25 | 1.28 | 1.30 | 1.32 | 1.33 | 1.34 | 1.35 | 1.36 | .50 |
| | .75 | 2.57 | 3.00 | 3.15 | 3.23 | 3.28 | 3.31 | 3.34 | 3.35 | 3.37 | 3.38 | 3.39 | 3.39 | .75 |
| | .90 | 8.53 | 9.00 | 9.16 | 9.24 | 9.29 | 9.33 | 9.35 | 9.37 | 9.38 | 9.39 | 9.40 | 9.41 | .90 |
| | .95 | 18.5 | 19.0 | 19.2 | 19.2 | 19.3 | 19.3 | 19.4 | 19.4 | 19.4 | 19.4 | 19.4 | 19.4 | .95 |
| | .975 | 38.5 | 39.0 | 39.2 | 39.2 | 39.3 | 39.3 | 39.4 | 39.4 | 39.4 | 39.4 | 39.4 | 39.4 | .975 |
| | .99 | 98.5 | 99.0 | 99.2 | 99.2 | 99.3 | 99.3 | 99.4 | 99.4 | 99.4 | 99.4 | 99.4 | 99.4 | .99 |
| | .995 | 198 | 199 | 199 | 199 | 199 | 199 | 199 | 199 | 199 | 199 | 199 | 199 | .995 |
| | .999 | 998 | 999 | 999 | 999 | 999 | 999 | 999 | 999 | 999 | 999 | 999 | 999 | .999 |
| | .9995 | $200^{1}$ | $200^{1}$ | $200^{1}$ | $200^{1}$ | $200^{1}$ | $200^{1}$ | $200^{1}$ | $200^{1}$ | $200^{1}$ | $200^{1}$ | $200^{1}$ | $200^{1}$ | .9995 |
| 3 | .0005 | $.0^{6}46$ | $.0^{3}50$ | $.0^{2}44$ | .012 | .023 | .033 | .043 | .052 | .060 | .067 | .074 | .079 | .0005 |
| | .001 | $.0^{5}19$ | $.0^{2}10$ | $.0^{2}71$ | .018 | .030 | .042 | .053 | .063 | .072 | .079 | .086 | .093 | .001 |
| | .005 | $.0^{4}46$ | $.0^{2}50$ | .021 | .041 | .060 | .077 | .092 | .104 | .115 | .124 | .132 | .138 | .005 |
| | .01 | $.0^{3}19$ | .010 | .034 | .060 | .083 | .102 | .118 | .132 | .143 | .153 | .161 | .168 | .01 |
| | .025 | $.0^{2}12$ | .026 | .065 | .100 | .129 | .152 | .170 | .185 | .197 | .207 | .216 | .224 | .025 |
| | .05 | $.0^{2}46$ | .052 | .108 | .152 | .185 | .210 | .230 | .246 | .259 | .270 | .279 | .287 | .05 |
| | .10 | .019 | .109 | .185 | .239 | .276 | .304 | .325 | .342 | .356 | .367 | .376 | .384 | .10 |
| | .25 | .122 | .317 | .424 | .489 | .531 | .561 | .582 | .600 | .613 | .624 | .633 | .641 | .25 |
| | .50 | .585 | .881 | 1.00 | 1.06 | 1.10 | 1.13 | 1.15 | 1.16 | 1.17 | 1.18 | 1.19 | 1.20 | .50 |
| | .75 | 2.02 | 2.28 | 2.36 | 2.39 | 2.41 | 2.42 | 2.43 | 2.44 | 2.44 | 2.44 | 2.45 | 2.45 | .75 |
| | .90 | 5.54 | 5.46 | 5.39 | 5.34 | 5.31 | 5.28 | 5.27 | 5.25 | 5.24 | 5.23 | 5.22 | 5.22 | .90 |
| | .95 | 10.1 | 9.55 | 9.28 | 9.12 | 9.01 | 8.94 | 8.89 | 8.85 | 8.81 | 8.79 | 8.76 | 8.74 | .95 |
| | .975 | 17.4 | 16.0 | 15.4 | 15.1 | 14.9 | 14.7 | 14.6 | 14.5 | 14.5 | 14.4 | 14.4 | 14.3 | .975 |
| | .99 | 34.1 | 30.8 | 29.5 | 28.7 | 28.2 | 27.9 | 27.7 | 27.5 | 27.3 | 27.2 | 27.1 | 27.1 | .99 |
| | .995 | 55.6 | 49.8 | 47.5 | 46.2 | 45.4 | 44.8 | 44.4 | 44.1 | 43.9 | 43.7 | 43.5 | 43.4 | .995 |
| | .999 | 167 | 149 | 141 | 137 | 135 | 133 | 132 | 131 | 130 | 129 | 129 | 128 | .999 |
| | .9995 | 266 | 237 | 225 | 218 | 214 | 211 | 209 | 208 | 207 | 206 | 204 | 204 | .9995 |

Read $.0^{3}56$ as .00056, $200^{1}$ as 2,000, $162^{4}$ as 1,620,000, and so on.

$\nu_1$, DEGREES OF FREEDOM FOR NUMERATOR

| Cum. prop. | 15 | 20 | 24 | 30 | 40 | 50 | 60 | 100 | 120 | 200 | 500 | ∞ | Cum. prop. | $\nu_2$, DEGREES OF FREEDOM FOR DENOMINATOR |
|---|---|---|---|---|---|---|---|---|---|---|---|---|---|---|
| .0005 | .051 | .058 | 062 | .066 | .069 | .072 | .074 | .077 | .078 | .080 | .081 | .083 | .0005 | 1 |
| .001 | .060 | .067 | .071 | .075 | .079 | .082 | .084 | .087 | .088 | .089 | .091 | .092 | .001 | |
| .005 | .093 | .101 | .105 | .109 | .113 | .116 | .118 | .121 | .122 | .124 | .126 | .127 | .005 | |
| .01 | .115 | .124 | .128 | .132 | .137 | .139 | .141 | .145 | .146 | .148 | .150 | .151 | .01 | |
| .025 | .161 | .170 | .175 | .180 | .184 | .187 | .189 | .193 | .194 | .196 | .198 | .199 | .025 | |
| .05 | .220 | .230 | .235 | .240 | .245 | .248 | .250 | .254 | .255 | .257 | .259 | .261 | .05 | |
| .10 | .325 | .336 | .342 | .347 | .353 | .356 | .358 | .362 | .364 | .366 | .368 | .370 | .10 | |
| .25 | .698 | .712 | .719 | .727 | .734 | .738 | .741 | .747 | .749 | .752 | .754 | .756 | .25 | |
| .50 | 2.09 | 2.12 | 2.13 | 2.15 | 2.16 | 2.17 | 2.17 | 2.18 | 2.18 | 2.19 | 2.19 | 2.20 | .50 | |
| .75 | 9.49 | 9.58 | 9.63 | 9.67 | 9.71 | 9.74 | 9.76 | 9.78 | 9.80 | 9.82 | 9.84 | 9.85 | .75 | |
| .90 | 61.2 | 61.7 | 62.0 | 62.3 | 62.5 | 62.7 | 62.8 | 63.0 | 63.1 | 63.2 | 63.3 | 63.3 | .90 | |
| .95 | 246 | 248 | 249 | 250 | 251 | 252 | 252 | 253 | 253 | 254 | 254 | 254 | .95 | |
| .975 | 985 | 993 | 997 | $100^{1}$ | $101^{1}$ | $101^{1}$ | $101^{1}$ | $101^{1}$ | $101^{1}$ | $102^{1}$ | $102^{1}$ | $102^{1}$ | .975 | |
| .99 | $616^{1}$ | $621^{1}$ | $623^{1}$ | $626^{1}$ | $629^{1}$ | $630^{1}$ | $631^{1}$ | $633^{1}$ | $634^{1}$ | $635^{1}$ | $636^{1}$ | $637^{1}$ | .99 | |
| .995 | $246^{2}$ | $248^{2}$ | $249^{2}$ | $250^{2}$ | $251^{2}$ | $252^{2}$ | $253^{2}$ | $253^{2}$ | $254^{2}$ | $254^{2}$ | $254^{2}$ | $255^{2}$ | .995 | |
| .999 | $616^{3}$ | $621^{3}$ | $623^{3}$ | $626^{3}$ | $629^{3}$ | $630^{3}$ | $631^{3}$ | $633^{3}$ | $634^{3}$ | $635^{3}$ | $636^{3}$ | $637^{3}$ | .999 | |
| .9995 | $246^{4}$ | $248^{4}$ | $249^{4}$ | $250^{4}$ | $251^{4}$ | $252^{4}$ | $252^{4}$ | $253^{4}$ | $253^{4}$ | $253^{4}$ | $254^{4}$ | $254^{4}$ | .9995 | |
| .0005 | .076 | .088 | .094 | .101 | .108 | .113 | .116 | .122 | .124 | .127 | .130 | .132 | .0005 | 2 |
| .001 | .088 | .100 | .107 | .114 | .121 | .126 | .129 | .135 | .137 | .140 | .143 | .145 | .001 | |
| .005 | .130 | .143 | .150 | .157 | .165 | .169 | .173 | .179 | .181 | .184 | .187 | .189 | .005 | |
| .01 | .157 | .171 | .178 | .186 | .193 | .198 | .201 | .207 | .209 | .212 | .215 | .217 | .01 | |
| .025 | .210 | .224 | .232 | .239 | .247 | .251 | .255 | .261 | .263 | .266 | .269 | .271 | .025 | |
| .05 | .272 | .286 | .294 | .302 | .309 | .314 | .317 | .324 | .326 | .329 | .332 | .334 | .05 | |
| .10 | .371 | .386 | .394 | .402 | .410 | .415 | .418 | .424 | .426 | .429 | .433 | .434 | .10 | |
| .25 | .657 | .672 | .680 | .689 | .697 | .702 | .705 | .711 | .713 | .716 | .719 | .721 | .25 | |
| .50 | 1.38 | 1.39 | 1.40 | 1.41 | 1.42 | 1.42 | 1.43 | 1.43 | 1.43 | 1.44 | 1.44 | 1.44 | .50 | |
| .75 | 3.41 | 3.43 | 3.43 | 3.44 | 3.45 | 3.45 | 3.46 | 3.47 | 3.47 | 3.48 | 3.48 | 3.48 | .75 | |
| .90 | 9.42 | 9.44 | 9.45 | 9.46 | 9.47 | 9.47 | 9.47 | 9.48 | 9.48 | 9.49 | 9.49 | 9.49 | .90 | |
| .95 | 19.4 | 19.4 | 19.5 | 19.5 | 19.5 | 19.5 | 19.5 | 19.5 | 19.5 | 19.5 | 19.5 | 19.5 | .95 | |
| .975 | 39.4 | 39.4 | 39.5 | 39.5 | 39.5 | 39.5 | 39.5 | 39.5 | 39.5 | 39.5 | 39.5 | 39.5 | .975 | |
| .99 | 99.4 | 99.4 | 99.5 | 99.5 | 99.5 | 99.5 | 99.5 | 99.5 | 99.5 | 99.5 | 99.5 | 99.5 | .99 | |
| .995 | 199 | 199 | 199 | 199 | 199 | 199 | 199 | 199 | 199 | 199 | 199 | 200 | .995 | |
| .999 | 999 | 999 | 999 | 999 | 999 | 999 | 999 | 999 | 999 | 999 | 999 | 999 | .999 | |
| .9995 | $200^{1}$ | $200^{1}$ | $200^{1}$ | $200^{1}$ | $200^{1}$ | $200^{1}$ | $200^{1}$ | $200^{1}$ | $200^{1}$ | $200^{1}$ | $200^{1}$ | $200^{1}$ | .9995 | |
| .0005 | .093 | .109 | .117 | .127 | .136 | .143 | .147 | .156 | .158 | .162 | .166 | .169 | .0005 | 3 |
| .001 | .107 | .123 | .132 | .142 | .152 | .158 | .162 | .171 | .173 | .177 | .181 | .184 | .001 | |
| .005 | .154 | .172 | .181 | .191 | .201 | .207 | .211 | .220 | .222 | .227 | .231 | .234 | .005 | |
| .01 | .185 | .203 | .212 | .222 | .232 | .238 | .242 | .251 | .253 | .258 | .262 | .264 | .01 | |
| .025 | .241 | .259 | .269 | .279 | .289 | .295 | .299 | .308 | .310 | .314 | .318 | .321 | .025 | |
| .05 | .304 | .323 | .332 | .342 | .352 | .358 | .363 | .370 | .373 | .377 | .382 | .384 | .05 | |
| .10 | .402 | .420 | .430 | .439 | .449 | .455 | .459 | .467 | .469 | .474 | .476 | .480 | .10 | |
| .25 | .658 | .675 | .684 | .693 | .702 | .708 | .711 | .719 | .721 | .724 | .728 | .730 | .25 | |
| .50 | 1.21 | 1.23 | 1.23 | 1.24 | 1.25 | 1.25 | 1.25 | 1.26 | 1.26 | 1.26 | 1.27 | 1.27 | .50 | |
| .75 | 2.46 | 2.46 | 2.46 | 2.47 | 2.47 | 2.47 | 2.47 | 2.47 | 2.47 | 2.47 | 2.47 | 2.47 | .75 | |
| .90 | 5.20 | 5.18 | 5.18 | 5.17 | 5.16 | 5.15 | 5.15 | 5.14 | 5.14 | 5.14 | 5.14 | 5.13 | .90 | |
| .95 | 8.70 | 8.66 | 8.63 | 8.62 | 8.59 | 8.58 | 8.57 | 8.55 | 8.55 | 8.54 | 8.53 | 8.53 | .95 | |
| .975 | 14.3 | 14.2 | 14.1 | 14.1 | 14.0 | 14.0 | 14.0 | 14.0 | 13.9 | 13.9 | 13.9 | 13.9 | .975 | |
| .99 | 26.9 | 26.7 | 26.6 | 26.5 | 26.4 | 26.4 | 26.3 | 26.2 | 26.2 | 26.2 | 26.1 | 26.1 | .99 | |
| .995 | 43.1 | 42.8 | 42.6 | 42.5 | 42.3 | 42.2 | 42.1 | 42.0 | 42.0 | 41.9 | 41.9 | 41.8 | .995 | |
| .999 | 127 | 126 | 126 | 125 | 125 | 125 | 124 | 124 | 124 | 124 | 124 | 123 | .999 | |
| .9995 | 203 | 201 | 200 | 199 | 199 | 198 | 198 | 197 | 197 | 197 | 196 | 196 | .9995 | |

## A-7c PERCENTILES OF THE $F(\nu_1, \nu_2)$ DISTRIBUTIONS (CONTINUED)

$\nu_1$, DEGREES OF FREEDOM FOR NUMERATOR

| $\nu_2$, degrees of freedom for denominator | Cum. prop. | 1 | 2 | 3 | 4 | 5 | 6 | 7 | 8 | 9 | 10 | 11 | 12 | Cum. prop. |
|---|---|---|---|---|---|---|---|---|---|---|---|---|---|---|
| **4** | .0005 | $.0^{6}44$ | $.0^{3}50$ | $.0^{2}46$ | .013 | .024 | .036 | .047 | .057 | .066 | .075 | .082 | .089 | .0005 |
| | .001 | $.0^{5}18$ | $.0^{2}10$ | $.0^{2}73$ | .019 | .032 | .046 | .058 | .069 | .079 | .089 | .097 | .104 | .001 |
| | .005 | $.0^{4}44$ | $.0^{2}50$ | .022 | .043 | .064 | .083 | .100 | .114 | .126 | .137 | .145 | .153 | .005 |
| | .01 | $.0^{3}18$ | .010 | .035 | .063 | .088 | .109 | .127 | .143 | .156 | .167 | .176 | .185 | .01 |
| | .025 | $.0^{2}11$ | .026 | .066 | .104 | .135 | .161 | .181 | .198 | .212 | .224 | .234 | .243 | .025 |
| | .05 | $.0^{2}44$ | .052 | .110 | .157 | .193 | .221 | .243 | .261 | .275 | .288 | .298 | .307 | .05 |
| | .10 | .018 | .108 | .187 | .243 | .284 | .314 | .338 | .356 | .371 | .384 | .394 | .403 | .10 |
| | .25 | .117 | .309 | .418 | .484 | .528 | .560 | .583 | .601 | .615 | .627 | .637 | .645 | .25 |
| | .50 | .549 | .828 | .941 | 1.00 | 1.04 | 1.06 | 1.08 | 1.09 | 1.10 | 1.11 | 1.12 | 1.13 | .50 |
| | .75 | 1.81 | 2.00 | 2.05 | 2.06 | 2.07 | 2.08 | 2.08 | 2.08 | 2.08 | 2.08 | 2.08 | 2.08 | .75 |
| | .90 | 4.54 | 4.32 | 4.19 | 4.11 | 4.05 | 4.01 | 3.98 | 3.95 | 3.94 | 3.92 | 3.91 | 3.90 | .90 |
| | .95 | 7.71 | 6.94 | 6.59 | 6.39 | 6.26 | 6.16 | 6.09 | 6.04 | 6.00 | 5.96 | 5.94 | 5.91 | .95 |
| | .975 | 12.2 | 10.6 | 9.98 | 9.60 | 9.36 | 9.20 | 9.07 | 8.98 | 8.90 | 8.84 | 8.79 | 8.75 | .975 |
| | .99 | 21.2 | 18.0 | 16.7 | 16.0 | 15.5 | 15.2 | 15.0 | 14.8 | 14.7 | 14.5 | 14.4 | 14.4 | .99 |
| | .995 | 31.3 | 26.3 | 24.3 | 23.2 | 22.5 | 22.0 | 21.6 | 21.4 | 21.1 | 21.0 | 20.8 | 20.7 | .995 |
| | .999 | 74.1 | 61.2 | 56.2 | 53.4 | 51.7 | 50.5 | 49.7 | 49.0 | 48.5 | 48.0 | 47.7 | 47.4 | .999 |
| | .9995 | 106 | 87.4 | 80.1 | 76.1 | 73.6 | 71.9 | 70.6 | 69.7 | 68.9 | 68.3 | 67.8 | 67.4 | .9995 |
| **5** | .0005 | $.0^{6}43$ | $.0^{3}50$ | $.0^{2}47$ | .014 | .025 | .038 | .050 | .061 | .070 | .081 | .089 | .096 | .0005 |
| | .001 | $.0^{5}17$ | $.0^{2}10$ | $.0^{2}75$ | .019 | .034 | .048 | .062 | .074 | .085 | .095 | .104 | .112 | .001 |
| | .005 | $.0^{4}43$ | $.0^{2}50$ | .022 | .045 | .067 | .087 | .105 | .120 | .134 | .146 | .156 | .165 | .005 |
| | .01 | $.0^{3}17$ | .010 | .035 | .064 | .091 | .114 | .134 | .151 | .165 | .177 | .188 | .197 | .01 |
| | .025 | $.0^{2}11$ | .025 | .067 | .107 | .140 | .167 | .189 | .208 | .223 | .236 | .248 | .257 | .025 |
| | .05 | $.0^{2}43$ | .052 | .111 | .160 | .198 | .228 | .252 | .271 | .287 | .301 | .313 | .322 | .05 |
| | .10 | .017 | .108 | .188 | .247 | .290 | .322 | .347 | .367 | .383 | .397 | .408 | .418 | .10 |
| | .25 | .113 | .305 | .415 | .483 | .528 | .560 | .584 | .604 | .618 | .631 | .641 | .650 | .25 |
| | .50 | .528 | .799 | .907 | .965 | 1.00 | 1.02 | 1.04 | 1.05 | 1.06 | 1.07 | 1.08 | 1.09 | .50 |
| | .75 | 1.69 | 1.85 | 1.88 | 1.89 | 1.89 | 1.89 | 1.89 | 1.89 | 1.89 | 1.89 | 1.89 | 1.89 | .75 |
| | .90 | 4.06 | 3.78 | 3.62 | 3.52 | 3.45 | 3.40 | 3.37 | 3.34 | 3.32 | 3.30 | 3.28 | 3.27 | .90 |
| | .95 | 6.61 | 5.79 | 5.41 | 5.19 | 5.05 | 4.95 | 4.88 | 4.82 | 4.77 | 4.74 | 4.71 | 4.68 | .95 |
| | .975 | 10.0 | 8.43 | 7.76 | 7.39 | 7.15 | 6.98 | 6.85 | 6.76 | 6.68 | 6.62 | 6.57 | 6.52 | .975 |
| | .99 | 16.3 | 13.3 | 12.1 | 11.4 | 11.0 | 10.7 | 10.5 | 10.3 | 10.2 | 10.1 | 9.96 | 9.89 | .99 |
| | .995 | 22.8 | 18.3 | 16.5 | 15.6 | 14.9 | 14.5 | 14.2 | 14.0 | 13.8 | 13.6 | 13.5 | 13.4 | .995 |
| | .999 | 47.2 | 37.1 | 33.2 | 31.1 | 29.7 | 28.8 | 28.2 | 27.6 | 27.2 | 26.9 | 26.6 | 26.4 | .999 |
| | .9995 | 63.6 | 49.8 | 44.4 | 41.5 | 39.7 | 38.5 | 37.6 | 36.9 | 36.4 | 35.9 | 35.6 | 35.2 | .9995 |
| **6** | .0005 | $.0^{6}43$ | $.0^{3}50$ | $.0^{2}47$ | .014 | .026 | .039 | .052 | .064 | .075 | .085 | .094 | .103 | .0005 |
| | .001 | $.0^{5}17$ | $.0^{2}10$ | $.0^{2}75$ | .020 | .035 | .050 | .064 | .078 | .090 | .101 | .111 | .119 | .001 |
| | .005 | $.0^{4}43$ | $.0^{2}50$ | .022 | .045 | .069 | .090 | .109 | .126 | .140 | .153 | .164 | .174 | .005 |
| | .01 | $.0^{3}17$ | .010 | .036 | .066 | .094 | .118 | .139 | .157 | .172 | .186 | .197 | .207 | .01 |
| | .025 | $.0^{2}11$ | .025 | .068 | .109 | .143 | .172 | .195 | .215 | .231 | .246 | .258 | .268 | .025 |
| | .05 | $.0^{2}43$ | .052 | .112 | .162 | .202 | .233 | .259 | .279 | .296 | .311 | .324 | .334 | .05 |
| | .10 | .017 | .107 | .189 | .249 | .294 | .327 | .354 | .375 | .392 | .406 | .418 | .429 | .10 |
| | .25 | .111 | .302 | .413 | .481 | .524 | .561 | .586 | .606 | .622 | .635 | .645 | .654 | .25 |
| | .50 | .515 | .780 | .886 | .942 | .977 | 1.00 | 1.02 | 1.03 | 1.04 | 1.05 | 1.05 | 1.06 | .50 |
| | .75 | 1.62 | 1.76 | 1.78 | 1.79 | 1.79 | 1.78 | 1.78 | 1.78 | 1.77 | 1.77 | 1.77 | 1.77 | .75 |
| | .90 | 3.78 | 3.46 | 3.29 | 3.18 | 3.11 | 3.05 | 3.01 | 2.98 | 2.96 | 2.94 | 2.92 | 2.90 | .90 |
| | .95 | 5.99 | 5.14 | 4.76 | 4.53 | 4.39 | 4.28 | 4.21 | 4.15 | 4.10 | 4.06 | 4.03 | 4.00 | .95 |
| | .975 | 8.81 | 7.26 | 6.60 | 6.23 | 5.99 | 5.82 | 5.70 | 5.60 | 5.52 | 5.46 | 5.41 | 5.37 | .975 |
| | .99 | 13.7 | 10.9 | 9.78 | 9.15 | 8.75 | 8.47 | 8.26 | 8.10 | 7.98 | 7.87 | 7.79 | 7.72 | .99 |
| | .995 | 18.6 | 14.5 | 12.9 | 12.0 | 11.5 | 11.1 | 10.8 | 10.6 | 10.4 | 10.2 | 10.1 | 10.0 | .995 |
| | .999 | 35.5 | 27.0 | 23.7 | 21.9 | 20.8 | 20.0 | 19.5 | 19.0 | 18.7 | 18.4 | 18.2 | 18.0 | .999 |
| | .9995 | 46.1 | 34.8 | 30.4 | 28.1 | 26.6 | 25.6 | 24.9 | 24.3 | 23.9 | 23.5 | 23.2 | 23.0 | .9995 |

$\nu_1$, DEGREES OF FREEDOM FOR NUMERATOR

| *Cum. prop.* | 15 | 20 | 24 | 30 | 40 | 50 | 60 | 100 | 120 | 200 | 500 | $\infty$ | *Cum. prop.* | $\nu_2$, DEGREES OF FREEDOM FOR DENOMINATOR |
|---|---|---|---|---|---|---|---|---|---|---|---|---|---|---|
| .0005 | .105 | .125 | .135 | .147 | .159 | .166 | .172 | .183 | .186 | .191 | .196 | .200 | .0005 | **4** |
| .001 | .121 | .141 | .152 | .163 | .176 | .183 | .188 | .200 | .202 | .208 | .213 | .217 | .001 | |
| .005 | .172 | .193 | .204 | .216 | .229 | .237 | .242 | .253 | .255 | .260 | .266 | .269 | .005 | |
| .01 | .204 | .226 | .237 | .249 | .261 | .269 | .274 | .285 | .287 | .293 | .298 | .301 | .01 | |
| .025 | .263 | .284 | .296 | .308 | .320 | .327 | .332 | .342 | .346 | .351 | .356 | .359 | .025 | |
| .05 | .327 | .349 | .360 | .372 | .384 | .391 | .396 | .407 | .409 | .413 | .418 | .422 | .05 | |
| .10 | .424 | .445 | .456 | .467 | .478 | .485 | .490 | .500 | .502 | .508 | .510 | .514 | .10 | |
| .25 | .664 | .683 | .692 | .702 | .712 | .718 | .722 | .731 | .733 | .737 | .740 | .743 | .25 | |
| .50 | 1.14 | 1.15 | 1.16 | 1.16 | 1.17 | 1.18 | 1.18 | 1.18 | 1.18 | 1.19 | 1.19 | 1.19 | .50 | |
| .75 | 2.08 | 2.08 | 2.08 | 2.08 | 2.08 | 2.08 | 2.08 | 2.08 | 2.08 | 2.08 | 2.08 | 2.08 | .75 | |
| .90 | 3.87 | 3.84 | 3.83 | 3.82 | 3.80 | 3.80 | 3.79 | 3.78 | 3.78 | 3.77 | 3.76 | 3.76 | .90 | |
| .95 | 5.86 | 5.80 | 5.77 | 5.75 | 5.72 | 5.70 | 5.69 | 5.66 | 5.66 | 5.65 | 5.64 | 5.63 | .95 | |
| .975 | 8.66 | 8.56 | 8.51 | 8.46 | 8.41 | 8.38 | 8.36 | 8.32 | 8.31 | 8.29 | 8.27 | 8.26 | .975 | |
| .99 | 14.2 | 14.0 | 13.9 | 13.8 | 13.7 | 13.7 | 13.7 | 13.6 | 13.6 | 13.5 | 13.5 | 13.5 | .99 | |
| .995 | 20.4 | 20.2 | 20.0 | 19.9 | 19.8 | 19.7 | 19.6 | 19.5 | 19.5 | 19.4 | 19.4 | 19.3 | .995 | |
| .999 | 46.8 | 46.1 | 45.8 | 45.4 | 45.1 | 44.9 | 44.7 | 44.5 | 44.4 | 44.3 | 44.1 | 44.0 | .999 | |
| .9995 | 66.5 | 65.5 | 65.1 | 64.6 | 64.1 | 63.8 | 63.6 | 63.2 | 63.1 | 62.9 | 62.7 | 62.6 | .9995 | |
| .0005 | .115 | .137 | .150 | .163 | .177 | .186 | .192 | .205 | .209 | .216 | .222 | .226 | .0005 | **5** |
| .001 | .132 | .155 | .167 | .181 | .195 | .204 | .210 | .223 | .227 | .233 | .239 | .244 | .001 | |
| .005 | .186 | .210 | .223 | .237 | .251 | .260 | .266 | .279 | .282 | .288 | .294 | .299 | .005 | |
| .01 | .219 | .244 | .257 | .270 | .285 | .293 | .299 | .312 | .315 | .322 | .328 | .331 | .01 | |
| .025 | .280 | .304 | .317 | .330 | .344 | .353 | .359 | .370 | .374 | .380 | .386 | .390 | .025 | |
| .05 | .345 | .369 | .382 | .395 | .408 | .417 | .422 | .432 | .437 | .442 | .448 | .452 | .05 | |
| .10 | .440 | .463 | .476 | .488 | .501 | .508 | .514 | .524 | .527 | .532 | .538 | .541 | .10 | |
| .25 | .669 | .690 | .700 | .711 | .722 | .728 | .732 | .741 | .743 | .748 | .752 | .755 | .25 | |
| .50 | 1.10 | 1.11 | 1.12 | 1.12 | 1.13 | 1.13 | 1.14 | 1.14 | 1.14 | 1.15 | 1.15 | 1.15 | .50 | |
| .75 | 1.89 | 1.88 | 1.88 | 1.88 | 1.88 | 1.88 | 1.87 | 1.87 | 1.87 | 1.87 | 1.87 | 1.87 | .75 | |
| .90 | 3.24 | 3.21 | 3.19 | 3.17 | 3.16 | 3.15 | 3.14 | 3.13 | 3.12 | 3.12 | 3.11 | 3.10 | .90 | |
| .95 | 4.62 | 4.56 | 4.53 | 4.50 | 4.46 | 4.44 | 4.43 | 4.41 | 4.40 | 4.39 | 4.37 | 4.36 | .95 | |
| .975 | 6.43 | 6.33 | 6.28 | 6.23 | 6.18 | 6.14 | 6.12 | 6.08 | 6.07 | 6.05 | 6.03 | 6.02 | .975 | |
| .99 | 9.72 | 9.55 | 9.47 | 9.38 | 9.29 | 9.24 | 9.20 | 9.13 | 9.11 | 9.08 | 9.04 | 9.02 | .99 | |
| .995 | 13.1 | 12.9 | 12.8 | 12.7 | 12.5 | 12.5 | 12.4 | 12.3 | 12.3 | 12.2 | 12.2 | 12.1 | .995 | |
| .999 | 25.9 | 25.4 | 25.1 | 24.9 | 24.6 | 24.4 | 24.3 | 24.1 | 24.1 | 23.9 | 23.8 | 23.8 | .999 | |
| .9995 | 34.6 | 33.9 | 33.5 | 33.1 | 32.7 | 32.5 | 32.3 | 32.1 | 32.0 | 31.8 | 31.7 | 31.6 | .9995 | |
| .0005 | .123 | .148 | .162 | .177 | .193 | .203 | .210 | .225 | .229 | .236 | .244 | .249 | .0005 | **6** |
| .001 | .141 | .166 | .180 | .195 | .211 | .222 | .229 | .243 | .247 | .255 | .262 | .267 | .001 | |
| .005 | .197 | .224 | .238 | .253 | .269 | .279 | .286 | .301 | .304 | .312 | .318 | .324 | .005 | |
| .01 | .232 | .258 | .273 | .288 | .304 | .313 | .321 | .334 | .338 | .346 | .352 | .357 | .01 | |
| .025 | .293 | .320 | .334 | .349 | .364 | .375 | .381 | .394 | .398 | .405 | .412 | .415 | .025 | |
| .05 | .358 | .385 | .399 | .413 | .428 | .437 | .444 | .457 | .460 | .467 | .472 | .476 | .05 | |
| .10 | .453 | .478 | .491 | .505 | .519 | .526 | .533 | .546 | .548 | .556 | .559 | .564 | .10 | |
| .25 | .675 | .696 | .707 | .718 | .729 | .736 | .741 | .751 | .753 | .758 | .762 | .765 | .25 | |
| .50 | 1.07 | 1.08 | 1.09 | 1.10 | 1.10 | 1.11 | 1.11 | 1.11 | 1.12 | 1.12 | 1.12 | 1.12 | .50 | |
| .75 | 1.76 | 1.76 | 1.75 | 1.75 | 1.75 | 1.75 | 1.74 | 1.74 | 1.74 | 1.74 | 1.74 | 1.74 | .75 | |
| .90 | 2.87 | 2.84 | 2.82 | 2.80 | 2.78 | 2.77 | 2.76 | 2.75 | 2.74 | 2.73 | 2.73 | 2.72 | .90 | |
| .95 | 3.94 | 3.87 | 3.84 | 3.81 | 3.77 | 3.75 | 3.74 | 3.71 | 3.70 | 3.69 | 3.68 | 3.67 | .95 | |
| .975 | 5.27 | 5.17 | 5.12 | 5.07 | 5.01 | 4.98 | 4.96 | 4.92 | 4.90 | 4.88 | 4.86 | 4.85 | .975 | |
| .99 | 7.56 | 7.40 | 7.31 | 7.23 | 7.14 | 7.09 | 7.06 | 6.99 | 6.97 | 6.93 | 6.90 | 6.88 | .99 | |
| .995 | 9.81 | 9.59 | 9.47 | 9.36 | 9.24 | 9.17 | 9.12 | 9.03 | 9.00 | 8.95 | 8.91 | 8.88 | .995 | |
| .999 | 17.6 | 17.1 | 16.9 | 16.7 | 16.4 | 16.3 | 16.2 | 16.0 | 16.0 | 15.9 | 15.8 | 15.7 | .999 | |
| .9995 | 22.4 | 21.9 | 21.7 | 21.4 | 21.1 | 20.9 | 20.7 | 20.5 | 20.4 | 20.3 | 20.2 | 20.1 | .9995 | |

## A-7c PERCENTILES OF THE $F(\nu_1, \nu_2)$ DISTRIBUTIONS (CONTINUED)

$\nu_1$, DEGREES OF FREEDOM FOR NUMERATOR

| $\nu_2$, DEGREES OF FREEDOM FOR DENOMINATOR | *Cum. prop.* | 1 | 2 | 3 | 4 | 5 | 6 | 7 | 8 | 9 | 10 | 11 | 12 | *Cum. prop.* |
|---|---|---|---|---|---|---|---|---|---|---|---|---|---|---|
| **7** | .0005 | $.0^{6}42$ | $.0^{3}50$ | $.0^{2}48$ | .014 | .027 | .040 | .053 | .066 | .078 | .088 | .099 | .108 | .0005 |
| | .001 | $.0^{5}17$ | $.0^{2}10$ | $.0^{2}76$ | .020 | .035 | .051 | .067 | .081 | .093 | .105 | .115 | .125 | .001 |
| | .005 | $.0^{4}42$ | $.0^{2}50$ | .023 | .046 | .070 | .093 | .113 | .130 | .145 | .159 | .171 | .181 | .005 |
| | .01 | $.0^{3}17$ | .010 | .036 | .067 | .096 | .121 | .143 | .162 | .178 | .192 | .205 | .216 | .01 |
| | .025 | $.0^{2}10$ | .025 | .068 | .110 | .146 | .176 | .200 | .221 | .238 | .253 | .266 | .277 | .025 |
| | .05 | $.0^{2}42$ | .052 | .113 | .164 | .205 | .238 | .264 | .286 | .304 | .319 | .332 | .343 | .05 |
| | .10 | .017 | .107 | .190 | .251 | .297 | .332 | .359 | .381 | .399 | .414 | .427 | .438 | .10 |
| | .25 | .110 | .300 | .412 | .481 | .528 | .562 | .588 | .608 | .624 | .637 | .649 | .658 | .25 |
| | .50 | .506 | .767 | .871 | .926 | .960 | .983 | 1.00 | 1.01 | 1.02 | 1.03 | 1.04 | 1.04 | .50 |
| | .75 | 1.57 | 1.70 | 1.72 | 1.72 | 1.71 | 1.71 | 1.70 | 1.70 | 1.69 | 1.69 | 1.69 | 1.68 | .75 |
| | .90 | 3.59 | 3.26 | 3.07 | 2.96 | 2.88 | 2.83 | 2.78 | 2.75 | 2.72 | 2.70 | 2.68 | 2.67 | .90 |
| | .95 | 5.59 | 4.74 | 4.35 | 4.12 | 3.97 | 3.87 | 3.79 | 3.73 | 3.68 | 3.64 | 3.60 | 3.57 | .95 |
| | .975 | 8.07 | 6.54 | 5.89 | 5.52 | 5.29 | 5.12 | 4.99 | 4.90 | 4.82 | 4.76 | 4.71 | 4.67 | .975 |
| | .99 | 12.2 | 9.55 | 8.45 | 7.85 | 7.46 | 7.19 | 6.99 | 6.84 | 6.72 | 6.62 | 6.54 | 6.47 | .99 |
| | .995 | 16.2 | 12.4 | 10.9 | 10.0 | 9.52 | 9.16 | 8.89 | 8.68 | 8.51 | 8.38 | 8.27 | 8.18 | .995 |
| | .999 | 29.2 | 21.7 | 18.8 | 17.2 | 16.2 | 15.5 | 15.0 | 14.6 | 14.3 | 14.1 | 13.9 | 13.7 | .999 |
| | .9995 | 37.0 | 27.2 | 23.5 | 21.4 | 20.2 | 19.3 | 18.7 | 18.2 | 17.8 | 17.5 | 17.2 | 17.0 | .9995 |
| **8** | .0005 | $.0^{6}42$ | $.0^{3}50$ | $.0^{2}48$ | .014 | .027 | .041 | .055 | .068 | .081 | .092 | .102 | .112 | .0005 |
| | .001 | $.0^{5}17$ | $.0^{2}10$ | $.0^{2}76$ | .020 | .036 | .053 | .068 | .083 | .096 | .109 | .120 | .130 | .001 |
| | .005 | $.0^{4}42$ | $.0^{2}50$ | .027 | .047 | .072 | .095 | .115 | .133 | .149 | .164 | .176 | .187 | .005 |
| | .01 | $.0^{3}17$ | .010 | .036 | .068 | .097 | .123 | .146 | .166 | .183 | .198 | .211 | .222 | .01 |
| | .025 | $.0^{2}10$ | .025 | .069 | .111 | .148 | .179 | .204 | .226 | .244 | .259 | .273 | .285 | .025 |
| | .05 | $.0^{2}42$ | .052 | .113 | .166 | .208 | .241 | .268 | .291 | .310 | .326 | .339 | .351 | .05 |
| | .10 | .017 | .107 | .190 | .253 | .299 | .335 | .363 | .386 | .405 | .421 | .435 | .445 | .10 |
| | .25 | .109 | .298 | .411 | .481 | .529 | .563 | .589 | .610 | .627 | .640 | .654 | .661 | .25 |
| | .50 | .499 | .757 | .860 | .915 | .948 | .971 | .988 | 1.00 | 1.01 | 1.02 | 1.02 | 1.03 | .50 |
| | .75 | 1.54 | 1.66 | 1.67 | 1.66 | 1.66 | 1.65 | 1.64 | 1.64 | 1.64 | 1.63 | 1.63 | 1.62 | .75 |
| | .90 | 3.46 | 3.11 | 2.92 | 2.81 | 2.73 | 2.67 | 2.62 | 2.59 | 2.56 | 2.54 | 2.52 | 2.50 | .90 |
| | .95 | 5.32 | 4.46 | 4.07 | 3.84 | 3.69 | 3.58 | 3.50 | 3.44 | 3.39 | 3.35 | 3.31 | 3.28 | .95 |
| | .975 | 7.57 | 6.06 | 5.42 | 5.05 | 4.82 | 4.65 | 4.53 | 4.43 | 4.36 | 4.30 | 4.24 | 4.20 | .975 |
| | .99 | 11.3 | 8.65 | 7.59 | 7.01 | 6.63 | 6.37 | 6.18 | 6.03 | 5.91 | 5.81 | 5.73 | 5.67 | .99 |
| | .995 | 14.7 | 11.0 | 9.60 | 8.81 | 8.30 | 7.95 | 7.69 | 7.50 | 7.34 | 7.21 | 7.10 | 7.01 | .995 |
| | .999 | 25.4 | 18.5 | 15.8 | 14.4 | 13.5 | 12.9 | 12.4 | 12.0 | 11.8 | 11.5 | 11.4 | 11.2 | .999 |
| | .9995 | 31.6 | 22.8 | 19.4 | 17.6 | 16.4 | 15.7 | 15.1 | 14.6 | 14.3 | 14.0 | 13.8 | 13.6 | .9995 |
| **9** | .0005 | $.0^{6}41$ | $.0^{3}50$ | $.0^{2}48$ | .015 | .027 | .042 | .056 | .070 | .083 | .094 | .105 | .115 | .0005 |
| | .001 | $.0^{5}17$ | $.0^{2}10$ | $.0^{2}77$ | .021 | .037 | .054 | .070 | .085 | .099 | .112 | .123 | .134 | .001 |
| | .005 | $.0^{4}42$ | $.0^{2}50$ | .023 | .047 | .073 | .096 | .117 | .136 | .153 | .168 | .181 | .192 | .005 |
| | .01 | $.0^{3}17$ | .010 | .037 | .068 | .098 | .125 | .149 | .169 | .187 | .202 | .216 | .228 | .01 |
| | .025 | $.0^{2}10$ | .025 | .069 | .112 | .150 | .181 | .207 | .230 | .248 | .265 | .279 | .291 | .025 |
| | .05 | $.0^{2}40$ | .052 | .113 | .167 | .210 | .244 | .272 | .296 | .315 | .331 | .345 | .358 | .05 |
| | .10 | .017 | .107 | .191 | .254 | .302 | .338 | .367 | .390 | .410 | .426 | .441 | .452 | .10 |
| | .25 | .108 | .297 | .410 | .480 | .529 | .564 | .591 | .612 | .629 | .643 | .654 | .664 | .25 |
| | .50 | .494 | .749 | .852 | .906 | .939 | .962 | .978 | .990 | 1.00 | 1.01 | 1.01 | 1.02 | .50 |
| | .75 | 1.51 | 1.62 | 1.63 | 1.63 | 1.62 | 1.61 | 1.60 | 1.60 | 1.59 | 1.59 | 1.58 | 1.58 | .75 |
| | .90 | 3.36 | 3.01 | 2.81 | 2.69 | 2.61 | 2.55 | 2.51 | 2.47 | 2.44 | 2.42 | 2.40 | 2.38 | .90 |
| | .95 | 5.12 | 4.26 | 3.86 | 3.63 | 3.48 | 3.37 | 3.29 | 3.23 | 3.18 | 3.14 | 3.10 | 3.07 | .95 |
| | .975 | 7.21 | 5.71 | 5.08 | 4.72 | 4.48 | 4.32 | 4.20 | 4.10 | 4.03 | 3.96 | 3.91 | 3.87 | .975 |
| | .99 | 10.6 | 8.02 | 6.99 | 6.42 | 6.06 | 5.80 | 5.61 | 5.47 | 5.35 | 5.26 | 5.18 | 5.11 | .99 |
| | .995 | 13.6 | 10.1 | 8.72 | 7.96 | 7.47 | 7.13 | 6.88 | 6.69 | 6.54 | 6.42 | 6.31 | 6.23 | .995 |
| | .999 | 22.9 | 16.4 | 13.9 | 12.6 | 11.7 | 11.1 | 10.7 | 10.4 | 10.1 | 9.89 | 9.71 | 9.57 | .999 |
| | .9995 | 28.0 | 19.9 | 16.8 | 15.1 | 14.1 | 13.3 | 12.8 | 12.4 | 12.1 | 11.8 | 11.6 | 11.4 | .9995 |

$\nu_1$, DEGREES OF FREEDOM FOR NUMERATOR

| Cum. prop. | 15 | 20 | 24 | 30 | 40 | 50 | 60 | 100 | 120 | 200 | 500 | ∞ | Cum. prop. | $\nu_2$, degrees of freedom for denominator |
|---|---|---|---|---|---|---|---|---|---|---|---|---|---|---|
| .0005 | .130 | .157 | .172 | .188 | .206 | .217 | .225 | .242 | .246 | .255 | .263 | .268 | .0005 | 7 |
| .001 | .148 | .176 | .191 | .208 | .225 | .237 | .245 | .261 | .266 | .274 | .282 | .288 | .001 | |
| .005 | .206 | .235 | .251 | .267 | .285 | .296 | .304 | .319 | .324 | .332 | .340 | .345 | .005 | |
| .01 | .241 | .270 | .286 | .303 | .320 | .331 | .339 | .355 | .358 | .366 | .373 | .379 | .01 | |
| .025 | .304 | .333 | .348 | .364 | .381 | .392 | .399 | .413 | .418 | .426 | .433 | .437 | .025 | |
| .05 | .369 | .398 | .413 | .428 | .445 | .455 | .461 | .476 | .479 | .485 | .493 | .498 | .05 | |
| .10 | .463 | .491 | .504 | .519 | .534 | .543 | .550 | .562 | .566 | .571 | .578 | .582 | .10 | |
| .25 | .679 | .702 | .713 | .725 | .737 | .745 | .749 | .760 | .762 | .767 | .772 | .775 | .25 | |
| .50 | 1.05 | 1.07 | 1.07 | 1.08 | 1.08 | 1.09 | 1.09 | 1.10 | 1.10 | 1.10 | 1.10 | 1.10 | .50 | |
| .75 | 1.68 | 1.67 | 1.67 | 1.66 | 1.66 | 1.66 | 1.65 | 1.65 | 1.65 | 1.65 | 1.65 | 1.65 | .75 | |
| .90 | 2.63 | 2.59 | 2.58 | 2.56 | 2.54 | 2.52 | 2.51 | 2.50 | 2.49 | 2.48 | 2.48 | 2.47 | .90 | |
| .95 | 3.51 | 3.44 | 3.41 | 3.38 | 3.34 | 3.32 | 3.30 | 3.27 | 3.27 | 3.25 | 3.24 | 3.23 | .95 | |
| .975 | 4.57 | 4.47 | 4.42 | 4.36 | 4.31 | 4.28 | 4.25 | 4.21 | 4.20 | 4.18 | 4.16 | 4.14 | .975 | |
| .99 | 6.31 | 6.16 | 6.07 | 5.99 | 5.91 | 5.86 | 5.82 | 5.75 | 5.74 | 5.70 | 5.67 | 5.65 | .99 | |
| .995 | 7.97 | 7.75 | 7.65 | 7.53 | 7.42 | 7.35 | 7.31 | 7.22 | 7.19 | 7.15 | 7.10 | 7.08 | .995 | |
| .999 | 13.3 | 12.9 | 12.7 | 12.5 | 12.3 | 12.2 | 12.1 | 11.9 | 11.9 | 11.8 | 11.7 | 11.7 | .999 | |
| .9995 | 16.5 | 16.0 | 15.7 | 15.5 | 15.2 | 15.1 | 15.0 | 14.7 | 14.7 | 14.6 | 14.5 | 14.4 | .9995 | |
| .0005 | .136 | .164 | .181 | .198 | .218 | .230 | .239 | .257 | .262 | .271 | .281 | .287 | .0005 | 8 |
| .001 | .155 | .184 | .200 | .218 | .238 | .250 | .259 | .277 | .282 | .292 | .300 | .306 | .001 | |
| .005 | .214 | .244 | .261 | .279 | .299 | .311 | .319 | .337 | .341 | .351 | .358 | .364 | .005 | |
| .01 | .250 | .281 | .297 | .315 | .334 | .346 | .354 | .372 | .376 | .385 | .392 | .398 | .01 | |
| .025 | .313 | .343 | .360 | .377 | .395 | .407 | .415 | .431 | .435 | .442 | .450 | .456 | .025 | |
| .05 | .379 | .409 | .425 | .441 | .459 | .469 | .477 | .493 | .496 | .505 | .510 | .516 | .05 | |
| .10 | .472 | .500 | .515 | .531 | .547 | .556 | .563 | .578 | .581 | .588 | .595 | .599 | .10 | |
| .25 | .684 | .707 | .718 | .730 | .743 | .751 | .756 | .767 | .769 | .775 | .780 | .783 | .25 | |
| .50 | 1.04 | 1.05 | 1.06 | 1.07 | 1.07 | 1.07 | 1.08 | 1.08 | 1.08 | 1.09 | 1.09 | 1.09 | .50 | |
| .75 | 1.62 | 1.61 | 1.60 | 1.60 | 1.59 | 1.59 | 1.59 | 1.58 | 1.58 | 1.58 | 1.58 | 1.58 | .75 | |
| .90 | 2.46 | 2.42 | 2.40 | 2.38 | 2.36 | 2.35 | 2.34 | 2.32 | 2.32 | 2.31 | 2.30 | 2.29 | .90 | |
| .95 | 3.22 | 3.15 | 3.12 | 3.08 | 3.04 | 3.02 | 3.01 | 2.97 | 2.97 | 2.95 | 2.94 | 2.93 | .95 | |
| .975 | 4.10 | 4.00 | 3.95 | 3.89 | 3.84 | 3.81 | 3.78 | 3.74 | 3.73 | 3.70 | 3.68 | 3.67 | .975 | |
| .99 | 5.52 | 5.36 | 5.28 | 5.20 | 5.12 | 5.07 | 5.03 | 4.96 | 4.95 | 4.91 | 4.88 | 4.86 | .99 | |
| .995 | 6.81 | 6.61 | 6.50 | 6.40 | 6.29 | 6.22 | 6.18 | 6.09 | 6.06 | 6.02 | 5.98 | 5.95 | .995 | |
| .999 | 10.8 | 10.5 | 10.3 | 10.1 | 9.92 | 9.80 | 9.73 | 9.57 | 9.54 | 9.46 | 9.39 | 9.34 | .999 | |
| .9995 | 13.1 | 12.7 | 12.5 | 12.2 | 12.0 | 11.8 | 11.8 | 11.6 | 11.5 | 11.4 | 11.4 | 11.3 | .9995 | |
| .0005 | .141 | .171 | .188 | .207 | .228 | .242 | .251 | .270 | .276 | .287 | .297 | .303 | .0005 | 9 |
| .001 | .160 | .191 | .208 | .228 | .249 | .262 | .271 | .291 | .296 | .307 | .316 | .323 | .001 | |
| .005 | .220 | .253 | .271 | .290 | .310 | .324 | .332 | .351 | .356 | .366 | .376 | .382 | .005 | |
| .01 | .257 | .289 | .307 | .326 | .346 | .358 | .368 | .386 | .391 | .400 | .410 | .415 | .01 | |
| .025 | .320 | .352 | .370 | .388 | .408 | .420 | .428 | .446 | .450 | .459 | .467 | .473 | .025 | |
| .05 | .386 | .418 | .435 | .452 | .471 | .483 | .490 | .508 | .510 | .518 | .526 | .532 | .05 | |
| .10 | .479 | .509 | .525 | .541 | .558 | .568 | .575 | .588 | .594 | .602 | .610 | .613 | .10 | |
| .25 | .687 | .711 | .723 | .736 | .749 | .757 | .762 | .773 | .776 | .782 | .787 | .791 | .25 | |
| .50 | 1.03 | 1.04 | 1.05 | 1.05 | 1.06 | 1.06 | 1.07 | 1.07 | 1.07 | 1.08 | 1.08 | 1.08 | .50 | |
| .75 | 1.57 | 1.56 | 1.56 | 1.55 | 1.55 | 1.54 | 1.54 | 1.53 | 1.53 | 1.53 | 1.53 | 1.53 | .75 | |
| .90 | 2.34 | 2.30 | 2.28 | 2.25 | 2.23 | 2.22 | 2.21 | 2.19 | 2.18 | 2.17 | 2.17 | 2.16 | .90 | |
| .95 | 3.01 | 2.94 | 2.90 | 2.86 | 2.83 | 2.80 | 2.79 | 2.76 | 2.75 | 2.73 | 2.72 | 2.71 | .95 | |
| .975 | 3.77 | 3.67 | 3.61 | 3.56 | 3.51 | 3.47 | 3.45 | 3.40 | 3.39 | 3.37 | 3.35 | 3.33 | .975 | |
| .99 | 4.96 | 4.81 | 4.73 | 4.65 | 4.57 | 4.52 | 4.48 | 4.42 | 4.40 | 4.36 | 4.33 | 4.31 | .99 | |
| .995 | 6.03 | 5.83 | 5.73 | 5.62 | 5.52 | 5.45 | 5.41 | 5.32 | 5.30 | 5.26 | 5.21 | 5.19 | .995 | |
| .999 | 9.24 | 8.90 | 8.72 | 8.55 | 8.37 | 8.26 | 8.19 | 8.04 | 8.00 | 7.93 | 7.86 | 7.81 | .999 | |
| .9995 | 11.0 | 10.6 | 10.4 | 10.2 | 9.94 | 9.80 | 9.71 | 9.53 | 9.49 | 9.40 | 9.32 | 9.26 | .9995 | |

$\nu_1$, DEGREES OF FREEDOM FOR NUMERATOR

| $\nu_2$, degrees of freedom for denominator | *Cum. prop.* | 1 | 2 | 3 | 4 | 5 | 6 | 7 | 8 | 9 | 10 | 11 | 12 | *Cum. prop.* |
|---|---|---|---|---|---|---|---|---|---|---|---|---|---|---|
| **10** | .0005 | $.0^{6}41$ | $.0^{3}50$ | $.0^{2}49$ | .015 | .028 | .043 | .057 | .071 | .085 | .097 | .108 | .119 | .0005 |
| | .001 | $.0^{5}17$ | $.0^{2}10$ | $.0^{2}77$ | .021 | .037 | .054 | .071 | .087 | .101 | .114 | .126 | .137 | .001 |
| | .005 | $.0^{4}41$ | $.0^{2}50$ | .023 | .048 | .073 | .098 | .119 | .139 | .156 | .171 | .185 | .197 | .005 |
| | .01 | $.0^{3}17$ | .010 | .037 | .069 | .100 | .127 | .151 | .172 | .190 | .206 | .220 | .233 | .01 |
| | .025 | $.0^{2}10$ | .025 | .069 | .113 | .151 | .183 | .210 | .233 | .252 | .269 | .283 | .296 | .025 |
| | .05 | $.0^{2}41$ | .052 | .114 | .168 | .211 | .246 | .275 | .299 | .319 | .336 | .351 | .363 | .05 |
| | .10 | .017 | .106 | .191 | .255 | .303 | .340 | .370 | .394 | .414 | .430 | .444 | .457 | .10 |
| | .25 | .107 | .296 | .409 | .480 | .529 | .565 | .592 | .613 | .631 | .645 | .657 | .667 | .25 |
| | .50 | .490 | .743 | .845 | .899 | .932 | .954 | .971 | .983 | .992 | 1.00 | 1.01 | 1.01 | .50 |
| | .75 | 1.49 | 1.60 | 1.60 | 1.59 | 1.59 | 1.58 | 1.57 | 1.56 | 1.56 | 1.55 | 1.55 | 1.54 | .75 |
| | .90 | 3.28 | 2.92 | 2.73 | 2.61 | 2.52 | 2.46 | 2.41 | 2.38 | 2.35 | 2.32 | 2.30 | 2.28 | .90 |
| | .95 | 4.96 | 4.10 | 3.71 | 3.48 | 3.33 | 3.22 | 3.14 | 3.07 | 3.02 | 2.98 | 2.94 | 2.91 | .95 |
| | .975 | 6.94 | 5.46 | 4.83 | 4.47 | 4.24 | 4.07 | 3.95 | 3.85 | 3.78 | 3.72 | 3.66 | 3.62 | .975 |
| | .99 | 10.0 | 7.56 | 6.55 | 5.99 | 5.64 | 5.39 | 5.20 | 5.06 | 4.94 | 4.85 | 4.77 | 4.71 | .99 |
| | .995 | 12.8 | 9.43 | 8.08 | 7.34 | 6.87 | 6.54 | 6.30 | 6.12 | 5.97 | 5.85 | 5.75 | 5.66 | .995 |
| | .999 | 21.0 | 14.9 | 12.6 | 11.3 | 10.5 | 9.92 | 9.52 | 9.20 | 8.96 | 8.75 | 8.58 | 8.44 | .999 |
| | .9995 | 25.5 | 17.9 | 15.0 | 13.4 | 12.4 | 11.8 | 11.3 | 10.9 | 10.6 | 10.3 | 10.1 | 9.93 | .9995 |
| **11** | .0005 | $.0^{6}41$ | $.0^{3}50$ | $.0^{2}49$ | .015 | .028 | .043 | .058 | .072 | .086 | .099 | .111 | .121 | .0005 |
| | .001 | $.0^{5}16$ | $.0^{2}10$ | $.0^{2}78$ | .021 | .038 | .055 | .072 | .088 | .103 | .116 | .129 | .140 | .001 |
| | .005 | $.0^{4}40$ | $.0^{2}50$ | .023 | .048 | .074 | .099 | .121 | .141 | .158 | .174 | .188 | .200 | .005 |
| | .01 | $.0^{3}16$ | .010 | .037 | .069 | .100 | .128 | .153 | .175 | .193 | .210 | .224 | .237 | .01 |
| | .025 | $.0^{2}10$ | .025 | .069 | .114 | .152 | .185 | .212 | .236 | .256 | .273 | .288 | .301 | .025 |
| | .05 | $.0^{2}41$ | .052 | .114 | .168 | .212 | .248 | .278 | .302 | .323 | .340 | .355 | .368 | .05 |
| | .10 | .017 | .106 | .192 | .256 | .305 | .342 | .373 | .397 | .417 | .435 | .448 | .461 | .10 |
| | .25 | .107 | .295 | .408 | .481 | .529 | .565 | .592 | .614 | .633 | .645 | .658 | .667 | .25 |
| | .50 | .486 | .739 | .840 | .893 | .926 | .948 | .964 | .977 | .986 | .994 | 1.00 | 1.01 | .50 |
| | .75 | 1.47 | 1.58 | 1.58 | 1.57 | 1.56 | 1.55 | 1.54 | 1.53 | 1.53 | 1.52 | 1.52 | 1.51 | .75 |
| | .90 | 3.23 | 2.86 | 2.66 | 2.54 | 2.45 | 2.39 | 2.34 | 2.30 | 2.27 | 2.25 | 2.23 | 2.21 | .90 |
| | .95 | 4.84 | 3.98 | 3.59 | 3.36 | 3.20 | 3.09 | 3.01 | 2.95 | 2.90 | 2.85 | 2.82 | 2.79 | .95 |
| | .975 | 6.72 | 5.26 | 4.63 | 4.28 | 4.04 | 3.88 | 3.76 | 3.66 | 3.59 | 3.53 | 3.47 | 3.43 | .975 |
| | .99 | 9.65 | 7.21 | 6.22 | 5.67 | 5.32 | 5.07 | 4.89 | 4.74 | 4.63 | 4.54 | 4.46 | 4.40 | .99 |
| | .995 | 12.2 | 8.91 | 7.60 | 6.88 | 6.42 | 6.10 | 5.86 | 5.68 | 5.54 | 5.42 | 5.32 | 5.24 | .995 |
| | .999 | 19.7 | 13.8 | 11.6 | 10.3 | 9.58 | 9.05 | 8.66 | 8.35 | 8.12 | 7.92 | 7.76 | 7.62 | .999 |
| | .9995 | 23.6 | 16.4 | 13.6 | 12.2 | 11.2 | 10.6 | 10.1 | 9.76 | 9.48 | 9.24 | 9.04 | 8.88 | .9995 |
| **12** | .0005 | $.0^{6}41$ | $.0^{3}50$ | $.0^{2}49$ | .015 | .028 | .044 | .058 | .073 | .087 | .101 | .113 | .124 | .0005 |
| | .001 | $.0^{5}16$ | $.0^{2}10$ | $.0^{2}78$ | .021 | .038 | .056 | .073 | .089 | .104 | .118 | .131 | .143 | .001 |
| | .005 | $.0^{4}39$ | $.0^{2}50$ | .023 | .048 | .075 | .100 | .122 | .143 | .161 | .177 | .191 | .204 | .005 |
| | .01 | $.0^{3}16$ | .010 | .037 | .070 | .101 | .130 | .155 | .176 | .196 | .212 | .227 | .241 | .01 |
| | .025 | $.0^{2}10$ | .025 | .070 | .114 | .153 | .186 | .214 | .238 | .259 | .276 | .292 | .305 | .025 |
| | .05 | $.0^{2}41$ | .052 | .114 | .169 | .214 | .250 | .280 | .305 | .325 | .343 | .358 | .372 | .05 |
| | .10 | .016 | .106 | .192 | .257 | .306 | .344 | .375 | .400 | .420 | .438 | .452 | .466 | .10 |
| | .25 | .106 | .295 | .408 | .480 | .530 | .566 | .594 | .616 | .633 | .649 | .662 | .671 | .25 |
| | .50 | .484 | .735 | .835 | .888 | .921 | .943 | .959 | .972 | .981 | .989 | .995 | 1.00 | .50 |
| | .75 | 1.46 | 1.56 | 1.56 | 1.55 | 1.54 | 1.53 | 1.52 | 1.51 | 1.51 | 1.50 | 1.50 | 1.49 | .75 |
| | .90 | 3.18 | 2.81 | 2.61 | 2.48 | 2.39 | 2.33 | 2.28 | 2.24 | 2.21 | 2.19 | 2.17 | 2.15 | .90 |
| | .95 | 4.75 | 3.89 | 3.49 | 3.26 | 3.11 | 3.00 | 2.91 | 2.85 | 2.80 | 2.75 | 2.72 | 2.69 | .95 |
| | .975 | 6.55 | 5.10 | 4.47 | 4.12 | 3.89 | 3.73 | 3.61 | 3.51 | 3.44 | 3.37 | 3.32 | 3.28 | .975 |
| | .99 | 9.33 | 6.93 | 5.95 | 5.41 | 5.06 | 4.82 | 4.64 | 4.50 | 4.39 | 4.30 | 4.22 | 4.16 | .99 |
| | .995 | 11.8 | 8.51 | 7.23 | 6.52 | 6.07 | 5.76 | 5.52 | 5.35 | 5.20 | 5.09 | 4.99 | 4.91 | .995 |
| | .999 | 18.6 | 13.0 | 10.8 | 9.63 | 8.89 | 8.38 | 8.00 | 7.71 | 7.48 | 7.29 | 7.14 | 7.01 | .999 |
| | .9995 | 22.2 | 15.3 | 12.7 | 11.2 | 10.4 | 9.74 | 9.28 | 8.94 | 8.66 | 8.43 | 8.24 | 8.08 | .9995 |

$\nu_1$, DEGREES OF FREEDOM FOR NUMERATOR

| *Cum. prop.* | 15 | 20 | 24 | 30 | 40 | 50 | 60 | 100 | 120 | 200 | 500 | ∞ | *Cum. prop.* | $\nu_2$, DEGREES OF FREEDOM FOR DENOMINATOR |
|---|---|---|---|---|---|---|---|---|---|---|---|---|---|---|
| .0005 | .145 | .177 | .195 | .215 | .238 | .251 | .262 | .282 | .288 | .299 | .311 | .319 | .0005 | **10** |
| .001 | .164 | .197 | .216 | .236 | .258 | .272 | .282 | .303 | .309 | .321 | .331 | .338 | .001 | |
| .005 | .226 | .260 | .279 | .299 | .321 | .334 | .344 | .365 | .370 | .380 | .391 | .397 | .005 | |
| .01 | .263 | .297 | .316 | .336 | .357 | .370 | .380 | .400 | .405 | .415 | .424 | .431 | .01 | |
| .025 | .327 | .360 | .379 | .398 | .419 | .431 | .441 | .459 | .464 | .474 | .483 | .488 | .025 | |
| .05 | .393 | .426 | .444 | .462 | .481 | .493 | .502 | .518 | .523 | .532 | .541 | .546 | .05 | |
| .10 | .486 | .516 | .532 | .549 | .567 | .578 | .586 | .602 | .605 | .614 | .621 | .625 | .10 | |
| .25 | .691 | .714 | .727 | .740 | .754 | .762 | .767 | .779 | .782 | .788 | .793 | .797 | .25 | |
| .50 | 1.02 | 1.03 | 1.04 | 1.05 | 1.05 | 1.06 | 1.06 | 1.06 | 1.06 | 1.07 | 1.07 | 1.07 | .50 | |
| .75 | 1.53 | 1.52 | 1.52 | 1.51 | 1.51 | 1.50 | 1.50 | 1.49 | 1.49 | 1.49 | 1.48 | 1.48 | .75 | |
| .90 | 2.24 | 2.20 | 2.18 | 2.16 | 2.13 | 2.12 | 2.11 | 2.09 | 2.08 | 2.07 | 2.06 | 2.06 | .90 | |
| .95 | 2.85 | 2.77 | 2.74 | 2.70 | 2.66 | 2.64 | 2.62 | 2.59 | 2.58 | 2.56 | 2.55 | 2.54 | .95 | |
| .975 | 3.52 | 3.42 | 3.37 | 3.31 | 3.26 | 3.22 | 3.20 | 3.15 | 3.14 | 3.12 | 3.09 | 3.08 | .975 | |
| .99 | 4.56 | 4.41 | 4.33 | 4.25 | 4.17 | 4.12 | 4.08 | 4.01 | 4.00 | 3.96 | 3.93 | 3.91 | .99 | |
| .995 | 5.47 | 5.27 | 5.17 | 5.07 | 4.97 | 4.90 | 4.86 | 4.77 | 4.75 | 4.71 | 4.67 | 4.64 | .995 | |
| .999 | 8.13 | 7.80 | 7.64 | 7.47 | 7.30 | 7.19 | 7.12 | 6.98 | 6.94 | 6.87 | 6.81 | 6.76 | .999 | |
| .9995 | 9.56 | 9.16 | 8.96 | 8.75 | 8.54 | 8.42 | 8.33 | 8.16 | 8.12 | 8.04 | 7.96 | 7.90 | .9995 | |
| .0005 | .148 | .182 | .201 | .222 | .246 | .261 | .271 | .293 | .299 | .312 | .324 | .331 | .0005 | **11** |
| .001 | .168 | .202 | .222 | .243 | .266 | .282 | .292 | .313 | .320 | .332 | .343 | .353 | .001 | |
| .005 | .231 | .266 | .286 | .308 | .330 | .345 | .355 | .376 | .382 | .394 | .403 | .412 | .005 | |
| .01 | .268 | .304 | .324 | .344 | .366 | .380 | .391 | .412 | .417 | .427 | .439 | .444 | .01 | |
| .025 | .332 | .368 | .386 | .407 | .429 | .442 | .450 | .472 | .476 | .485 | .495 | .503 | .025 | |
| .05 | .398 | .433 | .452 | .469 | .490 | .503 | .513 | .529 | .535 | .543 | .552 | .559 | .05 | |
| .10 | .490 | .524 | .541 | .559 | .578 | .588 | .595 | .614 | .617 | .625 | .633 | .637 | .10 | |
| .25 | .694 | .719 | .730 | .744 | .758 | .767 | .773 | .780 | .788 | .794 | .799 | .803 | .25 | |
| .50 | 1.02 | 1.03 | 1.03 | 1.04 | 1.05 | 1.05 | 1.05 | 1.06 | 1.06 | 1.06 | 1.06 | 1.06 | .50 | |
| .75 | 1.50 | 1.49 | 1.49 | 1.48 | 1.47 | 1.47 | 1.47 | 1.46 | 1.46 | 1.46 | 1.45 | 1.45 | .75 | |
| .90 | 2.17 | 2.12 | 2.10 | 2.08 | 2.05 | 2.04 | 2.03 | 2.00 | 2.00 | 1.99 | 1.98 | 1.97 | .90 | |
| .95 | 2.72 | 2.65 | 2.61 | 2.57 | 2.53 | 2.51 | 2.49 | 2.46 | 2.45 | 2.43 | 2.42 | 2.40 | .95 | |
| .975 | 3.33 | 3.23 | 3.17 | 3.12 | 3.06 | 3.03 | 3.00 | 2.96 | 2.94 | 2.92 | 2.90 | 2.88 | .975 | |
| .99 | 4.25 | 4.10 | 4.02 | 3.94 | 3.86 | 3.81 | 3.78 | 3.71 | 3.69 | 3.66 | 3.62 | 3.60 | .99 | |
| .995 | 5.05 | 4.86 | 4.76 | 4.65 | 4.55 | 4.49 | 4.45 | 4.36 | 4.34 | 4.29 | 4.25 | 4.23 | .995 | |
| .999 | 7.32 | 7.01 | 6.85 | 6.68 | 6.52 | 6.41 | 6.35 | 6.21 | 6.17 | 6.10 | 6.04 | 6.00 | .999 | |
| .9995 | 8.52 | 8.14 | 7.94 | 7.75 | 7.55 | 7.43 | 7.35 | 7.18 | 7.14 | 7.06 | 6.98 | 6.93 | .9995 | |
| .0005 | .152 | .186 | .206 | .228 | .253 | .269 | .280 | .305 | .311 | .323 | .337 | .345 | .0005 | **12** |
| .001 | .172 | .207 | .228 | .250 | .275 | .291 | .302 | .326 | .332 | .344 | .357 | .365 | .001 | |
| .005 | .235 | .272 | .292 | .315 | .339 | .355 | .365 | .388 | .393 | .405 | .417 | .424 | .005 | |
| .01 | .273 | .310 | .330 | .352 | .375 | .391 | .401 | .422 | .428 | .441 | .450 | .458 | .01 | |
| .025 | .337 | .374 | .394 | .416 | .437 | .450 | .461 | .481 | .487 | .498 | .508 | .514 | .025 | |
| .05 | .404 | .439 | .458 | .478 | .499 | .513 | .522 | .541 | .545 | .556 | .565 | .571 | .05 | |
| .10 | .496 | .528 | .546 | .564 | .583 | .595 | .604 | .621 | .625 | .633 | .641 | .647 | .10 | |
| .25 | .695 | .721 | .734 | .748 | .762 | .771 | .777 | .789 | .792 | .799 | .804 | .808 | .25 | |
| .50 | 1.01 | 1.02 | 1.03 | 1.03 | 1.04 | 1.04 | 1.05 | 1.05 | 1.05 | 1.05 | 1.06 | 1.06 | .50 | |
| .75 | 1.48 | 1.47 | 1.46 | 1.45 | 1.45 | 1.44 | 1.44 | 1.43 | 1.43 | 1.43 | 1.42 | 1.42 | .75 | |
| .90 | 2.11 | 2.06 | 2.04 | 2.01 | 1.99 | 1.97 | 1.96 | 1.94 | 1.93 | 1.92 | 1.91 | 1.90 | .90 | |
| .95 | 2.62 | 2.54 | 2.51 | 2.47 | 2.43 | 2.40 | 2.38 | 2.35 | 2.34 | 2.32 | 2.31 | 2.30 | .95 | |
| .975 | 3.18 | 3.07 | 3.02 | 2.96 | 2.91 | 2.87 | 2.85 | 2.80 | 2.79 | 2.76 | 2.74 | 2.72 | .975 | |
| .99 | 4.01 | 3.86 | 3.78 | 3.70 | 3.62 | 3.57 | 3.54 | 3.47 | 3.45 | 3.41 | 3.38 | 3.36 | .99 | |
| .995 | 4.72 | 4.53 | 4.43 | 4.33 | 4.23 | 4.17 | 4.12 | 4.04 | 4.01 | 3.97 | 3.93 | 3.90 | .995 | |
| .999 | 6.71 | 6.40 | 6.25 | 6.09 | 5.93 | 5.83 | 5.76 | 5.63 | 5.59 | 5.52 | 5.46 | 5.42 | .999 | |
| .9995 | 7.74 | 7.37 | 7.18 | 7.00 | 6.80 | 6.68 | 6.61 | 6.45 | 6.41 | 6.33 | 6.25 | 6.20 | .9995 | |

## A-7c PERCENTILES OF THE $F(\nu_1, \nu_2)$ DISTRIBUTIONS (CONTINUED)

$\nu_1$, DEGREES OF FREEDOM FOR NUMERATOR

| $\nu_2$, DEGREES OF FREEDOM FOR DENOMINATOR | *Cum. prop.* | 1 | 2 | 3 | 4 | 5 | 6 | 7 | 8 | 9 | 10 | 11 | 12 | *Cum. prop.* |
|---|---|---|---|---|---|---|---|---|---|---|---|---|---|---|
| **15** | .0005 | $.0^{6}41$ | $.0^{3}50$ | $.0^{2}49$ | .015 | .029 | .045 | .061 | .076 | .091 | .105 | .117 | .129 | .0005 |
| | .001 | $.0^{5}16$ | $.0^{2}10$ | $.0^{2}79$ | .021 | .039 | .057 | .075 | .092 | .108 | .123 | .137 | .149 | .001 |
| | .005 | $.0^{4}39$ | $.0^{2}50$ | .023 | .049 | .076 | .102 | .125 | .147 | .166 | .183 | .198 | .212 | .005 |
| | .01 | $.0^{3}16$ | .010 | .037 | .070 | .103 | .132 | .158 | .181 | .202 | .219 | .235 | .249 | .01 |
| | .025 | $.0^{2}10$ | .025 | .070 | .116 | .156 | .190 | .219 | .244 | .265 | .284 | .300 | .315 | .025 |
| | .05 | $.0^{2}41$ | .051 | .115 | .170 | .216 | .254 | .285 | .311 | .333 | .351 | .368 | .382 | .05 |
| | .10 | .016 | .106 | .192 | .258 | .309 | .348 | .380 | .406 | .427 | .446 | .461 | .475 | .10 |
| | .25 | .105 | .293 | .407 | .480 | .531 | .568 | .596 | .618 | .637 | .652 | .667 | .676 | .25 |
| | .50 | .478 | .726 | .826 | .878 | .911 | .933 | .948 | .960 | .970 | .977 | .984 | .989 | .50 |
| | .75 | 1.43 | 1.52 | 1.52 | 1.51 | 1.49 | 1.48 | 1.47 | 1.46 | 1.46 | 1.45 | 1.44 | 1.44 | .75 |
| | .90 | 3.07 | 2.70 | 2.49 | 2.36 | 2.27 | 2.21 | 2.16 | 2.12 | 2.09 | 2.06 | 2.04 | 2.02 | .90 |
| | .95 | 4.54 | 3.68 | 3.29 | 3.06 | 2.90 | 2.79 | 2.71 | 2.64 | 2.59 | 2.54 | 2.51 | 2.48 | .95 |
| | .975 | 6.20 | 4.76 | 4.15 | 3.80 | 3.58 | 3.41 | 3.29 | 3.20 | 3.12 | 3.06 | 3.01 | 2.96 | .975 |
| | .99 | 8.68 | 6.36 | 5.42 | 4.89 | 4.56 | 4.32 | 4.14 | 4.00 | 3.89 | 3.80 | 3.73 | 3.67 | .99 |
| | .995 | 10.8 | 7.70 | 6.48 | 5.80 | 5.37 | 5.07 | 4.85 | 4.67 | 4.54 | 4.42 | 4.33 | 4.25 | .995 |
| | .999 | 16.6 | 11.3 | 9.34 | 8.25 | 7.57 | 7.09 | 6.74 | 6.47 | 6.26 | 6.08 | 5.93 | 5.81 | .999 |
| | .9995 | 19.5 | 13.2 | 10.8 | 9.48 | 8.66 | 8.10 | 7.68 | 7.36 | 7.11 | 6.91 | 6.75 | 6.60 | .9995 |
| **20** | .0005 | $.0^{6}40$ | $.0^{3}50$ | $.0^{2}50$ | .015 | .029 | .046 | .063 | .079 | .094 | .109 | .123 | .136 | .0005 |
| | .001 | $.0^{5}16$ | $.0^{2}10$ | $.0^{2}79$ | .022 | .039 | .058 | .077 | .095 | .112 | .128 | .143 | .156 | .001 |
| | .005 | $.0^{4}39$ | $.0^{2}50$ | .023 | .050 | .077 | .104 | .129 | .151 | .171 | .190 | .206 | .221 | .005 |
| | .01 | $.0^{3}16$ | .010 | .037 | .071 | .105 | .135 | .162 | .187 | .208 | .227 | .244 | .259 | .01 |
| | .025 | $.0^{2}10$ | .025 | .071 | .117 | .158 | .193 | .224 | .250 | .273 | .292 | .310 | .325 | .025 |
| | .05 | $.0^{2}40$ | .051 | .115 | .172 | .219 | .258 | .290 | .318 | .340 | .360 | .377 | .393 | .05 |
| | .10 | .016 | .106 | .193 | .260 | .312 | .353 | .385 | .412 | .435 | .454 | .472 | .485 | .10 |
| | .25 | .104 | .292 | .407 | .480 | .531 | .569 | .598 | .622 | .641 | .656 | .671 | .681 | .25 |
| | .50 | .472 | .718 | .816 | .868 | .900 | .922 | .938 | .950 | .959 | .966 | .972 | .977 | .50 |
| | .75 | 1.40 | 1.49 | 1.48 | 1.47 | 1.45 | 1.44 | 1.43 | 1.42 | 1.41 | 1.40 | 1.39 | 1.39 | .75 |
| | .90 | 2.97 | 2.59 | 2.38 | 2.25 | 2.16 | 2.09 | 2.04 | 2.00 | 1.96 | 1.94 | 1.91 | 1.89 | .90 |
| | .95 | 4.35 | 3.49 | 3.10 | 2.87 | 2.71 | 2.60 | 2.51 | 2.45 | 2.39 | 2.35 | 2.31 | 2.28 | .95 |
| | .975 | 5.87 | 4.46 | 3.86 | 3.51 | 3.29 | 3.13 | 3.01 | 2.91 | 2.84 | 2.77 | 2.72 | 2.68 | .975 |
| | .99 | 8.10 | 5.85 | 4.94 | 4.43 | 4.10 | 3.87 | 3.70 | 3.56 | 3.46 | 3.37 | 3.29 | 3.23 | .99 |
| | .995 | 9.94 | 6.99 | 5.82 | 5.17 | 4.76 | 4.47 | 4.26 | 4.09 | 3.96 | 3.85 | 3.76 | 3.68 | .995 |
| | .999 | 14.8 | 9.95 | 8.10 | 7.10 | 6.46 | 6.02 | 5.69 | 5.44 | 5.24 | 5.08 | 4.94 | 4.82 | .999 |
| | .9995 | 17.2 | 11.4 | 9.20 | 8.02 | 7.28 | 6.76 | 6.38 | 6.08 | 5.85 | 5.66 | 5.51 | 5.38 | .9995 |
| **24** | .0005 | $.0^{6}40$ | $.0^{3}50$ | $.0^{2}50$ | .015 | .030 | .046 | .064 | .080 | .096 | .112 | .126 | .139 | .0005 |
| | .001 | $.0^{5}16$ | $.0^{2}10$ | $.0^{2}79$ | .022 | .040 | .059 | .079 | .097 | .115 | .131 | .146 | .160 | .001 |
| | .005 | $.0^{4}40$ | $.0^{2}50$ | .023 | .050 | .078 | .106 | .131 | .154 | .175 | .193 | .210 | .226 | .005 |
| | .01 | $.0^{3}16$ | .010 | .038 | .072 | .106 | .137 | .165 | .189 | .211 | .231 | .249 | .264 | .01 |
| | .025 | $.0^{2}10$ | .025 | .071 | .117 | .159 | .195 | .227 | .253 | .277 | .297 | .315 | .331 | .025 |
| | .05 | $.0^{2}40$ | .051 | .116 | .173 | .221 | .260 | .293 | .321 | .345 | .365 | .383 | .399 | .05 |
| | .10 | .016 | .106 | .193 | .261 | .313 | .355 | .388 | .416 | .439 | .459 | .476 | .491 | .10 |
| | .25 | .104 | .291 | .406 | .480 | .532 | .570 | .600 | .623 | .643 | .659 | .671 | .684 | .25 |
| | .50 | .469 | .714 | .812 | .863 | .895 | .917 | .932 | .944 | .953 | .961 | .967 | .972 | .50 |
| | .75 | 1.39 | 1.47 | 1.46 | 1.44 | 1.43 | 1.41 | 1.40 | 1.39 | 1.38 | 1.38 | 1.37 | 1.36 | .75 |
| | .90 | 2.93 | 2.54 | 2.33 | 2.19 | 2.10 | 2.04 | 1.98 | 1.94 | 1.91 | 1.88 | 1.85 | 1.83 | .90 |
| | .95 | 4.26 | 3.40 | 3.01 | 2.78 | 2.62 | 2.51 | 2.42 | 2.36 | 2.30 | 2.25 | 2.21 | 2.18 | .95 |
| | .975 | 5.72 | 4.32 | 3.72 | 3.38 | 3.15 | 2.99 | 2.87 | 2.78 | 2.70 | 2.64 | 2.59 | 2.54 | .975 |
| | .99 | 7.82 | 5.61 | 4.72 | 4.22 | 3.90 | 3.67 | 3.50 | 3.36 | 3.26 | 3.17 | 3.09 | 3.03 | .99 |
| | .995 | 9.55 | 6.66 | 5.52 | 4.89 | 4.49 | 4.20 | 3.99 | 3.83 | 3.69 | 3.59 | 3.50 | 3.42 | .995 |
| | .999 | 14.0 | 9.34 | 7.55 | 6.59 | 5.98 | 5.55 | 5.23 | 4.99 | 4.80 | 4.64 | 4.50 | 4.39 | .999 |
| | .9995 | 16.2 | 10.6 | 8.52 | 7.39 | 6.68 | 6.18 | 5.82 | 5.54 | 5.31 | 5.13 | 4.98 | 4.85 | .9995 |

$\nu_1$, DEGREES OF FREEDOM FOR NUMERATOR

| *Cum. prop.* | 15 | 20 | 24 | 30 | 40 | 50 | 60 | 100 | 120 | 200 | 500 | ∞ | *Cum. prop.* | $\nu_2$, DEGREES OF FREEDOM FOR DENOMINATOR |
|---|---|---|---|---|---|---|---|---|---|---|---|---|---|---|
| .0005 | .159 | .197 | .220 | .244 | .272 | .290 | .303 | .330 | .339 | .353 | .368 | .377 | .0005 | **15** |
| .001 | .181 | .219 | .242 | .266 | .294 | .313 | .325 | .352 | .360 | .375 | .388 | .398 | .001 | |
| .005 | .246 | .286 | .308 | .333 | .360 | .377 | .389 | .415 | .422 | .435 | .448 | .457 | .005 | |
| .01 | .284 | .324 | .346 | .370 | .397 | .413 | .425 | .450 | .456 | .469 | .483 | .490 | .01 | |
| .025 | .349 | .389 | .410 | .433 | .458 | .474 | .485 | .508 | .514 | .526 | .538 | .546 | .025 | |
| .05 | .416 | .454 | .474 | .496 | .519 | .535 | .545 | .565 | .571 | .581 | .592 | .600 | .05 | |
| .10 | .507 | .542 | .561 | .581 | .602 | .614 | .624 | .641 | .647 | .658 | .667 | .672 | .10 | |
| .25 | .701 | .728 | .742 | .757 | .772 | .782 | .788 | .802 | .805 | .812 | .818 | .822 | .25 | |
| .50 | 1.00 | 1.01 | 1.02 | 1.02 | 1.03 | 1.03 | 1.03 | 1.04 | 1.04 | 1.04 | 1.04 | 1.05 | .50 | |
| .75 | 1.43 | 1.41 | 1.41 | 1.40 | 1.39 | 1.39 | 1.38 | 1.38 | 1.37 | 1.37 | 1.36 | 1.36 | .75 | |
| .90 | 1.97 | 1.92 | 1.90 | 1.87 | 1.85 | 1.83 | 1.82 | 1.79 | 1.79 | 1.77 | 1.76 | 1.76 | .90 | |
| .95 | 2.40 | 2.33 | 2.39 | 2.25 | 2.20 | 2.18 | 2.16 | 2.12 | 2.11 | 2.10 | 2.08 | 2.07 | .95 | |
| .975 | 2.86 | 2.76 | 2.70 | 2.64 | 2.59 | 2.55 | 2.52 | 2.47 | 2.46 | 2.44 | 2.41 | 2.40 | .975 | |
| .99 | 3.52 | 3.37 | 3.29 | 3.21 | 3.13 | 3.08 | 3.05 | 2.98 | 2.96 | 2.92 | 2.89 | 2.87 | .99 | |
| .995 | 4.07 | 3.88 | 3.79 | 3.69 | 3.59 | 3.52 | 3.48 | 3.39 | 3.37 | 3.33 | 3.29 | 3.26 | .995 | |
| .999 | 5.54 | 5.25 | 5.10 | 4.95 | 4.80 | 4.70 | 4.64 | 4.51 | 4.47 | 4.41 | 4.35 | 4.31 | .999 | |
| .9995 | 6.27 | 5.93 | 5.75 | 5.58 | 5.40 | 5.29 | 5.21 | 5.06 | 5.02 | 4.94 | 4.87 | 4.83 | .9995 | |
| .0005 | .169 | .211 | .235 | .263 | .295 | .316 | .331 | .364 | .375 | .391 | .408 | .422 | .0005 | **20** |
| .001 | .191 | .233 | .258 | .286 | .318 | .339 | .354 | .386 | .395 | .413 | .429 | .441 | .001 | |
| .005 | .258 | .301 | .327 | .354 | .385 | .405 | .419 | .448 | .457 | .474 | .490 | .500 | .005 | |
| .01 | .297 | .340 | .365 | .392 | .422 | .441 | .455 | .483 | .491 | .508 | .521 | .532 | .01 | |
| .025 | .363 | .406 | .430 | .456 | .484 | .503 | .514 | .541 | .548 | .562 | .575 | .585 | .025 | |
| .05 | .430 | .471 | .493 | .518 | .544 | .562 | .572 | .595 | .603 | .617 | .629 | .637 | .05 | |
| .10 | .520 | .557 | .578 | .600 | .623 | .637 | .648 | .671 | .675 | .685 | .694 | .704 | .10 | |
| .25 | .708 | .736 | .751 | .767 | .784 | .794 | .801 | .816 | .820 | .827 | .835 | .840 | .25 | |
| .50 | .989 | 1.00 | 1.01 | 1.01 | 1.02 | 1.02 | 1.02 | 1.03 | 1.03 | 1.03 | 1.03 | 1.03 | .50 | |
| .75 | 1.37 | 1.36 | 1.35 | 1.34 | 1.33 | 1.33 | 1.32 | 1.31 | 1.31 | 1.30 | 1.30 | 1.29 | .75 | |
| .90 | 1.84 | 1.79 | 1.77 | 1.74 | 1.71 | 1.69 | 1.68 | 1.65 | 1.64 | 1.63 | 1.62 | 1.61 | .90 | |
| .95 | 2.20 | 2.12 | 2.08 | 2.04 | 1.99 | 1.97 | 1.95 | 1.91 | 1.90 | 1.88 | 1.86 | 1.84 | .95 | |
| .975 | 2.57 | 2.46 | 2.41 | 2.35 | 2.29 | 2.25 | 2.22 | 2.17 | 2.16 | 2.13 | 2.10 | 2.09 | .975 | |
| .99 | 3.09 | 2.94 | 2.86 | 2.78 | 2.69 | 2.64 | 2.61 | 2.54 | 2.52 | 2.48 | 2.44 | 2.42 | .99 | |
| .995 | 3.50 | 3.32 | 3.22 | 3.12 | 3.02 | 2.96 | 2.92 | 2.83 | 2.81 | 2.76 | 2.72 | 2.69 | .995 | |
| .999 | 4.56 | 4.29 | 4.15 | 4.01 | 3.86 | 3.77 | 3.70 | 3.58 | 3.54 | 3.48 | 3.42 | 3.38 | .999 | |
| .9995 | 5.07 | 4.75 | 4.58 | 4.42 | 4.24 | 4.15 | 4.07 | 3.93 | 3.90 | 3.82 | 3.75 | 3.70 | .9995 | |
| .0005 | .174 | .218 | .244 | .274 | .309 | .331 | .349 | .384 | .395 | .416 | .434 | .449 | .0005 | **24** |
| .001 | .196 | .241 | .268 | .298 | .332 | .354 | .371 | .405 | .417 | .437 | .455 | .469 | .001 | |
| .005 | .264 | .310 | .337 | .367 | .400 | .422 | .437 | .469 | .479 | .498 | .515 | .527 | .005 | |
| .01 | .304 | .350 | .376 | .405 | .437 | .459 | .473 | .505 | .513 | .529 | .546 | .558 | .01 | |
| .025 | .370 | .415 | .441 | .468 | .498 | .518 | .531 | .562 | .568 | .585 | .599 | .610 | .025 | |
| .05 | .437 | .480 | .504 | .530 | .558 | .575 | .588 | .613 | .622 | .637 | .649 | .659 | .05 | |
| .10 | .527 | .566 | .588 | .611 | .635 | .651 | .662 | .685 | .691 | .704 | .715 | .723 | .10 | |
| .25 | .712 | .741 | .757 | .773 | .791 | .802 | .809 | .825 | .829 | .837 | .844 | .850 | .25 | |
| .50 | .983 | .994 | 1.00 | 1.01 | 1.01 | 1.02 | 1.02 | 1.02 | 1.02 | 1.02 | 1.03 | 1.03 | .50 | |
| .75 | 1.35 | 1.33 | 1.32 | 1.31 | 1.30 | 1.29 | 1.29 | 1.28 | 1.28 | 1.27 | 1.27 | 1.26 | .75 | |
| .90 | 1.78 | 1.73 | 1.70 | 1.67 | 1.64 | 1.62 | 1.61 | 1.58 | 1.57 | 1.56 | 1.54 | 1.53 | .90 | |
| .95 | 2.11 | 2.03 | 1.98 | 1.94 | 1.89 | 1.86 | 1.84 | 1.80 | 1.79 | 1.77 | 1.75 | 1.73 | .95 | |
| .975 | 2.44 | 2.33 | 2.27 | 2.21 | 2.15 | 2.11 | 2.08 | 2.02 | 2.01 | 1.98 | 1.95 | 1.94 | .975 | |
| .99 | 2.89 | 2.74 | 2.66 | 2.58 | 2.49 | 2.44 | 2.40 | 2.33 | 2.31 | 2.27 | 2.24 | 2.21 | .99 | |
| .995 | 3.25 | 3.06 | 2.97 | 2.87 | 2.77 | 2.70 | 2.66 | 2.57 | 2.55 | 2.50 | 2.46 | 2.43 | .995 | |
| .999 | 4.14 | 3.87 | 3.74 | 3.59 | 3.45 | 3.35 | 3.29 | 3.16 | 3.14 | 3.07 | 3.01 | 2.97 | .999 | |
| .9995 | 4.55 | 4.25 | 4.09 | 3.93 | 3.76 | 3.66 | 3.59 | 3.44 | 3.41 | 3.33 | 3.27 | 3.22 | .9995 | |

$\nu_1$, DEGREES OF FREEDOM FOR NUMERATOR

| $\nu_2$, degrees of freedom for denominator | *Cum. prop.* | 1 | 2 | 3 | 4 | 5 | 6 | 7 | 8 | 9 | 10 | 11 | 12 | *Cum. prop.* |
|---|---|---|---|---|---|---|---|---|---|---|---|---|---|---|
| **30** | .0005 | .0$^6$40 | .0$^3$50 | .0$^2$50 | .015 | .030 | .047 | .065 | .082 | .098 | .114 | .129 | .143 | .0005 |
| | .001 | .0$^5$16 | .0$^2$10 | .0$^2$80 | .022 | .040 | .060 | .080 | .099 | .117 | .134 | .150 | .164 | .001 |
| | .005 | .0$^4$40 | .0$^2$50 | .024 | .050 | .079 | .107 | .133 | .156 | .178 | .197 | .215 | .231 | .005 |
| | .01 | .0$^3$16 | .010 | .038 | .072 | .107 | .138 | .167 | .192 | .215 | .235 | .254 | .270 | .01 |
| | .025 | .0$^2$10 | .025 | .071 | .118 | .161 | .197 | .229 | .257 | .281 | .302 | .321 | .337 | .025 |
| | .05 | .0$^2$40 | .051 | .116 | .174 | .222 | .263 | .296 | .325 | .349 | .370 | .389 | .406 | .05 |
| | .10 | .016 | .106 | .193 | .262 | .315 | .357 | .391 | .420 | .443 | .464 | .481 | .497 | .10 |
| | .25 | .103 | .290 | .406 | .480 | .532 | .571 | .601 | .625 | .645 | .661 | .676 | .688 | .25 |
| | .50 | .466 | .709 | .807 | .858 | .890 | .912 | .927 | .939 | .948 | .955 | .961 | .966 | .50 |
| | .75 | 1.38 | 1.45 | 1.44 | 1.42 | 1.41 | 1.39 | 1.38 | 1.37 | 1.36 | 1.35 | 1.35 | 1.34 | .75 |
| | .90 | 2.88 | 2.49 | 2.28 | 2.14 | 2.05 | 1.98 | 1.93 | 1.88 | 1.85 | 1.82 | 1.79 | 1.77 | .90 |
| | .95 | 4.17 | 3.32 | 2.92 | 2.69 | 2.53 | 2.42 | 2.33 | 2.27 | 2.21 | 2.16 | 2.13 | 2.09 | .95 |
| | .975 | 5.57 | 4.18 | 3.59 | 3.25 | 3.03 | 2.87 | 2.75 | 2.65 | 2.57 | 2.51 | 2.46 | 2.41 | .975 |
| | .99 | 7.56 | 5.39 | 4.51 | 4.02 | 3.70 | 3.47 | 3.30 | 3.17 | 3.07 | 2.98 | 2.91 | 2.84 | .99 |
| | .995 | 9.18 | 6.35 | 5.24 | 4.62 | 4.23 | 3.95 | 3.74 | 3.58 | 3.45 | 3.34 | 3.25 | 3.18 | .995 |
| | .999 | 13.3 | 8.77 | 7.05 | 6.12 | 5.53 | 5.12 | 4.82 | 4.58 | 4.39 | 4.24 | 4.11 | 4.00 | .999 |
| | .9995 | 15.2 | 9.90 | 7.90 | 6.82 | 6.14 | 5.66 | 5.31 | 5.04 | 4.82 | 4.65 | 4.51 | 4.38 | .9995 |
| **40** | .0005 | .0$^6$40 | .0$^3$50 | .0$^2$50 | .016 | .030 | .048 | .066 | .084 | .100 | .117 | .132 | .147 | .0005 |
| | .001 | .0$^5$16 | .0$^2$10 | .0$^2$80 | .022 | .042 | .061 | .081 | .101 | .119 | .137 | .153 | .169 | .001 |
| | .005 | .0$^4$40 | .0$^2$50 | .024 | .051 | .080 | .108 | .135 | .159 | .181 | .201 | .220 | .237 | .005 |
| | .01 | .0$^3$16 | .010 | .038 | .073 | .108 | .140 | .169 | .195 | .219 | .240 | .259 | .276 | .01 |
| | .025 | .0$^3$99 | .025 | .071 | .119 | .162 | .199 | .232 | .260 | .285 | .307 | .327 | .344 | .025 |
| | .05 | .0$^2$40 | .051 | .116 | .175 | .224 | .265 | .299 | .329 | .354 | .376 | .395 | .412 | .05 |
| | .10 | .016 | .106 | .194 | .263 | .317 | .360 | .394 | .424 | .448 | .469 | .488 | .504 | .10 |
| | .25 | .103 | .290 | .405 | .480 | .533 | .572 | .603 | .627 | .647 | .664 | .680 | .691 | .25 |
| | .50 | .463 | .705 | .802 | .854 | .885 | .907 | .922 | .934 | .943 | .950 | .956 | .961 | .50 |
| | .75 | 1.36 | 1.44 | 1.42 | 1.40 | 1.39 | 1.37 | 1.36 | 1.35 | 1.34 | 1.33 | 1.32 | 1.31 | .75 |
| | .90 | 2.84 | 2.44 | 2.23 | 2.09 | 2.00 | 1.93 | 1.87 | 1.83 | 1.79 | 1.76 | 1.73 | 1.71 | .90 |
| | .95 | 4.08 | 3.23 | 2.84 | 2.61 | 2.45 | 2.34 | 2.25 | 2.18 | 2.12 | 2.08 | 2.04 | 2.00 | .95 |
| | .975 | 5.42 | 4.05 | 3.46 | 3.13 | 2.90 | 2.74 | 2.62 | 2.53 | 2.45 | 2.39 | 2.33 | 2.29 | .975 |
| | .99 | 7.31 | 5.18 | 4.31 | 3.83 | 3.51 | 3.29 | 3.12 | 2.99 | 2.89 | 2.80 | 2.73 | 2.66 | .99 |
| | .995 | 8.83 | 6.07 | 4.98 | 4.37 | 3.99 | 3.71 | 3.51 | 3.35 | 3.22 | 3.12 | 3.03 | 2.95 | .995 |
| | .999 | 12.6 | 8.25 | 6.60 | 5.70 | 5.13 | 4.73 | 4.44 | 4.21 | 4.02 | 3.87 | 3.75 | 3.64 | .999 |
| | .9995 | 14.4 | 9.25 | 7.33 | 6.30 | 5.64 | 5.19 | 4.85 | 4.59 | 4.38 | 4.21 | 4.07 | 3.95 | .9995 |
| **60** | .0005 | .0$^6$40 | .0$^3$50 | .0$^2$51 | .016 | .031 | .048 | .067 | .085 | .103 | .120 | .136 | .152 | .0005 |
| | .001 | .0$^5$16 | .0$^2$10 | .0$^2$80 | .022 | .041 | .062 | .083 | .103 | .122 | .140 | .157 | .174 | .001 |
| | .005 | .0$^4$40 | .0$^2$50 | .024 | .051 | .081 | .110 | .137 | .162 | .185 | .206 | .225 | .243 | .005 |
| | .01 | .0$^3$16 | .010 | .038 | .073 | .109 | .142 | .172 | .199 | .223 | .245 | .265 | .283 | .01 |
| | .025 | .0$^3$99 | .025 | .071 | .120 | .163 | .202 | .235 | .264 | .290 | .313 | .333 | .351 | .025 |
| | .05 | .0$^2$40 | .051 | .116 | .176 | .226 | .267 | .303 | .333 | .359 | .382 | .402 | .419 | .05 |
| | .10 | .016 | .106 | .194 | .264 | .318 | .362 | .398 | .428 | .453 | .475 | .493 | .510 | .10 |
| | .25 | .102 | .289 | .405 | .480 | .534 | .573 | .604 | .629 | .650 | .667 | .680 | .695 | .25 |
| | .50 | .461 | .701 | .798 | .849 | .880 | .901 | .917 | .928 | .937 | .945 | .951 | .956 | .50 |
| | .75 | 1.35 | 1.42 | 1.41 | 1.38 | 1.37 | 1.35 | 1.33 | 1.32 | 1.31 | 1.30 | 1.29 | 1.29 | .75 |
| | .90 | 2.79 | 2.39 | 2.18 | 2.04 | 1.95 | 1.87 | 1.82 | 1.77 | 1.74 | 1.71 | 1.68 | 1.66 | .90 |
| | .95 | 4.00 | 3.15 | 2.76 | 2.53 | 2.37 | 2.25 | 2.17 | 2.10 | 2.04 | 1.99 | 1.95 | 1.92 | .95 |
| | .975 | 5.29 | 3.93 | 3.34 | 3.01 | 2.79 | 2.63 | 2.51 | 2.41 | 2.33 | 2.27 | 2.22 | 2.17 | .975 |
| | .99 | 7.08 | 4.98 | 4.13 | 3.65 | 3.34 | 3.12 | 2.95 | 2.82 | 2.72 | 2.63 | 2.56 | 2.50 | .99 |
| | .995 | 8.49 | 5.80 | 4.73 | 4.14 | 3.76 | 3.49 | 3.29 | 3.13 | 3.01 | 2.90 | 2.82 | 2.74 | .995 |
| | .999 | 12.0 | 7.76 | 6.17 | 5.31 | 4.76 | 4.37 | 4.09 | 3.87 | 3.69 | 3.54 | 3.43 | 3.31 | .999 |
| | .9995 | 13.6 | 8.65 | 6.81 | 5.82 | 5.20 | 4.76 | 4.44 | 4.18 | 3.98 | 3.82 | 3.69 | 3.57 | .9995 |

$\nu_1$, DEGREES OF FREEDOM FOR NUMERATOR

| *Cum. prop.* | 15 | 20 | 24 | 30 | 40 | 50 | 60 | 100 | 120 | 200 | 500 | ∞ | *Cum. prop.* | $\nu_2$, DEGREES OF FREEDOM FOR DENOMINATOR |
|---|---|---|---|---|---|---|---|---|---|---|---|---|---|---|
| .0005 | .179 | .226 | .254 | .287 | .325 | .350 | .369 | .410 | .420 | .444 | .467 | .483 | .0005 | **30** |
| .001 | .202 | .250 | .278 | .311 | .348 | .373 | .391 | .431 | .442 | .465 | .488 | .503 | .001 | |
| .005 | .271 | .320 | .349 | .381 | .416 | .441 | .457 | .495 | .504 | .524 | .543 | .559 | .005 | |
| .01 | .311 | .360 | .388 | .419 | .454 | .476 | .493 | .529 | .538 | .559 | .575 | .590 | .01 | |
| .025 | .378 | .426 | .453 | .482 | .515 | .535 | .551 | .585 | .592 | .610 | .625 | .639 | .025 | |
| .05 | .445 | .490 | .516 | .543 | .573 | .592 | .606 | .637 | .644 | .658 | .676 | .685 | .05 | |
| .10 | .534 | .575 | .598 | .623 | .649 | .667 | .678 | .704 | .710 | .725 | .735 | .746 | .10 | |
| .25 | .716 | .746 | .763 | .780 | .798 | .810 | .818 | .835 | .839 | .848 | .856 | .862 | .25 | |
| .50 | .978 | .989 | .994 | 1.00 | 1.01 | 1.01 | 1.01 | 1.02 | 1.02 | 1.02 | 1.02 | 1.02 | .50 | |
| .75 | 1.32 | 1.30 | 1.29 | 1.28 | 1.27 | 1.26 | 1.26 | 1.25 | 1.24 | 1.24 | 1.23 | 1.23 | .75 | |
| .90 | 1.72 | 1.67 | 1.64 | 1.61 | 1.57 | 1.55 | 1.54 | 1.51 | 1.50 | 1.48 | 1.47 | 1.46 | .90 | |
| .95 | 2.01 | 1.93 | 1.89 | 1.84 | 1.79 | 1.76 | 1.74 | 1.70 | 1.68 | 1.66 | 1.64 | 1.62 | .95 | |
| .975 | 2.31 | 2.20 | 2.14 | 2.07 | 2.01 | 1.97 | 1.94 | 1.88 | 1.87 | 1.84 | 1.81 | 1.79 | .975 | |
| .99 | 2.70 | 2.55 | 2.47 | 2.39 | 2.30 | 2.25 | 2.21 | 2.13 | 2.11 | 2.07 | 2.03 | 2.01 | .99 | |
| .995 | 3.01 | 2.82 | 2.73 | 2.63 | 2.52 | 2.46 | 2.42 | 2.32 | 2.30 | 2.25 | 2.21 | 2.18 | .995 | |
| .999 | 3.75 | 3.49 | 3.36 | 3.22 | 3.07 | 2.98 | 2.92 | 2.79 | 2.76 | 2.69 | 2.63 | 2.59 | .999 | |
| .9995 | 4.10 | 3.80 | 3.65 | 3.48 | 3.32 | 3.22 | 3.15 | 3.00 | 2.97 | 2.89 | 2.82 | 2.78 | .9995 | |
| .0005 | .185 | .236 | .266 | .301 | .343 | .373 | .393 | .441 | .453 | .480 | .504 | .525 | .0005 | **40** |
| .001 | .209 | .259 | .290 | .326 | .367 | .396 | .415 | .461 | .473 | .500 | .524 | .545 | .001 | |
| .005 | .279 | .331 | .362 | .396 | .436 | .463 | .481 | .524 | .534 | .559 | .581 | .599 | .005 | |
| .01 | .319 | .371 | .401 | .435 | .473 | .498 | .516 | .556 | .567 | .592 | .613 | .628 | .01 | |
| .025 | .387 | .437 | .466 | .498 | .533 | .556 | .573 | .610 | 620 | .641 | .662 | .674 | .025 | |
| .05 | .454 | .502 | .529 | .558 | .591 | .613 | .627 | .658 | .669 | .685 | .704 | .717 | .05 | |
| .10 | .542 | .585 | .609 | .636 | .664 | .683 | .696 | .724 | .731 | .747 | .762 | .772 | .10 | |
| .25 | .720 | .752 | .769 | .787 | .806 | .819 | .828 | .846 | .851 | .861 | .870 | .877 | .25 | |
| .50 | .972 | .983 | .989 | .994 | 1.00 | 1.00 | 1.01 | 1.01 | 1.01 | 1.01 | 1.02 | 1.02 | .50 | |
| .75 | 1.30 | 1.28 | 1.26 | 1.25 | 1.24 | 1.23 | 1.22 | 1.21 | 1.21 | 1.20 | 1.19 | 1.19 | .75 | |
| .90 | 1.66 | 1.61 | 1.57 | 1.54 | 1.51 | 1.48 | 1.47 | 1.43 | 1.42 | 1.41 | 1.39 | 1.38 | .90 | |
| .95 | 1.92 | 1.84 | 1.79 | 1.74 | 1.69 | 1.66 | 1.64 | 1.59 | 1.58 | 1.55 | 1.53 | 1.51 | .95 | |
| .975 | 2.18 | 2.07 | 2.01 | 1.94 | 1.88 | 1.83 | 1.80 | 1.74 | 1.72 | 1.69 | 1.66 | 1.64 | .975 | |
| .99 | 2.52 | 2.37 | 2.29 | 2.20 | 2.11 | 2.06 | 2.02 | 1.94 | 1.92 | 1.87 | 1.83 | 1.80 | .99 | |
| .995 | 2.78 | 2.60 | 2.50 | 2.40 | 2.30 | 2.23 | 2.18 | 2.09 | 2.06 | 2.01 | 1.96 | 1.93 | .995 | |
| .999 | 3.40 | 3.15 | 3.01 | 2.87 | 2.73 | 2.64 | 2.57 | 2.44 | 2.41 | 2.34 | 2.28 | 2.23 | .999 | |
| .9995 | 3.68 | 3.39 | 3.24 | 3.08 | 2.92 | 2.82 | 2.74 | 2.60 | 2.57 | 2.49 | 2.41 | 2.37 | .9995 | |
| .0005 | .192 | .246 | .278 | .318 | .365 | .398 | .421 | .478 | .493 | .527 | .561 | .585 | .0005 | **60** |
| .001 | .216 | .270 | .304 | .343 | .389 | .421 | .444 | .497 | .512 | .545 | .579 | .602 | .001 | |
| .005 | .287 | .343 | .376 | .414 | .458 | .488 | .510 | .559 | .572 | .602 | .633 | .652 | .005 | |
| .01 | .328 | .383 | .416 | .453 | .495 | .524 | .545 | .592 | .604 | .633 | .658 | .679 | .01 | |
| .025 | .396 | .450 | .481 | .515 | .555 | .581 | .600 | .641 | .654 | .680 | .704 | .720 | .025 | |
| .05 | .463 | .514 | .543 | .575 | .611 | .633 | .652 | .690 | .700 | .719 | .746 | .759 | .05 | |
| .10 | .550 | .596 | .622 | .650 | .682 | .703 | .717 | .750 | .758 | .776 | .793 | .806 | .10 | |
| .25 | .725 | .758 | .776 | .796 | .816 | .830 | .840 | .860 | .865 | .877 | .888 | .896 | .25 | |
| .50 | .967 | .978 | .983 | .989 | .994 | .998 | 1.00 | 1.00 | 1.01 | 1.01 | 1.01 | 1.01 | .50 | |
| .75 | 1.27 | 1.25 | 1.24 | 1.22 | 1.21 | 1.20 | 1.19 | 1.17 | 1.17 | 1.16 | 1.15 | 1.15 | .75 | |
| .90 | 1.60 | 1.54 | 1.51 | 1.48 | 1.44 | 1.41 | 1.40 | 1.36 | 1.35 | 1.33 | 1.31 | 1.29 | .90 | |
| .95 | 1.84 | 1.75 | 1.70 | 1.65 | 1.59 | 1.56 | 1.53 | 1.48 | 1.47 | 1.44 | 1.41 | 1.39 | .95 | |
| .975 | 2.06 | 1.94 | 1.88 | 1.82 | 1.74 | 1.70 | 1.67 | 1.60 | 1.58 | 1.54 | 1.51 | 1.48 | .975 | |
| .99 | 2.35 | 2.20 | 2.12 | 2.03 | 1.94 | 1.88 | 1.84 | 1.75 | 1.73 | 1.68 | 1.63 | 1.60 | .99 | |
| .995 | 2.57 | 2.39 | 2.29 | 2.19 | 2.08 | 2.01 | 1.96 | 1.86 | 1.83 | 1.78 | 1.73 | 1.69 | .995 | |
| .999 | 3.08 | 2.83 | 2.69 | 2.56 | 2.41 | 2.31 | 2.25 | 2.11 | 2.09 | 2.01 | 1.93 | 1.89 | .999 | |
| .9995 | 3.30 | 3.02 | 2.87 | 2.71 | 2.55 | 2.45 | 2.38 | 2.23 | 2.19 | 2.11 | 2.03 | 1.98 | .9995 | |

$\nu_1$, DEGREES OF FREEDOM FOR NUMERATOR

| $\nu_2$, DEGREES OF FREEDOM FOR DENOMINATOR | *Cum. prop.* | 1 | 2 | 3 | 4 | 5 | 6 | 7 | 8 | 9 | 10 | 11 | 12 | *Cum. prop.* |
|---|---|---|---|---|---|---|---|---|---|---|---|---|---|---|
| **120** | .0005 | $.0^{6}40$ | $.0^{3}50$ | $.0^{2}51$ | .016 | .031 | .049 | .067 | .087 | .105 | .123 | .140 | .156 | .0005 |
| | .001 | $.0^{5}16$ | $.0^{2}10$ | $.0^{2}81$ | .023 | .042 | .063 | .084 | .105 | .125 | .144 | .162 | .179 | .001 |
| | .005 | $.0^{4}39$ | $.0^{2}50$ | .024 | .051 | .081 | .111 | .139 | .165 | .189 | .211 | .230 | .249 | .005 |
| | .01 | $.0^{3}16$ | .010 | .038 | .074 | .110 | .143 | .174 | .202 | .227 | .250 | .271 | .290 | .01 |
| | .025 | $.0^{3}99$ | .025 | .072 | .120 | .165 | .204 | .238 | .268 | .295 | .318 | .340 | .359 | .025 |
| | .05 | $.0^{2}39$ | .051 | .117 | .177 | .227 | .270 | .306 | .337 | .364 | .388 | .408 | .427 | .05 |
| | .10 | .016 | .105 | .194 | .265 | .320 | .365 | .401 | .432 | .458 | .480 | .500 | .518 | .10 |
| | .25 | .102 | .288 | .405 | .481 | .534 | .574 | .606 | .631 | .652 | .670 | .685 | .699 | .25 |
| | .50 | .458 | .697 | .793 | .844 | .875 | .896 | .912 | .923 | .932 | .939 | .945 | .950 | .50 |
| | .75 | 1.34 | 1.40 | 1.39 | 1.37 | 1.35 | 1.33 | 1.31 | 1.30 | 1.29 | 1.28 | 1.27 | 1.26 | .75 |
| | .90 | 2.75 | 2.35 | 2.13 | 1.99 | 1.90 | 1.82 | 1.77 | 1.72 | 1.68 | 1.65 | 1.62 | 1.60 | .90 |
| | .95 | 3.92 | 3.07 | 2.68 | 2.45 | 2.29 | 2.18 | 2.09 | 2.02 | 1.96 | 1.91 | 1.87 | 1.83 | .95 |
| | .975 | 5.15 | 3.80 | 3.23 | 2.89 | 2.67 | 2.52 | 2.39 | 2.30 | 2.22 | 2.16 | 2.10 | 2.05 | .975 |
| | .99 | 6.85 | 4.79 | 3.95 | 3.48 | 3.17 | 2.96 | 2.79 | 2.66 | 2.56 | 2.47 | 2.40 | 2.34 | .99 |
| | .995 | 8.18 | 5.54 | 4.50 | 3.92 | 3.55 | 3.28 | 3.09 | 2.93 | 2.81 | 2.71 | 2.62 | 2.54 | .995 |
| | .999 | 11.4 | 7.32 | 5.79 | 4.95 | 4.42 | 4.04 | 3.77 | 3.55 | 3.38 | 3.24 | 3.12 | 3.02 | .999 |
| | .9995 | 12.8 | 8.10 | 6.34 | 5.39 | 4.79 | 4.37 | 4.07 | 3.82 | 3.63 | 3.47 | 3.34 | 3.22 | .9995 |
| **∞** | .0005 | $.0^{6}39$ | $.0^{3}50$ | $.0^{2}51$ | .016 | .032 | .050 | .069 | .088 | .108 | .127 | .144 | .161 | .0005 |
| | .001 | $.0^{5}16$ | $.0^{2}10$ | $.0^{2}81$ | .023 | .042 | .063 | .085 | .107 | .128 | .148 | .167 | .185 | .001 |
| | .005 | $.0^{4}39$ | $.0^{2}50$ | .024 | .052 | .082 | .113 | .141 | .168 | .193 | .216 | .236 | .256 | .005 |
| | .01 | $.0^{3}16$ | .010 | .038 | .074 | .111 | .145 | .177 | .206 | .232 | .256 | .278 | .298 | .01 |
| | .025 | $.0^{3}98$ | .025 | .072 | .121 | .166 | .206 | .241 | .272 | .300 | .325 | .347 | .367 | .025 |
| | .05 | $.0^{2}39$ | .051 | .117 | .178 | .229 | .273 | .310 | .342 | .369 | .394 | .417 | .436 | .05 |
| | .10 | .016 | .105 | .195 | .266 | .322 | .367 | .405 | .436 | .463 | .487 | .508 | .525 | .10 |
| | .25 | .102 | .288 | .404 | .481 | .535 | .576 | .608 | .634 | .655 | .674 | .690 | .703 | .25 |
| | .50 | .455 | .693 | .789 | .839 | .870 | .891 | .907 | .918 | .927 | .934 | .939 | .945 | .50 |
| | .75 | 1.32 | 1.39 | 1.37 | 1.35 | 1.33 | 1.31 | 1.29 | 1.28 | 1.27 | 1.25 | 1.24 | 1.24 | .75 |
| | .90 | 2.71 | 2.30 | 2.08 | 1.94 | 1.85 | 1.77 | 1.72 | 1.67 | 1.63 | 1.60 | 1.57 | 1.55 | .90 |
| | .95 | 3.84 | 3.00 | 2.60 | 2.37 | 2.21 | 2.10 | 2.01 | 1.94 | 1.88 | 1.83 | 1.79 | 1.75 | .95 |
| | .975 | 5.02 | 3.69 | 3.12 | 2.79 | 2.57 | 2.41 | 2.29 | 2.19 | 2.11 | 2.05 | 1.99 | 1.94 | .975 |
| | .99 | 6.63 | 4.61 | 3.78 | 3.32 | 3.02 | 2.80 | 2.64 | 2.51 | 2.41 | 2.32 | 2.25 | 2.18 | .99 |
| | .995 | 7.88 | 5.30 | 4.28 | 3.72 | 3.35 | 3.09 | 2.90 | 2.74 | 2.62 | 2.52 | 2.43 | 2.36 | .995 |
| | .999 | 10.8 | 6.91 | 5.42 | 4.62 | 4.10 | 3.74 | 3.47 | 3.27 | 3.10 | 2.96 | 2.84 | 2.74 | .999 |
| | .9995 | 12.1 | 7.60 | 5.91 | 5.00 | 4.42 | 4.02 | 3.72 | 3.48 | 3.30 | 3.14 | 3.02 | 2.90 | .9995 |

For sample sizes larger than, say, 30, a fairly good approximation to the $F$ distribution percentiles can be obtained from

$$\log F_\alpha(\nu_1, \nu_2) \approx \left(\frac{a}{\sqrt{h - b}}\right) - cg$$

where $h = 2\nu_1\nu_2/(\nu_1 + \nu_2)$, $g = (\nu_2 - \nu_1)/\nu_1\nu_2$, and $a$, $b$, $c$ are functions of $\alpha$ given below:

VALUES OF $\alpha$

| | $\alpha$ = .50 | .75 | .90 | .95 | .975 | .99 | .995 | .999 | .9995 |
|---|---|---|---|---|---|---|---|---|---|
| $a$ | 0 | .5859 | 1.1131 | 1.4287 | 1.7023 | 2.0206 | 2.2373 | 2.6841 | 2.8580 |
| $b$ | — | .58 | .77 | .95 | 1.14 | 1.40 | 1.61 | 2.09 | 2.30 |
| $c$ | .290 | .355 | .527 | .681 | .846 | 1.073 | 1.250 | 1.672 | 1.857 |

$\nu_1$, DEGREES OF FREEDOM FOR NUMERATOR

| Cum. prop. | 15 | 20 | 24 | 30 | 40 | 50 | 60 | 100 | 120 | 200 | 500 | ∞ | Cum. prop. | $\nu_2$, DEGREES OF FREEDOM FOR DENOMINATOR |
|---|---|---|---|---|---|---|---|---|---|---|---|---|---|---|
| .0005 | .199 | .256 | .293 | .338 | .390 | .429 | .458 | .524 | .543 | .578 | .614 | .676 | .0005 | **120** |
| .001 | .223 | .282 | .319 | .363 | .415 | .453 | .480 | .542 | .568 | .595 | .631 | .691 | .001 | |
| .005 | .297 | .356 | .393 | .434 | .484 | .520 | .545 | .605 | .623 | .661 | .702 | .733 | .005 | |
| .01 | .338 | .397 | .433 | .474 | .522 | .556 | .579 | .636 | .652 | .688 | .725 | .755 | .01 | |
| .025 | .406 | .464 | .498 | .536 | .580 | .611 | .633 | .684 | .698 | .729 | .762 | .789 | .025 | |
| .05 | .473 | .527 | .559 | .594 | .634 | .661 | .682 | .727 | .740 | .767 | .785 | .819 | .05 | |
| .10 | .560 | .609 | .636 | .667 | .702 | .726 | .742 | .781 | .791 | .815 | .838 | .855 | .10 | |
| .25 | .730 | .765 | .784 | .805 | .828 | .843 | .853 | .877 | .884 | .897 | .911 | .923 | .25 | |
| .50 | .961 | .972 | .978 | .983 | .989 | .992 | .994 | 1.00 | 1.00 | 1.00 | 1.01 | 1.01 | .50 | |
| .75 | 1.24 | 1.22 | 1.21 | 1.19 | 1.18 | 1.17 | 1.16 | 1.14 | 1.13 | 1.12 | 1.11 | 1.10 | .75 | |
| .90 | 1.55 | 1.48 | 1.45 | 1.41 | 1.37 | 1.34 | 1.32 | 1.27 | 1.26 | 1.24 | 1.21 | 1.19 | .90 | |
| .95 | 1.75 | 1.66 | 1.61 | 1.55 | 1.50 | 1.46 | 1.43 | 1.37 | 1.35 | 1.32 | 1.28 | 1.25 | .95 | |
| .975 | 1.95 | 1.82 | 1.76 | 1.69 | 1.61 | 1.56 | 1.53 | 1.45 | 1.43 | 1.39 | 1.34 | 1.31 | .975 | |
| .99 | 2.19 | 2.03 | 1.95 | 1.86 | 1.76 | 1.70 | 1.66 | 1.56 | 1.53 | 1.48 | 1.42 | 1.38 | .99 | |
| .995 | 2.37 | 2.19 | 2.09 | 1.98 | 1.87 | 1.80 | 1.75 | 1.64 | 1.61 | 1.54 | 1.48 | 1.43 | .995 | |
| .999 | 2.78 | 2.53 | 2.40 | 2.26 | 2.11 | 2.02 | 1.95 | 1.82 | 1.76 | 1.70 | 1.62 | 1.54 | .999 | |
| .9995 | 2.96 | 2.67 | 2.53 | 2.38 | 2.21 | 2.11 | 2.01 | 1.88 | 1.84 | 1.75 | 1.67 | 1.60 | .9995 | |
| .0005 | .207 | .270 | .311 | .360 | .422 | .469 | .505 | .599 | .624 | .704 | .804 | 1.00 | .0005 | **∞** |
| .001 | .232 | .296 | .338 | .386 | .448 | .493 | .527 | .617 | .649 | .719 | .819 | 1.00 | .001 | |
| .005 | .307 | .372 | .412 | .460 | .518 | .559 | .592 | .671 | .699 | .762 | .843 | 1.00 | .005 | |
| .01 | .349 | .413 | .452 | .499 | .554 | .595 | .625 | .699 | .724 | .782 | .858 | 1.00 | .01 | |
| .025 | .418 | .480 | .517 | .560 | .611 | .645 | .675 | .741 | .763 | .813 | .878 | 1.00 | .025 | |
| .05 | .484 | .543 | .577 | .617 | .663 | .694 | .720 | .781 | .797 | .840 | .896 | 1.00 | .05 | |
| .10 | .570 | .622 | .652 | .687 | .726 | .752 | .774 | .826 | .838 | .877 | .919 | 1.00 | .10 | |
| .25 | .736 | .773 | .793 | .816 | .842 | .860 | .872 | .901 | .910 | .932 | .957 | 1.00 | .25 | |
| .50 | .956 | .967 | .972 | .978 | .983 | .987 | .989 | .993 | .994 | .997 | .999 | 1.00 | .50 | |
| .75 | 1.22 | 1.19 | 1.18 | 1.16 | 1.14 | 1.13 | 1.12 | 1.09 | 1.08 | 1.07 | 1.04 | 1.00 | .75 | |
| .90 | 1.49 | 1.42 | 1.38 | 1.34 | 1.30 | 1.26 | 1.24 | 1.18 | 1.17 | 1.13 | 1.08 | 1.00 | .90 | |
| .95 | 1.67 | 1.57 | 1.52 | 1.46 | 1.39 | 1.35 | 1.32 | 1.24 | 1.22 | 1.17 | 1.11 | 1.00 | .95 | |
| .975 | 1.83 | 1.71 | 1.64 | 1.57 | 1.48 | 1.43 | 1.39 | 1.30 | 1.27 | 1.21 | 1.13 | 1.00 | .975 | |
| .99 | 2.04 | 1.88 | 1.79 | 1.70 | 1.59 | 1.52 | 1.47 | 1.36 | 1.32 | 1.25 | 1.15 | 1.00 | .99 | |
| .995 | 2.19 | 2.00 | 1.90 | 1.79 | 1.67 | 1.59 | 1.53 | 1.40 | 1.36 | 1.28 | 1.17 | 1.00 | .995 | |
| .999 | 2.51 | 2.27 | 2.13 | 1.99 | 1.84 | 1.73 | 1.66 | 1.49 | 1.45 | 1.34 | 1.21 | 1.00 | .999 | |
| .9995 | 2.65 | 2.37 | 2.22 | 2.07 | 1.91 | 1.79 | 1.71 | 1.53 | 1.48 | 1.36 | 1.22 | 1.00 | .9995 | |

The values given in this table are abstracted with permission from the following sources:

1. All values for $\nu_1, \nu_2$ equal to 50, 100, 200, 500 are from A. Hald, *Statistical Tables and Formulas*, John Wiley & Sons, Inc., New York, 1952.

2. For cumulative proportions .5, .75, .9, .95, .975, .99, .995 most of the values are from M. Merrington and C. M. Thompson, *Biometrika*, vol. 33 (1943), p. 73.

3. For cumulative proportions .999 the values are from C. Colcord and L. S. Deming, *Sankhyā*, vol. 2 (1936), p. 423.

4. For cum. prop. $= \alpha < .5$ the values are the reciprocals of values for $1 - \alpha$ (with $\nu_1$ and $\nu_2$ interchanged). The values in Merrington and Thompson and in Colcord and Deming are to five significant figures, and it is hoped (but not expected) that the reciprocals are correct as given. The values in Hald are to three significant figures, and the reciprocals are probably accurate within one to two digits in the third significant figure except for those values very close to unity, where they may be off four to five digits in the third significant figure.

5. Gaps remaining in the table after using the above sources were filled in by interpolation.

$$\alpha = \frac{(\nu_1/\nu_2)^{\frac{1}{2}\nu_1}}{\beta(\frac{1}{2}\nu_1, \frac{1}{2}\nu_2)} \int_{-\infty}^{F\alpha} F^{\frac{1}{2}\nu_1 - 1} \left(1 + \frac{\nu_1 F}{\nu_2}\right)^{-(\nu_1+\nu_2)/2} dF$$

## A-8a PERCENTILE ESTIMATES IN LARGE SAMPLES

### A-8*a*(1) Mean

| | Percentile estimate | *Eff.* |
|---|---|---|
| 1 | $P_{50}$ | .64 |
| 2 | $.5(P_{25} + P_{75})$ | .81 |
| 3 | $.3333(P_{17} + P_{50} + P_{83})$ | .88 |
| 4 | $.25(P_{12.5} + P_{37.5} + P_{62.5} + P_{87.5})$ | .91 |
| 5 | $.20(P_{10} + P_{30} + P_{50} + P_{70} + P_{90})$ | .93 |
| ... | . . . . . . . . . . . . . . . . . . . . . . . . . . . . | ... |
| 10 | $.10(P_{05} + P_{15} + P_{25} + P_{35} + P_{45} + P_{55} + P_{65} + P_{75} + P_{85} + P_{95})$ | .97 |

### A-8*a*(2) Standard deviation

| | Percentile estimate | *Eff.* |
|---|---|---|
| 2 | $.3388(P_{93} - P_{07})$ | .65 |
| 4 | $.1714(P_{97} + P_{85} - P_{15} - P_{03})$ | .80 |
| 6 | $.1180(P_{98} + P_{91} + P_{80} - P_{20} - P_{09} - P_{02})$ | .87 |
| 8 | $.0935(P_{98} + P_{93} + P_{86} + P_{77} - P_{23} - P_{14} - P_{07} - P_{02})$ | .90 |
| 10 | $.0739(P_{98.5} + P_{95} + P_{90} + P_{84} + P_{75} - P_{25} - P_{16} - P_{10} - P_{05} - P_{01.5})$ | .92 |

### A-8*a*(3) Mean and standard deviation

| | | EFFICIENCY | | |
|---|---|---|---|---|
| | *Percentiles* | *Mean* | *Std. dev.* | *K* |
| 2 | 15, 85 | .73 | .56 | .4824 |
| 4 | 05, 30, 70, 95 | .80 | .74 | .2305 |
| 6 | 05, 15, 40, 60, 85, 95 | .89 | .80 | .1704 |
| 8 | 03, 10, 25, 45, 55, 75, 90, 97 | .90 | .86 | .1262 |
| 10 | 03, 10, 20, 30, 50, 50, 70, 80, 90, 97 | .94 | .87 | .1104 |

## A-8b ORDER-STATISTIC ESTIMATES IN SMALL SAMPLES

### A-8*b*(1) Unbiased estimate of $\sigma$ using the range $w$ (variance to be multiplied by $\sigma^2$)

| *N* | $K_1w$ | *Variance* | *Eff.* | *N* | $K_1w$ | *Variance* | *Eff.* |
|---|---|---|---|---|---|---|---|
| 2 | $.886w$ | .571 | 1.000 | 11 | $.315w$ | .0616 | .831 |
| 3 | $.591w$ | .275 | .992 | 12 | $.307w$ | .0571 | .814 |
| 4 | $.486w$ | .183 | .975 | 13 | $.300w$ | .0533 | .797 |
| 5 | $.430w$ | .138 | .955 | 14 | $.294w$ | .0502 | .781 |
| 6 | $.395w$ | .112 | .933 | 15 | $.288w$ | .0474 | .766 |
| 7 | $.370w$ | .0949 | .911 | 16 | $.283w$ | .0451 | .751 |
| 8 | $.351w$ | .0829 | .890 | 17 | $.279w$ | .0430 | .738 |
| 9 | $.337w$ | .0740 | .869 | 18 | $.275w$ | .0412 | .725 |
| 10 | $.325w$ | .0671 | .850 | 19 | $.271w$ | .0395 | .712 |
| | | | | 20 | $.268w$ | .0381 | .700 |

**Percentiles of the distribution of $w/\sigma$**

| $N$ | $P_{0.1}$ | $P_{0.5}$ | $P_{01}$ | $P_{02.5}$ | $P_{05}$ | $P_{10}$ | $P_{90}$ | $P_{95}$ | $P_{97.5}$ | $P_{99}$ | $P_{99.5}$ | $P_{99.9}$ |
|---|---|---|---|---|---|---|---|---|---|---|---|---|
| 2 | .00 | .01 | .02 | .04 | .09 | .18 | 2.33 | 2.77 | 3.17 | 3.64 | 3.97 | 4.65 |
| 3 | .06 | .13 | .19 | .30 | .43 | .62 | 2.90 | 3.31 | 3.68 | 4.12 | 4.42 | 5.06 |
| 4 | .20 | .34 | .43 | .59 | .76 | .98 | 3.24 | 3.63 | 3.98 | 4.40 | 4.69 | 5.31 |
| 5 | .37 | .55 | .66 | .85 | 1.03 | 1.26 | 3.48 | 3.86 | 4.20 | 4.60 | 4.89 | 5.48 |
| 6 | .54 | .75 | .87 | 1.06 | 1.25 | 1.49 | 3.66 | 4.03 | 4.36 | 4.76 | 5.03 | 5.62 |
| 7 | .69 | .92 | 1.05 | 1.25 | 1.44 | 1.68 | 3.81 | 4.17 | 4.49 | 4.88 | 5.15 | 5.73 |
| 8 | .83 | 1.08 | 1.20 | 1.41 | 1.60 | 1.83 | 3.93 | 4.29 | 4.61 | 4.99 | 5.26 | 5.82 |
| 9 | .96 | 1.21 | 1.34 | 1.55 | 1.74 | 1.97 | 4.04 | 4.39 | 4.70 | 5.08 | 5.34 | 5.90 |
| 10 | 1.08 | 1.33 | 1.47 | 1.67 | 1.86 | 2.09 | 4.13 | 4.47 | 4.79 | 5.16 | 5.42 | 5.97 |
| 11 | 1.20 | 1.45 | 1.58 | 1.78 | 1.97 | 2.20 | 4.21 | 4.55 | 4.86 | 5.23 | 5.49 | 6.04 |
| 12 | 1.30 | 1.55 | 1.68 | 1.88 | 2.07 | 2.30 | 4.29 | 4.62 | 4.92 | 5.29 | 5.54 | 6.09 |
| 13 | 1.38 | 1.64 | 1.77 | 1.97 | 2.16 | 2.39 | 4.35 | 4.69 | 4.98 | 5.35 | 5.60 | 6.15 |
| 14 | 1.47 | 1.72 | 1.85 | 2.06 | 2.24 | 2.47 | 4.41 | 4.74 | 5.04 | 5.40 | 5.65 | 6.20 |
| 15 | 1.55 | 1.80 | 1.93 | 2.14 | 2.32 | 2.54 | 4.47 | 4.80 | 5.09 | 5.45 | 5.70 | 6.23 |
| 16 | 1.62 | 1.87 | 2.00 | 2.21 | 2.39 | 2.61 | 4.52 | 4.85 | 5.14 | 5.49 | 5.74 | 6.27 |
| 17 | 1.69 | 1.94 | 2.07 | 2.27 | 2.45 | 2.67 | 4.57 | 4.89 | 5.18 | 5.54 | 5.79 | 6.30 |
| 18 | 1.75 | 2.01 | 2.14 | 2.33 | 2.52 | 2.73 | 4.61 | 4.93 | 5.22 | 5.58 | 5.82 | 6.35 |
| 19 | 1.82 | 2.07 | 2.20 | 2.39 | 2.57 | 2.79 | 4.65 | 4.97 | 5.26 | 5.61 | 5.86 | 6.38 |
| 20 | 1.88 | 2.13 | 2.25 | 2.45 | 2.63 | 2.84 | 4.69 | 5.01 | 5.30 | 5.65 | 5.89 | 6.40 |

**Unbiased estimate of $\sigma$ based on $s$**

| $N$ | *Estimate* | *Variance* | $N$ | *Estimate* | *Variance* |
|---|---|---|---|---|---|
| 2 | $1.253s$ | $.571\sigma^2$ | 7 | $1.042s$ | $.0865\sigma^2$ |
| 3 | $1.128s$ | $.273\sigma^2$ | 8 | $1.036s$ | $.0738\sigma^2$ |
| 4 | $1.085s$ | $.178\sigma^2$ | 9 | $1.032s$ | $.0643\sigma^2$ |
| 5 | $1.064s$ | $.132\sigma^2$ | 10 | $1.028s$ | $.0570\sigma^2$ |
| 6 | $1.051s$ | $.104\sigma^2$ | $\infty$ | $\left[1 + \frac{1}{4(N-1)}\right]s$ | $\sigma^2/2N$ |

**A-8*b*(2) Mean deviation estimate of $\sigma$**

| $N$ | *Estimate* | *Eff.* |
|---|---|---|
| 2 | $.8862(X_2 - X_1)$ | 1.00 |
| 3 | $.5908(X_3 - X_1)$ | .99 |
| 4 | $.3770(X_4 + X_3 - X_2 - X_1)$ | .91 |
| 5 | $.3016(X_5 + X_4 - X_2 - X_1)$ | .94 |
| 6 | $.2369(X_6 + X_5 + X_4 - X_3 - X_2 - X_1)$ | .90 |
| 7 | $.2031(X_7 + X_6 + X_5 - X_3 - X_2 - X_1)$ | .92 |
| 8 | $.1723(X_8 + X_7 + X_6 + X_5 - X_4 - X_3 - X_2 - X_1)$ | .90 |
| 9 | $.1532(X_9 + X_8 + X_7 + X_6 - X_4 - X_3 - X_2 - X_1)$ | .91 |
| 10 | $.1353(X_{10} + X_9 + X_8 + X_7 + X_6 - X_5 - X_4 - X_3 - X_2 - X_1)$ | .89 |

### A-8b(3) Modified linear estimate of $\sigma$ (variance to be multiplied by $\sigma^2$)

| $N$ | *Estimate* | *Variance* | *Eff.* |
|---|---|---|---|
| 2 | $.8862(X_2 - X_1)$ | .571 | 1.000 |
| 3 | $.5908(X_3 - X_1)$ | .275 | .992 |
| 4 | $.4857(X_4 - X_1)$ | .183 | .975 |
| 5 | $.4299(X_5 - X_1)$ | .138 | .955 |
| 6 | $.2619(X_6 + X_5 - X_2 - X_1)$ | .109 | .957 |
| 7 | $.2370(X_7 + X_6 - X_2 - X_1)$ | .0895 | .967 |
| 8 | $.2197(X_8 + X_7 - X_2 - X_1)$ | .0761 | .970 |
| 9 | $.2068(X_9 + X_8 - X_2 - X_1)$ | .0664 | .968 |
| 10 | $.1968(X_{10} + X_9 - X_2 - X_1)$ | .0591 | .964 |
| 11 | $.1608(X_{11} + X_{10} + X_8 - X_4 - X_2 - X_1)$ | .0529 | .967 |
| 12 | $.1524(X_{12} + X_{11} + X_9 - X_4 - X_2 - X_1)$ | .0478 | .972 |
| 13 | $.1456(X_{13} + X_{12} + X_{10} - X_4 - X_2 - X_1)$ | .0436 | .975 |
| 14 | $.1399(X_{14} + X_{13} + X_{11} - X_4 - X_2 - X_1)$ | .0401 | .977 |
| 15 | $.1352(X_{15} + X_{14} + X_{12} - X_4 - X_2 - X_1)$ | .0372 | .977 |
| 16 | $.1311(X_{16} + X_{15} + X_{13} - X_4 - X_2 - X_1)$ | .0347 | .975 |
| 17 | $.1050(X_{17} + X_{16} + X_{15} + X_{13} - X_5 - X_3 - X_2 - X_1)$ | .0325 | .978 |
| 18 | $.1020(X_{18} + X_{17} + X_{16} + X_{14} - X_5 - X_3 - X_2 - X_1)$ | .0305 | .978 |
| 19 | $.09939(X_{19} + X_{18} + X_{17} + X_{15} - X_5 - X_3 - X_2 - X_1)$ | .0288 | .979 |
| 20 | $.09706(X_{20} + X_{19} + X_{18} + X_{16} - X_5 - X_3 - X_2 - X_1)$ | .0272 | .978 |

### A-8b(4) Several estimates of the mean (variance to be multiplied by $\sigma^2$)

| | MEDIAN | | MIDRANGE | | MEAN OF BEST TWO | | | $(X_2 + X_3 + \cdots + X_{N-1})/(N - 2)$ | |
|---|---|---|---|---|---|---|---|---|---|
| $N$ | *Var.* | *Eff.* | *Var.* | *Eff.* | *Statistic* | *Var.* | *Eff.* | *Var.* | *Eff.* |
| 2 | .500 | 1.000 | .500 | 1.000 | $\frac{1}{2}(X_1 + X_2)$ | .500 | 1.000 | | |
| 3 | .449 | .743 | .362 | .920 | $\frac{1}{2}(X_1 + X_3)$ | .362 | .920 | .449 | .743 |
| 4 | .298 | .838 | .298 | .838 | $\frac{1}{2}(X_2 + X_3)$ | .298 | .838 | .298 | .838 |
| 5 | .287 | .697 | .261 | .767 | $\frac{1}{2}(X_2 + X_4)$ | .231 | .867 | .227 | .881 |
| 6 | .215 | .776 | .236 | .706 | $\frac{1}{2}(X_2 + X_5)$ | .193 | .865 | .184 | .906 |
| 7 | .210 | .679 | .218 | .654 | $\frac{1}{2}(X_2 + X_6)$ | .168 | .849 | .155 | .922 |
| 8 | .168 | .743 | .205 | .610 | $\frac{1}{2}(X_3 + X_6)$ | .149 | .837 | .134 | .934 |
| 9 | .166 | .669 | .194 | .572 | $\frac{1}{2}(X_3 + X_7)$ | .132 | .843 | .118 | .942 |
| 10 | .138 | .723 | .186 | .539 | $\frac{1}{2}(X_3 + X_8)$ | .119 | .840 | .105 | .949 |
| 11 | .137 | .663 | .178 | .510 | $\frac{1}{2}(X_3 + X_9)$ | .109 | .832 | .0952 | .955 |
| 12 | .118 | .709 | .172 | .484 | $\frac{1}{2}(X_4 + X_9)$ | .100 | .831 | .0869 | .959 |
| 13 | .117 | .659 | .167 | .461 | $\frac{1}{2}(X_4 + X_{10})$ | .0924 | .833 | .0799 | .963 |
| 14 | .102 | .699 | .162 | .440 | $\frac{1}{2}(X_4 + X_{11})$ | .0860 | .830 | .0739 | .966 |
| 15 | .102 | .656 | .158 | .422 | $\frac{1}{2}(X_4 + X_{12})$ | .0808 | .825 | .0688 | .969 |
| 16 | .0904 | .692 | .154 | .392 | $\frac{1}{2}(X_5 + X_{12})$ | .0756 | .827 | .0644 | .971 |
| 17 | .0901 | .653 | .151 | .389 | $\frac{1}{2}(X_5 + X_{13})$ | .0711 | .827 | .0605 | .973 |
| 18 | .0810 | .686 | .148 | .375 | $\frac{1}{2}(X_5 + X_{14})$ | .0673 | .825 | .0570 | .975 |
| 19 | .0808 | .651 | .145 | .362 | $\frac{1}{2}(X_6 + X_{14})$ | .0640 | .823 | .0539 | .976 |
| 20 | .0734 | .681 | .143 | .350 | $\frac{1}{2}(X_6 + X_{15})$ | .0607 | .824 | .0511 | .978 |
| $\infty$ | $1.57/N$ | .637 | | .000 | $\frac{1}{2}(P_{25} + P_{75})$ | $1.24/N$ | .808 | | 1.000 |

### A-8b(5) Mean values of the order statistics

| $N$ | $X_1$ | $X_2$ | $X_3$ | $X_4$ | $X_5$ | $X_6$ | $X_7$ | $X_8$ | $X_9$ | $X_{10}$ |
|---|---|---|---|---|---|---|---|---|---|---|
| 2 | −.564 | .564 | | | | | | | | |
| 3 | −.846 | .000 | .846 | | | | | | | |
| 4 | −1.029 | −.297 | .297 | 1.029 | | | | | | |
| 5 | −1.163 | −.495 | .000 | .495 | 1.163 | | | | | |
| 6 | −1.267 | −.642 | −.202 | .202 | .642 | 1.267 | | | | |
| 7 | −1.352 | −.757 | −.353 | .000 | .353 | .757 | 1.352 | | | |
| 8 | −1.424 | −.852 | −.473 | −.153 | .153 | .473 | .852 | 1.424 | | |
| 9 | −1.485 | −.932 | −.572 | −.275 | .000 | .275 | .572 | .932 | 1.485 | |
| 10 | −1.539 | −1.001 | −.656 | −.376 | −.123 | .123 | .376 | .656 | 1.001 | 1.539 |
| 11 | −1.586 | −1.062 | −.729 | −.462 | −.225 | .000 | .225 | .462 | .729 | 1.062 |
| 12 | −1.629 | −1.116 | −.793 | −.537 | −.312 | −.103 | .103 | .312 | .537 | .793 |
| 13 | −1.668 | −1.164 | −.850 | −.603 | −.388 | −.191 | .000 | .191 | .388 | .603 |
| 14 | −1.703 | −1.208 | −.901 | −.662 | −.456 | −.267 | −.088 | .088 | .267 | .456 |
| 15 | −1.736 | −1.248 | −.948 | −.715 | −.516 | −.335 | −.165 | .000 | .165 | .335 |
| 16 | −1.766 | −1.285 | −.990 | −.763 | −.570 | −.396 | −.234 | −.077 | .077 | .234 |
| 17 | −1.794 | −1.319 | −1.029 | −.807 | −.619 | −.451 | −.295 | −.146 | .000 | .146 |
| 18 | −1.820 | −1.350 | −1.066 | −.848 | −.665 | −.502 | −.351 | −.208 | −.069 | .069 |
| 19 | −1.844 | −1.380 | −1.099 | −.886 | −.707 | −.548 | −.402 | −.264 | −.131 | .000 |
| 20 | −1.867 | −1.408 | −1.131 | −.921 | −.745 | −.590 | −.448 | −.315 | −.187 | −.062 |

Values given are in deviations from the mean in standard-deviation units. For sample sizes greater than 10 the mean values of $X_{11}$, $X_{12}$, . . . , can be obtained by symmetry. For example, for $N = 17$ the expected value of $X_{15}$ is the same except for sign as the expected value of $X_3$.

### A-8b(6) Best linear estimate of $\sigma$

| $N$ | *Estimate* | *Eff.* |
|---|---|---|
| 2 | $.8862(X_2 - X_1)$ | 1.000 |
| 3 | $.5908(X_3 - X_1)$ | .992 |
| 4 | $.4539(X_4 - X_1) + .1102(X_3 - X_2)$ | .989 |
| 5 | $.3724(X_5 - X_1) + .1352(X_4 - X_2)$ | .988 |
| 6 | $.3175(X_6 - X_1) + .1386(X_5 - X_2) + .0432(X_4 - X_3)$ | .988 |
| 7 | $.2778(X_7 - X_1) + .1351(X_6 - X_2) + .0625(X_5 - X_3)$ | .989 |
| 8 | $.2476(X_8 - X_1) + .1294(X_7 - X_2) + .0713(X_6 - X_3) + .0230(X_5 - X_4)$ | .989 |
| 9 | $.2237(X_9 - X_1) + .1233(X_8 - X_2) + .0751(X_7 - X_3) + .0360(X_6 - X_4)$ | .989 |
| 10 | $.2044(X_{10} - X_1) + .1172(X_9 - X_2) + .0763(X_8 - X_3) + .0436(X_7 - X_4) + .0142(X_6 - X_5)$ | .990 |

**A-8c(1) Percentiles for $\tau_1 = \frac{\bar{X} - \mu}{w}$**

| $N$ | $P_{95}$ | $P_{97.5}$ | $P_{99}$ | $P_{99.5}$ | $P_{99.9}$ | $P_{99.95}$ |
|---|---|---|---|---|---|---|
| 2 | 3.175 | 6.353 | 15.910 | 31.828 | 159.16 | 318.31 |
| 3 | .885 | 1.304 | 2.111 | 3.008 | 6.77 | 9.58 |
| 4 | .529 | .717 | 1.023 | 1.316 | 2.29 | 2.85 |
| 5 | .388 | .507 | .685 | .843 | 1.32 | 1.58 |
| 6 | .312 | .399 | .523 | .628 | .92 | 1.07 |
| 7 | .263 | .333 | .429 | .507 | .71 | .82 |
| 8 | .230 | .288 | .366 | .429 | .59 | .67 |
| 9 | .205 | .255 | .322 | .374 | .50 | .57 |
| 10 | .186 | .230 | .288 | .333 | .44 | .50 |
| 11 | .170 | .210 | .262 | .302 | .40 | .44 |
| 12 | .158 | .194 | .241 | .277 | .36 | .40 |
| 13 | .147 | .181 | .224 | .256 | .33 | .37 |
| 14 | .138 | .170 | .209 | .239 | .31 | .34 |
| 15 | .131 | .160 | .197 | .224 | .29 | .32 |
| 16 | .124 | .151 | .186 | .212 | .27 | .30 |
| 17 | .118 | .144 | .177 | .201 | .26 | .28 |
| 18 | .113 | .137 | .168 | .191 | .24 | .26 |
| 19 | .108 | .131 | .161 | .182 | .23 | .25 |
| 20 | .104 | .126 | .154 | .175 | .22 | .24 |
| $N$ | $-P_{05}$ | $-P_{02.5}$ | $-P_{01}$ | $-P_{0.5}$ | $-P_{0.1}$ | $-P_{0.05}$ |

When the table is read from the foot, the values are to be prefixed with a negative sign.

**A-8c(2) Percentiles for** $\tau_d = \dfrac{\bar{X}_1 - \bar{X}_2}{\frac{1}{2}(w_1 + w_2)}$

| $N_1 = N_2$ | $P_{95}$ | $P_{97.5}$ | $P_{99}$ | $P_{99.5}$ | $P_{99.9}$ | $P_{99.95}$ |
|---|---|---|---|---|---|---|
| 2 | 2.322 | 3.427 | 5.553 | 7.916 | 17.81 | 25.23 |
| 3 | .974 | 1.272 | 1.715 | 2.093 | 3.27 | 4.18 |
| 4 | .644 | .813 | 1.047 | 1.237 | 1.74 | 1.99 |
| 5 | .493 | .613 | .772 | .896 | 1.21 | 1.35 |
| 6 | .405 | .499 | .621 | .714 | .94 | 1.03 |
| 7 | .347 | .426 | .525 | .600 | .77 | .85 |
| 8 | .306 | .373 | .459 | .521 | .67 | .73 |
| 9 | .275 | .334 | .409 | .464 | .59 | .64 |
| 10 | .250 | .304 | .371 | .419 | .53 | .58 |
| 11 | .233 | .280 | .340 | .384 | .48 | .52 |
| 12 | .214 | .260 | .315 | .355 | .44 | .48 |
| 13 | .201 | .243 | .294 | .331 | .41 | .45 |
| 14 | .189 | .228 | .276 | .311 | .39 | .42 |
| 15 | .179 | .216 | .261 | .293 | .36 | .39 |
| 16 | .170 | .205 | .247 | .278 | .34 | .37 |
| 17 | .162 | .195 | .236 | .264 | .33 | .35 |
| 18 | .155 | .187 | .225 | .252 | .31 | .34 |
| 19 | .149 | .179 | .216 | .242 | .30 | .32 |
| 20 | .143 | .172 | .207 | .232 | .29 | .31 |
| $N_1 = N_2$ | $-P_{05}$ | $-P_{02.5}$ | $-P_{01}$ | $-P_{0.5}$ | $-P_{0.1}$ | $-P_{0.05}$ |

When the table is read from the foot, the tabled values are to be prefixed with a negative sign.

**A-8c(3) Percentiles for** $\tau_2 = \dfrac{\frac{1}{2}(X_1 + X_N) - \mu}{w}$

| $N$ | $P_{95}$ | $P_{97.5}$ | $P_{99}$ | $P_{99.5}$ |
|---|---|---|---|---|
| 2 | 3.16 | 6.35 | 15.91 | 31.83 |
| 3 | .90 | 1.30 | 2.11 | 3.02 |
| 4 | .55 | .74 | 1.04 | 1.37 |
| 5 | .42 | .52 | .71 | .85 |
| 6 | .35 | .43 | .56 | .66 |
| 7 | .30 | .37 | .47 | .55 |
| 8 | .26 | .33 | .42 | .47 |
| 9 | .24 | .30 | .38 | .42 |
| 10 | .22 | .27 | .35 | .39 |
| $N$ | $-P_{05}$ | $-P_{02.5}$ | $-P_{01}$ | $-P_{0.5}$ |

**Percentiles of the distribution of $w_1/w_2$, where $w_1$ is the range of $N_1$ observations and $w_2$ the range of $N_2$ observations from normal populations having equal variances**

$N_1$, SAMPLE SIZE FOR NUMERATOR

| $N_2$, sample size for denominator | *Cum. prop.* | 2 | 3 | 4 | 5 | 6 | 7 | 8 | 9 | 10 | 11 | 12 | 13 | 14 | 15 |
|---|---|---|---|---|---|---|---|---|---|---|---|---|---|---|---|
| 2 | .001 | $.0^{2}16$ | .043 | .121 | .200 | .269 | .326 | .375 | .416 | .452 | .483 | .511 | .536 | .558 | .578 |
| | .005 | $.0^{2}79$ | .096 | .210 | .308 | .388 | .452 | .505 | .550 | .589 | .623 | .653 | .680 | .704 | .726 |
| | .010 | .016 | .136 | .268 | .375 | .460 | .527 | .583 | .630 | .671 | .706 | .737 | .766 | .791 | .814 |
| | .025 | .039 | .217 | .376 | .495 | .587 | .661 | .721 | .772 | .816 | .854 | .888 | .919 | .946 | .972 |
| | .050 | .079 | .313 | .493 | .624 | .724 | .803 | .869 | .924 | .972 | 1.01 | 1.05 | 1.08 | 1.11 | 1.14 |
| | .100 | .158 | .461 | .668 | .814 | .926 | 1.01 | 1.09 | 1.15 | 1.20 | 1.25 | 1.29 | 1.33 | 1.37 | 1.40 |
| | .250 | .414 | .841 | 1.11 | 1.30 | 1.44 | 1.56 | 1.65 | 1.74 | 1.81 | 1.87 | 1.93 | 1.98 | 2.02 | 2.07 |
| | .500 | 1.00 | 1.65 | 2.06 | 2.36 | 2.59 | 2.77 | 2.93 | 3.06 | 3.18 | 3.28 | 3.37 | 3.45 | 3.53 | 3.60 |
| | .750 | 2.41 | 3.70 | 4.52 | 5.12 | 5.59 | 5.97 | 6.29 | 6.56 | 6.80 | 7.02 | 7.21 | 7.38 | 7.54 | 7.68 |
| | .900 | 6.31 | 9.50 | 11.6 | 13.1 | 14.3 | 15.2 | 16.0 | 16.7 | 17.3 | 17.8 | 18.3 | 18.8 | 19.2 | 19.5 |
| | .950 | 12.7 | 19.1 | 23.2 | 26.2 | 28.6 | 30.5 | 32.1 | 33.5 | 34.7 | 35.8 | 36.7 | 37.6 | 38.4 | 39.1 |
| | .975 | 25.4 | 38.2 | 46.4 | 52.5 | 57.2 | 61.0 | 64.2 | 67. | 69.4 | 71.6 | 73.5 | 75.3 | 76.9 | 78.3 |
| | .990 | 63.7 | 95.5 | 116 | 131 | 143 | 153 | 161 | 168 | 174 | 179 | 184 | 188 | 192 | 196 |
| | .995 | 127 | 191 | 232 | 263 | 286 | 305 | 321 | 335 | 347 | 358 | 368 | 376 | 384 | 392 |
| | .999 | 637 | 955 | 1,162 | 1,312 | 1,430 | 1,526 | 1,607 | 1,676 | 1,736 | 1,790 | 1,838 | 1,882 | 1,922 | 1,959 |
| 3 | .001 | $.0^{2}10$ | .031 | .095 | .163 | .224 | .276 | .321 | .359 | .393 | .422 | .448 | .471 | .493 | .512 |
| | .005 | $.0^{2}52$ | .071 | .164 | .249 | .319 | .377 | .426 | .467 | .503 | .534 | .562 | .587 | .609 | .630 |
| | .010 | .010 | .100 | .209 | .301 | .375 | .436 | .486 | .529 | .566 | .599 | .627 | .653 | .676 | .697 |
| | .025 | .026 | .160 | .290 | .392 | .472 | .536 | .590 | .635 | .674 | .709 | .739 | .766 | .791 | .813 |
| | .050 | .052 | .229 | .376 | .486 | .571 | .639 | .696 | .743 | .785 | .821 | .853 | .882 | .908 | .932 |
| | .100 | .105 | .332 | .497 | .617 | .709 | .782 | .842 | .894 | .939 | .978 | 1.01 | 1.04 | 1.07 | 1.10 |
| | .250 | .270 | .576 | .772 | .911 | 1.02 | 1.10 | 1.17 | 1.24 | 1.29 | 1.34 | 1.38 | 1.41 | 1.45 | 1.48 |
| | .500 | .605 | 1.00 | 1.25 | 1.42 | 1.56 | 1.67 | 1.76 | 1.84 | 1.91 | 1.97 | 2.03 | 2.08 | 2.12 | 2.16 |
| | .750 | 1.19 | 1.73 | 2.08 | 2.33 | 2.53 | 2.69 | 2.83 | 2.95 | 3.05 | 3.14 | 3.22 | 3.30 | 3.37 | 3.43 |
| | .900 | 2.17 | 3.01 | 3.55 | 3.96 | 4.27 | 4.53 | 4.75 | 4.94 | 5.10 | 5.25 | 5.38 | 5.50 | 5.62 | 5.72 |
| | .950 | 3.19 | 4.37 | 5.14 | 5.71 | 6.16 | 6.53 | 6.84 | 7.11 | 7.34 | 7.55 | 7.74 | 7.91 | 8.07 | 8.22 |
| | .975 | 4.61 | 6.27 | 7.35 | 8.16 | 8.79 | 9.31 | 9.76 | 10.1 | 10.5 | 10.8 | 11.0 | 11.3 | 11.5 | 11.7 |
| | .990 | 7.37 | 9.99 | 11.7 | 13.0 | 14.0 | 14.8 | 15.5 | 16.1 | 16.6 | 17.1 | 17.5 | 17.9 | 18.3 | 18.6 |
| | .995 | 10.5 | 14.2 | 16.6 | 18.4 | 19.8 | 21.0 | 22.0 | 22.8 | 23.6 | 24.2 | 24.8 | 25.4 | 25.9 | 26.3 |
| | .999 | 23.5 | 31.8 | 37.2 | 41.2 | 44.4 | 47.0 | 49.2 | 51.1 | 52.8 | 54.3 | 55.7 | 56.9 | 58.0 | 59.0 |

**Percentiles of the distribution of $w_1/w_2$, where $w_1$ is the range of $N_1$ observations and $w_2$ the range of $N_2$ observations from normal populations having equal variances**

$N_1$, SAMPLE SIZE FOR NUMERATOR

| $N_2$, SAMPLE SIZE FOR DENOMINATOR | *Cum. prop.* | 2 | 3 | 4 | 5 | 6 | 7 | 8 | 9 | 10 | 11 | 12 | 13 | 14 | 15 |
|---|---|---|---|---|---|---|---|---|---|---|---|---|---|---|---|
| 4 | .001 | $.0^{3}86$ | .027 | .083 | .145 | .201 | .251 | .293 | .330 | .362 | .390 | .415 | .438 | .458 | .477 |
| | .005 | $.0^{2}43$ | .060 | .143 | .221 | .286 | .340 | .386 | .426 | .460 | .490 | .516 | .540 | .561 | .581 |
| | .010 | $.0^{2}86$ | .085 | .182 | .266 | .335 | .392 | .440 | .480 | .515 | .546 | .573 | .598 | .620 | .640 |
| | .025 | .022 | .136 | .252 | .345 | .419 | .479 | .529 | .571 | .608 | .640 | .668 | .694 | .717 | .739 |
| | .050 | .043 | .194 | .325 | .425 | .503 | .566 | .618 | .663 | .701 | .734 | .764 | .791 | .815 | .838 |
| | .100 | .086 | .281 | .427 | .534 | .617 | .684 | .739 | .786 | .826 | .862 | .894 | .923 | .948 | .972 |
| | .250 | .221 | .480 | .649 | .769 | .861 | .935 | .998 | 1.05 | 1.10 | 1.14 | 1.17 | 1.21 | 1.24 | 1.26 |
| | .500 | .485 | .803 | 1.00 | 1.14 | 1.25 | 1.34 | 1.41 | 1.48 | 1.53 | 1.58 | 1.63 | 1.67 | 1.70 | 1.74 |
| | .750 | .903 | 1.30 | 1.54 | 1.72 | 1.86 | 1.98 | 2.07 | 2.16 | 2.23 | 2.29 | 2.35 | 2.41 | 2.45 | 2.50 |
| | .900 | 1.50 | 2.01 | 2.34 | 2.58 | 2.77 | 2.93 | 3.07 | 3.18 | 3.28 | 3.37 | 3.45 | 3.53 | 3.60 | 3.66 |
| | .950 | 2.03 | 2.66 | 3.07 | 3.38 | 3.62 | 3.82 | 3.99 | 4.14 | 4.26 | 4.38 | 4.48 | 4.58 | 4.66 | 4.74 |
| | .975 | 2.66 | 3.45 | 3.97 | 4.36 | 4.66 | 4.91 | 5.13 | 5.31 | 5.47 | 5.62 | 5.75 | 5.87 | 5.98 | 6.08 |
| | .990 | 3.72 | 4.79 | 5.49 | 6.01 | 6.42 | 6.76 | 7.06 | 7.31 | 7.53 | 7.73 | 7.91 | 8.07 | 8.22 | 8.36 |
| | .995 | 4.75 | 6.09 | 6.97 | 7.63 | 8 15 | 8.58 | 8.94 | 9.26 | 9.54 | 9.79 | 10.00 | 10.2 | 10.4 | 10.6 |
| | .999 | 8.25 | 10.5 | 12.0 | 13.1 | 14.0 | 14.8 | 15.4 | 15.9 | 16.4 | 16.8 | 17.20 | 17.6 | 17.9 | 18.2 |
| 5 | .001 | $.0^{3}76$ | .024 | .076 | .134 | .187 | .234 | .275 | .311 | .342 | .369 | .394 | .416 | .436 | .454 |
| | .005 | $.0^{2}38$ | .054 | .131 | .204 | .265 | .317 | .362 | .400 | .432 | .461 | .487 | .510 | .531 | .550 |
| | .010 | $.0^{2}76$ | .077 | .166 | .245 | .311 | .365 | .410 | .449 | .483 | .513 | .539 | .563 | .584 | .604 |
| | .025 | .019 | .123 | .230 | .317 | .387 | .444 | .491 | .532 | .567 | .598 | .625 | .650 | .672 | .692 |
| | .050 | .038 | .175 | .296 | .389 | .463 | .522 | .571 | .613 | .650 | .682 | .710 | .736 | .759 | .780 |
| | .100 | .077 | .253 | .387 | .487 | .564 | .626 | .678 | .722 | .760 | .794 | .823 | .850 | .874 | .897 |
| | .250 | .195 | .428 | .581 | .691 | .775 | .843 | .899 | .948 | .990 | 1.03 | 1.06 | 1.09 | 1.12 | 1.14 |
| | .500 | .424 | .703 | .876 | .000 | 1.10 | 1.17 | 1.24 | 1.29 | 1.34 | 1.39 | 1.42 | 1.46 | 1.49 | 1.52 |
| | .750 | .772 | 1.10 | 1.30 | 1.45 | 1.56 | 1.66 | 1.74 | 1.80 | 1.86 | 1.92 | 1.97 | 2.01 | 2.05 | 2.09 |
| | .900 | 1.23 | 1.62 | 1.87 | 2.05 | 2.20 | 2.32 | 2.42 | 2.51 | 2.58 | 2.65 | 2.71 | 2.77 | 2.82 | 2.87 |
| | .950 | 1.60 | 2.06 | 2.35 | 2.57 | 2.74 | 2.88 | 3.00 | 3.11 | 3.20 | 3.28 | 3.36 | 3.42 | 3.49 | 3.54 |
| | .975 | 2.02 | 2.55 | 2.90 | 3.16 | 3.36 | 3.53 | 3.67 | 3.80 | 3.91 | 4.01 | 4.10 | 4.18 | 4.25 | 4.32 |
| | .990 | 2.66 | 3.32 | 3.76 | 4.08 | 4.33 | 4.55 | 4.73 | 4.88 | 5.02 | 5.15 | 5.26 | 5.36 | 5.46 | 5.54 |
| | .995 | 3.24 | 4.02 | 4.53 | 4.91 | 5.22 | 5.47 | 5.69 | 5.87 | 6.04 | 6.19 | 6.32 | 6.44 | 6.55 | 6.66 |
| | .999 | 4.99 | 6.14 | 6.90 | 7.47 | 7.93 | 8.31 | 8.63 | 8.91 | 9.16 | 9.38 | 9.58 | 9.76 | 9.93 | 10.10 |

**Percentiles of the distribution of $w_1/w_2$, where $w_1$ is the range of $N_1$ observations and $w_2$ the range of $N_2$ observations from normal populations having equal variances**

$N_1$, SAMPLE SIZE FOR NUMERATOR

| $N_2$, SAMPLE SIZE FOR DENOMINATOR | *Cum. prop.* | 2 | 3 | 4 | 5 | 6 | 7 | 8 | 9 | 10 | 11 | 12 | 13 | 14 | 15 |
|---|---|---|---|---|---|---|---|---|---|---|---|---|---|---|---|
| 6 | .001 | .0$^3$70 | .023 | .071 | .126 | .177 | .223 | .262 | .297 | .327 | .354 | .378 | .400 | .419 | .437 |
| | .005 | .0$^2$35 | .050 | .123 | .192 | .251 | .301 | .344 | .381 | .413 | .441 | .466 | .489 | .509 | .527 |
| | .010 | .0$^2$70 | .072 | .156 | .231 | .293 | .346 | .390 | .428 | .460 | .489 | .515 | .538 | .558 | .578 |
| | .025 | .017 | .114 | .215 | .298 | .364 | .419 | .465 | .504 | .538 | .568 | .595 | .618 | .640 | .660 |
| | .050 | .035 | .162 | .276 | .365 | .435 | .492 | .539 | .580 | .615 | .646 | .672 | .697 | .719 | .740 |
| | .100 | .070 | .234 | .360 | .455 | .528 | .587 | .637 | .678 | .715 | .747 | .776 | .801 | .824 | .846 |
| | .250 | .179 | .395 | .537 | .640 | .719 | .782 | .835 | .881 | .921 | .955 | .986 | 1.01 | 1.04 | 1.06 |
| | .500 | .387 | .642 | .799 | .912 | 1.00 | 1.07 | 1.13 | 1.18 | 1.23 | 1.26 | 1.30 | 1.33 | 1.36 | 1.39 |
| | .750 | .694 | .983 | 1.16 | 1.29 | 1.39 | 1.47 | 1.54 | 1.60 | 1.65 | 1.70 | 1.74 | 1.78 | 1.82 | 1.85 |
| | .900 | 1.08 | 1.41 | 1.62 | 1.77 | 1.89 | 1.99 | 2.08 | 2.15 | 2.21 | 2.27 | 2.32 | 2.37 | 2.41 | 2.45 |
| | .950 | 1.38 | 1.75 | 1.99 | 2.16 | 2.30 | 2.41 | 2.51 | 2.59 | 2.67 | 2.73 | 2.79 | 2.85 | 2.90 | 2.94 |
| | .975 | 1.70 | 2.12 | 2.39 | 2.59 | 2.74 | 2.88 | 2.99 | 3.08 | 3.17 | 3.24 | 3.31 | 3.38 | 3.43 | 3.49 |
| | .990 | 2.18 | 2.66 | 2.98 | 3.22 | 3.41 | 3.56 | 3.70 | 3.81 | 3.92 | 4.01 | 4.09 | 4.17 | 4.24 | 4.30 |
| | .995 | 2.58 | 3.13 | 3.50 | 3.77 | 3.98 | 4.16 | 4.32 | 4.45 | 4.57 | 4.67 | 4.77 | 4.86 | 4.94 | 5.01 |
| | .999 | 3.72 | 4.47 | 4.97 | 5.34 | 5.64 | 5.88 | 6.09 | 6.28 | 6.44 | 6.59 | 6.72 | 6.84 | 6.95 | 7.06 |
| 7 | .001 | .0$^3$65 | .021 | .068 | .120 | .170 | .214 | .253 | .287 | .316 | .343 | .366 | .387 | .407 | .424 |
| | .005 | .0$^2$33 | .048 | .117 | .183 | .240 | .289 | .331 | .367 | .398 | .426 | .450 | .472 | .492 | .510 |
| | .010 | .0$^2$65 | .068 | .148 | .220 | .281 | .331 | .374 | .411 | .443 | .471 | .496 | .519 | .539 | .558 |
| | .025 | .016 | .107 | .204 | .283 | .348 | .401 | .446 | .484 | .517 | .546 | .572 | .595 | .616 | .635 |
| | .050 | .033 | .153 | .262 | .347 | .414 | .469 | .515 | .554 | .588 | .618 | .645 | .668 | .690 | .710 |
| | .100 | .066 | .221 | .341 | .431 | .502 | .559 | .606 | .647 | .682 | .713 | .740 | .765 | .787 | .808 |
| | .250 | .167 | .371 | .506 | .604 | .678 | .739 | .790 | .833 | .871 | .904 | .934 | .961 | .984 | 1.01 |
| | .500 | .361 | .599 | .746 | .852 | .934 | 1.00 | 1.06 | 1.10 | 1.14 | 1.18 | 1.21 | 1.24 | 1.27 | 1.30 |
| | .750 | .643 | .906 | 1.07 | 1.19 | 1.28 | 1.35 | 1.42 | 1.47 | 1.52 | 1.56 | 1.60 | 1.63 | 1.66 | 1.69 |
| | .900 | .985 | 1.28 | 1.46 | 1.60 | 1.70 | 1.79 | 1.86 | 1.93 | 1.98 | 2.03 | 2.08 | 2.12 | 2.16 | 2.19 |
| | .950 | 1.24 | 1.56 | 1.77 | 1.92 | 2.03 | 2.13 | 2.21 | 2.28 | 2.35 | 2.40 | 2.46 | 2.50 | 2.54 | 2.58 |
| | .975 | 1.51 | 1.86 | 2.09 | 2.25 | 2.38 | 2.49 | 2.59 | 2.67 | 2.74 | 2.80 | 2.86 | 2.91 | 2.96 | 3.00 |
| | .990 | 1.90 | 2.29 | 2.55 | 2.74 | 2.89 | 3.02 | 3.13 | 3.22 | 3.30 | 3.38 | 3.44 | 3.51 | 3.56 | 3.61 |
| | .995 | 2.21 | 2.65 | 2.94 | 3.15 | 3.32 | 3.46 | 3.58 | 3.69 | 3.78 | 3.86 | 3.94 | 4.01 | 4.07 | 4.13 |
| | .999 | 3.06 | 3.62 | 3.99 | 4.27 | 4.49 | 4.67 | 4.83 | 4.96 | 5.09 | 5.20 | 5.29 | 5.39 | 5.47 | 5.55 |

**Percentiles of the distribution of $w_1/w_2$, where $w_1$ is the range of $N_1$ observations and $w_2$ the range of $N_2$ observations from normal populations having equal variances**

| $N_2$, sample size for denominator | *Cum. prop.* | $N_1$, sample size for numerator: 2 | 3 | 4 | 5 | 6 | 7 | 8 | 9 | 10 | 11 | 12 | 13 | 14 | 15 |
|---|---|---|---|---|---|---|---|---|---|---|---|---|---|---|---|
| 8 | .001 | $.0^{3}62$ | .020 | .065 | .116 | .164 | .207 | .245 | .278 | .307 | .333 | .357 | .378 | .397 | .414 |
| | .005 | $.0^{2}31$ | .046 | .112 | .176 | .232 | .279 | .320 | .355 | .386 | .413 | .437 | .459 | .478 | .497 |
| | .010 | $.0^{2}62$ | .065 | .142 | .212 | .270 | .320 | .362 | .398 | .429 | .457 | .481 | .504 | .524 | .542 |
| | .025 | .016 | .103 | .195 | .272 | .335 | .387 | .430 | .467 | .500 | .528 | .553 | .576 | .597 | .615 |
| | .050 | .031 | .146 | .251 | .333 | .398 | .452 | .497 | .535 | .568 | .597 | .623 | .646 | .667 | .687 |
| | .100 | .062 | .211 | .326 | .413 | .481 | .537 | .583 | .622 | .656 | .686 | .713 | .737 | .759 | .779 |
| | .250 | .159 | .353 | .482 | .576 | .648 | .706 | .755 | .797 | .833 | .865 | .894 | .919 | .943 | .964 |
| | .500 | .342 | .567 | .707 | .807 | .885 | .947 | 1.00 | 1.04 | 1.08 | 1.12 | 1.15 | 1.18 | 1.20 | 1.23 |
| | .750 | .605 | .851 | 1.00 | 1.11 | 1.20 | 1.27 | 1.32 | 1.38 | 1.42 | 1.46 | 1.49 | 1.53 | 1.55 | 1.58 |
| | .900 | .919 | 1.19 | 1.35 | 1.47 | 1.57 | 1.65 | 1.72 | 1.77 | 1.82 | 1.87 | 1.91 | 1.95 | 1.98 | 2.01 |
| | .950 | 1.15 | 1.44 | 1.62 | 1.75 | 1.85 | 1.94 | 2.01 | 2.08 | 2.13 | 2.18 | 2.23 | 2.27 | 2.31 | 2.34 |
| | .975 | 1.39 | 1.69 | 1.89 | 2.03 | 2.15 | 2.24 | 2.32 | 2.39 | 2.46 | 2.51 | 2.56 | 2.61 | 2.65 | 2.69 |
| | .990 | 1.71 | 2.06 | 2.27 | 2.44 | 2.57 | 2.67 | 2.76 | 2.84 | 2.91 | 2.98 | 3.03 | 3.09 | 3.14 | 3.18 |
| | .995 | 1.98 | 2.35 | 2.59 | 2.76 | 2.91 | 3.02 | 3.13 | 3.21 | 3.29 | 3.36 | 3.42 | 3.48 | 3.53 | 3.58 |
| | .999 | 2.67 | 3.12 | 3.41 | 3.63 | 3.81 | 3.96 | 4.08 | 4.19 | 4.29 | 4.38 | 4.46 | 4.53 | 4.60 | 4.66 |
| 9 | .001 | $.0^{3}60$ | .020 | .063 | .112 | .159 | .201 | .239 | .271 | .300 | .326 | .349 | .369 | .388 | .405 |
| | .005 | $.0^{2}30$ | .044 | .108 | .170 | .225 | .271 | .311 | .346 | .376 | .403 | .427 | .448 | .468 | .485 |
| | .010 | $.0^{2}60$ | .062 | .137 | .205 | .262 | .310 | .352 | .387 | .418 | .445 | .469 | .491 | .511 | .529 |
| | .025 | .015 | .099 | .188 | .263 | .324 | .375 | .418 | .454 | .486 | .514 | .539 | .561 | .581 | .600 |
| | .050 | .030 | .141 | .242 | .322 | .386 | .438 | .481 | .519 | .551 | .580 | .605 | .628 | .649 | .668 |
| | .100 | .060 | .202 | .314 | .399 | .465 | .519 | .564 | .602 | .635 | .665 | .691 | .714 | .736 | .755 |
| | .250 | .152 | .339 | .464 | .554 | .624 | .680 | .727 | .767 | .803 | .833 | .861 | .887 | .909 | .929 |
| | .500 | .327 | .543 | .677 | .772 | .846 | .906 | .957 | 1.00 | 1.04 | 1.07 | 1.10 | 1.13 | 1.15 | 1.18 |
| | .750 | .576 | .809 | .952 | 1.05 | 1.13 | 1.20 | 1.25 | 1.30 | 1.34 | 1.38 | 1.41 | 1.44 | 1.47 | 1.50 |
| | .900 | .869 | 1.12 | 1.27 | 1.38 | 1.47 | 1.55 | 1.61 | 1.66 | 1.71 | 1.75 | 1.79 | 1.82 | 1.85 | 1.88 |
| | .950 | 1.08 | 1.34 | 1.51 | 1.63 | 1.72 | 1.80 | 1.87 | 1.93 | 1.98 | 2.02 | 2.07 | 2.10 | 2.14 | 2.17 |
| | .975 | 1.30 | 1.57 | 1.75 | 1.88 | 1.98 | 2.07 | 2.14 | 2.20 | 2.26 | 2.31 | 2.35 | 2.39 | 2.43 | 2.47 |
| | .990 | 1.59 | 1.89 | 2.08 | 2.23 | 2.34 | 2.43 | 2.51 | 2.58 | 2.65 | 2.70 | 2.75 | 2.80 | 2.84 | 2.88 |
| | .995 | 1.82 | 2.14 | 2.35 | 2.50 | 2.63 | 2.73 | 2.82 | 2.89 | 2.96 | 3.02 | 3.07 | 3.13 | 3.17 | 3.21 |
| | .999 | 2.40 | 2.78 | 3.03 | 3.22 | 3.37 | 3.49 | 3.59 | 3.69 | 3.77 | 3.84 | 3.91 | 3.97 | 4.03 | 4.08 |

**Percentiles of the distribution of $w_1/w_2$, where $w_1$ is the range of $N_1$ observations and $w_2$ the range of $N_2$ observations from normal populations having equal variances**

| $N_2$, SAMPLE SIZE FOR DENOMINATOR | *Cum. prop.* | $N_1$, SAMPLE SIZE FOR NUMERATOR: 2 | 3 | 4 | 5 | 6 | 7 | 8 | 9 | 10 | 11 | 12 | 13 | 14 | 15 |
|---|---|---|---|---|---|---|---|---|---|---|---|---|---|---|---|
| 10 | .001 | $.0^{3}58$ | .019 | .061 | .109 | .155 | .197 | .233 | .265 | .294 | .319 | .342 | .362 | .381 | .398 |
| | .005 | $.0^{2}29$ | .042 | .105 | .166 | .219 | .265 | .304 | .338 | .368 | .394 | .418 | .439 | .458 | .475 |
| | .010 | $.0^{2}58$ | .060 | .133 | .199 | .255 | .303 | .343 | .378 | .408 | .435 | .459 | .480 | .500 | .518 |
| | .025 | .014 | .096 | .183 | .256 | .316 | .365 | .407 | .443 | .474 | .502 | .526 | .548 | .568 | .586 |
| | .050 | .029 | .136 | .234 | .312 | .375 | .426 | .469 | .505 | .537 | .565 | .590 | .612 | .633 | .651 |
| | .100 | .058 | .196 | .305 | .387 | .452 | .504 | .548 | .586 | .618 | .647 | .672 | .695 | .716 | .736 |
| | .250 | .147 | .328 | .449 | .536 | .604 | .659 | .705 | .744 | .778 | .808 | .835 | .859 | .881 | .902 |
| | .500 | .315 | .523 | .652 | .744 | .816 | .874 | .922 | .964 | 1.00 | 1.03 | 1.06 | 1.09 | 1.11 | 1.13 |
| | .750 | .553 | .776 | .912 | 1.01 | 1.09 | 1.15 | 1.20 | 1.25 | 1.29 | 1.32 | 1.35 | 1.38 | 1.41 | 1.43 |
| | .900 | .830 | 1.06 | 1.21 | 1.32 | 1.40 | 1.47 | 1.52 | 1.57 | 1.62 | 1.66 | 1.69 | 1.72 | 1.75 | 1.78 |
| | .950 | 1.03 | 1.27 | 1.43 | 1.54 | 1.63 | 1.70 | 1.76 | 1.81 | 1.86 | 1.90 | 1.94 | 1.98 | 2.01 | 2.04 |
| | .975 | 1.23 | 1.48 | 1.64 | 1.76 | 1.86 | 1.93 | 2.00 | 2.06 | 2.11 | 2.15 | 2.20 | 2.23 | 2.27 | 2.30 |
| | .990 | 1.49 | 1.77 | 1.94 | 2.07 | 2.17 | 2.26 | 2.33 | 2.39 | 2.45 | 2.50 | 2.54 | 2.59 | 2.63 | 2.66 |
| | .995 | 1.70 | 1.99 | 2.17 | 2.31 | 2.42 | 2.51 | 2.59 | 2.66 | 2.72 | 2.77 | 2.82 | 2.87 | 2.91 | 2.95 |
| | .999 | 2.21 | 2.55 | 2.76 | 2.92 | 3.05 | 3.16 | 3.25 | 3.33 | 3.40 | 3.47 | 3.53 | 3.58 | 3.63 | 3.68 |
| 11 | .001 | $.0^{3}56$ | .018 | .059 | .107 | .152 | .192 | .228 | .260 | .288 | .313 | .336 | .356 | .374 | .391 |
| | .005 | $.0^{2}28$ | .041 | .102 | .162 | .214 | .259 | .298 | .331 | .360 | .387 | .410 | .431 | .450 | .467 |
| | .010 | $.0^{2}56$ | .058 | .129 | .194 | .249 | .296 | .336 | .370 | .400 | .426 | .450 | .471 | .491 | .508 |
| | .025 | .014 | .093 | .178 | .250 | .308 | .357 | .398 | .433 | .464 | .491 | .515 | .537 | .556 | .574 |
| | .050 | .028 | .132 | .228 | .305 | .366 | .416 | .458 | .494 | .525 | .553 | .577 | .600 | .620 | .638 |
| | .100 | .056 | .190 | .296 | .377 | .440 | .492 | .535 | .572 | .604 | .632 | .657 | .680 | .700 | .719 |
| | .250 | .142 | .318 | .436 | .521 | .588 | .641 | .686 | .724 | .757 | .787 | .813 | .837 | .858 | .878 |
| | .500 | .305 | .507 | .632 | .721 | .790 | .846 | .893 | .934 | .969 | 1.00 | 1.03 | 1.05 | 1.08 | 1.10 |
| | .750 | .535 | .749 | .880 | .974 | 1.05 | 1.11 | 1.16 | 1.20 | 1.24 | 1.27 | 1.30 | 1.33 | 1.35 | 1.38 |
| | .900 | .799 | 1.02 | 1.16 | 1.26 | 1.34 | 1.40 | 1.46 | 1.50 | 1.55 | 1.58 | 1.62 | 1.65 | 1.67 | 1.70 |
| | .950 | .986 | 1.22 | 1.36 | 1.47 | 1.55 | 1.62 | 1.67 | 1.72 | 1.77 | 1.81 | 1.84 | 1.88 | 1.91 | 1.93 |
| | .975 | 1.17 | 1.41 | 1.56 | 1.67 | 1.76 | 1.83 | 1.89 | 1.95 | 1.99 | 2.04 | 2.07 | 2.11 | 2.14 | 2.17 |
| | .990 | 1.42 | 1.67 | 1.83 | 1.95 | 2.04 | 2.12 | 2.19 | 2.25 | 2.30 | 2.34 | 2.39 | 2.42 | 2.46 | 2.49 |
| | .995 | 1.60 | 1.87 | 2.04 | 2.17 | 2.27 | 2.35 | 2.42 | 2.48 | 2.54 | 2.59 | 2.63 | 2.67 | 2.71 | 2.74 |
| | .999 | 2.07 | 2.37 | 2.56 | 2.71 | 2.82 | 2.92 | 3.00 | 3.07 | 3.13 | 3.19 | 3.24 | 3.29 | 3.34 | 3.38 |

**Percentiles of the distribution of $w_1/w_2$, where $w_1$ is the range of $N_1$ observations and $w_2$ the range of $N_2$ observations from normal populations having equal variances**

| $N_2$, sample size for denominator | *Cum. prop.* | $N_1$, sample size for numerator: 2 | 3 | 4 | 5 | 6 | 7 | 8 | 9 | 10 | 11 | 12 | 13 | 14 | 15 |
|---|---|---|---|---|---|---|---|---|---|---|---|---|---|---|---|
| 12 | .001 | $.0^354$ | .018 | .058 | .104 | .149 | .189 | .224 | .256 | .284 | .308 | .331 | .351 | .369 | .386 |
| | .005 | $.0^227$ | .040 | .100 | .158 | .210 | .254 | .292 | .325 | .354 | .380 | .403 | .424 | .442 | .460 |
| | .010 | $.0^254$ | .057 | .126 | .190 | .244 | .290 | .329 | .363 | .393 | .419 | .442 | .463 | .482 | .500 |
| | .025 | .014 | .091 | .174 | .244 | .302 | .350 | .390 | .425 | .455 | .482 | .506 | .527 | .547 | .564 |
| | .050 | .027 | .129 | .223 | .298 | .358 | .407 | .449 | .484 | .515 | .542 | .566 | .588 | .608 | .626 |
| | .100 | .055 | .186 | .290 | .368 | .430 | .481 | .523 | .560 | .591 | .619 | .644 | .666 | .686 | .705 |
| | .250 | .139 | .310 | .425 | .509 | .573 | .626 | .669 | .707 | .740 | .768 | .794 | .817 | .838 | .858 |
| | .500 | .297 | .493 | .615 | .702 | .769 | .823 | .869 | .908 | .942 | .973 | 1.00 | 1.02 | 1.05 | 1.07 |
| | .750 | .519 | .726 | .853 | .943 | 1.01 | 1.07 | 1.12 | 1.16 | 1.20 | 1.23 | 1.26 | 1.29 | 1.31 | 1.33 |
| | .900 | .773 | .987 | 1.12 | 1.21 | 1.29 | 1.35 | 1.40 | 1.45 | 1.49 | 1.52 | 1.55 | 1.58 | 1.61 | 1.63 |
| | .950 | .951 | 1.17 | 1.31 | 1.41 | 1.49 | 1.55 | 1.61 | 1.65 | 1.69 | 1.73 | 1.77 | 1.80 | 1.82 | 1.85 |
| | .975 | 1.13 | 1.35 | 1.50 | 1.60 | 1.68 | 1.75 | 1.81 | 1.86 | 1.90 | 1.94 | 1.98 | 2.01 | 2.04 | 2.07 |
| | .990 | 1.36 | 1.59 | 1.74 | 1.85 | 1.94 | 2.01 | 2.08 | 2.13 | 2.18 | 2.22 | 2.26 | 2.30 | 2.33 | 2.36 |
| | .995 | 1.53 | 1.78 | 1.94 | 2.05 | 2.14 | 2.22 | 2.29 | 2.34 | 2.39 | 2.44 | 2.48 | 2.52 | 2.55 | 2.59 |
| | .999 | 1.96 | 2.23 | 2.41 | 2.54 | 2.64 | 2.73 | 2.80 | 2.87 | 2.93 | 2.98 | 3.02 | 3.07 | 3.11 | 3.15 |
| 13 | .001 | $.0^353$ | .018 | .057 | .102 | .146 | .186 | .221 | .252 | .279 | .304 | .326 | .346 | .364 | .381 |
| | .005 | $.0^227$ | .039 | .098 | .155 | .206 | .250 | .287 | .320 | .349 | .374 | .397 | .417 | .436 | .453 |
| | .010 | $.0^253$ | .056 | .124 | .186 | .240 | .285 | .324 | .357 | .387 | .412 | .436 | .456 | .475 | .492 |
| | .025 | .013 | .089 | .170 | .239 | .296 | .343 | .383 | .418 | .448 | .474 | .498 | .519 | .538 | .556 |
| | .050 | .027 | .126 | .218 | .292 | .351 | .400 | .440 | .475 | .506 | .532 | .556 | .578 | .598 | .615 |
| | .100 | .053 | .182 | .283 | .361 | .422 | .472 | .513 | .549 | .580 | .608 | .632 | .654 | .674 | .692 |
| | .250 | .135 | .303 | .416 | .498 | .561 | .612 | .655 | .692 | .724 | .752 | .778 | .801 | .821 | .840 |
| | .500 | .289 | .481 | .600 | .685 | .750 | .803 | .848 | .886 | .920 | .949 | .976 | 1.00 | 1.02 | 1.04 |
| | .750 | .506 | .707 | .829 | .917 | .986 | 1.04 | 1.09 | 1.13 | 1.16 | 1.19 | 1.22 | 1.25 | 1.27 | 1.29 |
| | .900 | .750 | .958 | 1.08 | 1.18 | 1.25 | 1.31 | 1.36 | 1.40 | 1.44 | 1.47 | 1.50 | 1.53 | 1.55 | 1.58 |
| | .950 | .922 | 1.13 | 1.26 | 1.36 | 1.43 | 1.50 | 1.55 | 1.59 | 1.63 | 1.67 | 1.70 | 1.73 | 1.76 | 1.78 |
| | .975 | 1.09 | 1.30 | 1.44 | 1.54 | 1.62 | 1.68 | 1.74 | 1.78 | 1.82 | 1.86 | 1.90 | 1.93 | 1.96 | 1.98 |
| | .990 | 1.31 | 1.53 | 1.67 | 1.78 | 1.86 | 1.93 | 1.99 | 2.04 | 2.08 | 2.12 | 2.16 | 2.19 | 2.22 | 2.25 |
| | .995 | 1.47 | 1.70 | 1.85 | 1.96 | 2.05 | 2.12 | 2.18 | 2.23 | 2.28 | 2.32 | 2.36 | 2.40 | 2.43 | 2.46 |
| | .999 | 1.87 | 2.12 | 2.28 | 2.40 | 2.50 | 2.58 | 2.65 | 2.71 | 2.76 | 2.81 | 2.85 | 2.89 | 2.93 | 2.96 |

**Percentiles of the distribution of $w_1/w_2$, where $w_1$ is the range of $N_1$ observations and $w_2$ the range of $N_2$ observations from normal populations having equal variances**

| $N_2$, sample size for denominator | *Cum. prop.* | $N_1$, sample size for numerator: 2 | 3 | 4 | 5 | 6 | 7 | 8 | 9 | 10 | 11 | 12 | 13 | 14 | 15 |
|---|---|---|---|---|---|---|---|---|---|---|---|---|---|---|---|
| 14 | .001 | $.0^3 52$ | .017 | .056 | .101 | .144 | .183 | .217 | .248 | .275 | .300 | .322 | .341 | .359 | .376 |
| | .005 | $.0^2 26$ | .039 | .096 | .153 | .203 | .246 | .283 | .315 | .344 | .369 | .392 | .412 | .430 | .447 |
| | .010 | $.0^2 52$ | .055 | .122 | .183 | .236 | .281 | .319 | .352 | .381 | .407 | .429 | .450 | .469 | .486 |
| | .025 | .013 | .087 | .167 | .235 | .291 | .338 | .377 | .411 | .441 | .467 | .490 | .511 | .530 | .548 |
| | .050 | .026 | .124 | .214 | .287 | .345 | .393 | .433 | .468 | .498 | .524 | .548 | .569 | .589 | .606 |
| | .100 | .052 | .178 | .278 | .354 | .414 | .464 | .505 | .540 | .570 | .597 | .622 | .643 | .663 | .681 |
| | .250 | .133 | .297 | .407 | .488 | .550 | .601 | .643 | .679 | .711 | .738 | .763 | .786 | .806 | .825 |
| | .500 | .283 | .471 | .587 | .670 | .734 | .786 | .830 | .867 | .900 | .929 | .955 | .978 | 1.00 | 1.02 |
| | .750 | .494 | .690 | .810 | .895 | .961 | 1.02 | 1.06 | 1.10 | 1.13 | 1.16 | 1.19 | 1.22 | 1.24 | 1.26 |
| | .900 | .732 | .932 | 1.05 | 1.14 | 1.21 | 1.27 | 1.32 | 1.36 | 1.40 | 1.43 | 1.46 | 1.48 | 1.51 | 1.53 |
| | .950 | .897 | 1.10 | 1.23 | 1.32 | 1.39 | 1.45 | 1.50 | 1.54 | 1.58 | 1.61 | 1.64 | 1.67 | 1.70 | 1.72 |
| | .975 | 1.06 | 1.26 | 1.39 | 1.49 | 1.56 | 1.62 | 1.68 | 1.72 | 1.76 | 1.80 | 1.83 | 1.86 | 1.89 | 1.91 |
| | .990 | 1.26 | 1.48 | 1.61 | 1.71 | 1.79 | 1.85 | 1.91 | 1.96 | 2.00 | 2.04 | 2.07 | 2.10 | 2.13 | 2.16 |
| | .995 | 1.42 | 1.64 | 1.78 | 1.88 | 1.96 | 2.03 | 2.09 | 2.14 | 2.18 | 2.22 | 2.26 | 2.29 | 2.32 | 2.35 |
| | .999 | 1.79 | 2.03 | 2.18 | 2.29 | 2.38 | 2.46 | 2.52 | 2.58 | 2.63 | 2.67 | 2.71 | 2.75 | 2.78 | 2.81 |

**Percentiles of the distribution of $w_1/w_2$, where $w_1$ is the range of $N_1$ observations and $w_2$ the range of $N_2$ observations from normal populations having equal variances**

| $N_2$, sample size for denominator | Cum. prop. | $N_1$, sample size for numerator: 2 | 3 | 4 | 5 | 6 | 7 | 8 | 9 | 10 | 11 | 12 | 13 | 14 | 15 |
|---|---|---|---|---|---|---|---|---|---|---|---|---|---|---|---|
| 15 | .001 | $.0^351$ | .017 | .055 | .099 | .142 | .180 | .215 | .245 | .272 | .296 | .318 | .337 | .355 | .372 |
| | .005 | $.0^226$ | .038 | .094 | .150 | .199 | .242 | .279 | .311 | .339 | .364 | .387 | .407 | .425 | .442 |
| | .010 | $.0^251$ | .054 | .120 | .180 | .232 | .277 | .314 | .347 | .376 | .401 | .424 | .444 | .463 | .480 |
| | .025 | .013 | .085 | .165 | .232 | .287 | .333 | .372 | .405 | .435 | .461 | .484 | .504 | .523 | .541 |
| | .050 | .026 | .122 | .211 | .282 | .340 | .387 | .427 | .461 | .490 | .517 | .540 | .561 | .580 | .598 |
| | .100 | .051 | .175 | .273 | .348 | .408 | .456 | .497 | .531 | .561 | .588 | .612 | .633 | .653 | .671 |
| | .250 | .130 | .292 | .400 | .479 | .541 | .590 | .632 | .668 | .698 | .726 | .750 | .772 | .792 | .811 |
| | .500 | .278 | .462 | .576 | .657 | .720 | .771 | .814 | .850 | .883 | .911 | .936 | .960 | .981 | 1.00 |
| | .750 | .484 | .676 | .792 | .875 | .940 | .993 | 1.04 | 1.08 | 1.11 | 1.14 | 1.17 | 1.19 | 1.21 | 1.23 |
| | .900 | .715 | .910 | 1.03 | 1.11 | 1.18 | 1.24 | 1.28 | 1.32 | 1.36 | 1.39 | 1.42 | 1.44 | 1.47 | 1.49 |
| | .950 | .875 | 1.07 | 1.19 | 1.28 | 1.35 | 1.41 | 1.46 | 1.50 | 1.53 | 1.57 | 1.60 | 1.63 | 1.65 | 1.67 |
| | .975 | 1.03 | 1.23 | 1.35 | 1.44 | 1.52 | 1.57 | 1.63 | 1.67 | 1.71 | 1.74 | 1.77 | 1.80 | 1.83 | 1.85 |
| | .990 | 1.23 | 1.43 | 1.56 | 1.66 | 1.73 | 1.79 | 1.84 | 1.89 | 1.93 | 1.97 | 2.00 | 2.03 | 2.06 | 2.08 |
| | .995 | 1.38 | 1.59 | 1.72 | 1.82 | 1.90 | 1.96 | 2.01 | 2.06 | 2.10 | 2.14 | 2.18 | 2.21 | 2.24 | 2.26 |
| | .999 | 1.73 | 1.95 | 2.10 | 2.20 | 2.29 | 2.36 | 2.42 | 2.47 | 2.51 | 2.56 | 2.59 | 2.63 | 2.66 | 2.69 |

## A-8e CRITERIA FOR TESTING FOR EXTREME MEAN

| Statistic | No. of obs., $k$ | $P_{70}$ | $P_{80}$ | $P_{90}$ | $P_{95}$ | $P_{98}$ | $P_{99}$ | $P_{99.5}$ |
|---|---|---|---|---|---|---|---|---|
| $r_{10} = \dfrac{X_2 - X_1}{X_k - X_1}$ | 3 | .684 | .781 | .886 | .941 | .976 | .988 | .994 |
| | 4 | .471 | .560 | .679 | .765 | .846 | .889 | .926 |
| | 5 | .373 | .451 | .557 | .642 | .729 | .780 | .821 |
| | 6 | .318 | .386 | .482 | .560 | .644 | .698 | .740 |
| | 7 | .281 | .344 | .434 | .507 | .586 | .637 | .680 |
| $r_{11} = \dfrac{X_2 - X_1}{X_{k-1} - X_1}$ | 8 | .318 | .385 | .479 | .554 | .631 | .683 | .725 |
| | 9 | .288 | .352 | .441 | .512 | .587 | .635 | .677 |
| | 10 | .265 | .325 | .409 | .477 | .551 | .597 | .639 |
| $r_{21} = \dfrac{X_3 - X_1}{X_{k-1} - X_1}$ | 11 | .391 | .442 | .517 | .576 | .638 | .679 | .713 |
| | 12 | .370 | .419 | .490 | .546 | .605 | .642 | .675 |
| | 13 | .351 | .399 | .467 | .521 | .578 | .615 | .649 |
| $r_{22} = \dfrac{X_3 - X_1}{X_{k-2} - X_1}$ | 14 | .370 | .421 | .492 | .546 | .602 | .641 | .674 |
| | 15 | .353 | .402 | .472 | .525 | .579 | .616 | .647 |
| | 16 | .338 | .386 | .454 | .507 | .559 | .595 | .624 |
| | 17 | .325 | .373 | .438 | .490 | .542 | .577 | .605 |
| | 18 | .314 | .361 | .424 | .475 | .527 | .561 | .589 |
| | 19 | .304 | .350 | .412 | .462 | .514 | .547 | .575 |
| | 20 | .295 | .340 | .401 | .450 | .502 | .535 | .562 |
| | 21 | .287 | .331 | .391 | .440 | .491 | .524 | .551 |
| | 22 | .280 | .323 | .382 | .430 | .481 | .514 | .541 |
| | 23 | .274 | .316 | .374 | .421 | .472 | .505 | .532 |
| | 24 | .268 | .310 | .367 | .413 | .464 | .497 | .524 |
| | 25 | .262 | .304 | .360 | .406 | .457 | .489 | .516 |

## A-8f EFFICIENCY OF WINSORIZED MEAN AS AN ESTIMATE OF $\mu$ IN NORMAL SAMPLES, $N \leq 20$ AND FOUR LEVELS OF WINSORIZING

LEVEL OF WINSORIZING, $k$

| $N$ | 1 | 2 | 3 | 4 |
|---|---|---|---|---|
| 3 | .743 | | | |
| 4 | .838 | | | |
| 5 | .886 | .697 | | |
| 6 | .913 | .776 | | |
| 7 | .931 | .825 | .679 | |
| 8 | .943 | .858 | .743 | |
| 9 | .952 | .881 | .788 | .669 |
| 10 | .958 | .899 | .821 | .723 |
| 11 | .964 | .912 | .846 | .763 |
| 12 | .968 | .922 | .865 | .794 |
| 13 | .971 | .931 | .880 | .819 |
| 14 | .974 | .938 | .893 | .839 |
| 15 | .976 | .944 | .903 | .855 |
| 16 | .978 | .948 | .912 | .868 |
| 17 | .980 | .953 | .919 | .880 |
| 18 | .981 | .956 | .926 | .890 |
| 19 | .983 | .959 | .931 | .898 |
| 20 | .984 | .962 | .936 | .905 |

**Confidence coefficient .80**

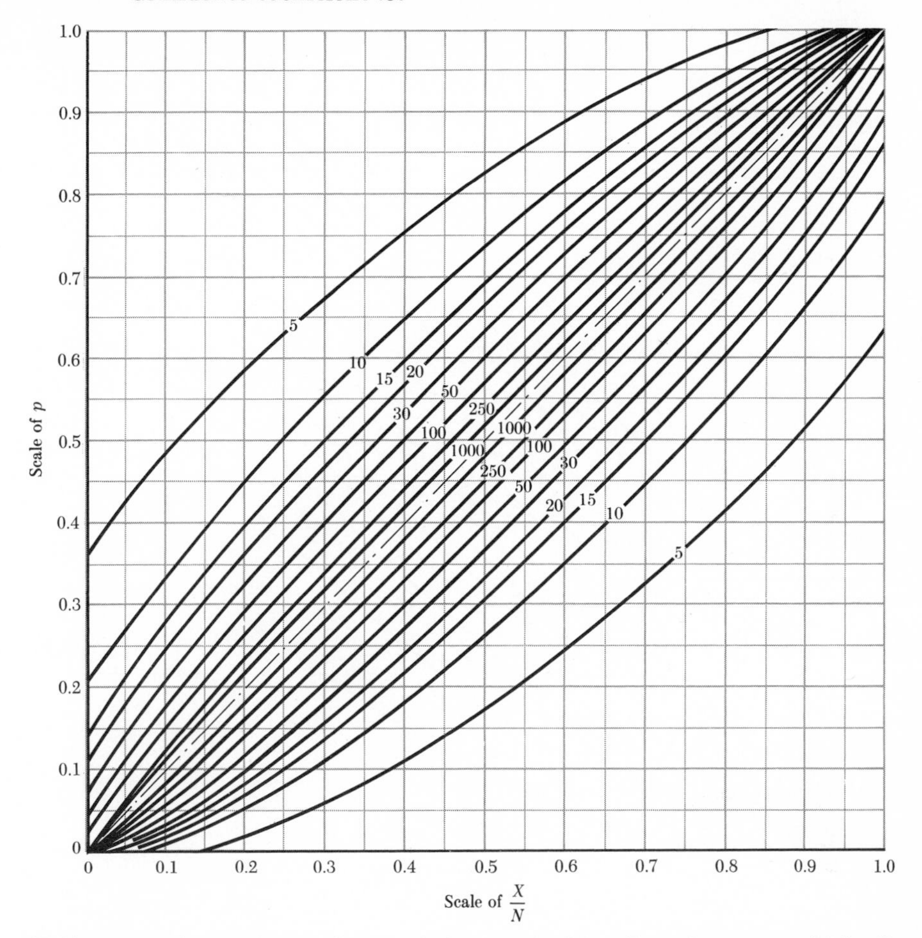


Table A-8$a$(1) is reproduced with permission from F. Mosteller, "On some useful 'inefficient' statistics," *Annals of Mathematical Statistics*, vol. 17 (1946), p. 377. Tables A-8$a$(2) and (3) are reproduced with permission from E. K. Yost, "Joint estimation of mean and standard deviation by percentiles," unpublished master's thesis, University of Oregon, Eugene, Ore., 1948. Table A-8$b$(1) is reproduced with permission from E. S. Pearson, "The probability integral of the range in samples of n observations from a normal population," *Biometrika*, vol. 32 (1942), p. 301. Tables A-8$b$(2) to (5) are reproduced from an unpublished manuscript, W. J. Dixon, University California, Los Angeles. Tables A-8$c$(1), (2) are reproduced with the permission of E. S. Pearson from E. Lord, "The use of the range in place of the standard deviation in the $t$ test," *Biometrika*, vol. 34 (1947), p. 41. Table A-8$c$(3) is reproduced with permission from J. E. Walsh, "On the range-midrange test and some tests with bounded significance levels," *Annals of Mathematical Statistics*, vol. 20 (1949), p. 257. Table A-8$d$ is reproduced from H. L. Harter, *Percentage Points of the Ratio of Two Ranges and Related Tables*, Aeronautical Research Laboratories, Wright-Patterson Air Force Base, Ohio, 1962 (ARL 62-378). Table A-8$e$ is reproduced from W. J. Dixon, "Processing data for outliers," *Biometrics*, vol. 9 (1953), p. 74. Table A-8$f$ is derived from W. J. Dixon, "Simplified estimation from censored normal samples," *Annals of Mathematical Statistics*, vol. 31 (1960), p. 385.

## A-9b CONFIDENCE BELTS FOR PROPORTIONS

**Confidence cofficient .90**

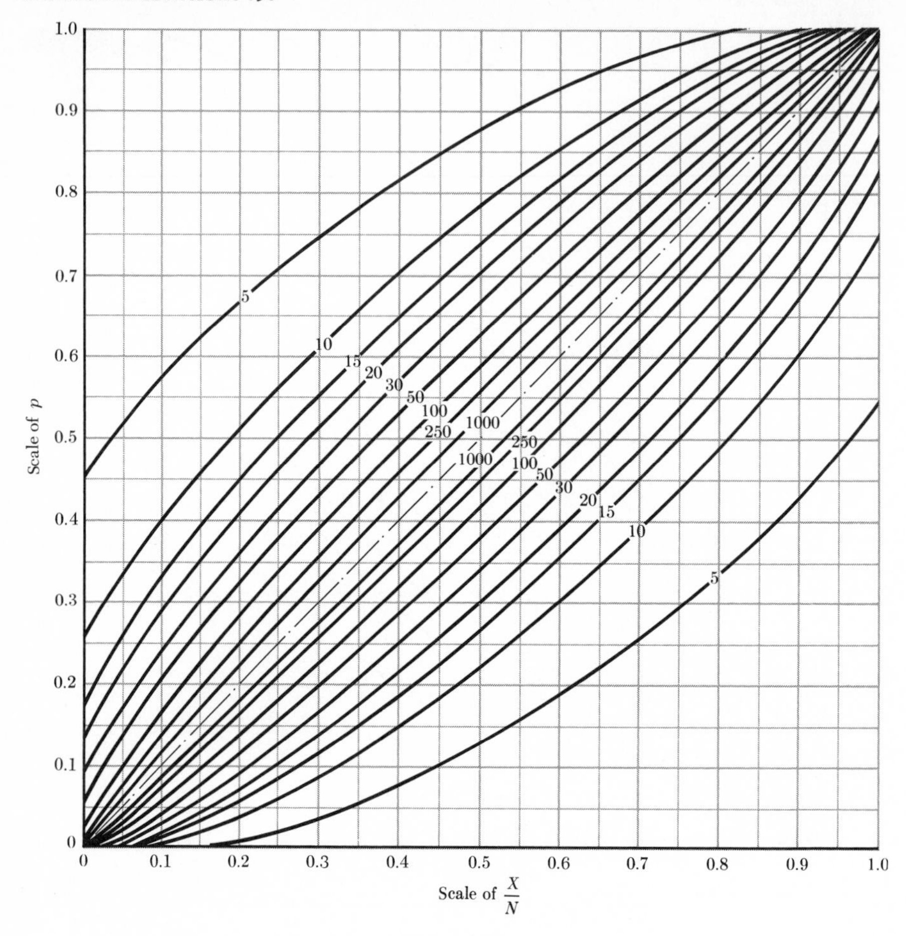

**Confidence coefficient .95**

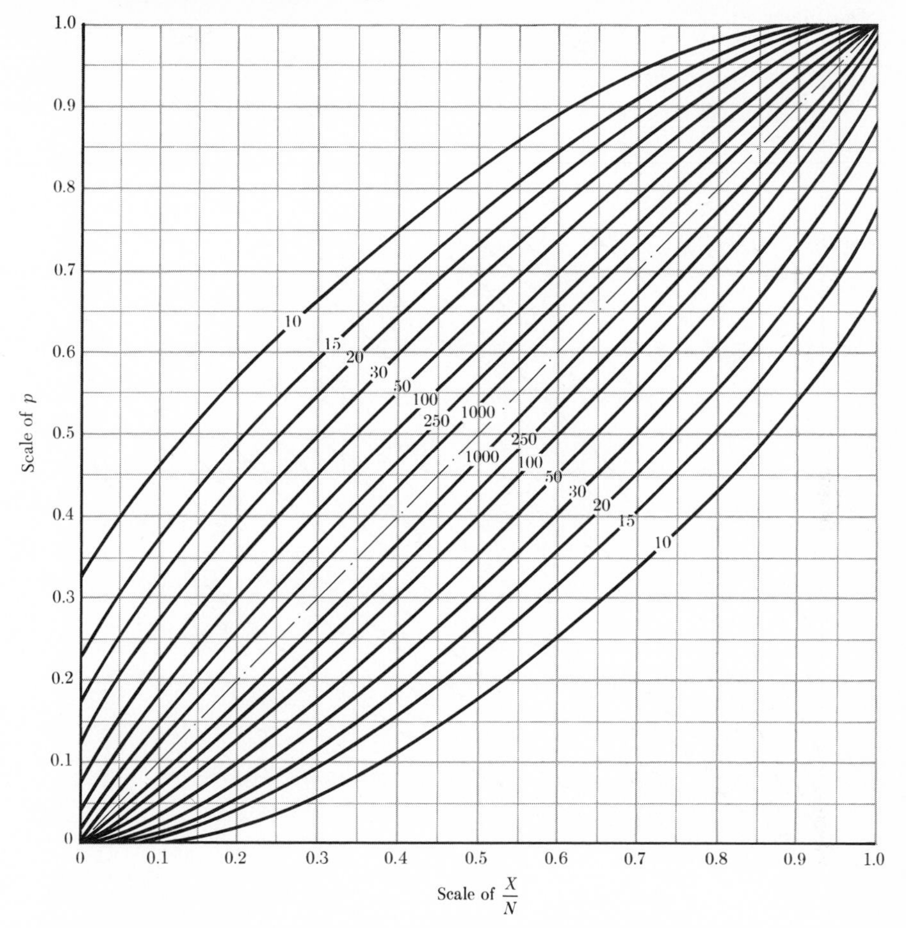


By permission of Prof. E. S. Pearson from C. J. Clopper, E. S. Pearson, "The use of confidence or fiducial limits illustrated in the case of the binomial," *Biometrika,* vol. 26 (1934), p. 404.

## A-9d CONFIDENCE BELTS FOR PROPORTIONS

**Confidence coefficient .99**

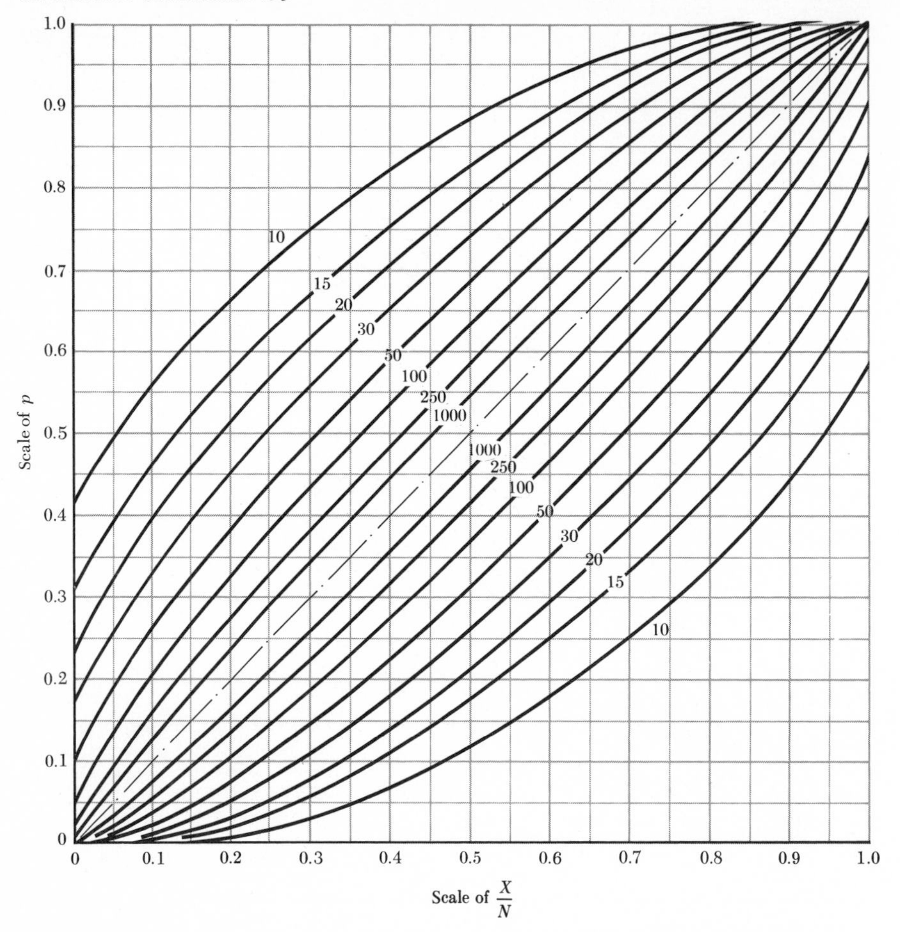


By permission of Prof. E. S. Pearson from C. J. Clopper, E. S. Pearson, "The use of confidence or fiducial limits illustrated in the case of the binomial," *Biometrika*, vol. 26 (1934), p. 404.

| $N$ | $S_1$ | $S_2$ | $X$ | Probability Obs. | Probability Other | Probability Total |
|---|---|---|---|---|---|---|
| 2 | 1 | 1 | 0 | .500 | .500 | 1.000 |
| 2 | 1 | 1 | 1 | .500 | .500 | 1.000 |
| 3 | 1 | 1 | 0 | .667 | .333 | 1.000 |
| 3 | 1 | 1 | 1 | .333 | .000 | .333 |
| 4 | 1 | 1 | 0 | .750 | .250 | 1.000 |
| 4 | 1 | 1 | 1 | .250 | .000 | .250 |
| 4 | 1 | 2 | 0 | .500 | .500 | 1.000 |
| 4 | 1 | 2 | 1 | .500 | .500 | 1.000 |
| 4 | 2 | 2 | 0 | .167 | .167 | .333 |
| 4 | 2 | 2 | 1 | 1.000 | .000 | 1.000 |
| 4 | 2 | 2 | 2 | .167 | .167 | .333 |
| 5 | 1 | 1 | 0 | .800 | .200 | 1.000 |
| 5 | 1 | 1 | 1 | .200 | .000 | .200 |
| 5 | 1 | 2 | 0 | .600 | .400 | 1.000 |
| 5 | 1 | 2 | 1 | .400 | .000 | .400 |
| 5 | 2 | 2 | 0 | .300 | .100 | .400 |
| 5 | 2 | 2 | 1 | .700 | .300 | 1.000 |
| 5 | 2 | 2 | 2 | .100 | .000 | .100 |
| 6 | 1 | 1 | 0 | .833 | .167 | 1.000 |
| 6 | 1 | 1 | 1 | .167 | .000 | .167 |
| 6 | 1 | 2 | 0 | .667 | .333 | 1.000 |
| 6 | 1 | 2 | 1 | .333 | .000 | .333 |
| 6 | 1 | 3 | 0 | .500 | .500 | 1.000 |
| 6 | 1 | 3 | 1 | .500 | .500 | 1.000 |
| 6 | 2 | 2 | 0 | .400 | .067 | .467 |
| 6 | 2 | 2 | 1 | .533 | .467 | 1.000 |
| 6 | 2 | 2 | 2 | .067 | .000 | .067 |
| 6 | 2 | 3 | 0 | .200 | .200 | .400 |
| 6 | 2 | 3 | 1 | 1.000 | .000 | 1.000 |
| 6 | 2 | 3 | 2 | .200 | .200 | .400 |
| 6 | 3 | 3 | 0 | .050 | .050 | .100 |
| 6 | 3 | 3 | 1 | .500 | .500 | 1.000 |
| 6 | 3 | 3 | 2 | .500 | .500 | 1.000 |
| 6 | 3 | 3 | 3 | .050 | .050 | .100 |
| 7 | 1 | 1 | 0 | .857 | .143 | 1.000 |
| 7 | 1 | 1 | 1 | .143 | .000 | .143 |
| 7 | 1 | 2 | 0 | .714 | .286 | 1.000 |
| 7 | 1 | 2 | 1 | .286 | .000 | .286 |
| 7 | 1 | 3 | 0 | .571 | .429 | 1.000 |
| 7 | 1 | 3 | 1 | .429 | .000 | .429 |
| 7 | 2 | 2 | 0 | .476 | .524 | 1.000 |
| 7 | 2 | 2 | 1 | .524 | .000 | .524 |
| 7 | 2 | 2 | 2 | .048 | .000 | .048 |
| 7 | 2 | 3 | 0 | .286 | .143 | .429 |
| 7 | 2 | 3 | 1 | .714 | .286 | 1.000 |
| 7 | 2 | 3 | 2 | .143 | .000 | .143 |
| 7 | 3 | 3 | 0 | .114 | .029 | .143 |
| 7 | 3 | 3 | 1 | .628 | .372 | 1.000 |
| 7 | 3 | 3 | 2 | .372 | .114 | .486 |
| 7 | 3 | 3 | 3 | .029 | .000 | .029 |
| 8 | 1 | 1 | 0 | .875 | .125 | 1.000 |
| 8 | 1 | 1 | 1 | .125 | .000 | .125 |
| 8 | 1 | 2 | 0 | .750 | .250 | 1.000 |
| 8 | 1 | 2 | 1 | .250 | .000 | .250 |
| 8 | 1 | 3 | 0 | .625 | .375 | 1.000 |
| 8 | 1 | 3 | 1 | .375 | .000 | .375 |
| 8 | 1 | 4 | 0 | .500 | .500 | 1.000 |
| 8 | 1 | 4 | 1 | .500 | .500 | 1.000 |
| 8 | 2 | 2 | 0 | .536 | .464 | 1.000 |
| 8 | 2 | 2 | 1 | .464 | .536 | 1.000 |
| 8 | 2 | 2 | 2 | .035 | .000 | .035 |
| 8 | 2 | 3 | 0 | .357 | .107 | .464 |
| 8 | 2 | 3 | 1 | .643 | .357 | 1.000 |
| 8 | 2 | 3 | 2 | .107 | .000 | .107 |
| 8 | 2 | 4 | 0 | .214 | .214 | .428 |
| 8 | 2 | 4 | 1 | 1.000 | .000 | 1.000 |
| 8 | 2 | 4 | 2 | .214 | .214 | .428 |
| 8 | 3 | 3 | 0 | .179 | .018 | .197 |
| 8 | 3 | 3 | 1 | .715 | .286 | 1.000 |
| 8 | 3 | 3 | 2 | .286 | .179 | .465 |
| 8 | 3 | 3 | 3 | .018 | .000 | .018 |
| 8 | 3 | 4 | 0 | .071 | .071 | .143 |
| 8 | 3 | 4 | 1 | .500 | .500 | 1.000 |
| 8 | 3 | 4 | 2 | .500 | .500 | 1.000 |
| 8 | 3 | 4 | 3 | .071 | .071 | .143 |
| 8 | 4 | 4 | 0 | .014 | .014 | .029 |
| 8 | 4 | 4 | 1 | .243 | .243 | .486 |
| 8 | 4 | 4 | 2 | 1.000 | .000 | 1.000 |
| 8 | 4 | 4 | 3 | .243 | .243 | .486 |
| 8 | 4 | 4 | 4 | .014 | .014 | .029 |
| 9 | 1 | 1 | 0 | .889 | .111 | 1.000 |
| 9 | 1 | 1 | 1 | .111 | .000 | .111 |
| 9 | 1 | 2 | 0 | .778 | .222 | 1.000 |
| 9 | 1 | 2 | 1 | .222 | .000 | .222 |
| 9 | 1 | 3 | 0 | .667 | .333 | 1.000 |
| 9 | 1 | 3 | 1 | .333 | .000 | .333 |
| 9 | 1 | 4 | 0 | .556 | .444 | 1.000 |
| 9 | 1 | 4 | 1 | .444 | .000 | .444 |
| 9 | 2 | 2 | 0 | .583 | .417 | 1.000 |
| 9 | 2 | 2 | 1 | .417 | .000 | .417 |
| 9 | 2 | 2 | 2 | .028 | .000 | .028 |
| 9 | 2 | 3 | 0 | .417 | .083 | .500 |
| 9 | 2 | 3 | 1 | .583 | .417 | 1.000 |
| 9 | 2 | 3 | 2 | .083 | .000 | .083 |
| 9 | 2 | 4 | 0 | .278 | .167 | .444 |
| 9 | 2 | 4 | 1 | .722 | .278 | 1.000 |
| 9 | 2 | 4 | 2 | .167 | .000 | .167 |
| 9 | 3 | 3 | 0 | .238 | .226 | .464 |
| 9 | 3 | 3 | 1 | 1.000 | .000 | 1.000 |
| 9 | 3 | 3 | 2 | .226 | .238 | .464 |
| 9 | 3 | 3 | 3 | .012 | .000 | .012 |
| 9 | 3 | 4 | 0 | .012 | .048 | .060 |
| 9 | 3 | 4 | 1 | .488 | .512 | 1.000 |
| 9 | 3 | 4 | 2 | .405 | .012 | .417 |
| 9 | 3 | 4 | 3 | .048 | .000 | .048 |
| 9 | 4 | 4 | 0 | .040 | .008 | .048 |
| 9 | 4 | 4 | 1 | .357 | .167 | .524 |
| 9 | 4 | 4 | 2 | .643 | .357 | 1.000 |
| 9 | 4 | 4 | 3 | .167 | .040 | .206 |
| 9 | 4 | 4 | 4 | .008 | .000 | .008 |
| 10 | 1 | 1 | 0 | .900 | .100 | 1.000 |
| 10 | 1 | 1 | 1 | .100 | .000 | .100 |
| 10 | 1 | 2 | 0 | .800 | .200 | 1.000 |
| 10 | 1 | 2 | 1 | .200 | .000 | .200 |
| 10 | 1 | 3 | 0 | .700 | .300 | 1.000 |
| 10 | 1 | 3 | 1 | .300 | .000 | .300 |
| 10 | 1 | 4 | 0 | .600 | .400 | 1.000 |
| 10 | 1 | 4 | 1 | .400 | .000 | .400 |
| 10 | 1 | 5 | 0 | .500 | .500 | 1.000 |
| 10 | 1 | 5 | 1 | .500 | .500 | 1.000 |
| 10 | 2 | 2 | 0 | .622 | .378 | 1.000 |
| 10 | 2 | 2 | 1 | .378 | .000 | .378 |
| 10 | 2 | 2 | 2 | .022 | .000 | .022 |
| 10 | 2 | 3 | 0 | .467 | .067 | .533 |
| 10 | 2 | 3 | 1 | .533 | .467 | 1.000 |
| 10 | 2 | 3 | 2 | .067 | .000 | .067 |
| 10 | 2 | 4 | 0 | .333 | .133 | .467 |
| 10 | 2 | 4 | 1 | .667 | .333 | 1.000 |
| 10 | 2 | 4 | 2 | .133 | .000 | .133 |
| 10 | 2 | 5 | 0 | .222 | .222 | .444 |
| 10 | 2 | 5 | 1 | .778 | .778 | 1.000 |
| 10 | 2 | 5 | 2 | .222 | .222 | .444 |
| 10 | 3 | 3 | 0 | .292 | .183 | .475 |
| 10 | 3 | 3 | 1 | .708 | .292 | 1.000 |
| 10 | 3 | 3 | 2 | .183 | .000 | .183 |
| 10 | 3 | 3 | 3 | .008 | .000 | .008 |
| 10 | 3 | 4 | 0 | .167 | .033 | .200 |
| 10 | 3 | 4 | 1 | .667 | .333 | 1.000 |

Cumulative probabilities are given for deviations in the observed direction from equality and for deviation of the same size or greater in the opposite direction. The total probabilities can be used for two-tail tests. Tables are extracted from more extensive tables prepared by Donald Goyette and M. Ray Mickey, Health Sciences Computing Facility, UCLA. In the probability columns "Obs." refers to the probability of a deviation as large or larger in the observed direction and "Other" refers to the probability of a deviation as large or larger in the opposite direction.

| | | |
|---|---|---|
| $X$ | — | $S_1$ |
| — | — | — |
| $S_2$ | — | $N$ |

$S_1$ is the smallest marginal total and $S_2$ is the next smallest; $X$ is the frequency in the cell corresponding to the two smallest totals.

# A-9e PROBABILITIES FOR FOURFOLD TABLES, $N \leq 15$ (CONTINUED)

| *N* | $S_1$ | $S_2$ | *X* | PROBABILITY *Obs.* | *Other* | *Total* |
|---|---|---|---|---|---|---|
| 10 | 3 | 4 | 2 | .333 | .167 | .500 |
| 10 | 3 | 4 | 3 | .033 | 0 | .033 |
| 10 | 3 | 5 | 0 | .083 | .083 | .167 |
| 10 | 3 | 5 | 1 | .500 | .500 | 1.000 |
| 10 | 3 | 5 | 2 | .500 | .500 | 1.000 |
| 10 | 3 | 5 | 3 | .083 | .083 | .167 |
| 10 | 4 | 4 | 0 | .071 | .005 | .076 |
| 10 | 4 | 4 | 1 | .452 | .119 | .571 |
| 10 | 4 | 4 | 2 | .548 | .452 | 1.000 |
| 10 | 4 | 4 | 3 | .119 | .071 | .190 |
| 10 | 4 | 4 | 4 | .005 | 0 | .005 |
| 10 | 4 | 5 | 0 | .024 | .024 | .048 |
| 10 | 4 | 5 | 1 | .262 | .262 | .524 |
| 10 | 4 | 5 | 2 | .738 | .738 | 1.000 |
| 10 | 4 | 5 | 3 | .262 | .262 | .524 |
| 10 | 4 | 5 | 4 | .024 | .024 | .048 |
| 10 | 5 | 5 | 0 | .004 | .004 | .008 |
| 10 | 5 | 5 | 1 | .103 | .103 | .206 |
| 10 | 5 | 5 | 2 | .500 | .500 | 1.000 |
| 10 | 5 | 5 | 3 | .500 | .500 | 1.000 |
| 10 | 5 | 5 | 4 | .103 | .103 | .206 |
| 10 | 5 | 5 | 5 | .004 | .004 | .008 |
| 11 | 1 | 1 | 0 | .909 | .091 | 1.000 |
| 11 | 1 | 1 | 1 | .091 | 0 | .091 |
| 11 | 1 | 2 | 0 | .818 | .182 | 1.000 |
| 11 | 1 | 2 | 1 | .182 | 0 | .182 |
| 11 | 1 | 3 | 0 | .727 | .273 | 1.000 |
| 11 | 1 | 3 | 1 | .273 | 0 | .273 |
| 11 | 1 | 4 | 0 | .636 | .364 | 1.000 |
| 11 | 1 | 4 | 1 | .364 | 0 | .364 |
| 11 | 1 | 5 | 0 | .545 | .455 | 1.000 |
| 11 | 1 | 5 | 1 | .455 | 0 | .455 |
| 11 | 2 | 2 | 0 | .655 | .345 | 1.000 |
| 11 | 2 | 2 | 1 | .345 | 0 | .345 |
| 11 | 2 | 2 | 2 | .018 | 0 | .018 |
| 11 | 2 | 3 | 0 | .509 | .055 | .564 |
| 11 | 2 | 3 | 1 | .491 | .509 | 1.000 |
| 11 | 2 | 3 | 2 | .055 | 0 | .055 |
| 11 | 2 | 4 | 0 | .382 | .109 | .491 |
| 11 | 2 | 4 | 1 | .618 | .382 | 1.000 |
| 11 | 2 | 4 | 2 | .109 | 0 | .109 |
| 11 | 2 | 5 | 0 | .273 | .182 | .455 |
| 11 | 2 | 5 | 1 | .727 | .273 | 1.000 |
| 11 | 2 | 5 | 2 | .182 | 0 | .182 |
| 11 | 3 | 3 | 0 | .339 | .152 | .491 |
| 11 | 3 | 3 | 1 | .661 | .339 | 1.000 |
| 11 | 3 | 3 | 2 | .152 | 0 | .152 |
| 11 | 3 | 3 | 3 | .006 | 0 | .006 |
| 11 | 3 | 4 | 0 | .212 | .024 | .236 |
| 11 | 3 | 4 | 1 | .721 | .279 | 1.000 |

| *N* | $S_1$ | $S_2$ | *X* | PROBABILITY *Obs.* | *Other* | *Total* |
|---|---|---|---|---|---|---|
| 11 | 3 | 4 | 2 | .279 | .212 | .491 |
| 11 | 3 | 4 | 3 | .024 | 0 | .024 |
| 11 | 3 | 5 | 0 | .121 | .061 | .182 |
| 11 | 3 | 5 | 1 | .576 | .424 | 1.000 |
| 11 | 3 | 5 | 2 | .424 | .121 | .545 |
| 11 | 3 | 5 | 3 | .061 | 0 | .061 |
| 11 | 4 | 4 | 0 | .106 | .088 | .194 |
| 11 | 4 | 4 | 1 | .530 | .470 | 1.000 |
| 11 | 4 | 4 | 2 | .470 | .106 | .576 |
| 11 | 4 | 4 | 3 | .088 | 0 | .088 |
| 11 | 4 | 4 | 4 | .003 | 0 | .003 |
| 11 | 4 | 5 | 0 | .045 | .015 | .061 |
| 11 | 4 | 5 | 1 | .348 | .197 | .545 |
| 11 | 4 | 5 | 2 | .652 | .348 | 1.000 |
| 11 | 4 | 5 | 3 | .197 | .045 | .242 |
| 11 | 4 | 5 | 4 | .015 | 0 | .015 |
| 11 | 5 | 5 | 0 | .013 | .002 | .015 |
| 11 | 5 | 5 | 1 | .175 | .067 | .242 |
| 11 | 5 | 5 | 2 | .608 | .392 | 1.000 |
| 11 | 5 | 5 | 3 | .392 | .175 | .567 |
| 11 | 5 | 5 | 4 | .067 | .013 | .080 |
| 11 | 5 | 5 | 5 | .002 | 0 | .002 |
| 12 | 1 | 1 | 0 | .917 | .083 | 1.000 |
| 12 | 1 | 1 | 1 | .083 | 0 | .083 |
| 12 | 1 | 2 | 0 | .833 | .167 | 1.000 |
| 12 | 1 | 2 | 1 | .167 | 0 | .167 |
| 12 | 1 | 3 | 0 | .750 | .250 | 1.000 |
| 12 | 1 | 3 | 1 | .250 | 0 | .250 |
| 12 | 1 | 4 | 0 | .667 | .333 | 1.000 |
| 12 | 1 | 4 | 1 | .333 | 0 | .333 |
| 12 | 1 | 5 | 0 | .583 | .417 | 1.000 |
| 12 | 1 | 5 | 1 | .417 | 0 | .417 |
| 12 | 1 | 6 | 0 | .500 | .500 | 1.000 |
| 12 | 1 | 6 | 1 | .500 | .500 | 1.000 |
| 12 | 2 | 2 | 0 | .682 | .318 | 1.000 |
| 12 | 2 | 2 | 1 | .318 | 0 | .318 |
| 12 | 2 | 2 | 2 | .015 | 0 | .015 |
| 12 | 2 | 3 | 0 | .545 | .455 | 1.000 |
| 12 | 2 | 3 | 1 | .455 | .545 | 1.000 |
| 12 | 2 | 3 | 2 | .045 | 0 | .045 |
| 12 | 2 | 4 | 0 | .424 | .091 | .515 |
| 12 | 2 | 4 | 1 | .576 | .424 | 1.000 |
| 12 | 2 | 4 | 2 | .091 | 0 | .091 |
| 12 | 2 | 5 | 0 | .318 | .152 | .470 |
| 12 | 2 | 5 | 1 | .682 | .318 | 1.000 |
| 12 | 2 | 5 | 2 | .152 | 0 | .152 |
| 12 | 2 | 6 | 0 | .227 | .227 | .455 |
| 12 | 2 | 6 | 1 | .773 | .773 | 1.000 |
| 12 | 2 | 6 | 2 | .227 | .227 | .455 |
| 12 | 3 | 3 | 0 | .382 | .127 | .509 |

| *N* | $S_1$ | $S_2$ | *X* | PROBABILITY *Obs.* | *Other* | *Total* |
|---|---|---|---|---|---|---|
| 12 | 3 | 3 | 1 | .618 | .382 | 1.000 |
| 12 | 3 | 3 | 2 | .127 | 0 | .127 |
| 12 | 3 | 3 | 3 | .005 | 0 | .005 |
| 12 | 3 | 4 | 0 | .255 | .236 | .491 |
| 12 | 3 | 4 | 1 | .764 | .745 | 1.000 |
| 12 | 3 | 4 | 2 | .236 | .255 | .491 |
| 12 | 3 | 4 | 3 | .018 | 0 | .018 |
| 12 | 3 | 5 | 0 | .159 | .045 | .205 |
| 12 | 3 | 5 | 1 | .636 | .364 | 1.000 |
| 12 | 3 | 5 | 2 | .364 | .159 | .523 |
| 12 | 3 | 5 | 3 | .045 | 0 | .045 |
| 12 | 3 | 6 | 0 | .091 | .091 | .182 |
| 12 | 3 | 6 | 1 | .500 | .500 | 1.000 |
| 12 | 3 | 6 | 2 | .500 | .500 | 1.000 |
| 12 | 3 | 6 | 3 | .091 | .091 | .182 |
| 12 | 4 | 4 | 0 | .141 | .067 | .208 |
| 12 | 4 | 4 | 1 | .594 | .406 | 1.000 |
| 12 | 4 | 4 | 2 | .406 | .141 | .547 |
| 12 | 4 | 4 | 3 | .067 | 0 | .067 |
| 12 | 4 | 4 | 4 | .002 | 0 | .002 |
| 12 | 4 | 5 | 0 | .071 | .010 | .081 |
| 12 | 4 | 5 | 1 | .424 | .152 | .576 |
| 12 | 4 | 5 | 2 | .576 | .424 | 1.000 |
| 12 | 4 | 5 | 3 | .152 | .071 | .222 |
| 12 | 4 | 5 | 4 | .010 | 0 | .010 |
| 12 | 4 | 6 | 0 | .030 | .030 | .061 |
| 12 | 4 | 6 | 1 | .273 | .273 | .545 |
| 12 | 4 | 6 | 2 | .727 | .727 | 1.000 |
| 12 | 4 | 6 | 3 | .273 | .273 | .545 |
| 12 | 4 | 6 | 4 | .030 | .030 | .061 |
| 12 | 5 | 5 | 0 | .027 | .001 | .028 |
| 12 | 5 | 5 | 1 | .247 | .045 | .293 |
| 12 | 5 | 5 | 2 | .689 | .311 | 1.000 |
| 12 | 5 | 5 | 3 | .311 | .247 | .558 |
| 12 | 5 | 5 | 4 | .045 | .027 | .072 |
| 12 | 5 | 5 | 5 | .001 | 0 | .001 |
| 12 | 5 | 6 | 0 | .008 | .008 | .015 |
| 12 | 5 | 6 | 1 | .121 | .121 | .242 |
| 12 | 5 | 6 | 2 | .500 | .500 | 1.000 |
| 12 | 5 | 6 | 3 | .500 | .500 | 1.000 |
| 12 | 5 | 6 | 4 | .121 | .121 | .242 |
| 12 | 5 | 6 | 5 | .008 | .008 | .015 |
| 12 | 6 | 6 | 0 | .001 | .001 | .002 |
| 12 | 6 | 6 | 1 | .040 | .040 | .080 |
| 12 | 6 | 6 | 2 | .284 | .284 | .567 |
| 12 | 6 | 6 | 3 | 1.000 | .000 | 1.000 |
| 12 | 6 | 6 | 4 | .284 | .284 | .567 |
| 12 | 6 | 6 | 5 | .040 | .040 | .080 |
| 12 | 6 | 6 | 6 | .001 | .001 | .002 |
| 13 | 1 | 1 | 0 | .923 | .077 | 1.000 |

| $N$ | $S_1$ | $S_2$ | $X$ | Probability Obs. | Probability Other | Probability Total |
|---|---|---|---|---|---|---|
| 13 | 1 | 1 | 1 | .077 | 0 | .077 |
| 13 | 1 | 2 | 0 | .846 | .154 | 1.000 |
| 13 | 1 | 2 | 1 | .154 | 0 | .154 |
| 13 | 1 | 3 | 0 | .769 | .231 | 1.000 |
| 13 | 1 | 3 | 1 | .231 | 0 | .231 |
| 13 | 1 | 4 | 0 | .692 | .308 | 1.000 |
| 13 | 1 | 4 | 1 | .308 | 0 | .308 |
| 13 | 1 | 5 | 0 | .615 | .385 | 1.000 |
| 13 | 1 | 5 | 1 | .385 | 0 | .385 |
| 13 | 1 | 6 | 0 | .538 | .462 | 1.000 |
| 13 | 1 | 6 | 1 | .462 | 0 | .462 |
| 13 | 2 | 2 | 0 | .705 | .295 | 1.000 |
| 13 | 2 | 2 | 1 | .295 | 0 | .295 |
| 13 | 2 | 2 | 2 | .013 | 0 | .013 |
| 13 | 2 | 3 | 0 | .577 | .423 | 1.000 |
| 13 | 2 | 3 | 1 | .423 | 0 | .423 |
| 13 | 2 | 3 | 2 | .038 | 0 | .038 |
| 13 | 2 | 4 | 0 | .462 | .077 | .538 |
| 13 | 2 | 4 | 1 | .538 | .462 | 1.000 |
| 13 | 2 | 4 | 2 | .077 | 0 | .077 |
| 13 | 2 | 5 | 0 | .359 | .128 | .487 |
| 13 | 2 | 5 | 1 | .641 | .359 | 1.000 |
| 13 | 2 | 5 | 2 | .128 | 0 | .128 |
| 13 | 2 | 6 | 0 | .269 | .192 | .462 |
| 13 | 2 | 6 | 1 | .731 | .269 | 1.000 |
| 13 | 2 | 6 | 2 | .192 | 0 | .192 |
| 13 | 3 | 3 | 0 | .420 | .108 | .528 |
| 13 | 3 | 3 | 1 | .580 | .420 | 1.000 |
| 13 | 3 | 3 | 2 | .108 | 0 | .108 |
| 13 | 3 | 3 | 3 | .003 | 0 | .003 |
| 13 | 3 | 4 | 0 | .294 | .203 | .497 |
| 13 | 3 | 4 | 1 | .706 | .294 | 1.000 |
| 13 | 3 | 4 | 2 | .203 | 0 | .203 |
| 13 | 3 | 4 | 3 | .014 | 0 | .014 |
| 13 | 3 | 5 | 0 | .196 | .035 | .231 |
| 13 | 3 | 5 | 1 | .685 | .315 | 1.000 |
| 13 | 3 | 5 | 2 | .315 | .196 | .510 |
| 13 | 3 | 5 | 3 | .035 | 0 | .035 |
| 13 | 3 | 6 | 0 | .122 | .070 | .192 |
| 13 | 3 | 6 | 1 | .563 | .437 | 1.000 |
| 13 | 3 | 6 | 2 | .437 | .122 | .559 |
| 13 | 3 | 6 | 3 | .070 | 0 | .070 |
| 13 | 4 | 4 | 0 | .176 | .052 | .228 |
| 13 | 4 | 4 | 1 | .646 | .354 | 1.000 |
| 13 | 4 | 4 | 2 | .354 | .176 | .530 |
| 13 | 4 | 4 | 3 | .052 | 0 | .052 |
| 13 | 4 | 4 | 4 | .001 | 0 | .001 |
| 13 | 4 | 5 | 0 | .098 | .007 | .105 |
| 13 | 4 | 5 | 1 | .490 | .119 | .608 |
| 13 | 4 | 5 | 2 | .510 | .490 | 1.000 |
| 13 | 4 | 5 | 3 | .119 | .098 | .217 |
| 13 | 4 | 5 | 4 | .007 | 0 | .007 |
| 13 | 4 | 6 | 0 | .049 | .021 | .070 |
| 13 | 4 | 6 | 1 | .343 | .217 | .559 |
| 13 | 4 | 6 | 2 | .657 | .343 | 1.000 |
| 13 | 4 | 6 | 3 | .217 | .049 | .266 |
| 13 | 4 | 6 | 4 | .021 | 0 | .021 |
| 13 | 5 | 5 | 0 | .044 | .032 | .075 |
| 13 | 5 | 5 | 1 | .315 | .249 | .565 |
| 13 | 5 | 5 | 2 | .685 | .315 | 1.000 |
| 13 | 5 | 5 | 3 | .249 | .044 | .293 |
| 13 | 5 | 5 | 4 | .032 | 0 | .032 |
| 13 | 5 | 5 | 5 | .001 | 0 | .001 |
| 13 | 5 | 6 | 0 | .016 | .005 | .021 |
| 13 | 5 | 6 | 1 | .179 | .086 | .266 |
| 13 | 5 | 6 | 2 | .587 | .413 | 1.000 |
| 13 | 5 | 6 | 3 | .413 | .179 | .592 |
| 13 | 5 | 6 | 4 | .086 | .016 | .103 |
| 13 | 5 | 6 | 5 | .005 | 0 | .005 |
| 13 | 6 | 6 | 0 | .004 | .001 | .005 |
| 13 | 6 | 6 | 1 | .078 | .025 | .103 |
| 13 | 6 | 6 | 2 | .383 | .209 | .592 |
| 13 | 6 | 6 | 3 | .617 | .383 | 1.000 |
| 13 | 6 | 6 | 4 | .209 | .078 | .286 |
| 13 | 6 | 6 | 5 | .025 | .004 | .029 |
| 13 | 6 | 6 | 6 | .001 | 0 | .001 |
| 14 | 1 | 1 | 0 | .929 | .071 | 1.000 |
| 14 | 1 | 1 | 1 | .071 | 0 | .071 |
| 14 | 1 | 2 | 0 | .857 | .143 | 1.000 |
| 14 | 1 | 2 | 1 | .143 | 0 | .143 |
| 14 | 1 | 3 | 0 | .786 | .214 | 1.000 |
| 14 | 1 | 3 | 1 | .214 | 0 | .214 |
| 14 | 1 | 4 | 0 | .714 | .286 | 1.000 |
| 14 | 1 | 4 | 1 | .286 | 0 | .286 |
| 14 | 1 | 5 | 0 | .643 | .357 | 1.000 |
| 14 | 1 | 5 | 1 | .357 | 0 | .357 |
| 14 | 1 | 6 | 0 | .571 | .429 | 1.000 |
| 14 | 1 | 6 | 1 | .429 | 0 | .429 |
| 14 | 1 | 7 | 0 | .500 | .500 | 1.000 |
| 14 | 1 | 7 | 1 | .500 | .500 | 1.000 |
| 14 | 2 | 2 | 0 | .725 | .275 | 1.000 |
| 14 | 2 | 2 | 1 | .275 | 0 | .275 |
| 14 | 2 | 2 | 2 | .011 | 0 | .011 |
| 14 | 2 | 3 | 0 | .604 | .396 | 1.000 |
| 14 | 2 | 3 | 1 | .396 | 0 | .396 |
| 14 | 2 | 3 | 2 | .033 | 0 | .033 |
| 14 | 2 | 4 | 0 | .495 | .066 | .560 |
| 14 | 2 | 4 | 1 | .505 | .495 | 1.000 |
| 14 | 2 | 4 | 2 | .066 | 0 | .066 |
| 14 | 2 | 5 | 0 | .396 | .110 | .505 |
| 14 | 2 | 5 | 1 | .604 | .396 | 1.000 |
| 14 | 2 | 5 | 2 | .110 | 0 | .110 |
| 14 | 2 | 6 | 0 | .308 | .165 | .473 |
| 14 | 2 | 6 | 1 | .692 | .308 | 1.000 |
| 14 | 2 | 6 | 2 | .165 | 0 | .165 |
| 14 | 2 | 7 | 0 | .231 | .231 | .462 |
| 14 | 2 | 7 | 1 | .769 | .769 | 1.000 |
| 14 | 2 | 7 | 2 | .231 | .231 | .462 |
| 14 | 3 | 3 | 0 | .453 | .093 | .547 |
| 14 | 3 | 3 | 1 | .547 | .453 | 1.000 |
| 14 | 3 | 3 | 2 | .093 | 0 | .093 |
| 14 | 3 | 3 | 3 | .003 | 0 | .003 |
| 14 | 3 | 4 | 0 | .330 | .176 | .505 |
| 14 | 3 | 4 | 1 | .670 | .330 | 1.000 |
| 14 | 3 | 4 | 2 | .176 | 0 | .176 |
| 14 | 3 | 4 | 3 | .011 | 0 | .011 |
| 14 | 3 | 5 | 0 | .231 | .027 | .258 |
| 14 | 3 | 5 | 1 | .725 | .275 | 1.000 |
| 14 | 3 | 5 | 2 | .275 | .231 | .505 |
| 14 | 3 | 5 | 3 | .027 | 0 | .027 |
| 14 | 3 | 6 | 0 | .154 | .055 | .209 |
| 14 | 3 | 6 | 1 | .615 | .385 | 1.000 |
| 14 | 3 | 6 | 2 | .385 | .154 | .538 |
| 14 | 3 | 6 | 3 | .055 | 0 | .055 |
| 14 | 3 | 7 | 0 | .096 | .096 | .192 |
| 14 | 3 | 7 | 1 | .500 | .500 | 1.000 |
| 14 | 3 | 7 | 2 | .500 | .500 | 1.000 |
| 14 | 3 | 7 | 3 | .096 | .096 | .192 |
| 13 | 4 | 4 | 0 | .210 | .041 | .251 |
| 14 | 4 | 4 | 1 | .689 | .311 | 1.000 |
| 14 | 4 | 4 | 2 | .311 | .210 | .520 |
| 14 | 4 | 4 | 3 | .041 | 0 | .041 |
| 14 | 4 | 4 | 4 | .001 | 0 | .001 |
| 14 | 4 | 5 | 0 | .126 | .095 | .221 |
| 14 | 4 | 5 | 1 | .545 | .455 | 1.000 |
| 14 | 4 | 5 | 2 | .455 | .126 | .580 |
| 14 | 4 | 5 | 3 | .095 | 0 | .095 |
| 14 | 4 | 5 | 4 | .005 | 0 | .005 |
| 14 | 4 | 6 | 0 | .070 | .015 | .085 |
| 14 | 4 | 6 | 1 | .406 | .175 | .580 |
| 14 | 4 | 6 | 2 | .594 | .406 | 1.000 |
| 14 | 4 | 6 | 3 | .175 | .070 | .245 |
| 14 | 4 | 6 | 4 | .015 | 0 | .015 |
| 14 | 4 | 7 | 0 | .035 | .035 | .070 |
| 14 | 4 | 7 | 1 | .280 | .280 | .559 |
| 14 | 4 | 7 | 2 | .720 | .720 | 1.000 |
| 14 | 4 | 7 | 3 | .280 | .280 | .559 |
| 14 | 4 | 7 | 4 | .035 | .035 | .070 |
| 14 | 5 | 5 | 0 | .063 | .023 | .086 |
| 14 | 5 | 5 | 1 | .378 | .203 | .580 |

# PROBABILITIES FOR FOURFOLD TABLES, $N \leq 15$ (CONTINUED)

| | | | | PROBABILITY | | |
|---|---|---|---|---|---|---|
| *N* | $S_1$ | $S_2$ | *X* | *Obs.* | *Other* | *Total* |
| 14 | 5 | 5 | 2 | .622 | .378 | 1.000 |
| 14 | 5 | 5 | 3 | .203 | .063 | .266 |
| 14 | 5 | 5 | 4 | .023 | 0 | .023 |
| 14 | 5 | 5 | 5 | .000 | 0 | .000 |
| 14 | 5 | 6 | 0 | .028 | .003 | .031 |
| 14 | 5 | 6 | 1 | .238 | .063 | .301 |
| 14 | 5 | 6 | 2 | .657 | .343 | 1.000 |
| 14 | 5 | 6 | 3 | .343 | .238 | .580 |
| 14 | 5 | 6 | 4 | .063 | .028 | .091 |
| 14 | 5 | 6 | 5 | .003 | 0 | .003 |
| 14 | 5 | 7 | 0 | .010 | .010 | .021 |
| 14 | 5 | 7 | 1 | .133 | .133 | .266 |
| 14 | 5 | 7 | 2 | .500 | .500 | 1.000 |
| 14 | 5 | 7 | 3 | .500 | .500 | 1.000 |
| 14 | 5 | 7 | 4 | .133 | .133 | .266 |
| 14 | 5 | 7 | 5 | .010 | .010 | .021 |
| 14 | 6 | 6 | 0 | .009 | .000 | .010 |
| 14 | 6 | 6 | 1 | .121 | .016 | .138 |
| 14 | 6 | 6 | 2 | .471 | .156 | .627 |
| 14 | 6 | 6 | 3 | .529 | .471 | 1.000 |
| 14 | 6 | 6 | 4 | .156 | .121 | .277 |
| 14 | 6 | 6 | 5 | .016 | .009 | .026 |
| 14 | 6 | 6 | 6 | .000 | 0 | .000 |
| 14 | 6 | 7 | 0 | .002 | .002 | .005 |
| 14 | 6 | 7 | 1 | .051 | .051 | .103 |
| 14 | 6 | 7 | 2 | .296 | .296 | .592 |
| 14 | 6 | 7 | 3 | .704 | .704 | 1.000 |
| 14 | 6 | 7 | 4 | .296 | .296 | .592 |
| 14 | 6 | 7 | 5 | .051 | .051 | .103 |
| 14 | 6 | 7 | 6 | .002 | .002 | .005 |
| 14 | 7 | 7 | 0 | .000 | .000 | .001 |
| 14 | 7 | 7 | 1 | .015 | .015 | .029 |
| 14 | 7 | 7 | 2 | .143 | .143 | .286 |
| 14 | 7 | 7 | 3 | .500 | .500 | 1.000 |
| 14 | 7 | 7 | 4 | .500 | .500 | 1.000 |
| 14 | 7 | 7 | 5 | .143 | .143 | .286 |
| 14 | 7 | 7 | 6 | .015 | .015 | .029 |
| 14 | 7 | 7 | 7 | .000 | .000 | .001 |
| 15 | 1 | 1 | 0 | .933 | .067 | 1.000 |
| 15 | 1 | 1 | 1 | .067 | 0 | .067 |
| 15 | 1 | 2 | 0 | .867 | .133 | 1.000 |
| 15 | 1 | 2 | 1 | .133 | 0 | .133 |
| 15 | 1 | 3 | 0 | .800 | .200 | 1.000 |
| 15 | 1 | 3 | 1 | .200 | 0 | .200 |
| 15 | 1 | 4 | 0 | .733 | .267 | 1.000 |
| 15 | 1 | 4 | 1 | .267 | 0 | .267 |
| 15 | 1 | 5 | 0 | .667 | .333 | 1.000 |
| 15 | 1 | 5 | 1 | .333 | 0 | .333 |
| 15 | 1 | 6 | 0 | .600 | .400 | 1.000 |
| 15 | 1 | 6 | 1 | .400 | 0 | .400 |

| | | | | PROBABILITY | | |
|---|---|---|---|---|---|---|
| *N* | $S_1$ | $S_2$ | *X* | *Obs.* | *Other* | *Total* |
| 15 | 1 | 7 | 0 | .533 | .467 | 1.000 |
| 15 | 1 | 7 | 1 | .467 | 0 | .467 |
| 15 | 2 | 2 | 0 | .743 | .257 | 1.000 |
| 15 | 2 | 2 | 1 | .257 | 0 | .257 |
| 15 | 2 | 2 | 2 | .010 | 0 | .010 |
| 15 | 2 | 3 | 0 | .629 | .371 | 1.000 |
| 15 | 2 | 3 | 1 | .371 | 0 | .371 |
| 15 | 2 | 3 | 2 | .029 | 0 | .029 |
| 15 | 2 | 4 | 0 | .524 | .057 | .581 |
| 15 | 2 | 4 | 1 | .476 | .524 | 1.000 |
| 15 | 2 | 4 | 2 | .057 | 0 | .057 |
| 15 | 2 | 5 | 0 | .429 | .095 | .524 |
| 15 | 2 | 5 | 1 | .571 | .429 | 1.000 |
| 15 | 2 | 5 | 2 | .095 | 0 | .095 |
| 15 | 2 | 6 | 0 | .343 | .143 | .486 |
| 15 | 2 | 6 | 1 | .657 | .343 | 1.000 |
| 15 | 2 | 6 | 2 | .143 | 0 | .143 |
| 15 | 2 | 7 | 0 | .267 | .200 | .467 |
| 15 | 2 | 7 | 1 | .733 | .267 | 1.000 |
| 15 | 2 | 7 | 2 | .200 | 0 | .200 |
| 15 | 3 | 3 | 0 | .484 | .081 | .565 |
| 15 | 3 | 3 | 1 | .516 | .484 | 1.000 |
| 15 | 3 | 3 | 2 | .081 | 0 | .081 |
| 15 | 3 | 3 | 3 | .002 | 0 | .002 |
| 15 | 3 | 4 | 0 | .363 | .154 | .516 |
| 15 | 3 | 4 | 1 | .637 | .363 | 1.000 |
| 15 | 3 | 4 | 2 | .154 | 0 | .154 |
| 15 | 3 | 4 | 3 | .009 | 0 | .009 |
| 15 | 3 | 5 | 0 | .264 | .242 | .505 |
| 15 | 3 | 5 | 1 | .758 | .736 | 1.000 |
| 15 | 3 | 5 | 2 | .242 | .264 | .505 |
| 15 | 3 | 5 | 3 | .022 | 0 | .022 |
| 15 | 3 | 6 | 0 | .185 | .044 | .229 |
| 15 | 3 | 6 | 1 | .659 | .341 | 1.000 |
| 15 | 3 | 6 | 2 | .341 | .185 | .525 |
| 15 | 3 | 6 | 3 | .044 | 0 | .044 |
| 15 | 3 | 7 | 0 | .123 | .077 | .200 |
| 15 | 3 | 7 | 1 | .554 | .446 | 1.000 |
| 15 | 3 | 7 | 2 | .446 | .123 | .569 |
| 15 | 3 | 7 | 3 | .077 | 0 | .077 |
| 15 | 4 | 4 | 0 | .242 | .033 | .275 |
| 15 | 4 | 4 | 1 | .725 | .275 | 1.000 |
| 15 | 4 | 4 | 2 | .275 | .242 | .516 |
| 15 | 4 | 4 | 3 | .033 | 0 | .033 |
| 15 | 4 | 4 | 4 | .001 | 0 | .001 |
| 15 | 4 | 5 | 0 | .154 | .077 | .231 |
| 15 | 4 | 5 | 1 | .593 | .407 | 1.000 |
| 15 | 4 | 5 | 2 | .407 | .154 | .560 |
| 15 | 4 | 5 | 3 | .077 | 0 | .077 |
| 15 | 4 | 5 | 4 | .004 | 0 | .004 |

| | | | | PROBABILITY | | |
|---|---|---|---|---|---|---|
| *N* | $S_1$ | $S_2$ | *X* | *Obs.* | *Other* | *Total* |
| 15 | 4 | 6 | 0 | .092 | .011 | .103 |
| 15 | 4 | 6 | 1 | .462 | .143 | .604 |
| 15 | 4 | 6 | 2 | .538 | .462 | 1.000 |
| 15 | 4 | 6 | 3 | .143 | .092 | .235 |
| 15 | 4 | 6 | 4 | .011 | 0 | .011 |
| 15 | 4 | 7 | 0 | .051 | .026 | .077 |
| 15 | 4 | 7 | 1 | .338 | .231 | .569 |
| 15 | 4 | 7 | 2 | .662 | .338 | 1.000 |
| 15 | 4 | 7 | 3 | .231 | .051 | .282 |
| 15 | 4 | 7 | 4 | .026 | 0 | .026 |
| 15 | 5 | 5 | 0 | .084 | .017 | .101 |
| 15 | 5 | 5 | 1 | .434 | .167 | .600 |
| 15 | 5 | 5 | 2 | .566 | .434 | 1.000 |
| 15 | 5 | 5 | 3 | .167 | .084 | .251 |
| 15 | 5 | 5 | 4 | .017 | 0 | .017 |
| 15 | 5 | 5 | 5 | .000 | 0 | .000 |
| 15 | 5 | 6 | 0 | .042 | .047 | .089 |
| 15 | 5 | 6 | 1 | .294 | .287 | .580 |
| 15 | 5 | 6 | 2 | .713 | .706 | 1.000 |
| 15 | 5 | 6 | 3 | .287 | .294 | .580 |
| 15 | 5 | 6 | 4 | .047 | .042 | .089 |
| 15 | 5 | 6 | 5 | .002 | 0 | .002 |
| 15 | 5 | 7 | 0 | .019 | .007 | .026 |
| 15 | 5 | 7 | 1 | .182 | .100 | .282 |
| 15 | 5 | 7 | 2 | .573 | .427 | 1.000 |
| 15 | 5 | 7 | 3 | .427 | .182 | .608 |
| 15 | 5 | 7 | 4 | .100 | .019 | .119 |
| 15 | 5 | 7 | 5 | .007 | 0 | .007 |
| 15 | 6 | 6 | 0 | .017 | .011 | .028 |
| 15 | 6 | 6 | 1 | .168 | .119 | .287 |
| 15 | 6 | 6 | 2 | .545 | .455 | 1.000 |
| 15 | 6 | 6 | 3 | .455 | .168 | .622 |
| 15 | 6 | 6 | 4 | .119 | .017 | .136 |
| 15 | 6 | 6 | 5 | .011 | 0 | .011 |
| 15 | 6 | 6 | 6 | .000 | 0 | .000 |
| 15 | 6 | 7 | 0 | .006 | .001 | .007 |
| 15 | 6 | 7 | 1 | .084 | .035 | .119 |
| 15 | 6 | 7 | 2 | .378 | .231 | .608 |
| 15 | 6 | 7 | 3 | .622 | .378 | 1.000 |
| 15 | 6 | 7 | 4 | .231 | .084 | .315 |
| 15 | 6 | 7 | 5 | .035 | .006 | .041 |
| 15 | 6 | 7 | 6 | .001 | 0 | .001 |
| 15 | 7 | 7 | 0 | .001 | .000 | .001 |
| 15 | 7 | 7 | 1 | .032 | .009 | .041 |
| 15 | 7 | 7 | 2 | .214 | .100 | .315 |
| 15 | 7 | 7 | 3 | .595 | .405 | 1.000 |
| 15 | 7 | 7 | 4 | .405 | .214 | .619 |
| 15 | 7 | 7 | 5 | .100 | .032 | .132 |
| 15 | 7 | 7 | 6 | .009 | .001 | .010 |
| 15 | 7 | 7 | 7 | .000 | 0 | .000 |

(Two-tail percentage points are given for the binomial for $p = .5$)

| $N$ | 1% | 5% | 10% | 25% | $N$ | 1% | 5% | 10% | 25% |
|---|---|---|---|---|---|---|---|---|---|
| 1 | | | | | 51 | 15 | 18 | 19 | 20 |
| 2 | | | | | 52 | 16 | 18 | 19 | 21 |
| 3 | | | | 0 | 53 | 16 | 18 | 20 | 21 |
| 4 | | | | 0 | 54 | 17 | 19 | 20 | 22 |
| 5 | | | 0 | 0 | 55 | 17 | 19 | 20 | 22 |
| 6 | | 0 | 0 | 1 | 56 | 17 | 20 | 21 | 23 |
| 7 | | 0 | 0 | 1 | 57 | 18 | 20 | 21 | 23 |
| 8 | 0 | 0 | 1 | 1 | 58 | 18 | 21 | 22 | 24 |
| 9 | 0 | 1 | 1 | 2 | 59 | 19 | 21 | 22 | 24 |
| 10 | 0 | 1 | 1 | 2 | 60 | 19 | 21 | 23 | 25 |
| 11 | 0 | 1 | 2 | 3 | 61 | 20 | 22 | 23 | 25 |
| 12 | 1 | 2 | 2 | 3 | 62 | 20 | 22 | 24 | 25 |
| 13 | 1 | 2 | 3 | 3 | 63 | 20 | 23 | 24 | 26 |
| 14 | 1 | 2 | 3 | 4 | 64 | 21 | 23 | 24 | 26 |
| 15 | 2 | 3 | 3 | 4 | 65 | 21 | 24 | 25 | 27 |
| 16 | 2 | 3 | 4 | 5 | 66 | 22 | 24 | 25 | 27 |
| 17 | 2 | 4 | 4 | 5 | 67 | 22 | 25 | 26 | 28 |
| 18 | 3 | 4 | 5 | 6 | 68 | 22 | 25 | 26 | 28 |
| 19 | 3 | 4 | 5 | 6 | 69 | 23 | 25 | 27 | 29 |
| 20 | 3 | 5 | 5 | 6 | 70 | 23 | 26 | 27 | 29 |
| 21 | 4 | 5 | 6 | 7 | 71 | 24 | 26 | 28 | 30 |
| 22 | 4 | 5 | 6 | 7 | 72 | 24 | 27 | 28 | 30 |
| 23 | 4 | 6 | 7 | 8 | 73 | 25 | 27 | 28 | 31 |
| 24 | 5 | 6 | 7 | 8 | 74 | 25 | 28 | 29 | 31 |
| 25 | 5 | 7 | 7 | 9 | 75 | 25 | 28 | 29 | 32 |
| 26 | 6 | 7 | 8 | 9 | 76 | 26 | 28 | 30 | 32 |
| 27 | 6 | 7 | 8 | 10 | 77 | 26 | 29 | 30 | 32 |
| 28 | 6 | 8 | 9 | 10 | 78 | 27 | 29 | 31 | 33 |
| 29 | 7 | 8 | 9 | 10 | 79 | 27 | 30 | 31 | 33 |
| 30 | 7 | 9 | 10 | 11 | 80 | 28 | 30 | 32 | 34 |
| 31 | 7 | 9 | 10 | 11 | 81 | 28 | 31 | 32 | 34 |
| 32 | 8 | 9 | 10 | 12 | 82 | 28 | 31 | 33 | 35 |
| 33 | 8 | 10 | 11 | 12 | 83 | 29 | 32 | 33 | 35 |
| 34 | 9 | 10 | 11 | 13 | 84 | 29 | 32 | 33 | 36 |
| 35 | 9 | 11 | 12 | 13 | 85 | 30 | 32 | 34 | 36 |
| 36 | 9 | 11 | 12 | 14 | 86 | 30 | 33 | 34 | 37 |
| 37 | 10 | 12 | 13 | 14 | 87 | 31 | 33 | 35 | 37 |
| 38 | 10 | 12 | 13 | 14 | 88 | 31 | 34 | 35 | 38 |
| 39 | 11 | 12 | 13 | 15 | 89 | 31 | 34 | 36 | 38 |
| 40 | 11 | 13 | 14 | 15 | 90 | 32 | 35 | 36 | 39 |
| 41 | 11 | 13 | 14 | 16 | 91 | 32 | 35 | 37 | 39 |
| 42 | 12 | 14 | 15 | 16 | 92 | 33 | 36 | 37 | 39 |
| 43 | 12 | 14 | 15 | 17 | 93 | 33 | 36 | 38 | 40 |
| 44 | 13 | 15 | 16 | 17 | 94 | 34 | 37 | 38 | 40 |
| 45 | 13 | 15 | 16 | 18 | 95 | 34 | 37 | 38 | 41 |
| 46 | 13 | 15 | 16 | 18 | 96 | 34 | 37 | 39 | 41 |
| 47 | 14 | 16 | 17 | 19 | 97 | 35 | 38 | 39 | 42 |
| 48 | 14 | 16 | 17 | 19 | 98 | 35 | 38 | 40 | 42 |
| 49 | 15 | 17 | 18 | 19 | 99 | 36 | 39 | 40 | 43 |
| 50 | 15 | 17 | 18 | 20 | 100 | 36 | 39 | 41 | 43 |

For values of $N$ larger than 100, approximate values of $r$ may be found by taking the nearest integer less than $(N - 1)/2 - k\sqrt{N + 1}$, where $k$ is 1.2879, 0.9800, 0.8224, 0.5752 for the 1, 5, 10, 25% values, respectively.

# A-10b PERCENTILES FOR THE BINOMIAL FOR $p = .5$

| X \ N | 10 | 11 | 12 | 13 | 14 | 15 | 16 | 17 | 18 | 19 | 20 | 21 | 22 |
|---|---|---|---|---|---|---|---|---|---|---|---|---|---|
| 0 | .001 | .000 | .000 | .000 | .000 | .000 | .000 | .000 | .000 | .000 | .000 | .000 | .000 |
| 1 | .011 | .006 | .003 | .002 | .001 | .000 | .000 | .000 | .000 | .000 | .000 | .000 | .000 |
| 2 | .055 | .033 | .019 | .011 | .006 | .004 | .002 | .001 | .001 | .000 | .000 | .000 | .000 |
| 3 | .172 | .113 | .073 | .046 | .029 | .018 | .011 | .006 | .004 | .002 | .001 | .001 | .000 |
| 4 | .377 | .274 | .194 | .133 | .090 | .059 | .038 | .025 | .015 | .010 | .006 | .004 | .002 |
| 5 | .623 | .500 | .387 | .291 | .212 | .151 | .105 | .072 | .048 | .032 | .021 | .013 | .008 |
| 6 | | | .613 | .500 | .395 | .304 | .227 | .166 | .119 | .084 | .058 | .039 | .026 |
| 7 | | | | | .605 | .500 | .402 | .315 | .240 | .180 | .132 | .095 | .067 |
| 8 | | | | | | | .598 | .500 | .407 | .324 | .252 | .192 | .143 |
| 9 | | | | | | | | | .593 | .500 | .412 | .332 | .262 |
| 10 | | | | | | | | | | | .588 | .500 | .416 |
| 11 | | | | | | | | | | | | | .584 |

| X \ N | 23 | 24 | 25 | 26 | 27 | 28 | 29 | 30 | 31 | 32 | 33 | 34 | 35 |
|---|---|---|---|---|---|---|---|---|---|---|---|---|---|
| 4 | .001 | .001 | .000 | .000 | .000 | .000 | .000 | .000 | .000 | .000 | .000 | .000 | .000 |
| 5 | .005 | .003 | .002 | .001 | .001 | .000 | .000 | .000 | .000 | .000 | .000 | .000 | .000 |
| 6 | .017 | .011 | .007 | .005 | .003 | .002 | .001 | .001 | .000 | .000 | .000 | .000 | .000 |
| 7 | .047 | .032 | .022 | .014 | .010 | .006 | .004 | .003 | .002 | .001 | .001 | .000 | .000 |
| 8 | .105 | .076 | .054 | .038 | .026 | .018 | .012 | .008 | .005 | .004 | .002 | .001 | .001 |
| 9 | .202 | .154 | .115 | .084 | .061 | .044 | .031 | .021 | .015 | .010 | .007 | .005 | .003 |
| 10 | .339 | .271 | .212 | .163 | .124 | .092 | .068 | .049 | .035 | .025 | .018 | .012 | .008 |
| 11 | .500 | .419 | .345 | .279 | .221 | .172 | .132 | .100 | .075 | .055 | .040 | .029 | .020 |
| 12 | | .581 | .500 | .423 | .351 | .286 | .229 | .181 | .141 | .108 | .081 | .061 | .045 |
| 13 | | | | .577 | .500 | .425 | .356 | .292 | .237 | .189 | .148 | .115 | .088 |
| 14 | | | | | | .575 | .500 | .428 | .360 | .298 | .243 | .196 | .155 |
| 15 | | | | | | | | .572 | .500 | .430 | .364 | .304 | .250 |
| 16 | | | | | | | | | | .570 | .500 | .432 | .368 |
| 17 | | | | | | | | | | | | .568 | .500 |
| 18 | | | | | | | | | | | | | |

| X \ N | 36 | 37 | 38 | 39 | 40 | 41 | 42 | 43 | 44 | 45 | 46 | 47 | 48 |
|---|---|---|---|---|---|---|---|---|---|---|---|---|---|
| 8 | .001 | .000 | .000 | .000 | .000 | .000 | .000 | .000 | .000 | .000 | .000 | .000 | .000 |
| 9 | .002 | .001 | .001 | .001 | .000 | .000 | .000 | .000 | .000 | .000 | .000 | .000 | .000 |
| 10 | .006 | .004 | .003 | .002 | .001 | .001 | .000 | .000 | .000 | .000 | .000 | .000 | .000 |
| 11 | .014 | .010 | .007 | .005 | .003 | .002 | .001 | .001 | .001 | .000 | .000 | .000 | .000 |
| 12 | .033 | .024 | .017 | .012 | .008 | .006 | .004 | .003 | .002 | .001 | .001 | .001 | .000 |
| 13 | .066 | .049 | .036 | .027 | .019 | .014 | .010 | .007 | .005 | .003 | .002 | .002 | .001 |
| 14 | .121 | .094 | .072 | .054 | .040 | .030 | .022 | .016 | .011 | .008 | .006 | .004 | .003 |
| 15 | .203 | .162 | .128 | .100 | .077 | .059 | .044 | .033 | .024 | .018 | .013 | .009 | .007 |
| 16 | .309 | .256 | .209 | .168 | .134 | .106 | .082 | .063 | .048 | .036 | .027 | .020 | .015 |
| 17 | .434 | .371 | .314 | .261 | .215 | .174 | .140 | .111 | .087 | .068 | .052 | .039 | .030 |
| 18 | .566 | .500 | .436 | .375 | .318 | .266 | .220 | .180 | .146 | .116 | .092 | .072 | .056 |
| 19 | | | .564 | .500 | .437 | .378 | .322 | .271 | .226 | .186 | .151 | .121 | .097 |
| 20 | | | | | .563 | .500 | .439 | .380 | .326 | .276 | .231 | .191 | .156 |
| 21 | | | | | | | .561 | .500 | .440 | .383 | .329 | .280 | .235 |
| 22 | | | | | | | | | .560 | .500 | .441 | .385 | .333 |
| 23 | | | | | | | | | | | .559 | .500 | .443 |
| 24 | | | | | | | | | | | | | .557 |

| X \ N | 49 | 50 | 51 | 52 | 53 | 54 | 55 | 56 | 57 | 58 | 59 | 60 | 61 |
|---|---|---|---|---|---|---|---|---|---|---|---|---|---|
| 13 | .001 | .000 | .000 | .000 | .000 | .000 | .000 | .000 | .000 | .000 | .000 | .000 | .000 |
| 14 | .002 | .001 | .001 | .001 | .000 | .000 | .000 | .000 | .000 | .000 | .000 | .000 | .000 |
| 15 | .005 | .003 | .002 | .002 | .001 | .001 | .001 | .000 | .000 | .000 | .000 | .000 | .000 |
| 16 | .011 | .008 | .005 | .004 | .003 | .002 | .001 | .001 | .001 | .000 | .000 | .000 | .000 |
| 17 | .022 | .016 | .012 | .009 | .006 | .005 | .003 | .002 | .002 | .001 | .001 | .001 | .000 |
| 18 | .043 | .032 | .024 | .018 | .014 | .010 | .007 | .005 | .004 | .003 | .002 | .001 | .001 |
| 19 | .076 | .059 | .046 | .035 | .027 | .020 | .015 | .011 | .008 | .006 | .004 | .003 | .002 |
| 20 | .126 | .101 | .080 | .063 | .049 | .038 | .029 | .022 | .017 | .012 | .009 | .007 | .005 |
| 21 | .196 | .161 | .131 | .106 | .084 | .067 | .052 | .041 | .031 | .024 | .018 | .014 | .010 |
| 22 | .284 | .240 | .201 | .166 | .136 | .110 | .089 | .070 | .056 | .043 | .034 | .026 | .020 |
| 23 | .388 | .336 | .288 | .244 | .205 | .170 | .140 | .114 | .092 | .074 | .059 | .046 | .036 |
| 24 | .500 | .444 | .390 | .339 | .292 | .248 | .209 | .175 | .145 | .119 | .096 | .078 | .062 |
| 25 | | .556 | .500 | .445 | .392 | .342 | .295 | .252 | .214 | .179 | .149 | .123 | .100 |
| 26 | | | | .555 | .500 | .446 | .394 | .344 | .298 | .256 | .217 | .183 | .153 |
| 27 | | | | | | .554 | .500 | .447 | .396 | .347 | .301 | .259 | .221 |
| 28 | | | | | | | | .553 | .500 | .448 | .397 | .349 | .304 |
| 29 | | | | | | | | | | .552 | .500 | .449 | .399 |
| 30 | | | | | | | | | | | | .551 | .500 |
| 31 | | | | | | | | | | | | | |

| X \ N | 62 | 63 | 64 | 65 | 66 | 67 | 68 | 69 | 70 | 71 | 72 | 73 | 74 |
|---|---|---|---|---|---|---|---|---|---|---|---|---|---|
| 18 | .001 | .000 | .000 | .000 | .000 | .000 | .000 | .000 | .000 | .000 | .000 | .000 | .000 |
| 19 | .002 | .001 | .001 | .001 | .000 | .000 | .000 | .000 | .000 | .000 | .000 | .000 | .000 |
| 20 | .004 | .003 | .002 | .001 | .001 | .001 | .000 | .000 | .000 | .000 | .000 | .000 | .000 |
| 21 | .008 | .006 | .004 | .003 | .002 | .002 | .001 | .001 | .001 | .000 | .000 | .000 | .000 |
| 22 | .015 | .011 | .008 | .006 | .005 | .003 | .002 | .002 | .001 | .001 | .001 | .000 | .000 |
| 23 | .028 | .021 | .016 | .012 | .009 | .007 | .005 | .004 | .003 | .002 | .001 | .001 | .001 |
| 24 | .049 | .038 | .030 | .023 | .018 | .014 | .010 | .008 | .006 | .004 | .003 | .002 | .002 |
| 25 | .081 | .065 | .052 | .041 | .032 | .025 | .019 | .015 | .011 | .008 | .006 | .005 | .004 |
| 26 | .126 | .104 | .084 | .068 | .054 | .043 | .034 | .027 | .021 | .016 | .012 | .009 | .007 |
| 27 | .187 | .157 | .130 | .107 | .088 | .071 | .057 | .046 | .036 | .028 | .022 | .017 | .013 |
| 28 | .263 | .225 | .191 | .161 | .134 | .111 | .091 | .074 | .060 | .048 | .038 | .030 | .024 |
| 29 | .352 | .307 | .266 | .229 | .195 | .164 | .137 | .114 | .094 | .077 | .062 | .050 | .040 |
| 30 | .450 | .401 | .354 | .310 | .269 | .232 | .198 | .168 | .141 | .118 | .097 | .080 | .065 |
| 31 | .550 | .500 | .450 | .402 | .356 | .313 | .272 | .235 | .201 | .171 | .144 | .121 | .100 |
| 32 | | | .550 | .500 | .451 | .404 | .358 | .315 | .275 | .238 | .205 | .175 | .148 |
| 33 | | | | | .549 | .500 | .452 | .405 | .360 | .318 | .278 | .241 | .208 |
| 34 | | | | | | | .548 | .500 | .452 | .406 | .362 | .320 | .281 |
| 35 | | | | | | | | | .548 | .500 | .453 | .408 | .364 |
| 36 | | | | | | | | | | | .547 | .500 | .454 |
| 37 | | | | | | | | | | | | | .546 |

| X \ N | 75 | 76 | 77 | 78 | 79 | 80 | 81 | 82 | 83 | 84 | 85 | 86 | 87 |
|---|---|---|---|---|---|---|---|---|---|---|---|---|---|
| 23 | .001 | .000 | .000 | .000 | .000 | .000 | .000 | .000 | .000 | .000 | .000 | .000 | .000 |
| 24 | .001 | .001 | .001 | .000 | .000 | .000 | .000 | .000 | .000 | .000 | .000 | .000 | .000 |
| 25 | .003 | .002 | .001 | .001 | .001 | .001 | .000 | .000 | .000 | .000 | .000 | .000 | .000 |
| 26 | .005 | .004 | .003 | .002 | .002 | .001 | .001 | .001 | .000 | .000 | .000 | .000 | .000 |
| 27 | .010 | .008 | .006 | .004 | .003 | .002 | .002 | .001 | .001 | .001 | .001 | .000 | .000 |
| 28 | .018 | .014 | .011 | .008 | .006 | .005 | .004 | .003 | .002 | .001 | .001 | .001 | .001 |
| 29 | .032 | .025 | .020 | .015 | .012 | .009 | .007 | .005 | .004 | .003 | .002 | .002 | .001 |
| 30 | .053 | .042 | .034 | .027 | .021 | .016 | .013 | .010 | .008 | .006 | .004 | .003 | .003 |
| 31 | .083 | .068 | .055 | .044 | .036 | .028 | .022 | .018 | .014 | .011 | .008 | .006 | .005 |
| 32 | .124 | .103 | .086 | .070 | .057 | .046 | .037 | .030 | .024 | .019 | .015 | .011 | .009 |
| 33 | .178 | .151 | .127 | .106 | .088 | .073 | .060 | .049 | .039 | .031 | .025 | .020 | .016 |
| 34 | .244 | .211 | .181 | .154 | .130 | .109 | .091 | .075 | .062 | .051 | .041 | .033 | .027 |
| 35 | .322 | .283 | .247 | .214 | .184 | .157 | .133 | .112 | .094 | .078 | .064 | .053 | .043 |
| 36 | .409 | .366 | .324 | .286 | .250 | .217 | .187 | .160 | .136 | .115 | .096 | .080 | .066 |
| 37 | .500 | .454 | .410 | .367 | .326 | .288 | .253 | .220 | .190 | .163 | .139 | .118 | .099 |
| 38 | | .546 | .500 | .455 | .411 | .369 | .328 | .291 | .255 | .223 | .193 | .166 | .142 |
| 39 | | | | .545 | .500 | .456 | .412 | .370 | .330 | .293 | .258 | .225 | .196 |
| 40 | | | | | | .544 | .500 | .456 | .413 | .372 | .332 | .295 | .260 |
| 41 | | | | | | | | .544 | .500 | .457 | .414 | .373 | .334 |
| 42 | | | | | | | | | | .543 | .500 | .457 | .415 |
| 43 | | | | | | | | | | | | .543 | .500 |

| X \ N | 88 | 89 | 90 | 91 | 92 | 93 | 94 | 95 | 96 | 97 | 98 | 99 | 100 |
|---|---|---|---|---|---|---|---|---|---|---|---|---|---|
| 29 | .001 | .001 | .000 | .000 | .000 | .000 | .000 | .000 | .000 | .000 | .000 | .000 | .000 |
| 30 | .002 | .001 | .001 | .001 | .001 | .000 | .000 | .000 | .000 | .000 | .000 | .000 | .000 |
| 31 | .004 | .003 | .002 | .002 | .001 | .001 | .001 | .000 | .000 | .000 | .000 | .000 | .000 |
| 32 | .007 | .005 | .004 | .003 | .002 | .002 | .001 | .001 | .001 | .001 | .000 | .000 | .000 |
| 33 | .012 | .010 | .007 | .006 | .004 | .003 | .003 | .002 | .001 | .001 | .001 | .001 | .000 |
| 34 | .021 | .017 | .013 | .010 | .008 | .006 | .005 | .004 | .003 | .002 | .002 | .001 | .001 |
| 35 | .035 | .028 | .022 | .018 | .014 | .011 | .009 | .007 | .005 | .004 | .003 | .002 | .002 |
| 36 | .055 | .045 | .036 | .029 | .024 | .019 | .015 | .012 | .009 | .007 | .006 | .004 | .003 |
| 37 | .083 | .069 | .057 | .046 | .038 | .031 | .025 | .020 | .016 | .012 | .010 | .008 | .006 |
| 38 | .120 | .102 | .085 | .071 | .059 | .048 | .039 | .032 | .026 | .021 | .017 | .013 | .010 |
| 39 | .169 | .145 | .123 | .104 | .087 | .073 | .061 | .050 | .041 | .034 | .027 | .022 | .018 |
| 40 | .228 | .198 | .171 | .147 | .126 | .107 | .090 | .075 | .063 | .052 | .043 | .035 | .028 |
| 41 | .297 | .263 | .230 | .201 | .174 | .150 | .128 | .109 | .092 | .077 | .065 | .054 | .044 |
| 42 | .375 | .336 | .299 | .265 | .233 | .203 | .177 | .152 | .131 | .111 | .094 | .080 | .067 |
| 43 | .458 | .416 | .376 | .338 | .301 | .267 | .235 | .206 | .179 | .155 | .133 | .114 | .097 |
| 44 | .542 | .500 | .458 | .417 | .377 | .339 | .303 | .269 | .238 | .208 | .182 | .157 | .136 |
| 45 | | | .542 | .500 | .459 | .418 | .379 | .341 | .305 | .271 | .240 | .211 | .184 |
| 46 | | | | | .541 | .500 | .459 | .419 | .380 | .342 | .307 | .273 | .242 |
| 47 | | | | | | | .541 | .500 | .459 | .420 | .381 | .344 | .309 |
| 48 | | | | | | | | | .541 | .500 | .460 | .420 | .382 |
| 49 | | | | | | | | | | | .540 | .500 | .460 |
| 50 | | | | | | | | | | | | | .540 |

Values are of $\alpha = P(X \text{ or less})$. For larger values of $X$ read the table by entering with $N - X$ and $1 - \alpha$. For example, if $N = 20$, the probability of observing $X = 13$ or less is $1 - .132 = .868$.

## A-11 DISTRIBUTION OF TOTAL NUMBER OF RUNS $u$ IN SAMPLES OF SIZE $(N_1,N_2)$

| $(N_1,N_2)$ \ $u$ | 2 | 3 | 4 | 5 | 6 | 7 | 8 | 9 | 10 | 11 | 12 | 13 | 14 | 15 | 16 | 17 | 18 | 19 | 20 |
|---|---|---|---|---|---|---|---|---|---|---|---|---|---|---|---|---|---|---|---|
| (2,3) | .200 | .500 | .900 | 1.000 | | | | | | | | | | | | | | | |
| (2,4) | .133 | .400 | .800 | 1.000 | | | | | | | | | | | | | | | |
| (2,5) | .095 | .333 | .714 | 1.000 | | | | | | | | | | | | | | | |
| (2,6) | .071 | .286 | .643 | 1.000 | | | | | | | | | | | | | | | |
| (2,7) | .056 | .250 | .583 | 1.000 | | | | | | | | | | | | | | | |
| (2,8) | .044 | .222 | .533 | 1.000 | | | | | | | | | | | | | | | |
| (2,9) | .036 | .200 | .491 | 1.000 | | | | | | | | | | | | | | | |
| (2,10) | .030 | .182 | .455 | 1.000 | | | | | | | | | | | | | | | |
| (3,3) | .100 | .300 | .700 | .900 | 1.000 | | | | | | | | | | | | | | |
| (3,4) | .057 | .200 | .543 | .800 | .971 | 1.000 | | | | | | | | | | | | | |
| (3,5) | .036 | .143 | .429 | .714 | .929 | 1.000 | | | | | | | | | | | | | |
| (3,6) | .024 | .107 | .345 | .643 | .881 | 1.000 | | | | | | | | | | | | | |
| (3,7) | .017 | .083 | .283 | .583 | .833 | 1.000 | | | | | | | | | | | | | |
| (3,8) | .012 | .067 | .236 | .533 | .788 | 1.000 | | | | | | | | | | | | | |
| (3,9) | .009 | .055 | .200 | .491 | .745 | 1.000 | | | | | | | | | | | | | |
| (3,10) | .007 | .045 | .171 | .455 | .706 | 1.000 | | | | | | | | | | | | | |
| (4,4) | .029 | .114 | .371 | .629 | .886 | .971 | 1.000 | | | | | | | | | | | | |
| (4,5) | .016 | .071 | .262 | .500 | .786 | .929 | .992 | 1.000 | | | | | | | | | | | |
| (4,6) | .010 | .048 | .190 | .405 | .690 | .881 | .976 | 1.000 | | | | | | | | | | | |
| (4,7) | .006 | .033 | .142 | .333 | .606 | .833 | .954 | 1.000 | | | | | | | | | | | |
| (4,8) | .004 | .024 | .109 | .279 | .533 | .788 | .929 | 1.000 | | | | | | | | | | | |
| (4,9) | .003 | .018 | .085 | .236 | .471 | .745 | .902 | 1.000 | | | | | | | | | | | |
| (4,10) | .002 | .014 | .068 | .203 | .419 | .706 | .874 | 1.000 | | | | | | | | | | | |
| (5,5) | .008 | .040 | .167 | .357 | .643 | .833 | .960 | .992 | 1.000 | | | | | | | | | | |
| (5,6) | .004 | .024 | .110 | .262 | .522 | .738 | .911 | .976 | .998 | 1.000 | | | | | | | | | |
| (5,7) | .003 | .015 | .076 | .197 | .424 | .652 | .854 | .955 | .992 | 1.000 | | | | | | | | | |
| (5,8) | .002 | .010 | .054 | .152 | .347 | .576 | .793 | .929 | .984 | 1.000 | | | | | | | | | |
| (5,9) | .001 | .007 | .039 | .119 | .287 | .510 | .734 | .902 | .972 | 1.000 | | | | | | | | | |
| (5,10) | .001 | .005 | .029 | .095 | .239 | .455 | .678 | .874 | .958 | 1.000 | | | | | | | | | |
| (6,6) | .002 | .013 | .067 | .175 | .392 | .608 | .825 | .933 | .987 | .998 | 1.000 | | | | | | | | |
| (6,7) | .001 | .008 | .043 | .121 | .296 | .500 | .733 | .879 | .966 | .992 | .999 | 1.000 | | | | | | | |
| (6,8) | .001 | .005 | .028 | .086 | .226 | .413 | .646 | .821 | .937 | .984 | .998 | 1.000 | | | | | | | |
| (6,9) | .000 | .003 | .019 | .063 | .175 | .343 | .566 | .762 | .902 | .972 | .994 | 1.000 | | | | | | | |
| (6,10) | .000 | .002 | .013 | .047 | .137 | .288 | .497 | .706 | .864 | .958 | .990 | 1.000 | | | | | | | |
| (7,7) | .001 | .004 | .025 | .078 | .209 | .383 | .617 | .791 | .922 | .975 | .996 | .999 | 1.000 | | | | | | |
| (7,8) | .000 | .002 | .015 | .051 | .149 | .296 | .514 | .704 | .867 | .949 | .988 | .998 | 1.000 | 1.000 | | | | | |
| (7,9) | .000 | .001 | .010 | .035 | .108 | .231 | .427 | .622 | .806 | .916 | .975 | .994 | .999 | 1.000 | | | | | |
| (7,10) | .000 | .001 | .006 | .024 | .080 | .182 | .355 | .549 | .743 | .879 | .957 | .990 | .998 | 1.000 | | | | | |
| (8,8) | .000 | .001 | .009 | .032 | .100 | .214 | .405 | .595 | .786 | .900 | .968 | .991 | .999 | 1.000 | 1.000 | | | | |
| (8,9) | .000 | .001 | .005 | .020 | .069 | .157 | .319 | .500 | .702 | .843 | .939 | .980 | .996 | .999 | 1.000 | 1.000 | | | |
| (8,10) | .000 | .000 | .003 | .013 | .048 | .117 | .251 | .419 | .621 | .782 | .903 | .964 | .990 | .998 | 1.000 | 1.000 | | | |
| (9,9) | .000 | .000 | .003 | .012 | .044 | .109 | .238 | .399 | .601 | .762 | .891 | .956 | .988 | .997 | 1.000 | 1.000 | 1.000 | | |
| (9,10) | .000 | .000 | .002 | .008 | .029 | .077 | .179 | .319 | .510 | .681 | .834 | .923 | .974 | .992 | .999 | 1.000 | 1.000 | 1.000 | |
| (10,10) | .000 | .000 | .001 | .004 | .019 | .051 | .128 | .242 | .414 | .586 | .758 | .872 | .949 | .981 | .996 | .999 | 1.000 | 1.000 | 1.000 |

By permission from C. Eisenhart and F. Swed, "Tables for testing randomness of grouping in a sequence of alternatives," *Annals of Mathematical Statistics*, vol. 14 (1943), p. 66.

The values listed on the previous page give the chance that $u$ or fewer runs will occur. For example, for two samples of size 4 the chance of three or fewer runs is .114. For sample sizes $N_1 = N_2$ larger than 10 the following table can be used. The columns headed $P_{0.5}$, $P_{01}$, $P_{02.5}$, $P_{05}$ give values of $u$ such that $u$ or fewer runs occur with chance less than the indicated percentage. For example, for $N_1 = N_2 = 12$ the chance of 8 or fewer runs is about .05. The columns headed $P_{95}$, $P_{97.5}$, $P_{99}$, $P_{99.5}$ give values of $u$ for which the chance of $u$ or more runs is less than 5, 2.5, 1, .5 per cent.

| $N_1 = N_2$ | $P_{0.5}$ | $P_{01}$ | $P_{02.5}$ | $P_{05}$ | $P_{95}$ | $P_{97.5}$ | $P_{99}$ | $P_{99.5}$ | *Mean* | *Var.* | *Std. dev.* |
|---|---|---|---|---|---|---|---|---|---|---|---|
| 11 | 5 | 6 | 7 | 7 | 16 | 16 | 17 | 18 | 12 | 5.24 | 2.29 |
| 12 | 6 | 7 | 7 | 8 | 17 | 18 | 18 | 19 | 13 | 5.74 | 2.40 |
| 13 | 7 | 7 | 8 | 9 | 18 | 19 | 20 | 20 | 14 | 6.24 | 2.50 |
| 14 | 7 | 8 | 9 | 10 | 19 | 20 | 21 | 22 | 15 | 6.74 | 2.60 |
| 15 | 8 | 9 | 10 | 11 | 20 | 21 | 22 | 23 | 16 | 7.24 | 2.69 |
| 16 | 9 | 10 | 11 | 11 | 22 | 22 | 23 | 24 | 17 | 7.74 | 2.78 |
| 17 | 10 | 10 | 11 | 12 | 23 | 24 | 25 | 25 | 18 | 8.24 | 2.87 |
| 18 | 10 | 11 | 12 | 13 | 24 | 25 | 26 | 27 | 19 | 8.74 | 2.96 |
| 19 | 11 | 12 | 13 | 14 | 25 | 26 | 27 | 28 | 20 | 9.24 | 3.04 |
| 20 | 12 | 13 | 14 | 15 | 26 | 27 | 28 | 29 | 21 | 9.74 | 3.12 |
| 25 | 16 | 17 | 18 | 19 | 32 | 33 | 34 | 35 | 26 | 12.24 | 3.50 |
| 30 | 20 | 21 | 22 | 24 | 37 | 39 | 40 | 41 | 31 | 14.75 | 3.84 |
| 35 | 24 | 25 | 27 | 28 | 43 | 44 | 46 | 47 | 36 | 17.25 | 4.15 |
| 40 | 29 | 30 | 31 | 33 | 48 | 50 | 51 | 52 | 41 | 19.75 | 4.44 |
| 45 | 33 | 34 | 36 | 37 | 54 | 55 | 57 | 58 | 46 | 22.25 | 4.72 |
| 50 | 37 | 38 | 40 | 42 | 59 | 61 | 63 | 64 | 51 | 24.75 | 4.97 |
| 55 | 42 | 43 | 45 | 46 | 65 | 66 | 68 | 69 | 56 | 27.25 | 5.22 |
| 60 | 46 | 47 | 49 | 51 | 70 | 72 | 74 | 75 | 61 | 29.75 | 5.45 |
| 65 | 50 | 52 | 54 | 56 | 75 | 77 | 79 | 81 | 66 | 32.25 | 5.68 |
| 70 | 55 | 56 | 58 | 60 | 81 | 83 | 85 | 86 | 71 | 34.75 | 5.89 |
| 75 | 59 | 61 | 63 | 65 | 86 | 88 | 90 | 92 | 76 | 37.25 | 6.10 |
| 80 | 64 | 65 | 68 | 70 | 91 | 93 | 96 | 97 | 81 | 39.75 | 6.30 |
| 85 | 68 | 70 | 72 | 74 | 97 | 99 | 101 | 103 | 86 | 42.25 | 6.50 |
| 90 | 73 | 74 | 77 | 79 | 102 | 104 | 107 | 108 | 91 | 44.75 | 6.69 |
| 95 | 77 | 79 | 82 | 84 | 107 | 109 | 112 | 114 | 96 | 47.25 | 6.87 |
| 100 | 82 | 84 | 86 | 88 | 113 | 115 | 117 | 119 | 101 | 49.75 | 7.05 |

The mean and variance of the sampling distribution of $u$ are

$$\frac{2N_1N_2}{N_1 + N_2} + 1 \quad \text{and} \quad \frac{2N_1N_2(2N_1N_2 - N_1 - N_2)}{(N_1 + N_2)^2(N_1 + N_2 - 1)}$$

respectively. For example, if $N_1 = N_2 = 20$ the mean of the distribution of $u$ is 21 and the variance is 9.74. For large values of $N_1$ and $N_2$ the distribution of $u$ is approximately normal; for example, for $N_1 = N_2 = 20$ approximate 97.5 and 2.5 percentiles are $21 + 1.96\sqrt{9.74} = 27.1$ and $21 - 1.96\sqrt{9.74} = 14.9$. The percentile approximation is improved if .5 is subtracted from the computed values.

## A-12a POWER CURVES FOR ONE- AND TWO-SIDED TESTS WHEN $\sigma$ IS KNOWN

For one-sided tests $d = \dfrac{\mu - \mu_0}{\sigma/\sqrt{N}}$ or $d = \dfrac{(\mu_1 - \mu_2)}{\sigma\sqrt{(1/N_1) + (1/N_2)}}$.

For two-sided tests $d = \dfrac{|\mu - \mu_0|}{\sigma/\sqrt{N}}$ or $d = \dfrac{|\mu_1 - \mu_2|}{\sigma\sqrt{(1/N_1) + (1/N_2)}}$.

Curve (*a*) is for two-sided $\alpha = .10$ tests, curve (*b*) is for one-sided $\alpha = .05$ tests, curve (*c*) is for two-sided $\alpha = .05$ tests, curve (*d*) is for one-sided $\alpha = .025$ tests, curve (*e*) is for two-sided $\alpha = .01$ tests, and curve (*f*) is for one-sided $\alpha = .005$ tests.

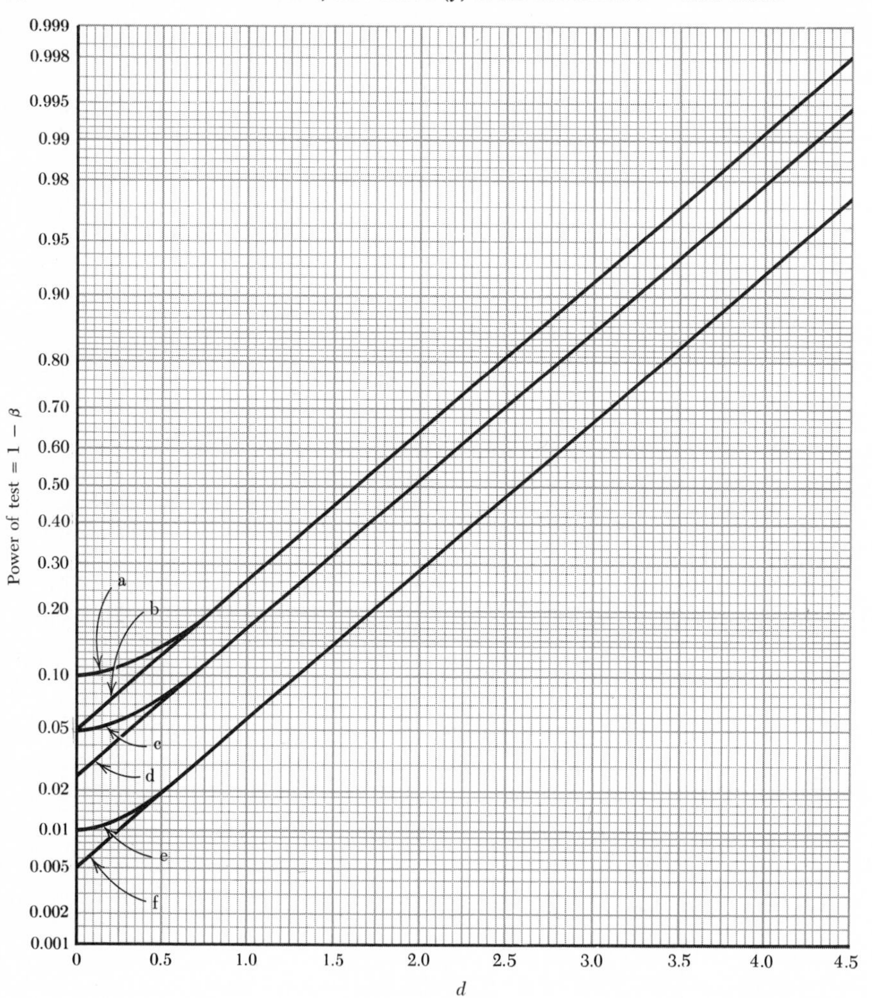

## VALUES OF $d = \frac{\mu - \mu_0}{\sigma/\sqrt{N}}$ FOR A SINGLE SAMPLE OR $d = \frac{\mu_1 - \mu_2}{\sigma\sqrt{1/N_1 + 1/N_2}}$ FOR TWO SAMPLES SUCH THAT A ONE-SIDED TEST AT THE $\alpha$ LEVEL OF SIGNIFICANCE HAS POWER $1 - \beta$

The values in the tables are approximate values of $|d|$ for two-sided tests at level of significance $2\alpha$.

$\alpha = .005$, ONE-SIDED TEST; $\alpha = .01$, TWO SIDED TEST

| $df$ \ $1-\beta$ | .1 | .2 | .3 | .4 | .5 | .6 | .7 | .8 | .9 | .95 | .975 | .99 | .995 |
|---|---|---|---|---|---|---|---|---|---|---|---|---|---|
| 4 | 1.99 | 2.73 | 3.29 | 3.78 | 4.25 | 4.73 | 5.25 | 5.88 | 6.77 | 7.52 | 8.18 | 8.98 | 9.51 |
| 5 | 1.83 | 2.48 | 2.97 | 3.39 | 3.79 | 4.20 | 4.64 | 5.17 | 5.91 | 6.53 | 7.08 | 7.73 | 8.18 |
| 6 | 1.72 | 2.33 | 2.78 | 3.16 | 3.53 | 3.90 | 4.30 | 4.77 | 5.43 | 5.99 | 6.47 | 7.05 | 7.43 |
| 7 | 1.65 | 2.23 | 2.65 | 3.01 | 3.36 | 3.71 | 4.08 | 4.52 | 5.13 | 5.65 | 6.09 | 6.62 | 6.98 |
| 8 | 1.60 | 2.16 | 2.56 | 2.91 | 3.24 | 3.57 | 3.92 | 4.34 | 4.92 | 5.41 | 5.84 | 6.33 | 6.67 |
| 9 | 1.56 | 2.10 | 2.50 | 2.83 | 3.15 | 3.47 | 3.81 | 4.21 | 4.77 | 5.24 | 5.65 | 6.12 | 6.45 |
| 10 | 1.53 | 2.06 | 2.45 | 2.78 | 3.09 | 3.40 | 3.73 | 4.12 | 4.67 | 5.12 | 5.51 | 5.98 | 6.29 |
| 12 | 1.49 | 2.00 | 2.37 | 2.69 | 2.99 | 3.28 | 3.60 | 3.98 | 4.50 | 4.94 | 5.31 | 5.75 | 6.05 |
| 16 | 1.44 | 1.93 | 2.28 | 2.59 | 2.87 | 3.16 | 3.46 | 3.82 | 4.32 | 4.73 | 5.09 | 5.50 | 5.79 |
| 24 | 1.39 | 1.86 | 2.20 | 2.49 | 2.77 | 3.04 | 3.33 | 3.68 | 4.15 | 4.54 | 4.89 | 5.28 | 5.55 |
| 36 | 1.36 | 1.82 | 2.15 | 2.44 | 2.70 | 2.97 | 3.25 | 3.59 | 4.05 | 4.43 | 4.76 | 5.15 | 5.41 |
| $\infty$ | 1.294 | 1.734 | 2.052 | 2.323 | 2.576 | 2.829 | 3.100 | 3.418 | 3.858 | 4.221 | 4.536 | 4.902 | 5.152 |

$\alpha = .0125$, ONE-SIDED TEST; $\alpha = .025$, TWO-SIDED TEST

| $df$ \ $1-\beta$ | .1 | .2 | .3 | .4 | .5 | .6 | .7 | .8 | .9 | .95 | .975 | .99 | .995 |
|---|---|---|---|---|---|---|---|---|---|---|---|---|---|
| 4 | 1.32 | 1.96 | 2.43 | 2.87 | 3.24 | 3.64 | 4.07 | 4.58 | 5.31 | 5.93 | 6.47 | 7.10 | 7.53 |
| 5 | 1.25 | 1.83 | 2.26 | 2.63 | 2.98 | 3.34 | 3.72 | 4.18 | 4.82 | 5.35 | 5.82 | 6.37 | 6.76 |
| 6 | 1.19 | 1.75 | 2.15 | 2.50 | 2.83 | 3.16 | 3.52 | 3.94 | 4.53 | 5.03 | 5.46 | 5.96 | 6.30 |
| 7 | 1.15 | 1.69 | 2.08 | 2.41 | 2.73 | 3.05 | 3.39 | 3.79 | 4.35 | 4.82 | 5.22 | 5.70 | 6.02 |
| 8 | 1.13 | 1.65 | 2.03 | 2.35 | 2.66 | 2.96 | 3.29 | 3.68 | 4.22 | 4.67 | 5.06 | 5.52 | 5.84 |
| 9 | 1.11 | 1.62 | 1.99 | 2.32 | 2.60 | 2.90 | 3.22 | 3.60 | 4.13 | 4.56 | 4.94 | 5.39 | 5.69 |
| 10 | 1.09 | 1.59 | 1.96 | 2.27 | 2.56 | 2.86 | 3.18 | 3.54 | 4.06 | 4.48 | 4.86 | 5.29 | 5.59 |
| 12 | 1.07 | 1.56 | 1.91 | 2.21 | 2.50 | 2.79 | 3.11 | 3.45 | 3.95 | 4.37 | 4.73 | 5.14 | 5.43 |
| 16 | 1.05 | 1.52 | 1.87 | 2.17 | 2.44 | 2.72 | 3.01 | 3.36 | 3.84 | 4.24 | 4.58 | 4.98 | 5.26 |
| 24 | 1.01 | 1.48 | 1.81 | 2.10 | 2.37 | 2.63 | 2.92 | 3.26 | 3.72 | 4.11 | 4.44 | 4.83 | 5.09 |
| 36 | .99 | 1.45 | 1.78 | 2.06 | 2.32 | 2.59 | 2.87 | 3.19 | 3.66 | 4.03 | 4.36 | 4.74 | 5.00 |
| $\infty$ | .959 | 1.399 | 1.717 | 1.988 | 2.241 | 2.494 | 2.765 | 3.083 | 3.523 | 3.886 | 4.201 | 4.567 | 4.817 |

$\alpha = .025$, ONE-SIDED TEST; $\alpha = .05$, TWO-SIDED TEST

| $df$ \ $1-\beta$ | .1 | .2 | .3 | .4 | .5 | .6 | .7 | .8 | .9 | .95 | .975 | .99 | .995 |
|---|---|---|---|---|---|---|---|---|---|---|---|---|---|
| 4 | .87 | 1.44 | 1.86 | 2.23 | 2.58 | 2.93 | 3.31 | 3.76 | 4.40 | 4.93 | 5.40 | 5.94 | 6.32 |
| 5 | .82 | 1.37 | 1.77 | 2.11 | 2.43 | 2.75 | 3.10 | 3.51 | 4.09 | 4.58 | 5.00 | 5.48 | 5.82 |
| 6 | .80 | 1.32 | 1.70 | 2.03 | 2.34 | 2.64 | 2.98 | 3.37 | 3.91 | 4.37 | 4.76 | 5.23 | 5.54 |
| 7 | .78 | 1.29 | 1.66 | 1.98 | 2.27 | 2.57 | 2.89 | 3.27 | 3.80 | 4.23 | 4.61 | 5.06 | 5.36 |
| 8 | .77 | 1.27 | 1.63 | 1.94 | 2.23 | 2.52 | 2.83 | 3.20 | 3.71 | 4.14 | 4.51 | 4.94 | 5.23 |
| 9 | .76 | 1.25 | 1.60 | 1.91 | 2.20 | 2.48 | 2.79 | 3.15 | 3.65 | 4.07 | 4.43 | 4.85 | 5.14 |
| 10 | .75 | 1.23 | 1.59 | 1.89 | 2.17 | 2.45 | 2.76 | 3.11 | 3.60 | 4.01 | 4.37 | 4.78 | 5.07 |
| 12 | .74 | 1.21 | 1.56 | 1.86 | 2.13 | 2.41 | 2.71 | 3.05 | 3.54 | 3.94 | 4.28 | 4.69 | 4.96 |
| 16 | .72 | 1.19 | 1.53 | 1.82 | 2.09 | 2.36 | 2.65 | 2.98 | 3.46 | 3.84 | 4.18 | 4.58 | 4.84 |
| 24 | .71 | 1.16 | 1.50 | 1.78 | 2.04 | 2.31 | 2.59 | 2.92 | 3.38 | 3.76 | 4.09 | 4.47 | 4.72 |
| 36 | .70 | 1.15 | 1.48 | 1.75 | 2.01 | 2.27 | 2.55 | 2.88 | 3.33 | 3.71 | 4.03 | 4.41 | 4.66 |
| $\infty$ | .678 | 1.118 | 1.436 | 1.707 | 1.960 | 2.213 | 2.484 | 2.802 | 3.242 | 3.605 | 3.920 | 4.286 | 4.536 |

$\alpha = .05$, ONE-SIDED TEST; $\alpha = .10$, TWO-SIDED TEST

| $df$ \ $1-\beta$ | .1 | .2 | .3 | .4 | .5 | .6 | .7 | .8 | .9 | .95 | .975 | .99 | .995 |
|---|---|---|---|---|---|---|---|---|---|---|---|---|---|
| 4 | .43 | .96 | 1.34 | 1.68 | 1.99 | 2.30 | 2.64 | 3.04 | 3.60 | 4.07 | 4.48 | 4.95 | 5.29 |
| 5 | .42 | .92 | 1.29 | 1.61 | 1.91 | 2.21 | 2.51 | 2.90 | 3.43 | 3.87 | 4.25 | 4.70 | 5.00 |
| 6 | .40 | .90 | 1.26 | 1.57 | 1.86 | 2.15 | 2.46 | 2.82 | 3.33 | 3.75 | 4.12 | 4.55 | 4.84 |
| 7 | .40 | .89 | 1.24 | 1.54 | 1.82 | 2.11 | 2.41 | 2.77 | 3.26 | 3.68 | 4.03 | 4.45 | 4.73 |
| 8 | .39 | .88 | 1.22 | 1.52 | 1.80 | 2.08 | 2.38 | 2.73 | 3.21 | 3.62 | 3.97 | 4.37 | 4.66 |
| 9 | .39 | .87 | 1.21 | 1.50 | 1.78 | 2.06 | 2.35 | 2.70 | 3.18 | 3.57 | 3.92 | 4.32 | 4.60 |
| 10 | .39 | .86 | 1.20 | 1.49 | 1.77 | 2.04 | 2.33 | 2.67 | 3.15 | 3.54 | 3.88 | 4.28 | 4.56 |
| 12 | .38 | .85 | 1.19 | 1.48 | 1.75 | 2.02 | 2.30 | 2.64 | 3.11 | 3.50 | 3.82 | 4.23 | 4.49 |
| 16 | .38 | .84 | 1.17 | 1.45 | 1.72 | 1.98 | 2.27 | 2.60 | 3.06 | 3.44 | 3.77 | 4.16 | 4.42 |
| 24 | .37 | .83 | 1.15 | 1.43 | 1.69 | 1.95 | 2.23 | 2.56 | 3.01 | 3.39 | 3.71 | 4.09 | 4.35 |
| 36 | .37 | .82 | 1.14 | 1.42 | 1.68 | 1.94 | 2.21 | 2.54 | 2.98 | 3.36 | 3.68 | 4.05 | 4.31 |
| $\infty$ | .363 | .803 | 1.121 | 1.392 | 1.645 | 1.898 | 2.169 | 2.487 | 2.927 | 3.290 | 3.605 | 3.971 | 4.221 |

## A-12c SAMPLE SIZE NEEDED TO ATTAIN POWER $1 - \beta$ AGAINST $d' = (\mu - \mu_0)/\sigma$ IN ONE-SAMPLE SINGLE-SIDED CASE AND AGAINST $d' = (\mu_1 - \mu_2)/(\sqrt{2}\,\sigma)$ IN TWO-SAMPLE SINGLE-SIDED CASE, $N_1 = N_2 = N$

Values given are $N = [(z_{1-\alpha} + z_{1-\beta})/d']^2$ rounded up to the next higher integer.

| $d'$ \ $1-\beta$ | ONE-SIDED TEST, $\alpha = .005$, $z_{.995} = 2.576$ | | | | | | ONE-SIDED TEST, $\alpha = .01$, $z_{.99} = 2.326$ | | | | | |
|---|---|---|---|---|---|---|---|---|---|---|---|---|
| | .30 | .50 | .70 | .90 | .95 | .99 | .30 | .50 | .70 | .90 | .95 | .99 |
| .1 | 422 | 664 | 961 | 1,489 | 1,782 | 2,403 | 325 | 542 | 813 | 1,302 | 1,577 | 2,165 |
| .2 | 106 | 166 | 241 | 373 | 446 | 601 | 82 | 136 | 204 | 326 | 395 | 542 |
| .3 | 47 | 74 | 107 | 166 | 198 | 267 | 37 | 61 | 91 | 145 | 176 | 241 |
| .4 | 27 | 42 | 61 | 94 | 112 | 151 | 21 | 34 | 51 | 82 | 99 | 136 |
| .5 | 17 | 27 | 39 | 60 | 72 | 97 | 13 | 22 | 33 | 53 | 64 | 87 |
| .6 | 12 | 19 | 27 | 42 | 50 | 67 | 10 | 16 | 23 | 37 | 44 | 61 |
| .7 | 9 | 14 | 20 | 31 | 37 | 50 | 7 | 12 | 17 | 27 | 33 | 45 |
| .8 | 7 | 11 | 16 | 24 | 28 | 38 | 6 | 9 | 13 | 21 | 25 | 34 |
| .9 | 6 | 9 | 12 | 19 | 22 | 30 | 5 | 7 | 11 | 17 | 20 | 27 |
| 1.0 | 5 | 7 | 10 | 15 | 18 | 25 | 4 | 6 | 9 | 14 | 16 | 22 |
| 1.2 | 3 | 5 | 7 | 11 | 13 | 17 | 3 | 4 | 6 | 10 | 11 | 16 |
| 1.4 | 3 | 4 | 5 | 8 | 10 | 13 | 2 | 3 | 5 | 7 | 9 | 12 |
| 1.6 | 2 | 3 | 4 | 6 | 7 | 10 | 2 | 3 | 4 | 6 | 7 | 9 |
| 1.8 | 2 | 3 | 3 | 5 | 6 | 8 | 2 | 2 | 3 | 5 | 5 | 7 |
| 2.0 | 2 | 2 | 3 | 4 | 5 | 7 | 1 | 2 | 3 | 4 | 4 | 6 |
| 2.2 | 1 | 2 | 2 | 4 | 4 | 5 | | 2 | 2 | 3 | 4 | 5 |
| 2.4 | | 2 | 2 | 3 | 4 | 5 | | 1 | 2 | 3 | 3 | 4 |
| 2.6 | | 1 | 2 | 3 | 3 | 4 | | | 2 | 2 | 3 | 4 |
| 2.8 | | | 2 | 2 | 3 | 4 | | | 2 | 2 | 3 | 3 |
| 3.0 | | | 2 | 2 | 2 | 3 | | | 1 | 2 | 2 | 3 |
| $(z_{1-\alpha} + z_{1-\beta})^2$ | 4.211 | 6.636 | 9.610 | 14.88 | 17.82 | 24.03 | 3.247 | 5.410 | 8.123 | 13.02 | 15.77 | 21.64 |

For $t$ test add 4 to each entry for the single-sample case and 2 for the two-sample case. (α = .005)

For $t$ test add 3 to each entry for the single-sample case and 2 for the two-sample case. (α = .01)

| $d'$ \ $1-\beta$ | ONE-SIDED TEST, $\alpha = .025$, $z_{.975} = 1.960$ | | | | | | ONE-SIDED TEST, $\alpha = .05$, $z_{.95} = 1.645$ | | | | | |
|---|---|---|---|---|---|---|---|---|---|---|---|---|
| | .30 | .50 | .70 | .90 | .95 | .99 | .30 | .50 | .70 | .90 | .95 | .99 |
| .1 | 207 | 385 | 618 | 1,052 | 1,300 | 1,837 | 126 | 271 | 471 | 857 | 1,083 | 1,577 |
| .2 | 52 | 97 | 155 | 263 | 325 | 460 | 32 | 68 | 118 | 215 | 271 | 395 |
| .3 | 23 | 43 | 69 | 117 | 145 | 205 | 14 | 31 | 53 | 96 | 121 | 176 |
| .4 | 13 | 25 | 39 | 66 | 82 | 115 | 8 | 17 | 30 | 54 | 68 | 99 |
| .5 | 9 | 16 | 25 | 43 | 52 | 74 | 6 | 11 | 19 | 35 | 44 | 64 |
| .6 | 6 | 11 | 18 | 30 | 37 | 52 | 4 | 8 | 14 | 24 | 31 | 44 |
| .7 | 5 | 8 | 13 | 22 | 27 | 38 | 3 | 6 | 10 | 18 | 23 | 33 |
| .8 | 4 | 7 | 10 | 17 | 21 | 29 | 2 | 5 | 8 | 14 | 17 | 25 |
| .9 | 3 | 5 | 8 | 13 | 17 | 23 | 2 | 4 | 6 | 11 | 14 | 20 |
| 1.0 | 3 | 4 | 7 | 11 | 13 | 19 | 2 | 3 | 5 | 9 | 11 | 16 |
| 1.2 | 2 | 3 | 5 | 8 | 10 | 13 | 1 | 2 | 4 | 6 | 8 | 11 |
| 1.4 | 2 | 2 | 4 | 6 | 7 | 10 | | 2 | 3 | 5 | 6 | 9 |
| 1.6 | 1 | 2 | 3 | 5 | 6 | 8 | | 2 | 2 | 4 | 5 | 7 |
| 1.8 | | 2 | 2 | 4 | 5 | 6 | | 1 | 2 | 3 | 4 | 5 |
| 2.0 | | 1 | 2 | 3 | 4 | 5 | | | 2 | 3 | 3 | 4 |
| 2.2 | | | 2 | 3 | 3 | 4 | | | 1 | 2 | 3 | 4 |
| 2.4 | | | 1 | 2 | 3 | 4 | | | | 2 | 2 | 3 |
| 2.6 | | | | 2 | 2 | 3 | | | | 2 | 2 | 3 |
| 2.8 | | | | 2 | 2 | 3 | | | | 2 | 2 | 3 |
| 3.0 | | | | 2 | 2 | 3 | | | | 1 | 2 | 2 |
| $(z_{1-\alpha} + z_{1-\beta})^2$ | 2.062 | 3.842 | 6.170 | 10.51 | 13.00 | 18.37 | 1.257 | 2.706 | 4.705 | 8.567 | 10.82 | 15.77 |

For $t$ test add 2 to each entry for the single-sample case and 1 for the two-sample case. (α = .025)

For $t$ test add 2 to each entry for the single-sample case and 1 for the two-sample case. (α = .05)

ALTERNATIVES $d' = (\mu - \mu_0)/\sigma$ FOR THE SINGLE-SAMPLE ONE-SIDED CASE WITH $N$ OBSERVATIONS AND $d' = (\mu_1 - \mu_2)/\sigma$ FOR THE TWO-SAMPLE CASE WITH $N_1 = N_2 = N$ OBSERVATIONS FOR WHICH THE POWER IS $1 - \beta$ USING A ONE-SIDED $100\alpha$ PER CENT LEVEL OF SIGNIFICANCE TEST

Values given are $(z_{1-\alpha} + z_{1-\beta})/\sqrt{N}$ for the one-sample case and $(z_{1-\alpha} + z_{1-\beta})\sqrt{2/N}$ for the two-sample case.

| | SINGLE-SAMPLE CASE, $\alpha = .05$ | | | | | | TWO-SAMPLE CASE, $\alpha = .05$ | | | | | |
|---|---|---|---|---|---|---|---|---|---|---|---|---|
| $N$ \ $1 - \beta$ | .30 | .50 | .70 | .90 | .95 | .99 | .30 | .50 | .70 | .90 | .95 | .99 |
| 5 | .50 | .73 | .97 | 1.30 | 1.47 | 1.77 | .70 | 1.04 | 1.37 | 1.85 | 2.08 | 2.51 |
| 10 | .35 | .52 | .68 | .92 | 1.04 | 1.25 | .50 | .73 | .97 | 1.30 | 1.47 | 1.77 |
| 20 | .25 | .36 | .48 | .65 | .73 | .88 | .35 | .52 | .68 | .92 | 1.04 | 1.25 |
| 40 | .17 | .26 | .34 | .46 | .52 | .62 | .25 | .36 | .48 | .65 | .73 | .88 |
| 60 | .14 | .21 | .28 | .37 | .42 | .51 | .20 | .30 | .39 | .53 | .60 | .72 |
| 80 | .12 | .18 | .24 | .32 | .36 | .44 | .17 | .26 | .34 | .46 | .52 | .62 |
| 100 | .11 | .16 | .21 | .29 | .32 | .39 | .15 | .23 | .30 | .41 | .46 | .56 |
| 150 | .09 | .13 | .17 | .23 | .26 | .32 | .12 | .18 | .25 | .33 | .37 | .45 |
| 200 | .07 | .11 | .15 | .20 | .23 | .28 | .11 | .16 | .21 | .29 | .32 | .39 |
| 300 | .06 | .09 | .12 | .16 | .18 | .22 | .09 | .13 | .17 | .23 | .26 | .32 |
| 400 | .05 | .08 | .10 | .14 | .16 | .19 | .07 | .11 | .15 | .20 | .23 | .28 |
| 500 | .05 | .07 | .09 | .13 | .14 | .17 | .07 | .10 | .13 | .18 | .20 | .25 |
| 600 | .04 | .06 | .08 | .11 | .13 | .16 | .06 | .09 | .12 | .16 | .18 | .22 |
| 800 | .03 | .05 | .07 | .10 | .11 | .14 | .05 | .08 | .10 | .14 | .16 | .19 |
| 1,000 | .035+ | .052 | .069 | .093 | .104 | .126 | .050 | .074 | .097 | .131 | .147 | .178 |
| 2,000 | .025+ | .037 | .049 | .065+ | .074 | .089 | .035+ | .052 | .069 | .093 | .104 | .126 |
| 5,000 | .016 | .023 | .031 | .041 | .047 | .056 | .022 | .033 | .043 | .059 | .066 | .079 |
| 10,000 | .011 | .016 | .022 | .030 | .033 | .040 | .016 | .023 | .031 | .041 | .047 | .056 |
| $z_{1-\alpha} + z_{1-\beta}$ | 1.121 | 1.645 | 2.169 | 2.927 | 3.290 | 3.971 | 1.585 | 2.326 | 3.067 | 4.139 | 4.653 | 5.616 |

## A-12d ALTERNATIVES $d' = (\mu - \mu_0)/\sigma$ FOR THE SINGLE-SAMPLE ONE-SIDED CASE WITH $N$ OBSERVATIONS AND $d' = (\mu_1 - \mu_2)/\sigma$ FOR THE TWO-SAMPLE CASE WITH $N_1 = N_2 = N$ OBSERVATIONS FOR WHICH THE POWER IS $1 - \beta$ USING A ONE-SIDED $100\alpha$ PER CENT LEVEL OF SIGNIFICANCE TEST (CONTINUED)

| | SINGLE-SAMPLE CASE, $\alpha = .025$ | | | | | | TWO-SAMPLE CASE, $\alpha = .025$ | | | | | |
|---|---|---|---|---|---|---|---|---|---|---|---|---|
| $N$ \ $1 - \beta$ | .30 | .50 | .70 | .90 | .95 | .99 | .30 | .50 | .70 | .90 | .95 | .99 |
| 5 | .64 | .87 | 1.11 | 1.44 | 1.61 | 1.91 | .90 | 1.23 | 1.57 | 2.05 | 2.28 | 2.71 |
| 10 | .45 | .61 | .78 | 1.02 | 1.14 | 1.35 | .64 | .87 | 1.11 | 1.44 | 1.61 | 1.91 |
| 20 | .32 | .43 | .55 | .72 | .80 | .95 | .45 | .61 | .78 | 1.02 | 1.14 | 1.35 |
| 40 | .22 | .30 | .39 | .51 | .57 | .67 | .32 | .43 | .55 | .72 | .80 | .95 |
| 60 | .18 | .25 | .32 | .41 | .46 | .55 | .26 | .35 | .45 | .59 | .65 | .78 |
| 80 | .16 | .21 | .27 | .36 | .40 | .47 | .22 | .30 | .39 | .51 | .57 | .67 |
| 100 | .14 | .19 | .24 | .32 | .36 | .42 | .20 | .27 | .35 | .45 | .55 | .60 |
| 150 | .11 | .16 | .20 | .26 | .29 | .34 | .16 | .22 | .28 | .37 | .41 | .49 |
| 200 | .10 | .13 | .17 | .22 | .25 | .30 | .14 | .19 | .24 | .32 | .36 | .42 |
| 300 | .08 | .11 | .14 | .18 | .20 | .24 | .11 | .16 | .20 | .26 | .29 | .34 |
| 400 | .07 | .09 | .12 | .16 | .18 | .21 | .10 | .13 | .17 | .22 | .25 | .30 |
| 500 | .06 | .08 | .11 | .14 | .16 | .19 | .09 | .12 | .15 | .20 | .24 | .27 |
| 600 | .05 | .08 | .10 | .13 | .14 | .17 | .08 | .11 | .14 | .18 | .20 | .24 |
| 800 | .05 | .06 | .08 | .11 | .12 | .15 | .07 | .09 | .12 | .16 | .18 | .21 |
| 1,000 | .045+ | .062 | .079 | .103 | .114 | .136 | .064 | .088 | .111 | .145− | .161 | .192 |
| 2,000 | .032 | .044 | .056 | .072 | .081 | .096 | .045+ | .062 | .079 | .103 | .114 | .136 |
| 5,000 | .020 | .028 | .035 | .046 | .051 | .061 | .029 | .039 | .050 | .065 | .072 | .086 |
| 10,000 | .014 | .020 | .025 | .032 | .036 | .043 | .020 | .028 | .035+ | .046 | .051 | .061 |
| $z_{1-\alpha} + z_{1-\beta}$ | 1.436 | 1.960 | 2.484 | 3.242 | 3.605 | 4.286 | 2.031 | 2.772 | 3.513 | 4.585 | 5.098 | 6.061 |

## ALTERNATIVES $d' = (\mu - \mu_0)/\sigma$ FOR THE SINGLE-SAMPLE ONE-SIDED CASE WITH $N$ OBSERVATIONS AND $d' = (\mu_1 - \mu_2)/\sigma$ FOR THE TWO-SAMPLE CASE WITH $N_1 = N_2 = N$ OBSERVATIONS FOR WHICH THE POWER IS $1 - \beta$ USING A ONE-SIDED $100\alpha$ PER CENT LEVEL OF SIGNIFICANCE TEST (CONTINUED)



| | SINGLE-SAMPLE CASE, $\alpha = .01$ | | | | | | TWO-SAMPLE CASE, $\alpha = .01$ | | | | | |
|---|---|---|---|---|---|---|---|---|---|---|---|---|
| $N$ \ $1 - \beta$ | .30 | .50 | .70 | .90 | .95 | .99 | .30 | .50 | .70 | .90 | .95 | .99 |
| 5 | .80 | 1.04 | 1.27 | 1.61 | 1.77 | 2.08 | 1.13 | 1.47 | 1.80 | 2.28 | 2.51 | 2.94 |
| 10 | .56 | .73 | .90 | 1.14 | 1.25 | 1.47 | .80 | 1.04 | 1.27 | 1.61 | 1.77 | 2.08 |
| 20 | .40 | .52 | .63 | .80 | .88 | 1.04 | .56 | .73 | .90 | 1.14 | 1.25 | 1.47 |
| 40 | .28 | .36 | .45 | .57 | .62 | .73 | .40 | .52 | .63 | .80 | .88 | 1.04 |
| 60 | .23 | .30 | .36 | .46 | .51 | .60 | .32 | .42 | .52 | .65 | .72 | .84 |
| 80 | .20 | .26 | .31 | .40 | .44 | .52 | .28 | .36 | .45 | .57 | .62 | .73 |
| 100 | .18 | .23 | .28 | .36 | .39 | .46 | .25 | .32 | .40 | .51 | .56 | .65 |
| 150 | .14 | .18 | .23 | .29 | .32 | .37 | .20 | .26 | .36 | .41 | .49 | .53 |
| 200 | .12 | .16 | .20 | .25 | .28 | .32 | .18 | .23 | .28 | .36 | .39 | .46 |
| 300 | .10 | .13 | .16 | .20 | .22 | .26 | .14 | .18 | .23 | .29 | .32 | .37 |
| 400 | .09 | .11 | .14 | .18 | .19 | .23 | .12 | .16 | .20 | .25 | .28 | .32 |
| 500 | .08 | .10 | .12 | .16 | .17 | .20 | .11 | .14 | .18 | .22 | .25 | .29 |
| 600 | .07 | .09 | .11 | .14 | .16 | .18 | .10 | .13 | .16 | .20 | .22 | .26 |
| 800 | .06 | .08 | .10 | .12 | .14 | .16 | .09 | .11 | .14 | .18 | .19 | .23 |
| 1,000 | .057 | .074 | .090 | .114 | .125+ | .147 | .081 | .104 | .127 | .161 | .178 | .208 |
| 2,000 | .040 | .052 | .064 | .081 | .089 | .104 | .057 | .074 | .090 | .114 | .126 | .147 |
| 5,000 | .025 | .033 | .040 | .051 | .056 | .066 | .036 | .047 | .057 | .072 | .079 | .093 |
| 10,000 | .018 | .023 | .029 | .036 | .040 | .047 | .025+ | .033 | .040 | .051 | .056 | .066 |
| $z_{1-\alpha} + z_{1-\beta}$ | 1.802 | 2.326 | 2.850 | 3.608 | 3.971 | 4.652 | 2.548 | 3.289 | 4.031 | 5.102 | 5.616 | 6.579 |

**A-12d ALTERNATIVES $d' = (\mu - \mu_0)/\sigma$ FOR THE SINGLE-SAMPLE ONE-SIDED CASE WITH $N$ OBSERVATIONS AND $d' = (\mu_1 - \mu_2)/\sigma$ FOR THE TWO-SAMPLE CASE WITH $N_1 = N_2 = N$ OBSERVATIONS FOR WHICH THE POWER IS $1 - \beta$ USING A ONE-SIDED $100\alpha$ PER CENT LEVEL OF SIGNIFICANCE TEST (CONTINUED)**

| | SINGLE-SAMPLE CASE, $\alpha = .005$ | | | | | | TWO-SAMPLE CASE, $\alpha = .005$ | | | | | |
|---|---|---|---|---|---|---|---|---|---|---|---|---|
| $N$ \ $1-\beta$ | .30 | .50 | .70 | .90 | .95 | .99 | .30 | .50 | .70 | .90 | .95 | .99 |
| 5 | .91 | 1.15 | 1.38 | 1.72 | 1.88 | 2.19 | 1.29 | 1.62 | 1.96 | 2.44 | 2.66 | 3.10 |
| 10 | .64 | .81 | .98 | 1.22 | 1.33 | 1.55 | .91 | 1.15 | 1.38 | 1.72 | 1.88 | 2.19 |
| 20 | .45 | .57 | .69 | .86 | .94 | 1.09 | .64 | .81 | .98 | 1.22 | 1.33 | 1.55 |
| 40 | .32 | .40 | .49 | .61 | .66 | .77 | .45 | .57 | .69 | .86 | .94 | 1.09 |
| 60 | .26 | .33 | .40 | .49 | .54 | .63 | .37 | .47 | .56 | .70 | .77 | .89 |
| 80 | .22 | .28 | .34 | .43 | .47 | .54 | .32 | .40 | .49 | .61 | .66 | .77 |
| 100 | .20 | .25 | .30 | .38 | .45 | .49 | .29 | .36 | .43 | .54 | .59 | .69 |
| 150 | .16 | .21 | .25 | .31 | .34 | .40 | .23 | .29 | .35 | .44 | .48 | .56 |
| 200 | .14 | .18 | .21 | .27 | .29 | .34 | .20 | .25 | .30 | .38 | .42 | .49 |
| 300 | .11 | .14 | .17 | .22 | .24 | .28 | .16 | .21 | .25 | .31 | .34 | .40 |
| 400 | .10 | .12 | .15 | .19 | .21 | .24 | .14 | .18 | .21 | .27 | .29 | .34 |
| 500 | .09 | .11 | .13 | .17 | .18 | .21 | .12 | .16 | .19 | .24 | .26 | .31 |
| 600 | .08 | .10 | .12 | .15 | .17 | .20 | .11 | .14 | .17 | .22 | .24 | .28 |
| 800 | .07 | .09 | .10 | .13 | .14 | .17 | .10 | .12 | .15 | .19 | .21 | .24 |
| 1,000 | .065− | .081 | .098 | .122 | .133 | .155+ | .092 | .115+ | .139 | .173 | .189 | .219 |
| 2,000 | .046 | .058 | .069 | .086 | .094 | .110 | .065− | .081 | .098 | .122 | .133 | .155+ |
| 5,000 | .029 | .036 | .044 | .055− | .060 | .069 | .041 | .052 | .062 | .077 | .084 | .098 |
| 10,000 | .021 | .026 | .031 | .039 | .042 | .049 | .029 | .036 | .044 | .055− | .060 | .069 |
| $z_{1-\alpha} + z_{1-\beta}$ | 2.052 | 2.576 | 3.100 | 3.858 | 4.221 | 4.902 | 2.902 | 3.643 | 4.384 | 5.456 | 5.969 | 6.932 |

Values are for the necessary sample size for the probability $1 - \alpha$ that the sample mean $\bar{X}$ of a random sample from a normally distributed population will fall between $\mu - k\sigma$ and $\mu + k\sigma$ (for most naturally occurring populations the restriction to normality is not required if the sample size is larger than 5). For example, for a probability $1 - \alpha = .99$ that $\bar{X}$ will fall between $\mu - .5\sigma$ and $\mu + .5\sigma$, a sample of size $N = 27$ would be required.

| $k$ | $P_{50}$ | $P_{60}$ | $P_{70}$ | $P_{80}$ | $P_{90}$ | $P_{95}$ | $P_{99}$ | $P_{99.9}$ |
|---|---|---|---|---|---|---|---|---|
| .01 | 4,543 | 7,090 | 10,733 | 16,436 | 27,061 | 38,416 | 66,357 | 108,306 |
| .05 | 182 | 284 | 430 | 658 | 1,083 | 1,537 | 2,655 | 4,333 |
| .10 | 46 | 71 | 108 | 165 | 271 | 385 | 664 | 1,084 |
| .15 | 21 | 32 | 48 | 74 | 121 | 171 | 295 | 482 |
| .20 | 12 | 18 | 27 | 42 | 68 | 97 | 166 | 271 |
| .25 | 8 | 12 | 18 | 27 | 44 | 62 | 107 | 174 |
| .30 | 6 | 8 | 12 | 19 | 31 | 43 | 74 | 121 |
| .40 | 3 | 5 | 7 | 11 | 17 | 25 | 42 | 68 |
| .50 | 2 | 3 | 5 | 7 | 11 | 16 | 27 | 44 |
| .60 | 2 | 2 | 3 | 5 | 8 | 11 | 19 | 31 |
| .70 | 1 | 2 | 3 | 4 | 6 | 8 | 14 | 23 |
| .80 | | 2 | 2 | 3 | 5 | 7 | 11 | 17 |
| .90 | | 1 | 2 | 3 | 4 | 5 | 9 | 14 |
| 1.00 | | | 2 | 2 | 3 | 4 | 7 | 11 |
| 1.25 | | | 1 | 2 | 3 | 2 | 5 | 7 |
| 1.50 | | | | 1 | 2 | 2 | 3 | 5 |
| 1.75 | | | | | 1 | 2 | 3 | 4 |
| 2.00 | | | | | | 1 | 2 | 3 |
| 2.50 | | | | | | | 2 | 2 |
| 3.00 | | | | | | | 1 | 2 |
| 3.25 | | | | | | | | 2 |
| 3.50 | | | | | | | | 1 |

## A-12f SAMPLE SIZES FOR TWO SAMPLE BINOMIAL TESTS

Required sample sizes for each of two samples in a test of $p_1 = p_2$ at level of significance $\alpha$ to have power $1 - \beta$ for the alternative $p_2 = p_1 + d$.

$\alpha = \beta = .01$

| $d$ \ $p_1$ | .05 | .10 | .15 | .20 | .25 | .30 | .35 | .40 | .45 | .50 |
|---|---|---|---|---|---|---|---|---|---|---|
| .05 | 1,275 | 1,968 | 2,574 | 3,093 | 3,526 | 3,873 | 4,132 | 4,306 | 4,392 | 4,392 |
| .10 | 424 | 586 | 727 | 846 | 944 | 1,019 | 1,073 | 1,106 | 1,117 | 1,106 |
| .15 | 231 | 299 | 357 | 405 | 443 | 472 | 491 | 501 | 501 | 491 |
| .20 | 152 | 188 | 217 | 242 | 261 | 274 | 283 | 285 | 283 | 274 |
| .25 | 110 | 131 | 149 | 163 | 173 | 180 | 183 | 183 | 180 | 173 |
| .30 | 85 | 98 | 109 | 117 | 123 | 127 | 128 | 127 | 123 | 117 |
| .35 | 68 | 76 | 84 | 89 | 92 | 94 | 94 | 92 | 89 | 84 |
| .40 | 55 | 61 | 66 | 70 | 72 | 72 | 72 | 70 | 66 | 61 |
| .45 | 46 | 50 | 54 | 56 | 57 | 57 | 56 | 54 | 50 | 46 |
| .50 | 39 | 42 | 44 | 45 | 46 | 45 | 44 | 42 | 39 | 35 |
| .55 | 33 | 35 | 37 | 37 | 37 | 37 | 35 | 33 | 30 | |
| .60 | 28 | 30 | 31 | 31 | 31 | 30 | 28 | 26 | | |
| .65 | 24 | 26 | 26 | 26 | 26 | 24 | 23 | | | |
| .70 | 21 | 22 | 22 | 22 | 21 | 20 | | | | |
| .75 | 18 | 19 | 19 | 18 | 17 | | | | | |
| .80 | 16 | 16 | 16 | 15 | | | | | | |
| .85 | 13 | 13 | 13 | | | | | | | |
| .90 | 11 | 11 | | | | | | | | |
| .95 | 9 | | | | | | | | | |

$\alpha = \beta = .05$

| $d$ \ $p_1$ | .05 | .10 | .15 | .20 | .25 | .30 | .35 | .40 | .45 | .50 |
|---|---|---|---|---|---|---|---|---|---|---|
| .05 | 676 | 1,023 | 1,327 | 1,587 | 1,803 | 1,977 | 2,107 | 2,193 | 2,236 | 2,236 |
| .10 | 231 | 313 | 383 | 443 | 492 | 530 | 557 | 573 | 578 | 573 |
| .15 | 128 | 162 | 191 | 215 | 235 | 249 | 259 | 264 | 264 | 259 |
| .20 | 86 | 103 | 118 | 131 | 140 | 147 | 151 | 152 | 151 | 147 |
| .25 | 63 | 73 | 82 | 89 | 94 | 98 | 100 | 100 | 98 | 94 |
| .30 | 49 | 55 | 61 | 65 | 68 | 70 | 71 | 70 | 68 | 65 |
| .35 | 39 | 44 | 47 | 50 | 52 | 53 | 53 | 52 | 50 | 47 |
| .40 | 32 | 35 | 38 | 40 | 41 | 41 | 41 | 40 | 38 | 35 |
| .45 | 27 | 29 | 31 | 32 | 33 | 33 | 32 | 31 | 29 | 27 |
| .50 | 23 | 25 | 26 | 27 | 27 | 27 | 26 | 25 | 23 | 21 |
| .55 | 20 | 21 | 22 | 22 | 22 | 22 | 21 | 20 | 18 | |
| .60 | 17 | 18 | 19 | 19 | 19 | 18 | 17 | 16 | | |
| .65 | 15 | 16 | 16 | 16 | 16 | 15 | 14 | | | |
| .70 | 13 | 14 | 14 | 14 | 13 | 13 | | | | |
| .75 | 12 | 12 | 12 | 12 | 11 | | | | | |
| .80 | 10 | 10 | 10 | 10 | | | | | | |
| .85 | 9 | 9 | 9 | | | | | | | |
| .90 | 8 | 8 | | | | | | | | |
| .95 | 7 | | | | | | | | | |

$\alpha = \beta = .10$

| $d$ \ $p_1$ | .05 | .10 | .15 | .20 | .25 | .30 | .35 | .40 | .45 | .50 |
|---|---|---|---|---|---|---|---|---|---|---|
| .05 | 440 | 652 | 836 | 994 | 1,126 | 1,231 | 1,310 | 1,363 | 1,389 | 1,389 |
| .10 | 155 | 205 | 248 | 284 | 314 | 337 | 353 | 363 | 366 | 363 |
| .15 | 87 | 108 | 126 | 141 | 153 | 161 | 167 | 170 | 170 | 167 |
| .20 | 59 | 70 | 79 | 87 | 83 | 97 | 99 | 100 | 99 | 97 |
| .25 | 44 | 50 | 56 | 60 | 63 | 65 | 66 | 66 | 65 | 63 |
| .30 | 34 | 38 | 42 | 44 | 46 | 47 | 48 | 47 | 46 | 44 |
| .35 | 28 | 31 | 33 | 34 | 36 | 36 | 36 | 36 | 34 | 33 |
| .40 | 23 | 25 | 27 | 28 | 28 | 28 | 28 | 28 | 27 | 25 |
| .45 | 20 | 21 | 22 | 23 | 23 | 23 | 23 | 22 | 21 | 20 |
| .50 | 17 | 18 | 19 | 19 | 19 | 19 | 19 | 18 | 17 | 16 |
| .55 | 15 | 15 | 16 | 16 | 16 | 15 | 16 | 15 | 15 | 14 |
| .60 | 13 | 13 | 14 | 14 | 14 | 13 | 13 | 12 | | |
| .65 | 11 | 12 | 12 | 12 | 12 | 11 | 11 | | | |
| .70 | 10 | 10 | 10 | 10 | 10 | 10 | | | | |
| .75 | 9 | 9 | 9 | 9 | 9 | | | | | |
| .80 | 8 | 8 | 8 | 8 | | | | | | |
| .85 | 7 | 7 | 7 | | | | | | | |
| .90 | 8 | 6 | | | | | | | | |
| .95 | 5 | | | | | | | | | |

$\alpha = \beta = .20$

| $d$ \ $p_1$ | .50 | .10 | .15 | .20 | .25 | .30 | .35 | .40 | .45 | .50 |
|---|---|---|---|---|---|---|---|---|---|---|
| .05 | 230 | 323 | 403 | 472 | 529 | 575 | 609 | 632 | 643 | 543 |
| .10 | 86 | 108 | 127 | 143 | 156 | 166 | 174 | 178 | 179 | 178 |
| .15 | 50 | 60 | 68 | 74 | 80 | 83 | 86 | 87 | 87 | 86 |
| .20 | 35 | 40 | 44 | 47 | 50 | 52 | 53 | 53 | 53 | 52 |
| .25 | 26 | 29 | 32 | 34 | 35 | 36 | 37 | 37 | 36 | 35 |
| .30 | 21 | 23 | 24 | 26 | 27 | 27 | 27 | 27 | 27 | 26 |
| .35 | 17 | 19 | 20 | 20 | 21 | 21 | 21 | 21 | 20 | 20 |
| .40 | 15 | 16 | 16 | 17 | 17 | 17 | 17 | 17 | 16 | 16 |
| .45 | 13 | 14 | 14 | 14 | 14 | 14 | 14 | 14 | 13 | 13 |
| .50 | 11 | 11 | 12 | 12 | 12 | 12 | 12 | 11 | 11 | 10 |
| .50 | 10 | 10 | 10 | 10 | 10 | 10 | 10 | 10 | 9 | |
| .60 | 9 | 9 | 9 | 9 | 9 | 9 | 9 | 8 | | |
| .65 | 8 | 8 | 8 | 8 | 8 | 8 | 7 | | | |
| .70 | 7 | 7 | 7 | 7 | 7 | 7 | | | | |
| .75 | 6 | 6 | 6 | 6 | 6 | | | | | |
| .80 | 8 | 6 | 6 | 6 | | | | | | |
| .85 | 5 | 5 | 5 | | | | | | | |
| .90 | 5 | 5 | | | | | | | | |
| .95 | 4 | | | | | | | | | |

# A-13 POWER OF THE ANALYSIS-OF-VARIANCE TEST

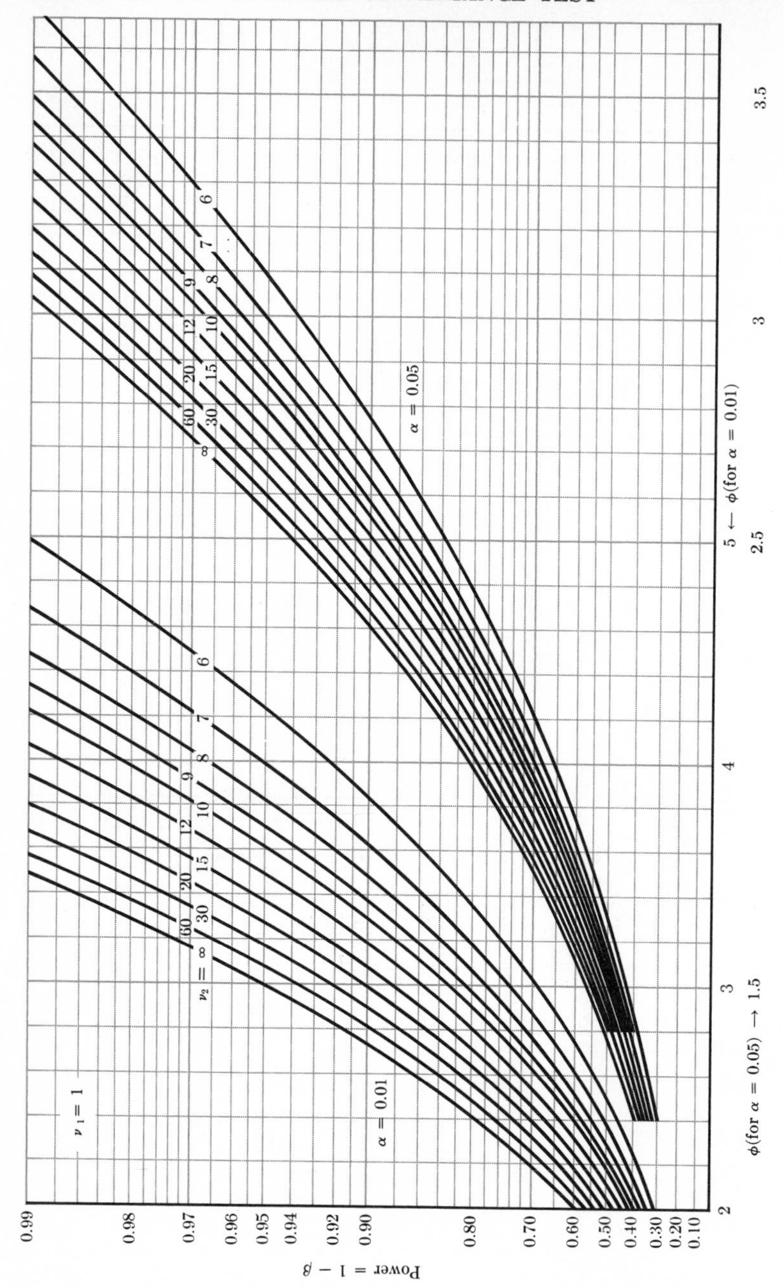

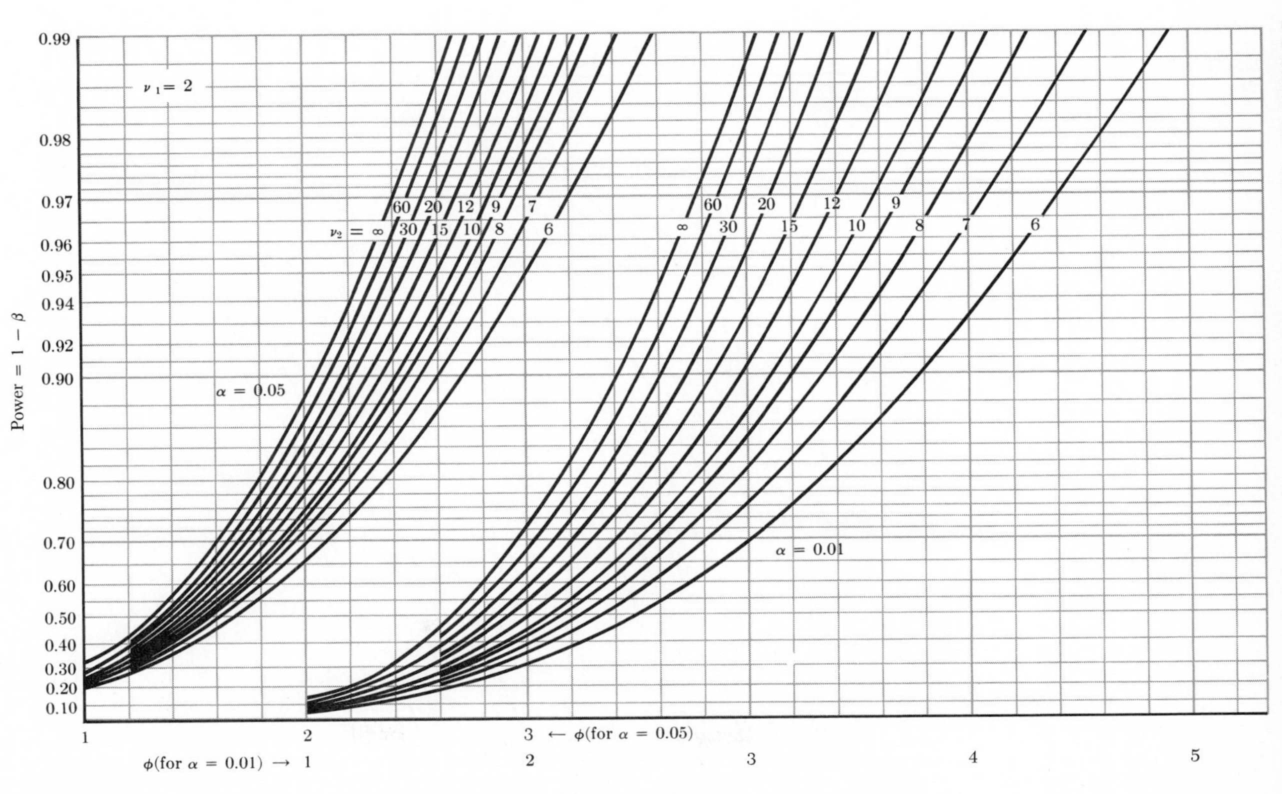

$\nu_1 = 2$
$\alpha = 0.05$
$\alpha = 0.01$
$\nu_2 = \infty$ 60 30 20 15 12 10 9 8 7 6
$\infty$ 60 30 20 15 12 10 9 8 7 6
Power = 1 − β
0.99 0.98 0.97 0.96 0.95 0.94 0.92 0.90 0.80 0.70 0.60 0.50 0.40 0.30 0.20 0.10
1 2 3 ← φ(for α = 0.05)
φ(for α = 0.01) → 1 2 3 4 5

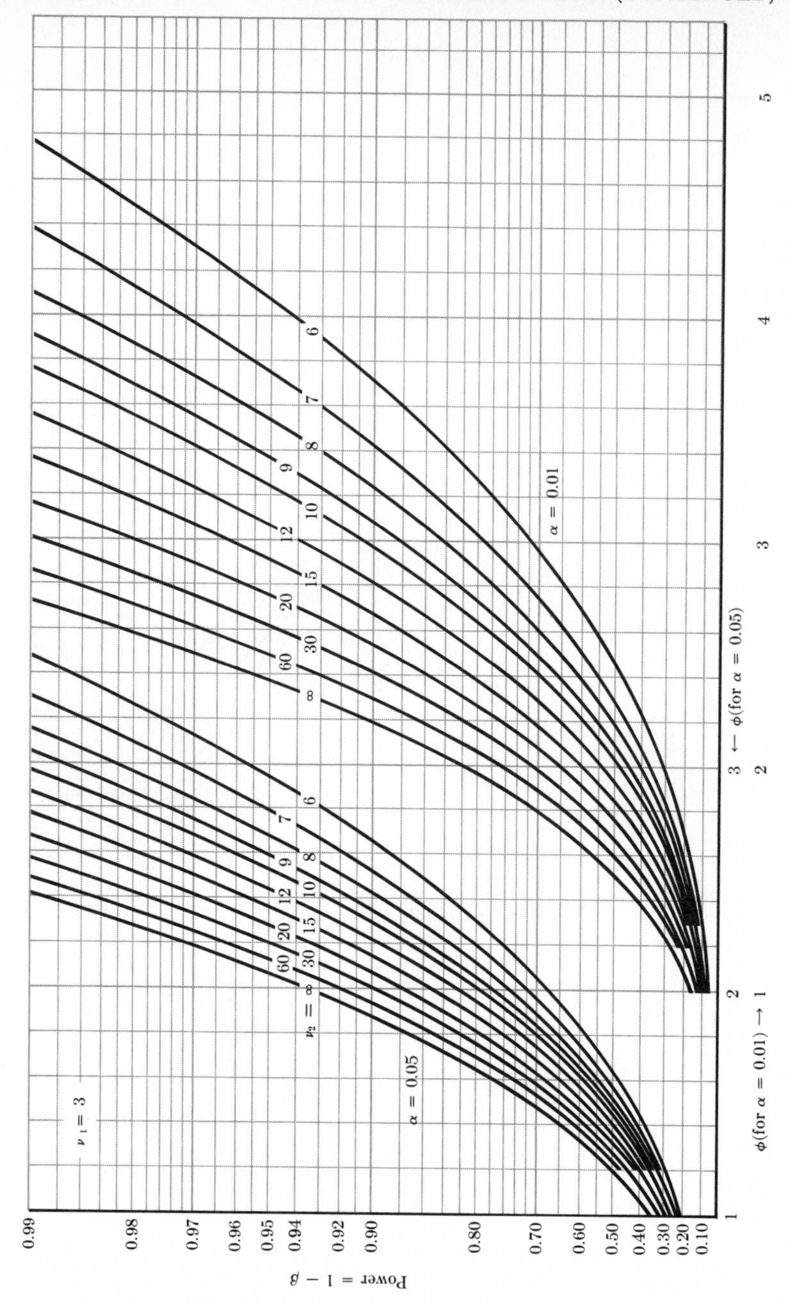

$\nu_1 = 3$
$\alpha = 0.05$
$\alpha = 0.01$
$\nu_2 = \infty$
60
30
20
15
12
10
9
8
7
6
$\infty$
Power = $1 - \beta$
0.99
0.98
0.97
0.96
0.95
0.94
0.92
0.90
0.80
0.70
0.60
0.50
0.40
0.30
0.20
0.10
$\phi$(for $\alpha$ = 0.01) $\rightarrow$ 1
3 $\leftarrow$ $\phi$(for $\alpha$ = 0.05)
1
2
3
4
5

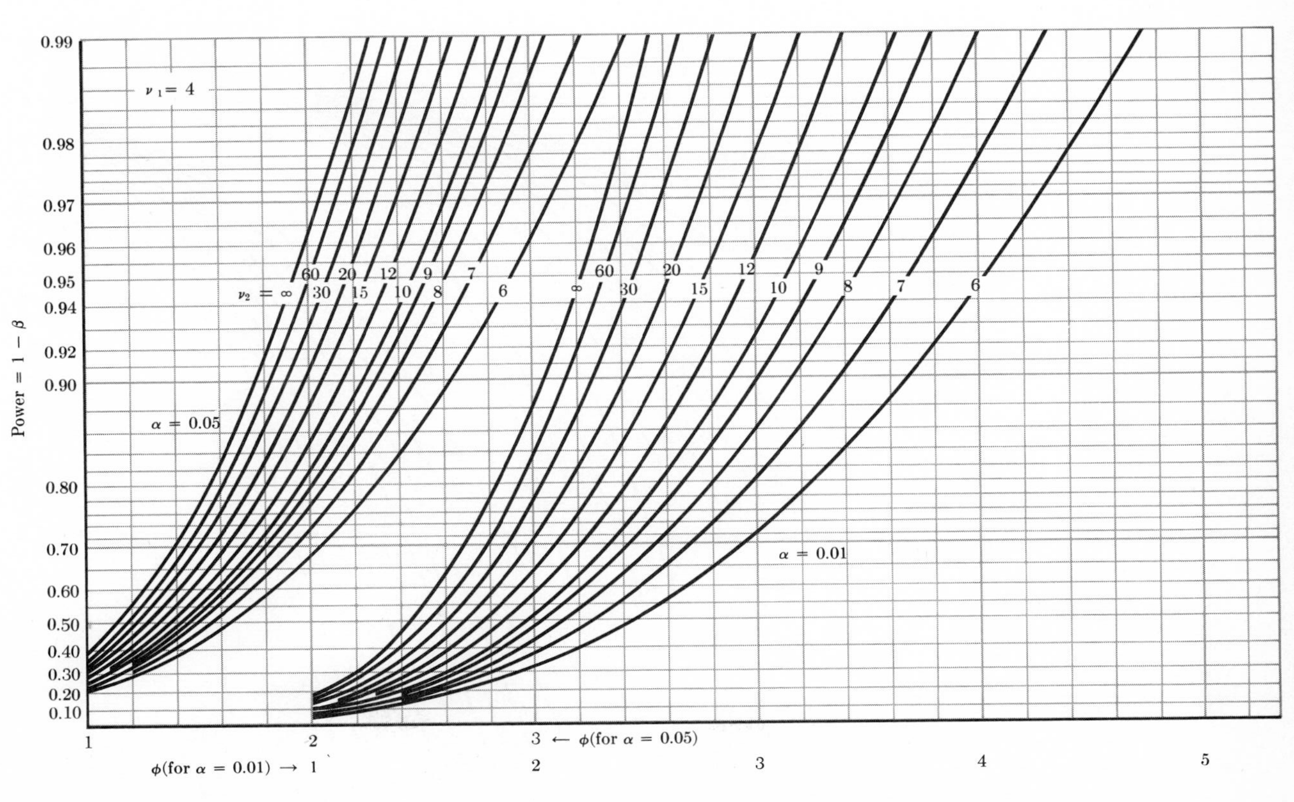
Power = 1 − β
0.99
0.98
0.97
0.96
0.95
0.94
0.92
0.90
0.80
0.70
0.60
0.50
0.40
0.30
0.20
0.10
ν1 = 4
α = 0.05
α = 0.01
ν2 = ∞ 60 30 20 15 12 10 9 8 7 6
∞ 60 30 20 15 12 10 9 8 7 6
1 2
3 ← φ(for α = 0.05)
φ(for α = 0.01) → 1 2 3 4 5

## A-13 POWER OF THE ANALYSIS-OF-VARIANCE TEST (CONTINUED)

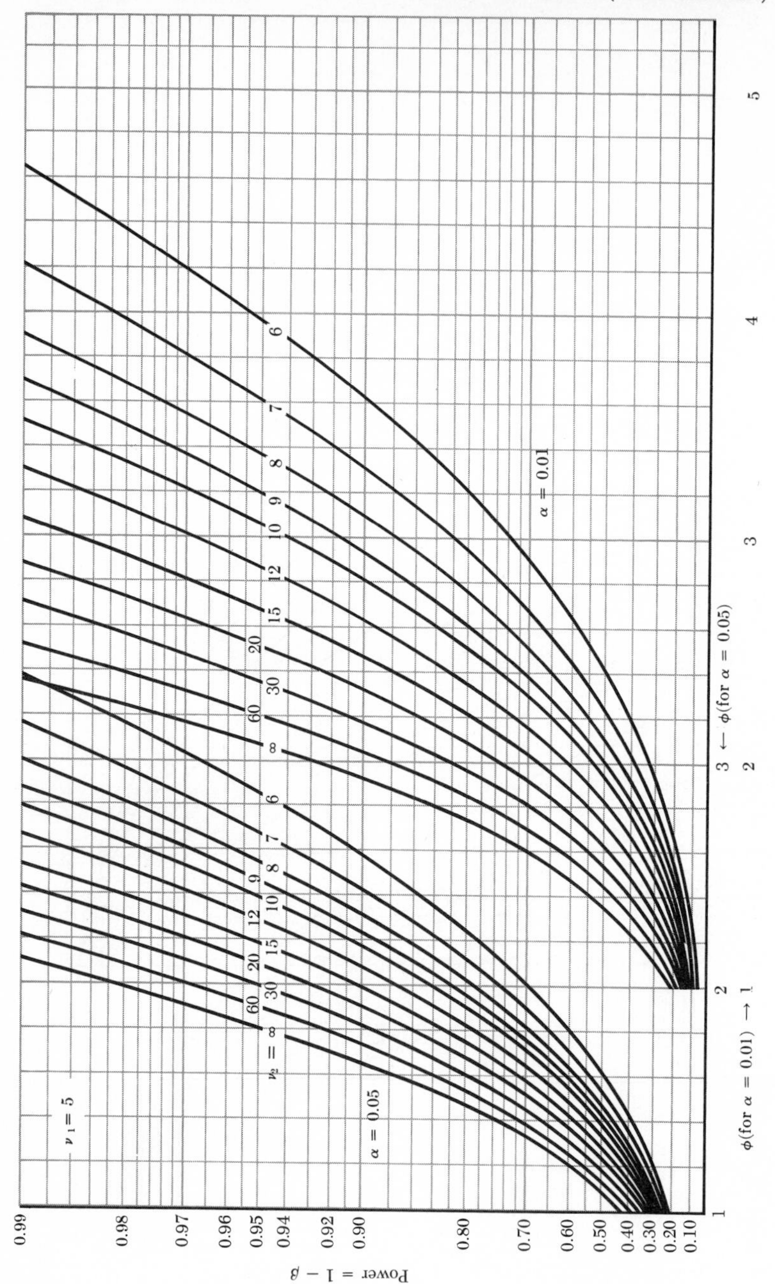

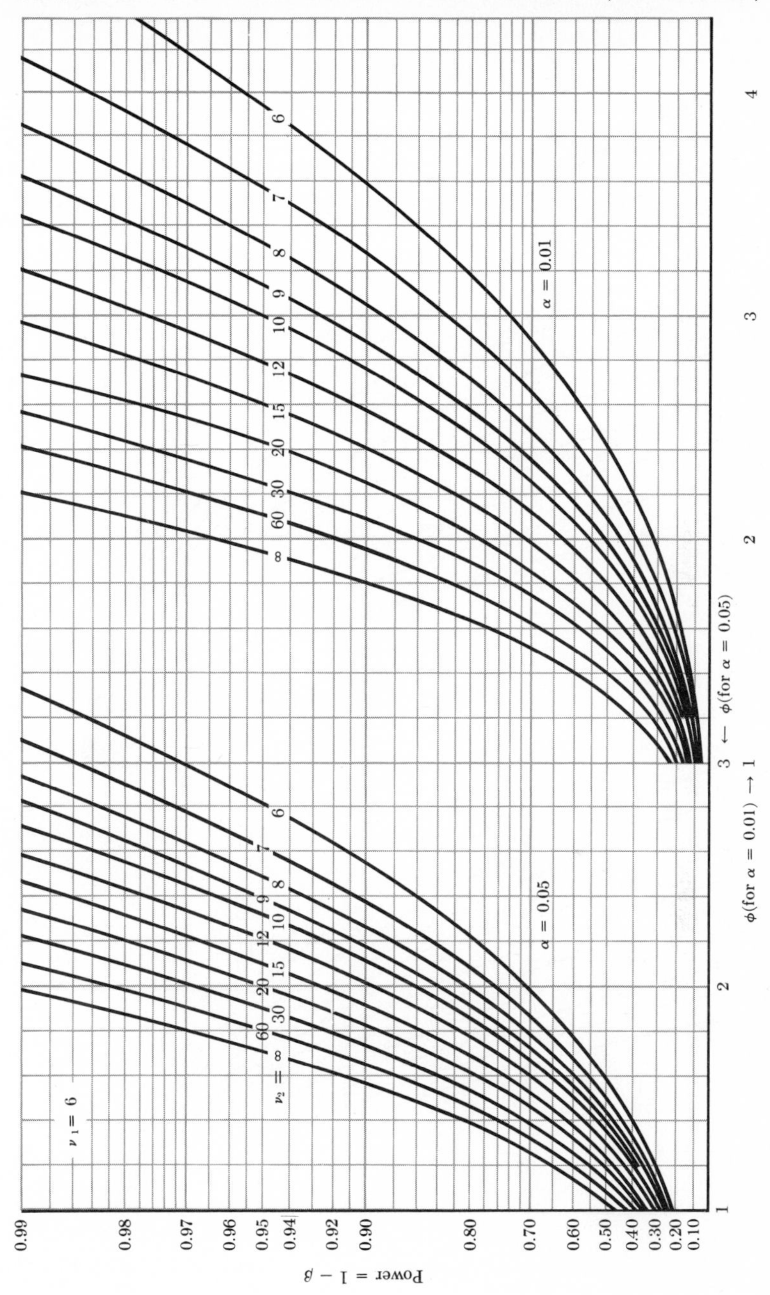

$\nu_1 = 6$
$\nu_2 = \infty$ 60 30 20 15 12 10 9 8 7 6
$\infty$ 60 30 20 15 12 10 9 8 7 6
$\alpha = 0.05$
$\alpha = 0.01$
Power = $1 - \beta$
0.99 0.98 0.97 0.96 0.95 0.94 0.92 0.90 0.80 0.70 0.60 0.50 0.40 0.30 0.20 0.10
1 2 3
2 3 4
$\phi$(for $\alpha$ = 0.01) → 1
3 ← $\phi$(for $\alpha$ = 0.05)

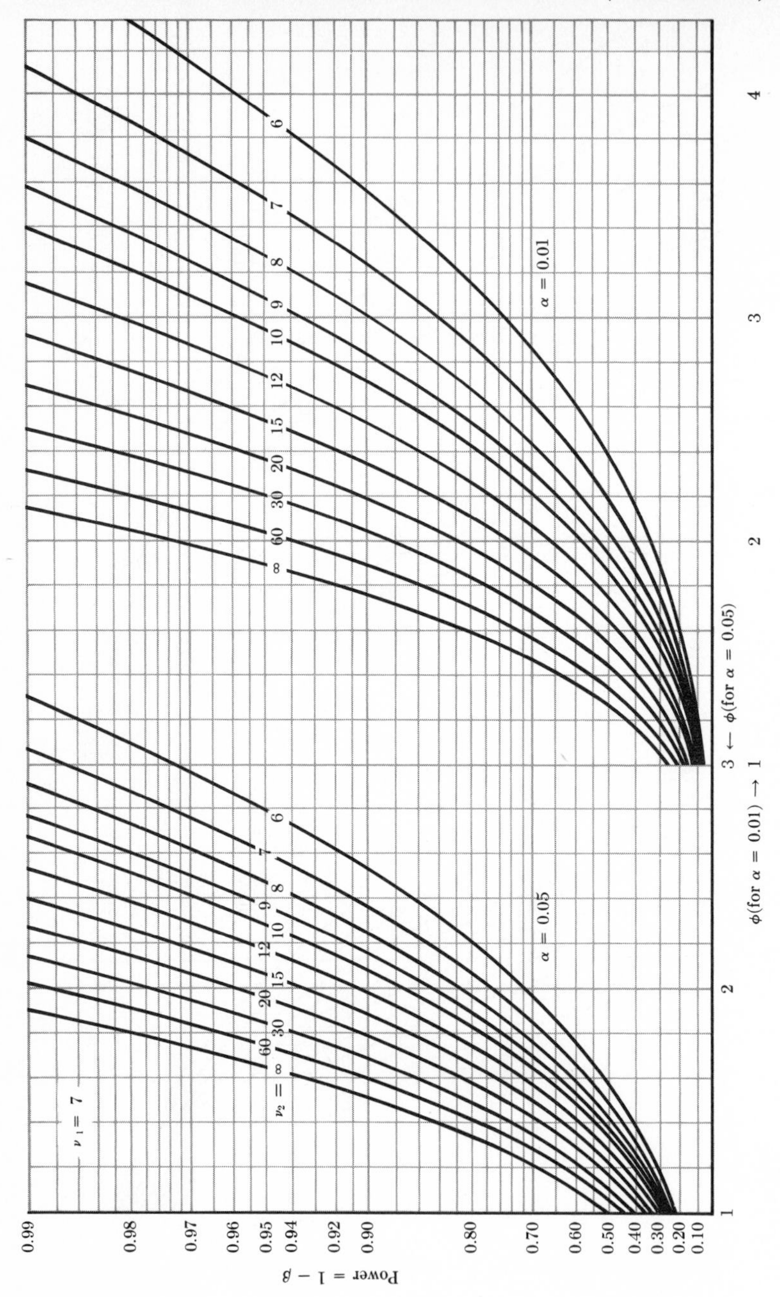
$\nu_1 = 7$
$\nu_2 =$ ∞ 60 30 20 15 12 10 9 8 7 6
∞ 60 30 20 15 12 10 9 8 7 6
$\alpha = 0.05$
$\alpha = 0.01$
Power = $1 - \beta$
0.99 0.98 0.97 0.96 0.95 0.94 0.92 0.90 0.80 0.70 0.60 0.50 0.40 0.30 0.20 0.10
$\phi$(for $\alpha = 0.01$) → 1 2 3
3 ← $\phi$(for $\alpha = 0.05$) 2 3 4

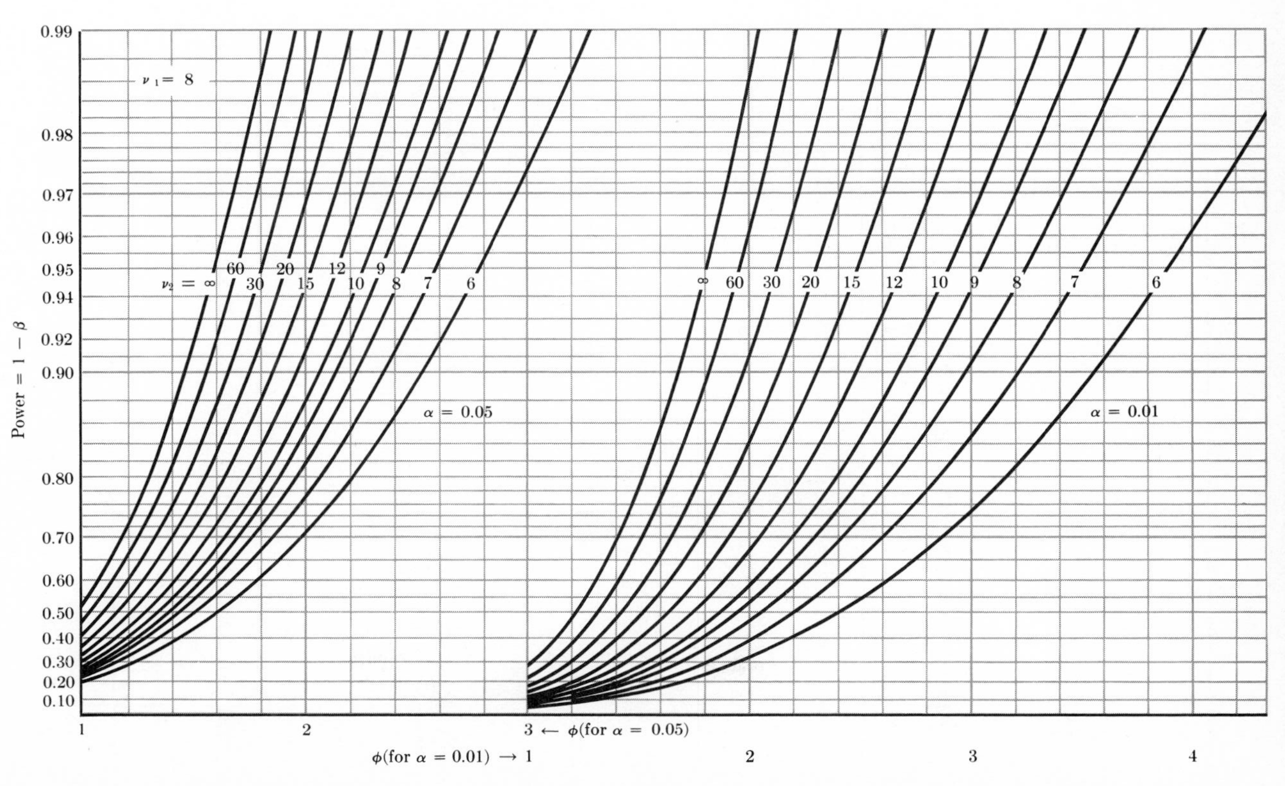
ν1 = 8
ν2 = ∞
60
30
20
15
12
10
9
8
7
6
α = 0.05
α = 0.01
Power = 1 − β
0.99
0.98
0.97
0.96
0.95
0.94
0.92
0.90
0.80
0.70
0.60
0.50
0.40
0.30
0.20
0.10
1
2
3 ← φ(for α = 0.05)
φ(for α = 0.01) → 1
2
3
4

## A-14a NATURAL LOGARITHMS (BASE $e$)

| | .0 | .1 | .2 | .3 | .4 | .5 | .6 | .7 | .8 | .9 |
|---|---|---|---|---|---|---|---|---|---|---|
| 1 | .0000 | .0953 | .1823 | .2624 | .3365 | .4055 | .4700 | .5306 | .5878 | .6419 |
| 2 | .6932 | .7419 | .7885 | .8329 | .8755 | .9163 | .9555 | .9933 | 1.0296 | 1.0647 |
| 3 | 1.0986 | 1.1314 | 1.1632 | 1.1939 | 1.2238 | 1.2528 | 1.2809 | 1.3083 | 1.3350 | 1.3610 |
| 4 | 1.3863 | 1.4110 | 1.4351 | 1.4586 | 1.4816 | 1.5041 | 1.5261 | 1.5476 | 1.5686 | 1.5892 |
| 5 | 1.6094 | 1.6292 | 1.6487 | 1.6677 | 1.6864 | 1.7048 | 1.7228 | 1.7405 | 1.7579 | 1.7750 |
| 6 | 1.7918 | 1.8083 | 1.8246 | 1.8406 | 1.8563 | 1.8718 | 1.8871 | 1.9021 | 1.9169 | 1.9315 |
| 7 | 1.9459 | 1.9601 | 1.9741 | 1.9879 | 2.0015 | 2.0149 | 2.0282 | 2.0412 | 2.0541 | 2.0669 |
| 8 | 2.0794 | 2.0919 | 2.1041 | 2.1163 | 2.1282 | 2.1401 | 2.1518 | 2.1633 | 2.1748 | 2.1861 |
| 9 | 2.1972 | 2.2083 | 2.2192 | 2.2300 | 2.2407 | 2.2513 | 2.2618 | 2.2721 | 2.2824 | 2.2925 |

## A-14b COMMON LOGARITHMS (BASE 10)

| | .0 | .1 | .2 | .3 | .4 | .5 | .6 | .7 | .8 | .9 |
|---|---|---|---|---|---|---|---|---|---|---|
| 1 | .0000 | .0414 | .0792 | .1139 | .1461 | .1761 | .2041 | .2304 | .2553 | .2788 |
| 2 | .3010 | .3222 | .3424 | .3617 | .3802 | .3979 | .4150 | .4314 | .4472 | .4624 |
| 3 | .4771 | .4914 | .5051 | .5185 | .5315 | .5441 | .5563 | .5682 | .5798 | .5911 |
| 4 | .6021 | .6128 | .6232 | .6335 | .6435 | .6532 | .6628 | .6721 | .6812 | .6902 |
| 5 | .6990 | .7076 | .7160 | .7243 | .7324 | .7404 | .7482 | .7559 | .7634 | .7709 |
| 6 | .7782 | .7853 | .7924 | .7993 | .8062 | .8129 | .8195 | .8261 | .8325 | .8388 |
| 7 | .8451 | .8513 | .8573 | .8633 | .8692 | .8751 | .8808 | .8865 | .8921 | .8976 |
| 8 | .9031 | .9085 | .9138 | .9191 | .9243 | .9294 | .9345 | .9395 | .9445 | .9494 |
| 9 | .9542 | .9590 | .9638 | .9685 | .9731 | .9777 | .9823 | .9868 | .9912 | .9956 |

To obtain the natural (or common) logarithm of a number larger than 9.9:
divide the number by 10 and add 2.3026 (or 1) to the logarithm obtained,
*or* divide the number by 100 and add 4.6052 (or 2) to the logarithm obtained,
*or* divide the number by 1000 and add 6.9078 (or 3) to the logarithm obtained, and so on.
To obtain the natural (or common) logarithm of a number smaller than 1:
multiply the number by 10 and subtract 2.3026 (or 1) from the logarithm,
*or* multiply the number by 100 and subtract 4.6052 (or 2) from the logarithm,
*or* multiply the number by 1000 and subtract 6.9078 (or 3) from the logarithm, and so on.
For example:

$\ln 34 = 1.2238 + 2.3026 = 3.5264$ $\qquad \log 34 = 5.315 + 1 = .15315$

$\ln .072 = 1.9741 - 4.6052 = -2.1311$ $\qquad \log .072 = .8573 - 2 = -1.1427$

To obtain a number from its logarithm, read the table in reverse. If the logarithm is not in the table, repeated addition or subtraction of 2.3026 for natural logarithms (or 1 for common logarithms) can be made until the number can be read from the table. For each addition the decimal is moved one place to the left or for each subtraction the decimal is moved one place to the right.

For example:

if $\log X = -2.1550$, we add 3, obtaining .8450, and read the value 7. Then, moving the decimal three places, we get $X = .0070$.

If $\ln X = 3.9513$, we subtract 2.3026, obtaining 1.6487, and read 5.2. Then, moving the decimal one place, we get $X = 52$.

## A-14c SPECIAL VALUES OF $\ln [(1 - \beta)/\alpha]$ FOR SEQUENTIAL TESTS

| $\alpha$ \ $\beta$ | .001 | .005 | .01 | .02 | .03 | .04 | .05 | .10 | .15 | .20 | .25 |
|---|---|---|---|---|---|---|---|---|---|---|---|
| .001 | 6.907 | 6.903 | 6.898 | 6.888 | 6.877 | 6.867 | 6.856 | 6.802 | 6.745 | 6.685 | 6.620 |
| .005 | 5.297 | 5.293 | 5.288 | 5.278 | 5.268 | 5.257 | 5.247 | 5.193 | 5.136 | 5.075 | 5.011 |
| .01 | 4.604 | 4.600 | 4.595 | 4.585 | 4.575 | 4.564 | 4.554 | 4.500 | 4.443 | 4.382 | 4.317 |
| .02 | 3.911 | 3.907 | 3.902 | 3.892 | 3.882 | 3.871 | 3.861 | 3.807 | 3.750 | 3.689 | 3.624 |
| .03 | 3.506 | 3.502 | 3.496 | 3.486 | 3.476 | 3.466 | 3.455 | 3.401 | 3.344 | 3.283 | 3.219 |
| .04 | 3.218 | 3.214 | 3.209 | 3.199 | 3.188 | 3.178 | 3.168 | 3.114 | 3.056 | 2.996 | 2.931 |
| .05 | 2.995 | 2.991 | 2.986 | 2.976 | 2.965 | 2.955 | 2.944 | 2.890 | 2.833 | 2.773 | 2.708 |
| .10 | 2.302 | 2.298 | 2.293 | 2.282 | 2.272 | 2.261 | 2.251 | 2.197 | 2.140 | 2.079 | 2.015 |
| .15 | 1.896 | 1.892 | 1.887 | 1.877 | 1.867 | 1.855 | 1.846 | 1.792 | 1.735 | 1.674 | 1.609 |
| .20 | 1.608 | 1.604 | 1.599 | 1.589 | 1.579 | 1.568 | 1.558 | 1.504 | 1.447 | 1.386 | 1.322 |
| .25 | 1.385 | 1.381 | 1.376 | 1.366 | 1.356 | 1.344 | 1.335 | 1.281 | 1.224 | 1.163 | 1.099 |

Note that $\ln [\beta/(1-\alpha)]$ is equal to $-\ln [(1-\alpha)/\beta]$, which can be read from the table by interchanging the arguments $\alpha$ and $\beta$ and prefixing a minus sign.

PROPORTION OF CASES LESS THAN OR EQUAL TO $X$

| $\lambda$ \ $X$ | 0 | 1 | 2 | 3 | 4 | 5 | 6 | 7 | 8 | 9 | 10 | 11 | 12 |
|---|---|---|---|---|---|---|---|---|---|---|---|---|---|
| .05 | .951 | .999 | 1.000 | | | | | | | | | | |
| .10 | .905 | .995 | 1.000 | | | | | | | | | | |
| .15 | .861 | .990 | .999 | 1.000 | | | | | | | | | |
| .20 | .819 | .982 | .999 | 1.000 | | | | | | | | | |
| .25 | .779 | .974 | .998 | 1.000 | | | | | | | | | |
| .30 | .741 | .963 | .996 | 1.000 | | | | | | | | | |
| .35 | .705 | .951 | .994 | 1.000 | | | | | | | | | |
| .40 | .670 | .938 | .992 | .999 | 1.000 | | | | | | | | |
| .45 | .638 | .925 | .989 | .999 | 1.000 | | | | | | | | |
| .50 | .607 | .910 | .986 | .998 | 1.000 | | | | | | | | |
| .55 | .577 | .894 | .982 | .998 | 1.000 | | | | | | | | |
| .60 | .549 | .878 | .977 | .997 | 1.000 | | | | | | | | |
| .65 | .522 | .861 | .972 | .996 | .999 | 1.000 | | | | | | | |
| .70 | .497 | .844 | .966 | .994 | .999 | 1.000 | | | | | | | |
| .75 | .472 | .827 | .959 | .993 | .999 | 1.000 | | | | | | | |
| .80 | .449 | .809 | .953 | .991 | .999 | 1.000 | | | | | | | |
| .85 | .427 | .791 | .945 | .989 | .998 | 1.000 | | | | | | | |
| .90 | .407 | .772 | .937 | .987 | .998 | 1.000 | | | | | | | |
| .95 | .387 | .754 | .929 | .984 | .997 | 1.000 | | | | | | | |
| 1.00 | .368 | .736 | .920 | .981 | .996 | .999 | 1.000 | | | | | | |
| 1.1 | .333 | .699 | .900 | .974 | .995 | .999 | 1.000 | | | | | | |
| 1.2 | .301 | .663 | .879 | .966 | .992 | .998 | 1.000 | | | | | | |
| 1.3 | .273 | .627 | .857 | .957 | .989 | .998 | 1.000 | | | | | | |
| 1.4 | .247 | .592 | .833 | .946 | .986 | .997 | .999 | 1.000 | | | | | |
| 1.5 | .223 | .558 | .809 | .934 | .981 | .996 | .999 | 1.000 | | | | | |
| 1.6 | .202 | .525 | .783 | .921 | .976 | .994 | .999 | 1.000 | | | | | |
| 1.7 | .183 | .493 | .757 | .907 | .970 | .992 | .998 | 1.000 | | | | | |
| 1.8 | .165 | .463 | .731 | .891 | .964 | .990 | .997 | .999 | 1.000 | | | | |
| 1.9 | .150 | .434 | .704 | .875 | .956 | .987 | .997 | .999 | 1.000 | | | | |
| 2.0 | .135 | .406 | .677 | .857 | .947 | .983 | .995 | .999 | 1.000 | | | | |
| 2.2 | .111 | .355 | .623 | .819 | .928 | .975 | .993 | .998 | 1.000 | | | | |
| 2.4 | .091 | .308 | .570 | .779 | .904 | .964 | .988 | .997 | .999 | 1.000 | | | |
| 2.6 | .074 | .267 | .518 | .736 | .877 | .951 | .983 | .995 | .999 | 1.000 | | | |
| 2.8 | .061 | .231 | .469 | .692 | .848 | .935 | .976 | .992 | .998 | .999 | 1.000 | | |
| 3.0 | .050 | .199 | .423 | .647 | .815 | .916 | .966 | .988 | .996 | .999 | 1.000 | | |
| 3.2 | .041 | .171 | .380 | .603 | .781 | .895 | .955 | .983 | .994 | .998 | 1.000 | | |
| 3.4 | .033 | .147 | .340 | .558 | .744 | .871 | .942 | .977 | .992 | .997 | .999 | 1.000 | |
| 3.6 | .027 | .126 | .303 | .515 | .706 | .844 | .927 | .969 | .988 | .996 | .999 | 1.000 | |
| 3.8 | .022 | .107 | .269 | .473 | .668 | .816 | .909 | .960 | .984 | .994 | .998 | .999 | 1.000 |
| 4.0 | .018 | .092 | .238 | .433 | .629 | .785 | .889 | .949 | .979 | .992 | .997 | .999 | 1.000 |

| N \ P | γ = 0.75 | | | | | γ = 0.90 | | | | | γ = 0.95 | | | | | γ = 0.99 | | | | |
|---|---|---|---|---|---|---|---|---|---|---|---|---|---|---|---|---|---|---|---|---|
| | 0.75 | 0.90 | 0.95 | 0.99 | 0.999 | 0.75 | 0.90 | 0.95 | 0.99 | 0.999 | 0.75 | 0.90 | 0.95 | 0.99 | 0.999 | 0.75 | 0.90 | 0.95 | 0.99 | 0.999 |
| 2 | 4.498 | 6.301 | 7.414 | 9.531 | 11.920 | 11.407 | 15.978 | 18.800 | 24.167 | 30.227 | 22.858 | 32.019 | 37.674 | 48.430 | 60.573 | 114.363 | 160.193 | 188.491 | 242.300 | 303.054 |
| 3 | 2.501 | 3.538 | 4.187 | 5.431 | 6.844 | 4.132 | 5.847 | 6.919 | 8.974 | 11.309 | 5.922 | 8.380 | 9.916 | 12.861 | 16.208 | 13.378 | 18.930 | 22.401 | 29.055 | 36.616 |
| 4 | 2.035 | 2.892 | 3.431 | 4.471 | 5.657 | 2.932 | 4.166 | 4.943 | 6.440 | 8.149 | 3.779 | 5.369 | 6.370 | 8.299 | 10.502 | 6.614 | 9.398 | 11.150 | 14.527 | 18.383 |
| 5 | 1.825 | 2.599 | 3.088 | 4.033 | 5.117 | 2.454 | 3.494 | 4.152 | 5.423 | 6.879 | 3.002 | 4.275 | 5.079 | 6.634 | 8.415 | 4.643 | 6.612 | 7.855 | 10.260 | 13.015 |
| 6 | 1.704 | 2.429 | 2.889 | 3.779 | 4.802 | 2.196 | 3.131 | 3.723 | 4.870 | 6.188 | 2.604 | 3.712 | 4.414 | 5.775 | 7.337 | 3.743 | 5.337 | 6.345 | 8.301 | 10.548 |
| 7 | 1.624 | 2.318 | 2.757 | 3.611 | 4.593 | 2.034 | 2.902 | 3.452 | 4.521 | 5.750 | 2.361 | 3.369 | 4.007 | 5.248 | 6.676 | 3.233 | 4.613 | 5.488 | 7.187 | 9.142 |
| 8 | 1.568 | 2.238 | 2.663 | 3.491 | 4.444 | 1.921 | 2.743 | 3.264 | 4.278 | 5.446 | 2.197 | 3.136 | 3.732 | 4.891 | 6.226 | 2.905 | 4.147 | 4.936 | 6.468 | 8.234 |
| 9 | 1.525 | 2.178 | 2.593 | 3.400 | 4.330 | 1.839 | 2.626 | 3.125 | 4.098 | 5.220 | 2.078 | 2.967 | 3.532 | 4.631 | 5.899 | 2.677 | 3.822 | 4.550 | 5.966 | 7.600 |
| 10 | 1.492 | 2.131 | 2.537 | 3.328 | 4.241 | 1.775 | 2.535 | 3.018 | 3.959 | 5.046 | 1.987 | 2.839 | 3.379 | 4.433 | 5.649 | 2.508 | 3.582 | 4.265 | 5.594 | 7.129 |
| 11 | 1.465 | 2.093 | 2.493 | 3.271 | 4.169 | 1.724 | 2.463 | 2.933 | 3.849 | 4.906 | 1.916 | 2.737 | 3.259 | 4.277 | 5.452 | 2.378 | 3.397 | 4.045 | 5.308 | 6.766 |
| 12 | 1.443 | 2.062 | 2.456 | 3.223 | 4.110 | 1.683 | 2.404 | 2.863 | 3.758 | 4.792 | 1.858 | 2.655 | 3.162 | 4.150 | 5.291 | 2.274 | 3.250 | 3.870 | 5.079 | 6.477 |
| 13 | 1.425 | 2.036 | 2.424 | 3.183 | 4.059 | 1.648 | 2.355 | 2.805 | 3.682 | 4.697 | 1.810 | 2.587 | 3.081 | 4.044 | 5.158 | 2.190 | 3.130 | 3.727 | 4.893 | 6.240 |
| 14 | 1.409 | 2.013 | 2.398 | 3.148 | 4.016 | 1.619 | 2.314 | 2.756 | 3.618 | 4.615 | 1.770 | 2.529 | 3.012 | 3.955 | 5.045 | 2.120 | 3.029 | 3.608 | 4.737 | 6.043 |
| 15 | 1.395 | 1.994 | 2.375 | 3.118 | 3.979 | 1.594 | 2.278 | 2.713 | 3.562 | 4.545 | 1.735 | 2.480 | 2.954 | 3.878 | 4.949 | 2.060 | 2.945 | 3.507 | 4.605 | 5.876 |
| 16 | 1.383 | 1.977 | 2.355 | 3.092 | 3.946 | 1.572 | 2.246 | 2.676 | 3.514 | 4.484 | 1.705 | 2.437 | 2.903 | 3.812 | 4.865 | 2.009 | 2.872 | 3.421 | 4.492 | 5.732 |
| 17 | 1.372 | 1.962 | 2.337 | 3.069 | 3.917 | 1.552 | 2.219 | 2.643 | 3.471 | 4.430 | 1.679 | 2.400 | 2.858 | 3.754 | 4.791 | 1.965 | 2.808 | 3.345 | 4.393 | 5.607 |
| 18 | 1.363 | 1.948 | 2.321 | 3.048 | 3.891 | 1.535 | 2.194 | 2.614 | 3.433 | 4.382 | 1.655 | 2.366 | 2.819 | 3.702 | 4.725 | 1.926 | 2.753 | 3.279 | 4.307 | 5.497 |
| 19 | 1.355 | 1.936 | 2.307 | 3.030 | 3.867 | 1.520 | 2.172 | 2.588 | 3.399 | 4.339 | 1.635 | 2.337 | 2.784 | 3.656 | 4.667 | 1.891 | 2.703 | 3.221 | 4.230 | 5.399 |
| 20 | 1.347 | 1.925 | 2.294 | 3.013 | 3.846 | 1.506 | 2.152 | 2.564 | 3.368 | 4.300 | 1.616 | 2.310 | 2.752 | 3.615 | 4.614 | 1.860 | 2.659 | 3.168 | 4.161 | 5.312 |
| 21 | 1.340 | 1.915 | 2.282 | 2.998 | 3.827 | 1.493 | 2.135 | 2.543 | 3.340 | 4.264 | 1.599 | 2.286 | 2.723 | 3.577 | 4.567 | 1.833 | 2.620 | 3.121 | 4.100 | 5.234 |
| 22 | 1.334 | 1.906 | 2.271 | 2.984 | 3.809 | 1.482 | 2.118 | 2.524 | 3.315 | 4.232 | 1.584 | 2.264 | 2.697 | 3.543 | 4.523 | 1.808 | 2.584 | 3.078 | 4.044 | 5.163 |
| 23 | 1.328 | 1.898 | 2.261 | 2.971 | 3.793 | 1.471 | 2.103 | 2.506 | 3.292 | 4.203 | 1.570 | 2.244 | 2.673 | 3.512 | 4.484 | 1.785 | 2.551 | 3.040 | 3.993 | 5.098 |
| 24 | 1.322 | 1.891 | 2.252 | 2.959 | 3.778 | 1.462 | 2.089 | 2.489 | 3.270 | 4.176 | 1.557 | 2.225 | 2.651 | 3.483 | 4.447 | 1.764 | 2.522 | 3.004 | 3.947 | 5.039 |
| 25 | 1.317 | 1.883 | 2.244 | 2.948 | 3.764 | 1.453 | 2.077 | 2.474 | 3.251 | 4.151 | 1.545 | 2.208 | 2.631 | 3.457 | 4.413 | 1.745 | 2.494 | 2.972 | 3.904 | 4.985 |
| 26 | 1.313 | 1.877 | 2.236 | 2.938 | 3.751 | 1.444 | 2.065 | 2.460 | 3.232 | 4.127 | 1.534 | 2.193 | 2.612 | 3.432 | 4.382 | 1.727 | 2.469 | 2.941 | 3.865 | 4.935 |
| 27 | 1.309 | 1.871 | 2.229 | 2.929 | 3.740 | 1.437 | 2.054 | 2.447 | 3.215 | 4.106 | 1.523 | 2.178 | 2.595 | 3.409 | 4.353 | 1.711 | 2.446 | 2.914 | 3.828 | 4.888 |

| | | | | | | | | | | | | | | | | | | | | |
|---|---|---|---|---|---|---|---|---|---|---|---|---|---|---|---|---|---|---|---|---|
| 30 | 1.297 | 1.855 | 2.210 | 2.904 | 3.708 | 1.417 | 2.025 | 2.413 | 3.170 | 4.049 | 1.497 | 2.140 | 2.549 | 3.350 | 4.278 | 1.668 | 2.385 | 2.841 | 3.733 | 4.768 |
| 35 | 1.283 | 1.834 | 2.185 | 2.871 | 3.667 | 1.390 | 1.988 | 2.368 | 3.112 | 3.974 | 1.462 | 2.090 | 2.490 | 3.272 | 4.179 | 1.613 | 2.306 | 2.748 | 3.611 | 4.611 |
| 40 | 1.271 | 1.818 | 2.166 | 2.846 | 3.635 | 1.370 | 1.959 | 2.334 | 3.066 | 3.917 | 1.435 | 2.052 | 2.445 | 3.213 | 4.104 | 1.571 | 2.247 | 2.677 | 3.518 | 4.493 |
| 45 | 1.262 | 1.805 | 2.150 | 2.826 | 3.609 | 1.354 | 1.935 | 2.306 | 3.030 | 3.871 | 1.414 | 2.021 | 2.408 | 3.165 | 4.042 | 1.539 | 2.200 | 2.621 | 3.444 | 4.399 |
| 50 | 1.255 | 1.794 | 2.138 | 2.809 | 3.588 | 1.340 | 1.916 | 2.284 | 3.001 | 3.833 | 1.396 | 1.996 | 2.379 | 3.126 | 3.993 | 1.512 | 2.162 | 2.576 | 3.385 | 4.323 |
| 55 | 1.249 | 1.785 | 2.127 | 2.795 | 3.571 | 1.329 | 1.901 | 2.265 | 2.976 | 3.801 | 1.382 | 1.976 | 2.354 | 3.094 | 3.951 | 1.490 | 2.130 | 2.538 | 3.335 | 4.260 |
| 60 | 1.243 | 1.778 | 2.118 | 2.784 | 3.556 | 1.320 | 1.887 | 2.248 | 2.955 | 3.774 | 1.369 | 1.958 | 2.333 | 3.066 | 3.916 | 1.471 | 2.103 | 2.506 | 3.293 | 4.206 |
| 65 | 1.239 | 1.771 | 2.110 | 2.773 | 3.543 | 1.312 | 1.875 | 2.235 | 2.937 | 3.751 | 1.359 | 1.943 | 2.315 | 3.042 | 3.886 | 1.455 | 2.080 | 2.478 | 3.257 | 4.160 |
| 70 | 1.235 | 1.765 | 2.104 | 2.764 | 3.531 | 1.304 | 1.865 | 2.222 | 2.920 | 3.730 | 1.349 | 1.929 | 2.299 | 3.021 | 3.859 | 1.440 | 2.060 | 2.454 | 3.225 | 4.120 |
| 75 | 1.231 | 1.760 | 2.098 | 2.757 | 3.521 | 1.298 | 1.856 | 2.211 | 2.906 | 3.712 | 1.341 | 1.917 | 2.285 | 3.002 | 3.835 | 1.428 | 2.042 | 2.433 | 3.197 | 4.084 |
| 80 | 1.228 | 1.756 | 2.092 | 2.749 | 3.512 | 1.292 | 1.848 | 2.202 | 2.894 | 3.696 | 1.334 | 1.907 | 2.272 | 2.986 | 3.814 | 1.417 | 2.026 | 2.414 | 3.173 | 4.053 |
| 85 | 1.225 | 1.752 | 2.087 | 2.743 | 3.504 | 1.287 | 1.841 | 2.193 | 2.882 | 3.682 | 1.327 | 1.897 | 2.261 | 2.971 | 3.795 | 1.407 | 2.012 | 2.397 | 3.150 | 4.024 |
| 90 | 1.223 | 1.748 | 2.083 | 2.737 | 3.497 | 1.283 | 1.834 | 2.185 | 2.872 | 3.669 | 1.321 | 1.889 | 2.251 | 2.958 | 3.778 | 1.398 | 1.999 | 2.382 | 3.130 | 3.999 |
| 95 | 1.220 | 1.745 | 2.079 | 2.732 | 3.490 | 1.278 | 1.828 | 2.178 | 2.863 | 3.657 | 1.315 | 1.881 | 2.241 | 2.945 | 3.763 | 1.390 | 1.987 | 2.368 | 3.112 | 3.976 |
| 100 | 1.218 | 1.742 | 2.075 | 2.727 | 3.484 | 1.275 | 1.822 | 2.172 | 2.854 | 3.646 | 1.311 | 1.874 | 2.233 | 2.934 | 3.748 | 1.383 | 1.977 | 2.355 | 3.096 | 3.954 |
| 110 | 1.214 | 1.736 | 2.069 | 2.719 | 3.473 | 1.268 | 1.813 | 2.160 | 2.839 | 3.626 | 1.302 | 1.861 | 2.218 | 2.915 | 3.723 | 1.369 | 1.958 | 2.333 | 3.066 | 3.917 |
| 120 | 1.211 | 1.732 | 2.063 | 2.712 | 3.464 | 1.262 | 1.804 | 2.150 | 2.826 | 3.610 | 1.294 | 1.850 | 2.205 | 2.898 | 3.702 | 1.358 | 1.942 | 2.314 | 3.041 | 3.885 |
| 130 | 1.208 | 1.728 | 2.059 | 2.705 | 3.456 | 1.257 | 1.797 | 2.141 | 2.814 | 3.595 | 1.288 | 1.841 | 2.194 | 2.883 | 3.683 | 1.349 | 1.928 | 2.298 | 3.019 | 3.857 |
| 140 | 1.206 | 1.724 | 2.054 | 2.700 | 3.449 | 1.252 | 1.791 | 2.134 | 2.804 | 3.582 | 1.282 | 1.833 | 2.184 | 2.870 | 3.666 | 1.340 | 1.916 | 2.283 | 3.000 | 3.833 |
| 150 | 1.204 | 1.721 | 2.051 | 2.695 | 3.443 | 1.248 | 1.785 | 2.127 | 2.795 | 3.571 | 1.277 | 1.825 | 2.175 | 2.859 | 3.652 | 1.332 | 1.905 | 2.270 | 2.983 | 3.811 |
| 160 | 1.202 | 1.718 | 2.047 | 2.691 | 3.437 | 1.245 | 1.780 | 2.121 | 2.787 | 3.561 | 1.272 | 1.819 | 2.167 | 2.848 | 3.638 | 1.326 | 1.896 | 2.259 | 2.968 | 3.792 |
| 170 | 1.200 | 1.716 | 2.044 | 2.687 | 3.432 | 1.242 | 1.775 | 2.116 | 2.780 | 3.552 | 1.268 | 1.813 | 2.160 | 2.839 | 3.627 | 1.320 | 1.887 | 2.248 | 2.955 | 3.774 |
| 180 | 1.198 | 1.713 | 2.042 | 2.683 | 3.427 | 1.239 | 1.771 | 2.111 | 2.774 | 3.543 | 1.264 | 1.808 | 2.154 | 2.831 | 3.616 | 1.314 | 1.879 | 2.239 | 2.942 | 3.759 |
| 190 | 1.197 | 1.711 | 2.039 | 2.680 | 3.423 | 1.236 | 1.767 | 2.106 | 2.768 | 3.536 | 1.261 | 1.803 | 2.148 | 2.823 | 3.606 | 1.309 | 1.872 | 2.230 | 2.931 | 3.744 |
| 200 | 1.195 | 1.709 | 2.037 | 2.677 | 3.419 | 1.234 | 1.764 | 2.102 | 2.762 | 3.529 | 1.258 | 1.798 | 2.143 | 2.816 | 3.597 | 1.304 | 1.865 | 2.222 | 2.921 | 3.731 |
| 250 | 1.190 | 1.702 | 2.028 | 2.665 | 3.404 | 1.224 | 1.750 | 2.085 | 2.740 | 3.501 | 1.245 | 1.780 | 2.121 | 2.788 | 3.561 | 1.286 | 1.839 | 2.191 | 2.880 | 3.678 |
| 300 | 1.186 | 1.696 | 2.021 | 2.656 | 3.393 | 1.217 | 1.740 | 2.073 | 2.725 | 3.481 | 1.236 | 1.767 | 2.106 | 2.767 | 3.535 | 1.273 | 1.820 | 2.169 | 2.850 | 3.641 |
| 400 | 1.181 | 1.688 | 2.012 | 2.644 | 3.378 | 1.207 | 1.726 | 2.057 | 2.703 | 3.453 | 1.223 | 1.749 | 2.084 | 2.739 | 3.499 | 1.255 | 1.794 | 2.138 | 2.809 | 3.589 |
| 500 | 1.177 | 1.683 | 2.006 | 2.636 | 3.368 | 1.201 | 1.717 | 2.046 | 2.689 | 3.434 | 1.215 | 1.737 | 2.070 | 2.721 | 3.475 | 1.243 | 1.777 | 2.117 | 2.783 | 3.555 |
| 600 | 1.175 | 1.680 | 2.002 | 2.631 | 3.360 | 1.196 | 1.710 | 2.038 | 2.678 | 3.421 | 1.209 | 1.729 | 2.060 | 2.707 | 3.458 | 1.234 | 1.764 | 2.102 | 2.763 | 3.530 |
| 700 | 1.173 | 1.677 | 1.998 | 2.626 | 3.355 | 1.192 | 1.705 | 2.032 | 2.670 | 3.411 | 1.204 | 1.722 | 2.052 | 2.697 | 3.445 | 1.227 | 1.755 | 2.091 | 2.748 | 3.511 |
| 800 | 1.171 | 1.675 | 1.996 | 2.623 | 3.350 | 1.189 | 1.701 | 2.027 | 2.663 | 3.402 | 1.201 | 1.717 | 2.046 | 2.688 | 3.434 | 1.222 | 1.747 | 2.082 | 2.736 | 3.495 |
| 900 | 1.170 | 1.673 | 1.993 | 2.620 | 3.347 | 1.187 | 1.697 | 2.023 | 2.658 | 3.396 | 1.198 | 1.712 | 2.040 | 2.682 | 3.426 | 1.218 | 1.741 | 2.075 | 2.726 | 3.483 |
| 1000 | 1.169 | 1.671 | 1.992 | 2.617 | 3.344 | 1.185 | 1.695 | 2.019 | 2.654 | 3.390 | 1.195 | 1.709 | 2.036 | 2.676 | 3.418 | 1.214 | 1.736 | 2.068 | 2.718 | 3.472 |
| ∞ | 1.150 | 1.645 | 1.960 | 2.576 | 3.291 | 1.150 | 1.645 | 1.960 | 2.576 | 3.291 | 1.150 | 1.645 | 1.960 | 2.576 | 3.291 | 1.150 | 1.645 | 1.960 | 2.576 | 3.291 |

By permission from C. Eisenhart, M. W. Hastay, W. A. Wallis, *Techniques of Statistical Analysis*, chap. 2. McGraw-Hill Book Company, New York, 1947.

# A-17 CRITICAL VALUES FOR COCHRAN'S TEST

Values given are for the statistic (largest $s^2$)/($\Sigma s_i^2$), where each of the $k$ values of $s^2$ has $\nu$ degrees of freedom.

PERCENTILE 95

| $k$ \ $\nu$ | 1 | 2 | 3 | 4 | 5 | 6 | 7 | 8 | 9 | 10 | 16 | 36 | 144 | $\infty$ |
|---|---|---|---|---|---|---|---|---|---|---|---|---|---|---|
| 2 | 0.9985 | 0.9750 | 0.9392 | 0.9057 | 0.8772 | 0.8534 | 0.8332 | 0.8159 | 0.8010 | 0.7880 | 0.7341 | 0.6602 | 0.5813 | 0.5000 |
| 3 | 0.9669 | 0.8709 | 0.7977 | 0.7457 | 0.7071 | 0.6771 | 0.6530 | 0.6333 | 0.6167 | 0.6025 | 0.5466 | 0.4748 | 0.4031 | 0.3333 |
| 4 | 0.9065 | 0.7679 | 0.6841 | 0.6287 | 0.5895 | 0.5598 | 0.5365 | 0.5175 | 0.5017 | 0.4884 | 0.4366 | 0.3720 | 0.3093 | 0.2500 |
| 5 | 0.8412 | 0.6838 | 0.5981 | 0.5441 | 0.5065 | 0.4783 | 0.4564 | 0.4387 | 0.4241 | 0.4118 | 0.3645 | 0.3066 | 0.2513 | 0.2000 |
| 6 | 0.7808 | 0.6161 | 0.5321 | 0.4803 | 0.4447 | 0.4184 | 0.3980 | 0.3817 | 0.3682 | 0.3568 | 0.3135 | 0.2612 | 0.2119 | 0.1667 |
| 7 | 0.7271 | 0.5612 | 0.4800 | 0.4307 | 0.3974 | 0.3726 | 0.3535 | 0.3384 | 0.3259 | 0.3154 | 0.2756 | 0.2278 | 0.1833 | 0.1429 |
| 8 | 0.6798 | 0.5157 | 0.4377 | 0.3910 | 0.3595 | 0.3362 | 0.3185 | 0.3043 | 0.2926 | 0.2829 | 0.2462 | 0.2022 | 0.1616 | 0.1250 |
| 9 | 0.6385 | 0.4775 | 0.4027 | 0.3584 | 0.3286 | 0.3067 | 0.2901 | 0.2768 | 0.2659 | 0.2568 | 0.2226 | 0.1820 | 0.1446 | 0.1111 |
| 10 | 0.6020 | 0.4450 | 0.3733 | 0.3311 | 0.3029 | 0.2823 | 0.2666 | 0.2541 | 0.2439 | 0.2353 | 0.2032 | 0.1655 | 0.1308 | 0.1000 |
| 12 | 0.5410 | 0.3924 | 0.3264 | 0.2880 | 0.2624 | 0.2439 | 0.2299 | 0.2187 | 0.2098 | 0.2020 | 0.1737 | 0.1403 | 0.1100 | 0.0833 |
| 15 | 0.4709 | 0.3346 | 0.2758 | 0.2419 | 0.2195 | 0.2034 | 0.1911 | 0.1815 | 0.1736 | 0.1671 | 0.1429 | 0.1144 | 0.0889 | 0.0667 |
| 20 | 0.3894 | 0.2705 | 0.2205 | 0.1921 | 0.1735 | 0.1602 | 0.1501 | 0.1422 | 0.1357 | 0.1303 | 0.1108 | 0.0879 | 0.0675 | 0.0500 |
| 24 | 0.3434 | 0.2354 | 0.1907 | 0.1656 | 0.1493 | 0.1374 | 0.1286 | 0.1216 | 0.1160 | 0.1113 | 0.0942 | 0.0743 | 0.0567 | 0.0417 |
| 30 | 0.2929 | 0.1980 | 0.1593 | 0.1377 | 0.1237 | 0.1137 | 0.1061 | 0.1002 | 0.0958 | 0.0921 | 0.0771 | 0.0604 | 0.0457 | 0.0333 |
| 40 | 0.2370 | 0.1576 | 0.1259 | 0.1082 | 0.0968 | 0.0887 | 0.0827 | 0.0780 | 0.0745 | 0.0713 | 0.0595 | 0.0462 | 0.0347 | 0.0250 |
| 60 | 0.1737 | 0.1131 | 0.0895 | 0.0765 | 0.0682 | 0.0623 | 0.0583 | 0.0552 | 0.0520 | 0.0497 | 0.0411 | 0.0316 | 0.0234 | 0.0167 |
| 120 | 0.0998 | 0.0632 | 0.0495 | 0.0419 | 0.0371 | 0.0337 | 0.0312 | 0.0292 | 0.0279 | 0.0266 | 0.0218 | 0.0165 | 0.0120 | 0.0083 |
| $\infty$ | 0 | 0 | 0 | 0 | 0 | 0 | 0 | 0 | 0 | 0 | 0 | 0 | 0 | 0 |

PERCENTILE 99

| k \ ν | 1 | 2 | 3 | 4 | 5 | 6 | 7 | 8 | 9 | 10 | 16 | 36 | 144 | ∞ |
|---|---|---|---|---|---|---|---|---|---|---|---|---|---|---|
| 2 | 0.9999 | 0.9950 | 0.9794 | 0.9586 | 0.9373 | 0.9172 | 0.8988 | 0.8823 | 0.8674 | 0.8539 | 0.7949 | 0.7067 | 0.6062 | 0.5000 |
| 3 | 0.9933 | 0.9423 | 0.8831 | 0.8335 | 0.7933 | 0.7606 | 0.7335 | 0.7107 | 0.6912 | 0.6743 | 0.6059 | 0.5153 | 0.4230 | 0.3333 |
| 4 | 0.9676 | 0.8643 | 0.7814 | 0.7212 | 0.6761 | 0.6410 | 0.6129 | 0.5897 | 0.5702 | 0.5536 | 0.4884 | 0.4057 | 0.3251 | 0.2500 |
| 5 | 0.9279 | 0.7885 | 0.6957 | 0.6329 | 0.5875 | 0.5531 | 0.5259 | 0.5037 | 0.4854 | 0.4697 | 0.4094 | 0.3351 | 0.2644 | 0.2000 |
| 6 | 0.8828 | 0.7218 | 0.6258 | 0.5635 | 0.5195 | 0.4866 | 0.4608 | 0.4401 | 0.4229 | 0.4084 | 0.3529 | 0.2858 | 0.2229 | 0.1667 |
| 7 | 0.8376 | 0.6644 | 0.5685 | 0.5080 | 0.4659 | 0.4347 | 0.4105 | 0.3911 | 0.3751 | 0.3616 | 0.3105 | 0.2494 | 0.1929 | 0.1429 |
| 8 | 0.7945 | 0.6152 | 0.5209 | 0.4627 | 0.4226 | 0.3932 | 0.3704 | 0.3522 | 0.3373 | 0.3248 | 0.2779 | 0.2214 | 0.1700 | 0.1250 |
| 9 | 0.7544 | 0.5727 | 0.4810 | 0.4251 | 0.3870 | 0.3592 | 0.3378 | 0.3207 | 0.3067 | 0.2950 | 0.2514 | 0.1992 | 0.1521 | 0.1111 |
| 10 | 0.7175 | 0.5358 | 0.4469 | 0.3934 | 0.3572 | 0.3308 | 0.3106 | 0.2945 | 0.2813 | 0.2704 | 0.2297 | 0.1811 | 0.1376 | 0.1000 |
| 12 | 0.6528 | 0.4751 | 0.3919 | 0.3428 | 0.3099 | 0.2861 | 0.2680 | 0.2535 | 0.2419 | 0.2320 | 0.1961 | 0.1535 | 0.1157 | 0.0833 |
| 15 | 0.5747 | 0.4069 | 0.3317 | 0.2882 | 0.2593 | 0.2386 | 0.2228 | 0.2104 | 0.2002 | 0.1918 | 0.1612 | 0.1251 | 0.0934 | 0.0667 |
| 20 | 0.4799 | 0.3297 | 0.2654 | 0.2288 | 0.2048 | 0.1877 | 0.1748 | 0.1646 | 0.1567 | 0.1501 | 0.1248 | 0.0960 | 0.0709 | 0.0500 |
| 24 | 0.4247 | 0.2871 | 0.2295 | 0.1970 | 0.1759 | 0.1608 | 0.1495 | 0.1406 | 0.1338 | 0.1283 | 0.1060 | 0.0810 | 0.0595 | 0.0417 |
| 30 | 0.3632 | 0.2412 | 0.1913 | 0.1635 | 0.1454 | 0.1327 | 0.1232 | 0.1157 | 0.1100 | 0.1054 | 0.0867 | 0.0658 | 0.0480 | 0.0333 |
| 40 | 0.2940 | 0.1915 | 0.1508 | 0.1281 | 0.1135 | 0.1033 | 0.0957 | 0.0898 | 0.0853 | 0.0816 | 0.0668 | 0.0503 | 0.0363 | 0.0250 |
| 60 | 0.2151 | 0.1371 | 0.1069 | 0.0902 | 0.0796 | 0.0722 | 0.0668 | 0.0625 | 0.0594 | 0.0567 | 0.0461 | 0.0344 | 0.0245 | 0.0167 |
| 120 | 0.1225 | 0.0759 | 0.0585 | 0.0489 | 0.0429 | 0.0387 | 0.0357 | 0.0334 | 0.0316 | 0.0302 | 0.0242 | 0.0178 | 0.0125 | 0.0083 |
| ∞ | 0 | 0 | 0 | 0 | 0 | 0 | 0 | 0 | 0 | 0 | 0 | 0 | 0 | 0 |

By permission from C. Eisenhart, M. W. Hastay, W. A. Wallis, *Techniques of Statistical Analysis*, chap. 15. McGraw-Hill Book Company, New York, 1947.

## A-18a PERCENTILES OF THE DISTRIBUTION OF $q = w/s$

$w$ is the range of $k$ observations, and $\nu$ is the number of degrees of freedom in an independent standard deviation $s$.

| $k$ | $\nu$ | $q_{.01}$ | $q_{.05}$ | $q_{.50}$ | $q_{.90}$ | $q_{.95}$ | $q_{.99}$ | $q_{.995}$ | $q_{.999}$ |
|---|---|---|---|---|---|---|---|---|---|
| 2 | 1 | .022 | .111 | 1.41 | 8.93 | 18.0 | 90.0 | 180. | 900. |
| | 2 | .020 | .100 | 1.16 | 4.13 | 6.09 | 14.0 | 19.9 | 44.7 |
| | 3 | .019 | .096 | 1.08 | 3.33 | 4.50 | 8.26 | 10.6 | 18.3 |
| | 4 | .019 | .094 | 1.05 | 3.02 | 3.93 | 6.51 | 7.92 | 12.2 |
| | 6 | .019 | .093 | 1.02 | 2.75 | 3.46 | 5.24 | 6.11 | 8.43 |
| | 8 | .018 | .092 | .999 | 2.63 | 3.26 | 4.75 | 5.42 | 7.13 |
| | 10 | .018 | .091 | .990 | 2.56 | 3.15 | 4.48 | 5.07 | 6.49 |
| | 12 | .018 | .091 | .984 | 2.52 | 3.08 | 4.32 | 4.85 | 6.11 |
| | 16 | .018 | .090 | .976 | 2.47 | 3.00 | 4.13 | 4.60 | 5.68 |
| | 20 | .018 | .090 | .972 | 2.44 | 2.95 | 4.02 | 4.46 | 5.44 |
| | 24 | .018 | .090 | .969 | 2.42 | 2.92 | 3.96 | 4.37 | 5.30 |
| | 30 | .018 | .089 | .966 | 2.40 | 2.89 | 3.89 | 4.29 | 5.16 |
| | 40 | .018 | .089 | .963 | 2.38 | 2.86 | 3.83 | 4.20 | 5.02 |
| | 60 | .018 | .089 | .960 | 2.36 | 2.83 | 3.76 | 4.12 | 4.89 |
| | 120 | .018 | .089 | .957 | 2.34 | 2.80 | 3.70 | 4.05 | 4.77 |
| | $\infty$ | .018 | .089 | .954 | 2.33 | 2.77 | 3.64 | 3.97 | 4.65 |
| 3 | 1 | .192 | .443 | 2.34 | 13.4 | 27.0 | 135. | 270. | 1,351. |
| | 2 | .191 | .437 | 1.91 | 5.73 | 8.33 | 19.0 | 27.0 | 60.4 |
| | 3 | .191 | .435 | 1.79 | 4.47 | 5.91 | 10.6 | 13.5 | 23.3 |
| | 4 | .191 | .434 | 1.74 | 3.98 | 5.04 | 8.12 | 9.81 | 15.0 |
| | 6 | .191 | .433 | 1.69 | 3.56 | 4.34 | 6.33 | 7.31 | 9.96 |
| | 8 | .191 | .433 | 1.66 | 3.37 | 4.04 | 5.64 | 6.37 | 8.25 |
| | 10 | .191 | .433 | 1.65 | 3.27 | 3.88 | 5.27 | 5.89 | 7.41 |
| | 12 | .191 | .432 | 1.64 | 3.20 | 3.77 | 5.05 | 5.60 | 6.92 |
| | 16 | .191 | .432 | 1.62 | 3.12 | 3.65 | 4.79 | 5.26 | 6.37 |
| | 20 | .191 | .432 | 1.62 | 3.08 | 3.58 | 4.64 | 5.07 | 6.07 |
| | 24 | .191 | .432 | 1.61 | 3.05 | 3.53 | 4.55 | 4.96 | 5.88 |
| | 30 | .191 | .432 | 1.61 | 3.02 | 3.49 | 4.46 | 4.84 | 5.70 |
| | 40 | .191 | .432 | 1.60 | 2.99 | 3.44 | 4.37 | 4.73 | 5.53 |
| | 60 | .191 | .432 | 1.60 | 2.96 | 3.40 | 4.29 | 4.63 | 5.37 |
| | 120 | .191 | .432 | 1.59 | 2.93 | 3.36 | 4.20 | 4.52 | 5.21 |
| | $\infty$ | .191 | .431 | 1.59 | 2.90 | 3.31 | 4.12 | 4.42 | 5.06 |
| 4 | 1 | .380 | .698 | 2.92 | 16.4 | 32.8 | 164. | 329. | 1,643. |
| | 2 | .399 | .718 | 2.38 | 6.77 | 9.80 | 22.3 | 31.6 | 70.8 |
| | 3 | .408 | .729 | 2.23 | 5.20 | 6.83 | 12.2 | 15.5 | 26.7 |
| | 4 | .414 | .735 | 2.16 | 4.59 | 5.76 | 9.17 | 11.1 | 16.8 |
| | 6 | .420 | .742 | 2.10 | 4.07 | 4.90 | 7.03 | 8.09 | 11.0 |
| | 8 | .423 | .746 | 2.07 | 3.83 | 4.53 | 6.20 | 6.98 | 8.98 |
| | 10 | .425 | .748 | 2.05 | 3.70 | 4.33 | 5.77 | 6.41 | 8.01 |
| | 12 | .426 | .750 | 2.04 | 3.62 | 4.20 | 5.50 | 6.07 | 7.44 |
| | 16 | .428 | .752 | 2.02 | 3.52 | 4.05 | 5.19 | 5.67 | 6.80 |
| | 20 | .429 | .754 | 2.01 | 3.46 | 3.96 | 5.02 | 5.46 | 6.45 |
| | 24 | .430 | .755 | 2.01 | 3.42 | 3.90 | 4.91 | 5.32 | 6.24 |
| | 30 | .431 | .756 | 2.00 | 3.39 | 3.85 | 4.80 | 5.18 | 6.03 |
| | 40 | .431 | .757 | 2.00 | 3.35 | 3.79 | 4.70 | 5.05 | 5.84 |
| | 60 | .432 | .758 | 1.99 | 3.31 | 3.74 | 4.60 | 4.93 | 5.65 |
| | 120 | .433 | .759 | 1.98 | 3.28 | 3.69 | 4.50 | 4.81 | 5.48 |
| | $\infty$ | .434 | .760 | 1.98 | 3.24 | 3.63 | 4.40 | 4.69 | 5.31 |

| $k$ | $\nu$ | $q_{.01}$ | $q_{.05}$ | $q_{.50}$ | $q_{.90}$ | $q_{.95}$ | $q_{.99}$ | $q_{.995}$ | $q_{.999}$ |
|---|---|---|---|---|---|---|---|---|---|
| 5 | 1 | .531 | .883 | 3.34 | 18.5 | 37.1 | 186. | 371. | 1,856. |
| | 2 | .576 | .930 | 2.71 | 7.54 | 10.9 | 24.7 | 35.0 | 78.4 |
| | 3 | .598 | .953 | 2.55 | 5.74 | 7.50 | 13.3 | 16.9 | 29.1 |
| | 4 | .611 | .968 | 2.47 | 5.04 | 6.29 | 9.96 | 12.0 | 18.2 |
| | 6 | .626 | .985 | 2.39 | 4.44 | 5.31 | 7.56 | 8.67 | 11.7 |
| | 8 | .635 | .995 | 2.36 | 4.17 | 4.89 | 6.63 | 7.44 | 9.52 |
| | 10 | .640 | 1.00 | 2.34 | 4.02 | 4.65 | 6.14 | 6.80 | 8.45 |
| | 12 | .644 | 1.01 | 2.32 | 3.92 | 4.51 | 5.84 | 6.42 | 7.82 |
| | 16 | .649 | 1.01 | 2.31 | 3.80 | 4.33 | 5.49 | 5.98 | 7.12 |
| | 20 | .652 | 1.02 | 2.30 | 3.74 | 4.23 | 5.29 | 5.73 | 6.74 |
| | 24 | .654 | 1.02 | 2.29 | 3.69 | 4.17 | 5.17 | 5.58 | 6.50 |
| | 30 | .656 | 1.02 | 2.28 | 3.65 | 4.10 | 5.05 | 5.43 | 6.28 |
| | 40 | .658 | 1.02 | 2.28 | 3.61 | 4.04 | 4.93 | 5.28 | 6.06 |
| | 60 | .660 | 1.02 | 2.27 | 3.56 | 3.98 | 4.82 | 5.15 | 3.86 |
| | 120 | .663 | 1.03 | 2.26 | 3.52 | 3.92 | 4.71 | 5.01 | 5.67 |
| | ∞ | .665 | 1.03 | 2.26 | 3.48 | 3.86 | 4.60 | 4.89 | 5.48 |
| 6 | 1 | .650 | 1.02 | 3.66 | 20.2 | 40.4 | 202. | 404. | 2,022. |
| | 2 | .718 | 1.09 | 2.97 | 8.14 | 11.7 | 26.6 | 37.7 | 84.5 |
| | 3 | .753 | 1.13 | 2.79 | 6.16 | 8.04 | 14.2 | 18.1 | 31.1 |
| | 4 | .775 | 1.15 | 2.70 | 5.39 | 6.71 | 10.6 | 12.7 | 19.3 |
| | 6 | .800 | 1.18 | 2.62 | 4.73 | 5.63 | 7.97 | 9.14 | 12.3 |
| | 8 | .815 | 1.19 | 2.58 | 4.43 | 5.17 | 6.96 | 7.80 | 9.96 |
| | 10 | .824 | 1.20 | 2.56 | 4.26 | 4.91 | 6.43 | 7.11 | 8.80 |
| | 12 | .831 | 1.21 | 2.55 | 4.16 | 4.75 | 6.10 | 6.69 | 8.13 |
| | 16 | .840 | 1.22 | 2.53 | 4.03 | 4.56 | 5.72 | 6.22 | 7.37 |
| | 20 | .845 | 1.23 | 2.52 | 3.95 | 4.45 | 5.51 | 5.95 | 6.97 |
| | 24 | .849 | 1.23 | 2.51 | 3.90 | 4.37 | 5.37 | 5.78 | 6.71 |
| | 30 | .853 | 1.24 | 2.50 | 3.85 | 4.30 | 5.24 | 5.62 | 6.47 |
| | 40 | .857 | 1.24 | 2.49 | 3.80 | 4.23 | 5.11 | 5.47 | 6.24 |
| | 60 | .861 | 1.24 | 2.49 | 3.76 | 4.16 | 5.00 | 5.32 | 6.02 |
| | 120 | .865 | 1.25 | 2.48 | 3.71 | 4.10 | 4.87 | 5.17 | 5.82 |
| | ∞ | .870 | 1.25 | 2.47 | 3.66 | 4.03 | 4.76 | 5.03 | 5.62 |
| 7 | 1 | .746 | 1.14 | 3.92 | 21.5 | 43.1 | 216. | 432. | 2,158. |
| | 2 | .835 | 1.22 | 3.18 | 8.63 | 12.4 | 28.2 | 40.0 | 89.5 |
| | 3 | .882 | 1.27 | 2.99 | 6.51 | 8.48 | 15.0 | 19.0 | 32.7 |
| | 4 | .911 | 1.30 | 2.90 | 5.68 | 7.05 | 11.1 | 13.4 | 20.3 |
| | 6 | .947 | 1.34 | 2.81 | 4.97 | 5.90 | 8.32 | 9.52 | 12.8 |
| | 8 | .967 | 1.36 | 2.77 | 4.65 | 5.40 | 7.24 | 8.10 | 10.3 |
| | 10 | .981 | 1.37 | 2.74 | 4.47 | 5.12 | 6.67 | 7.37 | 9.10 |
| | 12 | .991 | 1.38 | 2.73 | 4.35 | 4.95 | 6.32 | 6.92 | 8.38 |
| | 16 | 1.00 | 1.39 | 2.71 | 4.21 | 4.74 | 5.92 | 6.41 | 7.59 |
| | 20 | 1.01 | 1.40 | 2.69 | 4.12 | 4.62 | 5.69 | 6.13 | 7.15 |
| | 24 | 1.02 | 1.41 | 2.69 | 4.07 | 4.54 | 5.54 | 5.95 | 6.88 |
| | 30 | 1.02 | 1.41 | 2.68 | 4.02 | 4.46 | 5.40 | 5.78 | 6.63 |
| | 40 | 1.03 | 1.42 | 2.67 | 3.96 | 4.39 | 5.27 | 5.61 | 6.39 |
| | 60 | 1.04 | 1.43 | 2.66 | 3.91 | 4.31 | 5.13 | 5.45 | 6.16 |
| | 120 | 1.04 | 1.43 | 2.65 | 3.86 | 4.24 | 5.01 | 5.30 | 5.94 |
| | ∞ | 1.05 | 1.44 | 2.65 | 3.81 | 4.17 | 4.88 | 5.15 | 6.73 |

## A-18a PERCENTILES OF THE DISTRIBUTION OF $q = w/s$ (CONTINUED)

| $k$ | $\nu$ | $q_{.01}$ | $q_{.05}$ | $q_{.50}$ | $q_{.90}$ | $q_{.95}$ | $q_{.99}$ | $q_{.995}$ | $q_{.999}$ |
|---|---|---|---|---|---|---|---|---|---|
| 9 | 1 | .825 | 1.23 | 4.14 | 22.6 | 45.4 | 227. | 454. | 2,272. |
| | 2 | .932 | 1.33 | 3.36 | 9.05 | 13.0 | 29.5 | 41.8 | 93.7 |
| | 3 | .989 | 1.39 | 3.15 | 6.81 | 8.85 | 15.6 | 19.8 | 34.1 |
| | 4 | 1.03 | 1.42 | 3.06 | 5.93 | 7.35 | 11.6 | 13.9 | 21.0 |
| | 6 | 1.07 | 1.47 | 2.96 | 5.17 | 6.12 | 8.61 | 9.85 | 13.3 |
| | 8 | 1.10 | 1.49 | 2.92 | 4.83 | 5.60 | 7.47 | 8.35 | 10.6 |
| | 10 | 1.11 | 1.51 | 2.89 | 4.64 | 5.31 | 6.88 | 7.58 | 9.35 |
| | 12 | 1.13 | 1.53 | 2.88 | 4.51 | 5.12 | 6.51 | 7.12 | 8.60 |
| | 16 | 1.14 | 1.54 | 2.86 | 4.36 | 4.90 | 6.08 | 6.58 | 7.77 |
| | 20 | 1.16 | 1.55 | 2.84 | 4.27 | 4.77 | 5.84 | 6.29 | 7.31 |
| | 24 | 1.16 | 1.56 | 2.83 | 4.21 | 4.68 | 5.69 | 6.10 | 7.03 |
| | 30 | 1.17 | 1.57 | 2.83 | 4.16 | 4.60 | 5.54 | 5.91 | 6.76 |
| | 40 | 1.18 | 1.58 | 2.82 | 4.10 | 4.52 | 5.39 | 5.74 | 6.51 |
| | 60 | 1.19 | 1.58 | 2.81 | 4.04 | 4.44 | 5.25 | 5.57 | 6.27 |
| | 120 | 1.20 | 1.59 | 2.80 | 3.99 | 4.36 | 5.12 | 5.41 | 6.04 |
| | ∞ | 1.21 | 1.60 | 2.79 | 3.93 | 4.29 | 4.99 | 5.26 | 5.82 |
| 10 | 1 | .949 | 1.37 | 4.49 | 24.5 | 49.1 | 246. | 491. | 2,455. |
| | 2 | 1.08 | 1.50 | 3.65 | 9.73 | 14.0 | 31.7 | 44.9 | 101. |
| | 3 | 1.16 | 1.57 | 3.42 | 7.29 | 9.46 | 16.7 | 21.2 | 36.4 |
| | 4 | 1.21 | 1.62 | 3.31 | 6.33 | 7.83 | 12.3 | 14.7 | 22.3 |
| | 6 | 1.27 | 1.68 | 3.21 | 5.50 | 6.49 | 9.10 | 10.4 | 14.0 |
| | 8 | 1.31 | 1.71 | 3.17 | 5.13 | 5.92 | 7.86 | 8.78 | 11.2 |
| | 10 | 1.33 | 1.74 | 3.14 | 4.91 | 5.60 | 7.21 | 7.94 | 9.77 |
| | 12 | 1.35 | 1.76 | 3.12 | 4.78 | 5.40 | 6.81 | 7.44 | 8.96 |
| | 16 | 1.38 | 1.78 | 3.10 | 4.61 | 5.15 | 6.35 | 6.86 | 8.06 |
| | 20 | 1.39 | 1.79 | 3.08 | 4.51 | 5.01 | 6.09 | 6.54 | 7.58 |
| | 24 | 1.40 | 1.80 | 3.07 | 4.45 | 4.92 | 5.92 | 6.33 | 7.27 |
| | 30 | 1.41 | 1.81 | 3.06 | 4.38 | 4.82 | 5.76 | 6.14 | 6.98 |
| | 40 | 1.43 | 1.83 | 3.05 | 4.32 | 4.74 | 5.60 | 5.94 | 6.71 |
| | 60 | 1.44 | 1.84 | 3.04 | 4.25 | 4.65 | 5.45 | 5.76 | 6.45 |
| | 120 | 1.45 | 1.85 | 3.03 | 4.19 | 4.56 | 5.30 | 5.59 | 6.21 |
| | ∞ | 1.47 | 1.86 | 3.02 | 4.13 | 4.47 | 5.16 | 5.42 | 5.97 |
| 12 | 1 | 1.04 | 1.49 | 4.77 | 25.9 | 52.0 | 260. | 520. | 2,600. |
| | 2 | 1.20 | 1.63 | 3.87 | 10.3 | 14.8 | 33.4 | 47.3 | 106. |
| | 3 | 1.29 | 1.72 | 3.63 | 7.67 | 9.95 | 17.5 | 22.2 | 38.2 |
| | 4 | 1.35 | 1.77 | 3.52 | 6.65 | 8.21 | 12.8 | 15.4 | 23.4 |
| | 6 | 1.43 | 1.84 | 3.41 | 5.76 | 6.79 | 9.49 | 10.8 | 14.5 |
| | 8 | 1.47 | 1.88 | 3.36 | 5.36 | 6.18 | 8.18 | 9.12 | 11.6 |
| | 10 | 1.50 | 1.91 | 3.33 | 5.13 | 5.83 | 7.49 | 8.23 | 10.1 |
| | 12 | 1.53 | 1.93 | 3.31 | 4.99 | 5.62 | 7.06 | 7.70 | 9.25 |
| | 16 | 1.56 | 1.96 | 3.28 | 4.81 | 5.35 | 6.56 | 7.08 | 8.30 |
| | 20 | 1.58 | 1.98 | 3.27 | 4.70 | 5.20 | 6.29 | 6.74 | 7.79 |
| | 24 | 1.59 | 2.00 | 3.26 | 4.62 | 5.10 | 6.11 | 6.52 | 7.47 |
| | 30 | 1.61 | 2.01 | 3.25 | 4.56 | 5.00 | 5.93 | 6.31 | 7.16 |
| | 40 | 1.63 | 2.02 | 3.24 | 4.50 | 4.90 | 5.76 | 6.11 | 6.87 |
| | 60 | 1.64 | 2.04 | 3.22 | 4.42 | 4.81 | 5.60 | 5.91 | 6.60 |
| | 120 | 1.66 | 2.04 | 3.22 | 4.35 | 4.71 | 5.44 | 5.73 | 6.34 |
| | ∞ | 1.68 | 2.07 | 3.21 | 4.29 | 4.62 | 5.29 | 5.55 | 6.09 |

| $k$ | $\nu$ | $q_{.01}$ | $q_{.05}$ | $q_{.50}$ | $q_{.90}$ | $q_{.95}$ | $q_{.99}$ | $q_{.995}$ | $q_{.999}$ |
|---|---|---|---|---|---|---|---|---|---|
| 16 | 1 | 1.18 | 1.65 | 5.18 | 28.1 | 56.3 | 282. | 564. | 2,818. |
| | 2 | 1.37 | 1.83 | 4.20 | 11.1 | 16.0 | 36.0 | 51.0 | 114. |
| | 3 | 1.48 | 1.93 | 3.94 | 8.25 | 10.7 | 18.8 | 23.8 | 41.0 |
| | 4 | 1.56 | 2.00 | 3.82 | 7.13 | 8.79 | 13.7 | 16.5 | 24.9 |
| | 6 | 1.66 | 2.08 | 3.70 | 6.16 | 7.24 | 10.1 | 11.5 | 15.4 |
| | 8 | 1.72 | 2.14 | 3.65 | 5.72 | 6.57 | 8.66 | 9.64 | 12.2 |
| | 10 | 1.76 | 2.17 | 3.62 | 5.47 | 6.19 | 7.91 | 8.68 | 10.6 |
| | 12 | 1.79 | 2.20 | 3.59 | 5.31 | 5.95 | 7.44 | 8.10 | 9.71 |
| | 16 | 1.83 | 2.24 | 3.57 | 5.11 | 5.66 | 6.90 | 7.43 | 8.68 |
| | 20 | 1.86 | 2.26 | 3.55 | 4.99 | 5.49 | 6.59 | 7.05 | 8.12 |
| | 24 | 1.88 | 2.28 | 3.54 | 4.91 | 5.38 | 6.39 | 6.81 | 7.77 |
| | 30 | 1.90 | 2.30 | 3.53 | 4.83 | 5.27 | 6.20 | 6.58 | 7.44 |
| | 40 | 1.93 | 2.32 | 3.52 | 4.75 | 5.16 | 6.02 | 6.36 | 7.12 |
| | 60 | 1.95 | 2.34 | 3.51 | 4.68 | 5.06 | 5.84 | 6.15 | 6.82 |
| | 120 | 1.98 | 2.37 | 3.49 | 4.60 | 4.95 | 5.66 | 5.94 | 6.54 |
| | $\infty$ | 2.01 | 2.39 | 3.48 | 4.52 | 4.85 | 5.49 | 5.74 | 6.27 |
| 20 | 1 | 1.28 | 1.77 | 5.49 | 29.7 | 59.6 | 298. | 596. | 2,980. |
| | 2 | 1.50 | 1.97 | 4.45 | 11.7 | 16.8 | 38.0 | 53.7 | 120. |
| | 3 | 1.62 | 2.08 | 4.17 | 8.68 | 11.2 | 20.0 | 25.0 | 43.1 |
| | 4 | 1.71 | 2.16 | 4.04 | 7.50 | 9.23 | 14.4 | 17.3 | 26.1 |
| | 6 | 1.82 | 2.26 | 3.92 | 6.47 | 7.59 | 10.5 | 12.0 | 16.1 |
| | 8 | 1.89 | 2.32 | 3.86 | 6.00 | 6.87 | 9.03 | 10.0 | 12.7 |
| | 10 | 1.94 | 2.36 | 3.83 | 5.73 | 6.47 | 8.23 | 9.03 | 11.0 |
| | 12 | 1.98 | 2.40 | 3.81 | 5.55 | 6.21 | 7.73 | 8.41 | 10.1 |
| | 16 | 2.03 | 2.44 | 3.78 | 5.33 | 5.90 | 7.15 | 7.69 | 8.96 |
| | 20 | 2.07 | 2.47 | 3.76 | 5.21 | 5.71 | 6.82 | 7.29 | 8.37 |
| | 24 | 2.09 | 2.49 | 3.75 | 5.12 | 5.59 | 6.61 | 7.03 | 8.00 |
| | 30 | 2.12 | 2.52 | 3.74 | 5.03 | 5.48 | 6.41 | 6.79 | 7.65 |
| | 40 | 2.15 | 2.54 | 3.72 | 4.95 | 5.36 | 6.21 | 6.55 | 7.31 |
| | 60 | 2.18 | 2.57 | 3.71 | 4.86 | 5.24 | 6.02 | 6.32 | 7.00 |
| | 120 | 2.22 | 2.60 | 3.70 | 4.78 | 5.13 | 5.83 | 6.10 | 6.70 |
| | $\infty$ | 2.25 | 2.63 | 3.69 | 4.69 | 5.01 | 5.65 | 5.89 | 6.41 |
| 24 | 1 | 1.36 | 1.87 | 5.73 | 31.0 | 62.1 | 311. | 622. | 3,108. |
| | 2 | 1.59 | 2.07 | 4.65 | 12.2 | 17.5 | 39.5 | 55.9 | 125. |
| | 3 | 1.73 | 2.20 | 4.36 | 9.03 | 11.7 | 20.5 | 26.0 | 44.7 |
| | 4 | 1.82 | 2.28 | 4.22 | 7.79 | 9.58 | 14.9 | 18.0 | 27.1 |
| | 6 | 1.95 | 2.39 | 4.10 | 6.71 | 7.86 | 10.9 | 12.4 | 16.6 |
| | 8 | 2.03 | 2.46 | 4.03 | 6.21 | 7.11 | 9.32 | 10.4 | 13.1 |
| | 10 | 2.09 | 2.51 | 4.00 | 5.93 | 6.69 | 8.48 | 9.30 | 11.4 |
| | 12 | 2.13 | 2.55 | 3.97 | 5.74 | 6.41 | 7.96 | 8.65 | 10.3 |
| | 16 | 2.19 | 2.60 | 3.94 | 5.52 | 6.08 | 7.36 | 7.90 | 9.19 |
| | 20 | 2.23 | 2.63 | 3.92 | 5.38 | 5.89 | 7.01 | 7.48 | 8.57 |
| | 24 | 2.26 | 2.66 | 3.91 | 5.29 | 5.76 | 6.79 | 7.21 | 8.19 |
| | 30 | 2.29 | 2.68 | 3.90 | 5.20 | 5.64 | 6.57 | 6.95 | 7.82 |
| | 40 | 2.32 | 2.71 | 3.89 | 5.11 | 5.51 | 6.36 | 6.70 | 7.47 |
| | 60 | 2.36 | 2.74 | 3.87 | 5.02 | 5.39 | 6.16 | 6.46 | 7.13 |
| | 120 | 2.40 | 2.78 | 3.86 | 4.92 | 5.27 | 5.96 | 6.23 | 6.82 |
| | $\infty$ | 2.45 | 2.81 | 3.85 | 4.83 | 5.14 | 5.77 | 6.01 | 6.52 |

## A-18b PERCENTILES OF THE DISTRIBUTION OF $h = \max |z_i - \mu_i|/s$

The $z_1, \ldots, z_k$ are $k$ independent normal variables with means $\mu_i$ and variances $\sigma^2$, and $s^2$ is an independent estimate of $\sigma^2$ having a $\chi^2/df$ distribution with $\nu$ degrees of freedom.

| | $k = 1$ ($t$ DISTRIBUTION) | | | | $k = 2$ | | | | $k = 3$ | | | |
|---|---|---|---|---|---|---|---|---|---|---|---|---|
| $\nu$ | $h_{.90}$ | $h_{.95}$ | $h_{.975}$ | $h_{.99}$ | $h_{.90}$ | $h_{.95}$ | $h_{.975}$ | $h_{.99}$ | $h_{.90}$ | $h_{.95}$ | $h_{.975}$ | $h_{.99}$ |
| 6 | 1.94 | 2.45 | 2.97 | 3.71 | 2.39 | 2.92 | 3.47 | 4.27 | 2.64 | 3.19 | 3.77 | 4.61 |
| 8 | 1.86 | 2.31 | 2.75 | 3.36 | 2.26 | 2.72 | 3.18 | 3.81 | 2.49 | 2.96 | 3.43 | 4.08 |
| 10 | 1.81 | 2.23 | 2.63 | 3.17 | 2.19 | 2.61 | 3.02 | 3.57 | 2.41 | 2.83 | 3.24 | 3.80 |
| 12 | 1.78 | 2.18 | 2.56 | 3.05 | 2.15 | 2.54 | 2.92 | 3.42 | 2.36 | 2.75 | 3.13 | 3.63 |
| 16 | 1.78 | 2.12 | 2.47 | 2.92 | 2.10 | 2.46 | 2.80 | 3.24 | 2.29 | 2.65 | 2.99 | 3.43 |
| 20 | 1.72 | 2.09 | 2.42 | 2.85 | 2.07 | 2.41 | 2.74 | 3.15 | 2.26 | 2.59 | 2.92 | 3.32 |
| 24 | 1.71 | 2.06 | 2.39 | 2.80 | 2.04 | 2.38 | 2.69 | 3.09 | 2.23 | 2.56 | 2.87 | 3.25 |
| 30 | 1.70 | 2.04 | 2.36 | 2.75 | 2.02 | 2.35 | 2.65 | 3.03 | 2.21 | 2.52 | 2.82 | 3.18 |
| 45 | 1.68 | 2.01 | 2.32 | 2.69 | 2.00 | 2.31 | 2.60 | 2.95 | 2.17 | 2.48 | 2.75 | 3.10 |
| $\infty$ | 1.65 | 1.96 | 2.24 | 2.58 | 1.95 | 2.24 | 2.50 | 2.81 | 2.11 | 2.39 | 2.64 | 2.93 |

| | $k = 4$ | | | | $k = 5$ | | | | $k = 6$ | | | |
|---|---|---|---|---|---|---|---|---|---|---|---|---|
| $\nu$ | $h_{.90}$ | $h_{.95}$ | $h_{.975}$ | $h_{.99}$ | $h_{.90}$ | $h_{.95}$ | $h_{.975}$ | $h_{.99}$ | $h_{.90}$ | $h_{.95}$ | $h_{.975}$ | $h_{.99}$ |
| 6 | 2.82 | 3.39 | 3.99 | 4.86 | 2.96 | 3.54 | 4.16 | 5.05 | 3.07 | 3.66 | 4.29 | 5.20 |
| 8 | 2.66 | 3.13 | 3.61 | 4.27 | 2.78 | 3.26 | 3.75 | 4.42 | 2.88 | 3.36 | 3.86 | 4.55 |
| 10 | 2.56 | 2.98 | 3.40 | 3.97 | 2.68 | 3.10 | 3.53 | 4.10 | 2.77 | 3.20 | 3.63 | 4.20 |
| 12 | 2.50 | 2.89 | 3.28 | 3.78 | 2.61 | 3.00 | 3.39 | 3.90 | 2.70 | 3.10 | 3.48 | 3.99 |
| 16 | 2.43 | 2.78 | 3.13 | 3.57 | 2.53 | 2.89 | 3.23 | 3.67 | 2.62 | 2.97 | 3.31 | 3.75 |
| 20 | 2.39 | 2.72 | 3.04 | 3.45 | 2.49 | 2.82 | 3.14 | 3.54 | 2.57 | 2.90 | 3.21 | 3.62 |
| 24 | 2.36 | 2.68 | 2.98 | 3.37 | 2.46 | 2.77 | 3.08 | 3.46 | 2.53 | 2.85 | 3.15 | 3.53 |
| 30 | 2.33 | 2.64 | 2.93 | 3.29 | 2.43 | 2.73 | 3.02 | 3.38 | 2.50 | 2.81 | 3.09 | 3.45 |
| 45 | 2.30 | 2.59 | 2.86 | 3.20 | 2.39 | 2.68 | 2.94 | 3.28 | 2.46 | 2.75 | 3.01 | 3.34 |
| $\infty$ | 2.23 | 2.49 | 2.73 | 3.02 | 2.31 | 2.57 | 2.80 | 3.09 | 2.38 | 2.63 | 2.86 | 3.15 |

| | $k = 7$ | | | | $k = 8$ | | | | $k = 9$ | | | |
|---|---|---|---|---|---|---|---|---|---|---|---|---|
| $\nu$ | $h_{.90}$ | $h_{.95}$ | $h_{.975}$ | $h_{.99}$ | $h_{.90}$ | $h_{.95}$ | $h_{.975}$ | $h_{.99}$ | $h_{.90}$ | $h_{.95}$ | $h_{.975}$ | $h_{.99}$ |
| 6 | 3.17 | 3.77 | 4.41 | 5.33 | 3.25 | 3.86 | 4.51 | 5.45 | 3.32 | 3.94 | 4.59 | 5.55 |
| 8 | 2.96 | 3.45 | 3.95 | 4.65 | 3.04 | 3.53 | 4.04 | 4.74 | 3.10 | 3.60 | 4.11 | 4.82 |
| 10 | 2.85 | 3.28 | 3.71 | 4.29 | 2.92 | 3.35 | 3.79 | 4.37 | 2.98 | 3.41 | 3.85 | 4.44 |
| 12 | 2.78 | 3.17 | 3.56 | 4.08 | 2.84 | 3.24 | 3.63 | 4.15 | 2.90 | 3.29 | 3.69 | 4.21 |
| 16 | 2.69 | 3.04 | 3.38 | 3.82 | 2.75 | 3.10 | 3.44 | 3.88 | 2.80 | 3.15 | 3.49 | 3.94 |
| 20 | 2.63 | 2.96 | 3.28 | 3.68 | 2.69 | 3.02 | 3.33 | 3.74 | 2.74 | 3.07 | 3.38 | 3.79 |
| 24 | 2.60 | 2.91 | 3.21 | 3.56 | 2.66 | 2.97 | 3.27 | 3.64 | 2.70 | 3.02 | 3.31 | 3.69 |
| 30 | 2.57 | 2.87 | 3.15 | 3.50 | 2.62 | 2.92 | 3.20 | 3.55 | 2.67 | 2.96 | 3.24 | 3.60 |
| 45 | 2.52 | 2.80 | 3.07 | 3.39 | 2.57 | 2.85 | 3.11 | 3.44 | 2.62 | 2.89 | 3.16 | 3.48 |
| $\infty$ | 2.43 | 2.68 | 2.91 | 3.19 | 2.48 | 2.73 | 2.95 | 3.23 | 2.52 | 2.77 | 2.99 | 3.27 |

| | $k = 10$ | | | | $k = 11$ | | | | $k = 12$ | | | |
|---|---|---|---|---|---|---|---|---|---|---|---|---|
| $\nu$ | $h_{.90}$ | $h_{.95}$ | $h_{.975}$ | $h_{.99}$ | $h_{.90}$ | $h_{.95}$ | $h_{.975}$ | $h_{.99}$ | $h_{.90}$ | $h_{.95}$ | $h_{.975}$ | $h_{.99}$ |
| 6 | 3.38 | 4.01 | 4.67 | 5.64 | 3.44 | 4.07 | 4.74 | 5.72 | 3.49 | 4.13 | 4.81 | 5.79 |
| 8 | 3.16 | 3.66 | 4.18 | 4.89 | 3.21 | 3.71 | 4.23 | 4.96 | 3.26 | 3.76 | 4.29 | 5.02 |
| 10 | 3.03 | 3.47 | 3.91 | 4.50 | 3.08 | 3.52 | 3.96 | 4.56 | 3.12 | 3.56 | 4.01 | 4.61 |
| 12 | 2.95 | 3.35 | 3.74 | 4.26 | 2.99 | 3.39 | 3.79 | 4.31 | 3.03 | 3.43 | 3.83 | 4.36 |
| 16 | 2.85 | 3.20 | 3.54 | 3.99 | 2.89 | 3.24 | 3.58 | 4.03 | 2.93 | 3.28 | 3.62 | 4.07 |
| 20 | 2.79 | 3.11 | 3.43 | 3.83 | 2.83 | 3.15 | 3.47 | 3.87 | 2.86 | 3.19 | 3.50 | 3.91 |
| 24 | 2.75 | 3.06 | 3.36 | 3.73 | 2.79 | 3.10 | 3.39 | 3.77 | 2.82 | 3.13 | 3.43 | 3.80 |
| 30 | 2.71 | 3.01 | 3.28 | 3.64 | 2.75 | 3.04 | 3.32 | 3.67 | 2.78 | 3.08 | 3.35 | 3.70 |
| 45 | 2.66 | 2.93 | 3.20 | 3.52 | 2.69 | 2.97 | 3.23 | 3.55 | 2.73 | 3.00 | 3.26 | 3.58 |
| $\infty$ | 2.56 | 2.80 | 3.02 | 3.29 | 2.59 | 2.83 | 3.05 | 3.31 | 2.62 | 2.86 | 3.07 | 3.33 |

The percentiles listed cover the range $\alpha = .005$ to $.125$ for every sample size up to $N = 20$. Values $T_\alpha$ are such that the probability is $\alpha$ that the signed rank statistic is less than or equal to $T_\alpha$. The values $T_{1-\alpha}$ are such that the probability is $\alpha$ that $T$ is greater than or equal to $T_{1-\alpha}$.

| $T_\alpha$ | $T_{1-\alpha}$ | $\alpha$ |
|---|---|---|
| | $N = 1$ | |
| 0 | 1 | .500 |
| | $N = 2$ | |
| 0 | 3 | .250 |
| | $N = 3$ | |
| 0 | 6 | .125 |
| | $N = 4$ | |
| 0 | 10 | .062 |
| 1 | 9 | .125 |
| | $N = 5$ | |
| 0 | 15 | .031 |
| 1 | 14 | .062 |
| 2 | 13 | .094 |
| 3 | 12 | .156 |
| | $N = 6$ | |
| 0 | 21 | .016 |
| 1 | 20 | .031 |
| 2 | 19 | .047 |
| 3 | 18 | .078 |
| 4 | 17 | .109 |
| 5 | 16 | .156 |
| | $N = 7$ | |
| 0 | 28 | .008 |
| 1 | 27 | .016 |
| 2 | 26 | .023 |
| 3 | 25 | .039 |
| 4 | 24 | .055 |
| 5 | 23 | .078 |
| 6 | 22 | .109 |
| 7 | 21 | .148 |
| | $N = 8$ | |
| 0 | 36 | .004 |
| 1 | 35 | .008 |
| 2 | 34 | .012 |
| 3 | 33 | .020 |
| 4 | 32 | .027 |
| 5 | 31 | .039 |
| 6 | 30 | .055 |
| 7 | 29 | .074 |
| 8 | 28 | .098 |
| 9 | 27 | .125 |
| | $N = 9$ | |
| 1 | 44 | .004 |
| 2 | 43 | .006 |
| 3 | 42 | .010 |
| | $N = 9$ (*Cont.*) | |
| 4 | 41 | .014 |
| 5 | 40 | .020 |
| 6 | 39 | .027 |
| 7 | 38 | .037 |
| 8 | 37 | .049 |
| 9 | 36 | .064 |
| 10 | 35 | .082 |
| 11 | 34 | .102 |
| 12 | 33 | .125 |
| | $N = 10$ | |
| 3 | 52 | .005 |
| 4 | 51 | .007 |
| 5 | 50 | .010 |
| 6 | 49 | .014 |
| 7 | 48 | .019 |
| 8 | 47 | .024 |
| 9 | 46 | .032 |
| 10 | 45 | .042 |
| 11 | 44 | .053 |
| 12 | 43 | .065 |
| 13 | 42 | .080 |
| 14 | 41 | .097 |
| 15 | 40 | .116 |
| 16 | 39 | .138 |
| | $N = 11$ | |
| 5 | 61 | .005 |
| 6 | 60 | .007 |
| 7 | 59 | .009 |
| 8 | 58 | .012 |
| 9 | 57 | .016 |
| 10 | 56 | .021 |
| 11 | 55 | .027 |
| 12 | 54 | .034 |
| 13 | 53 | .042 |
| 14 | 52 | .051 |
| 15 | 51 | .062 |
| 16 | 50 | .074 |
| 17 | 49 | .087 |
| 18 | 48 | .103 |
| 19 | 47 | .120 |
| 20 | 46 | .139 |
| | $N = 12$ | |
| 7 | 71 | .005 |
| 8 | 70 | .006 |
| | $N = 12$ (*Cont.*) | |
| 9 | 69 | .008 |
| 10 | 68 | .010 |
| 11 | 67 | .013 |
| 12 | 66 | .017 |
| 13 | 65 | .021 |
| 14 | 64 | .026 |
| 15 | 63 | .032 |
| 16 | 62 | .039 |
| 17 | 61 | .046 |
| 18 | 60 | .055 |
| 19 | 59 | .065 |
| 20 | 58 | .076 |
| 21 | 57 | .088 |
| 22 | 56 | .102 |
| 23 | 55 | .117 |
| 24 | 54 | .133 |
| | $N = 13$ | |
| 9 | 82 | .004 |
| 10 | 81 | .005 |
| 11 | 80 | .007 |
| 12 | 79 | .009 |
| 13 | 78 | .011 |
| 14 | 77 | .013 |
| 15 | 76 | .016 |
| 16 | 75 | .020 |
| 17 | 74 | .024 |
| 18 | 73 | .029 |
| 19 | 72 | .034 |
| 20 | 71 | .040 |
| 21 | 70 | .047 |
| 22 | 69 | .055 |
| 23 | 68 | .064 |
| 24 | 67 | .073 |
| 25 | 66 | .084 |
| 26 | 65 | .095 |
| 27 | 64 | .108 |
| 28 | 63 | .122 |
| 29 | 62 | .137 |
| | $N = 14$ | |
| 12 | 93 | .004 |
| 13 | 92 | .005 |
| 14 | 91 | .007 |
| 15 | 90 | .008 |
| 16 | 89 | .010 |
| | $N = 14$ (*Cont.*) | |
| 17 | 88 | .012 |
| 18 | 87 | .015 |
| 19 | 86 | .018 |
| 20 | 85 | .021 |
| 21 | 84 | .025 |
| 22 | 83 | .029 |
| 23 | 82 | .034 |
| 24 | 81 | .039 |
| 25 | 80 | .045 |
| 26 | 79 | .052 |
| 27 | 78 | .059 |
| 28 | 77 | .068 |
| 29 | 76 | .077 |
| 30 | 75 | .086 |
| 31 | 74 | .097 |
| 32 | 73 | .108 |
| 33 | 72 | .121 |
| 34 | 71 | .134 |
| | $N = 15$ | |
| 15 | 105 | .004 |
| 16 | 104 | .005 |
| 17 | 103 | .006 |
| 18 | 102 | .008 |
| 19 | 101 | .009 |
| 20 | 100 | .011 |
| 21 | 99 | .013 |
| 22 | 98 | .015 |
| 23 | 97 | .018 |
| 24 | 96 | .021 |
| 25 | 95 | .024 |
| 26 | 94 | .028 |
| 27 | 93 | .032 |
| 28 | 92 | .036 |
| 29 | 91 | .042 |
| 30 | 90 | .047 |
| 31 | 89 | .053 |
| 32 | 88 | .060 |
| 33 | 87 | .068 |
| 34 | 86 | .076 |
| 35 | 85 | .084 |
| 36 | 84 | .094 |
| 37 | 83 | .104 |
| 38 | 82 | .115 |
| 39 | 81 | .126 |

| $T_\alpha$ | $T_{1-\alpha}$ | $\alpha$ | $T_\alpha$ | $T_{1-\alpha}$ | $\alpha$ | $T_\alpha$ | $T_{1-\alpha}$ | $\alpha$ | $T_\alpha$ | $T_{1-\alpha}$ | $\alpha$ |
|---|---|---|---|---|---|---|---|---|---|---|---|
| | $N = 16$ | | | $N = 17$ (*Cont.*) | | | $N = 18$ (*Cont.*) | | | $N = 19$ (*Cont.*) | |
| 19 | 117 | .005 | 36 | 117 | .028 | 51 | 120 | .071 | 64 | 126 | .113 |
| 20 | 116 | .005 | 37 | 116 | .032 | 52 | 119 | .077 | 65 | 125 | .121 |
| 21 | 115 | .007 | 38 | 115 | .036 | 53 | 118 | .084 | 66 | 124 | .129 |
| 22 | 114 | .008 | 39 | 114 | .040 | 54 | 117 | .091 | | $N = 20$ | |
| 23 | 113 | .009 | 40 | 113 | .044 | 55 | 116 | .098 | 37 | 173 | .005 |
| 24 | 112 | .011 | 41 | 112 | .049 | 56 | 115 | .106 | 38 | 172 | .005 |
| 25 | 111 | .012 | 42 | 111 | .054 | 57 | 114 | .114 | 39 | 171 | .006 |
| 26 | 110 | .014 | 43 | 110 | .060 | 58 | 113 | .123 | 40 | 170 | .007 |
| 27 | 109 | .017 | 44 | 109 | .066 | 59 | 112 | .132 | 41 | 169 | .008 |
| 28 | 108 | .019 | 45 | 108 | .073 | | $N = 19$ | | 42 | 168 | .009 |
| 29 | 107 | .022 | 46 | 107 | .080 | 32 | 158 | .005 | 43 | 167 | .010 |
| 30 | 106 | .025 | 47 | 106 | .087 | 33 | 157 | .005 | 44 | 166 | .011 |
| 31 | 105 | .029 | 48 | 105 | .095 | 34 | 156 | .006 | 45 | 165 | .012 |
| 32 | 104 | .033 | 49 | 104 | .103 | 35 | 155 | .007 | 46 | 164 | .013 |
| 33 | 103 | .037 | 50 | 103 | .112 | 36 | 154 | .008 | 47 | 163 | .015 |
| 34 | 102 | .042 | 51 | 102 | .122 | 37 | 153 | .009 | 48 | 162 | .016 |
| 35 | 101 | .047 | 52 | 101 | .132 | 38 | 152 | .010 | 49 | 161 | .018 |
| 36 | 100 | .052 | | $N = 18$ | | 39 | 151 | .011 | 50 | 160 | .020 |
| 37 | 99 | .058 | 27 | 144 | .004 | 40 | 150 | .013 | 51 | 159 | .022 |
| 38 | 98 | .065 | 28 | 143 | .005 | 41 | 149 | .014 | 52 | 158 | .024 |
| 39 | 97 | .072 | 29 | 142 | .006 | 42 | 148 | .016 | 53 | 157 | .027 |
| 40 | 96 | .080 | 30 | 141 | .007 | 43 | 147 | .018 | 54 | 156 | .029 |
| 41 | 95 | .088 | 31 | 140 | .008 | 44 | 146 | .020 | 55 | 155 | .032 |
| 42 | 94 | .096 | 32 | 139 | .009 | 45 | 145 | .022 | 56 | 154 | .035 |
| 43 | 93 | .106 | 33 | 138 | .010 | 46 | 144 | .025 | 57 | 153 | .038 |
| 44 | 92 | .116 | 34 | 137 | .012 | 47 | 143 | .027 | 58 | 152 | .041 |
| 45 | 91 | .126 | 35 | 136 | .013 | 48 | 142 | .030 | 59 | 151 | .045 |
| 46 | 90 | .137 | 36 | 135 | .015 | 49 | 141 | .033 | 60 | 150 | .049 |
| | $N = 17$ | | 37 | 134 | .017 | 50 | 140 | .036 | 61 | 149 | .053 |
| 23 | 130 | .005 | 38 | 133 | .019 | 51 | 139 | .040 | 62 | 148 | .057 |
| 24 | 129 | .005 | 39 | 132 | .022 | 52 | 138 | .044 | 63 | 147 | .062 |
| 25 | 128 | .006 | 40 | 131 | .024 | 53 | 137 | .048 | 64 | 146 | .066 |
| 26 | 127 | .007 | 41 | 130 | .027 | 54 | 136 | .052 | 65 | 145 | .071 |
| 27 | 126 | .009 | 42 | 129 | .030 | 55 | 135 | .057 | 66 | 144 | .077 |
| 28 | 125 | .010 | 43 | 128 | .033 | 56 | 134 | .062 | 67 | 143 | .082 |
| 29 | 124 | .012 | 44 | 127 | .037 | 57 | 133 | .067 | 68 | 142 | .088 |
| 30 | 123 | .013 | 45 | 126 | .041 | 58 | 132 | .072 | 69 | 141 | .095 |
| 31 | 122 | .015 | 46 | 125 | .045 | 59 | 131 | .078 | 70 | 140 | .101 |
| 32 | 121 | .017 | 47 | 124 | .049 | 60 | 130 | .084 | 71 | 139 | .108 |
| 33 | 120 | .020 | 48 | 123 | .054 | 61 | 129 | .091 | 72 | 138 | .115 |
| 34 | 119 | .022 | 49 | 122 | .059 | 62 | 128 | .098 | 73 | 137 | .123 |
| 35 | 118 | .025 | 50 | 121 | .065 | 63 | 127 | .105 | 74 | 136 | .131 |

The values of $T'_\alpha$, $T'_{1-\alpha}$, and $\alpha$ are such that if the $N_1$ and $N_2$ observations are chosen at random from the same population the chance that the rank sum $T'$ of the $N_1$ observations in the smaller sample is equal to or less than $T'_\alpha$ is $\alpha$ and the chance that $T'$ is equal to or greater than $T'_{1-\alpha}$ is $\alpha$. The sample sizes are shown in parentheses $(N_1,N_2)$.

| $T'_\alpha$ | $T'_{1-\alpha}$ | $\alpha$ |
|---|---|---|
| | (1,1) | |
| 1 | 2 | .500 |
| | (1,2) | |
| 1 | 3 | .333 |
| 2 | 2 | .667 |
| | (1,3) | |
| 1 | 4 | .250 |
| 2 | 3 | .500 |
| | (1,4) | |
| 1 | 5 | .200 |
| 2 | 4 | .400 |
| 3 | 3 | .600 |
| | (1,5) | |
| 1 | 6 | .167 |
| 2 | 5 | .333 |
| 3 | 4 | .500 |
| | (1,6) | |
| 1 | 7 | .143 |
| 2 | 6 | .286 |
| 3 | 5 | .428 |
| 4 | 4 | .571 |
| | (1,7) | |
| 1 | 8 | .125 |
| 2 | 7 | .250 |
| 3 | 6 | .375 |
| 4 | 5 | .500 |
| | (1,8) | |
| 1 | 9 | .111 |
| 2 | 8 | .222 |
| 3 | 7 | .333 |
| 4 | 6 | .444 |
| 5 | 5 | .556 |
| | (1,9) | |
| 1 | 10 | .100 |
| 2 | 9 | .200 |
| 3 | 8 | .300 |
| 4 | 7 | .400 |
| 5 | 6 | .500 |
| | (1,10) | |
| 1 | 11 | .091 |
| 2 | 10 | .182 |
| 3 | 9 | .273 |
| 4 | 8 | .364 |
| 5 | 7 | .455 |
| 6 | 6 | .545 |

| $T'_\alpha$ | $T'_{1-\alpha}$ | $\alpha$ |
|---|---|---|
| | (2,2) | |
| 3 | 7 | .167 |
| 4 | 6 | .333 |
| 5 | 5 | .667 |
| | (2,3) | |
| 3 | 9 | .100 |
| 4 | 8 | .200 |
| 5 | 7 | .400 |
| 6 | 6 | .600 |
| | (2,4) | |
| 3 | 11 | .067 |
| 4 | 10 | .133 |
| 5 | 9 | .267 |
| 6 | 8 | .400 |
| 7 | 7 | .600 |
| | (2,5) | |
| 3 | 13 | .047 |
| 4 | 12 | .095 |
| 5 | 11 | .190 |
| 6 | 10 | .286 |
| 7 | 9 | .429 |
| 8 | 8 | .571 |
| | (2,6) | |
| 3 | 15 | .036 |
| 4 | 14 | .071 |
| 5 | 13 | .143 |
| 6 | 12 | .214 |
| 7 | 11 | .321 |
| 8 | 10 | .429 |
| 9 | 9 | .571 |
| | (2,7) | |
| 3 | 17 | .028 |
| 4 | 16 | .056 |
| 5 | 15 | .111 |
| 6 | 14 | .167 |
| 7 | 13 | .250 |
| 8 | 12 | .333 |
| 9 | 11 | .444 |
| 10 | 10 | .556 |
| | (2,8) | |
| 3 | 19 | .022 |
| 4 | 18 | .044 |
| 5 | 17 | .089 |
| 6 | 16 | .133 |
| 7 | 15 | .200 |

| $T'_\alpha$ | $T'_{1-\alpha}$ | $\alpha$ |
|---|---|---|
| | (2,8) (*Cont.*) | |
| 8 | 14 | .267 |
| 9 | 13 | .356 |
| 10 | 12 | .444 |
| 11 | 11 | .556 |
| | (2,9) | |
| 3 | 21 | .018 |
| 4 | 20 | .036 |
| 5 | 19 | .073 |
| 6 | 18 | .109 |
| 7 | 17 | .164 |
| 8 | 16 | .218 |
| 9 | 15 | .291 |
| 10 | 14 | .364 |
| 11 | 13 | .455 |
| 12 | 12 | .545 |
| | (2,10) | |
| 3 | 23 | .015 |
| 4 | 22 | .030 |
| 5 | 21 | .061 |
| 6 | 20 | .091 |
| 7 | 19 | .136 |
| 8 | 18 | .182 |
| 9 | 17 | .242 |
| 10 | 16 | .303 |
| 11 | 15 | .379 |
| 12 | 14 | .455 |
| 13 | 13 | .545 |
| | (3,3) | |
| 6 | 15 | .050 |
| 7 | 14 | .100 |
| 8 | 13 | .200 |
| 9 | 12 | .350 |
| 10 | 11 | .500 |
| | (3,4) | |
| 6 | 18 | .028 |
| 7 | 17 | .057 |
| 8 | 16 | .114 |
| 9 | 15 | .200 |
| 10 | 14 | .314 |
| 11 | 13 | .429 |
| 12 | 12 | .571 |
| | (3,5) | |
| 6 | 21 | .018 |
| 7 | 20 | .036 |

| $T'_\alpha$ | $T'_{1-\alpha}$ | $\alpha$ |
|---|---|---|
| | (3,5) (*Cont.*) | |
| 8 | 19 | .071 |
| 9 | 18 | .125 |
| 10 | 17 | .196 |
| 11 | 16 | .286 |
| 12 | 15 | .393 |
| 13 | 14 | .500 |
| | (3,6) | |
| 6 | 24 | .012 |
| 7 | 23 | .024 |
| 8 | 22 | .048 |
| 9 | 21 | .083 |
| 10 | 20 | .131 |
| 11 | 19 | .190 |
| 12 | 18 | .274 |
| 13 | 17 | .357 |
| 14 | 16 | .452 |
| 15 | 15 | .548 |
| | (3,7) | |
| 6 | 27 | .008 |
| 7 | 26 | .017 |
| 8 | 25 | .033 |
| 9 | 24 | .058 |
| 10 | 23 | .092 |
| 11 | 22 | .133 |
| 12 | 21 | .192 |
| 13 | 20 | .258 |
| 14 | 19 | .333 |
| 15 | 18 | .417 |
| 16 | 17 | .500 |
| | (3,8) | |
| 6 | 30 | .006 |
| 7 | 29 | .012 |
| 8 | 28 | .024 |
| 9 | 27 | .042 |
| 10 | 26 | .067 |
| 11 | 25 | .097 |
| 12 | 24 | .139 |
| 13 | 23 | .188 |
| 14 | 22 | .248 |
| 15 | 21 | .315 |
| 16 | 20 | .387 |
| 17 | 19 | .461 |
| 18 | 18 | .539 |

| $T'_\alpha$ | $T'_{1-\alpha}$ | $\alpha$ |
|---|---|---|
| | (3,9) | |
| 6 | 33 | .005 |
| 7 | 32 | .009 |
| 8 | 31 | .018 |
| 9 | 30 | .032 |
| 10 | 29 | .050 |
| 11 | 28 | .073 |
| 12 | 27 | .105 |
| 13 | 26 | .141 |
| 14 | 25 | .186 |
| 15 | 24 | .241 |
| 16 | 23 | .300 |
| 17 | 22 | .363 |
| 18 | 21 | .432 |
| 19 | 20 | .500 |
| | (3,10) | |
| 6 | 36 | .003 |
| 7 | 35 | .007 |
| 8 | 34 | .014 |
| 9 | 33 | .024 |
| 10 | 32 | .038 |
| 11 | 31 | .056 |
| 12 | 30 | .080 |
| 13 | 29 | .108 |
| 14 | 28 | .143 |
| 15 | 27 | .185 |
| 16 | 26 | .234 |
| 17 | 25 | .287 |
| 18 | 24 | .346 |
| 19 | 23 | .406 |
| 20 | 22 | .469 |
| 21 | 21 | .531 |
| | (4,4) | |
| 10 | 26 | .014 |
| 11 | 25 | .029 |
| 12 | 24 | .057 |
| 13 | 23 | .100 |
| 14 | 22 | .171 |
| 15 | 21 | .243 |
| 16 | 20 | .343 |
| 17 | 19 | .443 |
| 18 | 18 | .557 |
| | (4,5) | |
| 10 | 30 | .008 |
| 11 | 29 | .016 |
| 12 | 28 | .032 |
| 13 | 27 | .056 |
| 14 | 26 | .095 |
| 15 | 25 | .143 |
| 16 | 24 | .206 |

| $T'_\alpha$ | $T'_{1-\alpha}$ | $\alpha$ |
|---|---|---|
| | (4,5) (*Cont.*) | |
| 17 | 23 | .278 |
| 18 | 22 | .365 |
| 19 | 21 | .452 |
| 20 | 20 | .548 |
| | (4,6) | |
| 10 | 34 | .005 |
| 11 | 33 | .010 |
| 12 | 32 | .019 |
| 13 | 31 | .033 |
| 14 | 30 | .057 |
| 15 | 29 | .086 |
| 16 | 28 | .129 |
| 17 | 27 | .176 |
| 18 | 26 | .238 |
| 19 | 25 | .305 |
| 20 | 24 | .381 |
| 21 | 23 | .457 |
| 22 | 22 | .545 |
| | (4,7) | |
| 10 | 38 | .003 |
| 11 | 37 | .006 |
| 12 | 36 | .012 |
| 13 | 35 | .021 |
| 14 | 34 | .036 |
| 15 | 33 | .055 |
| 16 | 32 | .082 |
| 17 | 31 | .115 |
| 18 | 30 | .158 |
| 19 | 29 | .206 |
| 20 | 28 | .264 |
| 21 | 27 | .324 |
| 22 | 26 | .394 |
| 23 | 25 | .464 |
| 24 | 24 | .538 |
| | (4,8) | |
| 10 | 42 | .002 |
| 11 | 41 | .004 |
| 12 | 40 | .008 |
| 13 | 39 | .014 |
| 14 | 38 | .024 |
| 15 | 37 | .036 |
| 16 | 36 | .055 |
| 17 | 35 | .077 |
| 18 | 34 | .107 |
| 19 | 33 | .141 |
| 20 | 32 | .184 |
| 21 | 31 | .230 |
| 22 | 30 | .285 |
| 23 | 29 | .341 |

| $T'_\alpha$ | $T'_{1-\alpha}$ | $\alpha$ |
|---|---|---|
| | (4,8) (*Cont.*) | |
| 24 | 28 | .404 |
| 25 | 27 | .467 |
| 26 | 26 | .533 |
| | (4,9) | |
| 10 | 46 | .001 |
| 11 | 45 | .003 |
| 12 | 44 | .006 |
| 13 | 43 | .010 |
| 14 | 42 | .017 |
| 15 | 41 | .025 |
| 16 | 40 | .038 |
| 17 | 39 | .053 |
| 18 | 38 | .074 |
| 19 | 37 | .099 |
| 20 | 36 | .130 |
| 21 | 35 | .165 |
| 22 | 34 | .207 |
| 23 | 33 | .252 |
| 24 | 32 | .302 |
| 25 | 31 | .355 |
| 26 | 30 | .413 |
| 27 | 29 | .470 |
| 28 | 28 | .530 |
| | (4,10) | |
| 10 | 50 | .001 |
| 11 | 49 | .002 |
| 12 | 48 | .004 |
| 13 | 47 | .007 |
| 14 | 46 | .012 |
| 15 | 45 | .018 |
| 16 | 44 | .026 |
| 17 | 43 | .038 |
| 18 | 42 | .053 |
| 19 | 41 | .071 |
| 20 | 40 | .094 |
| 21 | 39 | .120 |
| 22 | 38 | .152 |
| 23 | 37 | .187 |
| 24 | 36 | .227 |
| 25 | 35 | .270 |
| 26 | 34 | .318 |
| 27 | 33 | .367 |
| 28 | 32 | .420 |
| 29 | 31 | .473 |
| 30 | 30 | .527 |
| | (5,5) | |
| 15 | 40 | .004 |
| 16 | 39 | .008 |
| 17 | 38 | .016 |

| $T'_\alpha$ | $T'_{1-\alpha}$ | $\alpha$ |
|---|---|---|
| | (5,5) (*Cont.*) | |
| 18 | 37 | .028 |
| 19 | 36 | .048 |
| 20 | 35 | .075 |
| 21 | 34 | .111 |
| 22 | 33 | .155 |
| 23 | 32 | .210 |
| 24 | 31 | .274 |
| 25 | 30 | .345 |
| 26 | 29 | .421 |
| 27 | 28 | .500 |
| | (5,6) | |
| 15 | 45 | .002 |
| 16 | 44 | .004 |
| 17 | 43 | .009 |
| 18 | 42 | .015 |
| 19 | 41 | .026 |
| 20 | 40 | .041 |
| 21 | 39 | .063 |
| 22 | 38 | .089 |
| 23 | 37 | .123 |
| 24 | 36 | .165 |
| 25 | 35 | .214 |
| 26 | 34 | .268 |
| 27 | 33 | .331 |
| 28 | 32 | .396 |
| 29 | 31 | .465 |
| 30 | 30 | .535 |
| | (5,7) | |
| 15 | 50 | .001 |
| 16 | 49 | .003 |
| 17 | 48 | .005 |
| 18 | 47 | .009 |
| 19 | 46 | .015 |
| 20 | 45 | .024 |
| 21 | 44 | .037 |
| 22 | 43 | .053 |
| 23 | 42 | .074 |
| 24 | 41 | .101 |
| 25 | 40 | .134 |
| 26 | 39 | .172 |
| 27 | 38 | .216 |
| 28 | 37 | .265 |
| 29 | 36 | .319 |
| 30 | 35 | .378 |
| 31 | 34 | .438 |
| 32 | 33 | .500 |
| | (5,8) | |
| 15 | 55 | .001 |
| 16 | 54 | .002 |

(5,8) *(Cont.)*

| $T'_\alpha$ | $T'_{1-\alpha}$ | $\alpha$ |
|---|---|---|
| 17 | 53 | .003 |
| 18 | 52 | .005 |
| 19 | 51 | .009 |
| 20 | 50 | .015 |
| 21 | 49 | .023 |
| 22 | 48 | .033 |
| 23 | 47 | .047 |
| 24 | 46 | .064 |
| 25 | 45 | .085 |
| 26 | 44 | .111 |
| 27 | 43 | .142 |
| 28 | 42 | .177 |
| 29 | 41 | .217 |
| 30 | 40 | .262 |
| 31 | 39 | .311 |
| 32 | 38 | .362 |
| 33 | 37 | .416 |
| 34 | 36 | .472 |
| 35 | 35 | .528 |

(5,9)

| $T'_\alpha$ | $T'_{1-\alpha}$ | $\alpha$ |
|---|---|---|
| 15 | 60 | .000 |
| 16 | 59 | .001 |
| 17 | 58 | .002 |
| 18 | 57 | .003 |
| 19 | 56 | .006 |
| 20 | 55 | .009 |
| 21 | 54 | .014 |
| 22 | 53 | .021 |
| 23 | 52 | .030 |
| 24 | 51 | .041 |
| 25 | 50 | .056 |
| 26 | 49 | .073 |
| 27 | 48 | .095 |
| 28 | 47 | .120 |
| 29 | 46 | .149 |
| 30 | 45 | .182 |
| 31 | 44 | .219 |
| 32 | 43 | .259 |
| 33 | 42 | .303 |
| 34 | 41 | .350 |
| 35 | 40 | .399 |
| 36 | 39 | .449 |
| 37 | 38 | .500 |

(5,10)

| $T'_\alpha$ | $T'_{1-\alpha}$ | $\alpha$ |
|---|---|---|
| 15 | 65 | .000 |
| 16 | 64 | .001 |
| 17 | 63 | .001 |
| 18 | 62 | .002 |
| 19 | 61 | .004 |
| 20 | 60 | .006 |
| 21 | 59 | .010 |
| 22 | 58 | .014 |
| 23 | 57 | .020 |
| 24 | 56 | .028 |
| 25 | 55 | .038 |
| 26 | 54 | .050 |
| 27 | 53 | .065 |
| 28 | 52 | .082 |
| 29 | 51 | .103 |
| 30 | 50 | .127 |
| 31 | 49 | .155 |
| 32 | 48 | .185 |
| 33 | 47 | .220 |
| 34 | 46 | .257 |
| 35 | 45 | .297 |
| 36 | 44 | .339 |
| 37 | 43 | .384 |
| 38 | 42 | .430 |
| 39 | 41 | .477 |
| 40 | 40 | .523 |

(6,6)

| $T'_\alpha$ | $T'_{1-\alpha}$ | $\alpha$ |
|---|---|---|
| 21 | 57 | .001 |
| 22 | 56 | .002 |
| 23 | 55 | .004 |
| 24 | 54 | .008 |
| 25 | 53 | .013 |
| 26 | 52 | .021 |
| 27 | 51 | .032 |
| 28 | 50 | .047 |
| 29 | 49 | .066 |
| 30 | 48 | .090 |
| 31 | 47 | .120 |
| 32 | 46 | .155 |
| 33 | 45 | .197 |
| 34 | 44 | .242 |
| 35 | 43 | .294 |
| 36 | 42 | .350 |
| 37 | 41 | .409 |
| 38 | 40 | .469 |
| 39 | 39 | .531 |

(6,7)

| $T'_\alpha$ | $T'_{1-\alpha}$ | $\alpha$ |
|---|---|---|
| 21 | 63 | .001 |
| 22 | 62 | .001 |
| 23 | 61 | .002 |
| 24 | 60 | .004 |
| 25 | 59 | .007 |
| 26 | 58 | .011 |
| 27 | 57 | .017 |
| 28 | 56 | .026 |
| 29 | 55 | .037 |
| 30 | 54 | .051 |
| 31 | 53 | .069 |
| 32 | 52 | .090 |
| 33 | 51 | .117 |
| 34 | 50 | .147 |
| 35 | 49 | .183 |
| 36 | 48 | .223 |
| 37 | 47 | .267 |
| 38 | 46 | .314 |
| 39 | 45 | .365 |
| 40 | 44 | .418 |
| 41 | 43 | .473 |
| 42 | 42 | .527 |

(6,8)

| $T'_\alpha$ | $T'_{1-\alpha}$ | $\alpha$ |
|---|---|---|
| 21 | 69 | .000 |
| 22 | 68 | .001 |
| 23 | 67 | .001 |
| 24 | 66 | .002 |
| 25 | 65 | .004 |
| 26 | 64 | .006 |
| 27 | 63 | .010 |
| 28 | 62 | .015 |
| 29 | 61 | .021 |
| 30 | 60 | .030 |
| 31 | 59 | .041 |
| 32 | 58 | .054 |
| 33 | 57 | .071 |
| 34 | 56 | .091 |
| 35 | 55 | .114 |
| 36 | 54 | .141 |
| 37 | 53 | .172 |
| 38 | 52 | .207 |
| 39 | 51 | .245 |
| 40 | 50 | .286 |
| 41 | 49 | .331 |
| 42 | 48 | .377 |
| 43 | 47 | .426 |
| 44 | 46 | .475 |
| 45 | 45 | .525 |

(6,9)

| $T'_\alpha$ | $T'_{1-\alpha}$ | $\alpha$ |
|---|---|---|
| 21 | 75 | .000 |
| 22 | 74 | .000 |
| 23 | 73 | .001 |
| 24 | 72 | .001 |
| 25 | 71 | .002 |
| 26 | 70 | .004 |
| 27 | 69 | .006 |
| 28 | 68 | .009 |
| 29 | 67 | .013 |
| 30 | 66 | .018 |
| 31 | 65 | .025 |
| 32 | 64 | .033 |
| 33 | 63 | .044 |
| 34 | 62 | .057 |
| 35 | 61 | .072 |
| 36 | 60 | .091 |
| 37 | 59 | .112 |
| 38 | 58 | .136 |
| 39 | 57 | .164 |
| 40 | 56 | .194 |
| 41 | 55 | .228 |
| 42 | 54 | .264 |
| 43 | 53 | .303 |
| 44 | 52 | .344 |
| 45 | 51 | .388 |
| 46 | 50 | .432 |
| 47 | 49 | .477 |
| 48 | 48 | .523 |

(6,10)

| $T'_\alpha$ | $T'_{1-\alpha}$ | $\alpha$ |
|---|---|---|
| 21 | 81 | .000 |
| 22 | 80 | .000 |
| 23 | 79 | .000 |
| 24 | 78 | .001 |
| 25 | 77 | .001 |
| 26 | 76 | .002 |
| 27 | 75 | .004 |
| 28 | 74 | .005 |
| 29 | 73 | .008 |
| 30 | 72 | .011 |
| 31 | 71 | .016 |
| 32 | 70 | .021 |
| 33 | 69 | .028 |
| 34 | 68 | .036 |
| 35 | 67 | .047 |
| 36 | 66 | .059 |
| 37 | 65 | .074 |
| 38 | 64 | .090 |
| 39 | 63 | .110 |
| 40 | 62 | .132 |
| 41 | 61 | .157 |
| 42 | 60 | .184 |
| 43 | 59 | .214 |
| 44 | 58 | .246 |
| 45 | 57 | .281 |
| 46 | 56 | .318 |
| 47 | 55 | .356 |

| $T'_{\alpha}$ | $T'_{1-\alpha}$ | $\alpha$ |
|---|---|---|
| | (6,10) (*Cont.*) | |
| 48 | 54 | .396 |
| 49 | 53 | .437 |
| 50 | 52 | .479 |
| 51 | 51 | .521 |
| | (7,7) | |
| 28 | 77 | .000 |
| 29 | 76 | .001 |
| 30 | 75 | .001 |
| 31 | 74 | .002 |
| 32 | 73 | .003 |
| 33 | 72 | .006 |
| 34 | 71 | .009 |
| 35 | 70 | .013 |
| 36 | 69 | .019 |
| 37 | 68 | .027 |
| 38 | 67 | .036 |
| 39 | 66 | .049 |
| 40 | 65 | .064 |
| 41 | 64 | .082 |
| 42 | 63 | .104 |
| 43 | 62 | .130 |
| 44 | 61 | .159 |
| 45 | 60 | .191 |
| 46 | 59 | .228 |
| 47 | 58 | .267 |
| 48 | 57 | .310 |
| 49 | 56 | .355 |
| 50 | 55 | .402 |
| 51 | 54 | .451 |
| 52 | 53 | .500 |
| | (7,8) | |
| 28 | 84 | .000 |
| 29 | 83 | .000 |
| 30 | 82 | .001 |
| 31 | 81 | .001 |
| 32 | 80 | .002 |
| 33 | 79 | .003 |
| 34 | 78 | .005 |
| 35 | 77 | .007 |
| 36 | 76 | .010 |
| 37 | 75 | .014 |
| 38 | 74 | .020 |
| 39 | 73 | .027 |
| 40 | 72 | .036 |
| 41 | 71 | .047 |
| 42 | 70 | .060 |
| 43 | 69 | .076 |
| 44 | 68 | .095 |
| 45 | 67 | .116 |

| $T'_{\alpha}$ | $T'_{1-\alpha}$ | $\alpha$ |
|---|---|---|
| | (7,8) (*Cont.*) | |
| 46 | 66 | .140 |
| 47 | 65 | .168 |
| 48 | 64 | .198 |
| 49 | 63 | .232 |
| 50 | 62 | .268 |
| 51 | 61 | .306 |
| 52 | 60 | .347 |
| 53 | 59 | .389 |
| 54 | 58 | .433 |
| 55 | 57 | .478 |
| 56 | 56 | .522 |
| | (7,9) | |
| 28 | 91 | .000 |
| 29 | 90 | .000 |
| 30 | 89 | .000 |
| 31 | 88 | .001 |
| 32 | 87 | .001 |
| 33 | 86 | .002 |
| 34 | 85 | .003 |
| 35 | 84 | .004 |
| 36 | 83 | .006 |
| 37 | 82 | .008 |
| 38 | 81 | .011 |
| 39 | 80 | .016 |
| 40 | 79 | .021 |
| 41 | 78 | .027 |
| 42 | 77 | .036 |
| 43 | 76 | .045 |
| 44 | 75 | .057 |
| 45 | 74 | .071 |
| 46 | 73 | .087 |
| 47 | 72 | .105 |
| 48 | 71 | .126 |
| 49 | 70 | .150 |
| 50 | 69 | .175 |
| 51 | 68 | .204 |
| 52 | 67 | .235 |
| 53 | 66 | .268 |
| 54 | 65 | .303 |
| 55 | 64 | .340 |
| 56 | 63 | .379 |
| 57 | 62 | .419 |
| 58 | 61 | .459 |
| 59 | 60 | .500 |
| | (7,10) | |
| 28 | 98 | .000 |
| 29 | 97 | .000 |
| 30 | 96 | .000 |
| 31 | 95 | .000 |

| $T'_{\alpha}$ | $T'_{1-\alpha}$ | $\alpha$ |
|---|---|---|
| | (7,10) (*Cont.*) | |
| 32 | 94 | .001 |
| 33 | 93 | .001 |
| 34 | 92 | .001 |
| 35 | 91 | .002 |
| 36 | 90 | .003 |
| 37 | 89 | .005 |
| 38 | 88 | .007 |
| 39 | 87 | .009 |
| 40 | 86 | .012 |
| 41 | 85 | .017 |
| 42 | 84 | .022 |
| 43 | 83 | .028 |
| 44 | 82 | .035 |
| 45 | 81 | .044 |
| 46 | 80 | .054 |
| 47 | 79 | .067 |
| 48 | 78 | .081 |
| 49 | 77 | .097 |
| 50 | 76 | .115 |
| 51 | 75 | .135 |
| 52 | 74 | .157 |
| 53 | 73 | .182 |
| 54 | 72 | .209 |
| 55 | 71 | .237 |
| 56 | 70 | .268 |
| 57 | 69 | .300 |
| 58 | 68 | .335 |
| 59 | 67 | .370 |
| 60 | 66 | .406 |
| 61 | 65 | .443 |
| 62 | 64 | .481 |
| 63 | 63 | .519 |
| | (8,8) | |
| 36 | 100 | .000 |
| 37 | 99 | .000 |
| 38 | 98 | .000 |
| 39 | 97 | .001 |
| 40 | 96 | .001 |
| 41 | 95 | .001 |
| 42 | 94 | .002 |
| 43 | 93 | .003 |
| 44 | 92 | .005 |
| 45 | 91 | .007 |
| 46 | 90 | .010 |
| 47 | 89 | .014 |
| 48 | 88 | .019 |
| 49 | 87 | .025 |
| 50 | 86 | .032 |
| 51 | 85 | .041 |

| $T'_{\alpha}$ | $T'_{1-\alpha}$ | $\alpha$ |
|---|---|---|
| | (8,8) (*Cont.*) | |
| 52 | 84 | .052 |
| 53 | 83 | .065 |
| 54 | 82 | .080 |
| 55 | 81 | .097 |
| 56 | 80 | .117 |
| 57 | 79 | .139 |
| 58 | 78 | .164 |
| 59 | 77 | .191 |
| 60 | 76 | .221 |
| 61 | 75 | .253 |
| 62 | 74 | .287 |
| 63 | 73 | .323 |
| 64 | 72 | .360 |
| 65 | 71 | .399 |
| 66 | 70 | .439 |
| 67 | 69 | .480 |
| 68 | 68 | .520 |
| | (8,9) | |
| 36 | 108 | .000 |
| 40 | 104 | .000 |
| 41 | 103 | .001 |
| 42 | 102 | .001 |
| 43 | 101 | .002 |
| 44 | 100 | .003 |
| 45 | 99 | .004 |
| 46 | 98 | .006 |
| 47 | 97 | .008 |
| 48 | 96 | .010 |
| 49 | 95 | .014 |
| 50 | 94 | .018 |
| 51 | 93 | .023 |
| 52 | 92 | .030 |
| 53 | 91 | .037 |
| 54 | 90 | .046 |
| 55 | 89 | .057 |
| 56 | 88 | .069 |
| 57 | 87 | .084 |
| 58 | 86 | .100 |
| 59 | 85 | .118 |
| 60 | 84 | .138 |
| 61 | 83 | .161 |
| 62 | 82 | .185 |
| 63 | 81 | .212 |
| 64 | 80 | .240 |
| 65 | 79 | .271 |
| 66 | 78 | .303 |
| 67 | 77 | .336 |
| 68 | 76 | .371 |
| 69 | 75 | .407 |

| $T'_\alpha$ | $T'_{1-\alpha}$ | $\alpha$ | $T'_\alpha$ | $T'_{1-\alpha}$ | $\alpha$ | $T'_\alpha$ | $T'_{1-\alpha}$ | $\alpha$ | $T'_\alpha$ | $T'_{1-\alpha}$ | $\alpha$ |
|---|---|---|---|---|---|---|---|---|---|---|---|
| (8,9) (*Cont.*) | | | (9,9) | | | (9,10) (*Cont.*) | | | (10,10) (*Cont.*) | | |
| 70 | 74 | .444 | 45 | 126 | .000 | 54 | 126 | .001 | 65 | 145 | .001 |
| 71 | 73 | .481 | 50 | 121 | .000 | 55 | 125 | .001 | 66 | 144 | .001 |
| 72 | 72 | .519 | 51 | 120 | .001 | 56 | 124 | .002 | 67 | 143 | .001 |
| (8,10) | | | 52 | 119 | .001 | 57 | 123 | .003 | 68 | 142 | .002 |
| 36 | 116 | .000 | 53 | 118 | .001 | 58 | 122 | .004 | 69 | 141 | .003 |
| 41 | 111 | .000 | 54 | 117 | .002 | 59 | 121 | .005 | 70 | 140 | .003 |
| 42 | 110 | .001 | 55 | 116 | .003 | 60 | 120 | .007 | 71 | 139 | .004 |
| 43 | 109 | .001 | 56 | 115 | .004 | 61 | 119 | .009 | 72 | 138 | .006 |
| 44 | 108 | .002 | 57 | 114 | .005 | 62 | 118 | .011 | 73 | 137 | .007 |
| 45 | 107 | .002 | 58 | 113 | .007 | 63 | 117 | .014 | 74 | 136 | .009 |
| 46 | 106 | .003 | 59 | 112 | .009 | 64 | 116 | .017 | 75 | 135 | .012 |
| 47 | 105 | .004 | 60 | 111 | .012 | 65 | 115 | .022 | 76 | 134 | .014 |
| 48 | 104 | .006 | 61 | 110 | .016 | 66 | 114 | .027 | 77 | 133 | .018 |
| 49 | 103 | .008 | 62 | 109 | .020 | 67 | 113 | .033 | 78 | 132 | .022 |
| 50 | 102 | .010 | 63 | 108 | .025 | 68 | 112 | .039 | 79 | 131 | .026 |
| 51 | 101 | .013 | 64 | 107 | .031 | 69 | 111 | .047 | 80 | 130 | .032 |
| 52 | 100 | .017 | 65 | 106 | .039 | 70 | 110 | .056 | 81 | 129 | .038 |
| 53 | 99 | .022 | 66 | 105 | .047 | 71 | 109 | .067 | 82 | 128 | .045 |
| 54 | 98 | .027 | 67 | 104 | .057 | 72 | 108 | .078 | 83 | 127 | .053 |
| 55 | 97 | .034 | 68 | 103 | .068 | 73 | 107 | .091 | 84 | 126 | .062 |
| 56 | 96 | .042 | 69 | 102 | .081 | 74 | 106 | .106 | 85 | 125 | .072 |
| 57 | 95 | .051 | 70 | 101 | .095 | 75 | 105 | .121 | 86 | 124 | .083 |
| 58 | 94 | .061 | 71 | 100 | .111 | 76 | 104 | .139 | 87 | 123 | .095 |
| 59 | 93 | .073 | 72 | 99 | .129 | 77 | 103 | .158 | 88 | 122 | .109 |
| 60 | 92 | .086 | 73 | 98 | .149 | 78 | 102 | .178 | 89 | 121 | .124 |
| 61 | 91 | .102 | 74 | 97 | .170 | 79 | 101 | .200 | 90 | 120 | .140 |
| 62 | 90 | .118 | 75 | 96 | .193 | 80 | 100 | .223 | 91 | 119 | .157 |
| 63 | 89 | .137 | 76 | 95 | .218 | 81 | 99 | .248 | 92 | 118 | .176 |
| 64 | 88 | .158 | 77 | 94 | .245 | 82 | 98 | .274 | 93 | 117 | .197 |
| 65 | 87 | .180 | 78 | 93 | .273 | 83 | 97 | .302 | 94 | 116 | .218 |
| 66 | 86 | .204 | 79 | 92 | .302 | 84 | 96 | .330 | 95 | 115 | .241 |
| 67 | 85 | .230 | 80 | 91 | .333 | 85 | 95 | .360 | 96 | 114 | .264 |
| 68 | 84 | .257 | 81 | 90 | .365 | 86 | 94 | .390 | 97 | 113 | .289 |
| 69 | 83 | .286 | 82 | 89 | .398 | 87 | 93 | .421 | 98 | 112 | .315 |
| 70 | 82 | .317 | 83 | 88 | .432 | 88 | 92 | .452 | 99 | 111 | .342 |
| 71 | 81 | .348 | 84 | 87 | .466 | 89 | 91 | .484 | 100 | 110 | .370 |
| 72 | 80 | .381 | 85 | 86 | .500 | 90 | 90 | .516 | 101 | 109 | .398 |
| 73 | 79 | .414 | (9,10) | | | (10,10) | | | 102 | 108 | .427 |
| 74 | 78 | .448 | 45 | 135 | .000 | 55 | 155 | .000 | 103 | 107 | .456 |
| 75 | 77 | .483 | 52 | 128 | .000 | 63 | 147 | .000 | 104 | 106 | .485 |
| 76 | 76 | .517 | 53 | 127 | .001 | 64 | 146 | .001 | 105 | 105 | .515 |

For sample sizes greater than 10 the chance that the statistic $T'$ will be less than or equal to an integer $k$ is given approximately by the area under the standard normal curve to the left of

$$z = \frac{k + \frac{1}{2} - N_1(N_1 + N_2 + 1)/2}{\sqrt{N_1 N_2 (N_1 + N_2 + 1)/12}}$$

## A-21a PERCENTILES OF THE DISTRIBUTION OF $d$

| $N$ \ $1-\alpha$ | .80 | .85 | .90 | .95 | .99 |
|---|---|---|---|---|---|
| 5 | .45 | .47 | .51 | .56 | .67 |
| 10 | .32 | .34 | .37 | .41 | .49 |
| 20 | .23 | .25 | .26 | .29 | .35 |
| 25 | .21 | .22 | .24 | .26 | .32 |
| 30 | .19 | .20 | .22 | .24 | .29 |
| 35 | .18 | .19 | .20 | .23 | .27 |
| 40 | .17 | .18 | .19 | .21 | .25 |
| 45 | .16 | .17 | .18 | .20 | .24 |
| 50 | .15 | .16 | .17 | .19 | .23 |
| For larger values | $\frac{1.07}{\sqrt{N}}$ | $\frac{1.14}{\sqrt{N}}$ | $\frac{1.22}{\sqrt{N}}$ | $\frac{1.36}{\sqrt{N}}$ | $\frac{1.63}{\sqrt{N}}$ |

$\alpha$ is the chance that the maximum deviation between the cumulative distributions of the population and of the sample exceeds the value $d$ given in the table.

Exact chances that the maximum deviation exceeds most of the values 0, $1/N$, $2/N$, . . . , 1 for all $N$'s $\leq 100$ are given by Z. W. Birnbaum, "Numerical tabulation of the distribution of Kolmogorov's statistic for finite sample size," *Journal of the American Statistical Association*, vol. 47 (1952), p. 425.

For two samples of sizes $N_1$ and $N_2$, respectively, the sampling distribution of the maximum deviation between the two cumulative distribution curves has percentiles as follows: $P_{80} = 1.07\sqrt{(1/N_1) + (1/N_2)}$, $P_{85} = 1.14\sqrt{(1/N_1) + (1/N_2)}$,

$$P_{90} = 1.22\sqrt{(1/N_1) + (1/N_2)}$$

$P_{95} = 1.36\sqrt{(1/N_1) + (1/N_2)}$, $P_{99} = 1.63\sqrt{(1/N_1) + (1/N_2)}$. These percentiles are approximate for large samples, say, $N_1$ and $N_2$ each greater than about 30. Some exact chances are given by F. J. Massey, in the *Annals of Mathematical Statistics*, vol. 22 (1951), p. 125, and vol. 23 (1952), p. 435.

## A-21b REQUIRED SIZE OF A RANDOM SAMPLE SO THAT THERE IS PROBABILITY $1 - \alpha$ THAT THE SAMPLE CUMULATIVE DISTRIBUTION WILL BE WITHIN $\pm d$ UNITS OF THE POPULATION CUMULATIVE DISTRIBUTION

| $d$ \ $1-\alpha$ | .80 | .85 | .90 | .95 | .99 |
|---|---|---|---|---|---|
| .20 | 27 | 30 | 35 | 45 | 67 |
| .10 | 115 | 130 | 149 | 185 | 266 |
| .05 | 458 | 520 | 596 | 740 | 1,064 |
| .025 | 1,832 | 2,080 | 2,384 | 2,960 | 4,256 |
| .01 | 11,450 | 13,000 | 14,880 | 18,500 | 26,570 |
| For $d < .2$ | $(1.07/d)^2$ | $(1.14/d)^2$ | $(1.22/d)^2$ | $(1.36/d)^2$ | $(1.63/d)^2$ |

| | SET NUMBER | | | | | |
|---|---|---|---|---|---|---|
| | 1 | 2 | 3 | 4 | 5 | 6 |
| Color of figures | Black | Red | Blue | Green | Black | Black |
| Mean | 0 | 0 | +2 | 0 | +4 | 0 |
| Standard deviation | 1.715 | 1.715 | 1.715 | 3.470 | 1.715 | 6.055 |
| *Number on tag* | *Frequency* | | | | | |
| 10 | | | | 1 | | 10 |
| 9 | | | | 1 | 1 | 10 |
| 8 | | | | 1 | 3 | 10 |
| 7 | | | 1 | 3 | 10 | 10 |
| 6 | | | 3 | 5 | 23 | 10 |
| 5 | 1 | 1 | 10 | 8 | 39 | 10 |
| 4 | 3 | 3 | 23 | 12 | 48 | 10 |
| 3 | 10 | 10 | 39 | 16 | 39 | 10 |
| 2 | 23 | 23 | 48 | 20 | 23 | 10 |
| 1 | 39 | 39 | 39 | 22 | 10 | 10 |
| 0 | 48 | 48 | 23 | 23 | 3 | 10 |
| −1 | 39 | 39 | 10 | 22 | 1 | 10 |
| −2 | 23 | 23 | 3 | 20 | | 10 |
| −3 | 10 | 10 | 1 | 16 | | 10 |
| −4 | 3 | 3 | | 12 | | 10 |
| −5 | 1 | 1 | | 8 | | 10 |
| −6 | | | | 5 | | 10 |
| −7 | | | | 3 | | 10 |
| −8 | | | | 1 | | 10 |
| −9 | | | | 1 | | 10 |
| −10 | | | | 1 | | 10 |
| *Total* | 200 | 200 | 200 | 201 | 200 | 210 |

From E. G. Olds, L. A. Knowler, "Teaching statistical quality control for town and gown," *Journal of the American Statistical Association*, vol. 44 (1949), pp. 213–230.

# A-23 RANDOM NORMAL NUMBERS, $\mu = 2$ AND $\sigma = 1$

| 01 | 02 | 03 | 04 | 05 | 06 | 07 | 08 | 09 | 10 |
|---|---|---|---|---|---|---|---|---|---|
| 2.422 | 0.130 | 2.232 | 1.700 | 1.903 | 0.725 | 2.031 | 0.515 | −0.684 | 2.788 |
| 0.694 | 2.556 | 1.868 | 1.263 | 2.115 | 1.516 | 1.972 | 3.627 | 1.482 | 3.263 |
| 1.875 | 2.273 | 0.655 | 2.299 | 0.055 | 1.955 | −0.147 | 2.168 | 2.193 | 1.879 |
| 1.017 | 0.757 | 1.288 | 1.322 | 2.080 | 2.170 | 1.502 | 2.953 | 0.171 | 1.951 |
| 2.453 | 4.199 | 1.403 | 2.017 | 3.496 | 0.165 | 2.556 | 1.003 | 1.973 | 2.159 |
| 2.274 | 1.767 | 1.564 | 2.412 | 2.207 | 0.475 | 2.656 | 1.579 | 0.394 | 1.225 |
| 3.000 | 1.618 | 1.530 | 2.224 | 2.881 | 2.715 | 3.103 | 1.941 | 2.179 | 3.748 |
| 2.510 | 2.256 | 1.146 | 5.177 | 1.931 | 1.693 | 1.021 | 3.337 | 2.137 | 1.839 |
| 1.233 | 2.085 | 2.251 | 1.578 | 3.796 | 3.017 | 2.863 | 2.514 | 1.615 | 1.548 |
| 3.075 | 1.730 | 2.427 | 2.990 | 1.680 | 3.250 | 3.050 | 3.243 | 1.846 | 1.798 |
| 1.344 | −0.095 | 2.166 | 4.116 | 2.500 | 1.939 | 1.567 | 3.047 | 1.385 | −0.831 |
| 1.246 | 3.860 | 1.253 | 1.876 | 4.373 | 1.993 | 1.262 | 2.319 | 2.488 | 2.406 |
| 0.889 | 2.299 | 2.458 | 1.790 | 1.048 | 2.302 | 0.138 | 2.383 | 1.170 | 2.204 |
| 1.154 | 1.401 | 1.935 | 3.106 | 1.548 | −0.096 | 2.153 | 2.333 | 1.761 | 3.728 |
| 3.031 | 1.048 | 0.719 | 1.474 | 2.779 | 0.292 | 2.341 | 2.707 | 1.741 | 2.353 |
| 0.534 | 1.155 | 1.705 | 1.662 | 0.457 | 0.602 | 1.365 | 2.663 | 3.755 | 1.900 |
| 2.230 | 3.096 | 0.045 | 3.639 | 0.680 | 0.970 | 1.593 | 2.117 | 2.395 | 1.935 |
| 2.355 | 1.761 | 1.816 | 1.822 | 1.434 | 2.259 | 3.788 | 3.280 | 1.317 | 2.940 |
| 1.461 | 0.947 | 0.717 | 2.923 | 2.133 | 2.526 | 2.687 | 2.144 | 1.692 | 1.469 |
| 3.034 | 1.778 | 2.122 | 2.025 | 3.008 | 1.447 | −0.305 | 2.452 | 1.726 | 0.870 |
| 2.761 | 0.473 | 3.726 | 1.893 | 2.455 | 1.633 | 1.654 | 3.006 | 3.523 | 2.317 |
| 1.961 | 0.965 | 1.481 | 1.402 | 2.106 | 2.214 | 1.727 | 3.670 | 3.795 | 2.258 |
| 2.639 | 4.010 | 1.915 | 1.713 | 1.484 | 1.443 | 1.444 | 2.394 | 1.688 | 0.793 |
| 1.349 | 2.225 | 0.644 | 1.404 | 2.583 | 2.149 | 2.359 | 2.274 | 1.432 | 1.610 |
| 2.959 | 2.797 | 4.635 | 3.268 | 2.889 | 2.349 | 0.933 | 3.403 | 2.206 | −0.214 |
| 2.440 | 2.919 | 1.455 | 0.695 | 1.466 | 1.124 | 1.257 | 1.265 | 0.096 | 3.412 |
| 3.078 | 3.279 | 0.352 | 2.583 | 1.690 | 0.729 | 2.072 | 1.332 | 1.158 | 1.827 |
| 1.736 | 1.968 | 0.011 | 2.418 | 1.026 | 1.342 | 2.103 | 1.792 | 2.175 | 1.646 |
| 3.275 | 3.147 | 2.800 | 2.172 | 0.004 | 1.763 | 3.801 | 2.510 | 2.517 | −0.117 |
| 2.579 | 2.297 | 2.030 | 2.725 | 3.721 | 2.545 | 1.631 | −0.346 | −0.011 | 1.961 |
| 2.549 | 3.546 | 2.805 | 1.250 | 0.769 | 2.238 | 2.284 | 3.722 | 2.085 | 2.653 |
| 2.954 | 1.990 | 1.249 | 1.028 | 3.241 | 1.926 | 3.056 | 1.732 | 2.116 | 1.825 |
| 1.442 | 2.542 | 2.557 | 1.741 | 0.630 | 2.117 | 1.662 | 2.237 | −0.046 | 3.132 |
| 4.039 | 2.030 | 2.859 | 3.538 | 2.424 | 2.169 | 3.643 | 3.290 | 2.742 | 1.336 |
| 2.127 | 0.288 | 2.921 | 0.175 | 1.670 | 3.151 | 1.443 | 0.935 | 1.125 | 2.872 |
| 1.102 | 2.536 | 1.476 | 2.980 | 0.416 | 1.784 | 2.521 | 1.867 | 1.709 | 1.558 |
| 2.938 | 2.112 | 1.350 | 2.115 | 1.164 | 1.761 | 1.350 | 1.798 | 3.160 | 2.593 |
| 2.975 | 2.681 | 0.721 | 1.291 | 2.276 | 2.131 | 2.187 | 2.752 | 1.380 | 0.676 |
| 1.386 | 1.712 | 1.692 | 2.844 | 1.559 | 0.418 | 3.020 | 0.785 | 1.962 | 3.184 |
| 2.834 | 1.485 | 0.632 | 0.872 | 0.735 | 1.934 | 1.221 | 2.544 | 1.797 | 1.410 |
| 3.346 | 1.147 | 1.766 | 1.862 | 2.595 | 1.524 | 3.499 | 2.652 | 2.139 | 2.533 |
| 2.243 | 3.881 | 2.846 | 2.670 | 3.377 | 1.380 | 4.183 | 0.883 | 1.373 | 1.992 |
| 2.705 | 2.661 | 1.521 | 1.290 | 2.280 | 1.638 | 0.884 | 2.636 | 2.077 | 1.012 |
| 2.760 | 1.182 | 1.152 | 3.074 | 1.073 | 2.917 | 2.150 | 2.866 | 1.688 | 1.684 |
| 2.086 | 1.250 | 1.577 | 2.871 | 2.985 | 2.585 | 2.897 | 2.398 | 0.999 | 1.764 |
| 0.802 | 1.421 | 4.793 | 0.268 | 2.838 | 2.227 | 3.331 | 2.395 | 2.064 | 2.916 |
| 4.165 | 2.014 | 0.616 | 1.929 | 0.641 | 2.304 | 1.263 | 2.125 | 0.908 | 1.768 |
| 2.291 | 2.549 | 0.851 | 1.856 | 2.452 | 3.282 | 0.978 | 2.255 | 1.683 | 1.926 |
| 1.428 | 4.194 | 2.262 | 2.957 | 1.991 | 2.759 | 1.553 | 3.538 | 1.272 | 3.417 |
| 2.051 | 2.455 | 2.759 | 2.267 | 2.794 | 4.106 | 2.373 | 1.401 | 2.562 | 2.502 |

By permission from tables of the RAND Corporation.

| 11 | 12 | 13 | 14 | 15 | 16 | 17 | 18 | 19 | 20 |
|---|---|---|---|---|---|---|---|---|---|
| 1.911 | 0.626 | 2.289 | 1.628 | 1.638 | 2.676 | 0.900 | 1.685 | 1.605 | 1.366 |
| 3.196 | 2.979 | 2.447 | 2.099 | 1.273 | 2.733 | 2.653 | 2.219 | 1.318 | 3.129 |
| 0.398 | 2.304 | 1.019 | 0.363 | 1.286 | 2.428 | 0.677 | 1.684 | 1.267 | 0.651 |
| 1.228 | 2.134 | 0.300 | 1.785 | 2.547 | 1.566 | 2.545 | 2.428 | 1.702 | 2.276 |
| 1.190 | 3.020 | 0.954 | 2.907 | 2.916 | 1.279 | 3.403 | 2.698 | 1.629 | 1.448 |
| 0.953 | 2.127 | 1.723 | 2.302 | 1.474 | 0.826 | 1.644 | 2.035 | 2.359 | 2.930 |
| 1.479 | 1.956 | 1.280 | 1.722 | 0.938 | 0.922 | 2.734 | 3.484 | 1.659 | 2.789 |
| 1.509 | 0.952 | 1.258 | −0.864 | 1.620 | 1.789 | 2.931 | 2.616 | 1.622 | 1.566 |
| 0.627 | 2.404 | 0.571 | 2.940 | 2.705 | 1.709 | 2.404 | 1.456 | 2.486 | 2.869 |
| 1.923 | 2.765 | 2.422 | 1.725 | 1.009 | 2.372 | 1.925 | 1.083 | 3.314 | 1.961 |
| 2.760 | 2.633 | 3.011 | 2.277 | 1.539 | 0.873 | 2.379 | 2.610 | 1.635 | −0.625 |
| 2.009 | 3.204 | 1.114 | 2.269 | 0.912 | 0.831 | 2.485 | 2.076 | 1.230 | 3.607 |
| 0.876 | 1.124 | 2.137 | 1.448 | 1.236 | 1.699 | 1.408 | 1.454 | 2.018 | 1.514 |
| 1.430 | 1.920 | 2.969 | 1.518 | 1.543 | 1.509 | 4.071 | 3.444 | 0.907 | 2.478 |
| 3.422 | 2.307 | 2.919 | 1.833 | 1.792 | 3.090 | 2.212 | 0.814 | 1.661 | 0.865 |
| 3.304 | 1.292 | 1.863 | 2.785 | 1.666 | 0.323 | 2.384 | 3.133 | 3.393 | 2.814 |
| 2.329 | 2.671 | 3.353 | 1.166 | 1.016 | 3.036 | 2.024 | 1.439 | 2.203 | 1.128 |
| 1.402 | 1.964 | 1.505 | 1.746 | 1.912 | 1.202 | 0.595 | 0.527 | 1.881 | 3.456 |
| 2.274 | 1.209 | 1.450 | 2.241 | 1.678 | 1.565 | 2.746 | 2.149 | 1.829 | 1.520 |
| 1.205 | 0.531 | 2.975 | 3.024 | 3.357 | 2.558 | 1.450 | 2.192 | 1.665 | 3.373 |
| 2.462 | 1.328 | 1.301 | 3.312 | 1.959 | 2.010 | 2.482 | 1.530 | 1.909 | 3.171 |
| 0.227 | 3.166 | 1.989 | 2.976 | 2.188 | 1.399 | 1.407 | 2.610 | 1.903 | 0.624 |
| 2.142 | 2.926 | 1.634 | 1.940 | 0.785 | 2.331 | 1.663 | 0.847 | 2.533 | 1.166 |
| 2.558 | 0.903 | 0.082 | 1.299 | 2.366 | 2.554 | 1.948 | 1.055 | 1.559 | 1.787 |
| 0.818 | 3.174 | 0.123 | −1.149 | 1.606 | 2.118 | −0.044 | 0.022 | 0.866 | 2.336 |
| 3.083 | 2.287 | 2.379 | 2.909 | 2.520 | 0.708 | 1.600 | 0.790 | 1.751 | 2.480 |
| 2.517 | 1.470 | 2.621 | 0.880 | 1.931 | 1.495 | 1.943 | 1.868 | 2.048 | 3.879 |
| 2.594 | 1.571 | 1.218 | 2.346 | 2.267 | 0.946 | 2.840 | 1.753 | 2.237 | 0.687 |
| 0.411 | 0.760 | 1.114 | 1.842 | 1.756 | 3.951 | 2.110 | 2.251 | 2.116 | 1.042 |
| 2.853 | 3.054 | 2.421 | 2.418 | 1.542 | 2.070 | 0.641 | 0.753 | 1.040 | 0.702 |
| 1.262 | −0.591 | 1.320 | 2.049 | 2.705 | 3.826 | 3.272 | 1.054 | 2.494 | 2.050 |
| 0.540 | 1.678 | 2.534 | 1.944 | 1.939 | 2.544 | 1.582 | 1.333 | 1.895 | 1.746 |
| 2.381 | 2.968 | 1.656 | 3.152 | 1.730 | 3.927 | 3.183 | 3.211 | 3.765 | 2.035 |
| 2.225 | 1.420 | 1.334 | 1.923 | 1.664 | 1.939 | 0.680 | 2.785 | 1.569 | 1.701 |
| 1.953 | 2.779 | 2.584 | 2.228 | 0.221 | 1.378 | 1.381 | 2.209 | 2.979 | 2.906 |
| 3.413 | 2.229 | 2.976 | 2.535 | 3.589 | 0.615 | 3.425 | 1.187 | 2.748 | 0.906 |
| 1.610 | 2.376 | 2.086 | 0.610 | 2.532 | 3.083 | 1.332 | 1.776 | 0.407 | 0.721 |
| 0.984 | 2.243 | 2.939 | 1.704 | 2.277 | 1.026 | 1.879 | 0.405 | 1.003 | 0.755 |
| 1.808 | 2.362 | 1.717 | 0.831 | 2.160 | 0.546 | 2.686 | 1.924 | 1.756 | 1.829 |
| −0.766 | 3.529 | 2.361 | 0.955 | 2.148 | 1.104 | 0.541 | 1.460 | 1.840 | 1.579 |
| 2.643 | 2.051 | 1.384 | 2.229 | 2.952 | 2.203 | 0.765 | 4.381 | 1.611 | 1.936 |
| 1.952 | 2.752 | 3.588 | 2.481 | 1.911 | 3.753 | 1.428 | 3.223 | 1.873 | 2.034 |
| 2.590 | 2.306 | 3.280 | 1.664 | 2.281 | 1.443 | 2.024 | 2.126 | 3.250 | 1.384 |
| 0.622 | 2.617 | 1.969 | 2.231 | −0.079 | 0.768 | 2.547 | 1.365 | 1.163 | 1.280 |
| 0.433 | −0.560 | 3.292 | 1.987 | 1.065 | 2.766 | 1.425 | 0.846 | 2.520 | 0.981 |
| 3.146 | 3.323 | 1.713 | 1.887 | 2.010 | 1.277 | 0.491 | 2.489 | 1.503 | 1.974 |
| 4.021 | 1.744 | 0.598 | 2.954 | 2.633 | 1.960 | 1.539 | 2.393 | 4.012 | 3.356 |
| 1.188 | 0.450 | 2.958 | 1.177 | 1.482 | 1.090 | 1.671 | 3.021 | 0.386 | 3.560 |
| 1.211 | 2.575 | 0.158 | 3.124 | 3.632 | 2.647 | 3.029 | 3.526 | 2.237 | 0.671 |
| 3.750 | 2.362 | 1.407 | 0.642 | 1.274 | 1.632 | 2.378 | 2.601 | 0.003 | 1.261 |

## A-23 RANDOM NORMAL NUMBERS, $\mu = 2$ AND $\sigma = 1$ (CONTINUED)

| 21 | 22 | 23 | 24 | 25 | 26 | 27 | 28 | 29 | 30 |
|---|---|---|---|---|---|---|---|---|---|
| 1.707 | 2.089 | 1.315 | 0.278 | 3.045 | 2.968 | 1.396 | 1.534 | 2.365 | 2.746 |
| 1.113 | 1.779 | 1.935 | 0.971 | 4.024 | 0.847 | 1.382 | 2.342 | 2.110 | 0.316 |
| 1.847 | 0.547 | 3.697 | 1.250 | 1.586 | 2.036 | 2.924 | 0.585 | 0.456 | 2.859 |
| 2.713 | 2.761 | 1.664 | 2.461 | 2.158 | 3.453 | 2.078 | 1.113 | 1.769 | 1.263 |
| 0.676 | 0.432 | 2.667 | 2.515 | 1.369 | 3.196 | 2.979 | 2.447 | 2.099 | 1.273 |
| 2.167 | 1.828 | 2.867 | 1.178 | 2.078 | 1.500 | 2.622 | 2.341 | 1.504 | 2.468 |
| 3.445 | 3.323 | 2.558 | 1.789 | 1.595 | 1.191 | 1.175 | 2.872 | 1.257 | 1.062 |
| 1.284 | 3.180 | 3.315 | 1.210 | 1.842 | 3.384 | 2.942 | 2.550 | 0.727 | 1.736 |
| 2.135 | 2.590 | 2.533 | 1.635 | 1.983 | 0.614 | 0.377 | −0.663 | 0.427 | 2.445 |
| 2.944 | 2.043 | 2.220 | 1.987 | 2.859 | 3.029 | 2.091 | 1.052 | 1.532 | 1.956 |
| 3.654 | 2.333 | 1.468 | 3.126 | 1.241 | 2.936 | 1.557 | 2.020 | 1.423 | 2.701 |
| 0.821 | 1.542 | 2.365 | 2.199 | 3.479 | 3.111 | −0.107 | 1.644 | 1.337 | 1.442 |
| 2.483 | 2.583 | 2.075 | 1.026 | 0.668 | 2.281 | 1.566 | 1.255 | 2.020 | 1.135 |
| 0.715 | 1.384 | 2.080 | 2.542 | 2.368 | 0.019 | 2.906 | 2.325 | 2.175 | 5.197 |
| 4.638 | 2.662 | 1.012 | 2.941 | 1.336 | 0.574 | 3.034 | 2.937 | 2.553 | 0.174 |
| 2.327 | 2.152 | 3.057 | 2.077 | 2.321 | 0.861 | 2.892 | 1.394 | −0.556 | 1.459 |
| −0.082 | 0.676 | 3.038 | 2.470 | 1.394 | 2.131 | 1.262 | 3.207 | 1.810 | 0.322 |
| 2.051 | 1.576 | 2.087 | 3.030 | 2.030 | 2.827 | 2.183 | 1.182 | 1.507 | −0.042 |
| 2.438 | 0.924 | 1.699 | 0.477 | 2.449 | 2.540 | 1.620 | 2.509 | 2.347 | 3.022 |
| 2.284 | 2.159 | 2.975 | 3.268 | 0.484 | 1.862 | 1.676 | 1.449 | 2.475 | 2.556 |
| 0.872 | 0.474 | 2.213 | 3.602 | 3.244 | 3.078 | 1.376 | 2.612 | 2.421 | 1.014 |
| 2.236 | 1.963 | 1.839 | 1.598 | 2.195 | 2.680 | 2.228 | 1.107 | −0.661 | 1.041 |
| 2.425 | 2.412 | 1.500 | 2.278 | 2.328 | 2.102 | 2.087 | 3.098 | 2.697 | 0.765 |
| 1.511 | 2.431 | 1.434 | 1.558 | 1.020 | 2.864 | 0.871 | 2.523 | 1.878 | 1.370 |
| 1.600 | 2.040 | 2.993 | 0.873 | 0.568 | 2.703 | 2.578 | 1.515 | 3.627 | 2.097 |
| 3.076 | 1.939 | 0.682 | 3.085 | 2.877 | 2.696 | −0.771 | 2.560 | 1.954 | 0.999 |
| 2.593 | 1.610 | 2.800 | 2.456 | 0.226 | 3.575 | 1.435 | 2.170 | 1.165 | 3.506 |
| 1.362 | 2.727 | 2.145 | 2.023 | 0.509 | 0.336 | 2.045 | 0.375 | 1.010 | 2.316 |
| 1.603 | 2.783 | 0.682 | 2.108 | 2.031 | 0.854 | 2.028 | 2.357 | 0.722 | 1.562 |
| 1.908 | 1.635 | 2.009 | 1.203 | 1.775 | 2.868 | 1.949 | 1.391 | 1.151 | 1.352 |
| 3.486 | 0.507 | 2.322 | 1.204 | 2.434 | 1.720 | 1.804 | 2.235 | 2.439 | 1.492 |
| 2.029 | 1.352 | 3.629 | 2.076 | 1.587 | 0.891 | 3.029 | 1.242 | 0.014 | 4.019 |
| 2.894 | 1.688 | 0.657 | 1.800 | 2.943 | 1.373 | 1.269 | 3.411 | 1.316 | 2.405 |
| 0.965 | −0.028 | 1.904 | 2.241 | 2.563 | 1.149 | 2.375 | 1.386 | 1.562 | 2.882 |
| 2.191 | 2.133 | 2.676 | 0.229 | 2.319 | 1.114 | 3.197 | 2.588 | 3.163 | 2.423 |
| 2.115 | 2.418 | 2.741 | 1.839 | 2.416 | 1.452 | 0.319 | 0.853 | 2.774 | 0.929 |
| 1.120 | 2.126 | 0.773 | 0.798 | 3.436 | 2.374 | 2.173 | −0.333 | 2.004 | 1.765 |
| 3.524 | 0.008 | 3.260 | 1.109 | 4.111 | 2.474 | 2.482 | 2.416 | 0.832 | 4.059 |
| 0.103 | 2.774 | 2.056 | 2.463 | 0.383 | −0.962 | 2.458 | 2.388 | 1.556 | 1.088 |
| 1.573 | 2.519 | 2.153 | 3.188 | 1.618 | 2.477 | 2.185 | 1.851 | 0.498 | 2.066 |
| 1.138 | 3.032 | 2.390 | 2.436 | 2.655 | 1.484 | 2.378 | 3.166 | 2.531 | 2.082 |
| 2.665 | 2.960 | 2.518 | 1.940 | 0.026 | 2.570 | 2.703 | 2.592 | 3.094 | 1.862 |
| 1.397 | 1.859 | 2.208 | 2.559 | 1.749 | 0.624 | 0.074 | 1.398 | 0.996 | 2.910 |
| 1.875 | 2.250 | 0.183 | 2.214 | 1.356 | 4.282 | 2.370 | 3.006 | 1.413 | 2.412 |
| 3.195 | 2.671 | 1.918 | 3.305 | 3.722 | 1.372 | 2.564 | 2.106 | 1.871 | 1.792 |
| 1.464 | 2.055 | 3.045 | 2.367 | 1.992 | 0.919 | 3.006 | 2.713 | 4.049 | 4.618 |
| 3.328 | 1.781 | 2.565 | 1.304 | 2.041 | 1.597 | 0.225 | 2.309 | −0.558 | 2.504 |
| 2.804 | 3.606 | 1.858 | 3.028 | 2.456 | 1.730 | 1.430 | 3.405 | 0.474 | 2.222 |
| 2.590 | 1.641 | 3.857 | 2.582 | 2.594 | 1.933 | 3.341 | 1.002 | 2.704 | 1.341 |
| 2.980 | 0.601 | 1.595 | 2.248 | 2.381 | 1.911 | 0.626 | 2.289 | 1.628 | 1.638 |

| 31 | 32 | 33 | 34 | 35 | 36 | 37 | 38 | 39 | 40 |
|---|---|---|---|---|---|---|---|---|---|
| 3.355 | 0.073 | 3.139 | 2.472 | 1.825 | 0.296 | 1.685 | 3.401 | 1.820 | 1.428 |
| 1.086 | 1.955 | 2.529 | 2.503 | 1.687 | 1.754 | 4.138 | 2.394 | 0.303 | 3.776 |
| 2.367 | 1.525 | 2.625 | 1.789 | 0.991 | 4.127 | 0.915 | 3.023 | 1.377 | 3.435 |
| 0.248 | 0.749 | 3.697 | 4.166 | 2.544 | 1.620 | 3.217 | 1.083 | 1.907 | 2.951 |
| 1.694 | 0.258 | 1.836 | 1.953 | 1.853 | 3.590 | 3.604 | 1.907 | 1.995 | 2.468 |
| 1.546 | 1.255 | 2.856 | 3.221 | 2.397 | −0.010 | 2.169 | 2.781 | 3.001 | 2.536 |
| 1.266 | 2.089 | 2.974 | 1.305 | 2.376 | −0.475 | 1.792 | 1.546 | 0.583 | 1.214 |
| 0.713 | 2.473 | 1.381 | 1.750 | 1.064 | 3.744 | 2.470 | 1.004 | 2.155 | 2.332 |
| −0.001 | 1.600 | 2.166 | 0.561 | 0.898 | 2.587 | 0.580 | −0.461 | 0.954 | 1.364 |
| 3.406 | 2.207 | 2.110 | 1.522 | 3.923 | 1.379 | 3.613 | 3.379 | 2.716 | 2.796 |
| 1.432 | 1.651 | 1.584 | 3.649 | 2.485 | 2.820 | 2.948 | 2.626 | 1.763 | 3.329 |
| 1.541 | 1.154 | 4.311 | 2.354 | 2.257 | 1.262 | 2.304 | 2.178 | 1.657 | 2.126 |
| 2.216 | 3.505 | −0.056 | 1.332 | 0.980 | 1.675 | 1.850 | 2.487 | 2.051 | 1.433 |
| 1.602 | 2.225 | 2.949 | 3.945 | 3.753 | 3.855 | 2.769 | 0.760 | 2.095 | 1.419 |
| 2.211 | 1.804 | 2.642 | 0.975 | 1.646 | 2.552 | 2.291 | 1.277 | 2.341 | −0.219 |
| 3.006 | 2.279 | 1.097 | 3.473 | 0.919 | 2.535 | 2.459 | 3.934 | 1.826 | 1.587 |
| 2.520 | 2.468 | 2.156 | 2.438 | 1.625 | 1.604 | 1.628 | 1.139 | 2.608 | 2.643 |
| 2.666 | 4.058 | 2.805 | 3.069 | 0.945 | 2.533 | 2.761 | 1.140 | 2.604 | 1.627 |
| 2.852 | 0.570 | 3.920 | 1.572 | 2.924 | 2.135 | 1.558 | 2.604 | 2.191 | 2.529 |
| 2.014 | 2.825 | 3.502 | 2.006 | 1.879 | 3.304 | 1.538 | 0.906 | 3.125 | 1.009 |
| 1.540 | 0.444 | 1.541 | 1.850 | 1.793 | 2.284 | 1.890 | 3.091 | 2.293 | 2.491 |
| 1.190 | 2.087 | 2.159 | 1.157 | 2.314 | 1.753 | 0.722 | 2.447 | 2.124 | 2.927 |
| 0.741 | 3.411 | 1.689 | 1.945 | 0.286 | 3.288 | 1.390 | 0.240 | 1.448 | 2.768 |
| 2.169 | 1.937 | 2.261 | 0.766 | 2.075 | 0.457 | 2.031 | 0.831 | −0.009 | 4.316 |
| 0.979 | 1.935 | 2.232 | 1.812 | 3.290 | 2.031 | 3.222 | 2.520 | 4.105 | 0.705 |
| 1.405 | 0.166 | 0.137 | 3.246 | 4.142 | 2.808 | 2.526 | 2.687 | −0.627 | 2.023 |
| 0.923 | 2.287 | 1.164 | 0.732 | 0.736 | 0.892 | 2.633 | 2.107 | 1.260 | 0.615 |
| 1.529 | 3.188 | 2.153 | 3.828 | 3.610 | 1.654 | 2.596 | 0.957 | 1.479 | 1.497 |
| 0.781 | 3.562 | 3.633 | 0.889 | 0.832 | 2.068 | 2.103 | 3.360 | 1.686 | 1.538 |
| 2.153 | 2.125 | 1.930 | 3.161 | 2.931 | 1.941 | 3.108 | 1.732 | 4.296 | 1.830 |
| 3.204 | 3.945 | 0.682 | 4.165 | 2.419 | 0.565 | 1.637 | 1.931 | 1.092 | 2.482 |
| 2.154 | 1.889 | 1.391 | 1.690 | 1.356 | 2.560 | 1.784 | 1.041 | 2.808 | 0.576 |
| 3.391 | 2.602 | 2.496 | 2.177 | 1.564 | 1.781 | 0.302 | 2.499 | 1.501 | 1.410 |
| 3.266 | 2.051 | 1.958 | 0.979 | 2.454 | 1.438 | 2.098 | 3.208 | 2.374 | 2.710 |
| 1.842 | 0.513 | 1.736 | 2.878 | 1.893 | 1.614 | 2.775 | 1.060 | 1.508 | 1.197 |
| 2.799 | 0.757 | 0.625 | 3.336 | 2.268 | 1.418 | 1.616 | 2.363 | 0.751 | 2.138 |
| 0.104 | 3.564 | 0.681 | 1.231 | 2.527 | 0.172 | 1.331 | 0.991 | 3.570 | 1.382 |
| 2.232 | 3.514 | −0.433 | 2.932 | 3.245 | 2.778 | 2.196 | −0.326 | 1.034 | 1.889 |
| 4.201 | 3.351 | 1.761 | 1.957 | 1.342 | 3.575 | 3.216 | 1.335 | 1.527 | 0.812 |
| 1.046 | 1.646 | 1.363 | 1.051 | 4.600 | 3.209 | 3.041 | 3.234 | 2.034 | 0.682 |
| 2.874 | 1.663 | 2.591 | 1.396 | 1.052 | 1.068 | 2.226 | 3.048 | 1.906 | 2.755 |
| 1.389 | 2.966 | 2.846 | 2.410 | 1.663 | 3.620 | 2.151 | 2.036 | 3.733 | 1.462 |
| −0.144 | 1.641 | 1.693 | 1.599 | 2.704 | 3.083 | 1.387 | 0.593 | 1.191 | 2.707 |
| 2.177 | 0.829 | 2.094 | 1.737 | 1.625 | 1.766 | 1.415 | 2.238 | 0.549 | 1.887 |
| 2.595 | 2.094 | 2.851 | 1.175 | 0.425 | 2.242 | 1.477 | 3.237 | 2.614 | 1.226 |
| 1.655 | 3.804 | 0.607 | 1.958 | 4.251 | 1.457 | 3.369 | 2.077 | 1.511 | 1.458 |
| 2.601 | 2.255 | 1.787 | 1.136 | 2.912 | 3.060 | 2.562 | 3.137 | 3.248 | 1.382 |
| 2.308 | 2.422 | 3.081 | 2.185 | 1.963 | 3.855 | 2.389 | 4.057 | 2.428 | 3.054 |
| 1.196 | 4.160 | 2.841 | 1.550 | 0.919 | 1.884 | 1.911 | 1.386 | 2.607 | 1.625 |
| 0.843 | 1.330 | 1.678 | 2.198 | 1.398 | 0.709 | 1.810 | 2.269 | 4.242 | 0.777 |

# A-23 RANDOM NORMAL NUMBERS, $\mu = 2$ AND $\sigma = 1$ (CONTINUED)

| 41 | 42 | 43 | 44 | 45 | 46 | 47 | 48 | 49 | 50 |
|---|---|---|---|---|---|---|---|---|---|
| 1.017 | 2.773 | 3.278 | 2.557 | 1.003 | 4.181 | 0.946 | 3.464 | 1.945 | 2.929 |
| 0.723 | 0.781 | 1.546 | 1.649 | 2.723 | 4.542 | 1.819 | 0.511 | 2.580 | 3.707 |
| 1.681 | 1.200 | 0.335 | 3.391 | 2.382 | 3.080 | 0.685 | 1.924 | 1.085 | 1.459 |
| 0.622 | 0.742 | 2.495 | 1.860 | 1.145 | 2.040 | 2.103 | −0.256 | 0.976 | 2.414 |
| 1.815 | 2.061 | 2.092 | 2.089 | 2.281 | 2.377 | 1.821 | 1.760 | 2.515 | 1.898 |
| 4.334 | 1.662 | 0.044 | 1.363 | 0.681 | 2.111 | 2.443 | 1.603 | 1.803 | 2.149 |
| 0.863 | 2.642 | 5.436 | 0.332 | 2.847 | 1.466 | 3.031 | 1.571 | 3.024 | 1.988 |
| 2.414 | 1.988 | 2.666 | 0.867 | 1.589 | 2.192 | 3.027 | 2.257 | 1.868 | 1.927 |
| 1.505 | 2.364 | 0.762 | 1.955 | 1.888 | 1.845 | 3.180 | 2.618 | 1.730 | 1.603 |
| 3.048 | 2.037 | 2.759 | 2.609 | −0.042 | 1.215 | 1.292 | 1.582 | 1.522 | 2.097 |
| 2.347 | 4.816 | 1.535 | 1.367 | 0.385 | 2.013 | 2.557 | 2.041 | 3.070 | 1.934 |
| 2.637 | 2.563 | 1.892 | 2.131 | 0.191 | 2.484 | 1.788 | 2.762 | 2.166 | 1.211 |
| 4.176 | 2.393 | 1.075 | 3.911 | 0.959 | 2.438 | 3.201 | 1.810 | 2.049 | 3.476 |
| 0.814 | 1.055 | 0.395 | 2.185 | 1.741 | 2.742 | 2.228 | 2.151 | 1.997 | 3.302 |
| 2.972 | 3.710 | 4.682 | 4.813 | 0.468 | 2.311 | 2.382 | 0.810 | 0.155 | 1.685 |
| 3.210 | 2.294 | 1.751 | 2.719 | 3.103 | 2.459 | 3.656 | 4.862 | 3.724 | 3.457 |
| 4.647 | 2.777 | 2.450 | 4.247 | 3.151 | 0.197 | 3.602 | 1.754 | 1.739 | 2.646 |
| 2.398 | 2.318 | 1.071 | 4.416 | 1.063 | 1.568 | 3.057 | 0.985 | 3.425 | 1.924 |
| 2.846 | 1.300 | 1.631 | 2.344 | 1.073 | 1.049 | 2.743 | 0.365 | 1.949 | 0.378 |
| 2.654 | 1.044 | 4.907 | 3.688 | 2.752 | 2.365 | 2.083 | 1.669 | 2.538 | 2.617 |
| 2.522 | 2.231 | 1.380 | 1.734 | 2.419 | 1.313 | 2.226 | 2.524 | 2.073 | 3.032 |
| 0.711 | 1.460 | 1.175 | 2.244 | 0.929 | −0.091 | 1.096 | 2.061 | 3.099 | 2.186 |
| 3.372 | 3.769 | 0.942 | 3.646 | 2.481 | 1.554 | 3.715 | 1.193 | 1.956 | 1.735 |
| 2.854 | 1.464 | 3.607 | 2.428 | 1.384 | 1.977 | 1.504 | 0.492 | 2.102 | 2.624 |
| 1.851 | 0.855 | 2.913 | 2.684 | 3.043 | 2.595 | 1.803 | 1.303 | 0.233 | −0.429 |
| 0.851 | 0.943 | 2.635 | 1.671 | 0.778 | 2.899 | 2.145 | 2.747 | 1.342 | 2.313 |
| 2.348 | 2.970 | 1.982 | 3.217 | 1.025 | 2.626 | 2.164 | 2.568 | 1.080 | 2.814 |
| 2.284 | 2.458 | 3.307 | 0.374 | 1.370 | 2.631 | −0.649 | 1.111 | 1.296 | 2.403 |
| 0.983 | 2.360 | 1.880 | 4.331 | 3.672 | −0.018 | 3.053 | 2.068 | 2.051 | 2.506 |
| 3.603 | 1.047 | 1.433 | 3.600 | 2.465 | 2.472 | 1.190 | 3.504 | 2.205 | 0.793 |
| 1.809 | 3.479 | 1.013 | 3.249 | 3.934 | 1.432 | 2.893 | 1.707 | 3.498 | 1.429 |
| 1.277 | 2.925 | 2.783 | 1.597 | 2.619 | 2.000 | 1.513 | 2.888 | 1.421 | 0.639 |
| 0.303 | 3.879 | 2.063 | 2.132 | 2.682 | 2.316 | 1.718 | 2.201 | 4.431 | 3.085 |
| 2.498 | 3.072 | 3.567 | 2.302 | 3.157 | 1.860 | 2.802 | 2.098 | 0.902 | 1.945 |
| −0.542 | 0.666 | 3.987 | 2.668 | 2.360 | 2.762 | 1.351 | 2.835 | 2.972 | 1.796 |
| 1.640 | 2.193 | 0.976 | 1.777 | 1.383 | 3.100 | 1.663 | 1.609 | 2.503 | 1.596 |
| 2.248 | 1.911 | 0.620 | 2.295 | 1.884 | 3.421 | 1.086 | 2.085 | 1.861 | −0.191 |
| 1.900 | 0.623 | 3.047 | 1.127 | −0.199 | 3.653 | 0.976 | −0.088 | −0.966 | 2.626 |
| 1.536 | 0.718 | −0.513 | 2.675 | 3.145 | 1.838 | 1.609 | 0.857 | 1.854 | 3.839 |
| 2.503 | 3.434 | 2.290 | 2.397 | 1.162 | 1.932 | 2.626 | 0.816 | 1.229 | 1.753 |
| 1.142 | 1.628 | 1.783 | 2.148 | −0.105 | 3.072 | 2.312 | 3.666 | 2.784 | 2.102 |
| 1.877 | 3.107 | 0.960 | 1.363 | 1.139 | 1.135 | 2.370 | 4.245 | 3.284 | 0.188 |
| 3.632 | 2.586 | 1.531 | 1.613 | 1.645 | 0.963 | 4.596 | 1.979 | 2.649 | 1.435 |
| 4.072 | 0.554 | 1.319 | 2.224 | 1.879 | 1.806 | 1.606 | 3.049 | 0.099 | 2.996 |
| 1.564 | 1.624 | 1.014 | 1.414 | 1.796 | 1.244 | 1.712 | 2.319 | 2.166 | 2.727 |
| 2.876 | 0.772 | −0.646 | 1.254 | 3.797 | 1.827 | 2.039 | 4.280 | 2.208 | 1.842 |
| 2.833 | 3.289 | 1.977 | 1.568 | 2.582 | 0.198 | 1.190 | 2.708 | 3.264 | 2.876 |
| 1.108 | 2.332 | 1.546 | 0.872 | 4.085 | 2.583 | 3.384 | 3.934 | 3.073 | 1.029 |
| 2.644 | 1.766 | 1.846 | 3.098 | 2.757 | 3.840 | 2.353 | 3.384 | 2.716 | 2.435 |
| 2.105 | 1.828 | 1.889 | 0.854 | 2.878 | 1.569 | 4.151 | 0.821 | 2.818 | 2.038 |

| 01 | 02 | 03 | 04 | 05 | 06 | 07 | 08 | 09 | 10 |
|---|---|---|---|---|---|---|---|---|---|
| -0.221 | -0.540 | -0.701 | 5.511 | -2.404 | -0.987 | -0.158 | -0.578 | -1.893 | 0.854 |
| -2.454 | -2.816 | 0.580 | -1.068 | 1.010 | 1.209 | 2.234 | 3.224 | 3.750 | 1.285 |
| 0.089 | 0.418 | -0.421 | 2.448 | -0.279 | 1.916 | -3.166 | -0.773 | -0.818 | -1.411 |
| 0.931 | 1.345 | 3.164 | 0.019 | 0.767 | 0.439 | -3.412 | -0.982 | 0.520 | -0.473 |
| 0.361 | 0.794 | 0.120 | -0.347 | 2.785 | 0.980 | 1.003 | -1.796 | -1.778 | -0.783 |
| -0.559 | -2.111 | -3.396 | 4.236 | 2.764 | -1.990 | -0.060 | -2.488 | -0.503 | -4.406 |
| -4.816 | -1.369 | 1.856 | 0.383 | 0.016 | -2.144 | -0.187 | -1.561 | 1.441 | -2.246 |
| 0.784 | 0.607 | 0.663 | -0.764 | -1.395 | 1.738 | -2.055 | 2.962 | -1.616 | 1.326 |
| 2.576 | -3.024 | 0.191 | 1.084 | -3.698 | -3.031 | -0.517 | -0.179 | 0.681 | -0.719 |
| -1.232 | 1.234 | -0.046 | 1.338 | 1.726 | 1.448 | 2.216 | -1.662 | 2.188 | 2.308 |
| 2.129 | -1.936 | 3.381 | 1.319 | -3.131 | -1.037 | 1.191 | 1.449 | 0.690 | -0.251 |
| -2.753 | 1.049 | 1.616 | 1.232 | 2.910 | 0.389 | -3.766 | 2.044 | 1.459 | -0.002 |
| 0.071 | 1.869 | -5.827 | 0.866 | -1.191 | 2.508 | 1.552 | -1.052 | 1.914 | -0.274 |
| 0.507 | 0.595 | -0.202 | -0.775 | -1.732 | -2.771 | 0.049 | 5.221 | 3.059 | 0.015 |
| 0.384 | -1.574 | 1.414 | -0.789 | -1.263 | -0.470 | 0.020 | 1.489 | 0.497 | 1.316 |
| 1.688 | 4.311 | 2.305 | -5.632 | 1.776 | 1.540 | 0.208 | 0.611 | 0.810 | -1.241 |
| -0.045 | -1.563 | 3.687 | -0.160 | -0.101 | 1.838 | 1.590 | 1.222 | 0.377 | 2.069 |
| 2.516 | -1.339 | 0.956 | -1.285 | 0.301 | 3.739 | -3.320 | 0.183 | 0.993 | -4.678 |
| 0.536 | 1.965 | -0.580 | -0.307 | 1.564 | 0.163 | -2.239 | -2.460 | -2.003 | -1.609 |
| 0.775 | 1.427 | -0.626 | -1.134 | -3.109 | 1.652 | 2.331 | -0.188 | 2.137 | -1.316 |
| 0.964 | -3.740 | 1.995 | -1.349 | -1.068 | -0.172 | 1.907 | 5.515 | 0.863 | -1.018 |
| 2.597 | 2.328 | -0.722 | 3.057 | 1.632 | 0.655 | 0.972 | 1.401 | 1.840 | 4.508 |
| -1.343 | -2.859 | 0.903 | -0.631 | -2.810 | 3.345 | 1.997 | 0.356 | 1.215 | 0.501 |
| -1.097 | 0.798 | -1.057 | 3.880 | 2.321 | -1.677 | -3.746 | -1.125 | -1.090 | -1.972 |
| -0.977 | 0.225 | -0.004 | -0.513 | 3.613 | 1.030 | 2.349 | -1.278 | -1.301 | -5.159 |
| 2.421 | -1.732 | 2.170 | 0.451 | 1.013 | -0.912 | 0.615 | -0.532 | 1.453 | -2.155 |
| 0.921 | -0.932 | -2.511 | -0.164 | 0.154 | -0.004 | 0.516 | 2.240 | -0.020 | 4.432 |
| -1.854 | -3.192 | -3.633 | 0.067 | 3.709 | 0.560 | -0.156 | 0.964 | -2.618 | -0.718 |
| 2.457 | -0.566 | -1.439 | 0.194 | 1.440 | -1.568 | -2.407 | -1.356 | 0.849 | 0.801 |
| 1.143 | 0.212 | 4.088 | -0.832 | -0.361 | 0.303 | -2.984 | 1.378 | -0.649 | -2.399 |
| 1.184 | 1.622 | -1.896 | 0.026 | -2.163 | -1.683 | 3.778 | 3.585 | -3.853 | 2.352 |
| 3.842 | 1.179 | -0.987 | 0.498 | 2.348 | 3.263 | 1.924 | -4.421 | -0.680 | 2.129 |
| 1.930 | 0.114 | -6.145 | 0.737 | -0.353 | 2.478 | -2.104 | 0.020 | -2.250 | -1.096 |
| -0.174 | -0.403 | -1.539 | 1.740 | -1.293 | -1.922 | 1.228 | 1.433 | -2.659 | 2.923 |
| 3.017 | 2.409 | 1.876 | 4.534 | -0.539 | -1.534 | -0.847 | -0.107 | 0.796 | -1.257 |
| 2.632 | -2.417 | 0.136 | -1.155 | 4.277 | -1.035 | -0.968 | -0.400 | 1.393 | -0.858 |
| -1.589 | -2.199 | 0.776 | 0.821 | 3.237 | 4.810 | 1.012 | -4.102 | 1.088 | 1.958 |
| -1.308 | 0.561 | -0.882 | 4.041 | 1.923 | -0.717 | 0.599 | 1.705 | 0.241 | 2.677 |
| -1.881 | -1.808 | -2.767 | 0.426 | 0.234 | -4.060 | 1.036 | -1.657 | 0.471 | 1.753 |
| 2.869 | 4.397 | 1.986 | -2.123 | -0.065 | -0.705 | -0.968 | -2.265 | -1.979 | -1.114 |
| 0.252 | -1.324 | 0.302 | 2.426 | 0.710 | -1.454 | -0.319 | 2.277 | -0.971 | -3.217 |
| -0.833 | 1.653 | -2.738 | 2.856 | -0.789 | -0.873 | -0.809 | -1.538 | -1.334 | 2.289 |
| 0.276 | 0.020 | -0.162 | -1.720 | 0.048 | 0.401 | -2.073 | 2.430 | 2.776 | 1.174 |
| 0.742 | 3.058 | 1.994 | 3.090 | 0.170 | -0.789 | -2.526 | -0.980 | -1.331 | -1.834 |
| 0.770 | 1.419 | 4.391 | 1.502 | -2.856 | -3.648 | -1.179 | 1.556 | 3.176 | 2.613 |
| -2.208 | 1.766 | -0.282 | 3.051 | -1.734 | -0.032 | 1.234 | -0.626 | 1.052 | -3.146 |
| -1.161 | -0.803 | 5.530 | 2.219 | -0.371 | 1.372 | -1.649 | -2.059 | 1.456 | 0.677 |
| -4.276 | -0.196 | -1.456 | 0.139 | 0.094 | 2.367 | -1.902 | 1.123 | -1.222 | 0.323 |
| -1.615 | -0.140 | 0.697 | -0.647 | 1.289 | 1.416 | 0.811 | 0.523 | 1.406 | -1.022 |
| -3.831 | -0.105 | -2.271 | -3.207 | 0.539 | -1.010 | 2.646 | -1.985 | 0.347 | 0.712 |

By permission from tables of the RAND Corporation.

## A-24 RANDOM NORMAL NUMBERS, $\mu = 0$ AND $\sigma = 2$ (CONTINUED)

| 11 | 12 | 13 | 14 | 15 | 16 | 17 | 18 | 19 | 20 |
|---|---|---|---|---|---|---|---|---|---|
| −0.686 | 0.678 | −0.150 | −0.334 | 5.096 | 0.708 | 0.403 | 1.538 | 1.217 | −2.456 |
| 0.106 | −0.018 | 2.558 | −0.049 | −1.061 | −0.574 | −0.510 | −1.036 | −0.168 | 2.516 |
| −0.638 | 3.191 | −4.587 | −0.499 | −0.510 | −1.521 | 1.325 | −1.355 | −1.036 | −3.305 |
| −3.088 | −0.998 | 0.883 | 2.230 | 0.603 | 0.906 | −2.303 | 0.152 | 1.423 | 2.706 |
| 1.017 | −2.496 | −4.981 | 1.769 | −0.252 | 3.018 | 2.764 | 1.105 | −1.527 | −1.187 |
| 0.132 | −1.502 | −1.748 | −3.513 | 0.878 | −0.483 | −0.354 | −1.103 | −2.178 | 2.192 |
| 1.217 | −2.380 | −0.017 | −0.072 | −2.066 | 0.251 | 0.035 | 1.641 | −1.298 | 2.070 |
| 0.540 | −3.950 | 1.287 | −0.771 | −2.946 | −1.181 | −2.286 | −1.561 | 0.420 | −0.746 |
| −0.395 | −1.421 | −2.163 | 0.270 | −1.257 | 5.610 | 0.287 | −1.138 | −0.979 | −2.224 |
| −3.279 | −2.813 | 1.125 | −3.982 | −0.102 | 0.576 | −0.531 | −0.695 | −3.385 | −1.045 |
| 3.510 | 3.235 | 2.623 | −2.582 | 1.274 | 2.440 | 0.824 | 1.983 | −0.770 | 2.489 |
| −0.408 | 1.922 | 2.351 | −0.146 | −0.039 | −0.648 | −3.041 | 0.522 | −1.659 | −0.515 |
| −2.434 | 1.297 | 2.813 | −0.651 | 3.409 | 0.108 | 1.716 | −1.801 | 0.344 | 0.923 |
| 0.541 | 0.820 | −0.254 | 2.851 | −3.027 | 1.553 | 2.218 | 1.758 | 1.003 | −1.148 |
| −0.813 | 1.562 | 2.116 | −4.375 | −2.289 | 1.593 | 0.163 | −1.991 | 0.355 | 1.364 |
| −0.238 | −0.450 | 0.364 | 0.677 | −0.711 | −1.661 | 1.071 | 0.682 | 0.347 | 0.113 |
| −1.214 | −5.369 | 3.300 | 0.461 | 3.197 | −0.368 | −3.190 | 2.868 | −0.943 | 0.931 |
| 3.521 | 1.655 | 2.373 | 1.993 | −1.096 | 1.875 | −1.143 | −3.658 | 2.664 | 0.378 |
| −1.075 | −3.475 | −2.069 | −0.971 | −0.909 | −1.796 | −0.760 | −1.794 | −1.576 | 3.863 |
| −1.213 | −1.760 | 0.397 | −0.323 | 2.659 | 0.666 | −4.368 | 2.704 | 1.160 | −1.377 |
| 0.408 | 2.865 | 3.666 | −0.433 | −1.798 | −1.434 | −3.688 | 2.261 | −1.084 | 0.722 |
| −1.809 | −2.890 | −3.537 | −2.701 | 0.656 | 0.684 | 0.905 | 1.953 | 2.720 | −0.263 |
| −1.633 | 0.283 | −3.937 | −0.224 | −0.549 | 0.016 | −1.265 | −1.650 | −1.506 | 0.504 |
| −1.181 | 2.578 | 0.568 | 0.286 | 1.152 | −0.929 | −3.335 | 0.020 | 1.171 | 1.366 |
| 0.374 | 1.225 | −0.213 | −1.951 | 0.126 | −1.551 | −0.147 | 0.605 | 2.450 | −1.514 |
| −1.828 | −3.459 | 2.624 | 2.605 | 0.698 | −0.984 | −2.289 | −3.389 | 1.647 | −2.592 |
| −3.073 | 1.381 | 6.111 | 0.458 | −0.792 | −0.785 | −1.254 | 2.784 | −0.248 | −0.965 |
| −0.817 | −1.048 | −0.603 | −0.647 | −2.140 | 1.970 | 1.612 | −2.050 | 1.926 | −3.520 |
| −0.778 | −2.697 | 2.431 | −2.011 | 2.810 | 0.010 | 1.830 | −1.425 | 1.425 | −2.506 |
| 1.866 | 0.259 | −1.360 | 2.165 | 1.845 | −0.326 | 2.054 | −0.825 | 5.348 | −0.384 |
| 0.715 | −0.981 | −0.126 | 0.263 | −0.692 | −3.790 | −3.119 | −0.547 | −1.450 | −0.791 |
| 0.802 | −0.906 | −0.726 | 0.071 | 3.693 | −0.409 | 1.536 | −0.907 | 3.216 | −3.096 |
| −0.363 | 0.030 | −2.423 | −0.517 | −4.567 | 1.092 | −1.194 | 0.253 | −2.816 | 0.766 |
| 2.878 | −2.689 | 0.797 | 0.820 | 0.764 | 0.366 | −0.891 | −0.122 | 0.196 | 1.052 |
| −2.245 | 1.324 | 0.101 | 0.431 | −2.152 | 0.779 | −0.708 | 0.028 | 1.317 | −1.259 |
| −0.286 | 0.390 | 1.204 | −4.414 | −0.164 | −3.724 | 0.207 | 1.835 | 0.334 | 0.660 |
| −1.434 | −0.736 | −0.040 | 0.213 | 0.215 | −0.565 | 0.915 | −0.022 | 0.487 | −0.487 |
| 0.004 | −1.773 | −0.480 | 0.768 | −0.837 | −0.513 | 0.828 | 4.563 | 1.298 | 2.837 |
| 1.299 | −1.915 | 0.346 | 1.037 | −2.953 | −1.968 | −1.704 | 1.639 | 2.802 | −2.965 |
| −1.694 | 1.993 | 1.021 | −2.152 | 0.679 | −0.763 | 0.577 | 2.860 | 0.329 | −2.702 |
| −0.506 | 0.328 | −2.091 | −0.238 | 2.582 | −0.429 | 1.647 | −1.048 | −1.367 | 1.054 |
| 0.527 | −3.033 | −0.893 | −0.776 | −0.383 | −0.708 | −0.482 | −2.686 | 1.369 | 1.040 |
| 3.815 | −1.282 | 1.877 | −1.177 | 0.000 | −0.059 | 0.754 | 0.529 | −0.522 | 2.073 |
| 1.838 | −0.569 | 3.556 | −0.956 | −2.106 | 0.371 | 1.806 | 0.449 | −0.867 | 2.664 |
| 0.810 | 3.394 | 1.224 | −4.428 | 4.645 | −0.644 | −2.579 | −2.198 | 2.683 | 1.338 |
| 4.048 | −0.706 | 1.326 | 2.187 | 0.611 | 2.962 | −2.137 | −0.657 | −3.539 | −1.526 |
| 1.619 | −1.214 | −0.860 | −0.625 | −2.534 | 0.519 | −2.539 | 0.737 | −0.622 | −0.146 |
| 1.732 | −3.199 | 2.104 | 1.752 | 0.087 | −0.333 | 3.943 | 0.037 | 0.135 | 1.756 |
| −1.580 | −3.456 | 2.934 | 0.580 | 1.665 | 2.331 | −1.413 | −1.558 | 2.144 | 1.876 |
| 1.176 | −0.195 | 0.127 | 0.060 | −3.145 | −0.001 | 0.219 | −1.803 | −5.255 | 1.524 |

| 21 | 22 | 23 | 24 | 25 | 26 | 27 | 28 | 29 | 30 |
|---|---|---|---|---|---|---|---|---|---|
| −0.625 | −1.119 | 0.772 | −1.479 | 0.164 | 3.051 | 0.297 | 1.904 | 2.864 | 3.093 |
| 1.375 | 1.994 | 1.004 | −3.128 | −1.517 | −2.916 | 2.196 | 3.544 | −1.858 | −1.021 |
| −1.835 | 0.393 | −1.426 | −0.469 | −0.009 | −2.366 | 0.408 | −0.669 | −0.266 | −0.907 |
| 0.764 | −2.796 | −1.932 | −1.144 | −4.177 | −2.150 | 4.163 | 1.003 | −1.088 | −0.346 |
| −0.913 | −4.834 | 2.310 | −0.154 | −2.007 | −1.741 | −3.570 | 1.361 | −0.219 | −1.424 |
| −4.228 | 0.846 | −0.794 | −1.756 | 2.621 | 0.128 | −1.369 | 2.090 | −4.471 | 0.440 |
| −2.724 | 2.694 | −0.585 | 1.094 | 2.116 | −1.176 | 0.180 | 1.438 | 1.260 | −1.730 |
| 4.168 | −0.764 | −0.791 | −2.517 | −2.103 | 0.901 | 0.141 | 1.796 | −4.435 | 1.711 |
| 3.286 | 2.374 | 1.605 | −0.951 | −1.308 | −0.903 | 1.562 | 3.537 | 3.340 | −1.417 |
| 0.197 | 0.212 | 1.303 | −0.289 | 1.441 | −2.913 | −0.606 | −1.302 | 1.281 | 0.147 |
| 0.713 | −1.532 | −4.409 | −2.502 | −1.488 | 1.696 | −2.390 | 0.517 | −2.406 | −0.457 |
| 0.925 | −2.267 | 2.010 | −1.381 | −2.057 | 0.988 | −0.024 | −2.096 | 0.116 | 1.383 |
| −3.312 | 1.604 | 0.955 | −0.184 | 0.074 | −0.714 | 2.059 | −2.293 | 0.899 | −0.837 |
| 0.320 | −2.893 | −1.005 | 1.527 | −0.990 | 1.930 | −1.512 | 1.333 | 3.188 | −1.555 |
| 0.619 | −1.545 | 1.543 | −0.207 | −0.586 | 2.409 | −2.454 | −0.738 | −0.060 | −1.533 |
| 0.119 | −0.542 | −2.461 | −2.475 | −1.265 | −3.598 | 0.983 | −1.702 | −1.735 | −4.773 |
| 1.814 | −0.053 | −0.063 | −2.921 | 2.076 | −0.535 | 2.585 | −3.066 | −0.771 | 1.553 |
| −2.068 | 0.648 | 2.066 | 0.610 | −0.681 | 0.845 | 1.349 | 0.515 | −1.106 | −3.860 |
| −1.881 | −2.033 | 1.704 | 1.161 | 0.316 | 1.623 | −3.370 | −0.261 | −3.559 | −0.647 |
| −2.125 | 0.620 | −0.838 | 2.278 | 0.230 | 2.962 | 1.925 | −2.209 | −0.676 | 0.859 |
| 2.054 | −2.290 | 2.264 | 1.598 | 2.064 | −1.129 | −1.381 | −1.149 | −0.488 | 0.568 |
| −2.516 | −2.190 | −0.629 | 2.361 | 1.734 | 0.607 | 0.935 | 1.275 | 3.125 | −0.224 |
| −0.143 | −1.222 | 1.061 | −2.668 | −4.419 | 0.569 | 0.259 | −0.027 | 1.989 | 4.602 |
| −0.007 | 0.017 | −0.811 | −0.166 | 0.850 | 0.565 | 0.184 | −2.887 | 1.101 | 0.192 |
| −1.749 | 0.231 | −2.380 | −3.177 | −1.077 | 4.460 | 0.494 | 1.941 | −0.106 | 0.015 |
| 1.300 | −0.289 | −2.657 | −0.160 | −0.490 | −0.329 | 1.602 | −1.110 | 4.204 | 2.552 |
| 0.588 | −1.072 | 0.935 | −0.164 | 0.113 | 1.139 | −0.923 | −0.953 | 0.001 | −0.033 |
| 1.719 | −1.183 | −1.051 | −0.944 | 0.734 | 1.965 | 2.121 | 2.213 | 3.826 | −2.004 |
| 0.726 | 1.867 | 0.624 | 2.066 | −2.792 | −2.507 | −0.816 | −0.569 | 0.002 | −1.934 |
| 2.369 | −0.361 | 2.216 | −1.500 | −0.350 | −1.063 | −3.979 | −3.626 | −1.326 | −0.488 |
| 1.790 | −0.290 | 2.601 | 6.261 | −0.622 | −0.534 | 0.477 | 0.075 | 0.167 | −2.351 |
| 1.801 | −2.408 | 0.408 | −2.039 | 0.175 | 3.839 | 3.096 | −0.001 | 2.912 | −0.560 |
| 0.392 | 1.600 | −0.940 | −0.160 | −0.885 | −1.083 | −3.503 | 1.814 | −0.563 | −2.682 |
| −4.113 | −3.018 | 0.523 | −1.915 | −0.722 | −2.769 | 0.210 | −0.381 | −0.724 | −2.013 |
| −0.393 | −0.828 | −0.102 | −2.457 | 1.702 | 2.257 | −2.473 | −1.459 | −1.385 | −3.669 |
| 0.688 | −0.214 | 2.741 | 2.906 | −0.778 | 1.158 | 0.713 | 0.815 | −0.670 | 0.144 |
| 1.957 | 1.104 | 3.540 | 2.726 | −0.028 | −0.181 | −1.477 | −4.434 | 0.457 | 0.057 |
| 1.823 | −1.371 | −4.951 | 3.333 | 0.248 | 1.691 | 2.311 | −2.996 | 1.573 | 2.319 |
| −0.277 | 0.346 | −1.354 | 3.170 | 0.268 | 0.773 | 1.242 | 3.542 | 0.940 | −0.535 |
| −0.484 | 1.447 | −0.512 | −1.379 | −0.808 | 1.014 | 2.103 | 0.005 | −2.122 | 0.843 |
| 3.209 | −1.924 | −0.833 | 1.158 | 3.203 | −0.040 | −0.880 | −2.217 | 0.007 | 0.022 |
| 1.834 | 2.064 | −0.319 | 2.672 | 1.281 | 4.921 | 0.819 | 0.634 | −4.961 | −0.739 |
| −1.618 | 0.705 | 0.220 | −0.177 | −0.117 | −4.699 | 2.210 | 0.035 | 2.403 | −0.816 |
| 2.553 | 1.710 | −2.844 | −4.619 | −4.328 | 0.459 | −2.373 | −1.069 | 2.792 | 1.942 |
| −1.665 | 1.627 | 1.072 | −0.902 | 1.336 | 3.850 | −0.804 | 1.254 | −2.493 | −1.101 |
| −0.468 | 2.177 | 1.703 | 1.897 | 1.139 | −1.606 | −1.139 | −1.435 | 5.162 | −0.146 |
| −1.296 | −0.646 | 0.193 | 0.534 | −0.863 | −3.178 | 2.461 | −1.275 | 0.731 | 2.983 |
| −3.086 | −0.115 | 2.325 | 0.088 | 4.652 | 2.833 | −0.054 | −0.670 | −2.313 | −1.956 |
| −1.324 | 0.102 | 0.665 | 0.878 | −1.760 | −1.038 | 0.685 | −1.034 | 0.380 | −0.463 |
| 1.936 | 2.264 | 0.379 | 4.480 | −3.841 | −3.992 | −3.565 | 2.558 | −0.906 | −0.432 |

## A-24 RANDOM NORMAL NUMBERS, $\mu = 0$ AND $\sigma = 2$ (CONTINUED)

| 31 | 32 | 33 | 34 | 35 | 36 | 37 | 38 | 39 | 40 |
|---|---|---|---|---|---|---|---|---|---|
| −0.660 | −0.996 | −0.264 | −1.823 | 0.818 | −0.410 | −1.786 | 2.399 | 1.986 | 0.242 |
| 1.785 | −0.471 | 0.082 | −3.006 | −2.286 | −0.222 | 0.388 | −0.110 | −0.358 | −0.333 |
| 0.880 | 0.224 | 2.561 | 2.165 | 2.974 | 2.516 | −4.148 | −0.241 | −1.318 | −0.677 |
| 1.021 | 3.100 | −1.783 | −2.063 | −2.176 | 1.959 | 0.248 | 0.597 | 1.394 | 0.612 |
| −2.420 | 5.579 | 2.351 | 1.601 | 1.045 | −2.857 | 1.400 | 3.411 | −0.239 | −2.323 |
| −2.542 | −3.145 | −2.432 | −0.444 | −1.276 | −3.342 | 3.479 | 2.630 | −0.405 | 1.845 |
| −2.437 | 0.104 | 1.110 | 0.008 | −2.173 | 2.294 | −0.529 | 1.723 | −1.609 | 2.119 |
| 3.280 | −1.213 | −3.063 | 3.637 | −0.038 | 0.217 | −0.790 | −1.412 | −0.055 | −0.612 |
| 0.502 | −0.767 | −1.569 | 0.386 | −1.990 | −0.062 | 3.319 | −2.448 | −0.445 | 0.059 |
| −1.684 | 0.290 | 3.179 | 0.158 | 3.562 | −1.929 | −1.170 | 1.179 | 0.110 | 3.655 |
| 0.023 | 1.652 | 3.049 | −1.410 | −1.447 | −0.638 | −2.483 | 2.386 | −0.331 | 1.215 |
| −2.741 | −2.313 | −2.069 | −1.305 | −0.934 | −7.769 | 2.654 | −1.032 | 1.489 | 0.671 |
| −0.746 | 2.099 | −3.225 | 1.533 | −1.741 | −1.922 | 0.895 | −2.974 | −0.828 | 1.734 |
| −0.934 | −4.158 | −3.297 | −2.859 | −4.026 | 2.722 | −1.268 | 0.991 | −1.196 | −0.458 |
| −1.574 | 0.097 | 2.122 | −3.279 | −0.820 | 0.483 | 2.196 | 0.642 | −1.488 | 0.374 |
| 1.261 | −0.663 | 0.616 | −2.801 | 1.065 | 4.845 | 0.418 | −0.226 | 1.897 | 3.554 |
| 2.030 | 1.692 | 0.265 | 0.511 | −1.959 | 0.247 | −1.381 | −2.625 | 0.695 | −2.248 |
| −4.452 | 0.900 | −1.646 | 0.573 | 0.973 | −0.350 | 2.649 | 4.114 | 2.497 | 0.287 |
| −0.075 | −2.069 | −0.574 | 0.001 | −0.784 | −1.235 | −3.191 | 2.128 | 1.168 | −0.742 |
| 0.369 | 0.919 | −2.760 | 1.878 | −5.001 | −1.670 | 0.913 | −2.853 | 0.002 | 1.885 |
| 1.360 | −2.214 | −2.175 | 0.193 | 3.298 | −0.103 | 2.226 | 0.164 | −2.429 | −0.580 |
| −3.271 | 2.845 | −0.102 | −0.822 | −3.646 | 0.361 | −3.188 | −1.031 | 1.846 | −1.622 |
| −0.908 | −3.907 | 1.407 | 0.078 | 1.324 | 0.276 | −2.805 | 0.604 | 1.632 | 2.413 |
| −1.323 | 2.717 | −0.083 | −1.645 | 1.103 | −1.539 | −0.173 | 2.429 | −0.343 | 0.011 |
| 1.095 | −0.871 | 1.636 | 2.345 | −3.127 | 0.500 | 1.250 | −0.072 | −3.248 | 2.603 |
| −1.907 | −0.869 | −4.388 | −0.114 | 1.890 | 0.218 | 0.510 | −0.768 | −2.610 | 3.635 |
| 1.508 | −0.333 | −2.433 | 1.237 | −1.733 | −2.826 | −3.761 | −1.125 | 0.720 | −0.832 |
| 2.937 | 1.887 | −0.430 | −5.194 | 4.716 | −2.950 | −0.393 | −1.111 | 0.008 | −0.186 |
| −4.706 | −1.302 | −2.011 | −0.124 | −2.037 | 0.140 | 1.392 | −1.869 | −2.249 | −0.075 |
| 5.019 | −3.900 | 1.300 | −0.034 | −1.679 | −0.621 | 0.285 | 1.197 | −0.871 | −1.240 |
| −0.910 | −0.495 | 0.074 | 3.144 | −2.631 | −3.152 | 0.192 | −1.073 | 0.646 | 4.381 |
| 1.304 | −1.010 | −0.739 | −1.028 | 2.886 | −1.418 | 1.314 | 0.779 | −2.139 | 0.173 |
| 0.371 | −5.663 | −0.017 | −1.551 | −3.508 | 1.305 | −0.819 | −0.199 | −0.331 | −0.358 |
| −2.039 | −3.961 | 0.679 | 2.451 | −2.802 | 1.449 | −0.964 | −1.170 | 0.891 | 1.560 |
| −0.911 | 1.904 | 0.062 | −2.375 | −1.548 | −0.361 | 2.692 | 3.772 | 2.005 | 3.718 |
| −1.720 | 0.871 | 3.594 | 0.889 | 0.162 | 0.112 | −0.053 | −2.597 | −1.310 | −2.234 |
| −4.091 | 0.430 | 2.222 | −0.141 | 0.506 | 2.751 | −0.472 | −1.141 | 1.671 | −0.920 |
| −1.797 | −1.272 | 1.847 | 0.039 | 0.689 | −0.080 | 1.457 | −3.856 | 1.332 | −2.898 |
| −0.719 | 0.829 | 2.570 | 1.107 | −0.314 | −3.750 | 1.041 | −1.657 | −0.233 | 1.417 |
| −1.890 | 3.240 | 1.877 | 2.552 | 3.389 | 0.215 | 1.979 | −0.895 | −1.996 | 0.611 |
| −1.952 | −1.276 | −2.754 | −0.049 | −2.916 | 3.820 | 0.381 | 1.337 | 2.211 | 3.456 |
| 0.502 | 1.812 | −0.577 | 0.551 | −0.257 | 0.883 | 4.377 | −4.180 | −2.266 | −0.100 |
| 1.971 | 2.333 | −0.945 | 2.618 | 2.953 | −1.997 | 1.491 | −0.082 | 2.617 | 0.749 |
| 0.561 | −1.506 | 4.127 | 0.933 | −1.930 | 0.460 | −0.008 | 0.352 | −1.274 | 0.271 |
| −1.409 | −0.638 | 2.757 | 0.461 | 1.331 | 2.030 | −0.846 | −1.035 | −1.580 | −0.772 |
| −2.066 | 0.218 | 0.070 | −3.420 | 0.089 | −0.084 | 4.944 | −4.285 | 0.200 | 0.276 |
| −2.734 | 3.622 | −0.300 | 1.648 | 1.328 | 0.479 | −0.498 | 1.997 | 2.203 | 2.792 |
| −1.434 | 1.441 | 0.258 | −1.893 | −2.925 | −1.753 | 0.272 | 0.747 | −0.999 | −0.155 |
| −0.071 | −4.344 | −2.763 | 4.371 | 1.547 | 2.588 | 2.914 | 0.261 | 3.381 | 5.445 |
| 4.574 | 1.751 | 3.420 | −1.383 | 0.966 | −2.731 | 3.444 | 1.410 | 2.740 | −2.011 |

| 41 | 42 | 43 | 44 | 45 | 46 | 47 | 48 | 49 | 50 |
|---|---|---|---|---|---|---|---|---|---|
| −1.739 | 0.276 | 1.761 | 0.092 | 0.820 | 1.772 | −3.258 | 0.707 | −0.578 | −1.611 |
| −1.776 | −1.482 | 1.399 | 1.031 | −0.546 | −0.204 | 2.591 | 2.129 | 1.615 | 0.919 |
| −1.894 | 0.388 | 1.023 | −1.493 | 1.513 | 1.003 | 2.547 | −2.443 | −1.855 | 2.898 |
| −2.042 | 1.064 | −2.399 | −0.333 | −2.141 | −1.022 | −2.976 | −0.485 | 0.073 | −0.891 |
| 0.287 | 0.120 | −2.013 | 0.598 | 0.001 | 3.454 | 2.077 | −1.966 | −4.187 | 0.452 |
| 0.739 | −4.324 | 0.088 | 1.124 | 0.610 | 0.368 | 0.953 | −0.141 | −0.441 | 3.163 |
| 0.749 | 0.597 | 2.194 | −0.771 | 1.063 | 0.246 | 0.465 | −3.122 | −1.995 | 3.015 |
| −1.111 | −2.558 | 0.146 | −0.590 | −3.278 | 2.649 | −1.299 | −1.809 | 3.938 | −1.766 |
| −0.347 | 2.324 | 0.083 | −0.152 | −3.563 | −1.062 | 0.901 | 0.882 | 0.865 | 0.581 |
| −0.105 | 1.781 | −0.775 | −0.726 | −3.211 | −1.200 | −2.688 | 1.639 | −0.945 | −1.022 |
| 1.968 | 2.056 | −4.124 | −1.126 | −2.798 | −1.150 | −1.632 | −3.405 | 1.182 | 1.985 |
| 1.614 | −1.436 | −4.649 | −1.168 | 2.549 | 0.522 | −0.616 | 2.009 | −0.465 | 1.362 |
| −1.671 | −0.907 | −0.459 | 2.880 | 2.640 | −0.751 | −2.414 | −1.195 | −2.334 | −0.240 |
| −1.328 | 0.335 | −0.049 | −1.903 | 0.225 | −0.140 | −1.121 | 0.820 | 0.282 | 0.635 |
| −0.623 | −0.823 | 1.655 | 1.997 | −3.841 | 3.318 | 1.035 | 1.056 | 2.112 | 2.166 |
| −2.292 | −1.662 | 2.136 | −0.223 | 1.372 | −3.612 | −0.276 | −4.097 | −0.419 | −0.017 |
| 3.146 | 1.248 | 0.090 | −1.069 | −0.022 | 1.017 | −1.157 | 1.803 | −1.585 | −1.526 |
| 1.553 | −1.369 | 0.044 | 0.606 | −1.734 | −1.443 | −3.016 | −0.977 | 3.150 | 0.264 |
| −3.179 | 3.510 | −2.299 | 0.371 | 1.071 | −1.044 | −1.352 | 1.740 | 1.936 | 0.242 |
| 1.701 | −0.455 | 2.119 | −1.716 | −2.857 | −0.991 | −1.621 | 2.934 | 3.487 | 0.754 |
| −2.088 | 1.495 | 2.961 | −2.029 | −0.072 | −0.664 | 0.992 | 1.659 | 0.834 | −2.175 |
| −3.757 | 0.316 | 0.763 | −3.035 | 0.907 | −3.804 | −3.403 | 3.689 | 0.901 | −2.386 |
| −1.177 | −1.422 | −4.712 | 0.235 | −1.048 | −2.627 | 0.794 | −1.473 | 2.598 | −0.364 |
| −1.380 | −1.661 | −1.714 | −1.396 | 0.477 | 1.750 | −2.458 | −5.077 | −0.194 | 1.093 |
| −0.852 | 0.562 | −0.199 | 0.802 | 0.494 | −0.294 | 0.205 | 0.260 | −2.616 | 4.117 |
| 2.591 | 1.323 | 0.458 | 4.020 | −1.907 | −0.065 | −2.786 | 0.137 | 0.446 | 4.368 |
| −2.240 | 2.744 | 0.551 | −3.005 | −2.677 | 4.492 | 2.928 | 0.061 | −0.216 | 2.566 |
| −1.488 | −0.163 | −0.187 | 2.081 | −0.993 | 1.160 | 1.301 | −2.236 | 1.586 | 0.011 |
| 0.622 | −0.988 | −0.956 | −0.484 | −0.648 | −3.467 | −3.778 | 1.181 | 1.740 | 0.092 |
| −0.949 | −2.527 | −1.934 | 1.318 | 0.422 | 3.848 | 0.050 | −1.448 | 0.278 | 3.041 |
| −4.952 | 0.019 | 1.793 | 0.881 | 0.282 | 0.621 | 1.202 | −0.373 | 3.665 | 3.386 |
| −0.751 | 3.342 | 0.969 | 0.821 | 1.983 | −0.533 | −1.273 | −2.214 | −0.774 | −1.210 |
| 0.618 | −0.688 | −2.960 | −5.252 | −0.543 | 0.104 | −0.468 | −3.139 | 0.594 | −1.302 |
| 2.371 | 0.160 | 1.715 | 0.319 | 1.387 | 5.138 | 3.883 | −1.869 | −0.899 | −1.019 |
| −1.184 | 0.047 | 1.453 | −0.889 | −1.292 | 0.197 | −0.302 | −1.497 | −1.838 | −0.940 |
| −0.287 | 2.329 | 2.028 | −1.765 | 1.669 | −1.024 | 1.600 | 0.454 | 3.098 | 2.275 |
| 1.764 | −2.839 | −1.942 | 0.008 | 4.001 | 0.083 | −1.631 | 2.968 | −0.146 | −2.079 |
| 1.149 | −1.571 | 1.296 | 1.510 | −0.599 | 0.083 | −0.688 | 6.017 | 0.012 | −1.451 |
| 2.984 | −1.432 | −0.960 | −2.124 | 1.353 | 0.934 | 0.666 | 3.096 | 2.905 | −1.472 |
| −1.701 | −0.004 | 2.710 | 0.573 | 2.424 | −0.119 | −1.410 | 3.413 | −3.588 | 0.047 |
| −2.333 | 0.912 | −0.773 | −2.016 | 2.253 | 2.784 | 3.764 | 0.559 | 4.791 | 1.288 |
| −0.214 | 2.787 | 0.095 | −3.174 | 1.460 | 0.411 | 0.922 | −0.474 | 3.113 | −1.067 |
| 1.214 | 0.785 | −2.686 | 1.909 | −1.747 | −4.551 | 0.589 | −0.573 | −1.364 | −2.583 |
| 0.878 | 0.097 | 1.650 | 1.437 | −1.643 | −2.608 | 1.122 | 0.538 | 0.664 | −0.323 |
| −0.105 | −0.297 | 3.821 | 2.105 | 2.021 | −1.922 | 1.472 | 0.042 | 1.403 | 1.465 |
| −0.593 | 0.136 | 0.910 | −0.549 | −1.472 | 3.214 | −2.273 | 3.458 | 1.436 | 0.500 |
| 2.198 | 2.325 | −1.229 | −0.276 | 1.560 | −0.482 | −0.482 | 0.455 | −0.181 | 1.417 |
| 1.160 | 0.139 | 0.997 | −0.082 | −0.689 | 0.995 | −5.301 | 0.998 | 3.413 | −1.797 |
| 3.024 | −1.561 | 0.982 | −1.244 | 1.407 | −0.063 | −1.176 | 2.355 | 2.006 | −4.833 |
| 0.955 | 0.174 | −0.401 | 2.472 | 0.584 | 3.811 | 1.115 | 0.951 | −2.136 | −2.324 |

## A-25a CONFIDENCE INTERVALS FOR THE MEDIAN

If the observations are arranged in order of size $X_1 < X_2 < X_3 < \cdots < X_N$, then we are $100(1 - \alpha)$ per cent confident that the population median will be between $X_k$ and $X_{N-k+1}$, where $k$ and $\alpha$ are given below.

| $N$ | Largest $k$ | $\alpha \leq .05$ | Largest $k$ | $\alpha \leq .01$ | $N$ | Largest $k$ | $\alpha \leq .05$ | Largest $k$ | $\alpha \leq .01$ |
|---|---|---|---|---|---|---|---|---|---|
| 6 | 1 | .031 | | | 36 | 12 | .029 | 10 | .004 |
| 7 | 1 | .016 | | | 37 | 13 | .047 | 11 | .008 |
| 8 | 1 | .008 | 1 | .008 | 38 | 13 | .034 | 11 | .005 |
| 9 | 2 | .039 | 1 | .004 | 39 | 13 | .024 | 12 | .009 |
| 10 | 2 | .021 | 1 | .002 | 40 | 14 | .038 | 12 | .006 |
| 11 | 2 | .012 | 1 | .001 | 41 | 14 | .028 | 12 | .004 |
| 12 | 3 | .039 | 2 | .006 | 42 | 15 | .044 | 13 | .008 |
| 13 | 3 | .022 | 2 | .003 | 43 | 15 | .032 | 13 | .005 |
| 14 | 3 | .013 | 2 | .002 | 44 | 16 | .049 | 14 | .010 |
| 15 | 4 | .035 | 3 | .007 | 45 | 16 | .036 | 14 | .007 |
| 16 | 4 | .021 | 3 | .004 | 46 | 16 | .026 | 14 | .005 |
| 17 | 5 | .049 | 3 | .002 | 47 | 17 | .040 | 15 | .008 |
| 18 | 5 | .031 | 4 | .008 | 48 | 17 | .029 | 15 | .006 |
| 19 | 5 | .019 | 4 | .004 | 49 | 18 | .044 | 16 | .009 |
| 20 | 6 | .041 | 4 | .003 | 50 | 18 | .033 | 16 | .007 |
| 21 | 6 | .027 | 5 | .007 | 51 | 19 | .049 | 16 | .005 |
| 22 | 6 | .017 | 5 | .004 | 52 | 19 | .036 | 17 | .008 |
| 23 | 7 | .035 | 5 | .003 | 53 | 19 | .027 | 17 | .005 |
| 24 | 7 | .023 | 6 | .007 | 54 | 20 | .040 | 18 | .009 |
| 25 | 8 | .043 | 6 | .004 | 55 | 20 | .030 | 18 | .006 |
| 26 | 8 | .029 | 7 | .009 | 56 | 21 | .044 | 18 | .005 |
| 27 | 8 | .019 | 7 | .006 | 57 | 21 | .033 | 19 | .008 |
| 28 | 9 | .036 | 7 | .004 | 58 | 22 | .048 | 19 | .005 |
| 29 | 9 | .024 | 8 | .008 | 59 | 22 | .036 | 20 | .009 |
| 30 | 10 | .043 | 8 | .005 | 60 | 22 | .027 | 20 | .006 |
| 31 | 10 | .029 | 8 | .003 | 61 | 23 | .040 | 21 | .010 |
| 32 | 10 | .020 | 9 | .007 | 62 | 23 | .030 | 21 | .007 |
| 33 | 11 | .035 | 9 | .005 | 63 | 24 | .043 | 21 | .005 |
| 34 | 11 | .024 | 10 | .009 | 64 | 24 | .033 | 22 | .008 |
| 35 | 12 | .041 | 10 | .006 | 65 | 25 | .046 | 22 | .006 |

By permission of S. K. Banerjee from K. R. Nair, "Table of confidence interval for the median in samples from any continuous population," *Sankhya*, vol. 4 (1940), pp. 551–558.

$\gamma$ is the probability that an interval will cover a proportion $P$ of the population with a random sample of size $N$

### A-25b $P$ for interval between sample extremes

| $N$ \ $\gamma$ | .5 | .7 | .9 | .95 | .99 | .995 |
|---|---|---|---|---|---|---|
| 2 | .293 | .164 | .052 | .026 | .006 | .003 |
| 4 | .615 | .492 | .321 | .249 | .141 | .111 |
| 6 | .736 | .640 | .490 | .419 | .295 | .254 |
| 10 | .838 | .774 | .664 | .606 | .496 | .456 |
| 20 | .918 | .883 | .820 | .784 | .712 | .683 |
| 40 | .959 | .941 | .907 | .887 | .846 | .829 |
| 60 | .973 | .960 | .937 | .924 | .895 | .883 |
| 80 | .980 | .970 | .953 | .943 | .920 | .911 |
| 100 | .984 | .976 | .962 | .954 | .936 | .929 |
| 150 | .990 | .984 | .975 | .969 | .957 | .952 |
| 200 | .992 | .988 | .981 | .977 | .968 | .964 |
| 300 | .995 | .992 | .988 | .985 | .979 | .976 |
| 500 | .997 | .996 | .993 | .991 | .987 | .986 |
| 700 | .998 | .997 | .995 | .994 | .991 | .990 |
| 900 | .999 | .998 | .996 | .995 | .993 | .992 |
| 1,000 | .999 | .998 | .997 | .996 | .994 | .993 |

### A-25c $\gamma$ for interval between sample extremes

| $N$ \ $P$ | .5 | .7 | .9 | .95 | .99 |
|---|---|---|---|---|---|
| 2 | .250 | .090 | .010 | .003 | .000 |
| 4 | .688 | .348 | .052 | .014 | .001 |
| 6 | .891 | .580 | .114 | .033 | .001 |
| 10 | .989 | .851 | .264 | .086 | .004 |
| 20 | 1.000 | .992 | .608 | .264 | .017 |
| 40 | | 1.000 | .920 | .601 | .061 |
| 60 | | | .986 | .808 | .121 |
| 80 | | | .998 | .914 | .191 |
| 100 | | | 1.000 | .963 | .264 |
| 150 | | | | .996 | .443 |
| 200 | | | | 1.000 | .595 |
| 300 | | | | | .802 |
| 500 | | | | | .960 |
| 700 | | | | | .993 |
| 900 | | | | | .999 |
| 1,000 | | | | | 1.000 |

### A-25d $N$ for interval between sample extremes

| $P$ \ $\gamma$ | .50 | .70 | .90 | .95 | .99 | .995 |
|---|---|---|---|---|---|---|
| .995 | 336 | 488 | 777 | 947 | 1,325 | 1,483 |
| .99 | 168 | 244 | 388 | 473 | 662 | 740 |
| .95 | 34 | 49 | 77 | 93 | 130 | 146 |
| .90 | 17 | 24 | 38 | 46 | 64 | 72 |
| .85 | 11 | 16 | 25 | 30 | 42 | 47 |
| .80 | 9 | 12 | 18 | 22 | 31 | 34 |
| .75 | 7 | 10 | 15 | 18 | 24 | 27 |
| .70 | 6 | 8 | 12 | 14 | 20 | 22 |
| .60 | 4 | 6 | 9 | 10 | 14 | 16 |
| .50 | 3 | 5 | 7 | 8 | 11 | 12 |

### A-25e $N$ for interval below (above) the largest (smallest) sample value

| $P$ \ $\gamma$ | .50 | .70 | .95 | .99 | .995 |
|---|---|---|---|---|---|
| .995 | 139 | 241 | 598 | 919 | 1,379 |
| .990 | 69 | 120 | 299 | 459 | 688 |
| .950 | 14 | 24 | 59 | 90 | 135 |
| .90 | 7 | 12 | 29 | 44 | 66 |
| .85 | 5 | 8 | 19 | 29 | 43 |
| .80 | 4 | 6 | 14 | 21 | 31 |
| .75 | 3 | 5 | 11 | 17 | 25 |
| .70 | 2 | 4 | 9 | 13 | 20 |
| .60 | 2 | 3 | 6 | 10 | 14 |
| .50 | 1 | 2 | 5 | 7 | 10 |

# A-26 SOME ONE-SIDED AND SYMMETRICAL SIGNIFICANCE TESTS FOR $N \leq 10$

| | SIGNIFICANCE LEVEL OF TESTS, PER CENT | | SYMMETRICAL: ACCEPT $\mu \neq \mu_0$ IF EITHER | |
|---|---|---|---|---|
| $N$ | *One-sided* | *Symmetrical* | *One-sided: accept* $\mu < \mu_0$ *if* | *One-sided: accept* $\mu > \mu_0$ *if* |
| 4 | 6.2 | 12.5 | $X_4 < \mu_0$ | $X_1 > \mu_0$ |
| 5 | 6.2 | 12.5 | $\frac{1}{2}(X_4 + X_5) < \mu_0$ | $\frac{1}{2}(X_1 + X_2) > \mu_0$ |
| | 3.1 | 6.2 | $X_5 < \mu_0$ | $X_1 > \mu_0$ |
| 6 | 4.7 | 9.4 | $\max [X_5, \frac{1}{2}(X_4 + X_6)] < \mu_0$ | $\min [X_2, \frac{1}{2}(X_1 + X_3)] > \mu_0$ |
| | 3.1 | 6.2 | $\frac{1}{2}(X_5 + X_6) < \mu_0$ | $\frac{1}{2}(X_1 + X_2) > \mu_0$ |
| | 1.6 | 3.1 | $X_6 < \mu_0$ | $X_1 > \mu_0$ |
| 7 | 5.5 | 10.9 | $\max [X_5, \frac{1}{2}(X_4 + X_7)] < \mu_0$ | $\min [X_3, \frac{1}{2}(X_1 + X_4)] > \mu_0$ |
| | 2.3 | 4.7 | $\max [X_6, \frac{1}{2}(X_5 + X_7)] < \mu_0$ | $\min [X_2, \frac{1}{2}(X_1 + X_3)] > \mu_0$ |
| | 1.6 | 3.1 | $\frac{1}{2}(X_6 + X_7) < \mu_0$ | $\frac{1}{2}(X_1 + X_2) > \mu_0$ |
| | 0.8 | 1.6 | $X_7 < \mu_0$ | $X_1 > \mu_0$ |
| 8 | 4.3 | 8.6 | $\max [X_6, \frac{1}{2}(X_4 + X_8)] < \mu_0$ | $\min [X_3, \frac{1}{2}(X_1 + X_5)] > \mu_0$ |
| | 2.7 | 5.5 | $\max [X_6, \frac{1}{2}(X_5 + X_8)] < \mu_0$ | $\min [X_3, \frac{1}{2}(X_1 + X_4)] > \mu_0$ |
| | 1.2 | 2.3 | $\max [X_7, \frac{1}{2}(X_6 + X_8)] < \mu_0$ | $\min [X_2, \frac{1}{2}(X_1 + X_3)] > \mu_0$ |
| | 0.8 | 1.6 | $\frac{1}{2}(X_7 + X_8) < \mu_0$ | $\frac{1}{2}(X_1 + X_2) > \mu_0$ |
| | 0.4 | 0.8 | $X_8 < \mu_0$ | $X_1 > \mu_0$ |
| 9 | 5.1 | 10.2 | $\max [X_6, \frac{1}{2}(X_4 + X_9)] < \mu_0$ | $\min [X_4, \frac{1}{2}(X_1 + X_6)] > \mu_0$ |
| | 2.2 | 4.3 | $\max [X_7, \frac{1}{2}(X_5 + X_9)] < \mu_0$ | $\min [X_3, \frac{1}{2}(X_1 + X_5)] > \mu_0$ |
| | 1.0 | 2.0 | $\max [X_8, \frac{1}{2}(X_5 + X_9)] < \mu_0$ | $\min [X_2, \frac{1}{2}(X_1 + X_5)] > \mu_0$ |
| | 0.6 | 1.2 | $\max [X_8, \frac{1}{2}(X_7 + X_9)] < \mu_0$ | $\min [X_2, \frac{1}{2}(X_1 + X_3)] > \mu_0$ |
| | 0.4 | 0.8 | $\frac{1}{2}(X_8 + X_9) < \mu_0$ | $\frac{1}{2}(X_1 + X_2) > \mu_0$ |
| 10 | 5.6 | 11.1 | $\max [X_6, \frac{1}{2}(X_4 + X_{10})] < \mu_0$ | $\min [X_5, \frac{1}{2}(X_1 + X_7)] > \mu_0$ |
| | 2.5 | 5.1 | $\max [X_7, \frac{1}{2}(X_5 + X_{10})] < \mu_0$ | $\min [X_4, \frac{1}{2}(X_1 + X_6)] > \mu_0$ |
| | 1.1 | 2.1 | $\max [X_8, \frac{1}{2}(X_6 + X_{10})] < \mu_0$ | $\min [X_3, \frac{1}{2}(X_1 + X_5)] > \mu_0$ |
| | 0.5 | 1.0 | $\max [X_9, \frac{1}{2}(X_6 + X_{10})] < \mu_0$ | $\min [X_2, \frac{1}{2}(X_1 + X_5)] > \mu_0$ |

By permission from J. E. Walsh, "Applications of some significance tests for the median which are valid under very general conditions," *Journal of the American Statistical Association,* vol. 44 (1949), p. 342.

(CONFIDENCE COEFFICIENT .95)

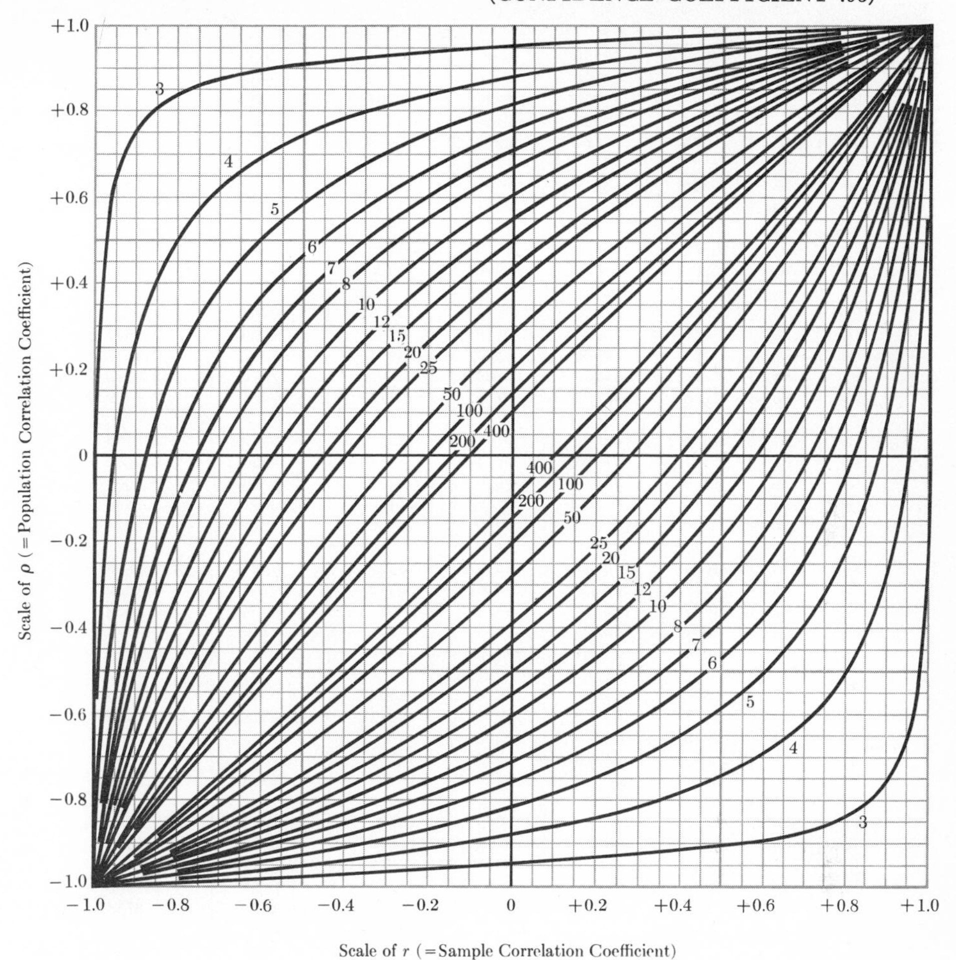


Scale of $r$ (= Sample Correlation Coefficient)

The numbers on the curves indicate sample size

By permission of Prof. E. S. Pearson from F. N. David, *Tables of the Ordinates and Probability Integral of the Distribution of the Correlation Coefficient in Small Samples*. The Biometrika Office, London.

## A-28 VARIOUS FUNCTIONS OF A PROPORTION $p$

| $p$ | $1-p$ | $p(1-p)$ | $\sqrt{p(1-p)}$ | $1-p^2$ | $1-(1-p)^2$ | $2 \arcsin \sqrt{p}$ | $2 \arcsin \sqrt{1-p}$ |
|---|---|---|---|---|---|---|---|
| .01 | .99 | .0099 | .09950 | .9999 | .0199 | .2003 | 2.9413 |
| .02 | .98 | .0196 | .14000 | .9996 | .0396 | .2838 | 2.8578 |
| .03 | .97 | .0291 | .17059 | .9991 | .0591 | .3482 | 2.7934 |
| .04 | .96 | .0384 | .19596 | .9984 | .0784 | .4027 | 2.7389 |
| .05 | .95 | .0475 | .21794 | .9975 | .0975 | .4510 | 2.6906 |
| .06 | .94 | .0564 | .23749 | .9964 | .1164 | .4949 | 2.6467 |
| .07 | .93 | .0651 | .25515 | .9951 | .1351 | .5355 | 2.6062 |
| .08 | .92 | .0736 | .27129 | .9936 | .1536 | .5735 | 2.5681 |
| .09 | .91 | .0819 | .28618 | .9919 | .1719 | .6094 | 2.5319 |
| .10 | .90 | .0900 | .30000 | .9900 | .1900 | .6435 | 2.4981 |
| .11 | .89 | .0979 | .31289 | .9879 | .2079 | .6761 | 2.4655 |
| .12 | .88 | .1056 | .32496 | .9856 | .2256 | .7075 | 2.4341 |
| .13 | .87 | .1131 | .33630 | .9831 | .2431 | .7377 | 2.4037 |
| .14 | .86 | .1204 | .34699 | .9804 | .2604 | .7670 | 2.3746 |
| .15 | .85 | .1275 | .35707 | .9775 | .2775 | .7954 | 2.3462 |
| .16 | .84 | .1344 | .36661 | .9744 | .2944 | .8230 | 2.3186 |
| .17 | .83 | .1411 | .37563 | .9711 | .3111 | .8500 | 2.2916 |
| .18 | .82 | .1476 | .38419 | .9676 | .3276 | .8763 | 2.2653 |
| .19 | .81 | .1539 | .39230 | .9639 | .3439 | .9021 | 2.2395 |
| .20 | .80 | .1600 | .40000 | .9600 | .3600 | .9273 | 2.2143 |
| .21 | .79 | .1659 | .40731 | .9559 | .3759 | .9521 | 2.1895 |
| .22 | .78 | .1716 | .41425 | .9516 | .3916 | .9764 | 2.1652 |
| .23 | .77 | .1771 | .42083 | .9471 | .4071 | 1.0004 | 2.1412 |
| .24 | .76 | .1824 | .42708 | .9424 | .4224 | 1.0239 | 2.1177 |
| .25 | .75 | .1875 | .43301 | .9375 | .4375 | 1.0472 | 2.0944 |
| .26 | .74 | .1924 | .43863 | .9324 | .4524 | 1.0701 | 2.0715 |
| .27 | .73 | .1971 | .44396 | .9271 | .4671 | 1.0928 | 2.0488 |
| .28 | .72 | .2016 | .44900 | .9216 | .4816 | 1.1152 | 2.0264 |
| .29 | .71 | .2059 | .45376 | .9159 | .4959 | 1.1374 | 2.0042 |
| .30 | .70 | .2100 | .45826 | .9100 | .5100 | 1.1593 | 1.9823 |
| .31 | .69 | .2139 | .46249 | .9039 | .5239 | 1.1810 | 1.9606 |
| .32 | .68 | .2176 | .46648 | .8976 | .5376 | 1.2025 | 1.9391 |
| .33 | .67 | .2211 | .47021 | .8911 | .5511 | 1.2239 | 1.9177 |
| .34 | .66 | .2244 | .47371 | .8844 | .5644 | 1.2451 | 1.8965 |
| .35 | .65 | .2275 | .47697 | .8775 | .5775 | 1.2661 | 1.8755 |
| .36 | .64 | .2304 | .48000 | .8704 | .5904 | 1.2870 | 1.8546 |
| .37 | .63 | .2331 | .48280 | .8631 | .6031 | 1.3078 | 1.8338 |
| .38 | .62 | .2356 | .48539 | .8556 | .6156 | 1.3284 | 1.8132 |
| .39 | .61 | .2379 | .48775 | .8479 | .6279 | 1.3490 | 1.7826 |
| .40 | .60 | .2400 | .48990 | .8400 | .6400 | 1.3694 | 1.7722 |
| .41 | .59 | .2419 | .49183 | .8319 | .6519 | 1.3898 | 1.7518 |
| .42 | .58 | .2436 | .49356 | .8236 | .6636 | 1.4101 | 1.7315 |
| .43 | .57 | .2451 | .49508 | .8151 | .6751 | 1.4303 | 1.7113 |
| .44 | .56 | .2464 | .49639 | .8064 | .6864 | 1.4505 | 1.6911 |
| .45 | .55 | .2475 | .49749 | .7975 | .6975 | 1.4706 | 1.6710 |
| .46 | .54 | .2484 | .49840 | .7884 | .7084 | 1.4907 | 1.6509 |
| .47 | .53 | .2491 | .49910 | .7791 | .7191 | 1.5108 | 1.6308 |
| .48 | .52 | .2496 | .49960 | .7696 | .7296 | 1.5308 | 1.6108 |
| .49 | .51 | .2499 | .49990 | .7599 | .7399 | 1.5508 | 1.5908 |
| .50 | .50 | .2500 | .50000 | .7500 | .7500 | 1.5708 | 1.5708 |

## BINOMIAL PROBABILITIES, $C(N,X)p^X(1-p)^{N-X}$ FOR $N \leq 10$ AND FOR VARIOUS VALUES OF $p$ A-29a

| N | X \ p | .01 | .05 | .10 | .15 | .20 | .25 | .30 | ⅓ | .35 | .40 | .45 | .50 |
|---|---|---|---|---|---|---|---|---|---|---|---|---|---|
| 2 | 0 | .9801 | .9025 | .8100 | .7225 | .6400 | .5625 | .4900 | .4444 | .4225 | .3600 | .3025 | .2500 |
| | 1 | .0198 | .0950 | .1800 | .2550 | .3200 | .3750 | .4200 | .4444 | .4550 | .4800 | .4950 | .5000 |
| | 2 | .0001 | .0025 | .0100 | .0225 | .0400 | .0625 | .0900 | .1111 | .1225 | .1600 | .2025 | .2500 |
| 3 | 0 | .9703 | .8574 | .7290 | .6141 | .5120 | .4219 | .3430 | .2963 | .2746 | .2160 | .1664 | .1250 |
| | 1 | .0294 | .1354 | .2430 | .3251 | .3840 | .4219 | .4410 | .4444 | .4436 | .4320 | .4084 | .3750 |
| | 2 | .0003 | .0071 | .0270 | .0574 | .0960 | .1406 | .1890 | .2222 | .2389 | .2880 | .3341 | .3750 |
| | 3 | .0000 | .0001 | .0010 | .0034 | .0080 | .0156 | .0270 | .0370 | .0429 | .0640 | .0911 | .1250 |
| 4 | 0 | .9606 | .8145 | .6561 | .5220 | .4096 | .3164 | .2401 | .1975 | .1785 | .1296 | .0915 | .0625 |
| | 1 | .0388 | .1715 | .2916 | .3685 | .4096 | .4219 | .4116 | .3951 | .3845 | .3456 | .2995 | .2500 |
| | 2 | .0006 | .0135 | .0486 | .0975 | .1536 | .2109 | .2646 | .2963 | .3105 | .3456 | .3675 | .3750 |
| | 3 | .0000 | .0005 | .0036 | .0115 | .0256 | .0469 | .0756 | .0988 | .1115 | .1536 | .2005 | .2500 |
| | 4 | .0000 | .0000 | .0001 | .0005 | .0016 | .0039 | .0081 | .0123 | .0150 | .0256 | .0410 | .0625 |
| 5 | 0 | .9510 | .7738 | .5905 | .4437 | .3277 | .2373 | .1681 | .1317 | .1160 | .0778 | .0503 | .0312 |
| | 1 | .0480 | .2036 | .3280 | .3915 | .4096 | .3955 | .3602 | .3292 | .3124 | .2592 | .2059 | .1562 |
| | 2 | .0010 | .0214 | .0729 | .1382 | .2048 | .2637 | .3087 | .3292 | .3364 | .3456 | .3369 | .3125 |
| | 3 | .0000 | .0011 | .0081 | .0244 | .0512 | .0879 | .1323 | .1646 | .1811 | .2304 | .2757 | .3125 |
| | 4 | .0000 | .0000 | .0004 | .0022 | .0064 | .0146 | .0284 | .0412 | .0488 | .0768 | .1128 | .1562 |
| | 5 | .0000 | .0000 | .0000 | .0001 | .0003 | .0010 | .0024 | .0041 | .0053 | .0102 | .0185 | .0312 |
| 6 | 0 | .9415 | .7351 | .5314 | .3771 | .2621 | .1780 | .1176 | .0878 | .0754 | .0467 | .0277 | .0156 |
| | 1 | .0571 | .2321 | .3543 | .3993 | .3932 | .3560 | .3025 | .2634 | .2437 | .1866 | .1359 | .0938 |
| | 2 | .0014 | .0305 | .0984 | .1762 | .2458 | .2966 | .3241 | .3292 | .3280 | .3110 | .2780 | .2344 |
| | 3 | .0000 | .0021 | .0146 | .0415 | .0819 | .1318 | .1852 | .2195 | .2355 | .2765 | .3032 | .3125 |
| | 4 | .0000 | .0001 | .0012 | .0055 | .0154 | .0330 | .0595 | .0823 | .0951 | .1382 | .1861 | .2344 |
| | 5 | .0000 | .0000 | .0001 | .0004 | .0015 | .0044 | .0102 | .0165 | .0205 | .0369 | .0609 | .0938 |
| | 6 | .0000 | .0000 | .0000 | .0000 | .0001 | .0002 | .0007 | .0014 | .0018 | .0041 | .0083 | .0156 |
| 7 | 0 | .9321 | .6983 | .4783 | .3206 | .2097 | .1335 | .0824 | .0585 | .0490 | .0280 | .0152 | .0078 |
| | 1 | .0659 | .2573 | .3720 | .3960 | .3670 | .3115 | .2471 | .2048 | .1848 | .1306 | .0872 | .0547 |
| | 2 | .0020 | .0406 | .1240 | .2097 | .2753 | .3115 | .3177 | .3073 | .2985 | .2613 | .2140 | .1641 |
| | 3 | .0000 | .0036 | .0230 | .0617 | .1147 | .1730 | .2269 | .2561 | .2679 | .2903 | .2918 | .2734 |
| | 4 | .0000 | .0002 | .0026 | .0109 | .0287 | .0577 | .0972 | .1280 | .1442 | .1935 | .2388 | .2734 |
| | 5 | .0000 | .0000 | .0002 | .0012 | .0043 | .0115 | .0250 | .0384 | .0466 | .0774 | .1172 | .1641 |
| | 6 | .0000 | .0000 | .0000 | .0001 | .0004 | .0013 | .0036 | .0064 | .0084 | .0172 | .0320 | .0547 |
| | 7 | .0000 | .0000 | .0000 | .0000 | .0000 | .0001 | .0002 | .0005 | .0006 | .0016 | .0037 | .0078 |
| 8 | 0 | .9227 | .6634 | .4305 | .2725 | .1678 | .1001 | .0576 | .0390 | .0319 | .0168 | .0084 | .0039 |
| | 1 | .0746 | .2793 | .3826 | .3847 | .3355 | .2670 | .1977 | .1561 | .1373 | .0896 | .0548 | .0312 |
| | 2 | .0026 | .0515 | .1488 | .2376 | .2936 | .3115 | .2965 | .2731 | .2587 | .2090 | .1569 | .1094 |
| | 3 | .0001 | .0054 | .0331 | .0839 | .1468 | .2076 | .2541 | .2731 | .2786 | .2787 | .2568 | .2188 |
| | 4 | .0000 | .0004 | .0046 | .0185 | .0459 | .0865 | .1361 | .1707 | .1875 | .2322 | .2627 | .2734 |
| | 5 | .0000 | .0000 | .0004 | .0026 | .0092 | .0231 | .0467 | .0683 | .0808 | .1239 | .1719 | .2188 |
| | 6 | .0000 | .0000 | .0000 | .0002 | .0011 | .0038 | .0100 | .0171 | .0217 | .0413 | .0703 | .1094 |
| | 7 | .0000 | .0000 | .0000 | .0000 | .0001 | .0004 | .0012 | .0024 | .0033 | .0079 | .0164 | .0312 |
| | 8 | .0000 | .0000 | .0000 | .0000 | .0000 | .0000 | .0001 | .0002 | .0002 | .0007 | .0017 | .0039 |
| 9 | 0 | .9135 | .6302 | .3874 | .2316 | .1342 | .0751 | .0404 | .0260 | .0207 | .0101 | .0046 | .0020 |
| | 1 | .0830 | .2985 | .3874 | .3679 | .3020 | .2253 | .1556 | .1171 | .1004 | .0605 | .0339 | .0176 |
| | 2 | .0034 | .0629 | .1722 | .2597 | .3020 | .3003 | .2668 | .2341 | .2162 | .1612 | .1110 | .0703 |
| | 3 | .0001 | .0077 | .0446 | .1069 | .1762 | .2336 | .2668 | .2731 | .2716 | .2508 | .2119 | .1641 |
| | 4 | .0000 | .0006 | .0074 | .0283 | .0661 | .1168 | .1715 | .2048 | .2194 | .2508 | .2600 | .2461 |
| | 5 | .0000 | .0000 | .0008 | .0050 | .0165 | .0389 | .0735 | .1024 | .1181 | .1672 | .2128 | .2461 |
| | 6 | .0000 | .0000 | .0001 | .0006 | .0028 | .0087 | .0210 | .0341 | .0424 | .0743 | .1160 | .1641 |
| | 7 | .0000 | .0000 | .0000 | .0000 | .0003 | .0012 | .0039 | .0073 | .0098 | .0212 | .0407 | .0703 |
| | 8 | .0000 | .0000 | .0000 | .0000 | .0000 | .0001 | .0004 | .0009 | .0013 | .0035 | .0083 | .0176 |
| | 9 | .0000 | .0000 | .0000 | .0000 | .0000 | .0000 | .0000 | .0001 | .0001 | .0003 | .0008 | .0020 |
| 10 | 0 | .9044 | .5987 | .3487 | .1969 | .1074 | .0563 | .0282 | .0173 | .0135 | .0060 | .0025 | .0010 |
| | 1 | .0914 | .3151 | .3874 | .3474 | .2684 | .1877 | .1211 | .0867 | .0725 | .0403 | .0207 | .0098 |
| | 2 | .0042 | .0746 | .1937 | .2759 | .3020 | .2816 | .2335 | .1951 | .1757 | .1209 | .0763 | .0439 |
| | 3 | .0001 | .0105 | .0574 | .1298 | .2013 | .2503 | .2668 | .2601 | .2522 | .2150 | .1665 | .1172 |
| | 4 | .0000 | .0010 | .0112 | .0401 | .0881 | .1460 | .2001 | .2276 | .2377 | .2508 | .2384 | .2051 |
| | 5 | .0000 | .0001 | .0015 | .0085 | .0264 | .0584 | .1029 | .1366 | .1536 | .2007 | .2340 | .2461 |
| | 6 | .0000 | .0000 | .0001 | .0012 | .0055 | .0162 | .0368 | .0569 | .0689 | .1115 | .1596 | .2051 |
| | 7 | .0000 | .0000 | .0000 | .0001 | .0008 | .0031 | .0090 | .0163 | .0212 | .0425 | .0746 | .1172 |
| | 8 | .0000 | .0000 | .0000 | .0000 | .0001 | .0004 | .0014 | .0030 | .0043 | .0106 | .0229 | .0439 |
| | 9 | .0000 | .0000 | .0000 | .0000 | .0000 | .0000 | .0001 | .0003 | .0005 | .0016 | .0042 | .0098 |
| | 10 | .0000 | .0000 | .0000 | .0000 | .0000 | .0000 | .0000 | .0000 | .0000 | .0001 | .0003 | .0010 |

## A-29b BINOMIAL COEFFICIENTS

| $N$ | $C(N,0)$ | $C(N,1)$ | $C(N,2)$ | $C(N,3)$ | $C(N,4)$ | $C(N,5)$ | $C(N,6)$ | $C(N,7)$ | $C(N,8)$ | $C(N,9)$ | $C(N,10)$ |
|---|---|---|---|---|---|---|---|---|---|---|---|
| 0 | 1 | | | | | | | | | | |
| 1 | 1 | 1 | | | | | | | | | |
| 2 | 1 | 2 | 1 | | | | | | | | |
| 3 | 1 | 3 | 3 | 1 | | | | | | | |
| 4 | 1 | 4 | 6 | 4 | 1 | | | | | | |
| 5 | 1 | 5 | 10 | 10 | 5 | 1 | | | | | |
| 6 | 1 | 6 | 15 | 20 | 15 | 6 | 1 | | | | |
| 7 | 1 | 7 | 21 | 35 | 35 | 21 | 7 | 1 | | | |
| 8 | 1 | 8 | 28 | 56 | 70 | 56 | 28 | 8 | 1 | | |
| 9 | 1 | 9 | 36 | 84 | 126 | 126 | 84 | 36 | 9 | 1 | |
| 10 | 1 | 10 | 45 | 120 | 210 | 252 | 210 | 120 | 45 | 10 | 1 |
| 11 | 1 | 11 | 55 | 165 | 330 | 462 | 462 | 330 | 165 | 55 | 11 |
| 12 | 1 | 12 | 66 | 220 | 495 | 792 | 924 | 792 | 495 | 220 | 66 |
| 13 | 1 | 13 | 78 | 286 | 715 | 1287 | 1716 | 1716 | 1287 | 715 | 286 |
| 14 | 1 | 14 | 91 | 364 | 1001 | 2002 | 3003 | 3432 | 3003 | 2002 | 1001 |
| 15 | 1 | 15 | 105 | 455 | 1365 | 3003 | 5005 | 6435 | 6435 | 5005 | 3003 |
| 16 | 1 | 16 | 120 | 560 | 1820 | 4368 | 8008 | 11440 | 12870 | 11440 | 8008 |
| 17 | 1 | 17 | 136 | 680 | 2380 | 6188 | 12376 | 19448 | 24310 | 24310 | 19448 |
| 18 | 1 | 18 | 153 | 816 | 3060 | 8568 | 18564 | 31824 | 43758 | 48620 | 43758 |
| 19 | 1 | 19 | 171 | 969 | 3876 | 11628 | 27132 | 50388 | 75582 | 92378 | 92378 |
| 20 | 1 | 20 | 190 | 1140 | 4845 | 15504 | 38760 | 77520 | 125970 | 167960 | 184756 |

## A-29c FACTORIALS

| $N$ | $N!$ | $N$ | $N!$ | $N$ | $N!$ | $N$ | $N!$ | $N$ | $N!$ |
|---|---|---|---|---|---|---|---|---|---|
| 1 | 1.0000 | 21 | 5.1091(19) | 41 | 3.3453(49) | 61 | 5.0758(83) | 81 | 5.7971(120) |
| 2 | 2.0000 | 22 | 1.1240(21) | 42 | 1.4050(51) | 62 | 3.1470(85) | 82 | 4.7536(122) |
| 3 | 6.0000 | 23 | 2.5852(22) | 43 | 6.0415(52) | 63 | 1.9826(87) | 83 | 3.9455(124) |
| 4 | 2.4000(1) | 24 | 6.2045(23) | 44 | 2.6583(54) | 64 | 1.2689(89) | 84 | 3.3142(126) |
| 5 | 1.2000(2) | 25 | 1.5511(25) | 45 | 1.1962(56) | 65 | 8.2477(90) | 85 | 2.8171(128) |
| 6 | 7.2000(2) | 26 | 4.0329(26) | 46 | 5.5026(57) | 66 | 5.4434(92) | 86 | 2.4227(130) |
| 7 | 5.0400(3) | 27 | 1.0889(28) | 47 | 2.5862(59) | 67 | 3.6471(94) | 87 | 2.1078(132) |
| 8 | 4.0320(4) | 28 | 3.0489(29) | 48 | 1.2414(61) | 68 | 2.4800(96) | 88 | 1.8548(134) |
| 9 | 3.6288(5) | 29 | 8.8418(30) | 49 | 6.0828(62) | 69 | 1.7112(98) | 89 | 1.6508(136) |
| 10 | 3.6288(6) | 30 | 2.6525(32) | 50 | 3.0414(64) | 70 | 1.1979(100) | 90 | 1.4857(138) |
| 11 | 3.9917(7) | 31 | 8.2228(33) | 51 | 1.5511(66) | 71 | 8.5048(101) | 91 | 1.3520(140) |
| 12 | 4.7900(8) | 32 | 2.6313(35) | 52 | 8.0658(67) | 72 | 6.1234(103) | 92 | 1.2438(142) |
| 13 | 6.2270(9) | 33 | 8.6833(36) | 53 | 4.2749(69) | 73 | 4.4701(105) | 93 | 1.1568(144) |
| 14 | 8.7178(10) | 34 | 2.9523(38) | 54 | 2.3084(71) | 74 | 3.3079(107) | 94 | 1.0874(146) |
| 15 | 1.3077(12) | 35 | 1.0333(40) | 55 | 1.2696(73) | 75 | 2.4809(109) | 95 | 1.0330(148) |
| 16 | 2.0923(13) | 36 | 3.7199(41) | 56 | 7.1100(74) | 76 | 1.8855(111) | 96 | 9.9168(149) |
| 17 | 3.5569(14) | 37 | 1.3764(43) | 57 | 4.0527(76) | 77 | 1.4518(113) | 97 | 9.6193(151) |
| 18 | 6.4024(15) | 38 | 5.2302(44) | 58 | 2.3506(78) | 78 | 1.1324(115) | 98 | 9.4269(153) |
| 19 | 1.2165(17) | 39 | 2.0398(46) | 59 | 1.3868(80) | 79 | 8.9462(116) | 99 | 9.3326(155) |
| 20 | 2.4329(18) | 40 | 8.1592(47) | 60 | 8.3210(81) | 80 | 7.1569(118) | 100 | 9.3326(157) |

The number in parentheses indicates the power of 10 by which the entry must be multiplied. For example, 8! = 40,320.

| $N$ | $r_{.95}$ | $r_{.975}$ | $r_{.99}$ | $r_{.995}$ | $r_{.9995}$ | $N$ | $r_{.95}$ | $r_{.975}$ | $r_{.99}$ | $r_{.995}$ | $r_{.9995}$ |
|---|---|---|---|---|---|---|---|---|---|---|---|
| 5 | .805 | .878 | .934 | .959 | .991 | 20 | .378 | .444 | .516 | .561 | .679 |
| 6 | .729 | .811 | .882 | .917 | .974 | 22 | .360 | .423 | .492 | .537 | .652 |
| 7 | .669 | .754 | .833 | .875 | .951 | 24 | .344 | .404 | .472 | .515 | .629 |
| 8 | .621 | .707 | .789 | .834 | .925 | 26 | .330 | .388 | .453 | .496 | .607 |
| 9 | .582 | .666 | .750 | .798 | .898 | 28 | .317 | .374 | .437 | .479 | .588 |
| 10 | .549 | .632 | .715 | .765 | .872 | 30 | .306 | .361 | .423 | .463 | .570 |
| 11 | .521 | .602 | .685 | .735 | .847 | 40 | .264 | .312 | .366 | .402 | .501 |
| 12 | .497 | .576 | .658 | .708 | .823 | 50 | .235 | .279 | .328 | .361 | .451 |
| 13 | .476 | .553 | .634 | .684 | .801 | 60 | .214 | .254 | .300 | .330 | .414 |
| 14 | .457 | .532 | .612 | .661 | .780 | 80 | .185 | .220 | .260 | .286 | .361 |
| 15 | .441 | .514 | .592 | .641 | .760 | 100 | .165 | .196 | .232 | .256 | .324 |
| 16 | .426 | .497 | .574 | .623 | .742 | 250 | .104 | .124 | .147 | .163 | .207 |
| 17 | .412 | .482 | .558 | .606 | .725 | 500 | .074 | .088 | .104 | .115 | .147 |
| 18 | .400 | .468 | .543 | .590 | .708 | 1000 | .052 | .062 | .074 | .081 | .104 |
| 19 | .389 | .456 | .529 | .575 | .693 | $\infty$ | 0 | 0 | 0 | 0 | 0 |
| $N$ | $-r_{.05}$ | $-r_{.025}$ | $-r_{.01}$ | $-r_{.005}$ | $-r_{.0005}$ | $N$ | $-r_{.05}$ | $-r_{.025}$ | $-r_{.01}$ | $-r_{.005}$ | $-r_{.0005}$ |

Percentiles of the sampling distribution of $r$ may be computed from percentiles of the sampling distribution of $t$ by the relation

$$r_\alpha = \frac{t_\alpha}{\sqrt{t_\alpha^2 + N - 2}}$$

where $t_\alpha$ is from Table A-5 for $N - 2$ degrees of freedom.

## VALUES OF $z = \frac{1}{2} \ln \frac{1 + r}{1 - r}$ **A-30b**

| $r$ | .00 | .01 | .02 | .03 | .04 | .05 | .06 | .07 | .08 | .09 |
|---|---|---|---|---|---|---|---|---|---|---|
| .0 | .00000 | .01000 | .02000 | .03001 | .04002 | .05004 | .06007 | .07012 | .08017 | .09024 |
| .1 | .10034 | .11045 | .12058 | .13074 | .14093 | .15114 | .16139 | .17167 | .18198 | .19234 |
| .2 | .20273 | .21317 | .22366 | .23419 | .24477 | .25541 | .26611 | .27686 | .28768 | .29857 |
| .3 | .30952 | .32055 | .33165 | .34283 | .35409 | .36544 | .37689 | .38842 | .40006 | .41180 |
| .4 | .42365 | .43561 | .44769 | .45990 | .47223 | .48470 | .49731 | .51007 | .52298 | .53606 |
| .5 | .54931 | .56273 | .57634 | .59014 | .60415 | .61838 | .63283 | .64752 | .66246 | .67767 |
| .6 | .69315 | .70892 | .72500 | .74142 | .75817 | .77530 | .79281 | .81074 | .82911 | .84795 |
| .7 | .86730 | .88718 | .90764 | .92873 | .95048 | .97295 | .99621 | 1.02033 | 1.04537 | 1.07143 |
| .8 | 1.09861 | 1.12703 | 1.15682 | 1.18813 | 1.22117 | 1.25615 | 1.29334 | 1.33308 | 1.37577 | 1.42192 |
| .9 | 1.47222 | 1.52752 | 1.58902 | 1.65839 | 1.73805 | 1.83178 | 1.94591 | 2.09229 | 2.29756 | 2.64665 |

For negative values of $r$ put a minus sign in front of the tabled numbers.

## A-30c CUMULATIVE DISTRIBUTION OF $\Sigma d_i^2$ AND THE RANK CORRELATION COEFFICIENT $r_s$ FOR INDEPENDENT OBSERVATIONS FROM THE SAME POPULATION



The value $\alpha$ is the probability that $\Sigma d_i^2$ is less than or equal to the tabled value (or that $r_s$ is greater than or equal to the tabled value). An observed value of $\Sigma d_i^2$ is located in the table heading and the corresponding values of $r_s$ and $\alpha$ are in the body of the table arranged by sample size $N$.

| $N$ | | VALUES OF $\Sigma d_i^2$ 0 | 2 | 4 | 6 | 8 | 10 | 12 | 14 | 16 | 18 | 20 | 22 | 24 |
|---|---|---|---|---|---|---|---|---|---|---|---|---|---|---|
| 2 | $\alpha$ | .500 | 1.000 | | | | | | | | | | | |
| | $r_s$ | 1.00 | −1.00 | | | | | | | | | | | |
| 3 | $\alpha$ | .167 | .500 | — | .833 | 1.000 | | | | | | | | |
| | $r_s$ | 1.00 | .50 | — | −.50 | −1.00 | | | | | | | | |
| 4 | $\alpha$ | .042 | .167 | .208 | .375 | .458 | .542 | .625 | .792 | .833 | .958 | 1.000 | | |
| | $r_s$ | 1.00 | .80 | .60 | .40 | .20 | .00 | −.20 | −.40 | −.60 | −.80 | −1.00 | | |
| 5 | $\alpha$ | .008 | .042 | .067 | .117 | .175 | .225 | .258 | .342 | .392 | .475 | .525 | .608 | .658 |
| | $r_s$ | 1.00 | .90 | .80 | .70 | .60 | .50 | .40 | .30 | .20 | .10 | .00 | −.10 | −.20 |
| 6 | $\alpha$ | .001 | .008 | .017 | .029 | .051 | .068 | .088 | .121 | .149 | .178 | .210 | .249 | .282 |
| | $r_s$ | 1.00 | .94 | .89 | .83 | .77 | .71 | .66 | .60 | .54 | .49 | .43 | .37 | .31 |
| 7 | $\alpha$ | .000 | .001 | .003 | .006 | .012 | .017 | .024 | .033 | .044 | .055 | .069 | .083 | .100 |
| | $r_s$ | 1.00 | .96 | .93 | .89 | .86 | .82 | .79 | .75 | .71 | .68 | .64 | .61 | .57 |
| 8 | $\alpha$ | .000 | .000 | .001 | .001 | .002 | .004 | .005 | .008 | .011 | .014 | .018 | .023 | .029 |
| | $r_s$ | 1.00 | .98 | .95 | .93 | .90 | .88 | .86 | .83 | .81 | .79 | .76 | .74 | .71 |
| 9 | $\alpha$ | .000 | .000 | .000 | .000 | .000 | .001 | .001 | .002 | .002 | .003 | .004 | .005 | .007 |
| | $r_s$ | 1.00 | .98 | .97 | .95 | .93 | .92 | .90 | .88 | .87 | .85 | .83 | .82 | .80 |
| 10 | $\alpha$ | .000 | .000 | .000 | .000 | .000 | .000 | .000 | .000 | .000 | .001 | .001 | .001 | .001 |
| | $r_s$ | 1.00 | .99 | .98 | .96 | .95 | .94 | .93 | .92 | .90 | .89 | .88 | .87 | .86 |

## CUMULATIVE DISTRIBUTION OF $\Sigma d_i^2$ AND THE RANK CORRELATION COEFFICIENT $r_s$ FOR INDEPENDENT OBSERVATIONS FROM THE SAME POPULATION (CONTINUED) A-30c

| $N$ | | 26 | 28 | 30 | 32 | 34 | 36 | 38 | 40 | 42 | 44 | 46 | 48 |
|---|---|---|---|---|---|---|---|---|---|---|---|---|---|
| 5 | $\alpha$ | .742 | .775 | .825 | .883 | .933 | .958 | .992 | 1.000 | | | | |
| | $r_s$ | −.30 | −.40 | −.50 | −.60 | −.70 | −.80 | −.90 | −1.00 | | | | |
| 6 | $\alpha$ | .329 | .357 | .401 | .460 | .500 | .540 | .599 | .643 | .671 | .718 | .751 | .790 |
| | $r_s$ | .26 | .20 | .14 | .09 | .03 | −.03 | −.09 | −.14 | −.20 | −.26 | −.31 | −.37 |
| 7 | $\alpha$ | .118 | .133 | .151 | .177 | .200 | .222 | .249 | .278 | .297 | .331 | .357 | .391 |
| | $r_s$ | .54 | .50 | .46 | .43 | .39 | .36 | .32 | .29 | .25 | .21 | .18 | .14 |
| 8 | $\alpha$ | .035 | .042 | .048 | .057 | .066 | .076 | .085 | .098 | .108 | .122 | .134 | .150 |
| | $r_s$ | .69 | .67 | .64 | .62 | .60 | .57 | .55 | .52 | .50 | .48 | .45 | .43 |
| 9 | $\alpha$ | .009 | .011 | .013 | .016 | .018 | .022 | .025 | .029 | .033 | .038 | .043 | .048 |
| | $r_s$ | .78 | .77 | .75 | .73 | .72 | .70 | .68 | .67 | .65 | .63 | .62 | .60 |
| 10 | $\alpha$ | .002 | .002 | .003 | .004 | .004 | .005 | .006 | .007 | .009 | .010 | .012 | .013 |
| | $r_s$ | .84 | .83 | .82 | .81 | .79 | .78 | .77 | .76 | .75 | .73 | .72 | .71 |

| $N$ | | 50 | 52 | 54 | 56 | 58 | 60 | 62 | 64 | 66 | 68 | 70 | 72 | 74 |
|---|---|---|---|---|---|---|---|---|---|---|---|---|---|---|
| 6 | $\alpha$ | .822 | .851 | .879 | .913 | .932 | .949 | .971 | .983 | .992 | .999 | 1.000 | | |
| | $r_s$ | −.43 | −.49 | −.54 | −.60 | −.66 | −.71 | −.77 | −.83 | −.89 | −.94 | −1.00 | | |
| 7 | $\alpha$ | .420 | .453 | .482 | .518 | .547 | .580 | .609 | .643 | .669 | .703 | .723 | .751 | .778 |
| | $r_s$ | .11 | .07 | .04 | .00 | −.04 | −.07 | −.11 | −.14 | −.18 | −.21 | −.25 | −.29 | −.32 |
| 8 | $\alpha$ | .163 | .180 | .195 | .214 | .231 | .250 | .268 | .291 | .310 | .332 | .352 | .376 | .397 |
| | $r_s$ | .40 | .38 | .36 | .33 | .31 | .29 | .26 | .24 | .21 | .19 | .17 | .14 | .12 |
| 9 | $\alpha$ | .054 | .060 | .066 | .074 | .081 | .089 | .097 | .106 | .115 | .125 | .135 | .146 | .156 |
| | $r_s$ | .58 | .57 | .55 | .53 | .52 | .50 | .48 | .47 | .45 | .43 | .42 | .40 | .38 |
| 10 | $\alpha$ | .015 | .017 | .019 | .022 | .024 | .027 | .030 | .033 | .037 | .040 | .044 | .048 | .052 |
| | $r_s$ | .70 | .68 | .67 | .66 | .65 | .64 | .62 | .61 | .60 | .59 | .58 | .56 | .55 |

## A-30c CUMULATIVE DISTRIBUTION OF $\Sigma d_i^2$ AND THE RANK CORRELATION COEFFICIENT $r_s$ FOR INDEPENDENT OBSERVATIONS FROM THE SAME POPULATION (CONTINUED)

The value $\alpha$ is the probability that $\Sigma d_i^2$ is less than or equal to the tabled value (or that $r_s$ is greater than or equal to the tabled value). An observed value of $\Sigma d_i^2$ is located in the table heading and the corresponding values of $r_s$ and $\alpha$ are in the body of the table arranged by sample size $N$.

| $N$ | | 76 | 78 | 80 | 82 | 84 | 86 | 88 | 90 | 92 | 94 | 96 | 98 | 100 |
|---|---|---|---|---|---|---|---|---|---|---|---|---|---|---|
| 7 | $\alpha$ | .802 | .823 | .849 | .867 | .882 | .900 | .917 | .931 | .945 | .956 | .967 | .976 | .983 |
| | $r_s$ | −.36 | −.39 | −.43 | −.46 | −.50 | −.54 | −.57 | −.61 | −.64 | −.68 | −.71 | −.75 | −.76 |
| 8 | $\alpha$ | .420 | .441 | .467 | .488 | .512 | .533 | .559 | .580 | .603 | .624 | .648 | .668 | .690 |
| | $r_s$ | .10 | .07 | .05 | .02 | .00 | −.02 | −.05 | −.07 | −.10 | −.12 | −.14 | −.17 | −.19 |
| 9 | $\alpha$ | .168 | .179 | .193 | .205 | .218 | .231 | .243 | .260 | .276 | .290 | .307 | .322 | .339 |
| | $r_s$ | .37 | .35 | .33 | .32 | .30 | .28 | .27 | .25 | .23 | .22 | .20 | .18 | .17 |
| 10 | $\alpha$ | .057 | .062 | .067 | .072 | .077 | .083 | .089 | .096 | .102 | .109 | .116 | .124 | .132 |
| | $r_s$ | .54 | .53 | .52 | .50 | .49 | .48 | .47 | .45 | .44 | .43 | .42 | .41 | .39 |

| $N$ | | 102 | 104 | 106 | 108 | 110 | 112 | 114 | 116 | 118 | 120 | 122 | 124 | 126 |
|---|---|---|---|---|---|---|---|---|---|---|---|---|---|---|
| 7 | $\alpha$ | .988 | .944 | .997 | .999 | 1.000 | 1.000 | | | | | | | |
| | $r_s$ | −.82 | −.86 | −.89 | −.93 | −.96 | −1.00 | | | | | | | |
| 8 | $\alpha$ | 709 | .732 | .750 | .769 | .786 | .805 | .820 | .837 | .850 | .866 | .878 | .892 | .902 |
| | $r_s$ | −.21 | −.24 | −.26 | −.29 | −.31 | −.33 | −.36 | −.38 | −.40 | −.43 | −.45 | −.48 | −.50 |
| 9 | $\alpha$ | .354 | .372 | .388 | .405 | .422 | .440 | .456 | .474 | .491 | .509 | .526 | .544 | .560 |
| | $r_s$ | .15 | .13 | .12 | .10 | .08 | .07 | .05 | .03 | .02 | .00 | −.02 | −.03 | −.05 |
| 10 | $\alpha$ | .139 | .148 | .156 | .165 | .174 | .184 | .193 | .203 | .214 | .224 | .235 | .246 | .257 |
| | $r_s$ | .38 | .37 | .36 | .35 | .33 | .32 | .31 | .30 | .29 | .27 | .26 | .25 | .24 |

| $N$ | | 128 | 130 | 132 | 134 | 136 | 138 | 140 | 142 | 144 | 146 | 148 | 150 | 152 |
|---|---|---|---|---|---|---|---|---|---|---|---|---|---|---|
| 8 | $\alpha$ | .915 | .924 | .934 | .943 | .952 | .958 | .965 | .971 | .977 | .982 | .986 | .989 | .992 |
| | $r_s$ | −.52 | −.55 | −.57 | −.60 | −.62 | −.64 | −.67 | −.69 | −.71 | −.74 | −.76 | −.79 | −.81 |
| 9 | $\alpha$ | .578 | .595 | .612 | .628 | .646 | .661 | .678 | .693 | .710 | .724 | .740 | .753 | .769 |
| | $r_s$ | −.07 | −.08 | −.10 | −.12 | −.13 | −.15 | −.17 | −.18 | −.20 | −.22 | −.23 | −.25 | −.27 |
| 10 | $\alpha$ | .268 | .280 | .292 | .304 | .316 | .328 | .341 | .354 | .367 | .379 | .393 | .406 | .419 |
| | $r_s$ | .22 | .21 | .20 | .19 | .18 | .16 | .15 | .14 | .13 | .12 | .10 | .09 | .08 |

# CUMULATIVE DISTRIBUTION OF $\Sigma d_i^2$ AND THE RANK CORRELATION COEFFICIENT $r_s$ FOR INDEPENDENT OBSERVATIONS FROM THE SAME POPULATION (CONTINUED)

| $N$ | | 154 | 156 | 158 | 160 | 162 | 164 | 166 | 168 | 170 | 172 | 174 | 176 | 178 |
|---|---|---|---|---|---|---|---|---|---|---|---|---|---|---|
| 8 | $\alpha$ | .995 | .996 | .998 | .999 | .999 | 1.000 | 1.000 | 1.000 | | | | | |
| | $r_s$ | −.83 | −.86 | −.88 | −.90 | −.93 | −.95 | −.98 | −1.00 | | | | | |
| 9 | $\alpha$ | .782 | .795 | .807 | .821 | .832 | .844 | .854 | .865 | .875 | .885 | .894 | .903 | .911 |
| | $r_s$ | −.28 | −.30 | −.32 | −.33 | −.35 | −.37 | −.38 | −.40 | −.42 | −.43 | −.45 | −.47 | −.48 |
| 10 | $\alpha$ | .433 | .446 | .459 | .473 | .486 | .500 | .514 | .527 | .541 | .554 | .567 | .581 | .594 |
| | $r_s$ | .07 | .05 | .04 | .03 | .02 | .01 | −.01 | −.02 | −.03 | −.04 | −.05 | −.07 | −.08 |

| $N$ | | 180 | 182 | 184 | 186 | 188 | 190 | 192 | 194 | 196 | 198 | 200 | 202 | 204 |
|---|---|---|---|---|---|---|---|---|---|---|---|---|---|---|
| 9 | $\alpha$ | .919 | .926 | .934 | .940 | .946 | .952 | .957 | .962 | .967 | .971 | .975 | .978 | .982 |
| | $r_s$ | −.50 | −.52 | −.53 | −.55 | −.57 | −.58 | −.60 | −.61 | −.63 | −.65 | −.67 | −.68 | −.70 |
| 10 | $\alpha$ | .607 | .621 | .633 | .646 | .659 | .672 | .684 | .696 | .708 | .720 | .732 | .743 | .754 |
| | $r_s$ | −.09 | −.10 | −.12 | −.13 | −.14 | −.15 | −.16 | −.18 | −.19 | −.20 | −.21 | −.22 | −.24 |

| $N$ | | 206 | 208 | 210 | 212 | 214 | 216 | 218 | 220 | 222 | 224 | 226 | 228 | 230 |
|---|---|---|---|---|---|---|---|---|---|---|---|---|---|---|
| 9 | $\alpha$ | .984 | .987 | .989 | .991 | .993 | .995 | .996 | .997 | .998 | .998 | .999 | .999 | 1.000 |
| | $r_s$ | −.72 | −.73 | −.75 | −.77 | −.78 | −.80 | −.82 | −.83 | −.85 | −.87 | −.88 | −.90 | −.92 |
| 10 | $\alpha$ | .765 | .776 | .786 | .797 | .807 | .816 | .826 | .835 | .844 | .852 | .861 | .868 | .876 |
| | $r_s$ | −.25 | −.26 | −.27 | −.28 | −.30 | −.31 | −.32 | −.33 | −.35 | −.36 | −.37 | −.38 | −.39 |

| $N$ | | 232 | 234 | 236 | 238 | 240 | 242 | 244 | 246 | 248 | 250 | 252 | 254 | 256 |
|---|---|---|---|---|---|---|---|---|---|---|---|---|---|---|
| 9 | $\alpha$ | 1.000 | 1.000 | 1.000 | 1.000 | 1.000 | | | | | | | | |
| | $r_s$ | −.93 | −.95 | −.97 | −.98 | −1.00 | | | | | | | | |
| 10 | $\alpha$ | .884 | .891 | .898 | .904 | .911 | .917 | .923 | .928 | .933 | .938 | .943 | .948 | .952 |
| | $r^s$ | −.41 | −.42 | −.43 | −.44 | −.45 | −.47 | −.48 | −.49 | −.50 | −.52 | −.53 | −.54 | −.55 |

## A-30c CUMULATIVE DISTRIBUTION OF $\Sigma d_i^2$ AND THE RANK CORRELATION COEFFICIENT $r_s$ FOR INDEPENDENT OBSERVATIONS FROM THE SAME POPULATION (CONTINUED)

The value $\alpha$ is the probability that $\Sigma d_i^2$ is less than or equal to the tabled value (or that $r_s$ is greater than or equal to the tabled value). An observed value of $\Sigma d_i^2$ is located in the table heading and the corresponding values of $r_s$ and $\alpha$ are in the body of the table arranged by sample size $N$.

| $N$ | | 258 | 260 | 262 | 264 | 266 | 268 | 270 | 272 | 274 | 276 | 278 | 280 | 282 |
|---|---|---|---|---|---|---|---|---|---|---|---|---|---|---|
| 10 | $\alpha$ | .956 | .960 | .963 | .967 | .970 | .973 | .976 | .978 | .981 | .983 | .985 | .987 | .988 |
| | $r_s$ | −.56 | −.58 | −.59 | −.60 | −.61 | −.62 | −.64 | −.65 | −.66 | −.67 | −.68 | −.70 | −.71 |

| $N$ | | 284 | 286 | 288 | 290 | 292 | 294 | 296 | 298 | 300 | 302 | 304 | 306 | 308 |
|---|---|---|---|---|---|---|---|---|---|---|---|---|---|---|
| 10 | $\alpha$ | .990 | .991 | .993 | .994 | .995 | .996 | .996 | .997 | .998 | .998 | .999 | .999 | .999 |
| | $r_s$ | −.72 | −.73 | −.75 | −.76 | −.77 | −.78 | −.79 | −.81 | −.82 | −.83 | −.84 | −.85 | −.87 |

| $N$ | | 310 | 312 | 314 | 316 | 318 | 320 | 322 | 324 | 326 | 328 | 330 |
|---|---|---|---|---|---|---|---|---|---|---|---|---|
| 10 | $\alpha$ | .999 | 1.000 | 1.000 | 1.000 | 1.000 | 1.000 | 1.000 | 1.000 | 1.000 | 1.000 | 1.000 |
| | $r_s$ | −.88 | −.89 | −.90 | −.92 | −.93 | −.94 | −.95 | −.96 | −.98 | −.99 | −1.00 |

# DISTRIBUTION OF QUADRANT SUM FOR CORNER TEST FOR ASSOCIATION FOR SAMPLES OF SIZE $2N$

The table gives $\alpha$, the chance that the quadrant sum equals or exceeds $S$.

| $S$ \ $2N$ | 6 | 8 | 10 | 14 | $\infty$ |
|---|---|---|---|---|---|
| 8 | .111 | .126 | .133 | .132 | .122 |
| 9 | .100 | .084 | .093 | .092 | .081 |
| 10 | .100 | .055 | .064 | .063 | .053 |
| 11 | .100 | .038 | .044 | .043 | .034 |
| 12 | .100 | .030 | .029 | .030 | .022 |
| 13 | .000 | .029 | .019 | .020 | .013 |
| 14 | | .029 | .013 | .014 | .008 |
| 15 | | .029 | .010 | .010 | .005 |
| 16 | | .029 | .008 | .007 | .003 |
| 17 | | .000 | .008 | .005 | .002 |
| 18 | | | .008 | .003 | .001 |
| 19 | | | .008 | .002 | .001 |
| 20 | | | .008 | .001 | .000 |
| 21 | | | .000 | .001 | .000 |
| 22 | | | | .001 | .000 |
| 23 | | | | .001 | .000 |

| | 0 | 1 | 2 | 3 | 4 | 5 | 6 | 7 | 8 | 9 |
|---|---|---|---|---|---|---|---|---|---|---|
| 1.00 | 1.00000 | .99900 | .99800 | .99701 | .99602 | .99502 | .99404 | .99305 | .99206 | .99108 |
| 1.01 | .99010 | .98912 | .98814 | .98717 | .98619 | .98522 | .98425 | .98328 | .98232 | .98135 |
| 1.02 | .98039 | .97943 | .97847 | .97752 | .97656 | .97561 | .97466 | .97371 | .97276 | .97182 |
| 1.03 | .97087 | .96993 | .96899 | .96805 | .96712 | .96618 | .96525 | .96432 | .96339 | .96246 |
| 1.04 | .96154 | .96061 | .95969 | .95877 | .95785 | .95694 | .95602 | .95511 | .95420 | .95329 |
| 1.05 | .95238 | .95147 | .95057 | .94967 | .94877 | .94787 | .94697 | .94607 | .94518 | .94429 |
| 1.06 | .94340 | .94251 | .94162 | .94073 | .93985 | .93897 | .93809 | .93721 | .93633 | .93545 |
| 1.07 | .93458 | .93371 | .93284 | .93197 | .93110 | .93023 | .92937 | .92851 | .92764 | .92678 |
| 1.08 | .92593 | .92507 | .92421 | .92336 | .92251 | .92166 | .92081 | .91996 | .91912 | .91827 |
| 1.09 | .91743 | .91659 | .91575 | .91491 | .91408 | .91324 | .91241 | .91158 | .91075 | .90992 |
| 1.1 | .90909 | .90090 | .89286 | .88496 | .87719 | .86957 | .86207 | .85470 | .84746 | .84034 |
| 1.2 | .83333 | .82645 | .81967 | .81301 | .80645 | .80000 | .79365 | .78740 | .78125 | .77519 |
| 1.3 | .76923 | .76336 | .75758 | .75188 | .74627 | .74074 | .73529 | .72993 | .72464 | .71942 |
| 1.4 | .71429 | .70922 | .70423 | .69930 | .69444 | .68966 | .68493 | .68027 | .67568 | .67114 |
| 1.5 | .66667 | .66225 | .65789 | .65359 | .64935 | .64516 | .64103 | .63694 | .63291 | .62893 |
| 1.6 | .62500 | .62112 | .61728 | .61350 | .60976 | .60606 | .60241 | .59880 | .59524 | .59172 |
| 1.7 | .58824 | .58480 | .58140 | .57803 | .57471 | .57143 | .56818 | .56497 | .56180 | .55866 |
| 1.8 | .55556 | .55249 | .54945 | .54645 | .54348 | .54054 | .53763 | .53476 | .53191 | .52910 |
| 1.9 | .52632 | .52356 | .52083 | .51813 | .51546 | .51282 | .51020 | .50761 | .50505 | .50251 |
| 2.0 | .50000 | .49751 | .49505 | .49261 | .49020 | .48780 | .48544 | .48309 | .48077 | .47847 |
| 2.1 | .47619 | .47393 | .47170 | .46948 | .46729 | .46512 | .46296 | .46083 | .45872 | .45662 |
| 2.2 | .45455 | .45249 | .45045 | .44843 | .44643 | .44444 | .44248 | .44053 | .43860 | .43668 |
| 2.3 | .43478 | .43290 | .43103 | .42918 | .42735 | .42553 | .42373 | .42194 | .42017 | .41841 |
| 2.4 | .41667 | .41494 | .41322 | .41152 | .40984 | .40816 | .40650 | .40486 | .40323 | .40161 |
| 2.5 | .40000 | .39841 | .39683 | .39526 | .39370 | .39216 | .39063 | .38911 | .38760 | .38610 |
| 2.6 | .38462 | .38314 | .38168 | .38023 | .37879 | .37736 | .37594 | .37453 | .37313 | .37175 |
| 2.7 | .37037 | .36900 | .36765 | .36630 | .36496 | .36364 | .36232 | .36101 | .35971 | .35842 |
| 2.8 | .35714 | .35587 | .35461 | .35336 | .35211 | .35088 | .34965 | .34843 | .34722 | .34602 |
| 2.9 | .34483 | .34364 | .34247 | .34130 | .34014 | .33898 | .33784 | .33670 | .33557 | .33445 |
| 3.0 | .33333 | .33223 | .33113 | .33003 | .32895 | .32787 | .32680 | .32573 | .32468 | .32362 |
| 3.1 | .32258 | .32154 | .32051 | .31949 | .31847 | .31746 | .31646 | .31546 | .31447 | .31348 |
| 3.2 | .31250 | .31153 | .31056 | .30960 | .30864 | .30769 | .30675 | .30581 | .30488 | .30395 |
| 3.3 | .30303 | .30211 | .30120 | .30030 | .29940 | .29851 | .29762 | .29674 | .29586 | .29499 |
| 3.4 | .29412 | .29326 | .29240 | .29155 | .29070 | .28986 | .28902 | .28818 | .28736 | .28653 |
| 3.5 | .28571 | .28490 | .28409 | .28329 | .28249 | .28169 | .28090 | .28011 | .27933 | .27855 |
| 3.6 | .27778 | .27701 | .27624 | .27548 | .27473 | .27397 | .27322 | .27248 | .27174 | .27100 |
| 3.7 | .27027 | .26954 | .26882 | .26810 | .26738 | .26667 | .26596 | .26525 | .26455 | .26385 |
| 3.8 | .26316 | .26247 | .26178 | .26110 | .26042 | .25974 | .25907 | .25840 | .25773 | .25707 |
| 3.9 | .25641 | .25575 | .25510 | .25445 | .25381 | .25316 | .25253 | .25189 | .25126 | .25063 |
| 4.0 | .25000 | .24938 | .24876 | .24814 | .24752 | .24691 | .24631 | .24570 | .24510 | .24450 |
| 4.1 | .24390 | .24331 | .24272 | .24213 | .24155 | .24096 | .24038 | .23981 | .23923 | .23866 |
| 4.2 | .23810 | .23753 | .23697 | .23641 | .23585 | .23529 | .23474 | .23419 | .23364 | .23310 |
| 4.3 | .23256 | .23202 | .23148 | .23095 | .23041 | .22989 | .22936 | .22883 | .22831 | .22779 |
| 4.4 | .22727 | .22676 | .22624 | .22573 | .22523 | .22472 | .22422 | .22371 | .22321 | .22272 |
| 4.5 | .22222 | .22173 | .22124 | .22075 | .22026 | .21978 | .21930 | .21882 | .21834 | .21786 |
| 4.6 | .21739 | .21692 | .21645 | .21598 | .21552 | .21505 | .21459 | .21413 | .21368 | .21322 |
| 4.7 | .21277 | .21231 | .21186 | .21142 | .21097 | .21053 | .21008 | .20964 | .20921 | .20877 |
| 4.8 | .20833 | .20790 | .20747 | .20704 | .20661 | .20619 | .20576 | .20534 | .20492 | .20450 |
| 4.9 | .20408 | .20367 | .20325 | .20284 | .20243 | .20202 | .20161 | .20121 | .20080 | .20040 |

| | 0 | 1 | 2 | 3 | 4 | 5 | 6 | 7 | 8 | 9 |
|---|---|---|---|---|---|---|---|---|---|---|
| 5.0 | .20000 | .19960 | .19920 | .19881 | .19841 | .19802 | .19763 | .19724 | .19685 | .19646 |
| 5.1 | .19608 | .19569 | .19531 | .19493 | .19455 | .19417 | .19380 | .19342 | .19305 | .19268 |
| 5.2 | .19231 | .19194 | .19157 | .19120 | .19084 | .19048 | .19011 | .18975 | .18939 | .18904 |
| 5.3 | .18868 | .18832 | .18797 | .18762 | .18727 | .18692 | .18657 | .18622 | .18587 | .18553 |
| 5.4 | .18519 | .18484 | .18450 | .18416 | .18382 | .18349 | .18315 | .18282 | .18248 | .18215 |
| 5.5 | .18182 | .18149 | .18116 | .18083 | .18051 | .18018 | .17986 | .17953 | .17921 | .17889 |
| 5.6 | .17857 | .17825 | .17794 | .17762 | .17731 | .17699 | .17668 | .17637 | .17606 | .17575 |
| 5.7 | .17544 | .17513 | .17483 | .17452 | .17422 | .17391 | .17361 | .17331 | .17301 | .17271 |
| 5.8 | .17241 | .17212 | .17182 | .17153 | .17123 | .17094 | .17065 | .17036 | .17007 | .16978 |
| 5.9 | .16949 | .16920 | .16892 | .16863 | .16835 | .16807 | .16779 | .16750 | .16722 | .16694 |
| 6.0 | .16667 | .16639 | .16611 | .16584 | .16556 | .16529 | .16502 | .16474 | .16447 | .16420 |
| 6.1 | .16393 | .16367 | .16340 | .16313 | .16287 | .16260 | .16234 | .16207 | .16181 | .16155 |
| 6.2 | .16129 | .16103 | .16077 | .16051 | .16026 | .16000 | .15974 | .15949 | .15924 | .15898 |
| 6.3 | .15873 | .15848 | .15823 | .15798 | .15773 | .15748 | .15723 | .15699 | .15674 | .15649 |
| 6.4 | .15625 | .15601 | .15576 | .15552 | .15528 | .15504 | .15480 | .15456 | .15432 | .15408 |
| 6.5 | .15385 | .15361 | .15337 | .15314 | .15291 | .15267 | .15244 | .15221 | .15198 | .15175 |
| 6.6 | .15152 | .15129 | .15106 | .15083 | .15060 | .15038 | .15015 | .14993 | .14970 | .14948 |
| 6.7 | .14925 | .14903 | .14881 | .14859 | .14837 | .14815 | .14793 | .14771 | .14749 | .14728 |
| 6.8 | .14706 | .14684 | .14663 | .14641 | .14620 | .14599 | .14577 | .14556 | .14535 | .14514 |
| 6.9 | .14493 | .14472 | .14451 | .14430 | .14409 | .14388 | .14368 | .14347 | .14327 | .14306 |
| 7.0 | .14286 | .14265 | .14245 | .14225 | .14205 | .14184 | .14164 | .14144 | .14124 | .14104 |
| 7.1 | .14085 | .14065 | .14045 | .14025 | .14006 | .13986 | .13966 | .13947 | .13928 | .13908 |
| 7.2 | .13889 | .13870 | .13850 | .13831 | .13812 | .13793 | .13774 | .13755 | .13736 | .13717 |
| 7.3 | .13699 | .13680 | .13661 | .13643 | .13624 | .13605 | .13587 | .13569 | .13550 | .13532 |
| 7.4 | .13514 | .13495 | .13477 | .13459 | .13441 | .13423 | .13405 | .13387 | .13369 | .13351 |
| 7.5 | .13333 | .13316 | .13298 | .13280 | .13263 | .13245 | .13228 | .13210 | .13193 | .13175 |
| 7.6 | .13158 | .13141 | .13123 | .13106 | .13089 | .13072 | .13055 | .13038 | .13021 | .13004 |
| 7.7 | .12987 | .12970 | .12953 | .12937 | .12920 | .12903 | .12887 | .12870 | .12853 | .12837 |
| 7.8 | .12821 | .12804 | .12788 | .12771 | .12755 | .12739 | .12723 | .12706 | .12690 | .12674 |
| 7.9 | .12658 | .12642 | .12626 | .12610 | .12594 | .12579 | .12563 | .12547 | .12531 | .12516 |
| 8.0 | .12500 | .12484 | .12469 | .12453 | .12438 | .12422 | .12407 | .12392 | .12376 | .12361 |
| 8.1 | .12346 | .12330 | .12315 | .12300 | .12285 | .12270 | .12255 | .12240 | .12225 | .12210 |
| 8.2 | .12195 | .12180 | .12165 | .12151 | .12136 | .12121 | .12107 | .12092 | .12077 | .12063 |
| 8.3 | .12048 | .12034 | .12019 | .12005 | .11990 | .11976 | .11962 | .11947 | .11933 | .11919 |
| 8.4 | .11905 | .11891 | .11876 | .11862 | .11848 | .11834 | .11820 | .11806 | .11792 | .11779 |
| 8.5 | .11765 | .11751 | .11737 | .11723 | .11710 | .11696 | .11682 | .11669 | .11655 | .11641 |
| 8.6 | .11628 | .11614 | .11601 | .11587 | .11574 | .11561 | .11547 | .11534 | .11521 | .11507 |
| 8.7 | .11494 | .11481 | .11468 | .11455 | .11442 | .11429 | .11416 | .11403 | .11390 | .11377 |
| 8.8 | .11364 | .11351 | .11338 | .11325 | .11312 | .11299 | .11287 | .11274 | .11261 | .11249 |
| 8.9 | .11236 | .11223 | .11211 | .11198 | .11186 | .11173 | .11161 | .11148 | .11136 | .11123 |
| 9.0 | .11111 | .11099 | .11086 | .11074 | .11062 | .11050 | .11038 | .11025 | .11013 | .11001 |
| 9.1 | .10989 | .10977 | .10965 | .10953 | .10941 | .10929 | .10917 | .10905 | .10893 | .10881 |
| 9.2 | .10870 | .10858 | .10846 | .10834 | .10823 | .10811 | .10799 | .10787 | .10776 | .10764 |
| 9.3 | .10753 | .10741 | .10730 | .10718 | .10707 | .10695 | .10684 | .10672 | .10661 | .10650 |
| 9.4 | .10638 | .10627 | .10616 | .10604 | .10593 | .10582 | .10571 | .10560 | .10549 | .10537 |
| 9.5 | .10526 | .10515 | .10504 | .10493 | .10482 | .10471 | .10460 | .10449 | .10438 | .10428 |
| 9.6 | .10417 | .10406 | .10395 | .10384 | .10373 | .10363 | .10352 | .10341 | .10331 | .10320 |
| 9.7 | .10309 | .10299 | .10288 | .10277 | .10267 | .10256 | .10246 | .10235 | .10225 | .10215 |
| 9.8 | .10204 | .10194 | .10183 | .10173 | .10163 | .10152 | .10142 | .10132 | .10121 | .10111 |
| 9.9 | .10101 | .10091 | .10081 | .10070 | .10060 | .10050 | .10040 | .10030 | .10020 | .10010 |

| $N$ | .00 | .01 | .02 | .03 | .04 | .05 | .06 | .07 | .08 | .09 |
|---|---|---|---|---|---|---|---|---|---|---|
| 1.0 | 1.000 | 1.020 | 1.040 | 1.061 | 1.082 | 1.103 | 1.124 | 1.145 | 1.166 | 1.188 |
| 1.1 | 1.210 | 1.232 | 1.254 | 1.277 | 1.300 | 1.323 | 1.346 | 1.369 | 1.392 | 1.416 |
| 1.2 | 1.440 | 1.464 | 1.488 | 1.513 | 1.538 | 1.563 | 1.588 | 1.613 | 1.638 | 1.664 |
| 1.3 | 1.690 | 1.716 | 1.742 | 1.769 | 1.796 | 1.823 | 1.850 | 1.877 | 1.904 | 1.932 |
| 1.4 | 1.960 | 1.988 | 2.016 | 2.045 | 2.074 | 2.103 | 2.132 | 2.161 | 2.190 | 2.220 |
| 1.5 | 2.250 | 2.280 | 2.310 | 2.341 | 2.372 | 2.403 | 2.434 | 2.465 | 2.496 | 2.528 |
| 1.6 | 2.560 | 2.592 | 2.624 | 2.657 | 2.690 | 2.723 | 2.756 | 2.789 | 2.822 | 2.856 |
| 1.7 | 2.890 | 2.924 | 2.958 | 2.993 | 3.028 | 3.063 | 3.098 | 3.133 | 3.168 | 3.204 |
| 1.8 | 3.240 | 3.276 | 3.312 | 3.349 | 3.386 | 3.423 | 3.460 | 3.497 | 3.534 | 3.572 |
| 1.9 | 3.610 | 3.648 | 3.686 | 3.725 | 3.764 | 3.803 | 3.842 | 3.881 | 3.920 | 3.960 |
| 2.0 | 4.000 | 4.040 | 4.080 | 4.121 | 4.162 | 4.203 | 4.244 | 4.285 | 4.326 | 4.368 |
| 2.1 | 4.410 | 4.452 | 4.494 | 4.537 | 4.580 | 4.623 | 4.666 | 4.709 | 4.752 | 4.796 |
| 2.2 | 4.840 | 4.884 | 4.928 | 4.973 | 5.018 | 5.063 | 5.108 | 5.153 | 5.198 | 5.244 |
| 2.3 | 5.290 | 5.336 | 5.382 | 5.429 | 5.476 | 5.523 | 5.570 | 5.617 | 5.664 | 5.712 |
| 2.4 | 5.760 | 5.808 | 5.856 | 5.905 | 5.954 | 6.003 | 6.052 | 6.101 | 6.150 | 6.200 |
| 2.5 | 6.250 | 6.300 | 6.350 | 6.401 | 6.452 | 6.503 | 6.554 | 6.605 | 6.656 | 6.708 |
| 2.6 | 6.760 | 6.812 | 6.864 | 6.917 | 6.970 | 7.023 | 7.076 | 7.129 | 7.182 | 7.236 |
| 2.7 | 7.290 | 7.344 | 7.398 | 7.453 | 7.508 | 7.563 | 7.618 | 7.673 | 7.728 | 7.784 |
| 2.8 | 7.840 | 7.896 | 7.952 | 8.009 | 8.066 | 8.123 | 8.180 | 8.237 | 8.294 | 8.352 |
| 2.9 | 8.410 | 8.468 | 8.526 | 8.585 | 8.644 | 8.703 | 8.762 | 8.821 | 8.880 | 8.940 |
| 3.0 | 9.000 | 9.060 | 9.120 | 9.181 | 9.242 | 9.303 | 9.364 | 9.425 | 9.486 | 9.548 |
| 3.1 | 9.610 | 9.672 | 9.734 | 9.797 | 9.860 | 9.923 | 9.986 | 10.05 | 10.11 | 10.18 |
| 3.2 | 10.24 | 10.30 | 10.37 | 10.43 | 10.50 | 10.56 | 10.63 | 10.69 | 10.76 | 10.82 |
| 3.3 | 10.89 | 10.96 | 11.02 | 11.09 | 11.16 | 11.22 | 11.29 | 11.36 | 11.42 | 11.49 |
| 3.4 | 11.56 | 11.63 | 11.70 | 11.76 | 11.83 | 11.90 | 11.97 | 12.04 | 12.11 | 12.18 |
| 3.5 | 12.25 | 12.32 | 12.39 | 12.46 | 12.53 | 12.60 | 12.67 | 12.74 | 12.82 | 12.89 |
| 3.6 | 12.96 | 13.03 | 13.10 | 13.18 | 13.25 | 13.32 | 13.40 | 13.47 | 13.54 | 13.62 |
| 3.7 | 13.69 | 13.76 | 13.84 | 13.91 | 13.99 | 14.06 | 14.14 | 14.21 | 14.29 | 14.36 |
| 3.8 | 14.44 | 14.52 | 14.59 | 14.67 | 14.75 | 14.82 | 14.90 | 14.98 | 15.05 | 15.13 |
| 3.9 | 15.21 | 15.29 | 15.37 | 15.44 | 15.52 | 15.60 | 15.68 | 15.76 | 15.84 | 15.92 |
| 4.0 | 16.00 | 16.08 | 16.16 | 16.24 | 16.32 | 16.40 | 16.48 | 16.56 | 16.65 | 16.73 |
| 4.1 | 16.81 | 16.89 | 16.97 | 17.06 | 17.14 | 17.22 | 17.31 | 17.39 | 17.47 | 17.56 |
| 4.2 | 17.64 | 17.72 | 17.81 | 17.89 | 17.98 | 18.06 | 18.15 | 18.23 | 18.32 | 18.40 |
| 4.3 | 18.49 | 18.58 | 18.66 | 18.75 | 18.84 | 18.92 | 19.01 | 19.10 | 19.18 | 19.27 |
| 4.4 | 19.36 | 19.45 | 19.54 | 19.62 | 19.71 | 19.80 | 19.89 | 19.98 | 20.07 | 20.16 |
| 4.5 | 20.25 | 20.34 | 20.43 | 20.52 | 20.61 | 20.70 | 20.79 | 20.88 | 20.98 | 21.07 |
| 4.6 | 21.16 | 21.25 | 21.34 | 21.44 | 21.53 | 21.62 | 21.72 | 21.81 | 21.90 | 22.00 |
| 4.7 | 22.09 | 22.18 | 22.28 | 22.37 | 22.47 | 22.56 | 22.66 | 22.75 | 22.85 | 22.94 |
| 4.8 | 23.04 | 23.14 | 23.23 | 23.33 | 23.43 | 23.52 | 23.62 | 23.72 | 23.81 | 23.91 |
| 4.9 | 24.01 | 24.11 | 24.21 | 24.30 | 24.40 | 24.50 | 24.60 | 24.70 | 24.80 | 24.90 |
| 5.0 | 25.00 | 25.10 | 25.20 | 25.30 | 25.40 | 25.50 | 25.60 | 25.70 | 25.81 | 25.91 |
| 5.1 | 26.01 | 26.11 | 26.21 | 26.32 | 26.42 | 26.52 | 26.63 | 26.73 | 26.83 | 26.94 |
| 5.2 | 27.04 | 27.14 | 27.25 | 27.35 | 27.46 | 27.56 | 27.67 | 27.77 | 27.88 | 27.98 |
| 5.3 | 28.09 | 28.20 | 28.30 | 28.41 | 28.52 | 28.62 | 28.73 | 28.84 | 28.94 | 29.05 |
| 5.4 | 29.16 | 29.27 | 29.38 | 29.48 | 29.59 | 29.70 | 29.81 | 29.92 | 30.03 | 30.14 |
| 5.5 | 30.25 | 30.36 | 30.47 | 30.58 | 30.69 | 30.80 | 30.91 | 31.02 | 31.14 | 31.25 |
| 5.6 | 31.36 | 31.47 | 31.58 | 31.70 | 31.81 | 31.92 | 32.04 | 32.15 | 32.26 | 32.38 |
| 5.7 | 32.49 | 32.60 | 32.72 | 32.83 | 32.95 | 33.06 | 33.18 | 33.29 | 33.41 | 33.52 |
| 5.8 | 33.64 | 33.76 | 33.87 | 33.99 | 34.11 | 34.22 | 34.34 | 34.46 | 34.57 | 34.69 |
| 5.9 | 34.81 | 34.93 | 35.05 | 35.16 | 35.28 | 35.40 | 35.52 | 35.64 | 35.76 | 35.88 |

| $N$ | .00 | .01 | .02 | .03 | .04 | .05 | .06 | .07 | .08 | .09 |
|---|---|---|---|---|---|---|---|---|---|---|
| 6.0 | 36.00 | 36.12 | 36.24 | 36.36 | 36.48 | 36.60 | 36.72 | 36.84 | 36.97 | 37.09 |
| 6.1 | 37.21 | 37.33 | 37.45 | 37.58 | 37.70 | 37.82 | 37.95 | 38.07 | 38.19 | 38.32 |
| 6.2 | 38.44 | 38.56 | 38.69 | 38.81 | 38.94 | 39.06 | 39.19 | 39.31 | 39.44 | 39.56 |
| 6.3 | 39.69 | 39.82 | 39.94 | 40.07 | 40.20 | 40.32 | 40.45 | 40.58 | 40.70 | 40.83 |
| 6.4 | 40.96 | 41.09 | 41.22 | 41.34 | 41.47 | 41.60 | 41.73 | 41.86 | 41.99 | 42.12 |
| 6.5 | 42.25 | 42.38 | 42.51 | 42.64 | 42.77 | 42.90 | 43.03 | 43.16 | 43.30 | 43.43 |
| 6.6 | 43.56 | 43.69 | 43.82 | 43.96 | 44.09 | 44.22 | 44.36 | 44.49 | 44.62 | 44.76 |
| 6.7 | 44.89 | 45.02 | 45.16 | 45.29 | 45.43 | 45.56 | 45.70 | 45.83 | 45.97 | 46.10 |
| 6.8 | 46.24 | 46.38 | 46.51 | 46.65 | 46.79 | 46.92 | 47.06 | 47.20 | 47.33 | 47.47 |
| 6.9 | 47.61 | 47.75 | 47.89 | 48.02 | 48.16 | 48.30 | 48.44 | 48.58 | 48.72 | 48.86 |
| 7.0 | 49.00 | 49.14 | 49.28 | 49.42 | 49.56 | 49.70 | 49.84 | 49.98 | 50.13 | 50.27 |
| 7.1 | 50.41 | 50.55 | 50.69 | 50.84 | 50.98 | 51.12 | 51.27 | 51.41 | 51.55 | 51.70 |
| 7.2 | 51.84 | 51.98 | 52.13 | 52.27 | 52.42 | 52.56 | 52.71 | 52.85 | 53.00 | 53.14 |
| 7.3 | 53.29 | 53.44 | 53.58 | 53.73 | 53.88 | 54.02 | 54.17 | 54.32 | 54.46 | 54.61 |
| 7.4 | 54.76 | 54.91 | 55.06 | 55.20 | 55.35 | 55.50 | 55.65 | 55.80 | 55.95 | 56.10 |
| 7.5 | 56.25 | 56.40 | 56.55 | 56.70 | 56.85 | 57.00 | 57.15 | 57.30 | 57.46 | 57.61 |
| 7.6 | 57.76 | 57.91 | 58.06 | 58.22 | 58.37 | 58.52 | 58.68 | 58.83 | 58.98 | 59.14 |
| 7.7 | 59.29 | 59.44 | 59.60 | 59.75 | 59.91 | 60.06 | 60.22 | 60.37 | 60.53 | 60.68 |
| 7.8 | 60.84 | 61.00 | 61.15 | 61.31 | 61.47 | 61.62 | 61.78 | 61.94 | 62.09 | 62.25 |
| 7.9 | 62.41 | 62.57 | 62.73 | 62.88 | 63.04 | 63.20 | 63.36 | 63.52 | 63.68 | 63.84 |
| 8.0 | 64.00 | 64.16 | 64.32 | 64.48 | 64.64 | 64.80 | 64.96 | 65.12 | 65.29 | 65.45 |
| 8.1 | 65.61 | 65.77 | 65.93 | 66.10 | 66.26 | 66.42 | 66.59 | 66.75 | 66.91 | 67.08 |
| 8.2 | 67.24 | 67.40 | 67.57 | 67.73 | 67.90 | 68.06 | 68.23 | 68.39 | 68.56 | 68.72 |
| 8.3 | 68.89 | 69.06 | 69.22 | 69.39 | 69.56 | 69.72 | 69.89 | 70.06 | 70.22 | 70.39 |
| 8.4 | 70.56 | 70.73 | 70.90 | 71.06 | 71.23 | 71.40 | 71.57 | 71.74 | 71.91 | 72.08 |
| 8.5 | 72.25 | 72.42 | 72.59 | 72.76 | 72.93 | 73.10 | 73.27 | 73.44 | 73.62 | 73.79 |
| 8.6 | 73.96 | 74.13 | 74.30 | 74.48 | 74.65 | 74.82 | 75.00 | 75.17 | 75.34 | 75.52 |
| 8.7 | 75.69 | 75.86 | 76.04 | 76.21 | 76.39 | 76.56 | 76.74 | 76.91 | 77.09 | 77.26 |
| 8.8 | 77.44 | 77.62 | 77.79 | 77.97 | 78.15 | 78.32 | 78.50 | 78.68 | 78.85 | 79.03 |
| 8.9 | 79.21 | 79.39 | 79.57 | 79.74 | 79.92 | 80.10 | 80.28 | 80.46 | 80.64 | 80.82 |
| 9.0 | 81.00 | 81.18 | 81.36 | 81.54 | 81.72 | 81.90 | 82.08 | 82.26 | 82.45 | 82.63 |
| 9.1 | 82.81 | 82.99 | 83.17 | 83.36 | 83.54 | 83.72 | 83.91 | 84.09 | 84.27 | 84.46 |
| 9.2 | 84.64 | 84.82 | 85.01 | 85.19 | 85.38 | 85.56 | 85.75 | 85.93 | 86.12 | 86.30 |
| 9.3 | 86.49 | 86.68 | 86.86 | 87.05 | 87.24 | 87.42 | 87.61 | 87.80 | 87.98 | 88.17 |
| 9.4 | 88.36 | 88.55 | 88.74 | 88.92 | 89.11 | 89.30 | 89.49 | 89.68 | 89.87 | 90.06 |
| 9.5 | 90.25 | 90.44 | 90.63 | 90.82 | 91.01 | 91.20 | 91.39 | 91.58 | 91.78 | 91.97 |
| 9.6 | 92.16 | 92.35 | 92.54 | 92.74 | 92.93 | 93.12 | 93.32 | 93.51 | 93.70 | 93.90 |
| 9.7 | 94.09 | 94.28 | 94.48 | 94.67 | 94.87 | 95.06 | 95.26 | 95.45 | 95.65 | 95.84 |
| 9.8 | 96.04 | 96.24 | 96.43 | 96.63 | 96.83 | 97.02 | 97.22 | 97.42 | 97.61 | 97.81 |
| 9.9 | 98.01 | 98.21 | 98.41 | 98.60 | 98.80 | 99.00 | 99.20 | 99.40 | 99.60 | 99.80 |

The square root of a number may be found by finding the first three places from this table, dividing the number by the three-place root and averaging the quotient and divisor. Repetition will give the square root to any desired accuracy. For example, $\sqrt{80.00}$ is just less than 8.95. First, we find $80.00/8.95 = 8.9385475$. Second, $(8.95 + 8.9385475)/2 = 8.9442738$, which is close to the more precise value of 8.9442719.

| | 0 | 1 | 2 | 3 | 4 | 5 | 6 | 7 | 8 | 9 |
|---|---|---|---|---|---|---|---|---|---|---|
| 1.0 | 1.0000 | 1.0050 | 1.0100 | 1.0149 | 1.0198 | 1.0247 | 1.0296 | 1.0344 | 1.0392 | 1.0440 |
| 10. | 3.1623 | 3.1780 | 3.1937 | 3.2094 | 3.2249 | 3.2404 | 3.2558 | 3.2711 | 3.2863 | 3.3015 |
| 1.1 | 1.0488 | 1.0536 | 1.0583 | 1.0630 | 1.0677 | 1.0724 | 1.0770 | 1.0817 | 1.0863 | 1.0909 |
| 11. | 3.3166 | 3.3317 | 3.3466 | 3.3615 | 3.3764 | 3.3912 | 3.4059 | 3.4205 | 3.4351 | 3.4496 |
| 1.2 | 1.0954 | 1.1000 | 1.1045 | 1.1091 | 1.1136 | 1.1180 | 1.1225 | 1.1269 | 1.1314 | 1.1358 |
| 12. | 3.4641 | 3.4785 | 3.4928 | 3.5071 | 3.5214 | 3.5355 | 3.5496 | 3.5637 | 3.5777 | 3.5917 |
| 1.3 | 1.1402 | 1.1446 | 1.1489 | 1.1533 | 1.1576 | 1.1619 | 1.1662 | 1.1705 | 1.1747 | 1.1790 |
| 13. | 3.6056 | 3.6194 | 3.6332 | 3.6469 | 3.6606 | 3.6742 | 3.6878 | 3.7014 | 3.7148 | 3.7283 |
| 1.4 | 1.1832 | 1.1874 | 1.1916 | 1.1958 | 1.2000 | 1.2042 | 1.2083 | 1.2124 | 1.2166 | 1.2207 |
| 14. | 3.7417 | 3.7550 | 3.7683 | 3.7815 | 3.7947 | 3.8079 | 3.8210 | 3.8341 | 3.8471 | 3.8601 |
| 1.5 | 1.2247 | 1.2288 | 1.2329 | 1.2369 | 1.2410 | 1.2450 | 1.2490 | 1.2530 | 1.2570 | 1.2610 |
| 15. | 3.8730 | 3.8859 | 3.8987 | 3.9115 | 3.9243 | 3.9370 | 3.9497 | 3.9623 | 3.9749 | 3.9875 |
| 1.6 | 1.2649 | 1.2689 | 1.2728 | 1.2767 | 1.2806 | 1.2845 | 1.2884 | 1.2923 | 1.2961 | 1.3000 |
| 16. | 4.0000 | 4.0125 | 4.0249 | 4.0373 | 4.0497 | 4.0620 | 4.0743 | 4.0866 | 4.0988 | 4.1110 |
| 1.7 | 1.3038 | 1.3077 | 1.3115 | 1.3153 | 1.3191 | 1.3229 | 1.3266 | 1.3304 | 1.3342 | 1.3379 |
| 17. | 4.1231 | 4.1352 | 4.1473 | 4.1593 | 4.1713 | 4.1833 | 4.1952 | 4.2071 | 4.2190 | 4.2308 |
| 1.8 | 1.3416 | 1.3454 | 1.3491 | 1.3528 | 1.3565 | 1.3601 | 1.3638 | 1.3675 | 1.3711 | 1.3748 |
| 18. | 4.2426 | 4.2544 | 4.2661 | 4.2778 | 4.2895 | 4.3012 | 4.3128 | 4.3243 | 4.3359 | 4.3474 |
| 1.9 | 1.3784 | 1.3820 | 1.3856 | 1.3892 | 1.3928 | 1.3964 | 1.4000 | 1.4036 | 1.4071 | 1.4107 |
| 19. | 4.3589 | 4.3704 | 4.3818 | 4.3932 | 4.4045 | 4.4159 | 4.4272 | 4.4385 | 4.4497 | 4.4609 |
| 2.0 | 1.4142 | 1.4177 | 1.4213 | 1.4248 | 1.4283 | 1.4318 | 1.4353 | 1.4387 | 1.4422 | 1.4457 |
| 20. | 4.4721 | 4.4833 | 4.4944 | 4.5056 | 4.5166 | 4.5277 | 4.5387 | 4.5497 | 4.5607 | 4.5717 |
| 2.1 | 1.4491 | 1.4526 | 1.4560 | 1.4595 | 1.4629 | 1.4663 | 1.4697 | 1.4731 | 1.4765 | 1.4799 |
| 21. | 4.5826 | 4.5935 | 4.6043 | 4 6152 | 4.6260 | 4.6368 | 4.6476 | 4.6583 | 4.6690 | 4.6797 |
| 2.2 | 1.4832 | 1.4866 | 1.4900 | 1.4933 | 1.4967 | 1.5000 | 1.5033 | 1.5067 | 1.5100 | 1.5133 |
| 22. | 4.6904 | 4.7011 | 4.7117 | 4.7223 | 4.7329 | 4.7434 | 4.7539 | 4.7645 | 4.7749 | 4.7854 |
| 2.3 | 1.5166 | 1.5199 | 1.5232 | 1.5264 | 1.5297 | 1.5330 | 1.5362 | 1.5395 | 1.5427 | 1.5460 |
| 23. | 4.7958 | 4.8062 | 4.8166 | 4.8270 | 4.8374 | 4.8477 | 4.8580 | 4.8683 | 4.8785 | 4.8888 |
| 2.4 | 1.5492 | 1.5524 | 1.5556 | 1.5588 | 1.5620 | 1.5652 | 1.5684 | 1.5716 | 1.5748 | 1.5780 |
| 24. | 4.8990 | 4.9092 | 4.9193 | 4.9295 | 4.9396 | 4.9497 | 4.9598 | 4.9699 | 4.9800 | 4.9900 |

| | 0 | 1 | 2 | 3 | 4 | 5 | 6 | 7 | 8 | 9 |
|---|---|---|---|---|---|---|---|---|---|---|
| 2.5 | 1.5811 | 1.5843 | 1.5875 | 1.5906 | 1.5937 | 1.5969 | 1.6000 | 1.6031 | 1.6062 | 1.6093 |
| 25. | 5.0000 | 5.0100 | 5.0200 | 5.0299 | 5.0398 | 5.0498 | 5.0596 | 5.0695 | 5.0794 | 5.0892 |
| 2.6 | 1.6125 | 1.6155 | 1.6186 | 1.6217 | 1.6248 | 1.6279 | 1.6310 | 1.6340 | 1.6371 | 1.6401 |
| 26. | 5.0990 | 5.1088 | 5.1186 | 5.1284 | 5.1381 | 5.1478 | 5.1575 | 5.1672 | 5.1769 | 5.1865 |
| 2.7 | 1.6432 | 1.6462 | 1.6492 | 1.6523 | 1.6553 | 1.6583 | 1.6613 | 1.6643 | 1.6673 | 1.6703 |
| 27. | 5.1962 | 5.2058 | 5.2154 | 5.2249 | 5.2345 | 5.2440 | 5.2536 | 5.2631 | 5.2726 | 5.2820 |
| 2.8 | 1.6733 | 1.6763 | 1.6793 | 1.6823 | 1.6852 | 1.6882 | 1.6912 | 1.6941 | 1.6971 | 1.7000 |
| 28. | 5.2915 | 5.3009 | 5.3104 | 5.3198 | 5.3292 | 5.3385 | 5.3479 | 5.3572 | 5.3666 | 5.3759 |
| 2.9 | 1.7029 | 1.7059 | 1.7088 | 1.7117 | 1.7146 | 1.7176 | 1.7205 | 1.7234 | 1.7263 | 1.7292 |
| 29. | 5.3852 | 5.3944 | 5.4037 | 5.4129 | 5.4222 | 5.4314 | 5.4406 | 5.4498 | 5.4589 | 5.4681 |
| 3.0 | 1.7321 | 1.7349 | 1.7378 | 1.7407 | 1.7436 | 1.7464 | 1.7493 | 1.7521 | 1.7550 | 1.7578 |
| 30. | 5.4772 | 5.4863 | 5.4955 | 5.5045 | 5.5136 | 5.5227 | 5.5317 | 5.5408 | 5.5498 | 5.5588 |
| 3.1 | 1.7607 | 1.7635 | 1.7664 | 1.7692 | 1.7720 | 1.7748 | 1.7776 | 1.7804 | 1.7833 | 1.7861 |
| 31. | 5.5678 | 5.5767 | 5.5857 | 5.5946 | 5.6036 | 5.6125 | 5.6214 | 5.6303 | 5.6391 | 5.6480 |
| 3.2 | 1.7889 | 1.7916 | 1.7944 | 1.7972 | 1.8000 | 1.8028 | 1.8055 | 1.8083 | 1.8111 | 1.8138 |
| 32. | 5.6569 | 5.6657 | 5.6745 | 5.6833 | 5.6921 | 5.7009 | 5.7096 | 5.7184 | 5.7271 | 5.7359 |
| 3.3 | 1.8166 | 1.8193 | 1.8221 | 1.8248 | 1.8276 | 1.8303 | 1.8330 | 1.8358 | 1.8385 | 1.8412 |
| 33. | 5.7446 | 5.7533 | 5.7619 | 5.7706 | 5.7793 | 5.7879 | 5.7966 | 5.8052 | 5.8138 | 5.8224 |
| 3.4 | 1.8439 | 1.8466 | 1.8493 | 1.8520 | 1.8547 | 1.8574 | 1.8601 | 1.8628 | 1.8655 | 1.8682 |
| 34. | 5.8310 | 5.8395 | 5.8481 | 5.8566 | 5.8652 | 5.8737 | 5.8822 | 5.8907 | 5.8992 | 5.9076 |
| 3.5 | 1.8708 | 1.8735 | 1.8762 | 1.8788 | 1.8815 | 1.8841 | 1.8868 | 1.8894 | 1.8921 | 1.8947 |
| 35. | 5.9161 | 5.9245 | 5.9330 | 5.9414 | 5.9498 | 5.9582 | 5.9666 | 5.9749 | 5.9833 | 5.9917 |
| 3.6 | 1.8974 | 1.9000 | 1.9026 | 1.9053 | 1.9079 | 1.9105 | 1.9131 | 1.9157 | 1.9183 | 1.9209 |
| 36. | 6.0000 | 6.0083 | 6.0166 | 6.0249 | 6.0332 | 6.0415 | 6.0498 | 6.0581 | 6.0663 | 6.0745 |
| 3.7 | 1.9235 | 1.9261 | 1.9287 | 1.9313 | 1.9339 | 1.9365 | 1.9391 | 1.9416 | 1.9442 | 1.9468 |
| 37. | 6.0828 | 6.0910 | 6.0992 | 6.1074 | 6.1156 | 6.1237 | 6.1319 | 6.1400 | 6.1482 | 6.1563 |
| 3.8 | 1.9494 | 1.9519 | 1.9545 | 1.9570 | 1.9596 | 1.9621 | 1.9647 | 1.9672 | 1.9698 | 1.9723 |
| 38. | 6.1644 | 6.1725 | 6.1806 | 6.1887 | 6.1968 | 6.2048 | 6.2129 | 6.2209 | 6.2290 | 6.2370 |
| 3.9 | 1.9748 | 1.9774 | 1.9799 | 1.9824 | 1.9849 | 1.9875 | 1.9900 | 1.9925 | 1.9950 | 1.9975 |
| 39. | 6.2450 | 6.2530 | 6.2610 | 6.2690 | 6.2769 | 6.2849 | 6.2929 | 6.3008 | 6.3087 | 6.3166 |

| | 0 | 1 | 2 | 3 | 4 | 5 | 6 | 7 | 8 | 9 |
|---|---|---|---|---|---|---|---|---|---|---|
| 4.0 | 2.0000 | 2.0025 | 2.0050 | 2.0075 | 2 0100 | 2.0125 | 2.0149 | 2.0174 | 2.0199 | 2.0224 |
| 40. | 6.3246 | 6.3325 | 6.3403 | 6.3482 | 6.3561 | 6.3640 | 6.3718 | 6.3797 | 6.3875 | 6.3953 |
| 4.1 | 2.0248 | 2.0273 | 2.0298 | 2.0322 | 2.0347 | 2.0372 | 2.0396 | 2.0421 | 2.0445 | 2.0469 |
| 41. | 6.4031 | 6.4109 | 6.4187 | 6.4265 | 6.4343 | 6.4420 | 6.4498 | 6.4576 | 6.4653 | 6.4730 |
| 4.2 | 2.0494 | 2.0518 | 2.0543 | 2.0567 | 2.0591 | 2.0616 | 2.0640 | 2.0664 | 2.0688 | 2.0712 |
| 42. | 6.4807 | 6.4885 | 6.4962 | 6.5038 | 6.5115 | 6.5192 | 6.5269 | 6.5345 | 6.5422 | 6.5498 |
| 4.3 | 2.0736 | 2.0761 | 2.0785 | 2.0809 | 2.0833 | 2.0857 | 2.0881 | 2.0905 | 2.0928 | 2.0952 |
| 43. | 6.5574 | 6.5651 | 6.5727 | 6.5803 | 6.5879 | 6.5955 | 6.6030 | 6.6106 | 6.6182 | 6.6257 |
| 4.4 | 2.0976 | 2.1000 | 2.1024 | 2.1048 | 2.1071 | 2.1095 | 2.1119 | 2.1142 | 2.1166 | 2.1190 |
| 44. | 6.6332 | 6.6408 | 6 6483 | 6.6558 | 6.6633 | 6.6708 | 6.6783 | 6.6858 | 6.6933 | 6.7007 |
| 4.5 | 2.1213 | 2.1237 | 2.1260 | 2.1284 | 2.1307 | 2.1331 | 2.1354 | 2.1378 | 2.1401 | 2.1424 |
| 45. | 6.7082 | 6.7157 | 6.7231 | 6.7305 | 6.7380 | 6.7454 | 6.7528 | 6.7602 | 6.7676 | 6.7750 |
| 4.6 | 2.1448 | 2.1471 | 2.1494 | 2.1517 | 2.1541 | 2.1564 | 2.1587 | 2.1610 | 2.1633 | 2.1656 |
| 46. | 6.7823 | 6.7897 | 6.7971 | 6.8044 | 6.8118 | 6.8191 | 6.8264 | 6.8337 | 6.8411 | 6.8484 |
| 4.7 | 2.1679 | 2.1703 | 2.1726 | 2.1749 | 2.1772 | 2.1794 | 2.1817 | 2.1840 | 2.1863 | 2.1886 |
| 47. | 6.8557 | 6.8629 | 6.8702 | 6.8775 | 6.8848 | 6.8920 | 6.8993 | 6.9065 | 6.9138 | 6.9210 |
| 4.8 | 2.1909 | 2.1932 | 2.1954 | 2.1977 | 2.2000 | 2.2023 | 2.2045 | 2.2068 | 2.2091 | 2.2113 |
| 48. | 6.9282 | 6.9354 | 6.9426 | 6.9498 | 6.9570 | 6.9642 | 6.9714 | 6.9785 | 6.9857 | 6.9929 |
| 4.9 | 2.2136 | 2.2159 | 2.2181 | 2.2204 | 2.2226 | 2.2249 | 2.2271 | 2.2293 | 2.2316 | 2.2338 |
| 49. | 7.0000 | 7.0071 | 7.0143 | 7.0214 | 7.0285 | 7.0356 | 7.0427 | 7.0498 | 7.0569 | 7.0640 |
| 5.0 | 2.2361 | 2.2383 | 2.2405 | 2.2428 | 2.2450 | 2.2472 | 2.2494 | 2.2517 | 2.2539 | 2.2561 |
| 50. | 7.0711 | 7.0781 | 7.0852 | 7.0922 | 7.0993 | 7.1063 | 7.1134 | 7.1204 | 7.1274 | 7.1344 |
| 5.1 | 2.2583 | 2.2605 | 2.2627 | 2.2650 | 2.2672 | 2.2694 | 2.2716 | 2.2738 | 2.2760 | 2.2782 |
| 51. | 7.1414 | 7.1484 | 7.1554 | 7.1624 | 7.1694 | 7.1764 | 7.1833 | 7.1903 | 7.1972 | 7.2042 |
| 5.2 | 2.2804 | 2.2825 | 2.2847 | 2.2869 | 2.2891 | 2.2913 | 2.2935 | 2.2956 | 2.2978 | 2.3000 |
| 52. | 7.2111 | 7.2180 | 7.2250 | 7.2319 | 7.2388 | 7.2457 | 7.2526 | 7.2595 | 7.2664 | 7.2732 |
| 5.3 | 2.3022 | 2.3043 | 2.3065 | 2.3087 | 2.3108 | 2.3130 | 2.3152 | 2.3173 | 2.3195 | 2.3216 |
| 53. | 7.2801 | 7.2870 | 7.2938 | 7.3007 | 7.3075 | 7.3144 | 7.3212 | 7.3280 | 7.3348 | 7.3417 |
| 5.4 | 2.3238 | 2.3259 | 2.3281 | 2.3302 | 2.3324 | 2.3345 | 2.3367 | 2.3388 | 2.3409 | 2.3431 |
| 54. | 7.3485 | 7.3553 | 7.3621 | 7.3689 | 7.3756 | 7.3824 | 7.3892 | 7.3959 | 7.4027 | 7.4095 |

| | 0 | 1 | 2 | 3 | 4 | 5 | 6 | 7 | 8 | 9 |
|---|---|---|---|---|---|---|---|---|---|---|
| 5.5 | 2.3452 | 2.3473 | 2.3495 | 2.3516 | 2.3537 | 2.3558 | 2.3580 | 2.3601 | 2.3622 | 2.3643 |
| 55. | 7.4162 | 7.4229 | 7.4297 | 7.4364 | 7.4431 | 7.4498 | 7.4565 | 7.4632 | 7.4699 | 7.4766 |
| 5.6 | 2.3664 | 2.3685 | 2.3707 | 2.3728 | 2.3749 | 2.3770 | 2.3791 | 2.3812 | 2.3833 | 2.3854 |
| 56. | 7.4833 | 7.4900 | 7.4967 | 7.5033 | 7.5100 | 7.5166 | 7.5233 | 7.5299 | 7.5366 | 7.5432 |
| 5.7 | 2.3875 | 2.3896 | 2.3917 | 2.3937 | 2.3958 | 2.3979 | 2.4000 | 2.4021 | 2.4042 | 2.4062 |
| 57. | 7.5498 | 7.5565 | 7.5631 | 7.5697 | 7.5763 | 7.5829 | 7.5895 | 7.5961 | 7.6026 | 7.6092 |
| 5.8 | 2.4083 | 2.4104 | 2.4125 | 2.4145 | 2.4166 | 2.4187 | 2.4207 | 2.4228 | 2.4249 | 2.4269 |
| 58. | 7.6158 | 7.6223 | 7.6289 | 7.6354 | 7.6420 | 7.6485 | 7.6551 | 7.6616 | 7.6681 | 7.6746 |
| 5.9 | 2.4290 | 2.4310 | 2.4331 | 2.4352 | 2.4372 | 2.4393 | 2.4413 | 2.4434 | 2.4454 | 2.4474 |
| 59. | 7.6811 | 7.6877 | 7.6942 | 7.7006 | 7.7071 | 7.7136 | 7.7201 | 7.7266 | 7.7330 | 7.7395 |
| 6.0 | 2.4495 | 2.4515 | 2.4536 | 2.4556 | 2.4576 | 2.4597 | 2.4617 | 2.4637 | 2.4658 | 2.4678 |
| 60. | 7.7460 | 7.7524 | 7.7589 | 7.7653 | 7.7717 | 7.7782 | 7.7846 | 7.7910 | 7.7974 | 7.8038 |
| 6.1 | 2.4698 | 2.4718 | 2.4739 | 2.4759 | 2.4779 | 2.4799 | 2.4819 | 2.4839 | 2.4860 | 2.4880 |
| 61. | 7.8102 | 7.8166 | 7.8230 | 7.8294 | 7.8358 | 7.8422 | 7.8486 | 7.8549 | 7.8613 | 7.8677 |
| 6.2 | 2.4900 | 2.4920 | 2.4940 | 2.4960 | 2.4980 | 2.5000 | 2.5020 | 2.5040 | 2.5060 | 2.5080 |
| 62. | 7.8740 | 7.8804 | 7.8867 | 7.8930 | 7.8994 | 7.9057 | 7.9120 | 7.9183 | 7.9246 | 7.9310 |
| 6.3 | 2.5100 | 2.5120 | 2.5140 | 2.5159 | 2.5179 | 2.5199 | 2.5219 | 2.5239 | 2.5259 | 2.5278 |
| 63. | 7.9373 | 7.9436 | 7.9498 | 7.9561 | 7.9624 | 7.9687 | 7.9750 | 7.9812 | 7.9875 | 7.9937 |
| 6.4 | 2.5298 | 2.5318 | 2.5338 | 2.5357 | 2.5377 | 2.5397 | 2.5417 | 2.5436 | 2.5456 | 2.5475 |
| 64. | 8.0000 | 8.0062 | 8.0125 | 8.0187 | 8.0250 | 8.0312 | 8.0374 | 8.0436 | 8.0498 | 8.0561 |
| 6.5 | 2.5495 | 2.5515 | 2.5534 | 2.5554 | 2.5573 | 2.5593 | 2.5612 | 2.5632 | 2.5652 | 2.5671 |
| 65. | 8.0623 | 8.0685 | 8.0747 | 8.0808 | 8.0870 | 8.0932 | 8.0994 | 8.1056 | 8.1117 | 8.1179 |
| 6.6 | 2.5690 | 2.5710 | 2.5729 | 2.5749 | 2.5768 | 2.5788 | 2.5807 | 2.5826 | 2.5846 | 2.5865 |
| 66. | 8.1240 | 8.1302 | 8.1363 | 8.1425 | 8.1486 | 8.1548 | 8.1609 | 8.1670 | 8.1731 | 8.1792 |
| 6.7 | 2.5884 | 2.5904 | 2.5923 | 2.5942 | 2.5962 | 2.5981 | 2.6000 | 2.6019 | 2.6038 | 2.6058 |
| 67. | 8.1854 | 8.1915 | 8.1976 | 8.2037 | 8.2098 | 8.2158 | 8.2219 | 8.2280 | 8.2341 | 8.2401 |
| 6.8 | 2.6077 | 2.6096 | 2.6115 | 2.6134 | 2.6153 | 2.6173 | 2.6192 | 2.6211 | 2.6230 | 2.6249 |
| 68. | 8.2462 | 8.2523 | 8.2583 | 8.2644 | 8.2704 | 8.2765 | 8.2825 | 8.2885 | 8.2946 | 8.3006 |
| 6.9 | 2.6268 | 2.6287 | 2.6306 | 2.6325 | 2.6344 | 2.6363 | 2.6382 | 2.6401 | 2.6420 | 2.6439 |
| 69. | 8.3066 | 8.3126 | 8.3187 | 8.3247 | 8.3307 | 8.3367 | 8.3427 | 8.3487 | 8.3546 | 8.3606 |

## A-33 SQUARE ROOTS (CONTINUED)

| | 0 | 1 | 2 | 3 | 4 | 5 | 6 | 7 | 8 | 9 |
|---|---|---|---|---|---|---|---|---|---|---|
| 7.0 | 2.6458 | 2.6476 | 2.6495 | 2.6514 | 2.6533 | 2.6552 | 2.6571 | 2.6589 | 2.6608 | 2.6627 |
| 70. | 8.3666 | 8.3726 | 8.3785 | 8.3845 | 8.3905 | 8.3964 | 8.4024 | 8.4083 | 8.4143 | 8.4202 |
| 7.1 | 2.6646 | 2.6665 | 2.6683 | 2.6702 | 2.6721 | 2.6739 | 2.6758 | 2.6777 | 2.6796 | 2.6814 |
| 71. | 8.4261 | 8.4321 | 8.4380 | 8.4439 | 8.4499 | 8.4558 | 8.4617 | 8.4676 | 8.4735 | 8.4794 |
| 7.2 | 2.6833 | 2.6851 | 2.6870 | 2.6889 | 2.6907 | 2.6926 | 2.6944 | 2.6963 | 2.6981 | 2.7000 |
| 72. | 8.4853 | 8.4912 | 8.4971 | 8.5029 | 8.5088 | 8.5147 | 8.5206 | 8.5264 | 8.5323 | 8.5381 |
| 7.3 | 2.7019 | 2.7037 | 2.7055 | 2.7074 | 2.7092 | 2.7111 | 2.7129 | 2.7148 | 2.7166 | 2.7185 |
| 73. | 8.5440 | 8.5499 | 8.5557 | 8.5615 | 8.5674 | 8.5732 | 8.5790 | 8.5849 | 8.5907 | 8.5965 |
| 7.4 | 2.7203 | 2.7221 | 2.7240 | 2.7258 | 2.7276 | 2.7295 | 2.7313 | 2.7331 | 2.7350 | 2.7368 |
| 74. | 8.6023 | 8.6081 | 8.6139 | 8.6197 | 8.6255 | 8.6313 | 8.6371 | 8.6429 | 8.6487 | 8.6545 |
| 7.5 | 2.7386 | 2.7404 | 2.7423 | 2.7441 | 2.7459 | 2.7477 | 2.7495 | 2.7514 | 2.7532 | 2.7550 |
| 75. | 8.6603 | 8.6660 | 8.6718 | 8.6776 | 8.6833 | 8.6891 | 8.6948 | 8.7006 | 8.7063 | 8.7121 |
| 7.6 | 2.7568 | 2.7586 | 2.7604 | 2.7622 | 2.7641 | 2.7659 | 2.7677 | 2.7695 | 2.7713 | 2.7731 |
| 76. | 8.7178 | 8.7235 | 8.7293 | 8.7350 | 8.7407 | 8.7464 | 8.7521 | 8.7579 | 8.7636 | 8.7693 |
| 7.7 | 2.7749 | 2.7767 | 2.7785 | 2.7803 | 2.7821 | 2.7839 | 2.7857 | 2.7875 | 2.7893 | 2.7911 |
| 77. | 8.7750 | 8.7807 | 8.7864 | 8.7920 | 8.7977 | 8.8034 | 8.8091 | 8.8148 | 8.8204 | 8.8261 |
| 7.8 | 2.7928 | 2.7946 | 2.7964 | 2.7982 | 2.8000 | 2.8018 | 2.8036 | 2.8054 | 2.8071 | 2.8089 |
| 78. | 8.8318 | 8.8374 | 8.8431 | 8.8487 | 8.8544 | 8.8600 | 8.8657 | 8.8713 | 8.8769 | 8.8826 |
| 7.9 | 2.8107 | 2.8125 | 2.8142 | 2.8160 | 2.8178 | 2.8196 | 2.8213 | 2.8231 | 2.8249 | 2.8267 |
| 79. | 8.8882 | 8.8938 | 8.8994 | 8.9051 | 8.9107 | 8.9163 | 8.9219 | 8.9275 | 8.9331 | 8.9387 |
| 8.0 | 2.8284 | 2.8302 | 2.8320 | 2.8337 | 2.8355 | 2.8373 | 2.8390 | 2.8408 | 2.8425 | 2.8443 |
| 80. | 8.9443 | 8.9499 | 8.9554 | 8.9610 | 8.9666 | 8.9722 | 8.9778 | 8.9833 | 8.9889 | 8.9944 |
| 8.1 | 2.8460 | 2.8478 | 2.8496 | 2.8513 | 2.8531 | 2.8548 | 2.8566 | 2.8583 | 2.8601 | 2.8618 |
| 81. | 9.0000 | 9.0056 | 9.0111 | 9.0167 | 9.0222 | 9.0277 | 9.0333 | 9.0388 | 9.0443 | 9.0499 |
| 8.2 | 2.8636 | 2.8653 | 2.8671 | 2.8688 | 2.8705 | 2.8723 | 2.8740 | 2.8758 | 2.8775 | 2.8792 |
| 82. | 9.0554 | 9.0609 | 9.0664 | 9.0719 | 9.0774 | 9.0830 | 9.0885 | 9.0940 | 9.0995 | 9.1049 |
| 8.3 | 2.8810 | 2.8827 | 2.8844 | 2.8862 | 2.8879 | 2.8896 | 2.8914 | 2.8931 | 2.8948 | 2.8965 |
| 83. | 9.1104 | 9.1159 | 9.1214 | 9.1269 | 9.1324 | 9.1378 | 9.1433 | 9.1488 | 9.1542 | 9.1597 |
| 8.4 | 2.8983 | 2.9000 | 2.9017 | 2.9034 | 2.9052 | 2.9069 | 2.9086 | 2.9103 | 2.9120 | 2.9138 |
| 84. | 9.1652 | 9.1706 | 9.1761 | 9.1815 | 9.1869 | 9.1924 | 9.1978 | 9.2033 | 9.2087 | 9.2141 |

| | 0 | 1 | 2 | 3 | 4 | 5 | 6 | 7 | 8 | 9 |
|---|---|---|---|---|---|---|---|---|---|---|
| 8.5 | 2.9155 | 2.9172 | 2.9189 | 2.9206 | 2.9223 | 2.9240 | 2.9257 | 2.9275 | 2.9292 | 2.9309 |
| 85. | 9.2195 | 9.2250 | 9.2304 | 9.2358 | 9.2412 | 9.2466 | 9.2520 | 9.2574 | 9.2628 | 9.2682 |
| 8.6 | 2.9326 | 2.9343 | 2.9360 | 2.9377 | 2.9394 | 2.9411 | 2.9428 | 2.9445 | 2.9462 | 2.9479 |
| 86. | 9.2736 | 9.2790 | 9.2844 | 9.2898 | 9.2952 | 9.3005 | 9.3059 | 9.3113 | 9.3167 | 9.3220 |
| 8.7 | 2.9496 | 2.9513 | 2.9530 | 2.9547 | 2.9563 | 2.9580 | 2.9597 | 2.9614 | 2.9631 | 2.9648 |
| 87. | 9.3274 | 9.3327 | 9.3381 | 9.3434 | 9.3488 | 9.3541 | 9.3595 | 9.3648 | 9.3702 | 9.3755 |
| 8.8 | 2.9665 | 2.9682 | 2.9698 | 2.9715 | 2.9732 | 2.9749 | 2.9766 | 2.9783 | 2.9799 | 2.9816 |
| 88. | 9.3808 | 9.3862 | 9.3915 | 9.3968 | 9.4021 | 9.4074 | 9.4128 | 9.4181 | 9.4234 | 9.4287 |
| 8.9 | 2.9833 | 2.9850 | 2.9866 | 2.9883 | 2.9900 | 2.9917 | 2.9933 | 2.9950 | 2.9967 | 2.9983 |
| 89. | 9.4340 | 9.4393 | 9.4446 | 9.4499 | 9.4552 | 9.4604 | 9.4657 | 9.4710 | 9.4763 | 9.4816 |
| 9.0 | 3.0000 | 3.0017 | 3.0033 | 3.0050 | 3.0067 | 3.0083 | 3.0100 | 3.0116 | 3.0133 | 3.0150 |
| 90. | 9.4868 | 9.4921 | 9.4974 | 9.5026 | 9.5079 | 9.5131 | 9.5184 | 9.5237 | 9.5289 | 9.5341 |
| 9.1 | 3.0166 | 3.0183 | 3.0199 | 3.0216 | 3.0232 | 3.0249 | 3.0265 | 3.0282 | 3.0299 | 3.0315 |
| 91. | 9.5394 | 9.5446 | 9.5499 | 9.5551 | 9.5603 | 9.5656 | 9.5708 | 9.5760 | 9.5812 | 9.5864 |
| 9.2 | 3.0332 | 3.0348 | 3.0364 | 3.0381 | 3.0397 | 3.0414 | 3.0430 | 3.0447 | 3.0463 | 3.0480 |
| 92. | 9.5917 | 9.5969 | 9.6021 | 9.6073 | 9.6125 | 9.6177 | 9.6229 | 9.6281 | 9.6333 | 9.6385 |
| 9.3 | 3.0496 | 3.0512 | 3.0529 | 3.0545 | 3.0561 | 3.0578 | 3.0594 | 3.0610 | 3.0627 | 3.0643 |
| 93. | 9.6437 | 9.6488 | 9.6540 | 9.6592 | 9.6644 | 9.6695 | 9.6747 | 9.6799 | 9.6850 | 9.6902 |
| 9.4 | 3.0659 | 3.0676 | 3.0692 | 3.0708 | 3.0725 | 3.0741 | 3.0757 | 3.0773 | 3.0790 | 3.0806 |
| 94. | 9.6954 | 9.7005 | 9.7057 | 9.7108 | 9.7160 | 9.7211 | 9.7263 | 9.7314 | 9.7365 | 9.7417 |
| 9.5 | 3.0822 | 3.0838 | 3.0854 | 3.0871 | 3.0887 | 3.0903 | 3.0919 | 3.0935 | 3.0952 | 3.0968 |
| 95. | 9.7468 | 9.7519 | 9.7570 | 9.7622 | 9.7673 | 9.7724 | 9.7775 | 9.7826 | 9.7877 | 9.7929 |
| 9.6 | 3.0984 | 3.1000 | 3.1016 | 3.1032 | 3.1048 | 3.1064 | 3.1081 | 3.1097 | 3.1113 | 3.1129 |
| 96. | 9.7980 | 9.8031 | 9.8082 | 9.8133 | 9.8184 | 9.8234 | 9.8285 | 9.8336 | 9.8387 | 9.8438 |
| 9.7 | 3.1145 | 3.1161 | 3.1177 | 3.1193 | 3.1209 | 3.1225 | 3.1241 | 3.1257 | 3.1273 | 3.1289 |
| 97. | 9.8489 | 9.8539 | 9.8590 | 9.8641 | 9.8691 | 9.8742 | 9.8793 | 9.8843 | 9.8894 | 9.8944 |
| 9.8 | 3.1305 | 3.1321 | 3.1337 | 3.1353 | 3.1369 | 3.1385 | 3.1401 | 3.1417 | 3.1432 | 3.1448 |
| 98. | 9.8995 | 9.9045 | 9.9096 | 9.9146 | 9.9197 | 9.9247 | 9.9298 | 9.9348 | 9.9398 | 9.9448 |
| 9.9 | 3.1464 | 3.1480 | 3.1496 | 3.1512 | 3.1528 | 3.1544 | 3.1559 | 3.1575 | 3.1591 | 3.1607 |
| 99. | 9.9499 | 9.9549 | 9.9599 | 9.9649 | 9.9700 | 9.9750 | 9.9800 | 9.9850 | 9.9900 | 9.9950 |

# Answers

## CHAPTER TWO

**1.** (*a*) meas., cont.; (*b*) meas., cont.; (*c*) disc.; (*d*) disc.; (*e*) disc.; (*f*) disc.; (*g*) meas., cont.; (*h*) disc.; (*i*) meas., cont.; (*j*) disc.; (*k*) meas., cont.; (*l*) disc.; (*m*) scale, cont.; (*n*) meas., cont.; (*o*) meas., cont.

**2.** 74 per cent.

**6 and 7**

| | DATA IN TABLE 2-2*a* | | DATA IN TABLE 2-2*b* | |
|---|---|---|---|---|
| *Age* | *Frequency* | *Cumulative per cent* | *Frequency* | *Cumulative per cent* |
| 70–74 | 0 | 100 | 1 | 100 |
| 65–69 | 4 | 100 | 1 | 99 |
| 60–64 | 6 | 96 | 5 | 98 |
| 55–59 | 11 | 90 | 6 | 93 |
| 50–54 | 14 | 79 | 13 | 87 |
| 45–49 | 9 | 65 | 14 | 74 |
| 40–44 | 22 | 56 | 13 | 60 |
| 35–39 | 11 | 34 | 9 | 47 |
| 30–34 | 13 | 23 | 12 | 38 |
| 25–29 | 8 | 10 | 24 | 26 |
| 20–24 | 2 | 2 | 2 | 2 |
| | 100 | | 100 | |

**8** and **9**

| *Systolic blood pressure* | DATA IN TABLE 2-2*a* *Frequency* | DATA IN TABLE 2-2*a* *Cumulative per cent* | DATA IN TABLE 2-2*b* *Frequency* | DATA IN TABLE 2-2*b* *Cumulative per cent* |
|---|---|---|---|---|
| 190–199 | 1 | 100 | 2 | 100 |
| 180–189 | 0 | 99 | 0 | 98 |
| 170–179 | 0 | 99 | 0 | 98 |
| 160–169 | 4 | 99 | 3 | 98 |
| 150–159 | 2 | 95 | 3 | 95 |
| 140–149 | 5 | 93 | 5 | 92 |
| 130–139 | 18 | 88 | 13 | 87 |
| 120–129 | 25 | 70 | 32 | 74 |
| 110–119 | 32 | 45 | 32 | 42 |
| 100–109 | 10 | 13 | 10 | 10 |
| 90– 99 | 3 | 3 | 0 | 0 |
| | 100 | | 100 | |

**10** and **11**

| *Diastolic blood pressure* | DATA IN TABLE 2-2a *Frequency* | DATA IN TABLE 2-2a *Cumulative per cent* | DATA IN TABLE 2-2*b* *Frequency* | DATA IN TABLE 2-2*b* *Cumulative per cent* |
|---|---|---|---|---|
| 110–114 | 4 | 100 | 1 | 100 |
| 105–109 | 1 | 96 | 0 | 99 |
| 100–104 | 2 | 95 | 4 | 99 |
| 95– 99 | 1 | 93 | 1 | 95 |
| 90– 94 | 18 | 92 | 24 | 94 |
| 85– 89 | 8 | 74 | 6 | 70 |
| 80– 84 | 34 | 66 | 32 | 64 |
| 75– 79 | 11 | 32 | 11 | 32 |
| 70– 74 | 16 | 21 | 14 | 21 |
| 65– 69 | 2 | 5 | 5 | 7 |
| 60– 64 | 2 | 3 | 2 | 2 |
| 55– 59 | 1 | 1 | 0 | 0 |
| | 100 | | 100 | |

**12 and 13**

| Blood cholesterol | DATA IN TABLE 2-2*a* Frequency | DATA IN TABLE 2-2*a* Cumulative per cent | DATA IN TABLE 2-2*b* Frequency | DATA IN TABLE 2-2*b* Cumulative per cent |
|---|---|---|---|---|
| 500–524 | 1 | 100 | 0 | 100 |
| 475–499 | 0 | 99 | 0 | 100 |
| 450–474 | 1 | 99 | 1 | 100 |
| 425–449 | 1 | 98 | 0 | 99 |
| 400–424 | 1 | 97 | 4 | 99 |
| 375–399 | 7 | 96 | 1 | 95 |
| 350–374 | 10 | 89 | 5 | 94 |
| 325–349 | 15 | 79 | 6 | 89 |
| 300–324 | 16 | 64 | 7 | 83 |
| 275–299 | 13 | 48 | 12 | 76 |
| 250–274 | 25 | 35 | 20 | 64 |
| 225–249 | 7 | 10 | 13 | 44 |
| 200–224 | 2 | 3 | 13 | 31 |
| 175–199 | 1 | 1 | 10 | 18 |
| 150–174 | 0 | 0 | 7 | 8 |
| 125–149 | 0 | 0 | 1 | 1 |
| | 100 | | 100 | |

**14 and 15**

| Height | DATA IN TABLE 2-2*a* Frequency | DATA IN TABLE 2-2*a* Cumulative per cent | DATA IN TABLE 2-2*b* Frequency | DATA IN TABLE 2-2*b* Cumulative per cent |
|---|---|---|---|---|
| 74 | 4 | 100 | 2 | 100 |
| 73 | 3 | 96 | 6 | 98 |
| 72 | 2 | 93 | 7 | 92 |
| 71 | 6 | 91 | 16 | 85 |
| 70 | 12 | 85 | 14 | 69 |
| 69 | 18 | 73 | 9 | 55 |
| 68 | 16 | 55 | 10 | 46 |
| 67 | 17 | 39 | 18 | 36 |
| 66 | 11 | 22 | 10 | 18 |
| 65 | 6 | 11 | 5 | 8 |
| 64 | 2 | 5 | 2 | 3 |
| 63 | 2 | 3 | 1 | 1 |
| 62 | 1 | 1 | 0 | 0 |
| | 100 | | 100 | |

**16** and **17**

| *Weight* | DATA IN TABLE 2-2*a* *Frequency* | *Cumulative per cent* | DATA IN TABLE 2-2*b* *Frequency* | *Cumulative per cent* |
|---|---|---|---|---|
| 251–265 | 0 | 100 | 1 | 100 |
| 236–250 | 2 | 100 | 1 | 99 |
| 221–235 | 1 | 98 | 1 | 98 |
| 206–220 | 3 | 97 | 3 | 97 |
| 191–205 | 6 | 94 | 4 | 94 |
| 176–190 | 17 | 88 | 24 | 90 |
| 161–175 | 25 | 71 | 21 | 66 |
| 146–160 | 23 | 46 | 21 | 45 |
| 131–145 | 16 | 23 | 21 | 24 |
| 116–130 | 5 | 7 | 3 | 3 |
| 101–115 | 2 | 2 | 0 | 0 |
| | 100 | | 100 | |

**18** and **19**

| *Pulse rate* | DATA IN TABLE 2-2*a* *Frequency* | *Cumulative per cent* | DATA IN TABLE 2-2*b* *Frequency* | *Cumulative per cent* |
|---|---|---|---|---|
| 105–109 | 0 | 100 | 2 | 100 |
| 100–104 | 2 | 100 | 3 | 98 |
| 95– 99 | 3 | 98 | 2 | 95 |
| 90– 94 | 3 | 95 | 1 | 93 |
| 85– 89 | 4 | 92 | 5 | 92 |
| 80– 84 | 27 | 88 | 28 | 87 |
| 75– 79 | 10 | 61 | 9 | 59 |
| 70– 74 | 24 | 51 | 26 | 50 |
| 65– 69 | 9 | 27 | 8 | 24 |
| 60– 64 | 17 | 18 | 16 | 16 |
| 55– 59 | 0 | 1 | 0 | 0 |
| 50– 54 | 1 | 1 | 0 | 0 |
| | 100 | | 100 | |

**27.** A cumulative distribution presupposes a natural ordering of the observed variable. Here, for example, we do not have an ordering of "patient's choice" and "no private physician."

**Summary computations from Table 2-2b and Table 2-2a**

| | *Age* | | *Systolic B.P.* | | *Diastolic B.P.* | | *Cholesterol* | | *Height* | | *Weight* | |
|---|---|---|---|---|---|---|---|---|---|---|---|---|
| | $\Sigma X_i$ | $\Sigma X_i^2$ | $\Sigma X_i$ | $\Sigma X_i^2$ | $\Sigma X_i$ | $\Sigma X_i^2$ | $\Sigma X_i$ | $\Sigma X_i^2$ | $\Sigma X_i$ | $\Sigma X_i^2$ | $\Sigma X_i$ | $\Sigma X_i^2$ |
| Table 2-2*b* | | | | | | | | | | | | |
| 1st–10 | 400 | 16,810 | 1,204 | 145,378 | 834 | 69,764 | 2,845 | 852,501 | 697 | 48,619 | 1,804 | 336,084 |
| 2nd–10 | 461 | 24,413 | 1,412 | 208,094 | 879 | 79,077 | 3,103 | 1,017,049 | 669 | 44,783 | 1,570 | 249,990 |
| 3rd–10 | 452 | 21,862 | 1,240 | 156,200 | 827 | 68,929 | 3,250 | 1,104,352 | 685 | 46,963 | 1,622 | 265,876 |
| 4th–10 | 387 | 15,601 | 1,168 | 138,454 | 810 | 66,550 | 2,758 | 785,842 | 680 | 46,302 | 1,617 | 264,941 |
| 5th–10 | 407 | 18,747 | 1,219 | 150,641 | 785 | 62,325 | 2,204 | 533,004 | 690 | 47,674 | 1,618 | 266,622 |
| 6th–10 | 442 | 20,450 | 1,194 | 144,156 | 804 | 65,616 | 2,685 | 732,625 | 701 | 49,195 | 1,789 | 331,761 |
| 7th–10 | 346 | 12,740 | 1,155 | 133,723 | 769 | 59,677 | 2,288 | 538,002 | 689 | 47,537 | 1,582 | 253,932 |
| 8th–10 | 369 | 14,319 | 1,152 | 134,070 | 765 | 59,325 | 2,199 | 489,595 | 696 | 48,496 | 1,544 | 241,274 |
| 9th–10 | 411 | 17,449 | 1,234 | 153,404 | 856 | 73,566 | 2,442 | 622,692 | 697 | 48,641 | 1,780 | 320,304 |
| 10th–10 | 416 | 18,538 | 1,219 | 151,429 | 818 | 68,022 | 2,375 | 584,461 | 685 | 46,997 | 1,672 | 283,898 |
| 1st–20 | 861 | 41,223 | 2,616 | 353,472 | 1,713 | 148,841 | 5,948 | 1,869,550 | 1,366 | 93,402 | 3,374 | 586,074 |
| 2nd–20 | 839 | 37,463 | 2,408 | 294,654 | 1,637 | 135,479 | 6,008 | 1,890,194 | 1,365 | 93,265 | 3,239 | 530,817 |
| 3rd–20 | 849 | 39,197 | 2,413 | 294,797 | 1,589 | 127,941 | 4,889 | 1,265,629 | 1,391 | 96,869 | 3,407 | 598,383 |
| 4th–20 | 715 | 27,059 | 2,307 | 267,793 | 1,534 | 119,002 | 4,487 | 1,027,597 | 1,385 | 96,033 | 3,126 | 495,206 |
| 5th–20 | 827 | 35,987 | 2,453 | 304,833 | 1,674 | 141,588 | 4,817 | 1,207,153 | 1,382 | 95,638 | 3,452 | 604,202 |
| 1st–50 | 2,107 | 97,433 | 6,243 | 798,767 | 4,135 | 346,645 | 14,160 | 4,292,748 | 3,421 | 234,341 | 8,231 | 1,383,513 |
| 2nd–50 | 1,984 | 83,496 | 5,954 | 716,782 | 4,012 | 326,206 | 11,989 | 2,967,375 | 3,468 | 240,866 | 8,367 | 1,431,169 |
| all 100 | 4,091 | 180,929 | 12,197 | 1,515,549 | 8,147 | 672,851 | 26,149 | 7,260,123 | 6,889 | 475,207 | 16,598 | 2,814,682 |
| Table 2-2*a* | | | | | | | | | | | | |
| 1st–50 | 2,106 | 93,556 | 6,025 | 739,879 | 4,051 | 333,827 | 15,881 | 5,255,135 | 3,447 | 237,877 | 8,429 | 1,447,043 |
| 2nd–50 | 2,315 | 114,793 | 6,105 | 759,097 | 4,120 | 344,584 | 14,992 | 4,614,116 | 3,379 | 228,663 | 8,011 | 1,319,553 |
| all 100 | 4,421 | 208,349 | 12,130 | 1,498,976 | 8,171 | 678,411 | 30,873 | 9,839,251 | 6,826 | 466,540 | 16,440 | 2,766,596 |

## CHAPTER THREE

| | $\bar{X}$ | $s^2$ | $s$ |
|---|---|---|---|
| **1.** | 57.8 | 365.8 | 19.1 |
| **2.** (a) | 49.0 | 277.3 | 16.7 |
| (b) | 50.3 | 222.8 | 14.9 |
| **3.** G | 44.5 | 4.3 | 2.1 |
| B | 46.0 | 4.3 | 2.1 |
| **4.** | 44.8 | 128.0 | 11.3 |
| **5.** | 41.3 | 141.5 | 11.9 |
| **6.** | 124.6 | 270.5 | 16.4 |
| **7.** | 125.6 | 280.4 | 16.7 |
| **8.** | 83.6 | 109.9 | 10.5 |
| **9.** | 83.4 | 88.9 | 9.4 |
| **10.** | 308.0 | 3,086.0 | 55.5 |
| **11.** | 262.2 | 4,413.0 | 66.4 |
| **12.** | 68.3 | 6.0 | 2.5 |
| **13.** | 68.9 | 6.3 | 2.5 |
| **14.** | 164.6 | 661.9 | 25.7 |

| | $\bar{X}$ | $s^2$ | $s$ |
|---|---|---|---|
| **15.** | 166.1 | 621.5 | 24.9 |
| **16.** | 75.9 | 99.9 | 10.0 |
| **17.** | 76.8 | 110.1 | 10.5 |
| **18.** | 12.3 | 16.7 | 4.1 |
| **19.** | 2.9 | 3.5 | 1.9 |
| **20.** | 2.1 | 2.3 | 1.5 |
| **21.** P | 6.7 | 7.5 | 2.7 |
| C | 6.5 | 6.9 | 2.6 |
| **22.** | 293.0 | 4,212.0 | 64.9 |
| **23.** | 1,842.0 | 2,383,653.0 | 1,544.0 |
| **24.** | 98.6 | 8.8 | 3.0 |
| **28.** | 16.0 | .02 | .12 |
| **29.** | 4.64 | 1.40 | 1.19 |
| **31.** | 5.70 | 3.50 | 1.87 |
| **32.** $Z = 616$ | | | |

## CHAPTER FOUR

**3.** 0; .05; 1.

**4.** 69.7; .70; 3.46.

**5.** .5 and .25/10 = .025.

**7.** $\sigma_{\bar{X}} = .022$ for $N = 2{,}000$; $\sigma_{\bar{X}} = .011$ for $N = 8{,}000$.

**8.** $\mu = \frac{1}{3}$, $\sigma = \sqrt{2}/3$; $\sigma_{\bar{X}} = \sqrt{2}/15$; $N = 222{,}223$.

**9.** .04

**10.** $N = 36$.

**11.** Var $(\bar{X}) = .1712$; without replacement $\sigma_{\bar{X}} = .414$, with replacement $\sigma_{\bar{X}} = .424$.

**12.** Increase to $4N$, $16N$, $64N$, and $256N$, respectively, where $N$ is the original sample size.

## CHAPTER FIVE

**1.** (a) .8413, (b) .9772, (c) .0228, (d) −1.96, (e) 0, (f) 1.645, (g) 2.576.

**2.** $P_{10} = -1.282$, $P_{25} = -.674$, $P_{75} = .674$, $P_{97.5} = 1.96$.

**3.** (a) .1151; (b) .7698; (c) .4772; (d) .0808; (e) 38.23; (f) 21.77, 38.23.

**4.** (a) .441; (b) 12.43, 82.77; (c) $P_{10} = 26.83$, $P_{30} = 39.11$, $P_{99} = 85.28$; (d) $z = 1$ for 63.8, $z = -.5$ for 39.5.

**5.** (a) .8413; (b) 96.1, 103.9; (c) 94.8, 105.2.

**6.** .6826; .0228; .0228; 60.27, 75.73; 70.02; 66.43.

**7.** .0228 over 90; .1587 below 75.

**8.** 266,000 before 35 days; 227,000 after 46 days.

**9.** $P_{10} = 23.8$, $P_{50} = 43$, $P_{75} = 53.1$, $P_{90} = 62.2$, $P_{99} = 77.9$.

**15.** $\mu = \frac{2}{3}$, $\sigma^2 = \frac{2}{9}$.

**16.** .1151 for fewer than 10; .0228 for fewer than 8; .7698 for 10 to 15.

**17.** $z = 6$, so the chance is greater than the last value read from Table A-3, that is, 0.999994 for $z = 5$; .77.

## CHAPTER SIX

**1.** $4{,}604 < \mu < 4{,}996$ for 95 per cent confidence interval; $4{,}542 < \mu < 5{,}058$ for 99 per cent confidence interval.

**2.** $126.1 < \mu < 133.9$.

**3.** $480.4 < \mu < 519.6$.

**4.** $63.42 < \mu < 64.98$ for 90 per cent confidence interval; $62.64 < \mu < 65.76$ for 99.9 per cent confidence interval.

**5.** $N = 97$; $N = 956$.

**6.** $N = 96$.

**7.** $4{,}593.6 < \mu < 5{,}006.4$; it is normally distributed.

**8.** $1.0019 < \mu < 1.0061$.

**9.** $\bar{X} = 10.4$ per cent, $s = 1.17$ per cent; $9.72 < \mu < 11.08$ for 90 per cent confidence interval.

**10.** For boys $\bar{X} = 45.96$, $s = 2.08$; $45.37 < \mu < 46.55$ for 95 per cent confidence interval. For girls $\bar{X} = 44.46$, $s = 2.06$; $43.87 < \mu < 45.05$ for 95 per cent confidence interval. No.

**11.**
2. (a) $49 \pm 1.97 \times 16.7/\sqrt{2{,}096}$
   (b) $50.3 \pm 1.9 \times 14.9/\sqrt{459}$
3. $44.5 \pm 2.01 \times 2.1/\sqrt{50}$
   $46 \pm 2.01 \times 2.1/\sqrt{50}$
4. $44.8 \pm 1.99 \times 11.3/\sqrt{100}$
5. $41.3 \pm 1.99 \times 11.9/\sqrt{100}$
6. $124.6 \pm 1.99 \times 16.4/\sqrt{100}$
7. $125.6 \pm 1.99 \times 16.7/\sqrt{100}$
8. $83.6 \pm 1.99 \times 10.5/\sqrt{100}$
9. $83.4 \pm 1.99 \times 9.4/\sqrt{100}$
10. $308 \pm 1.99 \times 55.5/\sqrt{100}$
11. $262.2 \pm 1.99 \times 66.4/\sqrt{100}$
12. $68.3 \pm 1.99 \times 2.5/\sqrt{100}$
13. $68.9 \pm 1.99 \times 2.5/\sqrt{100}$
14. $164.6 \pm 1.99 \times 25.7/\sqrt{100}$
15. $166.1 \pm 1.99 \times 24.9/\sqrt{100}$
16. $75.9 \pm 1.99 \times 10/\sqrt{100}$
17. $76.8 \pm 1.99 \times 10.5/\sqrt{100}$
18. $12.3 \pm 2.005 \times 4.1/\sqrt{55}$
19. $2.9 \pm 1.97 \times 1.9/\sqrt{246}$
20. $2.1 \pm 1.97 \times 1.5/\sqrt{646}$
21. $P$ $6.7 \pm 1.97 \times 2.7/\sqrt{434}$
    $C$ $6.5 \pm 1.97 \times 2.6/\sqrt{407}$
22. $293 \pm 1.97 \times 64.9/\sqrt{1{,}502}$
23. $1{,}842.2 \pm 1.97 \times 1{,}543.9/\sqrt{3{,}200}$
24. $98.6 \pm 2.571 \times 3/\sqrt{5}$.

**12.** $\bar{X} < 96.14$ and $\bar{X} > 103.86$.

**13.** $z = 3.33$; reject at $\alpha = .05$, .01, and .001.

**14.** Reject if $\bar{X} < 48.63$ or $\bar{X} > 51.37$.

**15.** (a) $z = 2.50$, reject; (b) $z = -2.50$, accept; (c) $z = 2.50$, accept; (d) $z = 1.20$, accept; (e) $z = 1$, accept.

**16.** $H$: $\mu \le 1$, $\alpha = .05$, reject if $t > t_{.95}(9) = 1.833$; $t = 4.219$, so we reject.

**17.** $H$: $\mu = 10$, $\alpha = .05$, reject if $t < -2.262$ or $t > 2.262$; $t = 1.081$, so we accept ($\bar{X} = 10.4$, $s = 1.17$).

**18.** $H$: $\mu = 0$, $\alpha = .01$, reject if $t < -3.250$ or $t > 3.250$; $t = 7.133$, so we reject ($\bar{X} = 3$, $s = 1.33$).

**19.** (*a*) Reject if $\bar{X} < 135.1$ or $\bar{X} > 144.9$; (*b*) $z_1 = -5.96$, $z_2 = -2.04$, so $\beta = .0207$.

**20.** (*a*) For $N = 100$ reject if $\bar{X} < 66.23$ or $\bar{X} > 67.77$, for $N = 25$ reject if $\bar{X} < 65.45$ or $\bar{X} > 68.55$. (*b*) For example, for $N = 100$, if $\mu = 68$, $z_1 = (66.23 - 68)/.3 = -5.90$, $z_2 = (67.77 - 68)/.3 = -.77$, so $\beta = .22$. (*c*) Approximately .05 for $\mu = 67.25$ and .48 for $\mu = 67.75$.

**21.** (*a*) $z_1 = -11.96$, $z_2 = -8.04$, so $\beta < .2$; (*b*) $N = 19$.

**22.** (*a*) $H: \mu \geq 24$, $\alpha = .05$, reject if $t < -2.13$, 4 degrees of freedom; $t = 1.173$, so we accept the mean as 24 or larger ($\bar{X} = 25.6$, $s = 3.05$). Thus the manufacturer could not safely claim that the mean is less than 24. (*b*) The chance of rejecting when the mean is actually greater than or equal to 24 is $\alpha = .05$. In doing so, he would advertise falsely. (*c*) By accepting the hypothesis when $\mu < 24$. At this point in the text the student has no way to compute $\beta$ ($\sigma$ not known). In making an error of this type the manufacturer loses an opportunity to promote his product appropriately.

**23.** (*a*) $H: \mu \leq 20$, $\alpha = .05$, reject if $t > 2.132$, 4 degrees of freedom; $t = -.932$, so we accept the mean as 20 or less ($\bar{X} = 19.2$, $s = 1.92$). Thus the laboratory could not safely say that the manufacturer's claim is correct. (*b*) By rejecting when $\mu \leq 20$, $\alpha = .05$. In doing so, the manufacturer would be unjustly supported. (*c*) By accepting the hypothesis when $\mu > 20$. With an error of this type satisfactory batteries would not be recognized.

## CHAPTER SEVEN

**1.** $H: \mu \geq 150$, alternative $\mu < 150$, $\alpha = .05$, reject if $t < t_{.05}(49) = -1.677$; $t = -3.536$, so we reject. Individual weight changes would give better information.

**2.** $H: \mu \leq 65$, alternative $\mu > 65$, $\alpha = .05$, reject if $t > t_{.95}(99) = 1.662$; $t = -1.95$, so we accept.

**3.** For first 10 observations $H: \mu = 110$, $\alpha = .05$, reject if $t < t_{.025}(9) = -2.262$ or if $t > t_{.975}(9) = 2.262$; with $\bar{X} = 120.4$, $s = 6.80$, $t = 4.836$, so we reject. For first 20 observations $H: \mu = 110$, $\alpha = .05$, reject if $t < t_{.025}(19) = -2.093$ or if $t > t_{.975}(19) = 2.093$; with $\bar{X} = 130.8$, $s = 24.39$, $t = 3.814$, so we reject.

**4.** The standard error of the mean is $\sigma_{\bar{X}} = \sqrt{p(1-p)/N}$; for example, $p = .4$, $N = 100$, $\sigma_{\bar{X}} = .049$.

**5.** $N = 2{,}500$ (use $p = .5$).

**6.** $N = 900$.

**7.** $.554 < p < .646$.

**8.** $.179 < p < .221$.

**9.** $N = 13{,}526$.

**10.** At $\alpha = .01$, for $p = .50$, $\bar{X} < .436$ and $\bar{X} > .564$; for $p = \frac{1}{3}$, $\bar{X} < .273$ and $\bar{X} > .394$; for $p = .25$, $\bar{X} < .194$ and $\bar{X} > .306$. At $\alpha = .05$, for $p \geq .50$, $\bar{X} < .459$, and for $p \leq .50$, $\bar{X} > .541$; for $p \geq \frac{1}{3}$, $\bar{X} < .294$, and for $p \leq \frac{1}{3}$, $\bar{X} > .372$; for $p \geq .25$, $\bar{X} < .214$, and for $p \leq .25$, $\bar{X} > .286$.

**11.** Yes; critical region is $\bar{X} < .302$ and $\bar{X} > .364$.

**12.** $198.6 < \sigma^2 < 593.6$.

**13.** $69.29 < \mu < 70.11$; $2.63 < \sigma^2 < 4.90$.

**14.** $N = 141$.

**15.** (*a*) $s^2 < 37.2$, $s^2 > 200$; (*b*) $s^2 > 20.3$; (*c*) $s^2 < 6.94$, $s^2 > 39.80$; (*d*) $s^2 < 20.25$, $s^2 > 41.70$; (*e*) $s^2 < 0.011$.

**16.** $H_1$: $\sigma^2 = 20$, $\alpha = .05$, reject if $s^2/\sigma^2 < \chi^2_{.025}/df(24) = .517$ or if $s^2/\sigma^2 > \chi^2_{.975}/df(24) = 1.64$; $s^2/\sigma^2 = .63$, so we accept $H_1$. $H_2$: $\sigma^2 \geq 20$, $\alpha = .05$, reject if $s^2/\sigma^2 < \chi^2_{.05}/df(24) = .577$; $s^2/\sigma^2 = .63$, so we accept $H_2$.

**17.** (*a*) .78, (*b*) .55, (*c*) .75, (*d*) .04, (*e*) $1.00^-$.

**18.** $N = 11$.

## CHAPTER EIGHT

**1.** 16.7, 28.7, .941, 1.06, 1, .351, 8.75.

**2.** They are equal.

**3.** $F_\alpha(N_1,N_2)F_{1-\alpha}(N_2,N_1) = 1$; they are reciprocals.

**4.** (*a*) $F_{.025}(15,15) = .349$, $F_{.975}(15,15) = 2.86$; $F = 3.13$, so we reject $H$.
(*b*) $F_{.005}(40,12) = .339$, $F_{.995}(40,12) = 4.23$; $F = 2.47$, so we accept $H$.
(*c*) $F_{.01}(59,119) = .58$, $F_{.99}(59,119) = 1.66$; $F = .47$, so we reject $H$.

**5.** $H$: $\sigma_1^2 = \sigma_2^2$, $F_{.025}(49,39) = .556$, $F_{.975}(49,39) = 1.83$; $F = .694$, so we accept $H$.

**6.** $H$: $\sigma_1^2 = \sigma_2^2$, $F_{.05}(9,9) = .315$, $F_{.95}(9,9) = 3.18$; $F = 1.10$ ($s_1^2 = 24.18$, $s_2^2 = 22.01$), so we accept $H$.

**7.** $H$: $\sigma_A^2 = \sigma_B^2$, $F_{.05}(7,7) = .264$, $F_{.95}(7,7) = 3.79$; $F = .649$, so we accept $H$.

**8.** (*a*) .75, (*b*) .85, (*c*) .25.

**9.** Reject if $F < .200$ or if $F > 4.99$ at $\alpha = .05$; $\beta = .855$, $1 - \beta = .145$.

**10.** $12.1 < \sigma^2 < 27.8$.

**11.** $21.6 < \sigma^2 < 36.3$.

**12.** $s_p^2 = 5.67$, $s^2 = 73.27$.

**13.** $H$: $\mu_c \geq \mu_p$, reject if $t > t_{.95}(18) = 1.734$; $t = -2.5$, so we accept. However, the hypothesis of equality of variances would be rejected [$F = 4.77$ compared with $F_{.95}(9,9) = 3.18$], and the analysis might follow that presented at the end of Sec. 8-4: reject if $t > t_{.95}(14.2) = 1.76$. The recomputed $t$ is also $-2.5$, so we accept the hypothesis.

**14.** $H$: $\mu_1 = \mu_2$, $t_{.025}(37) = -2.027$, $t_{.975}(37) = 2.027$, $s_p = 7.15$; $t = -3.05$, so we reject.

**15.** For uniformity we test $H$: $\sigma_1^2 = \sigma_2^2$, $F_{.025}(8,8) = .226$, $F_{.975}(8,8) = 4.43$; $F = 1.69$, so we accept. For difference in means we test $H$: $\mu_1 = \mu_2$, $t_{.025}(16) = -2.12$, $t_{.975}(16) = 2.12$, $s_p = .0575$; $t = 9.56$, so we reject.

**16.** $H$: $\sigma_1^2 = \sigma_2^2$, $F_{.025}(49,49) = .565$, $F_{.975}(49,49) = 1.77$; $F = 4.253/4.325 = .98$, so we accept. $H$: $\mu_1 = \mu_2$, $t_{.025}(98) = -1.99$, $t_{.975}(98) = 1.99$, $s_p = 2.071$; $t = -3.62$, so we reject. $.81 < \mu_2 - \mu_1 < 2.19$ for 90 per cent confidence interval.

**17.** In Prob. 5 we accepted $H$: $\sigma_1^2 = \sigma_2^2$. $H$: $\mu_1 = \mu_2$, $t_{.025}(88) = -1.99$, $t_{.975}(88) = 1.99$. $s_p = 2{,}733$, $t = -3.45$, so we reject.

**18.** $N = 72$.

**19.** $H$: $\mu_1 = \mu_2$. $t_{.025}(9) = -2.262$, $t_{.975}(9) = 2.262$, $\bar{d} = \bar{X}_1 - \bar{X}_2 = -2.8$; $t = -6.73$, so we reject. $2.04 < \mu_2 - \mu_1 < 3.56$ for 90 per cent confidence interval.

**20.** (*a*) $H$: $\mu_P \leq \mu_N$, $t_{.95}(6) = 1.943$; with paired observations $t = 6.95$, so we reject. There is evidence that pollination gives a higher mean yield of seed. (*b*) By rejecting the hypothesis when the mean yield of seed from pollinated plants is actually not greater than the mean yield of seed from nonpollinated plants. The consequence would be using the pollination when it actually does not increase the yield. (*c*) By accepting the hypothesis when pollination actually does increase the yield. The consequence is a loss of potential increased yield. (*d*) $.35 < \mu_1 - \mu_2 < .63$.

## CHAPTER NINE

**1.** $N = 14$.

**2.** $N = 157$ if median is used. Using the mean of 100 observations costs $15.00, while using the median of 157 costs $15.70.

**3.** $\bar{X}$ for 163 observations.

**4.** $.5(P_{25} + P_{75})$.

**5.** From Table A-8*a*, estimate of $\mu$, 136.7. $E = .88$.

**6.** (*a*) 69.5, (*b*) 69.6, (*c*) 10.7.

**7.** (*a*) 4.7, (*b*) 4.8, (*c*) 5.2.

**8.** For $A$, median $= 1.23$, $M_r = 1.24$, $w = .18$, estimate of $\sigma$, .061; For $B$, median $= .98$, $M_r = .975$, $w = .17$, estimate of $\sigma$, .057.

**9.** Estimate of $\mu$, 13.2, estimate of $\sigma$, 1.96; $N = 650$.

**10.** $P_{15}, P_{40}, P_{60}, P_{85}$; estimate of $\mu$, 99.25; estimate of $\sigma$, $.3040(P_{95} - P_{05}) = 30.40$. $P_{05}, P_{30}, P_{70}, P_{95}$; estimate of $\mu$, 100, estimate of $\sigma$, 30.43. To estimate $\mu$ from all 19, sum and divide by 19; estimate of $\sigma$, $.0739[(P_{95} - P_{05}) + \cdots + (P_{55} - P_{45})]$ [it is only coincidental that this factor .0739 is the same as in Table A-8*a*(2)].

**11.** Estimate of $\sigma$, 28.80; .65; yes, but they are not the most efficient ones.

**12.** $H$: $\mu_2 \leq \mu_1$, $\alpha = .05$, reject if $\tau_d < -.493$; $\tau_d = -.27$, so we accept.

**13.** $H$: $\sigma_1^2 = \sigma_2^2$, $\alpha = .10$, reject if $F < .505$ or $F > 1.98$; $F = w_1/w_2 = 1.5$, so we accept.

**14.** $.062 < \mu_c - \mu_p < .578$ for 95 per cent confidence interval. $H$: $\mu_p \leq \mu_c$, $\alpha = .05$, reject if $\tau_d < -.250$; $\tau_d = .377$, so we accept.

**15.** With $\tau_1$, $9.71 < \mu < 11.09$ for 95 per cent confidence interval; $H$: $\mu = 10$, $\alpha = .05$, reject if $\tau_1 < -.230$ or if $\tau_1 > .230$; $\tau_1 = .133$, so we accept. With $\tau_2$, $9.69 < \mu < 11.31$ for 95 per cent confidence interval; $H$: $\mu = 10$, $\alpha = .05$, reject if $\tau_2 < -.27$ or if $\tau_2 > .27$; $\tau_2 = .167$, so we accept. With $t$, $9.56 < \mu < 11.24$ for 95 per cent confidence interval; $H$: $\mu = 10$, $\alpha = .05$, reject if $t < -2.262$ or if $t > 2.262$; $t = 1.08$, so we accept.

**16.** With $\tau_1$, $1.88 < \mu < 3.72$ for 95 per cent confidence interval; $H$: $\mu = 0$, $\alpha = .05$, reject if $\tau_1 < -.230$ or if $\tau_1 > .230$; $\tau_1 = .70$, so we reject. With $\tau_2$, $2.92 < \mu < 5.08$ for 95 per cent confidence interval; $H$: $\mu = 0$, $\alpha = .05$, reject if $\tau_2 < -.27$ or if $\tau_2 > .27$; $\tau_2 = 1$, so we reject. Pairing is necessary to eliminate the effects caused by differences in locality.

**17.** (*a*) $40.284 < \mu < 40.360$, (*b*) $40.283 < \mu < 40.361$, (*c*) $40.278 < \mu < 40.362$.

**18.** $-.081 < \mu < .055$.

**19.** (*a*) For $N = 10$, $-.035 < \mu < .053$; for $N = 40$, $-.003 < \mu < .041$; (*b*) For $N = 10$, $.8846 < \sigma^2 < 1.2307$; for $N = 40$, $.8821 < \sigma^2 < 1.2271$.

**20.** $6.19 < \mu < 14.61$.

**21.** Approximately 85 per cent.

## CHAPTER TEN

**1**

| | *Sum of squares* | *Degrees of freedom* | *Mean square* | *F ratio* |
|---|---|---|---|---|
| Means | 30.920 | 2 | 15.46 | .032 |
| Within | 30,704.838 | 63 | 487.38 | |
| *Total* | 30,735.758 | 65 | 472.85 | |

$H$: $\mu_1 = \mu_2 = \mu_3$, $\alpha = .05$, $F_{.95}(2,63) = 3.15$; $F = .032$, so we accept.

**2**

| | *Sum of squares* | *Degrees of freedom* | *Mean square* | *F ratio* |
|---|---|---|---|---|
| Means | 349.6 | 3 | 116.5 | 1.18 |
| Within | 1,967.7 | 20 | 98.4 | |
| *Total* | 2,317.3 | 23 | 100.8 | |

$H$: all means are equal, $\alpha = .05$, $F_{.95}(3,20) = 3.10$; $F = 1.18$, so we accept.

**3**

| | *Sum of squares* | *Degrees of freedom* | *Mean square* | *F ratio* |
|---|---|---|---|---|
| Means | 11,467 | 2 | 5,733 | 3.17 |
| Within | 537,500 | 297 | 1,810 | |
| *Total* | 548,967 | 299 | | |

$H: \mu_{\text{I}} = \mu_{\text{II}} = \mu_{\text{III}}$, $\alpha = .05$, $F_{.95}(2,297) = 3.00$; $F = 3.17$, so we reject.

**4.** $H: \mu_1 = \mu_2$, $\alpha = .05$, $F_{.95}(1,14) = 4.60$.

| | *Sum of squares* | *Degrees of freedom* | *Mean square* | *F ratio* |
|---|---|---|---|---|
| Means | 25 | 1 | 25 | 4.32 |
| Within | 81 | 14 | 5.79 | |
| *Total* | 106 | 15 | | |

$F = 4.32$, so we accept.

**5**

| $\bar{X}_{A\cdot}$ | $\bar{X}_{B\cdot}$ | $\bar{X}_{C\cdot}$ | *Confidence limits* | *Population contrast* |
|---|---|---|---|---|
| 1 | −1 | 0 | −26.08, 2.08 | $\mu_A - \mu_B$ |
| 1 | 0 | −1 | −12.08, 16.08 | $\mu_A - \mu_C$ |
| 0 | 1 | −1 | − .08, 28.08 | $\mu_B - \mu_C$ |

$s_p = 42.54$, $q_{.95}(3,297) = 3.31$, $n = 100$.

**6.** $H: \mu_1 = \mu_2$, $q_{.95}(2,14) = 3.03$. $q = 2.939$, so we accept; $(q/\sqrt{2})^2 = 4.32$. $F_{.95}(1,14) = t^2_{.975}(14) = q^2_{.95}(2,14)/2 = 4.60$. This is an illustration of the fact that for two groups, each of size $n$, $F_{1-\alpha}(1,df) = [t_{1-\alpha/2}(df)]^2 = [q_{1-\alpha}(2,df)/\sqrt{2}]^2$, and that results of a $t$ test, analysis of variance, and a $q$ test always agree.

**7.** $5.002 < \mu_2 - \mu_1 < 12.998$, $4.002 < \mu_4 - \mu_2 < 11.998$
$-.998 < \mu_1 - \mu_3 < 6.998$, $16.002 < \mu_5 - \mu_2 < 23.998$
$13.002 < \mu_4 - \mu_1 < 20.998$, $16.002 < \mu_4 - \mu_3 < 23.998$
$25.002 < \mu_5 - \mu_1 < 32.998$, $28.002 < \mu_5 - \mu_3 < 35.998$
$8.002 < \mu_2 - \mu_3 < 15.998$, $8.002 < \mu_5 - \mu_4 < 15.998$

**8.** $H$: all $\mu_i$ are equal, $\alpha = .05$, $F_{.95}(3,96) = 2.71$.

(*a*) Systolic blood pressure

**From Table 2-2a**

| | *Sum of squares* | *Degrees of freedom* | *Mean square* | *F ratio* |
|---|---|---|---|---|
| Means | 4,094.363 | 3 | 1,364.788 | 5.57 |
| Within | 23,512.492 | 96 | 244.922 | |
| *Total* | 27,606.856 | 99 | | |

$F = 5.57$, so we reject.

**From Table 2-2b**

| | *Sum of squares* | *Degrees of freedom* | *Mean square* | *F ratio* |
|---|---|---|---|---|
| Means | 11,734.262 | 3 | 3,911.42 | 23.3 |
| Within | 16,146.523 | 96 | 168.19 | |
| *Total* | 27,880.785 | 99 | | |

$F = 23.3$, so we reject.

(*b*) Diastolic blood pressure

**From Table 2-2a**

| | *Sum of squares* | *Degrees of freedom* | *Mean square* | *F ratio* |
|---|---|---|---|---|
| Means | 1,874.504 | 3 | 624.83 | 6.75 |
| Within | 8,883.984 | 96 | 92.54 | |
| *Total* | 10,758.488 | 99 | | |

$F = 6.75$, so we reject.

**From Table 2-2b**

| | *Sum of squares* | *Degrees of freedom* | *Mean square* | *F ratio* |
|---|---|---|---|---|
| Means | 1,808.051 | 3 | 602.68 | 7.92 |
| Within | 7,306.758 | 96 | 76.11 | |
| *Total* | 9,114.809 | 99 | | |

$F = 7.92$, so we reject.

(*c*) Cholesterol

**From Table 2-2a**

| | *Sum of squares* | *Degrees of freedom* | *Mean square* | *F ratio* |
|---|---|---|---|---|
| Means | 30,403.250 | 3 | 10,134.4 | 3.51 |
| Within | 277,423.875 | 96 | 2,889.8 | |
| *Total* | 307,827.125 | 99 | | |

$F = 3.51$, so we reject.

**From Table 2-2b**

| | Sum of squares | Degrees of freedom | Mean square | F ratio |
|---|---|---|---|---|
| Means | 91,691.125 | 3 | 30,563.71 | 8.87 |
| Within | 330,726.938 | 96 | 3,445.07 | |
| *Total* | 422,418.063 | 99 | | |

$F = 8.87$, so we reject.

**9**

| | Sum of squares | Degrees of freedom | Mean square | F ratio |
|---|---|---|---|---|
| Means | 21,306.2 | 3 | 7,102.07 | 136 |
| Within | 836.0 | 16 | 52.25 | |
| *Total* | 22,142.2 | 19 | | |

$H$: $\mu_A = \mu_B = \mu_C = \mu_D$, $\alpha = .05$, $F_{.95}(3,16) = 3.24$; $F = 136$, so we reject. $s_p = 7.23$, $q_{.95}(4,16) = 4.05$. 95 per cent simultaneous confidence limits for all pairs of means are as follows:

$-25.49 < \mu_A - \mu_B < .69$
$-95.89 < \mu_A - \mu_C < -69.71$
$-63.89 < \mu_A - \mu_D < -33.71$
$-83.49 < \mu_B - \mu_C < -57.31$
$-51.49 < \mu_B - \mu_D < -25.31$
$18.91 < \mu_C - \mu_D < 45.09$

| | Sum of squares | Degrees of freedom | Mean square | F ratio |
|---|---|---|---|---|
| Means | 19.845 | 1 | 19.845 | 3.22 |
| Within | 1,221.030 | 198 | 6.167 | |
| *Total* | 1,240.875 | 199 | | |

$H$: $\mu_a = \mu_b$, $\alpha = .05$, $F_{.95}(1,198) = 3.84$, $\bar{X}_a = 68.26$, $\bar{X}_b = 68.89$; $F = 3.22$, so we accept.

**11**

| | Sum of squares | Degrees of freedom | Mean square | F ratio |
|---|---|---|---|---|
| Means | 124.820 | 1 | 124.8 | .20 |
| Within | 123,605.960 | 198 | 624.3 | |
| *Total* | 123,730.780 | 199 | | |

$H$: $\mu_a = \mu_b$, $\alpha = .05$, $F_{.95}(1,198) = 3.84$, $\bar{X}_a = 164.40$, $\bar{X}_b = 165.98$; $F = .20$, so we accept.

12. $H: \mu_1 = \mu_2 = \mu_3 = \mu_4$, $\alpha = .05$, $F_{.95}(3,96) = 2.71$.

(a) Systolic blood pressure

**From Table 2-2a**

$\bar{X}_1 = 121.73$ $\bar{X}_2 = 116.41$ $\bar{X}_3 = 125.82$ $\bar{X}_4 = 130.04$

$n_1 = 11$ $n_2 = 54$ $n_3 = 11$ $n_4 = 24$

| | *Sum of squares* | *Degrees of freedom* | *Mean square* | *F ratio* |
|---|---|---|---|---|
| Means | 3,353.19 | 3 | 1,117.7 | 4.42 |
| Within | 24,253.75 | 96 | 252.6 | |
| *Total* | 27,606.94 | 99 | | |

$F = 4.42$, so we reject.

**From Table 2-2b**

$\bar{X}_1 = 118.20$ $\bar{X}_2 = 116.66$ $\bar{X}_3 = 152.25$ $\bar{X}_4 = 129.50$

$n_1 = 10$ $n_2 = 64$ $n_3 = 8$ $n_4 = 18$

| | *Sum of squares* | *Degrees of freedom* | *Mean square* | *F ratio* |
|---|---|---|---|---|
| Means | 10,304.871 | 3 | 3,434.96 | 18.8 |
| Within | 17,576.023 | 96 | 183.08 | |
| *Total* | 27,880.894 | 99 | | |

$F = 18.8$, so we reject.

(b) Diastolic blood pressure

**From Table 2-2a**

$\bar{X}_1 = 79.36$ $\bar{X}_2 = 79.09$ $\bar{X}_3 = 86.64$ $\bar{X}_4 = 86.42$

$n_1 = 11$ $n_2 = 54$ $n_3 = 11$ $n_4 = 24$

| | *Sum of squares* | *Degrees of freedom* | *Mean square* | *F ratio* |
|---|---|---|---|---|
| Means | 1,229.127 | 3 | 409.71 | 4.13 |
| Within | 9,529.441 | 96 | 99.26 | |
| *Total* | 10,758.566 | 99 | | |

$F = 4.13$, so we reject.

**From Table 2-2b**

$\bar{X}_1 = 78.70$ $\bar{X}_2 = 79.33$ $\bar{X}_3 = 94.50$ $\bar{X}_4 = 84.83$

$n_1 = 10$ $n_2 = 64$ $n_3 = 8$ $n_4 = 18$

| | *Sum of squares* | *Degrees of freedom* | *Mean square* | *F ratio* |
|---|---|---|---|---|
| Means | 1,932.207 | 3 | 644.07 | 8.61 |
| Within | 7,182.691 | 96 | 74.82 | |
| *Total* | 9,114.898 | 99 | | |

$F = 8.61$, so we reject.

(*c*) Cholesterol

**From Table 2-2a**

$\bar{X}_1 = 290.91$ $\bar{X}_2 = 304.56$ $\bar{X}_3 = 313.91$ $\bar{X}_4 = 323.92$

$n_1 = 11$ $n_2 = 54$ $n_3 = 11$ $n_4 = 24$

| | *Sum of squares* | *Degrees of freedom* | *Mean square* | *F ratio* |
|---|---|---|---|---|
| Means | 10,264.262 | 3 | 3,421.42 | 1.10 |
| Within | 297,563.812 | 96 | 3,099.62 | |
| *Total* | 307,828.064 | 99 | | |

$F = 1.10$, so we accept.

**From Table 2-2b**

$\bar{X}_1 = 274.099$ $\bar{X}_2 = 242.641$ $\bar{X}_3 = 322.625$ $\bar{X}_4 = 294.333$

$n_1 = 10$ $n_2 = 64$ $n_3 = 8$ $n_4 = 18$

| | *Sum of squares* | *Degrees of freedom* | *Mean square* | *F ratio* |
|---|---|---|---|---|
| Means | 73,645.312 | 3 | 24,548.44 | 6.76 |
| Within | 348,774.188 | 96 | 3,633.06 | |
| *Total* | 422,419.500 | 99 | | |

$F = 6.76$, so we reject.

(*d*) Gradient

**From Table 2-2a**

$\bar{X}_1 = 42.36$ $\bar{X}_2 = 37.31$ $\bar{X}_3 = 39.18$ $\bar{X}_4 = 43.62$

$n_1 = 11$ $n_2 = 54$ $n_3 = 11$ $n_4 = 24$

| | *Sum of squares* | *Degrees of freedom* | *Mean square* | *F ratio* |
|---|---|---|---|---|
| Means | 756.739 | 3 | 252.246 | 2.74 |
| Within | 8,831.445 | 96 | 91.994 | |
| *Total* | 9,588.184 | 99 | | |

$F = 2.74$, so we reject.

**From Table 2-2b**

$\bar{X}_1 = 39.50$ $\bar{X}_2 = 37.33$ $\bar{X}_3 = 57.75$ $\bar{X}_4 = 44.67$

$n_1 = 10$ $n_2 = 64$ $n_3 = 8$ $n_4 = 18$

| | *Sum of squares* | *Degrees of freedom* | *Mean square* | *F ratio* |
|---|---|---|---|---|
| Means | 3,346.889 | 3 | 1,115.63 | 11.8 |
| Within | 9,100.106 | 96 | 94.79 | |
| *Total* | 12,446.992 | 99 | | |

$F = 11.8$, so we reject.

With the results shown in (a) for systolic blood pressure for Table 2-2b we have the following contrasts:

| *Population contrast* | $a_1$ | $a_2$ | $a_3$ | $a_4$ |
|---|---|---|---|---|
| $\mu_3 - \dfrac{\mu_1 + \mu_2 + \mu_4}{3}$ | $-\frac{1}{3}$ | $-\frac{1}{3}$ | 1 | $-\frac{1}{3}$ |
| $\mu_4 - \dfrac{\mu_2 + \mu_1}{2}$ | $-\frac{1}{2}$ | $-\frac{1}{2}$ | 0 | 1 |
| $\mu_1 - \mu_2$ | 1 | $-1$ | 0 | 0 |
| $\dfrac{\mu_3 + \mu_2}{2} - \dfrac{\mu_4 + \mu_1}{2}$ | $-\frac{1}{2}$ | $\frac{1}{2}$ | $\frac{1}{2}$ | $-\frac{1}{2}$ |

95 per cent confidence intervals and conclusions are as follows:

(16.157,45.437); reject the hypothesis that $\mu_3$ is equal to average of $\mu_1$, $\mu_2$, and $\mu_4$.

(.855,23.285); reject the hypothesis that $\mu_4$ is equal to average of the difference between $\mu_2$ and $\mu_1$.

(−11.578,14.658); accept the hypothesis that $\mu_1 = \mu_2$.

(.107,21.103); reject the hypothesis that the mean of $\mu_3$ and $\mu_2$ is equal to the mean of $\mu_4$ and $\mu_1$.

**13.** $H$: $\mu_1 = \mu_2 = \mu_3 = \mu_4$, $\alpha = .05$, $F_{.95}(3,96) = 2.71$.

(a) Systolic blood pressure

**From Table 2-2a**

$\bar{X}_1 = 115.05$ $\quad\bar{X}_2 = 126.07$ $\quad\bar{X}_3 = 123.83$ $\quad\bar{X}_4 = 129.86$

$n_1 = 41$ $\quad n_2 = 29$ $\quad n_3 = 23$ $\quad n_4 = 7$

| | *Sum of squares* | *Degrees of freedom* | *Mean square* | *F ratio* |
|---|---|---|---|---|
| Means | 2,921.072 | 3 | 973.69 | 3.79 |
| Within | 24,685.840 | 96 | 257.14 | |
| *Total* | 27,606.910 | 99 | | |

$F = 3.79$, so we reject.

**From Table 2-2b**

$\bar{X}_1 = 115.63$ $\quad\bar{X}_2 = 118.89$ $\quad\bar{X}_3 = 136.86$ $\quad\bar{X}_4 = 136.08$

$n_1 = 46$ $\quad n_2 = 28$ $\quad n_3 = 14$ $\quad n_4 = 12$

| | *Sum of squares* | *Degrees of freedom* | *Mean square* | *F ratio* |
|---|---|---|---|---|
| Means | 7,606.879 | 3 | 2,535.63 | 12 |
| Within | 20,274.004 | 96 | 211.19 | |
| *Total* | 27,880.883 | 99 | | |

$F = 12.0$, so we reject.

(b) Diastolic blood pressure

**From Table 2-2a**

$\bar{X}_1 = 77.54$ $\bar{X}_2 = 84.24$ $\bar{X}_3 = 83.78$ $\bar{X}_4 = 88.86$

$n_1 = 41$ $n_2 = 29$ $n_3 = 23$ $n_4 = 7$

| | *Sum of squares* | *Degrees of freedom* | *Mean square* | *F ratio* |
|---|---|---|---|---|
| Means | 1,356.320 | 3 | 452.11 | 4.62 |
| Within | 9,402.262 | 96 | 97.94 | |
| *Total* | 10,758.578 | 99 | | |

$F = 4.62$, so we reject.

**From Table 2-2b**

$\bar{X}_1 = 76.67$ $\bar{X}_2 = 83.46$ $\bar{X}_3 = 87.43$ $\bar{X}_4 = 88.25$

$n_1 = 46$ $n_2 = 28$ $n_3 = 14$ $n_4 = 12$

| | *Sum of squares* | *Degrees of freedom* | *Mean square* | *F ratio* |
|---|---|---|---|---|
| Means | 2,218.163 | 3 | 739.39 | 10.3 |
| Within | 6,896.738 | 96 | 71.84 | |
| *Total* | 9,114.901 | 99 | | |

$F = 10.3$, so we reject.

(c) Cholesterol

**From Table 2-2a**

$\bar{X}_1 = 304.512$ $\bar{X}_2 = 304.379$ $\bar{X}_3 = 320.304$ $\bar{X}_4 = 313.428$

$n_1 = 41$ $n_2 = 29$ $n_3 = 23$ $n_4 = 7$

| | *Sum of squares* | *Degrees of freedom* | *Mean square* | *F ratio* |
|---|---|---|---|---|
| Means | 4,513.945 | 3 | 1,504.6 | .48 |
| Within | 303,314.750 | 96 | 3,159.5 | |
| *Total* | 307,828.695 | 99 | | |

$F = .48$, so we accept.

**From Table 2-2b**

$\bar{X}_1 = 237.33$ $\bar{X}_2 = 262.61$ $\bar{X}_3 = 334.50$ $\bar{X}_4 = 266.33$

$n_1 = 46$ $n_2 = 28$ $n_3 = 14$ $n_4 = 12$

| | *Sum of squares* | *Degrees of freedom* | *Mean square* | *F ratio* |
|---|---|---|---|---|
| Means | 101,801.562 | 3 | 33,933.9 | 10.2 |
| Within | 320,617.812 | 96 | 3,339.8 | |
| *Total* | 422,419.375 | 99 | | |

$F = 10.2$, so we reject.

(*d*) Gradient

**From Table 2-2a**

$\bar{X}_1 = 37.51$ $\bar{X}_2 = 41.83$ $\bar{X}_3 = 40.04$ $\bar{X}_4 = 41$

$n_1 = 41$ $n_2 = 29$ $n_3 = 23$ $n_4 = 7$

| | *Sum of squares* | *Degrees of freedom* | *Mean square* | *F ratio* |
|---|---|---|---|---|
| Means | 340.849 | 3 | 113.62 | 1.18 |
| Within | 9,247.325 | 96 | 96.32 | |
| *Total* | 9,588.172 | 99 | | |

$F = 1.18$, so we accept.

**From Table 2-2b**

$\bar{X}_1 = 38.96$ $\bar{X}_2 = 35.43$ $\bar{X}_3 = 49.45$ $\bar{X}_4 = 47.83$

$n_1 = 46$ $n_2 = 28$ $n_3 = 14$ $n_4 = 12$

| | *Sum of squares* | *Degrees of freedom* | *Mean square* | *F ratio* |
|---|---|---|---|---|
| Means | 2,591.132 | 3 | 863.71 | 8.41 |
| Within | 9,855.852 | 96 | 102.66 | |
| *Total* | 12,446.981 | 99 | | |

$F = 8.41$, so we reject.

With the results shown in (*c*) for cholesterol for the data of Table 2-2*b* we have the following contrasts:

| *Population contrast* | $a_1$ | $a_2$ | $a_3$ | $a_4$ |
|---|---|---|---|---|
| $\mu_4 - \dfrac{\mu_1 + \mu_2}{2}$ | $-\frac{1}{2}$ | $\frac{1}{2}$ | 0 | 1 |
| $\mu_3 - \dfrac{\mu_1 + \mu_2 + \mu_4}{3}$ | $-\frac{1}{3}$ | $-\frac{1}{3}$ | 1 | $-\frac{1}{3}$ |
| $\dfrac{\mu_2 + \mu_4}{2} - \mu_1$ | $-1$ | $\frac{1}{2}$ | 0 | $\frac{1}{2}$ |
| $\mu_4 - \mu_1$ | $-1$ | 0 | 0 | 1 |

95 per cent confidence intervals and conclusions are as follows:

$(-35.143, 67.863)$; accept the hypothesis that $\mu$ equals the average of $\mu_1$ and $\mu_2$.

$(30.456, 127.702)$; reject the hypothesis that $\mu_3$ equals the average of $\mu_1$, $\mu_2$, and $\mu_4$.

$(-10.254, 64.534)$; accept the hypothesis that $\mu_1$ equals the average of $\mu_2$ and $\mu_4$.

$(-24.412, 82.412)$; accept the hypothesis that $\mu_4$ equals $\mu_1$.

**14.** $H$: $\mu_1 = \mu_2 = \mu_3 = \mu_4$, $\alpha = .05$, $F_{.95}(3,96) = 2.71$.

(*a*) Weight

| | *Sum of squares* | *Degrees of freedom* | *Mean square* | *F ratio* |
|---|---|---|---|---|
| Means | 10,070.856 | 3 | 3,356.95 | 5.99 |
| Within | 53,788.953 | 96 | 560.30 | |
| *Total* | 63,859.809 | 99 | | |

$F = 5.99$, so we reject.

(*b*) Systolic blood pressure

| | *Sum of squares* | *Degrees of freedom* | *Mean square* | *F ratio* |
|---|---|---|---|---|
| Means | 757.059 | 3 | 252.35 | .90 |
| Within | 26,849.758 | 96 | 279.68 | |
| *Total* | 27,606.816 | 99 | | |

$F = .90$, so we accept.

**15**

| | *Sum of squares* | *Degrees of freedom* | *Mean square* | *F ratio* |
|---|---|---|---|---|
| Means | 4,402.08 | 3 | 1,467.36 | 13.6 |
| Within | 864.58 | 8 | 108.07 | |
| *Total* | 5,266.66 | 11 | | |

$H: \mu_1 = \mu_2 = \mu_3 = \mu_4$; $\alpha = .05$, $F_{.95}(3,8) = 4.07$; $F = 13.6$, so we reject.

**16.** $q_{.95}(3,57) = 3.40$, $s_p/\sqrt{n} = .9579$; for 95 per cent confidence limits
$-5.06 < \mu_1 - \mu_2 < 1.46$, $-5.16 < \mu_2 - \mu_3 < 1.36$,
$-3.21 < (\mu_1 + \mu_3)/2 - \mu_2 < 3.31$.

**17**

| | *Sum of squares* | *Degrees of freedom* | *Mean square* | *F ratio* |
|---|---|---|---|---|
| Rows | 1.305 | 3 | .435 | 0.74 |
| Columns | .466 | 3 | .155 | 0.26 |
| Residual | 5.308 | 9 | .589 | |
| *Total* | 7.079 | 15 | | |

$H_1$: row effects $= 0$, $H_2$: column effects $= 0$, $\alpha = .05$, $F_{.95}(3,9) = 3.86$, both hypotheses are accepted.

**18.** $H_1$: row effects $= 0$, $H_2$: column effects $= 0$, $H_3$: interaction $= 0$.

| | *Sum of squares* | *Degrees of freedom* | *Mean square* | *F ratio* |
|---|---|---|---|---|
| Row | 190.1 | 3 | 63.37 | |
| Column | 200.2 | 4 | 50.05 | |
| Interaction | 110.1 | 12 | 9.17 | 1.81 |
| *Subtotal* | 500.4 | 19 | | |
| Within | 304.2 | 60 | 5.07 | |
| *Total* | 804.6 | 79 | | |

$F_{.95}(12,60) = 1.92$; $F = 1.81$, so we accept $H_3$ and pool.

| | Sum of squares | Degrees of freedom | Mean square | F ratio |
|---|---|---|---|---|
| Row | 190.1 | 3 | 63.37 | 11.02 |
| Column | 200.2 | 4 | 50.05 | 8.70 |
| Residual | 414.3 | 72 | 5.75 | |
| *Total* | 804.6 | 79 | | |

$F_{.95}(3,72) = 2.74$, so we reject $H_1$; $F_{.95}(4,72) = 2.51$, so we reject $H_2$.

**19**

| | Sum of squares | Degrees of freedom | Mean square | F ratio |
|---|---|---|---|---|
| Column | 62.12 | 4 | 15.53 | 34.13 |
| Row | 6.77 | 2 | 3.39 | 7.45 |
| Interaction | 84.55 | 8 | 10.57 | 23.23 |
| *Subtotal* | 153.44 | 14 | | |
| Within | 13.66 | 30 | .455 | |
| *Total* | 167.10 | 44 | | |

$H_1$: row effects = 0, $H_2$: column effects = 0, $H_3$: interaction = 0, $\alpha = .05$; $F_{.95}(8,30) = 2.27$, so we reject $H_3$; $F_{.95}(2,30) = 3.32$, so we reject $H_2$; $F_{.95}(4,30) = 2.69$, so we reject $H_1$.

**20.** (*a*) Cholesterol

**From Table 2-2a**

| AGE | HEIGHT ≤67 | 68–69 | ≥70 | *Row means* |
|---|---|---|---|---|
| ≤39 | 1. 243<br>2. 386<br>3. 281<br>4. 360<br>5. 248<br>$\bar{X}_{11\cdot} = 303.60$ | 1. 284<br>2. 302<br>3. 358<br>4. 253<br>5. 268<br>$\bar{X}_{21\cdot} = 293.00$ | 1. 240<br>2. 314<br>3. 245<br>4. 312<br>5. 251<br>272.40 | 289.67 |
| 40–49 | 1. 273<br>2. 285<br>3. 243<br>4. 378<br>5. 332<br>302.20 | 1. 279<br>2. 298<br>3. 273<br>4. 302<br>5. 394<br>309.20 | 1. 254<br>2. 369<br>3. 333<br>4. 279<br>5. 341<br>315.20 | 308.87 |
| ≥50 | 1. 384<br>2. 310<br>3. 337<br>4. 428<br>5. 370<br>365.80 | 1. 315<br>2. 367<br>3. 302<br>4. 474<br>5. 334<br>358.40 | 1. 307<br>2. 250<br>3. 341<br>4. 336<br>5. 243<br>295.40 | 339.87 |
| *Column means* | 323.87 | 320.20 | 294.33 | |

| | Sum of squares | Degrees of freedom | Mean square |
|---|---|---|---|
| Rows (age) | 19,248.3514 | 2 | 9,624.2 |
| Columns (height) | 7,773.7240 | 2 | 3,886.9 |
| Interaction | 10,133.1797 | 4 | 2,533.3 |
| *Subtotal* | 37,155.2551 | 8 | |
| Within | 95,151.5626 | 36 | 2,643.1 |
| *Total* | 132,306.8177 | 44 | |

For $\alpha$ = .05 the only significant finding is for differences in cholesterol associated with age.

**From Table 2-2b**

| AGE | HEIGHT ≤67 | HEIGHT 68–70 | HEIGHT ≥71 | Row means |
|---|---|---|---|---|
| ≤29 | 1. 135<br>2. 260<br>3. 235<br>4. 252<br>5. 156<br>$\bar{X}_{11\cdot}$ = 207.60 | 1. 252<br>2. 269<br>3. 235<br>4. 386<br>5. 352<br>$\bar{X}_{21\cdot}$ = 298.80 | 1. 222<br>2. 178<br>3. 277<br>4. 195<br>5. 206<br>215.60 | 240.67 |
| 30–44 | 1. 302<br>2. 344<br>3. 260<br>4. 216<br>5. 239<br>272.20 | 1. 403<br>2. 294<br>3. 319<br>4. 208<br>5. 290<br>302.80 | 1. 312<br>2. 264<br>3. 336<br>4. 282<br>5. 214<br>281.60 | 285.53 |
| ≥45 | 1. 370<br>2. 229<br>3. 244<br>4. 353<br>5. 453<br>364.60 | 1. 311<br>2. 269<br>3. 311<br>4. 420<br>5. 253<br>312.80 | 1. 286<br>2. 277<br>3. 353<br>4. 259<br>5. 218<br>278.60 | 318.67 |
| *Column means* | 281.47 | 304.80 | 258.60 | |

| | Sum of squares | Degrees of freedom | Mean square |
|---|---|---|---|
| Rows (age) | 45,974.0638 | 2 | 22,987. |
| Columns (height) | 16,008.7675 | 2 | 8,004. |
| Interaction | 30,702.2969 | 4 | 7,676. |
| *Subtotal* | 92,685.1282 | 8 | |
| Within | 126,110.5000 | 36 | 3,503. |
| *Total* | 218,795.6282 | 44 | |

$H$: interaction effects are zero, $\alpha = .05$, $F_{.95}(4,36) = 2.64$. $F = 2.19$, so we accept. Since 2.19 is greater than $2F_{.50}(4,36)$, we do not pool. $H$: row (age) effects are zero, $\alpha = .05$, $F_{.95}(2,36) = 3.27$; $F = 6.57$, so we reject. $H$: column (height) effects are zero, $\alpha = .05$, $F_{.95}(2,36) = 3.27$; $F = 2.29$, so we accept. We conclude that there is evidence that cholesterol varies in mean value with age but there is not evidence that it varies with height.

(*b*) Systolic blood pressure

**From Table 2-2a**

| AGE | HEIGHT $\leq 67$ | 68–69 | $\geq 70$ | *Row means* |
|---|---|---|---|---|
| $\leq 29$ | 1. 115<br>2. 120<br>3. 108<br>4. 124<br>5. 106<br>$\bar{X}_{11.} = 114.60$ | 1. 100<br>2. 132<br>3. 114<br>4. 100<br>5. 120<br>$\bar{X}_{21.} = 113.20$ | 1. 110<br>2. 120<br>3. 138<br>4. 110<br>5. 114<br>118.40 | 115.40 |
| 30–44 | 1. 130<br>2. 110<br>3. 134<br>4. 138<br>5. 100<br>122.40 | 1. 114<br>2. 130<br>3. 120<br>4. 115<br>5. 112<br>118.20 | 1. 124<br>2. 120<br>3. 150<br>4. 120<br>5. 120<br>126.80 | 122.47 |
| $\geq 45$ | 1. 110<br>2. 120<br>3. 120<br>4. 141<br>5. 130<br>124.20 | 1. 190<br>2. 130<br>3. 148<br>4. 130<br>5. 90<br>137.60 | 1. 120<br>2. 130<br>3. 140<br>4. 100<br>5. 140<br>126.00 | 129.27 |
| *Column means* | 120.40 | 123.00 | 123.73 | |

| | *Sum of squares* | *Degrees of freedom* | *Mean square* |
|---|---|---|---|
| Rows (age) | 1,442.3113 | 2 | 721.16 |
| Columns (height) | 92.0444 | 2 | 46.02 |
| Interaction | 694.1873 | 4 | 173.55 |
| *Subtotal* | 2,228.5430 | 8 | |
| Within | 10,423.9570 | 36 | 289.55 |
| *Total* | 12,652.5000 | 44 | |

After pooling, the residual mean square is 278, and neither row or column effects are significant.

**From Table 2-2b**

| AGE | HEIGHT ≤67 | 68–70 | ≥71 | Row means |
|---|---|---|---|---|
| ≤29 | 1. 130<br>2. 115<br>3. 120<br>4. 110<br>5. 120<br>$\bar{X}_{11.} = 119.00$ | 1. 112<br>2. 110<br>3. 110<br>4. 120<br>5. 120<br>$\bar{X}_{21.} = 114.40$ | 1. 120<br>2. 130<br>3. 110<br>4. 110<br>5. 108<br>115.60 | 116.33 |
| 30–44 | 1. 112<br>2. 110<br>3. 120<br>4. 100<br>5. 125<br>113.40 | 1. 115<br>2. 122<br>3. 120<br>4. 100<br>5. 100<br>111.40 | 1. 125<br>2. 120<br>3. 120<br>4. 134<br>5. 115<br>122.8 | 115.87 |
| ≥45 | 1. 160<br>2. 140<br>3. 190<br>4. 190<br>5. 155<br>167.00 | 1. 110<br>2. 120<br>3. 120<br>4. 130<br>5. 110<br>118.00 | 1. 130<br>2. 110<br>3. 110<br>4. 148<br>5. 130<br>125.6 | 136.87 |
| *Column means* | 133.13 | 114.60 | 121.33 | |

| | *Sum of squares* | *Degrees of freedom* | *Mean square* |
|---|---|---|---|
| Rows (age) | 4,314.1789 | 2 | 2,157.1 |
| Columns (height) | 2,640.3115 | 2 | 1,320.2 |
| Interaction | 4,741.6797 | 4 | 1,185.4 |
| *Subtotal* | 11,696.1703 | 8 | |
| Within | 4,998.7852 | 36 | 138.9 |
| *Total* | 16,694.9555 | 44 | |

$H$: interaction effects are zero, $\alpha = .05$, $F_{.95}(4,36) = 2.640$; $F = 8.54$, so we reject. We do not pool the interaction sum of squares with the "within." $H$: Row (age) effects are zero, $\alpha = .05$, $F_{.95}(2,36) = 3.27$; $F = 15.53$, so we reject. $H$: column (height) effects are zero, $\alpha = .05$, $F_{.95}(2,36) = 3.27$; $F = 9.51$, so we reject. We conclude that any one of the effects considered alone is significant. However, if we wish to test that various effects are significant at the 5 per cent level simultaneously, we form contrasts such as ($\bar{X}_{r_1}$ = mean of row 1, etc.)

$$\bar{X}_{r_1} - \tfrac{1}{2}\bar{X}_{r_2} - \tfrac{1}{2}\bar{X}_{r_3} \quad \text{and} \quad \bar{X}_{c_1} - \tfrac{1}{2}\bar{X}_{c_2} - \tfrac{1}{2}\bar{X}_{c_3}$$

and test with the $q$ statistic.

21

| | *Sum of squares* | *Degrees of freedom* | *Mean square* | *F ratio* |
|---|---|---|---|---|
| Means | 26,133.33 | 2 | 13,066.67 | 2,268 |
| Within | 3,438.72 | 597 | 5.76 | |
| *Total* | 29,572.05 | 599 | | |

$H$: $\mu_1 = \mu_2 = \mu_3$, $\alpha = .05$, $F_{.95}(2, \infty) = 3$. $F = 2{,}268$, so we reject. $\bar{X}_s = \bar{X}_1/10 + 4\bar{X}_2/10 + 5\bar{X}_3/10 = 11$. Assuming $\sigma_1^2 = \sigma_2^2 = \sigma_3^2 = \sigma^2$, we estimate $\sigma^2$ by $s_p^2 = 5.76$. The variance of the sampling distribution of $\bar{X}_s$ is

$$\frac{1}{100}\frac{\sigma^2}{200} + \frac{16}{100}\frac{\sigma^2}{200} + \frac{25}{100}\frac{\sigma^2}{200} = .0021\sigma^2$$

which is now estimated to be $.0021s_p^2 = .0021 \times 5.76 = .0121$, and the standard error of $\bar{X}_s$ is $\sqrt{.0121} = .11$. Given that $\bar{X}_s$ has a normal sampling distribution, 95 per cent confidence limits for the population mean $\mu$ are $11 - 1.96 \times .11 < \mu < 11 + 1.96 \times .11$.

**22.** $\bar{X}_s = (8\bar{X}_1 + 6\bar{X}_2 + 6\bar{X}_3 + 5\bar{X}_4)/25 = 12.4$. The variance of $\bar{X}_s$ is $(64 + 36 + 36 + 25)\sigma^2/(625 \times 100)$, which, if $\sigma = 10$, is .258. Thus a 95 per cent confidence interval for the population mean $\mu$ is $12.4 - 1.96 \times .51 < \mu < 12.4 + 1.96 \times .51$. A representative sample would have 128 freshmen, 96 sophomores, 96 juniors, and 80 seniors. Then $\bar{X}_s = (8\bar{X}_1 + 6\bar{X}_2 + 6\bar{X}_3 + 5\bar{X}_4)/25$ would have variance equal to $(\frac{64}{128} + \frac{36}{96} + \frac{36}{96} + \frac{25}{80})\sigma^2/625$, which for $\sigma = 10$ is equal to .250. This is slightly less than the .258 value for the equal-sample-size case.

## CHAPTER ELEVEN

**1.** (*b*) $\bar{X} = 1$, $\bar{Y} = 1.67$, $b = 1.0001$; $\bar{Y}_x = 1.67 + 1(X - 1)$

**2.** (*b*) $\bar{X} = 68.26$, $\bar{Y} = 164.4$, $s_x^2 = 6.0327$, $s_y^2 = 645.0505$, $b = 4.4098$, $s_{y \cdot x}^2 = 533.12$; $\bar{Y}_x = -136.59 + 4.4098X = 164.4 + 4.4098(X - 68.26)$. (*c*) 95 per cent confidence limits are $159.805 < A < 168.995$, $2.53 < B < 6.29$, $410.208 < \sigma_{y \cdot x}^2 < 721.$, $142.4 < \mu_{y \cdot 65} < 157.7$.

**3.** (*b*) $\Sigma X_i = 6{,}889$, $\Sigma Y_i = 16{,}598$, $\Sigma X_i^2 = 475{,}207$, $\Sigma Y_i^2 = 2{,}814{,}692$, $\Sigma X_i Y_i = 1{,}145{,}780$. (*c*) $\bar{Y}_x = 165.98 + 3.7575(X - 68.89) = -92.87 + 3.7575X$. (*d*) 95 per cent limits are $161.443 < A < 170.512$, $1.941 < B < 5.574$, $400. < \sigma_{y \cdot x}^2 < 703.$, $143.0 < \mu_{y \cdot 65} < 156.8$.

**4.** $\bar{X} = 76.95$, $\bar{Y} = 23.85$, $s_x = 13.01$, $s_y = 6.68$, $b = .39$, $s_{y \cdot x} = 4.40$; $\bar{Y}_x = 23.85 + .39(X - 76.95)$.

| | *Sum of squares* | *Degrees of freedom* | *Mean square* | *F ratio* |
|---|---|---|---|---|
| Within | 1,251 | 76 | 16.5 | |
| Regression | 2,094 | 1 | | |
| About regression | 295 | 4 | 73.7 | 4.5 |
| *Subtotal* | 2,389 | 5 | | |

$H$: all means are on a straight line, $\alpha = .05$, $F_{.95}(4,76) = 2.5$; $F = 4.5$, so we reject linearity. Inspection of the mean values suggests a curvilinear relationship. In fact, even though linearity is rejected, the estimated line may be of some value, since the standard deviation of the yield is reduced by one-third when it is used.

**5.** (*b*) $\bar{X} = 32.97$, $\bar{Y} = 2{,}264.33$, $s_x = 13$, $s_y = 879$, $r = .979$, $s_{y \cdot x} = 186.4$, $b = 66.2$, $b' = .0145$; (*d*) $\bar{Y}_{50} = 3{,}392$: 95 per cent limits are $3{,}060 < \mu_{y \cdot 50} < 3{,}725$.

**6.** $b = .3848$: (*a*) $\bar{Y}_x = 12.8 + .3848(X - 16)$, (*b*) $\bar{Y}_{17} = 13.184$, (*c*) $12.96 < \mu_{y \cdot 17} < 13.41$, (*d*) $10.3 < Y_{17} < 16.1$.

**7.** (*a*) $\bar{X} = 108.75$, $\bar{Y} = 54.3$, $s_x = 16.42$, $s_y = 14.09$, $r = .528$, $b = .454$, $b' = .617$. (*b*) $H$: $\rho = 0$, $\alpha = .05$, reject if $z\sqrt{N-3} > 1.96$ or if $z\sqrt{N-3} < -1.96$; $z\sqrt{N-3} = 5.8$, so we reject.
(*c*) 95 per cent limits are $108.3 < \mu_{x \cdot 60} < 116.3$, $42.15 < \mu_{y \cdot 90} < 49.13$.

**8.** (*b*) $\bar{X} = 94.11$, $\bar{Y} = 6.95$, $s_x = 12.8$, $s_y = .835$, $r = .806$, $s_{y \cdot x} = 2.49$, $b = .0526$, $b' = 12.35$. (*d*) To compute a tolerance interval we need a specified value of $X$. For instance, if $X = 100$, then $7.26 \pm .30$ has 90 per cent chance of covering 50 per cent of the population with $X = 100$.

**9.** (*b*) $\bar{X} = 44.21$, $\bar{Y} = 121.30$, $s_x^2 = 130.268$, $s_y^2 = 278.858$, $b = .502$, $s^2_{y \cdot x} = 248.616$; $\bar{Y}_x = 121.3 + .502(X - 44.21) = 99.107 + .502X$. (*c*) 95 per cent confidence limits are $118.162 < A < 124.438$, $.2258 < B < .7782$, $191.243 < \sigma^2_{y \cdot x} < 336.422$, $(99.106 + .502X) \pm .2762\sqrt{128.9653 + (X - 44.21)^2}$.

**10.** (*b*) $\Sigma X_i = 4{,}091$, $\Sigma Y_i = 12{,}197$, $\Sigma X_i^2 = 180{,}929$, $\Sigma Y_i^2 = 1{,}515{,}549$, $\Sigma X_i Y_i = 509{,}487$; (*c*) $\bar{Y}_x = 121.97 + .7745(X - 40.91) = 90.285 + .7745X$; (*d*) 95 per cent limits are $119.15 < A < 124.79$, $.5320 < B < 1.0170$, $155 < \sigma^2_{y \cdot x} < 273$, $125.43 < \mu_{y \cdot 50} < 132.59$.

**11.** (*b*) $\bar{X} = 164.40$, $\bar{Y} = 121.30$, $s_x^2 = 645.06$, $s_y^2 = 278.87$, $b = .17101$, $s^2_{y \cdot x} = 262.66$.

**12.** (*b*) $\Sigma X_i = 16{,}598$, $\Sigma Y_i = 12{,}197$, $\Sigma X_i^2 = 2{,}814{,}682$, $\Sigma Y_i^2 = 1{,}515{,}549$, $\Sigma X_i Y_i = 2{,}028{,}925$; (*c*) $\bar{Y}_x = 121.97 + .0745(X - 165.98) = 109.554 + .0748X$; (*d*) 95 per cent limits are $118.63 < A < 125.31$, $-.0617 < B < .2112$, $216 < \sigma^2_{y \cdot x} < 378$, $117 < \mu_{y \cdot 150} < 125$.

**13.** (*b*) $b_1 = 1$, $b_2 = 1$, $b_3 = .5454$; for $A$, $\bar{Y}_x = 9.25 + 1(X - 2)$, for $B$, $\bar{Y}_x = 11.75 + 1(X - 3.75)$, for $C$, $\bar{Y}_x = 6.5 + .5454(X - 1.75)$. (*c*) $\alpha = .05$, $t_{.975}(4) = 2.776$, reject if $t > 2.776$ or if $t < -2.776$. For $A$, $b_1 - b_2 = 0$; $t = 0$, so we accept. For $B$, $b_1 - b_3 = .4546$; $t = .33$, so we accept. For $C$, $b_2 - b_3 = .4546$; $t = .43$, so we accept.

**14.** (*a*) $100b_0 + 4{,}421b_1 + 16{,}440b_2 = 12{,}130$,
$4{,}421b_0 + 208{,}349b_1 + 726{,}526b_2 = 542{,}735$,
$16{,}440b_0 + 726{,}526b_1 + 2{,}766{,}596b_2 = 2{,}005{,}093$. (*b*) $b_0 = 70.47139$, $b_1 = .50536$, $b_2 = .17328$; $\bar{Y}_x = 70.47139 + .50536X_1 + .17328X_2$.

## CHAPTER TWELVE

**1**

| | *Degrees of freedom* | $\Sigma x^2$ | $\Sigma xy$ | $\Sigma y^2$ | *Degrees of freedom* | $\Sigma y'^2$ | *Mean square* |
|---|---|---|---|---|---|---|---|
| Among means | 2 | 21.1667 | 40.0833 | 100.1667 | 2 | 45.29 | 22.64 |
| Within groups | 9 | 129.7500 | 106.2500 | 96.5000 | 8 | 9.49 | 1.19 |
| *Total* | 11 | 150.9167 | 146.3333 | 196.6667 | 10 | 54.78 | |

$H$: there is no difference in the $Y$ means of the three groups after adjustment for the $X$ covariate values, $\alpha = .05$, $F_{.95}(2.8) = 4.46$. $F = 19.1$, so we reject.

**2**

| | Degrees of freedom | $\Sigma x^2$ | $\Sigma xy$ | $\Sigma y^2$ | Degrees of freedom | $\Sigma y'^2$ | Mean square |
|---|---|---|---|---|---|---|---|
| Among means | 1 | 6.125 | 21.875 | 78.125 | 1 | 43.49 | 43.49 |
| Within groups | 6 | 64.750 | 50.750 | 47.75 | 5 | 7.97 | 1.59 |
| *Total* | 7 | 70.875 | 72.625 | 125.875 | 6 | 51.46 | |

$H$: there is no difference in the $Y$ means of the two groups after adjustment for the $X$ covariate values, $\alpha = .05$, $F_{.95}(1,5) = 6.61$. $F = 27.3$, so we reject. Note that the entries in the above table compares to those which would have been made in Tables 11-5 and 11-6 in Chap. 11.

**3**

| | Degrees of freedom | $\Sigma x^2$ | $\Sigma xy$ | $\Sigma y^2$ | Degrees of freedom | $\Sigma y'^2$ | Mean square |
|---|---|---|---|---|---|---|---|
| Among means | 1 | 6.125 | 8.750 | 12.500 | 1 | .492 | .492 |
| Within groups | 6 | 4.750 | 4.750 | 11.500 | 5 | 6.750 | 1.350 |
| *Total* | 7 | 10.875 | 13.500 | 24.000 | 6 | 7.242 | |

$H$: there is no difference in the $Y$ means of the two groups after adjustment for the $X$ covariate values, $\alpha = .05$, $F_{.95}(1,5) = 6.61$. $F = .36$, so we accept. Again, the entries in this table correspond to those in Tables 11-5 and 11-6.

**4.** (*a*) $H$: the mean systolic blood pressure is the same for the four age groups after adjusting for weight, $\alpha = .05$, $F_{.95}(3,35) = 2.88$.

**From Table 2-2a**

| | Degrees of freedom | $\Sigma x^2$ | $\Sigma xy$ | $\Sigma y^2$ | Degrees of freedom | $\Sigma y'^2$ | Mean square |
|---|---|---|---|---|---|---|---|
| Among means | 3 | 1,970.000 | −1217.376 | 1,959.500 | 3 | 2,657.4 | 885.8 |
| Within groups | 36 | 24,389.000 | 6683.188 | 11,094.125 | 35 | 9,262.8 | 264.6 |
| *Total* | 39 | 26,359.000 | 5465.812 | 13,053.625 | 38 | 11,920.2 | |

$F = 3.35$, so we reject.

**From Table 2-2b**

| | Degrees of freedom | $\Sigma x^2$ | $\Sigma xy$ | $\Sigma y^2$ | Degrees of freedom | $\Sigma y'^2$ | Mean square |
|---|---|---|---|---|---|---|---|
| Among means | 3 | 30.000 | 293.625 | 6,155.312 | 3 | 6,154.9 | 2,051.6 |
| Within groups | 36 | 24,546.000 | −126.812 | 11,402.688 | 35 | 11,402.0 | 325.8 |
| *Total* | 39 | 24,576.000 | 166.813 | 17,558.000 | 38 | 17,556.9 | |

$F = 6.3$, so we reject.

(*b*) $H$: the mean diastolic blood pressure is the same for the four age groups after adjusting for weight, $\alpha = .05$, $F_{.95}(3,35) = 2.88$.

**From Table 2-2a**

| | Degrees of freedom | $\Sigma x^2$ | $\Sigma xy$ | $\Sigma y^2$ | Degrees of freedom | $\Sigma y'^2$ | Mean square |
|---|---|---|---|---|---|---|---|
| Among means | 3 | 1,970.000 | −568.125 | 837.625 | 3 | 1,062.6 | 354.2 |
| Within groups | 36 | 24,389.000 | 4,231.125 | 4,073.187 | 35 | 3,339.2 | 95.4 |
| *Total* | 39 | 26,359.000 | 3,663.000 | 4,910.812 | 38 | 4,401.8 | |

$F = 3.71$, so we reject.

**From Table 2-2b**

| | Degrees of freedom | $\Sigma x^2$ | $\Sigma xy$ | $\Sigma y^2$ | Degrees of freedom | $\Sigma y'^2$ | Mean square |
|---|---|---|---|---|---|---|---|
| Among means | 3 | 30.000 | 136.312 | 762.188 | 3 | 741.54 | 247.18 |
| Within groups | 36 | 24,546.000 | 1,807.688 | 2,857.188 | 35 | 2,724.06 | 77.83 |
| *Total* | 39 | 24,576.000 | 1,944.000 | 3,619.375 | 38 | 3,465.60 | |

$F = 3.18$, so we reject.

(*c*) $H$: the mean blood cholesterol is the same for the four age groups after adjusting for weight, $\alpha = .05$, $F_{.95}(3,35) = 2.88$.

**From Table 2-2a**

| | Degrees of freedom | $\Sigma x^2$ | $\Sigma xy$ | $\Sigma y^2$ | Degrees of freedom | $\Sigma y'^2$ | Mean square |
|---|---|---|---|---|---|---|---|
| Among means | 3 | 1,970.000 | −385.000 | 28,292.000 | 3 | 28,217 | 9,406 |
| Within groups | 36 | 24,389.000 | −4,698.000 | 89,995.000 | 35 | 89,090 | 2,545 |
| *Total* | 39 | 26,359.000 | −5,083.000 | 118,287.000 | 38 | 117,307 | |

$F = 3.70$, so we reject.

**From Table 2-2b**

| | Degrees of freedom | $\Sigma x^2$ | $\Sigma xy$ | $\Sigma y^2$ | Degrees of freedom | $\Sigma y'^2$ | Mean square |
|---|---|---|---|---|---|---|---|
| Among means | 3 | 30.000 | 653.000 | 31,886.000 | 3 | 31,825. | 10,608. |
| Within groups | 36 | 24,546.000 | 824.000 | 154,272.000 | 35 | 154,244. | 4,407. |
| *Total* | 39 | 24,576.000 | 1,477.000 | 186,158.000 | 38 | 186,069. | |

$F = 2.41$, so we accept.

**5.** (*a*) $H$: the mean systolic blood pressure is the same for the four age groups after adjusting for height, $\alpha = .05$, $F_{.95}(3,35) \doteq 2.88$.

**From Table 2-2a**

| | Degrees of freedom | $\Sigma x^2$ | $\Sigma xy$ | $\Sigma y^2$ | Degrees of freedom | $\Sigma y'^2$ | Mean square |
|---|---|---|---|---|---|---|---|
| Among means | 3 | 20.188 | −145.812 | 1,959.500 | 3 | 2,263.219 | 754.406 |
| Within groups | 36 | 227.625 | 303.062 | 11,094.125 | 35 | 10,690.621 | 305.446 |
| *Total* | 39 | 247.813 | 157.250 | 13,053.625 | 38 | 12,953.840 | |

$F = 2.47$, so we accept.

**From Table 2-2b**

| | Degrees of freedom | $\Sigma x^2$ | $\Sigma xy$ | $\Sigma y^2$ | Degrees of freedom | $\Sigma y'^2$ | Mean square |
|---|---|---|---|---|---|---|---|
| Among means | 3 | 3 | −88.562 | 6,155.312 | 3 | 5,745.973 | 1,915.324 |
| Within groups | 36 | 210 | −465.438 | 11,402.688 | 35 | 10,371.106 | 296.317 |
| *Total* | 39 | 213 | −554.000 | 17,558.000 | 38 | 16,117.078 | |

$F = 6.464$, so we reject.

(*b*) $H$: The mean diastolic blood pressure is the same for the four age groups after adjusting for height. $\alpha = .05$, $F_{.95}(3,35) = 2.88$.

**From Table 2-2a**

| | Degrees of freedom | $\Sigma x^2$ | $\Sigma xy$ | $\Sigma y^2$ | Degrees of freedom | $\Sigma y'^2$ | Mean square |
|---|---|---|---|---|---|---|---|
| Among means | 3 | 20.188 | −122.500 | 837.625 | 3 | 941.108 | 313.703 |
| Within groups | 36 | 227.625 | 157.000 | 4,073.187 | 35 | 3,964.900 | 113.283 |
| *Total* | 39 | 247.812 | 34.500 | 4,910.812 | 38 | 4,906.008 | |

$F = 2.769$, so we accept.

**From Table 2-2b**

| | Degrees of freedom | $\Sigma x^2$ | $\Sigma xy$ | $\Sigma y^2$ | Degrees of freedom | $\Sigma y'^2$ | Mean square |
|---|---|---|---|---|---|---|---|
| Among means | 3 | 3.000 | −25.250 | 762.188 | 3 | 727.453 | 242.48 |
| Within groups | 36 | 210.000 | −139.375 | 2,857.187 | 35 | 2,764.686 | 78.99 |
| *Total* | 39 | 213.000 | −164.625 | 3,619.375 | 38 | 3,492.138 | |

$F = 3.07$, so we reject.

(*c*) $H$: The mean blood cholesterol is the same for the four age groups after adjusting for height. $\alpha = .05$, $F_{.95}(3,35) = 2.88$.

**From Table 2-2a**

| | Degrees of freedom | $\Sigma x^2$ | $\Sigma xy$ | $\Sigma y^2$ | Degrees of freedom | $\Sigma y'^2$ | Mean square |
|---|---|---|---|---|---|---|---|
| Among means | 3 | 20.188 | −295.188 | 28,292.000 | 3 | 27,089.813 | 9,030 |
| Within groups | 36 | 227.625 | −378.562 | 89,995.000 | 35 | 89,365.375 | 2,553 |
| *Total* | 39 | 247.813 | −673.750 | 118,287.000 | 38 | 116,455.188 | |

$F = 3.53$, so we reject.

**From Table 2-2b**

| | Degrees of freedom | $\Sigma x^2$ | $\Sigma xy$ | $\Sigma y^2$ | Degrees of freedom | $\Sigma y'^2$ | Mean square |
|---|---|---|---|---|---|---|---|
| Among means | 3 | 3.000 | −129.563 | 31,886.000 | 3 | 31,551.813 | 10,517 |
| Within groups | 36 | 210.000 | −212.437 | 154,272.000 | 35 | 154,057.063 | 4,402 |
| *Total* | 39 | 213.000 | −342.000 | 186,158.000 | 38 | 185,608.875 | |

$F = 2.39$, so we accept.

**6.** (*a*) $H$: the mean systolic blood pressure is the same for the four groups after adjusting for weight, $\alpha = .05$.

**From Table 2-2a, $n = 10$**

| | Degrees of freedom | $\Sigma x^2$ | $\Sigma xy$ | $\Sigma y^2$ | Degrees of freedom | $\Sigma y'^2$ | Mean square |
|---|---|---|---|---|---|---|---|
| Among means | 3 | 8,273.000 | −133.938 | 656.500 | 3 | 1,457.559 | 485.9 |
| Within groups | 36 | 25,137.000 | 8,622.938 | 13,647.938 | 35 | 10,689.945 | 305.4 |
| *Total* | 39 | 33,410.000 | 8,489.000 | 14,304.438 | 38 | 12,147.504 | |

$F_{.95}(3,35) \doteq 2.88$; $F = 1.59$, so we accept.

**From Table 2-2b, $n = 8$**

| | Degrees of freedom | $\Sigma x^2$ | $\Sigma xy$ | $\Sigma y^2$ | Degrees of freedom | $\Sigma y'^2$ | Mean square |
|---|---|---|---|---|---|---|---|
| Among means | 3 | 3,376.750 | 1,172.000 | 5,870.500 | 3 | 5,918.680 | 1,973 |
| Within groups | 28 | 19,141.250 | −977.000 | 8,399.000 | 27 | 8,349.129 | 309 |
| *Total* | 31 | 22,518.000 | 195.000 | 14,269.500 | 30 | 14,267.809 | |

$F_{.95}(3,27) \doteq 2.94$; $F = 6.38$, so we reject.

(*b*) $H$: the mean blood cholesterol is the same for the four groups after adjusting for weight, $\alpha = .05$.

**From Table 2-2a, $n = 10$**

| | Degrees of freedom | $\Sigma x^2$ | $\Sigma xy$ | $\Sigma y^2$ | Degrees of freedom | $\Sigma y'^2$ | Mean square |
|---|---|---|---|---|---|---|---|
| Among means | 3 | 8,273 | −2,295 | 23,133 | 3 | 22,537.812 | 7,513 |
| Within groups | 36 | 25,137 | −4,922 | 106,745 | 35 | 105,781.188 | 3,022 |
| *Total* | 39 | 33,410 | −7,217 | 129,878 | 38 | 128,319.000 | |

$F_{.95}(3,35) \doteq 2.88$; $F = 2.49$, so we accept.

**From Table 2-2b, $n = 8$**

| | Degrees of freedom | $\Sigma x^2$ | $\Sigma xy$ | $\Sigma y^2$ | Degrees of freedom | $\Sigma y'^2$ | Mean square |
|---|---|---|---|---|---|---|---|
| Among means | 3 | 3,376.75 | 4,711 | 21,239 | 3 | 23,228.875 | 7,743 |
| Within means | 28 | 19,141.25 | −6,355 | 138,966 | 27 | 136,856.062 | 5,069 |
| *Total* | 31 | 22,518.00 | −1,644 | 160,205 | 30 | 160,084.938 | |

$F_{.95}(3,27) \doteq 2.94$; $F = 1.53$, so we accept.

**7**

| | Degrees of freedom | $\Sigma x^2$ | $\Sigma xy$ | $\Sigma y^2$ | Degrees of freedom | $\Sigma y'^2$ | Mean square |
|---|---|---|---|---|---|---|---|
| Among means | 2 | .5864 | 17.19 | 504 | 2 | 66.8 | 33.4 |
| Within groups | 45 | 7.7096 | 140.69 | 13,309 | 44 | 10,741.6 | 244.1 |
| *Total* | 47 | 8.2960 | 157.88 | 13,813 | 46 | 10,808.4 | |

$H$: there is no difference in final examination scores after adjustment for grade-point averages, $\alpha = .05$, $F_{.95}(2,44) = 3.21$. $F = .14$, so we accept $H$.

**8**

| | Degrees of freedom | $\Sigma x^2$ | $\Sigma xy$ | $\Sigma y^2$ | Degrees of freedom | $\Sigma y'^2$ | Mean square |
|---|---|---|---|---|---|---|---|
| Among means | 3 | 7,009.75 | −1,376.12 | 2826.69 | 3 | 3548.79 | 1182.9 |
| Within groups | 12 | 4,674.00 | 1,862.75 | 2675.75 | 11 | 1933.38 | 175.8 |
| *Total* | 15 | 11,683.75 | 486.62 | 5502.44 | 14 | 5482.17 | |

$H_1$: $\mu_A = \mu_B = \mu_C = \mu_D$, $\alpha = .05$, $F_{.95}(3,12) = 3.49$. Using unadjusted $y$ values, we find

$$F = \frac{2,826.69/3}{2,675.75/12} = 4.23$$

Hence we reject $H_1$. $H_2$: $\mu_A = \mu_B = \mu_C = \mu_D$, after adjustment for quantity of food, $\alpha = .05$. By analysis of covariance methods, we find

$$F = \frac{1,182.9}{175.8} = 6.73$$

Hence we reject $H_2$.

**9.** The plot of the points suggests that the regression lines of the three groups may be parallel but not linear.

| | Degrees of freedom | $\Sigma x^2$ | $\Sigma xy$ | $\Sigma y^2$ | Degrees of freedom | $\Sigma y'^2$ | Mean square |
|---|---|---|---|---|---|---|---|
| Among means | 2 | 0 | 0 | 34.67 | 2 | 34.67 | 17.33 |
| Within groups | 9 | 15.00 | −6.00 | 30.00 | 8 | 27.60 | 3.45 |
| *Total* | 11 | 15.00 | −6.00 | 64.67 | 10 | 62.27 | |

$H_1$: the mean $Y$ is the same for the three groups when adjusted for the $X$-covariate values, given parallel regression lines, $\alpha = .05$, $F_{.95}(2,8) = 4.46$; $F = 5.02$, so we reject $H_1$.

We proceed to investigate the cause of this significant $F$, using the steps following Table 12-6. We have already done step 1. We obtain $S_1 = 27.6$, $S_2 = 0$, $S_3 = 34.67$, $S_4 = 0$.

In step 2 we test for $H_2$: a common regression line can be used for all the observations, $\alpha = .05$. $F = 1.884$, $F_{.95}(4,6) = 4.53$, so we accept $H_2$. Since test 1 was significant, we shall proceed with the other tests even though test 2 was not significant. In step 2($a$) we test $H_3$: the slopes of the regression lines within the groups are the same; $F = 0$, so we accept $H_3$. In step 2($b$) we test $H_4$: the regression for means is linear; $F = 10.0$, $F_{.95}(1,8) = 5.32$, so we reject $H_4$. In step 2($c$) we test $H_5$: the average slope within groups ($b_w$) is the same as the slope of the regression line for group means ($b_m$); $F = 0$, so we accept.

This analysis accepts equal slopes, and in fact the least-squares lines are exactly parallel. The heights of the least-squares lines are different with the $F = 5.02$, which is significant. The possible curvature of individual regression lines (curves) is disregarded in the analysis, but the curvature in the values of the mean $Y$ is indicated by the significant $F = 10.0$.

## CHAPTER THIRTEEN

**1.** $H$: proportions in 9:3:3:1 ratio, $\alpha = 0.05$, $\chi^2_{.95}(3) = 7.81$; $\chi^2 = 1.91$, so we accept.

**2.** $H$: the die is balanced, $\alpha = 0.05$, $\chi^2_{.95}(5) = 11.07$; $\chi^2 = 6.40$, so we accept.

**3.** Fewer than 38 or more than 62 heads.

**4.** $H$: chance of even = chance of odd, $\alpha = 0.05$, $\chi^2_{.95}(1) = 3.84$; $\chi^2 = 5.76$, so we reject.

**5.** $\chi^2 = (10 - 16)^2/16 + (65 - 68)^2/68 + (25 - 16)^2/16 = 7.44$, which is significant at the 5 per cent level.

**6.** We assume that these data represent a random sample from all families. $H$: distributions of number of children for the three categories are the same, reject if $\chi^2 > \chi^2_{.95}(6) = 12.59$; $\chi^2 = 41.4$, so we reject.

**7.** $H$: $p_1 - p_2 = 0$, $\alpha = 0.05$, $\chi^2_{.95}(1) = 3.84$; $\chi^2 = 14.13$, so we reject. This could also be analyzed by the method in Sec. 13-6.

**8.** $H$: the grade distribution is the same for each instructor, $\alpha = 0.05$, $\chi^2_{.95}(8) = 15.51$; $\chi^2 = 8.8$, so we accept.

**9.** $H$: control and treated groups have the same distribution, $\alpha = .05$, $\chi^2_{.95}(2) = 5.99$; $\chi^2 = 6.34$, so we reject.

**10.** $H$: $p_1 = p_2$, $\alpha = 0.05$, $\chi^2_{.95}(1) = 3.84$; $\chi^2 = 2.4$, so we accept.

**11.** No record of number of cars built or sold during the periods is given. For example, if manufacturer $A$ built only 33, 127, and 256 cars, respectively, during the given periods, the interpretation of the results would be quite different than if he had built several thousand cars each period. If these data represent a random sample from all cars manufactured exactly 10 years ago, then the $\chi^2$ statistic can be used to test the hypothesis that the distributions, of lengths of life without overhaul, are the same for all three manufacturers. It can be seen from only a few terms that the observed chi-square will be considerably larger than $\chi^2_{.95}(10) = 18.31$.

**12.** $\chi^2_{.95}(2) = 5.99$. The computed $\chi^2 = 6.0$, so we reject the hypothesis of equal level.

**13.** $\chi^2_{.95}(3) = 7.81$. The computed $\chi^2 = 20.5$, so we reject the hypothesis of equal proportions. It is possible that the different interviewers were measuring different symptoms.

**14.** $\chi^2_{.95}(2) = 5.99$. The computed $\chi^2 = 10.2$, so we reject the hypothesis of equal proportions.

**15.** $\chi^2_{.95}(10) = 18.31$. The computed $\chi^2 = 30.5$ and we reject.

**16.** $\chi^2_{.95}(1) = 3.84$. The computed chi-square with continuity correction is

$$\chi^2 = \frac{(450 - 200 - 50)^2 100}{50 \times 50 \times 15 \times 85} = 1.2$$

which is not cause to reject the hypothesis of equal proportions.

**17.** $\chi^2_{.95}(1) = 3.84$. The computed chi-square with continuity correction is $\chi^2 = 29$ and we reject the hypothesis.

**18.** $N = 10$, $S_1 = 4$, $S_2 = 5$, $X = 1$, and we read in Table A-9*e* that the probability of having as large or larger value of $\chi^2$ is .524.

**20.** $\bar{X} = 0.70$ and $.48 < p < 0.87$ is a 90 per cent confidence interval for $p$.

**21.** Rejecting if the number of correct guesses is 0, 1, 9, or 10 in 10 tries gives $\alpha = 0.0216$. Rejecting with 0, 1, 2, 8, 9, or 10 correct guesses gives $\alpha = 0.1094$.

**22.** From Table A-15, $P(0) = 0.135$, $P(0 \text{ or } 1 \text{ or } 2) = 0.677$ and $P(4 \text{ or more}) = 1 - 0.857 = 0.143$.

**23.** (*b*) $.35 < p < .45$, (*c*) $.4 \pm 1.645\sqrt{.4 \times (.6)/250}$.

**37.** (*a*) $H: p = .03$, $\alpha = .05$, reject if $z > 1.96$ or if $z < -1.96$; $z = -10.8$, so we reject. 95 per cent confidence interval $.09 < p < .19$. (*b*) $H: p_1 = p_2$, $\alpha = .05$, reject if $z > 1.96$ or if $z < -1.96$; $z = 3.9$, so we reject. 99 per cent confidence interval $.03 < p_1 - p_2 < .15$. (*c*) $H: p_1 = p_2 = p_3 = p_4 = p_5$, $\alpha = .05$, $\chi^2_{.95}(4) = 9.49$; $\chi^2 = 36$, so we reject.

## CHAPTER FOURTEEN

**1.** Since the hypothesis has been rejected, a $\beta$ error may have been made. The 99 per cent is the chance that a true hypothesis will be accepted and *does not* indicate the proportion of rejected hypotheses which are false.

**2.** An "alternative" is selected because of its pertinence to the problem under investigation, and the power of the test is computed to ascertain the value of carrying out the study.

**3.** These graphs are simplified if they are drawn on normal-probability paper (see Table A-12*a*).

**4.** 87 per cent from Table A-12*a*, 99.8 per cent from Table A-4; $N = 31$.

**5.** $d = 1.67$, and we read from Table A-12*b* that $1 - \beta$ is between .3 and .4.

**6.** Using $2.8\sigma/\sqrt{N} = 3$ and putting $\sigma = 8$, we read from Table A-12*b* that $N = 56$.

**7.** $H: \mu \geq 105$, $\alpha = .10$; if $\mu = 100$, then $\beta = .10$; reject if $t = (\bar{X} - 105)/s/\sqrt{N} < t_{.10}(N-1)$. Using the normal approximation, we put $t_{.10} = -1.28$ and $s = \sigma = 16$. For $\beta = .10$ we ask that $(\bar{X} - 100)/\sigma/\sqrt{N} = t_{.90} = 1.28$. This gives $5 = 2.56 \times 16/\sqrt{N}$, or $N = 68$.

**8.** If $\mu = 67$, then $d = 1.05$, and we read from Table A-12*b* that the power is between .1 and .2. If $\mu = 64.5$, then $d = 1.58$, and we read from Table A-12*b* that the power is almost .3.

**9.** $d = .316$, and we read from Table A-12*b* that the power is less than .1.

**10.** $d = 5\sqrt{20}/\sqrt{500} = 1$, and Table A-12*b* gives $1 - \beta$ as approximately .25.

**11.** (*a*) A two-sample $z$ test with $N_1 = N_2 = 100$ has $d = .2/.30\sqrt{.01 + .01} = 5.6$; the selection of two heights, or any number of heights, essentially pairs the observations, so $d = .2/.15\sqrt{.04} = 6.7$, giving a smaller $\beta$. (*b*) $5.6 = .2/(.15\sqrt{2/N})$ gives $N = 35$.

**12.** $d' = .6/.3 = 2$, and from Table 12*c* we read $N = 3$ for a two-sided $\alpha = .05$ test if $\sigma$ was known or $N = 5$ for a $t$ test.

**13.** For over 20 years nonsmokers have a mean decrease of .8 liters and smokers have a mean decrease of .4 liters, giving $d' = .4/.3 = 1.33$; Table 12*c* for a one-sided $\alpha = .05$ test gives $N_1 = N_2 = 7$ if $\sigma$ is known or $N_1 = N_2 = 8$ if a $t$ test is used.

**14.** (*a*) 5 per cent. (*b*) $N = 100$. With a one-sided $\alpha = .05$ test of $H: p = p_0 = .20$, we reject if $\bar{X} < .20 - 1.645\sqrt{.2 \times .8/100} = .134$; if $p = .10$, we compute $z = (.134 - .10)/\sqrt{.1 \times .9/100} = 1.13$ and read from Table A-4 that $1 - \beta = .87$. (*c*) 1 per cent. (*d*) For $\alpha = .01$ the critical region is $\bar{X} < .20 - 2.326\sqrt{.2 \times .8/300} = .107$, and for $p = .10$ we compute $z = (.107 - .10)/\sqrt{.1 \times .9/100} = .23$ and read from Table A-4 that $1 - \beta = .59$. (*e*) For $\alpha = .05$, $N = 300$; at 5 per cent we reject if $\bar{X} < .20 - 1.645\sqrt{.2 \times .8/300} = .162$, giving $z = (.162 - .10)/\sqrt{.1 \times .9/300} = 3.58$, and $1 - \beta$ is approximately .9999; for $\alpha = .01$ we reject if $\bar{X} < .20 - 2.326\sqrt{.2 \times .8/300} = .183$, giving $z = (.183 - .10)/\sqrt{.1 \times .9/300} = 4.79$ and $1 - \beta$ is greater than .99999. (*f*) The value $N = 109$ approximately satisfies $.20 - 1.645\sqrt{.2 \times .8/N} = .10 + 1.282\sqrt{.1 \times .9/N}$ (see also Table A-12*e*).

**15.** $p_0 = .00001$, $\alpha = .05$ (one-sided), $p_1 = .000005$, $\beta = .05$, giving $.000010 - 1.645\sqrt{.000010 \times .999990/N} = .000005 + 1.645\sqrt{.000005 \times .999995/N}$, or $.000005\sqrt{N} = 1.645(.0031 + .0022)$, or $\sqrt{N} = 1{,}744$, or $N = 3{,}043{,}000$.

**16.** (*a*) 1 per cent. (*b*) $\bar{\mu} = \mu_1 + 5 = \mu_2 - 5$, $k = 2$, $\sigma = 10$, giving

$$\phi^2 = \frac{(5^2 + 5^2)/2}{100/n} = \frac{n}{4}$$

For $n = 20$, $\phi^2 = 5$, $\phi = 2.2$, and we read from Table A-13, with $\nu_1 = 1$ and $\nu_2 = 38$, that $1 - \beta = .66$ ($\alpha = .01$ curves). (*c*) 5 per cent; again, $\phi = 2.2$, and the $\alpha = .05$ curves give $1 - \beta = .86$. (*d*) For $n = 40$, $\phi^2 = 10$, $\phi = 3.1$, and we read from Table A-13, with $\nu_1 = 1$ and $\nu_2 = 78$, that $1 - \beta = .96$ for $\alpha = .01$ and $1 - \beta$ is greater than .99 for $\alpha = .05$. (*e*) We see from the answers in (*b*) and (*d*) above that the required $n$ is between 20 and 40. Trial and error gives $n = 32$.

**17.** $\phi = \sqrt{5}$, $\nu_1 = 3$, $\nu_2 = 76$, power $= .035$.

**18.** $\phi^2 = .154n$ and trial and error in Table A-13 gives $n$ as approximately 17.

## CHAPTER FIFTEEN

**1.** (*a*)

| | CELL SUMS $\Sigma X_i$ | | | | | | |
|---|---|---|---|---|---|---|---|
| | 66 | 52 | 78 | 60 | *Contrast* | $\Sigma a_i^2$ | *Mean square* |
| | CONTRAST COEFFICIENTS | | | | | | |
| Column | 1 | −1 | 1 | −1 | 32 | 4 | 64 |
| Row | 1 | 1 | −1 | −1 | −20 | 4 | 25 |
| Interaction | 1 | −1 | −1 | 1 | −4 | 4 | 1 |

(*b*)

| | *Sum of squares* | *Degrees of freedom* | *Mean square* | *F ratio* |
|---|---|---|---|---|
| Column | 64 | 1 | 64 | 48.1 |
| Row | 25 | 1 | 25 | 18.8 |
| Interaction | 1 | 1 | 1 | .75 |
| Residual | 16 | 12 | 1.33 | |
| *Total* | 106 | 15 | | |

$F_{.95}(1,12) = 4.75$; for rows $F = 18.8$, so we reject; for columns $F = 48.1$, so we reject.

(*c*) Between-means sum of squares = 90.

**2.** (*a*)

| | OBSERVATIONS $X_i$ | | | | | | | | | | |
|---|---|---|---|---|---|---|---|---|---|---|---|
| | 3.5 | 4.0 | 4.6 | 4.8 | 2.9 | 3.1 | 5.0 | 5.4 | *Contrast* | $\Sigma a_i^2$ | *Mean square* |
| | CONTRAST COEFFICIENTS | | | | | | | | | | |
| Column | 1 | 1 | −1 | −1 | 1 | 1 | −1 | −1 | −6.3 | 8 | 4.96125 |
| Row | 1 | 1 | 1 | 1 | −1 | −1 | −1 | −1 | .5 | 8 | .03125 |
| Interaction | 1 | 1 | −1 | −1 | −1 | −1 | 1 | 1 | 2.5 | 8 | .78125 |

(*b*)

| | *Sum of squares* | *Degrees of freedom* | *Mean square* | *F ratio* |
|---|---|---|---|---|
| Column | 4.96125 | 1 | 4.961 | 81 |
| Row | .03125 | 1 | .031 | .51 |
| Interaction | .78125 | 1 | .781 | 12.8 |
| Residual | .245 | 4 | .061 | |
| *Total* | 6.01875 | 7 | | |

$F_{.95}(1,4) = 7.71$; for column $F = 81.0$, so we reject; for rows $F = .51$, so we accept; for interaction $F = 12.8$, so we reject.

(*c*) Between-means sum of squares = 5.773.

**3.** (*a*)

| | CELL SUMS $\Sigma X_i$ | | | | | | |
|---|---|---|---|---|---|---|---|
| | 11 | 16 | 19 | 13 | *Contrast* | $\Sigma a_i^2$ | *Mean square* |
| | CONTRAST COEFFICIENTS | | | | | | |
| Linear | −3 | −1 | 1 | 3 | 9 | 20 | 1.35 |
| Quadratic | 1 | −1 | −1 | 1 | −11 | 4 | 10.08 |
| Cubic | −1 | 3 | −3 | 1 | −7 | 20 | .82 |

(*b*) Between-means sum of squares is 12.25.
(*c*)

| | *Sum of squares* | *Degrees of freedom* | *Mean square* | *F ratio* |
|---|---|---|---|---|
| Linear effect | 1.35 | 1 | 1.35 | .52 |
| Quadratic effect | 10.08 | 1 | 10.08 | 3.91 |
| Cubic effect | .82 | 1 | .82 | .32 |
| Residual | 20.67 | 8 | 2.58 | |
| *Total* | 32.92 | 11 | | |

$F_{.95}(1,8) = 5.32$.

(*e*) On a parabola. Although not large enough to be significant, at the 5 per cent level the value $F = 3.91$ is consistent with a parabolic appearance.

**4.**

| | CELL SUMS $\Sigma X_{ij}$ | | | | | | | | |
|---|---|---|---|---|---|---|---|---|---|
| | 16 | 7 | 15 | 25 | 20 | 25 | *Contrast* | $\Sigma a_i^2$ | *Mean square* |
| | CONTRAST COEFFICIENTS | | | | | | | | |
| Row | 1 | 1 | 1 | −1 | −1 | −1 | −32 | 6 | 56.889 |
| Column 1 | 1 | 0 | −1 | 1 | 0 | −1 | 1 | 4 | .083 |
| Column 2 | 1 | −2 | 1 | 1 | −2 | 1 | 27 | 12 | 20.250 |
| Interaction 1 | 1 | 0 | −1 | −1 | 0 | 1 | 1 | 4 | .083 |
| Interaction 2 | 1 | −2 | 1 | −1 | 2 | −1 | 7 | 12 | 1.361 |

**5.**

| | CELL SUMS | | | | | |
|---|---|---|---|---|---|---|
| | 36 | 43 | 21 | *Contrast* | $\Sigma a_i^2$ | *Mean square* |
| | CONTRAST COEFFICIENTS | | | | | |
| Difference in radiation | 1 | −1 | 0 | −7 | 2 | 4.900 |
| Difference in radiation and placebo | 1 | 1 | −2 | 37 | 6 | 45.633 |

$F_{.95}(1,12) = 4.75$.

| | *Sum of squares* | *Degrees of freedom* | *Mean square* | *F ratio* |
|---|---|---|---|---|
| Radiation type | 4.9 | 1 | 4.9 | .551 |
| Radiation vs. placebo | 45.633 | 1 | 45.633 | 5.13 |
| Residual | 106.8 | 12 | 8.9 | |
| *Total* | 157.333 | 14 | | |

For radiation type $F = .551$, so we accept; for radiation versus placebo $F = 5.13$, so we reject.

**6.** (*a*)

| | *Sum of squares* | *Degrees of freedom* | *Mean square* | *F ratio* |
|---|---|---|---|---|
| Concentration effect | 1.304 | 3 | .435 | .737 |
| Composition effect | .466 | 3 | .155 | .263 |
| Residual | 5.311 | 9 | .590 | |
| *Total* | 7.081 | 15 | | |

$F_{.95}(3,9) = 3.86$; for concentration $F = .737$, so we accept; for composition $F = .263$, so we accept.

(*b*)

| | OBSERVATIONS $X_i$ | | | | | | |
|---|---|---|---|---|---|---|---|
| | 28.80 | 29.12 | 29.76 | 30.56 | *Contrast* | $\Sigma a_i^2$ | *Mean square* |
| | CONTRAST COEFFICIENTS | | | | | | |
| Linear | −3 | −1 | 1 | 3 | 5.92 | 20 | 1.7523 |
| Quadratic | 1 | −1 | −1 | 1 | .48 | 4 | .0576 |
| Cubic | −1 | 3 | −3 | 1 | −.16 | 20 | .0013 |

(*c*)

| | *Sum of squares* | *Degrees of freedom* | *Mean square* | *F ratio* |
|---|---|---|---|---|
| Linear effect | 1.7523 | 1 | 1.7523 | 2.97 |
| Quadratic effect | .0576 | 1 | .0576 | .097 |
| Cubic effect | .0013 | 1 | .0013 | .002 |

$F_{.95}(1,9) = 5.12$; for cubic effect $F = .002$, so we accept no cubic term; for quadratic effect $F = .097$, so we accept no quadratic term; for linear effect $F = 2.97$, so we accept a constant.

(*d*) Between-means sum of squares for all the "all-sucrose" observations is 1.8112.

**7.** (*a*) Coefficients within each function, except (1) sum to 0; $\Sigma a_i b_i = 0$ for any pair of functions.

(*b*) $\Sigma a_i^2 = 8$ in each function:

| | LINEAR FUNCTIONS | | | | | | | |
|---|---|---|---|---|---|---|---|---|
| | 1 | 2 | 3 | 4 | 5 | 6 | 7 | 8 |
| *Contrast* | 22.1 | −4.9 | 4.1 | 3.1 | −.9 | −.3 | −.9 | 1.7 |
| *Mean square* | 61.05125 | 3.00125 | 2.10125 | 1.20125 | .10125 | .01125 | .10125 | .36125 |

(*d*) 95 per cent confidence intervals are $.052 < \sigma^2 < 1.19$,
$-4.9/8 - 2.776\sqrt{.144/8} < R < -4.9/8 + 2.776\sqrt{.144/8}$,
$4.1/8 - 2.776\sqrt{.144/8} < C < 4.1/8 + 2.776\sqrt{.144/8}$,
$3.1/8 - 2.776\sqrt{.144/8} < L < 3.1/8 + 2.776\sqrt{.144/8}$.

**8.** $\bar{D} = D_0 - .301$, $\Sigma(D_i - \bar{D})^2 = .906$, $\bar{Y} = 7.8$, $\Sigma(Y_i - \bar{Y})^2 = 50.4$, $\Sigma(D - \bar{D})(Y - \bar{Y}) = 6.02$, $b = 6.645 s^2_{x \cdot y} = .80$, $t_{.975}(13) = 2.160$.

| | Sum of squares | Degrees of freedom | Mean square | F ratio |
|---|---|---|---|---|
| Linear | 40 | 1 | 40 | 52.15 |
| Quadratic | 1.2 | 1 | 1.2 | 1.56 |
| Remainder | 9.2 | 12 | .767 | |
| *Total* | 50.4 | 14 | | |

$F_{.95}(1,12) = 4.75$; for quadratic $F = 1.56$, so we accept the hypothesis that there is no quadratic effect; for linear $F = 52.15$, so we reject the hypothesis that there is no linear effect. 95 per cent confidence interval for log dosage with mean $y = 9$ is $-.2014 < X - D_0 < -.0023$.

**9.** (*a*) $\bar{D} = D_0 - .602$, $\bar{Y} = .70804$, $\Sigma(D_i - \bar{D})^2 = .90601$, $\Sigma(Y_i - \bar{Y})^2 = .32203$, $\Sigma(D - \bar{D})(Y - \bar{Y}) = .53621$, $b = .59184$, $s^2_{x \cdot y} = .001558$, $df = 3$.

(*b*) 95 per cent confidence interval for log 7.1 is
$.3777 < X - D_0 < .6497[-.4228 < \log(X - D_0) < -.1873]$.

(*d*)

| | Sum of squares | Degrees of freedom | Mean square | F ratio |
|---|---|---|---|---|
| Linear | 50.176 | 1 | 50.176 | 178.56 |
| Quadratic | 1.143 | 1 | 1.143 | 4.07 |
| Remainder | .561 | 2 | .281 | |
| *Total* | 51.880 | 4 | | |

$F_{.95}(1,2) = 18.5$ for quadratic $F = 4.07$, so we accept the hypothesis that there is no quadratic effect; for linear $F = 178.56$, so we reject the hypothesis that there is no linear effect.

**10.** (*a*) Measurements on each man form a population.

(*b*)

| | Sum of squares | Degrees of freedom | Mean square | Estimate |
|---|---|---|---|---|
| Men | 7.357 | 4 | $s_M^2 = 1.839$ | $\sigma^2 + 7\sigma_M^2$ |
| Within | 1.666 | 30 | $s_p^2 = .056$ | $\sigma^2$ |
| *Total* | 9.023 | 34 | | |

Estimate of $\mu$, 6.41, estimate of $\sigma_m^2$, .255, estimate of $\sigma^2$, .056.

(*c*) Seven columns (days) and five rows (men) each consist of samples from larger groups of categories:

| | Sum of squares | Degrees of freedom | Mean square | Estimate |
|---|---|---|---|---|
| Men | 7.357 | 4 | 1.8393 | $\sigma^2 + 7\sigma_r^2$ |
| Days | 1.463 | 6 | .2438 | $\sigma^2 + 5\sigma_c^2$ |
| Remainder | .203 | 24 | .0085 | $\sigma^2$ |
| *Total* | 9.023 | 34 | | |

**11.** (*a*) Measuring 100 objects once each gives variance 101, while measuring 10 objects 10 times each gives variance 1,001.

(*b*) Given that $n_i = n$ values of $n = 3$, $k = 8$ results in a cost of \$104 with variance $\sigma_m^2/k + (\sigma^2/k^2)\Sigma(1/n_i) = 1{,}254$. For $n = 1$ and $k = 9$, the cost is \$99, with variance 1,122.

**12.**

| | Sum of squares | Degrees of freedom | Mean square | F ratio |
|---|---|---|---|---|
| Row | 1.5 | 3 | .5 | .23 |
| Column | 1.5 | 3 | .5 | .23 |
| Fertilizer | 90 | 3 | 30 | 13.8 |
| Residual | 13 | 6 | 2.167 | |
| *Total* | 106.0 | 15 | | |

$F_{.95}(3,6) = 4.76$; for rows $F = .23$, so we accept; for columns $F = .23$, so we accept; for fertilizer $F = 13.8$, so we reject.

**13.** (*a*) Simultaneous confidence intervals for population means: $1 - \alpha = .95$, $s_p^2 = 26.01$, $df = 121$, $k = 5$, $h_{1-\alpha} = 2.63$.

| | $\bar{X}_{1.}$ 63 | $\bar{X}_{2.}$ 72 | $\bar{X}_{3.}$ 60 | $\bar{X}_{4.}$ 80 | $\bar{X}_{5.}$ 92 | $\Sigma c_i \bar{X}_{i.} \pm h_{1-\alpha} s_p \sqrt{\frac{\Sigma c_i^2}{n}}$ | Estimate of mean of population |
|---|---|---|---|---|---|---|---|
| 1 | 1 | 0 | 0 | 0 | 0 | 63 ± 2.68 | 1 |
| 2 | 0 | 1 | 0 | 0 | 0 | 72 ± 2.68 | 2 |
| 3 | 0 | 0 | 1 | 0 | 0 | 60 ± 2.68 | 3 |
| 4 | 0 | 0 | 0 | 1 | 0 | 80 ± 2.68 | 4 |
| 5 | 0 | 0 | 0 | 0 | 1 | 92 ± 2.68 | 5 |

(*b*) 95 per cent confidence interval $60.97 < \mu_1 < 65.03$.

**14.** (*a*) Simultaneous confidence intervals for population means: $1 - \alpha = .99$, $s_p^2 = 52.25$, $df = 16$, $k = 4$, $h_{1-\alpha} = 3.57$.

| | $\bar{X}_{1.}$ 140.8 | $\bar{X}_{2.}$ 153.2 | $\bar{X}_{3.}$ 223.6 | $\bar{X}_{4.}$ 191.6 | $\Sigma c_i \bar{X}_i \pm h_{1-\alpha} s_p \sqrt{\frac{\Sigma c_i^2}{n}}$ |
|---|---|---|---|---|---|
| 1 | 1 | 0 | 0 | 0 | 140.8 ± 11.54 |
| 2 | 0 | 1 | 0 | 0 | 153.2 ± 11.54 |
| 3 | 0 | 0 | 1 | 0 | 223.6 ± 11.54 |
| 4 | 0 | 0 | 0 | 1 | 191.6 ± 11.54 |

(b) 99 per cent confidence interval $131.4 < \mu_1 < 150.2$.

**15.** Simultaneous confidence intervals for population means: $1 - \alpha = .95$, $s_p^2 = 1.33$, $df = 2$, $q_{.95}(4,12) = 4.2$, $k = 3$, $h_{1-\alpha} = 2.75$, $n = 4$.

| | $\bar{X}_{11\cdot}$ 16.5 | $\bar{X}_{21\cdot}$ 13.0 | $\bar{X}_{12\cdot}$ 19.5 | $\bar{X}_{22\cdot}$ 15.0 | FOR $q$ DISTRIBUTION $\Sigma c_{ij}\bar{X}_{ij\cdot} \pm \frac{q_{1-\alpha}s_p}{\sqrt{n}}$ | FOR $h$ DISTRIBUTION $\Sigma c_{ij}\bar{X}_{ij\cdot} \pm h_{1-\alpha}s_p\sqrt{\frac{\Sigma c_{ij}^2}{n}}$ |
|---|---|---|---|---|---|---|
| 1 | $\frac{1}{2}$ | $\frac{1}{2}$ | $-\frac{1}{2}$ | $-\frac{1}{2}$ | $-2.5 \pm 2.42$ | $-2.5 \pm 1.59$ |
| 2 | $\frac{1}{2}$ | $-\frac{1}{2}$ | $\frac{1}{2}$ | $-\frac{1}{2}$ | $4 \pm 2.42$ | $4 \pm 1.59$ |
| 3 | $\frac{1}{2}$ | $-\frac{1}{2}$ | $-\frac{1}{2}$ | $\frac{1}{2}$ | $-.5 \pm 2.42$ | $-.5 \pm 1.59$ |

**16.** (a) To show that $b$ and $\bar{Y}$ are orthogonal

| | $Y_1$ | $Y_2$ | $Y_3$ | $\cdots$ | $Y_N$ |
|---|---|---|---|---|---|
| | CONTRAST COEFFICIENTS | | | | |
| $a_i$ | $\frac{1}{N}$ | $\frac{1}{N}$ | $\frac{1}{N}$ | $\cdots$ | $\frac{1}{N}$ |
| $c_i$ | $\frac{X_1 - \bar{X}}{D}$ | $\frac{X_2 - \bar{X}}{D}$ | $\frac{X_3 - \bar{X}}{D}$ | $\cdots$ | $\frac{X_N - \bar{X}}{D}$ |

Where $a_i$ are coefficients for $\bar{Y}$, $c_i$ for $b$ and $D = \Sigma(X_i - \bar{X})^2$. Then $\Sigma a_i = 1$, $\Sigma c_i = 0$, $\Sigma a_i c_i = 0$.

(b) Simultaneous confidence intervals are obtained with $k = N$ and $n = 1$, and $s^2_{y\cdot x}$ has $N - 2$ degrees of freedom.

| | CONFIDENCE INTERVALS $\Sigma d_i Y_i \pm h_{1-\alpha}s_{y\cdot x}\sqrt{\Sigma d_i^2/n}$ |
|---|---|
| $\mu$ | $\bar{Y} \pm h_{1-\alpha}s_{y\cdot x}/\sqrt{N}$ |
| $B$ | $b \pm h_{1-\alpha}s_{y\cdot x}/\sqrt{\Sigma(X_i - \bar{X})^2}$ |

$d_i = a_i$ for $\bar{Y}$ and $d_i = c_i$ for $b$.

(c) $1 - \alpha = .95$, $df = 10$, $k = 2$, $h_{1-\alpha} = 2.61$, $s^2_{y\cdot x} = 92.52$, $\Sigma(X_i - \bar{X})^2 = 171.67$. Simultaneous 95 per cent confidence intervals are

$$141.67 - \frac{2.61\sqrt{92.52}}{\sqrt{12}} < \mu < 141.67 + \frac{2.61\sqrt{92.52}}{\sqrt{12}}$$

$$5.029 - \frac{2.61\sqrt{92.52}}{\sqrt{171.67}} < B < 5.029 + \frac{2.61\sqrt{92.52}}{\sqrt{171.67}}$$

**17.** (*a*) Two orthogonal contrasts for treatments

| Treatment | Contrast coefficients | Contrast | $\Sigma a_i^2$ | Mean square |
|---|---|---|---|---|
| 1 | 1 0 −1<br>−1 1 0<br>0 −1 1 | −.872 | 6 | .1267 |
| 2 | 1 −2 1<br>1 1 −2<br>−2 1 1 | −2.502 | 18 | .3478 |

(*b*) Two orthogonal contrasts for rows

| Row | Contrast coefficients | Contrast | $\Sigma a_i^2$ | Mean square |
|---|---|---|---|---|
| 1 | 1 −1 0<br>1 −1 0<br>1 −1 0 | .453 | 6 | .0342 |
| 2 | 1 1 −2<br>1 1 −2<br>1 1 −2 | 1.293 | 18 | .0929 |

## CHAPTER SIXTEEN

**2.** With $df = 99$, $\chi^2/df$ has 2.5 percentile and 97.5 percentile equal to .740 and 1.30, respectively, so 95 per cent confidence limits $\sigma^2$ are $\frac{1.21 \times 4}{1.30} < \sigma^2 < \frac{1.21 \times 4}{.740}$.
$t_{.975}(99) = 1.99$, and 95 per cent confidence limits for the mean are
$4.1 - 1.99(1.1/\sqrt{100}) < \mu < 4.1 + 1.99(1.1/\sqrt{100})$.

**3.** $\lambda = 1$:

| $X$ | 0 | 1 | 2 | 3 | 4 | 5 | 6 |
|---|---|---|---|---|---|---|---|
| $y = \sqrt{X} + \sqrt{X+1}$ | 1 | 2.41 | 3.15 | 3.73 | 4.24 | 4.69 | 5.1 |
| *Frequency* | .368 | .368 | .184 | .061 | .015 | .003 | .001 |

**4.**

| *Rank* | 1 | 2 | 3 | 4 | 5 | 6 | 7 |
|---|---|---|---|---|---|---|---|
| *T-score* | 65 | 58 | 54 | 50 | 46 | 42 | 35 |

**5.**

| *Test score* | 95 | 90 | 85 | 80 | 75 | 70 | 65 | 60 |
|---|---|---|---|---|---|---|---|---|
| *T-score* | 71 | 65 | 60 | 56 | 53 | 47 | 42 | 34 |

$u_y = 2.144$, $\sigma_y = .9425$.

**7.** Considering the ninth pair for rejection, we reject with $\alpha = .05$ if $r_{11} > .477$; $r_{11} = 20/24 = .8$, so we reject.

**8.** First level 95 per cent confidence interval $16.130 - .158 < \mu < 16.130 + .158$; second level 95 per cent confidence interval $16.132 - .184 < \mu < 16.132 + .184$.

## CHAPTER SEVENTEEN

**1.** $H$: $\mu_1 - \mu_2 = 0$, $\alpha = .05$, reject if $r \leq 1$; $r = 1$, so we reject. The major difference between the results of the $t$ and sign tests is due to the ninth pair, which should be rejected as extreme if the $t$ test is used.

**2.** $\alpha = .05$, reject if $r \leq 5$; $r = 6$, so we accept.

**3.** Only temperatures 1 and 2 are significant with $\alpha = .05$.

**4.** With $N = 15$, 10 positive and 5 negative is not significant with $\alpha \leq .25$.

**5.** Eight minus signs out of 18 is not significant at $\alpha = .05$.

**6.** $H$: differences have 0 median, $\alpha = .05$. (*a*) $N = 41$ (9 ties), reject if $r \leq 13$. $r = 19$, so accept. (*b*) $N = 50$, reject if $r \leq 17$. $r = 19$, so accept.

**7.** $H$: equal medians, $N = 10$, $\alpha = .048$, reject if $T$ is $\leq T_{.024} = 8$ or if $T$ is $\geq T_{.976} = 47$; $T = 45$, so we accept.

**8.** Depending on ties, there are 7 to 11 runs, none of which would be significant at the $\alpha = .05$ level.

**9.** With 13.05 as the median, there are six runs, which is significant at the $\alpha = .05$ level.

**10.** $\alpha = .05$, reject if $u \leq 14$ or $u > 27$; $u = 14$, so we reject the hypothesis of random arrangement (sequence of days is vertical).

**11.** $T' = 22.5$, which is not significant at the $\alpha = .05$ level.

**12.** $R_1 = 17$, $R_2 = 38$, $R_3 = 65$, $R_4 = 90$, $H = 17.3$. This is larger than $\chi^2_{.995}(3) = 12.84$, so we reject the hypothesis of four identically distributed populations.

**13.** $N = (1.63/.002)^2 = 664{,}000$.

**14.** Add and subtract .21 from the sample cumulative distribution.

**15.** Add and subtract .26 from the sample cumulative distribution.

**16.** Reject the hypothesis that the two population cumulatives coincide if the maximum deviation is greater than $1.36\sqrt{\frac{2}{20}} = .42$. The sample cumulatives differed by .4, so we accept the hypothesis.

**17.** $N = 4{,}444$ from Chebyshev's inequality and $N = 664$ from the normality of $\bar{X}$.

**18.** Reading Table A-25*d* with $\gamma = .90$, $P = .95$, we get $N = 77$.

**19.** From Table A-25*c* we get $\gamma = .920$ for $N = 40$, $P = .90$.

**20.** For $N = 10$ we read from Table A-25*a* that $X_2 = 56$ and $X_9 = 63$ are endpoints of a 97.9 per cent confidence interval for the population median.

**21.** For $N = 40$ we read from Table A-25*a* that $X_{14} = 22$ and $X_{27} = 28$ are endpoints of a 96.2 per cent confidence interval for the population median.

**22.** $H$: the rankings are independent, $\alpha = .05$, reject if $r_s \leq -.65$ or if $r_s \geq .65$; $r_s = .636$, so we reject.

**23.** For the data in Prob. 11-5, $r_s = .967$; Table A-30*a* gives the 97.5 percentile to be .361, so we reject the hypothesis of independence.

**24.** Chi-square corrected for continuity is $\chi^2 = .2$, so we accept the hypothesis of equality of distributions.

**25.** Chi-square corrected for continuity is $\chi^2 = 16.2$, so we reject equality of distribution.

**26.** $H$: scores at different times of day have the same distribution, $\chi^2_{.95}(16) = 11.07$; $\chi^2 = 10.5$ using the 15th, 34th, and 49th observations as cutting points. Accept.

## CHAPTER EIGHTEEN

**1.** $d_m \ln (.3/.1) + g_m \ln (.7/.9) = \ln (.9/.05)$ and
$d_m \ln (.3/.1) + g_m \ln (.7/.9) = \ln (.10/.95)$.

**2.** $d_m \ln (.5/.3) + g_m \ln (.5/.7) = \ln (.9/.1)$ and
$d_m \ln (.5/.3) + g_m \ln (.5/.7) = \ln (.1/.9)$.

**3.** $d_m \ln (.5/.2) + g_m \ln (.5/.8) = \ln (.9/.1)$ and
$d_m \ln (.5/.2) + g_m \ln (.5/.8) = \ln (.1/.9)$.

**4.** $p_0 = .04$, $d = .01$, $p_1 = .08$, $\beta = .05$, so
$d_m \ln (.08/.04) + g_m \ln (.92/.96) = \ln (.95/.01)$ and
$d_m \ln (.08/.04) + g_m \ln (.92/.96) = \ln (.05/.99)$.

**5.** With the methods in Chaps. 13 and 14, $N = 30$, $= .09$, reject if $\bar{X} = .19$; $\bar{X} = .23$, so we reject. With a sequential test, the lines are
$1.099d_m - .251g_m = 2.890$ and $1.099d_m - .251g_m = -2.251$. After 10 observations the value $p = .3$ is accepted.

**6.** (*a*) Equations of the lines are $\Sigma X_i = 60m + 58.9$ and $\Sigma X_i = 60m - 58.9$. The average sample number for $(\mu_1 + \mu_2)/2$ is 8.67. (*b*) Equations of the lines are $\Sigma X_i = 60m + 91.9$ and $\Sigma X_i = 60m - 91.9$. The average sample number for $(\mu_1 + \mu_2)/2$ is 21.1.

**7.** Equations of the curves are $\Sigma(X_i - \mu)^2 = 1{,}059 + 145.8m$ and
$\Sigma(X_i - \mu)^2 = -1{,}059 + 145.8m$. The average sample number corresponding to $\sigma^2 = D$ is 26.4.

## CHAPTER NINETEEN

**2.** (*a*) 3, (*b*) .107, (*c*) $2.96 < \mu < 3.04$, (*d*) $2.85 < P_{20} < 2.97$, (*e*) $P_{20} = 18.5$, $19.5 < P_{50} < 21$.

**3.** (*a*) $N' = 8$, $N = 6$, and the estimate of $P_{50}$ is $.602 + .831 \times .301 = .852$; (*b*) $\text{Log}_{10} X = .852$, and Table A-14*b* gives $X$ as approximately 7.1.

## CHAPTER TWENTY

**1.** $C(90,4)/C(100,4)$; $C(10,4)/C(100,4)$.

**2.** $C(2,1)C(38,2)/C(40,3)$; etc.

**3.** .362.

**4.** .1.

**6.** $(.95)^{10}$; $10(.5)(.95)^9, \cdots, (.05)^{10}$.

**7.** .001; $P$ (all recover in 20 cases given $p = .9$) $= (.9)^{20} = .1216$.

**8.** $C(100,40)(.5)^{100} = .0108$.

**9.** (*a*) $(.5)^5$, (*b*) $5(.5)^5$, (*c*) $10(.5)^5$, (*d*) $10(.5)^5$, (*e*) $5(.5)^5$, (*f*) $(.5)^5$.

**10.** The man must go a total of five blocks north and five blocks south; therefore the problem is equivalent to the probability of exactly five heads in tossing a true coin, or $C(10,5)(.5)^{10}$.

**11.** $C(3,2)(.15)^2(.85)$.

**12.** $6(.5)^6 + (.5)^6$.

**13.** $C(10,8)(.5)^{10}$; $C(10,8)(.5)^{10} + C(10,9)(.5)^{10} + C(10,10)(.5)^{10} = .0547$.

**14.** $C(10,8)(.2)^2(.8)^8 + C(10,9)(.2)(.8)^9 + (.8)^{10} = .6778$.

**16.** $(.25)^3$; $10(.5)^8$; $(.5)^5$; $30(.5)^8$; $(.5)^3$; $10(.5)^5$; $(.5)^5$; 0.

**18.** $6(.75)^2\,(.25)^2$.

**20.** None, 0, 1, 2.

**21.** .9790, .9981, which agree with the values of .021 and .002 recorded in Table A-25.

**23.** $\frac{2}{16}, \frac{6}{16}, \frac{6}{16}, \frac{2}{16}; \frac{2}{6}$.

**25.** All are density functions in that $f(x)$ is never negative and sums to 1.

**26.** (*a*) $C(10,x)C(190, 20 - x)/C(200,20)$, (*b*) $C(30,x)(.5)^{30}$, (*c*) $C(12,x)(.5)^{12}$, (*d*) $C(50,x)(.10)^x(.90)^{50-x}$, (*e*) $C(35,x)(.75)^x(.25)^{35-x}$

**32.** $(1 - 1/20{,}000)^{10{,}000}$, which is approximately equal to $(.5)^0e^{-.5}/0! = e^{-.5} = .607$ (Table A-15).

**33.** $e^{-.5} = .607$; $(.607)^5$.

**34.** $\mu = 180$, $\sigma = 360/\sqrt{12}$.

**36.** $\frac{20}{360} = .055$; $(.945)^{10}$.

**37.** .150, 1.76, 4.10.

**39**

| $X$ | $q_x$ | $l_x$ | $d_x$ |
|---|---|---|---|
| 0–1 | .60 | 100 | 60 |
| 1–2 | .50 | 40 | 20 |
| 2–3 | .40 | 20 | 8 |
| 3–4 | .30 | 12 | 4 |
| 4–5 | .30 | 8 | 3 |
| 5–6 | .30 | 5 | 2 |
| 6–7 | .30 | 3 | 1 |
| 7–8 | .30 | 2 | 1 |
| 8–9 | .30 | 1 | 0 |
| 9–10 | .30 | 1 | 1 |

$$e_2 = \frac{8 \times .5 + 4 \times 1.5 + 3 \times 2.5 + 2 \times 3.5 + 1 \times 4.5 + 1 \times 5.5 + 1 \times 7.5}{20} = 2.12$$

**40**

| $X$ | $q_x$ | $l_x$ | $d_x$ |
|---|---|---|---|
| 0–1 | .6 | 100,000 | 60,000 |
| 1–2 | .6 | 40,000 | 24,000 |
| 2–3 | .6 | 16,000 | 9,600 |
| 3–4 | .6 | 6,400 | 4,940 |
| 4–5 | .6 | 2,560 | 1,536 |
| 5–6 | .6 | 1,024 | 614 |
| 6–7 | .6 | 410 | 246 |
| 7–8 | .6 | 164 | 98 |
| 8–9 | .6 | 66 | 40 |
| 9–10 | .6 | 26 | 16 |
| 10–11 | .6 | 10 | 6 |
| 11–12 | .6 | 4 | 2 |
| 12–13 | .6 | 2 | 1 |
| 13–14 | .6 | 1 | 1 |
| 14–15 | .6 | 0 | |

# Index

# List of tables